MODERN CONSTRUCTION ENVELOPES

Systems for architectural design and prototyping

THIRD EDITION

ANDREW WATTS

newtecnic
building.engineering.technology

Newtecnic is an international world leader in the engineering design and architectural design of complex, highly ambitious construction projects and advanced building envelope systems.

The company undertakes the engineering and architectural design of building structures, facades and MEP installations in partnership with leading international developers and large contractors. Newtecnic also works on some projects with signature architects.

In partnership with the Engineering Departments of Cambridge University and the University of Houston. Newtecnic's R&D team analyses, develops, tests, validates and specifies new building products, technologies and methods.

DESIGN TO PROTOTYPE 4

Design to prototype 6
Comparison of systems 16
Preliminary specification for inventory models 22
Prototyping 44
BIM to fabrication 50

METAL WALLS 54

01 Sheet metal 56
02 Profiled metal cladding 66
03 Composite panels 76
04 Metal rainscreens 86
05 Mesh screens 96
06 Louvre screens 106

GLASS WALLS 11

07 Stick glazing 11
08 Unitised glazing 12
09 Clamped glazing 13
10 Bolt fixed glazing 14
11 Glass blocks 15
12 Steel windows 16
13 Aluminium windows 17
14 Timber windows 18

METAL ROOFS 316

26 Metal standing seam 318
27 Profiled metal sheet 328
28 Composite panels 338
29 Rainscreens 348
30 Metal canopies 358

GLASS ROOFS 368

31 Silicone-sealed glazing and rooflights 370
32 Bolt fixed glazing: small scale rooflights 380
33 Bolt fixed glazing: large scale rooflights 390
34 Bolt fixed glazing: twin wall roofs 400
35 Bonded glass rooflights 410

CONCRETE ROOFS 41

36 Concealed membrane 412
37 Exposed membrane 422
38 Planted 432

CONTENTS

NCRETE WALLS 198

5 Cast in situ / cast-in-place 200
6 Storey height precast panels 210
7 Small precast/GRC cladding panels 220

MASONRY WALLS 230

18 Masonry cavity walls brick 232
19 Masonry cavity walls block 242
20 Stone cladding 252
21 Terracotta rainscreens 262

POLYMER WALLS 272

22 Polymer-based cladding sealed panels 274
23 Rainscreens 284

TIMBER WALLS 294

24 Cladding the timber frame 296
25 Cladding panels and rainscreens 306

MBER ROOFS 442

39 Flat roofs: Bitumen-based sheet membranes 444
40 Pitched roofs: Tiles and sheet 454

POLYMER ROOFS 474

41 GRP rooflights 476
42 GRP panels and shells 486

FABRIC ROOFS 496

43 ETFE cushions 498
44 Single membrane roof 508

END 518

Index 518
Authorship 520
Bibliography 520

3D printing of facade components in site laboratory
Facade Engineer: Newtecnic

DESIGN TO PROTOTYPE

- Design to prototype
- Comparison of systems
- Preliminary specification for inventory models
- Prototyping
- BIM to fabrication

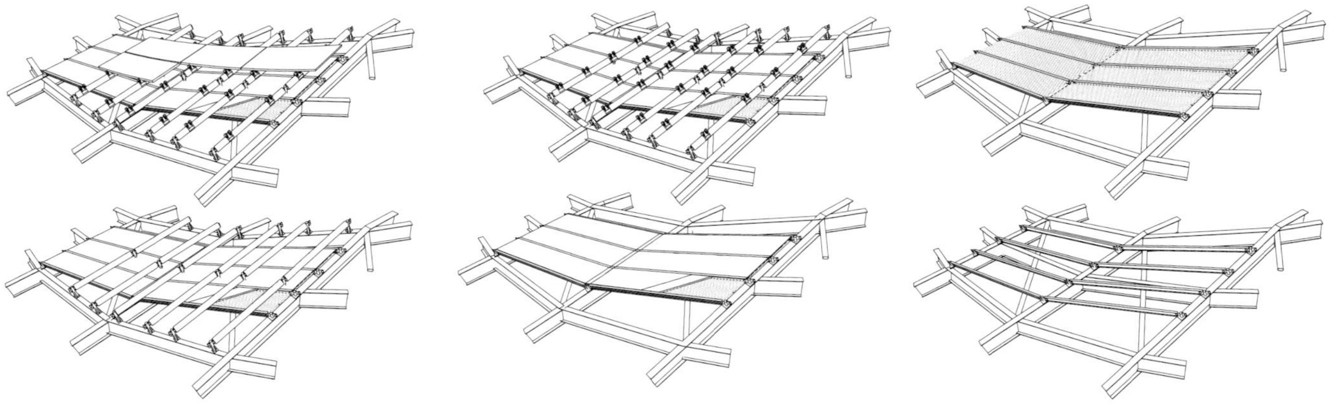

a | Build-up of a facade system showing the different layers required to provide supporting structure, thermal envelope, waterproofing, secondary structural support to facade panels, fixing brackets and facade panels

Modern Construction Series

The series is based around the Modern Construction Handbook (MCH). Topics from the Walls and Roofs chapters from the MCH are expanded in this book to provide more comprehensive information about each envelope technology. Books in the series organise system technologies based on the primary material used, highlighting characteristics of the technologies based on key design aspects. The series currently comprises the Modern Construction Handbook - providing an overview of system design; Modern Construction Envelopes - which provides detail of current facade technologies; and Modern Construction Case Studies - giving examples of the possible integrated applications of construction technologies.

Aims of this book

Modern Construction Envelopes combines the earlier Modern Construction Facades and Modern Construction Roofs into a single book with updated illustrations and third angle projections of envelope systems. This book is a tool for students and designers to apply the existing technologies, described here, in the design of facades for specific projects. It broadens knowledge beyond that provided in the Walls and Roofs chapters of the Modern Construction Handbook, describing the principles of the main facade types used today and illustrating this through typical generic details. The 12 chapters present envelope technologies from the standpoint of the primary material used in their construction, from metal to glass, concrete, masonry, plastics and timber. These technologies provide the information required to design project-specific envelope systems which are a combination of current technologies, based on the individual technology characteristics and performance aspects. The information in this book allows the designer to select and apply the technologies in a way that is flexible; it does not explicitly guide or restrict their application, and which can be applied to projects internationally.

New content for 3rd edition

Previous editions of Modern Construction Envelopes focused on individual components and assemblies and connections between components to form facade and roof assemblies. The earlier editions also showed how each envelope system interfaces with systems in other materials. This 3rd edition shows how to extend the system to include, for example, glazed panels within an opaque system rather than continue to show details where two dissimilar systems interface with one another. In addition, this edition re-focusses the content of drawings and renders specifically as prototypes, typical bays and inventories.

Design to Prototype

This 3rd edition of the MCE begins with a new section, Design to Prototype, which provides a 'user guide' to the systems for facades and roofs described in this book. This new section provides preliminary information to allow the designer to proceed from the hand drawn sketch to drawings for the fabrication of a facade or roof prototype which can be made during the design stages of any building project.

Each chapter begins with an image of a high profile example project where Newtecnic was the facade engineer. Each example illustrates, for each material for both facade and roof, what can be achieved when developed from the basis of the 2D drawings and renders shown in that chapter. Not all thermal breaks are shown; these vary as a result of climate and as a result of project-specific requirements.

Arrangement of information per topic

Each of the 44 envelope types is described on 10 pages, starting with a set of generic 3rd angle projection drawings which can be used as a basis for the design of a prototype. The following 6 pages show 2D details from which are generated views of a typical bay and can be used to inform the design of the prototype. Dimensions on 3rd angle drawings and 2D details refer to the prototype only and are used for general information about orders of magnitude, not specific sizes, as these vary between specific design applications.

3rd angle projection drawings

3rd angle projection of example prototypes can be used as a template for a prototype prepared by a facade contractor prior to competitive tender, or bidding stage. Prototypes can also be prepared after the award of contract as a benchmark visual mock-up. The third angle projection drawings of prototypes are supported by 2D details, which also have some typical dimensions added. These dimensions can be used only for preliminary guidance in the early stages of design development of drawings and models in the preparation of prototypes.

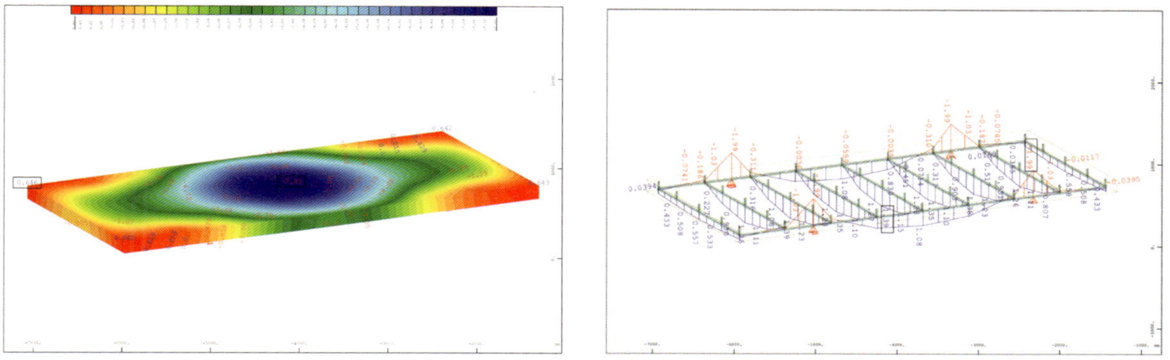

b | Finite element analysis of a typical facade panel

Prototype models
Information for the prototype provided is described as views of a 3D model, which are set out in third angle projection to describe a typical assembly that brings together specific aspects of that envelope system, such as the ability to form a fold or integrate a glazed panel. These third angle protection drawings are dimensioned to give an order of magnitude of component sizes that would be expected for a typical assembly.

Typical bay models
Views of a typical bay are used to show how a given facade system can be combined economically with its closest corresponding roof system, and vice versa. Views of these models are used for comparisons between their weight, volume, solar and thermal performance as set out in this chapter.

Typical bays are combined into renders of inventory models set out on the last two pages of any topic. These inventory models extending over two floors are usually shown with specific components and assemblies within each envelope type, linked to the material specification pages found within the introduction. Renders of the inventory model show how facade or roof assemblies are broken down into their individual components which are linked to material standards, as set out in this chapter.

Inventory models
The inventory pages at the end of each topic are linked to standards and codes for building materials through tags attached to individual components and assemblies. Components in each assembly are linked to this chapter. The inventory pages of each envelope system show how a typical bay is combined both horizontally and vertically to form a continuous assembly.

The role of 2D drawings
The renders of prototypes, typical bays and inventory models are supported by 2D drawings arranged around them in 3D space in order to link the 2D and 3D information together. The supporting structures shown are indicative only, and are typically shown constructed from the same primary material as the corresponding envelope system.

The drawn and rendered information has been developed to reflect the increasing use of BIM (building information modelling) and 3D printing techniques. This book provides information for those engaged with these current techniques. The use of 3rd angle projection and 2D detail drawings is a key part of the 2nd edition of Modern Construction Case Studies, a companion volume in the Modern Construction series. The current technologies set out in this book are explored further in their application as emerging technologies.

Qualifying comments
The building techniques discussed and the examples shown are designed to last for an extended period with a relatively high performance. Consequently, buildings for exhibitions and for temporary use are excluded. For addressing an international readership, references to building regulations and codes of practice have specifically not been included. This book explains the principles of accepted building techniques currently in use. Building codes throughout the world are undergoing increased harmonisation because of greater economic and intellectual globalisation. Building components and assemblies from many different countries are often used in a single building. Since building codes are written to protect users of buildings by providing for their health and safety, good construction practice will always uphold these codes as well as assist their advancement. The components, assemblies and details shown in this book describe many of the building techniques used by the building industry today, but this book does not necessarily endorse or justify their use since techniques in building are in a continual state of change and development. All details shown aim to demonstrate continuity in thermal insulation and waterproofing, together with two defences against rainwater penetration. Where specific items are not clearly present on drawings, these principles should still be followed.

From design concept to system design
Envelope systems can be developed from a sketch concept to a physical prototype through the combination of current technologies using computer-based 3D modelling and structural analysis (MCE_8, a and b). This combination of technologies is formed by comparing the envelope technologies set out in this book based on their inherent characteristics and material properties. The assembly of technologies

c | Digital model of a physical prototype

is then evaluated by computer-based analysis, which is also validated through hand-based calculation. The system design is then progressed through iterations of computer-based modelling and analysis towards a final design, optimised for the project requirements. The system is developed for efficient manufacture, with fabrication data produced directly from the integrated computer model (c), ensuring efficiency and precision. This approach allows the design team to provide information on the system design which can be directly inputted into machinery used by the fabricator or manufacturer (MCE_10, e).

The ability of the design team to provide these files for direct use in fabrication ensures a high level of detail while avoiding the need for generating a new set of detailed design proposals. If a dialogue between designer and fabricator can be established during the design stage, fabricators are then able to provide input on the system design and suggest more economic methods of manufacture. This also prevents the manufacturer being required to solve design issues, at a later stage, that could have been addressed and resolved at a much earlier stage; further time is then saved in the procurement time of the prototype (d) and the manufacture of the system.

There is a certain continuity between traditionally constructed pre-industrial buildings and the current technologies which are used in assemblies to form mass customised systems. Assemblies and their components are once again becoming simplified to generate a system which is economic, feasible and sustainable through local manufacture. There is a return from maximising off-site prefabrication toward local and on-site fabrication, though in a more controlled environment. This reduces the environmental impact of the project through reduced transportation of cumbersome prefabricated components which are not easily tessellated to provide an efficient volume of material, unlike raw materials which can usually be packed densely for transportation.

While the building envelope systems presented in this book are mostly described through rectilinear building forms, they can equally be applied to complex geometries (MCE_10, e). This is achieved through a global approach to the envelope system design described in the following section, aiming to design a single system which can be applied to the entire envelope. Mass customisation techniques are used in the design and manufacture process to achieve the desired complex form through a simplified assembly of current technologies.

An example is described here where Newtecnic worked with the United Arab Emirates' based contractor, Arabian Profiles, on an external envelope of complex geometry (MCE_14, k). The project was a cultural centre in Azerbaijan, which involved the design of a single envelope system for the project incorporating 16,000 cladding panels. These panels were either single-curved or double-curved to provide a continuously curved surface made from a combination of GRC (glass reinforced concrete) and GRP (glass reinforced polymer) types (MCE_10, e). The GRC is incorporated mainly on surfaces that are to be walked upon in the plaza spaces around the building, while the GRP is used for the upper cladding panels which are lighter in weight but have a comparable colour and surface finish.

The panels were designed and manufactured to create the complex geometry intended without the need for flat or facetted panels, while remaining economic in their method of manufacture. This outcome was achieved by translating the digital 3D model describing the geometry into individual panels, outputting the data in a form that could be used in flexible moulding tables to fabricate the panels using a method of mass customisation (MCE_11, f). For each panel, the three-dimensional form was described by a grid of points linked by regular curves that create the single curved forms with non-rectilinear edges. The flexible moulding tables use this data input to create visually complex forms without the need for hand-made components with their associated labour intensity and fabrication tolerances.

The use of a flexible moulding table allows complex panel shapes formed in single-curved geometry to be manufactured quickly and economically to a high standard. Digitally controlled devices are used to adjust the shape of the panel based on data provided by the 3D system model. Information for edge returns of the panels, used to stiffen panels at their edges, were provided by 'developed', or unrolled, shapes offset from the curved shapes in the 3D model. This avoidance of purely hand-crafted techniques ensures that the manufacture of systems for complex buildings can be applied to large-scale building envelopes.

d | Physical prototype built from digital model

Current and emerging technologies

Envelopes for ambitious projects are realised through simplification; not of the overall building design, but of the system concept and design, to ensure the project is feasible. An envelope system is formed from an assembly of existing and emerging technologies which are integrated using first principles of design as well as optimisation for simplicity of manufacture and assembly. The construction of increasingly complex envelopes has been made possible through both the new application of existing technologies, and the significant advance of new technologies and construction techniques. Envelope systems can be created through a combination of existing technologies into an adaptable facade or roof system optimised for manufacture and cost. By considering the manufacture and assembly of the system from an early stage, the design is shaped by the constraints and opportunities accordingly, providing confidence in the potential of the project to succeed. This is reaffirmed through iterations and detailed design based on the initial simplified system concept, providing a full system design which can be easily adopted by the contractor.

Traditional design approach and limitations

Traditional design approaches for facade and roof systems can be very restrictive, as design is generally based on a range of publications such as architectural monographs, codes of practice and standards. These primary sources are used by the various collaborators in the design process to inform the system design, based on the application of building technologies for generic applications of rectilinear buildings rather than the basic principles which govern them. This confines the use of system technologies to their previous applications, which is also commonly strictly divided between walls and roofs.

The standard design method also provides fewer opportunities for the generation of adaptable system designs and optimisation which would lead to a reduction in construction time, material quantities and cost. One example of this restriction is the difficulty in finding a single facade system which is suitable for use throughout a complex geometry project. In a complex geometry application, multiple systems are required which create interfaces within the building facade where these different systems meet. Interfaces can be extremely complex and difficult to design in advance, resulting in laborious on-site fabrication to resolve the interfaces, making it difficult to achieve high performance and high quality of construction across the whole facade.

The existing sources of information focus on the application of technologies to specific materials or building applications, grouping technologies in terms of building type or material type. This approach allows the corresponding current technologies for standard systems to be compared easily. However, it is often not possible to extract the driving principles behind the technology from an examination of the application in buildings alone. Also, the principles and generalised guidelines which made the technology applicable for a certain facade application may not be applicable to another facade system; an issue highlighted by Aksamija in their paper on the design of high performance facades [4]. These sources do not provide guidance to evaluate the suitability of a technology and to validate it for a design application; therefore, the use of these technologies in alternative system types is restricted.

In addition, technical sheets of current technologies from manufacturers, fabricators and suppliers are mainly used as the basis for calculations during the detailed design of a facade system. The early stages of an architectural and structural engineering design are much less specific in the choice of materials and systems for structure and facades. This late introduction of the specific facade technology into the design process creates a large disadvantage in terms of maximising system efficiency. A key fact which has been presented in multiple publications, such as those of Aksamija [4] and Ubbelohde [5], highlights that the building performance and associated construction costs are determined in the early stages of design. Demonstrating that technical design needs to be integrated early in the process, to exploit opportunities of increased facade performance, is difficult to apply due to the accepted methods of standard practice [5].

Architects or designers, for example, use technical publications to make a selection of a component or a system based on parameters of performance, cost and aesthetic. Most current architectural design tasks also make use of standard climate-based approaches, targeting the sustainability of the system by minimising the calculated environmental impact factors. This neglects the great advance in digital design and technological tools for the modern architectural industry,

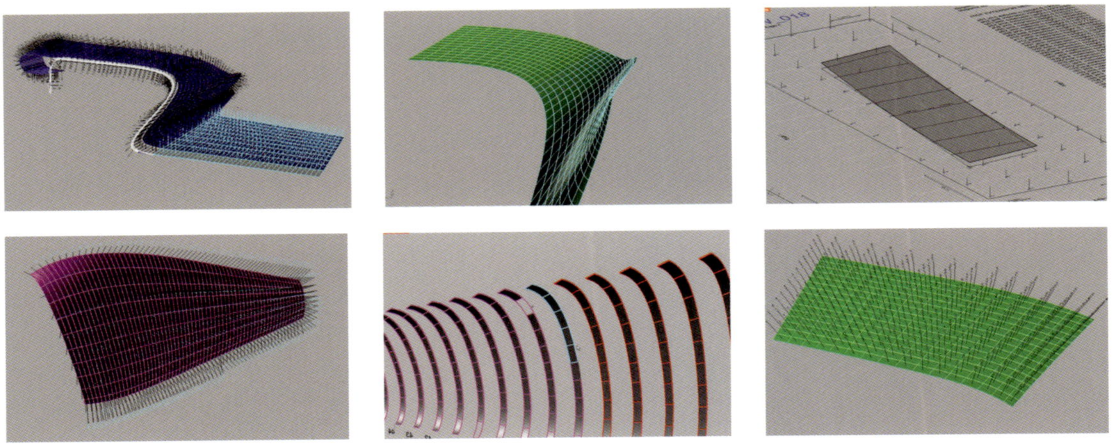

e | Panelisation for manufacture, using 3-D parametric software.

which provides new opportunities for both structural engineers and architectural engineers to optimise systems on a project-specific basis. These technological advancements are initiating an expansion of standard design methodologies, a need which was identified in Mendler's guidebook to sustainable design published back in 2000 [3]. This expansion in standard design approaches is also accompanied by the introduction of new ways of thinking in facade design, which is fuelling innovation in new system designs.

The traditional method generates a "loose fit" design which is typically inefficient and requires an unnecessarily complex assembly. This approach also ignores the need to understand the combined behaviour of components and assemblies in order to allow the system performance to be optimised in relation to the project requirements. Therefore, to achieve high performance of the building, the traditional design delivery of envelope design must be reconsidered. As Ubbelohde explains in a case study paper for high performance facades, the creation of these systems requires the need for "breaking [the] mould of conventional consulting" [5]. This is particularly important in the design of complex building envelopes, which requires an in-depth understanding of each component and how they can be combined successfully to meet the system requirements of a given facade or complete envelope.

It is clear that the traditional design method is inefficient for facade system design, especially in relation to complex geometry buildings. A new approach to design is required to optimise these systems. "We should not expect to produce fundamentally new buildings using the same traditional design process" [3].

Newtecnic's methodology

This following text describes a methodology for complex facade design using a product development style approach. The new approach removes the geometric and technological restrictions of design generated by the traditional method, centred around a change in mindset of designers and engineers to approach projects in a "product development" style. In this method existing products and technologies are used as inspiration, but they do not limit the development of the new products. The product is developed from an initial creative concept; this is commonly based on the concept geometry proposed by the architects, which can be realised through the integrated assembly of existing or emerging technologies. The design is then tested, analysed and refined in an iterative development process to create the final product. This process involves creativity backed by basic engineering principles, which is validated through in-depth research and physical testing. The building of a body of knowledge is the starting point; this involves research into the project, the constraints and opportunities of the envelope form, existing and emerging technologies which could be used and the potential design constraints. This body of knowledge is then used to inform the specific system design approach, which allows the project to be freed from the restrictions and uncertainties of standard design methods. The system design is formed from a combination of existing and emerging technologies, chosen based on the first principles which govern them. Rigorous modelling, analysis and testing is used to ensure the required system performance and quality is met. The system is analysed at both a global and local scale, performed in parallel to take into account the interrelationship of the two scales, allowing the global behaviour to inform the local analysis and vice-versa. The process is then repeated to optimise the design and respond to changes in the project through a series of iterations.

Not conforming to traditional design methods based on design standards and codes provides greater freedom during the design process. This approach enables innovative ways of using materials and technologies to be explored as part of the design process, creating opportunities for increased efficiency in the performance and reduced costs of the system design. This is particularly critical for ambitious building projects, as it allows them to be achieved in a reasonable timescale and always within budget.

The background research of the engineering design of a facade aims to develop a full understanding of the project and its constraints and opportunities. Research is conducted into emerging materials and technologies which could be used to improve the efficiency and performance of the facade design. In addition, the research provides examples of existing technologies and assemblies to demonstrate the feasibility of each assembly independently. This early stage research involves a large investment of time and money to ensure that both the project, and its associated performance requirements, are well

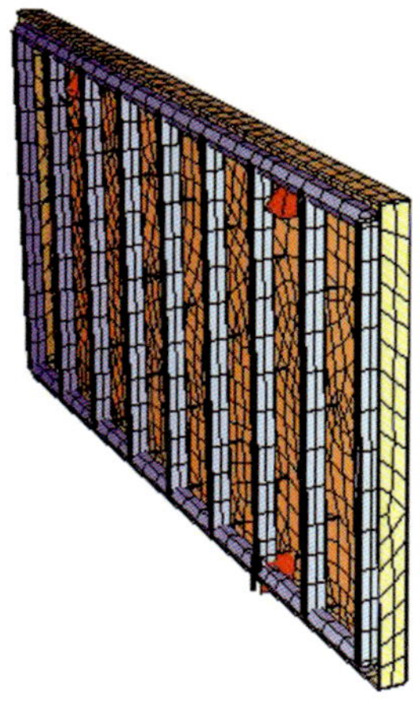

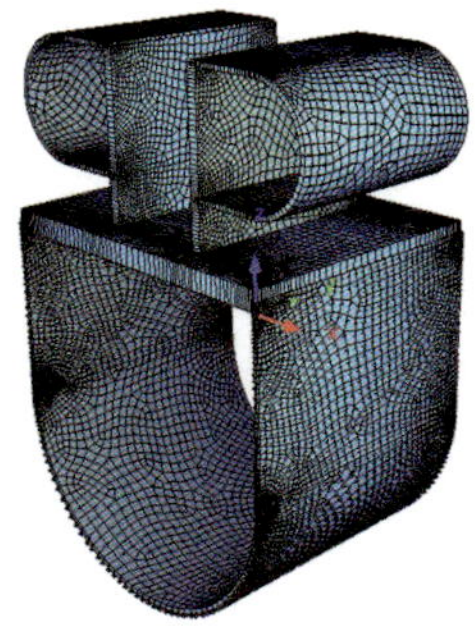

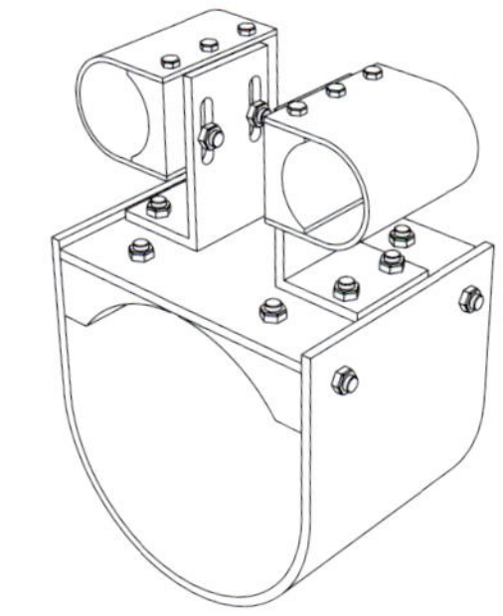

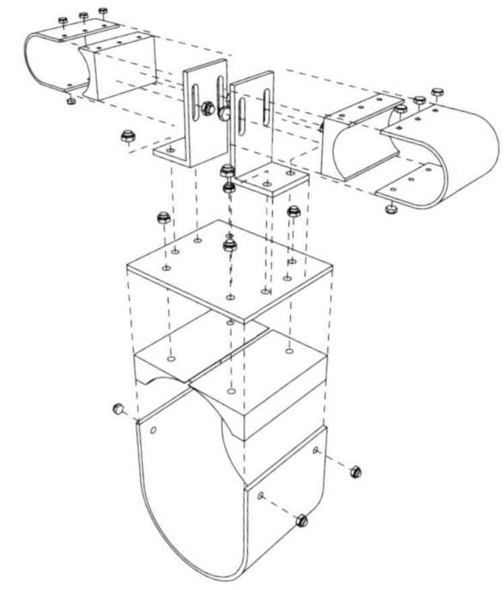

f | Structural analysis of the facade panel and support bracket. Analysis of deflection and stresses at an early stage of the design allows for the overall depth of a complex facade build-up to be established before the design of the building has progressed to the detailed stage.

understood. This allows specific design options to be considered in relation to one another, rather than making it difficult to draw comparisons between them.

Basic principles underpin the design of high-performance building systems, regardless of geometric complexity or external constraints; the interdependent governing aspects are considered in parallel and balanced to achieve an optimised design. Simple rectilinear systems can reach required efficiency through the design of components based on standard guidance and previous application. This approach is not sufficient for complex geometry designs as the behaviour of the system at different orientations changes, given that most systems are commonly designed to work either vertically or horizontally. Therefore, for a system to have a high performance across the complete envelope, the system must be designed from first principles. This approach is supported by computer-based analysis and physical testing. Background research of new and existing technologies is undertaken in partnership with leading universities, providing a wealth of knowledge surrounding the basic principles and behaviour of the technology. For emerging technologies, experimental analysis and physical testing is conducted to provide understanding of its inherent properties and behaviour, providing a firm basis for successfully integrating the technology within system designs.

The basic design principles are governed by the physics of envelope design, which are well understood in the academic setting but often overlooked within the industry. Building physics provides a robust platform from which to develop new design concepts as well to interpret published data from both manufacturers and technical papers [10] [11]. When comparing these publications, it is then possible to extrapolate general principles informing the existing technologies, thus enabling the building engineer to design using a given technology, without relying on specific products or project specific solutions. Basic principles of building systems commonly include predictable and direct load transfer, to allow for the accurate prediction of load path in the structural analysis. There is also the principle of heat transfer which must be considered to avoid thermal bridging which can lead to interstitial condensation. The "rain screen" principle is seeing increased use in facade system technologies; this principle uses an external layer of panels to protect the waterproof membrane layer and insulation underneath from the external elements such as UV radiation, wind pressures and direct rain and debris [6]. Core design principles such as sustainability, buildability and durability have fewer scientific definitions, but often have a large impact on the technical design of the system. In the proposed approach, these aspects are closely linked and focus on producing a system which exploits the various properties of the materials and technologies used. Simplicity is the key aspect in terms of promoting sustainability, achieved through a final product which does not require specialist equipment, allowing local manufacturers and contractors to successfully compete for the project. The simplified design, when combined with durable high-quality materials, results in a more robust system that can last over 60 years before components need to be replaced, compared to the regular design life of 25 years for a facade system.

The methodology is centered around an understanding of the underlying principles of each current technology informed by technical data sheets, engineering knowledge and experience gained from previous projects. Once the system requirements and basic principles are established, the initial concept is developed through design and analysis, taking into consideration published data and underlying principles to inform this iterative process. The design is then compared again with the precedents proposed and design requirements defined initially. This aim is achieved through the creation of 3D assembly sketches and the calculation of key performance parameters, such as structural and thermal performance. The behaviour and interaction of the individual components is also interrogated, focussing on the relationship between economic use of material, constraints of fabrication and installation. This method allows the priorities of the design development to be determined, informing the necessary analysis, prototyping and testing. Full-scale mock ups and physical testing are then undertaken to investigate performance of the system, which then informs any required design changes.

A project-specific approach allows for greater adaptability of the facade system design [6]. The different criteria of the design can be set as 'parameters' which can be linked to the process of modelling the building forms digitally in a parametric design software. Working parametrically allows the designer to establish what is 'fixed' and what will be 'variable' in the design development; this approach allows a digital design method

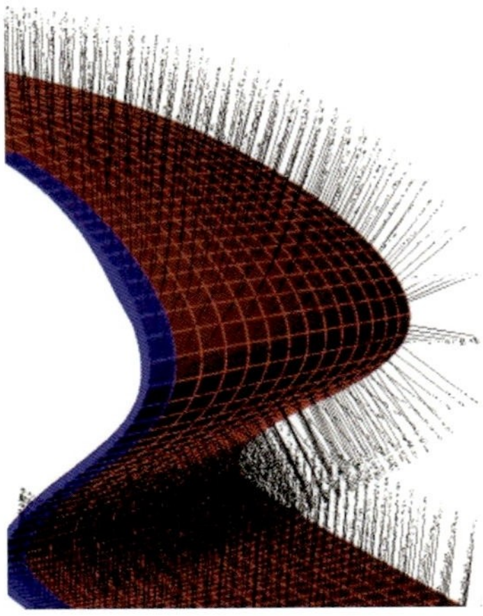

g | Panelisation of geometry for manufacture.

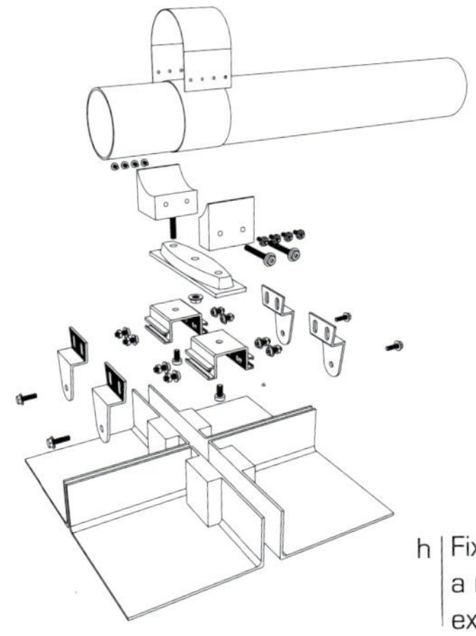

h | Fixing detail for panels using a minimum set of cast and extruded components.

to evolve through an engineering-based method of iteration, rather than starting the digital design model again each time a new option is explored. Parametric modelling provides an interrelationship between parts of the building as well as the urban space surrounding it; different assemblies, component sizings and arrangements can be explored through parametric modelling, enabling the optimisation of complex system arrangements. Through this method many more iterations can be investigated and explored than are possible by hand using 2D CAD, or even by conventional modelling. The use of parametric design further increases this adaptability through parametric 3D modelling and global analysis of the system. This flexible approach is performed alongside detailed analysis of each component and assembly. Using this method, changes to the design can be quickly implemented in the modelling and analysis process, enabling the impact of the design change on the complete design to be determined. Changes in design specification, or design constraints, can then be assessed and accommodated without causing delays to the project program.

Parametric modelling is applied to all the system components in parallel, taking into consideration their interaction through connections and load transfer. The process is then iterated to refine the design of each component. As a result, design decisions can be made quickly, based on the results from successive design developments, bringing the benefit of increasing the efficiency of both the design process and the resulting facade system.

For complex geometry envelopes, the division of the external skin to form a panelised envelope system is a process with great potential for optimising both the panel shapes and sizes. These factors must be balanced to create a feasible map of panels which is optimised for manufacture and ease of installation. Panelisation is the automated computer process of dividing an external envelope surface into a series of rectilinear shapes, which are commonly quadrilaterals. This is an important part of complex geometry facade design, as it allows the facade system surface to be divided into panels with a limited number of different geometries. The procedure involves the combination of scripts and advanced programming skills to create a parametric panel layout model. The model is easily adjustable and therefore allows multiple panel arrangements to be investigated quickly by adjusting the parametric inputs of the model. In addition, this makes the design able to quickly adapt to change, avoiding long project delays and providing a robust design process.

On completion of the project, the system and design processes are then analysed to determine the best practices of the processes used, and the knowledge gained. This is documented in the form of published and internal documents, allowing the system and design process to be implemented in future projects. Published work documenting this design process undertaken by Newtecnic includes design guidance books [6] [7], design research papers [8] and research testing papers [9]. This stage also provides an opportunity to identify gaps in current knowledge where there is a need for further analysis, research or testing.

The design methodology involves a more holistic approach, which, unlike methods described in existing technical publications, focuses on creating solutions based on both a greater knowledge base and engineering first principles. This approach provides facade system solutions which meet the expectations and needs of the client, whilst increasing the value of the deliverable product. The increasing integration of components, described in *Modern Construction Case Studies*, is achieved through high level early stage design, built on a large, researched knowledge base and thorough performance analysis to inform design decisions. The method allows fewer components to be used in the facade assembly and reduces the complexity of assembly tasks, reducing project cost and environmental impact. A robust design can be successfully provided through the method described, using basic principles and parametric design. The resulting system has the ability to quickly adapt to changes at any point during the project, without causing significant project delays and additional costs. The proposed 'product development' style method generates innovation and aids the development of a facade system design through a creative and unrestrictive approach to design. There are significant advantages of both the design approach and the resulting outputs which include economic benefits, enhanced sustainability and improved build quality. These are highlighted in the case studies of novel facade systems, in which a single simplified system was developed for use throughout the projects outlined above.

j | Triangulated roof structure to which envelope system is fixed.

Assembly of current technologies

Through the method described above, facade systems are created and developed based on an assembly of current technologies, which evolve as new emerging technologies. Current tecnologies are also developed and integrated further when they are needed to become more efficient. The components assembled to create the system are refined and simplified; this allows redundant components to be eliminated and increases the feasibility and sustainability of the system. The simplified component geometries can be produced without the need for expensive specialised fabrication equipment, which enables local manufacturers to successfully bid for the work, keeping the fabrication local and reducing the transport emissions associated with the project.

Striving for increased simplicity in system assembly also provides a 'visual integrity' to the construction, along with advantages of structural and material efficiency. The approach allows an integration to be created between facade systems informed by the characteristics and physical properties specific to the technologies used, developing a more efficient design with reduced system weight and assembly time. Integrated cassette panels provide an example of the integration of different facade layer technologies, incorporating thermal and waterproofing performance. Demonstrating the ability to reduce the number of different panels required in a system decreases the installation time on site and shifts the fabrication to specific construction labs, where a more controlled environment can provide high quality fabrication.

Interfaces between the facade system and the primary structure as well as between different facade assemblies are one of the most critical parts of the design. Interfaces are created where the different systems adopted for various areas of the envelope meet. These interfaces are highly complex to design and to optimise for high performance. For this reason, the associated detailing of the interface is often neglected in the design process; a solution is then developed during the construction stage. This inevitably results in either a reduction in the envelope system's performance or an inefficient design, as well as increased complex onsite fabrication. Using a single integrated system across the entire envelope eliminates interfaces between different building systems, which are often an expensive and slow-to-construct part of a continuous high-performance envelope. For a single system to be applicable across the entire envelope it must be highly adjustable and be able to perform highly at any orientation. This is achieved through an assembly of technologies created using the design methodology previously described, involving a highly integrated workflow with an investment in initial research and testing. An understanding of the basic principles of the technologies developed enables them to be applied in a way which is adaptable to different orientations and geometries, allowing them to perform effectively across the full facade. The systems are formed mostly from an assembly of current technology components, where the novelty of the system is in the combination of these technologies. Current technologies are used where appropriate to reduce cost and development time [12], as there is no benefit in creating new technologies where they are not required. Emerging technologies are then developed where it is deemed advantageous to improve the system's efficiency.

The primary advantages of a single system that uses a combination of current technologies are as follows:

Elimination of interfaces – as previously explained there are many issues related to the inclusion of system interfaces in building envelopes, including complex fabrication on site and reduced performance of envelope. The elimination of interfaces is beneficial in providing a higher performance envelope with minimal on-site fabrication.

Simplification of the assembly – Through design optimisation the number of different components used is reduced. This process also increases certainty in the design, through the use of global and local analysis alongside physical mock-ups and testing.

Integration of components – Integrated panels demonstrate the ability to reduce the number of different panels required in a system. This reduces the installation time on site and shifts the fabrication to specific construction labs, where a more controlled environment can provide high quality fabrication.

Reduced assembly time – The use of a single system across the envelope also simplifies whole fabrication and assembly processes through less complex coordination, providing less opportunity for error. Using a reduced number of components and increased simplicity of system design, allows the assembly time and associated cost to be reduced.

Single visual expression – The use of a single system and panel type enables a single continuous visual expression to be achieved. This is

k | Facades during completion.

particularly important in complex forms when a constant visual flow of the envelope is required.

Reduction of costs – As a further benefit of the advantages mentioned above, such as the reduction in components and assembly time, the cost of the project is reduced. The simplicity and robustness of the system design resulting from the described design methodology reduces the cost associated with risk.

Emerging technologies

During the past 20 years, significant progress has been made in architectural engineering and building technology. Facade design has played a primary role in this regard, with increased research in this area which has led to great developments in emerging technologies. An example of this is composite materials which are being increasingly applied to facade systems. Composites include glass-fibre reinforced concrete (GRC), glass-fibre reinforced polymer (GRP), fibre reinforced polymer (FRP), and ultra high-performance concrete (UHPC) [1]. Emerging technologies such as these allow building technology to move beyond the limitations of the physical properties of conventional construction materials such as concrete, timber, steel and aluminium alloys to create increasingly efficient systems.

System designs can begin to take advantage of the possibilities created by these emerging material technologies, producing building envelopes with double-curved geometry using emerging composite materials to form both facade panels and solar shading devices. These emerging technologies, however, give rise to new engineering design challenges within the design process. To enable the technologies both current and emerging to be successfully combined to form a facade system, the product development style approach previously described must be employed. In this approach there is a significant investment in the initial design stages, undertaking research testing and analysis to understand the underlying principles of the technology. The increased understanding and knowledge gained then informs the system design, enabling the creation of systems which are highly integrated and optimised for manufacture, weight and cost.

The development of emerging technologies is not limited to materials development; it can be seen in the facade system as a whole, including the interface between its different parts and the primary structure [2]. Several types of connection have been developed with the aim of meeting the requirements of both the architectural engineering and the structural engineering aspects of complex geometry systems. Flexibility in design, adaptability, simplicity and robustness of the fixing system are some of the key criteria considered during design of connection systems in facades [1]. The selection of the most suitable connection system is based on the structural requirements combined with on-site adjustability and cost. In the case of complex geometry envelopes where a single panelised system is employed, a single bracket design can be developed with the required adjustability to accommodate all the panel geometries and orientations incorporated in the envelope system. By minimising the number of connection types used, the assembly complexity and associated time and cost are reduced.

Example approaches are presented in the following paragraphs to provide inspiration for routes through which the current technologies presented in this book can be implemented and developed alongside emerging technologies, using project-specific approaches and methods of mass customisation to inform building design projects.

The first approach is a 'masterclass' in current technology, where an existing technology is utilised with a focus on optimising the current technology, to create a higher performing system. An example of this can be found in clamped glazing used in large-scale glazed facades, a system which provides a fully glazed facade across several floors vertically, with an intermediate supporting structure which is barely visible. Such designs push the boundaries of a current technology to achieve increased performance in order to allow the visual architectural intent of transparency to be bolder in a large-scale glazed wall. The increase in technical performance is achieved through structural efficiency, waterproofing and thermal insulation. This approach of maximising the performance of an existing technology is a relatively low-risk method of achieving increased performance with an integrated system which enables the desired architectural intent to be created.

k | Facades during completion.

Another approach is a building design which points to a new technology but is constructed using existing technologies; projects which provide inspiration and direction for the development of future new technologies. The designs are bold and push the boundaries of the existing facade technologies, creating an opening for the development of an emerging technology. An example of this is in complex curved glazed roofs consisting of small geometrically distinct panels. This points to the development of a digital panelisation tool to enable a glazed envelope formed as a complex geometry to be fabricated as separate panels. These digital tools allow the system to form a continuous watertight envelope, formed of prefabricated glazed panels which are fixed together with varying joint widths.

The third approach is one in which the development of an emerging technology is used to meet the requirements of a particular project. As previously mentioned, this is typically how the development of new envelope technologies is achieved. This approach could be a specific material, a construction method or a design tool that adds value to the envelope system. The approach generally presents a higher risk of increased cost over a current technology, as a larger initial investment is required for the development of the new technology for the design, analysis and testing required to demonstrate the feasibility and performance of a new technology. An example of this approach type can be seen in full height glazing combined with a GRP-clad structural frame, where the structure is fully integrated with the glazed wall system, significantly reducing the depth of the combined envelope and structure. This is achieved by coordinating the design of the envelope and the structure to ensure both are high performing and feasible when combined to form the final system. In this case the architectural intent could be realised in a way that also provided great structural and material efficiency.

The three different approaches show that development can either occur incrementally or in large steps, each method of development carries a certain level of risk, with the final case of developing and implementing emerging technologies carrying the greatest risk. All the approaches, however, follow the same thinking communicated in this book, of forming building systems from a collection of current technologies which are evaluated on their characteristics and performance capabilities.

References

[1] Paparo I., Overend M., Micoli F. (2015) 'Performance-based material selection and design for free form building envelopes', *Advanced Building Skins*

[2] Watts A. (2013) 'Modern Construction Handbook', Ambra | V, 3rd Edition, Vienna

[3] Narsalay R., Sen A., Light D. (2017) 'Feb 17: How manufacturers can get faster, more flexible, cheaper', *Harvard Business Review*

[4] Aksammija, A. (2016) 'Design methods for sustainable, high-performance building facades', *Advances in Building Energy Research*, 10(2), pp. 240-262.

[5] Ubbelohde, S et al., (2004) 'A case study in integrated design: Modelling for high-performance facades', *ASHRAE*

[6] Watts A. (2013) 'Modern Construction Handbook', Ambra | V, 3rd Edition, Vienna

[7] Watts A. (2016) 'Modern Construction Case Studies', Birkhäuser, Switzerland

[8] Paparo I., Overend M., Micoli F. (2015) 'Performance-based material selection and design for free form building envelopes', *Advanced Building Skins*

[9] Dona M., Overend M., (2017) 'Fatigue performance of a connection for GRC cladding panels', *Proceedings of the 8th International Conference on Advanced Composites in Construction* (ACIC 2017)

[10] Standards and Codes: BS EN ISO 10456:2007, BS EN ISO 6946:2007, BS EN ISO 10211:2007, BS EN ISO 14683:2007, BS EN ISO 12631:2012, Eurocode 1, Eurocode 2, Eurocode 3, Eurocode 4, Eurocode 8, Eurocode 9.

[11] MCRMA, (1995) 'Technical paper No 9: Composite roof and Wall cladding panel design guide', *Metal Cladding and Roofing Manufacturers Association*

[12] Mendler S., Odell W. (2000) 'The HOK Guidebook to Sustainable Design', John Wiley & Sons., New York

A primary objective of the Modern Construction Envelopes is to provide a comparative analysis of different facade technologies used for building envelopes, in relation to the climate and environment where they have been implemented on each project. The 44 systems illustrated in the book have been compared in the tables and graphs below in terms of the environmental and structural performance of their building envelope.

	Typical bay	Facade system*	Roof system
01	Typical Bay 01	Standing seam metal	Standing seam roof
02	Typical Bay 02	Profiled metal	Profiled roof
03	Typical Bay 03	Composite metal	Composite roof (with rainscreen)
04	Typical Bay 04	Rainscreen	Rainscreen roof
05	Typical Bay 05	Metal mesh	Canopy roof
06	Typical Bay 06	Louvres	Membrane roof
07	Typical Bay 07	Stick glazing	Louvres on roof, exposed membrane
08	Typical Bay 08	Unitised glazing	Silicone sealed roof
09	Typical Bay 09	Clamped glazing	Timber flat roof
10	Typical Bay 10	Bolted wall	Bonded rooflight
11	Typical Bay 11	Glass block	Concealled membrane roof
12	Typical Bay 12	Steel glazing window wall	Silicone sealed roof lights
13	Typical Bay 13	Aluminium glazing winodw wall	Silicone sealed rooflight
14	Typical Bay 14	Timber window wall glazing	Concealed timber roof
15	Typical Bay 15	In situ wall	Concealed membrane roof
16	Typical Bay 16	Storey height precast	Concealed membrane roof
17	Typical Bay 17	GRC cladding wall	Concealed membrane roof
18	Typical Bay 18	Masonry cavity brick	Tiled roof
19	Typical Bay 19	Masonry block wall	Tiled roof
20	Typical Bay 20	Stone cladding wall	Concealed membrane roof
21	Typical Bay 21	Terracotta wall	Concealed membrane roof
22	Typical Bay 22	Timber wall	Tiled roof
23	Typical Bay 23	Cladding panel	Tiled roof
24	Typical Bay 24	Sealed GRP wall	Composite panel roof
25	Typical Bay 25	GRP rainscreen wall	GRP rooflight

*refer to pages describing the structural design for each project for illustrations of the facade build-up

Weight of facade vs. facade zone

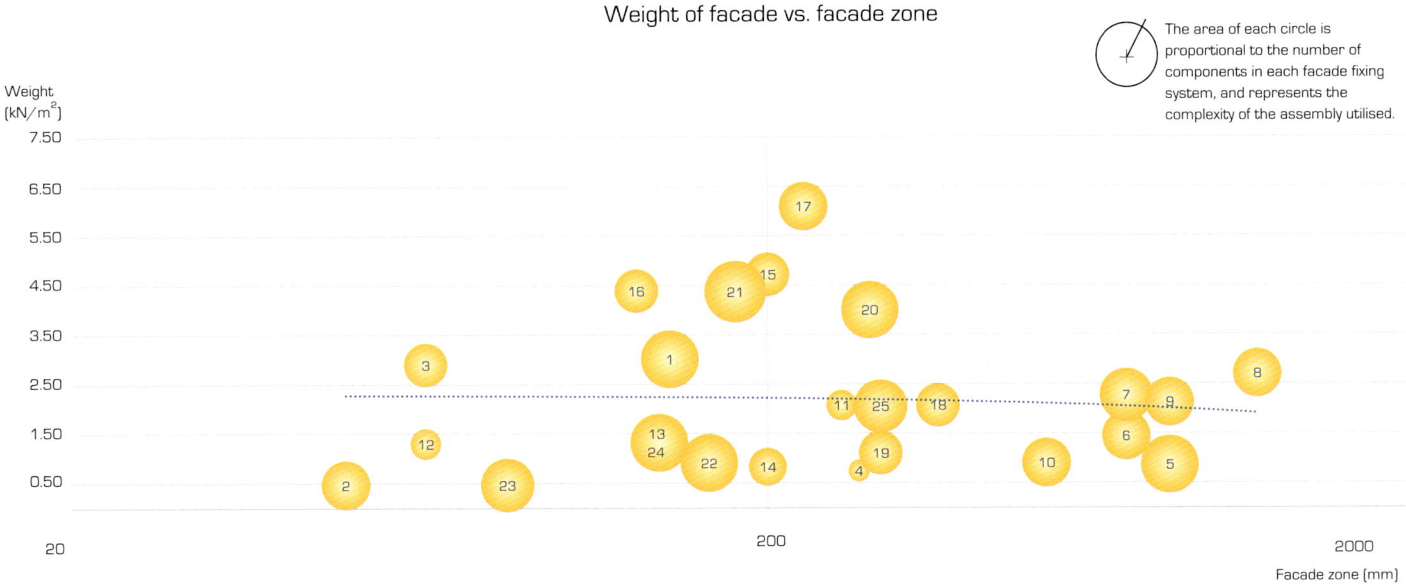

The numerical results used for the comparison have been obtained from the analysis performed on each system. Each facade technology, designed to suit all project conditions, has been assessed on a representative typical bay in order to compare facade systems. The numerical values provided in this book are for comparison between systems, not for their specific values, which are approximate only.

Facade zone	Secondary structure type	Weight of secondary structure (kN/m²)	Total weight of facade, including secondary structure (kN/m²)	
145mm	Steel brace frame	2.43	3	01
50mm	Steel portal frame	0.21	0.46	02
65mm	Steel gridshell	0.3	2.89	03
270mm	Steel space frame	0.55	0.74	04
750mm	Steel brace frame	0.27	0.86	05
650mm	Concrete brace frame	0.83	1.44	06
650mm	Concrete braced frame	1.22	2.26	07
1000mm	Steel brace frame	1.65	2.7	08
750mm	Steel braced frame	1.24	2.12	09
500mm	Steel gridshell	0	0.9	10
255mm	Concrete brace frame	0.11	2.06	11
65mm	Cconcrete braced frame	1.08	1.29	12
140mm	Concrete brace frame	0.86	1.37	13
200mm	Timber brace frame	0	0.82	14
200mm	Concrete shell	1.83	4.71	15
130mm	Concrete space frame	3.87	4.38	16
225mm	Concrete shell	4.25	6.09	17
350mm	Timber framed structure	0	2.06	18
290mm	Timber framed structure	0.08	1.1	19
280mm	Concrete brace frame	0.51	3.99	20
180mm	Concrete brace frame	3.71	4.36	21
165mm	Timber framed roof	0	0.9	22
85mm	Timber Frame	0.18	0.46	23
140mm	Steel gridshell	0.15	1.31	24
290mm	Steel grid shell	1.25	2.04	25

Weight of facade vs. weight of secondary structure

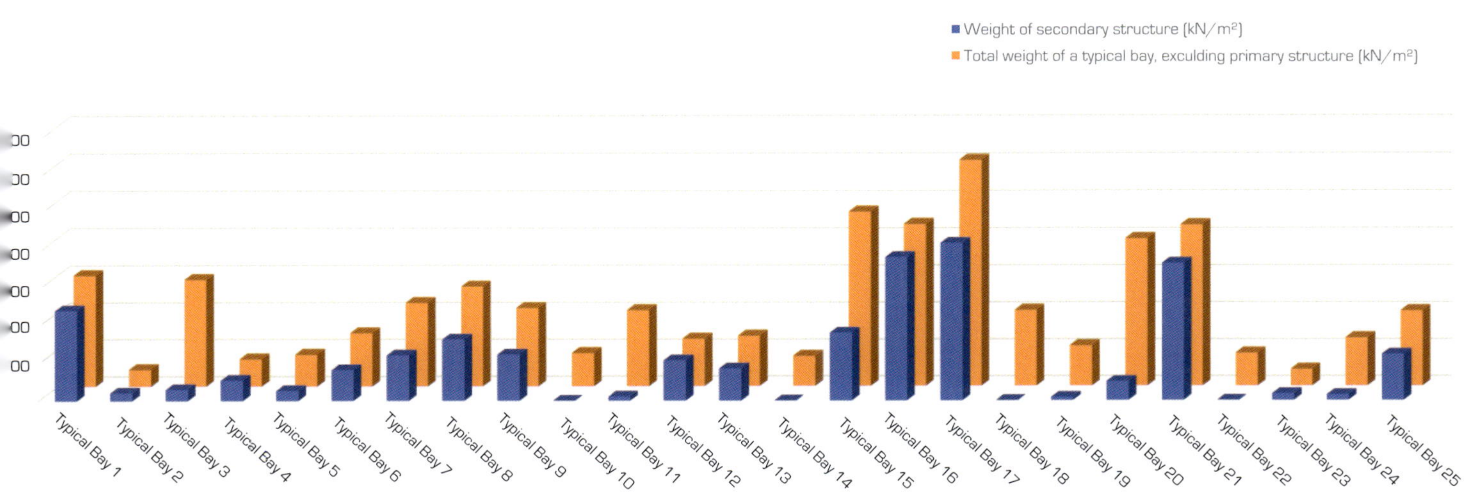

	Typical bay	Facade system*	Roof system
26	Typical Bay 26	Standing seam roof	Standing seam wall
27	Typical Bay 27	Profiled metal	Profiled roof
28	Typical Bay 28	Sealed GRP wall	Composite panel roof
29	Typical Bay 29	Rainscreen	Rainscreen roof
30	Typical Bay 30	Metal mesh	Canopy roof
31	Typical Bay 31	Steel glazing window wall	Silicone sealed roof lights
32	Typical Bay 32	Glass small rooflight	Stick glazing
33	Typical Bay 33	Glass large rooflight	Clamped glazing
34	Typical Bay 34	Bolted wall	Bonded rooflight
35	Typical Bay 35	In-situ wall	Concealed membrane roof
36	Typical Bay 36	Concrete exposed membrane	Concrete exposed membrane
37	Typical Bay 37	Planted roof	Storey-height precast
38	Typical Bay 38	Timber window wall glazing	Concealed timber roof
39	Typical Bay 39	Cladding panel	Tiled roof
40	Typical Bay 40	Timber pitched slate	Stone cladding
41	Typical Bay 41	GRP rainscreen wall	GRP rooflight
42	Typical Bay 42	GRP rainscreen wall	GRP rainscreen wall
43	Typical Bay 43	Polymer rainscreen wall	ETFE roof
44	Typical Bay 44	N/A	Single membrane roof

Weight roof vs. facade zone

Facade zone	Secondary structure type	Weight of secondary structure (kN/m²)	Total weight of facade, including secondary structure (kN/m²)	
140mm	Steel brace frame	0.31	0.60	26
190mm	Steel portal frame	0.04	0.52	27
140mm	Steel gridshell	0.28	0.73	28
180mm	Steel space frame	2.48	2.57	29
470mm	Steel brace frame	0.12	0.54	30
65mm	Concrete braced frame	0.23	0.31	31
220mm	Not shown	0.04	0.12	32
620mm	Concrete slab	0.27	0.38	33
800mm	Steel gridshell	0.08	0.19	34
200mm	Concrete shell	2.02	2.87	35
200mm	In-situ concrete	2.77	3.06	36
650mm	Timber framed roof	1.06	1.67	37
275mm	Timber brace frame	0.31	0.83	38
160mm	Timber Frame	0.11	0.80	39
120mm	Concrete wall	0.14	1.18	40
150mm	Steel grid shell	1.16	1.52	41
35mm	GRP truss	0.12	0.19	42
400-1400 mm	Steel braced frame	4.41	4.78	43
30mm	Steel cable	0.74	0.90	44

Weight of roof vs. weight of secondary structure

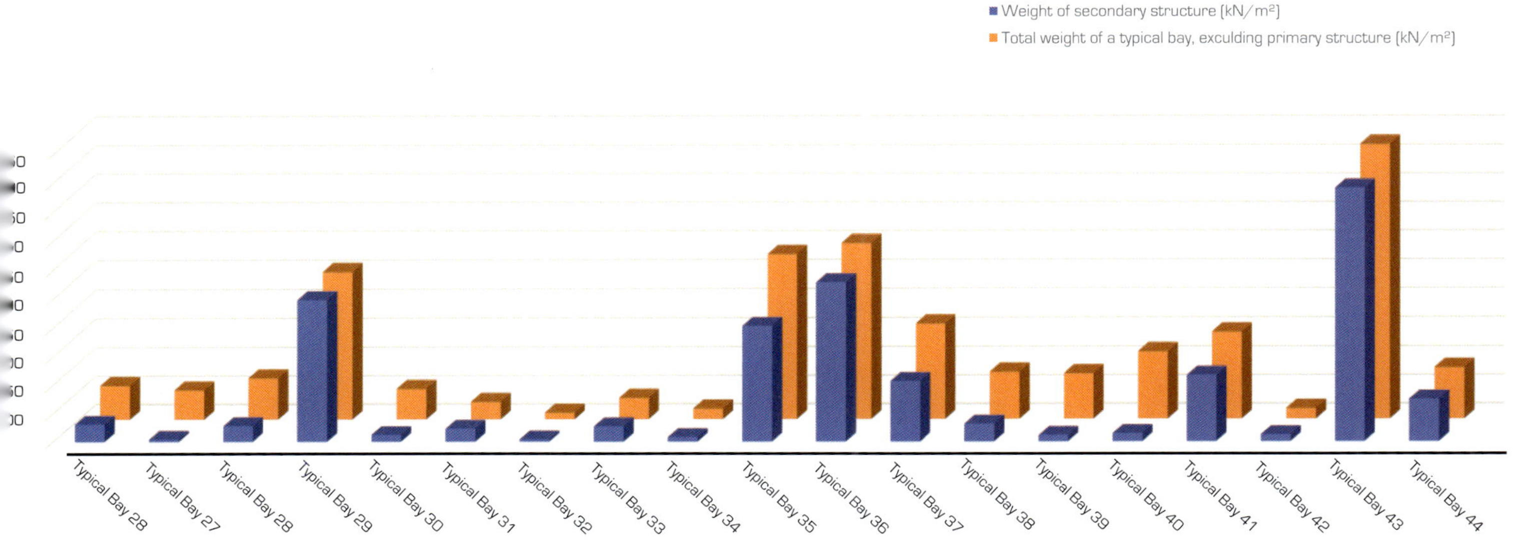

	Facade System name	U-value of system envelope (W/m²K)	Linear thermal bridging effect for system typical detail (W/mK)	Annual cumulative radiation with shading system (kWh/m²)	Reduction in annual solar gain by shading system (%)
01	Typical Bay 01	0.41	0.07	470	12
02	Typical Bay 02	0.51	0.04	315	2
03	Typical Bay 03	0.91	0.02	490	5
04	Typical Bay 04	0.32	0.02	610	4
05	Typical Bay 05	0.71	0.05	610	23
06	Typical Bay 06	1.20	0.04	405	21
07	Typical Bay 07	1.20	0.04	535	13
08	Typical Bay 08	1.20	0.04	310	6
09	Typical Bay 09	1.60	0.04	520	24
10	Typical Bay 10	1.40	0.04	645	5
11	Typical Bay 11	1.50	0.04	645	10
12	Typical Bay 12	1.50	0.04	790	11
13	Typical Bay 13	1.02	0.07	490	9
14	Typical Bay 14	1.60	0.04	395	11
15	Typical Bay 15	0.89	0.18	580	16
16	Typical Bay 16	0.90	0.29	580	15
17	Typical Bay 17	0.61	0.07	530	15
18	Typical Bay 18	0.45	0.07	670	8
19	Typical Bay 19	2.21	0.33	480	16
20	Typical Bay 20	0.70	0.03	655	12
21	Typical Bay 21	0.36	0.06	675	7
22	Typical Bay 22	0.45	0.10	555	12
23	Typical Bay 23	0.77	0.05	565	2
24	Typical Bay 24	1.63	0.35	550	9
25	Typical Bay 25	0.31	0.03	390	1

Effectiveness of external solar shading systems

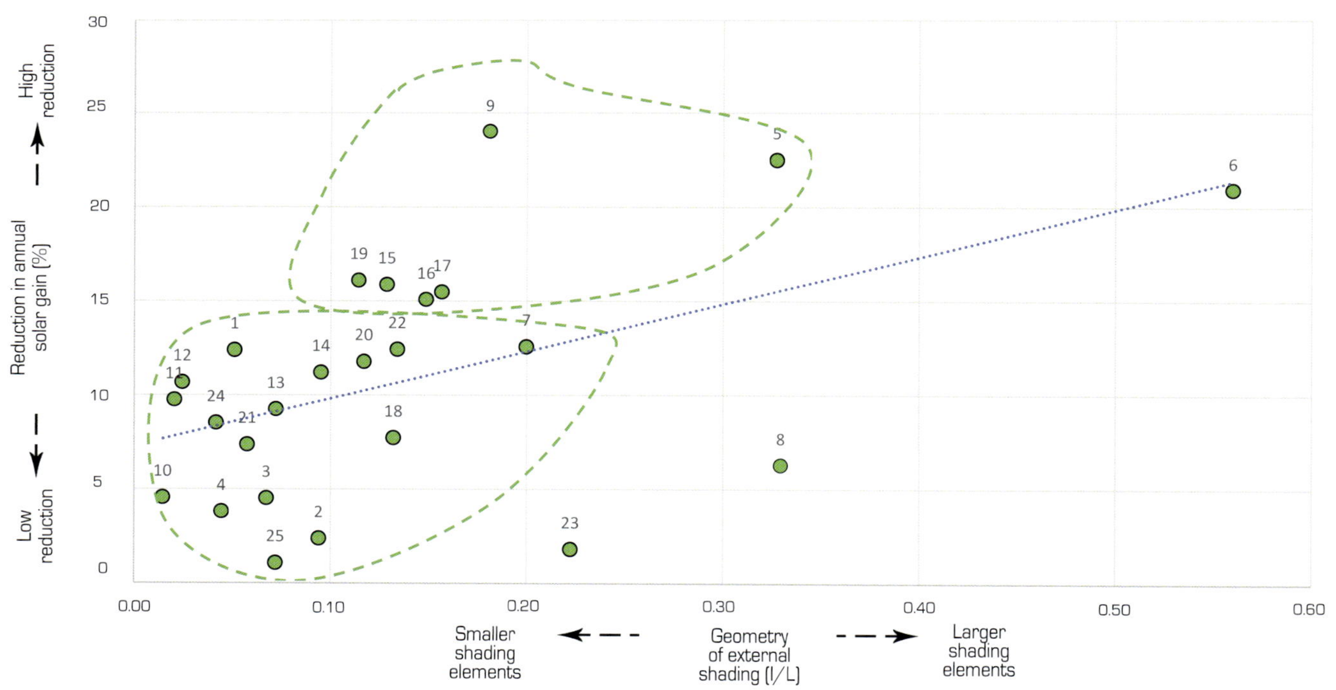

The linear relationship between the geometry of external shading and the reduction in annual solar gain illustrates the expected effectiveness of external shading systems and allows to identify areas on the graph that represent shading systems with high effectiveness.

	Roof system name	U-value of system envelope (W/m²K)	Linear thermal bridging effect for system typical detail (W/mK)	Insulation thickness (mm)	Annual cumulative radiation (kWh/m²)
26	Typical Bay 26	0.23	0.03	170	1015
27	Typical Bay 27	0.52	0.02	80	620
28	Typical Bay 28	1.00	0.08	40	1125
29	Typical Bay 29	0.54	0.05	80	750
30	Typical Bay 30	3.00	0.08	10	490
31	Typical Bay 31	1.60	0.04	16	1145
32	Typical Bay 32	1.40	0.04	16	1070
33	Typical Bay 33	1.50	0.04	16	870
34	Typical Bay 34	1.45	0.05	16	1150
35	Typical Bay 35	1.18	0.45	50	500
36	Typical Bay 36	1.23	0.85	50	850
37	Typical Bay 37	0.36	0.07	100	435
38	Typical Bay 36	0.21	0.03	180	1135
39	Typical Bay 39	0.96	0.08	25	995
40	Typical Bay 40	0.49	0.07	90	840
41	Typical Bay 41	0.86	0.03	100	1315
42	Typical Bay 42	1.45	0.34	20	890
43	Typical Bay 43	1.80	0.40	20	116
44	Typical Bay 44	6.00	1.00	N/A	870

U-value of system envelope vs thickness of insulation layer

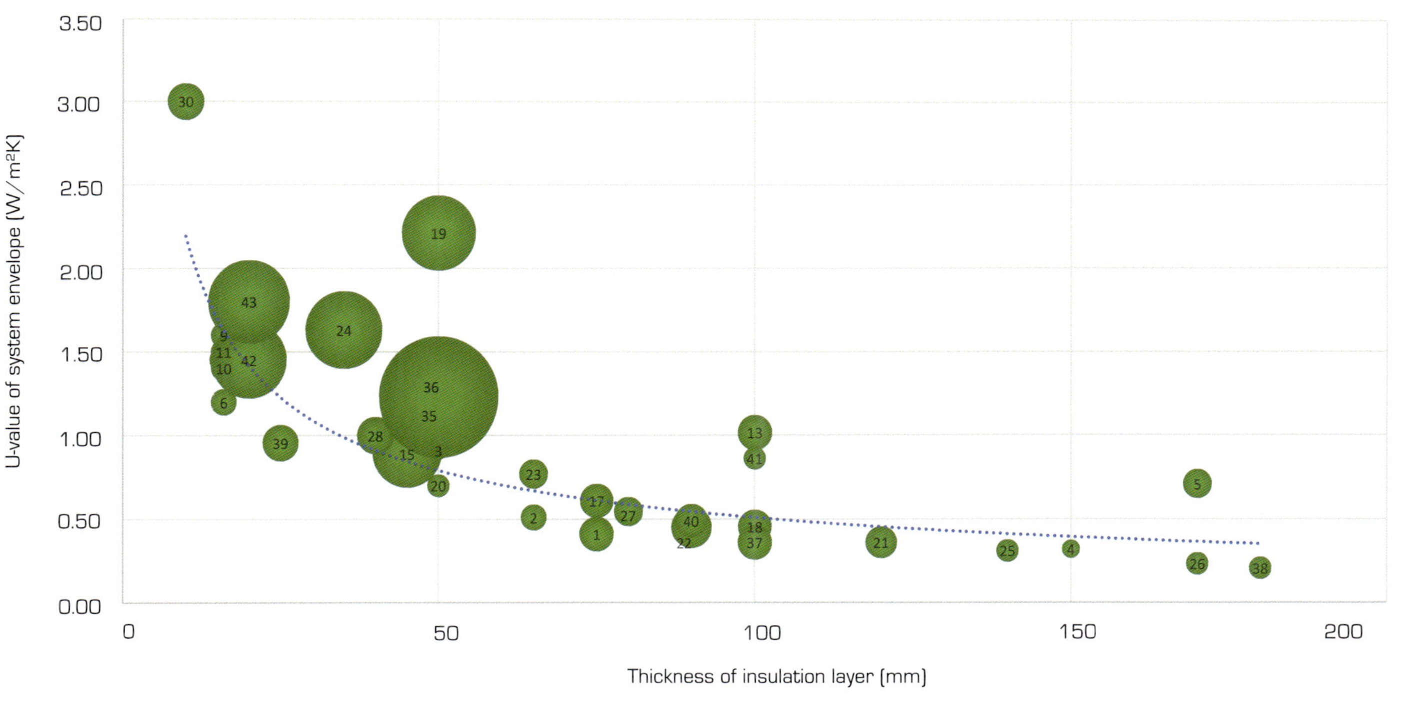

The relationship of inverse proportionality between U-value of each envelope system and thickness of the main insulating layer is illustrated in the graph, which also shows the potential effects of thermal bridging within more complex system assemblies.

This section sets out preliminary specification information for the main materials referred to elsewhere in this book. The standards and norms can be used in early stage design work only, with a view to deriving preliminary costs for project-specific use of the facade and roof systems set out in this book.

MAT01 - 1.0	**MAT01 – ULTRA HIGH PERFORMANCE CONCRETE**
MAT01 - 1.1	Shape: Produce UHPC units straight or curved to follow building lines as indicated on drawings.
MAT01 - 1.2	Cement: To EN 197: Part 1, CEM 1 Portland cement (ASTM C150/C150M)
MAT01 - 1.3	Sand: Silica sand to UHPC 'Specification for the manufacture, curing and testing of UHPC products', clause 16.13.2.1.
MAT01 - 1.4	Mixing Water
MAT01 - 1.4.1	To EN 1008 (ASTM C1602/C1602M).
MAT01 - 1.4.2	Water shall be clean and free of impurities that may affect colour, stability, setting or strength of panels.
MAT01 - 1.5	Admixtures
MAT01 - 1.5.1	Submit proposals of proposed types.
MAT01 - 1.5.2	Calcium chloride based admixtures shall not be used where UHPC contains cast-in steel components or fixings.
MAT01 - 1.6	Pigments
MAT01 - 1.6.1	Powder pigments or dispersions to EN 12878 (ASTM D1208).
MAT01 - 1.6.2	Pigments shall have proven successful use in UHPC cladding.
MAT01 - 1.6.3	Surface Coatings shall not be visible once applied and cured. The coating shall be water, oil and graffiti resistant. Panels shall be resilient to cleaning with the cleaning process demonstrated by the supplier on the performance mock-up, such that the cleaned surface shall not affect panel pigmentation, texture or surface finish. Proposals and samples to be provided for visual inspections and for acceptance by the Project Architect. The surface coating is expected to maintain its performance for cycles of 15 years through the lifespan of the building. Surface coating (spray) is to be applied to the front and to the sides of the panel - on the external/top surface of the panels.
MAT01 - 1.6.4	Mix Design: Mix design shall be determined by the Contractor in accordance with UHPC 'Specification for the manufacture, curing and testing of UHPC products', clause 15.8 in Specification Part 2 Performance Requirement, for acceptance by the Facade Consultant.
MAT01 – 1.7	Top Coating to be 4mm thick, cementitious finish. GRC and UHPC coating to be the same and colour of coating to match. GRC and UHPC coating to be confirmed by Project Architect and Facade Consultant. The process of application of coating on UHPC and GRC consists of preparing the surface and assess if an intercoat is needed. To avoid bonding problems and differential stress distributions, it is essential that the surface (of GRC and/or UHPC) is prepared before applying the coating. The surface should be free from chemical contaminants. Usually, the surface is smoothed by sandblasting and the coating should be applied without delay after surface preparation. To avoid adhesion problems between the substrate and the coating, an intercoat can be used, where necessary. This allows to avoid cracks and achieve visual appearance requirements.
MAT01 – 1.8	Materials Properties
MAT01 - 1.8.1	The following mechanical properties have been assumed for UHPC for the structural calculations in the development of the concept for the cladding panels: - Characteristic compressive strength fck: 150MPa - Characteristic limit of elasticity under tension fctfk: 9.0MPa - Young's modulus: 50GPa - Density: 25kN/m³ The contractor is to confirm the properties above for UHPC, and submit a full set of structural calculations based on the specific mechanical properties achieved by the materials they propose.
MAT02 - 1.0	**MAT02 – GLASS FIBRE REINFORCED CONCRETE**
MAT02 - 1.1	Shape: Produce GRC units straight or curved to follow building lines as indicated on drawings.
MAT02 - 1.2	Cement: To EN 197: Part 1, CEM 1 Portland cement (ASTM C150/C150M).
MAT02 - 1.3	Sand: Silica sand to GRCA 'Specification for the manufacture, curing and testing of GRC products', clause 16.13.2.1.
MAT02 - 1.4	Glass Fibre:
MAT02 - 1.4.1	Type: Alkali-resistant continuous filament fibre.

MAT02 - 1.4.2	Compliance: In accordance with GRCA 'Specification for alkali resistant glassfibre rovings and chopped strands for reinforcement of cements and concretes'.
MAT02 - 1.4.1	Content shall be determined in accordance with EN 1170 (ASTM C947).
MAT02 - 1.5	Mixing Water
MAT02 - 1.5.1	To EN 1008 (ASTM C1602/C1602M).
MAT02 - 1.5.2	Water shall be clean and free of impurities that may affect colour, stability, setting or strength of panels.
MAT02 - 2.6	Admixtures
MAT02 – 1.6.1	Submit proposals of proposed types.
MAT02 - 1.6.2	Calcium chloride based admixtures shall not be used where GRC contains cast-in steel components or fixings.
MAT02 - 1.7	Pigments
MAT02 - 1.7.1	Powder pigments or dispersions to EN 12878 (ASTM D1208).
MAT02 - 1.7.2	Pigments shall have proven successful use in GRC cladding.
MAT02 - 1.8	Surface Coatings shall not be visible once applied and cured. The coating shall be water, oil and graffiti resistant. Panels shall be resilient to cleaning with the cleaning process demonstrated by the supplier on the performance mock-up, such that the cleaned surface shall not affect panel pigmentation, texture or surface finish. Proposals and samples to be provided for visual inspections and for acceptance by the Project Architect. The surface coating is expected to maintain its performance for cycles of 15 years through the lifespan of the building. Surface coating (spray) is to be applied to the front and to the sides of the panel - on the external/top surface of the panels.
MAT02 - 1.9	Mix Design: Mix design shall be determined by the Contractor in accordance with GRCA 'Specification for the manufacture, curing and testing of GRC products', clause 15.8, for acceptance by the Project Architect.
MAT02 - 1.10	Top Coating to be 4mm thick, cementitious finish. GRC and UHPC coating to be the same and to be confirmed by Project Architect and Facade Consultant.
MAT02 - 1.11	Material Properties
MAT02 - 1.11.1	The following mechanical properties have been assumed for GRC for the structural calculations in the development of the concept for the cladding panels: - Dry density: 2200 kg.m^3 - Expected L.O.P: 8 MPa - Expected M.O.R: 21 MPa - Young's modulus: 19 GPa - Thermal expansion coefficient: 15x10-6 K-1 The contractor is to confirm the properties above for GRC, and submit a full set of structural calculations based on the specific mechanical properties achieved by the materials they propose.
MAT03 - 1.0	**MAT03 – STRUCTURAL STEEL WORKS – MILD STEEL**
MAT03 - 1.1	Grades of metals, section dimensions and properties: according to EN 10025 and relevant European standards and codes. When not specified, select grades and sections appropriate for the purpose (the minimum grade adopted shall comply with the minimum technical requirement).
MAT03 - 1.2	Prefinished metal: May be used if methods of fabrication do not damage or alter appearance of finish, and finish is adequately protected.
MAT03 - 1.3	Fasteners: according to EN ISO 898-1, EN 20898-2 (ASTM F2329, ASTM F2329) and relevant European standards and codes. Unless specified otherwise, of same metal as component being fastened, with matching coating or finish.
MAT03 - 1.4	Tubes: To EN 10210 (ASTM A450/A450M).
MAT03 - 1.4.1	Seamless circular: To EN 10297-1 (ASTM A423. Also to ASTM A519).
MAT03 - 1.4.2	Seamless cold drawn: To EN 10305-1 (ASTM A179/A179M).
MAT03 - 1.4.3	Welded and cold sized square and rectangular: To EN 10305-5 (ASTM A847/A847M).
MAT03 - 1.4.4	Welded circular: To EN 10296-1 (ASTM A1016).
MAT03 - 1.4.5	Welded cold drawn: To EN 10305-2 (ASTM A512).
MAT03 - 1.4.6	Welded cold sized: To EN 10305-3 (ASTM A847/A847M).
MAT03 - 1.5	Steel Long and Flat Products
MAT03 - 1.5.1	Hot rolled structural steels (excluding structural hollow sections and tubes): To EN 10025-1 (ASTM A700).
MAT03 - 1.5.2	Fine grain steels, including special steels: To EN 10025-3 and -4 (ASTM A1039 and ASTM A700).

MAT03 - 1.5.3	Steels with improved atmospheric corrosion resistance: To EN 10025-5 (ASTM A1039 and ASTM A700).
MAT03 - 1.6	Steel Plate, Sheet and Strip
MAT03 - 1.6.1	Plates and wide flats, high yield strength steel: To EN 10025-6 (ASTM A1039 and ASTM A700).
MAT03 - 1.7	Hot Rolled Steel Plate, Sheet and Strip
MAT03 - 1.7.1	Flat products, high yield strength for cold forming: To EN 10149-1, -2 and -3 (ASTM A666 and ASTM A700).
MAT03 - 1.7.2	Carbon steel sheet and strip for cold forming: To EN 10111 (ASTM A1039 and ASTM A700).
MAT03 - 1.7.3	Narrow strip, formable steel and steel for general engineering purposes: To EN 10029 (ASTM A568/A568M).
MAT03 - 1.8	Cold Rolled Steel Plate, Sheet and Strip
MAT03 - 1.8.1	Steel sections: To EN 10162 (A568/A568M).
MAT03 - 1.8.2	Flat products, high yield strength micro-alloyed steels for cold forming: To EN 10268 (ASTM A505 and ASTM A700).
MAT03 - 1.8.3	Carbon steel flat products for cold forming: To EN 10130 and EN 10131 (ASTM A700, ASTM A568/A568M, ASTM A109 and ASTM A918).
MAT03 - 1.8.4	Uncoated carbon steel narrow strip for cold forming: To EN 10139 and EN 10140 (ASTM A700).
MAT03 - 1.8.5	Narrow strip steel for structure: To EN 10132-1, -2, and -3 (ASTM A700).
MAT03 - 1.8.6	Carbon steel flat products for vitreous enamelling: To EN 10209 (ASTM A700).
MAT03 - 1.9	Coated Steel Flat Products
MAT03 - 1.9.1	Hot dip zinc coated carbon steel sheet and strip for cold forming: To EN 10346 and EN 10143 (ASTM A755).
MAT03 - 1.9.2	Hot dip zinc coated structural steel sheet and strip: To EN 10143 and EN 10346 (ASTM A700, ASTM A568/A568M, ASTM A123/A123M, ASTM A755 and ASTM A653).
MAT03 - 1.9.3	Hot dip zinc-aluminium (za) coated sheet and strip: To EN 10346 (ASTM A700, ASTM A123/A123M and ASTM A653).
MAT03 - 1.9.4	Hot dip aluminium-zinc (az) coated sheet and strip: To EN 10346 (ASTM A700, ASTM A123/A123M, ASTM A653).
MAT03 - 1.9.5	Organic coated flat products: To EN 10169 (ASTM A700).
MAT03 - 1.10	Steel Structural Hollow Sections (SHS)
MAT03 - 1.10.1	Non alloy and fine grain steels, hot finished: To EN 10210-1 and -2 (ASTM A568/A568M, ASTM A700).
MAT03 - 1.10.2	Non-alloy and fine grain steels, cold formed welded: To EN 10219-2 (ASTM A618/A618M).
MAT03 - 1.10.3	Weather resistant steels, hot finished: To EN 10210, EN ISO 14713 (ASTM A123/A123M and ASTM A53).
MAT03 - 1.11	Other Steel Sections
MAT03 - 1.11.1	Equal flange tees: To EN 10055 (ASTM A961/A961M).
MAT03 - 1.11.2	Wire, carbon steel for structure: To EN 10264 (ASTM A1000/A1000M).
MAT03 - 1.11.3	Wire and wire products, general: To EN 10218-2 (ASTM A1007).
MAT03 - 1.12	Quality and class of materials
MAT03 - 1.12.1	Minimum requirement: Grade: S355JR (ASTM A1007). The steel grade used for all structural sections shall be S355JR (ASTM A1007), unless instructions to the contrary are given on the drawings and/or use of a different steel grade is explicitly justified by calculations. Refer to relevant EN 10025 and European standards. (ASTM A572/A572M and ASTM A1039).
MAT03 - 1.12.2	Unless use of a different steel quality is explicitly justified by calculations, the steel quality used for all structural sections shall be Class C for external exposed steelwork and Class B for internal or external insulated steelwork. The steel quality used for tie rods and other elements subjected to permanent tension shall be as prescribed by the producer.
MAT03 - 1.12.3	Unless explicit justification to the contrary is provided by calculations, for all steel plates of thickness greater than 15mm and subject to through-thickness tensile loads, the material shall be of quality Class B or Class C and shall be subject to through-thickness tensile tests. The through-thickness quality required shall be calculated according to EN 1993-1-10 (ANSI/AISC 360-10).
MAT03 - 1.13	Welding:
MAT03 - 1.13.1	Type and class of weld: Weld type and weld class shall conform to EN ISO 9692-1 and current European standards (ASTM A269/A269M).

MAT03 - 1.13.2	Weld metal: Weld metals shall conform to EN ISO 14555, EN ISO 13918 and current European standards (ASTM A488/A488M). Welding consumables and procedures shall ensure that mechanical properties for the deposited weld metal, fusion line and heat affected zone are not less than the minimal specified for the parent metal.
MAT03 - 1.13.3	Welding procedures:
MAT03 - 1.13.3.1	All welders shall be approval tested to EN 287-1 or equivalent (ASTM A488 / A488M). Welder testing shall be witnessed and certificates endorsed by an independent Inspection Authority.
MAT03 - 1.13.3.2	Welding procedures shall conform to EN 288-3 or equivalent (D1.1/D1.1M:2010 Structural Welding Code-Steel and ASTM A488/A488M), and shall be approved by an independent Inspection Authority.
MAT03 - 1.13.3.3	The Facade Contractor shall store, handle and use welding consumables in accordance with the manufacturer's recommendations.
MAT03 - 1.13.3.4	The Facade Contractor shall show the welding method on the fabrication drawings for approval by the Project Architect.
MAT03 - 1.13.3.5	Charpy V-notch impact tests shall be included, according to EN 288: Part 3. Samples shall be taken from the base metal, HAZ/ fusion line and weld metal (ASTM A370).
MAT03 - 1.13.3.6	Welding procedures and sequencing shall take due account of plate shrinkage and shall ensure that distortion is reduced to a minimum. Refer to EN ISO 5817 (ASTM A488/A488M and D1.1/D1.1M:2010).
MAT03 - 1.13.4	Partial penetration butt welds: Partial penetration butt welds shall not be allowed unless they have been subject of prior approval on the part of the Project Architect.
MAT03 - 1.13.5	Butt welds: Where possible, butt welds shall be made with the use of run-on and run-off plates to ensure full throat thickness at the ends. Butt welds shall comply with the following requirements:
MAT03 - 1.13.5.1	Material specification for run-on and run-off plates shall be identical to that of the material being welded.
MAT03 - 1.13.5.2	Run-on and run-off plates shall be prepared in the same manner as the parts being joined.
MAT03 - 1.13.5.3	After completion of the welding process, the plates are to be removed by cutting.
MAT03 - 1.13.5.4	The surfaces affected by cutting are to be ground smooth and inspected for cracks.
MAT03 - 1.13.6	Weld testing and Checking: Welded material shall be subjected to testing in compliance with EN 288-3 and current European standards (ASTM A488/A488M).
MAT03 - 1.13.7	Scope of weld examination:
MAT03 - 1.13.7.1	All welds shall be subjected to visual examination in compliance with EN 1090-2. Welds carried out on site shall be subjected to dye penetrant testing in compliance with EN ISO 3452-1 (AWS B1.11:2000 and ASTM E1417/E1417M).
MAT03 - 1.13.7.2	Butt welds shall be subjected to ultrasonic examination to EN ISO 17640 (ASTM E273) or radiographic examination to EN ISO 17636-2 (ASTM E1032), and magnetic particle test to EN ISO 23278 (ISO 23278 and ASTM E709) or dye penetrant test to EN 1090-2 (ASTM E1417/E1417M).
MAT03 - 1.13.7.3	Partial penetration butt welds shall be subjected to ultrasonic examination and magnetic particle or dye penetrant testing on a randomly-determined 20% of the total joint length.
MAT03 - 1.13.7.4	10% of fillet welds shall be subjected to dye penetrant or magnetic particle testing. Not less than 15% of welded assembly surfaces and edges shall be subjected to dye penetrant or magnetic particle testing.
MAT03 - 1.13.7.5	Welded assembly plates subject to tensile loading perpendicular to the plane of the plate, and therefore at risk of through-thickness lamellar tearing, shall be subjected to ultrasonic examination across the welded zone. A written examination report shall be prepared and made available to the Project Architect for all welds executed within the scope of the present contract.
MAT03 - 1.13.8	Weld defects
MAT03 - 1.13.8.1	All weld defects are to be reported.
MAT03 - 1.13.8.2	Weld defect acceptance criteria shall be set in accordance with current National & European Standards or in accordance with the following criteria where the latter are more onerous.
MAT03 - 1.13.9	Visual checking
MAT03 - 1.13.9.1	There shall be no evidence of tearing or incomplete fusion.
MAT03 - 1.13.9.2	The length and size of welds shall in no case be less than that indicated on the plans.
MAT03 - 1.13.9.3	The root separation for angle welds shall not exceed 1.0mm total, and should not average above 0.5mm.
MAT03 - 1.13.10	Ultrasonic / Radiographic testing: No defect shall be present within a distance equal to twice the throat thickness, measured from the start or finish of a weld run.
MAT03 - 1.13.11	Dye penetrant or magnetic particle testing

MAT03 - 1.13.11.1	There shall be no evidence of cracks, tears or lack of fusion. Any undercut shall be intermittent and not greater than 5% of the thickness of the material or 0.5mm deep, whichever is the lesser.
MAT03 - 1.13.11.2	The sum of the diameters of piping or porosity shall not exceed 10mm total in any 25mm of weld length, nor 20mm total in any 300mm of weld length.
MAT03 - 1.13.11.3	The maximum length of a flaw may not exceed 2/3 of the effective throat thickness of the weld, or 20mm, whichever is the smaller.
MAT03 - 1.13.11.4	The flaw is to be further than three times the larger of its own width or length from the end of the weld or from an adjacent flaw.
MAT03 - 1.13.11.5	Any perceived sub-surface flaw shall be ground back as necessary to expose its full size, type and shape. Any inconclusive result shall be treated as a positive indication of a weld defect unless re-evaluation by the same method or by an alternative technique, subsequent to surface dressing, proves the contrary.
MAT03 - 1.13.12	Site welds:
MAT03 - 1.13.12.1	At the time of site-welding there shall be no paint, other than prefabrication primer, on the elements to be welded. Any paint located within 50mm of the weld shall be ground off prior to welding, to ensure that the weld is not contaminated by the paint.
MAT03 - 1.13.12.2	Site weld area shall be prepared to EN 1011 (ASTM A488/A488M). Other painted areas next to the weld areas shall be suitably protected during site-welding operations. Paint damaged by site-welding operations shall be repaired in accordance with the prescriptions of this standard, manufactures recommendations and the applicable Standards.
MAT03 - 1.13.12.3	Temporary protection shall be applied and removed before welding if this is necessary to ensure that rusting does not occur to a level beyond that allowed by this document.
MAT03 - 1.14	Rods and bars in tension
MAT03 - 1.14.1	The threading for steel rods and bars in tension shall present an improved fatigue profile according to ISO standards.
MAT03 - 1.14.2	Whenever possible, the threads shall be rolled. Changes of cross-section within the tension elements shall be designed so that excessive stress concentrations are not induced at these changes of cross-section, and this shall be demonstrated by testing.
MAT03 - 1.14.3	All rod and bar systems should be used in accordance with the local structural design codes. The commonly used systems (Mc Alloy, Pfeifer for example) are originally designed and tested by the producers in accordance with the requirements of EN 1993-1 Eurocode 3 (IBC).
MAT03 - 1.14.4	Checks shall be carried out during the fabrication process to demonstrate the straightness of the tension elements between adjacent points of restraint.
MAT04 - 1.0	**MAT04 – ANTI CORROSION PROTECTION FOR MILD STEEL: PAINT AND GALVANISATION FINISHES**
MAT04 - 1.1	Contractual Samples
MAT04 - 1.1.1	Before commencement of the works, the Contractor shall provide data sheets for all cleaning, preparation and paint materials he proposes to use, together with evidence to demonstrate that the proposed materials comply with current European standards and with the present document.
MAT04 - 1.1.2	The Contractor shall submit two sets of 3 no. A4 samples to indicate the colour and surface finish of each of the protective systems he proposes to use. The samples will be checked for compliance with the visual requirements of this specification by the Project Architect. Samples which are accepted as complying with this specification will form the performance standards for checking the colour and finish of all production components.
MAT04 - 1.2	Preparation for paint application
MAT04 - 1.2.1	All paint for factory or site application shall be supplied in sealed containers of not more than 25 litres capacity, and shall be clearly labelled with manufacturer's name, instructions for storage and application, colour and batch number and expiry date.
MAT04 - 1.2.2	Paint shall be carefully prepared for use in strict accordance with the manufacturer's instructions and with standard good practice. Paint shall not be used after the expiry date shown on the container. Surface preparation shall comply with the recommendations of EN ISO 8501-1 (ASTM D6386) and EN ISO 8504-3 (ASTM D2200).
MAT04 - 1.2.3	All welding slag, sputter, flux, residues, oxides, fumes, snags and strikes shall be removed and all sharp arrases and burrs ground smooth before cleaning and application of the protective coating system.
MAT04 - 1.2.4	Blast-cleaning where required shall be carried out in adequate lighting and atmospheric conditions; no blast cleaning shall be carried out when the temperature of the steelwork, is below the dew point.
MAT04 - 1.2.5	Abrasives shall be free from contamination. The minimum quality requirement shall be surface grade Sa 2.5 to ISO 8501-1 (ASTM D6386). The blast-cleaned surface must be vacuum cleaned before any coating is applied.
MAT04 - 1.2.6	Surfaces on which rust bloom has started to appear after treatment shall be re-blast cleaned prior to application of primer.

MAT04 - 1.2.7	Where surface preparation or painting is to be carried out on site, the Contractor shall adequately protect surrounding areas and adjacent structures from damage or marking, to the satisfaction of the Project Architect.
MAT04 - 1.2.8	The Contractor shall prepare and protect all connections, including fasteners, items of basketry and other small pieces fabricated separately to the main steelwork, to an equivalent standard to the adjacent steel unless noted otherwise in the drawings.
MAT04 - 1.3	Paint application
MAT04 - 1.3.1	The manufacturer instructions for use shall be strictly applied, with a painting that has to be adapted to the high relative humidity encountered in the project location.
MAT04 - 1.3.2	Coating shall not be carried out when: The ambient temperature is below 5°C. The surfaces are wet. Atmospheric conditions exist which are liable to generate condensation on surfaces either immediately before or during coating application.
MAT04 - 1.3.3	All surfaces shall be clean, dry and dust-free immediately prior to coating application. For each application, all weld runs, angles and edges shall be stripe-coated first. Minimum over-coating period for the primer shall be observed prior to over coating with the next coat over the whole area.
MAT04 - 1.3.4	At bolted connections, clearance at holes shall be maintained and runs through holes avoided.
MAT04 - 1.3.5	Temporary bolts shall be fitted in any threaded fittings and removed before the coating has set. Unless otherwise directed, all metal fittings and attachments capable of easy and safe removal shall be removed. They shall be cleaned and re-fixed after the finish coat is dry and hard. Where they cannot be removed, they shall be protected before surface preparations are begun.
MAT04 - 1.3.6	At all junctions of base metal with existing coatings, the coatings shall be feathered out over 100mm to provide a smooth graduated profile before any new coatings are applied.
MAT04 - 1.3.7	Unless application over partial area is approved by the paint manufacturers and the Project Architect, any coat to a member shall be applied over the whole surface in one continuous film operation to give the total dry thickness over the steel required by this specification. Wet thickness gauges shall be used continually by the applicator to ensure that the dry film thickness requirements will be met. The intervals between coats shall be in strict accordance with the recommendations of the manufacturer of the paint system. No subsequent coat shall be applied until the previous coat has dried and cured to the correct degree.
MAT04 - 1.3.8	If the total dry film thickness is less than specified, additional coats of finish paint shall be applied until the specified total dry film thickness is obtained. Total dry film thickness shall be measured in accordance with EN ISO 2808. An additional finishing coat shall be applied at places (such as exposed edges or plates) where corrosive conditions will be exceptionally sever, and over protruding or irregular features such as bolt heads and welds. In the event of any foreign materials being found embedded in the paint film, the affected areas shall be re-blasted and the complete paint system reapplied. All paint shall be applied in such a manner as to ensure continuous covering film, free from issues, tears, runs, sags etc. to give a finish of uniform colour and texture acceptable to the Project Architect. The applied coatings shall be inspected to verify that the coatings are free of blisters, pinholes, dry spray, mudcracking and other defects that may affect the performance or quality of the coating system. The Project Architect may direct the brushing out of any runs or sags, or the removal and repair of the paint.
MAT04 - 1.4	Minimum requirements for paint systems. The minimum requirements for paint systems shall be as follows:
MAT04 - 1.4.1	Paint system following galvanization, for external application:
MAT04 - 1.4.1.1	Etch primer or approved equivalent.
MAT04 - 1.4.1.2	Undercoat: 2 pack high-build epoxy with micaceous iron oxide to EN ISO 12944 (ASTM D2803) by airless spray to nominal dry film thickness of 100 microns.
MAT04 - 1.4.1.1.3	Finish coat: 2 pack polyurethane finish, by brush to nominal dry film thickness of 50 microns.
MAT04 - 1.4.2	Paint system for internal application in normally dry environment:
MAT04 - 1.4.2.1	Blast-cleaning to surface quality Sa 2.5
MAT04 - 1.4.2.2	Primer: Modified alkyd zinc phosphate primer to EN ISO 12944 or approved European equivalent (ASTM D823), by airless spray to nominal dry film thickness of 100 microns.
MAT04 - 1.4.2.3	Undercoat: Alkyd-based undercoat to EN ISO 12944 (ASTM D823), by brush to nominal dry film thickness of 35 microns.
MAT04 - 1.4.2.4	Finish coat: Alkyd-based finish coat to EN ISO 12944 (ASTM D823), by brush to nominal dry film thickness of 35 microns.
MAT04 - 1.5	Paint finish

MAT04 - 1.5.1	The visual requirements for paint finish described in the present document shall apply to all paint protected surfaces, including made good surfaces.
MAT04 - 1.5.2	The Contractor shall propose to the Project Architect for approval a suitable paint complex from a single manufacturer including primer, intermediate build coats and topcoat to comply with exposure conditions. The nature of all surface preparations shall be specified. All paints and coatings shall be used without thinning or adulteration other than in strict accordance with the manufacturer's recommended application method.
MAT04 - 1.5.3	The topcoat shall be a polyurethane paint. The colour shall be a RAL number, designation to be confirmed by the Project Architect.
MAT04 - 1.5.4	Where fabrication of steelwork involves welding, partial treatment of the steelwork prior to fabrication may only be undertaken when the primer has a welding certificate for the relevant dry film thickness. Any painted zones which are damaged during fabrication, transport or site works shall be submitted to the Project Architect for inspection and choice of remedial action. Where making good has been permitted by the Project Architect, the repair procedure shall be submitted for approval prior to execution and the end result shall show no discernible change in surface finish or colour between the original and the made good areas.
MAT04 - 1.6	Guarantees for paint systems
MAT04 - 1.6.1	Batch conformity:
MAT04 - 1.6.1.1	All products and cleaning solvents for any particular protective coating system are to be obtained from a single manufacturer and shall be certified by that manufacturer to provide a compatible paint system. The Contractor shall stockpile sufficient materials from one batch to complete the whole of the works. The Contractor and the manufacturer shall jointly guarantee the paint system proposed by submitting written approval of acceptance of responsibility on the paint finish.
MAT04 - 1.6.1.2	Conformity of paint system with exposure conditions: The paint system shall be appropriate to the exposure conditions and shall as a minimum comply with the requirements of EN ISO 12944 [ASTM D1014] for the following exposure classifications:
MAT04 - 1.6.1.2.1	Painted surfaces subject to external exposure, rain-protected or otherwise: "Exterior Exposed Polluted [Coastal]" classification, with typical time to first maintenance of 10–20 years.
MAT04 - 1.6.1.2.2	Painted surfaces subject to internal exposure, rain-protected: "Interior of buildings, normally dry" classification, reference SD2, with typical time to first maintenance of 10–20 years.
MAT04 - 1.6.2	Minimum performance guarantees for paint systems shall be at least the following:
MAT04 - 1.6.2.1	10 years for steel corrosion
MAT04 - 1.6.2.2	10 years for paint scaling
MAT04 - 1.6.2.3	10 years for colour degradation.
MAT04 - 1.6.2.4	10 years for aspect.
MAT04 - 1.6.2.5	If any pitting, scaling or rusting occurs before the end of the guarantee life of the works, the Contractor shall repair wherever necessary at his own expense.
MAT04 - 1.7	Compatibility tests for paint systems:
MAT04 - 1.7.1	Before commencement of the works, the Contractor shall supply, at his own expense, results of the following tests to provide evidence of compatibility:
MAT04 - 1.7.1.1	Compatibility between primer paint and all the metallic surfaces.
MAT04 - 1.7.1.2	Compatibility between different paint layers.
MAT04 - 1.7.1.3	Compatibility between finish paint and all materials in contact with it.
MAT04 - 1.7.1.4	Compatibility between all paint layers and any proposed rectification products.
MAT04 - 1.7.2	Tests carried out shall prove evidence of chemical and physical compatibility without degradation of durability, of corrosion protection or of colour. The test results shall be certified by an Approved External Inspecting Authority.
MAT04 - 1.8	Checking Procedures for painting works
MAT04 - 1.8.1	Self-checking by the Contractor: The Contractor shall carry out automatic checks on the following:
MAT04 - 1.8.1.1	Conformity with current National & European standards for all products used.
MAT04 - 1.8.1.2	Quality of the abrasive used for surface preparation.
MAT04 - 1.8.1.3	Characteristics of the compressed air used with the abrasive.
MAT04 - 1.8.1.4	Environmental temperature and hygrometry during coating application.
MAT04 - 1.8.1.5	Delay between two layers being painted.
MAT04 - 1.8.1.6	Thickness of the layers.
MAT04 - 1.8.1.7	Adhesion between successive layers.

MAT04 - 1.8.2	Test reports shall be written up and submitted to the Project Architect on demand, after completion of the painting works.
MAT04 - 1.8.3	Independent checking of paint thickness: At each step of the paint application process, the thickness and adhesion of the paint layers shall be checked by an Approved External Inspecting Authority. The cost of this inspection shall be borne by the Contractor. The rate and number of measurements shall be submitted to the Project Architect for approval. The location of measurement points shall be chosen by the Inspecting Authority. Checks shall be carried out on the whole steelwork paint package for:
MAT04 - 1.8.3.1	Thickness of the primer film.
MAT04 - 1.8.3.2	Thickness of the build complex before application of the finish film.
MAT04 - 1.8.3.3	Total thickness of the paint complex.
MAT04 - 1.8.3.4	Adhesion of the primer on the substrate surface.
MAT04 - 1.8.3.5	Adhesion of the finish film and intermediate build layers.
MAT04 - 1.8.3.6	Adhesion of the whole paint complex.
MAT04 - 1.8.4	Independent checking of paint thickness: The checking of film thickness shall be carried out by non-destructive procedures. At any point on the painted surface, the thickness of the coating shall be greater than or equal to the minimum thickness specified prior to commencement of the painting works. In case of doubt, more precise tests shall be carried out. If these tests shall prove the presence of a defect in quality, the paint coating complex shall be entirely removed and re-applied.
MAT04 - 1.8.5	Checking by the Project Architect: The Contractor shall provide full access on site to the Project Architect for checking the paint at every stage of the construction. The Project Architect shall require all coated surfaces to be washed down and cleaned to his satisfaction prior to final inspection of the site-applied coatings and repairs. If the coatings are proven to be defective, they shall be entirely removed and completely redone in compliance with the approved paint specification.
MAT04 - 1.9	Rectification of defects of paint systems
MAT04 - 1.9.1	Any coated surfaces damaged at any stage of the works (including bolt heads etc) must be repaired. The affected surfaces must be thoroughly cleaned to remove contamination in accordance with EN ISO 8501 (ASTM A780/A780M), without polishing the steel surface. Coating edges must be firm and feathered.
MAT04 - 1.9.2	All surrounding areas shall be cleaned as necessary to present a sound surface for local over-painting.
MAT04 - 1.9.3	Site application of paint repairs shall be carefully carried out to provide as neat an appearance as possible. The full specified DFT of the system shall be reinstated at the damaged zones; this will require multiple site-applied coats. All touch-up areas shall appear in a solid colour, homogeneous with the adjacent undamaged area. Procedures and materials used in the repair process must be approved by the paint manufacturer prior to use.
MAT04 - 1.9.4	Written approval from the Project Architect is necessary prior to any rectification of paint defects. Rectifications shall only be admitted for local and superficial defects.
MAT04 - 1.9.5	If rectification is not practicable, the entire element shall be cleaned and repainted on site or in shop. The rectification processes shall conform to the procedure approved by the Project Architect.
MAT04 – 1.10	Galvanizing
MAT04 – 1.10.1	Galvanizing shall be hot dip galvanizing and shall be carried out in accordance with EN ISO 1461 "Hot dip galvanized coating on fabricated iron and steel – specifications and test methods", and also with EN ISO 12944 "Paints and varnishes – Corrosion protection of steel structures by protective paint systems" (ASTM D823).
MAT04 – 1.10.2	The thickness of the zinc layer deposit shall be at least 600 g/m², or 85 microns minimum thickness for all steelwork elements (visible and concealed). Thickness shall not exceed 150 microns. Coating thickness shall be justified by testing certificates.
MAT04 – 1.10.3	All steel which shall be subject to hot-dip galvanization shall be supplied as galvanizable steel with controlled Silicium and phosphorous content. In particular, the Silicium content of steel which shall be subject to hot-dip galvanization must be coherent with the requirements of EN ISO 1461. Grit-blasting, if used, shall be carefully controlled to ensure that the subsequent surface finish of the steel does not generate excessive zinc build-up. Residues, oxides, fumes, snags and strikes shall be removed and all sharp arises and burrs ground smooth before cleaning. Rust, mill scale and dirt shall be removed and the steel surface shall be thoroughly cleaned and degreased with detergent or steam, rinsed and dried prior to galvanizing. The surface shall be pickled or blast-cleaned and flash-pickled to suit the required coating thickness. Steel surfaces shall not be more heavily pitted or rusted than the level specified as Grade C of ISO 8501-1 (ASTM D6386).
MAT04 – 1.10.4	The Contractor shall ensure that the galvanizing system used shall be sufficient to achieve the design performance specification. Galvanized surfaces which are to receive a subsequent paint finish shall be etch-primed or pre-treated with a suitable mordant wash system in accordance with the recommendations of the paint system manufacturer.

MAT04 - 1.10.5	Damaged areas shall be repaired with a suitable zinc-rich paint to give protection equal to that of undamaged surfaces, overlapping adjacent areas by 20mm. The damaged areas shall be adequately cleaned prepared for application of the zinc-rich paint in accordance with the recommendations of the paint system manufacturer.
MAT04 - 1.10.6	Prepare material and galvanize in accordance with EN ISO 8501-1 (ASTM D6386).
MAT04 - 1.10.7	Finish is to be a compatible, durable paint finish to galvanizing or minimum 10 year warranty paint system.
MAT04 - 1.10.8	Paint colour to any exposed structural steelwork to be selected from Metallic range, as confirmed by the Project Architect.
MAT04 - 1.10.9	All elements are to be pre-drilled, cut and welded prior to galvanizing.
MAT04 - 1.11	Tests and controls: All necessary tests and checks shall be carried out to demonstrate material and structural compliance with the standards and with this specification.
MAT04 - 1.11.1	Structural Steel – Paint Finish – External:
MAT04 - 1.11.1.1	All external visible structural steel is to be hot dip galvanized and painted.
MAT04 - 1.11.1.2	Surface treatment suitable for high quality architectural paint coating for visible steelwork.
MAT04 - 1.11.1.3	Degrease, wash and dry, then sweep blast galvanized surface with a non-metallic abrasive to impart a suitable surface roughness before priming. Prepare surface in accordance with EN ISO 8501-1 (ASTM D6386).
MAT04 - 1.11.1.4	Where steel is to be painted, surface preparation is to be to abrasive blast class 2.5 to EN ISO 8504-3 (ASTM D7055).
MAT04 - 1.11.1.5	Pressure wash and leave overnight to ensure no more than 70 mg/m² of chlorides.
MAT04 - 1.11.1.6	Painting system for steel to comply with EN ISO 8501-1 (ASTM D6386). Paint system must be capable of being re-coated in-situ to facilitate future maintenance and to allow touching up of construction damage.
MAT04 - 1.11.1.7	Paint colour: to be selected and confirmed by the Project Architect.
MAT04 - 1.11.2	Structural Steel – Paint Finish – Internal:
MAT04 - 1.11.2.1	All internal visible steelwork to be coated with suitable metal primer and powder coat finished.
MAT04 - 1.11.2.2	Grind all welds and finish to a smooth even surface prior to painting.
MAT04 - 1.11.2.3	Prepare all steel surfaces in accordance with EN ISO 8501-1 (ASTM D6386) prior to painting.
MAT04 - 1.11.2.4	Paint colour: to be selected and confirmed by the Project Architect.
MAT05 - 1.0	**MAT05 – STAINLESS STEEL**
MAT05 - 1.1	Material grade and quality
MAT05 - 1.1.1	Chemical composition and physical properties: To EN 10088-1 (ASTM A276/A276M).
MAT05 - 1.1.2	Sheet, strip and plate: To EN 10088-2 (ASTM A480/A480M).
MAT05 - 1.1.3	Semi-finished products bars, rods and sections: To EN 10088-3 (ASTM A276/A276M).
MAT05 - 1.1.4	Wire: To EN 1088-3 (ASTM A555/A555M).
MAT05 - 1.1.5	Tubes: To EN 10217 (ASTM A691/A691M).
MAT05 - 1.1.5.1	Welded circular: To EN 10296-2 (ASTM A268/A268M).
MAT05 - 1.1.5.2	Seamless circular: To EN 10297-2 (ASTM A268/A268M).
MAT05 - 1.1.6	Stainless Steels as referred to in this section are Austenitic Stainless Steels.
MAT05 - 1.1.7	All used stainless Steels shall have a high welding capacity. The welding procedure shall not require any specific heat treatment neither before nor after the welding process.
MAT05 - 1.1.8	Stainless steel shall be at least of grade 1.4547 (316/316L) unless noted otherwise, refer to EN 10088-3 (ASTM A240/A240M).
MAT05 - 1.1.9	Minimal required mechanical properties shall be:
MAT05 - 1.1.9.1	Yield Strength at 0.2% 300 MPa
MAT05 - 1.1.9.2	Ultimate tensile strength 650-850 MPa
MAT05 - 1.1.9.3	Notch Impact Strength at -20°C (KCV) 40 J
MAT05 - 1.2	Surface treatment
MAT05 - 1.2.1	All visible surfaces shall be properly discalced, cleaned and free of ferrous particles.
MAT05 - 1.2.2	All visible surfaces shall receive a chemical passivating treatment to ensure corrosion resistance. The type of passivating treatment shall be chosen so as not to lower the stainless steel grade in any way.
MAT05 - 1.2.3	All external and welded surfaces shall be free of cracks and discontinuities to avoid interstitial and inter-crystalline corrosion.

MAT05 - 1.2.4	All visible welds on stainless steel components shall be ground flat, unless otherwise stated on the drawings or in this specification.
MAT05 - 1.2.5	All visible surfaces of stainless steel castings and profiles shall receive a non-directional brushed finish, unless otherwise stated. For each component type, and prior to commencement of any fabrication run, the Facade Contractor shall provide samples of finishes for approval by the Project Architect.
MAT05 - 1.2.6	All visible surfaces of stainless steel sheet, plate and bars shall receive a non-directional brushed finish, unless otherwise stated. For each component type, and prior to commencement of any fabrication run, the Facade Contractor shall provide samples of finishes for approval by the Project Architect.
MAT05 - 1.2.7	Any alternative finishes which the Facade Contractor may wish to propose are subject to prior approval from the Client.
MAT05 - 1.3	Plate protection
MAT05 - 1.3.1	Stainless-steel plates and sheets shall be protected in the workshop and during handling with removable peels or plastic coverings against shocks or welding sparks. This protection shall be removed at the end of the fabrication process for thorough cleaning and local re-passivating as required. After cleaning, another removable protection shall be applied; this protection shall remain in place during transport, handling and installation on site.
MAT05 - 1.3.2	All prefabricated stainless steel components, including bars, profiles and castings, shall receive a removable protection prior to assembly on site to ensure that the stainless steel is not contaminated by adjacent works on site.
MAT05 - 1.3.3	At the end of the works, all stainless steel elements and components shall be stripped of their protective coverings and cleaned with appropriate products.
MAT05 - 1.4	Welding
MAT05 - 1.4.1	Type and class of weld: Weld type and weld class shall conform to EN ISO 9692-1 and current European standards (ASTM A269 / A269M).
MAT05 - 1.4.2	Stainless steel welding: Weld stainless steel shall conform to EN ISO 14555, EN ISO 13918 and current European standards (AWS D1.2). Welding consumables and procedures shall ensure that mechanical properties for the deposited weld metal, fusion line and heat affected zone are not less than the minimal specified for the parent metal.
MAT05 - 1.4.2.1	Samples shall be provided to the Project Architect for benchmarking of visual weld quality and approval of passivating method. Fabrication processes shall conform to the standards set by the pre-approved samples.
MAT05 - 1.4.2.2	TIG or MIG welding process shall be chosen on the basis of parent metal chemical composition and control of metal deposition. The proportion of ferrite in the weld metal shall be the same as that in the parent metal. Alloy separation shall be avoided. Welding materials shall conform to EN ISO 14555, EN ISO 13918 and current European standards (AWS D1.2).
MAT05 - 1.4.2.3	Systematic passivating of the weld and heat-affected zone shall be undertaken immediately after welding.
MAT05 - 1.4.2.4	The welding sequences and locations of any temporary restraints shall be designed to minimize weld distortion, taking due account of the increased coefficient of thermal expansion and the reduced conductivity of stainless steel. Refer to EN ISO 5817 (AWS D1.2).
MAT05 - 1.4.2.5	Stainless steel to carbon steel welds shall not be allowed unless they have received prior approval from the Project Architect, and only if compatibility of weld deposit metal has been demonstrated.
MAT05 - 1.4.3	Welding procedures
MAT05 - 1.4.3.1	All welders shall be approval tested to EN 287-1 (AWS D1.2) or equivalent. Welder testing shall be witnessed and certificates endorsed by an independent Inspection Authority.
MAT05 - 1.4.3.2	Welding procedures shall conform to EN 288-3 (AWS D1.2) or equivalent, and shall be approved by an independent Inspection Authority.
MAT05 - 1.4.3.3	The Facade Contractor shall store, handle and use welding consumables in accordance with the manufacturer's recommendations.
MAT05 - 1.4.3.4	The Facade Contractor shall show the welding method on the fabrication drawings for approval by the Project Architect.
MAT05 - 1.4.3.5	Charpy V-notch impact tests shall be included, according to EN 288: Part 3 (ASTM E23). Samples shall be taken from the base metal; HAZ/ fusion line and weld metal.
MAT05 - 1.4.3.6	Welding procedures and sequencing shall take due account of plate shrinkage and shall ensure that distortion is reduced to a minimum. Refer to EN ISO 5817 (AWS D1.2).
MAT05 - 1.4.4	Partial penetration butt welds: Partial penetration butt welds shall not be allowed unless they have been subject of prior approval on the part of the Project Architect.
MAT05 - 1.4.5	Butt welds: Where possible, butt welds shall be made with the use of run-on and run-off plates to ensure full throat thickness at the ends. Butt welds shall comply with the following requirements:

MAT05 - 1.4.5.1	Material specification for run-on and run-off plates shall be identical to that of the material being welded.
MAT05 - 1.4.5.2	Run-on and run-off plates shall be prepared in the same manner as the parts being joined.
MAT05 - 1.4.5.3	After completion of the welding process, the plates are to be removed by cutting.
MAT05 - 1.4.5.4	The surfaces affected by cutting are to be ground smooth and inspected for cracks.
MAT05 - 1.4.6	Weld testing and Checking: Welded material shall be subjected to testing in compliance with EN 288-3 and current European standards (ASTM A488/A488M).
MAT05 - 1.4.7	Scope of weld examination
MAT05 - 1.4.7.1	All welds shall be subjected to visual examination in compliance with EN 1090-2 (AWS B1.11:2000). Welds carried out on site shall be subjected to dye penetrant testing in compliance with EN ISO 3452-1 (ASTM E1417/E1417M).
MAT05 - 1.4.7.2	Butt welds shall be subjected to ultrasonic examination to EN ISO 17640 (ASTM E273) or radiographic examination to EN ISO 17636-2 (ASTM E709), and magnetic particle test to EN ISO 23278 (ASTM E1417) or dye penetrant test to EN 1090-2 (E1417M).
MAT05 - 1.4.7.3	Partial penetration butt welds shall be subjected to ultrasonic examination and magnetic particle or dye penetrant testing on a randomly-determined 20% of the total joint length.
MAT05 - 1.4.7.4	10% of fillet welds shall be subjected to dye penetrant or magnetic particle testing. Not less than 15% of welded assembly surfaces and edges shall be subjected to dye penetrant or magnetic particle testing.
MAT05 - 1.4.7.5	Welded assembly plates subject to tensile loading perpendicular to the plane of the plate, and therefore at risk of through-thickness lamellar tearing, shall be subjected to ultrasonic examination across the welded zone. A written examination report shall be prepared and made available to the Project Architect for all welds executed within the scope of the present contract.
MAT05 - 1.4.8	Weld defects
MAT05 - 1.4.8.1	All weld defects are to be reported.
MAT05 - 1.4.8.2	Weld defect acceptance criteria shall be set in accordance with current National & European Standards or in accordance with the following criteria where the latter are more onerous.
MAT05 - 1.4.9	Visual checking
MAT05 - 1.4.9.1	There shall be no evidence of tearing or incomplete fusion.
MAT05 - 1.4.9.2	The length and size of welds shall in no case be less than that indicated on the plans.
MAT05 - 1.4.9.3	The root separation for angle welds shall not exceed 1.0mm total, and should not average above 0.5mm.
MAT05 - 1.4.10	Ultrasonic / Radiographic testing: No defect shall be present within a distance equal to twice the throat thickness, measured from the start or finish of a weld run.
MAT05 - 1.4.11	Dye penetrant or magnetic particle testing
MAT05 - 1.4.11.1	There shall be no evidence of cracks, tears or lack of fusion. Any undercut shall be intermittent and not greater than 5% of the thickness of the material or 0.5mm deep, whichever is the lesser.
MAT05 - 1.4.11.2	The sum of the diameters of piping or porosity shall not exceed 10mm total in any 25mm of weld length, nor 20mm total in any 300mm of weld length.
MAT05 - 1.4.11.3	The maximum length of a flaw may not exceed 2/3 of the effective throat thickness of the weld, or 20mm, whichever is the smaller.
MAT05 - 1.4.11.4	The flaw is to be further than three times the larger of its own width or length from the end of the weld or from an adjacent flaw.
MAT05 - 1.4.11.5	Any perceived sub-surface flaw shall be ground back as necessary to expose its full size, type and shape. Any inconclusive result shall be treated as a positive indication of a weld defect unless re-evaluation by the same method or by an alternative technique, subsequent to surface dressing, proves the contrary.
MAT05 - 1.4.12	Site welds
MAT05 - 1.4.12.1	At the time of site-welding there shall be no paint, other than prefabrication primer, on the elements to be welded. Any paint located within 50mm of the weld shall be ground off prior to welding, to ensure that the weld is not contaminated by the paint.
MAT05 - 1.4.12.2	Site weld area shall be prepared to EN 1011 (AWS D1.2). Other painted areas next to the weld areas shall be suitably protected during site-welding operations. Paint damaged by site-welding operations shall be repaired in accordance with the prescriptions of this standard, manufactures recommendations and the applicable Standards.
MAT05 - 1.4.12.3	Temporary protection shall be applied and removed before welding if this is necessary to ensure that rusting does not occur to a level beyond that allowed by this document.

MAT05 - 1.4.13	Stainless Steel Fabrication
MAT05 - 1.4.13.1	Guillotining or punching: Do not use for metal thicknesses greater than 10 mm.
MAT05 - 1.4.13.2	Thermal cutting:
MAT05 - 1.4.13.3	Carbonation in the heat affected zone: Remove, after cutting.
MAT05 - 1.4.13.4	Bending
MAT05 - 1.4.13.5	Plates or bars: Cold bending radius not less than material thickness.
MAT05 - 1.4.13.6	Tubes: Cold bending radius not less than 2 x tube diameter.
MAT05 - 1.4.13.7	Welding: In addition to general welding requirements:
MAT05 - 1.4.13.8	Protect adjacent surfaces from weld spatter.
MAT05 - 1.4.13.9	Pickle all welds before post fabrication treatments.
MAT05 - 1.4.13.10	Protection: Provide protection to fabricated components during transit and on site.
MAT06 - 1.0	**MAT06 – CONNECTIONS AND FIXINGS**
MAT06 - 1.1	Materials for Connections:
MAT06 - 1.1.1	Ordinary Bolts: Shop and site bolts and nuts shall be at least to the following grades Grade 8.8 or Grade 10.9. Their dimensions, tolerance and thread length shall conform to EN ISO 898-1.
MAT06 - 1.1.2	High Strength Friction Grip (HSFG) Bolts: Shop and site HSFG bolts shall be at least of grade 8.8. Their dimensions, tolerance, thread length and the HSFG bolt assemblies considered as a whole shall conform to EN ISO 898-1 (ASTM F593).
MAT06 - 1.1.3	Stainless Steel Bolts: Stainless Steel Bolts and their assemblies shall conform to international standards EN ISO 898-1 (ASTM F593), or to equivalent standards. Shop and site stainless steel bolts and nuts shall be at least to the following grades: 1.4547 (316/316L), class A4. Bolts, pins and screws shall be delivered normalized and hardened. Washers and accessories shall be of the same alloy type and surface finish as the bolts.
MAT06 - 1.2	Assemblies
MAT06 - 1.2.1	General: Bolted assemblies shall be configured as shown on the drawings. The locations of bolt heads and nuts shall be scrupulously positioned and aligned as noted on the drawings.
MAT06 - 1.2.2	Holes: Holes for bolts shall be drilled, punched or punched and reamed. The holes shall not have ragged edges. Use of blowtorches is not allowed. Ordinary bolt holes shall conform to following tolerances:
MAT06 - 1.2.2.1	Diameter less or equal to bolt diameter plus 1mm for bolts of diameter less than or equal to 12mm.
MAT06 - 1.2.2.2	Diameter less or equal to bolt diameter plus 2mm for bolts of diameter less than or equal to 24mm.
MAT06 - 1.2.2.3	Diameter less or equal to bolt diameter plus 3mm for bolts of diameter over 36mm.
MAT06 - 1.2.3	HSFG bolt holes shall conform to current European standards. Punched out holes, whether reamed or not, shall not be accepted. Holes in hollow sections shall be sealed to avoid water penetration.
MAT06 - 1.2.4	Length of bolt shanks: The length of the bolt shank shall be such that, after tightening, at least three clear threads show above the nut and at least one thread plus the thread run out is clear between the nut and the unthreaded shank of the bolt.
MAT06 - 1.2.5	Combination of different bolt grades: Combinations of different bolt grades shall not be accepted within a single assembly. For the works covered by the present contract, the use of bolts of the same diameter but of different grades shall not be allowed in the workshop or on site.
MAT06 - 1.2.6	Washers: A flat washer shall be provided under each nut. Tapered washers shall be used wherever the angle measured between the bearing surface and the plan perpendicular to the bolt axis is equal to or greater than 3°.
MAT06 - 1.2.7	Assembly: Bolts shall be inserted effortlessly and then tightened to draw connected parts firmly together. Tightening shall conform to current standards.
MAT06 - 1.2.8	Locking nuts: All assemblies which may undergo load reversal or be subject to vibrations shall be made secure against unlocking of the threads. The locking will be achieved by the use of locking nuts, spring washers, gluing or any other approved method.
MAT06 - 1.2.9	Slotted holes: Where slotted holes are provided for movement connections, the joint shall be checked to ensure it remains free to move.
MAT06 - 1.2.9.1	The method for making bolted movement connections shall be the following:
MAT06 - 1.2.9.2	The slotted hole shall be wider than the un-slotted hole.
MAT06 - 1.2.9.3	A shouldered bolt shall be used, with a spring washer under the head and the shoulder bearing on the faying surface of the un-slotted hole.

MAT06 - 1.2.9.4	A flat washer shall be provided under the nut and the nut tightened onto the un-slotted member.
MAT06 - 1.2.9.5	The washer shall be sized to resist in bending the load applied from the bolt head across the span of the slotted or oversized hole.
MAT06 - 1.3	High strength friction grip fasteners
MAT06 - 1.3.1	The use of high strength friction grip bolts shall comply with EN 14399 (ASTM F3125). Load-indicating washers shall be used to confirm preload, in accordance with the manufacturer's recommendations and in accordance with EN 14399-9 (ASTM A325). If after tightening, a bolt or nut slackens off for any reason, the complete bolt assembly shall be discarded and shall not be re-used anywhere else in the works.
MAT06 - 1.3.2	Surface preparation for friction grip bolting: The Contractor shall check that all the faying surfaces of friction grip joints are free of distortion, deformities or contaminants which may reduce the slip factor below the stated design value. Deformed surfaces, other than the faying surfaces, shall be machined flat. The faying surfaces shall not be machined.
MAT06 - 1.3.3	All mill-scale shall be removed from the faying surfaces of friction grip joints. The faying surfaces shall be cleaned of dirt, grease or any other contaminants and then wire brushed.
MAT06 - 1.3.4	Paint shall not be applied to friction grip interfaces. The faying surfaces are to be masked, if necessary, to prevent rusting beyond Rust grade C of ISO 8501-1 (ASTM D6386).
MAT06 - 1.3.5	Tests shall be carried out to determine the slip factor onto stainless steel faying surfaces. If galvanizing or other metal coatings have been applied to the faying surfaces, the Contractor shall provide evidence or carry out tests to demonstrate that a slip factor not less than the design value is maintained. In all cases where one or both faying surfaces are in stainless steel, tests shall be carried out after machining and surface preparation to verify the slip factor of the assembly, in accordance with National or European standards.
MAT06 - 1.3.6	Checking of HSFG bolts HSFG: Following complete assembly, and prior to any site painting, all HSFG bolts shall be checked to ascertain that the minimum shank tension has been obtained and that appropriate hardened washers have been fitted in accordance with the requirements of EN 1993-1-8 (ASTM A490M).
MAT06 - 1.4	Surface protection
MAT06 - 1.4.1	All bolted joints shall be sealed against the ingress of water. Before site painting commences, gaps at joints shall be plugged with compatible filler. Care shall be taken that water is not sealed within the joint. Surface finish shall be identical to that of the assembled elements.
MAT06 - 1.4.2	Ordinary and HSFG Bolts: Plated, galvanized or sherardized bolt assemblies shall be primed with a compatible etch primer or shall be treated with a mordant solution prior to over-coating. The preparation and coating process shall conform to current European standards. Unless otherwise specified by the Project Architect, all exposed bolt assemblies shall be painted and all hidden bolt assemblies shall be sherardized. The coating thickness provided shall be 50 microns, or 375g/m^2. Special care shall be taken with the corrosion protection of bolts of grade 10.9 or above, in order to avoid their fragilisation by hydrogen cracking. The origin of these bolts has to be checked; and the Contractor has to submit the material sheet with relevant specifications and information.
MAT06 - 1.4.3	Stainless Steel Bolts: Stainless steel bolts shall not be assembled against mild steel or aluminium, unless an insulating plastic washer is provided or the contact zone is durably protected by painting or by sealing of threads.
MAT06 - 1.5	Fixings to bolted assemblies
MAT06 - 1.5.1	Cast-In Anchors and Channels:
MAT06 - 1.5.1.1	Facade fixings to new structure; use cast-in anchors and channels equal to HALFEN or approved fabricated cast-in plate.
MAT06 - 1.5.1.2	Cast-ins shall be minimum stainless steel grade 1.4547 (316/316L).
MAT06 - 1.5.1.3	Site welds to facade brackets shall be de-slagged and minimum 2 coat cold galvanised after welding.
MAT06 - 1.5.2	Fixings, Screws, Rivets and Washers
MAT06 - 1.5.2.1	Screws: minimum stainless steel grade 1.4547 (316/316L).
MAT06 - 1.5.2.2	Nuts, bolts & washers: minimum stainless steel grade 1.4547 (316/316L).
MAT06 - 1.5.2.3	Pop rivets (if required): aluminium grade ISO 17057 1999 or stainless steel (minimum grade 1.4547 (316/316L) recommended) with aluminium or stainless steel mandrels.
MAT06 - 1.5.2.4	Masonry anchors (expanding or chemical): minimum Stainless steel grade 1.4401; Masonry anchors installed inboard of the air seal may be galvanised or plated steel. Only recognised brands such as HILTI, Ramset and Powers may be used.
MAT06 - 1.5.2.5	Ensure all fixings are concealed. Exposed fixings are unacceptable unless approved by the Project Architect.
MAT06 - 1.5.2.6	All fixings shall be vibration proof under dead loads, live loads, wind loads and thermal movements.

MAT06 - 1.6	Fixings. This section refers to fixings generally. Fixings described in relation to specific materials take precedence.
MAT06 - 1.6.1	Fixings, within the framing components, shall not be visible unless indicated on the Design Drawings.
MAT06 - 1.6.2	Fixings shall be to the acceptance of the Building Control Officer and in accordance with the Structural Engineer's documentation.
MAT06 - 1.6.3	Fixings shall be inherently corrosion resistant or fully protected to prevent corrosion.
MAT06 - 1.6.4	All fixings shall be installed in accordance with the manufacturer's recommended procedures.
MAT06 - 1.6.5	Fasteners generally.
MAT06 - 1.6.5.1	Materials: To have:
MAT06 - 1.6.5.1.1	Bimetallic corrosion resistance appropriate to items being fixed.
MAT06 - 1.6.5.1.1.2	Atmospheric corrosion resistance appropriate to fixing location.
MAT06 - 1.6.5.2	Appearance: Submit samples on request.
MAT06 - 1.6.6	Packings
MAT06 - 1.6.6.1	Materials: Non compressible, corrosion proof.
MAT06 - 1.6.6.2	Area of packings: Sufficient to transfer loads.
MAT06 - 1.6.7	Masonry Fixings
MAT06 - 1.6.7.1	Light duty: Plugs and screws.
MAT06 - 1.6.7.2	Heavy duty: Expansion anchors or chemical anchors.
MAT06 - 1.6.8	Plugs. Proprietary types to suit substrate, loads to be supported and conditions expected in use.
MAT06 - 1.6.9	Anchors
MAT06 - 1.6.9.1	Types:
MAT06 - 1.6.9.1.1	Expansion: For use in substrate strong enough to resist forces generated by expansion of anchor.
MAT06 - 1.6.9.1.2	Adhesive or chemical: • For use in substrate where expansion of anchor would fracture substrate. • For use in irregular substrate where expansion anchors cannot transfer load on anchor.
MAT06 - 1.6.9.1.3	Cavity: For use where the anchor is retained by toggles of the plug locking onto the inside face of the cavity.
MAT06 - 1.6.10	Miscellaneous Screws
MAT06 - 1.6.10.1	Type: To suit the fixing requirement of the components and substrate.
MAT06 - 1.6.10.2	Pattern: Self-tapping, metallic drive screws, or power driven screws.
MAT06 - 1.6.10.3	Washers and screw cups: Where required to be of same material as screw.
MAT06 - 1.6.11	Fixing Generally
MAT06 - 1.6.11.1	Integrity of supported components: Select types, sizes, quantities and spacings of fixings, fasteners and packings to retain supported components without distortion or loss of support.
MAT06 - 1.6.11.2	Components, substrates, fixings and fasteners of dissimilar metals: Isolate with washers/ sleeves to avoid bimetallic corrosion.
MAT06 - 1.6.11.3	Appearance: Fixings to be in straight lines at regular centres.
MAT06 - 1.6.12	Fixing Through Finishes
MAT06 - 1.6.12.1	Penetration of fasteners and plugs into substrate: To achieve a secure fixing.
MAT06 - 1.6.13	Fixing Packings
MAT06 - 1.6.13.1	Function: To take up tolerances and prevent distortion of materials and components.
MAT06 - 1.6.13.2	Limits: Do not use packings beyond thicknesses recommended by fixings and fasteners Manufacturer.
MAT06 - 1.6.13.3	Locations: Not within zones to be filled with sealant.
MAT06 - 1.6.14	Using Powder Actuated Fixing Systems
MAT06 - 1.6.14.1	Powder actuated fixing tools: To EN 15895.
MAT06 - 1.6.14.2	Operatives: Trained and certified as competent by tool manufacturer.
MAT06 - 1.6.15	Fixings, Bolts, Anchors, Screws, Rivets, Shear Pins, Nuts, Washers and Other Components
MAT06 - 1.6.15.1	Mild steel in accordance with EN ISO 4016, EN ISO 4034, EN ISO 898-1 and EN ISO 898 (ASTM A449).
MAT06 - 1.6.15.2	Austenitic stainless steel complying with EN ISO 3506-1 -2 (ASTM A370).
MAT06 - 1.6.15.3	Aluminium alloy of appropriate grade complying with EN 754-3, -5, EN 755-1, -9, EN 573- 3, EN 515 and EN 12020-1 -2 (ASTM B210).

MAT07 - 1.0	MAT07 – ALUMINIUM
MAT07 - 1.1	Designations: • If the Contractor chooses an alloy other than those listed below and not listed in a recognised code then a certificate of properties shall be obtained from the billet maker. The data shall include all mechanical properties required for design, including the yield in compression. This data shall be used in calculations. • Designation system, chemical composition and forms: To EN 573-1, -2, -3 and -5 (ASTM B210) • Temper designations: To EN 515 (ASTM B210).
MAT07 - 1.2	Sheet, strip and plate: To EN 485-1 to -4
MAT07 - 1.3	Cold drawn rods, bars and tubes: To EN 754-1 and -2 (ASTM B211M).
MAT07 - 1.4	Extruded rods, bars, tubes and profiles: To EN 755-1 and -2 (ASTM B221M).
MAT07 - 1.5	Drawn wire: To EN 1301-1, -2 and -3 (ASTM B221M).
MAT07 - 1.6	Rivet, bolt and screw stock: To BS 1473 (ASTM F468).
MAT07 - 1.7	Structural sections: To EN 15088 (ASTM B308 / B308M).
MAT07 - 1.8	Fabrication generally
MAT07 - 1.8.1	Finished components: Rigid and free from distortion, cracks, burrs and sharp arrises.
MAT07 - 1.8.2	Moving parts: Free moving without binding.
MAT07 - 1.8.3	Corner junctions of identical sections: Mitre.
MAT07 - 1.9	Cold Formed Work
MAT07 - 1.9.1	Profiles: Accurate, with straight arrises.
MAT07 - 1.10	Adhesive Bonding
MAT07 - 1.10.1	Preparation of surfaces of metals to receive adhesives:
MAT07 - 1.10.1.1	Degrease.
MAT07 - 1.10.1.2	Abrade mechanically or chemically etch.
MAT07 - 1.10.1.3	Prime: To suit adhesive.
MAT07 - 1.10.1.4	Adhesive bond: Form under pressure.
MAT07 - 1.11	Welding
MAT07 - 1.11.1	Welding procedures:
MAT07 - 1.11.1.1	Method and standard: To be agreed with the Project Architect.
MAT07 - 1.11.1.2	Welding Procedure Specification (WPS): To be agreed with the Project Architect.
MAT07 - 1.11.2	Preparation:
MAT07 - 1.11.2.1	Joint preparation: Clean thoroughly.
MAT07 - 1.11.2.2	Surfaces of materials that will be self-finished and visible in the completed work: protect from weld splatter.
MAT07 - 1.11.3	Jointing:
MAT07 - 1.11.3.1	Joints: Fully bond parent and filler metal throughout with no inclusions, holes, porosity or cracks.
MAT07 - 1.11.3.2	Dissimilar metals: To be agreed with the Project Architect.
MAT07 - 1.11.3.3	Strength requirements: Welds to achieve design loads.
MAT07 - 1.11.3.4	Heat straightening: To be agreed with the Project Architect.
MAT07 - 1.11.3.5	Complex assemblies: Agree priority for welding members to minimize distortion caused by subsequent welds.
MAT07 - 1.11.3.6	Tack welds: Use only for temporary attachment.
MAT07 - 1.11.3.7	Jigs: Provide to support and restrain members during welding.
MAT07 - 1.11.3.8	Filler plates: To be agreed with the Project Architect.
MAT07 - 1.11.3.9	Lap joints: Minimum 5 x metal thickness or 25 mm, whichever is greater.
MAT07 - 1.11.3.10	Weld terminations: Clean and sound.
MAT07 - 1.12	Brazing
MAT07 - 1.12.1	Standard: To EN 14324 (ASTM F1076)
MAT07 - 1.12.1	Testing:
MAT07 - 1.12.1	Destructive testing: To EN 12797 (ASTM E1495). Non-destructive testing: To EN 12799 (E1495M)
MAT07 - 1.13	Testing:
MAT07 - 1.13.1	Testing standard: To be agreed with the Project Architect.
MAT07 - 1.13.1	Welding records and test results: Submit 2 copies.

MAT07 - 1.14	Finishing Welded And Brazed Joints Visible In Complete Work
MAT07 - 1.14.1	Standard: To EN ISO 8501-3 (ASTM D7396).
MAT07 - 1.14.1.1	Preparation grade: To be agreed with the Project Architect.
MAT07 - 1.14.2	Butt joints: Smooth, and flush with adjacent surfaces.
MAT07 - 1.14.3	Fillet joints: Neat.
MAT07 - 1.14.4	Grinding: Grind smooth where indicated on drawings.
MAT07 - 1.15	Preparation For Application Of Coatings
MAT07 - 1.15.1	General: Complete fabrication, and drill fixing holes before applying coatings.
MAT07 - 1.15.2	Paint, grease, flux, rust, burrs and sharp arises: Remove.
MAT07 - 1.16	Liquid Organic Coating For Aluminium Alloy Components
MAT07 - 1.16.1	Standard: To EN 12206 (ASTM B449 - 93(2015)).
MAT07 - 1.17	Zinc And Cadmium Plating Of Iron And Steel Surfaces
MAT07 - 1.17.1	Zinc plating: To EN ISO 2081 (ASTM B633).
MAT07 - 1.17.2	Cadmium plating: To EN ISO 2082 (ASTM B766)
MAT07 - 1.18	Chromium Plating
MAT07 - 1.18.1	Standard: To EN ISO 1456 (ASTM B650).
MAT07 - 1.19	Galvanizing
MAT07 - 1.19.1	Standard: To EN ISO 1461 (ASTM A123 / A123M)
MAT07 - 1.19.2	Preparation:
MAT07 - 1.19.2.1	Vent and drain holes: Provide in accordance with EN ISO 14713-1 and -2 (ASTM G1). Seal after sections have been drained and cooled.
MAT07 - 1.19.2.2	Components subjected to cold working stresses: Heat treat to relieve stresses before galvanizing.
MAT07 - 1.19.2.3	Welding slag: Remove.
MAT07 - 1.19.2.4	Component cleaning: To EN ISO 8501-3 (ASTM D7396)
MAT07 - 1.19.2.5	Grade: To be agreed with the Project Architect.
MAT07 - 1.20	Documentation
MAT07 - 1.20.1	Submit:
MAT07 - 1.20.1.1	Manufacturer's maintenance instructions.
MAT07 - 1.20.1.2	Guarantees, warranties, test certificates, record schedules and log books.
MAT07 - 1.20.2	Completion
MAT07 - 1.20.2.1	Protection: Remove.
MAT07 - 1.20.2.2	Cleaning and maintenance: Carry out in accordance with procedures detailed in fabricators' guarantees.
MAT07 - 1.21	Extruded profiles
MAT07 - 1.21.1	The aluminium alloy used for the manufacture of extruded profiles shall be grade 6063 T6 or equivalent.
MAT07 - 1.21.2	The exact alloy designation, together with its strength and weld ability characteristics and its surface finish and anodizing properties, shall be submitted to the Project Architect for approval prior to fabrication.
MAT07 - 1.21.3	The minimum wall thickness of all extruded structural profiles shall not be less than 18/10 mm.
MAT07 - 1.21.4	The minimum wall thickness of all extruded cover plates shall not be less than 15/10 mm.
MAT07 - 1.21.5	Extrusion tolerances shall conform to relevant European standards.
MAT07 - 1.21.5.1	Alloy 6063-T6 (typical extrusions) or similar.
MAT07 - 1.21.5.2	Alloy 6061-T6 (brackets).
MAT07 - 1.21.5.3	Minimum wall thickness 2.5mm for structural elements.
MAT07 - 1.21.5.4	minimum wall thickness 1.2mm for beads, etc.
MAT07 - 1.22	Welding of aluminium plate and extrusions
MAT07 - 1.22.1	All welding shall be carried out and tested in compliance with current European Standards. The Contractor shall demonstrate that the quantity of testing and the acceptance criteria retained for the interpretation of these tests are suited to the end-use of the welded assembly.
MAT07 - 1.22.2	The alloys to be welded shall be pre-selected for their weld ability. The exact alloy designation shall be submitted to the Project Architect for approval prior to fabrication.
MAT07 - 1.22.3	Loss of material strength due to the welding process shall be clearly stated and allowed for in design, either by a local increase in material thickness, or by positioning of the welded section in a zone where design stress levels are sufficiently reduced.

MAT07 - 1.22.4	All necessary care shall be taken during the weld preparation & welding process to ensure that straightness tolerances are met.
MAT07 - 1.22.5	Adequate allowance shall be made for localized thermal expansion during the welding process. The sequence of welding shall take all necessary care to avoid development of locked-in stresses due to inhibited thermal expansion.
MAT07 - 1.22.6	All visible weld beads shall be removed prior to application of surface finish and welds shall be ground flush.
MAT07 - 1.22.7	All weld testing certificates shall be submitted to the Project Architect for approval prior to installation.
MAT07 - 1.23	Surface finish
MAT07 - 1.23.1	Fluorocarbon finish: Aluminium surfaces exposed to view under outdoor circumstances shall receive a factory applied, oven-cured, two coat, one bake Valspar Mica Flake finish based on Kynar 500 fluoropolymer resin. The primer coat shall be an inhibitive primer with dry film thickness averaging 10 microns to 15 microns.
MAT07 - 1.23.2	The finish colour Mica Flake topcoat shall have a minimum dry film thickness of 30 microns. The total minimum dry film thickness of the system shall be 40 microns. Recessed, unexposed areas such as inside corners and channels are to be visually covered to the extent possible.
MAT07 - 1.23.3	Aluminium Surfaces Exposed to Indoor View:
MAT07 - 1.23.3.1	All aluminium surfaces exposed to view under indoor circumstances shall receive a factory applied, oven-cured powder coating.
MAT07 - 1.23.3.2	The dry film thickness shall be minimum 60-80 microns. Recessed, unexposed areas such as inside corners and channels are to be visually covered to the extent possible.
MAT07 - 1.23.4	Finish shall be applied by an applicator licensed by the formulator and applied in strict conformance with their specification for cleaning, priming, finish coat application and quality control. Applicator shall provide written certification of applicator license to Project Architect/Consultant by a formulator for approval prior to application of the finish.
MAT07 - 1.23.5	Colour variation between adjacent parts of the same finish shall be imperceptible to the naked eye under normal daylight conditions. The sub-contractor shall submit range samples that define the maximum variation of colour that can be anticipated in the work for approval by the Project Architect.
MAT07 - 1.23.6	The sub-contractor shall establish and submit for approval a quality control program to assure compliance with the specified requirements. The program shall include documented procedures, processes, testing, etc. Sub-contractor shall maintain complete certified inspection, testing and process records. Said records shall be available to the Project Architect and Consultant. No finishing shall be performed prior to approval of this quality control program. The sub-contractor shall not ship any material that has not been inspected and tested in the prescribed manner and does not fall within the prescribed colour range, or has been rejected by the Project Architect.
MAT07 - 1.23.7	Field touch-up of painted aluminium components shall only be performed with written approval from the Project Architect. Only an 'air dry' system is allowed, and field touch-up materials and procedures shall be performed in strict accordance with paint manufacturer's written instructions.
MAT07 - 1.23.7.1	Damaged material as determined by the Project Architect shall be replaced with new material.
MAT07 - 1.23.7.1	Sub-contractor shall perform a sample repair as instructed by the Project Architect and shall not proceed until the sample is approved for workmanship and colour uniformity.
MAT07 - 1.24	Anodizing
MAT07 - 1.24.1	Selected anodizer: Submit details before commencement of anodizing, including:
MAT07 - 1.24.1.1	Name and contact details.
MAT07 - 1.24.1.2	Details of accreditation schemes.
MAT07 - 1.24.1.3	Technical data of product including current certificates.
MAT07 - 1.24.2	Working Procedures
MAT07 - 1.24.2.1	Standard: To BS 3987 (ASTM B580) for anodic coatings on wrought aluminium.
MAT07 - 1.24.3	Anodizer Requirements
MAT07 - 1.24.3.1	Processing:
MAT07 - 1.24.3.2	Approved: By the Aluminium Finishing Association.
MAT07 - 1.24.3.3	Certified: To EN ISO 9001 (ASTM E1212).
MAT07 - 1.24.3.4	Anodizing plant: Each anodizer to use only one plant.
MAT07 - 1.24.4	Guarantees
MAT07 - 1.24.4.1	Anodizer guarantees: Submit sample copies before commencement of anodizing.
MAT07 - 1.24.4.2	Project specific guarantees: Submit signed copies on completion of work.
MAT07 - 1.24.4.3	Guarantees to cover:

MAT07 - 1.24.4.4	Life expectancy.
MAT07 - 1.24.4.5	Colour: Opacity and consistency.
MAT07 - 1.24.4.6	Texture: Gloss, satin or matt.
MAT07 - 1.24.4.7	Quality of coating.
MAT07 - 1.24.5	Control Samples
MAT07 - 1.24.5.1	Sequence: Prior to ordering materials for the works, obtain approval of appearance for:
MAT07 - 1.24.5.2	Anodic coated samples: Showing colour and texture variation.
MAT07 - 1.24.5.3	Fabrication samples: Showing joint assembly, how anodic coating is affected and how cut metal edges are finished and protected.
MAT07 - 1.24.6	Independent Inspection At Plant
MAT07 - 1.24.6.1	Requirement: Contractors/ suppliers of the following designated components must commission an approved Independent Inspection Authority to carry out acceptance inspections to confirm that anodic coating application complies with this specification.
MAT07 - 1.24.6.2	Designated components: As required by the Project Architect.
MAT07 - 1.24.6.3	Acceptance inspections: Carry out for each variation of colour and finish of each component work package at anodizer's plant prior to any fabrication of units, in accordance with the following:
MAT07 - 1.24.6.4	Where three of more production runs are required for application of coatings, not less than three acceptance inspections must be carried out in accordance with BS 6001-1 (ASTM E223), general inspection level 2, with an acceptance quality limit of 1%.
MAT07 - 1.24.6.5	Where less than three production runs are required for application of coatings, one acceptance inspection must be carried out in accordance with BS 6001-2 (ISO 2859-2), with a limiting quality of 5% where the probability of acceptance is 10%.
MAT07 - 1.24.6.6	Components failing inspection: Reprocess or replace and re-inspect.
MAT07 - 1.24.6.7	Inspection reports: Independent Inspection Authority must submit copies.
MAT07 - 1.24.7	Quality Assurance System
MAT07 - 1.24.7.1	Requirement: Powder and coating application to the following designated components is to be tested and approved in accordance with the Qualanod system.
MAT07 - 1.24.7.2	Designated components: As required by the Project Architect.
MAT07 - 1.24.8	Component Design
MAT07 - 1.24.8.1	Condition of components to be anodized:
MAT07 - 1.24.8.2	To comply with relevant recommendations of BS 4479-1, and -5 (ASTM D3933) of suitable size to fit plant capacity.
MAT07 - 1.24.9	Pre-treatment
MAT07 - 1.24.9.1	Condition of components to be anodized:
MAT07 - 1.24.9.2	Free from corrosion and damage.
MAT07 - 1.24.9.3	Suitable for and compatible with the pre-treatment and anodizing process.
MAT07 - 1.24.9.4	Process: In accordance with the specification requirements for the finish.
MAT07 - 1.24.10	Extent Of Anodic Coatings
MAT07 - 1.24.10.1	Application: To visible component surfaces, and concealed surfaces requiring protection. Coated surfaces will be deemed 'significant surfaces' for relevant BS 3987 performance requirements.
MAT07 - 1.24.11	Application Of Anodic Coatings
MAT07 - 1.24.11.1	Surfaces to receive anodic coatings: Clean.
MAT07 - 1.24.11.2	Commencement of anodic coating: To be continuous from pre-treatment.
MAT07 - 1.24.11.3	Jig points: To be agreed. Not on visible areas of anodic coated components.
MAT07 - 1.24.11.4	Use of touch-up paint: Not acceptable.
MAT07 - 1.24.12	Performance And Appearance Of Anodic Coatings
MAT07 - 1.24.12.1	Standard: To BS 3987 (ASTM B580).
MAT07 - 1.24.12.2	Visual inspection after anodizing: Significant surfaces to be free from visible coating/ defects when viewed from a distance of not less than 5 m for external and 3 m for internal applications.
MAT07 - 1.24.13	Fabrication
MAT07 - 1.24.11.1	Units may be assembled:
MAT07 - 1.24.11.2	Before anodizing, providing sufficient drainage holes are included in components to fully drain components.
MAT07 - 1.24.11.3	From components anodized after cutting to size.

MAT07 - 1.24.11.4	Where approved, from components anodized before cutting to size.
MAT07 - 1.24.11.5	Exposure of uncoated background metal: Not acceptable.
MAT07 - 1.24.11.6	Assembly sealants: Compatible with anodic coatings. Obtain approval of colour if sealants are visible after fabrication.
MAT07 - 1.24.14	Damaged Components - Repair/ Replacement
MAT07 - 1.24.14.1	Before delivery to site: Check all components for damage to anodic coatings. Replace damaged components.
MAT07 - 1.24.14.2	Site damage: Submit proposals for repair or replacement.
MAT07 - 1.24.15	Protection
MAT07 - 1.24.15.1	Anodic coated surfaces of components: Protect from damage during handling and installation, or by subsequent site operations.
MAT07 - 1.24.15.2	Protective coverings: Must be:
MAT07 - 1.24.15.3	Resistant to weather conditions.
MAT07 - 1.24.15.4	Partially removable to suit building in and access to fixing points.
MAT07 - 1.24.15.5	Protective tapes in contact with anodizing to be:
MAT07 - 1.24.15.6	Low tack, self adhesive and light in colour.
MAT07 - 1.24.15.7	Applied and removed in accordance with tape and anodizers recommendations.
MAT07 - 1.24.15.7	Inspection of protection: Carry out weekly. Promptly repair any deterioration or deficiency.
MAT07 - 1.24.16	Documentation
MAT07 - 1.24.16.1	Submit the following information for each batch of anodic coated components:
MAT07 - 1.24.16.2	Supplier.
MAT07 - 1.24.16.3	Trade name.
MAT07 - 1.24.16.4	Colour (if required).
MAT07 - 1.24.16.5	Batch and reference number.
MAT07 - 1.24.16.6	Statutory requirements.
MAT07 - 1.24.17	Completion
MAT07 - 1.24.17.1	Protection: Remove.
MAT07 - 1.24.17.2	Cleaning and maintenance of anodic coatings: Carry out in accordance with procedures detailed in anodizer's guarantees.
MAT08 - 1.0	**MAT08 – GLAZING**
MAT08 - 1.1	Glazing panels as described shall be integral to the glazed systems described in other parts of this document and as indicated on the Design Drawings to provide a consistent design irrespective of material.
MAT08 - 1.2	Certain glazing panels shall be used for both vertical and inclined conditions. Make any modifications necessary for use in all intended conditions, whilst maintaining the visual requirements. Laminated glass shall be used for single glazed units and the inner/ lower panel of multiple glazed units for inclined/ overhead conditions.
MAT08 - 1.3	Where edges of glazing are exposed, that is, bonded and unframed, they shall be suitable for open jointing or weather sealing jointing as described and indicated on the Design Drawings. Propose the treatment to the visible glass edge for acceptance by the Project Architect.
MAT08 - 1.4	Where applicable, include a solid black applied finish to the perimeter of the glazing to conceal the silicone bonding and frame behind, as accepted by the Project Architect through sampling.
MAT08 - 1.5	Spacer bars to double glazed units shall be black.
MAT08 - 1.6	Select all glazing materials and systems to comply with the performance requirements.
MAT08 - 1.7	The final glass thickness, type, make-up, type of breakage and containment in the various locations shall be determined, paying full attention to the safety requirements. Final selection of glass thickness, together with type and location of interlayer and coatings shall be specified by Security Consultant.
MAT08 - 1.8	References to applied coatings to glass such as but not necessarily limited to the low emissivity.
MAT08 - 1.9	Any high performance or acoustic qualities shall not affect the appearance / colour of the glass beyond what is accepted by the supervisor sample.
MAT08 - 1.10	When indicated on the drawings herein finishes as described will be applied to glass panels.
MAT08 - 1.11	The marking of visibility glazing should be included to comply with the requirements of this document and samples for validation of textures and colours must be submitted to the Project Architect.
MAT09 - 1.0	**MAT09 – MEMBRANES**
MAT09 - 1.1	Waterproof Membrane:

MAT09 - 1.1.1	Membrane shall be inert, durable, impermeable, rot-proof, vermin-proof and not degradable by moisture and extreme temperatures, and manufactured from ethylene propylenediene M-class rubber (EPDM) or acceptable equivalent.
MAT09 - 1.1.2	Membranes shall be certified to an internationally recognised certificate to the acceptance of the Project Architect.
MAT09 - 1.1.3	Membranes shall be of limited combustibility and achieve Euroclass Class B propagation and surface spread of flame, unless the Detailed Design of the system protects the membrane from the need for such requirements.
MAT09 - 1.2	Breather Membrane
MAT09 - 1.2.1	Membrane shall be water resistant, water vapour permeable and manufactured from non-woven, spunbonded high density polyethylene, or acceptable equivalent.
MAT09 - 1.2.2	Membranes shall be certified to an internationally recognised certificate to the acceptance of the Project Architect.
MAT09 - 1.2.3	Vapour resistance of the breather membrane shall not be greater than that of the substrate it is protecting.
MAT09 - 1.2.4	Membranes shall be of limited combustibility and achieve Euroclass Class B propagation and surface spread of flame, unless the Detailed Design of the system protects the membrane from the need for such requirements.
MAT09 - 1.3	Vapour Control Layers/ Air Leakage Barrier (Vapour Barrier)
MAT09 - 1.3.1	High performance reinforced membranes of metal foil or plastics protected both sides by rigid facings/ linings to meet the performance criteria specified. Foil adhered to plasterboard shall not be accepted.
MAT09 - 1.3.2	The product shall also perform as an air barrier.
MAT09 - 1.3.3	Membranes shall be certified to an internationally recognised and in acceptance of the Project Architect.
MAT09 - 1.3.4	Vapour resistance of the material shall be correct for the specific application.
MAT09 - 1.3.5	Include all necessary tapes, seals and accessories to provide a fully vapour and air sealed system at joints, perimeter conditions and penetrations.
MAT10 - 1.0	**MAT10 – THERMAL INSULATION**
MAT10 - 1.1	Materials
MAT10 - 1.1.1	Mineral fibre products shall comply with EN 13162 (ASTM C553).
MAT10 - 1.1.2	Cellular glass (CG) products shall comply with EN 13167 (C552).
MAT10 - 1.2	Particular Requirements
MAT10 - 1.2.1	Insulation behind cladding elements shall be inert, durable, rot-proof and vermin-proof and not be degradable by moisture, extreme temperatures or water vapour, unless the Detailed Design of the system protects the insulation from the need for such requirements.
MAT10 - 1.2.2	The insulation shall not bulge, sag, delaminate or detach during its installation or in its installed position during the life of the works.
MAT10 - 1.2.3	All combustible foam products shall be fire resistant modified. Fire rating requirements to be as defined in Fire Strategy Report from Fire and Life Safety Consultants.
MAT10 - 1.2.4	Insulation shall be a minimum of Euroclass B2 and must conform to all relevant standards.
MAT10 - 1.2.5	Insulation shall have zero Ozone Depleting Potential (ODP), be CFC and HFC free and have a Global Warming Potential (GWP) of less than five. Refer also to sections 10.1 - 10.3.
MAT10 - 1.2.6	Expanded polystyrene (EPS) shall not be permitted within the works.
MAT11 - 1.0	**MAT11 – GASKETS**
MAT11 - 1.1	Gaskets shall be made of either Ethylene Propylene material (EPDM) or of Silicone.
MAT12 - 1.0	**MAT12 – ADHESIVES**
MAT12 - 1.1	Determine suitable adhesives to achieve the requirements of this document.
MAT12 - 1.2	Adhesives for timber shall be to EN 204 or to the manufacturer's specification, as applicable.
MAT12 - 1.3	Adhesives shall be compatible with the proposed finishes and any preservative/ fire retardant treatments.
MAT13 - 1.0	**MAT13 – SEALANTS**
MAT13 - 1.1	Sealant products shall be used in accordance with the system manufacturer's recommendations, to suit the service conditions.
MAT13 - 1.2	Sealant shall not leak or bleed causing any discolouration or staining.
MAT13 - 1.3	Sealants shall not compromise the integrity of the works.
MAT13 - 1.4	Apply as a continuous bead unless specified or recommended otherwise by the manufacturer.
MAT13 - 1.5	Joints

MAT13 - 1.5.1	Primer, backing strip, bond breaker: Types recommended by sealant manufacturer.
MAT13 - 1.6	Suitability Of Joints
MAT13 - 1.6.1	Pre-sealing checks:
MAT13 - 1.6.1.1	Joint dimensions: Within limits specified for the sealant.
MAT13 - 1.6.1.2	Substrate quality: Surfaces regular, undamaged and stable.
MAT13 - 1.6.2	Joints not fit to receive sealant: To be agreed with the Project Architect.
MAT13 - 1.7	Preparing Joints
MAT13 - 1.7.1	Surfaces to which sealant must adhere:
MAT13 - 1.7.1.1	Remove temporary coatings, tapes, loosely adhering material, dust, oil, grease, surface water and contaminants that may affect bond.
MAT13 - 1.7.1.2	Clean using materials and methods recommended by sealant manufacturer.
MAT13 - 1.8	Applying Sealants
MAT13 - 1.8.1	Substrate: Dry (unless recommended otherwise) and unaffected by frost.
MAT13 - 1.8.2	Environmental conditions: Do not dry or raise temperature of joints by heating.
MAT13 - 1.8.3	Sealant application: Fill joints completely and neatly, ensuring firm adhesion to substrates.
MAT13 - 1.8.4	Sealant profiles:
MAT13 - 1.8.4.1	Butt and lap joints: Slightly concave.
MAT13 - 1.8.4.2	Fillet joints: Flat or slightly convex.
MAT13 - 1.8.5	Protection: Protect finished joints from contamination or damage until sealant has cured.
MAT14 - 1.0	**MAT14 – TIMBER**
MAT14 - 1.1	Strength and stiffness parameters shall be determined on the basis of tests for the types of action effects to which the material will be subjected to, or on the basis of comparisons with similar timber species and grades, or wood-based materials, or on well-established relations between the different properties.
MAT14 - 1.2	Timber members shall comply with EN 14081 - 1 (NDS - 2015).
MAT14 - 1.3	Glued laminated timber members shall comply with EN 14080 (NDS - 2015).
MAT14 - 1.4	Laminated veneer lumber structural members shall comply with EN 14374 (NDS - 2015).
MAT14 - 1.5	Adhesives
MAT14 - 1.5.1	Classification of structural wood adhesives to follow EN 301 (D7374 - 08).
MAT14 - 1.5.2	Adhesives to be tested following EN 302 (D3930 - 08).
MAT14 - 1.6	Metal fasteners shall comply with EN 14592 and metal connetors shall comply with EN 14545.
MAT15 - 1.0	**MAT15 – REINFORCED CONCRETE**
MAT15 - 1.1	Concrete mixes should be specified in accordance with BS 8500 - 1 (ACI 318 - 14).
MAT15 - 1.2	Aggregates should be of a grade which ensures adequate durability of the concrete.
MAT15 - 1.3	Admixtures are permitted in accordance with EN 206 - 1 (ASTM C494 / C494M - 17).
MAT15 - 1.4	Concrete cover to reinforcement should be adequate not only for exposure condition but also, where necessary, to resist fire.
MAT15 - 1.5	Testing of concrete in situ shall comply with EN 12504 (ASTM C42-136).
MAT15 - 1.6	Where the ends of slabs are cast monolithically with concrete members, reinforcement should be provided in accordance with EN 1992 - 1 - 1 to avoid developing surface cracking over the supports.
MAT16 - 1.0	**MAT16 – BRICK**
MAT16 - 1.1	Quality test
MAT16 - 1.1.1	All test cubes shall be made and tested in accordance with the requirements of BS4551.
MAT16 - 1.1.2	7 and 28-day mean compressive strength of tested specimens shall attain the strength specified.
MAT16 - 1.2	All bricks shall have a minimum average crushing strength of not less than 20 MPa (2900 psi).
MAT16 - 1.3	To be the best of their respective kind; hard square cornered and sound in accordance with BS 4729.
MAT16 - 1.4	The minimum bearing for lintels onto blockwork shall be 300mm (11 13/16 inch).
MAT16 - 1.5	Metal embedded in external solid walls shall be grade 18/8 austenitic stainless steel.
MAT16 - 1.6	Reinforcement: Where required in brickwork shall be approved.
MAT16- 1.7	Sealants used in brickwork joints shall be polysulphide stored in accordance with the manufacturer's instructions

MAT16 - 1.8	Color and texture of bricks delivered shall be approved by the Architect
MAT16 - 1.9	Unloaded by hand or machine onto a reasonably dry and level site and not tipped from vehicles
MAT17 - 1.0	**MAT17 – CONCRETE BLOCKWORK**
MAT17 - 1.1	To be carried out in accordance with BS 5628 (ASTM C170).
MAT17 - 1.2	Hard square cornered and sound in accordance with BS 6073 (ASTM C150).
MAT17 - 1.3	Mixing, molding and curing shall be in accordance with BS 4551 (ASTM C55 - 17).
MAT17 - 1.4	Evidence of quality control to be produced from suppliers as referred in BS 3921 (ASTM C62).
MAT17 - 1.5	Unloaded by hand or machine onto a reasonably dry and level site and not tipped from vehicles.
MAT17 - 1.6	Any pigment used in mortar must conform to the requirements of BS 1014. Pigment containing 'carbon black' will not be permitted.
MAT17 - 1.7	Metal embedded in external solid walls shall be grade 18/8 austenitic stainless steel.
MAT17 - 1.8	The minimum bearing for lintels onto blockwork shall be 300mm (11 13/16 inch).
MAT18 - 1.0	**MAT18 – GLASS FIBER REINFORCED POLYMER**
MAT18 - 1.1	The fabricator design composite systems to achieve a minimum performance level for interior and exterior applications consistent with a Class A per IBC Section 803 (ASTM E84).
MAT18 - 1.2	Screening process to be used with Cone Calorimeter piror to fabrication (ASTM E1354).
MAT18 - 1.3	Detailling of anchorages, connections, and joints must allow for differential dimensional changes of FRP components and the supporting primary structure due to thermal effects or differential deflection.
MAT18 - 1.4	Prefabricated stiffeners may by FRP components or metal members that are placed on the composite panels or embedded in the structure during panel fabrication.
MAT18 - 1.5	Reinforcement content shall be +-10% of the specified fibre volume fraction.
MAT18 - 1.6	Maximum permissible warpage of one corner out of the plane of the other three shall be 5 mm/m (1/8inch/ft).
MAT19 - 1.0	**MAT19 – ETFE**
MAT19 - 1.1	ETFE Film
MAT19 - 1.1.1	Number and thickness of layers to be determined according to thermal and structurral specifications.
MAT19 - 1.1.2	Color and printing of the ETFE shall be applied to respond to the required solar specifications.
MAT19 - 1.2	Cable and end fitness
MAT19 - 1.2.1	All cables in contact with the membrane shall be white PVC coated.
MAT19 - 1.2.2	All Structural wire rope cables except stainless steel, shall conform to the latest revision of ASTM A603.
MAT19 - 1.3	Aluminum Clamping system
MAT19 - 1.3.1	Structural aluminum clamping system shall be Aluminum Association alloy 6061-T6 or alloy 6063-T6 or EN AW-6060-T66.
MAT19 - 1.3.2	Bent plates shall be formed and then heat -treated to T6or T66, as applicable.
MAT19 - 1.3.3	All Structural "U Straps" shall be stainless steel type ALSI 316.
MAT19 - 1.4	ETFE system supporting structure shall not rely on the ETFE film for structural stability.
MAT19 - 1.5	Fire Performance: Burning Characteristic (ASTM E84), Class A
MAT20 - 1.0	**MAT20 – PTFE / GLASS FIBRE CLOTH/ MEMBRANE**
MAT20 - 1.1	X-ray radiographic examinations can be used for detection of macroscopic flaws.
MAT20 - 1.2	Specific gravity can be readily measured by water-displacement and gradient-tube techniques as per ASTM D792 or D1505.
MAT20 - 1.3	Tensile strength to achieve a minimum of 19.3 MPa (2800 psi) following ASTM D638.
MAT20 - 1.4	Sintering temperature exceed the melting point of PTFE degreeC (648 degreeF) and range from 360 to 380 degreeC (680 to 716 degreeF).
MAT20 - 1.5	Lab tests shall be conducted at 23+-2 degreeC (73.4 +- 4 degreeF). Since the resin does not absorb water, the maintenance of constant humidity during testing is not required (ASTM D4745 - 06).
MAT20 - 1.6	Use bolting load less than 6.9 MPa (1000 psi) for unconfined gaskets.

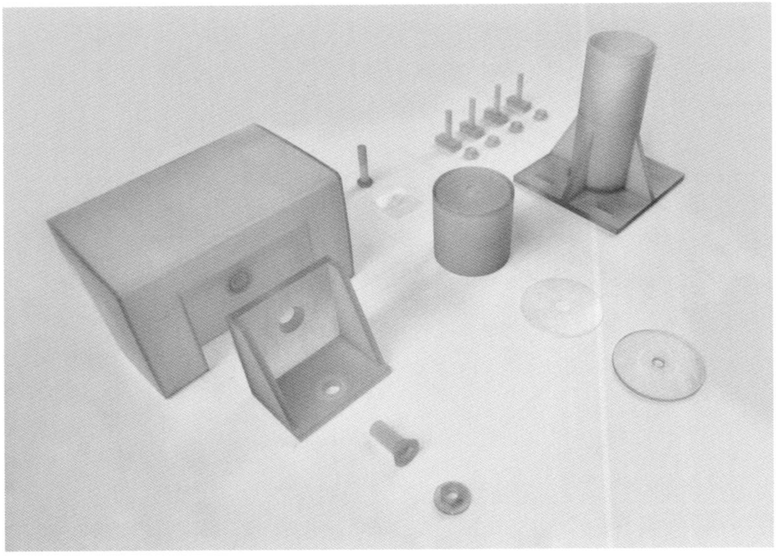

a | Elements assembly

b | Assembled bracket

Prototyping is a critical process during the design stage as it represents the first step in turning a concept into a physical output. Through a prototype the designer can test a number of physical, visual and performance qualities of the design during its development. These range from general massing of building blocks to testing material qualities, and the design of assemblies.

There are many benefits of prototyping during design. Small scale modelling is affordable and can be done in-house by the design team. Models are generally used as effective tools for presentations to clients or other design team members, as they can be used to effectively explain design solutions to a non-technical audience, or to other members of the design team who do not share the same specialist knowledge. Prototypes also allow the designer to design bespoke assemblies and test concepts before approaching manufacturers, in order to be able to convey the specific requirements of the design developed.

Working with physical models at an early stage in the design process can help to identify design issues that might not be evident from a two dimensional drawing, or even a 3D digital model, particularly if the prototype contains a series of components within a complex assembly, which can also include movable parts. Being able to identify issues at an early stage reduces the risk of major changes later in the design and construction process, where the impact on cost would be higher. The purpose of building prototypes during the design process is to allow the designer to iterate more quickly on a design concept and gather feedback more effectively from the various parties involved in the design.

Communication across the design team, and between designers and contractors, is critical, especially in the case of large and complex projects. Being able to see a physical model at either 1:1 or reduced scale allows others to immediately understand the design concept and in turn be able to provide feedback more effectively.

Prototypes, especially 3D prints can be an effective tool to interact with fabircators and explore construction tolerances and adjustabiltiy required for construction and installation of building systems. Prototypes built to test construction assemblies can assist in defining critical design constraints, such as joint widths in facade cladding where installation techniques need to be considered. Building models at 1:1 scale gives a realistic perspective of scaling and the space available to operate the tools required to fix components in place.

Prototyping can also be used to quickly compare design variations but also to explain the design either to the client or other members of the design team through direct comparisons. For example, through communicating a strategy to a stake holder by comparing one assembly technique that may lead to an apparent design cost in fabrication but aims to save time and cost during installation and reduce the risk of delays to project; another strategy would be cheap to fabricate but risky, time consuming and potentially dangerous to install, it becomes much easier to grasp the viability when physically attempting both options.

Prototyping is a very informative process for many aspects of design development. With assemblies especially, a prototype can be used to confirm both the design and specification of a component. The exact number of parts required be easily be determined and the sizing of those parts can be confirmed.

The relationship of components can be effectively demonstrated through physical modelling and any allowances for tolerances can be tested, which is particularly useful if they are designed to work in three axis.

Prototypes can be used to test the relationship and installation processes of one component with another to check the feasibility of the assembly.

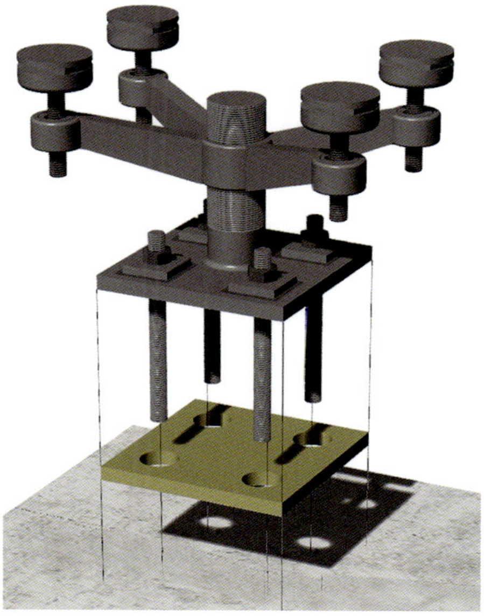

c | Digital model of the bracket

d | Element model of the bracket system

e | Bracket's prototype based on the digital model

f | Image of the final product: steel glazing bracket

There are different types of models which play different roles in the design process from massing studies, to assessing visual appearances and assemblies. The three most common forms of prototyping in the design process are visual, functional, and manufacturing prototypes which supply a combination of both visual and functional qualities to test.

Visual Prototypes

Visual prototypes are typically built for early stage design massing studies and overall site models, as well as detailed publicity and presentation models that aim to express the architectural intent of the project.

The primary aim of visual mock-ups is to represent the overall design of a project through building scale, as well as the basic geometrical form in relation to the site and the individual spaces that make up the whole. They become design tools such as site models and basic building mass models, which can be used for defining site organisation, and building placement, and orientation.

They are useful for organising both large multi-building concepts on a single site, as well for arranging and orienting single buildings on a site. For sites with stricter constraints or pre-defined building placement, these models can be used for understanding the context, and available visual and orientation options of internal spaces. As well as producing prototype models of a building on site, it can also be a valuable exercise to prototype the key spaces that make up the building.

These models allow for quick iterations and tests of spatial compilations before defining the overall form and geometrical detail of the design. If the project is a public building, these models can be used to determine the entrances and exits, their locations relative to the site context, and how the general flow of people will work traversing to and from the building as well as through it. This is an important consideration for constructing an accessible building.

Another type of visual prototyping which focuses on the aesthetic qualities of a design, are presentation models. These models are usually quite detailed and aim to give an impression of the desired final design intent both materialistic and geometrically. They bring to life the design for the client and allow a further iteration of internal spatial organisation as well as representation of the material qualities and systems chosen. These models can either be static and purely visual or also have a dual function to describe the construction sequence through the way they are assembled.

Visual prototypes can be made quickly with low budget materials as a way to express a concept or design. The typical materials used for producing visual prototypes depend on the purpose of the model. Card is the most common for spatial modelling and is used for its rigidity. Models can also be made out of acrylics and plastic sheet, for transparency purposes to either represent windows or represent physical boundaries without affecting visual boundaries. Other modelling materials include metal and wood depending on the desired material qualities and finishes required.

The techniques for making massing models are generally by hand as table top models, in workshops, or with the aid of a laser cutting machine either in office or outsourced. They tend to be made from flat two dimensional components due to the low material and skill requirements, and represent the first step in fabrication considerations during the design process. The objective of these models is to be made quickly and economically. This is particularly useful for buildings of complex geometries, that do not necessarily have clearly defined wall and roof systems and blur the boundaries of interface.

For the more detailed visually driven models, they will take a longer time to make and be more costly depending on the designer's available resources, but can be used throughout the design process as a reference point, presentation and marketing tool.

a | Digital model showing shingled glazing system supported on steel tubes

b | Prototype of the shingled glazing system

c | Fixing bracket - 3D digital model

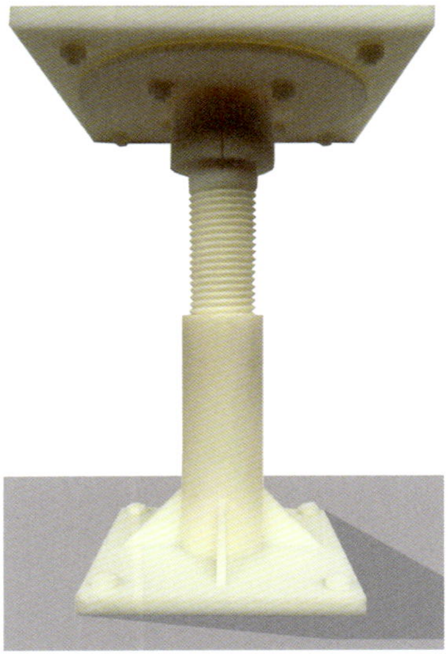

d | Fixing bracket - 3D printed prototype

e | Fixing bracket - Manufactured prototype

Functional prototypes

Functional prototypes become useful and valuable during the detailed design stages, and are used as the first step to test design concepts in a physical form and progress those designs from conceptual to constructible. The information in this book is aligned to allow the designer in making an early stage prototype of this category.

When designing building systems it is important to consider the assembly process, and take into account construction tolerances and feasibility of installation through 1:1 scale models for components and relative scaled models for full building bay installation sequencing.

The models can be used as a tool to confirm assembly methods as well as to a tool to describe to the client and other design team members what the final assembly method would be. This process saves time on developing complex animated sequences, and is also more useful than a series of static images. Having physical models made up of a set of components allows for discussion and experimentation as well as explanation. The purpose of these models is to explain how systems go together by physically demonstrating the process. As these models are interactive, they also allow the client to be directly involved and understand the processes.

Once a final design prototype is produced for a building element such as a fixing bracket the exact number of components can be scheduled, quantified and added to the specification. This begins the verification process of the design for construction.

Functional prototypes are the first test in communication between the design team and fabricators by gaining an understanding of techniques and any manufacturing constraints that may be of concern for a particular design proposal.

Details

1 | Envelope panels
2 | Brackets
3 | Waterproofing

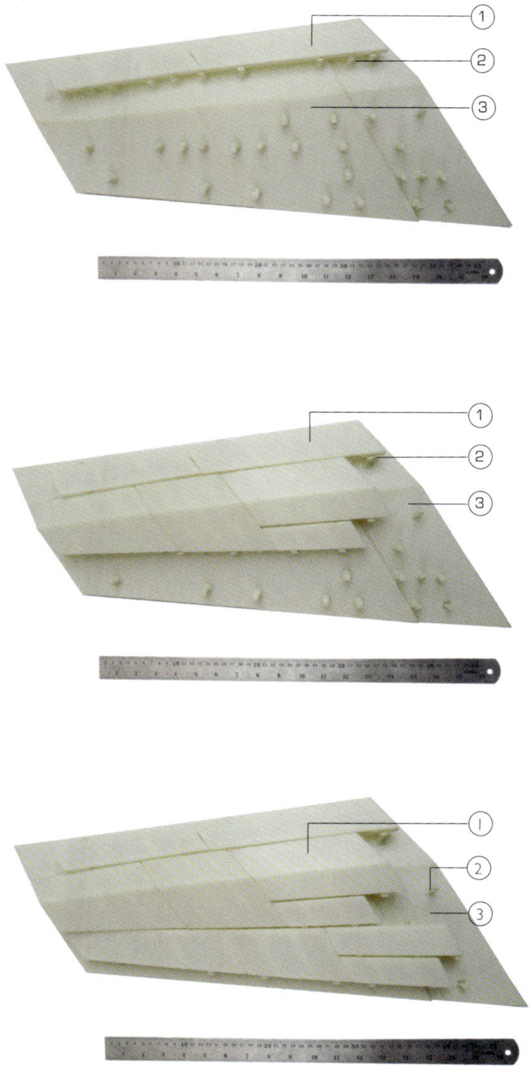

f | Prototype for system assembly

g | Facade design prototype: solution B

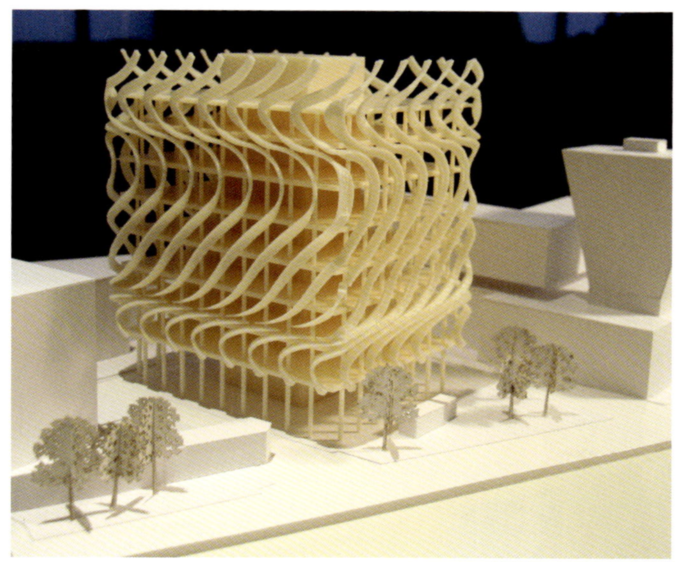

h | Facade design prototype: solution D

Rapid prototyping

One of the most common methods of producing functional mock-ups, especially for bespoke and complex designs is through the technique of rapid prototyping where three dimensional digital models are translated directly into physical models. One of the most common forms of rapid prototyping used in the design industry is 3D printing. This type of printing removes the process of translating a 3D model into a series of 2D parts to then be rebuilt as a physical 3D model, and creates not only complex forms that would be extremely difficult to achieve by hand let alone accurately, but also kinetic models where components are interlinked with each other.

There are currently many applications of 3D printing most commonly with plastics, but also using metal, ceramics and composite materials such as concrete. Depending on the resolution required for accuracy, different 3D printing techniques can be used. The early popular technique of FDM (fused deposition modelling), which involves a less precise process of extruding plastics is very economical for quick basic studies, but is restrictive in form and resolution. A common technique is photopolymerisation which produces a solid from a liquid using light to cure the material which is built up in ultra-thin layers. If using this method with support materials, complex geometrical applications become limitless, and the size and detail of components become more precise. The layering technique can be used to inform the orientation of a model during printing add structural integrity.

3D printing provides an accurate representation of a digital 3D model with a negligible fabrication tolerance. They provide quick solutions but are not representative of actual fabrication techniques, however they can be used in construction process to create casts and molds for fabrication. Printing in plastic for design remains a preference as it provides a lightweight model than can be easily lifted, explored and taken to meetings.

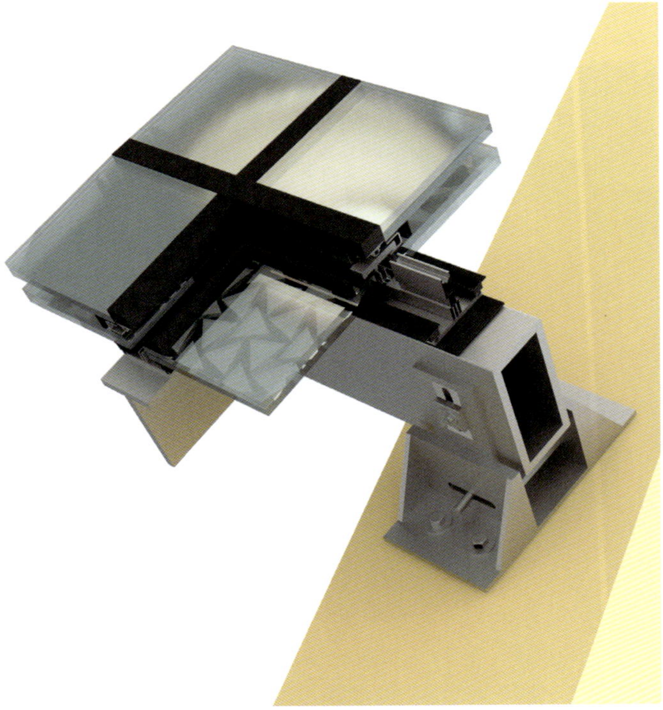

a | Standard toggle-fixed glazing prototype

b | Steel framed glazing prototype

Manufacturing Prototypes

Manufacturing prototypes are the next step in physical modelling during the design stages. These models aim to describe both the function and aesthetics of the developed building systems. They should represent the same manufacturing techniques as the intended fabrication of the final components. Through this process they aim to verify the assembly strategy and final quality achieved, to establish a benchmark for the design. Visually they can be used to test the quality of material finishes of the main system itself, but also any seals or fixings that are also visible, and add to the overall aesthetic of a system.

There are a number of qualities that can be assessed from manufacturer fabricated prototypes that would not be described during the massing and concept modelling processes. The main progression from visual and functional prototypes is to test fabrication processes with the actual materials. When 3D printing, all components are considered equal, and although appear to be an exact representation, they only describe the ideal scenario of an application. Although at this stage manufacture prototypes usually represent typical system conditions, they can also be used to solve complex designs.

When working with the specified materials, the manufacturing prototypes are used to describe system connections and interfaces. By using actual materials and fabrication techniques, the models become more realistic by taking into account not only the design tolerances but also manufacturing tolerances that may have an affect on connections. This includes how components are sealed, and how interface connections are dealt with. Welding metals, for example, may require a larger interface area than has been designed for.

Waterproofing and insulation strategies can be developed and tested to see how they perform within the system, and more importantly at the system interfaces. These prototypes can be used to test how waterproofing is fixed or bonded to the insulation, and highlight any weak points within the system or potential risks of leakage. These issues can only be picked up through physical modelling and testing of systems.

The design team can develop the specification by confirming the desired material finishes based on project requirements both visually for colours, textures and edges, but also performance-based on the required design life, as well as testing the assembly strategy. Composite materials need to be tested for their mix of strength and surface finish. The models can be used as tests for how the precision of a design in digital media can be recreated and to what level of accuracy. This is particularly useful for envelope systems to assess the visual appearance of panel joints, both in alignment and how any fixings or other visible components look, as well as in testing structural connections.

Working with manufacturers and fabricators during the design stages can help keep costs on budget and avoid tender returns that significantly exceed the budget. For complex geometries and bespoke designs, this is very important to initially understand whether systems can even be fabricated or complex shapes can be rationalised. Working with manufacturers allows the design team to understand what technologies are currently available, how techniques are chosen and applied, and highlight any critical fabrication issues or limitations that may affect the design development. It is also important to speak to multiple manufacturers to collect a range of design solutions and issues to understand what the common fabrication approaches are, especialy when dealing with bespoke systems.

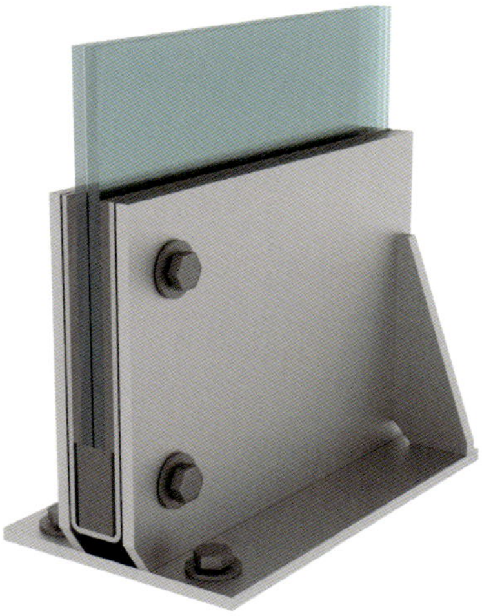

c | Glass balustrade prototype - digital model

d | Glass balustrade prototype – manufactured prototype

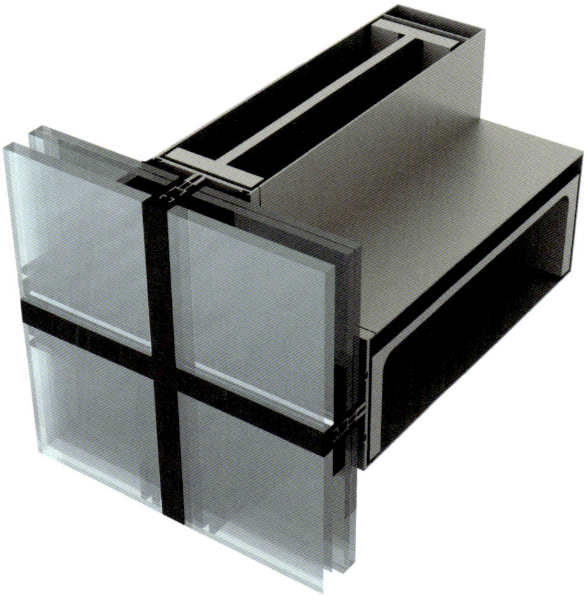

e | Toggle-fixed glazing on steel structure - digital model

f | Toggle-fixed glazing on steel structure – manufactured prototype

g | Unitised aluminium glazing system – digital model

h | Unitised aluminium glazing system – manufactured prototype

a | Building

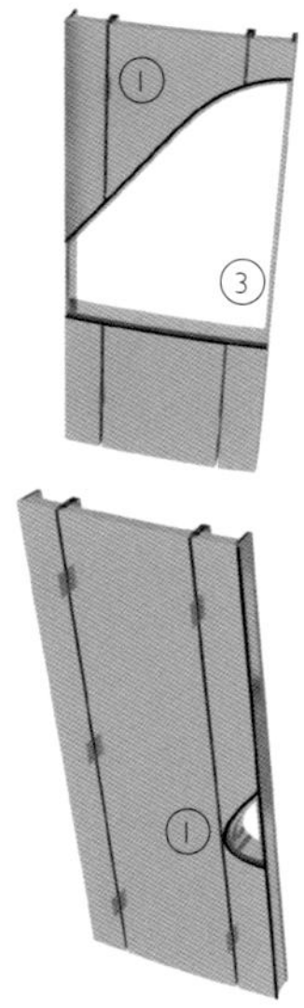

b | Components to be fabricated

c | Fabrication process

d | Physical mock-up

BIM and prototyping at Newtecnic

At Newtecnic, a building information model (BIM) is a 3D digital representation of all the components that comprise the structure, facades, MEP installations and interior construction. The model comprises components which can be identified individually through data attributes and parametric rules for each object in the model.

A BIM model is made of combinations of components and assemblies which are grouped in hierarchies which represent the way in which their fabrication and installation will be undertaken. Some components or assemblies have a generic geometry which is representative only of a more complex set of components, position in space and attributes within. Other components and assemblies are modelled with geometric precision if their forms are required to be part of a geometry-based model. However, all components and assemblies are defined by parameters such as key dimensions or geometrical properties, and their position in 3D space in relation to other components and assemblies. This approach allows the effects of geometry changes to specific components or assemblies to be communicated as a modification to related objects. Changes in geometry of a single facade panel, for example, affects not just the adjacent panels, but all the components to which that panel is fixed. In order to manage evolution of the design, a set of rules is established at an early stage of the project to ensure that modifications can be deployed rapidly across an entire building information model. Changes and modifications are used to develop the model, but they present few opportunities for a fresh start using the same model. Each model has its own rules and parameters; BIM can be used as a primary tool to increase efficiency, productivity and to reduce cost more than to can be used as a tool for continuous design changes.

Each model element carries specific attributes that provide information on quantities and weights of materials, geometry properties, material specification, unique identification numbers and thermal performance. These properties can be output to generate bills of quantities, schedules and specification documents. Additional BIM tools allow for the extraction of 2D views from a building information model for both rapid prototyping and the production of 2D drawings. These different 'viewpoints' within the model are set to suit the specific outputs, from visual communication of design drawings to third angle projection drawings for physical prototyping.

At Newtecnic, building information modelling is a collaborative way of working with fabricators at an early stage of the project, because it holds different design engineering and architectural layout information within the same model. Both architecture-based geometries and spaces are integrated with engineering design assemblies. As a result of the hierarchical structure of Newtecnic BIM models, which function as a series of interconnected files, the coordination between design engineers and fabricators allows for the easy transfer of components and assemblies, in which the model is like a machine assembly. This is different from traditional architectural outputs from BIM, which are based on the overlay of plans, sections and elevations, arranged in parallel planes. The possibility of including construction sequences, cost estimations and project life cycle information makes the Newtecnic BIM model the ideal platform for the project management of the construction of the building. Depending on the level of information required to be exchanged between Newtecnic and fabricator, four levels of model outputs can be applied to a building information model:

- Level 0: 3D CAD models and associated 2D drawings and text, with paper-based or electronic exchange of information. The models are based on the use of lines (vectors) rather than surfaces, so lines define the edges of surfaces in most instances.
- Level 1: 3D CAD models and associated 2D drawings, with the introduction of spatial coordination, a hierarchy of vector (line) based models and associated digital formats. This may include 2D information and 3D information such as visualisations or concept development models. These models are generally shared between different disciplines in a single project team.
- Level 2: A managed 3D environment set within a hierarchy of models, which data attached. Models are created separately for each discipline.

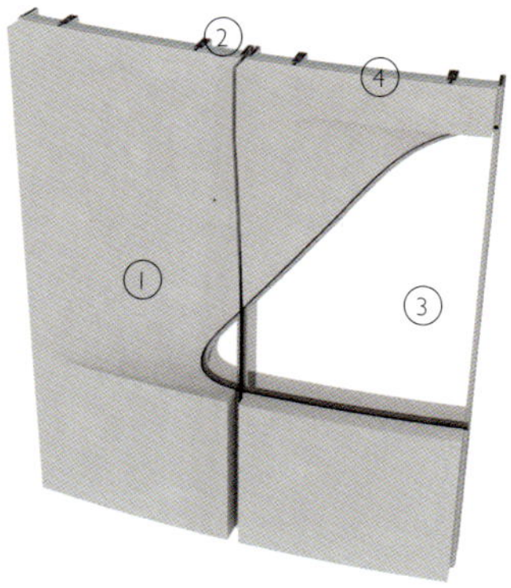

e | Digital model of a GRC cladding mock-up for unitised facade panels

Details

1 | GRC panel
2 | Supporting steel structure
3 | Glazing
4 | Insulation
5 | Steel bracket

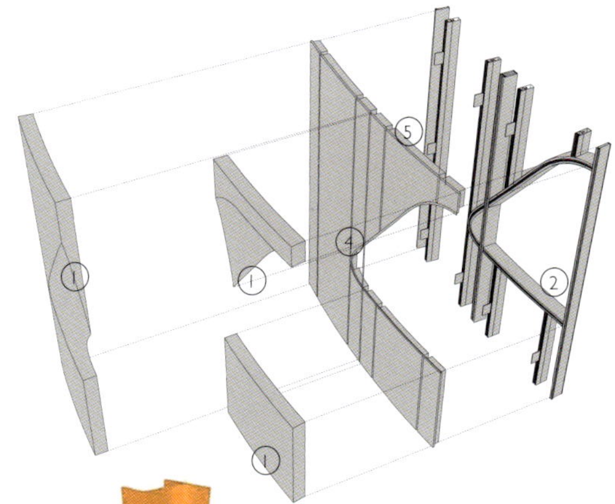

f | Digital model: exploded view of GRC cladding panel mock-up

Level of Development 100 | Level of Development 200 | Level of Development 300 | Level of Development 400 | Level of Development 500

g | Diagram of the levels of development for BIM modelling components

These separate models are assembled to form a single hierarchical model but maintain their independence as stand-alone assemblies. Data may include construction sequencing and associated cost information.
- Level 3: A single collaborative project model with construction sequencing, associated costs and project life cycle information. Although all Newtecnic models are built on cloud servers, this model type is required to start its life on the cloud server, with its future position in a much larger model hierarchy already in place.

The BIM execution plan at Newtecnic

At Newtecnic, a BIM execution plan is specific to any given project and is not a generic plan that can be used across different projects. The BIM execution plan defines the appropriate uses for BIM outputs within a given building design project. In addition, the plan is aligned with the method of describing the detailed design through documentation of drawings, 3D models, specifications, prototypes, descriptions and movies, such as the construction sequence. Once the plan is created, the team of Newtecnic design engineers can monitor their progress against this plan to gain the maximum benefits of coordination from the building information model. The aim in developing this structured procedure is to stimulate planning and direct communication by the project team during the early phases of a project.

The main content in a Newtecnic building information model is as follows:
- Model type across the design engineering disciplines;
- Required level of detail across the design engineering disciplines;
- Purpose of each model across the design engineering disciplines;
- Engineering design discipline responsible for creation of specific models;
- Schedule of dates for specific outputs from models that affect other models (such as geometry);
- Naming of components and assemblies across the design engineering disciplines;
- Use of BIM tools across the design engineering disciplines;
- coordination and clash detection across the design engineering disciplines.

A key term used across different BIM execution plans used in building design is the 'Level of Development' of a given building information model. The Level of Development (LOD) allows the inherent content and the specific uses of models to be set out in written form across the stages design development and construction of a given building project. At Newtecnic, the use of specific 'levels of development' provides a project-specific standard to describe the expected content of model outputs at different stages of the design development and execution of the project. This method enables consistency in communication and execution by facilitating the detailed definition of BIM stage completion times and associated outputs.

At Newtecnic, Levels of Development of a BIM model vary between countries and between projects. As result, the following standards are applied:
- LOD 100: The component or assembly within the model may be represented within the model only by a symbol or simple geometric shape. Information related to components and assemblies, such as unit costs and weight, can be derived from other models.
- LOD 200: The component or assembly is graphically represented within the model as a generic system, object, or assembly with approximate quantities, size, shape, location, and orientation. Text-based information is attached to the component or assembly.
- LOD 300: The component or assembly is graphically represented within the model as either a specific system, or as objects with quantity, size, shape, location, and orientation. Text-based information is attached to the component or assembly.
- LOD 350: In addition to the component or assembly being graphically represented within the model as a specific system or object with quantity, size, shape, orientation, the model contains interfaces between building systems provided by different engineering design disciplines. Text-based information is still attached to the component or assembly.
- LOD 400: The component or assembly is graphically represented within the model as a specific system or object with size, shape, location, quantity, and orientation with additional information from which fabrication and installation drawings can be prepared. Text-based information may also be attached to the component or assembly.

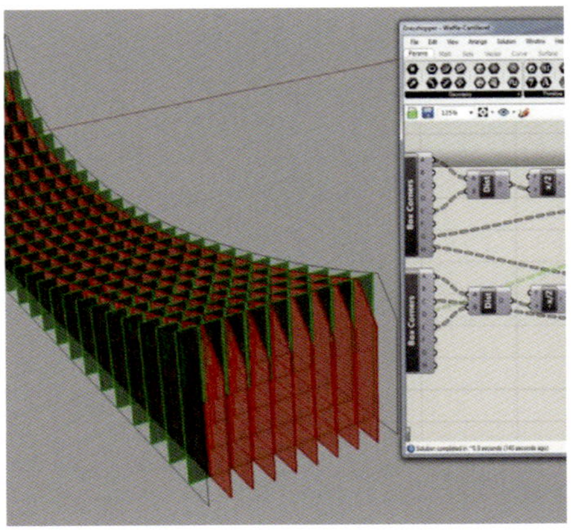
a | Digital 3d model using parametric scripting

b | Construction of a mould

c | Assembly of moulds

- LOD 500: The component or assembly is a field-verified representation in terms of size, shape, location, quantity, and orientation. Text-based information may also be attached to the component or assembly.

The use of BIM in design stages at Newtecnic

Building information modelling allows a project team to make specific design changes to a project during the design or documentation process. However, the underlying parameters of the project are set from the project spatial and organisational program from which the design work begins. This approach gives the engineering design team more time to add value to the design by reducing the number and physical weight of design assemblies in order to reduce cost and increase productivity during the fabrication and construction phases. Whenever a design change is made, the effects of that change can be updated rapidly within the building information model with the benefit of maintaining coordination of components, assemblies and building elements, throughout the period of the development of the project, from initial design to installation. This approach allows the design team to add value faster than with traditional 2D CAD and 3D modelling, because the complete building is constructed digitally long before any fabrication takes place.

A recent development in building information models allows integration with structural analysis and environmental analysis. A further essential benefit is the use of a software user interface called a 'dashboard' where effects of the change of one design parameter can be identified by the resulting change in another parameter. An obvious example is the effect of the arrangement of building volumes, assemblies and components on construction time, cost and associated productivity. Issues of productivity must be checked for safety in both the workshop and the construction site.

In the construction phase, building information modelling makes available concurrent information on construction schedule and cost. Newtecnic design engineers update the building information model for the production of estimates and detailed construction planning. The model is developed in a way that allows a fast update of 2D drawings, supporting specifications and associated schedules. At Newtecnic, building information models are required to use 'clash avoidance' tools that identify the exact location of discrepancies as they appear in the model.

The use of BIM in emerging construction-based tools

Building information models are increasingly able to interact with new tools for construction tasks that increase productivity, reduce waste, and increase site safety. These tools are as follows:

Construction Sequence Simulation

Simulating and animating construction sequences, linked to BIM, are growing in popularity as building projects become increasingly complex in the way they mix and combine functions. Protected public spaces that need to account for pedestrians and transport systems are increasingly being simulated, along with the buildings themselves. Putting them into context helps people better understand how to build harmoniously in tight spaces. Real-world simulations that include noise and air quality impacts, delivery schedules, make worksite management simpler by allowing stakeholders to see both the big picture, and the fine details, are being brought to life by these simulations.

On-site Construction Labs

While off-site manufacturing can be the right method for mass-produced building components, on-site construction labs, as devised by Newtecnic, are ideal for fabricating mass-customised parts from building information models. Bringing raw materials, to be made into components on-site, reduces transportation costs and enables a high level of manufacturing flexibility. Portable, very high-tech production facilities can use data, gathered from LiDAR equipped drones to fine- tune manufacturing so that each component perfectly fits the as-built structure. The movement towards 3D printing is very well accommodated by the use of construction labs which capitalise on the growing trend towards industry-wide adoption of BIM.

Training Workforces Locally

Knowledge transfer from building information models, to workforces in remote locations, can be facilitated through design engineering that ensures construction and installation complexity is greatly simplified. That means very ambitious structures can be built by local workers who, when provided with the right project specific information and training, can achieve world class performance.

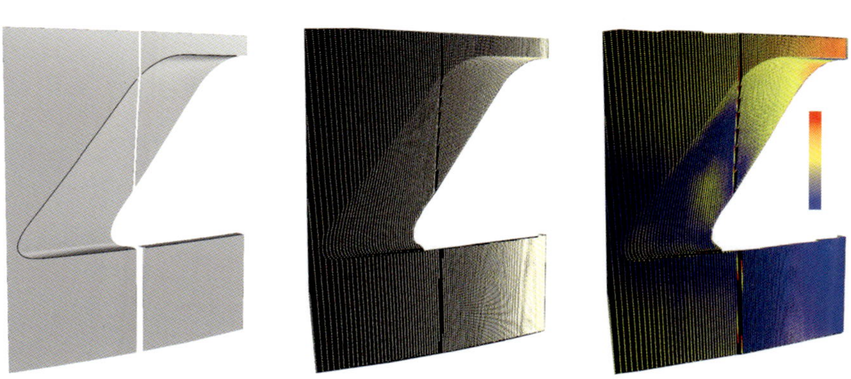

d | 3D scanning for deviations

e | CNC milling a mould for casting

Reducing Crane Use

Crane usage will be reduced by deploying specialist machinery and robotised systems able to more efficiently and safely carry, and potentially autonomously install, large building components. This trend is especially relevant to building refurbishment where, on low but wide buildings with continuous wall and roof facades, it may not be possible to use cranes to lift components into place. The application of BIM in refurbishment work, and associated site access by crane, presents big opportunities for the use of BIM.

Digital User Manuals

Buildings are becoming smarter and more like consumer products such as smart-phones, that can be upgraded throughout their lives. And just like consumer products, smart buildings require digital user manuals that can be continuously updated. Combining BIM data with engineering information, means that current and future occupants and owners can make the best use of resources, adapting and developing buildings and their interconnected machines and systems to make them future-resilient.

Robots and Cobots

Humans have always shared workspace with machines on construction sites. The development of robots, drones and cobots, that is, robots that can work together with people, to automate tasks and help humans with heavy lifting and repetitive work, is underway. This trend, made possible through BIM, will accompany advancements in fully or semi-autonomous robotics.

Inspection of building facades by drones will become safer and more thorough than using top-slung cradles because it will allow operators better views from the comfort of an office with an accompanying building information model. It will not be long before drones and robots go beyond inspection by also delivering components for installation. And because the engineering that makes this possible is being devised now; in less than a decade those parts will increasingly be installed using robots and cobots.

Light-weighting

The weight of a building is increasingly becoming important; each extra kilo requires more energy and resources to manufacture, transport and assemble, as well as to heat, cool, clean and maintain after construction. Immediate and substantial long-term saving can be made when weight is reduced. Therefore, the trend to produce precise weight calculations, such as are made for Newtecnic projects using BIM, will extend across the industry so that the immediate and long-term, consequential and extended costs can be accurately calculated.

Cloud

The major and growing trend for BIM-based cloud operation in construction will expand over the coming months and years as its benefits become increasingly apparent. Working on the cloud means there is only one set of building data that is shared and used by all stakeholders. Information cannot be siloed or hidden. Cloud operation using 3D digital twins of buildings, their components and their construction methods, lets everyone better understand the project and their role in it. Anyone can see the big picture and it myriad details. That leads to fewer mistakes, better quality, and collaboration that avoids confrontation and disputes.

Waste Reduction

European Commission has reported that around 30% of all waste in Europe is generated by the construction industry. The trend for engineering to find sustainable solutions to this problem is leading directly to increased efficiency, and productivity through eliminating wasted materials and time. BIM is the primary tool available for this task.

Generative Design

When design is freed from traditional industry practices, shapes and components can be based on first principles of engineering design, which can be created 'generatively'. This means that rather than being designed by a single person, geometry is based on functional requirements. In many cases the shapes that are generated have never been seen before, yet they are suited to purpose. This trend is often the starting point for designers to adapt these shapes and to be inspired to develop new types of facade and detailing that can be manufactured using 3D printing.

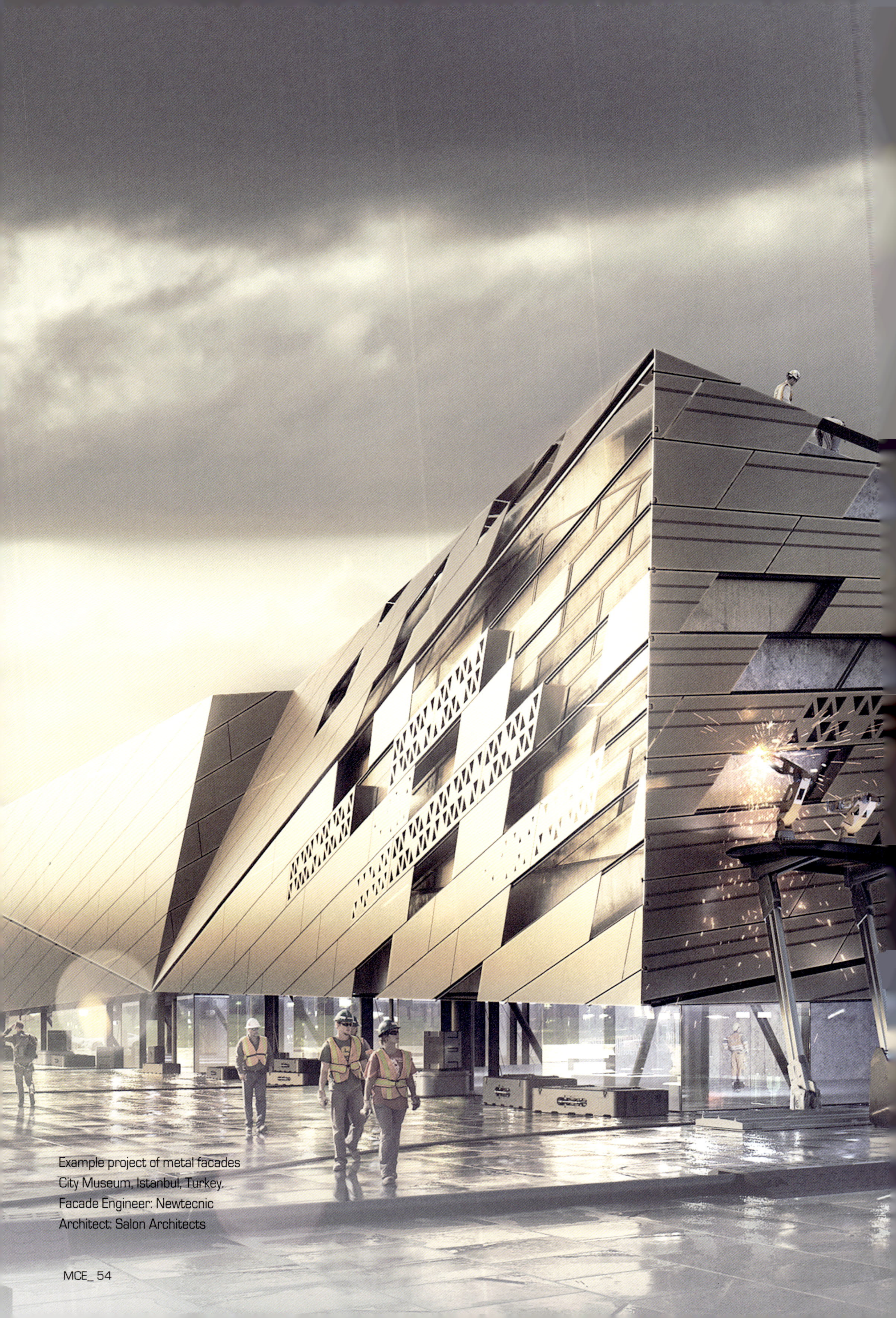

Example project of metal facades
City Museum, Istanbul, Turkey.
Facade Engineer: Newtecnic
Architect: Salon Architects

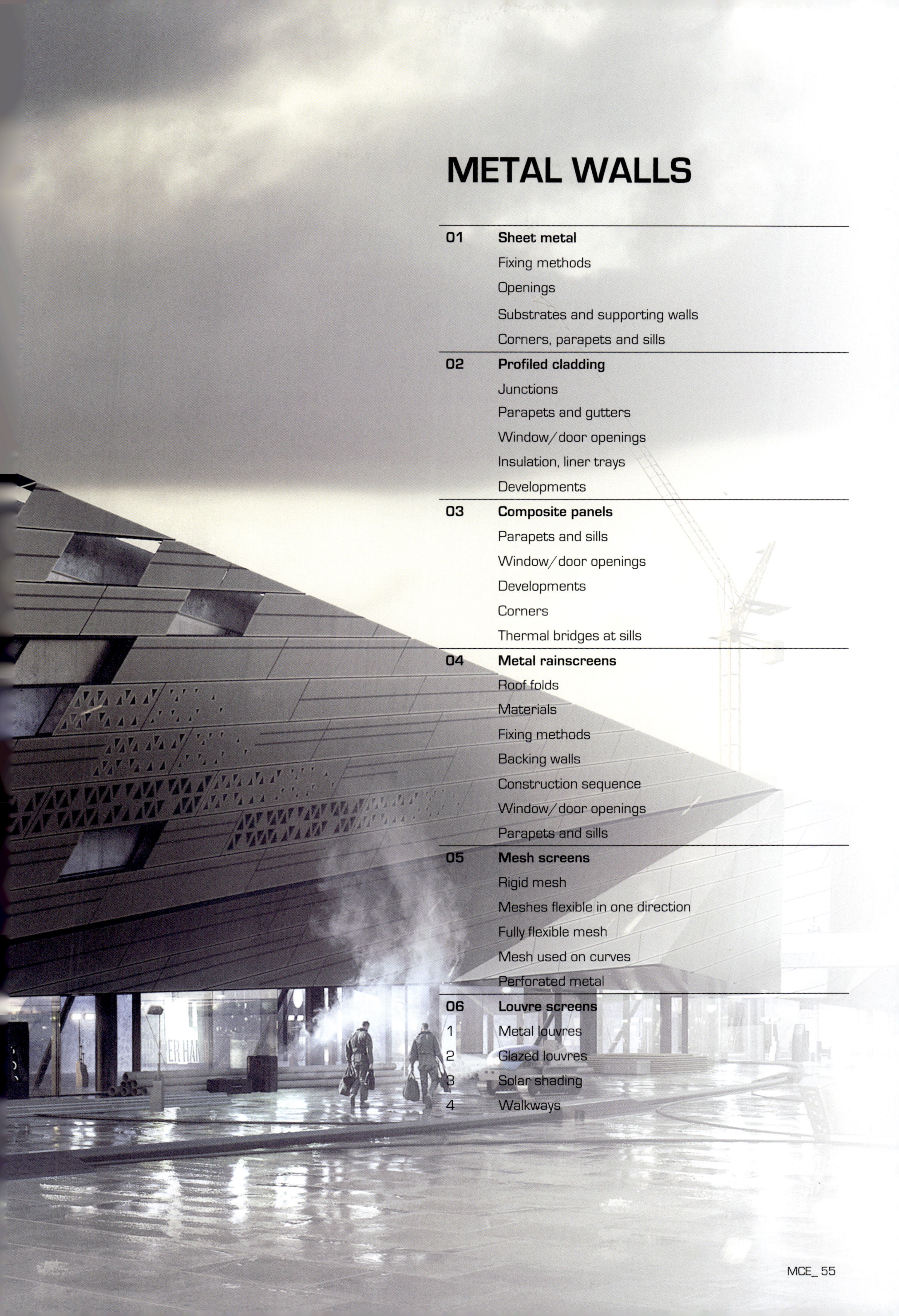

METAL WALLS

01	**Sheet metal**
	Fixing methods
	Openings
	Substrates and supporting walls
	Corners, parapets and sills
02	**Profiled cladding**
	Junctions
	Parapets and gutters
	Window/door openings
	Insulation, liner trays
	Developments
03	**Composite panels**
	Parapets and sills
	Window/door openings
	Developments
	Corners
	Thermal bridges at sills
04	**Metal rainscreens**
	Roof folds
	Materials
	Fixing methods
	Backing walls
	Construction sequence
	Window/door openings
	Parapets and sills
05	**Mesh screens**
	Rigid mesh
	Meshes flexible in one direction
	Fully flexible mesh
	Mesh used on curves
	Perforated metal
06	**Louvre screens**
1	Metal louvres
2	Glazed louvres
3	Solar shading
4	Walkways

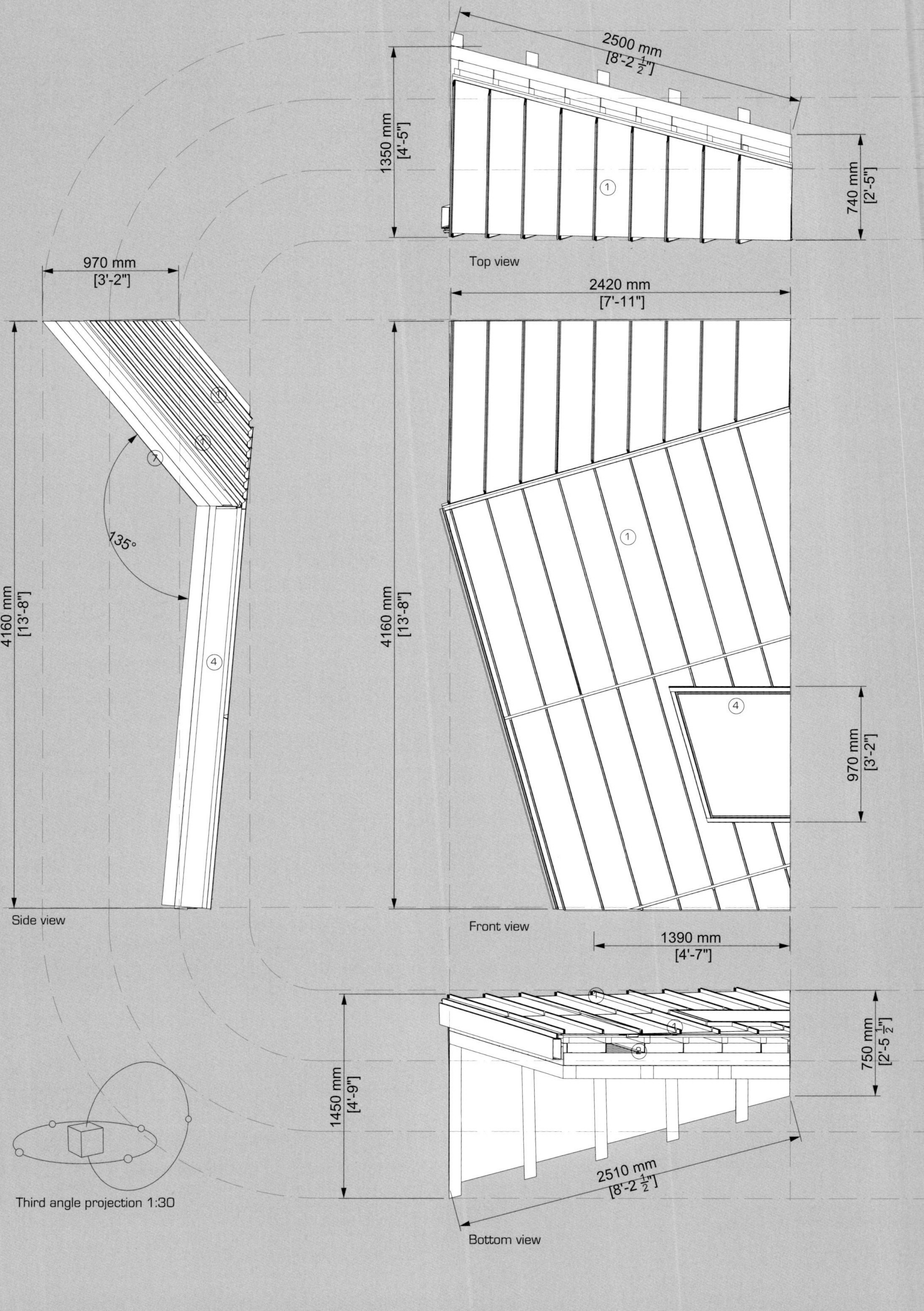
2500 mm
[8'-2 1/2"]
1350 mm
[4'-5"]
740 mm
[2'-5"]
Top view
970 mm
[3'-2"]
2420 mm
[7'-11"]
135°
4160 mm
[13'-8"]
4160 mm
[13'-8"]
970 mm
[3'-2"]
Side view
Front view
1390 mm
[4'-7"]
1450 mm
[4'-9"]
750 mm
[2'-5 1/2"]
2510 mm
[8'-2 1/2"]
Third angle projection 1:30
Bottom view

Details

1 | Folded metal sheet
2 | Fixing battens
3 | Standing seam joints
4 | Window frame
5 | Waterproof membrane, typically bitumen-based paint
6 | Internal finish
7 | Metal clips fixed at centres
8 | Timber window sill

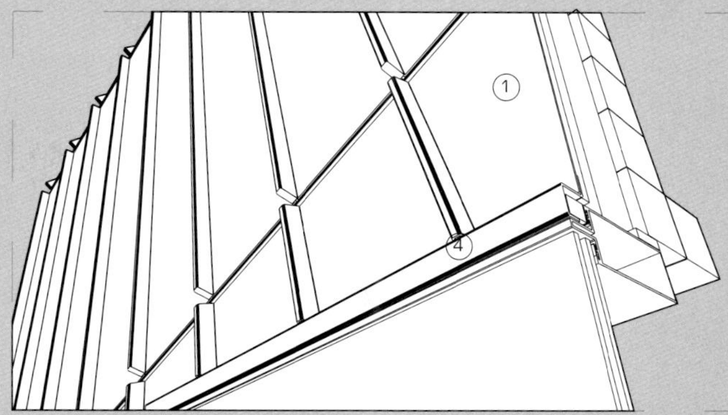

3D view of the assembly

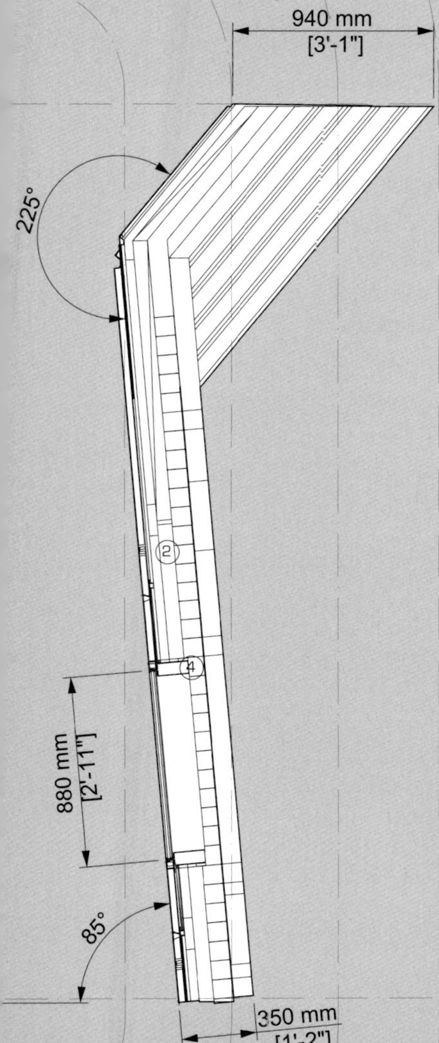

Side view

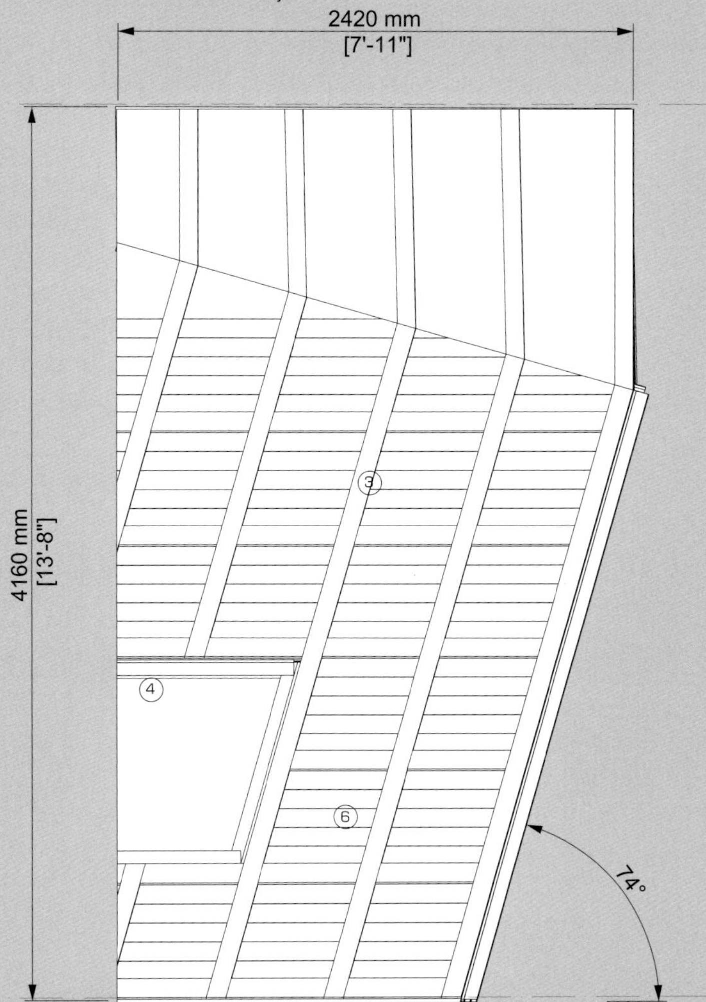

Back view

9 | Substrate in plywood or timber board
10 | Folded metal coping
11 | Thermal insulation
12 | Backing wall - timber/metal frame with plywood facing
13 | Vapour barrier
14 | Ventilated metal drip
15 | Structural concrete wall

3D view of the assembly

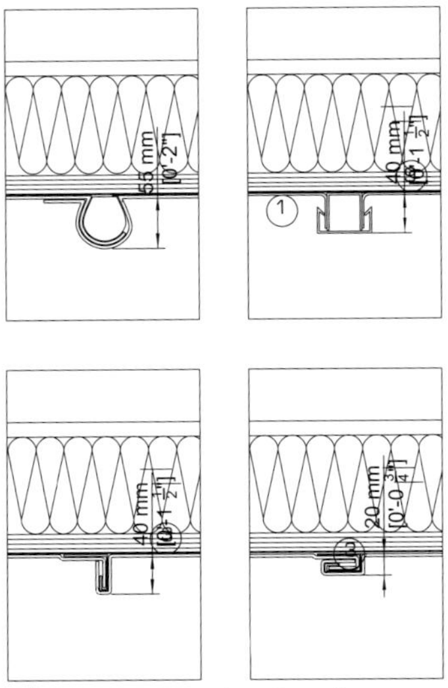

a | Horizontal section. Standing seam joints

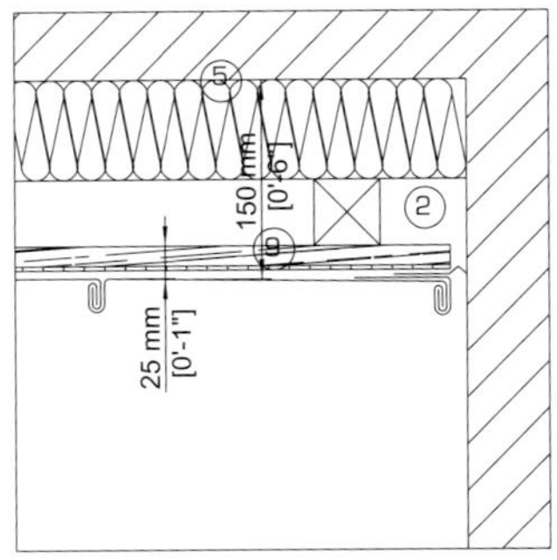

b | Horizontal section. Internal corners

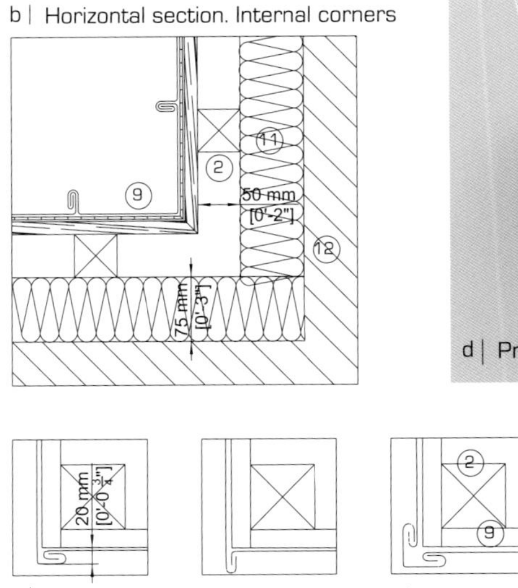

c | Horizontal section. Joints at external corners

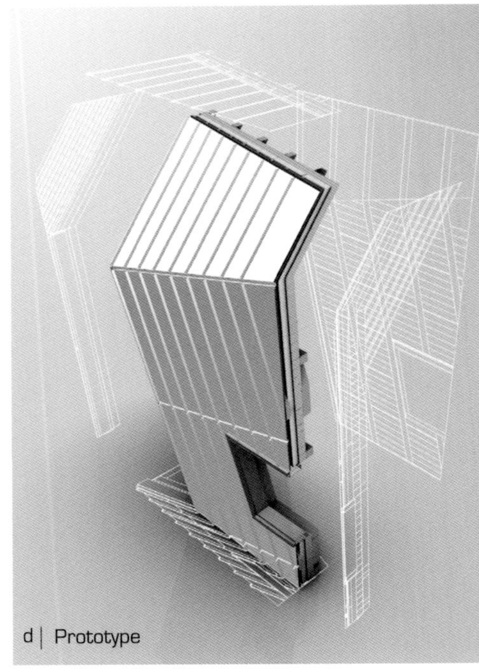

d | Prototype

Details

1 | Folded metal sheet
2 | Fixing battens
3 | Standing seam joints
4 | Window frame
5 | Waterproof membrane, typically bitumen-based paint
6 | Internal finish

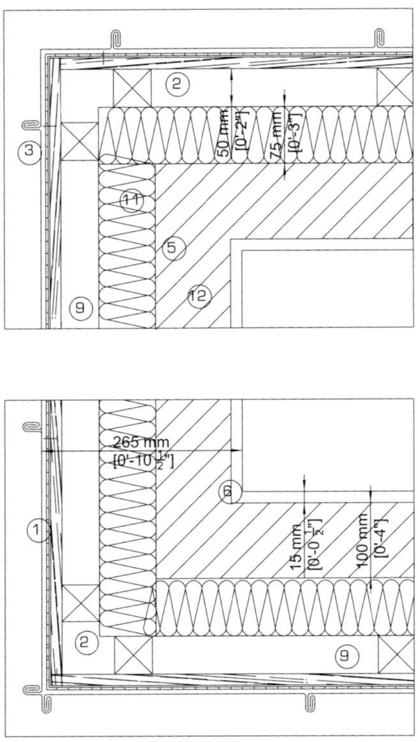

e | Horizontal section. External corner

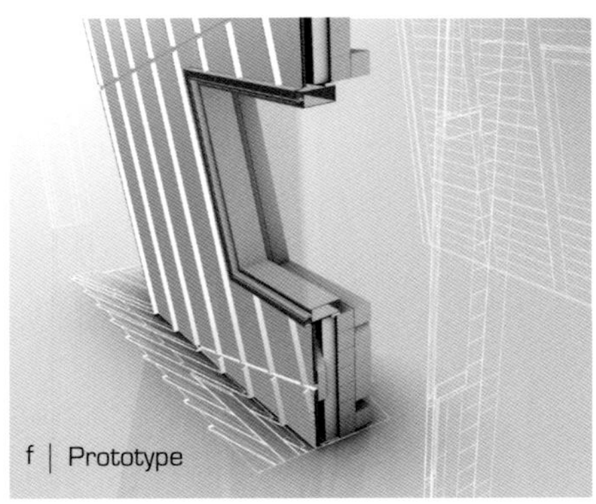

f | Prototype

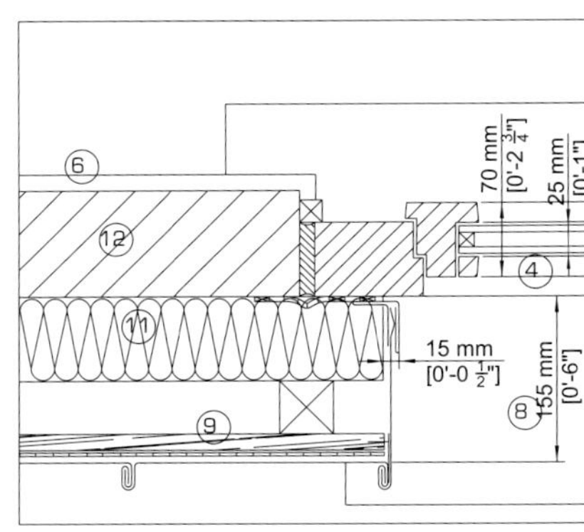

g | Horizontal section. Window jamb

Sheet metal is used for the rich surface textures that can be achieved with relatively soft materials applied to a continuous supporting substrate. This method does not provide the sharp lines and flat surfaces associated with rainscreen panels or composite panel construction. The most common metals used are copper, lead, zinc. More recently, stainless steel has come into use, but primarily as a roofing material. Copper sheet is a ductile material, but not as malleable as lead. Its characteristic green patina when fully weathered gives a consistent appearance. Lead sheet is extremely durable, and its softness allows it to be formed over complex geometries and panels with a high amount of surface relief. Zinc is durable, though more brittle than copper but is susceptible to corrosion from its underside if not ventilated. Stainless steel is a very durable material, but it still has an uneven surface when laid that provides a richness of reflection. The main disadvantage of working in stainless steel is its hardness, making it difficult to work when forming folds in jointing.

Fixing methods

There are three fixing methods for continuously supported sheet metal walls: continuous sheet, lapped tiles and recessed joints.

Continuous sheets are laid in varying widths with standing seams (a) in vertical joints that run continuously from top to bottom of a wall. This gives the facade a characteristic striped appearance with strong shadows across the standing seam joints in sunlight (f). The sheet metal is fixed on the horizontal joints with flattened seams that allow rainwater to drain off easily. Horizontal joints are at distances to suit the visual appearance of the design but 12.0 metres (39ft 4in) to 17.0 metres (55ft 9in) is the maximum depending on the metal used. Vertical joints align with the edges of windows and door openings. Horizontal joints are usually staggered to form a pattern rather than try to achieve a continuous straight line which is difficult to keep completely straight and horizontal. This is because horizontal joints are broken between each vertical seam.

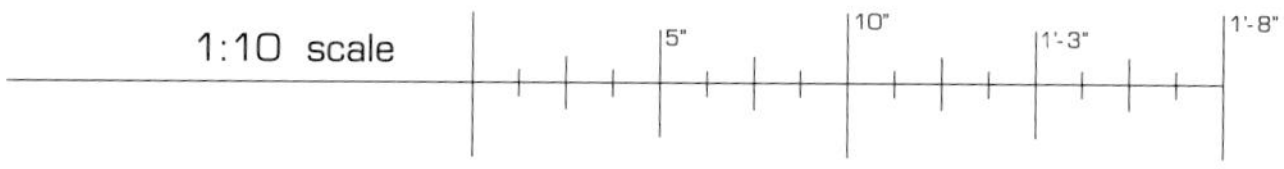

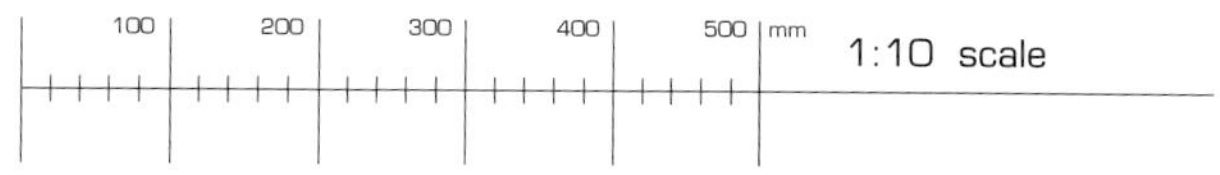

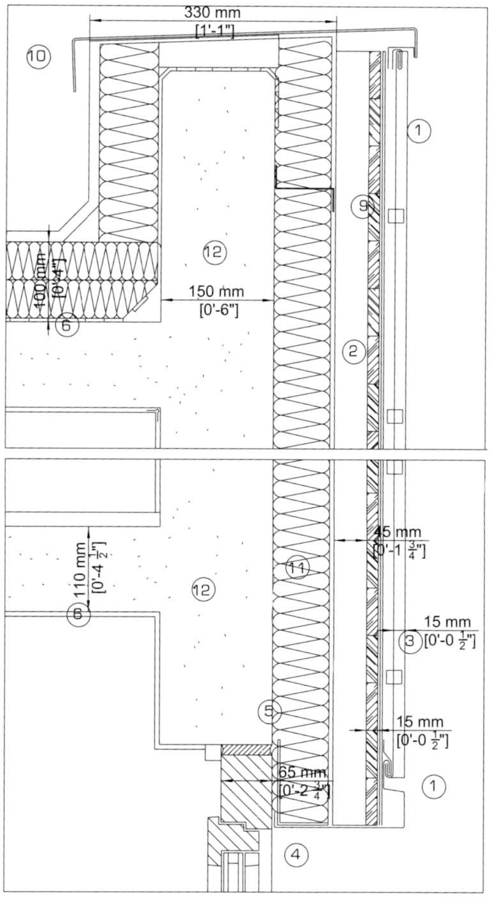

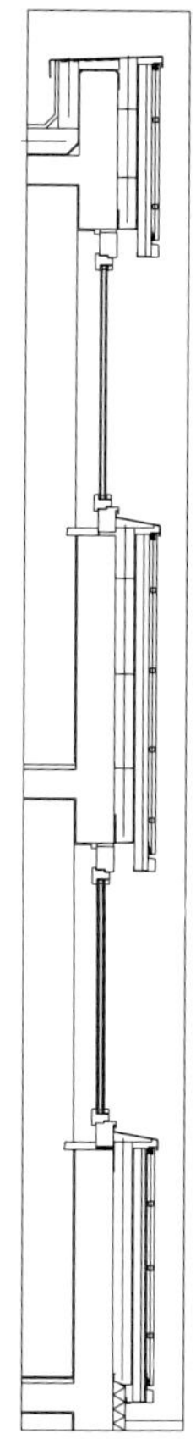
h | Vertical section

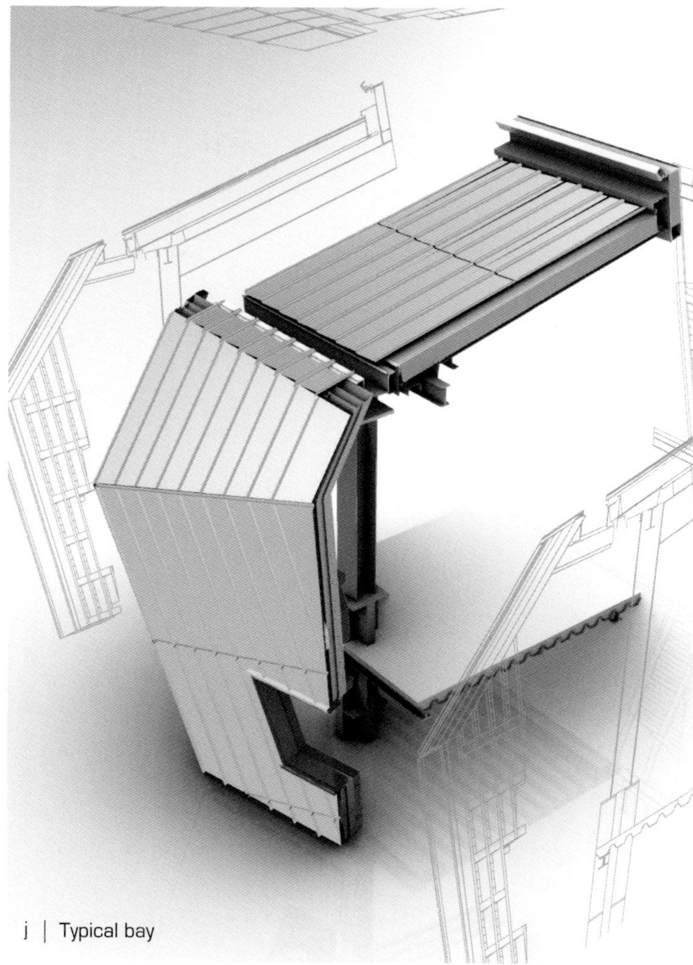
j | Typical bay

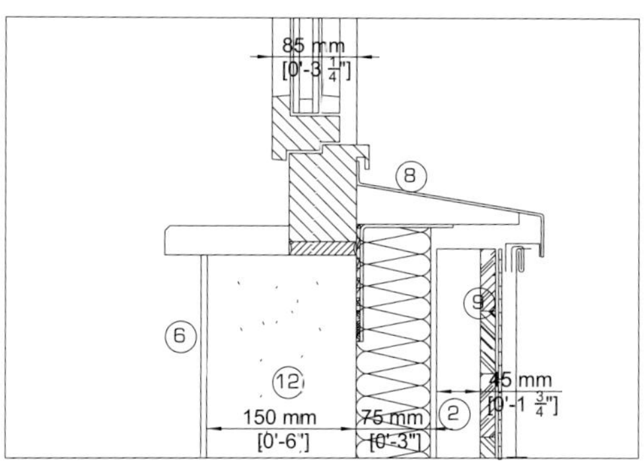

k | Vertical section. Parapet and window

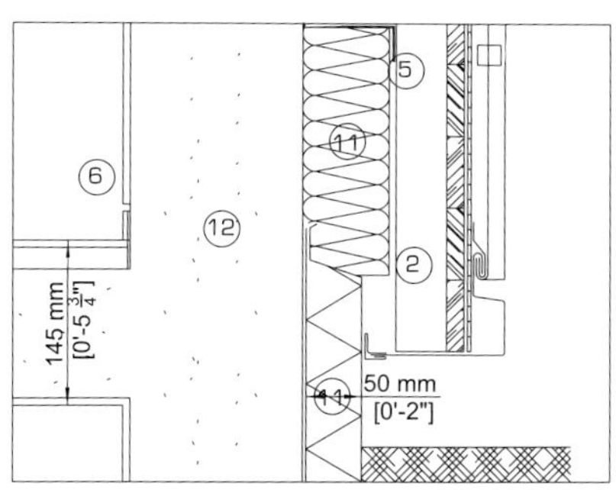

m| Vertical section. Ground level sill

7 | Metal clips fixed at centres
8 | Timber window sill
9 | Substrate in plywood or timber board
10 | Folded metal coping
11 | Thermal insulation
12 | Backing wall - timber/metal frame with plywood facing
13 | Vapour barrier
14 | Ventilated metal drip
15 | Structural concrete wall

Lapped tiles are made approximately 450mm to 600mm (18inx24in) square and are set in either horizontal and vertical edges or at a 45° angle (MCE_60, c). Other angles can be used but are harder to co-ordinate with the edges of corners and openings. Window and door openings are usually enclosed in a metal strip around the reveal of the opening, with a shadow gap or projecting corner detail. Tiles are not folded into openings due to the complexity of jointing and the difficulty of getting them to fold neatly at half way across the panels. Shadows across the surface of a tiled wall have small strong lines which give a very textured appearance to the facade. Tiles are lapped on four sides to give a continuous watertight joint on all edges.

Recessed joints are formed in sheet metal laid over a specially formed substrate to produce recessed lines, which are usually horizontal (f). The material is occasionally recessed on four sides and set on a plywood background with projecting panels formed by the plywood. However, this technique is seldom used any more due to the increased use of metal rainscreen panels with their advantage of flatness of panel and close fixing tolerances, providing crisp lines at joints which are more difficult to achieve in a sheet metal continuously supported on a profiled background. However, this technique may find favour again due to its rich surface texture. Drips are incorporated at horizontal recessed joints to avoid staining occurring as a result of dirt being washed off the flat surfaces of these joints.

Openings

Where vertical jointed sheet metal is used, window and door openings are usually positioned so that a joint falls on the edge of an opening. This gives a clean, co-ordinated appearance to the facade, but openings that are set deliberately 'off grid' from the vertical joints also look visually dramatic (b). A recent development has been to use sheet metal with vertical joints inclined at angles up to 45º from the vertical so that they

a | Prototpye

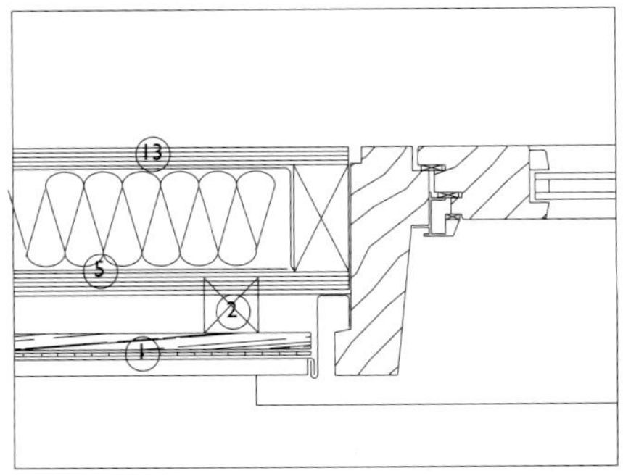

b | Horizontal section. Window jamb

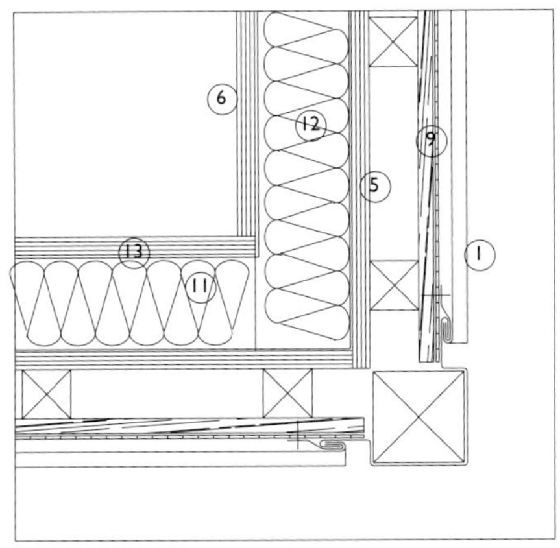

c | Horizontal section. External corner with batten

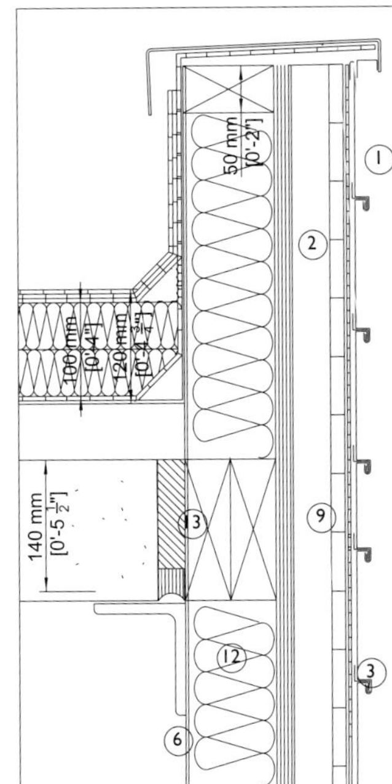

d | Vertical section. Parapet

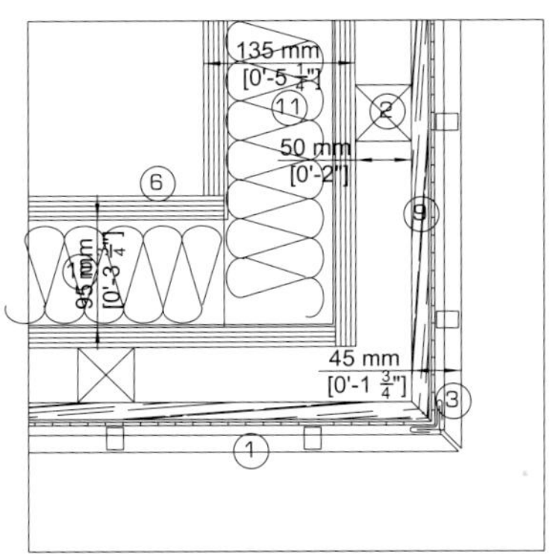

e | Horizontal section. External corner with metal fold

Details

1 | Folded metal sheet
2 | Fixing battens
3 | Standing seam joints
4 | Window frame
5 | Waterproof mebrane, typically bitumen-based paint
6 | Internal finish

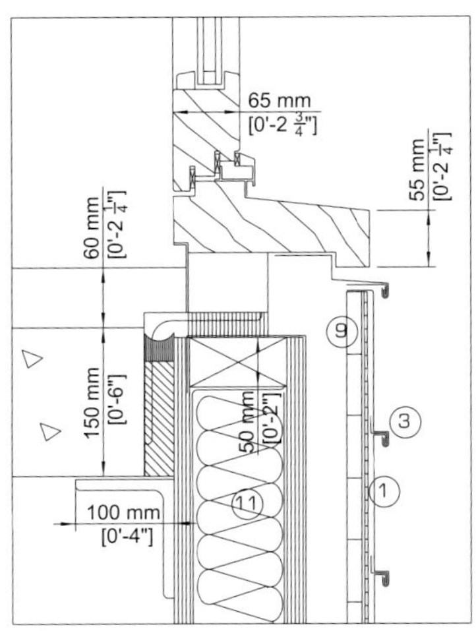

f | Vertical section. Window sill

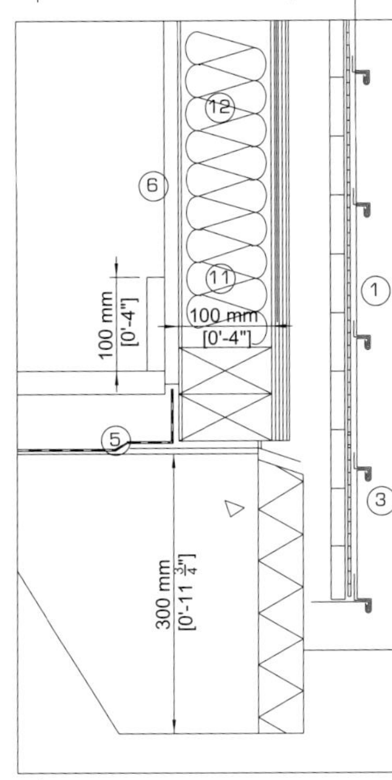

g | Vertical section. Ground level sill

contrast with rectilinear window openings, giving the sheet metal the appearance of a continuous 'non-gridded' texture across a complete facade. For all orientations of sheet metal, window and door openings have separate metal sheets forming reveals on all sides. Although this can result in some awkward pieces of metal to form junctions around windows, in practice they are practical and economic to form on site due to the site-based nature of fixing metal sheet. Sheet metal cladding is ideally suited to the complex junctions associated with non-rectilinear geometries. Attempts to make the material appear too regular can produce disappointing results, particularly where a pure rectilinear grid is attempted. In this instance, metal rainscreen panels would probably be more suitable.

Windows and doors are glazed with almost any available technique, but the ever-increasing use of double glazed units both to conserve energy and avoid condensation on the window or door surface has led to thermally broken sections being used very commonly. Window frames are often clad in the same sheet metal as is used in the adjacent facade, but this is expensive since the metal cladding will always be a decorative finish to a window that is designed for use without such a finish. The usual alternative is to use either a polyester powder coated or PVDF paint finish on aluminium to match the colour of the adjacent metal, or use a different material such as timber windows. A paint finish is obviously much easier to match if the sheet metal finish is pre-patinated (pre-weathered) so that its final colour will be very similar to the colour of the metal when installed. This is much more difficult in unweathered metals. The use of galvanised steel windows and doors with zinc is not so common due to the increased performance of paint coatings. However, galvanised finishes are ever-increasing in their durability and may eventually be used as a durable finish for window frames.

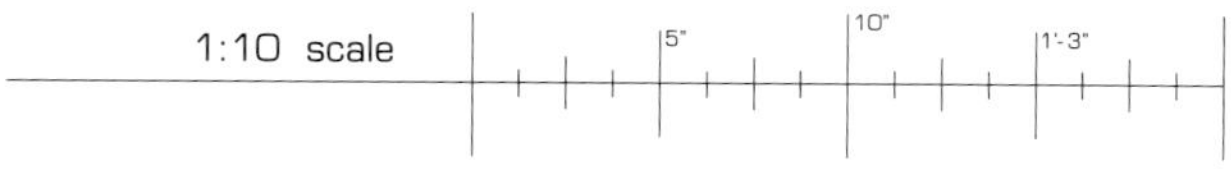

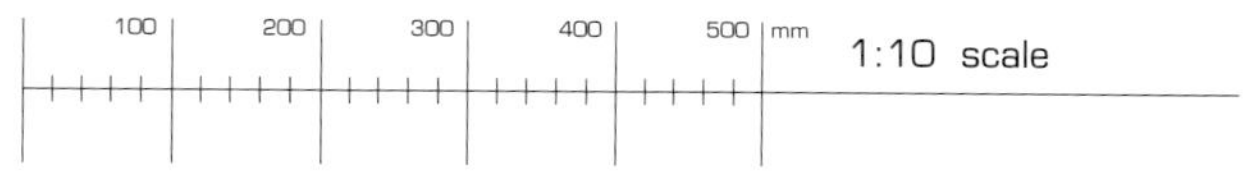

h | Typical bay

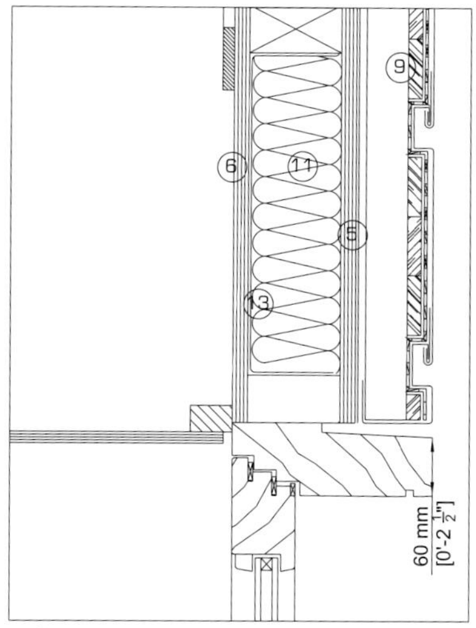

j | Vertical section. Window head

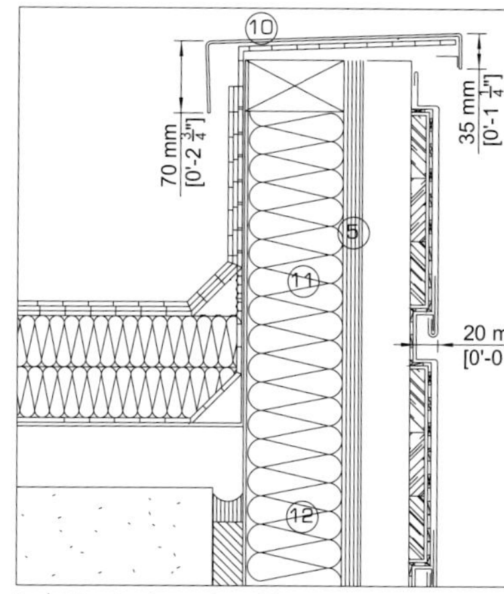

k | Vertical section. Parapet

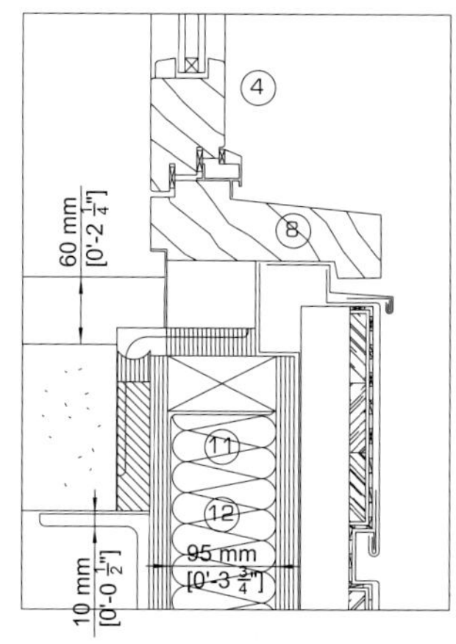

7 | Metal clips fixed at centres
8 | Timber window sill
9 | Substrate in plywood or timber board
10 | Folded metal coping
11 | Thermal insulation
12 | Backing wall - timber/metal frame with plywood facing
13 | Vapour barrier
14 | Ventilated metal drip
15 | Structural concrete wall

m | Horizontal section. External corner with metal fold

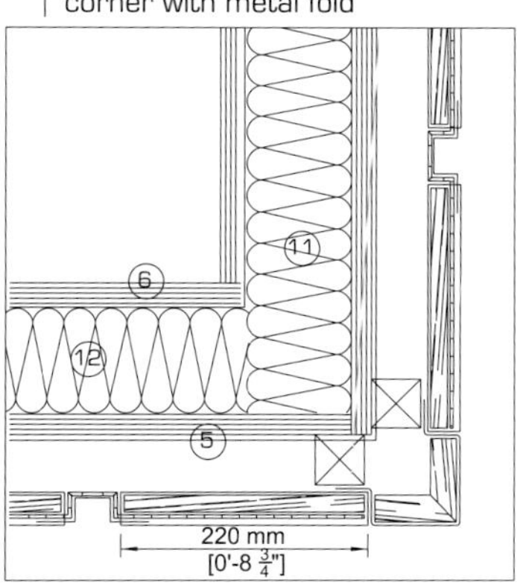

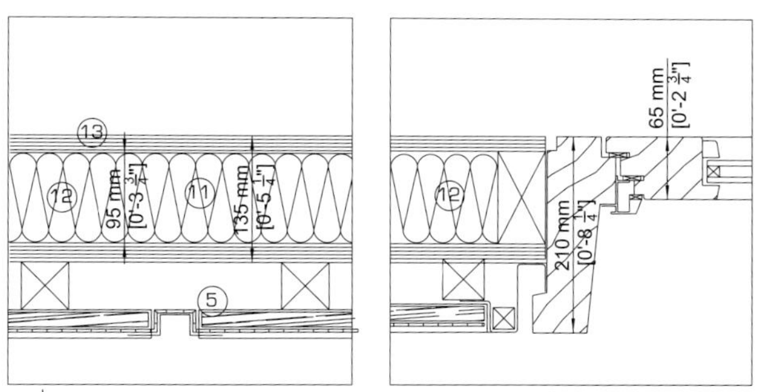

n | Horizontal section. Window jamb

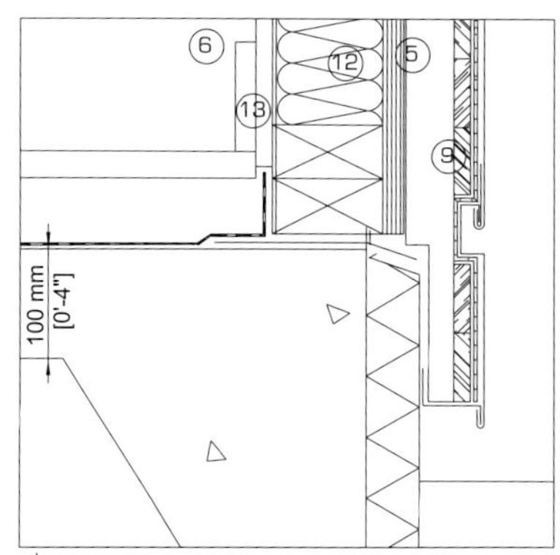

p | Vertical section. Ground level sill

Buildings clad partly in sheet metal are beginning to use large-scale glazed openings using a completely different system such as bolt fixed glazing. Whereas these two systems previously seemed incompatible, with sheet metal as an economic system and bolt fixed glazing as an expensive system, the two are now used together increasingly where a deliberate contrast of surface texture is sought. While bolt fixed glazing has a smooth, continuous surface uninterrupted by visible framing, sheet metal has joints in a direction at 400mm (1ft 4in) to 600mm (2ft) centres, with a comparatively uneven surface finish.

Substrates and supporting walls

Sheet metals can be laid directly onto a substrate, typically plywood (m), with the exception of zinc, which needs ventilation on its interior face to avoid corrosion. Plywood is preferred for its durability, since if it becomes wet before a repair can be undertaken, the material can dry out without being damaged. Other materials such as particle boards are not resistant to moisture penetration and so are not used. Timber boards are used but are usually more expensive, as is profiled metal sheet. Where timber framing is used for the wall construction, the timber substrate forms an integral part of the external wall, providing diaphragm stiffness in the frame. Profiled metal sheet is increasingly used as a substrate for zinc, since zinc is more rigid than other metals such as copper or lead. Profiled metal sheet can span the gap between the peaks of the cladding, while providing a ventilated zone behind to avoid corrosion of the zinc. The addition of a ventilation mat provides a full gap between the zinc and the profiled metal.

Sheet metal is increasingly fixed to walls constructed in a wide range of materials: timber frames, precast concrete, concrete block and lightweight steel frames made in cold formed sections.

a | Prototpye

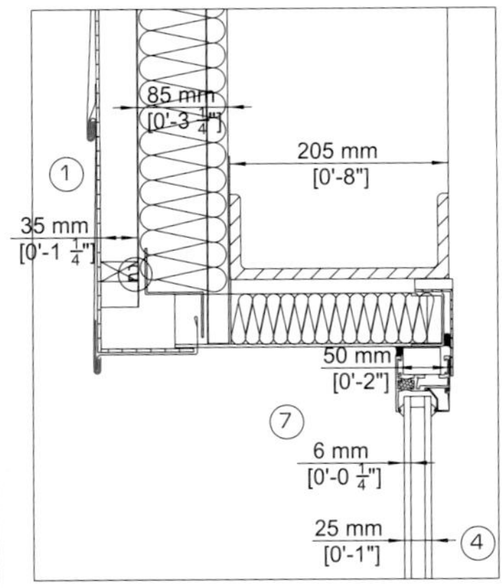

b | Vertical section. Window head

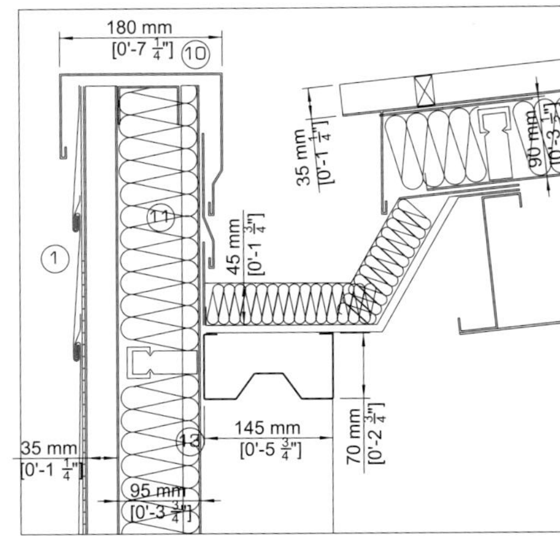

c | Vertical section. Parapet

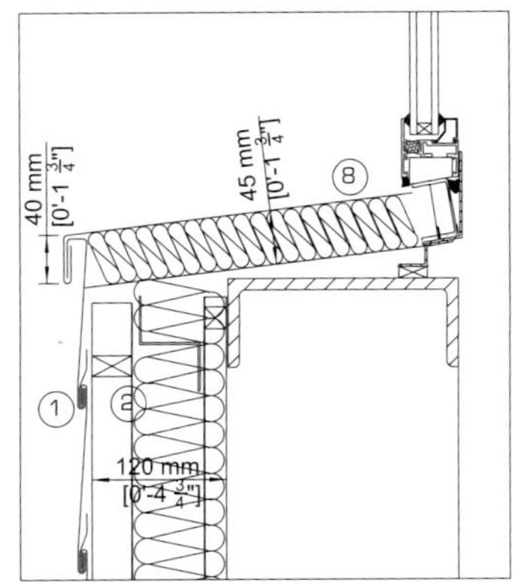

d | Vertical section. Window sill

9 | Substrate in plywood or timber board
10 | Folded metal coping
11 | Thermal insulation
12 | Backing wall - timber/metal frame with plywood facing
13 | Vapour barrier
14 | Ventilated metal drip
15 | Structural concrete wall

e | Vertical section. Ground level sill

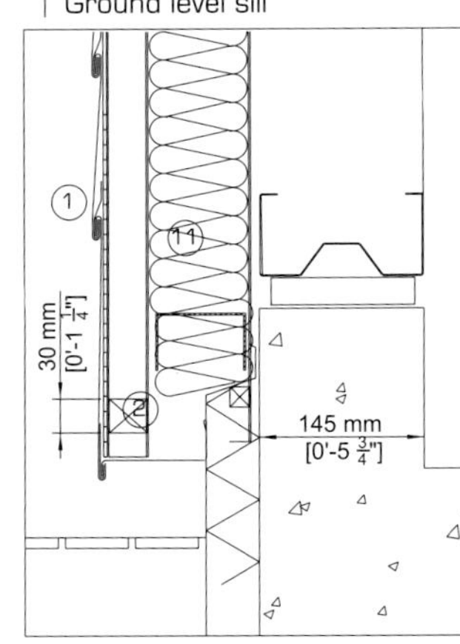

Details

1 | Folded metal sheet
2 | Fixing battens
3 | Standing seam joints
4 | Window frame
5 | Waterproof membrane, typically bitumen-based paint
6 | Internal finish
7 | Metal clips fixed at centres
8 | Timber window sill

In Type 1, timber frames use sheet metal as a cladding in a fairly traditional form of construction (MCE_61, h), or as infill panels to a timber, concrete or steel frame. The overall cross section of the wall remains thin due to the inclusion of thermal insulation within the frame rather than on an outside face (MCE_61, n). A vapour barrier is needed on the warm-in-winter side (in a temperate climate) to avoid vapour reaching the insulation from inside. The vapour barrier is needed in the same place in Type 3, where pressed steel, or 'light gauge' steel sections are used (c). The all-metal construction of Type 3 is undergoing refinement for use in housing, where almost all its components can be either recycled or unbolted and modified with the same kit of parts during its lifetime. Its flatness of appearance combined with small-scale standing seam joints make it ideal for a sealed metal cladding where profiled metal sheet or composite panels have too 'industrial' an appearance. In Type 2, thermal insulation is set on the outside of the concrete structure in order to use its thermal mass as well as to keep the structure at as even a temperature as possible (a). The metal cladding is then set forward of the insulation. A new development is the use of profiled metal cladding as a substrate in Type 4. Where zinc is used the void formed by the profiled sheet provides a ventilation zone without the use of timber (k). A plastic-based drainage mat is set between the zinc and the profiled sheet to complete the ventilation.

Corners, parapets and sills

Sheet metal can be joined at corners in facades with either recessed or projecting coverstrips. The covers need a timber or plywood support under them to provide rigidity. Corners for vertically set metal sheet can also be formed by setting standing seam joints at the corner, or close to the corner on either side of the edge (k). With tiled sheets, corners usually wrap around, ignoring the corner like a continuous pattern folded around the corner (k). Alternatively, corners can have coverstrips that break the pattern from one facade to another. There

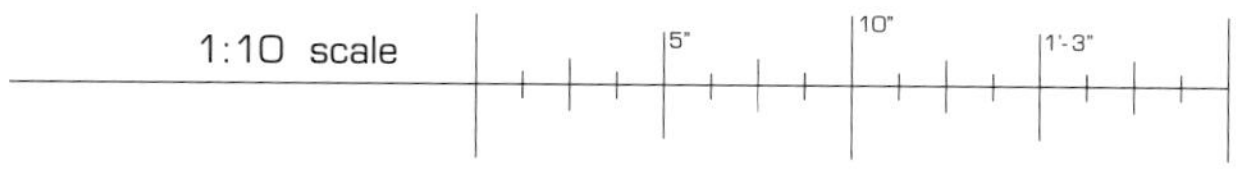

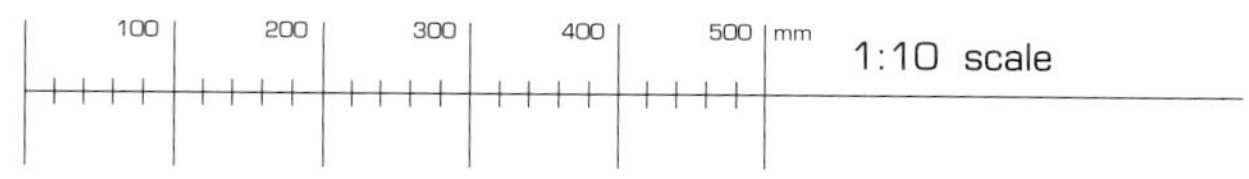

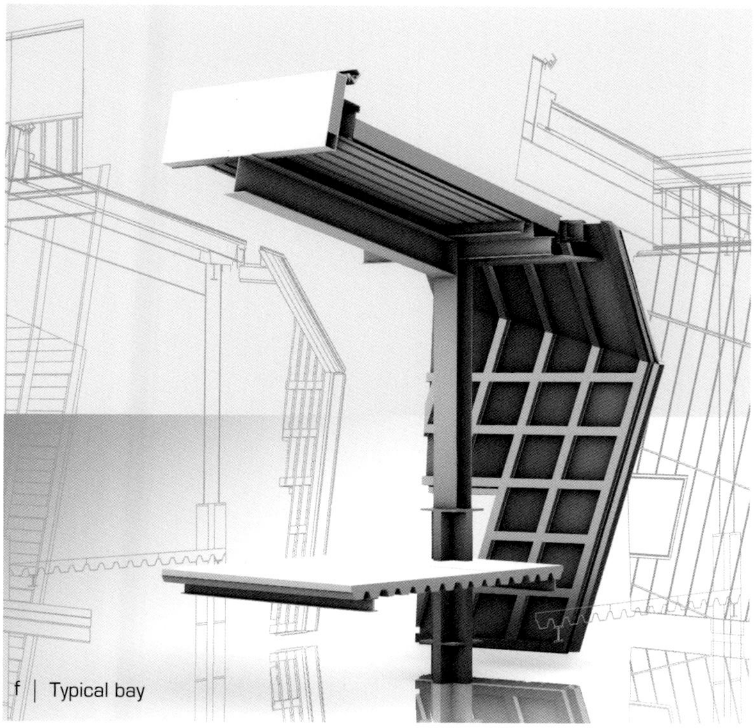

f | Typical bay

g | Vertical section. Window jamb

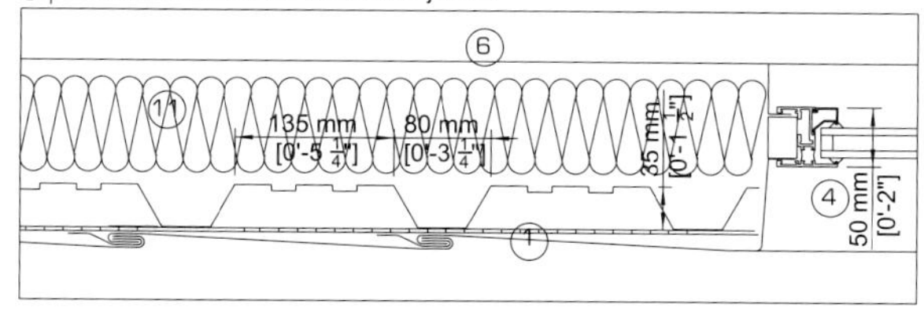

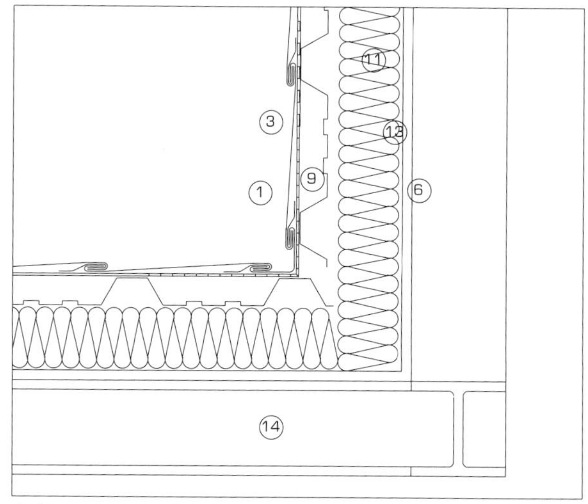

h | Horizontal section. Internal corner with joint

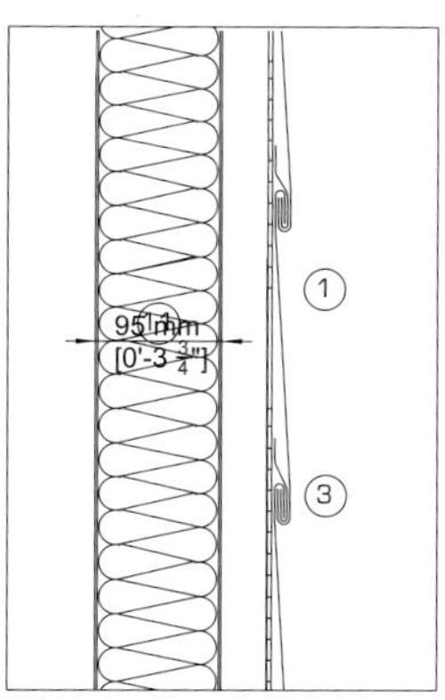

j | Vertical section. Wall

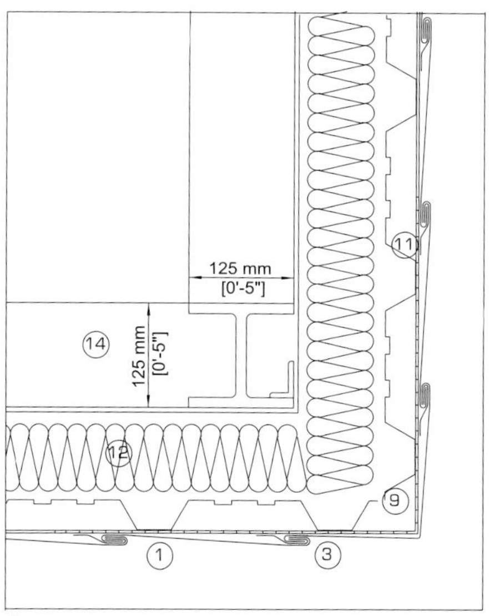

k | Horizontal section. External corner with metal fold

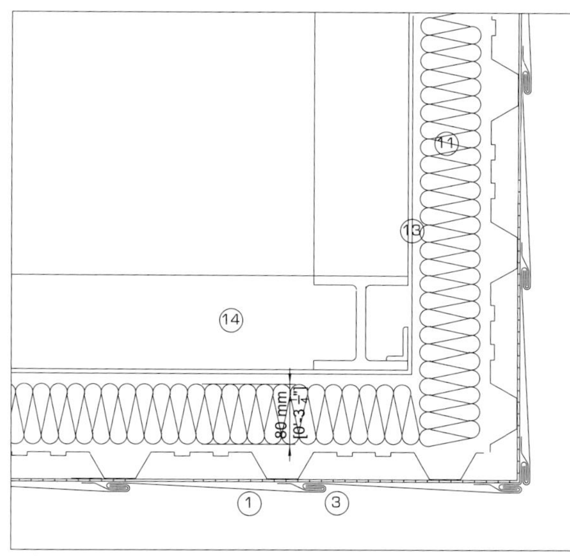

m | Horizontal section. External corner with joint

is an increased use of pressed metal clips and rails, as used in profiled metal cladding, to support sheet metal substrates, typically plywood. Clips and rails are made as proprietary systems which can be fixed to the backing wall quickly and easily adjusted to correct vertical or inclined alignment.

An advantage of sheet metal is that parapet copings and sill drips at the base of walls at windows can be formed in the same material with an identical finish. This is unlike many other metal cladding systems, where extruded aluminium or pressed steel or aluminium are most commonly used for parapets and sills. The ability to form metal on site in junctions of sheet metal walls with parapet copings is used either to form a recessed joint, which allows the standing seam joint in the cladding to be tapered down to the line of the coping, or a projecting parapet coping which allows the standing seam to butt up to the underside of the coping (c). With either solution, an undercloak flashing or waterproof layer is needed underneath the coping to provide additional waterproofing.

Sills are formed in a similar way, but with projecting or flush drips to throw water clear of the base of the cladding (d). Where unweathered metal is used, care should be taken that rainwater runoff from oxidising metal does not stain paved surfaces at the base of the wall. Slot drains or gravel edges can be used both to provide drainage and avoid visible staining. Drips are often reinforced with a steel or aluminium angle to create a strong, straight edge. Compatibility between the cladding material and support material must be ensured to avoid bimetallic corrosion. Where the void behind metal cladding is used for ventilation, parapets and ills are used to introduce fresh air. Insect mesh is introduced within the joint, but its presence does not alter air flow rates significantly.

01 Sheet metal

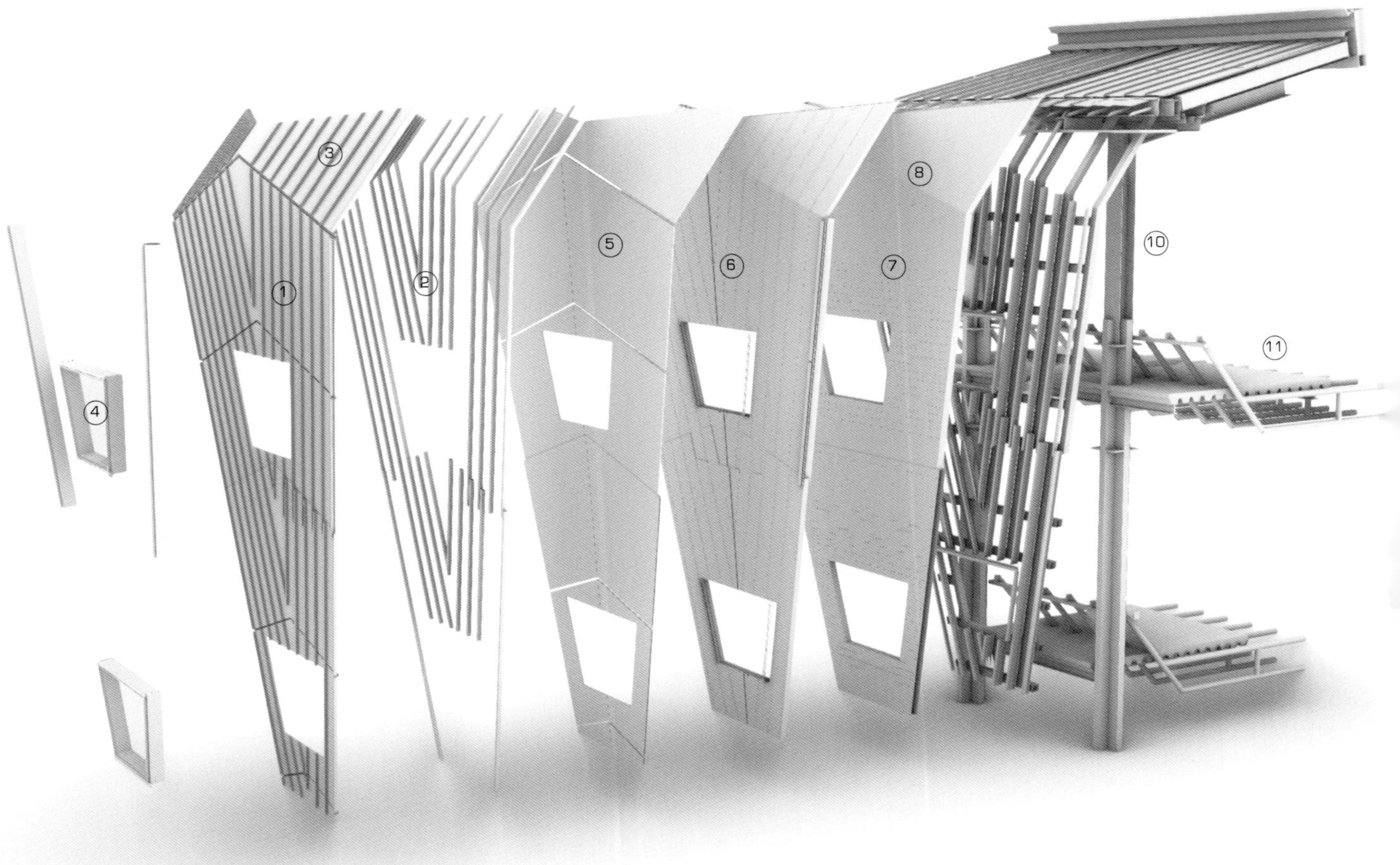

a | 3D exploded view showing sheet metal facade fixed to steel frame

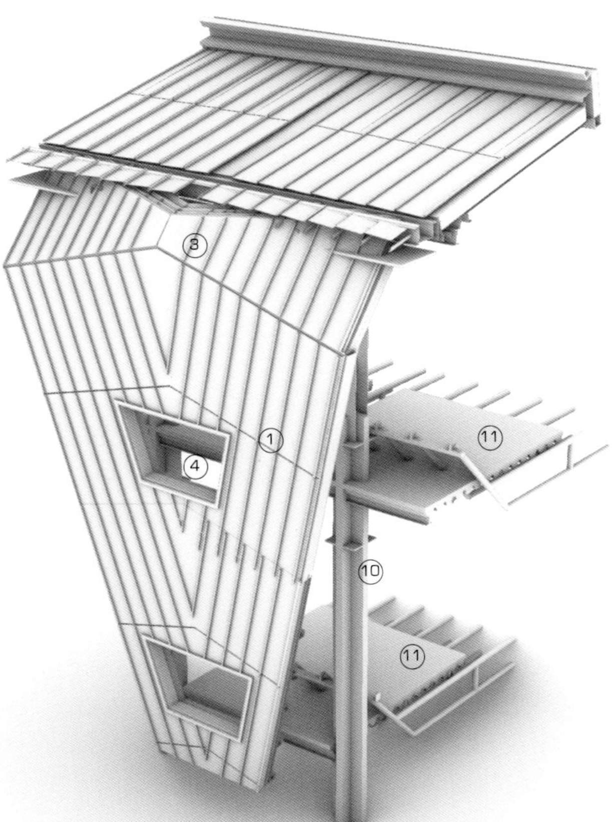

b | 3D view showing sheet metal facade fixed to steel frame

Details

1 | Folded metal sheet
2 | Fixing battens
3 | Standing seam joints
4 | Window frame
5 | Waterproof membrane
6 | Thermal insulation
7 | Vapour barrier
8 | Backing wall
9 | Internal finish
10 | Primary structure
11 | Floor slab

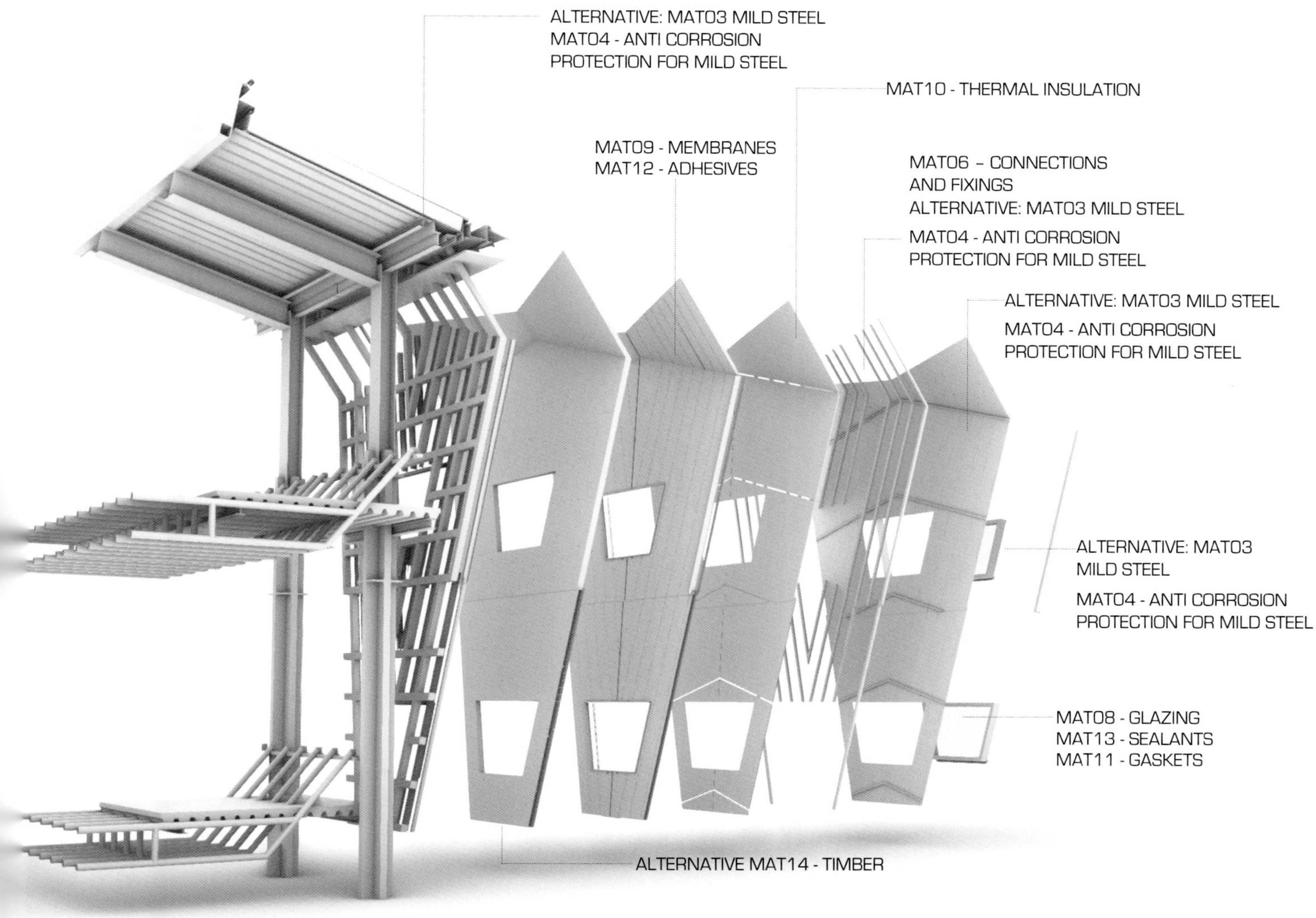

c | 3D exploded view of metal sheet facade components with material specification references

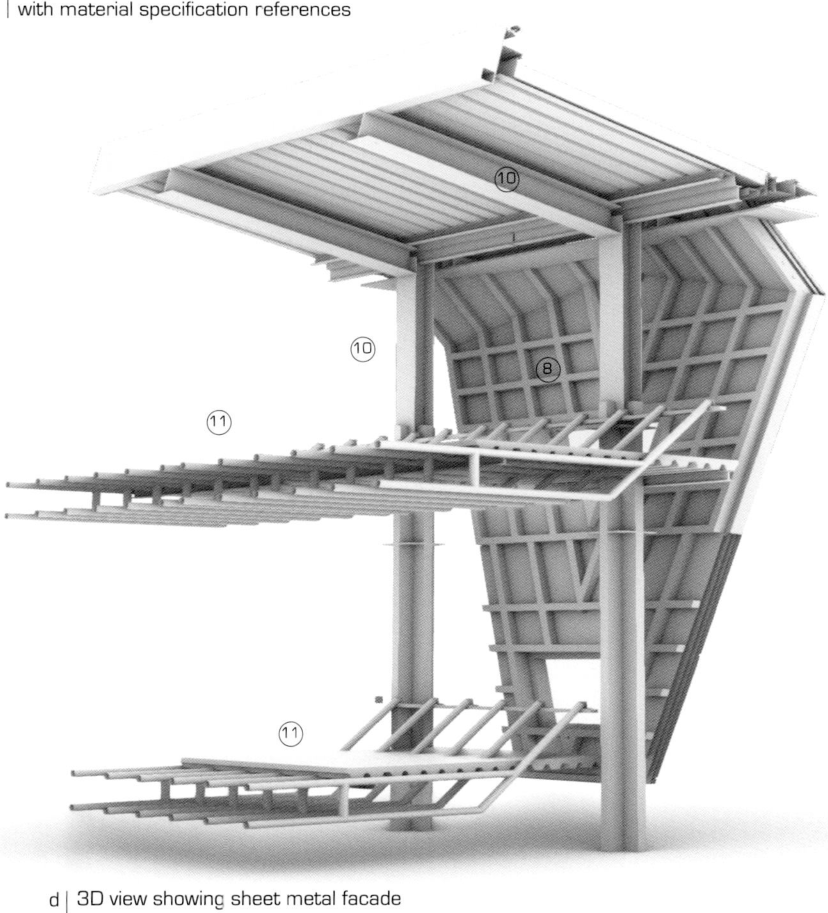

d | 3D view showing sheet metal facade fixed to steel frame

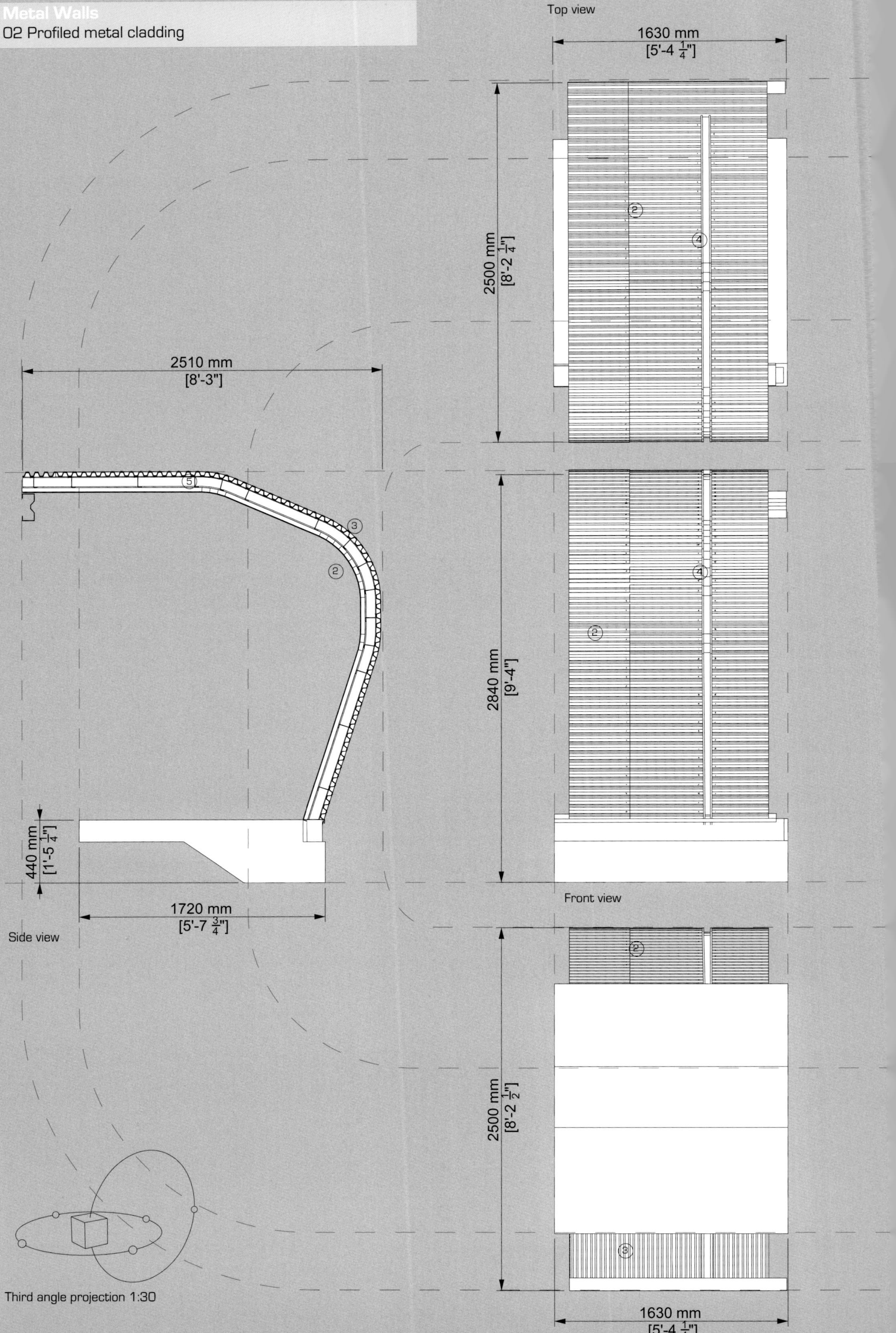

Third angle projection 1:30

Details

1 | Metal cover strip set
2 | Horizontally set profiled sheet
3 | Vertically set profiled sheet
4 | Z section steel fixing rails
5 | Thermal insulation
6 | Backing wall, typically timber/metal frame with plywood facing and waterproof membrane, or concrete block
7 | Vapour barrier

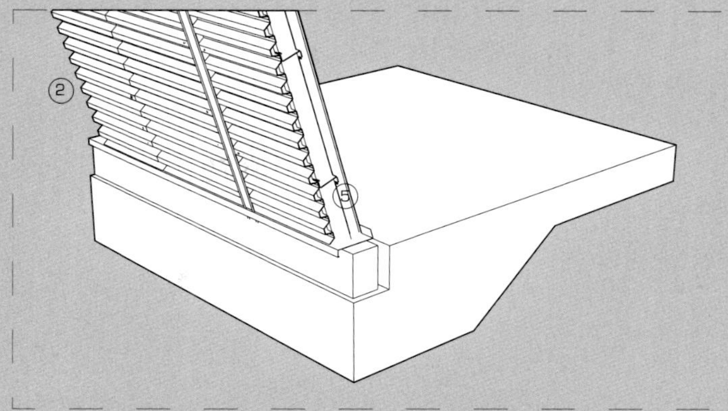

3D view of the assembly

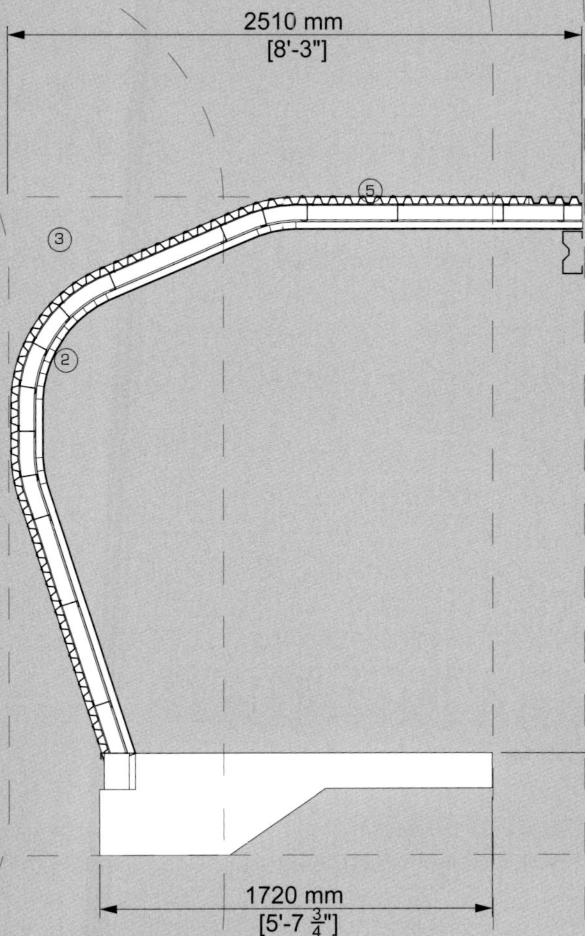

Side view

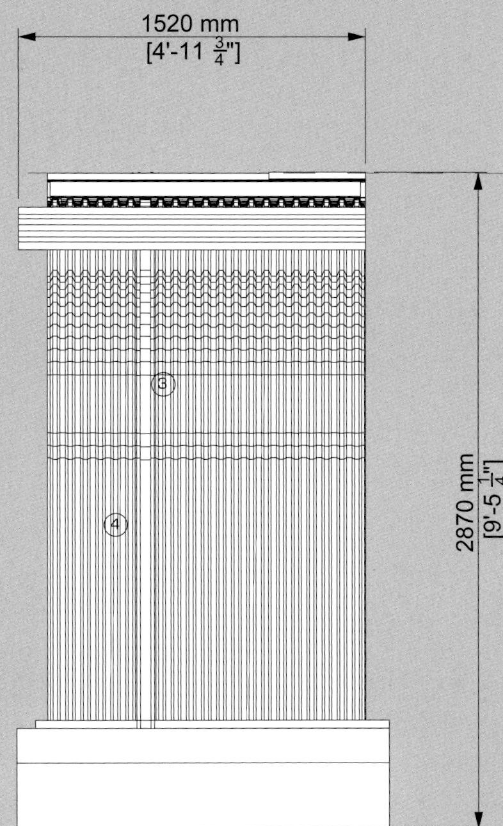

Back view

8 | Internal finish
9 | Roof, typically profiled metal sheet
10 | Curved eaves profile
11 | Concealed gutter
12 | Exposed gutter
13 | Supporting structural frame
14 | Structural slab
15 | Window frame
16 | Metal trim to window
17 | Metal parapet coping

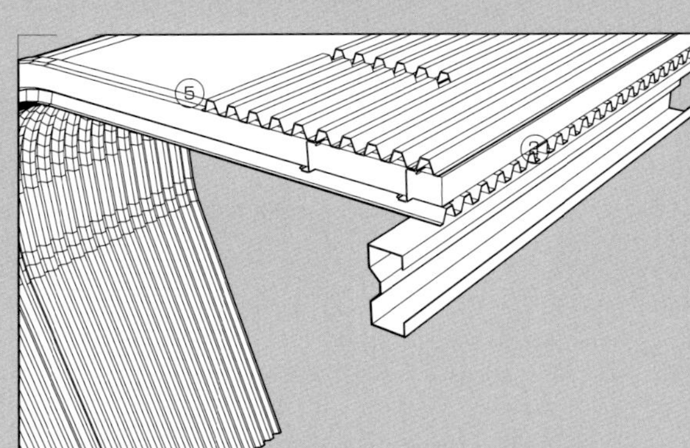

3D view of the assembly

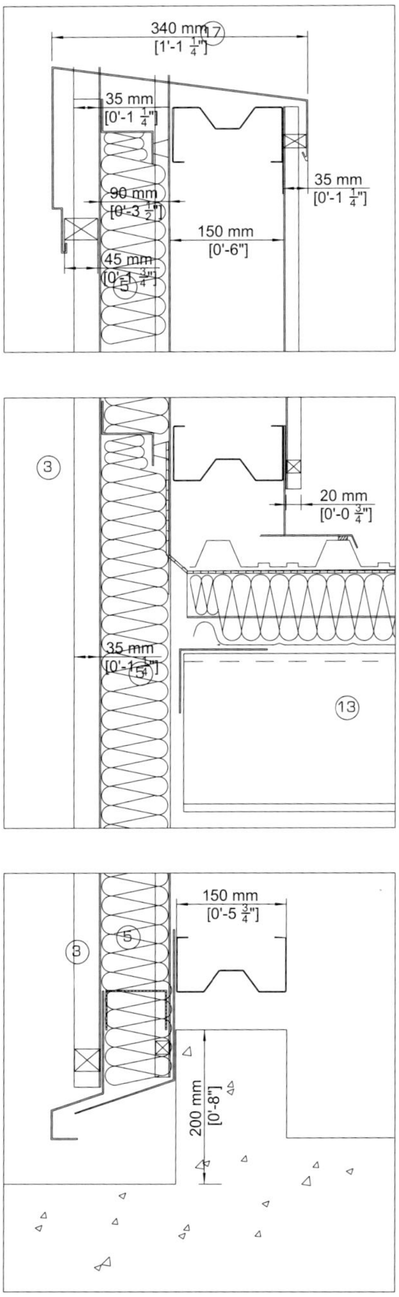

a| Vertical section. Parapet and ground level sill

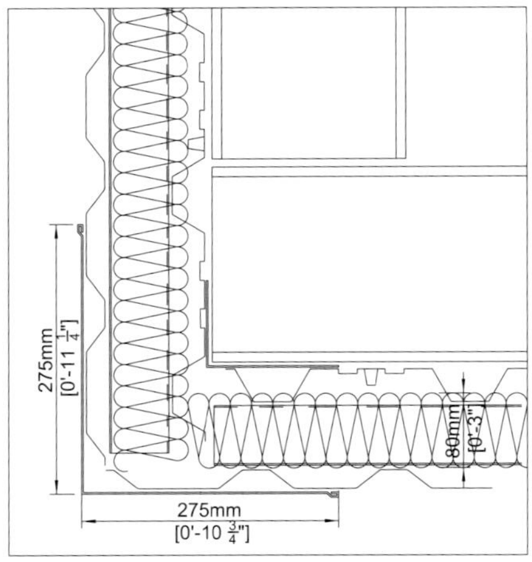

b| Horizontal section. External corner

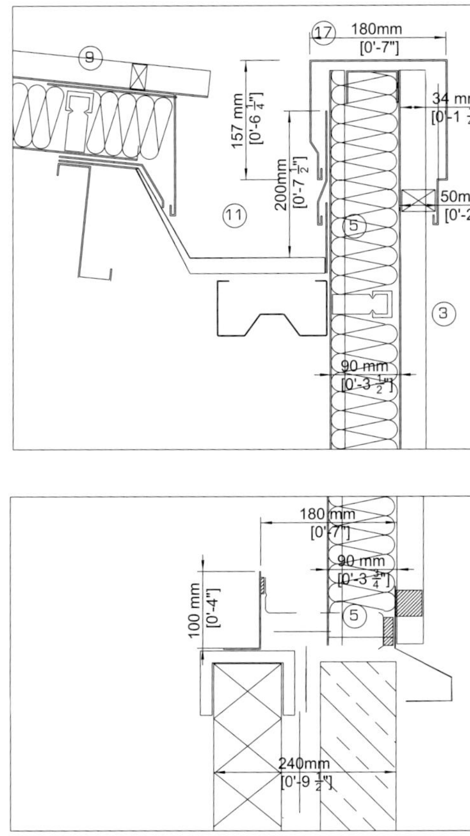

c| Vertical section. Parapet and sill

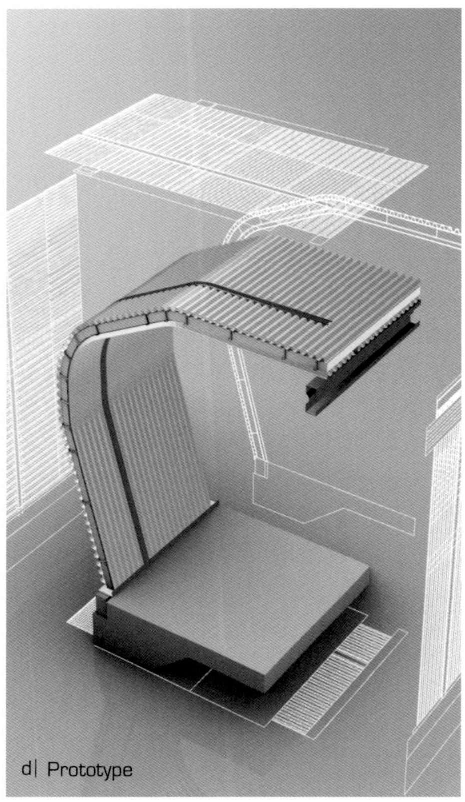

d| Prototype

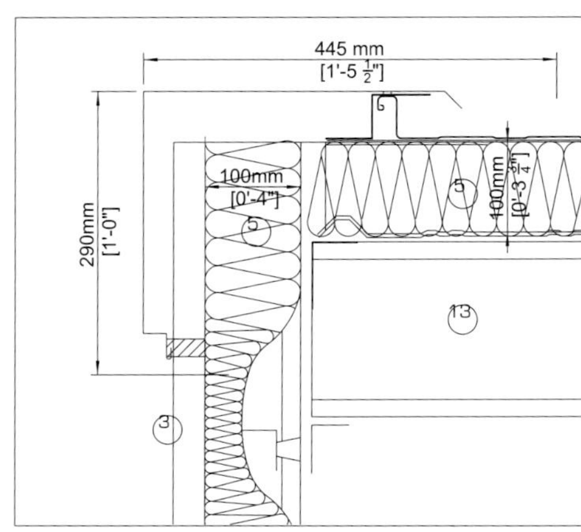

e| Vertical section. Roof junction at gable end

An advantage of profiled metal cladding is that it can be easily integrated with a similar system used for cladding the roof. Also, small areas of roof can be easily accommodated as steps within the facade with simple junctions between vertical and shallow pitch roofs. At the junction with the top of the roof an undercloak flashing is used to ensure water running off the wall is sent on down the roof and not into the joint at the base (f). At the bottom of the roof, either an exposed gutter or a concealed parapet gutter is used to collect the rainwater (c). Very small areas of roof can be drained without a gutter by projecting the roof beyond the cladding, allowing the rainwater to be thrown clear of the cladding and avoiding staining of the wall below. The effect of throwing water clear of the building needs to be integrated within the overall design.

Profiled metal cladding is most commonly used in large single-storey buildings such as factories or warehouses where it spans vertically from ground to roof without the need for additional support (MCE_73, f). This makes it a very economical solution for enclosing these building types. Although profiled metal cladding is used mainly for industrial buildings in conjunction with a portal frame in either steel or concrete, it provides an economic cladding system for larger-framed structures. The material can be set either vertically or horizontally to suit the design.

Horizontally set cladding is used where its strong horizontal lines are used for linear emphasis. Like vertically set cladding, the profiled sheet is supported at 3.0 to 5.0 metre (10ft to 16ft 6in) centres by posts or structural columns (MCE_73, f). This direction allows the material to enclose a building with a curved section. A useful aspect of profiled metal sheeting is its ability to be curved in one direction. This makes it an ideal cladding material for buildings with a curved vertical section (d). Slight irregularities in the surface finish or setting out of the curve are concealed by the profile itself. Polished stainless steel has been

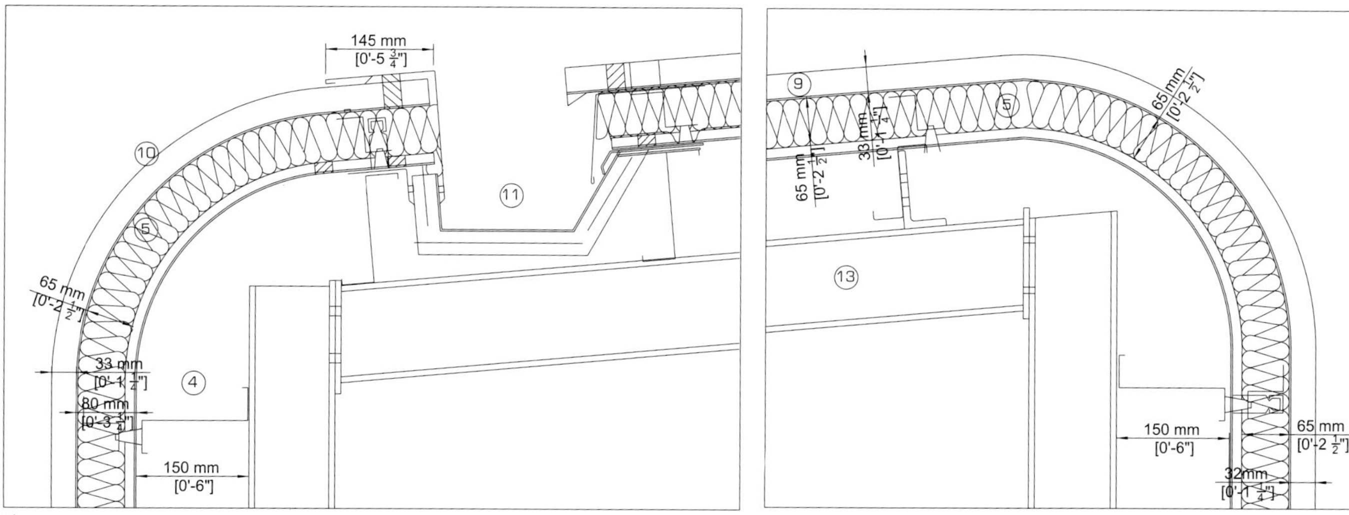

f | Vertical section. Curved eaves with parapet

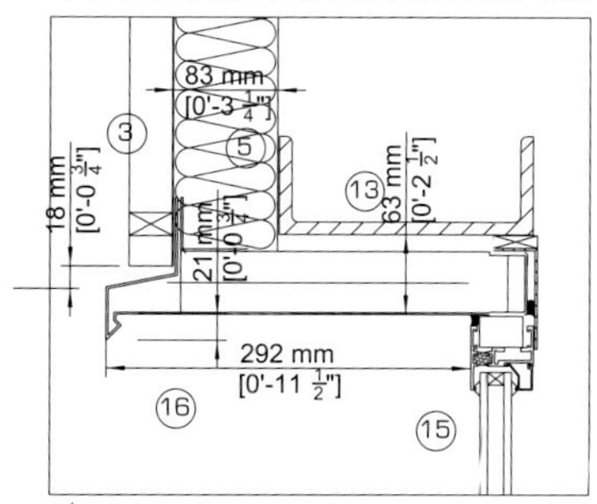

g | Vertical section. Window

Details

1 | Metal cover strip set
2 | Horizontally set profiled sheet
3 | Vertically set profiled sheet
4 | Z section steel fixing rails
5 | Thermal insulation
6 | Backing wall, typically timber/metal frame with plywood facing and waterproof membrane, or concrete block
7 | Vapour barrier
8 | Internal finish
9 | Roof, typically profiled metal sheet
10 | Curved eaves profile
11 | Concealed gutter
12 | Exposed gutter
13 | Supporting structural frame
14 | Structural slab
15 | Window frame
16 | Metal trim to window
17 | Metal parapet coping

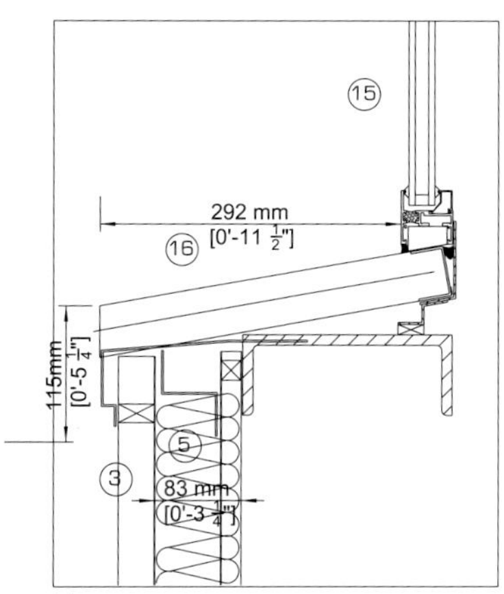

h | Vertical section. Roof junction at gable end

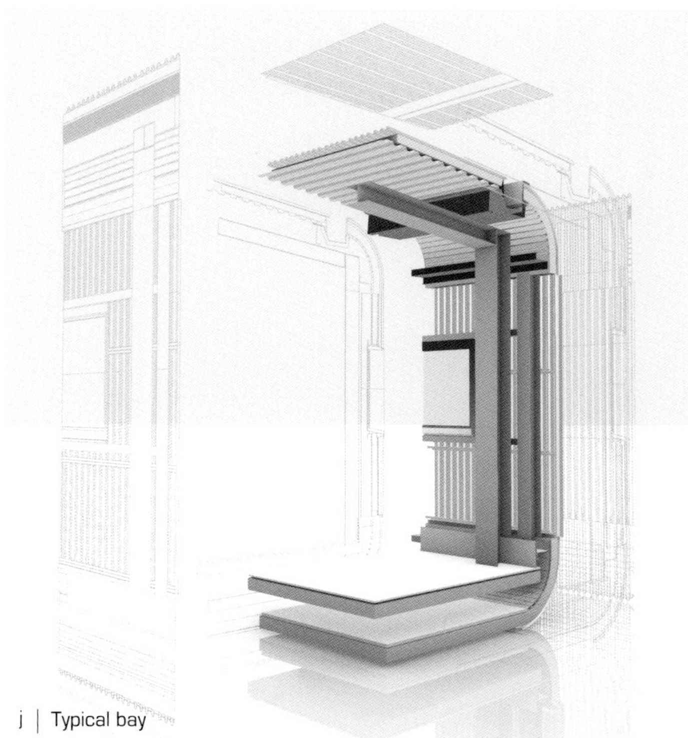

j | Typical bay

used for horizontally set cladding in public buildings where its high cost is balanced by longer durability than coated aluminium or coated mild steel types.

Vertically set sheeting has horizontal cladding rails at 3.0 metre to 5.0 metre (10ft to 16ft 6in) centres, depending on the floor height. In buildings of more than one storey, an inner lining to the wall is usually added since the gap created between the cladding and the floor slab is difficult to seal economically at floor level in a way that will allow people to walk on it. The additional inner lining may extend up to 1.0 metres (3ft 3in) above the finished floor level and may be either a metal lining tray forming part of the proprietary cladding system, or be a concrete block work wall around 100mm (4in) thick. A smoke seal or fire barrier may be required between the floors enclosed by the cladding, but this is very much dependent upon its particular application. Although horizontal rails can be set at wide centres, additional rails may be needed either to accommodate windows and doors or to increase the stiffness of the wall without using a much deeper profile, which would also increase stiffness.

Junctions

When laid vertically, sheets are joined by lapping them by around 150mm (6in) at vertical joints. Horizontal joints are also lapped with the upper sheet set over the lower one in the traditional manner. When laid horizontally, horizontal joints are formed with laps as when laid vertically, but horizontal joints are not usually lapped in the same way. This is mainly because it is difficult to form a continuous straight line in a joint that moves in and out with the shape of the profile. Instead, a recessed top hat section or projecting coverplate is used. The profiled sheet is butted up to the C-shaped section and sealed with silicone or mastic. The same principle is used for a projecting coverplate.

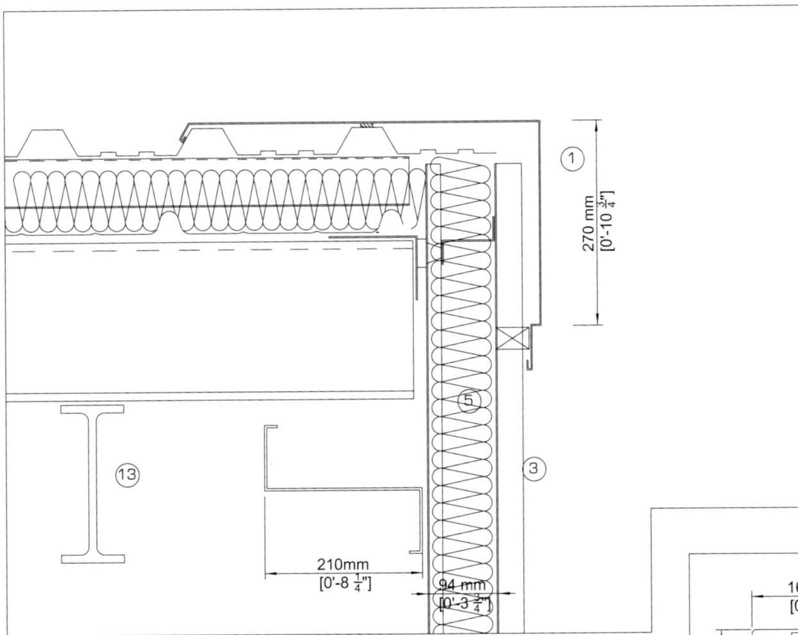

a| Vertical section. Roof junction at gable end

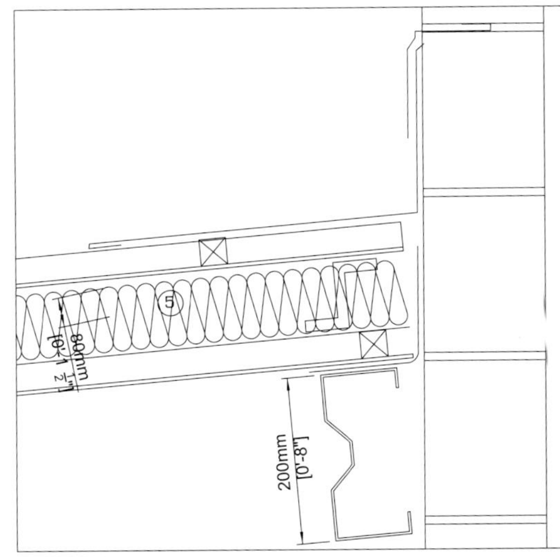

b| Vertical section. Roof junction at abutting wall

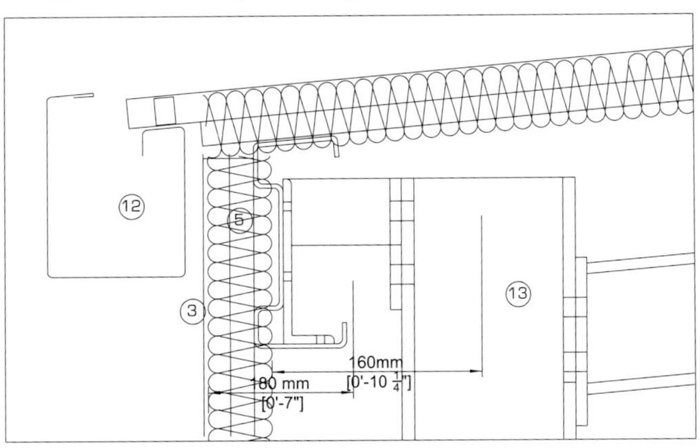

c| Vertical section. External gutter

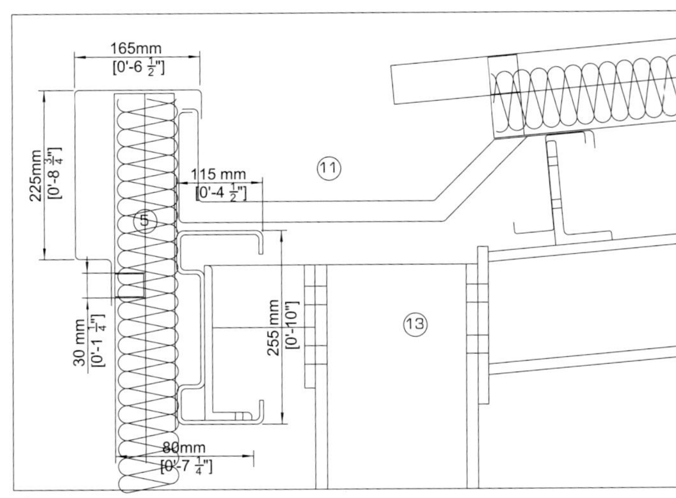

d| Vertical section. Roof parapet

Details

1 | Metal cover strip set
2 | Horizontally set profiled sheet
3 | Vertically set profiled sheet
4 | Z section steel fixing rails
5 | Thermal insulation
6 | Backing wall, typically timber/ metal frame with plywood facing and waterproof membrane, or concrete block
7 | Vapour barrier
8 | Internal finish
9 | Roof, typically profiled metal sheet
10 | Curved eaves profile
11 | Concealed gutter
12 | Exposed gutter
13 | Supporting structural frame
14 | Structural slab
15 | Window frame
16 | Metal trim to window
17 | Metal parapet coping

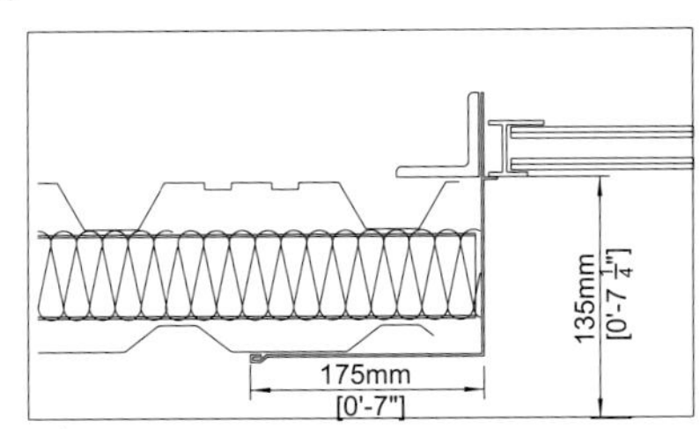

e| Horizontal section. Door opening

Corners are treated in a similar way. Corners to vertically and horizontally set cladding use projecting or recessed coverstrips. The profiled sheets that meet are lapped, however, to provide a weathertight seal and the coverplate provides both an additional seal along a potentially vulnerable joint as well as a crisp line to the corner (MCE_68, b). Regardless of sheet orientation, edging and jointing pieces are clearly visible, making them an important part of the design. Whereas profiled sheet can be lapped to give a continuous appearance on a large area of facade, the edging and jointing pieces of parapets, sills and corners are clearly visible (MCE_72, a). The visual impact of these junctions can be reduced with recessed joints. The use of curved eaves sheets (MCE_79, f) and curved (in plan) corner sheets was developed to avoid the need for visible corner pieces. 90º corner sheets are now available from some manufacturers that can be lapped smoothly over adjacent profiled metal sheets.

Parapets and gutters

Parapets are usually formed by either projecting the profiled sheet above the roof line in order to completely conceal the roof, which is often of the same material in the case of industrial buildings (MCE_68, a). Alternatively, a low parapet is formed at the level of the intersection of wall and roof, with a recessed gutter set immediately behind the parapet (MCE_72, a). A variation on this latter solution is to use curved eaves to give the idea of complete continuity between walls and roof with only a recessed gutter creating a line between the two. The recessed gutter in any of these configurations is useful when a pitched roof is used. On the gable elevation, the parapet can remain the same height while the roof rises and falls independently of the continuing line of the parapet on all sides. Curved eaves have mitred corner panels to allow a curved profile to be used continuously around a building (MCE_69, f).

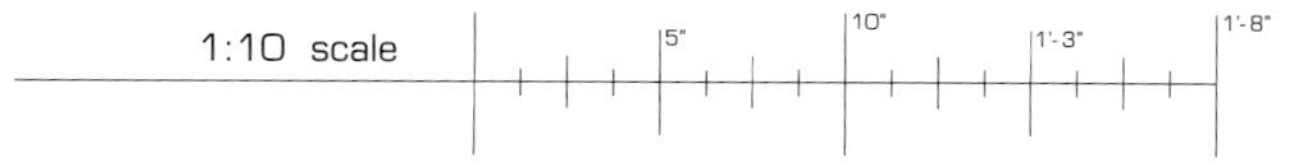

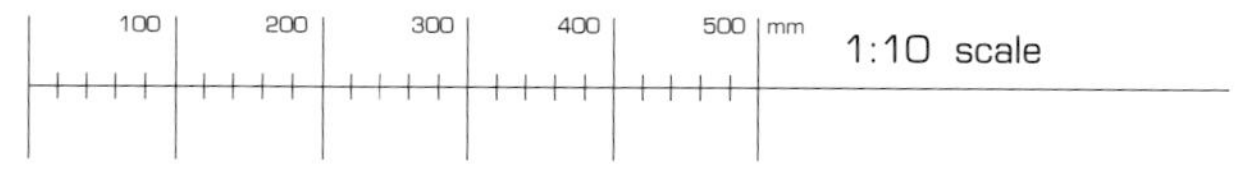

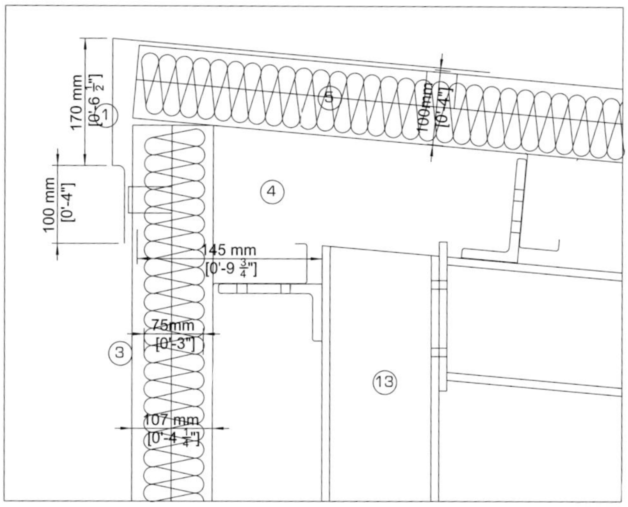

f | Vertical section. Roof junction at gable end

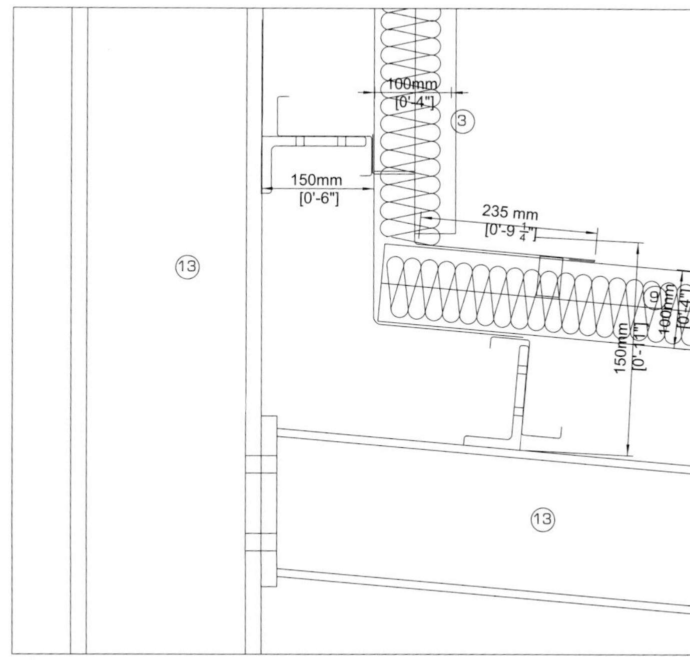

g | Vertical section. Junction with roof

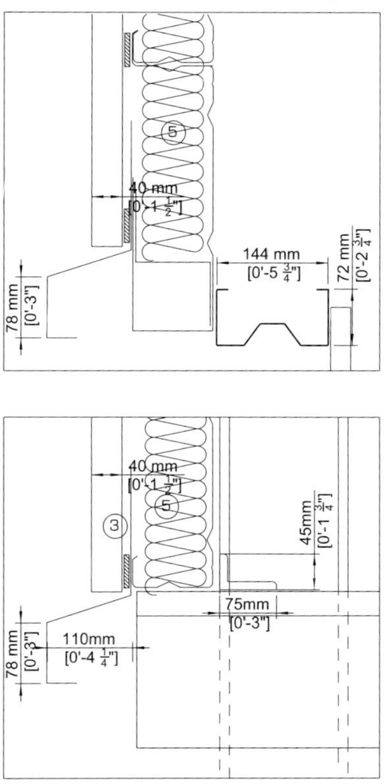

h | Vertical section. Door openings at head

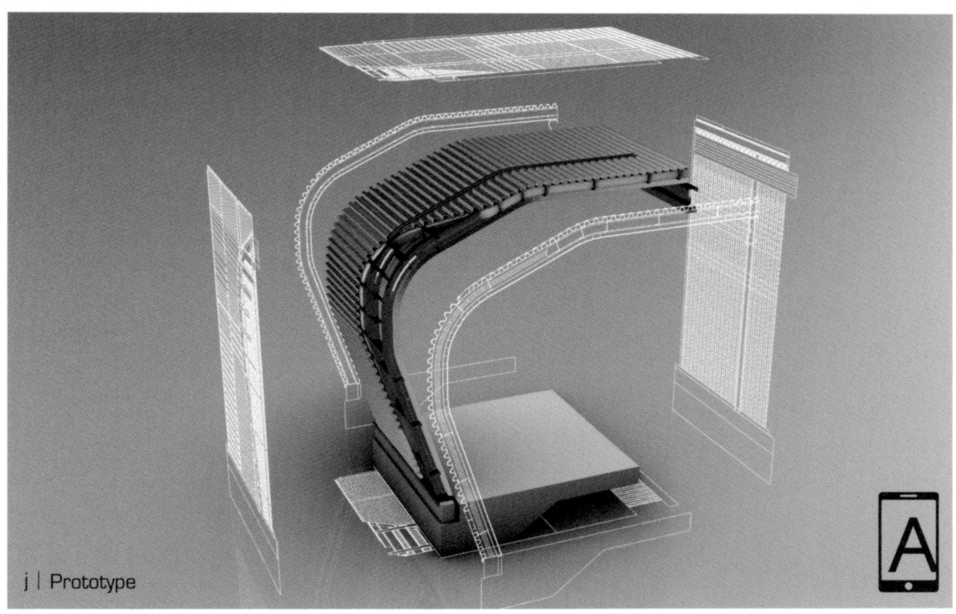

j | Prototype

Visible gutters are fixed on the outside face of the cladding. The roof projects over the top of the cladding in order to drain rainwater into the gutter, resulting in the roof visually projecting forward of the wall (c), unlike a parapet gutter. An advantage of this method is that rainwater is kept outside the building, avoiding the need to run vertical rainwater pipes within a building, then running rainwater back out through the foundations below ground level. Since gutters are needed only at the base of roof slopes, they are often not needed on all facades, giving an uneven appearance to the building. A solution to making gutters work on all facades is to design a hipped roof that drains equally into all gutters, but this can complicate roof design. Gutters require the support of brackets back to primary structure in order to support the weight of water when in use. The supporting brackets usually need to penetrate the cladding, requiring seals around the penetrations in order to make them weathertight. If the roof construction is required to be ventilated then the depth of the gutter will increase if the roof is intended to be hidden from view. Deep gutters have a strong visual presence on the facade.

Insulation and liner trays

Although profiled metal sheet is capable of long vertical spans, the thermal insulation and internal finish material require additional support. The insulation cannot be fixed directly to the metal sheet without being bonded to it. Fixing brackets to the profiled sheet would involve penetrating the sheet, creating a possible point for water ingress. Welding a support bracket would be both expensive and would easily distort the surface of the cladding. Bonding the insulation to the liner would be the next practical method, but this is done as a composite panel, which has constraints and is dealt with in the next section.

Flexible insulation quilt is fixed to intermediary sheeting rails that are also used to support an inner metal lining sheet. Sheeting rails are

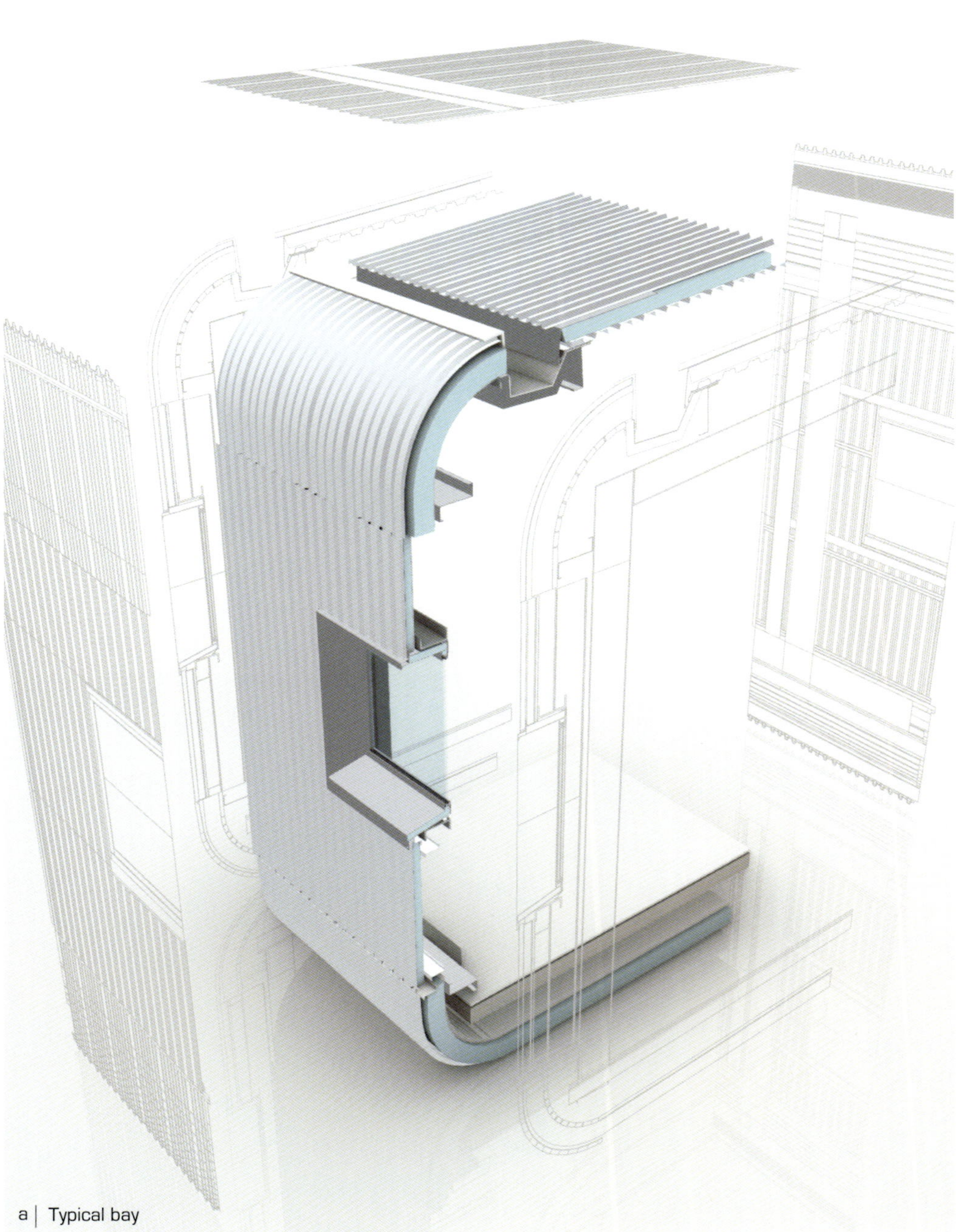

a | Typical bay

Details

1 | Metal cover strip set
2 | Horizontally set profiled sheet
3 | Vertically set profiled sheet
4 | Z section steel fixing rails
5 | Thermal insulation
6 | Backing wall, typically timber/metal frame with plywood facing and waterproof membrane, or concrete block
7 | Vapour barrier
8 | Internal finish
9 | Roof, typically profiled metal sheet
10 | Curved eaves profile
11 | Concealed gutter
12 | Exposed gutter
13 | Supporting structural frame
14 | Structural slab
15 | Window frame
16 | Metal trim to window
17 | Metal parapet coping

b | Prototype

made from pressed steel sections. Since the lining sheets are usually flat to create a smooth finish within the building, they do not span very far and require sheeting rails set at close centres. The rails can be used to give additional rigidity to the outer profiled sheet, but this requires penetrating the sheet with screw fixings which are sealed from the outside with plastic caps and washers.

An inner lining tray can also be formed from the same metal profiled sheet, as used in warehouse buildings where a smooth inner wall finish is not needed. Some intermediary sheeting rails are still required to support the thermal insulation. A more economic form of lining wall that does not interfere with the outer profiled cladding is concrete block work. In this instance, closed cell thermal insulation is fixed to the outside face of the wall.

Window and door openings

The reveals for windows and doors are formed in flat metal sheet, usually the same metal and same colour as the profiled sheeting. In practice the colour matching can be difficult if the coating (usually polyester powder coating or PVDF) is applied in different workshops or by different coating applicators. Contrasting colours are sometimes chosen for this reason. This is also true of window sections, which are usually supplied predated by a different manufacturer. Close co-ordination is needed between contractors to ensure a consistent colour throughout the project. An alternative approach is to reduce reveals to a small depth and use a colour for the windows different from that of the adjacent cladding. For example, with a silver metallic finish for cladding, a darker grey might be used for window frames without creating any contrast between the two colours used.

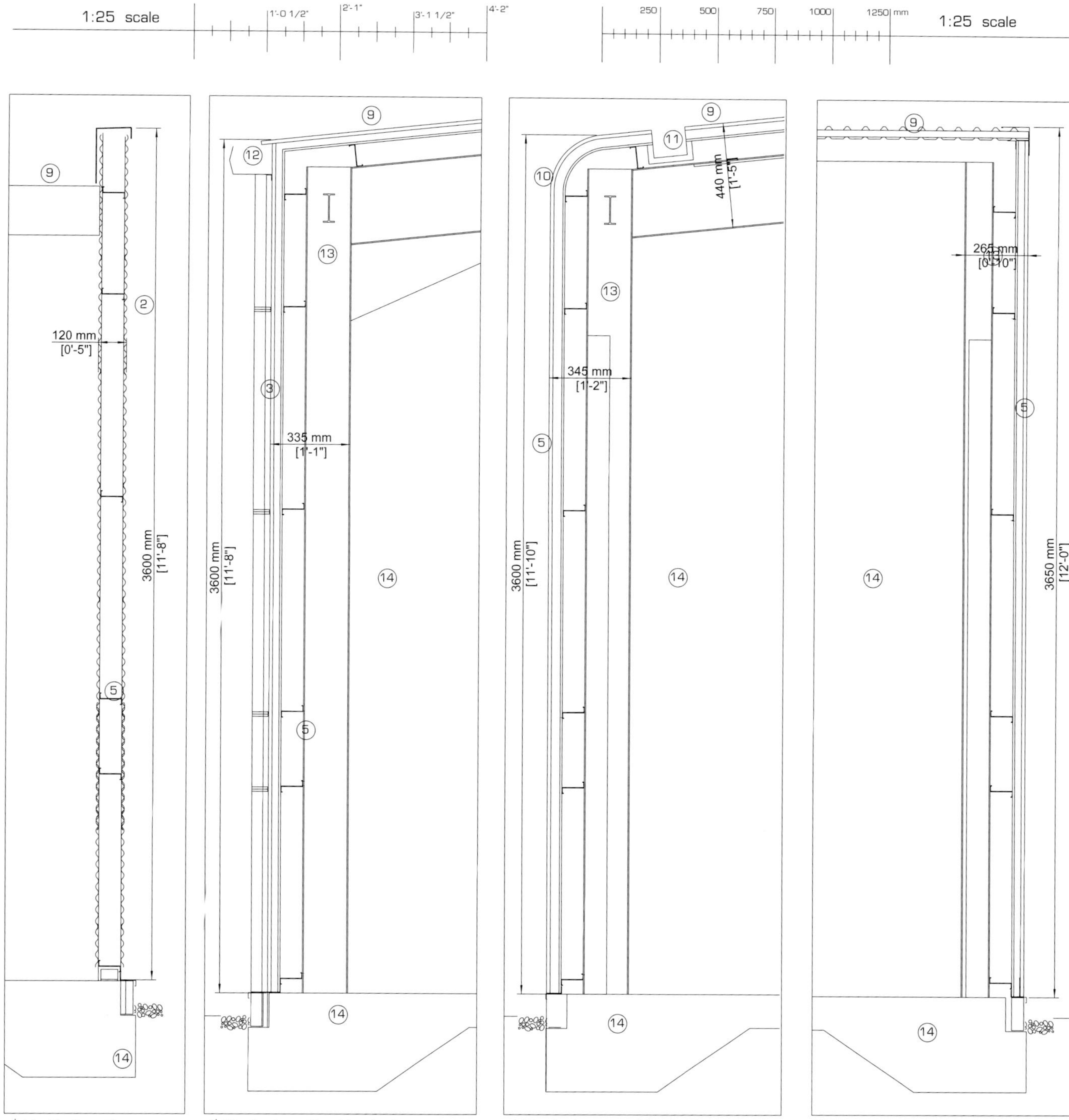

c | Vertical section. Horizontally-set cladding

d | Vertical section. Vertically-set cladding with gutter

e | Vertical section. Vertically-set cladding with curved eaves

f | Vertical section. Vertically-set cladding with gable end

Sills are formed in pressed metal which is inclined to drain water from its horizontal surface and has a projecting drip to avoid dirt washed off the sill running onto the cladding below, which would cause staining (a). Some drips have rising edges at the sides to avoid water running off at the sides that causes streaking in lines below the edges of the openings. Sills at ground level or at the base of the cladding are either flush or projecting, to suit visual requirements. As with sheet metal cladding, the sill is usually reinforced both to ensure it lies in a straight line and to protect it from accidental damage (MCE_69, h).

Developments

The range of profile types is steadily increasing, with wider, deeper profiles that were originally designed for use as roof decking being used as wall cladding. However, some of the interlocking types used on roofs are not suited to cladding since the standing seam joint, which is not designed to be tightly folded together, does not work when set in the vertical plane. This principle is also true of wall cladding types which are lapped and are not suited to use in roofs, where the seam is not high enough to be submerged under water during rain. A recent development has been the use of flat metal rainscreen panels fixed directly to a profiled sheet. This provides a smooth finish visually to the outside face of the cladding, while maintaining the economy and structural efficiency of the profiled sheet. Although the outer metal panel is fixed to the profiled sheet with screws or rivets that penetrate it, the pin jointed rainscreen configuration protects the fixings from the worst effects of windblown rain.

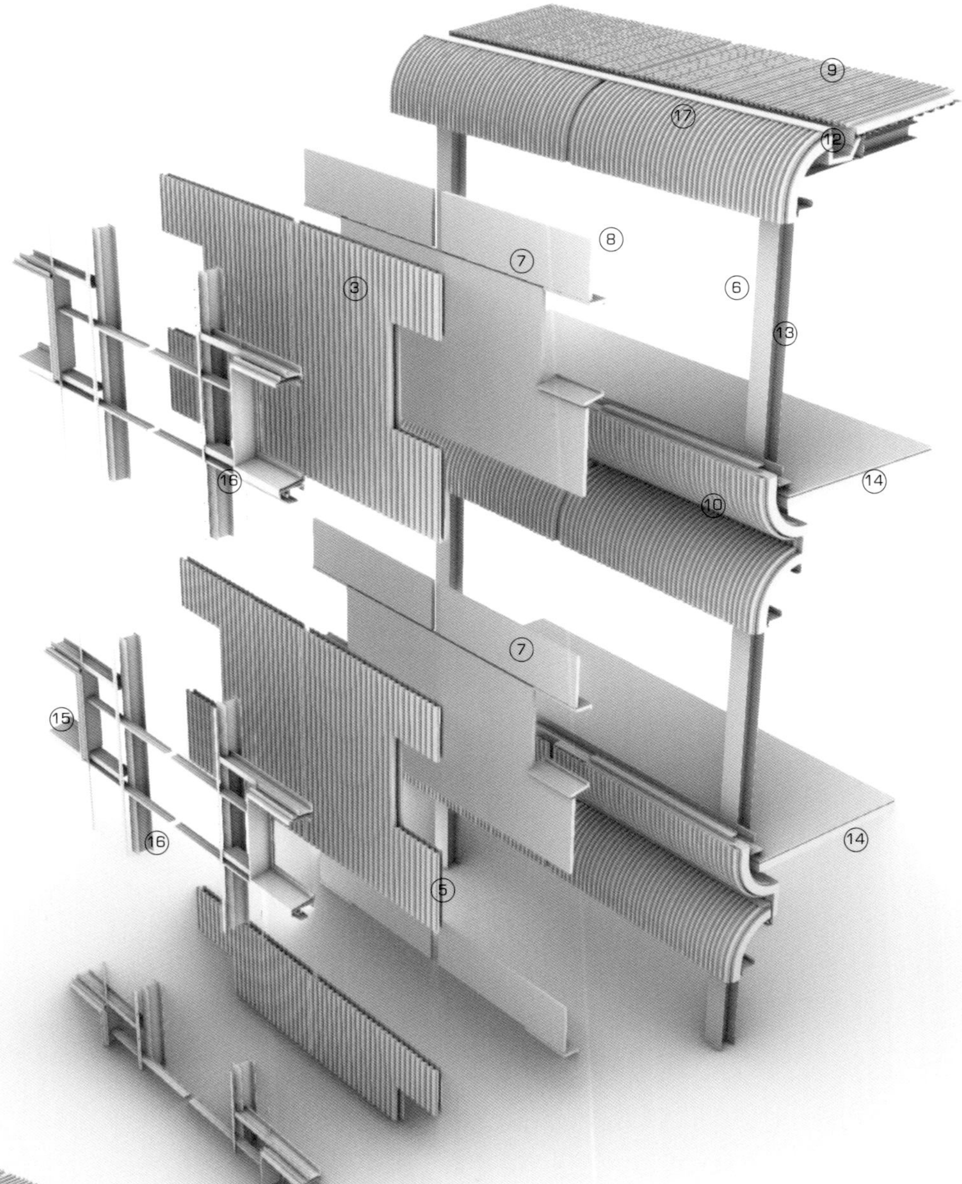

a | 3D exploded view showing profiled metal cladding facade fixed to steel frame

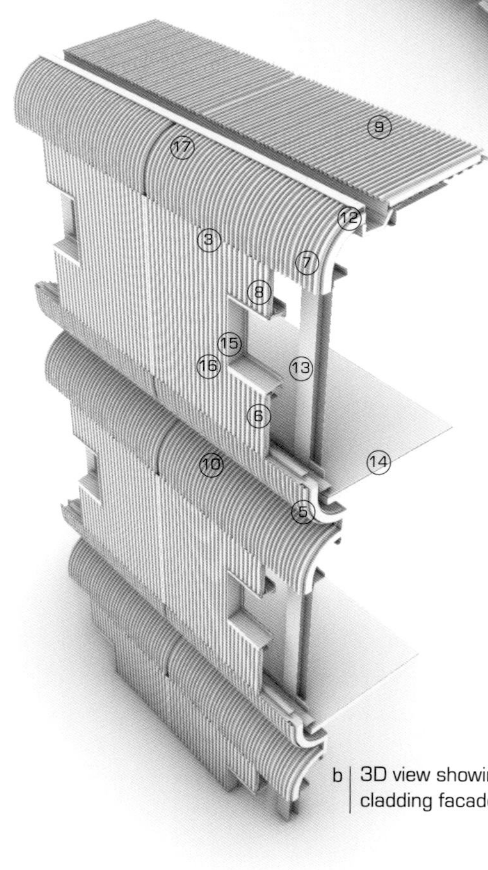

b | 3D view showing profiled metal cladding facade fixed to steel frame

Details

1 | Metal cover strip set
2 | Horizontally set profiled sheet
3 | Vertically set profiled sheet
4 | Z section steel fixing rails
5 | Thermal insulation
6 | Backing wall, typically timber/metal frame with plywood facing and waterproof membrane, or concrete block
7 | Vapour barrier

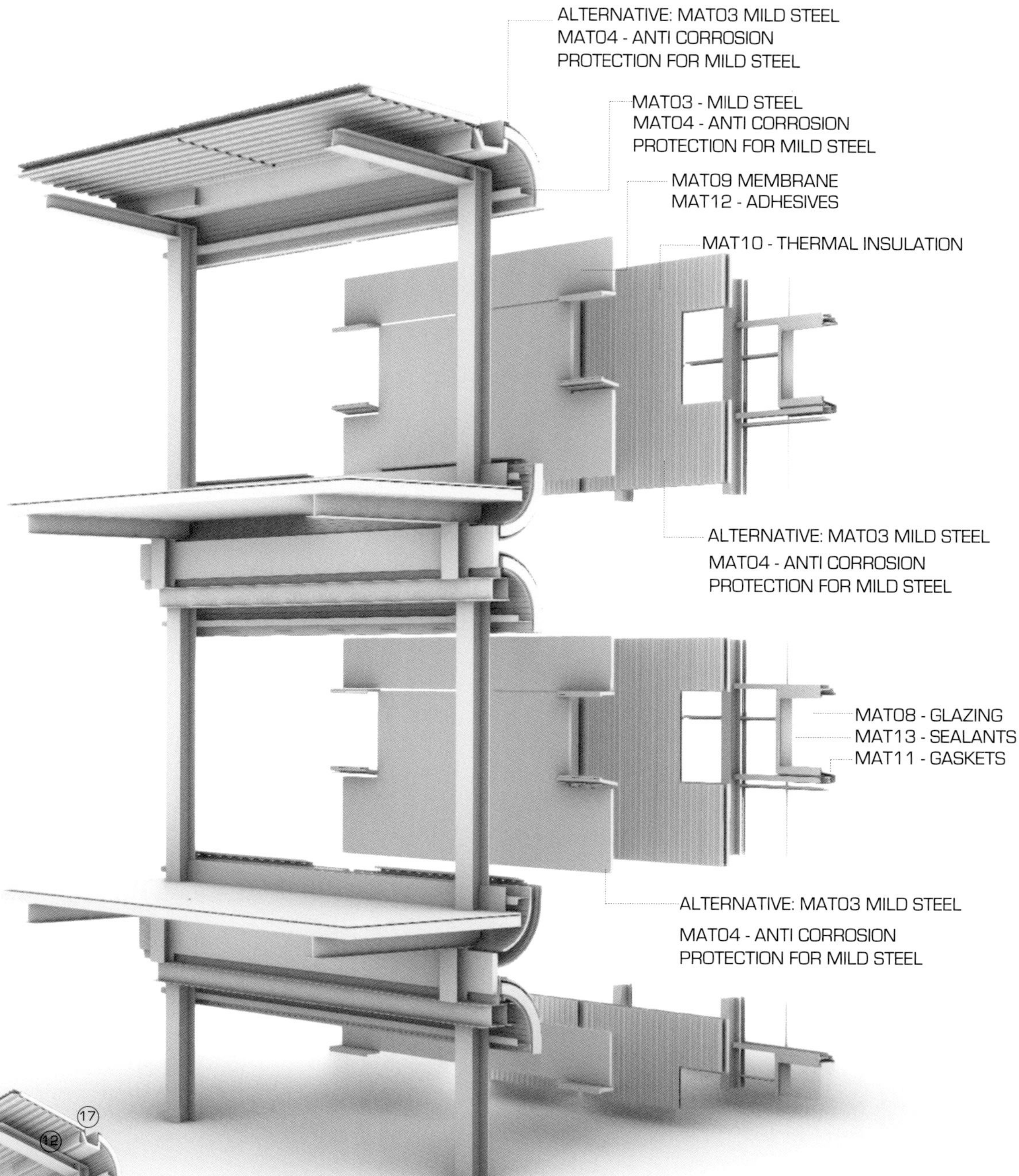

c | 3D exploded view of profiled metal cladding facade components with material specification references

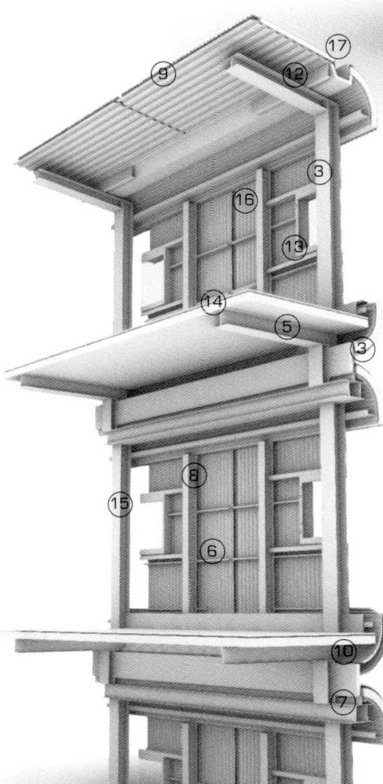

8 | Internal finish
9 | Roof, typically profiled metal sheet
10 | Curved eaves profile
11 | Concealed gutter
12 | Exposed gutter
13 | Supporting structural frame
14 | Structural slab
15 | Window frame
16 | Metal trim to window
17 | Metal parapet coping

d | 3D view showing profiled metal cladding facade fixed to steel frame

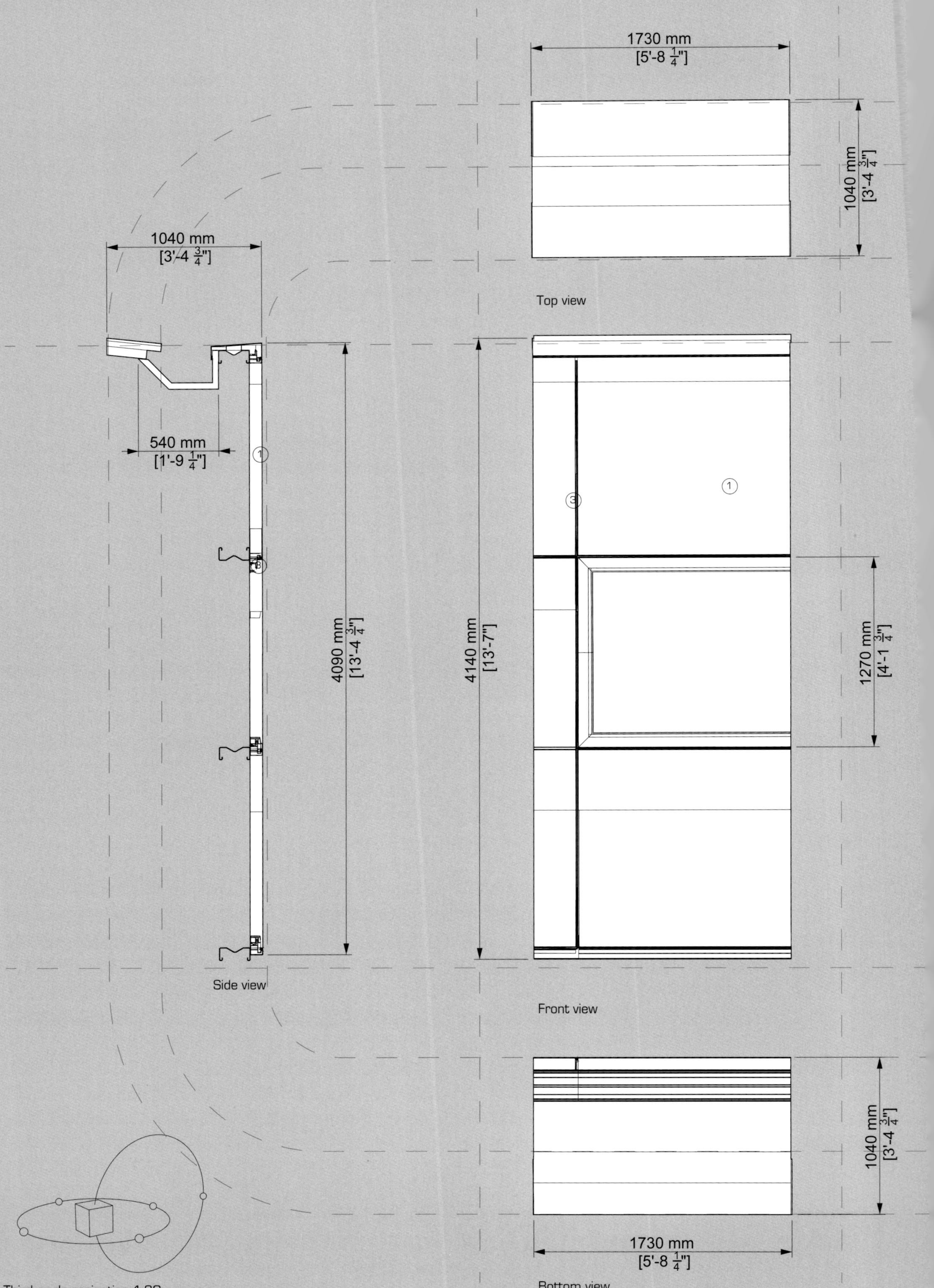

Third angle projection 1:30

Details

1 | Vertically set composite panel
2 | Horizontally set composite panel
3 | Silicone-based seal
4 | Outer metal facing
5 | Inner metal facing
6 | Inner insulation core
7 | Metal capping
8 | Concealed fixing
9 | Supporting structure
10 | 4-way interlocking composite panel

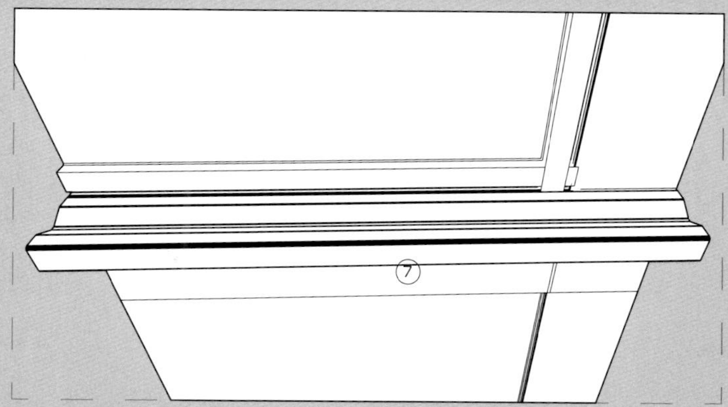

3D view of the assembly

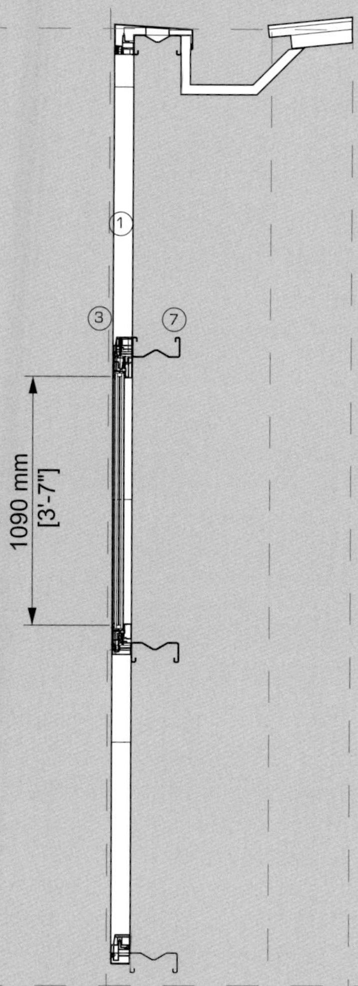

Side view

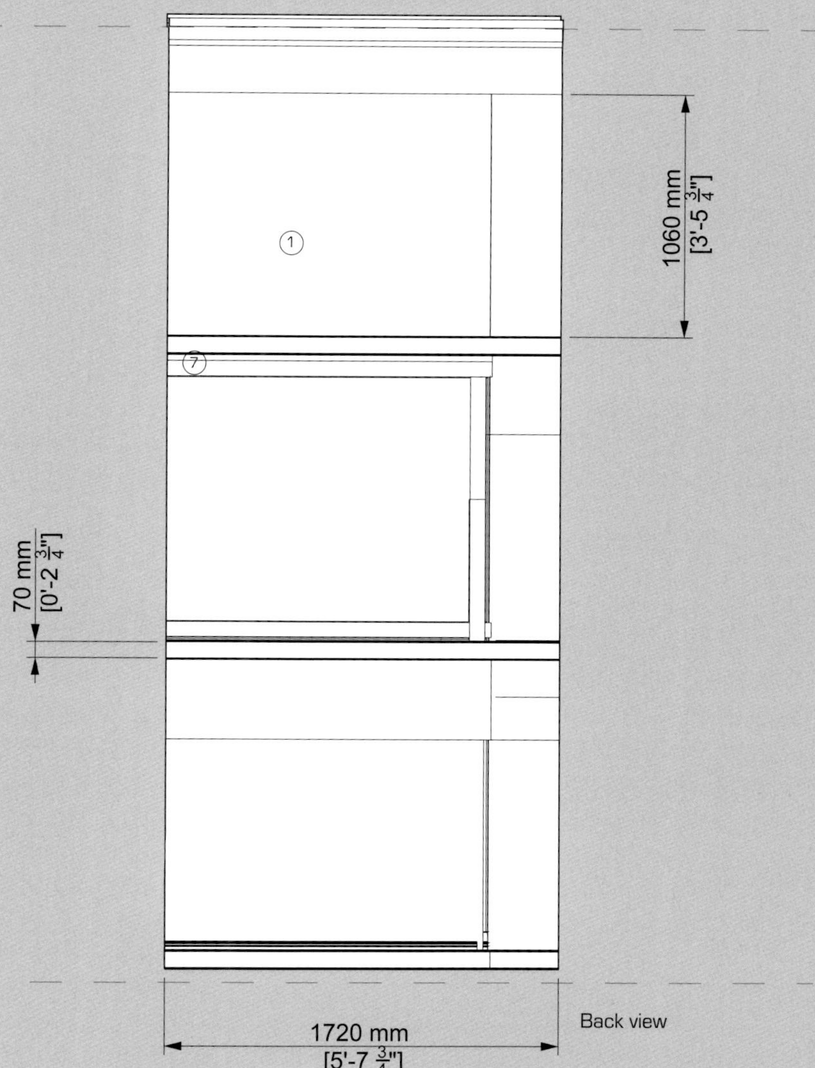

Back view

11 | Window frame
12 | Sectional roller shutter formed from composite panels
13 | Roof construction, composite panels are shown
14 | Metal trim
15 | Exposed gutter
16 | Concealed gutter
17 | Metal parapet coping
18 | Stick glazed curtain walling
19 | Door frame

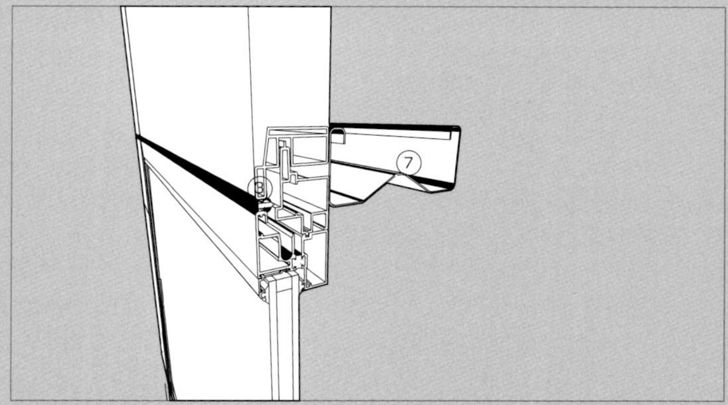

3D view of the assembly

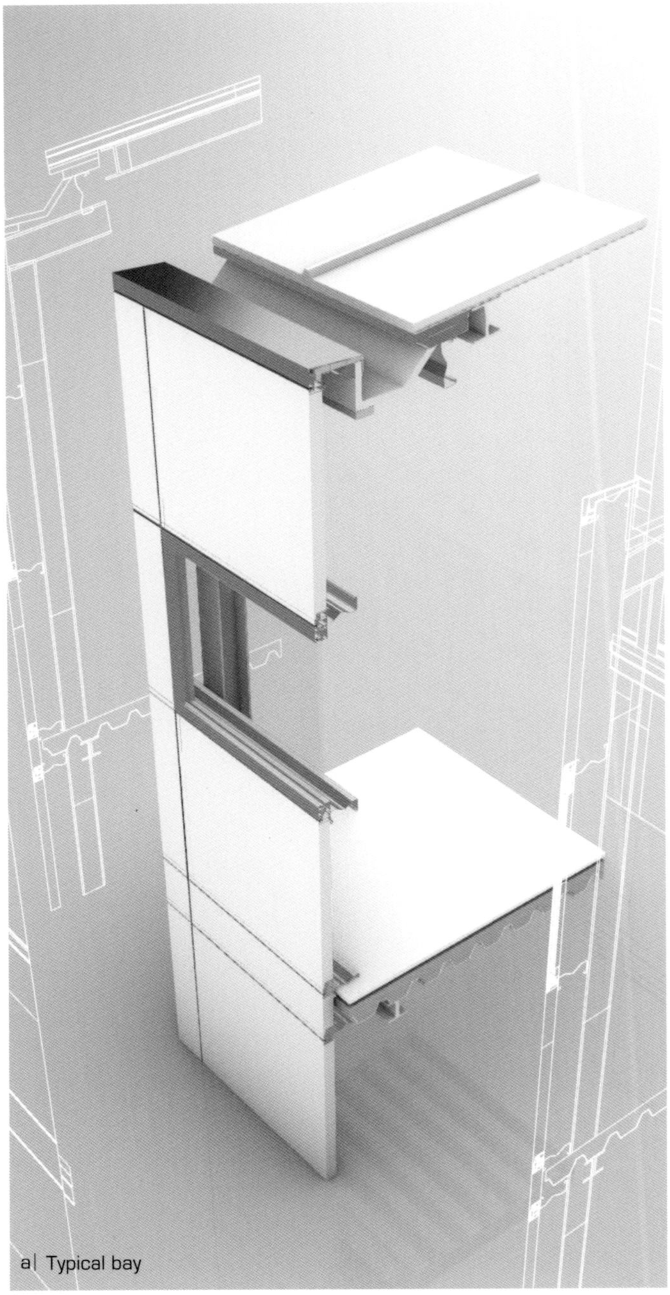

a| Typical bay

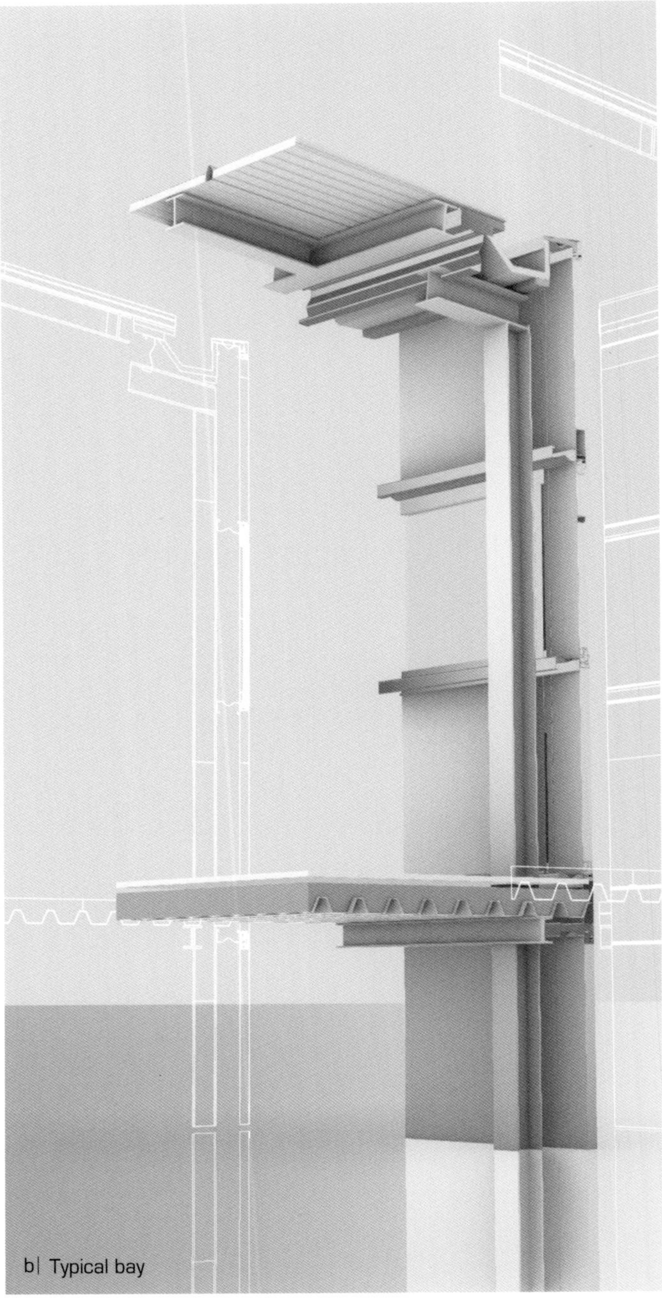

b| Typical bay

Composite metal panels require fewer components than for the 'kit of parts' used in the assembly of profiled metal cladding. Like profiled metal, panels are set either vertically or horizontally (a). Some panels interlock on two sides, while others interlock on four sides. Four-sided panels require no separate interface components for jointing but it is more difficult later to remove a damaged panel.

Horizontally set composite panels can be easily integrated with ribbon windows and suit building facades covering several floors. Panels are stacked one above the other with their vertical joints closed by rubber-based gaskets, recessed channel sections in aluminium, or projecting coverstrips in aluminium (MCE_84, a). Panels are fixed back to the primary structure or on a secondary steel frame, typically box sections (b), fixed to the sides of floor slabs if columns are spaced too far apart or columns are not positioned on the edge of floor slabs (MCE_80, b).

Where windows are used in a facade, additional support is needed to frame the opening (b). This is because windows are not supported by the composite panels except where made specifically as part of a proprietary system. In practice, windows are usually supplied by a specialist manufacturer.

The steel support framing is set on the face of the slab, making it easier to take up deflections in floor slabs. However, as is the case with curtain walling, the gap between composite panel and floor slab needs filling with a smoke seal or fire barrier. The floor finish usually has a metal angle to close off the gap at slab level and at the soffit level below. Four-sided interlocking panels use the same principle for fixing on all four sides. This also makes it easier to integrate windows within the system since a window panel is locked in like any other panel.

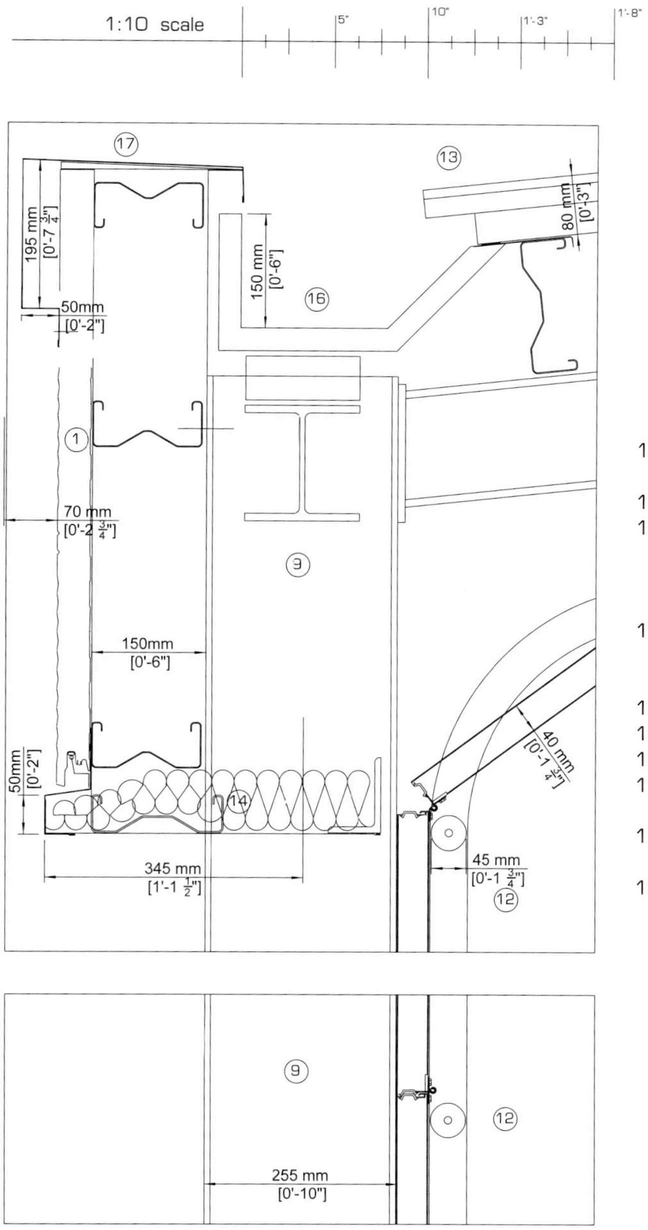

c | Vertical section. Parapet and door opening. Sectional roller shutter door shown, typical in industrial applications

Details

1 Vertically set composite panel
2 Horizontally set composite panel
3 Silicone-based seal
4 Outer metal facing
5 Inner metal facing
6 Inner insulation core
7 Metal capping
8 Concealed fixing
9 Supporting structure
10 4-way interlocking composite panel
11 Window frame
12 Sectional roller shutter formed from composite panels
13 Roof construction, composite panels are shown
14 Metal trim
15 Exposed gutter
16 Concealed gutter
17 Metal parapet coping
18 Stick glazed curtain walling
19 Door frame

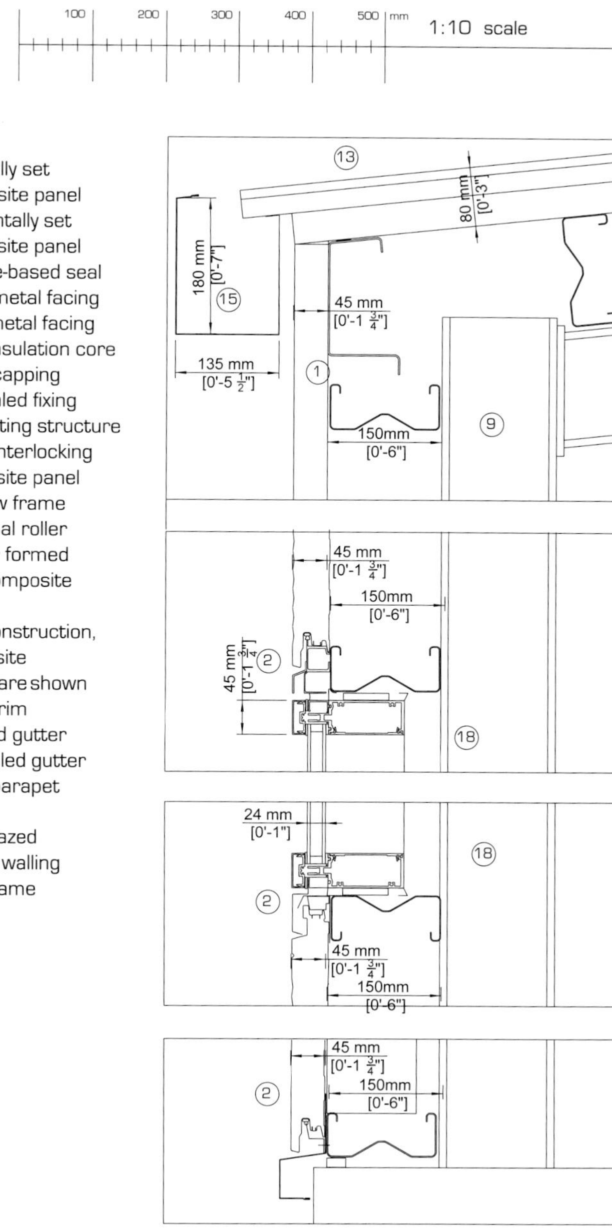

d | Vertical section. Parapet and sill with curtain walling-type window set into cladding, typical in industrial applications

Vertically set composite panels are more common in single-storey applications, but multi-storey applications are used increasingly. Panels are interlocked at vertical joints, while horizontal joints are formed by using a sill-type detail similar to that used in a transition from vertical panel to low pitched roof. A sill in extruded or folded aluminium or steel (depending on which metal is used for the composite panel faces) is used. The front of the trim projects beyond the face of the cladding to throw water clear and prevent staining to panels below (c). The back of the drip projects up the back of the upper panel to prevent water from penetrating the joint. Panels are also supported on either an interlocking frame, or occasionally they span between columns if panels are stiff enough to span unassisted. An additional method of fixing panels is to position them between floor slabs spanning from floor to ceiling when used as part of a rainscreen system. Panels sit on the floor slab with their outer face flush with the edge of the slab. The outer rainscreen is set forward of the composite panel, concealing both the panels and the edge of the floor slab.

Interlocking vertically set panels are of several types, unlike horizontally set types, which have a stepped joint to avoid rainwater penetration. The most common type for vertical joints is also a stepped joint with a recess on the outer face. An alternative is to have projecting nibs on the side of the panel to which a coverplate is fixed over the gap between the two panels. Rubber-based seals are set into the depths of all these joint types. Another joint is a C-shaped channel profile which interlocks with the profile of the adjacent panel. The outside face of the panel has a slightly projecting edge instead of a recessed joint in the stepped joint types.

All these panel types use jointing methods that avoid thermal bridges. Drips often penetrate from outside but the low condensation risk is assessed during the design stage.

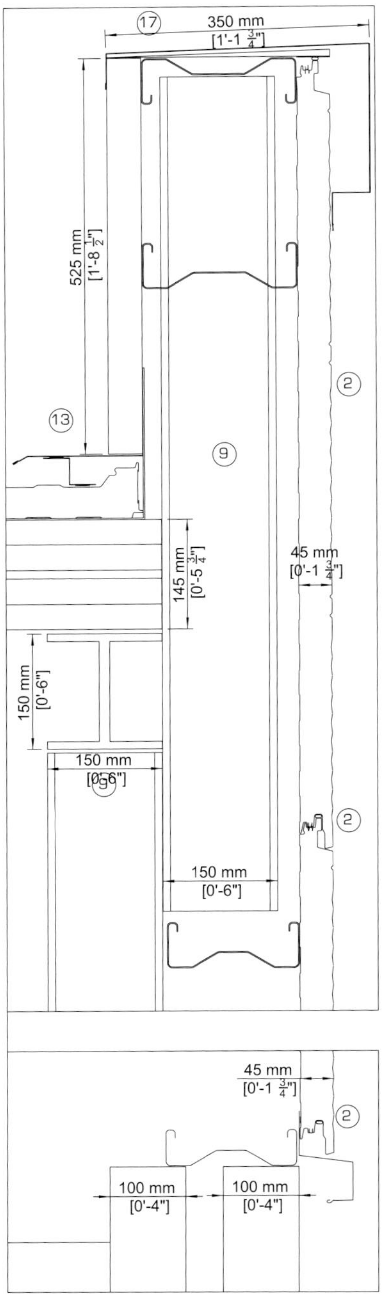

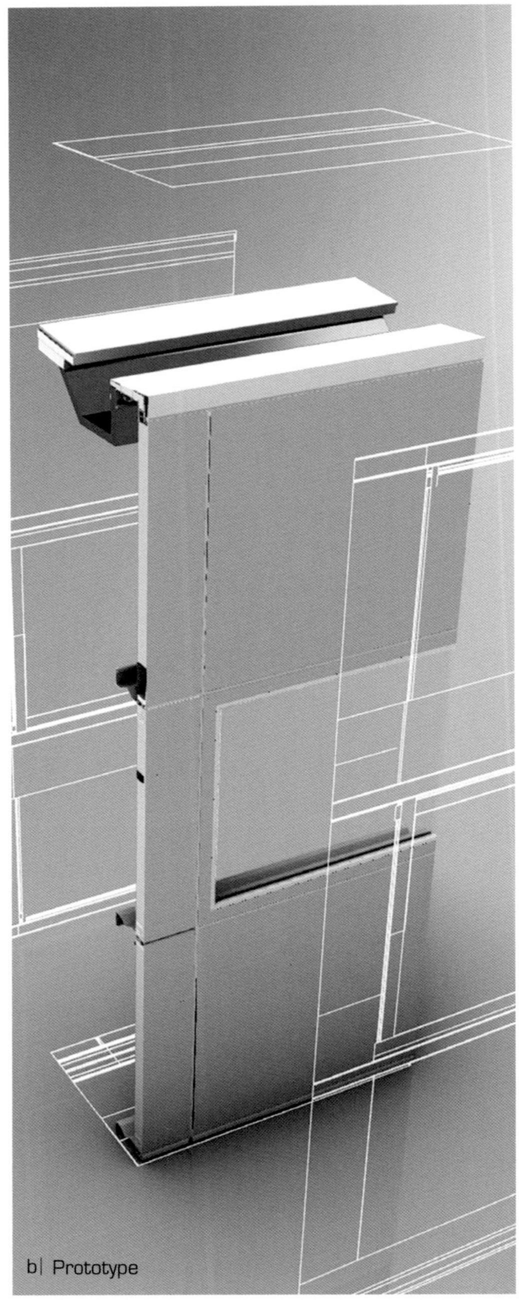
b| Prototype

c| Vertical section. Curved parapet and cill with aluminuim window set into reveal in cladding

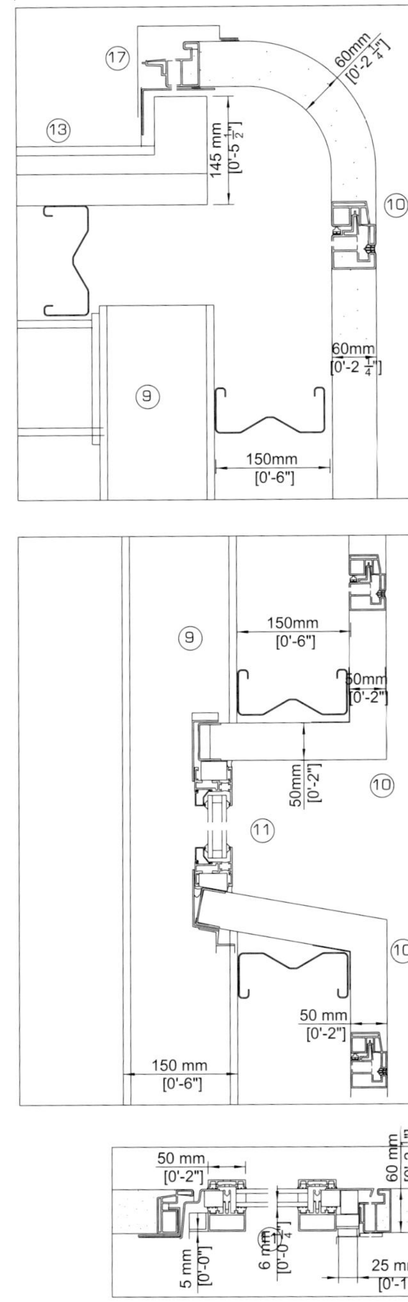

a| Vertical section. Parapet and ground level sill

Parapets and sills

Parapet copings and drips at ground level can also be made as composite panels, forming an integrated part of a proprietary system (e). This can be an advantage when seeking a seamless effect across a facade, but sets limitations on the variety of junctions (in terms of width and height) at parapet and base due to the need for repetition in specially made panels. A large number of different panel types cannot be produced economically for a single project. Although folded metal flashings and extrusions can appear more visually intrusive than a specially made composite panel (MCE_78, a), this method is far more flexible for dealing with varied parapet and ground level junctions, particularly for complex junctions.

Like fully supported sheet metal facades, parapets can be set either level with the roof, to create a continuous smooth envelope, or be stepped to allow the roof profile to be concealed. The same principles apply as those for continuously supported sheet metal parapets and sills.

Windows and door openings

There are two methods of creating an opening in a composite panel for both horizontally and vertically set panels. The first method is a special reveal panel that interlocks with the composite panels (MCE_83, h). With vertically set panels the window interlocks into vertical joints and has coverstrips on horizontal joints. Where horizontally set panels are used, the window interlocks with the horizontal joints and has coverstrips applied on the vertical joints. The second method is to use metal sheet to form a reveal. A single sheet of metal, 1200 or 1500mm wide coil (4ft or 5ft), is fixed to a light gauge steel frame which is insulated (g). The inner framed wall has a vapour barrier and an inner metal sheet, usually matching the adjacent inner face of the panels.

1:10 scale

d | Vertical section. Integrated parapet and windows, all flush with face of composite panels

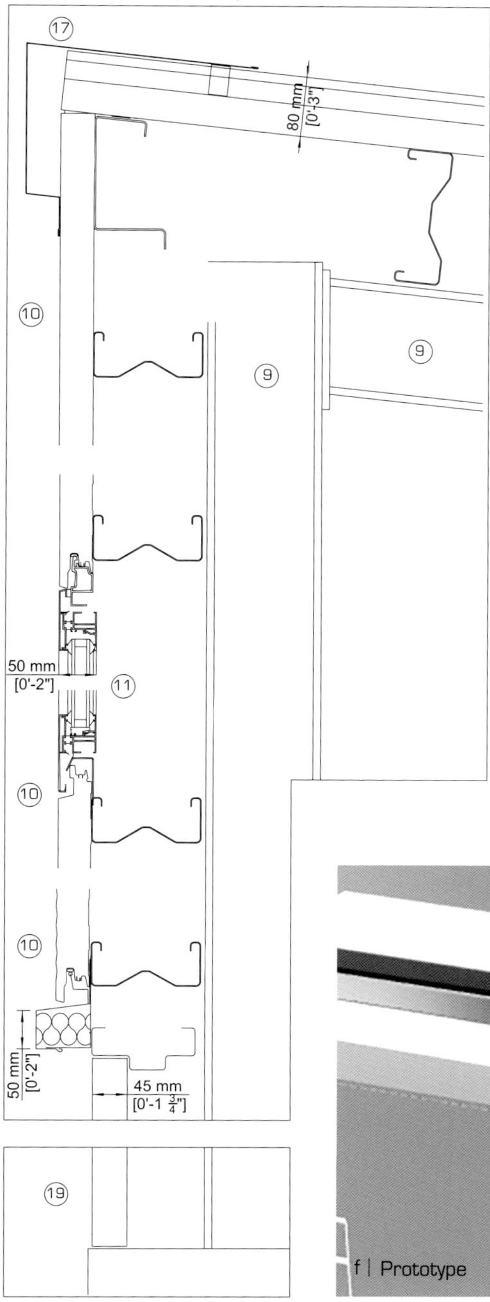

1:10 scale

e | Vertical section. Junction with roof and door opening

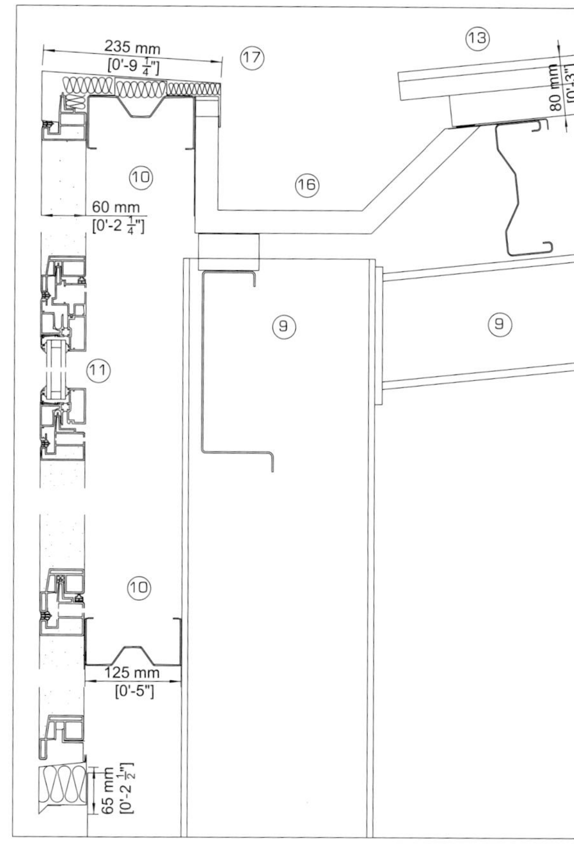

Details

1 Vertically set composite panel
2 Horizontally set composite panel
3 Silicone-based seal
4 Outer metal facing
5 Inner metal facing
6 Inner insulation core
7 Metal capping
8 Concealed fixing
9 Supporting structure
10 4-way interlocking composite panel
11 Window frame
12 Sectional roller shutter formed from composite panels
13 Roof construction, composite panels are shown
14 Metal trim
15 Exposed gutter
16 Concealed gutter
17 Metal parapet coping
18 Stick glazed curtain walling
19 Door frame

f | Prototype

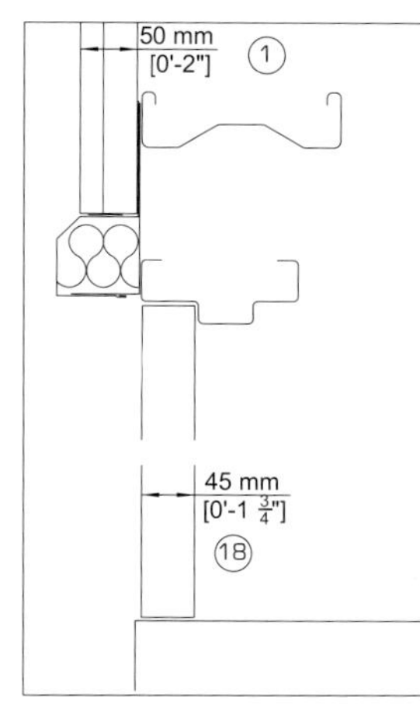

Heads and sills of openings are formed in the same way with either a purpose-made corner panel (the sill is an inclined surface) or with sheet metal and thermal insulation. However, the lightweight metal frame is not usually fixed back to the composite in order to avoid penetrating either the outer or inner face of the panels. Instead it is fixed back to a floor slab or to the primary structure.

Since the inclusion of reveals is not really in the nature of composite panel detailing, and the wall has little depth, the additional frame is also used to support the window, typically aluminium framed with a thermal break and double glazed units (MCE_78, a). More often, windows and doors are fixed with the outer face of the glass or door aligned with the outer face of the adjacent composite panels. An alternative form of glazing to windows and doors is to use curtain walling. Thermally broken stick systems are fixed directly against the adjacent composite panels with a seal set against the primary structure forming the opening (typically vertical sheeting rails). If horizontally set panels are used, then the curtain walling can be sealed against adjacent composite panels with a vertical cover strip used for all vertical joints.

Where many windows are required at a particular height on a facade, as when providing light into an upper floor, it is common to create a continuous ribbon of windows to avoid small infills of composite panels between windows. The continuous line of windows is fixed back to a secondary frame of steel box sections which may be exposed in the building or be concealed behind an inner finish such as a plasterboard lining (MCE_85, c). Continuous windows can also be glazed into horizontally set panels interlocking on two sides, or four-sided panels. This can avoid the need for additional support framing. In this instance the windows are braced back in the same way as the composite panels.

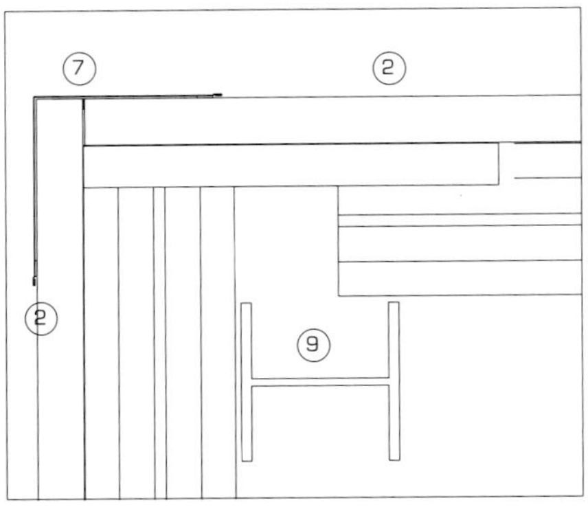

a | Horizontal sections. Corner details

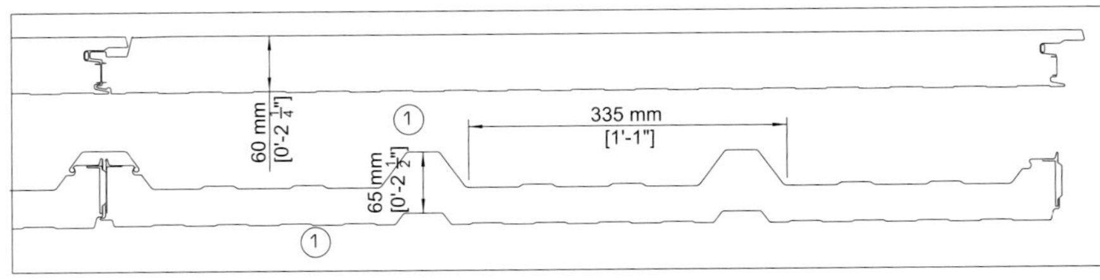

b | Horizontal section. Typical joint between panels spanning horizontally with interlocking joints on horizontal edges

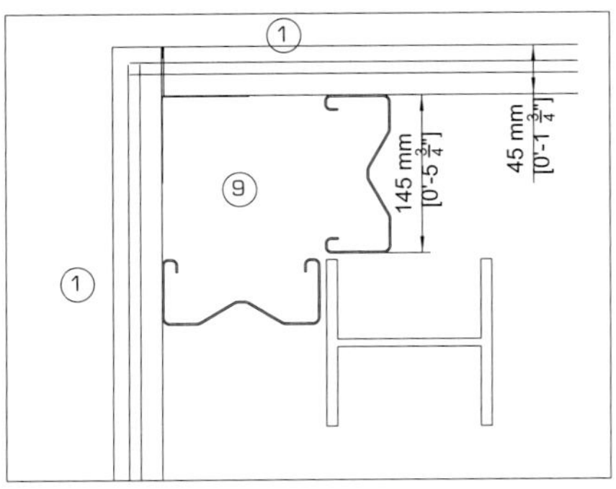

c | Horizontal section. Door jamb with metal sheet forming panel in door reveal

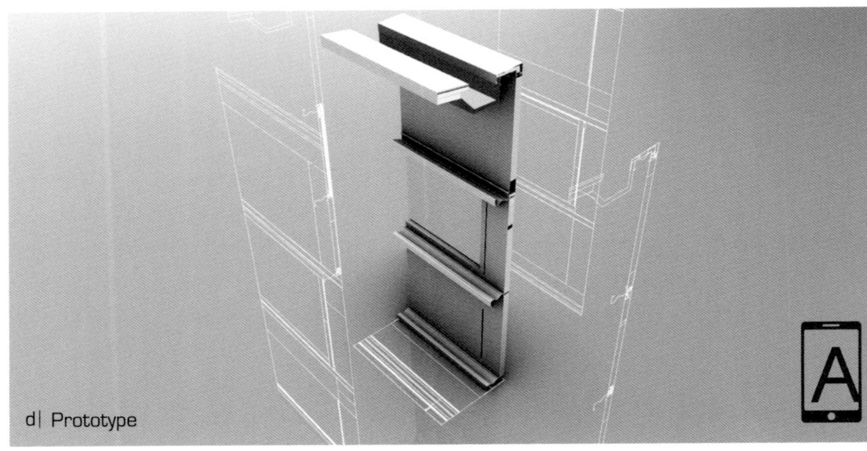

d | Prototype

e | Horizontal section. Panels spanning horizontally with recessed cover seal on vertical edges

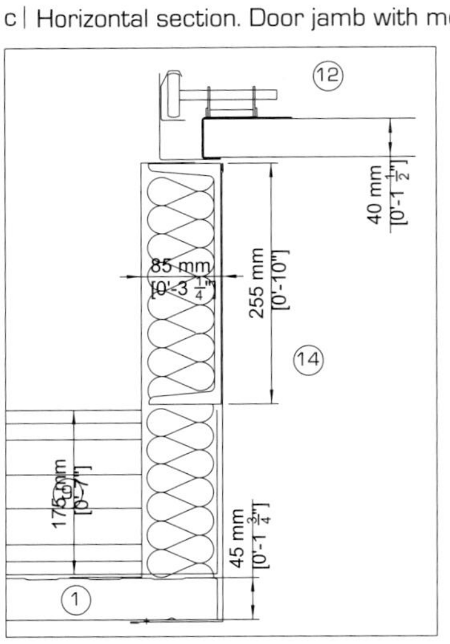

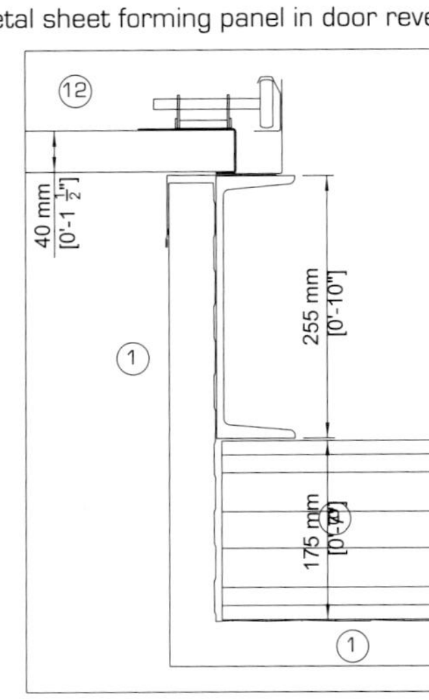

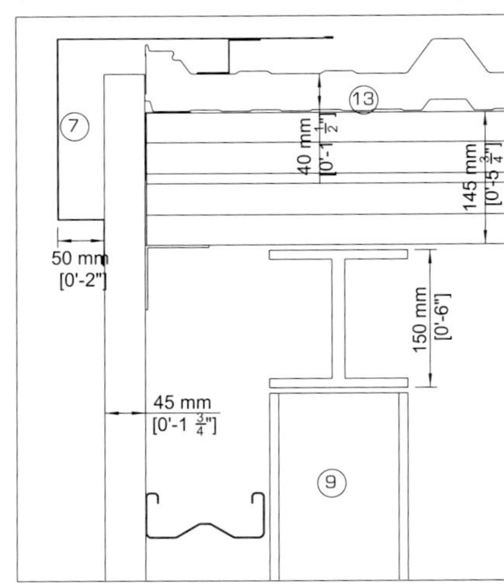

Developments

The use of composite metal panels is increasing into building types beyond industrial buildings, into office buildings and sports facilities. In the case of office buildings, they provide an economic spandrel panel where curtain walling is not always effective. In sports facilities, composite panels enclose large indoor spaces with a durable, crisply-made and relatively economic cladding system. Although the appearance of the external face is smooth and gridded, the supporting structure is visible and usually set on the inside face to avoid penetrations to the outside through the joints between panels. If the exposed structure is enclosed with an economic lining wall concealing the structure, this additional element can add considerable cost to the cladding, making it much less economic.

For this reason, a supporting structure that is designed to be seen, such as tubular steel posts, is increasingly used. The composite panels span between steel posts or trusses with little or no interlocking supporting structure (b). In order to keep the supporting structure as visually elegant as possible, trusses or posts are spaced as far apart as possible. This has led to panels getting longer, with a maximum length currently around 15 metres (49ft). Some proprietary systems include edges to panels which are deeper, making the continuous vertical joints and horizontal joints more rigid, allowing them to span greater distances, and thus reducing the amount of visible supporting structure needed.

Increasingly, window openings need not be dictated by the direction in which panels are laid. Horizontally set panels do not have windows arranged horizontally. Transitions between window openings and composite panels are becoming more economic with standard extrusions and rubber-based seals. This is evermore the case with four-sided, interlocking panels, where window panels and metal panels are fixed in

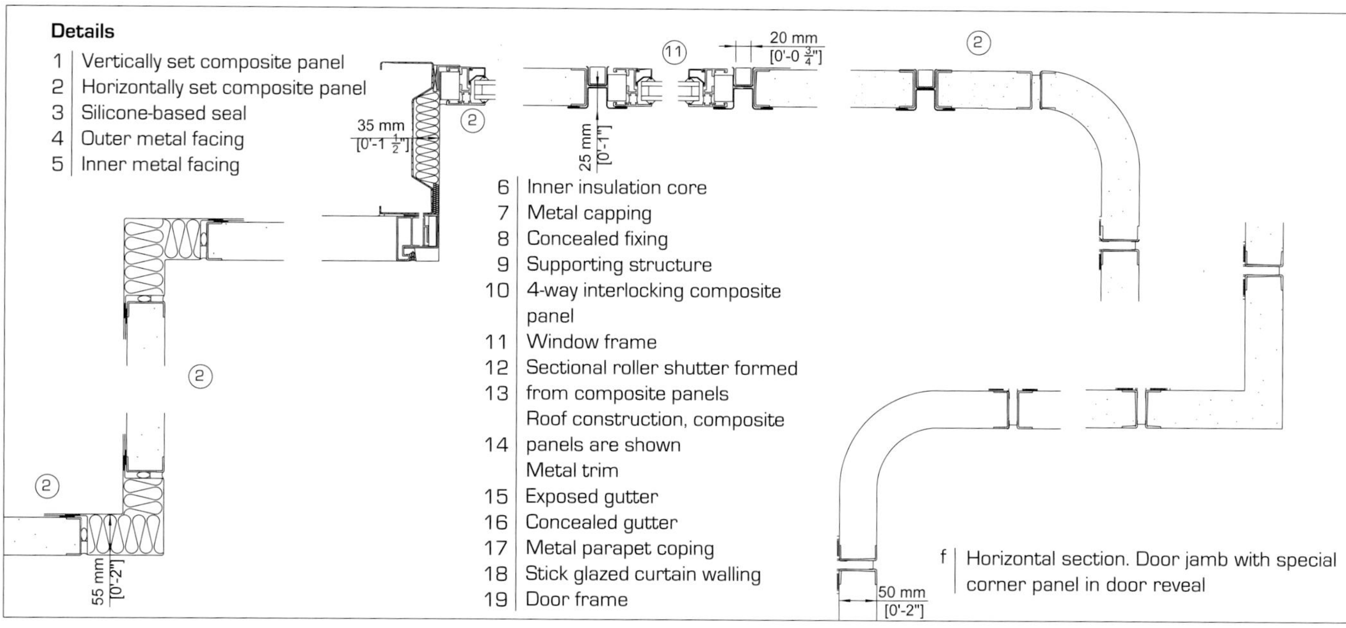

f | Horizontal section. Door jamb with special corner panel in door reveal

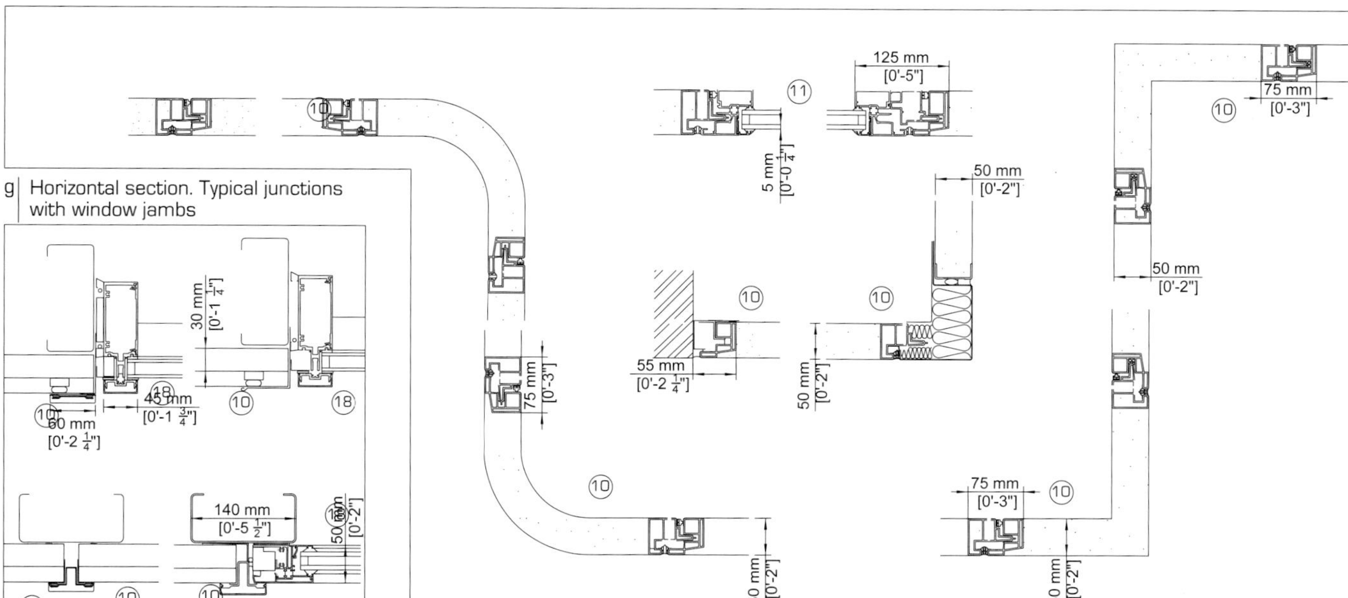

g | Horizontal section. Typical junctions with window jambs

h | Horizontal section. Panels interlocking on 4 sides, spanning either vertically or horizontally

the same way (h). Increasingly, irregular facade grids are being developed in designs to create a richer mix of panel sizes in visual patchwork of different sizes of panels.

Corners

Composite panels are connected at corners by one of two methods. Either specially made corner panels (h are used (typically why 90º is standard) or a coverstrip is added to cover the junction where the panels meet. Corner panels are more suited to vertically set arrangements, though panels for horizontally-laid panels are sometimes used for visual effect (h). Where corner coverstrips are used, their appearance resembles that of prof led metal cladding, which can give a facade an overall framed appearance. Metal trims at the parapet, base and corners can give this appearance. For this reason, the special corner panels and parapet panels are increasingly used.

Thermal bridges at sills

A weakness in composite panel systems has been the use of pressed metal sections or aluminium extrusions that pass from outside to inside without a thermal break. This is being remedied by the use of insulated sills, made in the manner of composite panels. This reduces the thermal bridge, in some cases a break in the section from outside to inside can be formed by turning the metal sill into the injected foam or polystyrene in the same way as a composite panel.

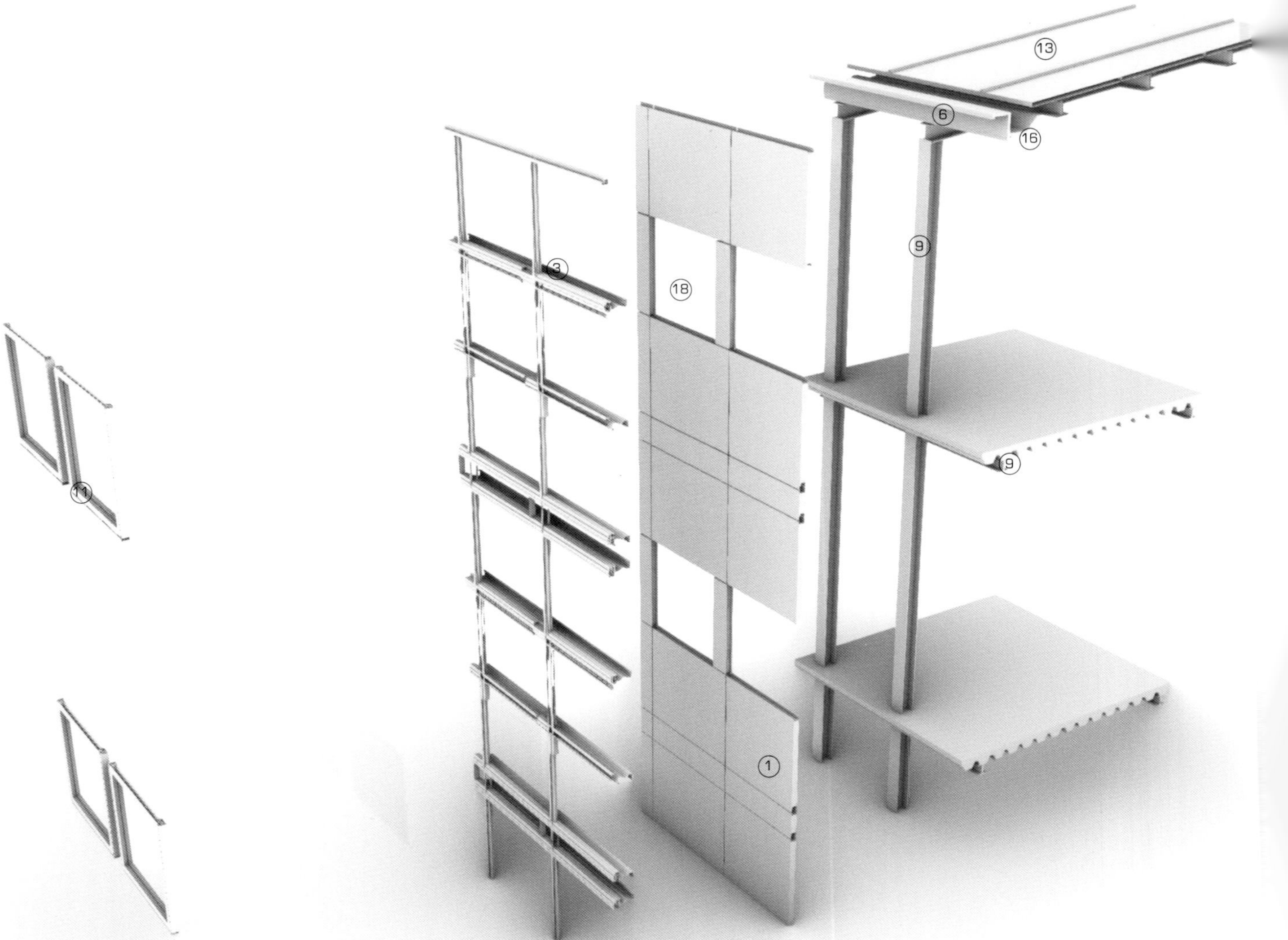

a | 3D exploded view showing composite panel facade fixed to steel frame

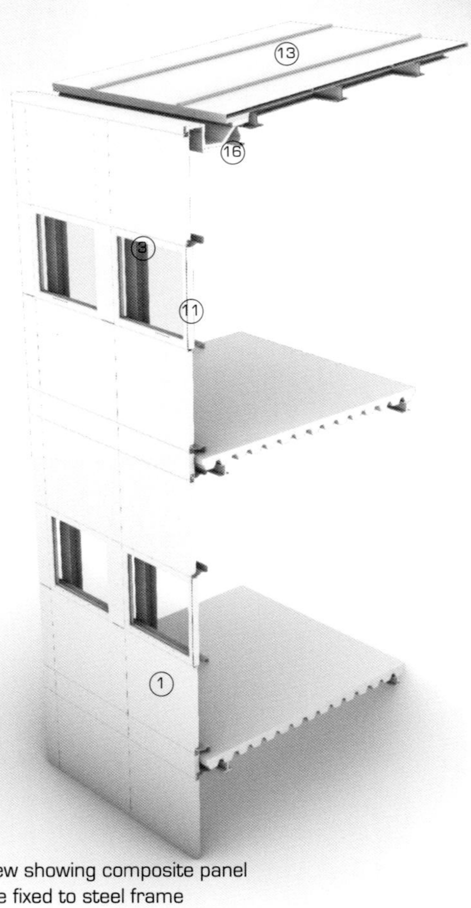

b | 3D view showing composite panel facade fixed to steel frame

Details

1 | Vertically set composite panel
2 | Horizontally set composite panel
3 | Silicone-based seal
4 | Outer metal facing
5 | Inner metal facing
6 | Inner insulation core
7 | Metal capping
8 | Concealed fixing
9 | Supporting structure
10 | 4-way interlocking composite panel

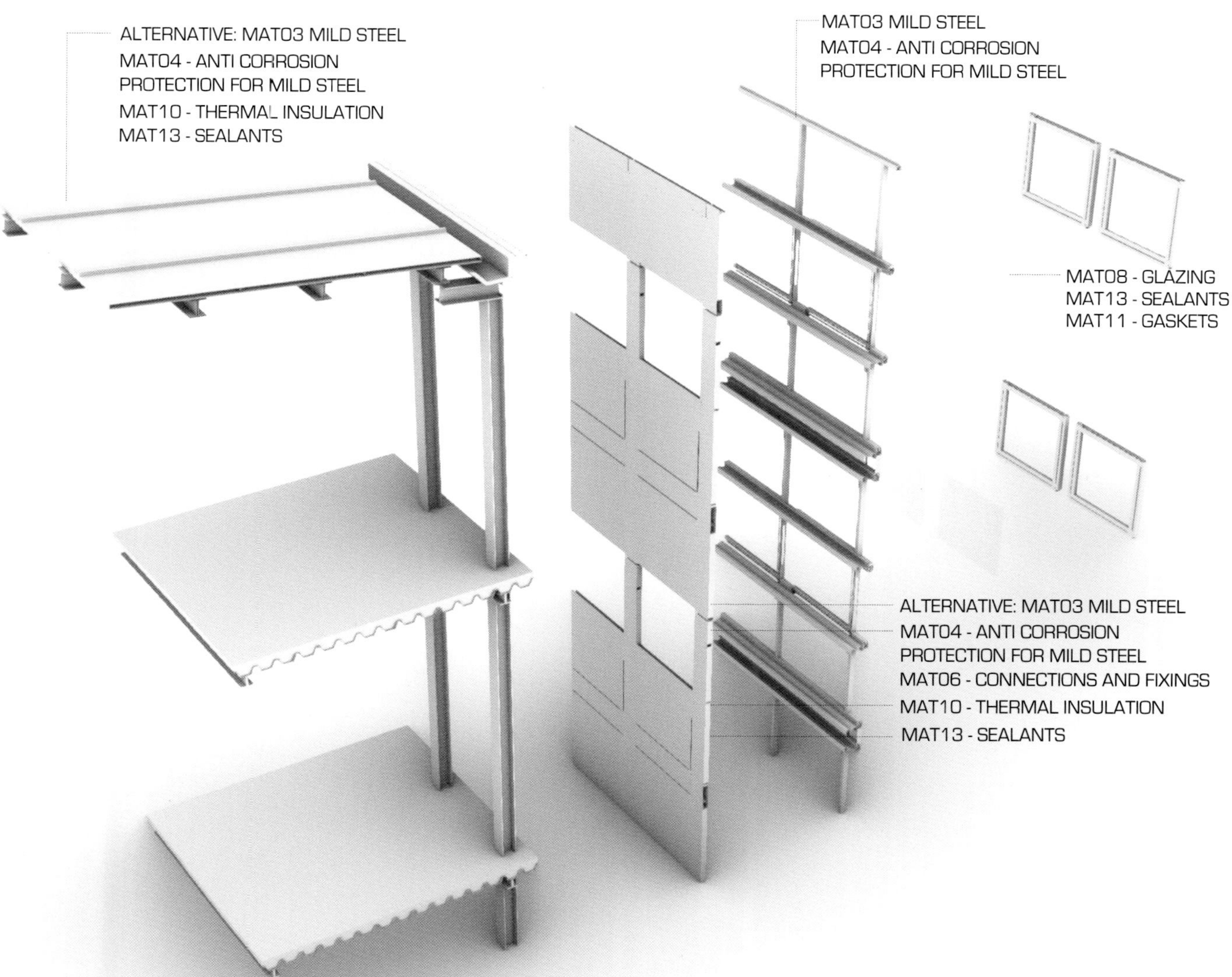

c | 3D exploded view of composite panel facade components with material specification references

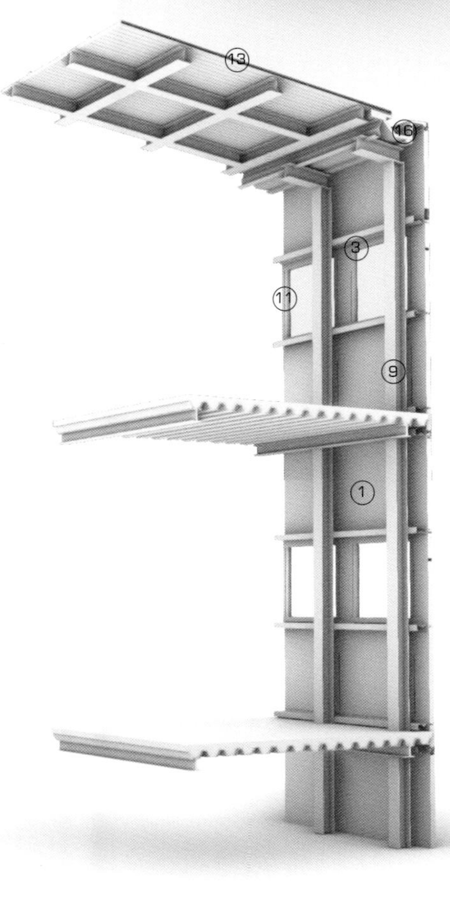

11 | Window frame
12 | Sectional roller shutter formed from composite panels
13 | Roof construction, composite panels are shown
14 | Metal trim
15 | Exposed gutter
16 | Concealed gutter
17 | Metal parapet coping
18 | Stick glazed curtain walling
19 | Door frame

d | 3D view showing composite panel facade fixed to steel frame

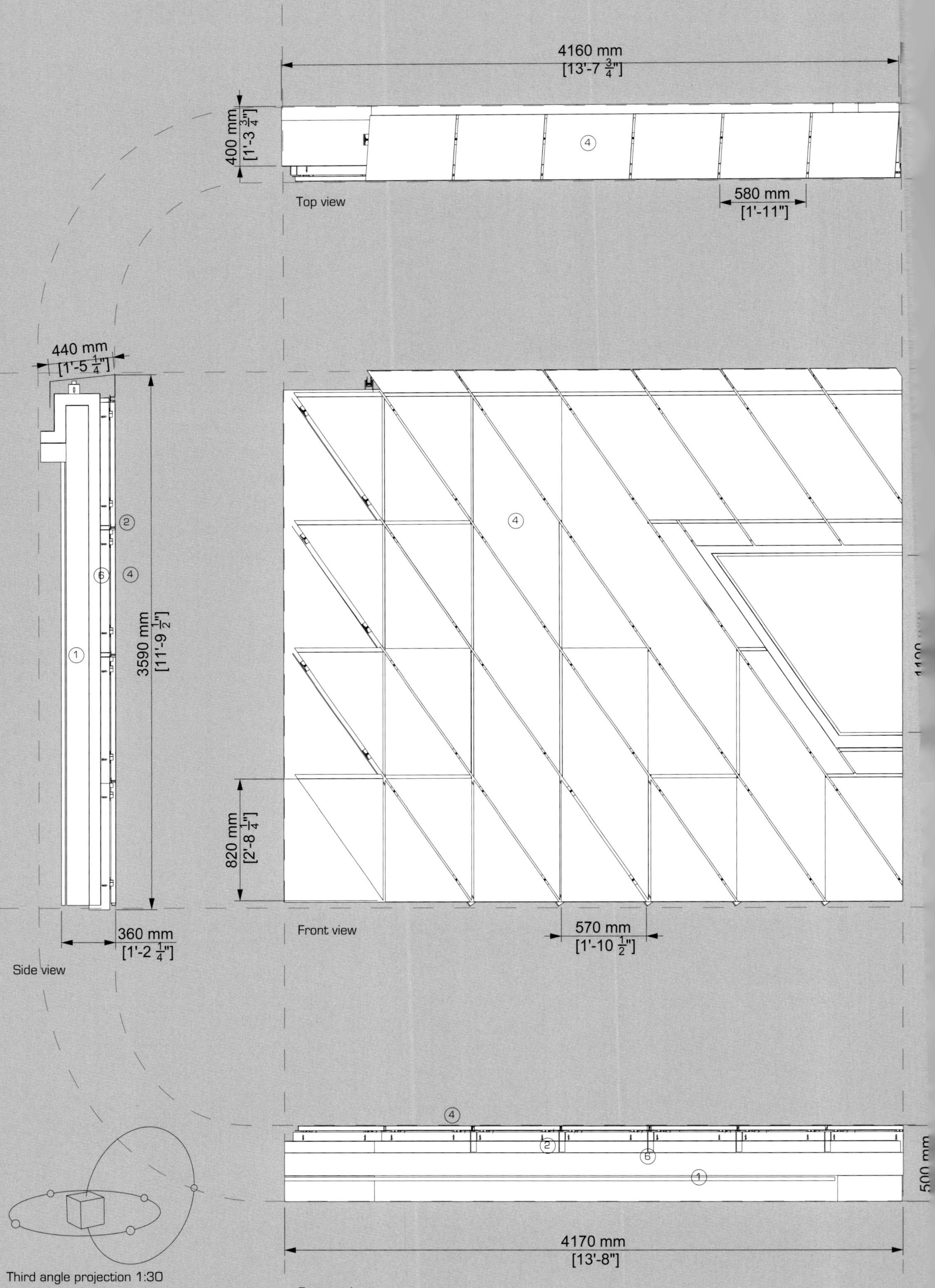

Third angle projection 1:30

Details

1 | Backing wall or structural wall supporting rainscreen
2 | Supporting frame
3 | Support bracket
4 | Metal rainscreen panel
5 | Open joint
6 | Closed cell thermal insulation

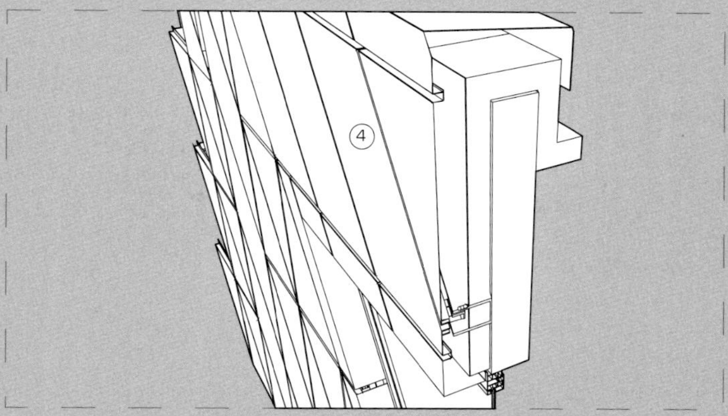

3D view of the assembly

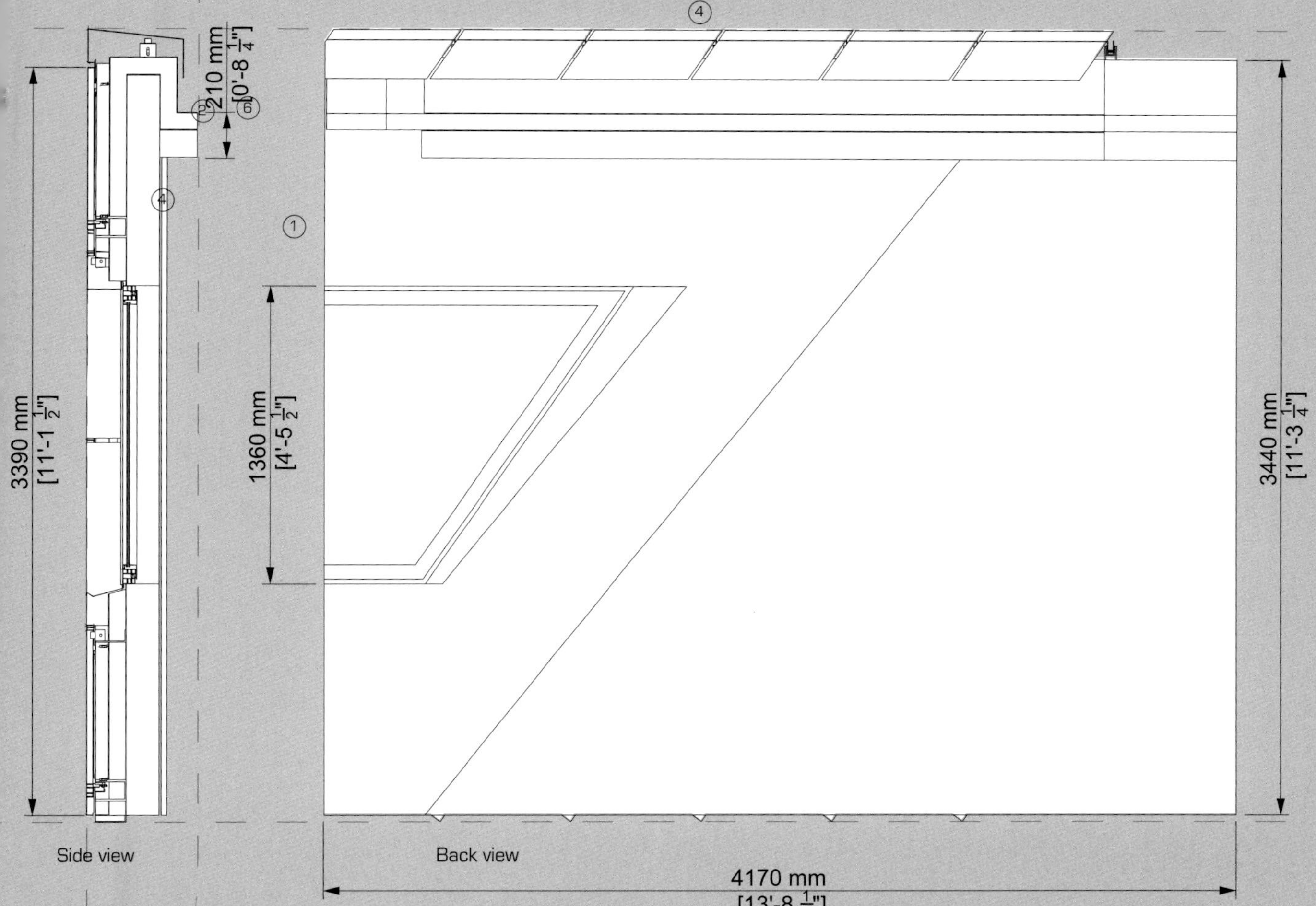

Side view

Back view

7 | Waterproof membrane
8 | Internal finish
9 | Supporting structure
10 | Pressed metal sill
11 | Pressed metal coping
12 | Continuity of waterproofing layers of wall and roof

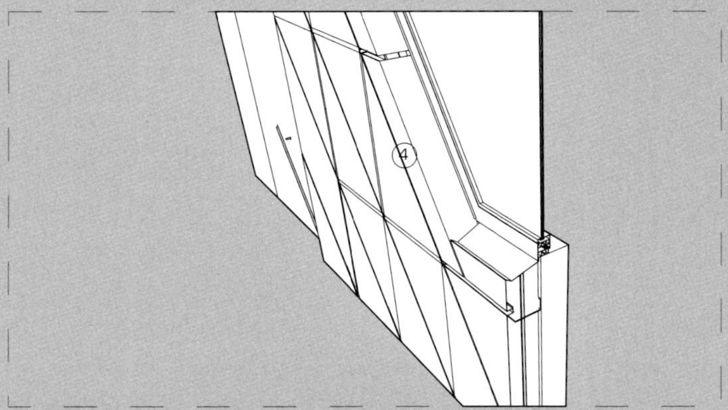

3D view of the assembly

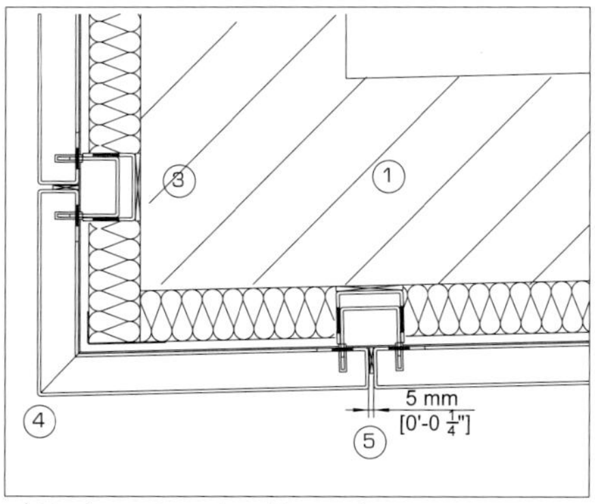

a | Horizontal section. External corner

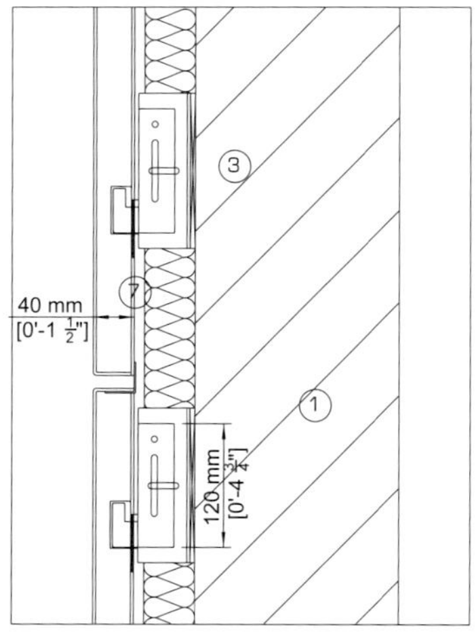

b | Vertical section. Parapet and junction with ground slab

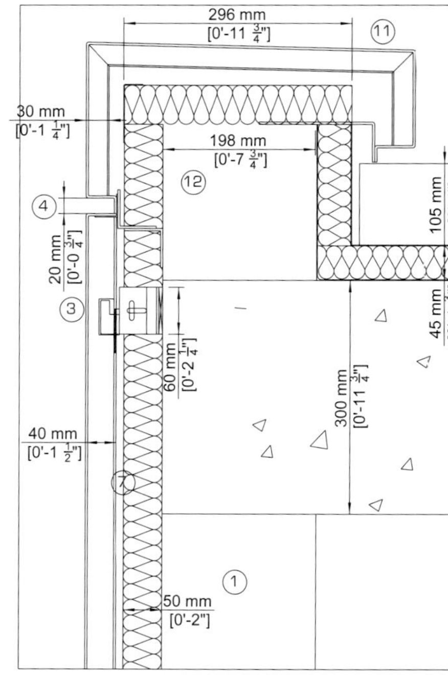

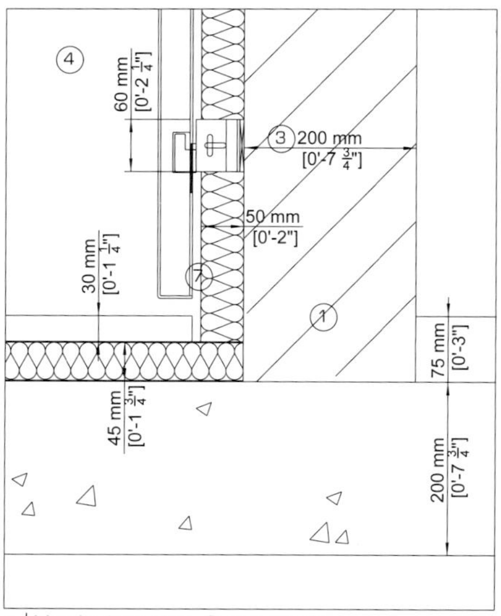

c | Vertical section. Ground level

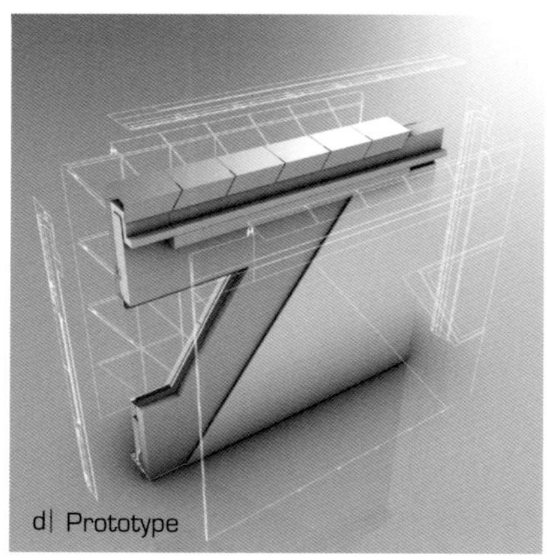

d | Prototype

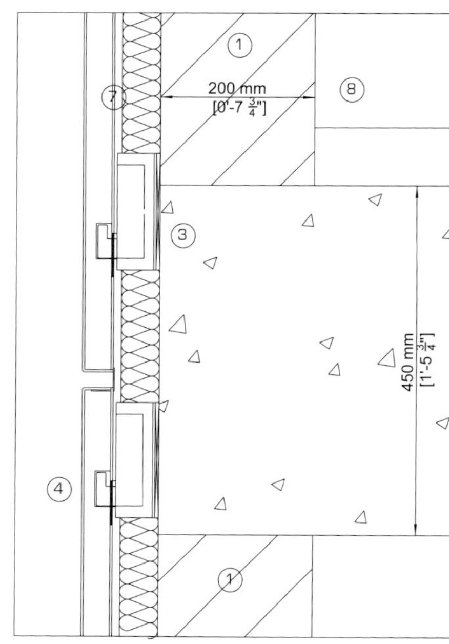

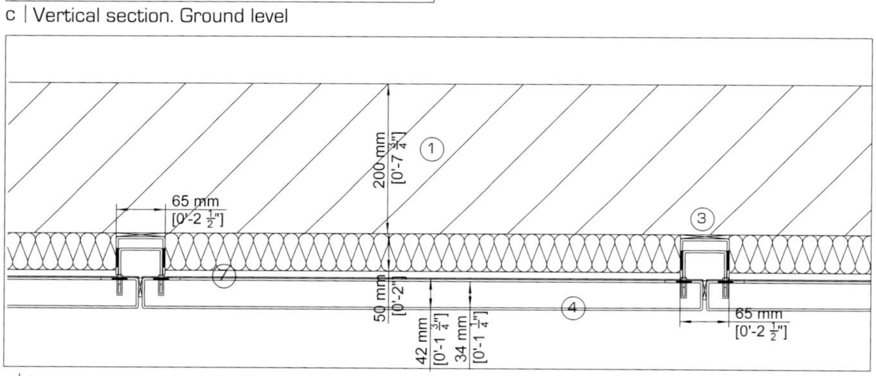

c | Horizontal section. Panel junction

A much wider range of materials is used in rainscreens than was the case five years ago. Copper and zinc have the advantage of being easier to form than steel or aluminium, particularly where site-based construction methods are preferred for either economy or in dealing with complex or curved geometries. This method of fixing rainscreens avoids the need for off-site fabrication of a large number of different panel types with different curved geometries. They can be made economically on-site.

The 'tiling' or 'shingling' of panels in copper has been developed from sheet metal cladding. This departs from metal rainscreen designs in that the surface appearance has a deliberately uneven texture that emphasises the oil-canning effect that gives the appearance of a layered, tiled surface. The move away from the emphasis on flat metal panels in rainscreen construction includes an increased use of profiled and curved metal panels. A major advantage is that fixings can be concealed by semi-interlocking panels in the manner of sheet metal cladding or traditional roof coverings.

There has been a recent increased use of semi-lapped assemblies that conceal the void behind the metal cladding. These usually have visible fixings at panel joints but allow the joints between panels to be less in shadow than is the case with other fixing methods. There has been a gradual development of rainscreens as visual screens rather than as weather-excluding panels. For example, perforated metal screens in mild steel or aluminium are used to create both modelling to a facade and solar shading set forward of glazed walling. In such designs the back of the rainscreen panel is visually as important as the outer visible face where the panel is seen through glazed openings in a facade. Fixings for such rainscreen panels often have screws and bolts that are set into the fixing rather than having projecting threaded bolt and exposed nuts.

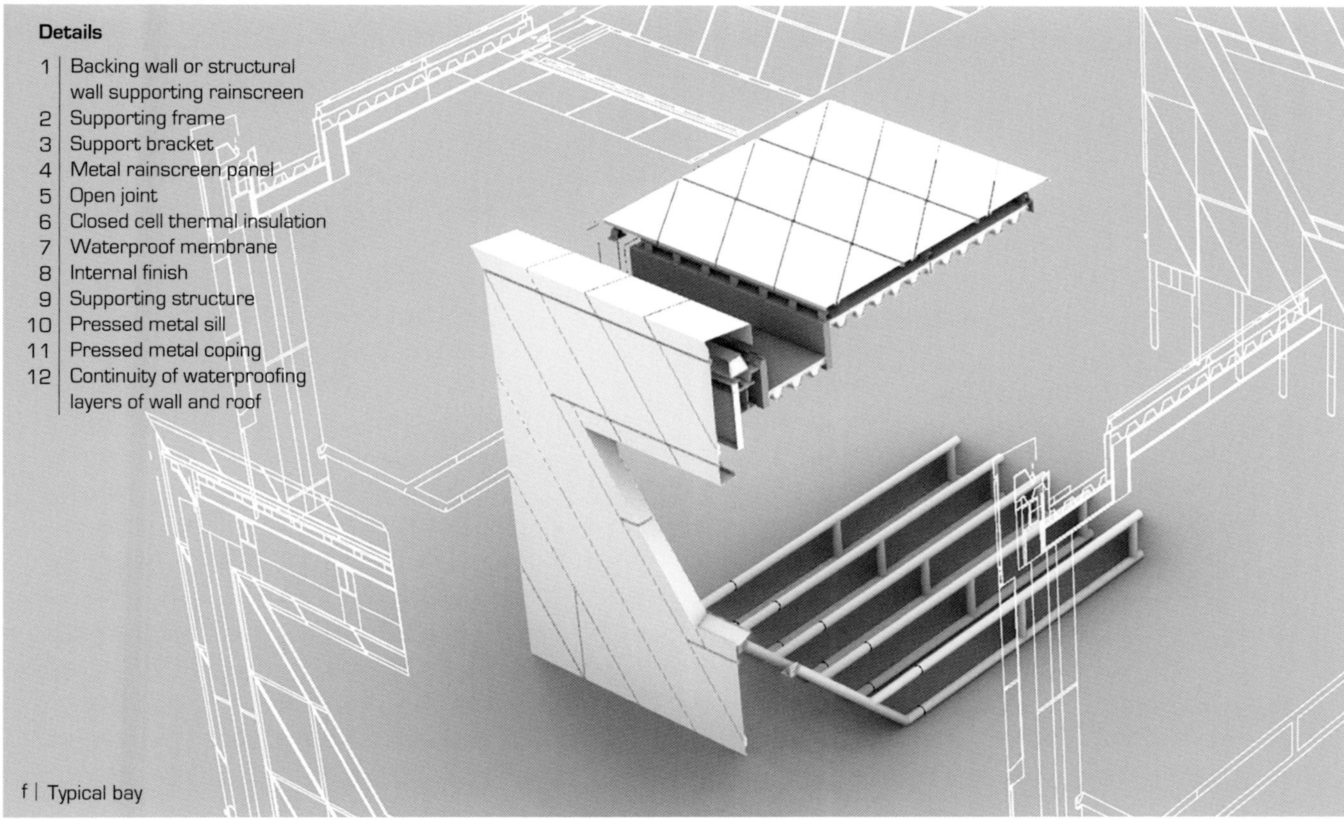

f | Typical bay

g | Vertical section. Ground level

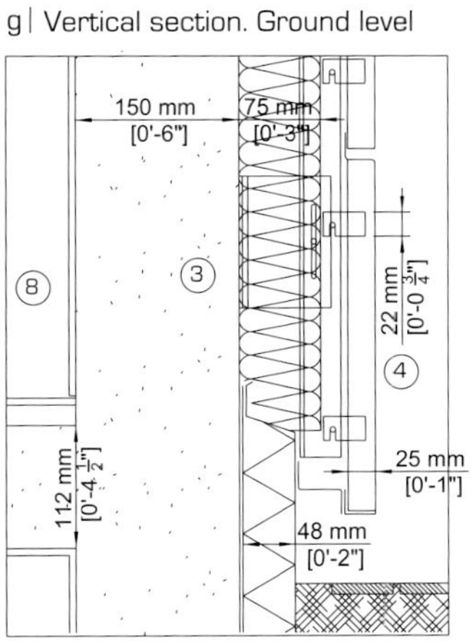

h | Horizontal section. External corner

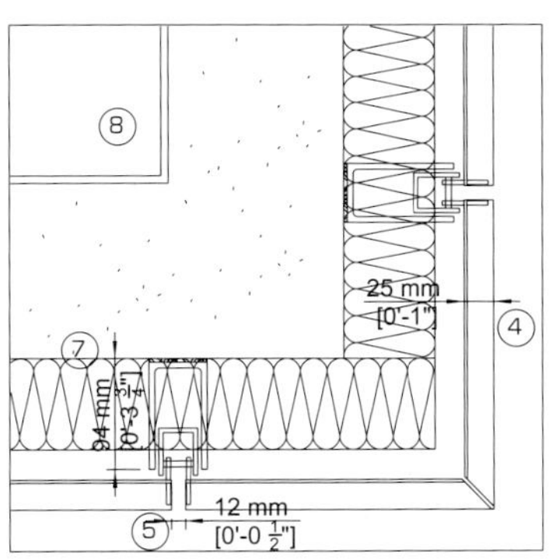

j | Horizontal section. Internal corner with joint at corner

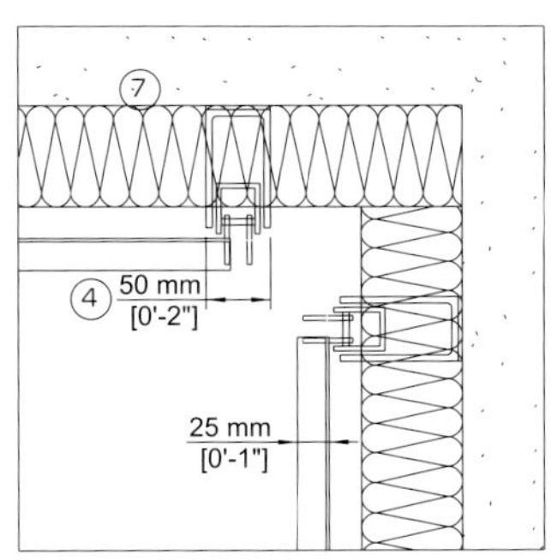

k | Horizontal section. Internal corner with jointa away from corner

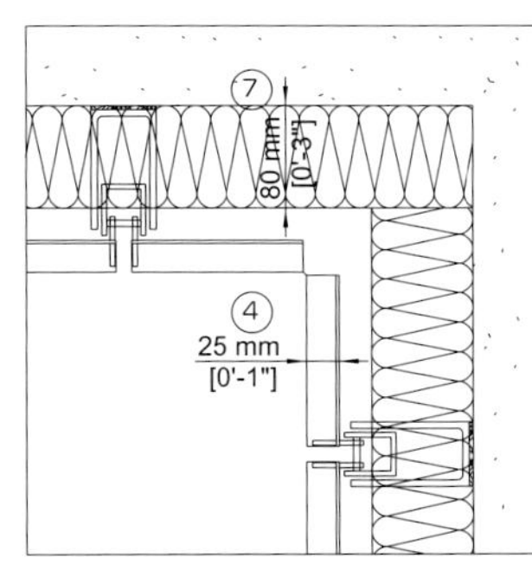

Materials

Rainscreen panels can have a flatness that is difficult to achieve with other methods. Panel flatness is achieved by either use of composites such as proprietary laminates where two sheets of aluminium are bonded on both sides of an inner core sheet of plastic of thickness 3-5mm (0.118in to 0.2in) or by a minimum 3mm (0.118in) thick aluminium sheet, or approximately 1mm thick steel sheet, depending on panel size (MCE_91, f). Honeycomb panels are also being used. A metal honeycomb layer, about 5mm (0.2in) thick, is bonded to thin metal sheets on either side. Aluminium is most commonly used. One of the outer sheets is factory paint coated from the rolled coil from which it is cut, giving the material a high level of colour consistency over large areas of panel.

Fixing methods

The three main fixing types used for metal rainscreens are (Type 1) visible point fixed (MCE_90, b and MCE_91, f); Type 2 are horizontal or vertical rails with partially exposed brackets (hung panels) (MCE_90, b) and Type 3, vertical and horizontal rails with partially interlocking panels and concealed fixings (MCE_90, c).

The choice of fixing method is often determined by what is seen through the joint from the outside. If dark shadows are sought at the joint then the backing wall should have a consistent dark colour. In this case, short lengths of bracket may be sufficient to support the panels, since they will not be visible. If the backing wall is likely to be visible through the open joints, such as if the backing wall is clad in exposed polystyrene insulation board, then the joints will need to be screened by a continuous channel.

a| Typical bay

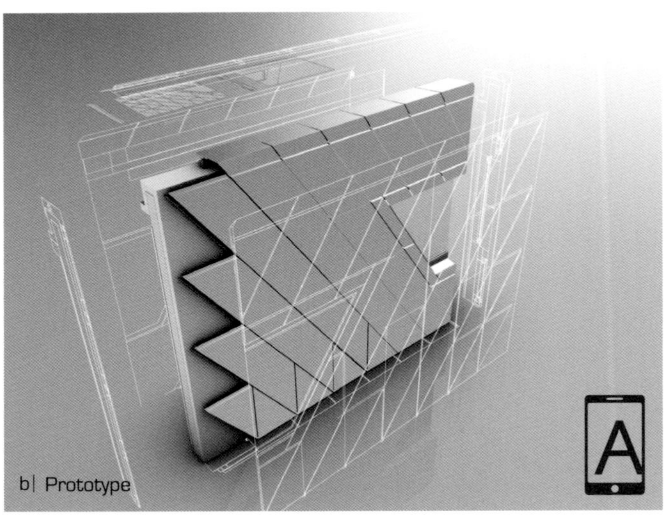

b| Prototype

Details

1 | Backing wall or structural wall supporting rainscreen
2 | Supporting frame
3 | Support bracket
4 | Metal rainscreen panel
5 | Open joint
6 | Closed cell thermal insulation
7 | Waterproof membrane
8 | Internal finish
9 | Supporting structure
10 | Pressed metal sill
11 | Pressed metal coping
12 | Continuity of waterproofing layers of wall and roof

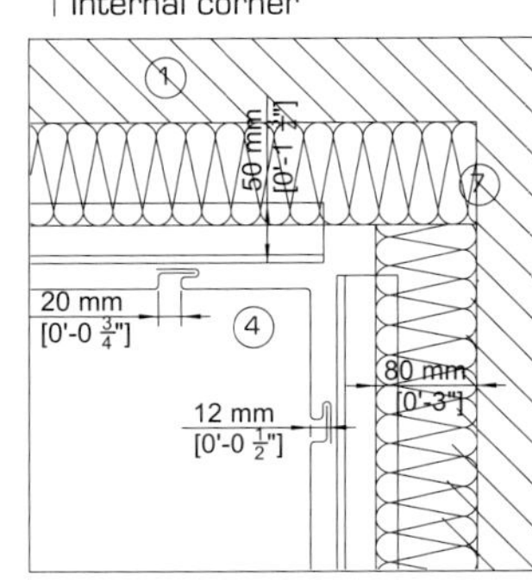

c| Horizontal section. Internal corner

Panels which are point fixed have either countersunk screws set flush with the panel face, or dome-headed screws which make a visible feature of them. The screws are fixed into rails set to suit their position. Vertical rails are often preferred since water can easily drain down them.

Hung panels are hooked onto supporting brackets. Panels are fixed by cutting slots into the sides of the rainscreen panel during manufacture and hooking them onto dowels projecting from C-shaped brackets. These brackets are in turn secured to vertical rails set at vertical joints between panels. The rails also act as screens to close off views into the cavity. Horizontal joints are formed by an upstand formed in the top or bottom edge of a panel.

Semi-interlocking panels are fixed by screwing the top of the panel to horizontal rails to suit the orientation of the panels. The adjacent panel is lapped into the panel next to it, both to secure it and conceal the fixing (MCE_92, f). The joint in the other direction is either fixed with a similar semi-interlocking edge to form a tiled appearance, or has a cover strip.

Unlike masonry-based rainscreens such as terracotta or stone, metal rainscreen panels are lightweight in comparison and need to be mechanically fixed at a minimum of one or two points, usually at the bottom of the panel if the panel is hung from the top. Fixing screws are usually applied at the joint, unless it is fixed with exposed fixings through the panel itself. This means that part of the bracket is usually visible. Short lengths of brackets then become visible and need to be incorporated as a visible part of the design.

Backing walls

Supporting walls to rainscreen panels are usually concrete block, which allows supporting panels to be fixed at any point across its surface (MCE_93, n), or framed, where rainscreen fixings are secured to the

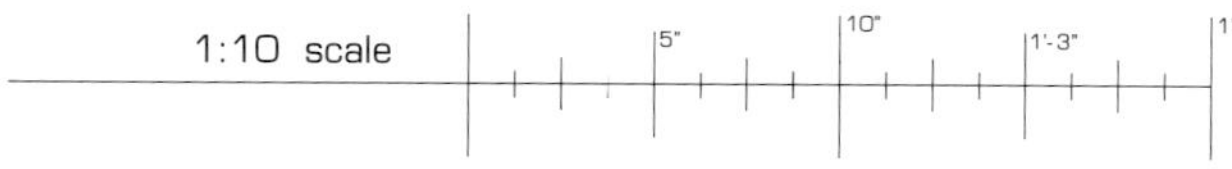

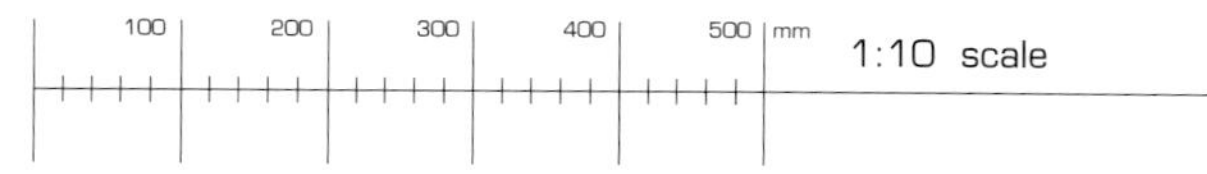

g | Vertical section. Parapet

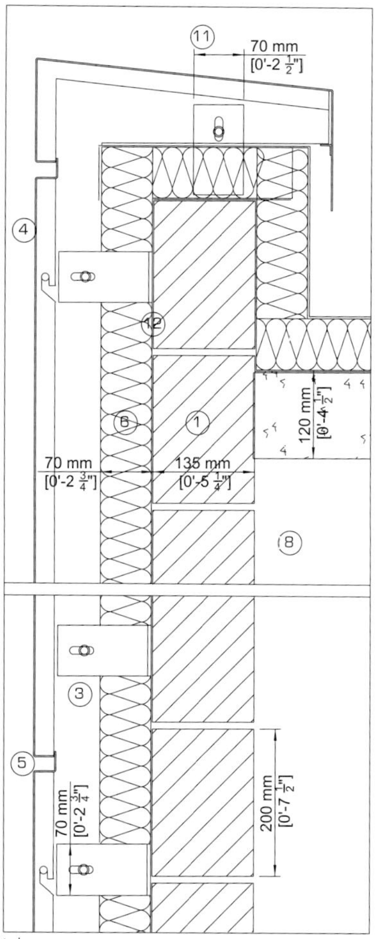

h | Vertical section. Window

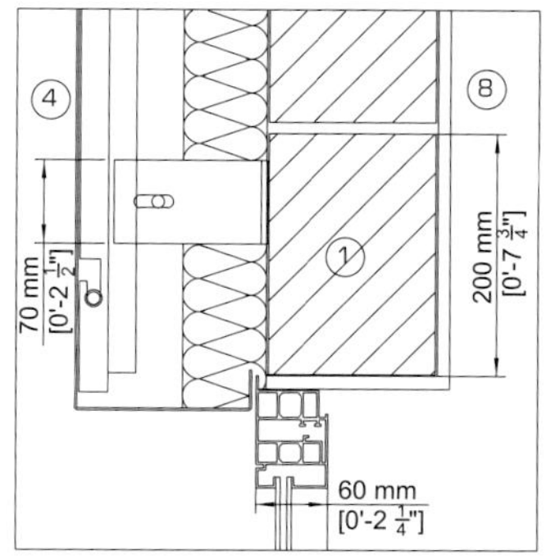

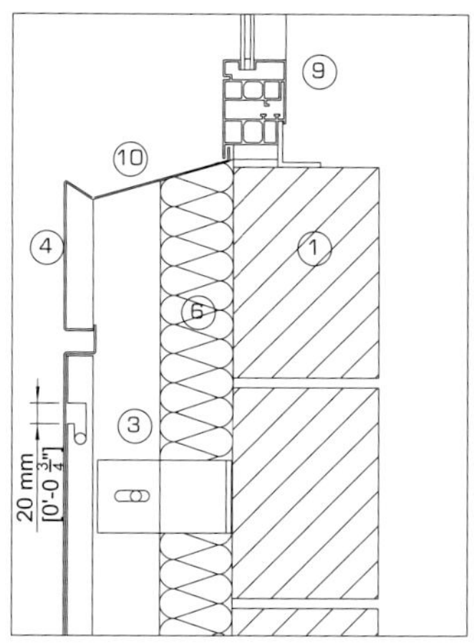

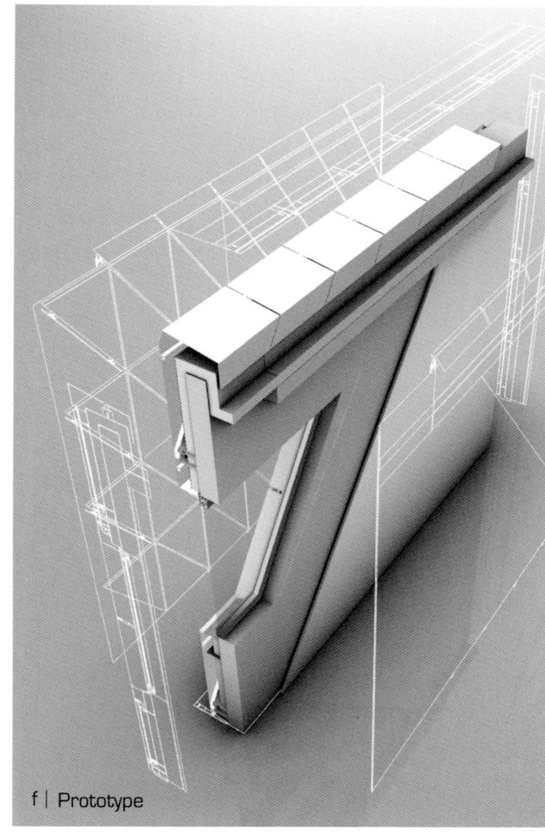
f | Prototype

k | Horizontal section. External corner and window jamb

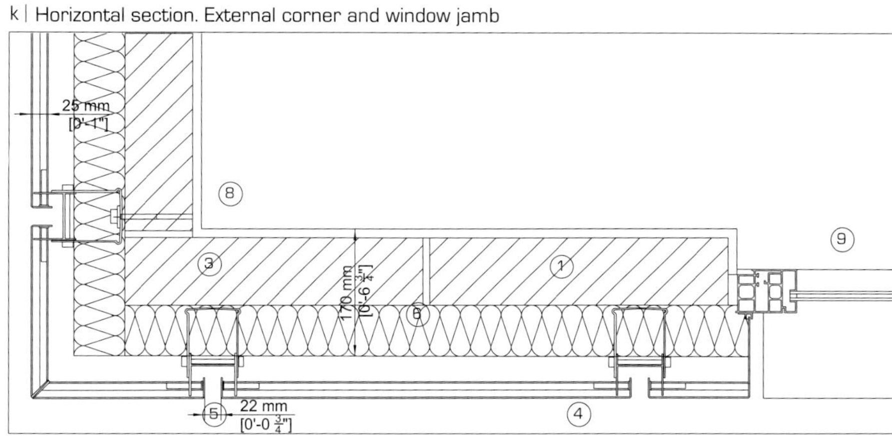

j | Horizontal section. External corner

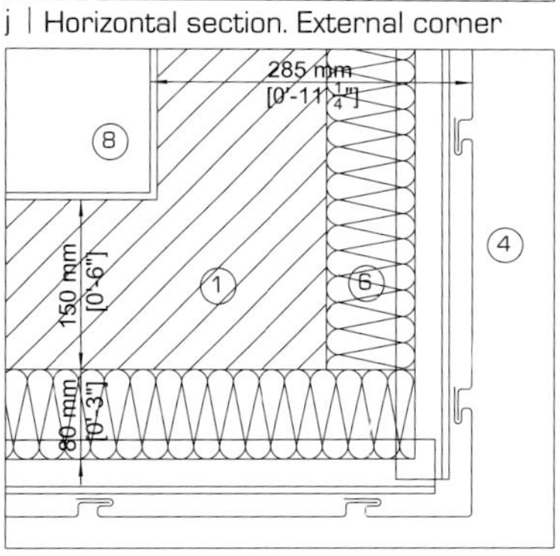

framing members rather than the outer skin of the backing wall. In some cases, if the outer skin is thick enough, say 6mm (0.25in) aluminium sheet, rainscreens can be fixed directly to the sheet material rather than the frame. If a lightweight backing wall cannot accept additional loads from rainscreen cladding onto it (as with composite panels), then support rails span from floor to floor as posts.

With concrete block backing walls, the thermal insulation is usually set on the outside face in order to keep the structure either warm or cool, depending on the geographical location. The waterproof layer is set directly on the outside face of the concrete (MCE_94, a). The thermal insulation used is closed cell-type in order for it not to absorb water which would drastically reduce its performance. The insulation is also used to protect the waterproof membrane but this makes it necessary for support brackets to be fixed through the insulation to the supporting wall behind. Sometimes holes have to be cut in the insulation, which reduces its effectiveness, but it is always better if the fixings for support rails can be fixed at the same time as the insulation in order to co-ordinate them and avoid later cutting of the insulation.

With lightweight backing walls in timber or pressed steel, thermal insulation is set within the frame. A waterproofing layer is set on the inside (warm in winter) face. An internal finish layer is then set in front of this vapour barrier. Rainscreen fixings for support rails are fixed directly to the outer waterproofing layer using sealing washers that avoid leaks through the fixing point. The framed backing wall is designed to receive fixing brackets at points which transfer loads down to the primary structure. The use of framed backing walls with rainscreens makes it necessary to co-ordinate the two elements of construction during the design rather than during the construction.

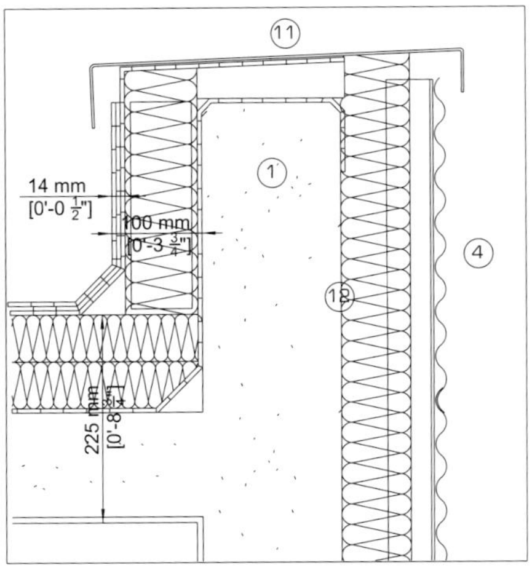

a | Vertical section. Parapet and sill at ground level with panels set horizontally

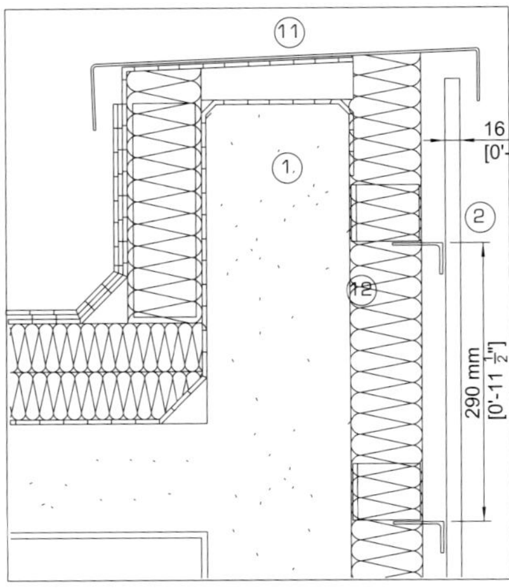

b | Vertical section. Parapet with panels set vertically

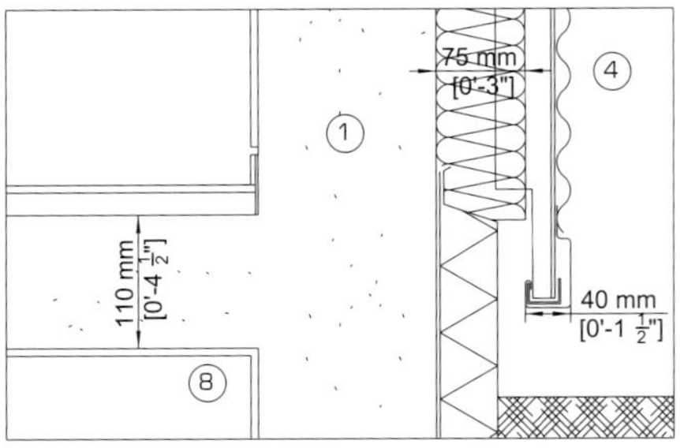

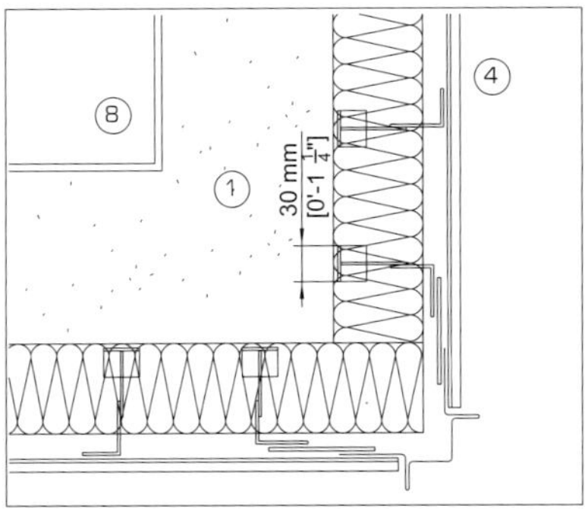

c | Vertical sections. Sill at ground level with panels set vertically

d | Horizontal section. External corner

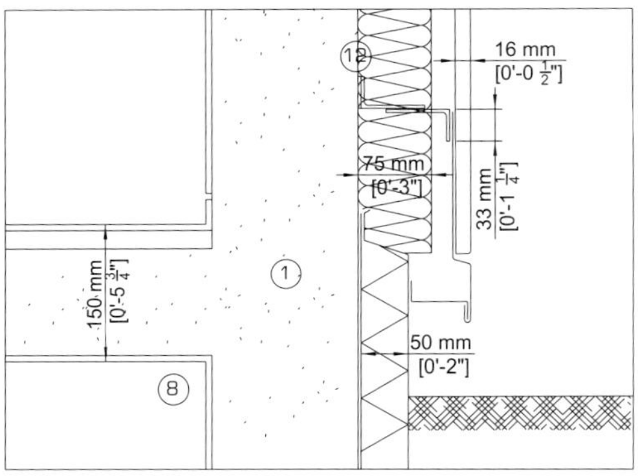

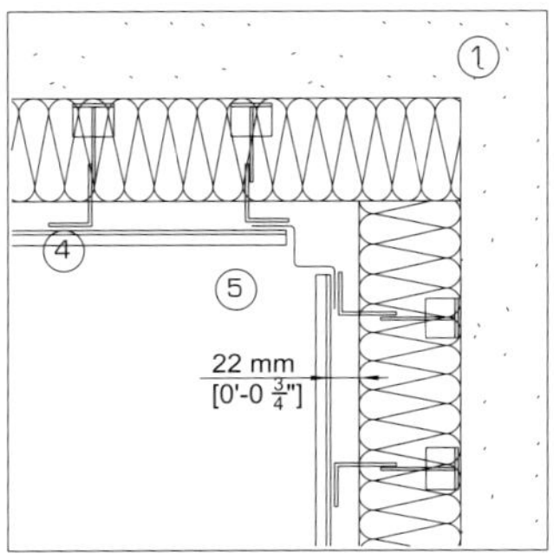

e | Horizontal section. Internal corner

f | Prototype

Construction sequence

An essential aspect of rainscreen construction is the sequence in which the various elements of the backing walls, windows, thermal insulation, waterproofing layer and rainscreen panel are brought together (MCE_94, a). Although the rainscreen principle is very effective and often very economic, its effectiveness can be reduced if seals are not properly applied or insulation is damaged because elements are assembled on site in the wrong order. Typically, windows are sealed against the backing wall before the rainscreen panels are set in place. An advantage of this form of construction is that the external wall can be made waterproof before the rainscreen panels are fixed.

A typical construction sequence is to build the backing wall first and set the waterproofing layer and insulation in place. Windows are then set into the backing wall and are sealed against its waterproofing layer. Thermal insulation, if set outside the line of the waterproofing layer (as in in-situ concrete, precast concrete or concrete block) makes it easier to set the thermal insulation on the backing wall after the windows and doors have been fixed. Support rails for the rainscreen panels are then fixed to the backing wall, followed by the metal panels themselves (MCE_94, a). Panels are usually fixed in horizontal rows from the bottom up so that corner panels and panels at window openings can be fixed from the outside from the top of the panel. Metal panels can then be set in a correct alignment with the windows in terms of their position and in setting the required joint width. The open jointed nature of the construction usually dictates that the rainscreen panel is set in a way that avoids a view through the joint to the backing wall beyond.

Window and door openings

Because window and door openings are usually set into an opening before the rainscreen panels are set, reveals are sealed with either individual rainscreen panels or with sheet metal trims similar to those

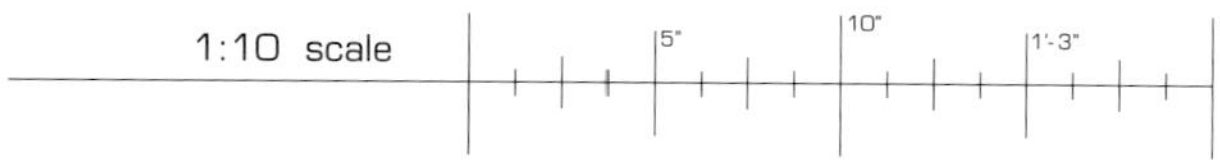

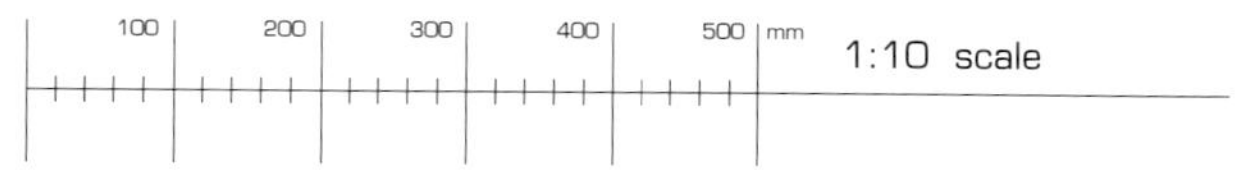

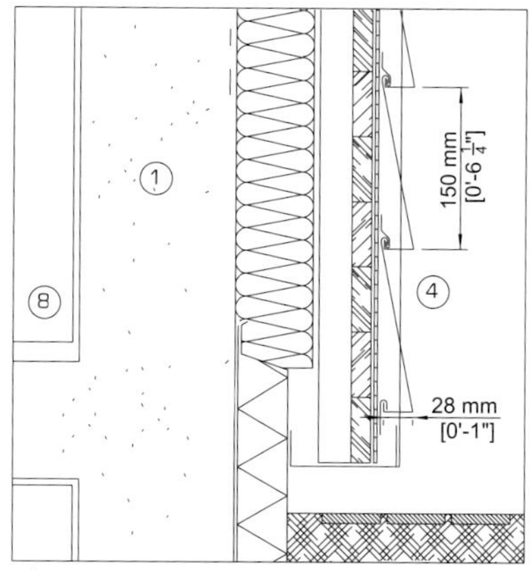

h | Vertical section. Sill at ground level

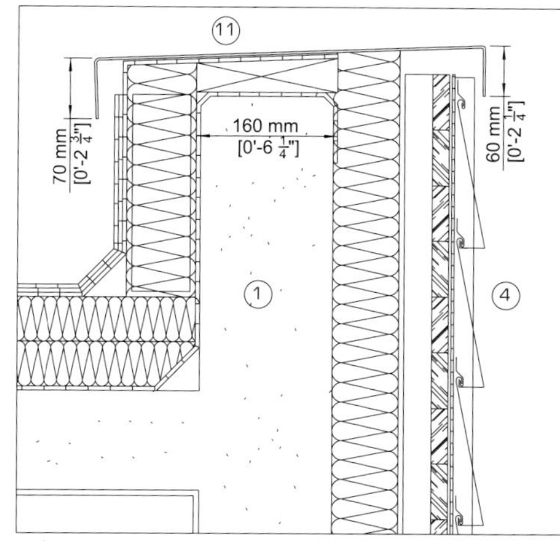

j | Vertical section. Parapet

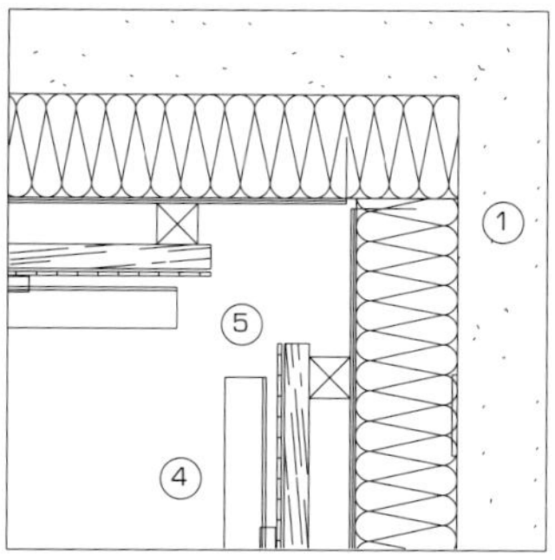

g | Horizontal section. Internal corner

Details

1	Backing wall or structural wall supporting rainscreen
2	Supporting frame
3	Support bracket
4	Metal rainscreen panel
5	Open joint
6	Closed cell thermal insulation
7	Waterproof membrane
8	Internal finish
9	Supporting structure
10	Pressed metal sill
11	Pressed metal coping
12	Continuity of waterproofing layers of wall and roof

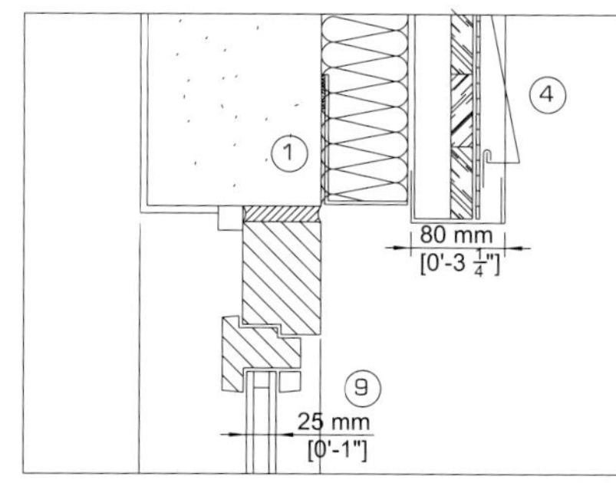

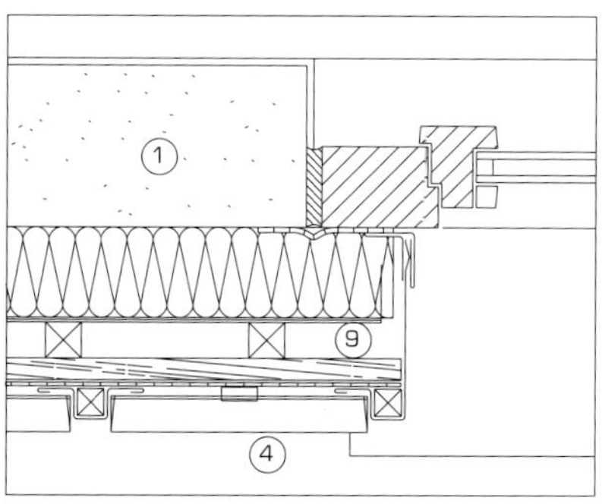

k | Horizontal section. Window jamb

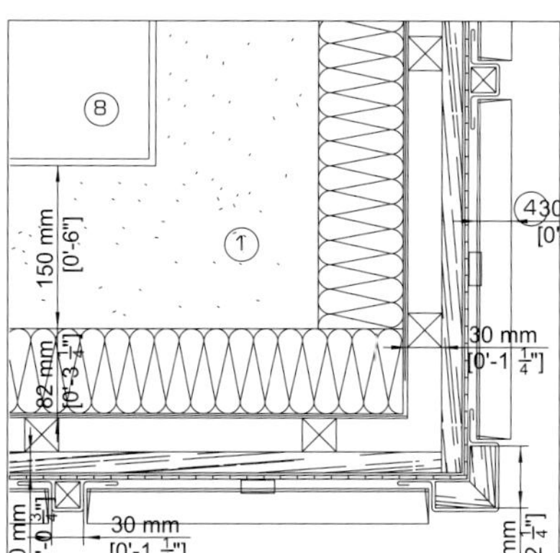

m | Horizontal section. External corner

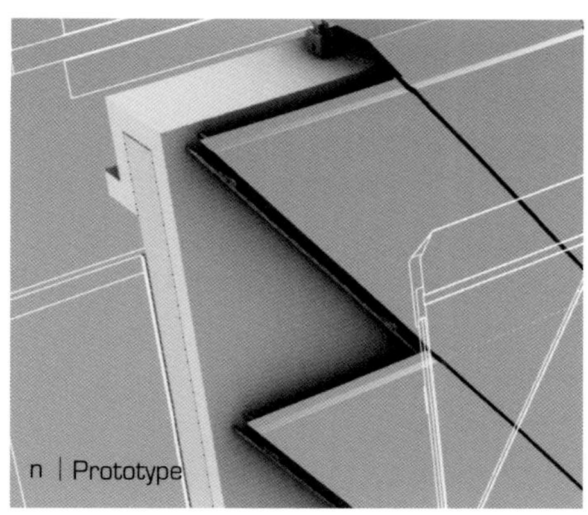

n | Prototype

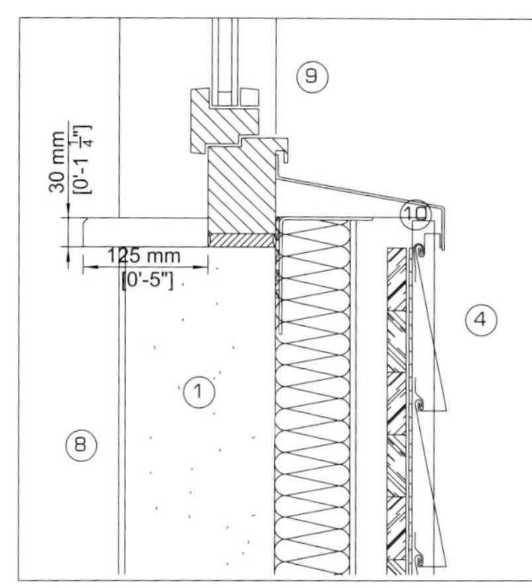

p | Vertical section. Window

used for sheet metal construction. Unlike sheet metal construction however, a gap is usually maintained between the trim and the window in order to maintain the joint principle. Similarly gaps between reveal trims and adjacent wall panels are also separated by an open joint (k). Since windows and doors are sealed against the waterproofing layer behind the rainscreen panel rather than to the panel itself, these separations are straightforward to achieve. These open joints around openings are made in a way that conceals the waterproof layer behind, thus protecting it from both accidental damage from building users and from the possible effects from sunlight heating up the membrane or attacking it with UV light. Windows and doors are usually provided with an additional wider frame or trim at their edges in order to allow thermal insulation in the cavity to provide continuity at the opening and allow the rainscreen panel to lap against it (f).

Parapets and sills

Parapets have an open joint between the parapet flashing and the panel below but horizontal joints between flashings are usually closed to protect the parapet from accidental damage to the waterproofing undercloak beneath or from the harmful effects of sunlight from above acting on its horizontal surface (n). Joints between flashings may be recessed to match the visual appearance of the rainscreen panels by providing a shadow or may be lapped in the manner of fully supported sheet metal cladding. The waterproofing layer will form a continuous seal with the adjacent roofing membrane. Sills at the base of the wall are detailed in a similar way with metal being continuous but with joints either recessed or lapped.

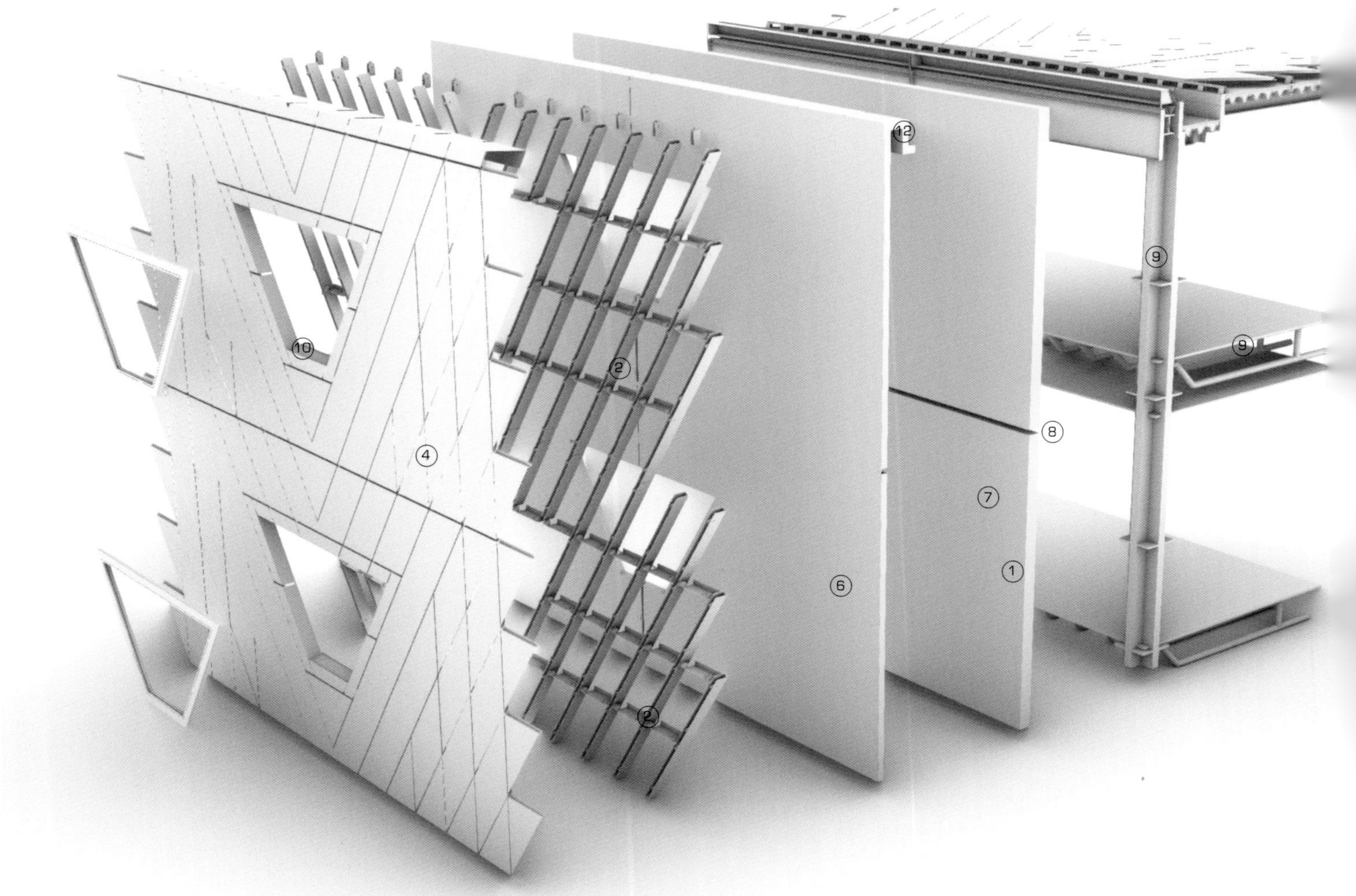

a | 3D exploded view showing metal rainscreen facade fixed to steel frame

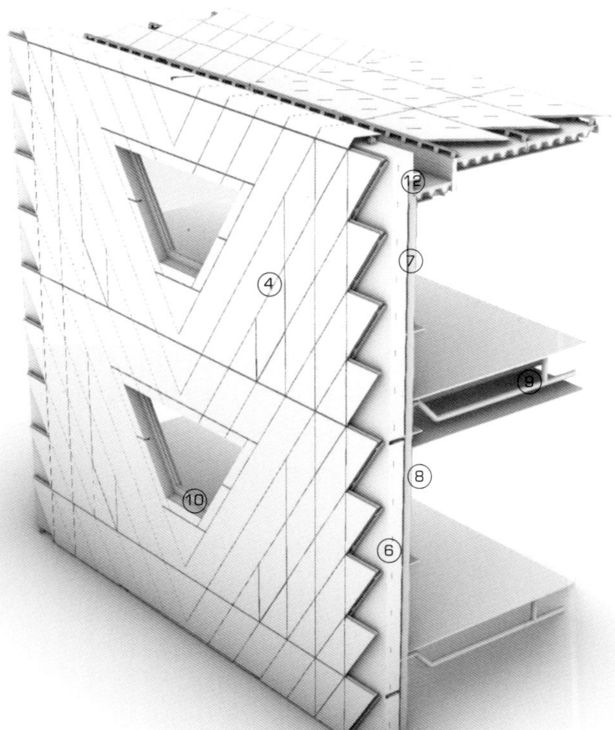

b | 3D view showing metal rainscreen facade fixed to steel frame

Details

1 | Backing wall or structural wall supporting rainscreen
2 | Supporting frame
3 | Support bracket
4 | Metal rainscreen panel
5 | Open joint
6 | Closed cell thermal insulation

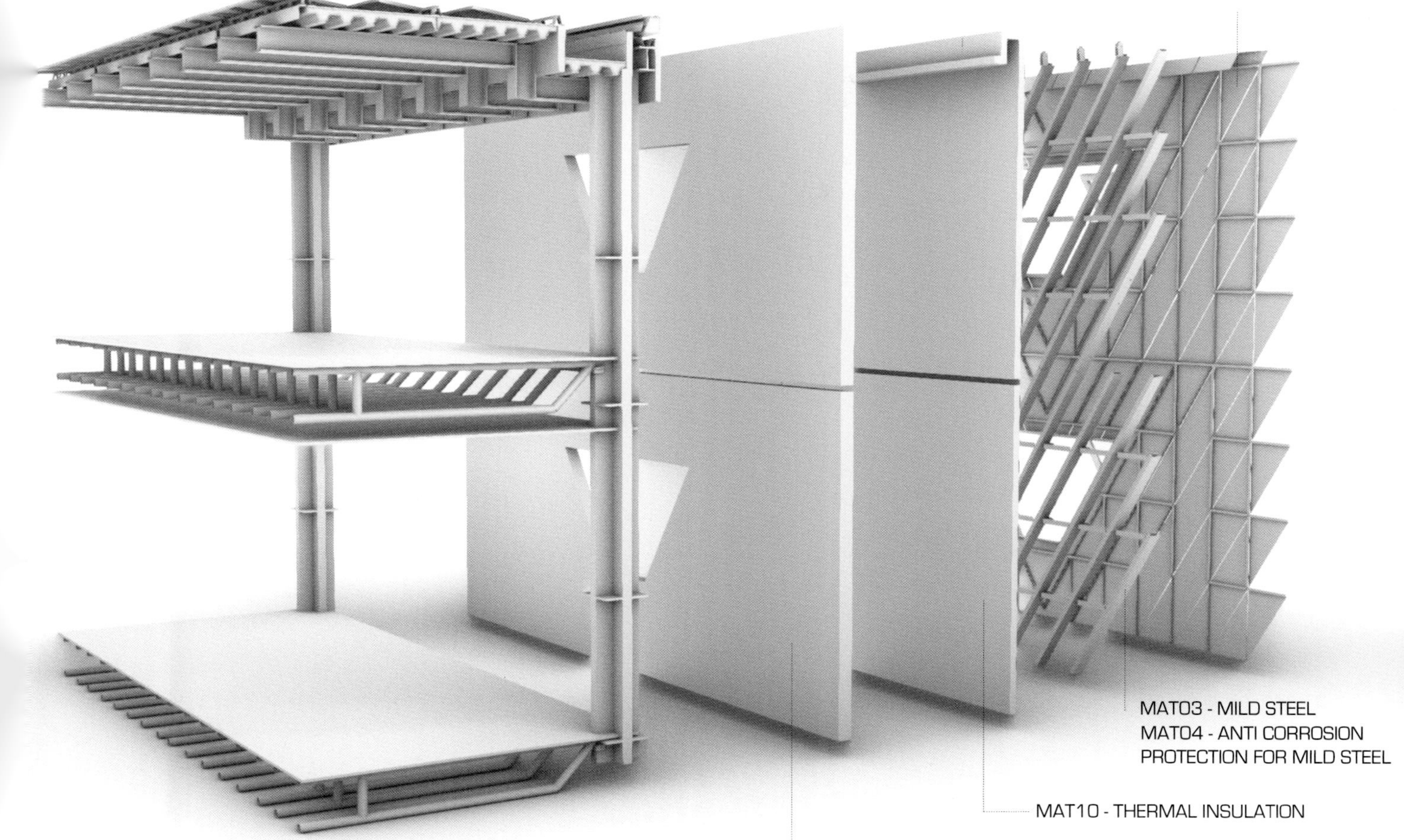

c | 3D exploded view of metal rainscreen facade components with material specification references

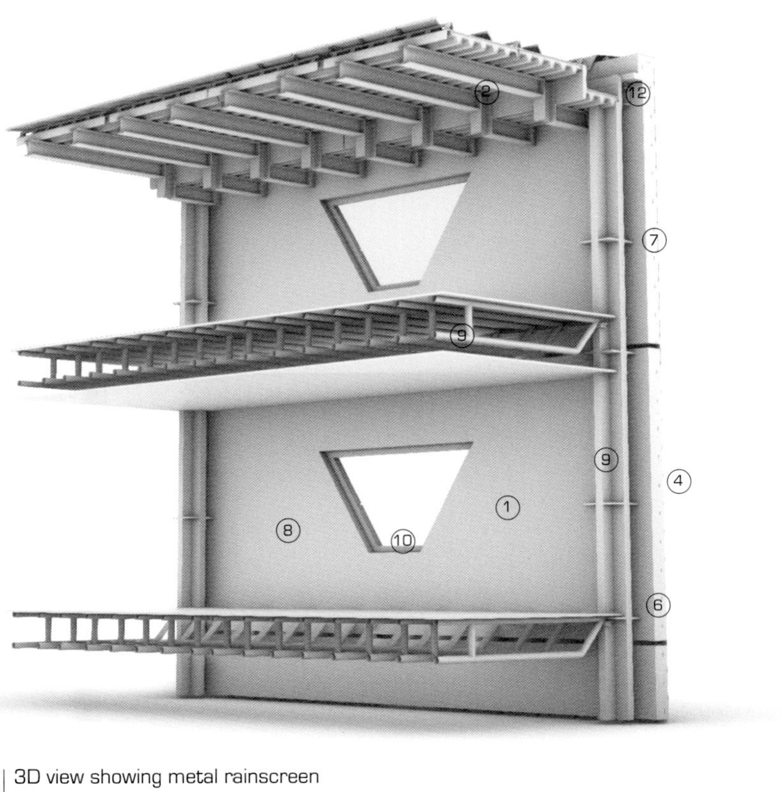

7 | Waterproof membrane
8 | Internal finish
9 | Supporting structure
10 | Pressed metal sill
11 | Pressed metal coping
12 | Continuity of waterproofing layers of wall and roof

d | 3D view showing metal rainscreen facade fixed to steel frame

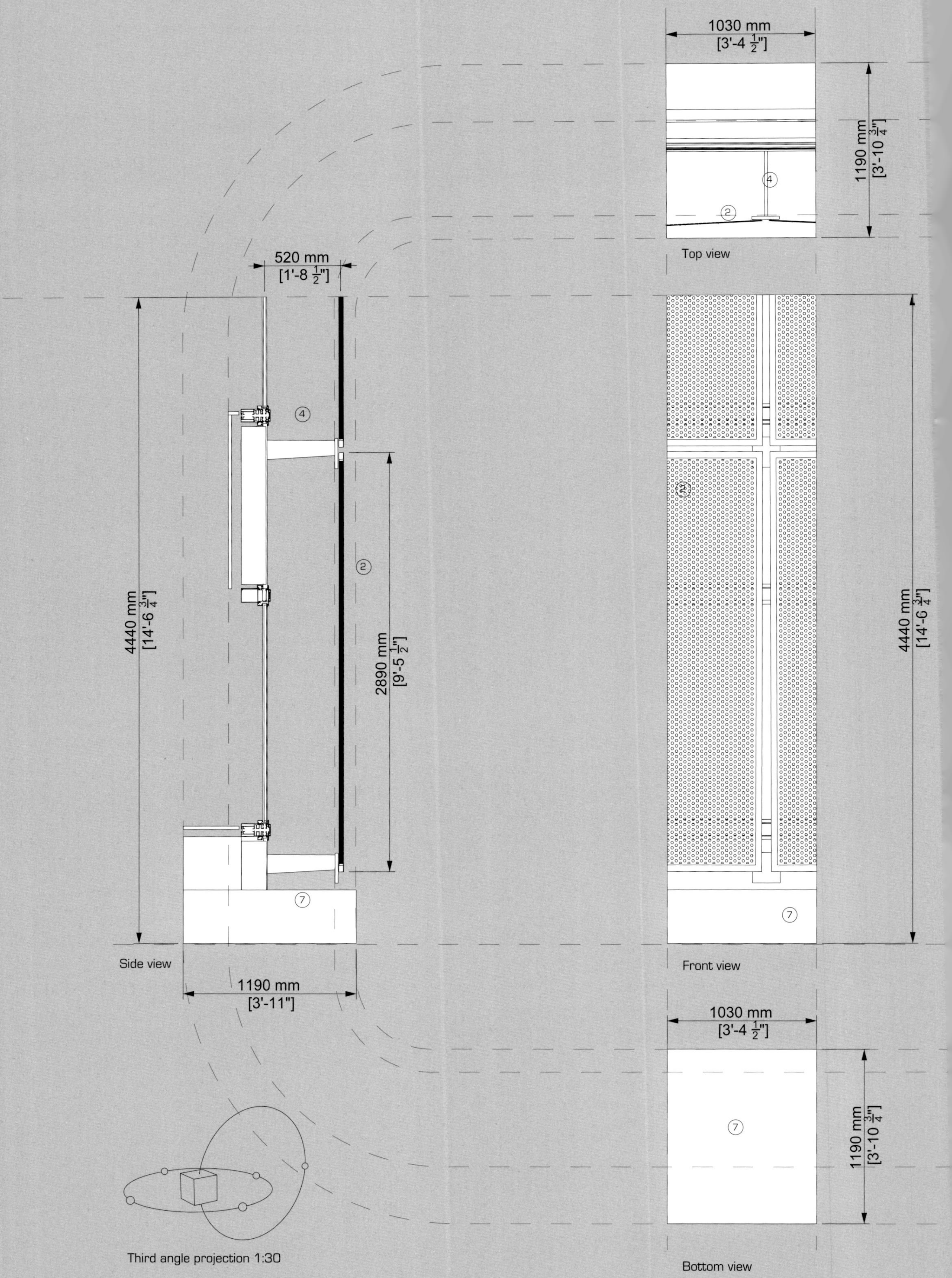

Third angle projection 1:30

Details

1 Metal support edge frame
2 Stainless steel mesh
3 Stainless steel spring
4 Metal fixing bracket
5 Metal support rod
6 Fixing bolt to tension mesh
7 Floor slab or backing wall
8 Adjacent curtain wall

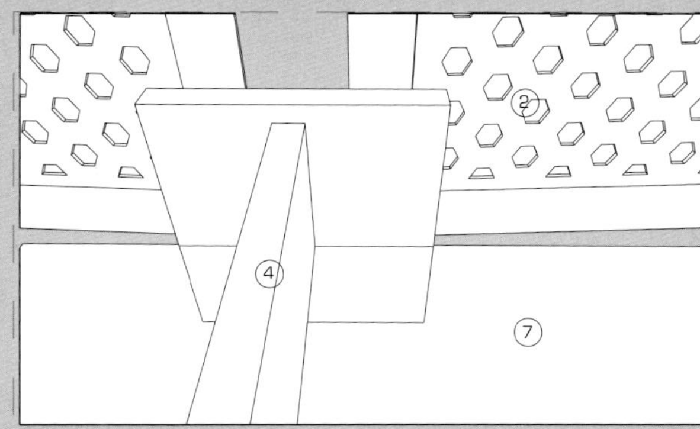
3D view of the assembly

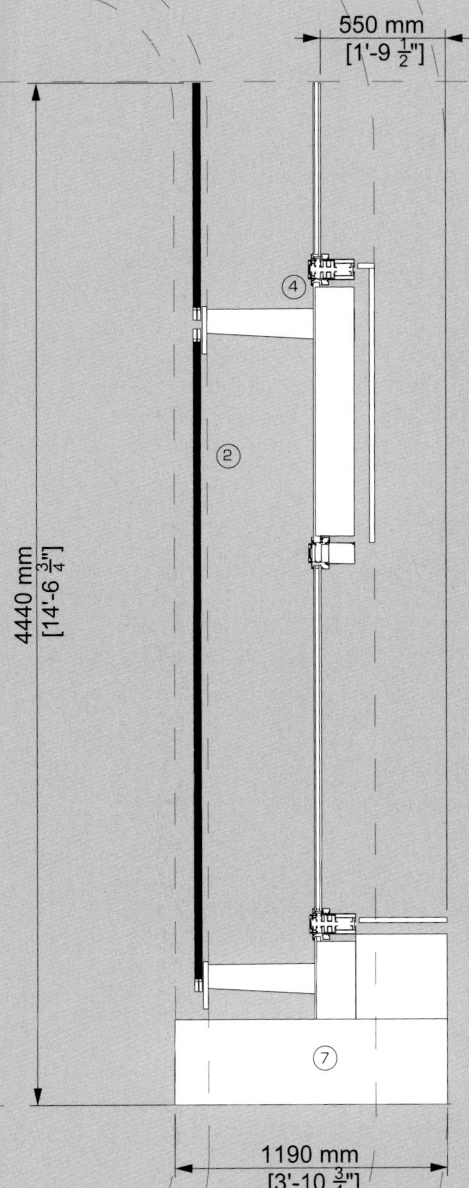

Side view

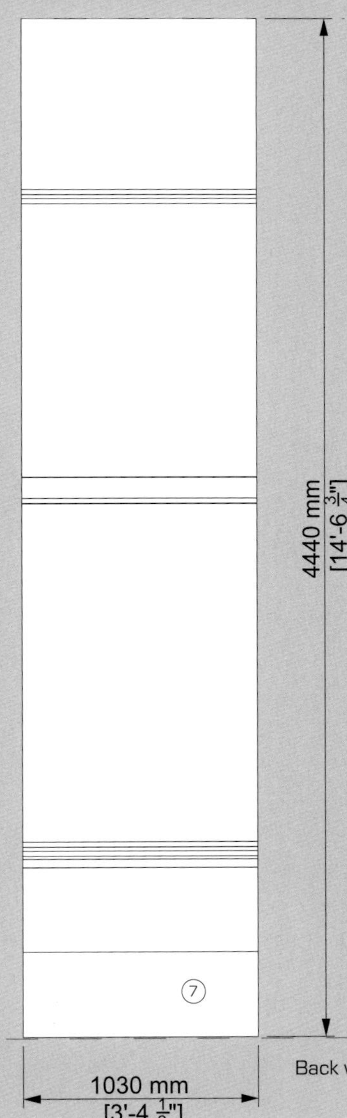

Back view

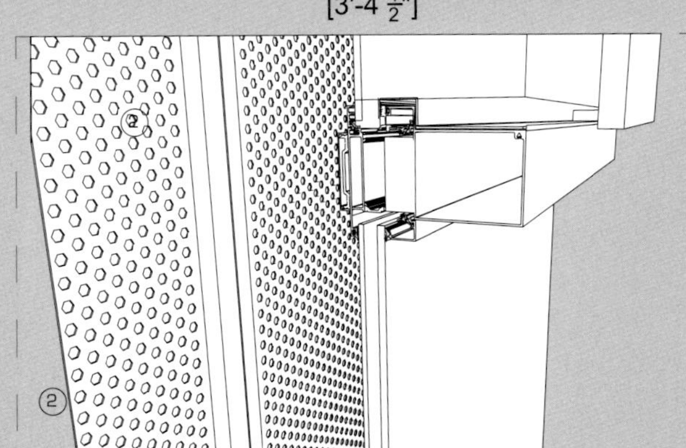
3D view of the assembly

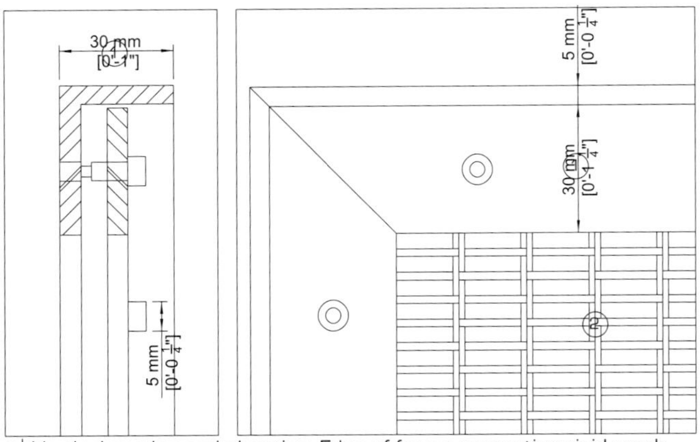

a | Vertical section and elevation. Edge of frame supporting rigid mesh

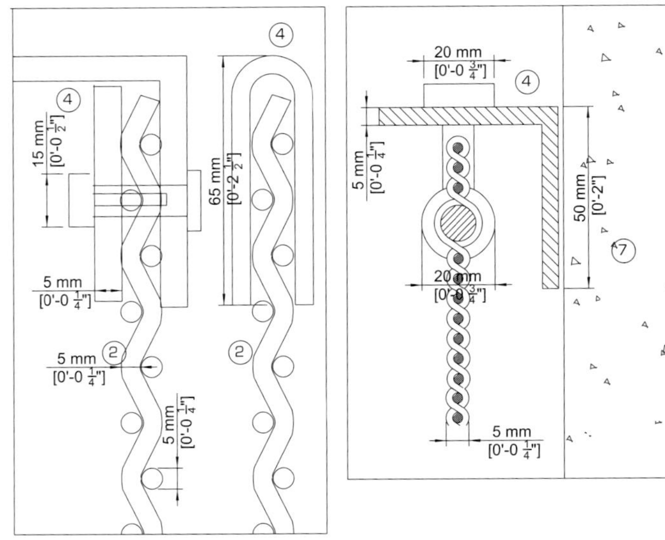

b | Vertical section and elevation. Corner of frame supporting rigid mesh

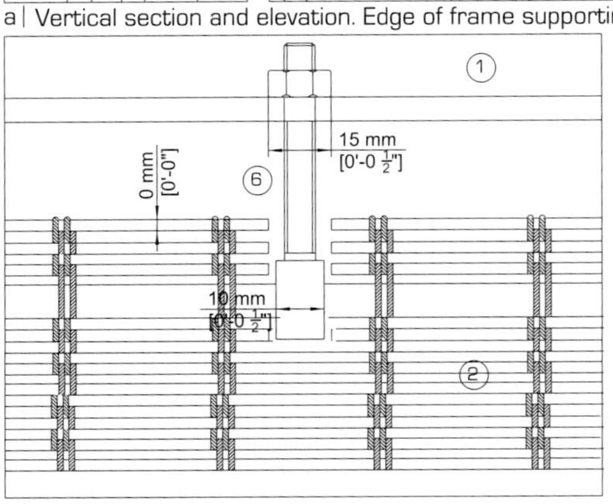

c | Vertical section and elevation.

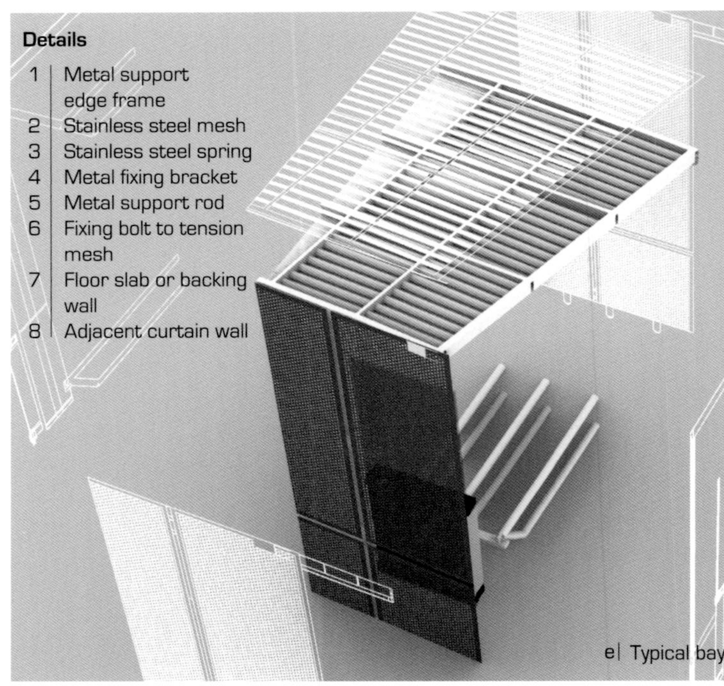

e | Typical bay

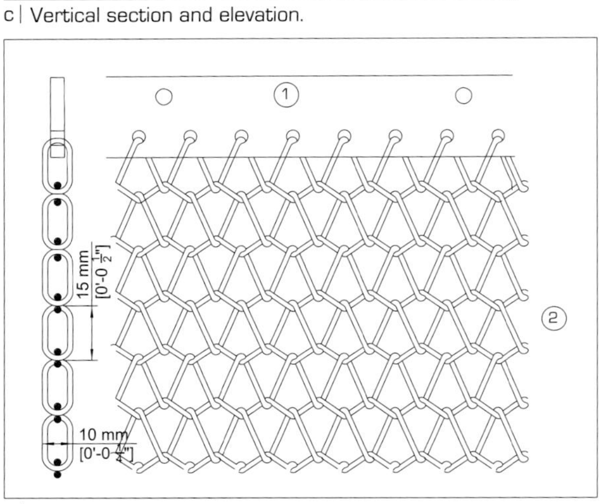

d | Vertical section and elevation. Edge of support for mesh flexible in one direction.

Stainless steel mesh screens have been introduced into mainstream building construction only in the last ten years. This metal is preferred for mesh screens due to its durability and weather resistance when used externally. Their appearance as a textile rather than as a sheet material has led to the use of mesh as 'wraps' to facades in a similar way, in visual terms, to rainscreens. Their purpose is often to provide a smooth, textile-like surface across a wall that can conceal a variety of different facade elements immediately behind it. It has also found favour in car park design, where open mesh decks are given homogeneity with a woven mesh screen. Their varying levels of translucency can be exploited, both in daylight by providing depth to a facade that gives some privacy to building users, and with a night time glow across its surface generated by electrical light within the building.

Meshes are of three essential types: rigid mesh made from rods (MCE_104, b), mesh flexible in one direction made as woven wire with rods in one direction and wire in the opposite direction (MCE_100, b), and mesh that is flexible in two directions which is made from woven wire only.

Rigid mesh

Rigid mesh is made in relatively small sheet sizes and is suitable mainly for balustrades or areas of facades where the material can be supported on a visible frame. It is also used as external solar shading where its lightness in weight allows it to be moved in a motorised system. The material is made in both mild steel and stainless steel but mild steel requires painting. Polyester powder coating is the most common finish. Rigid mesh cannot be tensioned and so it is clamped in a frame at its edges.

This material is usually made in relatively small panels of around 1800 x 1500mm (6ft x 5ft). Stainless steel bars are woven in two directions,

k | Vertical section and elevation. Rigid mesh pattern

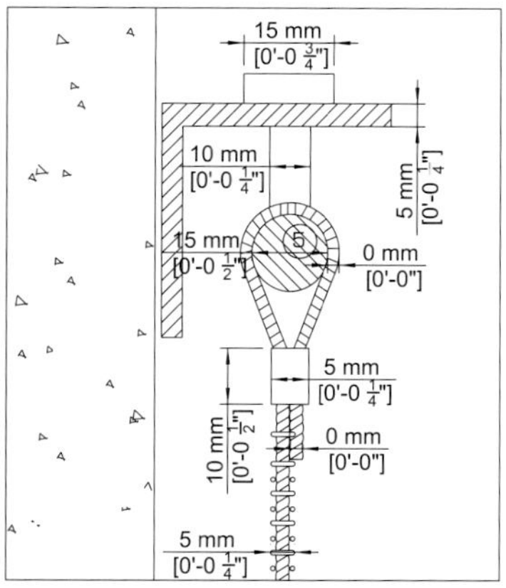

f | Vertical section and elevation. Edge of frame supporting mesh flexible in one direction. The mesh is fixed to a rod which is tensioned on a nut and bolt secured to a metal angle.

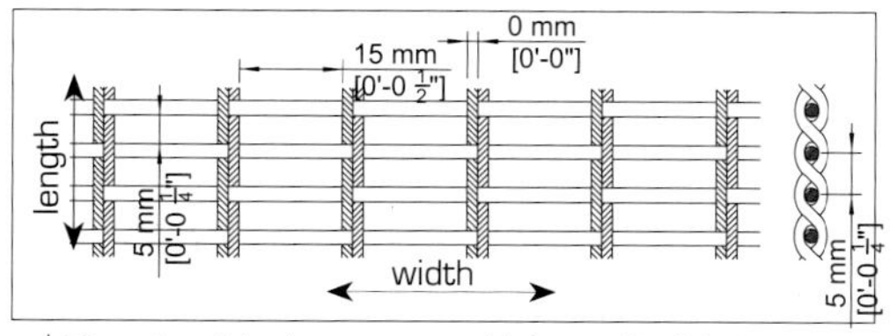

g | Elevation. Mesh patterns which are flexible in one direction.

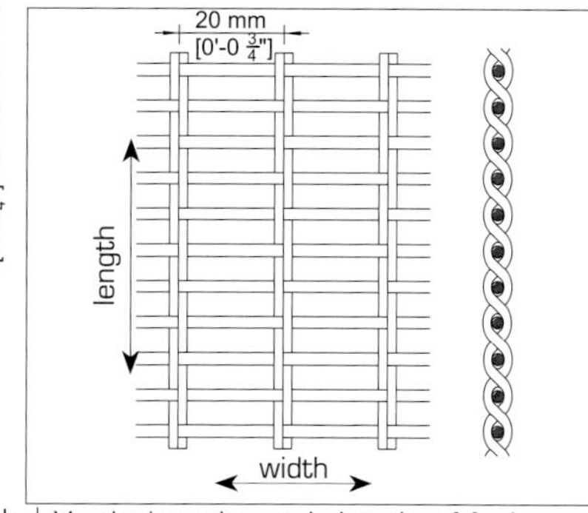

h | Vertical section and elevation. Mesh patterns which are flexible in one direction.

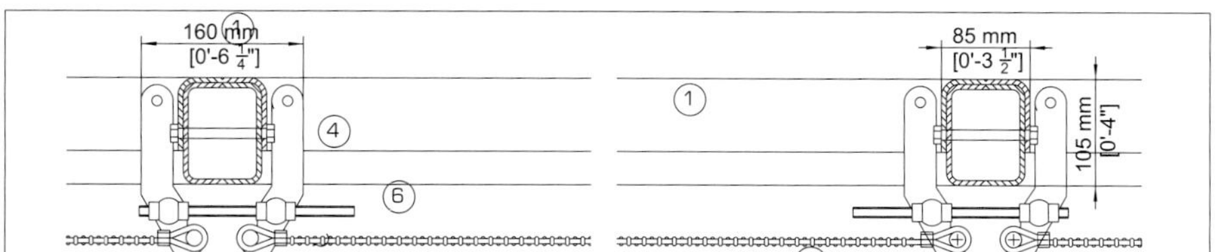

j | Horizontal section. Flexible mesh flexible supported on vertical edges with tensioning mechanism

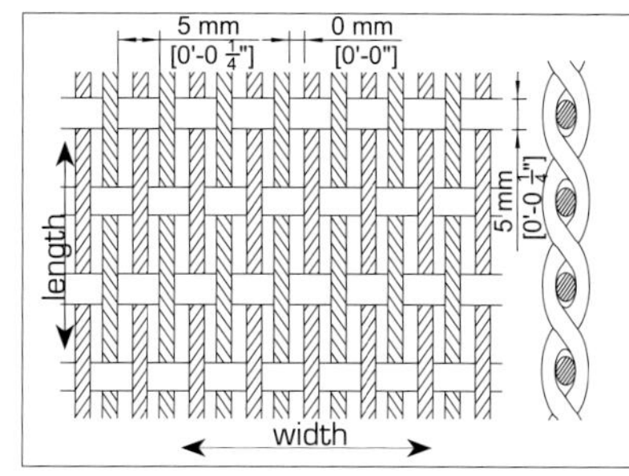

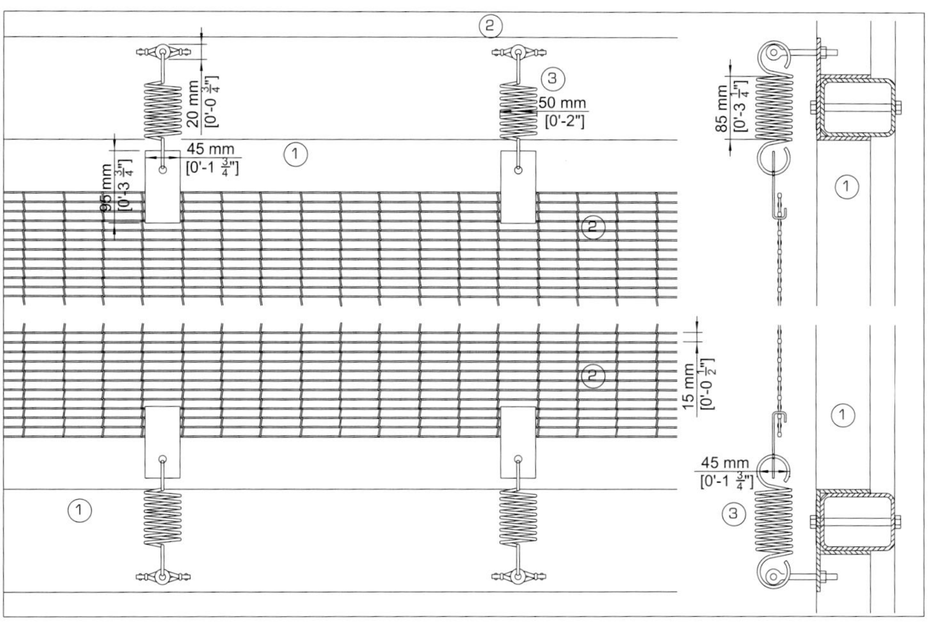

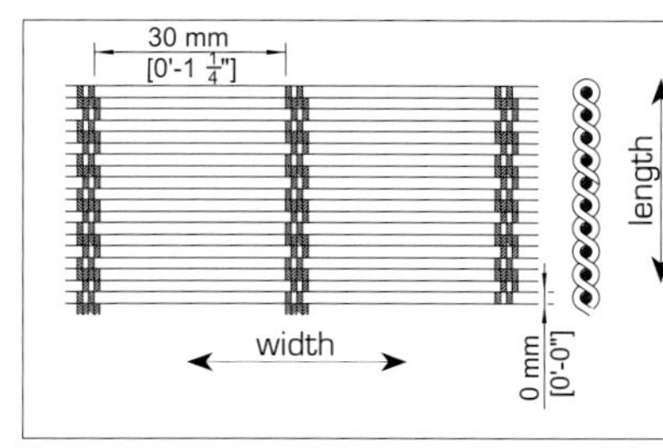

k | Vertical section and elevation. The mesh is fixed to patch fittings which are tensioned by springs. The springs are supported by a carrier frame.

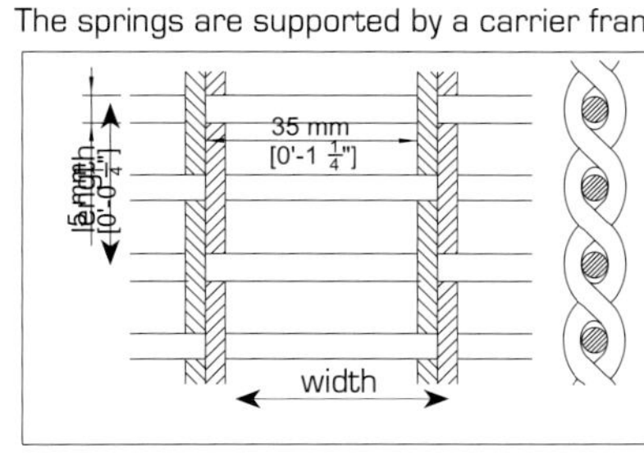

giving the material a stiffness comparable to aluminium sheet but with a surface texture much more undulating than perforated metal sheet. Rigid mesh can provide up to around 50% shading. Rod thickness is typically 1.5mm (0.054in) diameter woven to form openings of around 6mm x 2mm (0.25in x 0.08in). Because they cannot be tensioned, rigid meshes are either held in a continuous edge frame or supported at points in the manner of bolt fixed glazing. They are most commonly used in balustrading where rigidity of material is an essential requirement. The economic nature of the material allows it to be used as both balustrading and full-height screening as part of the same design. When used as a balustrade, the exposed edges of the material are held captive in a protective edging if not fixed into a full supporting frame. This avoids injury to building users. A folded flat sheet or a pair of flats are commonly used.

Meshes flexible in one direction

These meshes are made with rigid stainless steel rod woven in one direction and stainless steel cable in the other direction. An advantage presented by the cables is that they can be tensioned at each end to provide a large continuous flat area of translucent metal.

Most of these meshes come in a maximum width of around 7500mm (25ft). Since the material is made as a continuous run, it can be made in very long lengths, making it ideal for use in a single run of material from top to bottom of a facade without joints. In terms of transparency, the material can vary from around 25% light transmission to 65% depending on the weave. The amount of light transmission can be varied by increasing the thickness and frequency of cables. The distance between rods cannot be varied by reducing the thickness of the cable, allowing it to be more tightly woven, but more cables are usually introduced to compensate for the loss of strength in the cable when

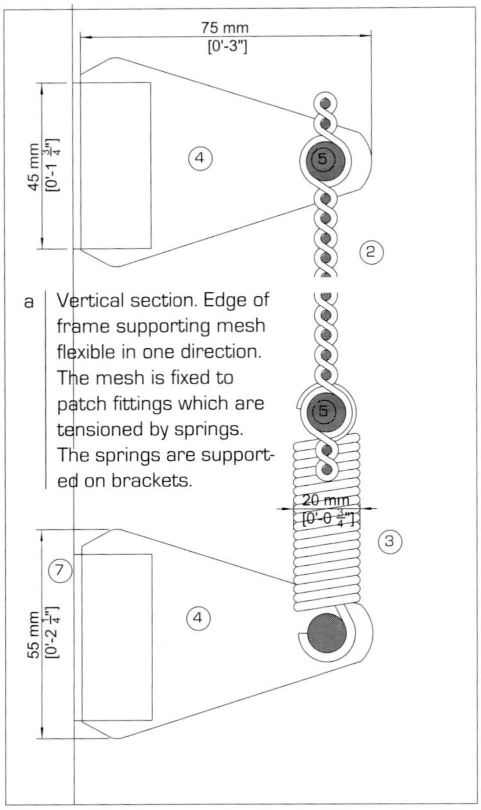

a | Vertical section. Edge of frame supporting mesh flexible in one direction. The mesh is fixed to patch fittings which are tensioned by springs. The springs are supported on brackets.

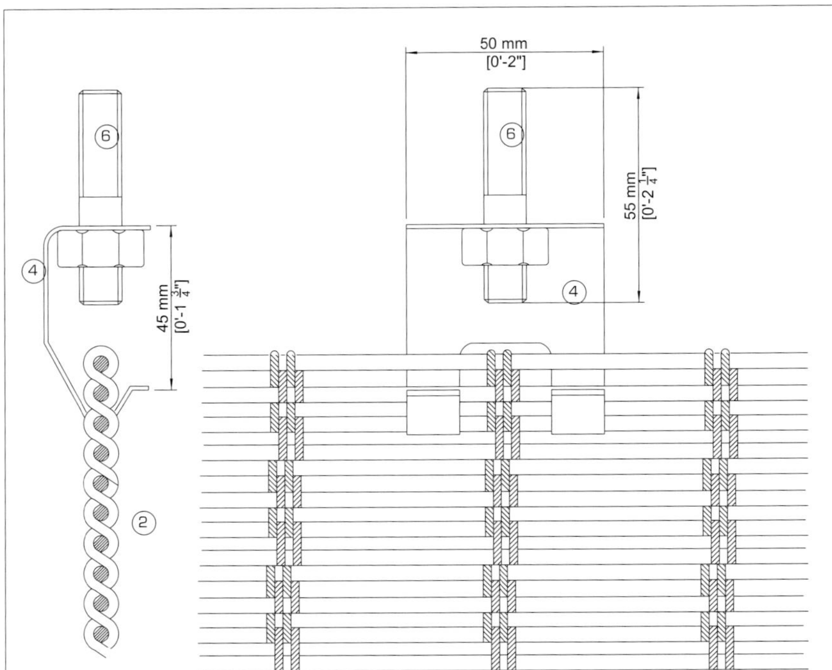

b | Vertical sections and elevations. Edge of frame supporting mesh flexible in one direction. The mesh is fixed to a rod which is tensioned on a nut and bolt secured to a metal angle.

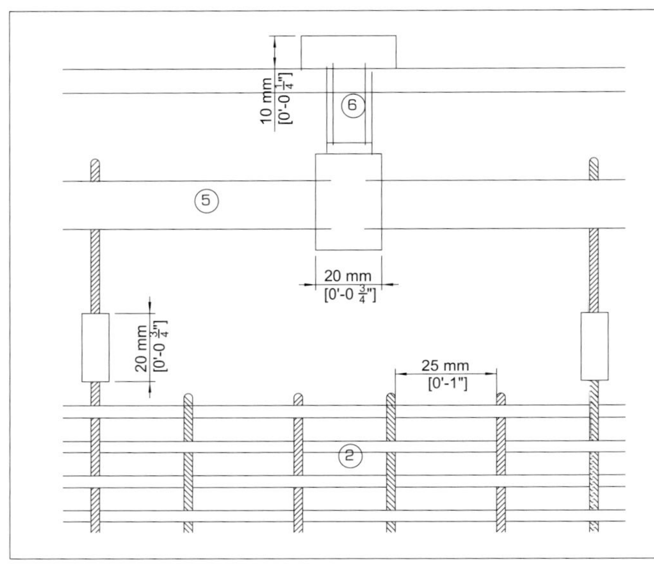

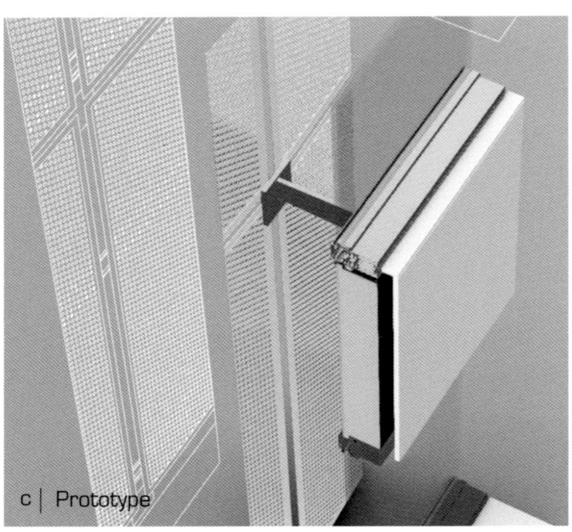

c | Prototype

tensioned. Cable thickness can vary from 2.0mm to 2.5mm (0.08in to 0.1in) diameter. Rod thicknesses can vary from 2.0mm (0.08in) up to 4mm (0.15in) diameter. The weave pattern can vary from 4mm x 10mm (0.18in x 0.37in) to 4mm x 100mm (0.18in x 4in), giving very different visual effects from dense to very open. In addition, varying densities can be woven into a single panel, or length, of material.

Meshes flexible in one direction are fixed by tensioning the cables at each end. The cable is usually set vertically to avoid sag associated with horizontal laying. The cables are looped in a secure loop at each end around a rod or bar. One end is fixed while the other is tensioned by springs set at intervals along the length of the horizontal bar (b). Springs are usually set at the bottom so that the mesh is first hung, then secured and tensioned at the bottom. Lateral stability to the mesh over a long run of the material is provided either by bars woven or fixed into the mesh, or by point fixings. The point fixings comprise discs set either side of the mesh to hold the material in place. A bolt runs through an opening in the mesh between the two discs, which is secured back to the supporting structure, typically a floor slab or backing wall. Meshes will span 2.0 to 2.5 metres (6ft 6in to 8ft) vertically between points of lateral restraint. Adjacent sheets can be fixed together by using a bracket with two bolts in the manner of bolt fixed glazing or by lapping the mesh panels and using a single bolt in the manner of a popper on textile cloths as used in denim jeans, for example. Applications of mesh using wide strips are increasingly common, in widths from 5 to 7metres, (16ft to 23ft) hung from continuous rods and restrained back to a frame at 1.0 to 1.5 metre (3ft 3in to 5ft) centres.

Fully flexible mesh

This material is made as a woven-wire cloth or as a crimped wire panel (MCE_102, a). Wires are crimped down the length of the material and straight wires across its width. The woven cloth type is manufactured

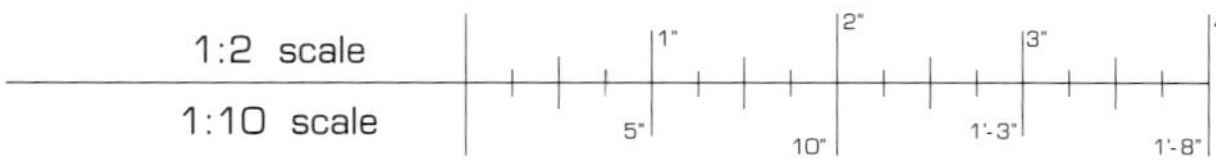

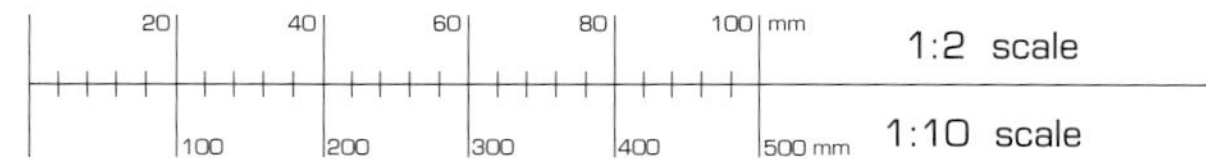

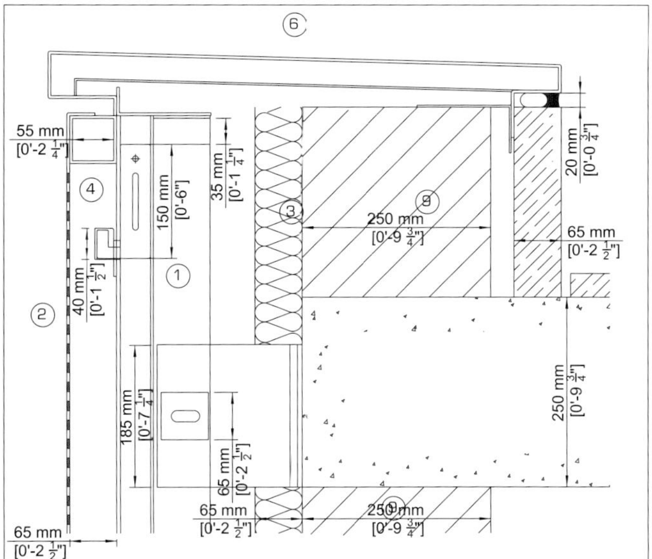

d | Vertical section. Parapet of perforated metal screen.

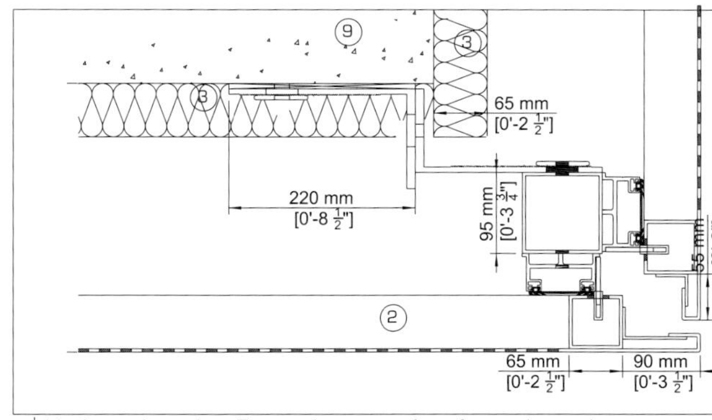

e | Horizontal section. External corner of perforated metal screen.

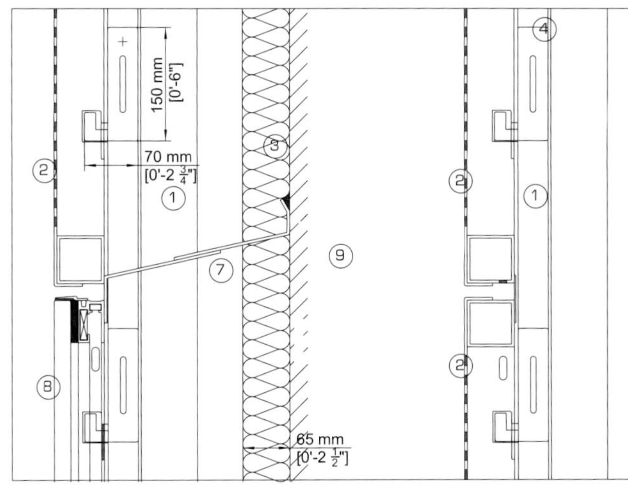

g | Vertical sections. Panel junction of perforated metal screen.

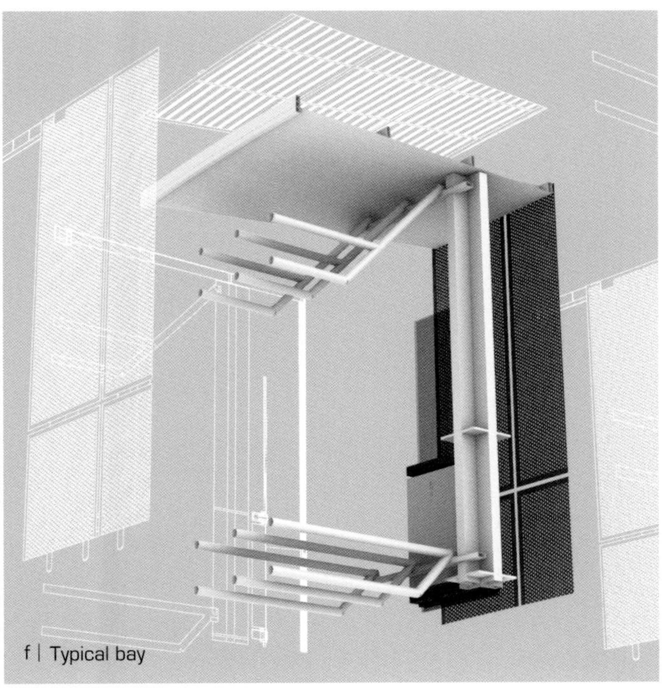

f | Typical bay

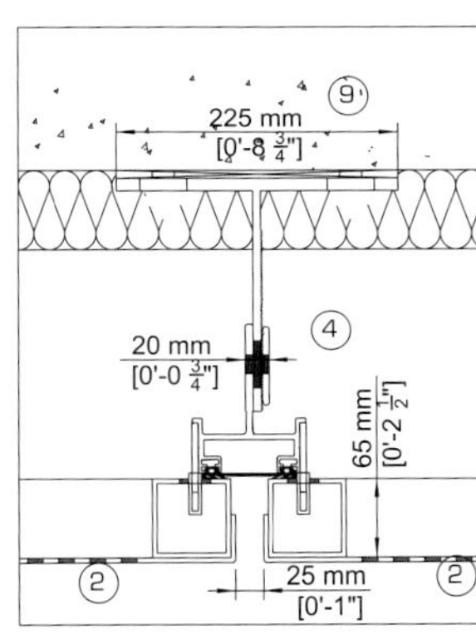

h | Horizontal section. Metal screen junction.

Details for perforated metal

1 | Metal support frame
2 | Perforated steel sheet
3 | Thermal insulation
4 | Metal fixing bracket
5 | Fixing bolt
6 | Pressed metal parapet coping
7 | Cavity tray
8 | Adjacent glazing
9 | Backing wall

primarily for small solar shading screens. It is also used for balustrades in continuous long lengths and as vertically- or horizontally set bands of solar shading material. Large panels are not interrupted by joint lines. The material is fixed by tensioning it vertically. Closely woven meshes have a light transmission of between 1% and 5%, while crimped wire screens vary from 25% to 50%. The tightly woven types are made in widths from around 1800mm to 2400mm (6ft to 8ft) and are made in very long lengths. The more open weaves have an appearance similar to those with cables, using straight rod in one direction and weaving rod in the other direction around the flat rod. These open weave types are made in widths of around 6000mm (20ft) though these sizes are difficult to use in facade panels since the support grid of around 2000mm (6ft 6in) is usually needed to restrain the material. Since the material is made from rod, metals other than stainless steel can be used, though typically copper and bronze are the most common alternatives. They are less rigid than stainless steel but can produce quite dramatic visual effects. A grid of 3mm x 1.5mm (0.125in x 0.0625in) is common in this more open weave material.

Mesh used on curves

Regular mesh that is rigid in one direction is difficult to curve, as it is suited to flat, rectilinear designs. Curves can be formed over lengths of 2-3 metres (6ft 6in to 10ft) by setting out the cables on a curve top and bottom which forces the thin rods in the opposite direction to the curve. However, meshes are being developed which can take up curves more easily for complex geometries. Instead of using cables, loops of stainless steel strip are woven in loops between rows of rod. This allows the rods to be bent around a form or bowed out by brackets, while allowing the loops to be individually stretched between each row of rods. Variations on this type of mesh are set to grow over the next ten years.

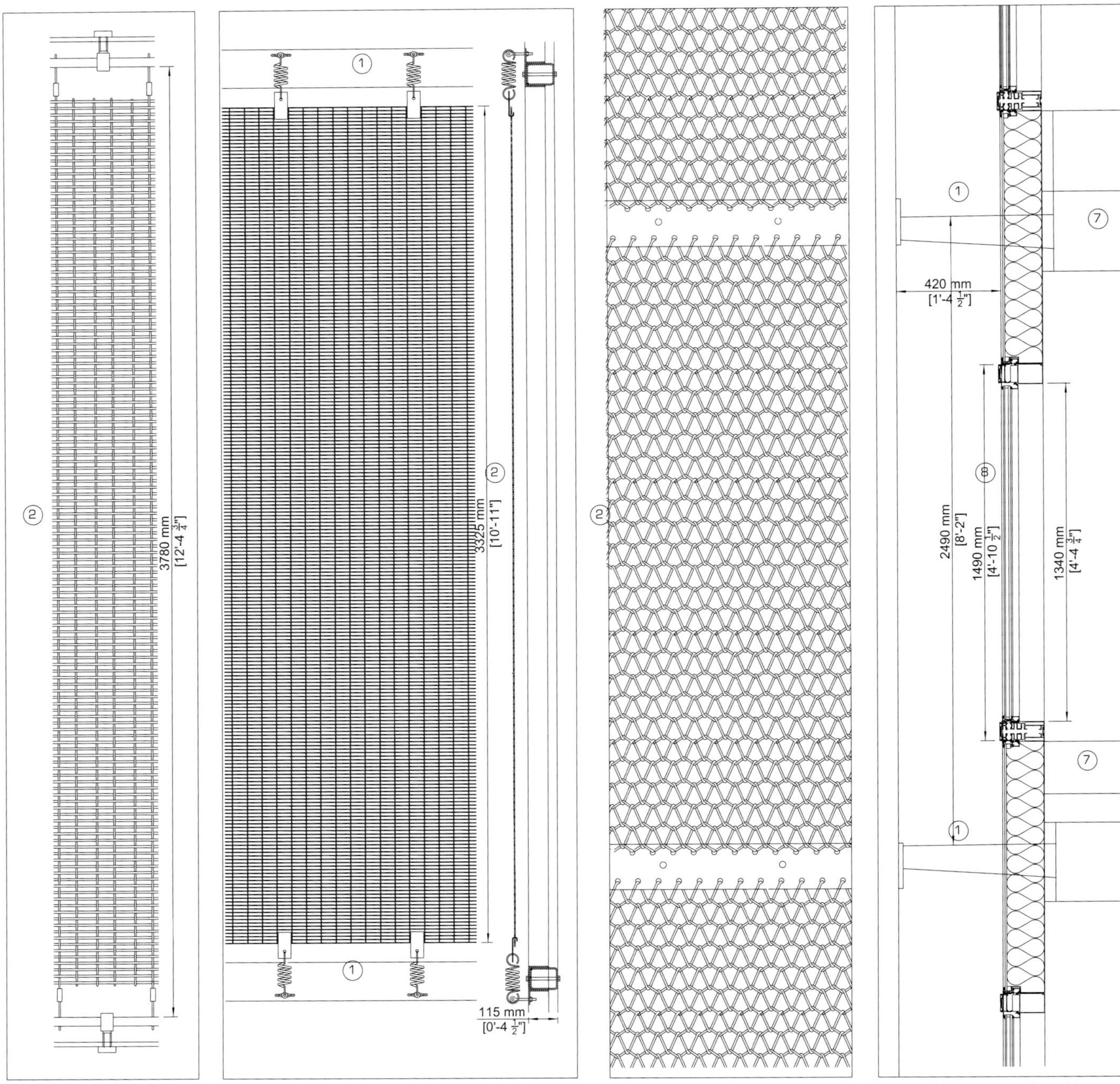

a | Vertical section and elevation. Edge of frame supporting mesh flexible in one direction. The mesh is fixed to patch fittings which are tensioned by springs. The springs are supported on brackets.

Perforated metal

Non-rectilinear or irregular geometric shapes in flat or curved form can more easily be achieved in perforated metal. Although panel sizes are much smaller than tensioned mesh, a greater range of forms can currently be made more economically. Perforated metal in each mild steel (paint or polyester powder coated) and aluminium (polyester powder coated or PVDF coated) are used. Both materials are manufactured with perforations of different shapes and percentages of perforations. Circular holes are the most commonly used as they are straightforward to manufacture (MCE_104, a). They are also able to have a closely controlled percentage of perforation by varying both the size of the holes and their proximity. This makes the material very useful if a precise shading coefficient (percentage of solar shading) or light transmission is specified for a facade. By varying both the hole diameter and the centres of the holes, different visual effects of transparency can be achieved. Squares and various decorative motifs are also made but with less control on precise perforation percentages.

Steel and aluminium sheet are commonly available in sizes up to around 3 metres x 2.5 metres (10ft x 8ft) in 3mm (0.118mm) thick sheet allowing panels to be reasonably large, depending on wind load considerations. In general, the higher the percentage of perforation, the lower the wind load on the perforated metal panel. Perforated metal panels are usually fixed back to an edge frame made from angle or profile in the same material (MCE_104, a). The edges of the metal are usually not perforated in order to conceal the frame behind. The increased use of water jet cutting machines allows for a much greater control of the extent of pattern on a sheet. Perforated metal panels

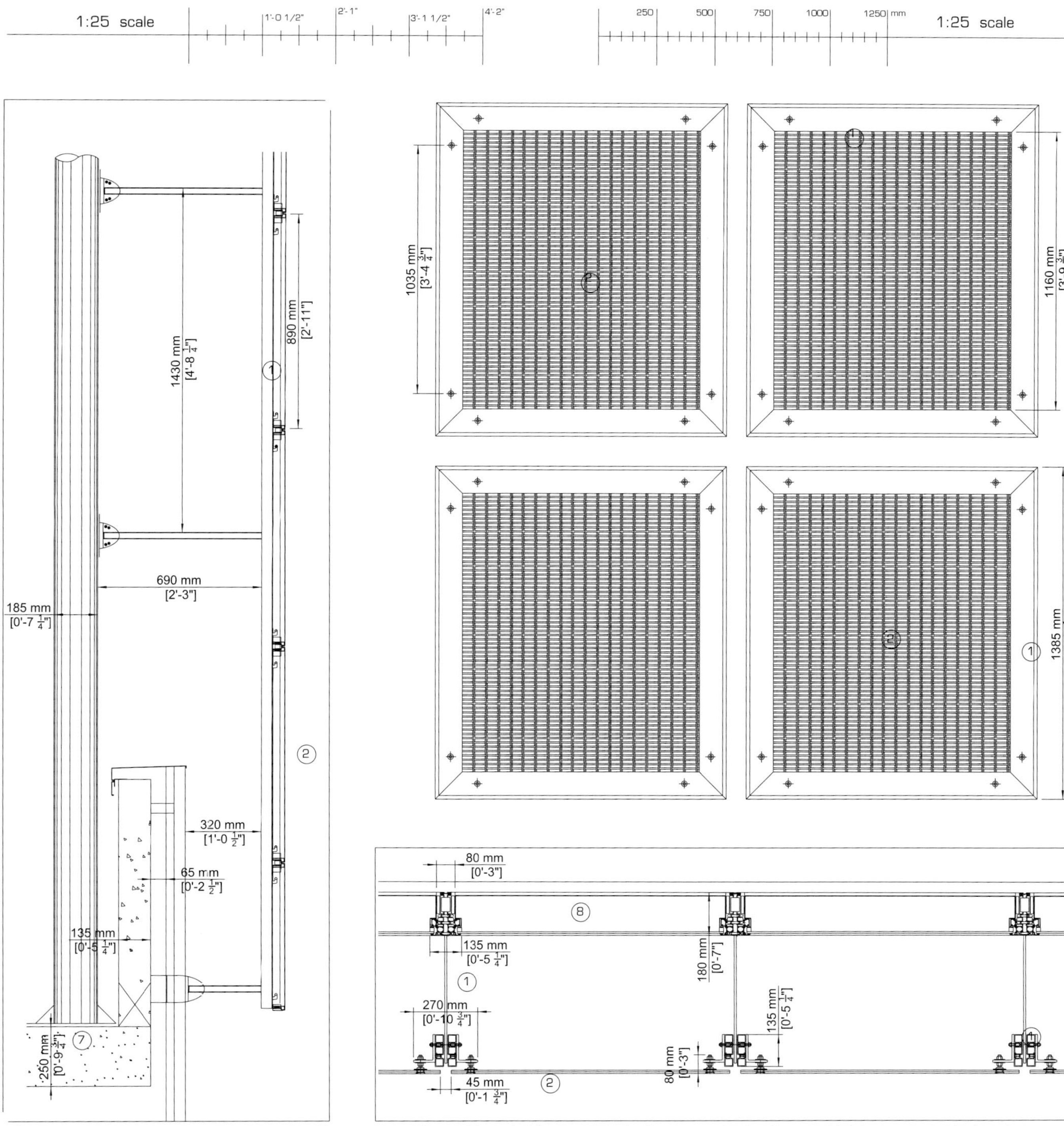

b | Horizontal, vertical section and elevation. Frame supporting rigid mesh. The mesh is fixed to a metal edge frame which is supported by brackets.

are then fixed back to the primary structure with a variety of hanging brackets, ties and struts to suit the design. The supporting structure is visually very refined if visible through the metal, particularly at night if the panel assembly is visible from lighting within the building. A series of forked pin connections or cast moment connectors are increasingly being used, with tapered tubular framing members or box sections with complex sections increasingly being the norm (a). Because the metal framing is exposed to the effects of weather, a high specification paint is used for steel, and either polyester powder coating or PVDF coatings are used for aluminium. Anodising has become more popular in recent years but requires very close control in the factory to avoid visible colour differences between adjacent anodised panels.

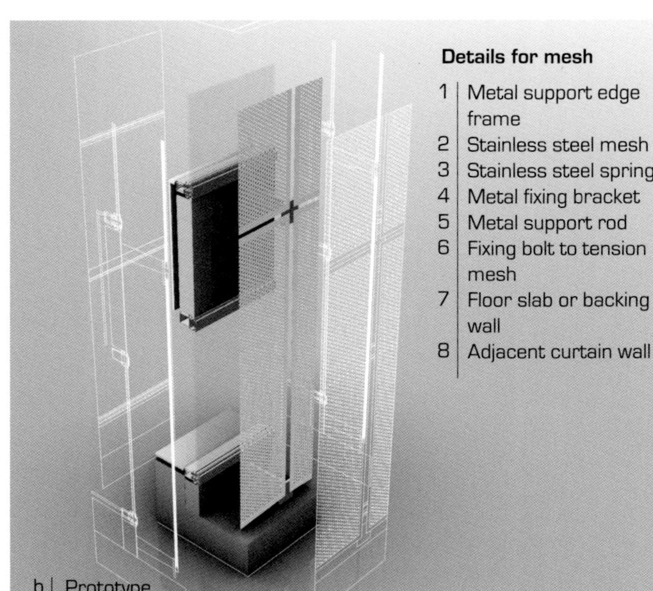

b | Prototype

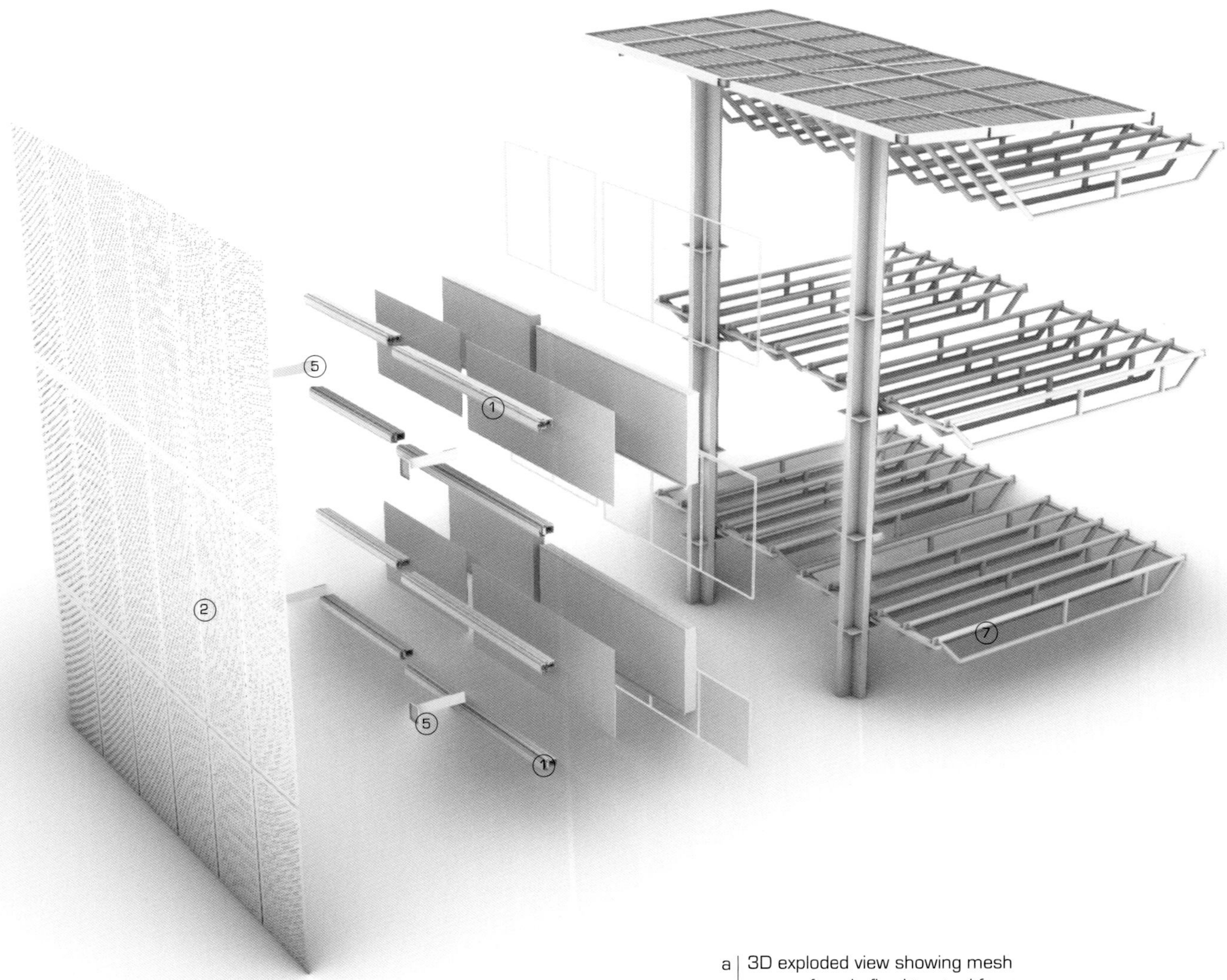

a | 3D exploded view showing mesh screen facade fixed to steel frame

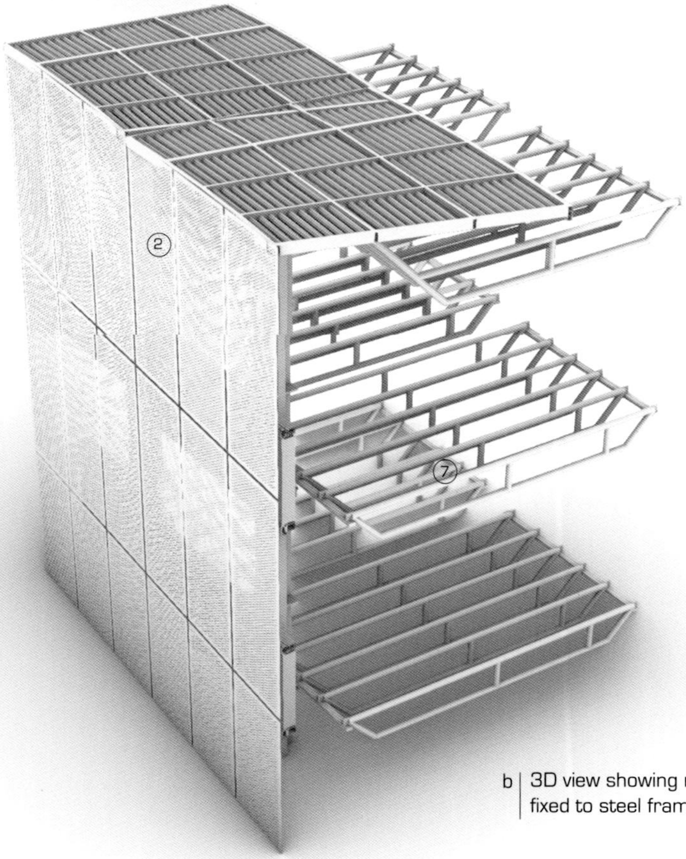

b | 3D view showing mesh screen facade fixed to steel frame

Details

1 | Metal support edge frame
2 | Stainless steel mesh
3 | Stainless steel spring
4 | Metal fixing bracket
5 | Metal support rod
6 | Fixing bolt to tension mesh
7 | Floor slab or backing wall
8 | Adjacent curtain wall

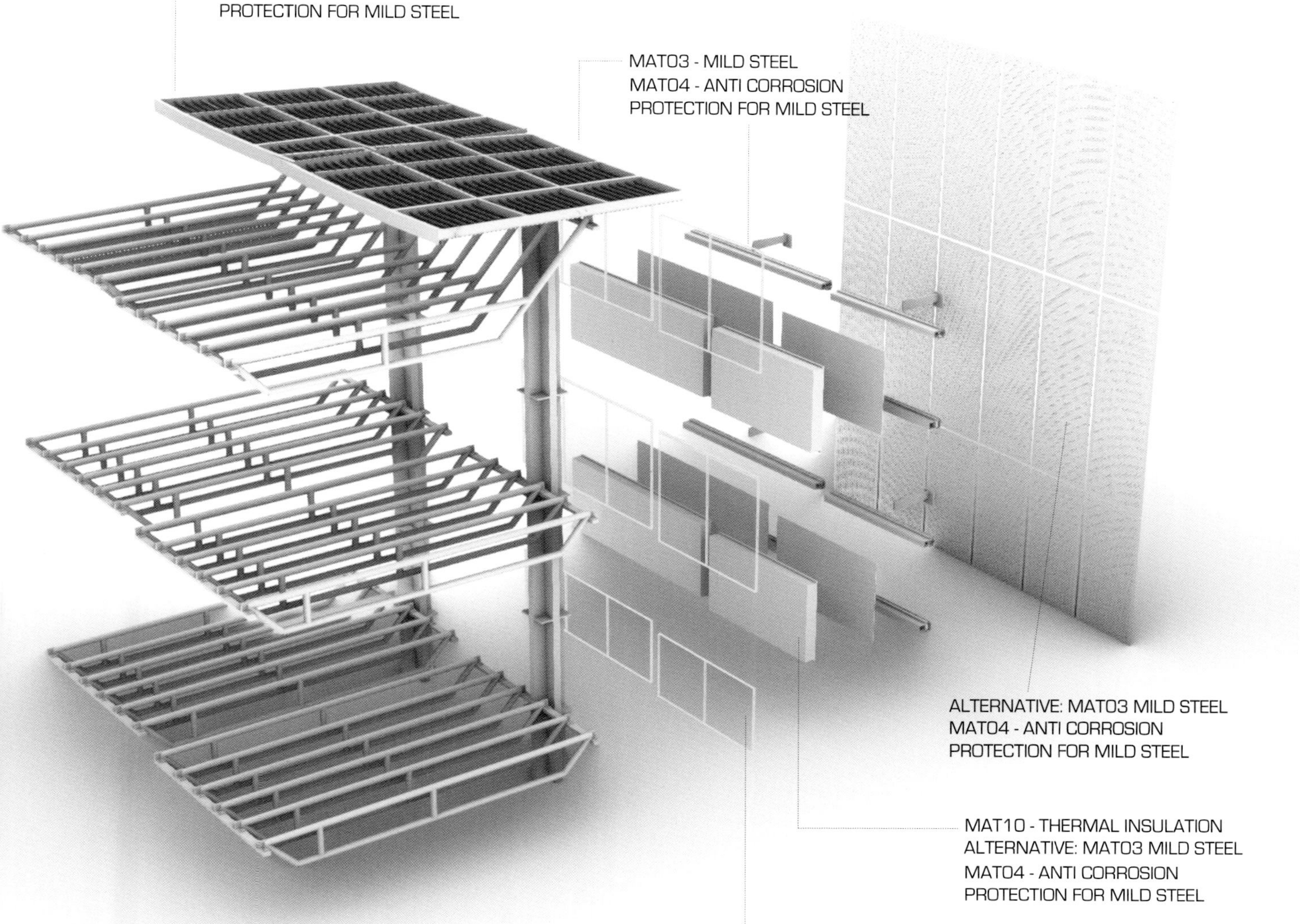

c | 3D exploded view of mesh screen facade components with material specification references

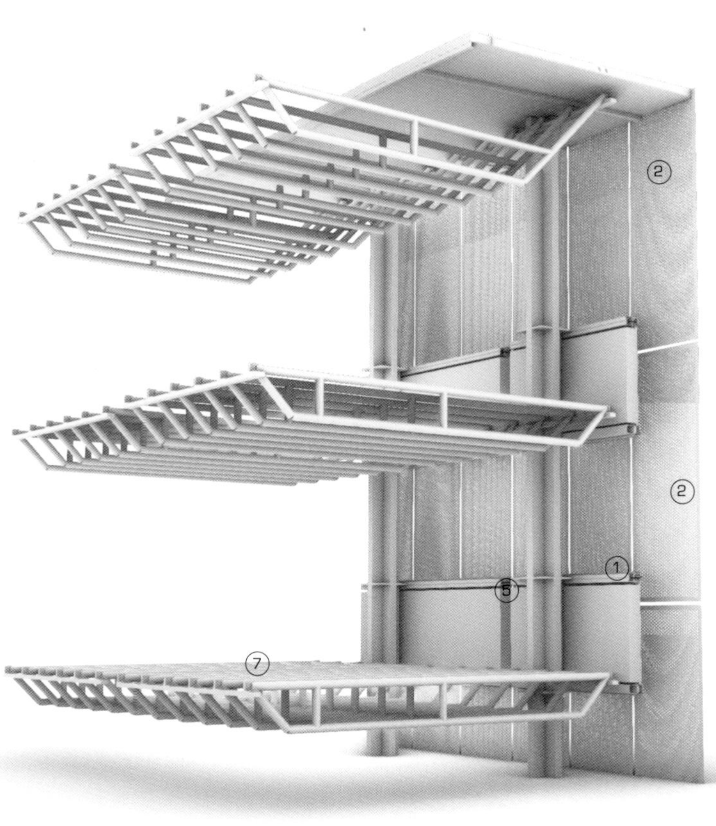

d | 3D view showing mesh screen facade fixed to steel frame

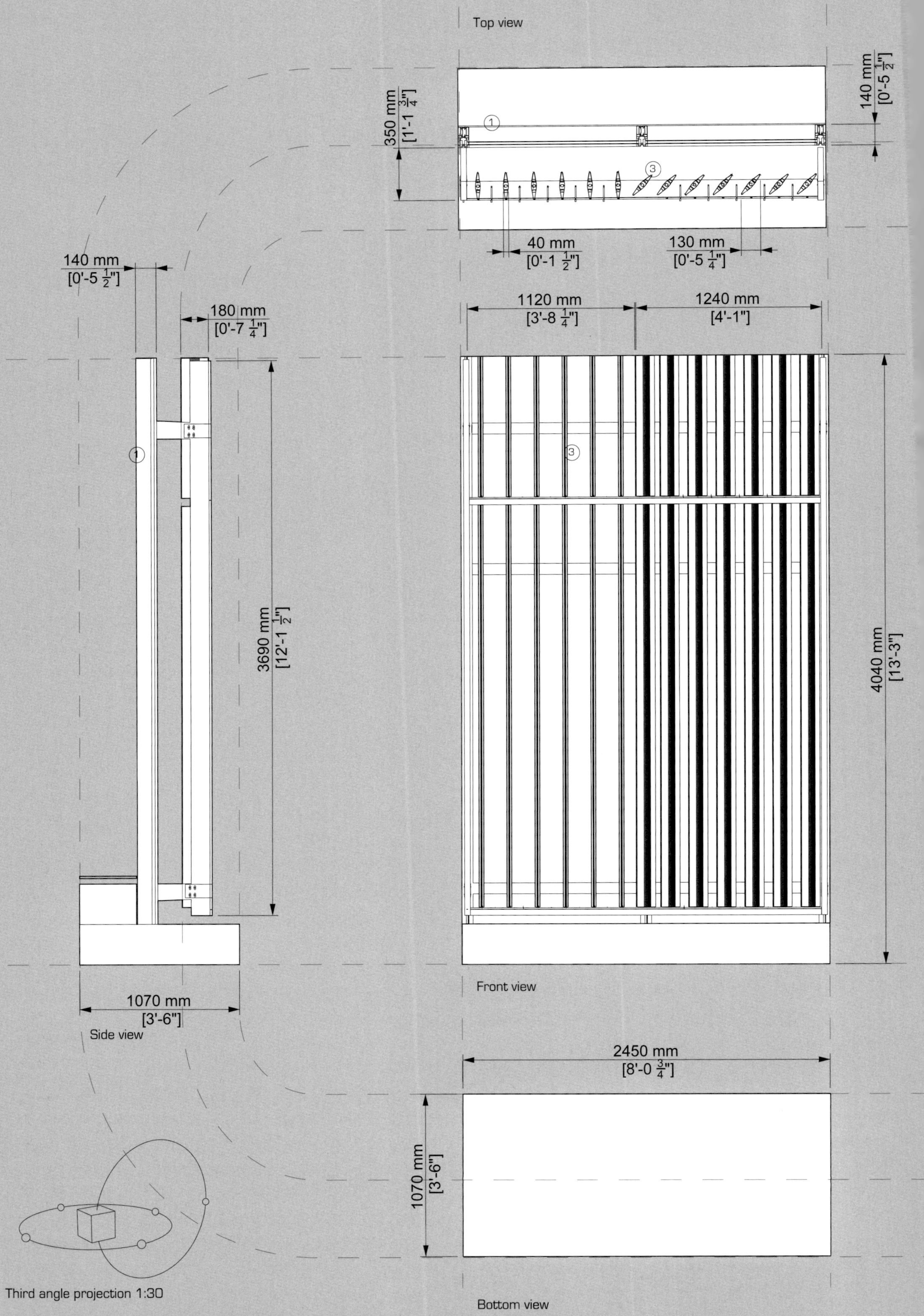

Third angle projection 1:30

Details

1 | Extruded aluminium frame
2 | Stick curtain walling-type carrier system
3 | Extruded aluminium louvre blades
4 | Backing wall
5 | Louvre panel door
6 | Thermal insulation

3D view of the assembly

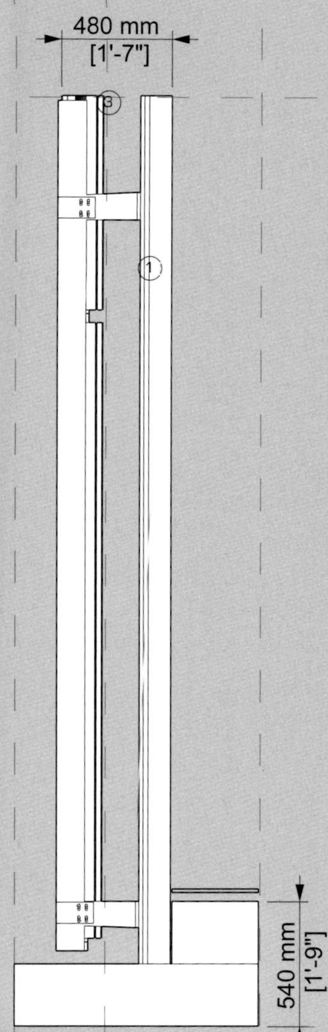

Side view

Back view

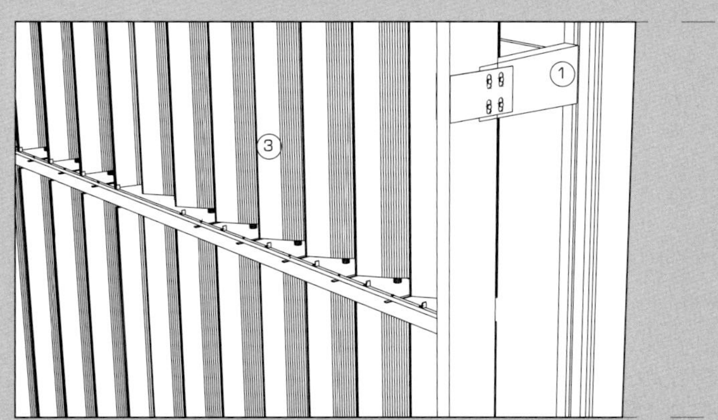

3D view of the assembly

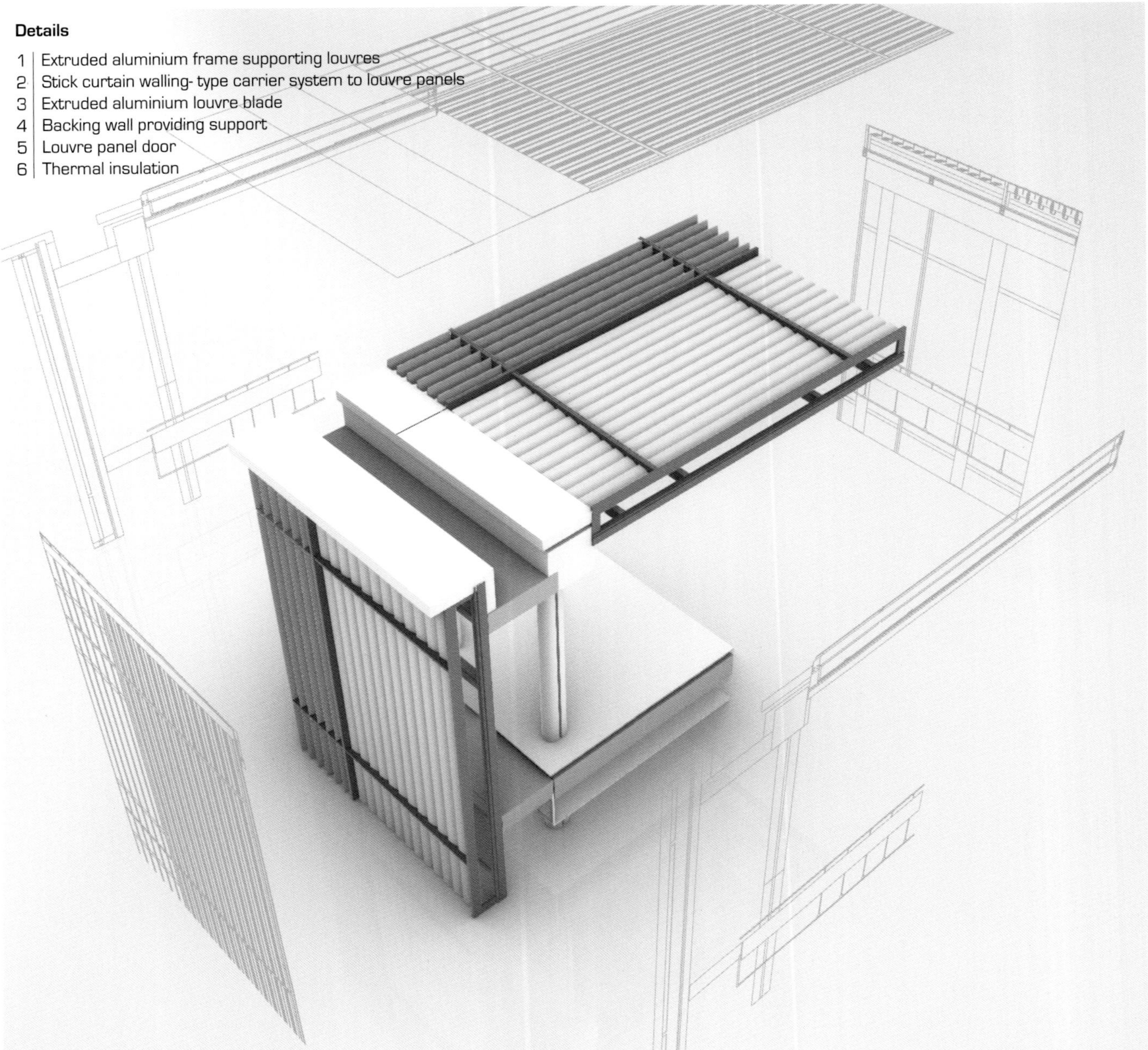

a| Typical bay

Metal louvres are typically used for two purposes: as weather-resisting screens to naturally ventilated spaces such as plant rooms, as terminations to the ends of air-handling ducts where they penetrate the external wall, or alternatively as solar shading on the outside of facades in front of glazed walls or windows. Glazed louvres are also used to provide natural ventilation to winter garden spaces where they also provide light, as in high level clerestorey glazing or in glazed walls in winter gardens in housing. Glass louvres are generally adjustable. Metal louvres are used for air handling ducts or plant rooms, and can be in single, double or triple bank depending on the amount of weather resistance required.

Metal louvres

Louvre panels can be set either horizontally or vertically. Horizontal louvres have inclined blades mounted in a frame of the same material. Some louvres throw the water clear of the edge of the blade at the front. Others, where more weather protection is needed, have either a drainage channel at the front or are made as a V-shape in order to drain water away into the sides of the frame where it is ejected to the outside at the sill. Open gauge mesh is usually set at the back of the louvre panel to prevent the passage of birds. Louvres without drainage channels are used in sheltered areas and also where rain that is blown through the louvre will not damage the building fabric. Drained louvres are used where more weather protection is needed, usually where exposure to the weather is more severe and where water penetration through the panel must be reduced. These drained louvres allow a little less air movement than the standard types, and more louvre area is usually provided to compensate for this. The use of a double bank louvre ensures that water blown over the top of the outer blade will run down the face of the inner blade and be drained away. The free area from the louvres is around 50%.

Vertically set louvres have blades set at an angle (in plan) such that water drains down the face of the louvres and is drained away at the bottom of the

b | Horizontal sections. Corners of vertically-set extruded aluminium louvres

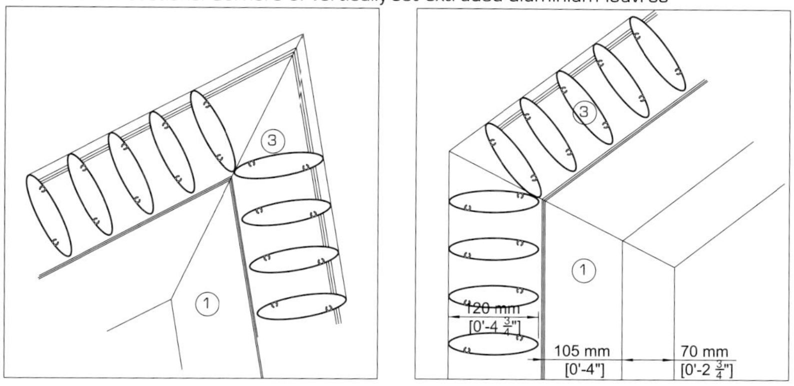

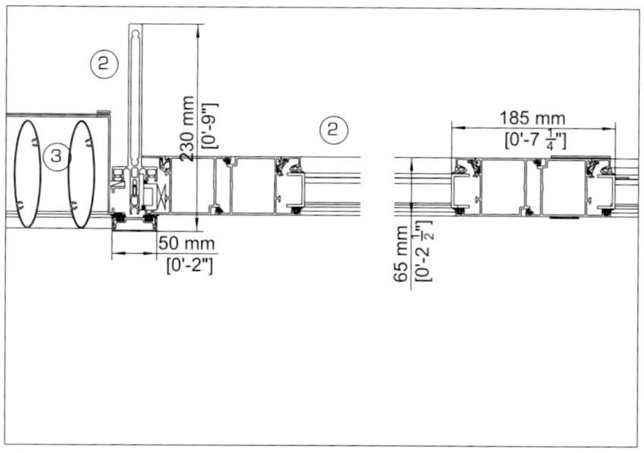

c | Horizontal section. Vertically-set extruded aluminium louvres. Panel to panel junction and panel to door junction

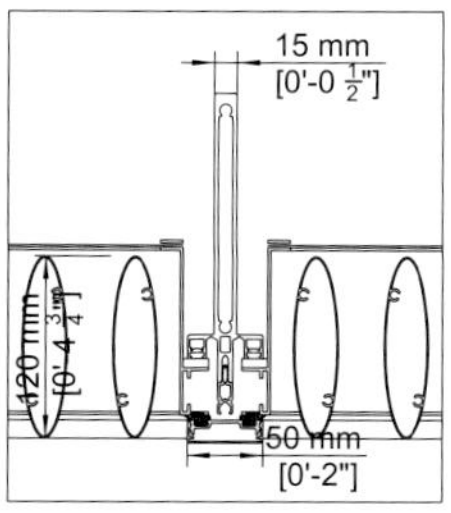

d | Vertical section. Vertically-set extruded aluminium louvres

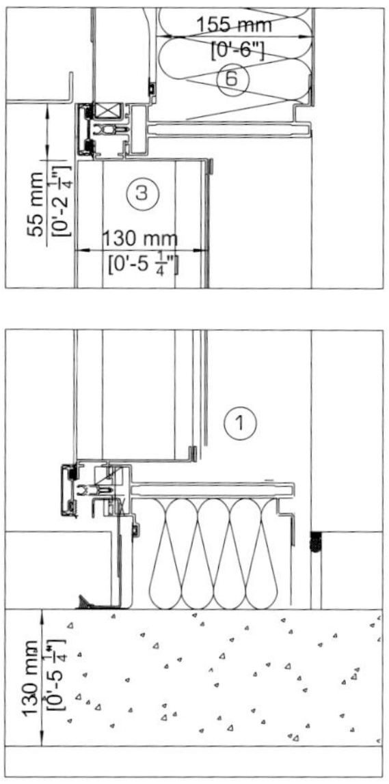

e | Vertical section. Louvred door

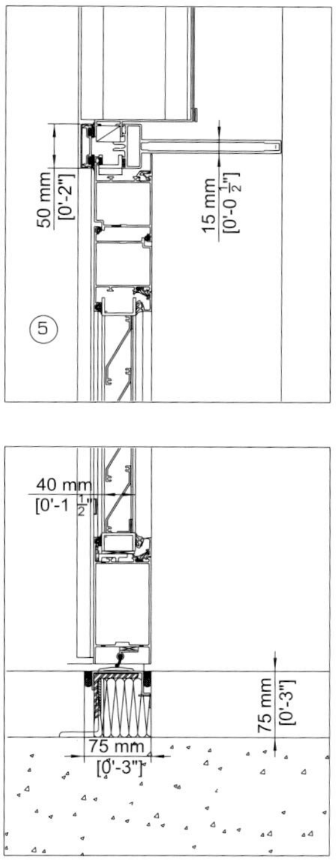

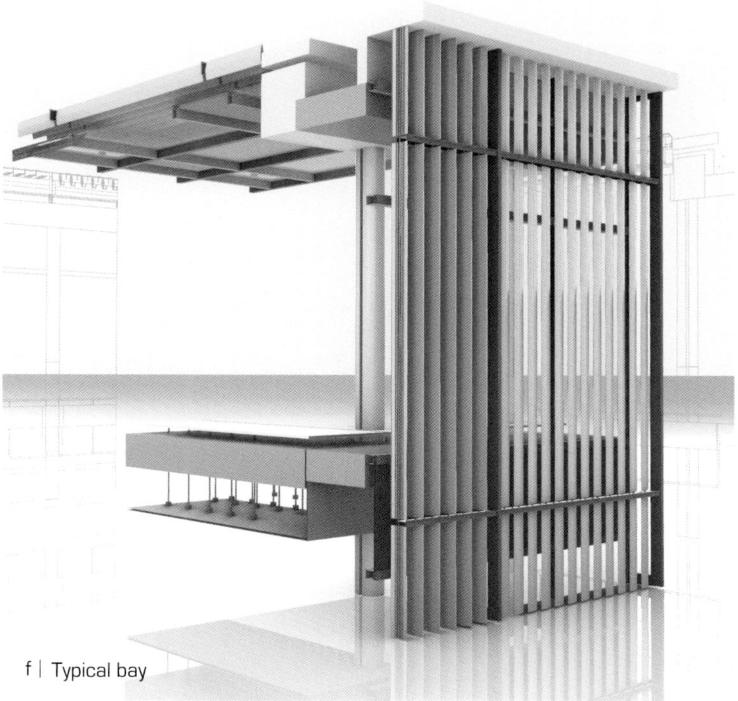

f | Typical bay

frame (a). They are deeper than the simplest horizontal types and perform in a similar way to the drained horizontal types (MCE_111, f). Their depth also ensures that views through the louvre panels are severely reduced. Profiles for vertical louvres can be both V-shaped and elliptical (MCE_111, e). The free area is also around 50%. Panels are assembled in a factory for coupling together on site. Typical panel sizes are around 1.5 metres x 2.5 metres (5ft x 8ft) for both vertical and horizontally set louvres. Larger panels can be fabricated, but transportation to site becomes more difficult. Louvred panels can be arranged in heights up to around 4 metres (13ft) with extruded aluminium mullions. Above this height, additional steel posts are required behind panel joints to provide both support and lateral restraint.

Horizontally set blades are set 50mm apart. Stiffener bars are set at centres from 1000mm (3ft 3in) to 1500mm (3ft) vertical centres, depending on blade size and material thickness. These bars are not visible directly from the outside but can be seen from below if louvres are set at an angle that allows views through them. Louvre blades are fixed to the stiffeners with extruded aluminium clips to stiffen the blades (MCE_111, d). Corners of frames to louvre panels are mitred and screwed or welded. The edge has different profiles to suit being glazed into adjacent curtain walling, and has either recessed joints between panels or a wide frame to suit the type of additional support needed for the complete louvred panel width.

Louvre panels can also be set at inclined angles, but resistance to rain penetration is significantly reduced on inclined walls that can be seen from below. Louvres can be set upside down to avoid views through the screen, but they provide no weather protection. In both cases, an additional vertical louvre is set behind to exclude rain.

Doors are made in the same way as louvre panels but have stiffer frames to suit movement. They are designed with louvre panels and are often the same size in order to conceal their appearance.

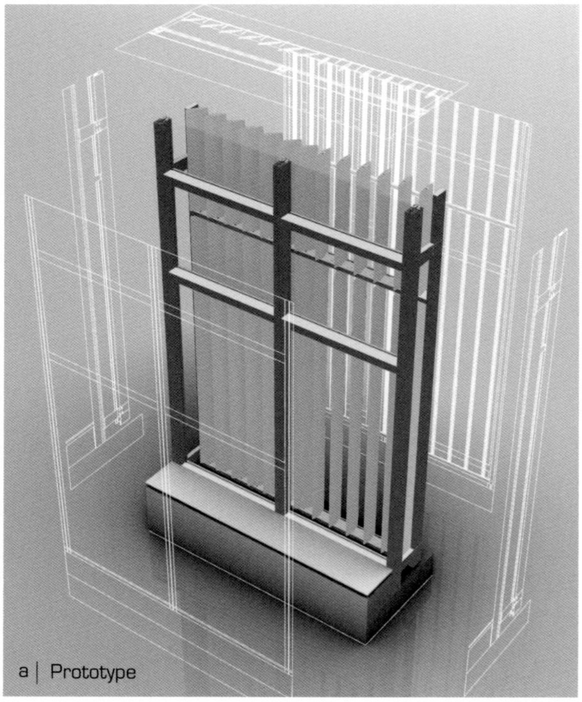

a | Prototype

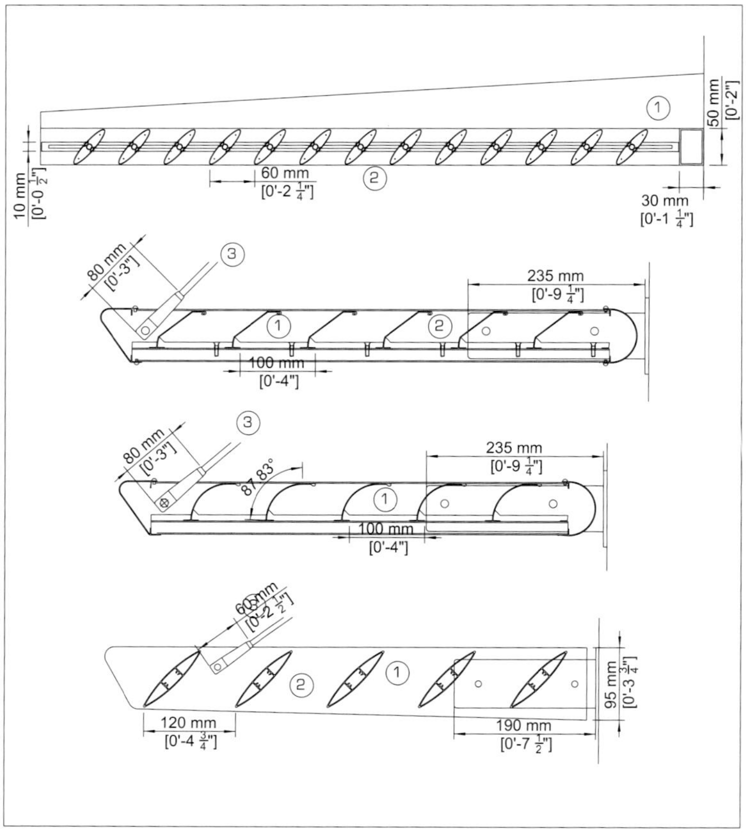

b | Vertical section. Typical louvred solar shading canopies

Details for canopies

1 | Metal support frame
2 | Extruded aluminium louvre blades
3 | Supporting rods or tubes
4 | Extruded aluminium decking for maintenance
5 | Balustrade

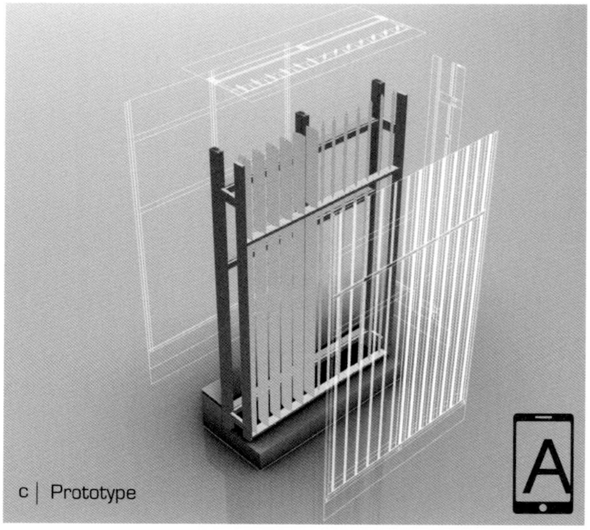

c | Prototype

Details

1 | Extruded aluminium frame supporting louvres
2 | Stick curtain walling- type carrier system to louvre panels
3 | Extruded aluminium louvre blade
4 | Backing wall providing support
5 | Louvre panel door
6 | Thermal insulation

Sandtrap louvres are designed for use in sandy and dusty conditions to prevent most airborne sand and dust from passing through the louvre. They consist of vertically set C-shaped aluminium profiles which interlock to provide a continuous barrier. Air passing through the louvre passes round the interlocking profiles while sand is trapped by the inner profile and falls to the bottom of the frame. The sill is inclined to ensure that sand falls out of the bottom of the panel at the front. Sandtrap louvres remove most of the sand and dust particles before the air reaches the filters on mechanical ventilation equipment. Insect screen is also usually provided and this has little effect on air flow rates. Maximum sizes are similar to those of other louvre types at 1.5 x 2.0 metres (5ft x 6ft 6in), though as with all louvre types, larger panels can be specially made.

Solar shading

Metal louvres are also used as solar shading on glazed facades. Louvres are located either in panels set away from the external wall or horizontally projecting from the facade as cantilevered panels (b). Louvres are set either horizontally or vertically to suit the protection needed from varying sun angles. Vertically set louvres are positioned forward of the glass, usually a minimum of 600mm (2ft) to allow a person to pass between the external wall and louvres for maintenance access and cleaning. The shading devices are fixed to vertical posts or mullions which usually coincide with the module of the glazing behind. Horizontally set panels consist of fixed louvres, usually inclined at 45º to the vertical in order to maximise the shading effect. The louvres are fixed to mild steel channels or T-sections which are in turn fixed to support brackets projecting through the external wall (MCE_114, a). Where glazed curtain walling is used, the support bracket goes through the mullion (vertical member) where it can be properly sealed against water penetration. Louvres for both vertical and horizontal shading usually span much further than those used in 1200mm (4ft) wide opening louvre panels, and need to be more rigid as a result. Panels project a maximum of around 1000mm (3ft 3in) without additional diagonal tie rods to prop the panels.

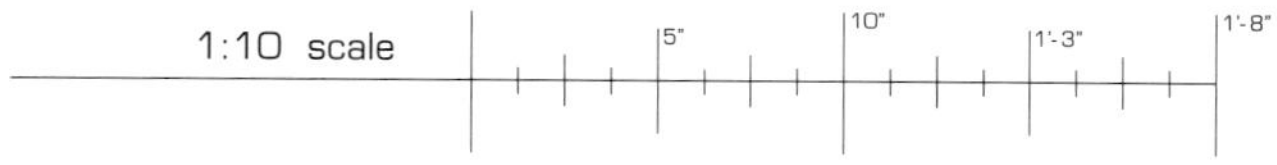

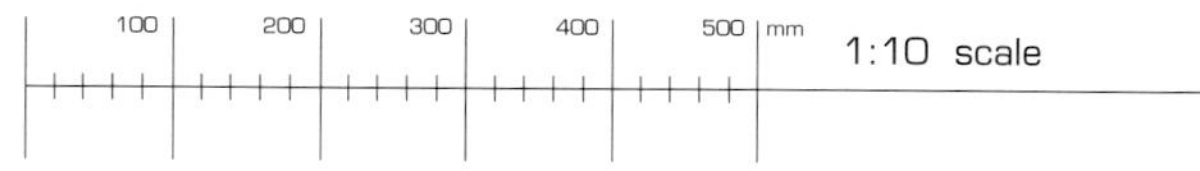

d | Vertical section. Horizontally-set extruded aluminium louvres

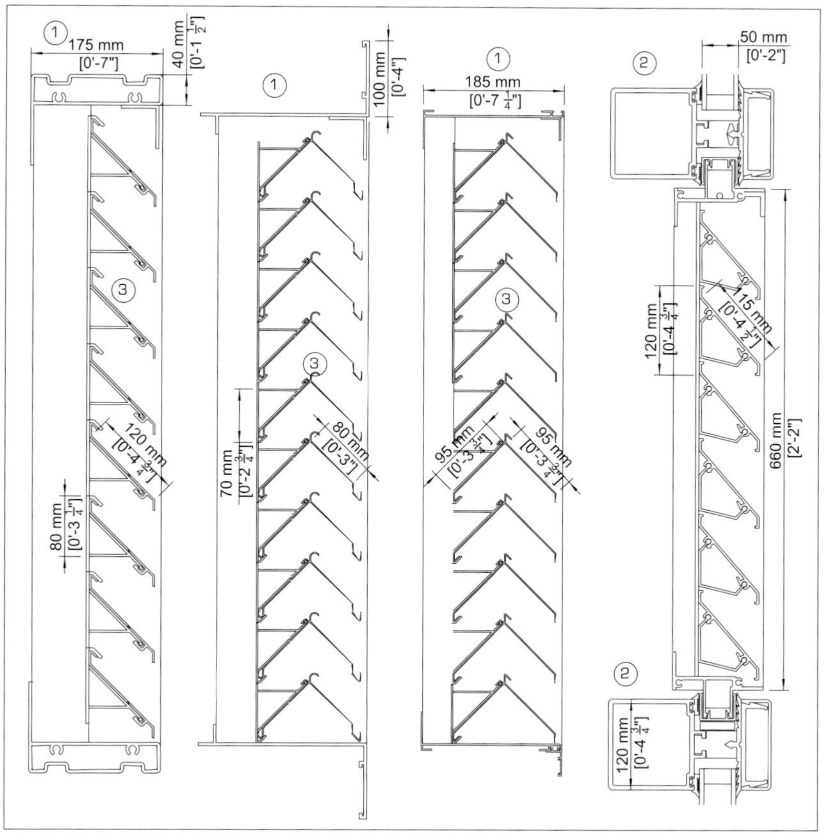

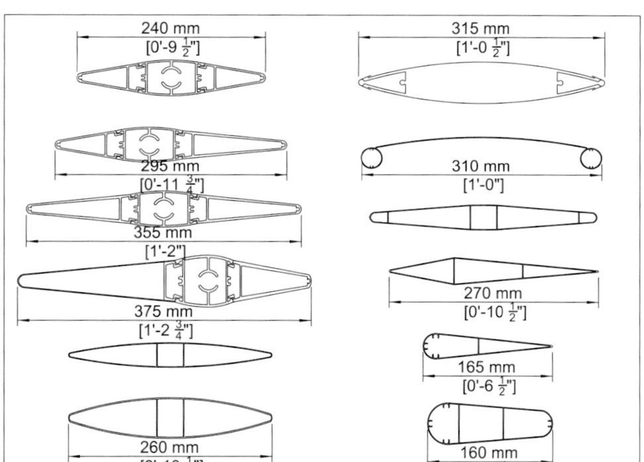

e | Horizontal section. Typical extruded solar shading louvre sections

f | Prototype

Most proprietary systems can reach 2000mm (6ft 6in) with an additional diagonal brace.

Elliptical and aerofoil-shaped aluminium profiles are most commonly used since their shape is seen outs de the building and from inside. More traditional Z-shaped louvre profiles are also used. An advantage of horizontally set louvres over projecting canopies is that they do not require rainwater drainage that is needed for a continuous horizontal surface. Small louvre blades are made as a single extrusion, but larger louvres, up to around 500mm (1ft 8in) in width, are made from an extruded aluminium core to which curved or flat aluminium sections, usually 3mm (0.118mm) thick, are fixed. The ends of profiles are fitted with extruded aluminium end caps both for visual reasons and to protect the inside surfaces from corrosion. Aluminium louvres are finished in either PVDF or polyester powder coating. Blades are usually fixed at their ends from the centre of the extrusion at a single point at each end.

Where louvres are motorised, they are fixed at single pivot points to vertical posts. Link rods are set into the support post joining each blade in order to operate a set of louvres from a single motor. Pivots have nylon bushes fitted to avoid long term rattle and noise from the moving parts.

Vertically set louvres use the same variety of section profiles and are set on transom (horizontal) sections that are connected back to mullions or directly back to the external wall. Vertical louvre sections using an elliptical or aerofoil profile can span up to around 3000mm (10ft) vertically without need for additional stiffeners.

Glazed louvres

These have traditionally been poor excluders of air (high air infiltration rates) but they have improved significantly in recent years. The amount of opening can be closely controlled. They consist of glass blades held in an aluminium frame or in aluminium clips, secured on pivots into an edge frame made

a| Typical bay

from extruded aluminium profiles. Some clips holding the glass in place are made from polypropylene rather than aluminium, to avoid rattling between the two components. The most recent development has been in thin aluminium clips and the use of bolt fixed glazing to hold the glass in place.

Louvre glass blades are usually made in laminated glass for safety, though float glass is sometimes used for small-scale applications. Solar control glasses are often used to match adjacent areas of glazing. Single or double glazed units are available. Glass louvre panels are either hand operated by a wire cable-type winding handle or are electrically operated with rods as used in metal louvres. Electrically operated units are generally 1500mm (5ft) high to suit the length of the rods. Panels are coupled vertically or horizontally to form a large-scale screen of panels. They can also be glazed into curtain walling systems. Maximum sizes are bigger than metal louvres in 2400mm x 2400mm (8ft x 8ft) approximately, but maximum length of ventilator unit is around 1200mm (4ft). Automatic opening types are used as smoke vents in the event of fire. They provide around 70% free area when fully open.

Banks of louvre blades can also be made to tilt to suit varying ventilation requirements, such as in buildings with a high degree of natural ventilation. Panels can open and close from rain and wind sensors forming part of the BMS (building management system). Panel sizes are similar to those of fixed panels, that is, 1.5 x 2.5 metres (5ft x 8ft). They have a single rod set into the edge frame that connects all the blades together and which is pushed up and down by a separate motor or is connected to a manual winding rod.

Where double glazed units are used, they are typically in 24mm thick overall units of 4/16/4 (glass/cavity/glass). The outer 4mm is slightly thicker if a laminated glass is used for safety. Like single glazed louvres, the maximum length of a panel is 1200mm (4ft), but two are commonly joined to provide an overall maximum panel length of 2400mm (8ft). The thicker frames give

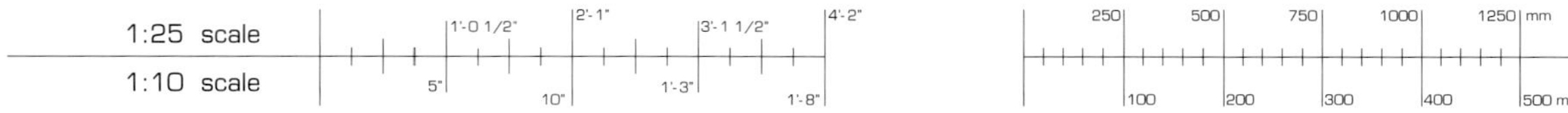

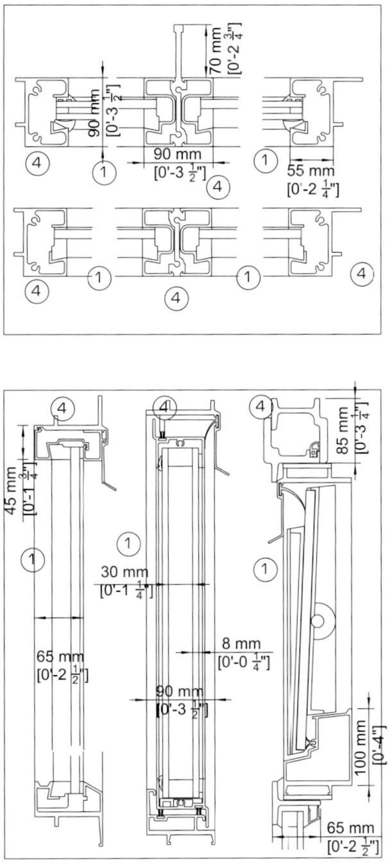

b | Horizontal and vertical sections. Glazed louvres

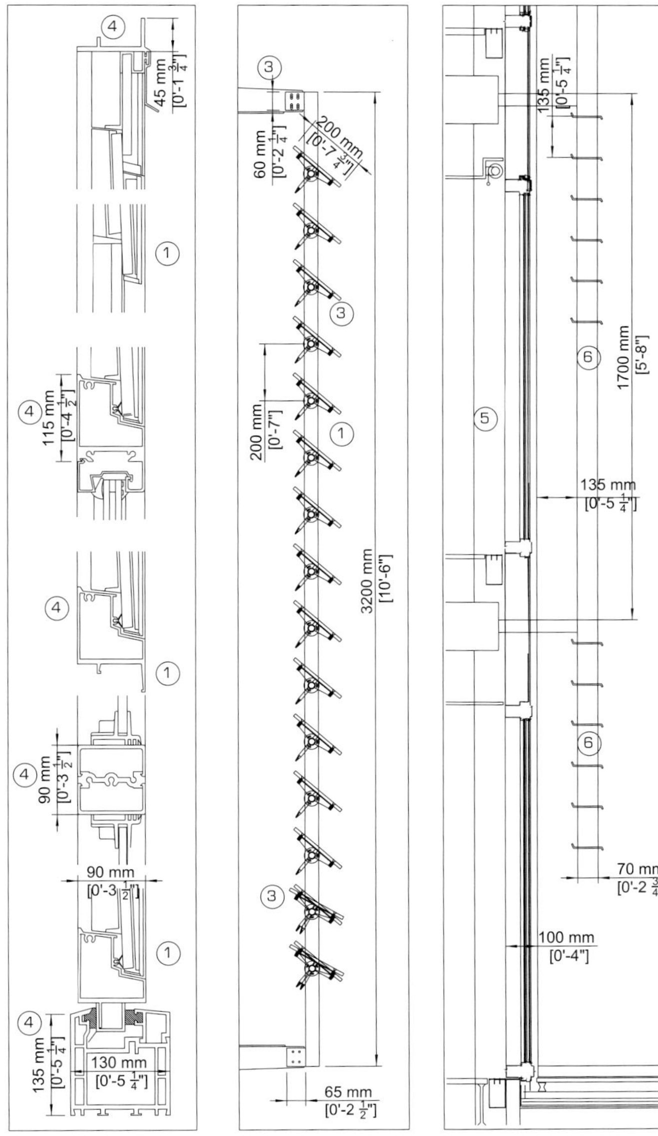

c | Horizontal and vertical section. Fixed horizontally-set louvres providing solar shading to glazed facade

Details

1 | Extruded aluminium frame supporting louvres
2 | Stick curtain walling- type carrier system to louvre panels
3 | Extruded aluminium louvre blade
4 | Backing wall providing support
5 | Louvre panel door
6 | Thermal insulation

d | Horizontal sections. Glazed louvres

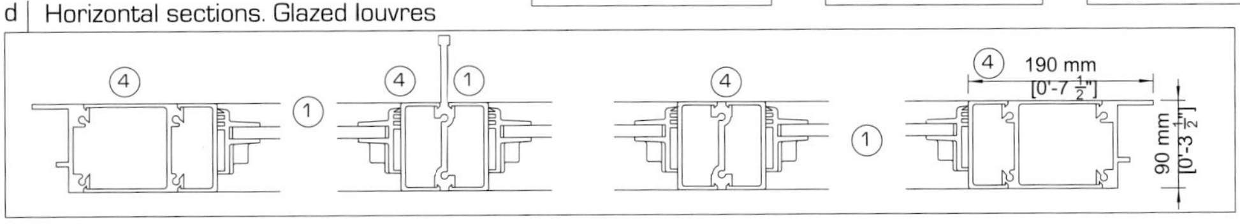

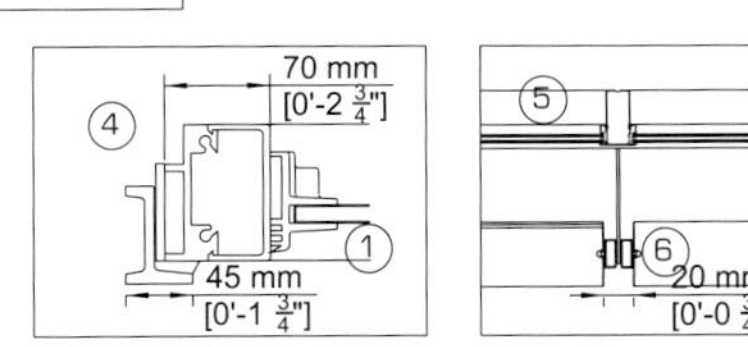

less free area for ventilation of around 50%. Although the glass is insulated for reasons of energy conservation, frames are not yet thermally broken. Condensation risk when windows are closed is assessed for each application.

Walkways

Horizontally set louvres can also be used as maintenance walkways if they are made sufficiently rigid. T-section aluminium profiles are used, and have a serrated top to provide an anti-slip surface. A fall arrest system is used to secure maintenance personnel to the walkway. This consists of a continuous cable or tube fixed to a convenient point along the walkway. A maintenance person in a harness is then linked to the continuous cable or tube by a secure line. The aluminium T-sections are fixed to steel I-sections or channels which span between column supports adjacent to the external wall, typically at around 7500mm (24ft 6in) centres. The main sections supporting the T-sections are fixed to stainless steel or aluminium brackets that project through the external wall from the edge of the floor slab.

A recent development in metal louvre design is in movable types. Louvres are set on a moving rack which allows them to be moved from open to closed, using perforated aluminium sheet to form louvres. When closed, the louvres create a translucent screen with 20% to 50% light transmission (depending on the degree of perforation in the steel) to 100% light transmission when open. This allows a glazed wall to deal with changing sun angles at different times of day and different times of year through a change in the angle of the blade only. This system is used on both vertical and horizontal planes, for example on large glass facades as well as projecting areas of horizontal glazing within the facade. The use of more complex metal profiles where blades interlock, together with different perforated metals and their connection to a BMS (building management system) is sure to develop significant over the next ten years. Such controls can reduce energy consumption within a building by reducing the amount of mechanical cooling needed as well as control glare from direct sunlight.

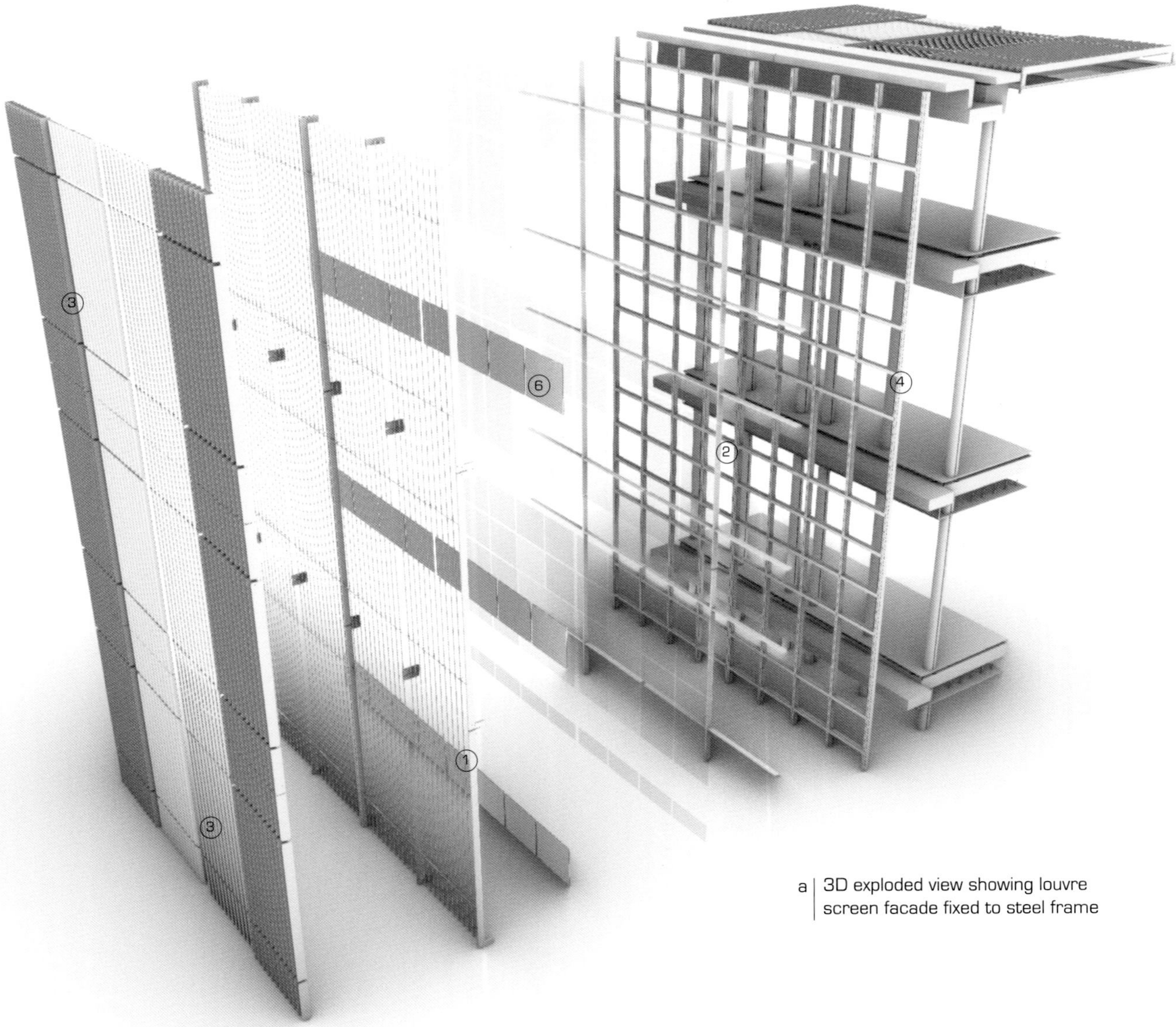

a | 3D exploded view showing louvre screen facade fixed to steel frame

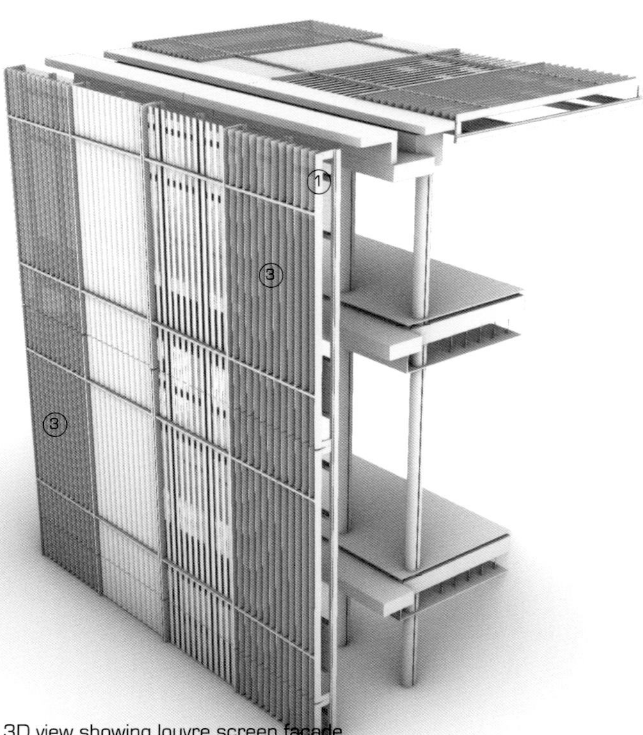

b | 3D view showing louvre screen facade fixed to steel frame

Details

1 | Extruded aluminium frame supporting louvres
2 | Stick curtain walling- type carrier system to louvre panels
3 | Extruded aluminium louvre blade
4 | Backing wall providing support
5 | Louvre panel door
6 | Thermal insulation

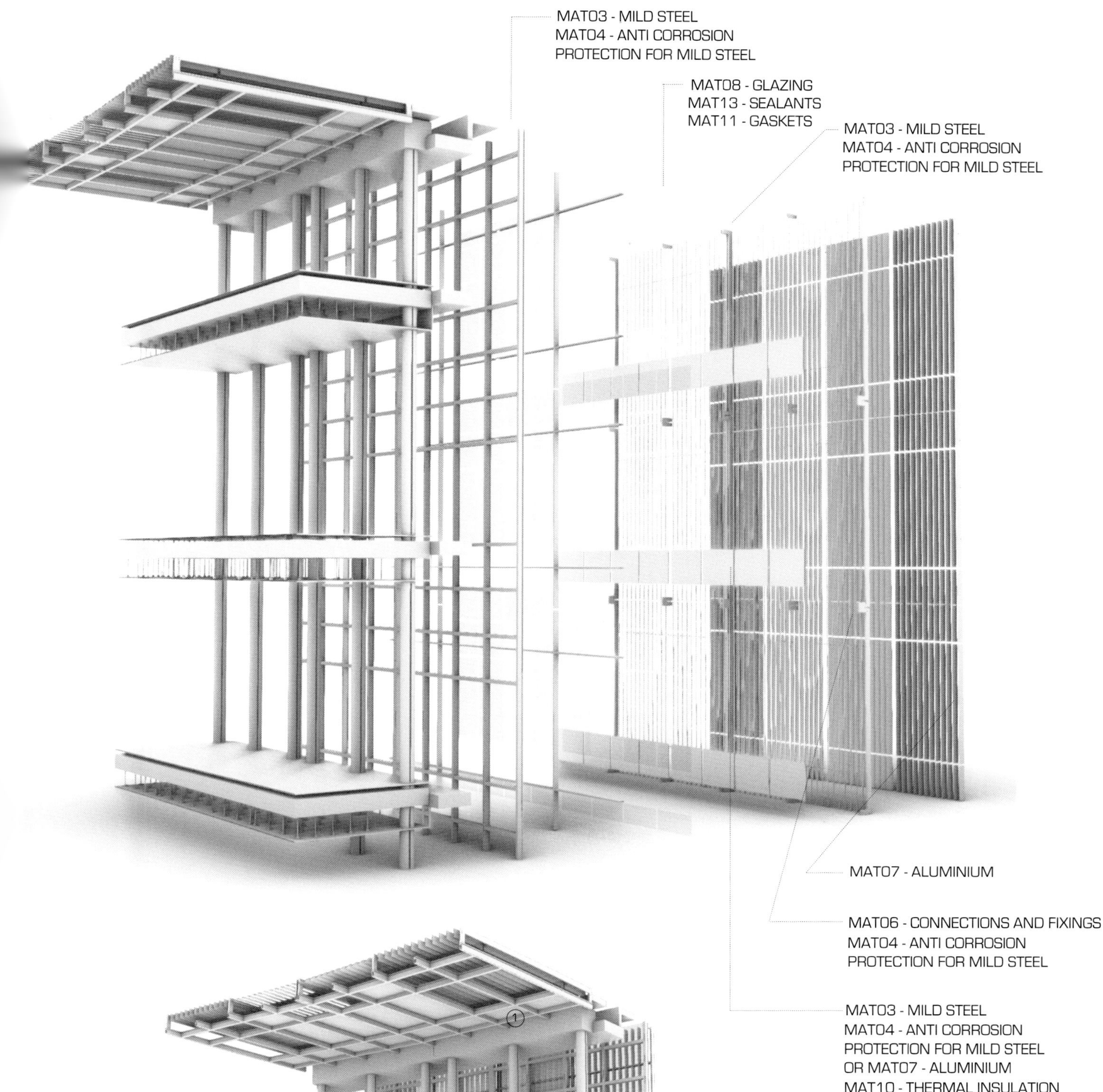

c | 3D exploded view of louvre screen facade components with material specification references

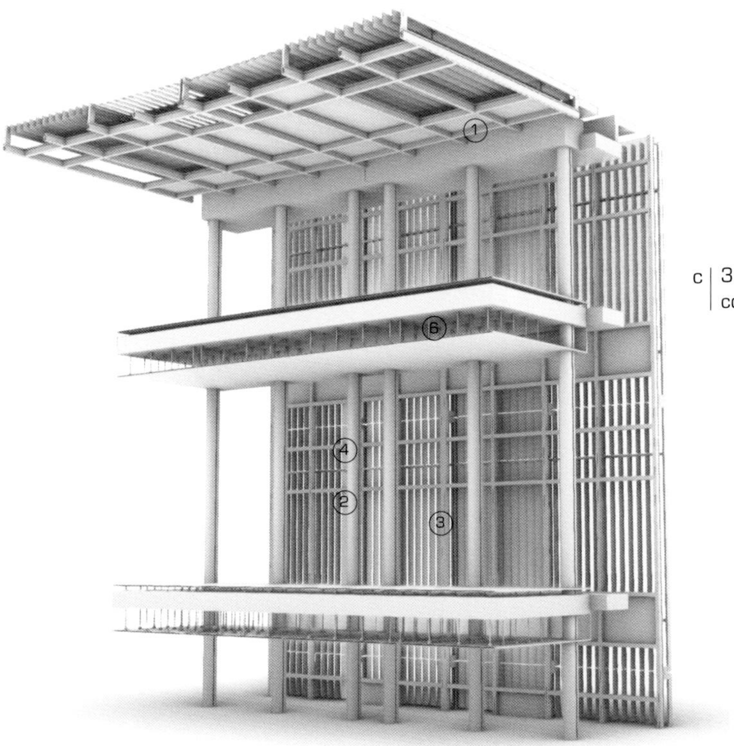

d | 3D view showing louvre screen facade fixed to steel frame

Example project of glazed facades
Evolution Tower, Moscow, Russia
Facade Engineer to RMJM: Newtecnic
Architect: RMJM, Philipp Nikandrov

GLASS WALLS

07 **Stick glazing:**
System assembly
Framing profiles
Opening lights
Corners
Spandrel panels

08 **Unitised glazing**
Jointing panels
Opening lights
Corner panels, parapets and sills
Silicone-bonded glazing

09 **Clamped glazing**
Patch plate glazing
Clamped glazing
Opaque glazing
Sealing clamped glazing

10 **Bolt fixed glazing**
Support methods
Bottom-supported glazing
Top-hung glazing
Corners
Seals and interfaces

11 **Glass blocks**
Fixing glass blocks
Support frames and walls
Cast glass channels

12 **Steel windows**
Small-scale glazing
Large-scale glazing

13 **Aluminium windows**
Windows in openings
Window walls
Composite windows

14 **Timber windows**
Window walls
Window design
Windows in openings

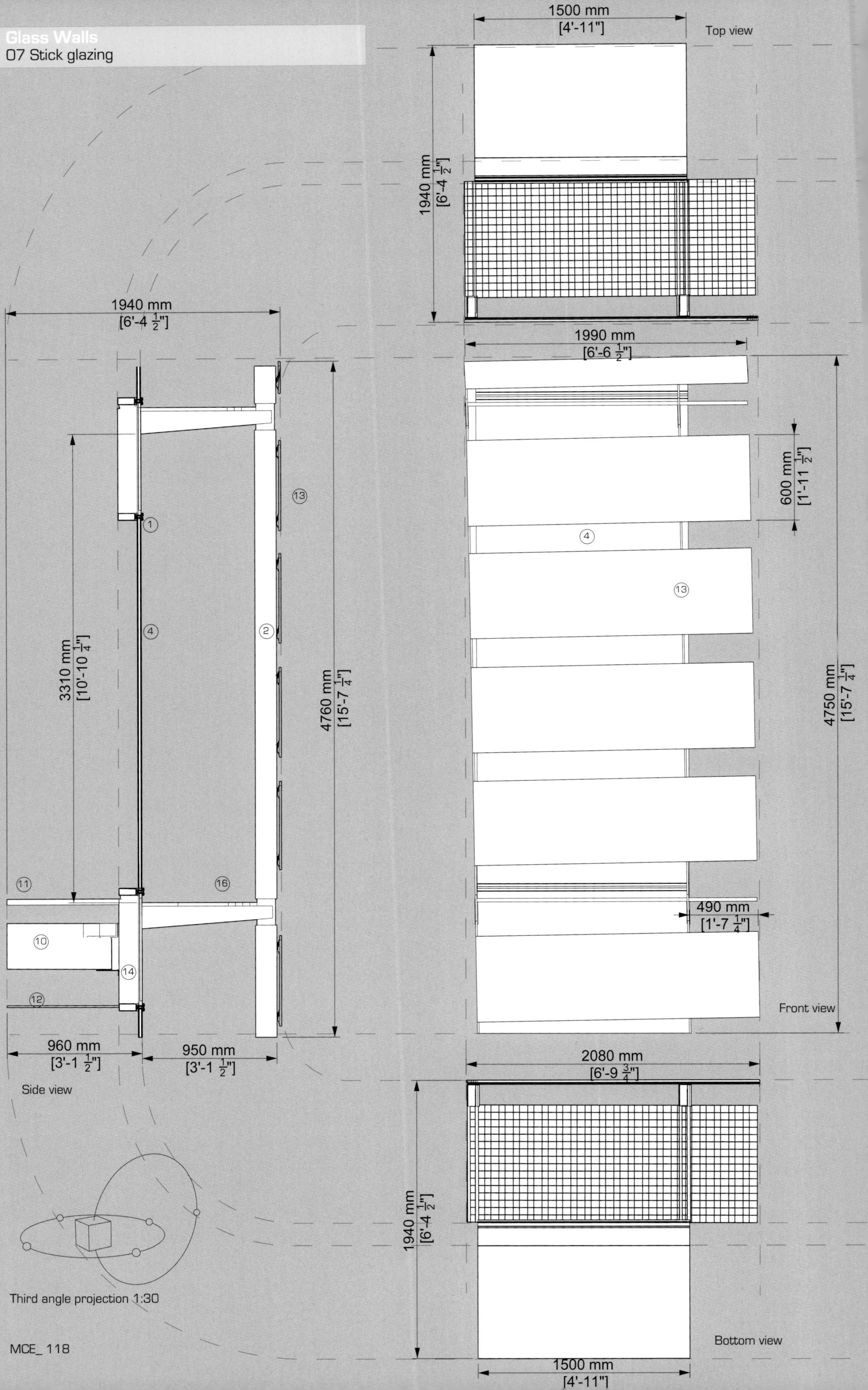

Third angle projection 1:30

3D view of the assembly

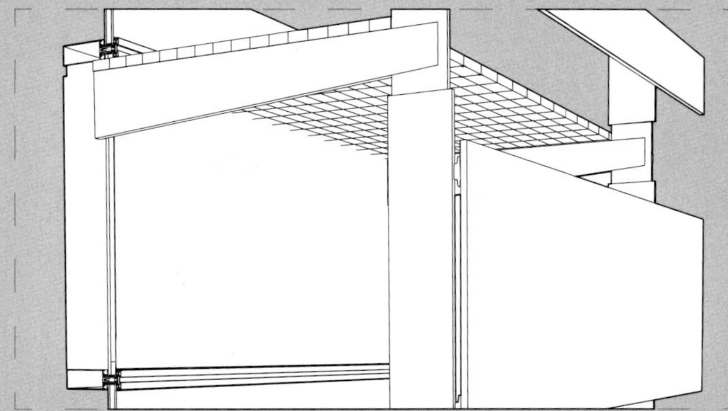

Details

1 | Extruded aluminium transom
2 | Extruded aluminium mullion
3 | Fixing bracket
4 | Single-glazed or double-glazed unit to suit application
5 | Pressure plate
6 | Rubber-based seal
7 | Thermal break
8 | Metal-faced or opaque glass-faced insulated panel
9 | Cover cap

Side view

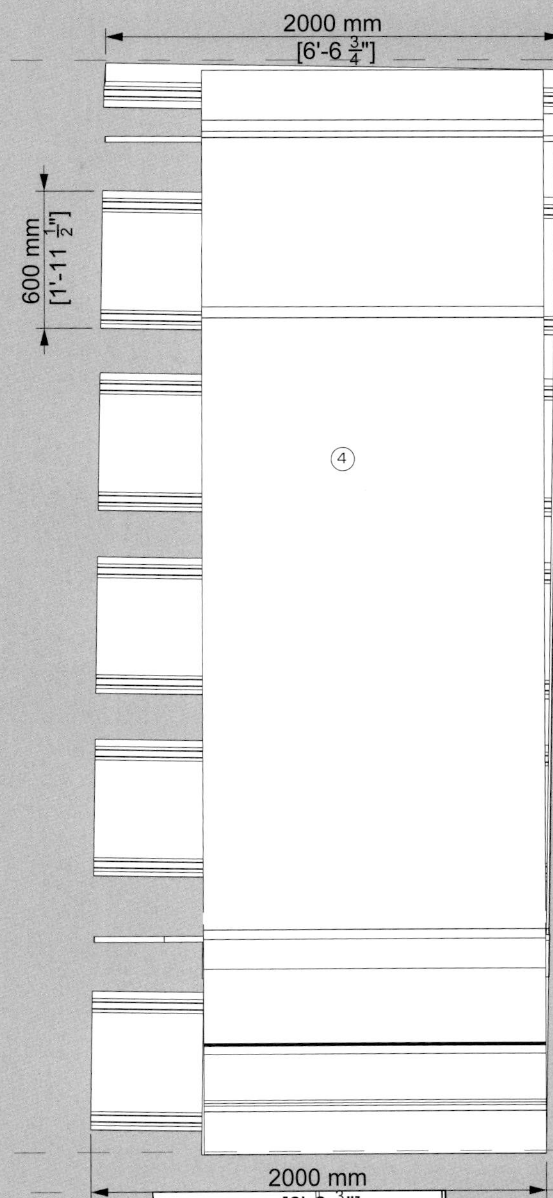

Back view

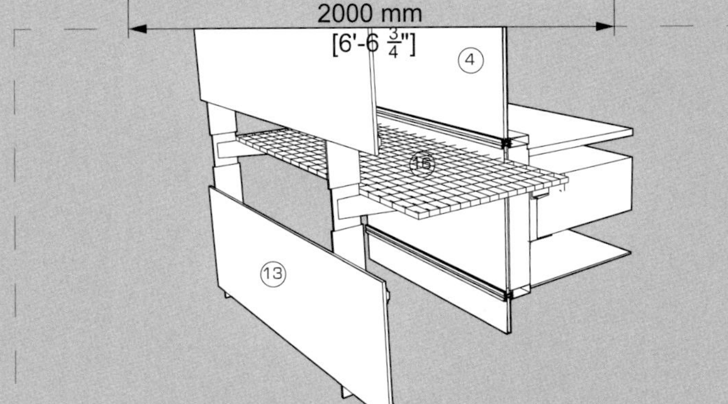

3D view of the assembly

10 | Floor slab
11 | Floor finish
12 | Ceiling finish
13 | Outer glazed screen providing solar shading
14 | Thermal insulation
15 | Metal sheet seal
16 | Maintenance access deck
17 | Window glazed in curtain walling
18 | Metal honeycomb panel
19 | Slot in mullion to receive fixing bracket for external screens, etc.
20 | Steel hollow section
21 | Adjacent wall. Metal rainscreen shown
22 | Smoke seal

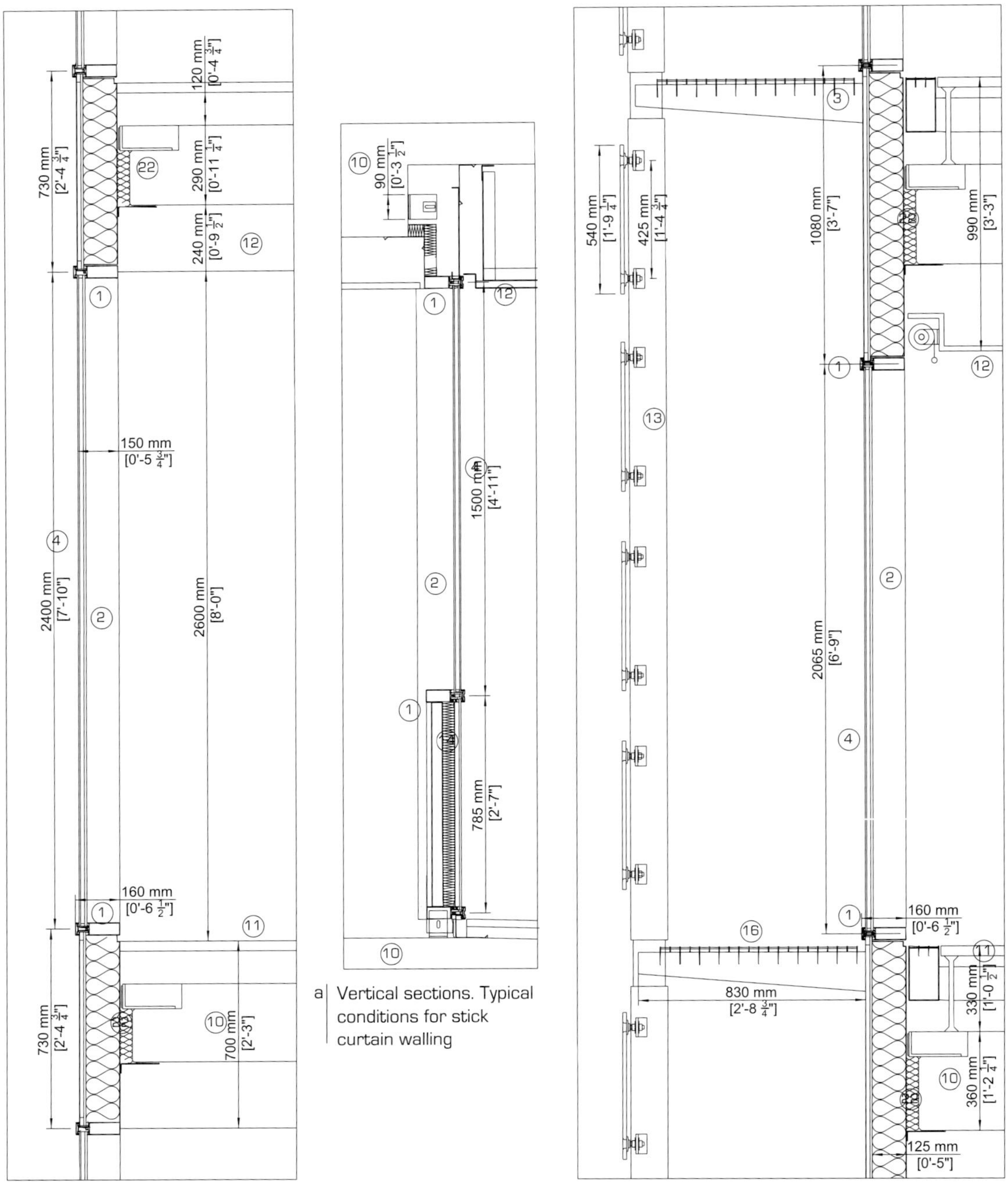

a Vertical sections. Typical conditions for stick curtain walling

Details
1 Extruded aluminium transom
2 Extruded aluminium mullion
3 Fixing bracket
4 Single-glazed or double-glazed unit to suit application
5 Pressure plate
6 Rubber-based seal
7 Thermal break
8 Metal-faced or opaque glass-faced insulated panel
9 Cover cap
10 Floor slab
11 Floor finish
12 Ceiling finish
13 Outer glazed screen providing solar shading
14 Thermal insulation
15 Metal sheet seal
16 Maintenance access deck
17 Window glazed in curtain walling
18 Metal honeycomb panel
19 Slot in mullion to receive fixing bracket for external screens, etc.
20 Steel hollow section
21 Adjacent wall. Metal rain screen shown
22 Smoke seal

Comparison with unitised glazing

Framed glazing systems are of two types: stick and unitised. Stick systems are assembled mainly on site, while unitised systems are assembled in a factory. Stick systems are well adapted for non-modular construction. While unitised systems require a repetition in panel sizes in order to keep the types to a small number in order to remain economic (MCE_123, e), stick systems allow a high degree of freedom in module size and in facade design. Mullions and transoms do not need to be continuous; glazing bars can set out in staggered grids and can be changed easily from smaller grid sizes to larger ones. Complex geometries can be taken up much more easily by stick systems than by unitised systems. Stick systems are often preferred for low-rise building, where scaffolding is used, but the increased use of mast climbers (moving platforms) is making stick systems viable for taller buildings (10 to 20 storeys) where unitised glazing would otherwise have been used. The increased dependence on mast climbers is in part due to the need to be independent of site cranes, which are increasingly needed to service the construction of the primary structure often being built at the same time, though usually several floors higher or several bays away from the glazed walling.

For low-rise buildings, or where there is a high degree of variation in the facade module, stick glazing is often preferred, since the wall is assembled in place on site rather than in a workshop, making it very economic when compared to unitised glazing. Although off-site fabrication saves time and can be of higher quality, it is often more expensive. Sometimes mullions and transoms are pre-assembled into carrier frames that are lifted into place and fixed without glass. This 'semi-unitised' approach can save time on construction where there is some degree of repetition.

A criticism of stick systems has traditionally been of their poorer quality of assembly when compared to unitised glazing, but this is much less the case today. However, bringing all the components together at the site - of

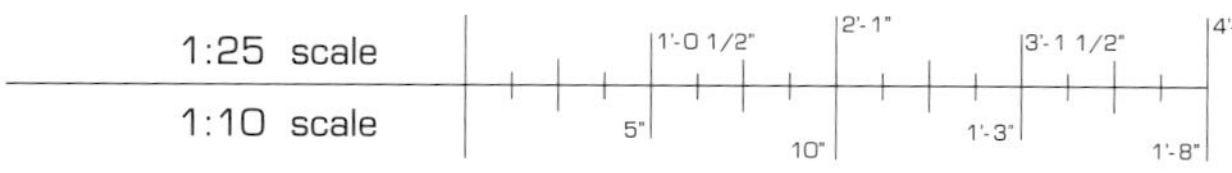

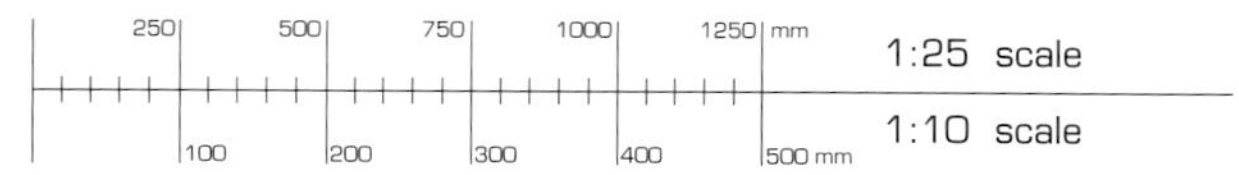

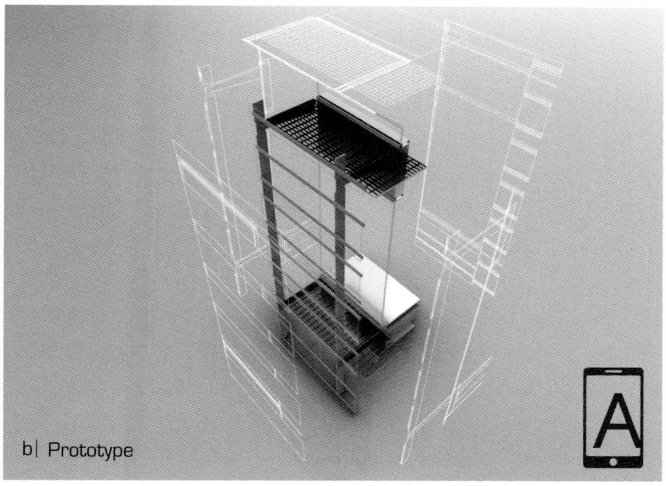

b| Prototype

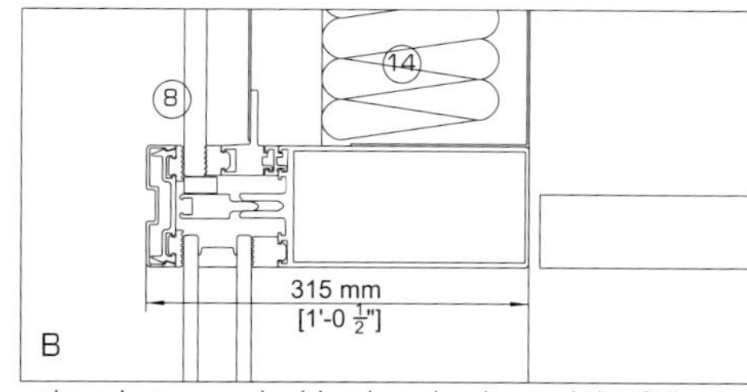

c| Vertical sections. Junctions between double glazed units and the following:

A | Another double glazed unit
B | Glass spandrel panel
C | Metal honeycomb panel
D | Inward opening window

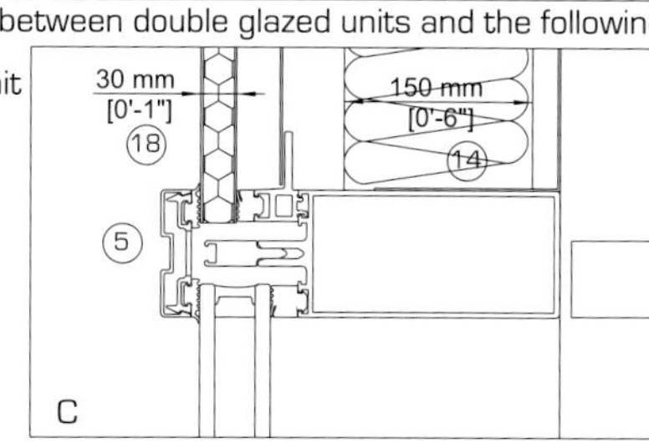

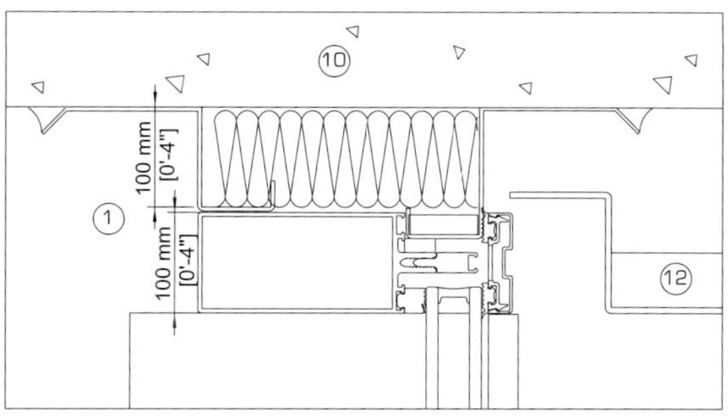

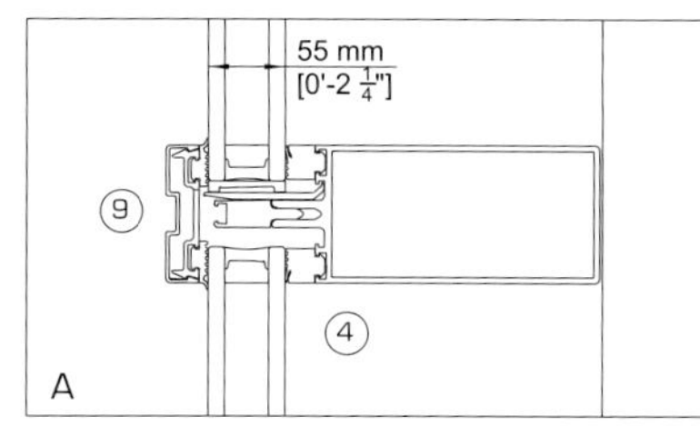

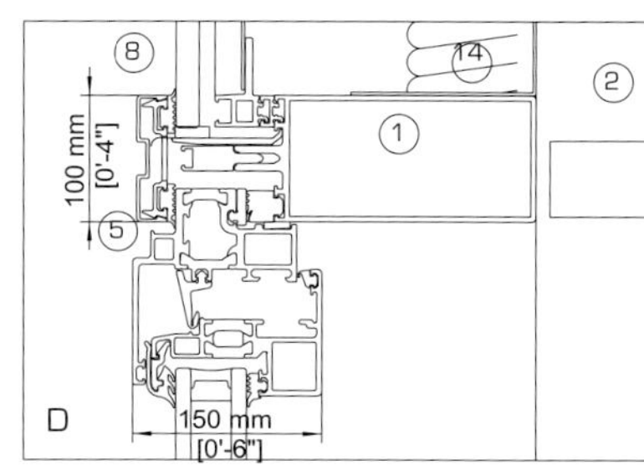

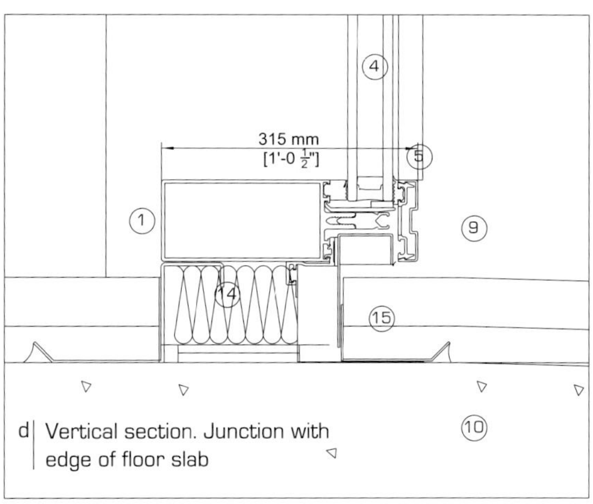

d| Vertical section. Junction with edge of floor slab

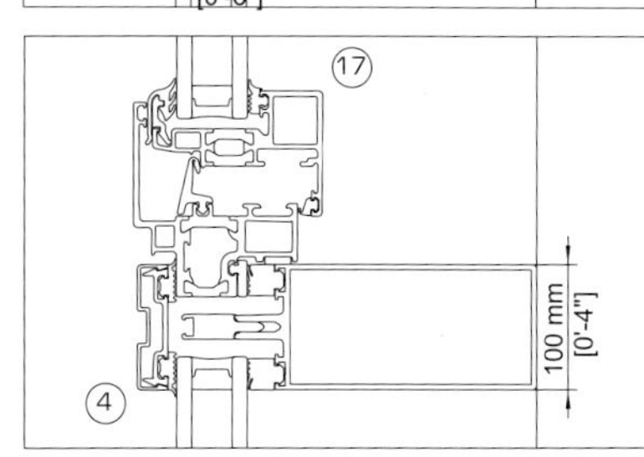

double or single glazed units, aluminium profiles, rubber-based gaskets and seals, folded metal flashings and copings, involves a much higher dependence on site-based work to achieve the quality of construction of the factory-based unitised systems.

System assembly

An essential aspect of stick systems is that they should be drained and ventilated in order to avoid water being drawn through the rubber-based seals into the building. Rainwater penetrating the outer seal is drained away in a ventilated zone that provides pressure equalisation between outside and inside the system (c). Pressure equalisation avoids water being drawn into the system by a pressure difference between two chambers, resulting in water being drawn through a joint. Any rainwater entering this zone is drained away to the outside forward of the inner seal forming a second line of defence against air and water infiltration. Most stick systems now provide a full thermal break through the aluminium profiles rather than the partial thermal breaks provided on previous systems. This reduces condensation risk on the inside face of the framing in temperate climates. In hot and humid climates, the condensation will occur harmlessly on the outside face of the framing if the interior is cooled by mechanical ventilation. In all climates, a thermal break improves the U-value of the glazed wall, thus reducing energy consumption within the building for either heating or cooling.

Stick glazing is assembled mainly on site. Mullions (verticals) are fixed to floor slabs with transoms (horizontals) spanning between the mullions to which they are fixed (b). These framing members can be pre-assembled into 'ladders' on the ground and lifted into place by crane in order to reduce time on site. The glass is set in place and the pressure plates fixed through thermal breaks back to the carrier frame of mullions and transoms (MCE_122, a). Decorative cappings are usually applied to the pressure plates to conceal them as well as the self tapping screws holding the

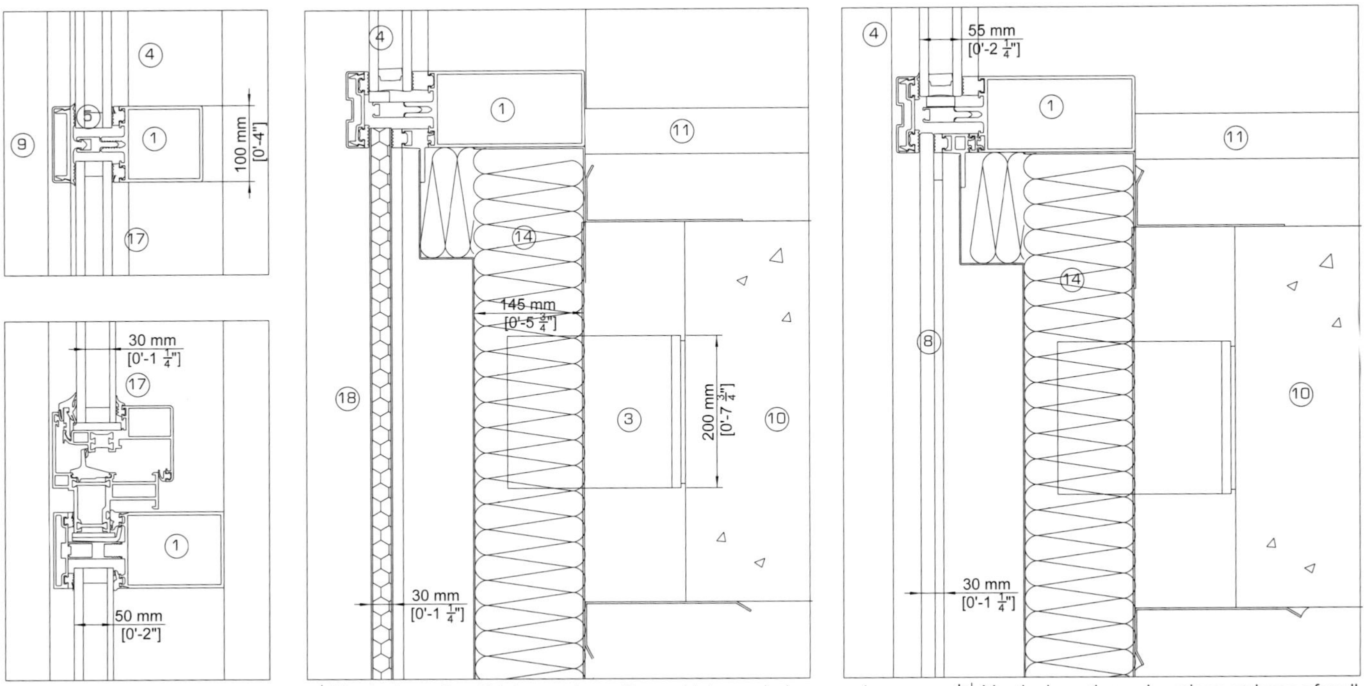

a| Vertical section. Stick curtain walling set into a window opening

b| Vertical sections. Junctions at base of wall

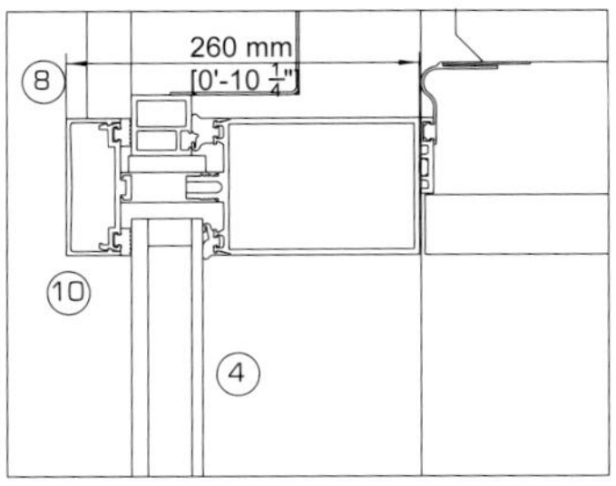

c| Vertical sections.

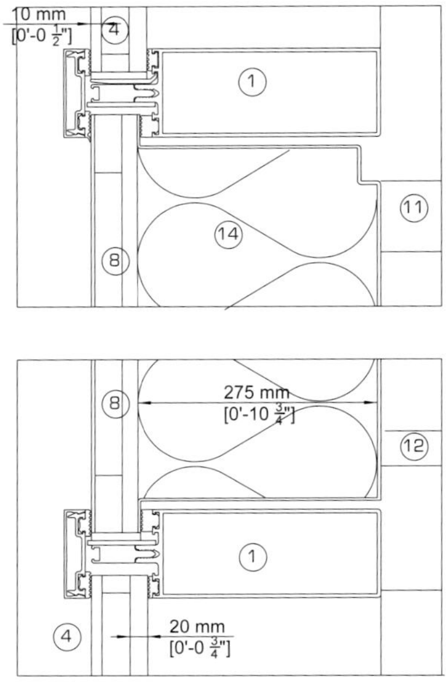

Details

1 | Extruded aluminium transom
2 | Extruded aluminium mullion
3 | Fixing bracket
4 | Single-glazed or double-glazed unit to suit application
5 | Pressure plate
6 | Rubber-based seal
7 | Thermal break
8 | Metal-faced or opaque glass-faced insulated panel
9 | Cover cap
10 | Floor slab
11 | Floor slab
12 | Ceiling finish
13 | Outer glazed screen providing solar shading

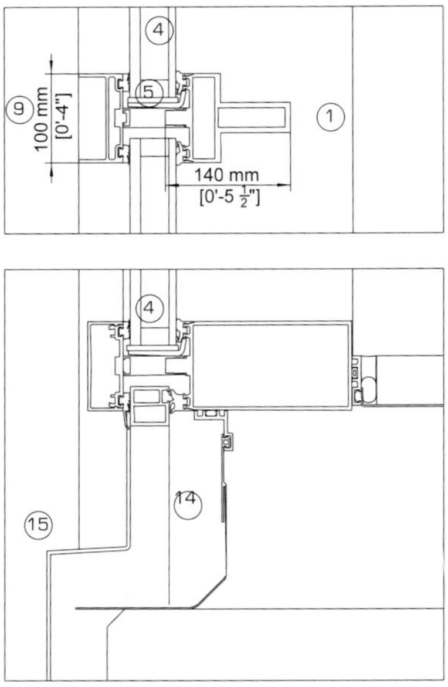

glass in place. Sometimes the cappings are omitted, but care must be taken to ensure that the screws are properly aligned and that the pressure plates are continuous and jointed carefully.

Stick systems are fixed from floor to floor at either side of the floor slab or on top (e). The design of the connection is dependent mainly on the method of forming the edge of the floor construction on floor depths or uses of the adjacent floor zone, and any additional elements such as brackets for solar shading or maintenance walkways which penetrate the stick glazing. The glazing is either hung from each floor level or sits on each floor level. Hanging is usually preferred but constraints imposed by the structure of the building may require the glazing to be supported at the base of each mullion. Mullions are linked by spigots so that the restrained end of one mullion fits into the supported end of the other mullion.

Because glazed walling is set forward of the edge of the slab, the gap between the two is closed with a floor finish. Seals between adjacent floors are designed as either smoke seals or as flame barriers. Smoke seals comprise a mineral wool or glass wool barrier, held in place and sealed with galvanised steel sheet above and below. Flame barriers, usually of one hour or 90-minute rating are provided by a specially protected spandrel panel at the level of the floor zone. The spandrel panel forming the barrier is fixed directly to the floor slab as well as forming part of the curtain wall system (a). This ensures that the spandrel remains in place long after the adjacent curtain walling has collapsed during a fire. The spandrel panel itself is usually protected with a fire resistant board as part of the barrier.

Where glazing is inclined significantly from the vertical, two-edge glazing is used to allow the water to run down unimpeded with transoms. This is a standard method for glazed roofs, where mullions are capped, but transoms running across the width of the slope are sealed with a flush silicone joint.

14 | Thermal insulation
15 | Metal sheet seal
16 | Maintenance access deck
17 | Window glazed in curtain walling
18 | Metal honeycomb panel
19 | Slot in mullion to receive fixing bracket for external screens, etc.
20 | Steel hollow section
21 | Adjacent wall. Metal rainscreen shown
22 | Smoke seal

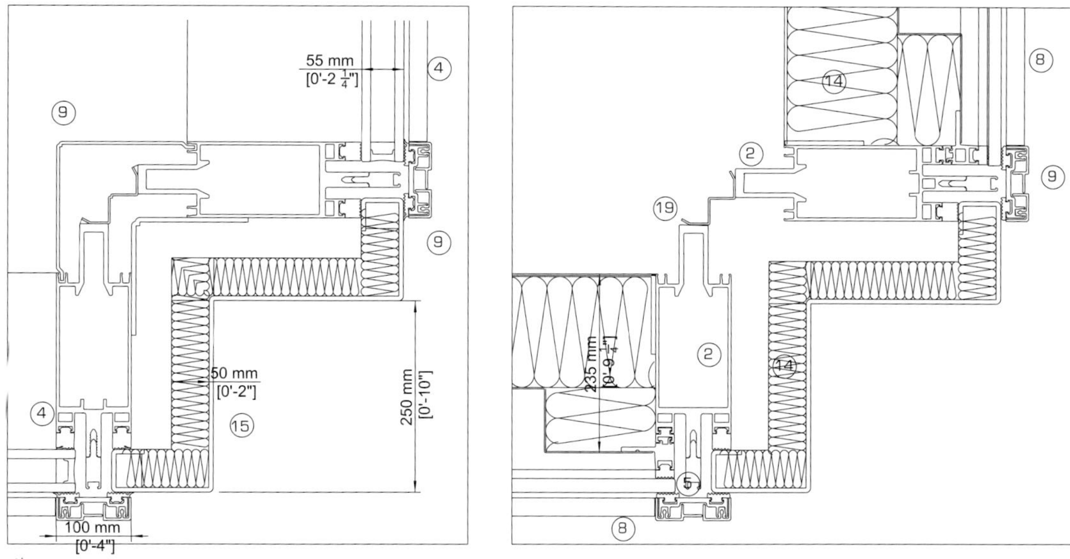

d| Horizontal sections.

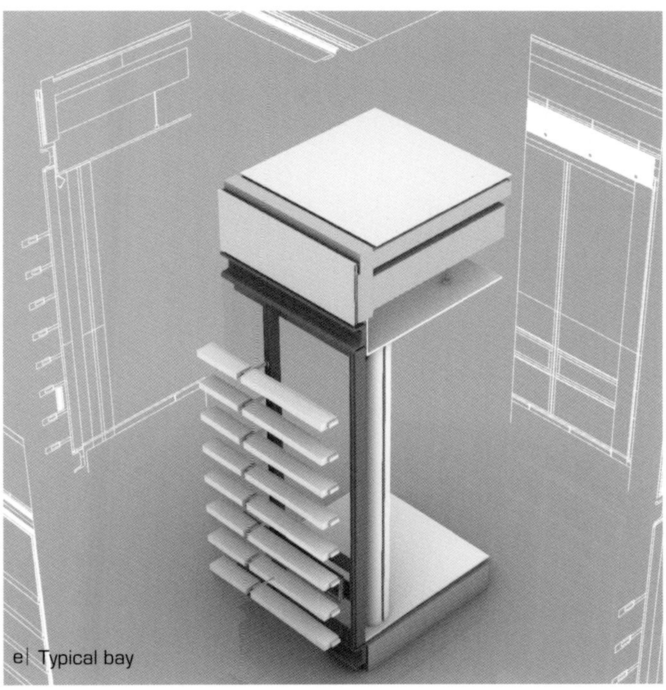

e| Typical bay

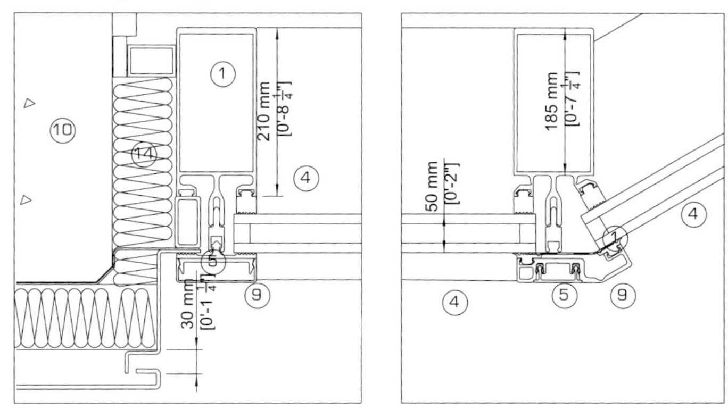

f | Horizontal sections.

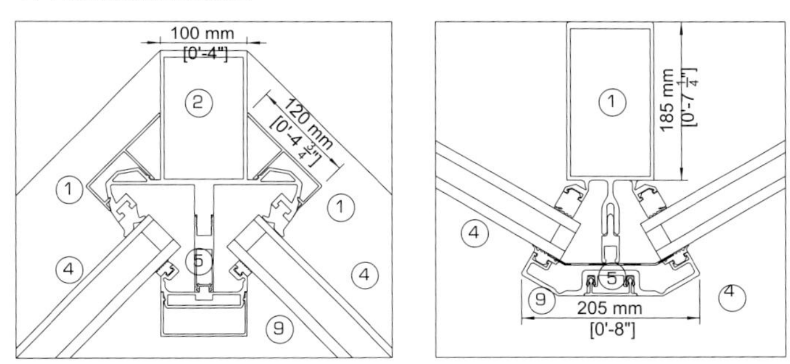

Framing profiles

Carrier frames are made from a wide variety of sections. The shape can usually be adapted for a specific project. The most common types are rectangular box sections that use the full width of the joint, and narrower T-shaped and I-shaped sections that are narrower than the joint width. Although the depth of the profile is determined by structural requirements, the overall shape can be adapted to suit other needs, such as incorporating roller blind guides into the sides of the vertical profiles. Mullions and transoms are often not of the same depth for structural reasons, but can be made so if required to support blinds, for example, or for visual reasons.

Where framing members have long vertical or horizontal spans, mild steel sections are sometimes used instead of a rectangular box section in aluminium, particularly if the equivalent aluminium section would be visually too deep or if the glazing system is fixed back to a steel frame forming part of the primary structure. The front part of the extrusion containing the seals is fixed directly to a mild steel box or T-section. The glazing is then fixed to this extrusion in the conventional way (MCE_124, a). If the steel frame has variable or complex curves then the rubber-based seals are set onto the steel frame without an extrusion with a technique commonly used in glazed roof structures. The glass is fixed with pressure plates directly to the steelwork with a rubber-based profile set between the glass and supporting steel.

Corners

Both internal and external corners are formed by glazing a folded aluminium strip into the mullion on each side of the corner. Alternatively, a mullion set at 45º is used to give a thin edge to the facade, with a joint width similar to that of joints elsewhere on the facade. Some manufacturers provide interlocking mullions for use at corners to allow varying angles at corners on a single building with a constant abutting of mullions on the inside face.

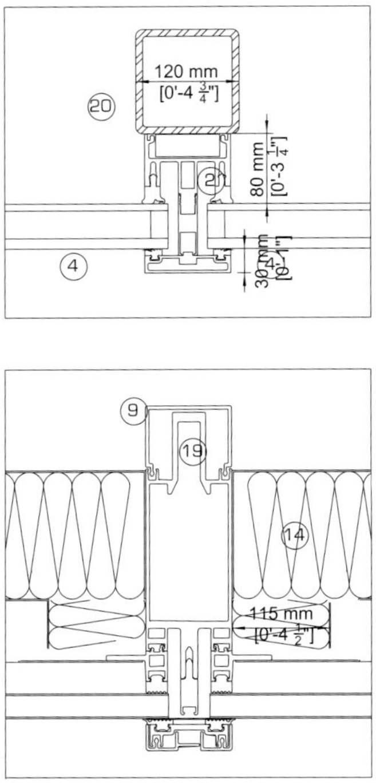

a| Horizontal sections.

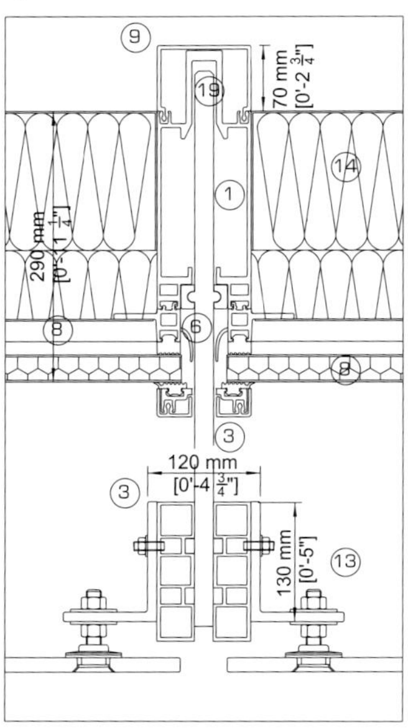

Details

1 | Extruded aluminium transom
2 | Extruded aluminium mullion
3 | Fixing bracket
4 | Single-glazed or double-glazed unit to suit application
5 | Pressure plate
6 | Rubber-based seal
7 | Thermal break
8 | Metal-faced or opaque glass-faced insulated panel
9 | Cover cap
10 | Floor slab
11 | Floor finish
12 | Ceiling finish
13 | Outer glazed screen providing solar shading
14 | Thermal insulation
15 | Metal sheet seal
16 | Maintenance access deck
17 | Window glazed in curtain walling

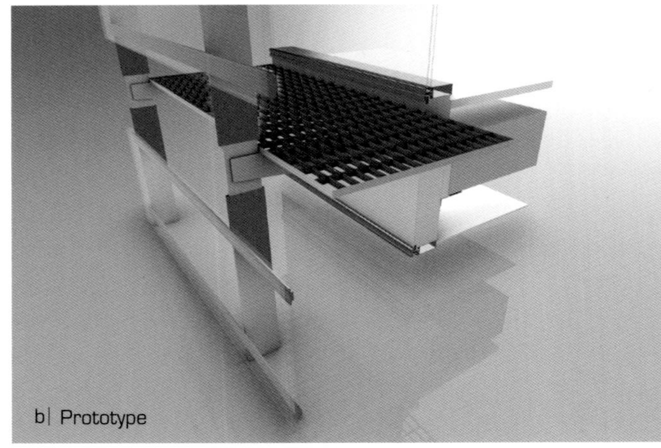

b| Prototype

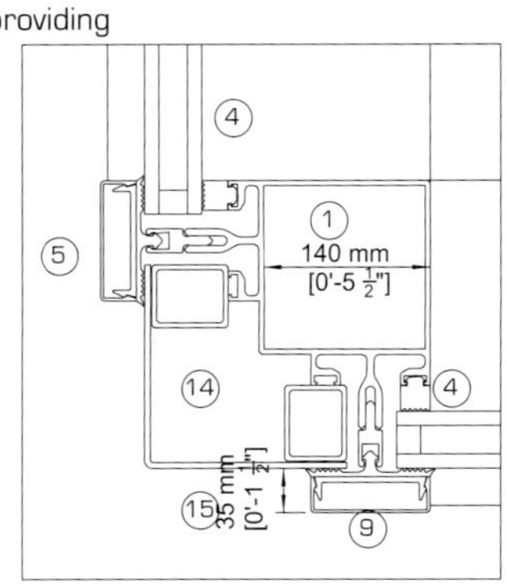

b| Horizontal sections. External corner

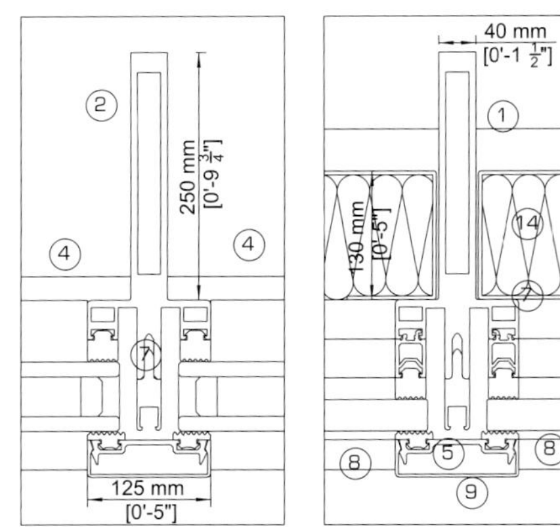

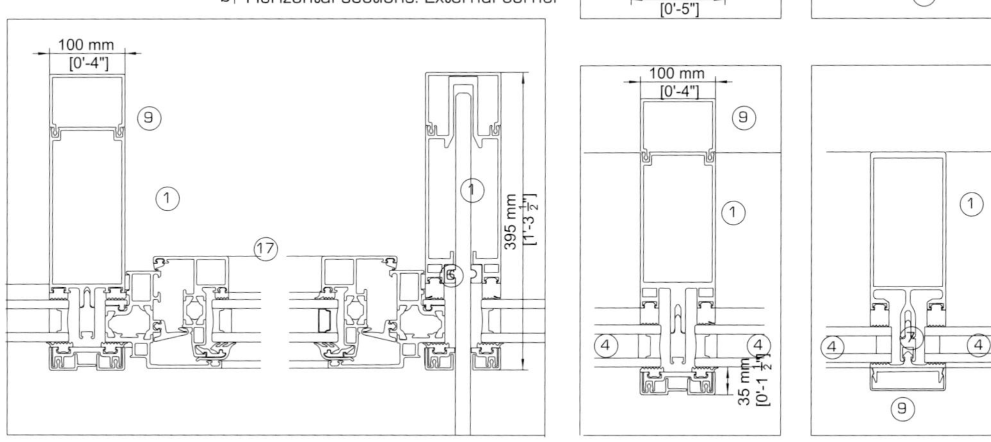

Opening lights

Windows, doors and smoke vents are fixed into stick glazing as items with their own frame rather than using the curtain wall framing directly. This is because the opening light usually needs its own drainage and weatherproofing profiles and seals. This gives the appearance of a thicker frame around opening lights than around the adjacent fixed lights.

Drips and seals form a part of the secondary frame in the same way as if the light were glazed into any other form of construction such as an opening in a masonry wall. The opening light is glazed into curtain walling around its edge with a thinner frame that corresponds to the thickness of the glazed units, allowing it to be fixed in place with the same technique as if it were a double glazed unit. These edge nibs are positioned on the frame to the opening light in a way that ensures that the glass of the opening light is in the same plane as the adjacent glass (MCE_121, c).

Parapets, sills and penetrations

Parapet copings are glazed into the curtain walling with a pressure plate in the same way as adjacent glazing. Copings usually project to align with the face of the cappings on the pressure plates or slightly forward of them to protect the glazing beneath from the vertical movement of maintenance cradles. Copings slope inwards towards a gutter rather than sloping forwards to avoid dirt accumulating on the top of the coping being washed down the facade by rain. This also avoids the need for the coping to project beyond the face of the glazing and form a drip. It should be noted that projecting drips on copings is still good practice in masonry construction, where an impervious coping washes rainwater onto a permeable masonry material beneath, causing staining. Sills at the base of stick glazing are formed with a pressed aluminium profile with a clip at the bottom which provides a clean line at the bottom of the wall, and provides rigidity to the profile (d). Where sill drips can be seen from below, the

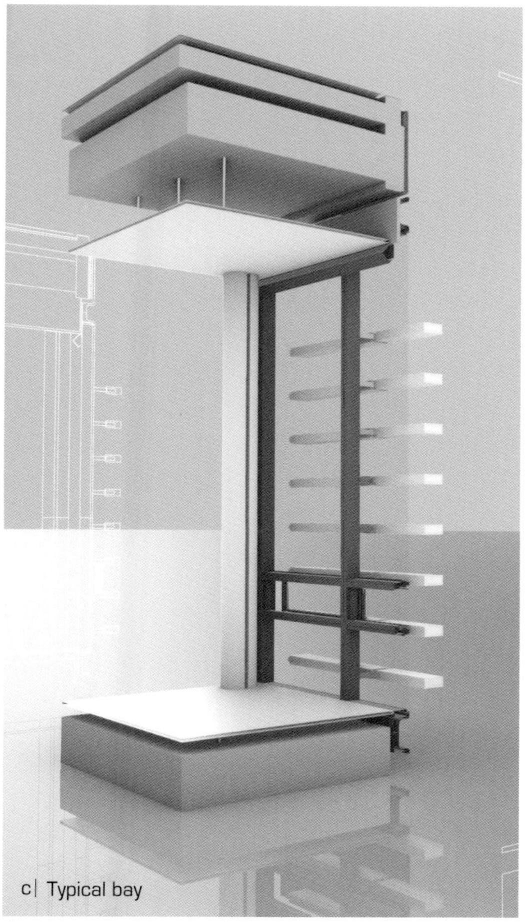
c| Typical bay

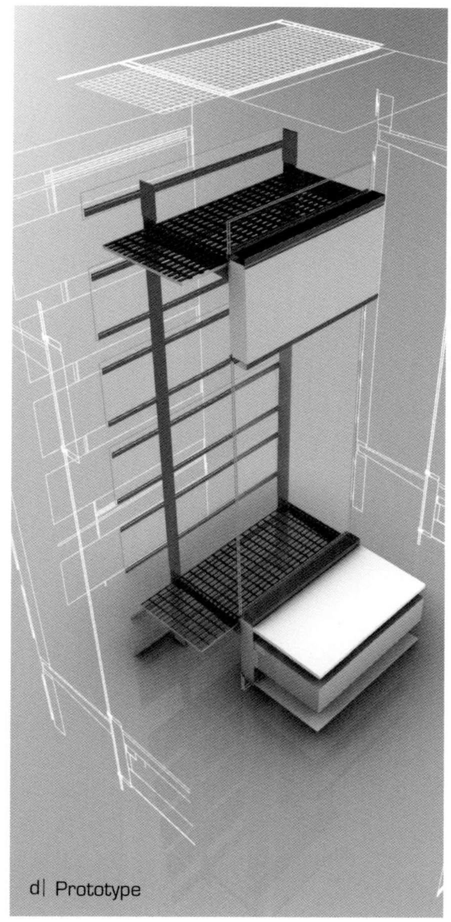
d| Prototype

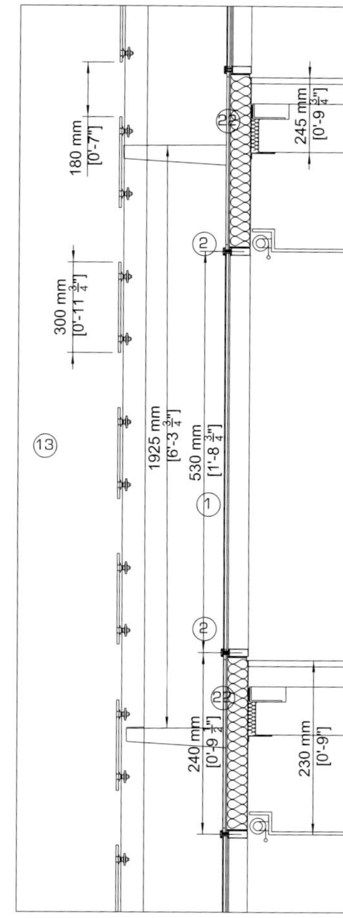

e| Horizontal and vertical section.
Stick system with additional outer glazed screen to provide solar shading or as part of a twin wall facade

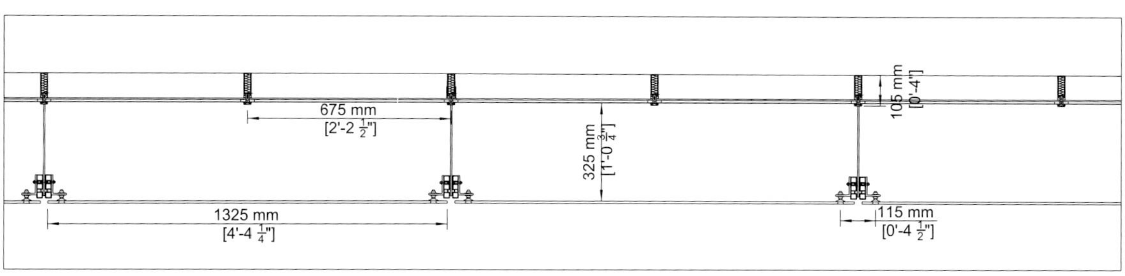

18 | Metal honeycomb panel
19 | Slot in mullion to receive fixing bracket for external screens, etc.
20 | Steel hollow section wall
21 | Adjacent wall. Metal rain screen shown
22 | Smoke seal

folded edge provides a smooth painted edge which is also protected from weather corrosion.

Where stick glazing abuts an adjacent area of wall in a different material, a blocking profile is glazed into the edge of the glazing profile, faced with an EPDM foil. The foil is then bonded to the adjacent wall. Where glazing abuts metal rainscreens, which is a common combination, the edge of the metal return to the rainscreen can also be glazed into the edge mullion or transom (MCE_123, f). Where a projecting sill is required at the base of the wall, for example, a projecting aluminium sill is glazed into the curtain walling at the top end, and projects down over the adjacent wall in a manner to suit the detailing of the wall beneath.

Spandrel panels

Spandrel panels are made either as a continuous sealed panel, draining in the same way as the glass panels, or as a ventilated box (MCE_122, b). Where metal is used, spandrels can be formed as trays glazed into the framing with insulation between. Glazed spandrels are made either with rigid insulation bonded to the back of laminated glass or on a sheet of laminated glass with a ventilated void behind both to cool the glass and avoid either a visual read-through of the insulation from the outside or any risk of the insulation delaminating from the glass. When glass is used it is made opaque (with either method) by screen printing, etching or a combination of both.

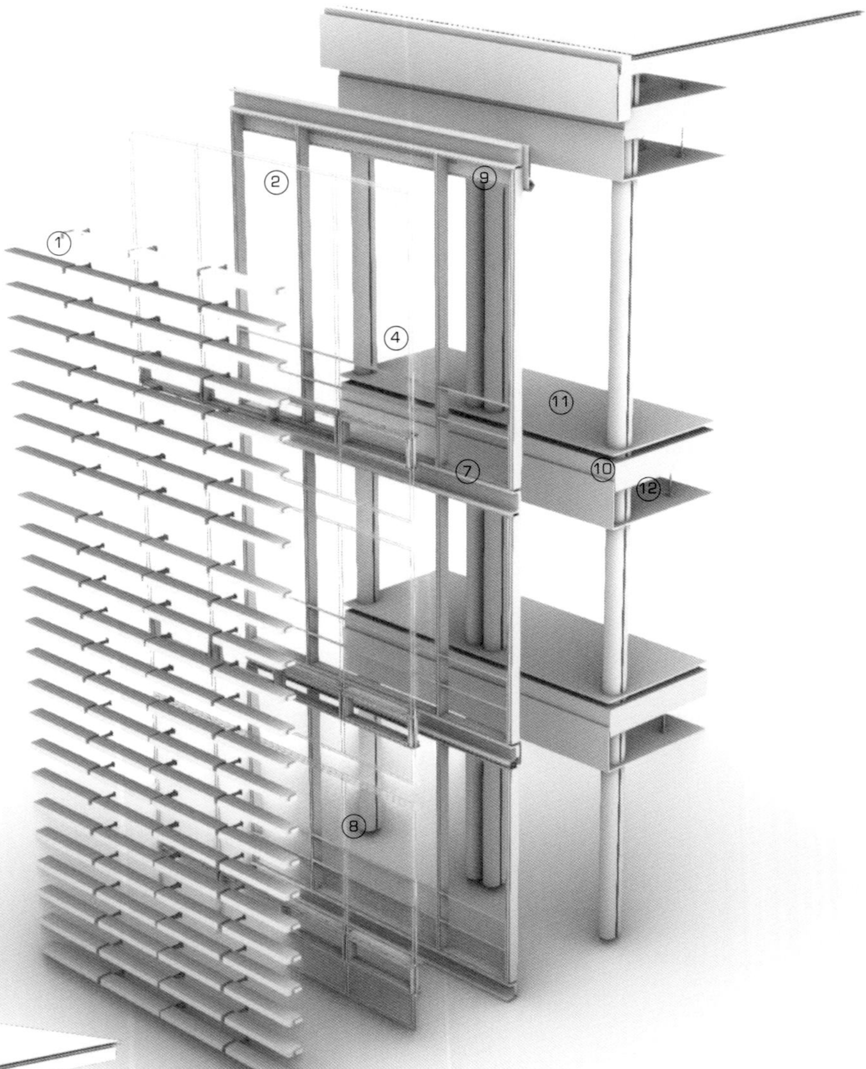

a | 3D exploded view showing stick glazed facade fixed to steel frame

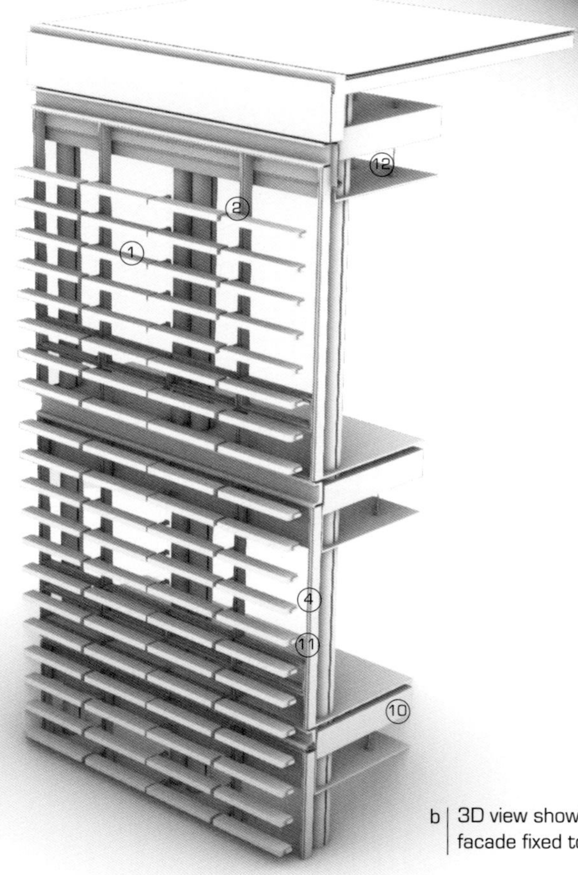

b | 3D view showing stick glazed facade fixed to steel frame

Details

1 | Extruded aluminium transom
2 | Extruded aluminium mullion
3 | Fixing bracket
4 | Single-glazed or double-glazed unit to suit application
5 | Pressure plate
6 | Rubber-based seal
7 | Thermal break
8 | Metal-faced or opaque glass-faced insulated panel
9 | Cover cap

c | 3D exploded view of stick glazed facade components with material specificification references

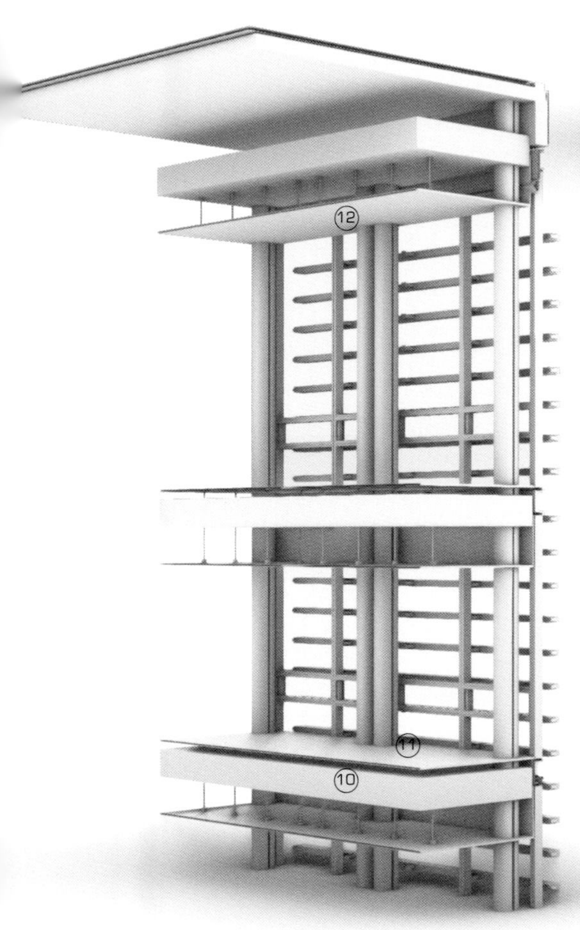

10 | Floor slab
11 | Floor finish
12 | Ceiling finish
13 | Outer glazed screen providing solar shading
14 | Thermal insulation
15 | Metal sheet seal
16 | Maintenance access deck
17 | Window glazed in curtain walling
18 | Metal honeycomb panel
19 | Slot in mullion to receive fixing bracket for external screens, etc.
20 | Steel hollow section
21 | Adjacent wall. Metal rainscreen shown
22 | Smoke seal

d | 3D view showing stick glazed facade fixed to steel frame

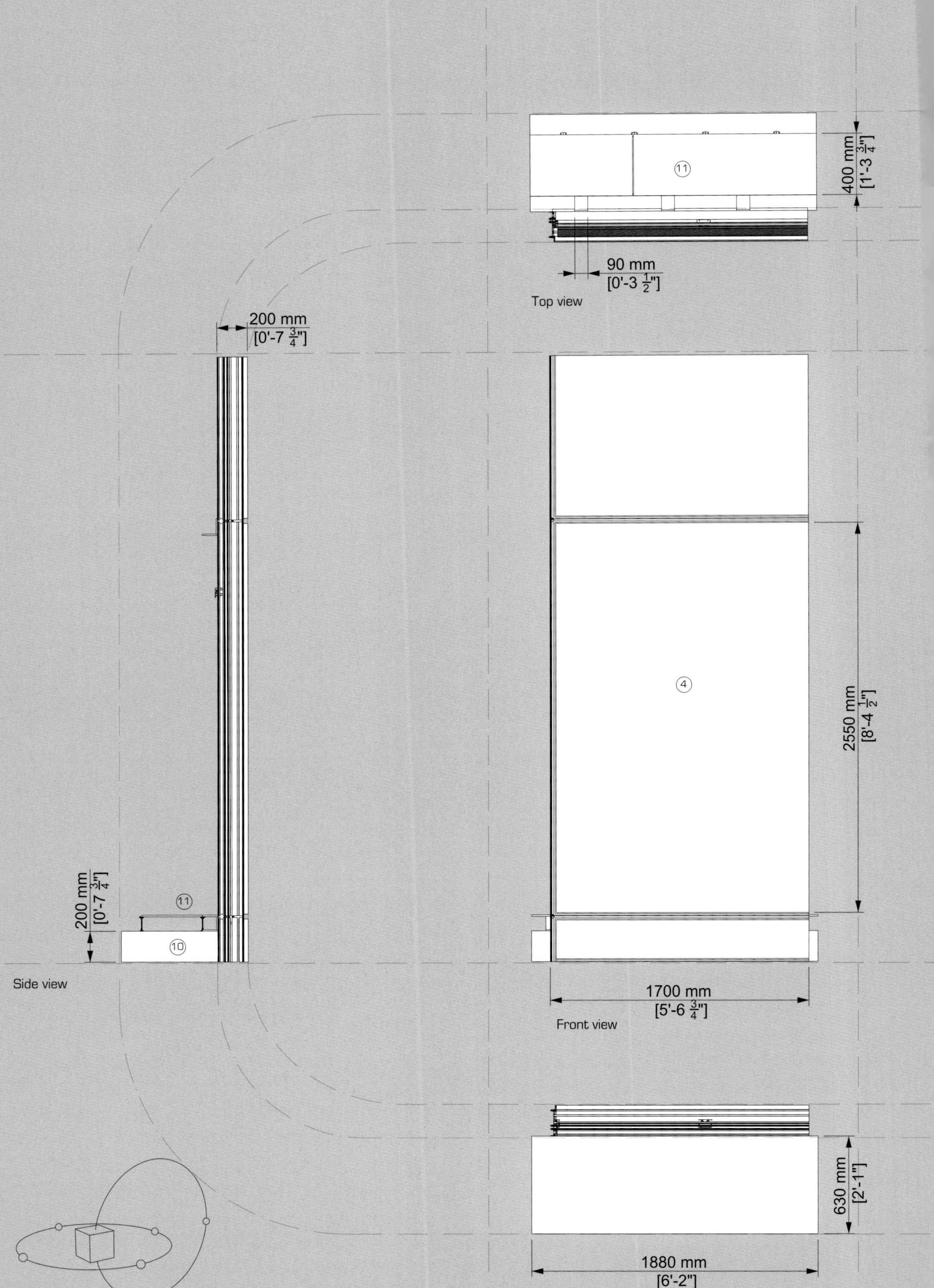

Third angle projection 1:30

Details

1 | Interlocking transom
2 | Plain transom
3 | Interlocking mullion
4 | Single-glazed or double-glazed unit to suit application
5 | Pressure plate
6 | Rubber-based seal
7 | Thermal break
8 | Metal parapet coping
9 | Metal-faced or opaque glass-faced insulated panels

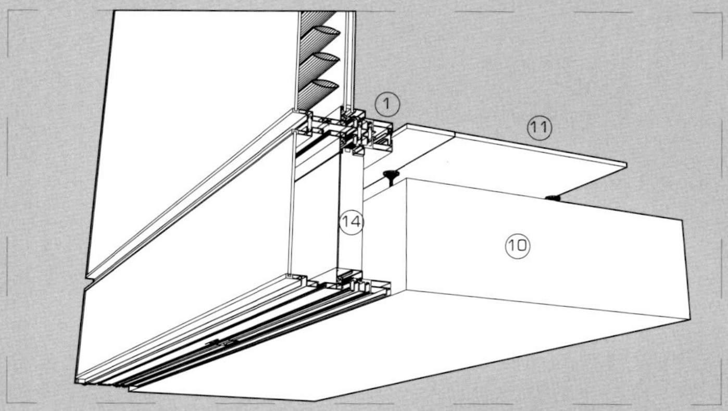

3D view of the assembly

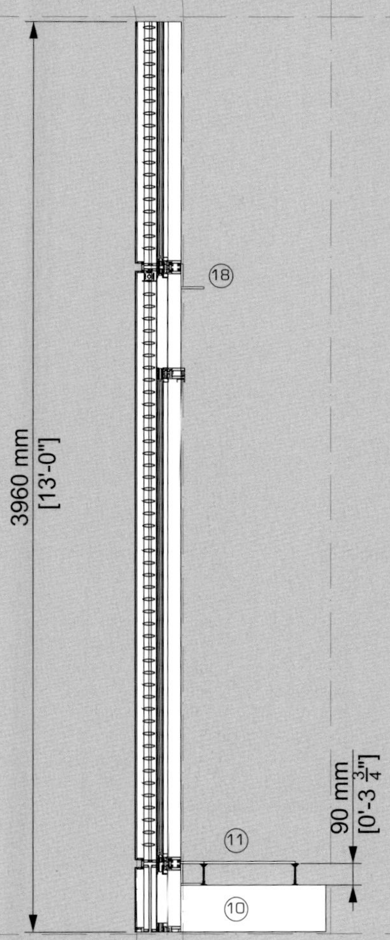

Side view

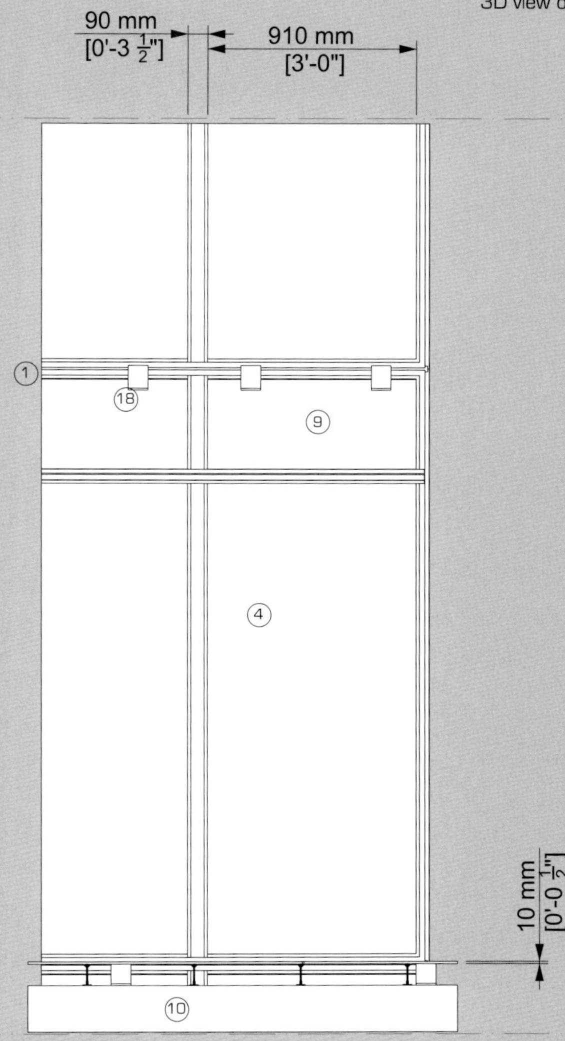

Back view

10 | Floor slab
11 | Floor finish
12 | Ceiling finish
13 | Outer screen providing solar shading
14 | Thermal insulation
15 | Sheet metal seal
16 | Cover cap
17 | Smoke seal
18 | Support bracket

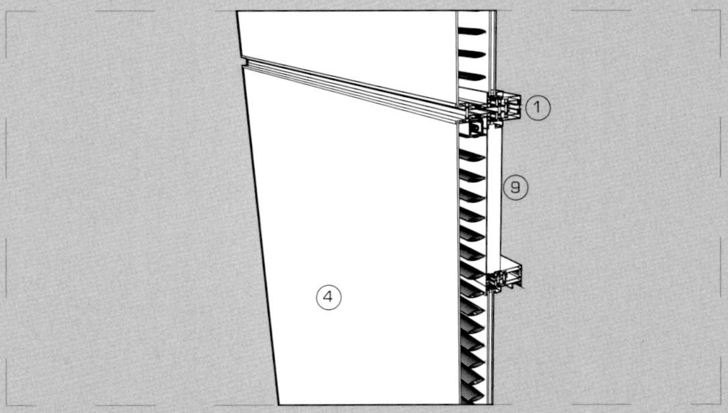

3D view of the assembly

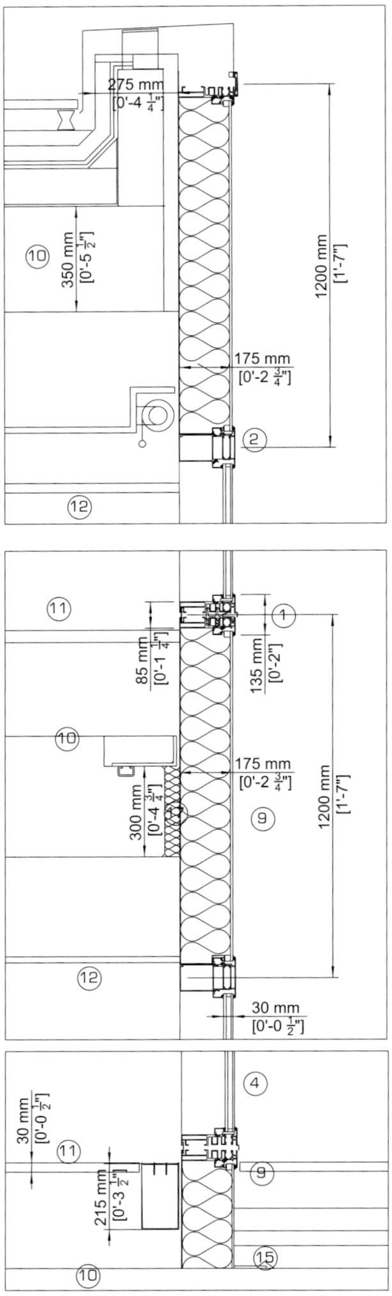

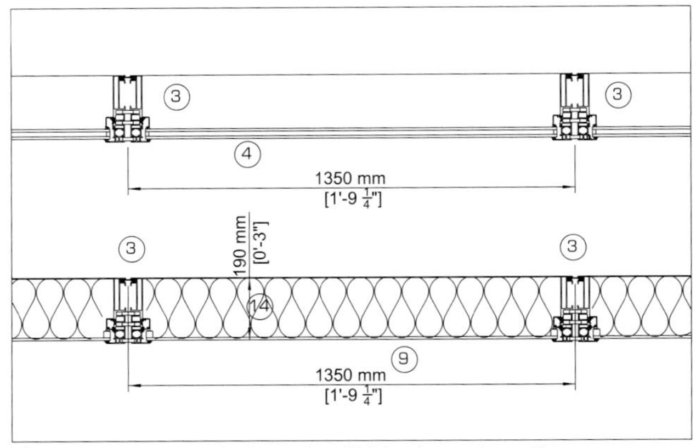

a| Horizontal and vertical section. Unitised curtain walling

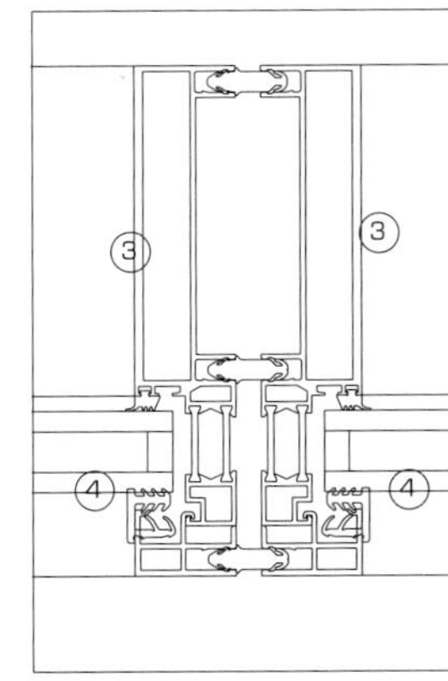

b| Horizontal section. Fully-unitised glazing

Details

1 | Interlocking transom
2 | Plain transom
3 | Interlocking mullion
4 | Single-glazed or double-glazed unit to suit application
5 | Pressure plate
6 | Rubber-based seal
7 | Thermal break

c| Horizontal sections. Semi-unitised glazing

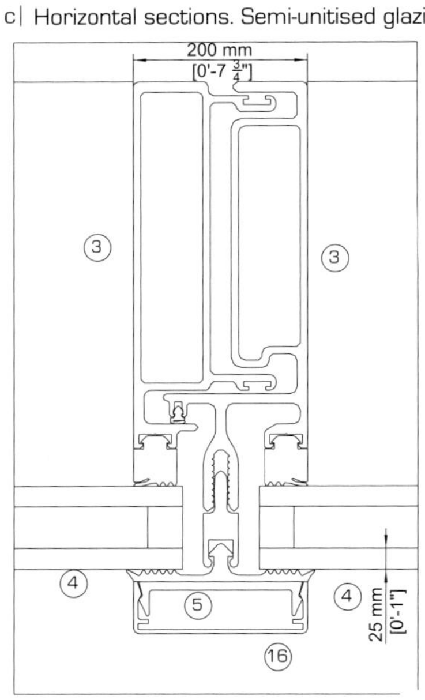

d| Horizontal section. Fully-unitised glazing

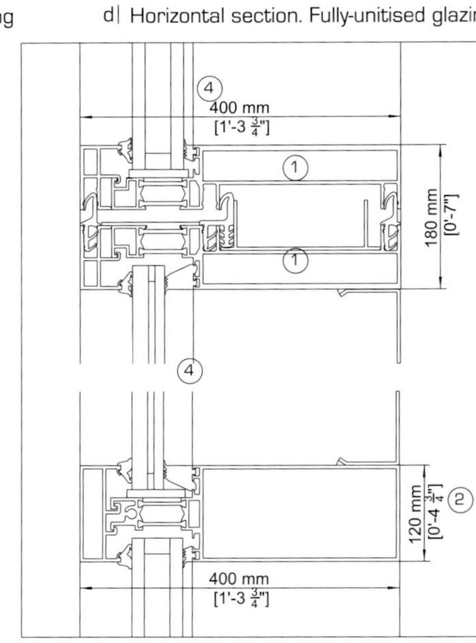

The main advantages of unitised glazing in construction are of speed of installation on site and of quality control of assembly workshop conditions over those on site. For taller buildings, for those over five or six storeys, scaffolding becomes much less practical, and so working from inside the building on a floor slab is a safer and faster method. Unitised panels are assembled and glazed in the factory. On site they are secured by brackets fixed to the floor slab and set side by side, then rising floor by floor, usually from the bottom of the building upwards. Panels are made either as panels set side by side, so that they can be replaced as complete panels if accidentally damaged, or be semi-interlocking where the glass is replaced without moving the panel. The fully unitised type has wider site lines (overall frame width) at around 80mm (3in), than the semi-interlocking type at around 65mm (2.5in). A comparable stick system would have a joint width of around 50mm 2in). From the visual point of view, the wider site lines of unitised glazing over stick glazing is the main disadvantage of this system.

Unitised glazing is much less suited to tall glazed walls, visually, that have heights greater than around four metres between floor slabs, that is, greater than a single panel height. The framing can appear to be visually very strong, and instead frameless construction is often preferred as either bolt fixed or clamped glazing. Its most common use is in modules of 1200mm or 1500mm wide (4ft or 5ft) where it suits the internal planning of office buildings, but storey height panels 3 metres wide are used. Because of the cost of setting up jigs and tools in the factory, unitised glazing is suited to use with a small number of panel types to create a high level of repetition. This suits very modular facades with relatively few panel types. Glazed units are fixed either with pressure plates, or are silicone-bonded to a light frame around the edge and then mechanically fixed to the frame of the panel.

Glazed units are fitted either externally or internally depending on the glass replacement method. When glazed externally, mullions and tran-

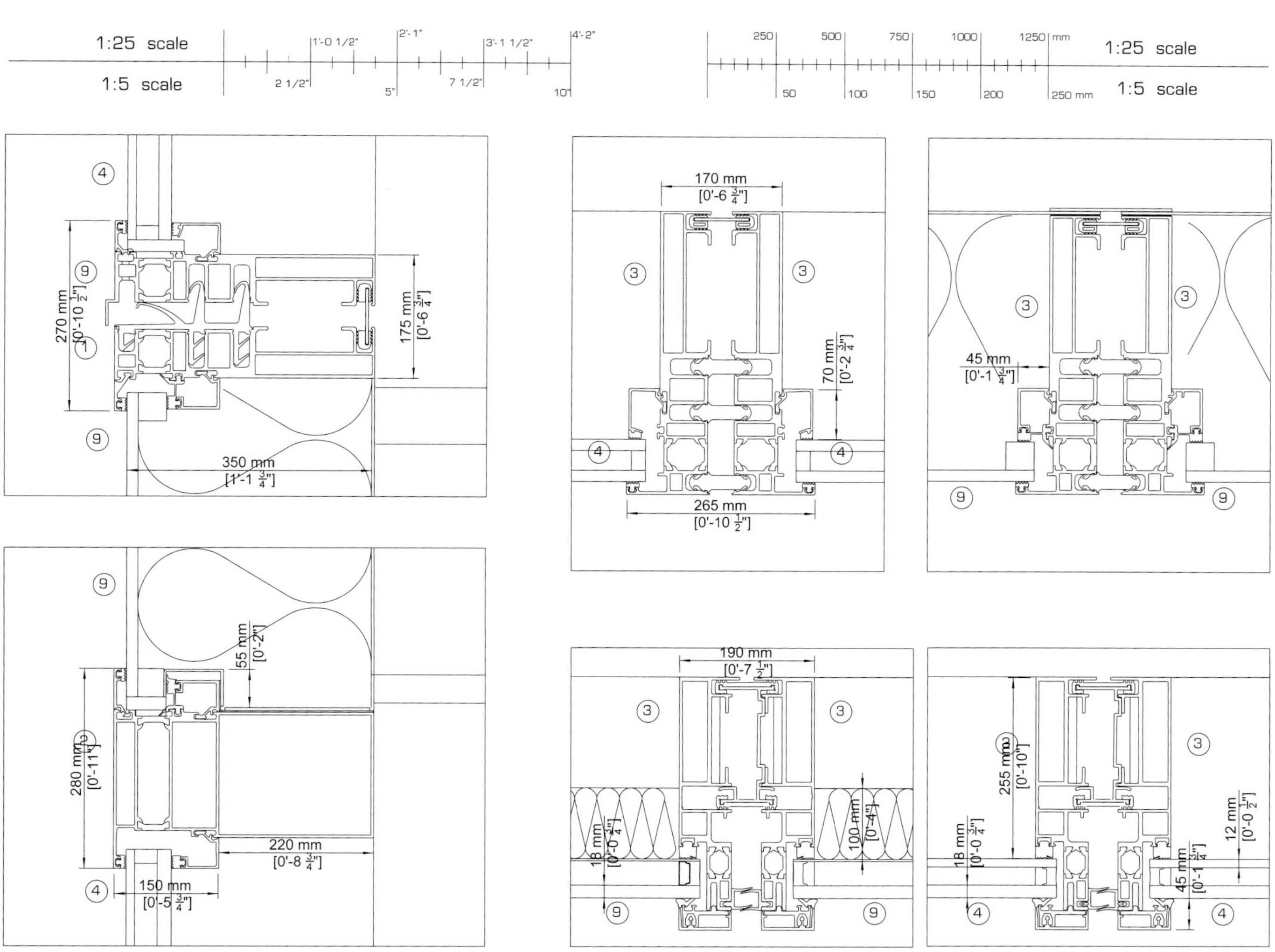

e | Vertical sections. Junction with slab edge and parapet. System used in temperate climates.

8 | Metal parapet coping
9 | Metal-faced or opaque glass-faced insulated panels
10 | Floor slab
11 | Floor finish
12 | Ceiling finish
13 | Outer screen providing solar shading
14 | Thermal insulation
15 | Sheet metal seal
16 | Cover cap
17 | Smoke seal
18 | Support bracket

f | Prototype

soms are adapted to have a clip on the outside to form a pressure plate (b). Where glazed units are internally glazed, clips are used which project from the sides of the mullion/transom section. This results in internally glazed sections usually being wider than those glazed externally. When glazed externally, a maintenance cradle is used with the capacity to lift glass and allow the panel to be lifted in place. This method allows glass to be replaced where it is either impractical to carry the glass through the building or where glass sizes are too large to enter stairwells or lifts. When internally glazed, glass units are carried up either in a lift or by stairs and are installed entirely from the floor plate.

Jointing panels

Most panels are made one storey-height, but larger panels are made to span two floors and one bay wide or one storey high and several bays wide. Those larger panels are usually done where installation time is critical. Like stick glazing, panels are either hung from the top or are supported at the bottom. Instead of a straightforward spigot between storey height lengths of mullion, panels are tied together with a 'stack' joint (d), so-named from the concept of panels being stacked above one another. The complete bottom horizontal edge of the panel intersects with the top edge of the panel below. This joint comprises two lines of defence against air and water infiltration, and a further air seal at the interior face of the panels. In stick glazing, seals are provided by a set of pressure plates applied when the carrier frame and glass are already in place. In unitised glazing, the waterproofing is mostly fixed to the panel before installation and has to work when panels are simply slotted into place. An outer line of defence is provided by rubber-based baffles set on each panel so that they press together to form a seal. An aluminium drip profile is sometimes added to the outside of this as a first barrier against wind, but allowing water to drip out again behind this profile. Any water passing through this profile is stopped in a pressure equalised chamber that stops water getting any further. This is

b| Prototype

a| Horizontal and vertical sections. Junction with slab edge and parapet. System used in hot and humid countries, mostly Middle East and South East Asia.

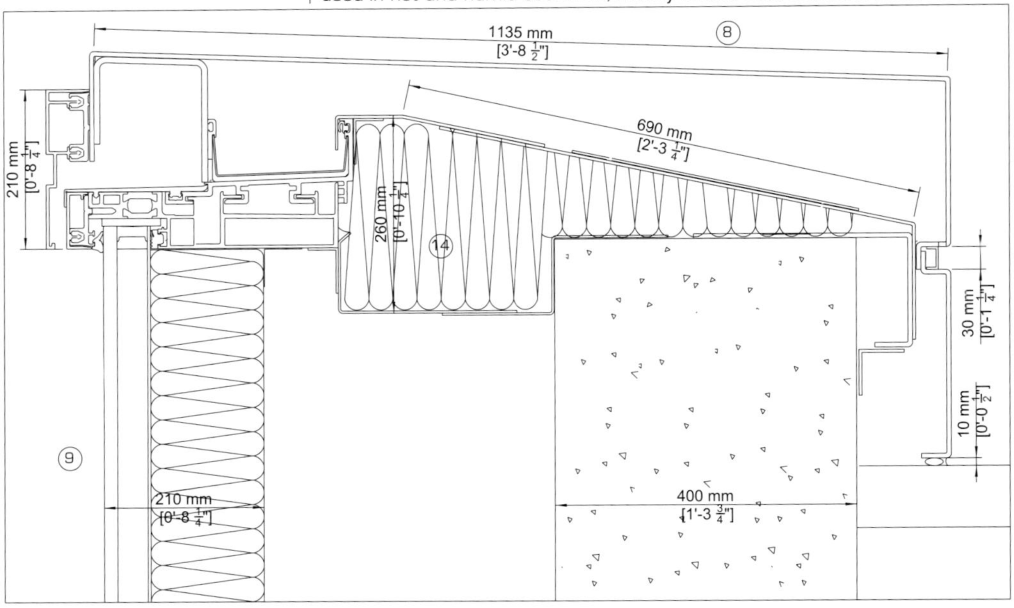

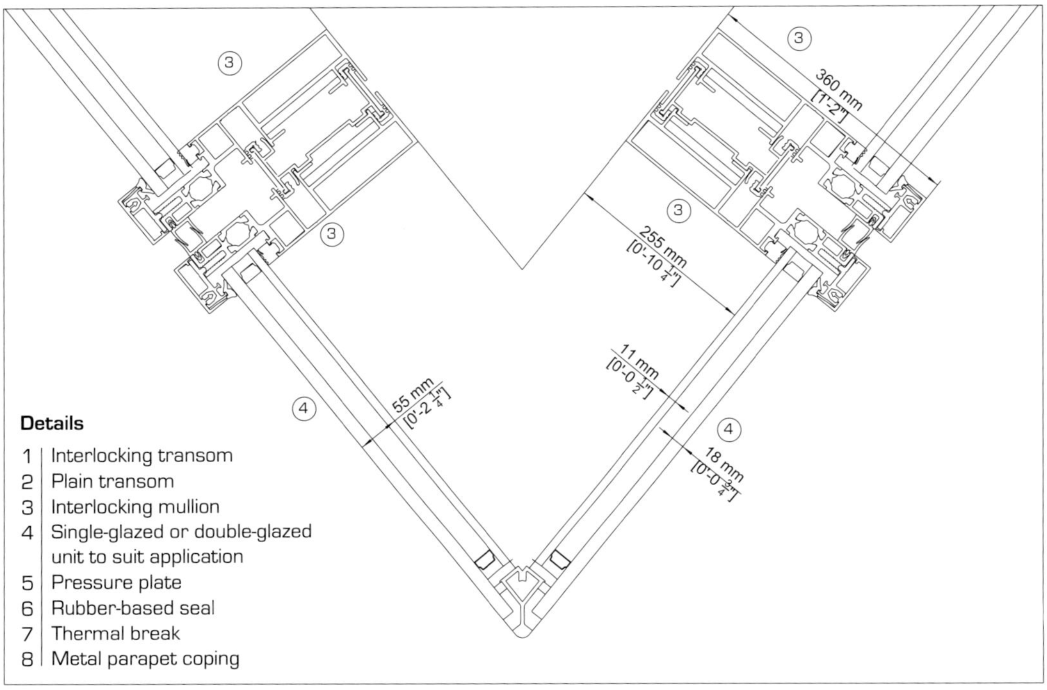

Details

1 | Interlocking transom
2 | Plain transom
3 | Interlocking mullion
4 | Single-glazed or double-glazed unit to suit application
5 | Pressure plate
6 | Rubber-based seal
7 | Thermal break
8 | Metal parapet coping

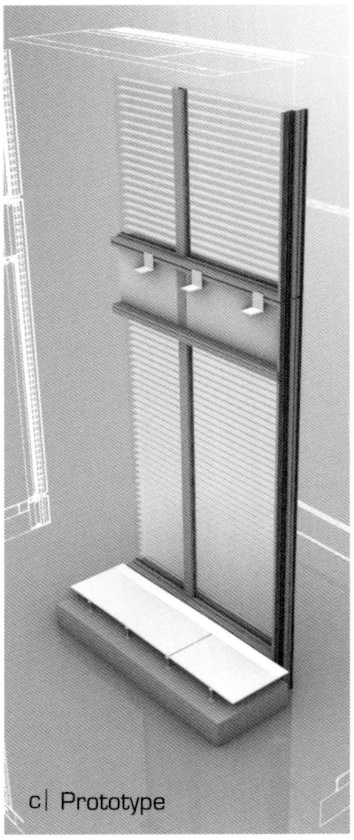

c| Prototype

drained through the forward baffle and out of the front of the panel. At the back of the joint is an air seal. In warm humid countries, this rear chamber is usually ventilated to the outside as well as to release condensation that inevitably forms in the full depth of the profile. In this detail, the performance of the inner air seal is critical to the success of the system.

The vertical joint between two panels is formed as a vertical continuity of horizontal joints. The forward baffle forms an outer seal, the inner seal is lapped onto the horizontal joint, and the inner air seal is continuous with its horizontal counterpart.

Both horizontal and vertical aluminium profiles have a thermal break set near the external face to avoid thermal bridging. However, since external air is allowed deep into the joint, thermal calculations are undertaken to check that the dewpoint falls in the pressure equalised drainage cavity. Spandrel panels are formed in the same way as with stick glazing with metal or glass panels. The transom dividing spandrel and glazing beneath is formed as one division within the unitised panel, though it can sometimes be at the position of the stack joint if the panels are seated on the fixing bracket at slab level rather than hanging from it.

Opening lights

Opening lights such as smoke vents are formed within each panel with a secondary frame inserted within the edge frame of the panel. Doors are usually made as separate items that are glazed into the adjacent frame, unlike stick glazing where doors are glazed into the stick system, making the overall widths greater. However, air infiltration rates for doors are usually half that of unitised glazing at around 300 Pascals for doors when compared with 600 Pascals for unitised glazing. Stick glazing provides similar levels of air infiltration. The poorer performance of traditional doors can sometimes lead to the use of window profiles as

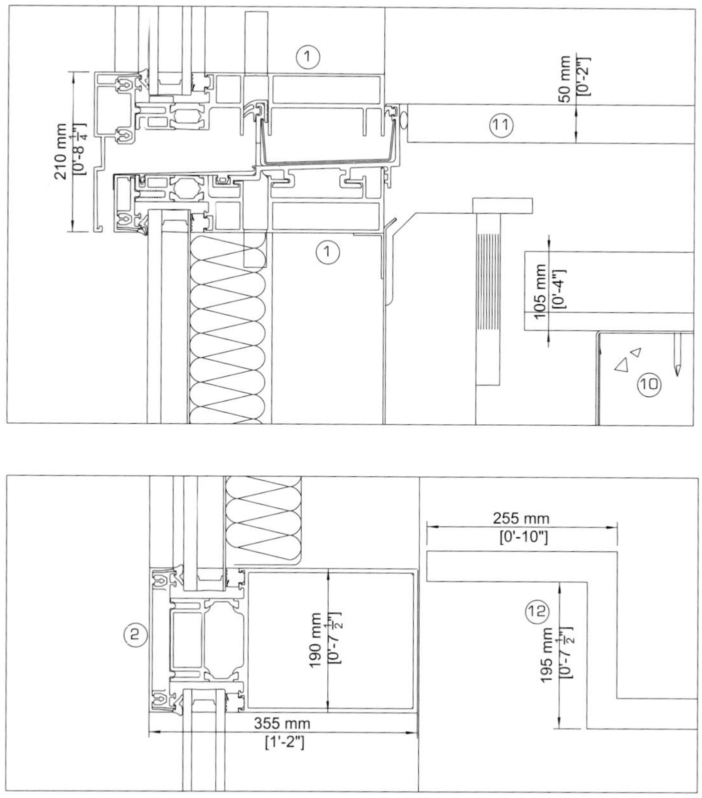

d| Horizontal section. External sun shading glass fin shown

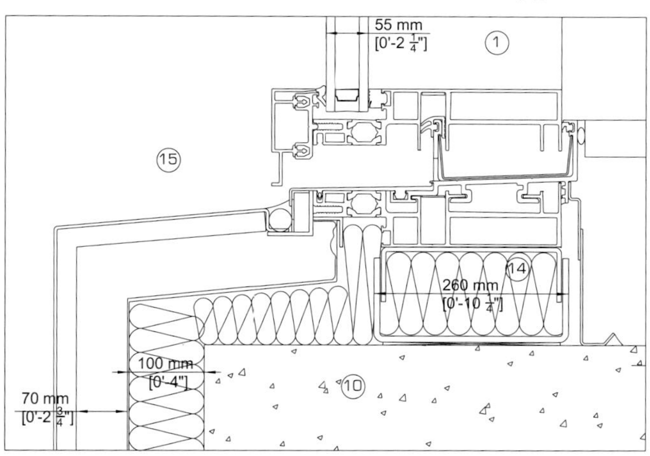

9 | Metal-faced or opaque glass-faced insulated panels
10 | Floor slab
11 | Floor finish
12 | Ceiling finish
13 | Outer screen providing solar shading
14 | Thermal insulation
15 | Sheet metal seal
16 | Cover cap
17 | Smoke seal
18 | Support bracket

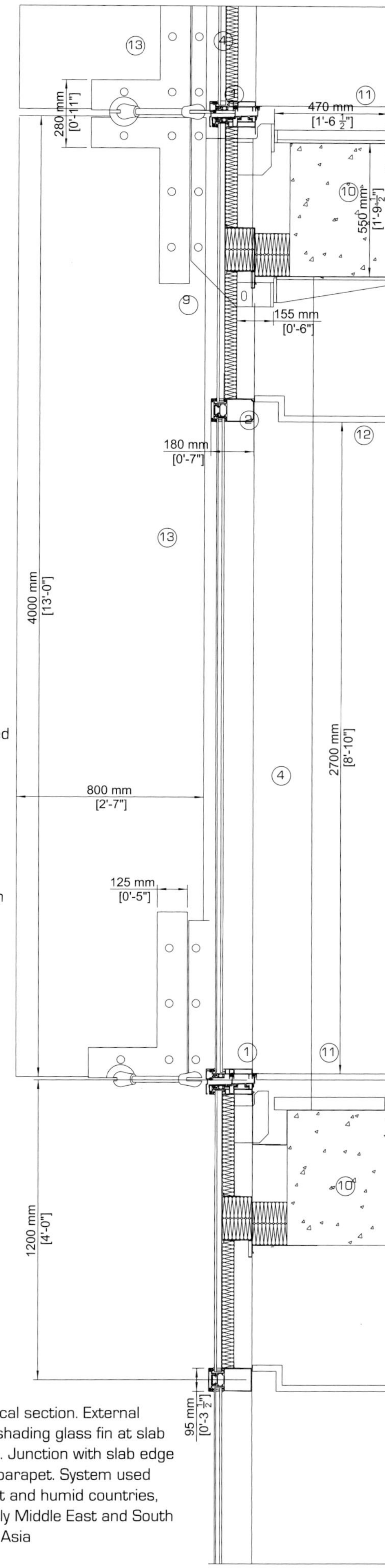

e| Vertical section. External sun shading glass fin at slab edge. Junction with slab edge and parapet. System used in hot and humid countries, mainly Middle East and South East Asia

doors since they have much lower air infiltration rates, some matching those of curtain walling.

Corner panels, parapets and sills

An advantage of unitised glazing is its ability to have frameless glazed corners, which is difficult to achieve in stick glazing, since the sealing joint between the two glazed units has to be achieved on site, where curing of the silicone is much more difficult than in the factory. Corner units with frameless corners are made by supporting the glass on three sides in a frame and bonding the two sides together. Usually a small square aluminium section, 30mm x 30mm (1.2in x 1.2in) approximately to match the thickness of the double glazed units, is used to support the corner of the panel when it is lifted by crane. This makes the corner of the panel slightly less all-glass than can be achieved by a double glazed bay window for example, with a fully glazed corner. In such a bay window the glass can be joined without the need for an

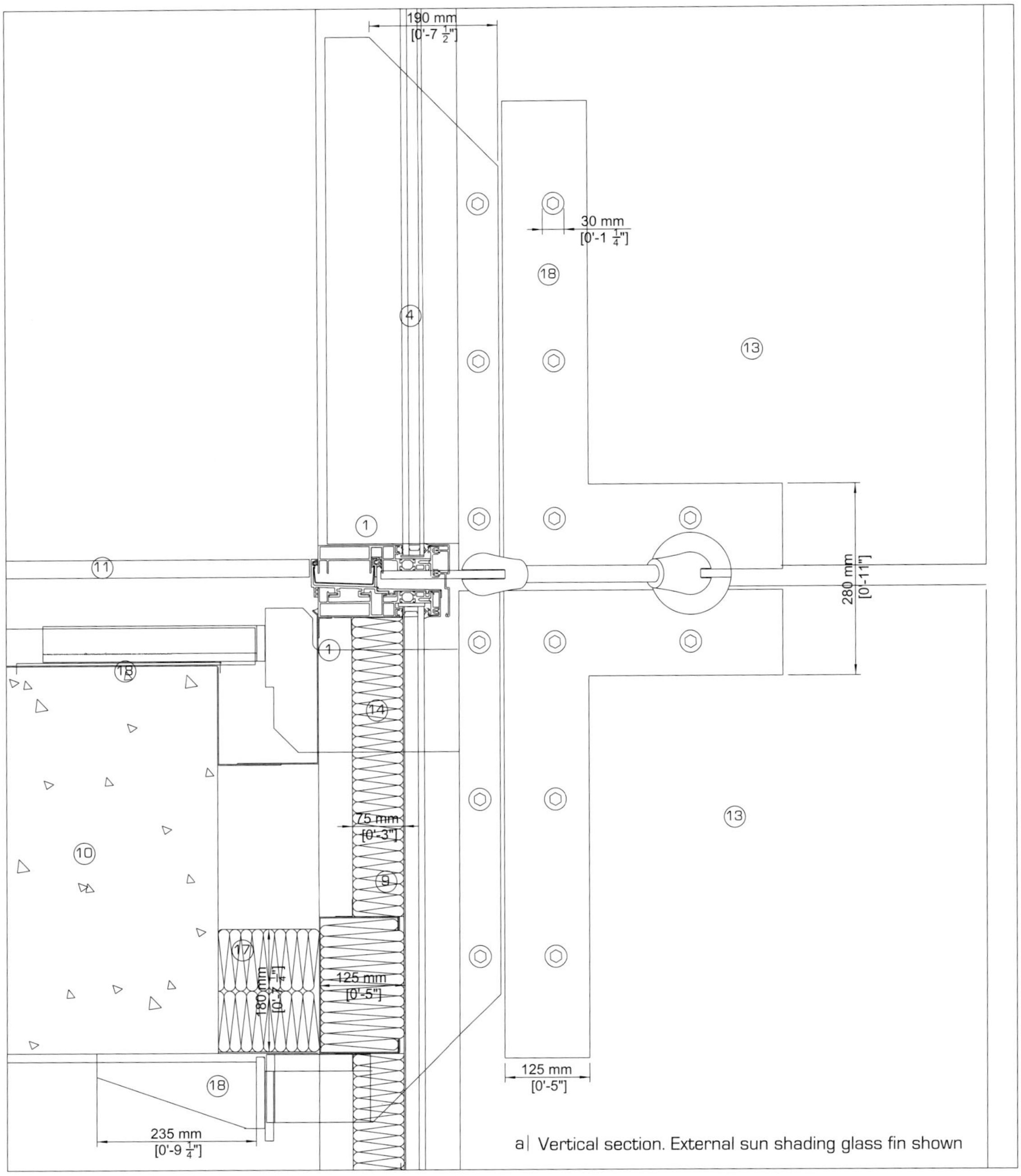

a| Vertical section. External sun shading glass fin shown

Details

1 | Interlocking transom
2 | Plain transom
3 | Interlocking mullion
4 | Single-glazed or double-glazed unit to suit application
5 | Pressure plate
6 | Rubber-based seal
7 | Thermal break
8 | Metal parapet coping
9 | Metal-faced or opaque glass-faced insulated panels
10 | Floor slab
11 | Floor finish
12 | Ceiling finish
13 | Outer screen providing solar shading
14 | Thermal insulation
15 | Sheet metal seal
16 | Cover cap
17 | Smoke seal
18 | Support bracket

additional aluminium post, using either a stepped or mitred joint. In a fully glazed corner in a unitised panel, the appearance of the corner is still around 40mm (1.5in), since the visible edging to the double glazed unit visually widens the joint to around 40mm. Panels can be formed either as equal corners, with up to around 1500mm (5ft) from each corner to a short corner of 1500mm (5ft) on one leg and 300mm (1ft) on the other leg. As the size of the corner panel increases, it becomes more difficult to lift and set in place, and consequently becomes more expensive. A corner unit with a framing member at the corner can be made either with a frame visible, or visible only from the inside, with a glass-to-glass junction as described in the paragraph. Sometimes the joint occurs on the corner of two flat panels, with either a mitred corner or a square corner with an infill piece.

Both parapets and sills are formed in a similar way to those used on stick systems, but are formed as separate panels unlike stick systems which are glazed into the top transom at roof level (MCE_133, h). Sills are formed in pressed metal which is inclined to drain water away from the glazing (MCE_133, d).

Penetrations

Penetrations through glazing for brackets, usually fixed at floor level to fix them back to the floor slab, occur at joints between panels through mullions or transoms. This is because the bracket can be far more easily sealed at this point, where the glazing system is drained and ventilated internally, than through spandrel panels, where it is difficult to apply a seal and where no mechanical pressure can be applied to the spandrel panel since there is no framing member behind.

Silicone-bonded glazing

A development over the past ten years has been in silicone-bonded glazing. Although this is used in unitised systems, it is also used partially

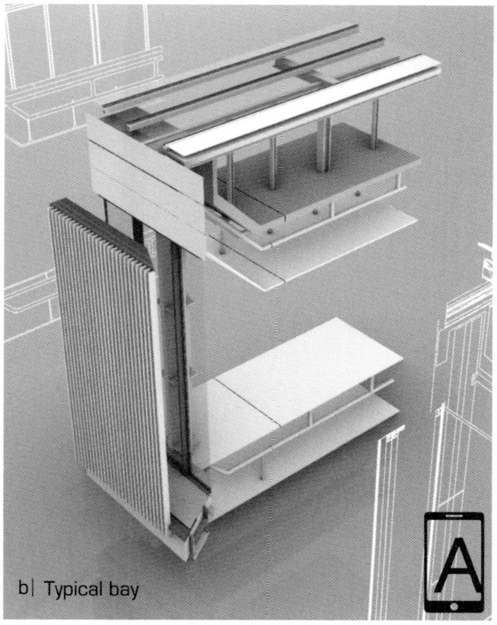

b| Typical bay

c| Typical bay

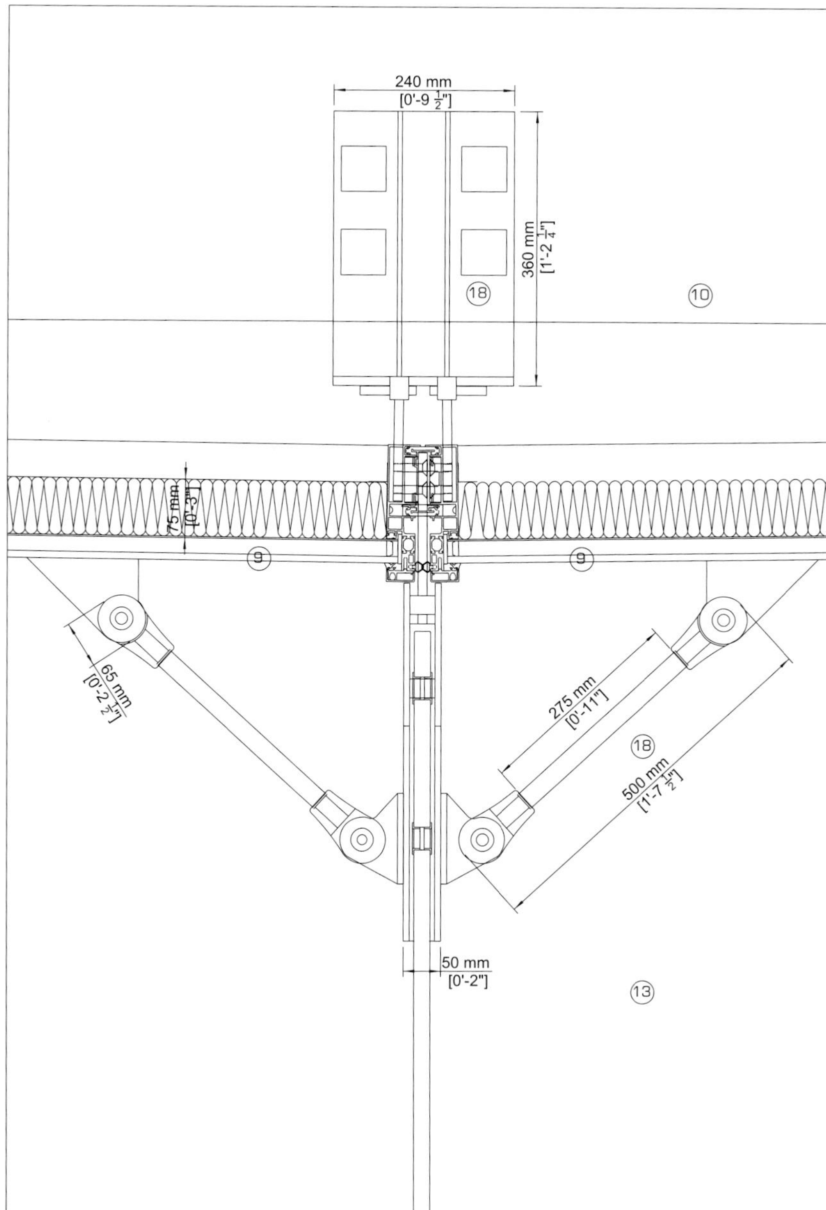

d| Vertical section. External sun shading glass fin shown

in stick glazing. The glass is restrained on two vertical sides by pressure plates and two horizontal sides by silicone. Alternatively, the glass is bonded on all four sides by silicone. This method avoids the need for visible cappings for mullions and/or transoms which are often around 1.5 times wider than an equivalent stick system. The use of silicone allows the joint between glass units to be flush rather than using a pressure plate which is set in front of the glass. This allows the facade to have no visible framing on the outside. Silicone is also used in the joint since this material cannot be combined with an EPDM, such as neoprene, used in seals in pressure plate systems.

The glass units are bonded to aluminium profiles to form a light frame at the edge of the glass. The glass is then mechanically fixed with screws to the carrier frame on site. This technique is increasingly used in unitised systems where damaged glass can be removed without the need to remove the complete panel. Although silicone-bonded glazing allows a facade to appear from the outside as a continuous glass surface interrupted only by smooth and narrow joints, with 20mm (0.78in) rather than 50mm (2in) minimum with pressure plates, the opaque zone across the frame is the same overall width as with pressure plates. This is because the area behind the double glazed units is made opaque with screen printing in order to conceal the width of the carrier frame behind.

Some manufacturers bond a light aluminium frame around the edge of the frame within the depth of the double glazed unit, fixing the glass with short lengths of pressure plate in the conventional manner, then applying sealant into the joint. Sometimes the glass is secured by additional short clips along the horizontal edges to increase the safety of the system, but building codes around the world differ in this requirement.

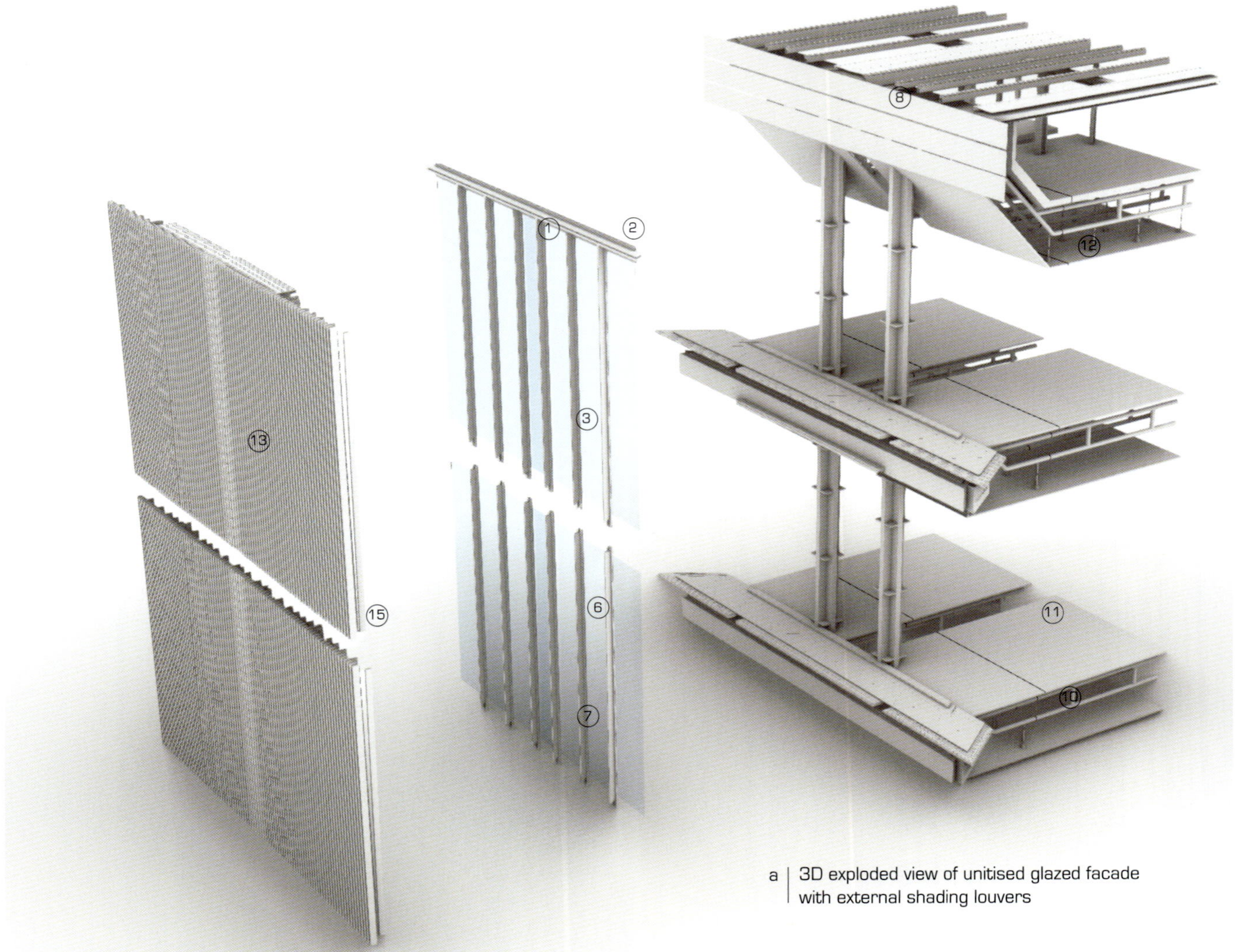

a | 3D exploded view of unitised glazed facade with external shading louvers

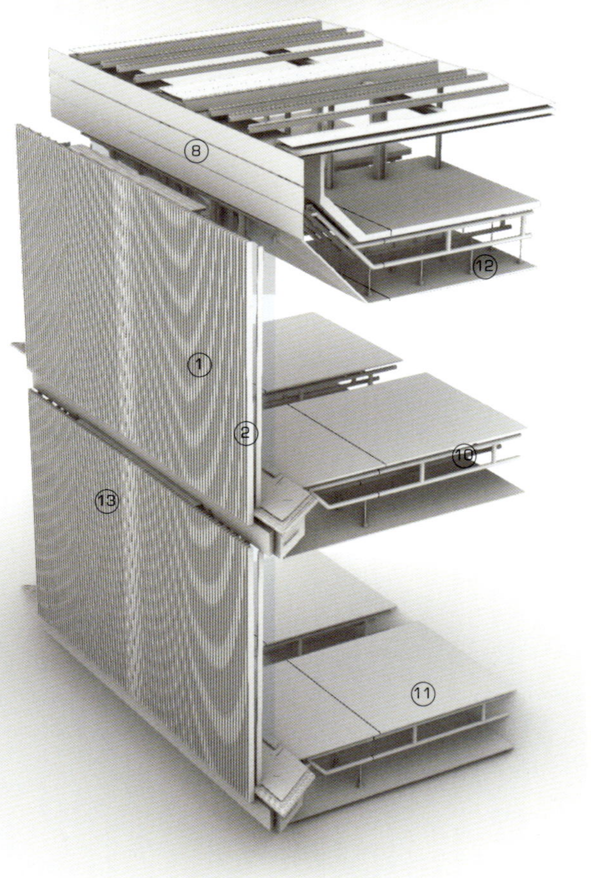

b | 3D view of unitised glazed facade with external shading louvers

Details

1 | Interlocking transom
2 | Plain transom
3 | Interlocking mullion
4 | Single-glazed or double-glazed unit to suit application
5 | Pressure plate
6 | Rubber-based seal
7 | Thermal break
8 | Metal parapet coping

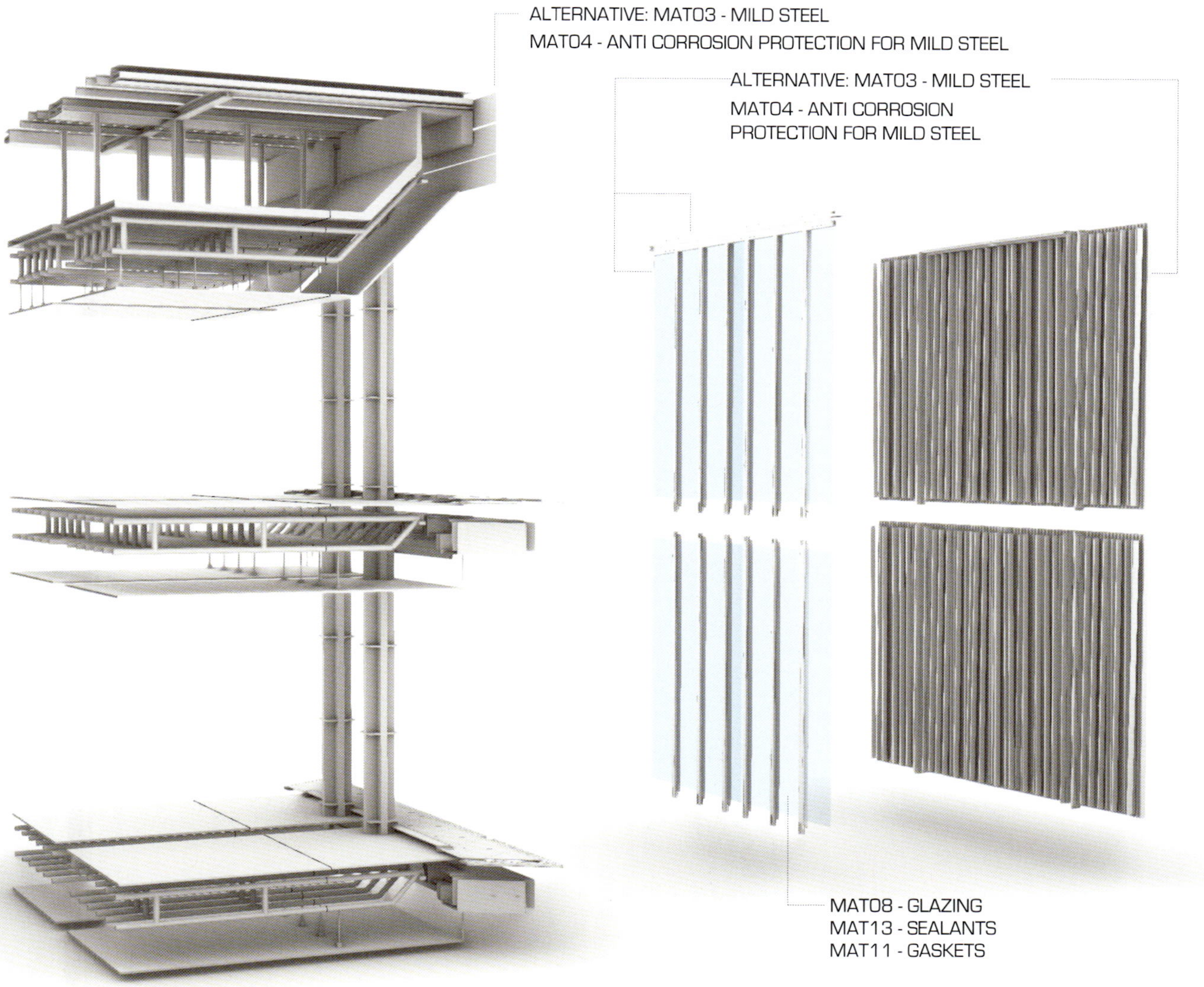

c | 3D exploded view of unitised glazed facade with external shading louver components with material specification references

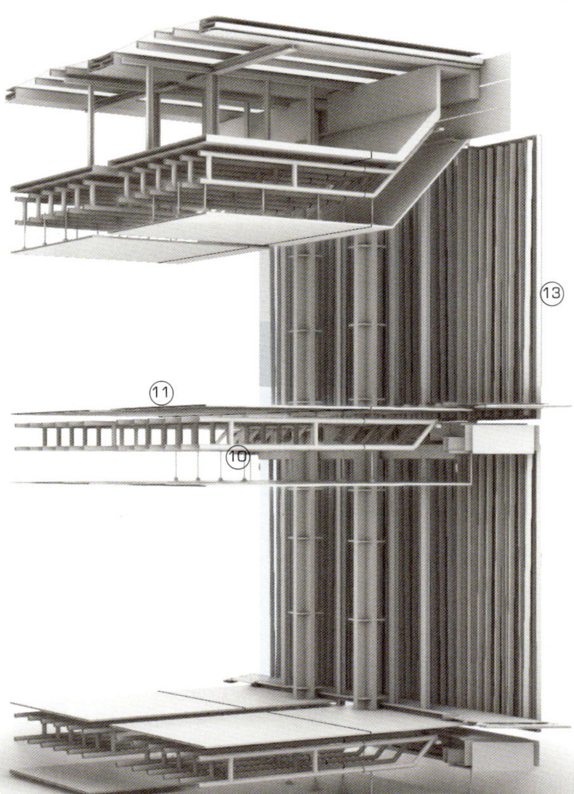

9 | Metal-faced or opaque glass-faced insulated panels
10 | Floor truss
11 | Floor finish
12 | Ceiling finish
13 | Outer screen providing solar shading
14 | Thermal insulation
15 | Sheet metal seal
16 | Cover cap
17 | Smoke seal
18 | Support bracket

d | 3D view of unitised glazed facade with external shading louvers

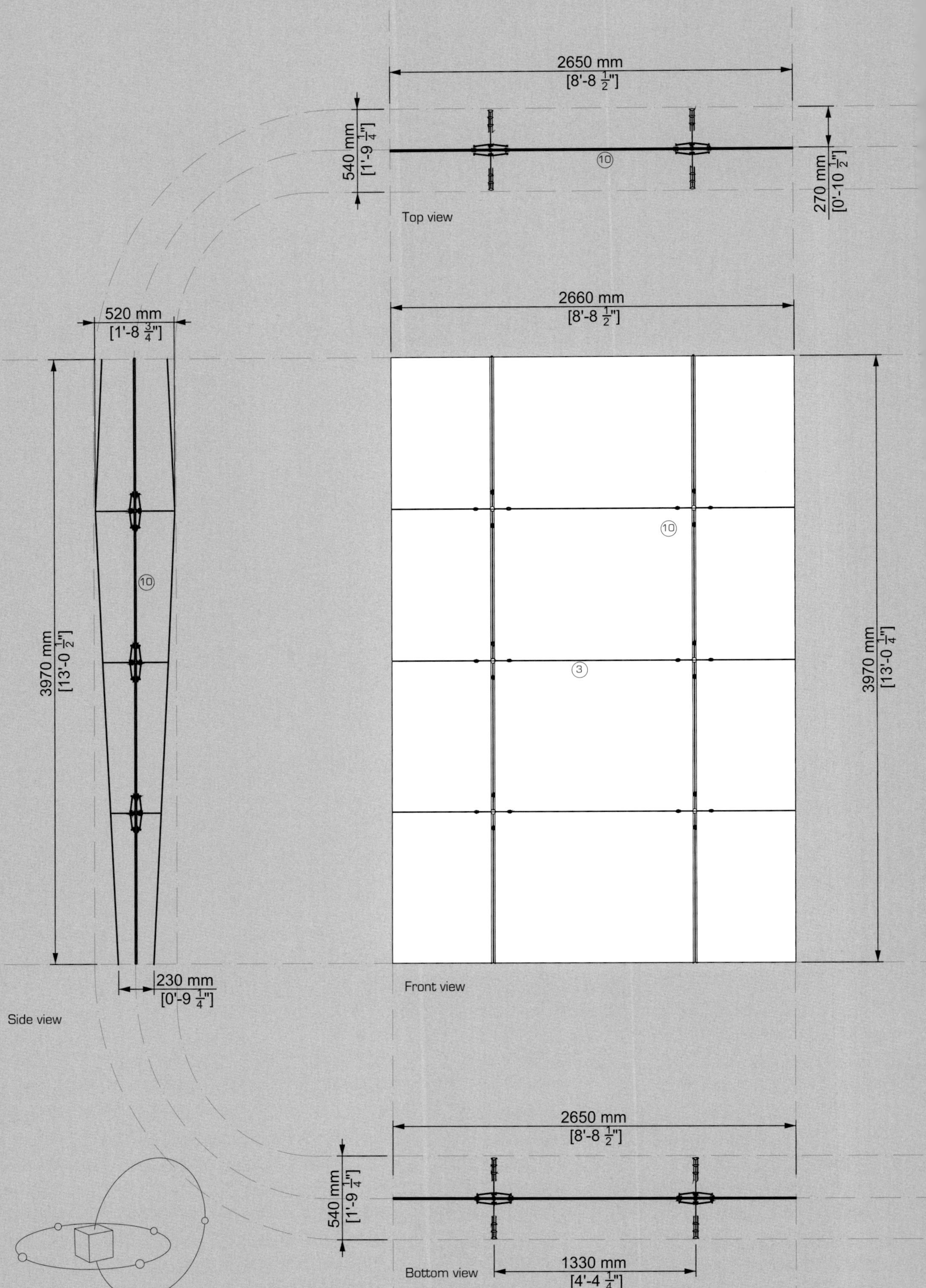

Third angle projection 1:30

Details

1 | Stainless steel patch plates
2 | Single-glazed or double-glazed unit to suit application
3 | Silicone seal
4 | Glass fin
5 | Support bracket

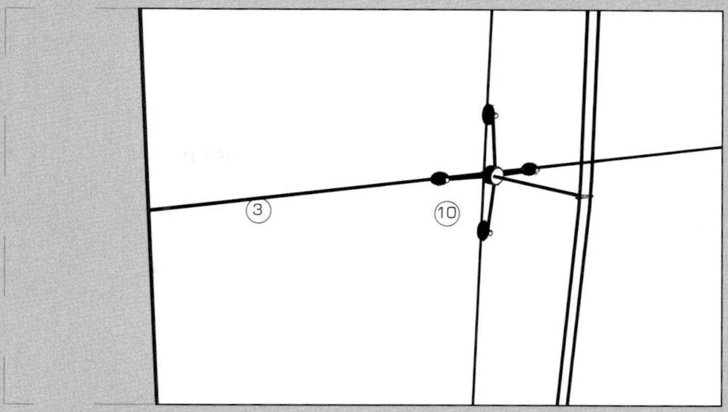

3D view of the assembly

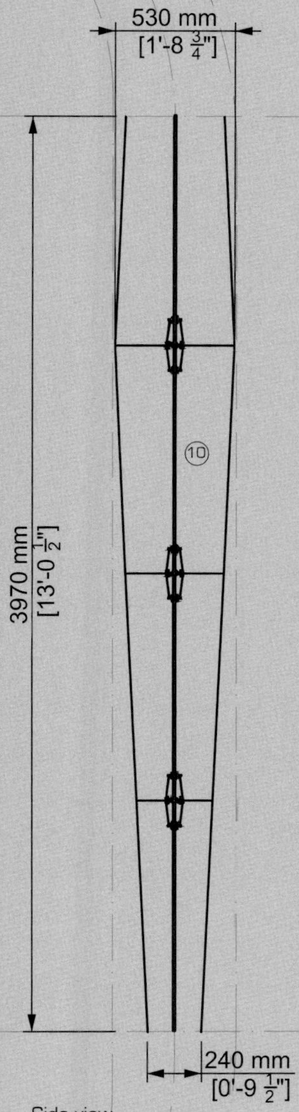

Side view

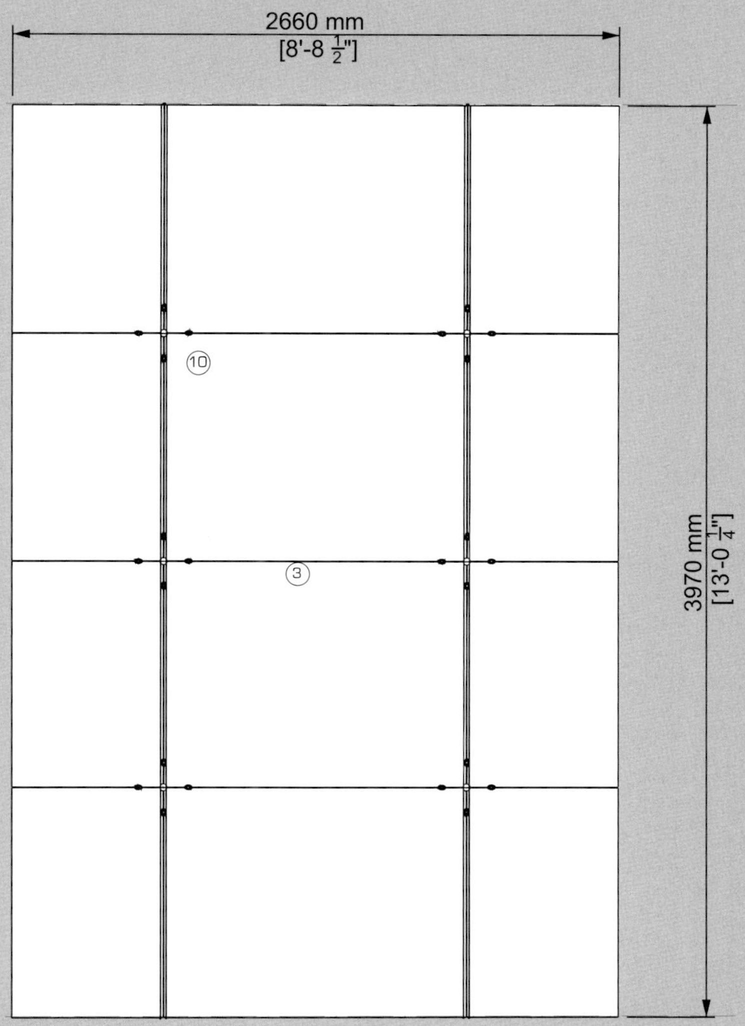

Back view

6 | Floor slab
7 | Ceiling finish
8 | Floor finish
9 | Access ladder
10 | Clamped glazed wall
11 | Inner framed curtain walling
12 | Fixing bolt
13 | Maintenance access deck

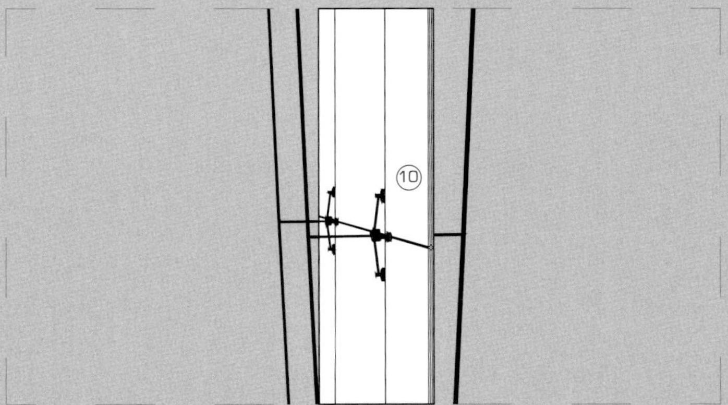

3D view of the assembly

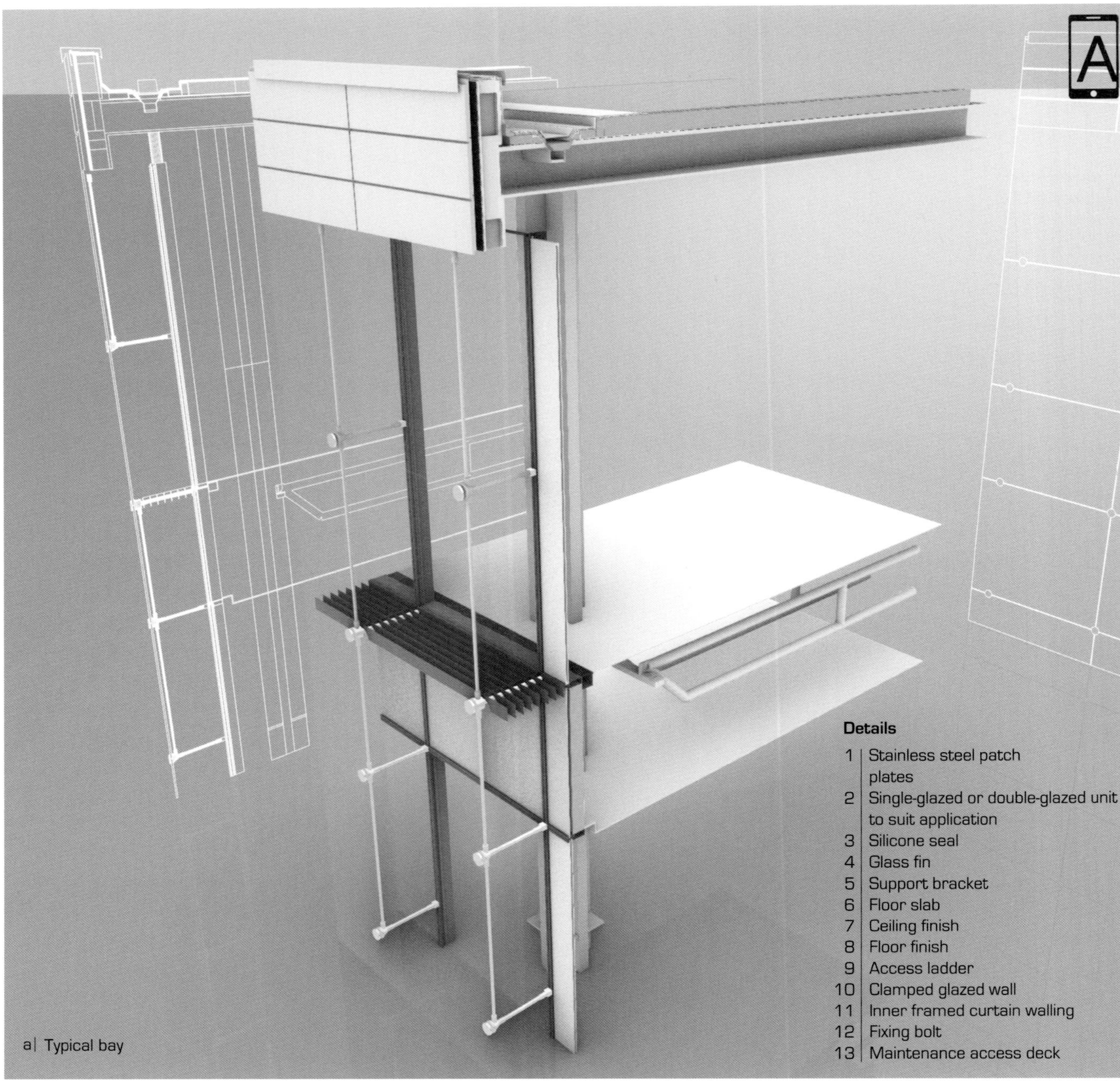

a | Typical bay

This glazing method is suited primarily to single glazing but double glazed examples are being constructed that avoid some of the wide joints resulting from the visible black edges of double glazed units. The sealant between double glazed units is also usually black to match the unit edges.

Comparison with bolt fixed glazing

As a method of frameless glazing, clamped glazing is more economical than bolt fixed glazing. Whereas bolt fixed glazing requires the drilling of glass, clamped glazing does not, as the glass is fixed with patches or clamps that pass through the joint between the glass sheets or double glazed units. The ability of clamp brackets to be simple and easily made allows the glass to be supported at different angles to one another in a tiled, non-planar manner. Glass has been lapped in the manner of traditional patent glazing and tiled in the manner of wood shingles, where glass is lapped on two edges to give a rich, undulating texture across glass facades. Because the facade is more visually vibrant, the fittings themselves can also project from the facade to allow less expensive stainless steel angle-type brackets to be used. These contrast both in the appearance and higher costs associated with bolt fixed glazing. A disadvantage of clamped glazing is that glass thicknesses are usually thicker than those in bolt fixed glazing, where the distances between fixings are reduced (reducing the span of the glass) by setting them into the material.

This combination of easily fabricated, easily modified brackets and fixing through the joint allows a geometrically complex facade to be fixed back to a rectilinear, economic supporting structure. This contrasts with the need to repeat expensive bolt fixings where there is little possibility of changing the geometry to allow for different fixing positions of the glass.

Clamped glazing is increasingly used in rainscreen configuration with open joints or open, lapped joints with an accessible area behind, as

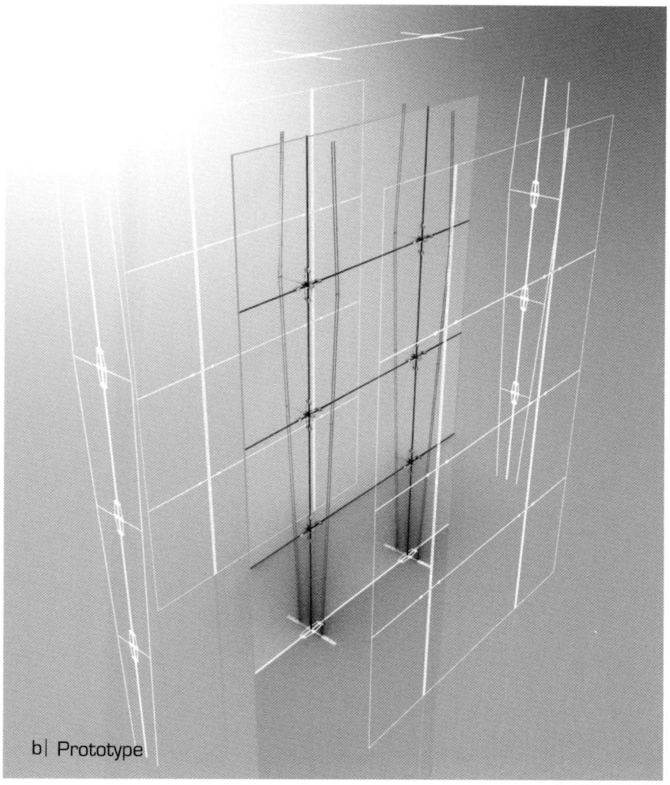

b| Prototype

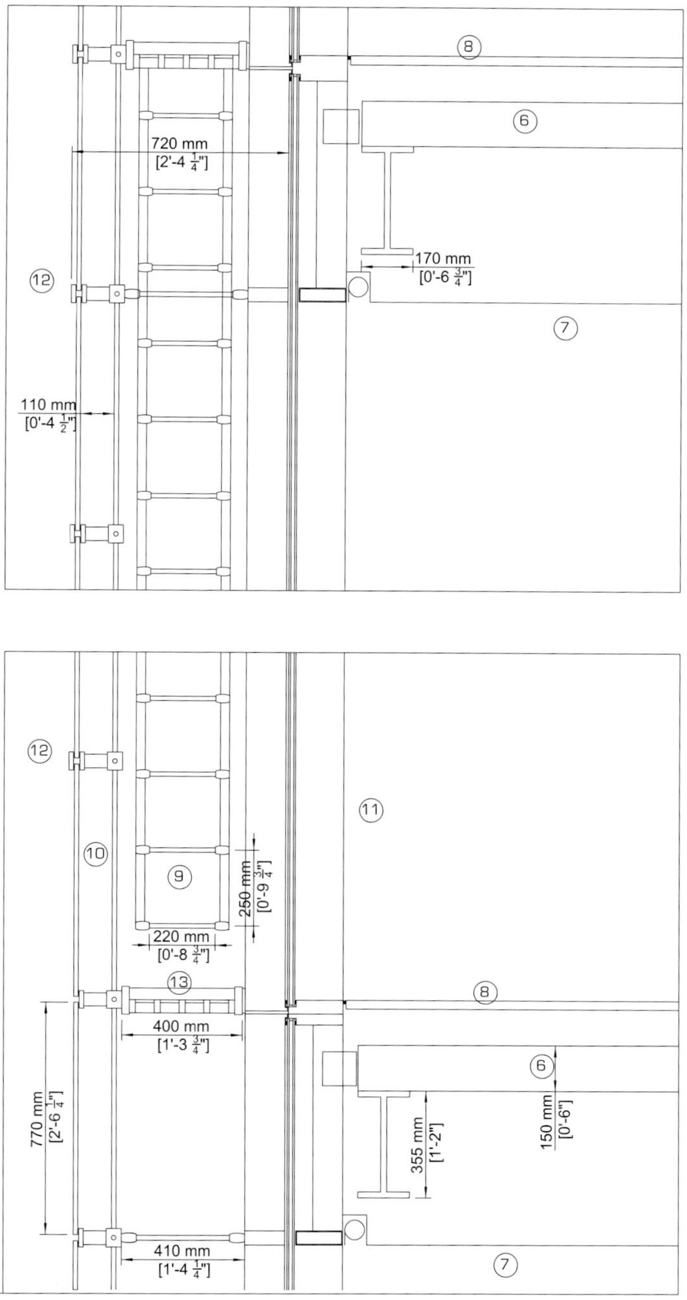

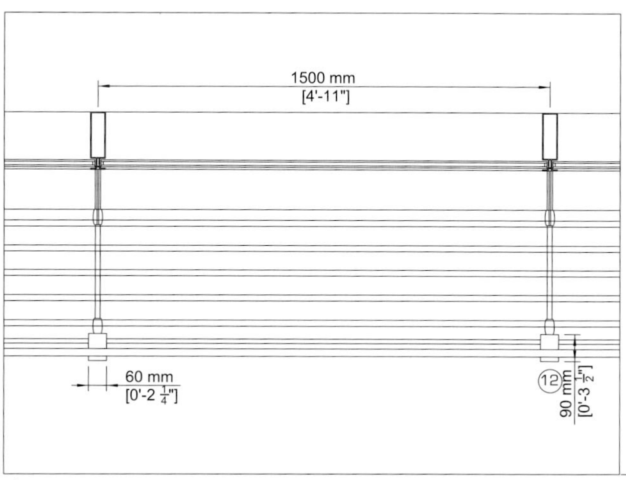

c| Horizontal and vertical sections. Twin wall construction with outer wall in clamped glazing

used in 'twin wall' glazed facades (a). The outer screen acts primarily as a weather barrier, allowing the inner wall to have opening windows where they would otherwise not be possible, as in tall buildings or buildings in areas of high ambient noise. The rainscreen principle has been developed further to avoid fixing clamps directly to the glass, but instead to silicone bond an edge frame around the glass and fix the frames back to the supporting structure with clamps. This silicone-bonded variation of clamped glazing is very useful in opaque or translucent glazing in a full rainscreen to an opaque backing wall. Joints between the glass can be sealed with rubber-based strips to avoid dirt getting into the void behind the glass and staining the back of the glass. The advantage of this system is in clamping the glass by means of a secondary frame that avoids penetrating an outer seal, thus making it suitable for glass rainscreen walls without visible framing and without the need for cleaning the inner face of the glass sheets.

Patch plate glazing

This method uses angles and plates that are bolted through the glass rather than through the joints. This is an earlier form of frameless glazing and was the forerunner of bolt fixed glazing. Although cheaper than bolt fixed glazing, it is restricted to single glazing. Glass sheets are fixed at their corners with a patch fitting that connects four sheets together. Like framed glazing in either stick or unitised systems, the glazing is either hung from the top with brackets or continuous stainless steel angles fixed back to the primary structure, or seated on support angles at its base. Patch fittings are not usually at the top or base of a wall. Instead, continuous angles or a glazing channel is visually preferred both top and bottom in order to provide a weathertight seal. Clamped angles are used where they are concealed beneath floor finishes. Glazing channels are used either for convenience or where the channel is visible at floor level. Where glazed balustrades are used, these clamps are not sufficient for cantilevered glass; that is, where the glass is fixed

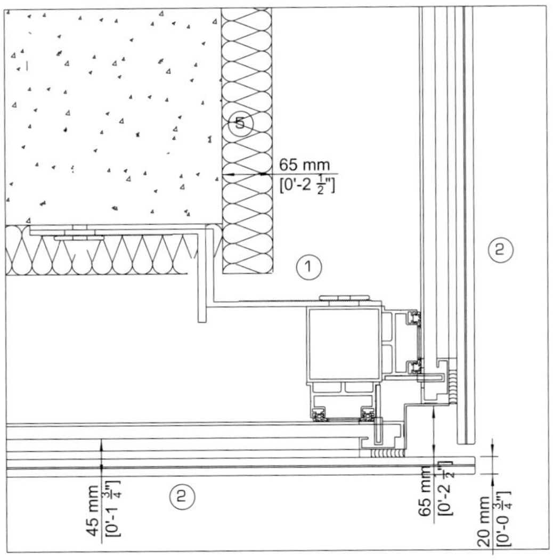

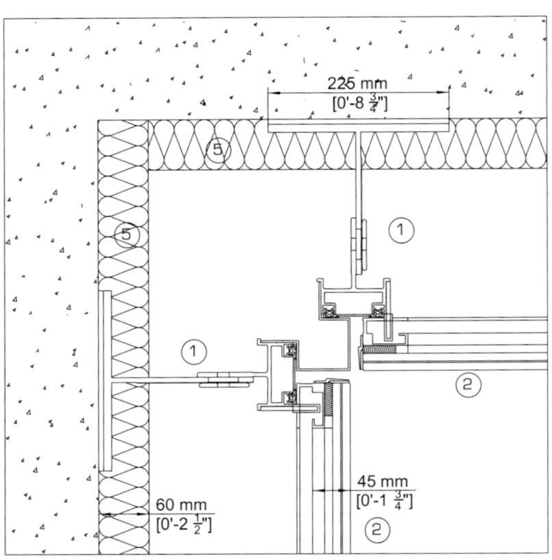

a | Horizontal and vertical section. Clear double glazed units with lapped joints supported by rods or cables

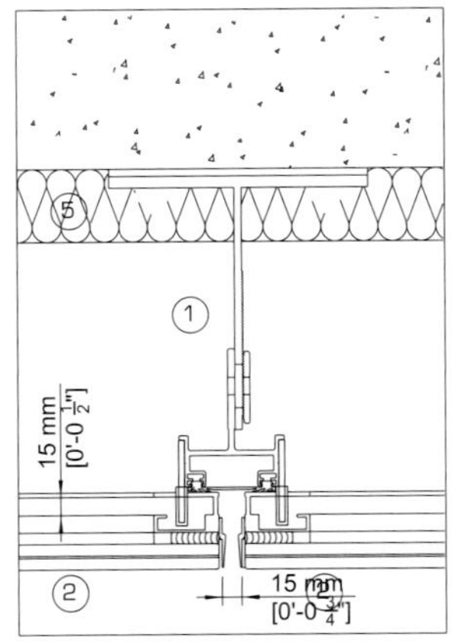

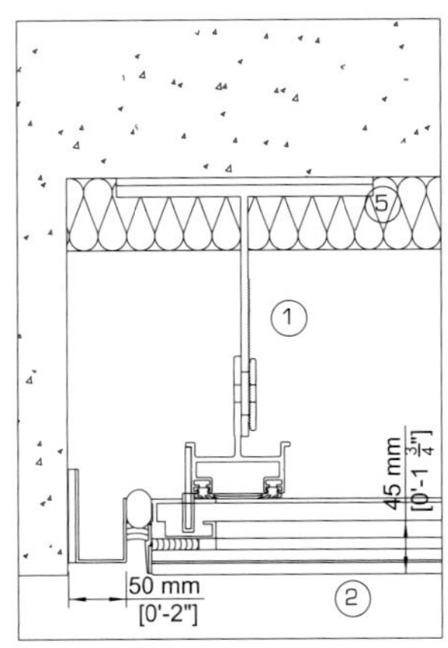

b | Horizontal and vertical section. Clear double glazed units with lapped joints supported by rods or cables

c | Vertical section. Opaque clamped glazed panels. Junctions at ground level, movement joint and panel-to-panel conditions

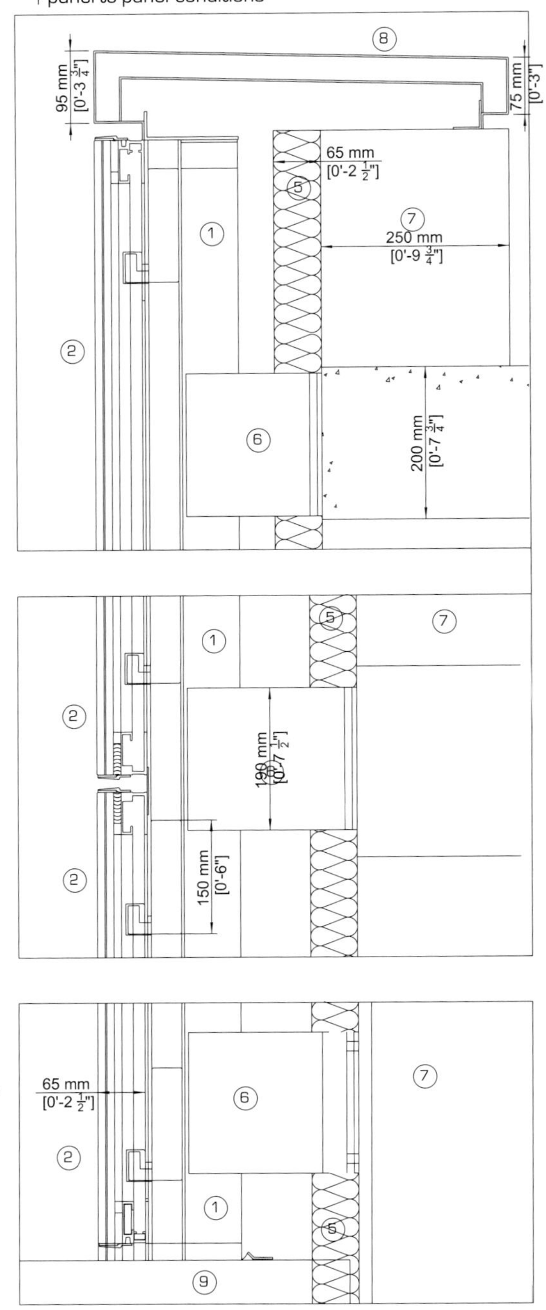

Details

1 | Extruded aluminium mullions
2 | Opaque glazing
3 | Clamp fixing
4 | Floor deck
5 | Thermal insulation
6 | Fixing bracket
7 | Backing wall
8 | Metal parapet coping
9 | Ground level

only at floor level and not restrained at the level of the handrail. Doors are fixed using patch plates where the door is hung, and are seated on pivots with a floor spring when the door is supported at floor level. Where patch plate glazing is hung from the top, doors are also hung in order to accommodate movement in the supporting structure. Similarly, if the glazed wall is supported at its base then the door should also be supported at floor level. If the glazing is hung and the door is supported differently at its base, then a difference in movement between wall and door will either damage the door or prevent it from working properly.

The use of bolts through the glass makes the size of each patch plate relatively big compared with other clamping methods, with up to four bolts per fixing in elevation. This method is often used with glass fins that provide stability in an all-glass wall. The brackets securing the fins can have from four to six bolts per bracket, giving a very strong appearance. Since stainless steel brackets and bolts are required for durability (mild steel is not sufficiently durable), these brackets are finished as polished to avoid surface tarnishing. The use of this polished finish gives this glazing a very particular appearance which is difficult to adapt to something lighter. Although the glazed wall itself is very transparent, the patch fittings have a strong visual presence.

Clamped glazing

This method avoids drilling through the glass by passing through the joint between glass sheets or units rather than by drilling holes in the glass itself. This allows double glazed units to be used, since plates either side of a unit can be clamped at its edge strip.

The disc or square plate either side of the joint is secured together and fixed back to a supporting cable, rod or tube structure (a). This method is more economic than the patch plate type, but still requires comparatively large discs to secure the glass. The use of discs has

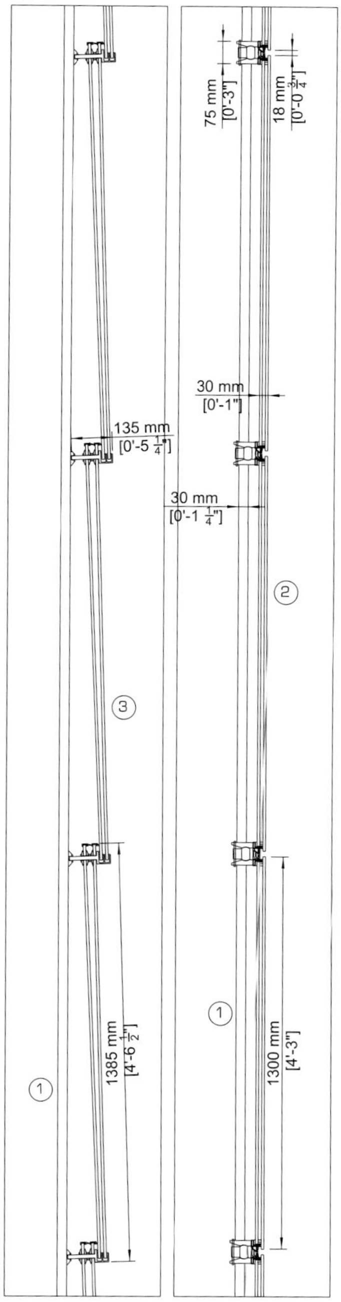

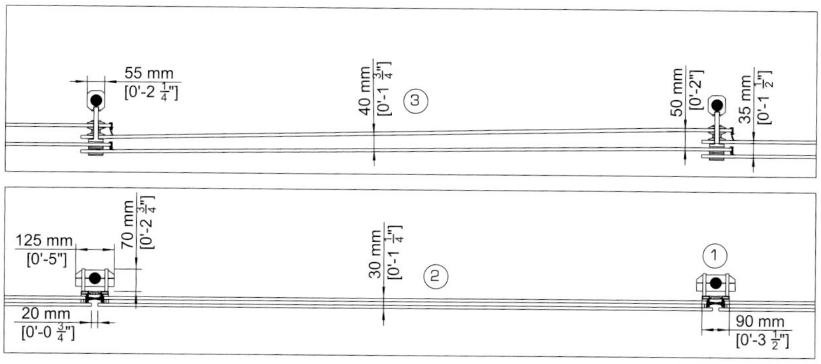

d| Horizontal and vertical sections. Panels in lapped arrangement and in vertically-aligned arrangement

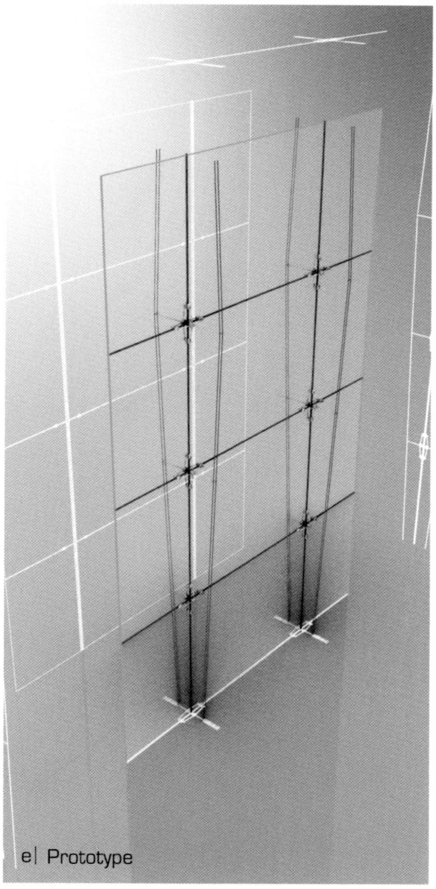

e| Prototype

Details

1 | Vertical cable support
2 | Vertically-aligned glazing
3 | Inclined and lapped glazing
4 | Extruded silicone gasket

f | Horizontal section. Opaque clamped glazed panels. Junctions at corners, panel-to-panel and junction with adjacent wall

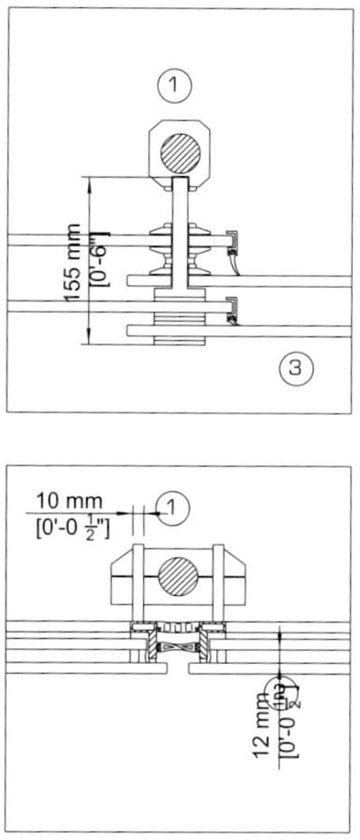

g | Vertical section. Opaque clamped glazed panels. Junctions at panel-to-panel and parapet

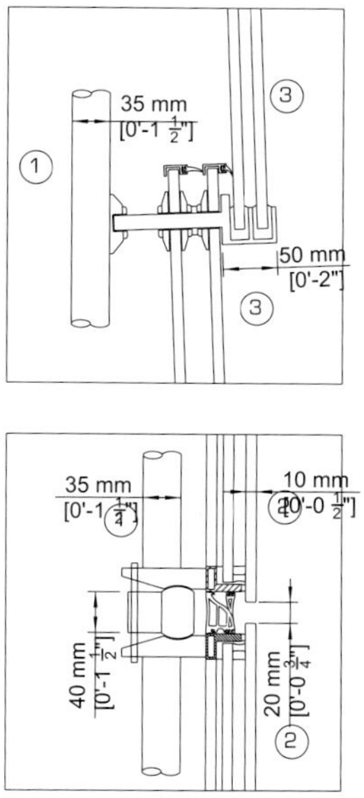

been combined with stick glazing, where the discs are used to secure the glass to stick glazed aluminium extrusions. Glass is set into rubber-based gaskets fixed to the carrier frame and discs are used to hold the glass in place. The gap between the double glazed units is filled with silicone sealant. This has the advantage of being a drained and pressure-equalised (ventilated) system.

The glass is fixed back to either cables or rods if the glass is hung, or back to a tube if the glass is bottom-supported. If top-hung, cables or thin rods are hung from the top and tensioned at the bottom (MCE_145, c). The glass is fixed to the cables with discs, usually at the edges of the glass to reduce the number of cables to a minimum. If wide sheets of glass are used, then the glass may be supported in from the edge. The use of cables provides a very thin wall, but with deflections that are much higher than a framed wall. Maximum deflections can reach up to 600mm to 800m (2ft to 2ft 8in) for walls around 20 metres (65ft) high, but this can be safely accommodated within the structural design. The usual difficulty is in accommodating doors which can deflect only tiny amounts in order to function normally. Doors are usually set into a separate frame and do not form part of the cable structure. If the glazing is supported at its base, then supporting steel tubes provide much more stiffness in the glazed wall, but with a more visible supporting structure. An advantage of such a supporting structure is that the glass can be fixed with a wide range of different geometries. Lapped glazing or 'fishscale' or 'shingled' glazing is possible. Glass is held in a bracket that can support the bottom of the glass in a different position from the top of the glass below. With double glazed units, the glass can be drilled to combine the advantage of clamped glazing with the seals provided by extruded silicone gasket 'flippers' that exclude rainwater without being punctured by the brackets, which pass through holes in the glass beneath. Such brackets are 'shoes' which usually support the glass away from corners to avoid the corner of the glass being

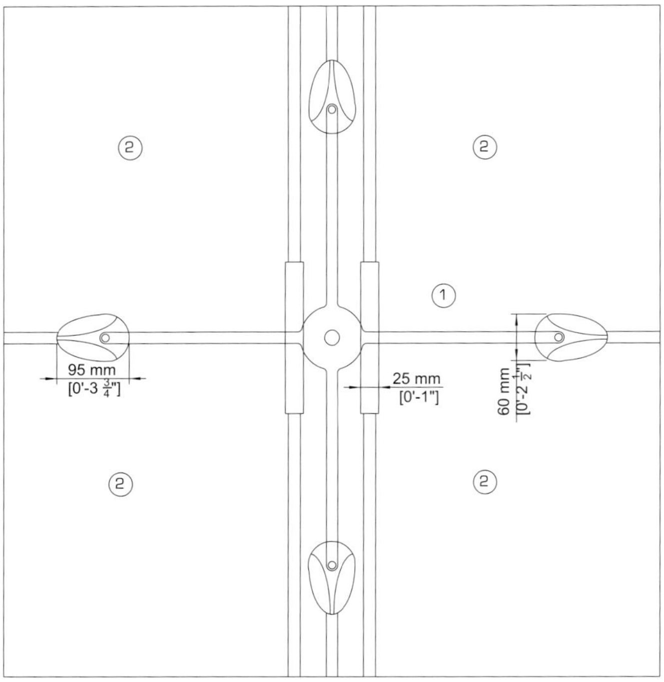

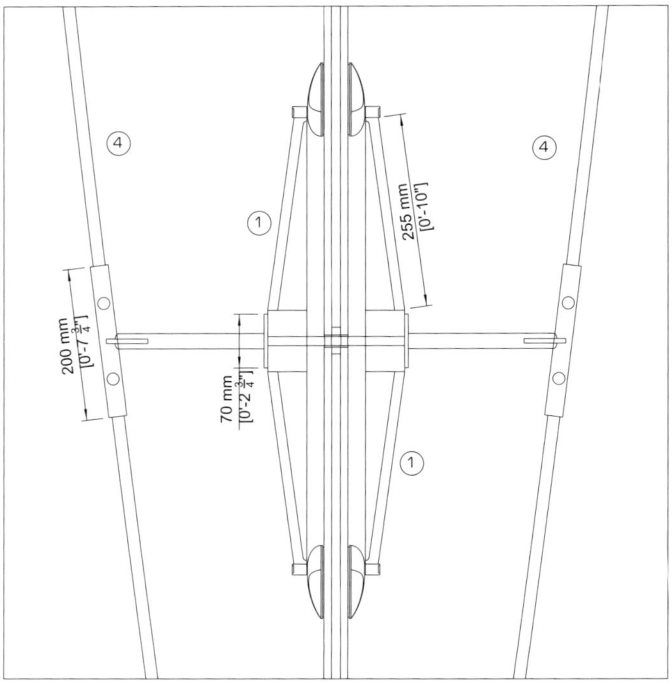

a| Horizontal, vertical section and elevation. Clamped glazing with cast steel clamps held in place by forming part of a continuous truss.

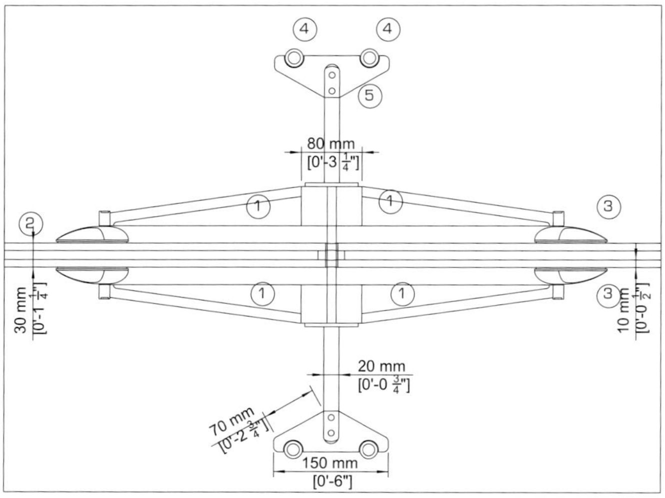

b| Prototype

unevenly supported. The glass is not structurally supported along its vertical edges as it is easier to sit them in the shoes. Sometimes clips along vertical joints are used for lateral restraint, but this depends very much on the specific design. An alternative method of fixing the glass is to use a silicone-bonded frame around the glass and fix that to cables or steel tubes. The joint between the glass is sealed with extruded silicone gaskets that can provide two lines of defence against windblown rain instead of the single line provided by site-applied silicone sealant. The use of extruded silicone allows the seal to be fixed to the unit in the factory, allowing the wall to be built in the manner of unitised glazing without the need for site scaffolding. Units are fixed from mast climbers or from scaffolding.

Opaque glazing

This system comprises single glazing bonded back to aluminium frames which are clamped to a supporting structure, typically a concrete blockwork wall or hollow terracotta block. Clamps are fixed between joints from the front as short lengths of pressure plate. Alternatively, panels may be hooked onto rainscreen-type support brackets before being clamped in place. Unlike rainscreen construction, the joints between panels are sealed with an extruded profile or rubber-based seal onto which the glass is fixed. Parapets are partially ventilated to avoid most of the water ingress, but allow some ventilation. The bottom is also ventilated, allowing moisture to find its way out. Opaque glazing can be easily integrated into adjacent areas of clear glazing by setting the aluminium extrusion away from the edge and fixing the edge of the glass into the adjacent glazing system.

Sealing clamped glazing

With all forms of clamped glazing, the movement or deflection of any fixed edges must be compatible with that of the general support system. This is particularly important in the case of cable-supported glaz-

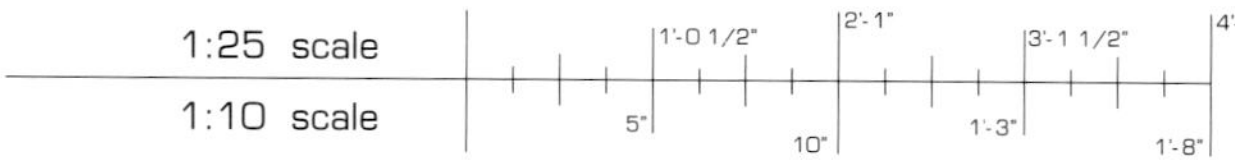

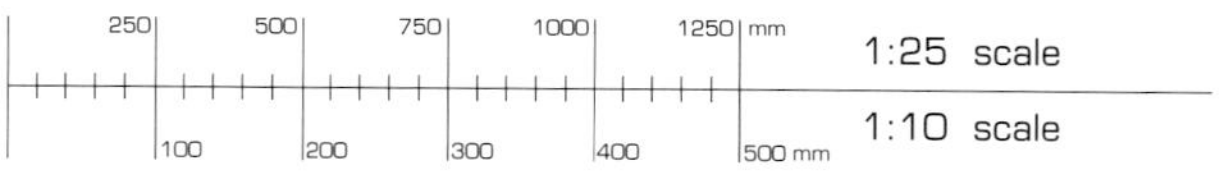

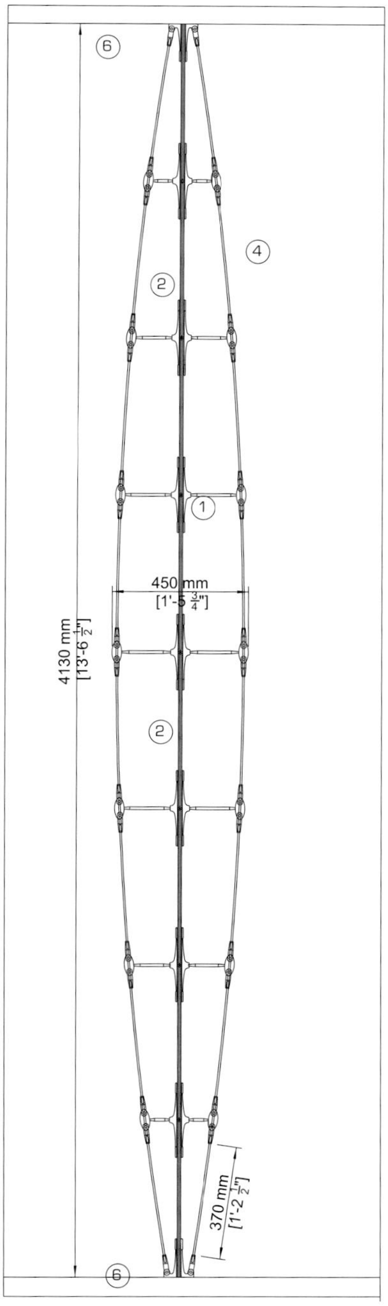

c| Horizontal and vertical section. Clamped glazing with cast steel clamps held in place by forming part of a continuous truss.

Details

1 | Cast metal fixing bracket
2 | Single glazed or double glazed unit to suit application
3 | Cast or machined metal clamp
4 | Stainless steel cable
5 | Metal connector
6 | Floor slab

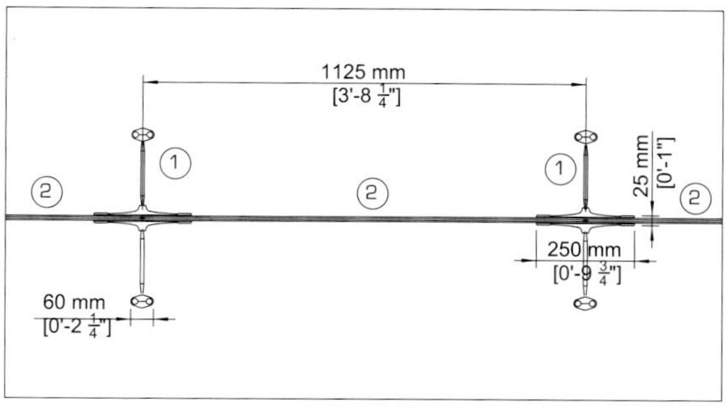

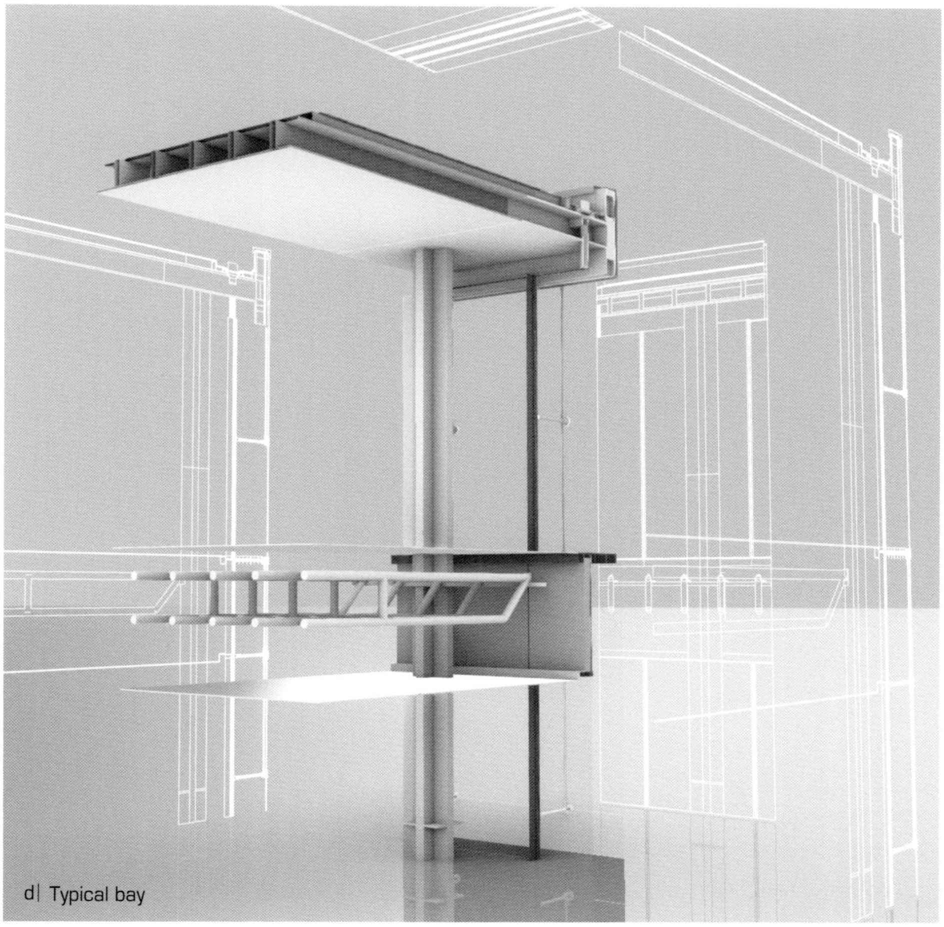
d| Typical bay

ing, where high deflections associated with the cables must be integrated with the small amounts of movement when the glass at the edge of the wall is fixed onto an adjacent form of construction such as a reinforced concrete wall. The edge channel or fixings must allow the glass to rotate within it to take up movement from deflection in a cable supporting the other end of the glass. Seals are usually formed by glazing channels or steel angles that provide a weathertight seal. High amounts of movement can be taken up at junctions with a flexible metal strip that is silicone-bonded to the edge of the glass to provide a flexible seal.

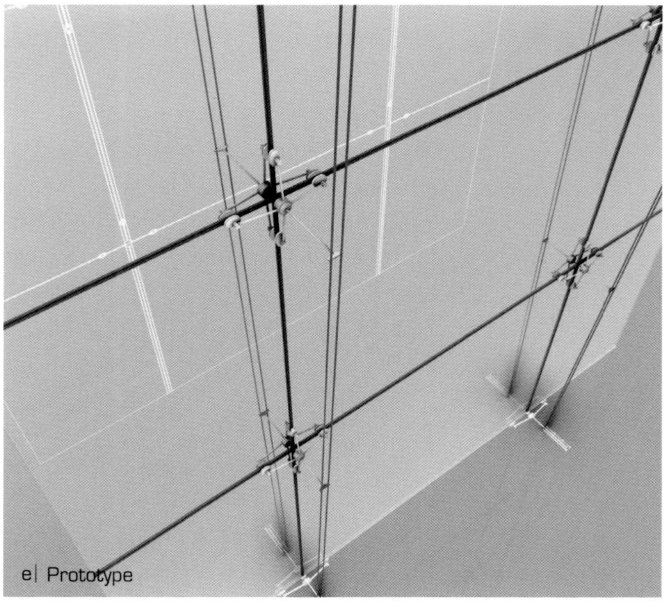
e| Prototype

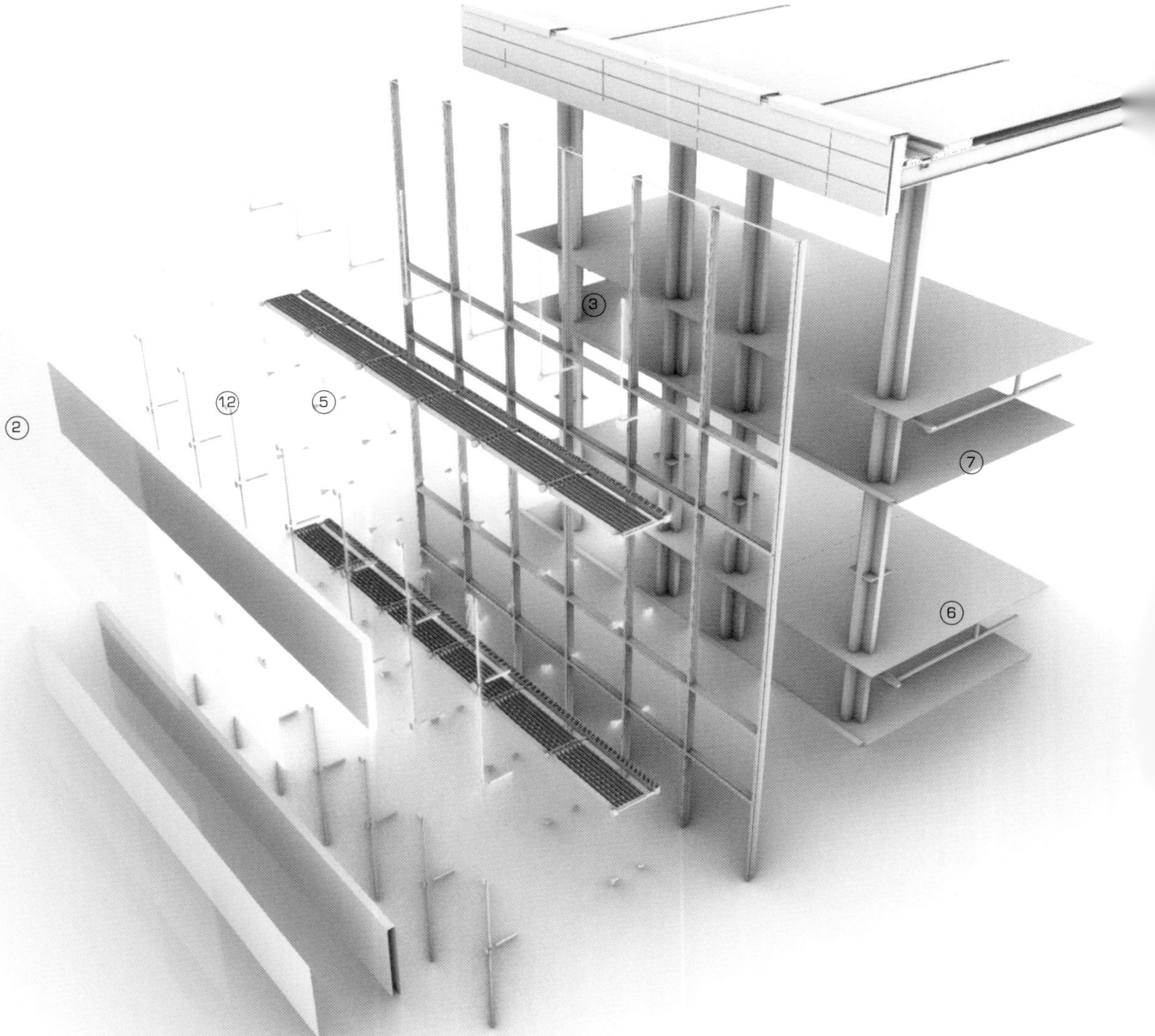

a | 3D exploded view showing clamped glazed facade fixed to steel frame

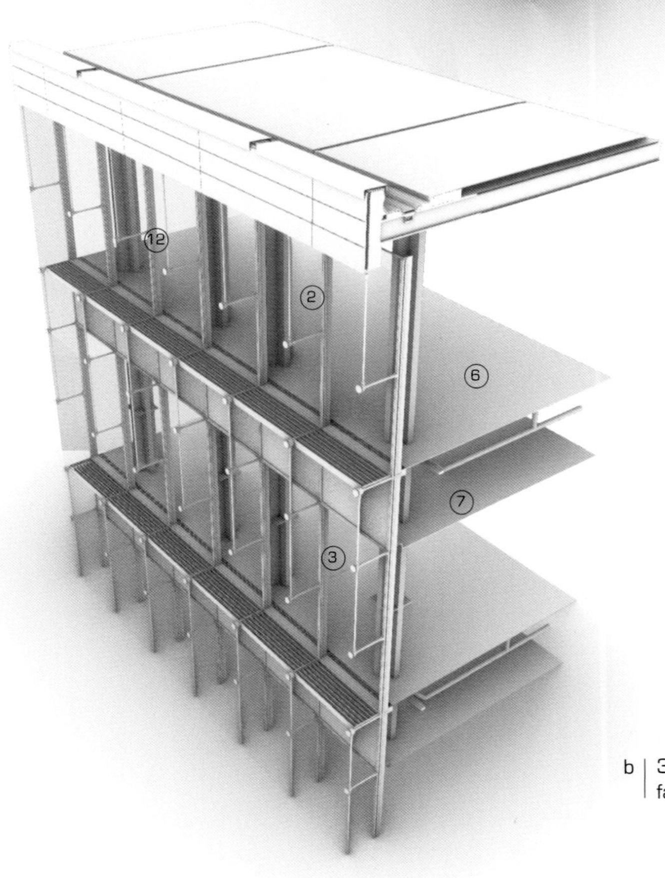

Details

1 | Stainless steel patch plates
2 | Single-glazed or double-glazed unit to suit application
3 | Silicone seal
4 | Glass fin
5 | Support bracket

b | 3D view showing clamped glazed facade fixed to steel frame

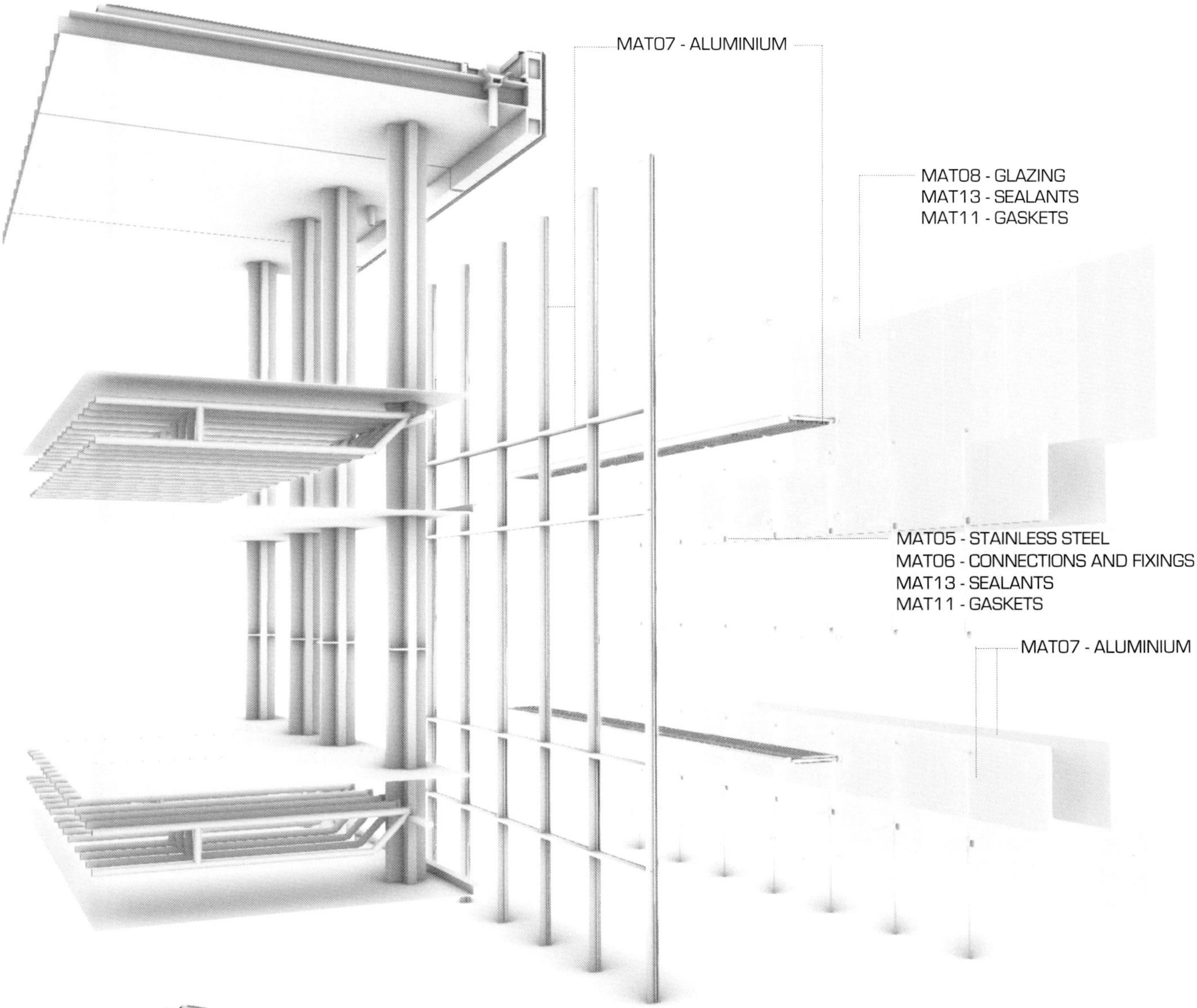

c | 3D exploded view of clamped glazed facade components with material specification references

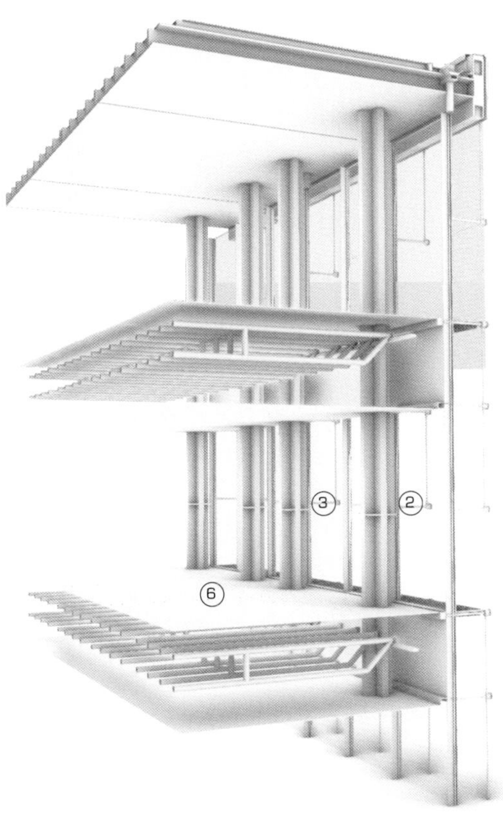

6 | Floor deck
7 | Ceiling finish
8 | Floor finish
9 | Access ladder
10 | Clamped glazed wall
11 | Inner framed curtain walling
12 | Fixing bolt
13 | Maintenance access deck

d | 3D view showing clamped glazed facade fixed to steel frame

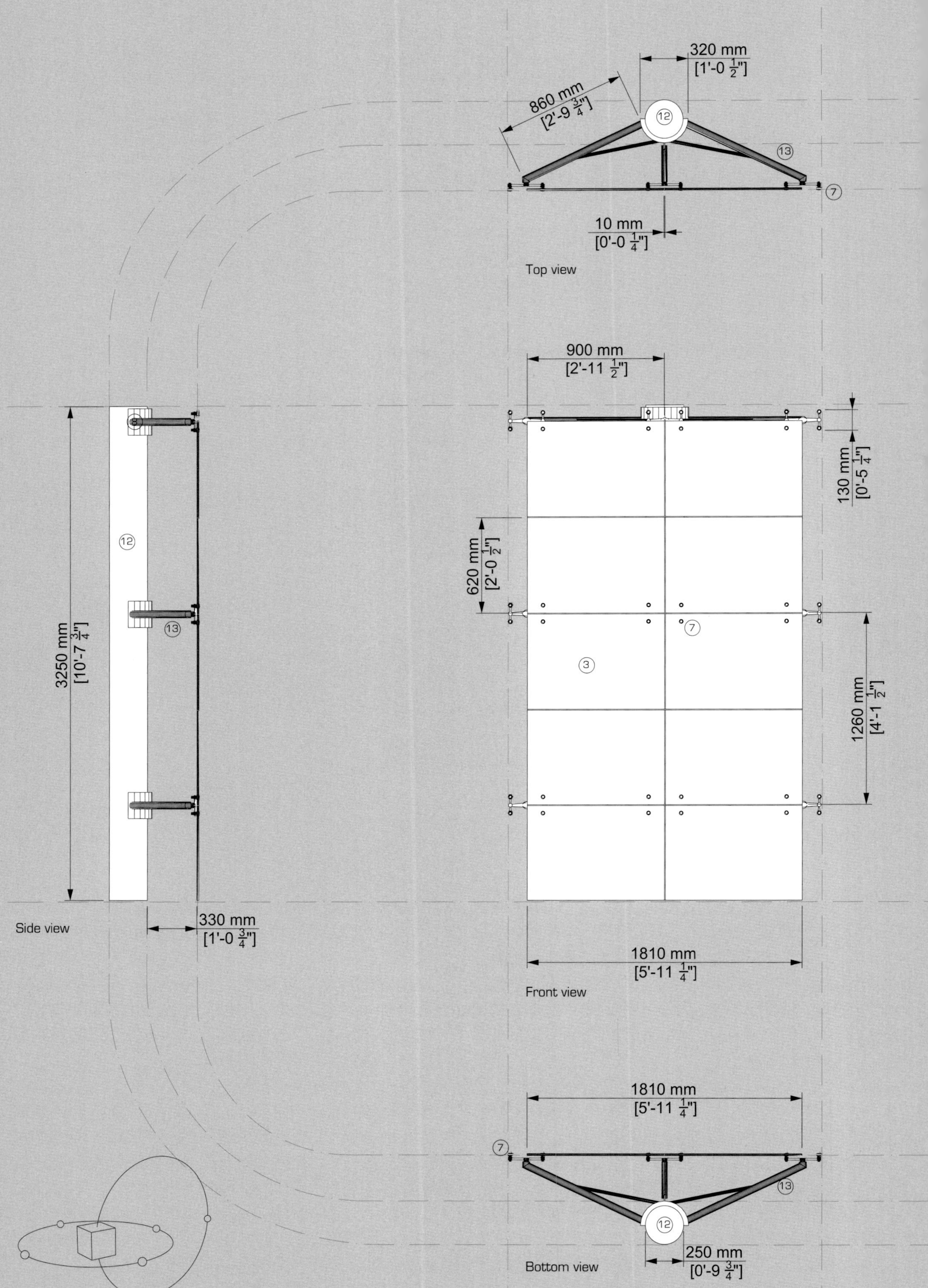

Third angle projection 1:30

Details

1 | Cast steel connector
2 | Mild steel or stainless steel angle bracket
3 | Single-glazed or double-glazed unit to suit application
4 | Outer silicone seal with inner rubber-based extruded seal
5 | Insulated panel
6 | Bolt-based cable end
7 | Stainless steel bolt fixing

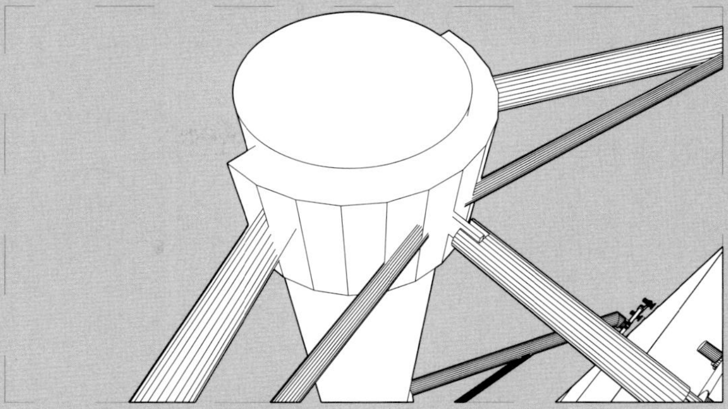

3D view of the assembly

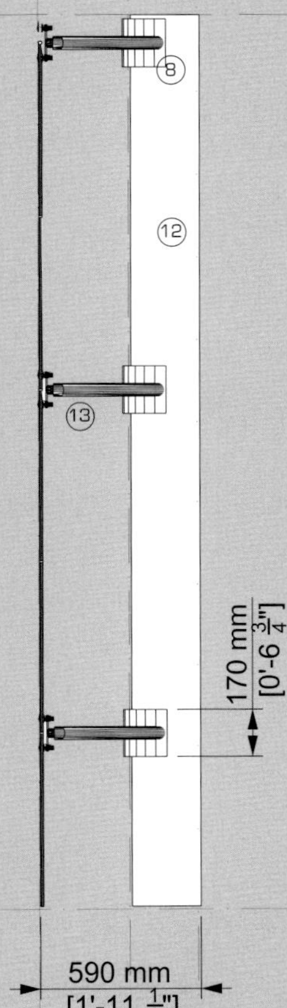

Side view

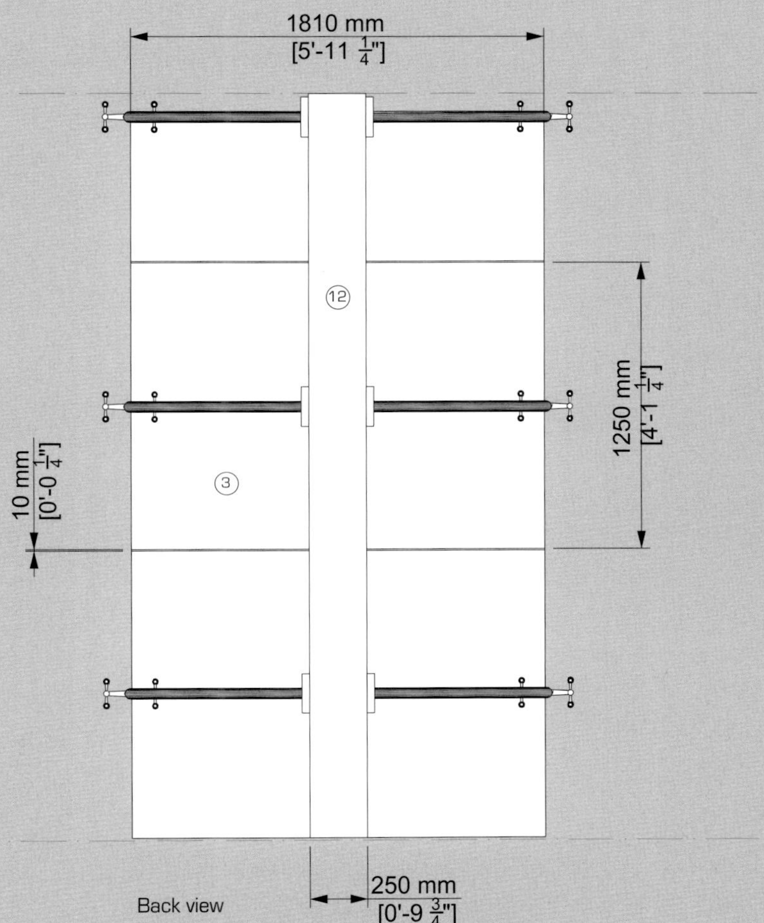

Back view

8 | Steel connector fixed to steel tube
9 | Stainless steel cable
10 | Floor slab/structural wall
11 | Glazing channel at floor level
12 | Structural column. Concrete shown
13 | Steel arm for lateral support
14 | Steel rod

3D view of the assembly

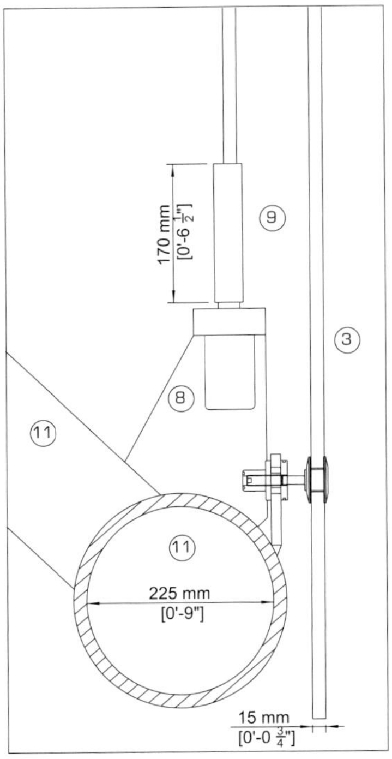

a | Elevation. Typical fixing arrangement for bolt fixed panels

Details

1 | Cast steel connector
2 | Mild steel or stainless steel angle bracket
3 | Single-glazed or double-glazed unit to suit application
4 | Outer silicone seal with inner rubber-based extruded seal
5 | Insulated panel
6 | Bolt-based cable end
7 | Stainless steel bolt fixing
8 | Steel connector fixed to steel tube
9 | Stainless steel cable
10 | Floor slab/ structural wall

c | Elevation. Typical layout of panels where cantilevered support brackets are used

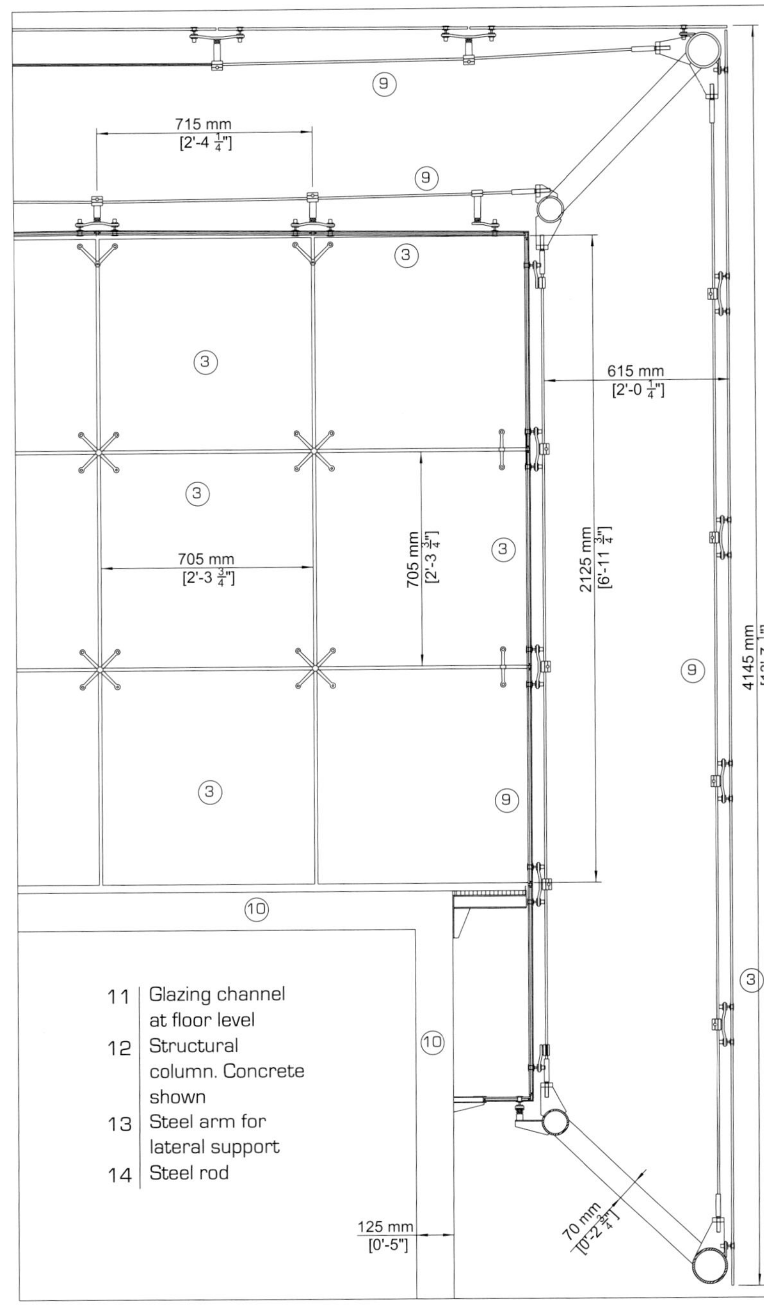

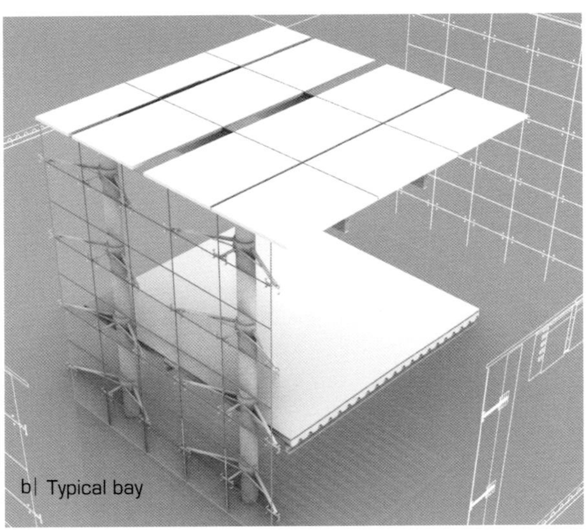

b | Typical bay

Like clamped glazing, bolt fixed glazing has a high degree of transparency provided by the bolts, which replace the need for metal framing around the edges of the glass sheet or double glazed units. One of the main advantages of bolt fixed glazing is the small size of the fixings when compared to the larger and more frequent clamped fixings. This system can also be double glazed much more easily since a spacer can be introduced into double glazed units during manufacture, away from the edge of the glass, where it also has the advantage of reducing the glass span, thus minimising glass thickness. However, the sight lines associated with double glazed units are almost the same as those for stick glazing, with the black line around the edge of two adjacent units, plus a 20mm (0.78in) average joint in silicone giving an overall 50mm (2in) thick joint width in a dark colour. Since the edges of double glazed units are made opaque to conceal the edging strip, translucent silicone is not used in the joint between two opaque edges. Instead, a dark coloured silicone is used, usually black.

Support methods

In common with some other glazed wall types, bolt fixed glazing is either hung from the top of the unit or is supported at the bottom. When hung from the top, bolt fixings are secured to stainless steel cables or rods fixed at the top and tensioned at the bottom (a). Fixing tolerances are provided in three directions: vertically, horizontally and laterally. Because the joints between the glass units or sheets are visible, and need to be well aligned, and due to the fact that the glass is drilled during its manufacture of the glass, fixing tolerances have to be provided by the bolt fixing and its connection to the supporting bracket.

The bolt fixings themselves vary in complexity from economic types where a threaded bolt is secured with nuts to a fixing bracket, to sophisticated types where all threads are concealed behind sleeves to give a very smooth appearance. All bolt types allow rotation to occur between the glass and the fixing, usually to a maximum of 12 degrees. A ball

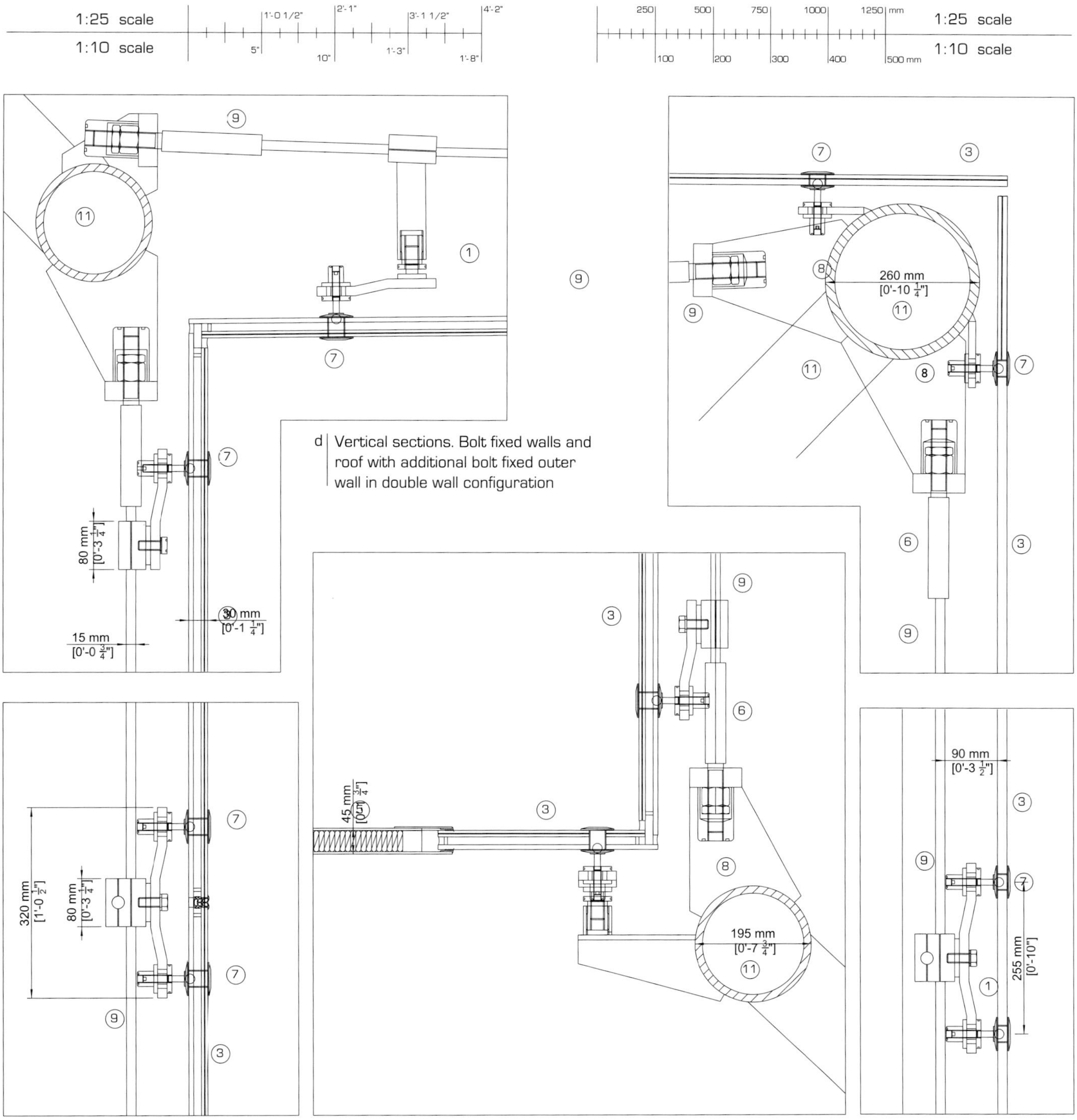

d | Vertical sections. Bolt fixed walls and roof with additional bolt fixed outer wall in double wall configuration

joint is housed where the fixing intersects with the glass for this purpose. Any sleeve that is set over the threaded rod behind the disc is positioned so that it does not interfere with the free movement of the ball joint. Support brackets to bolts are either cast or are machined and welded from plate. The choice will depend largely on shape and the number of brackets needed, since castings are economic only in large numbers cast from a single mould. Castings can be formed to take up complex geometries that look very cumbersome in welded plate but are more expensive and time-consuming to produce, except in large quantities. The choice of painted mild steel or stainless steel (polished or brushed) for the support bracket is very much a visual decision, particularly where a large amount of welding is required. If the welding is not performed to the highest standards then the results can be visually disappointing. However, the bolts themselves are always in stainless steel. Junctions between mild steel and stainless steel are isolated to avoid the strong electrolytic corrosion that occurs between these two metals.

Construction tolerances between supporting structure and glass panels are accommodated between the glass and the bolt fixing, between the bolt fixing and the support bracket or 'spider' and between the support bracket and the supporting column or truss. The gap between glass bolt and fixing is to allow three of the typical four bolt fixings on a glass unit to move freely, while the fourth unit is clamped tight without damaging the glass at the edge of the hole. If adjustment is provided between bolt fixing and support bracket in order to take up dimensional differences between glass and supporting structure then this results in different joint widths between glass units, with slightly uneven corners where four glass units meet. This method does, however, allow the support spider to be set in a fixed relationship with the supporting truss or column. If this adjustment is instead formed between spider and primary structure then the glass remains in perfect alignment, but the alignment between spider and column/truss will vary. The choice of where to position the

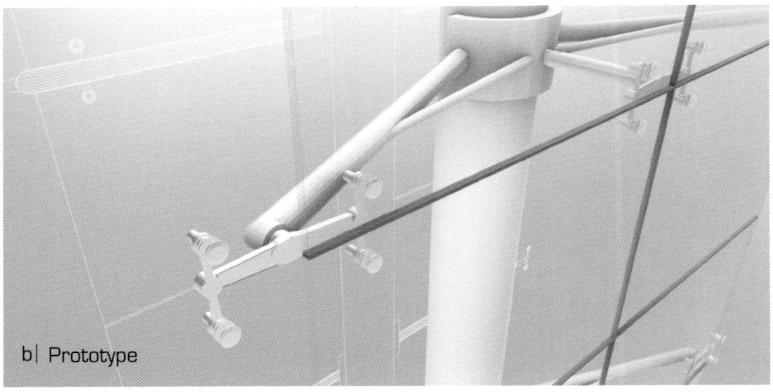

b| Prototype

b| Vertical sections and elevation. Typical details from vertical section shown on page 151

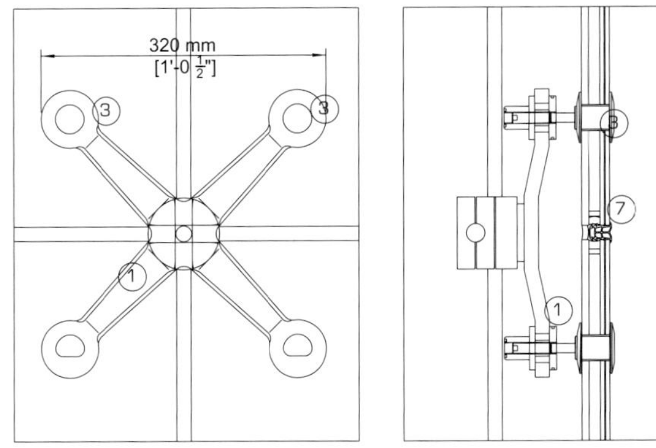

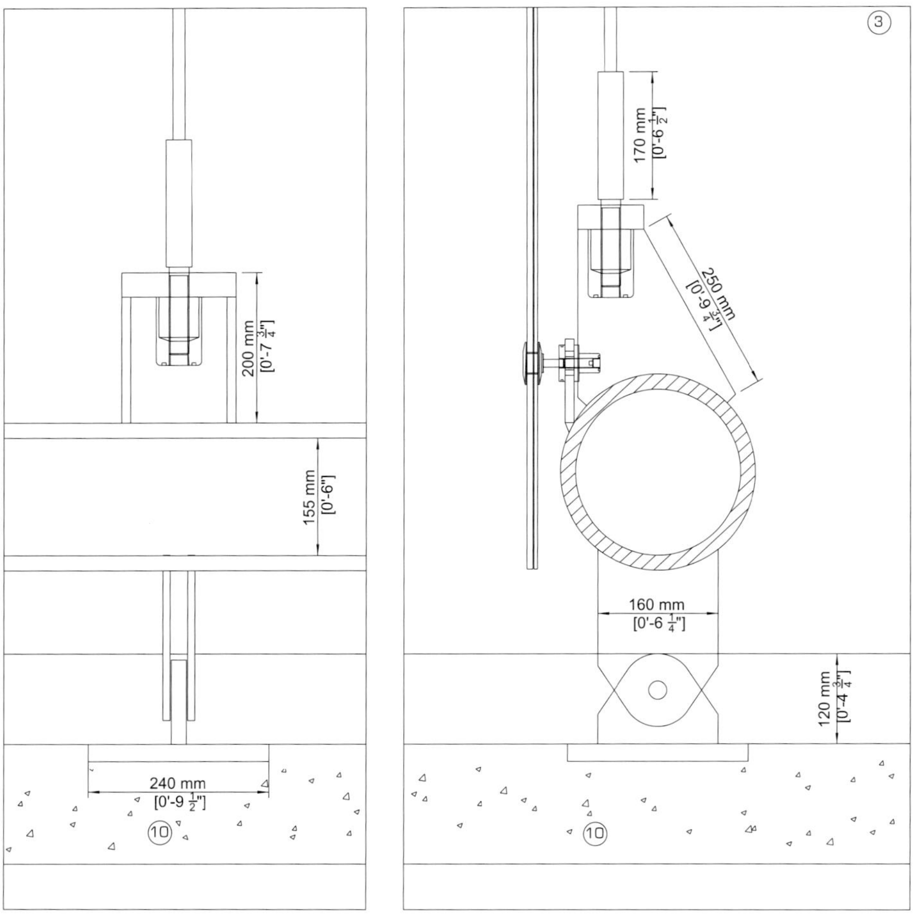

Details ③ ⑭

1 | Cast steel connector
2 | Mild steel or stainless steel angle bracket
3 | Single-glazed or double-glazed unit to suit application
4 | Outer silicone seal with inner rubber-based extruded seal
5 | Insulated panel
6 | Bolt-based cable end
7 | Stainless steel bolt fixing

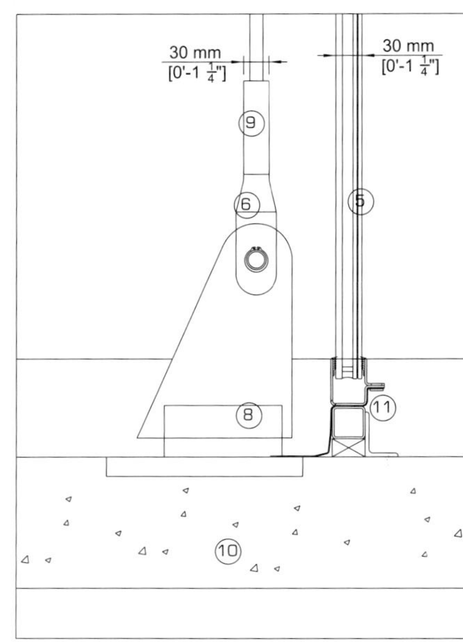

dimensional adjustment varies between designs and is largely based on visual preference.

Because all components in bolt fixed glazing are visible, both from inside and outside, the choice of fixing bolts and screws is visually very significant, as is the extent of visible thread on the bolt fixing itself. Countersunk bolts and pig-nose bolts (those with two small holes on their face to provide points for tightening) are often preferred to the more common hexagonal head types. Bolts with Allen keys (a square hole cut into a circular bolt head tightened with a socket wrench) are also preferred, as are smooth circular washers with small holes for tightening drilled into their edges.

Bottom-supported glazing

Glass supported at its base can be fixed by a variety of methods. Glass units are most commonly supported with glass fins or steel posts. When glass fins are used, the glass is stacked from its base, with the dead load passing partly to the fins, and partly through the bolt fixings to the glass below, depending on the design. The glass fins serve primarily to stiffen the glass wall and resist wind loads. This method is very much a development of the patch plate system described in the previous topic in order to provide maximum transparency with a minimum of metal supporting structure.

The glass units and fins are clamped at their base to the floor slab. Because the clamping plates are big, secured by four to six bolts which pass through the glass, they are often concealed below the finished floor level. The floor slab supporting the glazed wall is either stepped down to accommodate the base clamps or forms part of a raised floor zone. With either method, the pocket formed can also be used to accommodate convector heating used to reduce down-draughts in temperate climates. Above the base level, glass units are fixed with bolt fixings that transfer loads from the top glass to lower ones.

c | Horizontal and vertical section. Typical details from vertical section shown on page 151

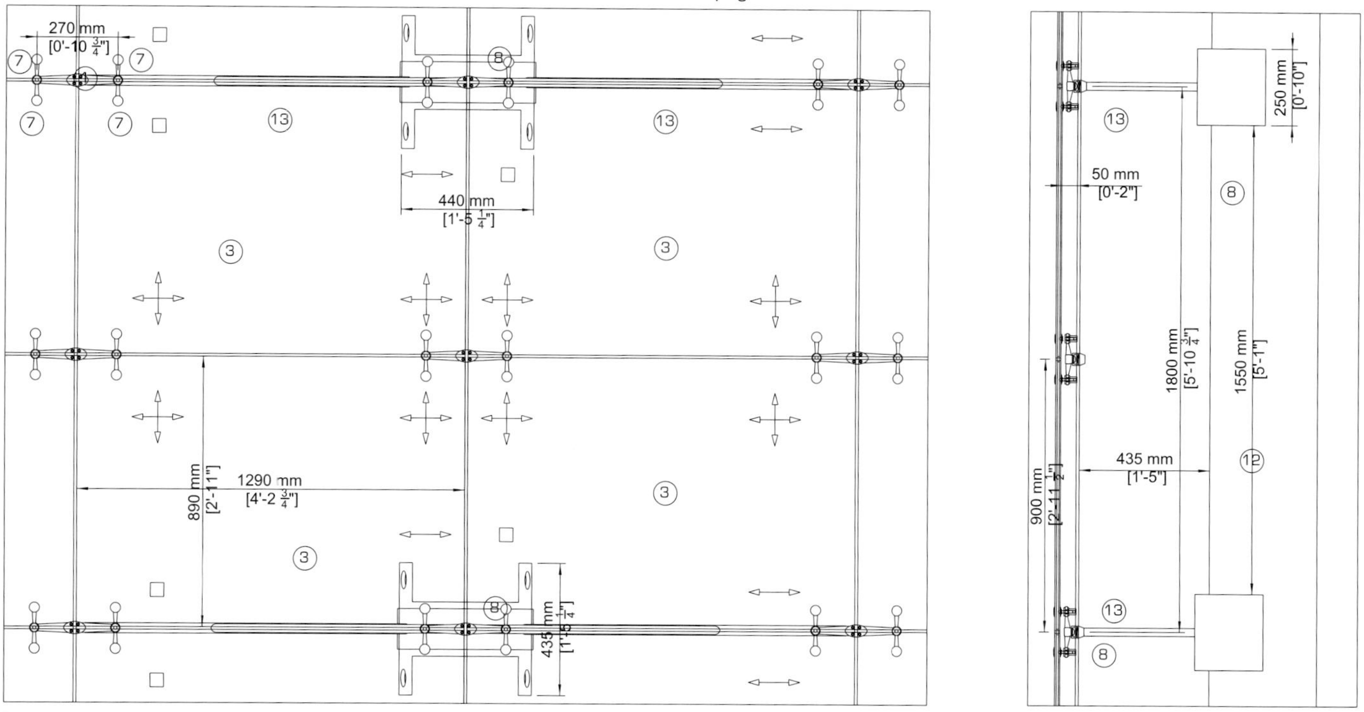

Key to fixing conditions typical bolt fixed wall

Glass fixed

Glass movable in one direction

Glass free to move

8 | Steel connector fixed to steel tube
9 | Stainless steel cable
10 | Floor slab/structural wall
11 | Glazing channel at floor level
12 | Structural column. Concrete shown
13 | Steel arm for lateral support
14 | Steel rod

d | Elevation. Bolt fixing or 'rotule' in cross-shaped design

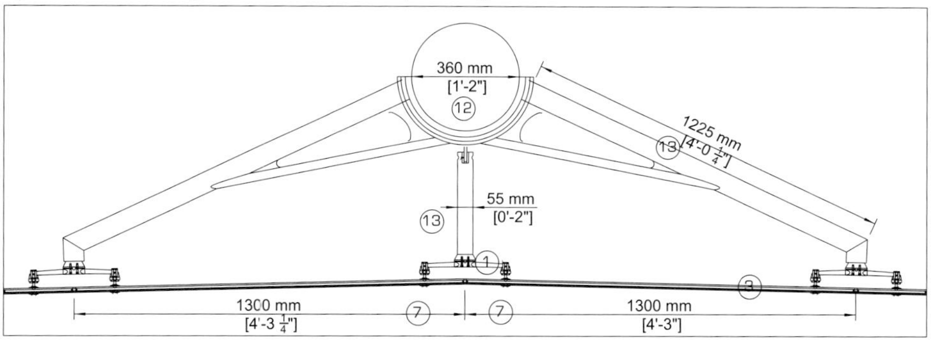

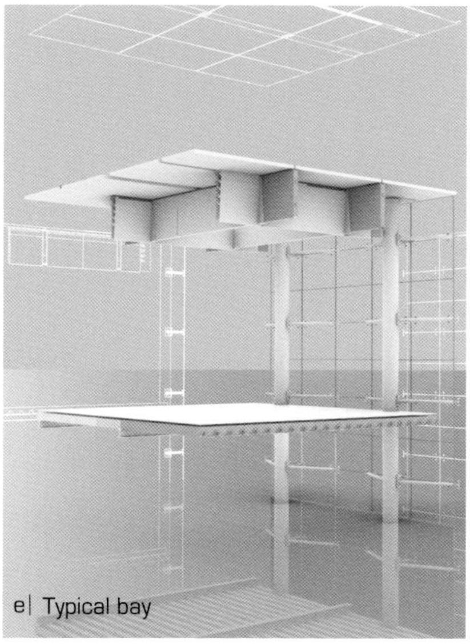

e | Typical bay

Structural posts, usually in mild steel, are an alternative method of supporting the glazing without introducing a visually intrusive element, though concrete columns are sometimes used where they form a necessary part of the primary structure. This method has the advantage of allowing each glass to be individually supported. In generic examples, the glass is fixed back to a steel post at each vertical joint. Lateral restraint is provided by vertical or horizontal wind trusses in either steel tube or cable (b). Cable is preferred for its visual lightness. The size (diameter) of the supporting posts or trusses can be reduced by adding outrigger brackets to posts set at wider centres. This reduces the number of support posts by two thirds, but each post will necessarily be bigger. The outriggers can also be reduced to become lateral restraints only by introducing small diameter posts, typically around 10mm (0.4in), to take the vertical load.

In all cases the bolts supporting the glass are fixed back to steel columns with brackets or to outriggers with similar brackets. The general shape of the bracket is determined both by the position of the holes in relation to the edge as well as by the method of fixing the bracket back to its supporting structure. The positions of the holes are usually so as to reduce the span of the glass in order to keep it within an economic thickness, typically 12mm (0.5in). An essential aspect of bracket design is in accommodating fixing tolerances. Since the supporting structure is erected before the glass, tolerances are introduced into the bolt and bracket fixing to take up the differences.

Top-hung glazing

A supporting structure to bolt fixed glazing can be made visually very lightweight by the use of cables or rods hung from the top and tensioned at the base (MCE_155, d). Stainless steel cables or painted mild steel rods are used. The choice is usually made for visual rea-

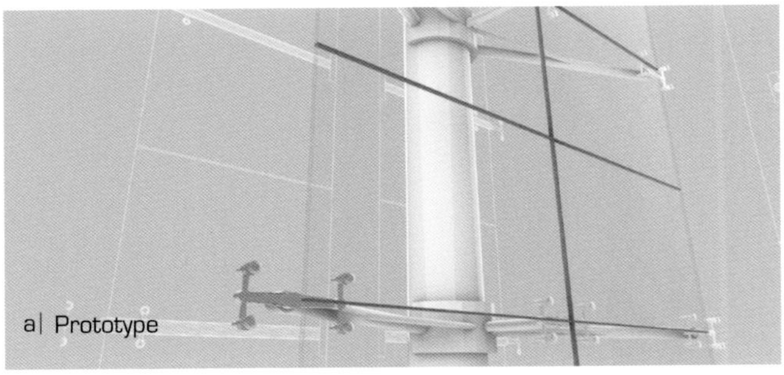

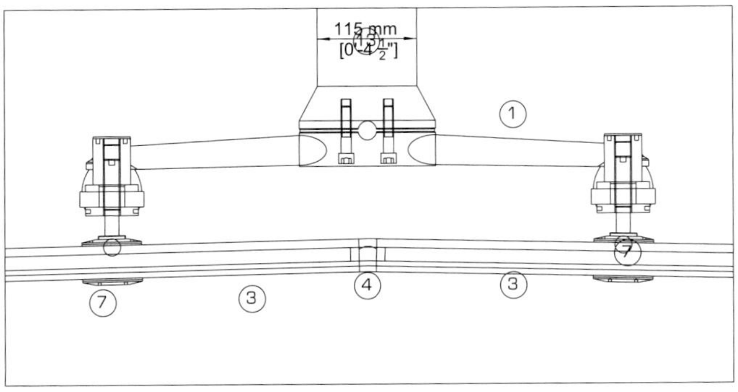

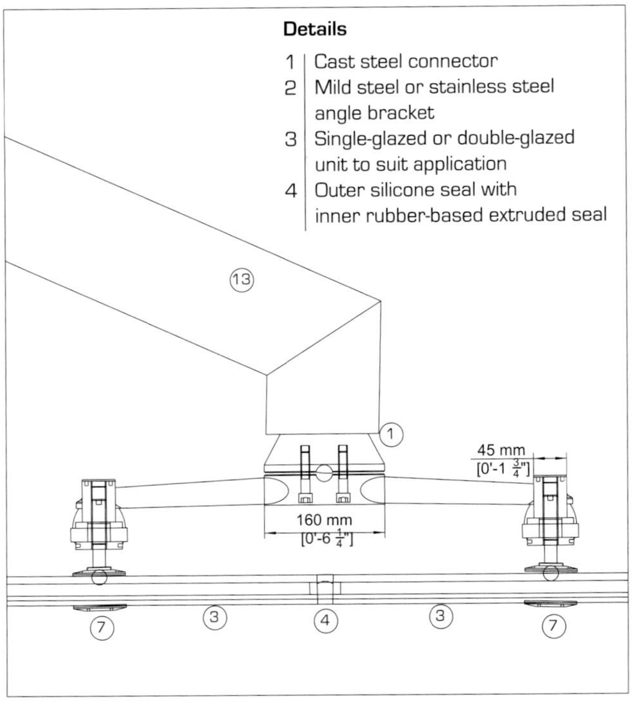

b| Vertical sections and elevation. Every second glass panel is supported by cantilevered arms spanning from a post. Remaining glass panels are fixed to the bolt fixing or 'rotule'

c| Vertical sections and elevation. 'H' style fixing connection

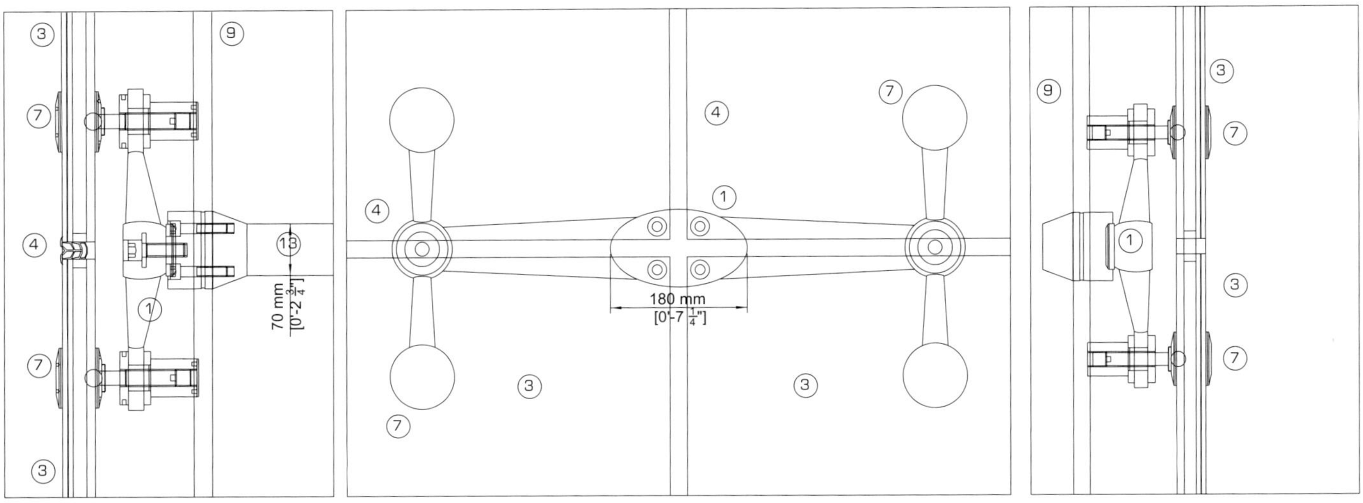

sons. However, like clamped glazing this method is associated with high deflections under wind load. This can be accepted either in the design (some cable-assisted walls deflect up to 800mm/2ft 8in at their centre) or be restrained with cable trusses or outriggers, typically at floor levels. Bolt fixings are secured with brackets that are secured directly to the cable or rod. They usually fit together in two halves when secured to cable (MCE_152, b). Where rod is used, the material is usually in lengths that connect into a fixing made as a single piece.

Again, the mixture of stainless steel and mild steel requires separators to avoid bimetallic corrosion, caused by contact between different metals in the presence of rainwater. While mild steel rod can be painted to match other parts of the construction such as adjacent supporting structure, care must be taken during assembly on site to ensure that factory-coated components and assemblies are not damaged. Where stainless steel is used instead of either mild steel rod or stainless steel cable, the diameter of the rod will be bigger than either of the other two options.

Corners

Corners are made by either introducing supporting structure at the corner or alternatively by cantilevering or fixing the glass at the corner and pinning the glass together with two bolts forming a corner bracket (MCE_151, d). Where structural support is provided at a corner, a special bracket is usually required, though manufacturers are increasingly providing standard corner brackets as part of their systems. Variations of corner condition are usually as economic as one another. Where double glazed units are used, care is taken to avoid exposed edges of glass where they are vulnerable to damage, usually from cleaning equipment.

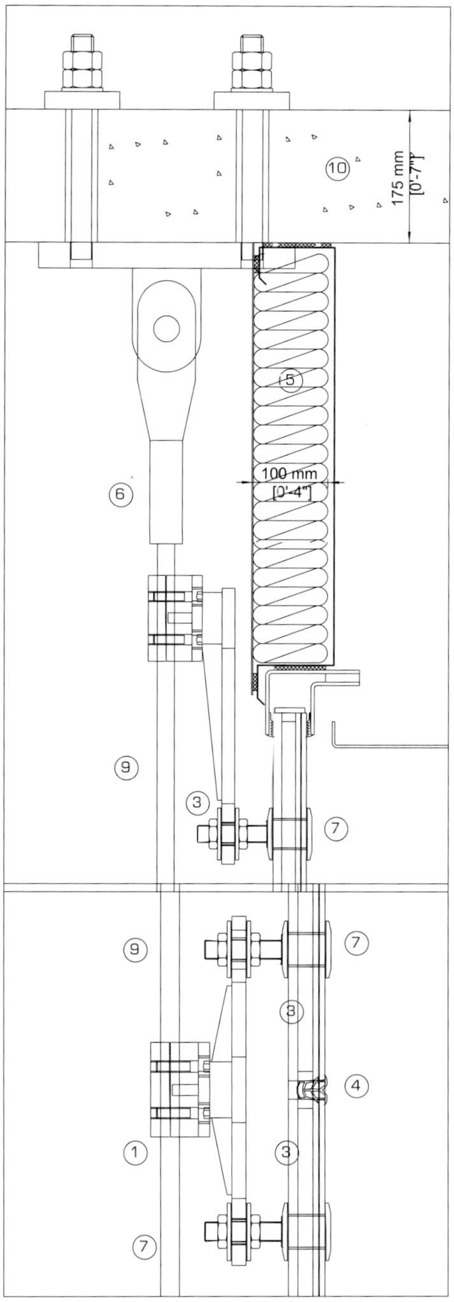

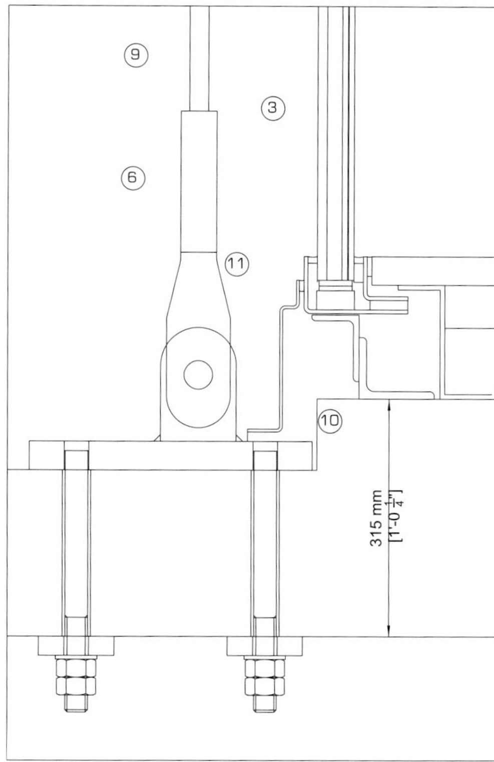

d | Vertical sections. Panel to panel and junction with floor slab

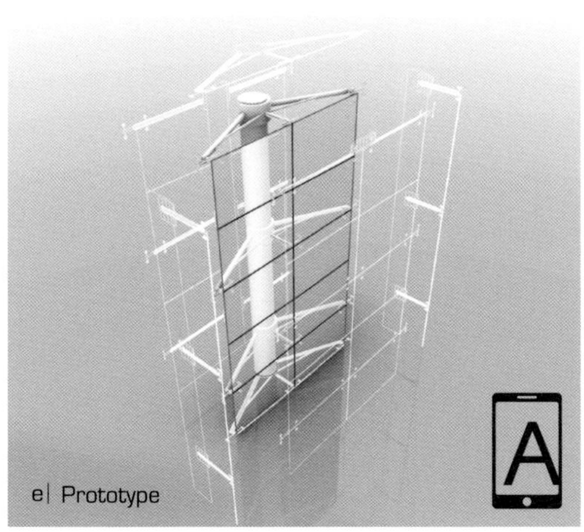

e | Prototype

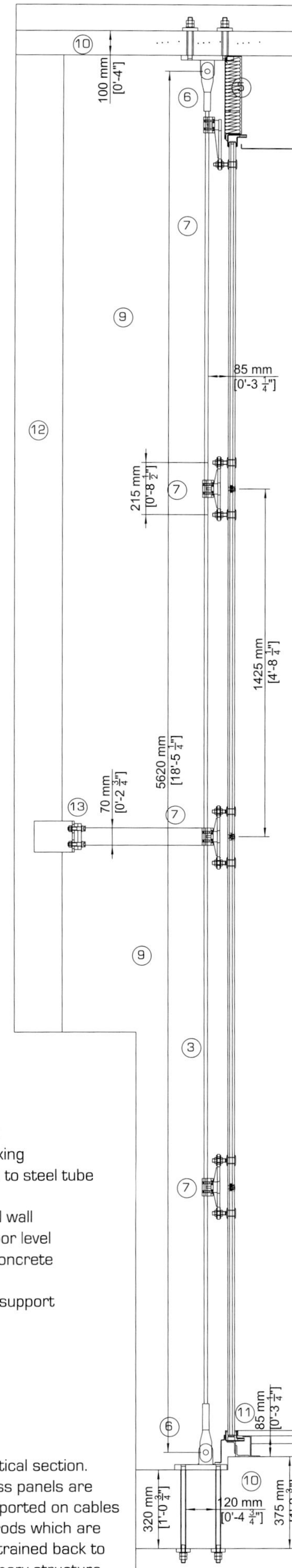

f | Vertical section. Glass panels are supported on cables or rods which are restrained back to primary structure

Seals and interfaces

Like clamped glazing, bolt fixed glazing has essentially a single barrier to water penetration. The workmanship of the silicone sealant on site is critical to its successful application. In common with clamped glazing, silicone extrusions are slowly being introduced to make installation easier (MCE_156, a). The depth of double glazed units (often around 30mm/1.2in) allows a small drainage channel to be introduced. This information can be found in the section on clamped glazing.

Where bolt fixed glazing interfaces with adjacent glazing systems, the edge of the glass is usually glazed-in directly to those systems. Where it meets a wall type in a non-glazed material, the bolt fixed glazing is usually terminated by a glazing channel or metal angles that are fixed to the adjacent wall. The detail is formed in a way that conceals the angles, leaving only the glass visible. Where bolt fixings are used at the interface with another material, half brackets are used.

5 | Insulated panel
6 | Bolt-based cable end
7 | Stainless steel bolt fixing
8 | Steel connector fixed to steel tube
9 | Stainless steel cable
10 | Floor slab/structural wall
11 | Glazing channel at floor level
12 | Structural column. Concrete shown
13 | Steel arm for lateral support
14 | Steel rod

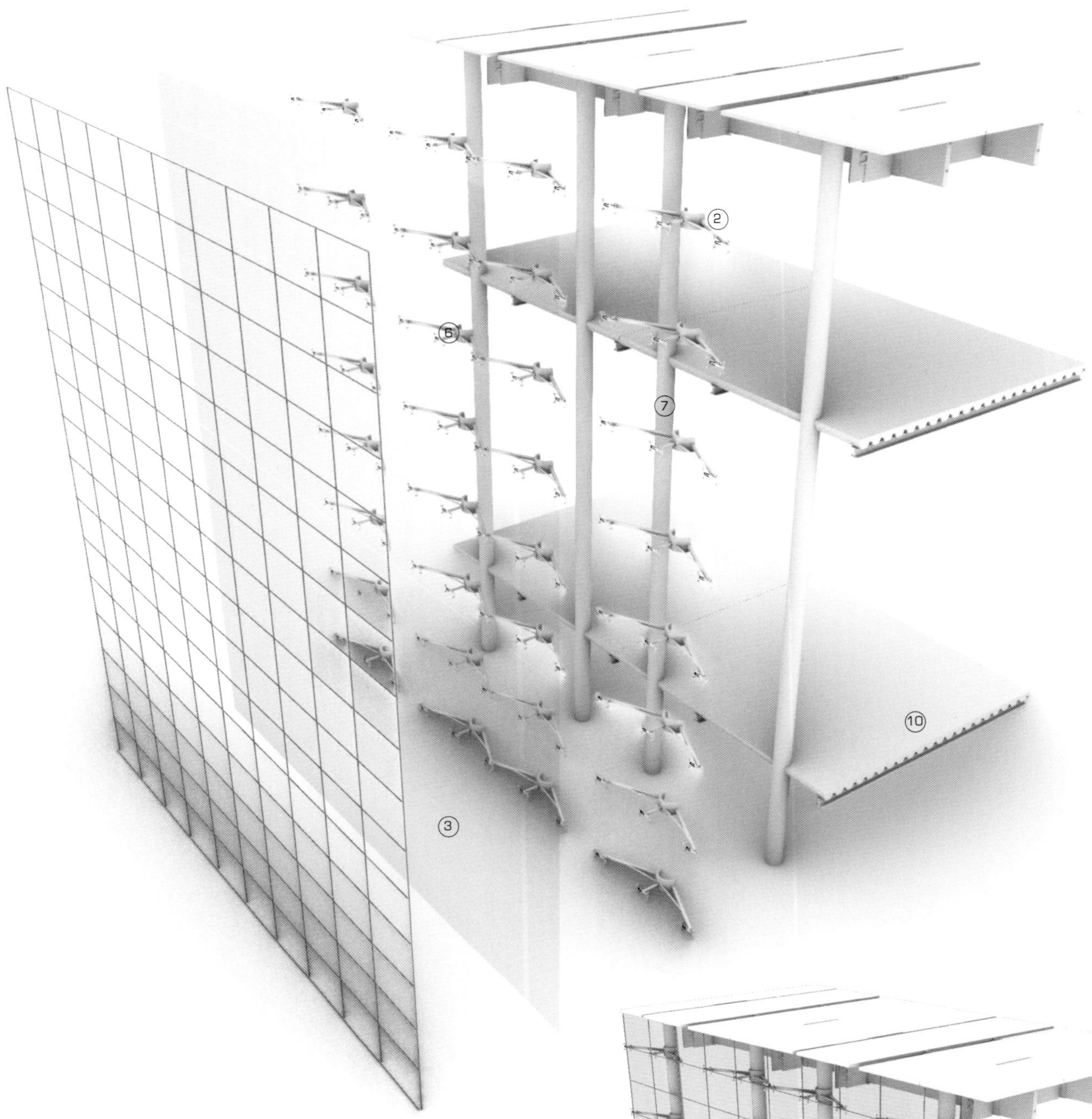

a | 3D exploded view showing bolt fixed glazed facade to steel frame

Details

1 | Cast steel connector
2 | Mild steel or stainless steel angle bracket
3 | Single-glazed or double-glazed unit to suit application
4 | Outer silicone seal with inner rubber-based extruded seal
5 | Insulated panel
6 | Bolt-based cable end
7 | Stainless steel bolt fixing

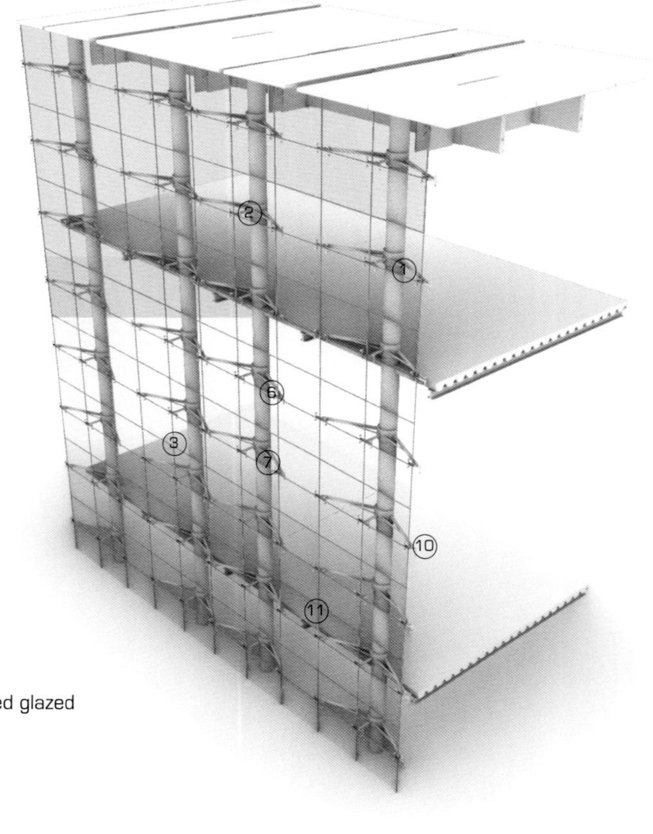

b | 3D view showing bolt fixed glazed facade to steel frame

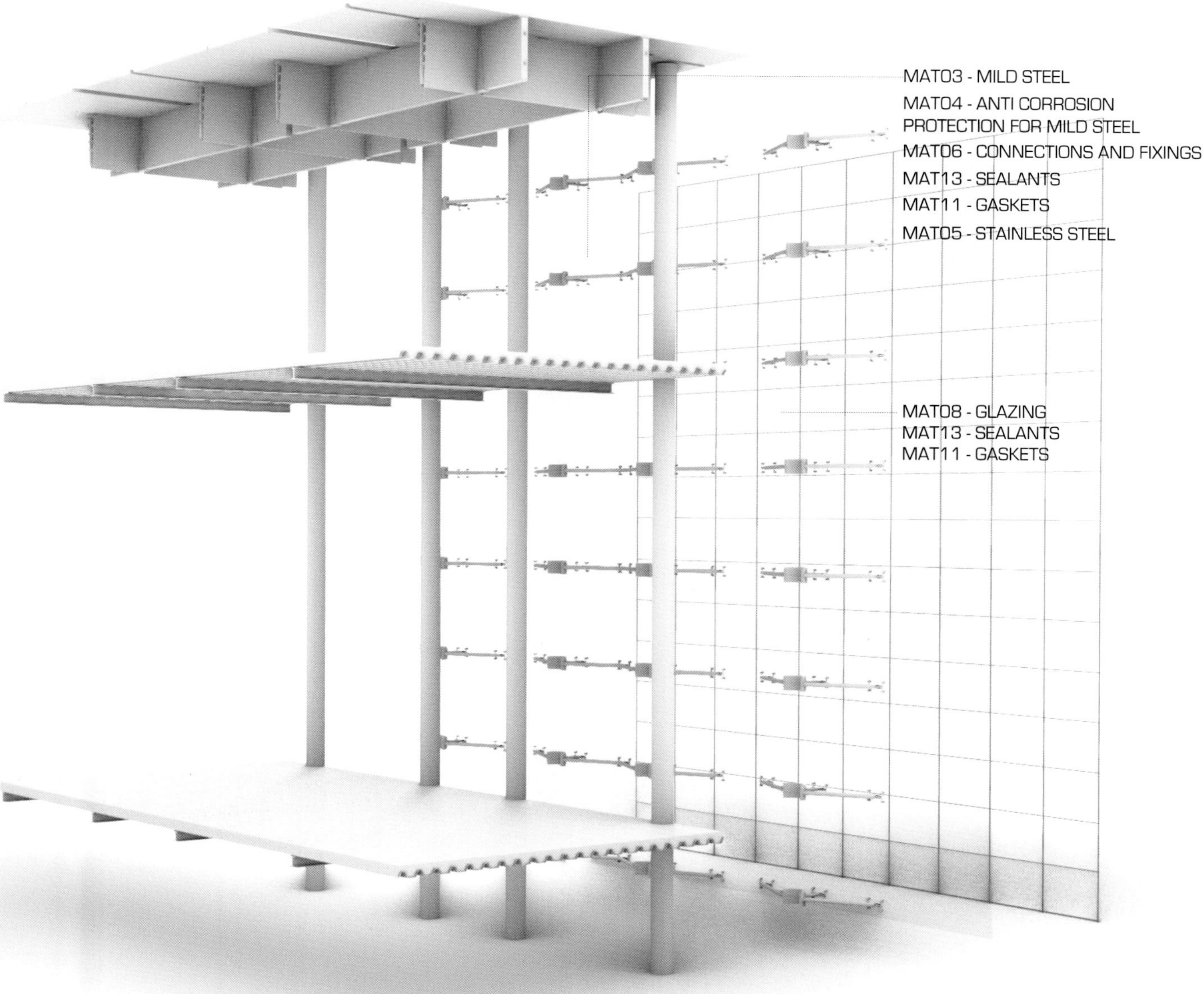

c | 3D exploded view showing bolt fixed glazed facade components with material specification references

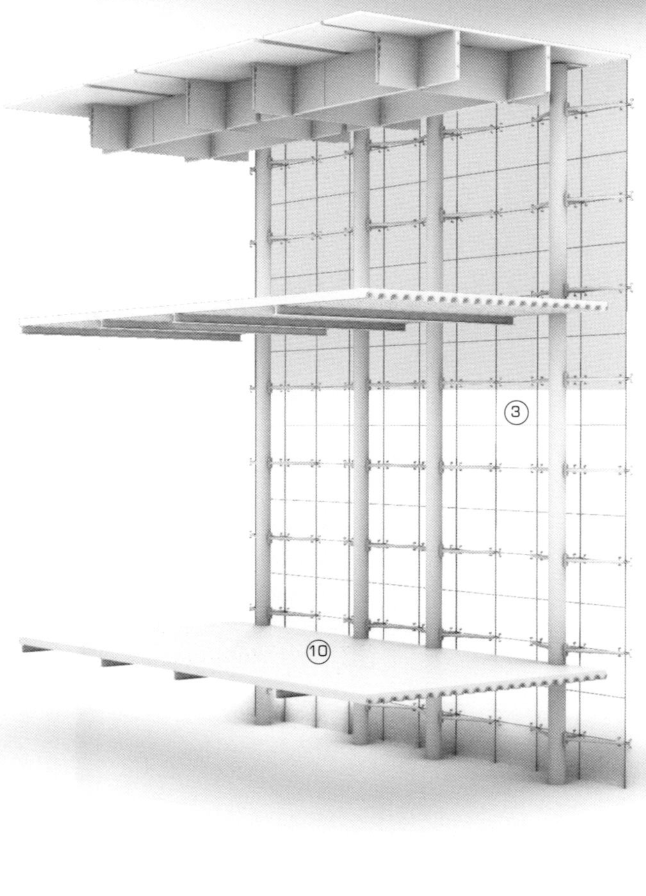

8 | Steel connector fixed to steel tube
9 | Stainless steel cable
10 | Floor slab/structural wall
11 | Glazing channel at floor level
12 | Structural column. Concrete shown
13 | Steel arm for lateral support
14 | Steel rod

d | 3D view showing bolt fixed glazed facade to steel frame

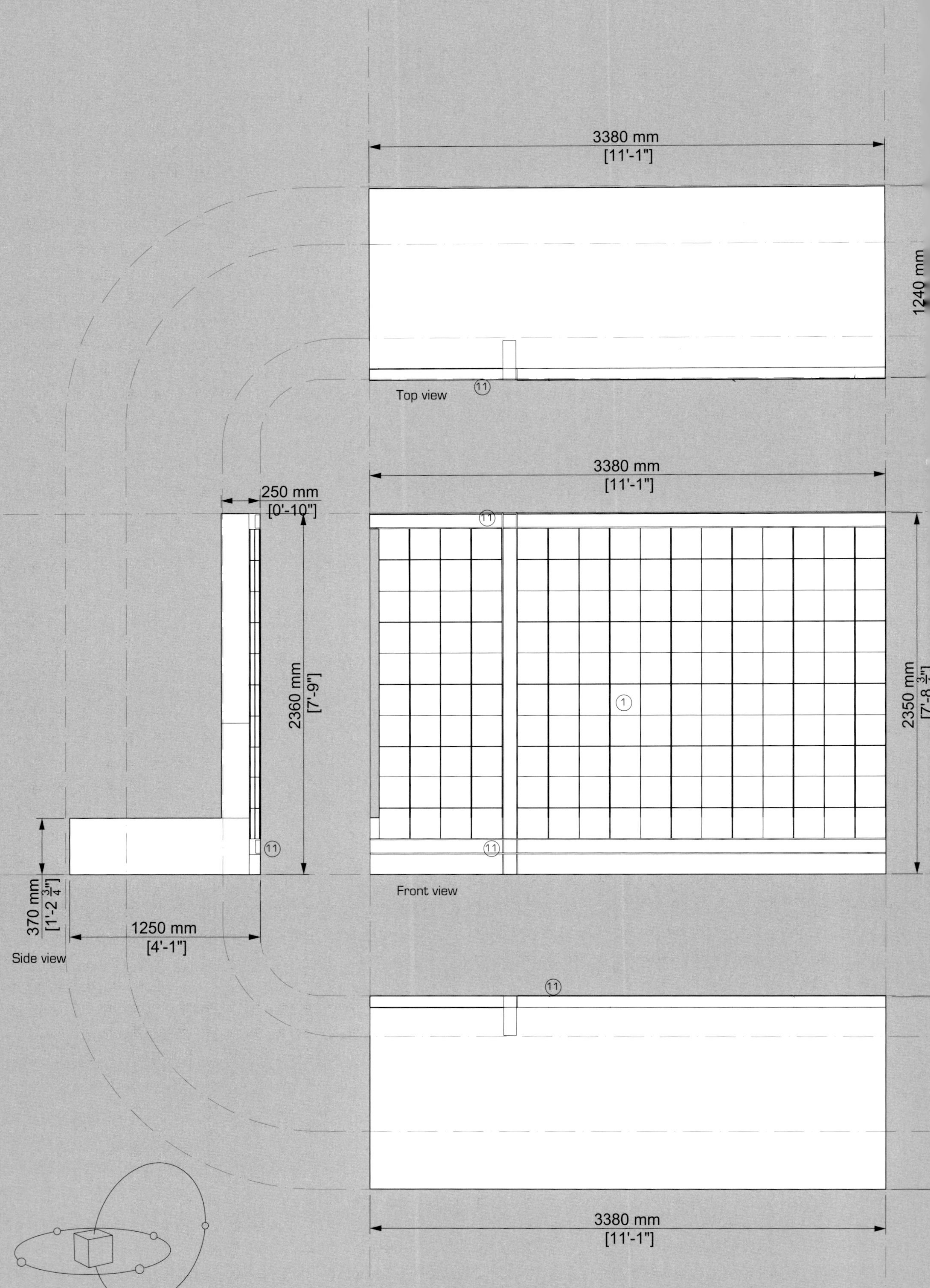

Third angle projection 1:30

Details

1 | Glass block
2 | Bedding reinforcement
3 | Bedding compound, mortar or silicone-based bond
4 | Adjacent blockwork wall with rainscreen
5 | Adjacent concrete wall
6 | Adjacent light gauge steel-framed wall
7 | Reinforced concrete frame
8 | Steel box section
9 | 90° corner block
10 | 45° corner block

3D view of the assembly

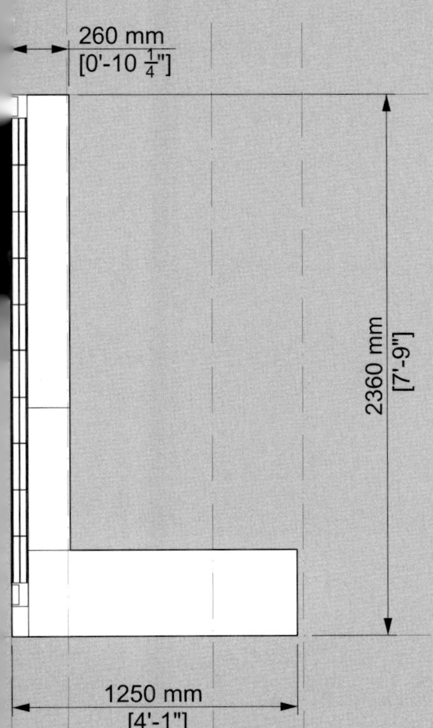

ide view

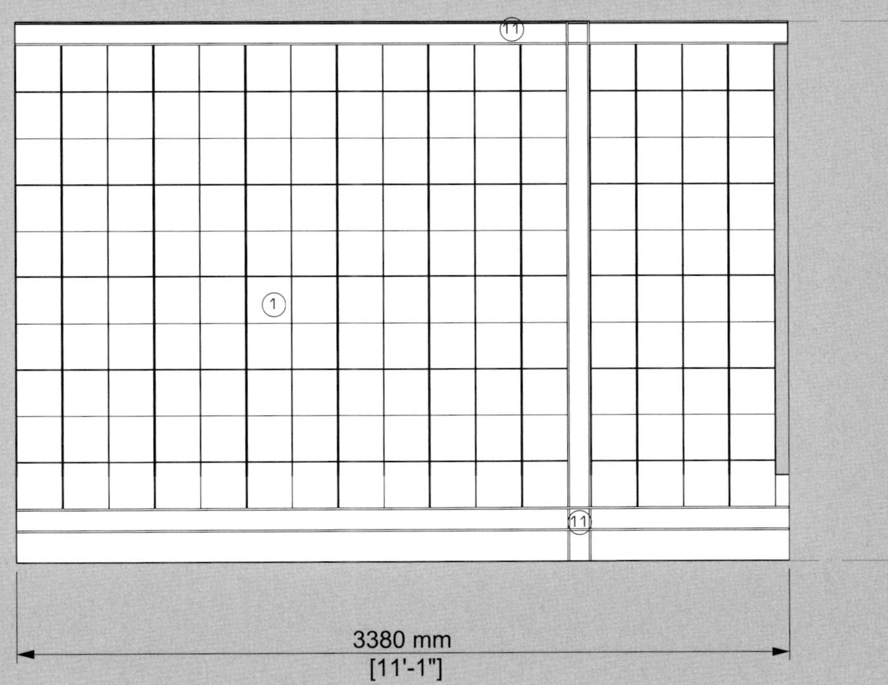

Back view

11 | Steel I-section, usually specially fabricated
12 | Steel angle
13 | Inside
14 | Outside
15 | Floor deck
16 | Steel T-section
17 | Adjacent brick cavity wall construction
18 | Thermal insulation
19 | Extruded aluminium sections providing support
20 | Adjacent reinforced concrete wall

3D view of the assembly

a| Prototype

Details

1 | Glass block
2 | Bedding reinforcement
3 | Bedding compound, mortar or silicone-based bond
4 | Adjacent blockwork wall with rainscreen

b| Horizontal section. Glass blocks supported by steel angles and box sections

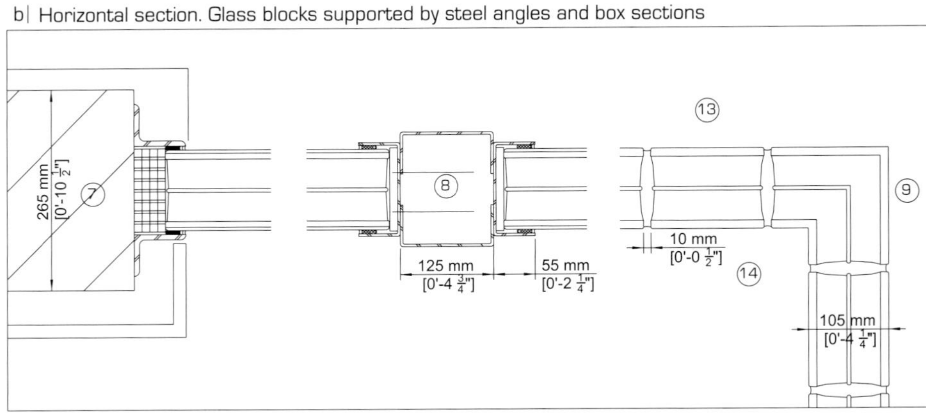

c| Horizontal and vertical sections. Glass blocks supported by welded sections and box sections

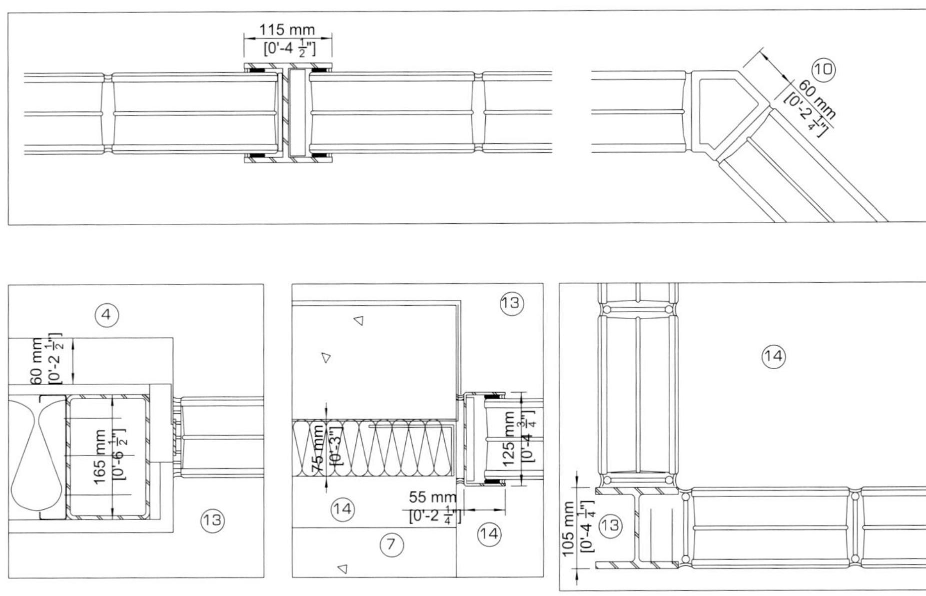

d| Horizontal section. Glass blocks supported by T-sections

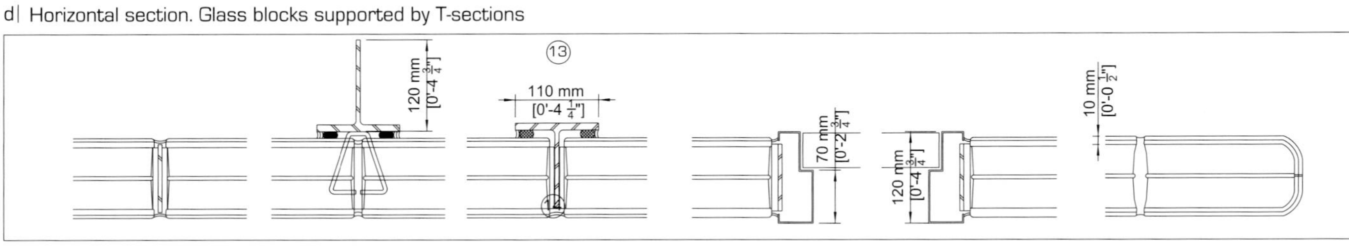

Glass blocks are used to form a robust, translucent and fire resisting glazed wall construction. Glass blocks are made as either solid or hollow components, but the hollow type has better thermal and acoustic insulation due to the partially air-evacuated void within and is most commonly used in wall construction.

Glass blocks are made by pressing together two half blocks of glass which are held at high temperature. The half blocks are made as either pressed or cast shapes. As the block cools after manufacture, a partial vacuum is created as the air pressure drops with temperature.

The most common sizes of glass block for walls are as follows:
190mm x 190mm thick (metric size)
8 x 8 x 4in thick (imperial size)
150mm x 150mm (nominal 6 x 6in)
200mm x 200mm (nominal 8 x 8in)
200mm x 100mm (nominal 8 x 4in)
300mm x 300mm (nominal 12 x 12in)
Typical thickness 100mm (nominal 4in)

General properties of glass blocks are as follows:
Thermal conductivity approximately 0.88 W/m^2 ^{0}C (0.51 Btu/hrft^{0}F)
Light transmission = 75% for hollow block

Although glass blocks are set in courses in the manner of masonry construction, they are used as a non-loadbearing material. Blocks are arranged in a rectilinear grid or 'stack bond' of continuous vertical and horizontal joints that give this material more the appearance of glazing than of masonry construction (g). Recently there has been a development in design by setting glass blocks forward of floor slabs instead of being seated on them, in the manner of glazed curtain walling. With this method, a stiffening frame is set within the joints between blocks, with

1:10 scale 5" 10" 1'-3" 1'-8"

100 200 300 400 500 mm 1:10 scale

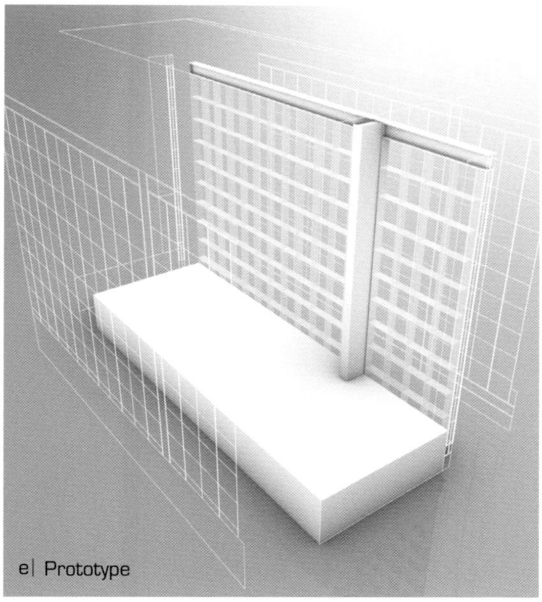
e| Prototype

5 | Adjacent concrete wall
6 | Adjacent light gauge steel-framed wall
7 | Reinforced concrete
8 | frame
9 | Steel box section
10 | 90° corner block
11 | 45° corner block
Steel I-section, usually
12 | specially fabricated
13 | Steel angle
14 | Inside
15 | Outside
16 | Floor deck
17 | Steel T-section
Adjacent brick cavity
18 | wall construction
19 | Thermal insulation
Extruded aluminium sections providing support
20 | Adjacent reinforced concrete wall

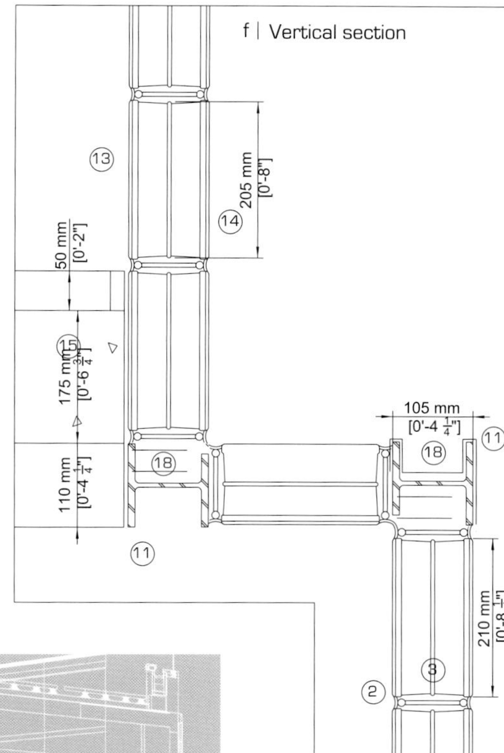

f | Vertical section

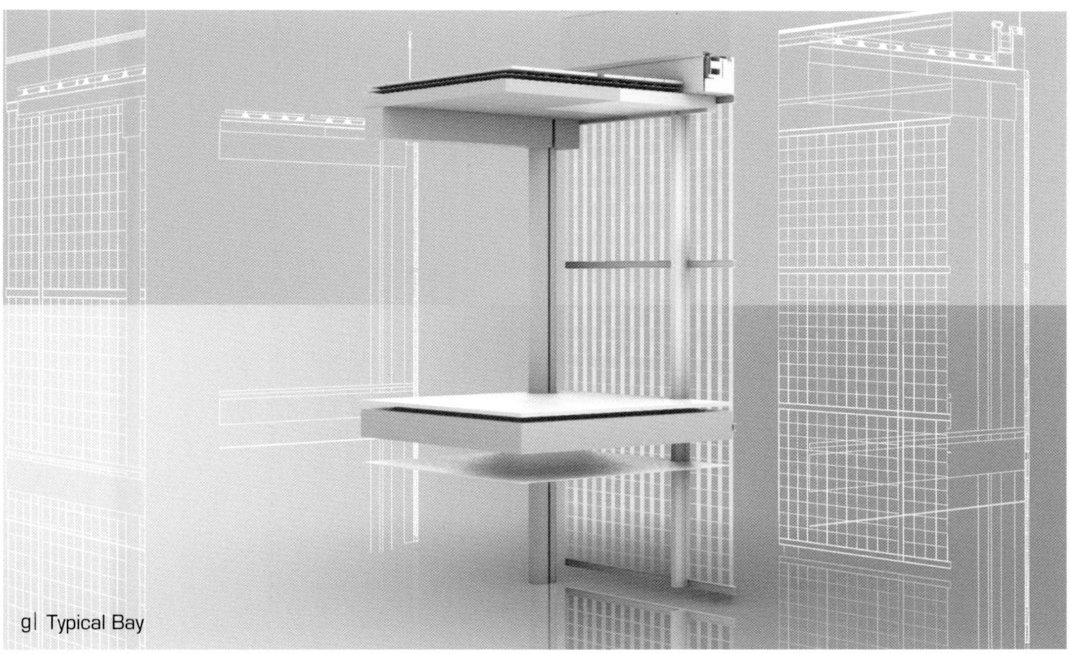
g| Typical Bay

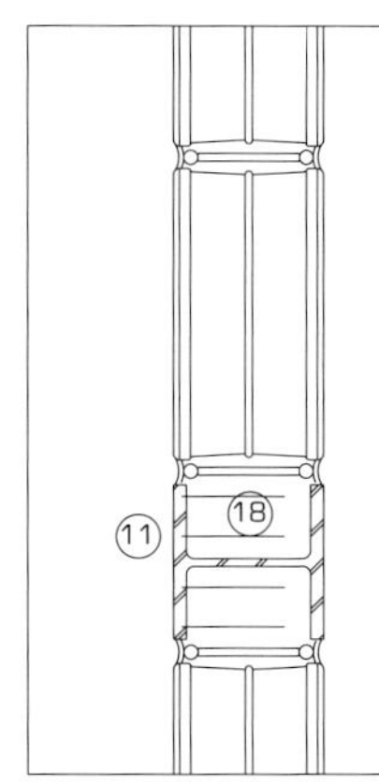

additional supports set away from the glass blocks visually to lighten their appearance (f).

Glass blocks are most commonly either set into concrete or masonry backing walls, or are set into a reinforced concrete or steel support frame. For non-fire rated construction, maximum panel sizes vary from around 3600 x 3600mm (12 x 12ft) to around 4500 x 4500mm (15ft x 15ft) depending on block thickness. This overall area can be adapted to suit a maximum height of around 6000mm (20ft) and a maximum width of around 7500mm (25ft).

When used as fire resisting construction, glass blocks can reach 90 minutes fire integrity; but above 60 minutes fire resisting construction, a metal channel restraint is used in the jointing for its greater reliability in fire over mortar or silicone-type materials. One hour fire-rated panels of glass block can be made to a maximum size of approximately 3000 x 3000mm (9ft x 9ft) with a maximum of 4000mm (12ft) in either height or width, depending on individual building codes. Thermal insulation of glass block is poor compared to double glazed units, with some condensation risk on the inside face in temperate climates. For this reason, blocks are often used with concrete and steel frames that do not have thermal insulation, often where the interior face of the wall is ventilated but not heated, as in semi-external circulation areas. An ideal environment for glass blocks is in regions which have a warm temperate climate for most of the year, where thermal insulation is not required.

Fixing glass blocks

Blocks are fixed in different ways. When supported on reinforced concrete floor slabs, they are bedded in mortar on all joints, with a flexible joint (sometimes with an additional aluminium edge frame) at the top and sides to allow for structural movement (MCE_162, a). Blocks can

a | Vertical sections. Glass blocks fixed into window frames in masonry walls

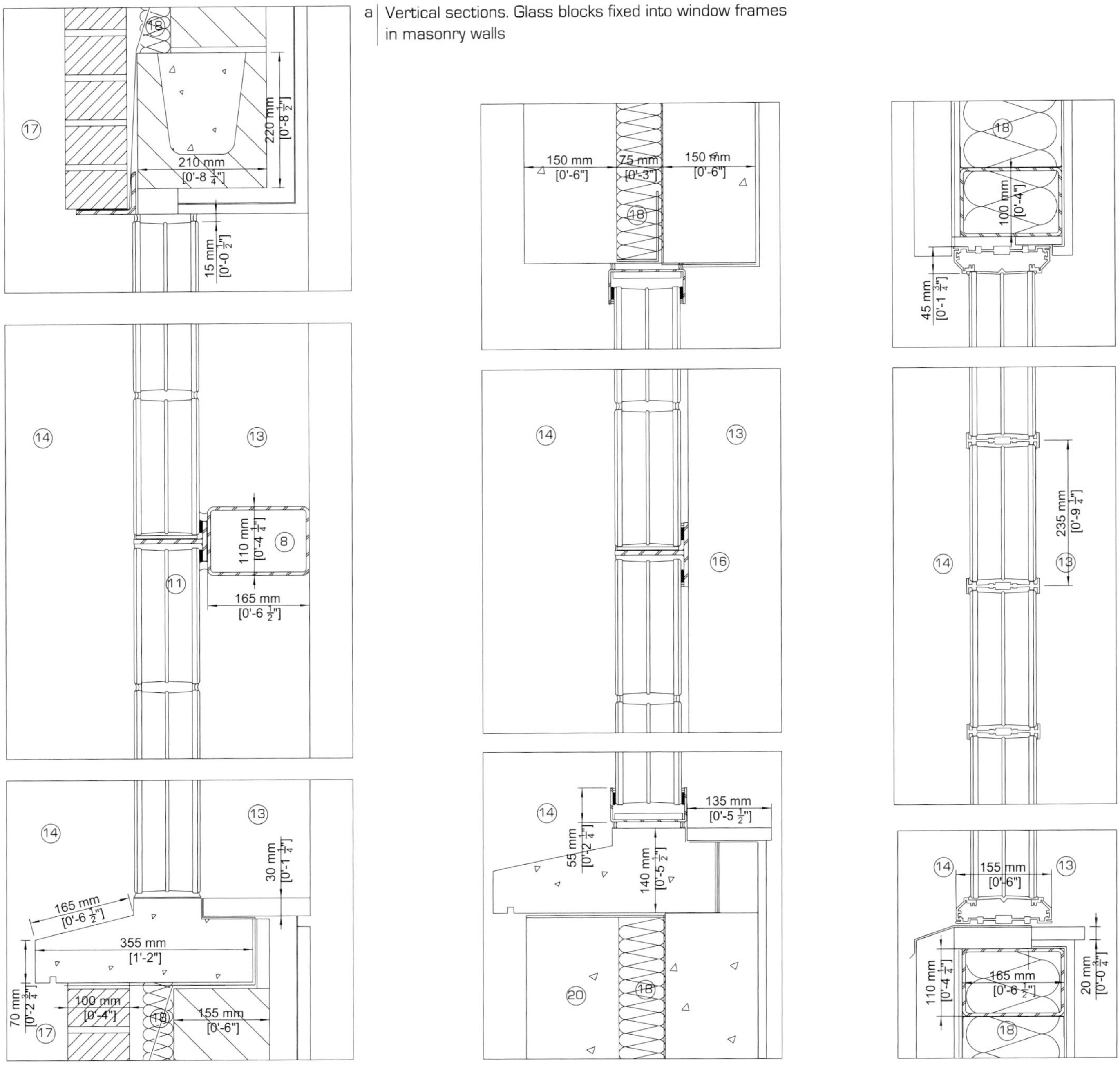

also be glued together with bonding silicone, using a sealing silicone on the outside faces. Blocks can also be fixed together with extruded aluminium clips that hold all the blocks together. Angle-shaped clips are used to fix the glass block panels to the structural opening without the use of a perimeter frame, the clips allowing for structural movement between the panel of blocks and the structural opening. The outside faces are also sealed with silicone.

Support frames and walls

In a generic example, glass blocks are fixed in panels to a maximum of around 3500mm x 3500mm (11ft 6in x 11ft 6in) square, in storey height frames, though half-storey frames can also be used to reduce the thickness of the block as well as the need for additional bed reinforcement. Blocks are set on the edge of the floor slab. Block courses are laid on a bed of mortar with the head of each panel of glass blocks allowing for the structural deflection of floor slabs. A reinforced concrete floor slab spanning 7500mm (25ft) between columns can have a deflection of around 20mm (0.8 in) at its centre. This deflection is allowed for the head of the block with an aluminium channel restraint used at the perimeter to hold the glass blocks in place but allow the slab above to deflect under load (MCE_161, f).

Glass blocks can be laid out in visually continuous lengths, but are divided up with steel T-sections at 2.0 - 3.0 metre (6ft 6in - 10ft) centres to provide the lateral stability of the panels. The additional T-section results in a wider joint than those used elsewhere in the panel, but in a long run of glass blocks this is barely visible from the outside, though visible on the inside (a).

Curved panels have the advantage of inherent structural stability but minimum radius dimensions for different block sizes as follows:

150mm x 150mm blocks:1200mm min. radius.

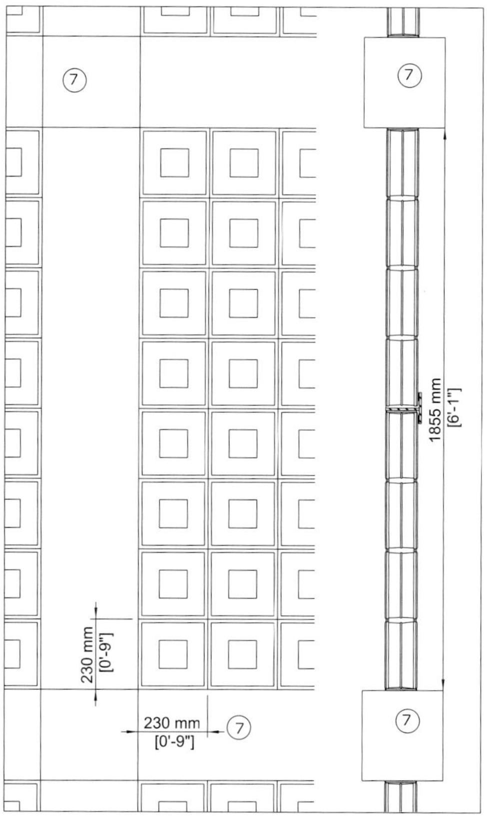

b | Elevations, vertical sections. Glass blocks fixed into window frames in masonry walls

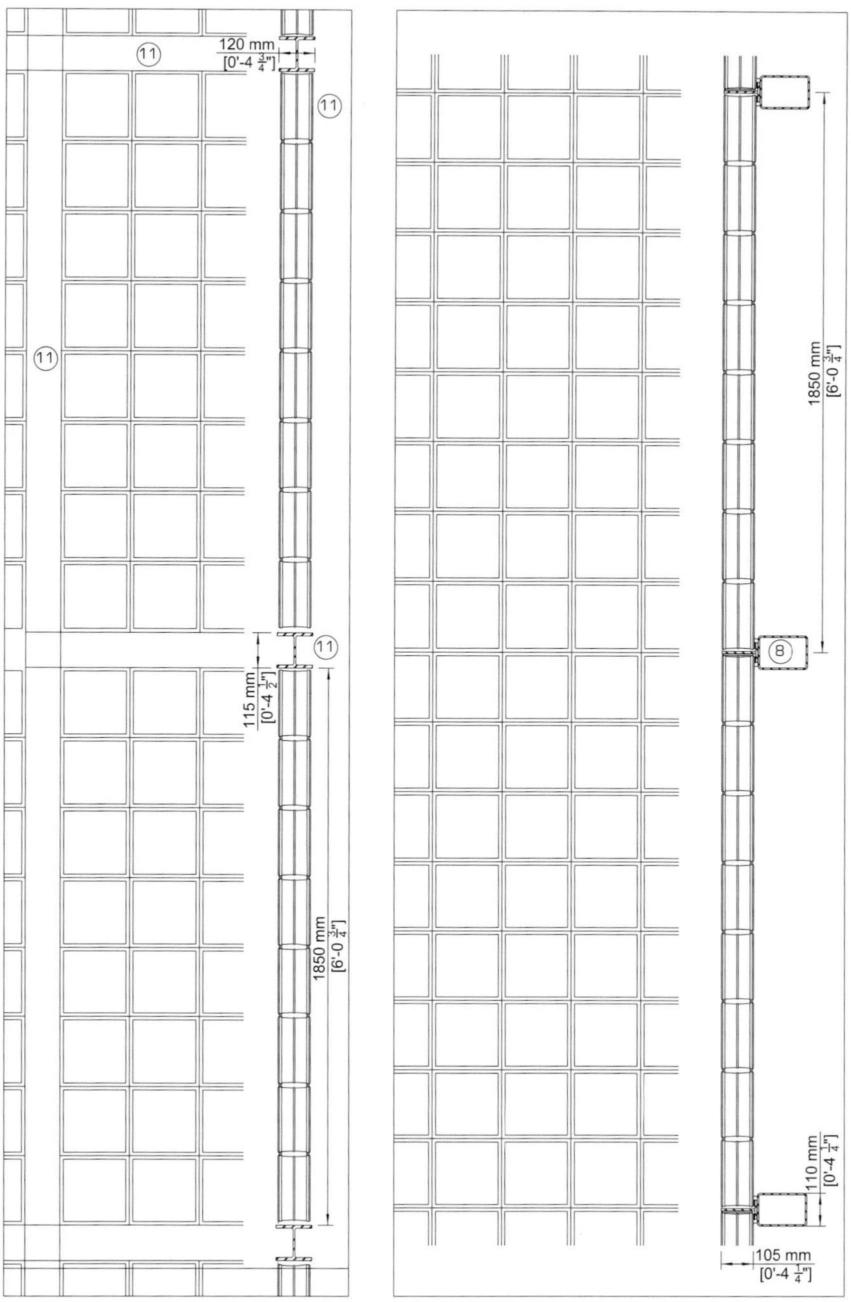

Details

1 | Glass block
2 | Bedding reinforcement
3 | Bedding compound, mortar or silicone-based bond
4 | Adjacent blockwork wall with rainscreen
5 | Adjacent concrete wall
6 | Adjacent light gauge steel-framed wall
7 | Reinforced concrete frame
8 | Steel box section
9 | 90° corner block
10 | 45° corner block
11 | Steel I-section, usually specially fabricated
12 | Steel angle
13 | Inside
14 | Outside
15 | Floor deck
16 | Steel T-section
17 | Adjacent brick cavity wall construction
18 | Thermal insulation
19 | Extruded aluminium sections providing support
20 | Adjacent reinforced concrete wall

(6in x 6in blocks: 48in min. radius)
200mm x 200mm blocks: 1600mm min. radius
(8in x 8in blocks: 65in min. radius)
300mm x3 00mm blocks: 2500mm min. radius
(12in x 12in blocks: 98in min. radius)

Movement joints are set at the junction of straight and curved panels. These joints usually serve also as structural joints for the blocks.

In alternative generic examples, glass blocks are set in openings in a reinforced concrete wall. Openings in the wall have to be formed in close tolerances in order to have even edges if the concrete is to be left exposed. Where openings cannot be formed so accurately, the glass blocks are set into an aluminium edge frame that laps over the concrete or forms a shadow gap around the opening. If the concrete wall is clad in a different material, such as metal or masonry rainscreen panels, then the cladding can be adjusted to meet the glass blocks neatly without the makeup piece in the opening being visible.

Steel support frames

Glass blocks can also be supported on steel frames. These have the advantage of making the glass blocks appear to have a continuous surface uninterrupted by a supporting structure.

In other cases, steel T-sections support the blocks for modest spans, which can be increased from 1.5 metres to 3.0 metres (5ft to 10ft) with the use of a box section set behind it to form a continuous supporting frame both vertically and horizontally. With both options, the frame is partially visible from outside; as a shadow line in daylight, and as a strong frame when lit internally at night from outside.

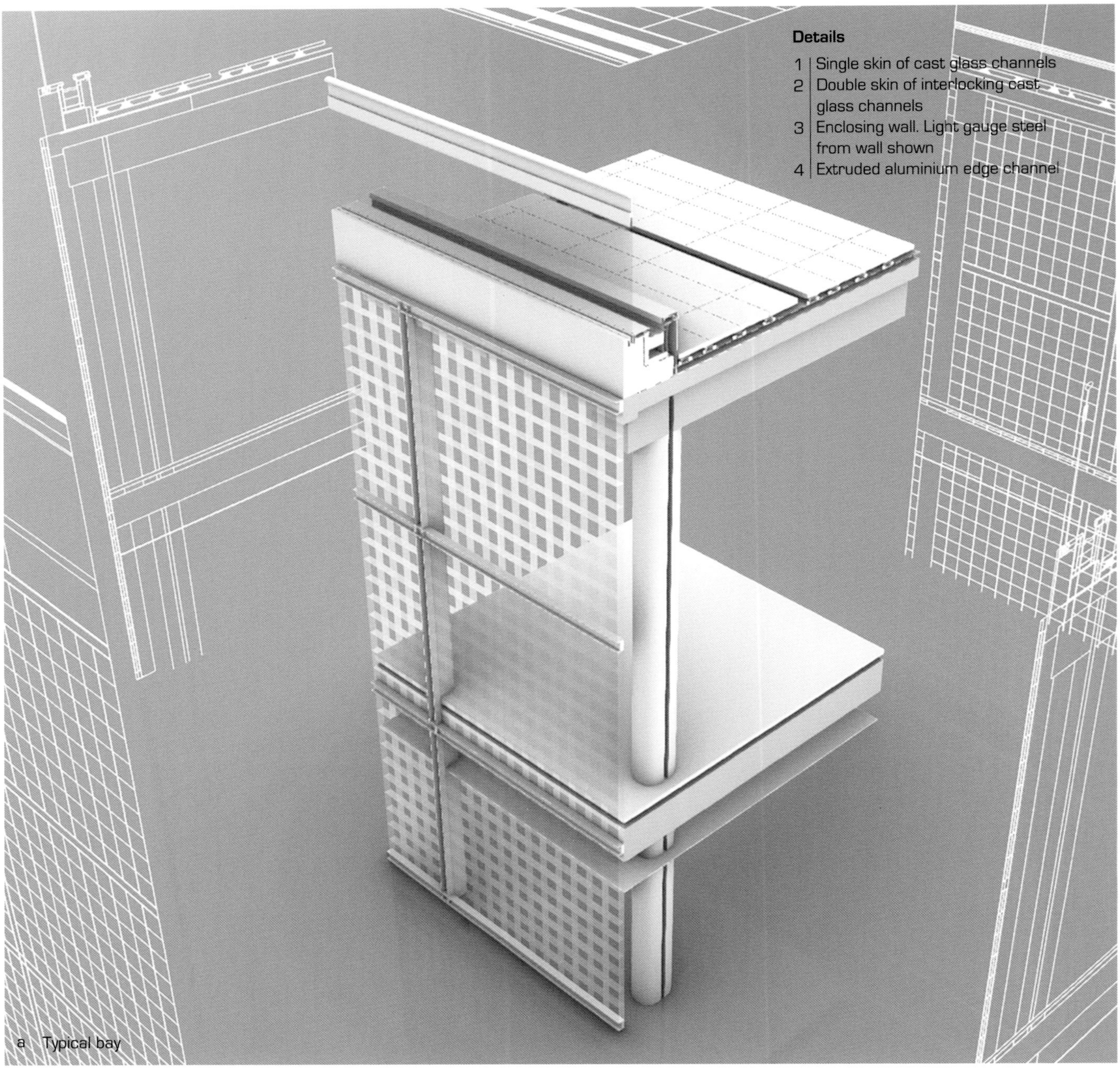

a Typical bay

A complete frame of rolled steel sections can support glass blocks by setting the blocks directly onto the steels. This method requires close dimensional co-ordination of steel size and block width in order for the blocks to be aligned on the outer face to drain water down the face of the steel without the need for a drip, which is visually obtrusive.

A recent development in glass block design has been to insert a steel support frame into the joints between individual blocks (MCE_160, c). This is made more economic by using the largest 300 x 300mm (12in x 12in) blocks and adapting their edges in order to conceal the steel box section within. This concealed steel frame can be supported by brackets back to secondary vertical posts, such as tubes. Alternatively, the concealed steel frame can be made to hang or sit on a supporting structure in the manner of curtain walling, using steel posts to provide lateral stability throughout its height. Glass blocks are bonded together with silicone, which allows for the higher amounts of structural movement associated with steel supports than those in reinforced concrete structures. The continuous surface of blocks, uninterrupted by floor slabs or concrete beams, does not require any drips to throw water clear of these surfaces, giving an overall smooth appearance to the glass block wall.

Masonry and timber framed window openings

Glass blocks cannot be bedded directly into masonry walls, such as concrete blocks, since any movement at joints between blocks will result in a crack in the nearest glass blocks. For this reason, when glass blocks are held in masonry openings, they can be set directly onto a concrete sill if one is used. Sometimes the concrete sill can form a complete reinforced concrete frame, around 50-75mm (2in-3in) wide, in which the glass blocks are set. Openings in timber framed walls and openings in cavity wall construction (typically brick outer leaf, air gap and inner block/timber wall) use an extruded aluminium frame to hold the

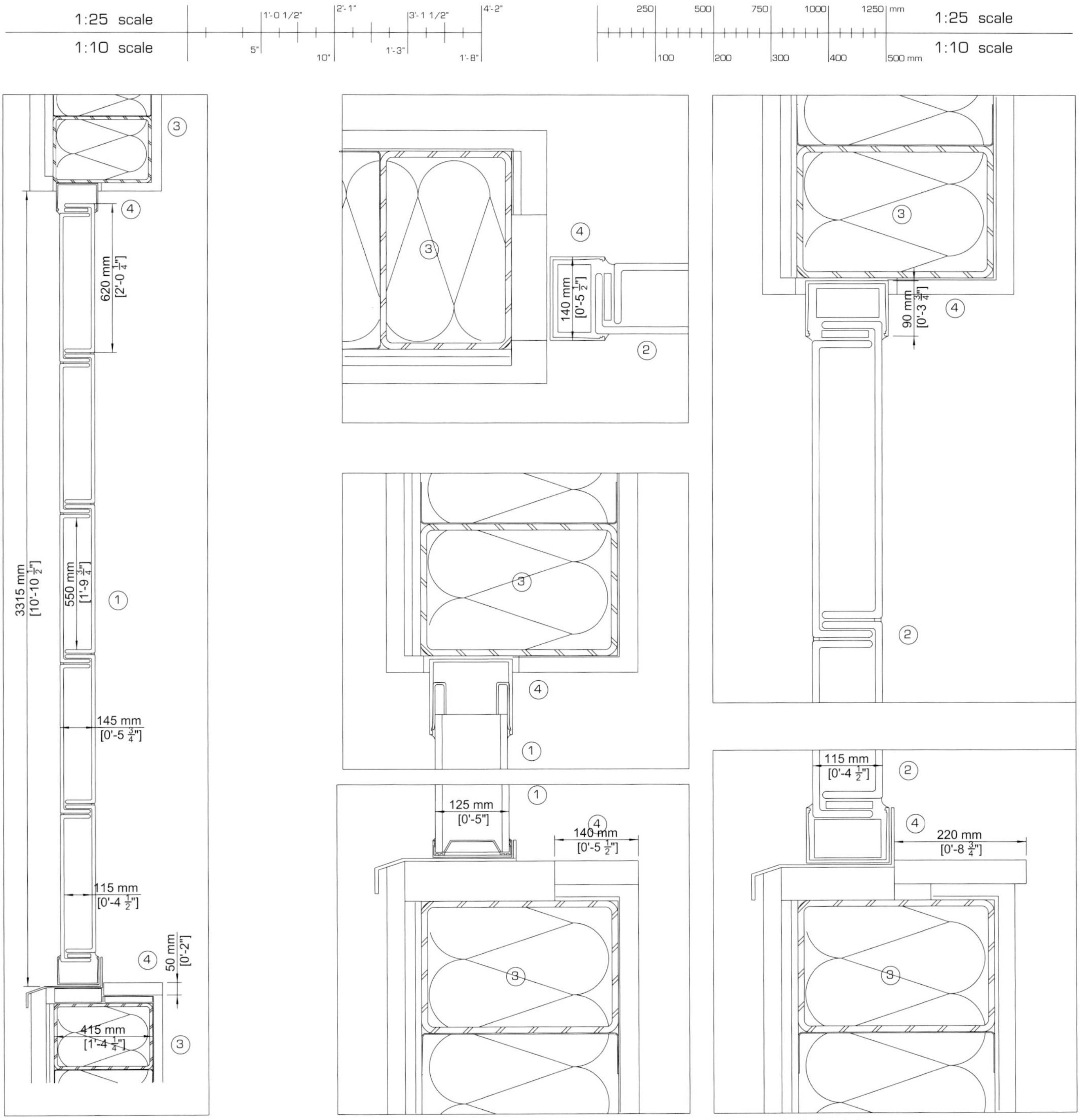

b | Horizontal and vertical sections. Cast glass channels fixed into a light gauge steel framed wall, typically extruded aluminium

glass blocks in place and which is usually polyester powder coated. The aluminium frames are thermally broken, fixed to the masonry opening, usually with screws. Similar aluminium sections are used between some joints to provide additional support if required. Where glass blocks are set into an opening in a timber frame, the aluminium profile edge can be combined with a timber sill to give a continuous timber appearance both internally and externally.

Cast glass channels

The effect of translucency provided by thick glass can also be provided by cast glass channels. These are like half blocks in section which are made in lengths of up to around 2.5 metres (1ft 6in). Most are around 250mm wide x 60mm deep (10in x 2.5in), with glass around 6-7mm (0.25in) thick. Planks can be set both vertically and horizontally, but vertical applications are the most common as they are far easier to fix. When set horizontally, glass planks cannot sit on top of one another; each plank is supported individually at its end to provide a waterproof seal between planks that can accommodate thermal movement (b).

An advantage of cast glass channels or 'planks' over glass blocks is that they are self-supporting, with the ability to span their full length of around 2.5 metres (8ft). The glass is held in place with thermally broken aluminium extrusions at its ends (b), using sections similar to those used for glass blocks in window openings. The extrusions are anodised or polyester powder coated to suit the design. Light transmission is around 85%, reducing to 70% if two channels are interlocked to form a wall of double thickness to improve thermal insulation and increased sound reduction through the wall (b). A double thickness of interlocking channels can provide a sound reduction similar to that of a double glazed unit in a glazed wall of around 40dB, with thermal insulation also similar to that of a double glazed unit at around 2.0 W/m²K if a hard low emissivity (low e) coating is applied to the inner face of the outer

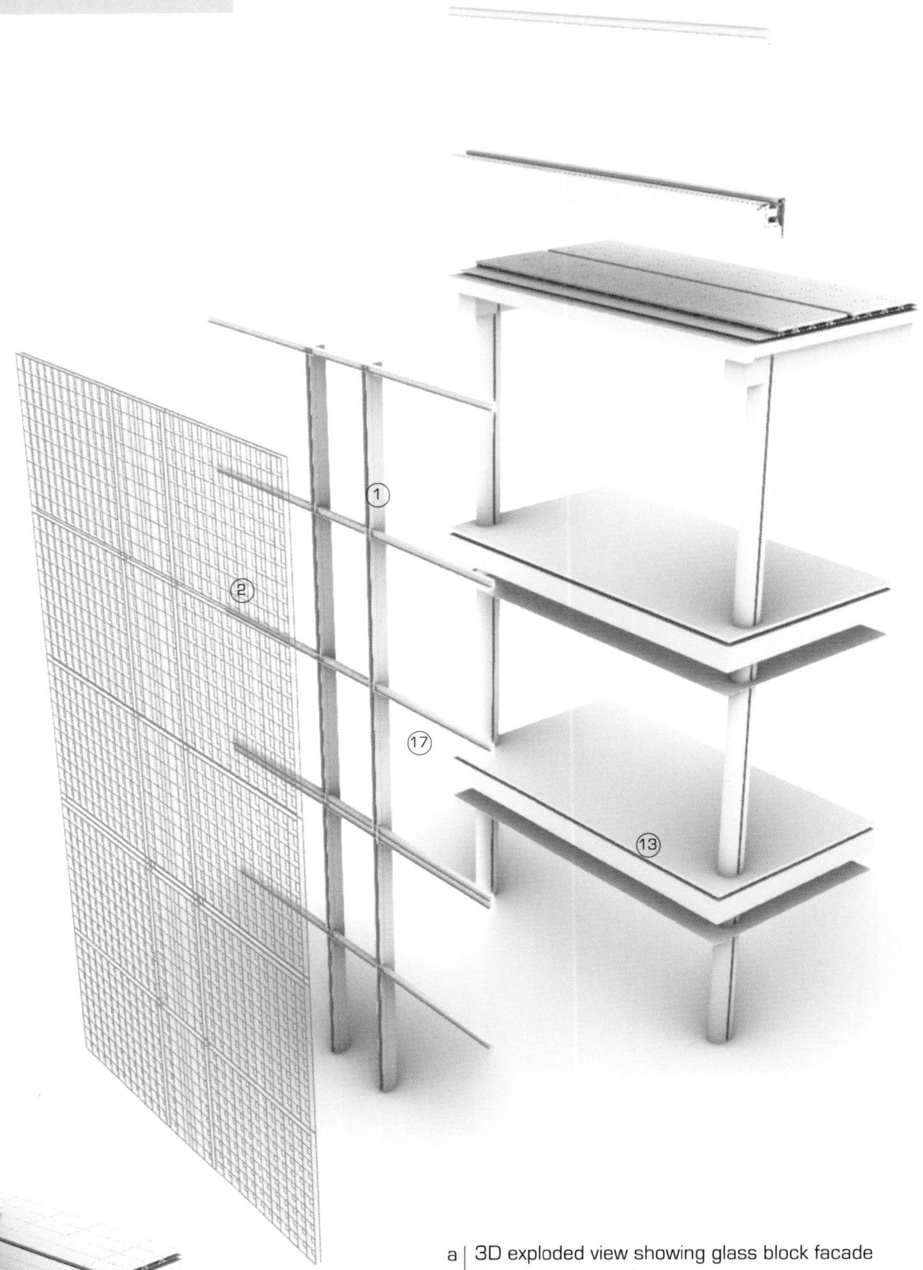

a | 3D exploded view showing glass block facade fixed to concrete frame

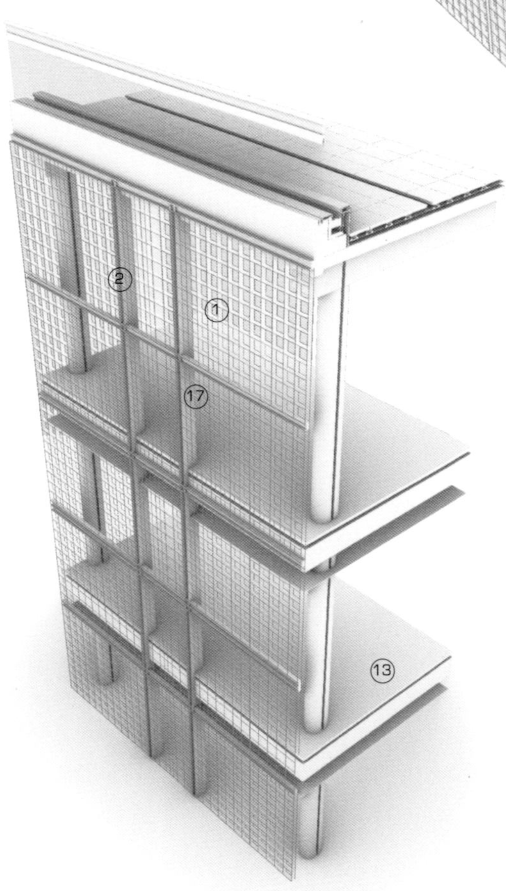

b | 3D view showing glass block facade fixed to concrete frame

channel. An advantage of cast glass channel glazing is that it is much cheaper than a conventional glazed wall system. This coating changes the surface appearance slightly. A hard coating allows the channels to be assembled on site. The higher performance soft coatings are required to be sealed as double glazed units in the factory, which is not possible with cast glass channels as their ends are open until installed in the aluminium edge framing already described.

Cast glass channels are sealed together with silicone, with translucent white being the most commonly used colour, to match the glass as closely as possible. Unlike glass blocks, cast glass channels can provide only a very limited fire resistance of around 30 minutes when the length of the plank is restricted to around 2.5 metres (8ft). The glass is reinforced with a grid of wires, giving the appearance of traditional wired glass, and joints use a fire retardant silicone.

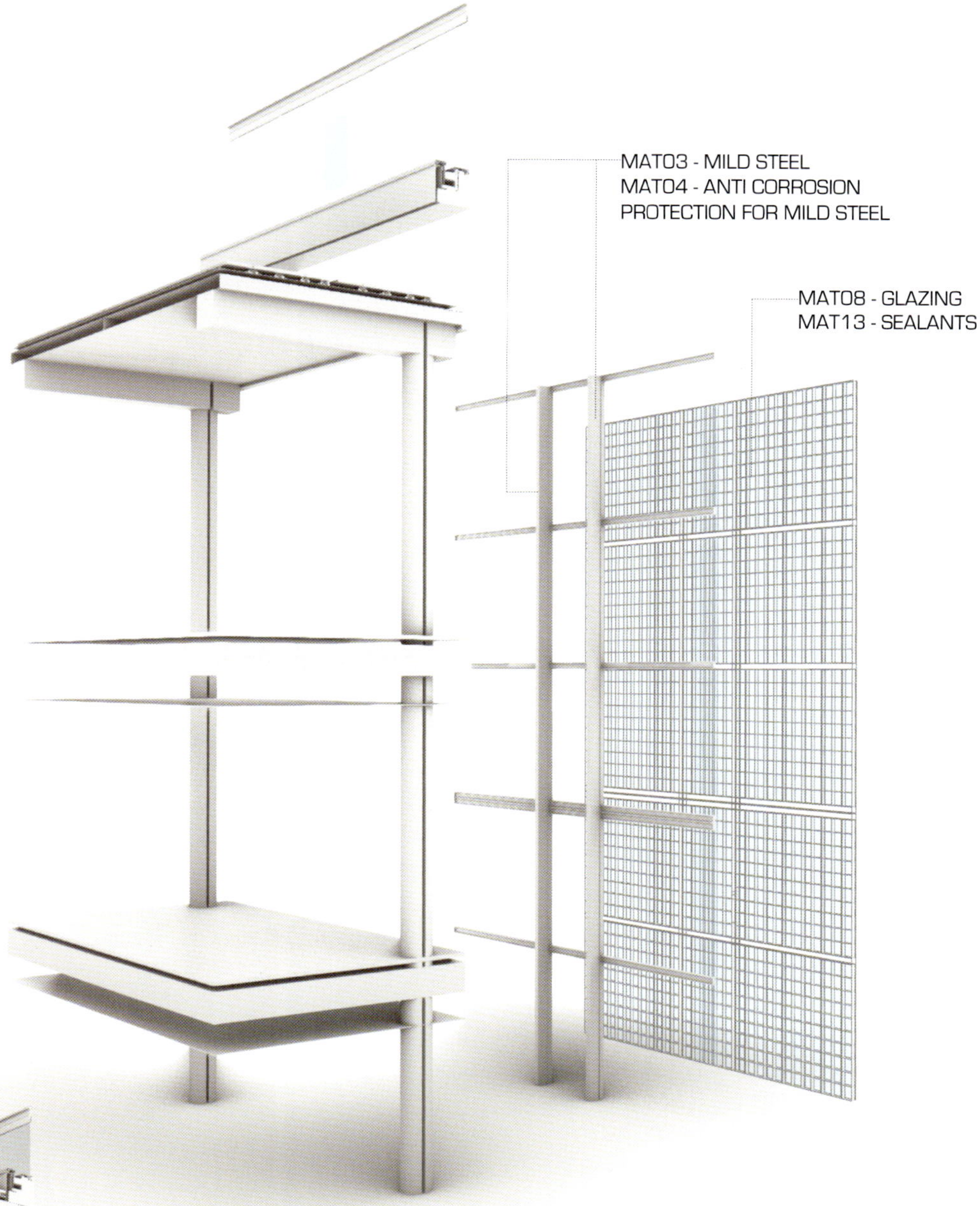

c | 3D exploded view of glass block facade components with material specification references

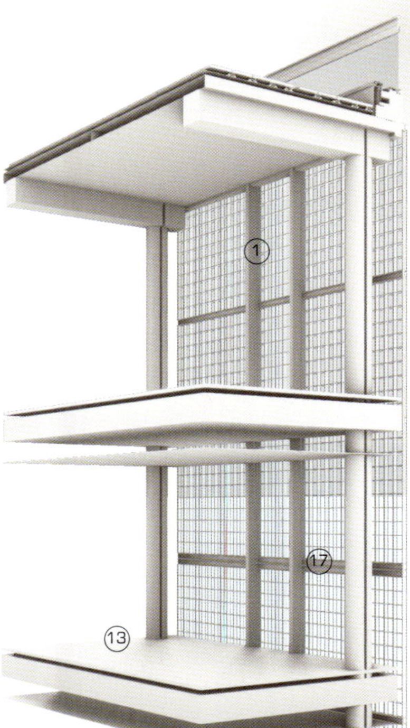

Details

1 | Glass block
2 | Bedding reinforcement
3 | Bedding compound, mortar or silicone-based bond
4 | Adjacent blockwork wall with rainscreen
5 | Adjacent concrete wall
6 | Adjacent light gauge steel-framed wall
7 | Reinforced concrete frame
8 | Steel box section
9 | 90° corner block
10 | 45° corner block
11 | Steel I-section, usually specially fabricated
12 | Steel angle
13 | Floor deck
14 | Steel T-section
15 | Adjacent brick cavity wall construction
16 | Thermal insulation
17 | Extruded aluminium sections providing support
18 | Adjacent reinforced concrete wall

d | 3D view showing glass block facade fixed to concrete frame

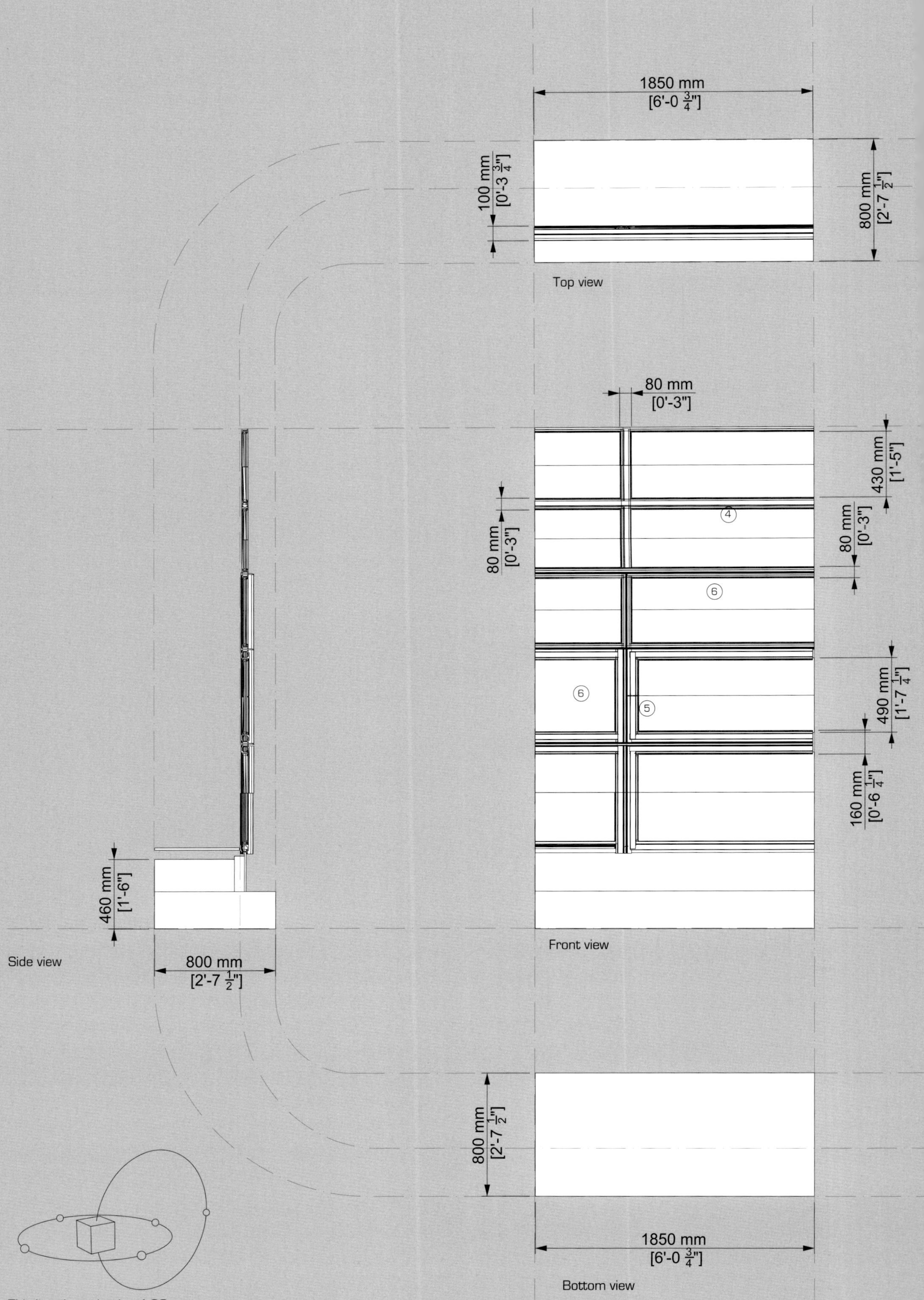

Third angle projection 1:30

Details

1 | Outside
2 | Inside
3 | Steel supporting structure
4 | Transom
5 | Mullion
6 | Single-glazed or double-glazed unit to suit application
7 | Fixing bead
8 | Fixing lug
9 | Projecting transom
10 | Rubber-based seal
11 | Fixed light

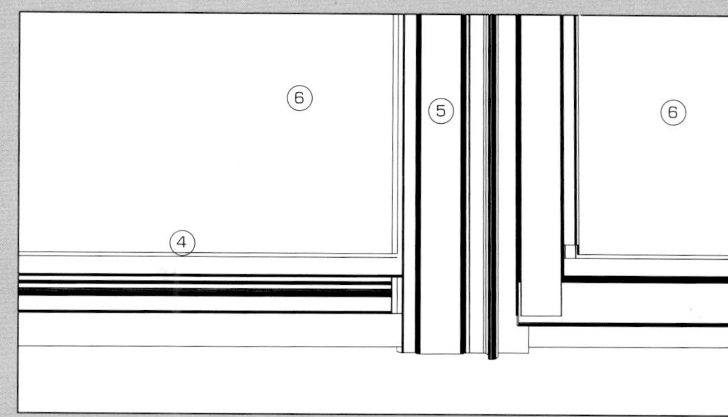

3D view of the assembly

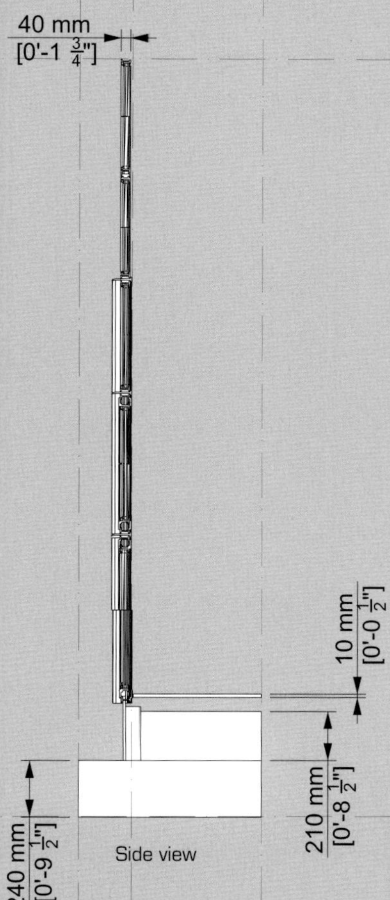

Side view

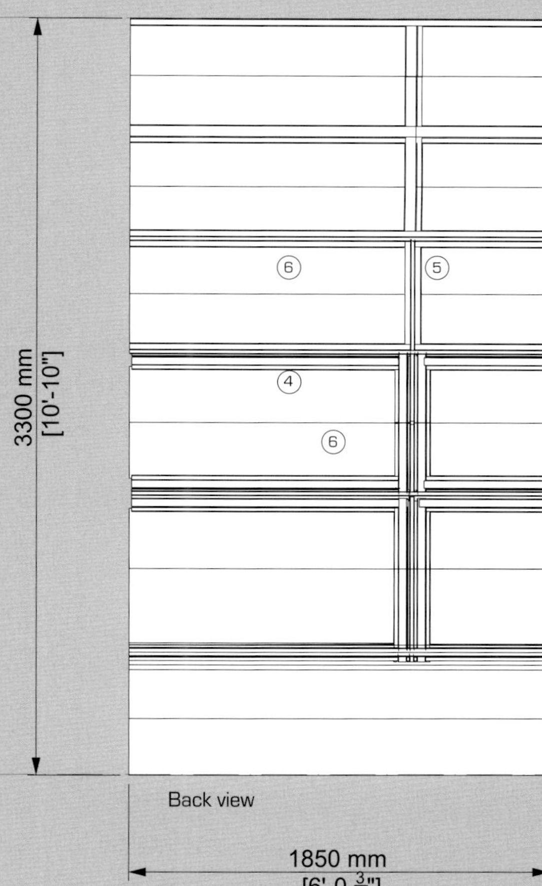

Back view

12 | Inward opening light
13 | Outward opening light
14 | Window sill
15 | Condensation tray
16 | Damp proof course (DPC)
17 | Internal finish
18 | Drip
19 | Packing
20 | Aluminium clip to secure double-glazed unit
21 | Steel sill
22 | Meeting stile

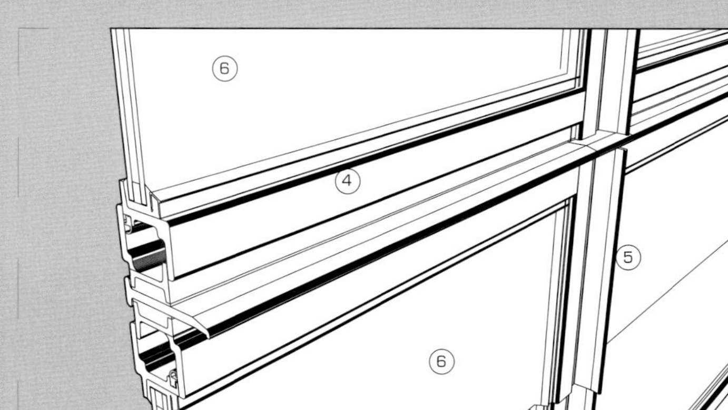

3D view of the assembly

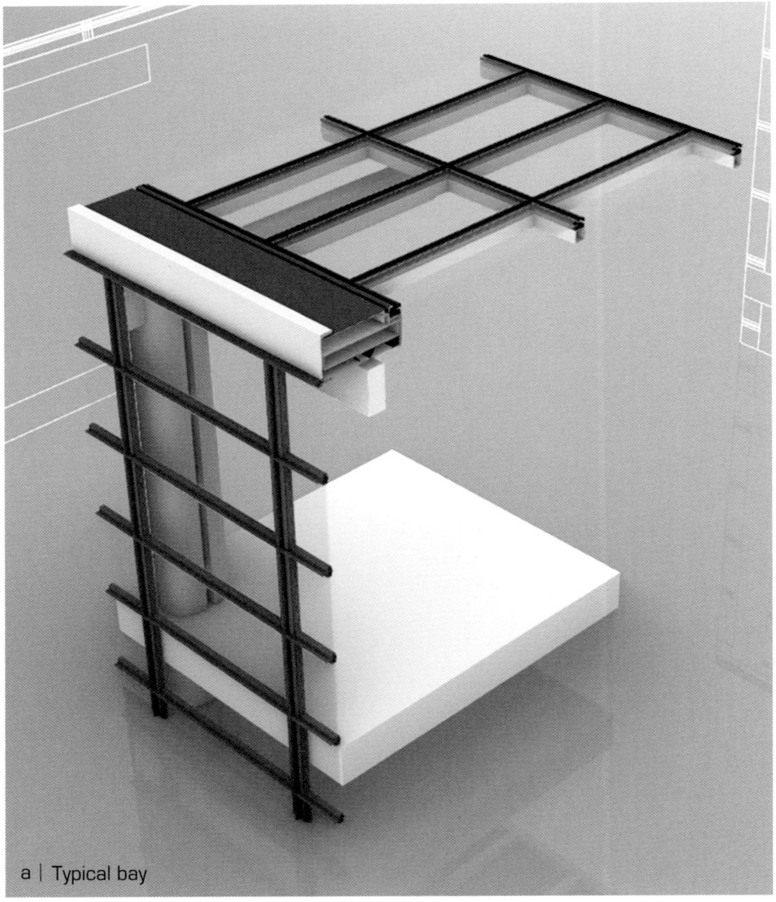

a | Typical bay

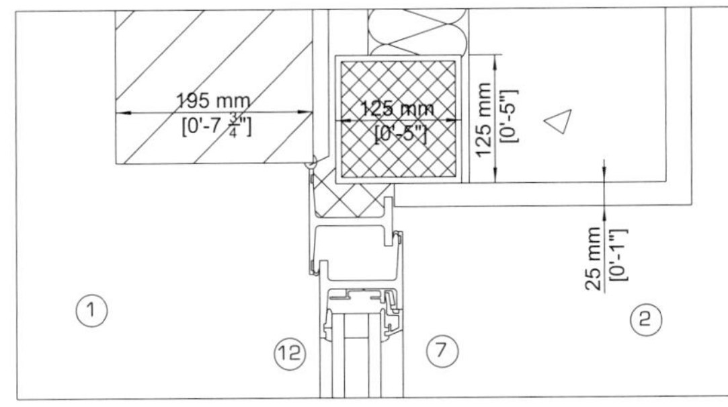

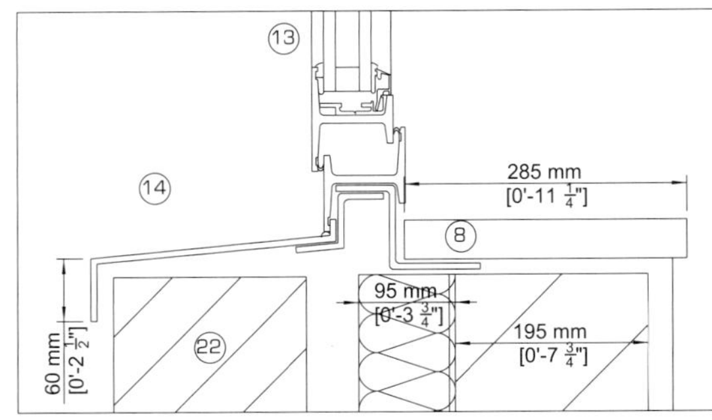

b | Horizontal and vertical sections. Rolled steel frames with double glazed units

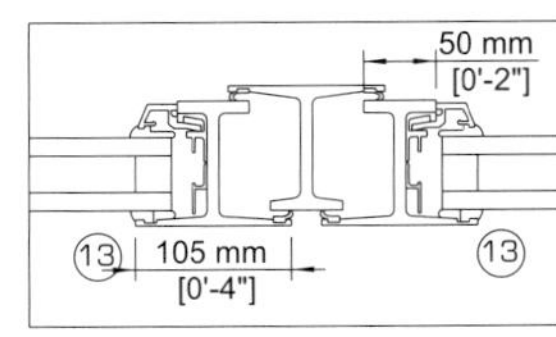

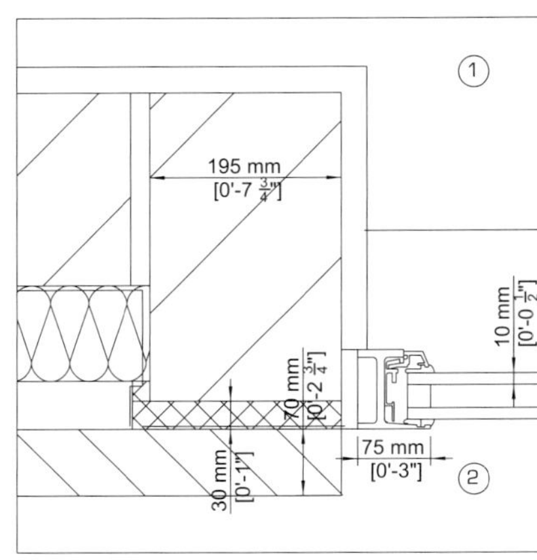

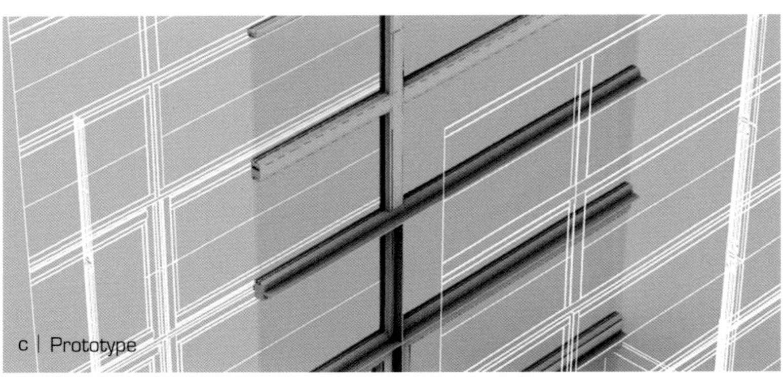

c | Prototype

Steel-framed windows are a robust window construction that is often used as a fire resistant glazing. Steel glazing was developed primarily for single glazing, where the main advantage was their thin sight lines when compared to those in aluminium. The thin, rigid, rolled steel sections were also more economic than their aluminium counterparts. The rolled sections can accommodate double glazed units of most thicknesses, but are not thermally broken. These rolled sections are used with small window sizes up to a maximum of around 3000 x 1800mm (10ft x 6ft), and a minimum of 250mm x 400mm (10in x 16in) with double glazed units. Fully glazed walls made from window sections can be made using this relatively small window size. The 'window wall' is stiffened with a frame formed from integral steel fins that pass through the window. These narrow window sections cannot incorporate a thermal break.

Window walls with larger panel sizes can be formed in larger pressed steel sections and rolled hollow sections rather than the smaller G-shaped or T-shaped sections. The sight lines of rolled steel sections are very similar to those in aluminium, but pressed steel sections are wider and deeper in section than the rolled types. However, pressed sections have the advantage of being able to be formed into larger-scale walls, as well as being able to incorporate a thermal break. Thermal breaks are held in place by folding the material tightly over the ends of the polymer material, which is currently too difficult to do economically in thin hot rolled sections.

The most common use of steel windows is in fire resistant glazing, typically where glazed walls enclose a fire resisting compartment. The fire integrity is usually between 30 minutes and 1 hour, but 2 hour integrity can be achieved when used with fire resistant glasses. Although steel

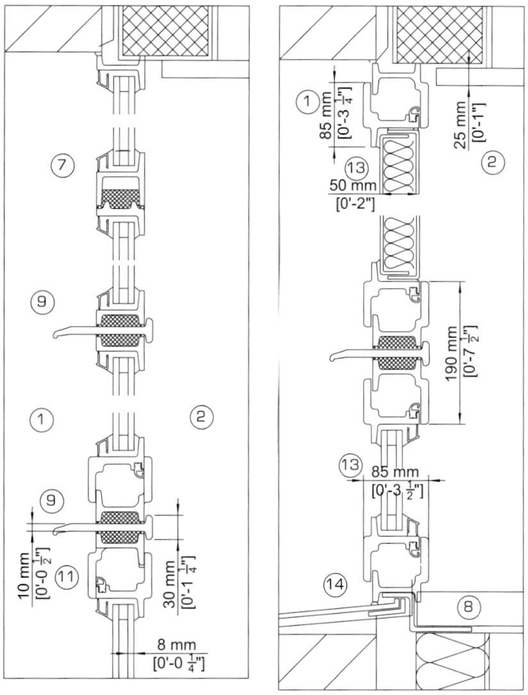

d| Horizontal section. Rolled steel frames glazed internally

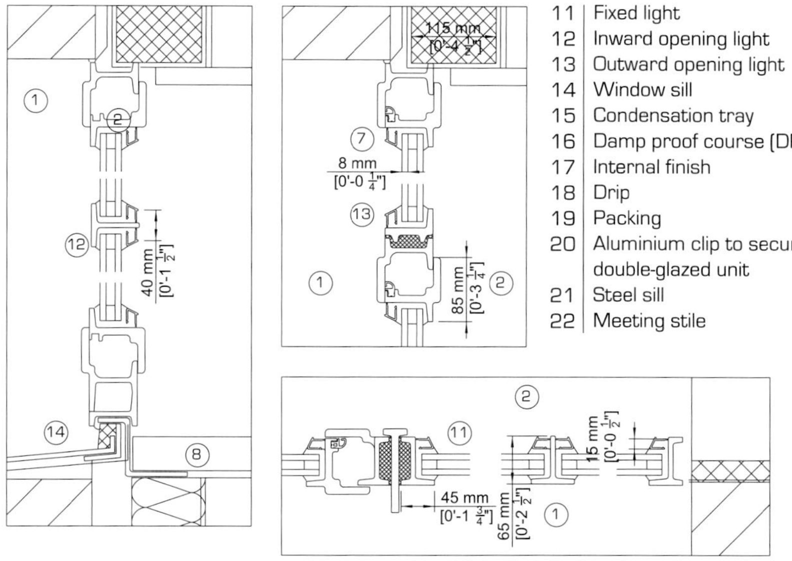

11 | Fixed light
12 | Inward opening light
13 | Outward opening light
14 | Window sill
15 | Condensation tray
16 | Damp proof course (DPC)
17 | Internal finish
18 | Drip
19 | Packing
20 | Aluminium clip to secure double-glazed unit
21 | Steel sill
22 | Meeting stile

f | Horizontal and vertical sections. Rolled steel frames glazed internally and externally

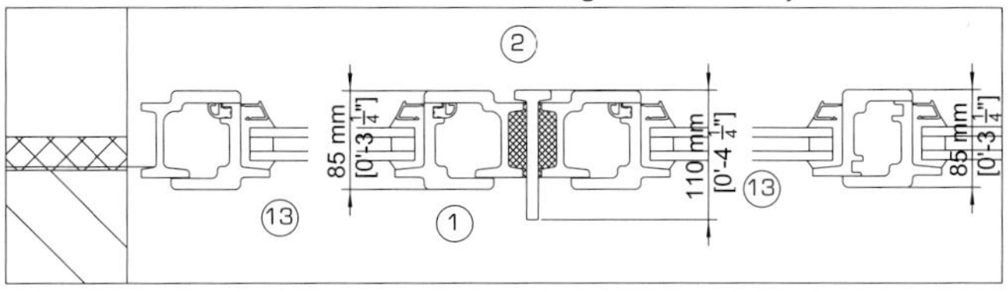

e| Horizontal section. Rolled steel frames glazed externally

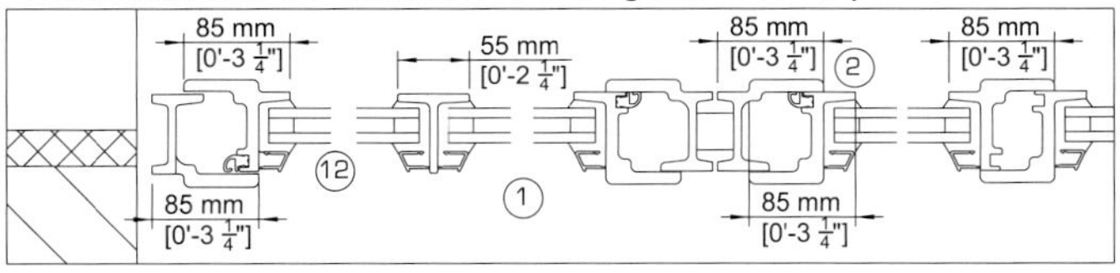

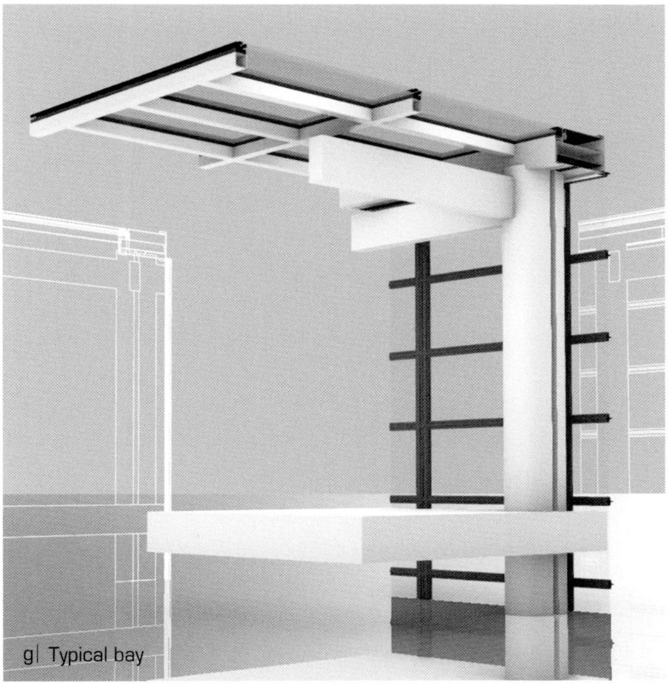

g| Typical bay

Details

1 | Outside
2 | Inside
3 | Steel supporting structure
4 | Transom
5 | Mullion
6 | Single-glazed or double-glazed unit to suit application
7 | Fixing bead
8 | Fixing lug
9 | Projecting transom
10 | Rubber-based seal

glazing provides structural integrity during a fire, it does not provide thermal insulation to counter the heat generated by fire. This is dealt with either by providing sprinklers that drench the wall to keep it cool, or else the wall is positioned in a way that building users do not come into direct contact. Areas of fire resisting glazing can be combined with areas of unprotected glass without changing the outward appearance of the glazing. This is a big advantage where a consistency of appearance is required over a continuous facade with different requirements for fire protection in different areas of the same facade.

Small-scale glazing

Fixed and opening lights framed in sizes of up to 3000mm x 1800mm (10ft x 6ft) use a standard single section which is used for both horizontal and vertical glazing sections as well as for opening lights. Where opening lights occur, the profile used for the window frame is different, as is the profile for the opening light itself. Also, profiles are different for inward and opening lights. Unlike aluminium framed glazed walling, where a single extrusion is used throughout, with opening lights glazed into the system, small-scale steel glazing requires a family of different profiles to create a single glazed wall. For this reason, manufacturers offer a small range of steel window types in order to remain economic. Because sections are rolled rather than extruded, there is very little possibility of creating a new profile for a specific project. Single glazed panes of glass were traditionally fixed with steel pins and glazing putty like traditional window frames, but the practice has given way to thin steel channels and angles. More recently, aluminium extrusions are used which are clipped into place. The extrusions are both more reliable in the long term and far easier to replace, particularly in large-scale glazed walls.

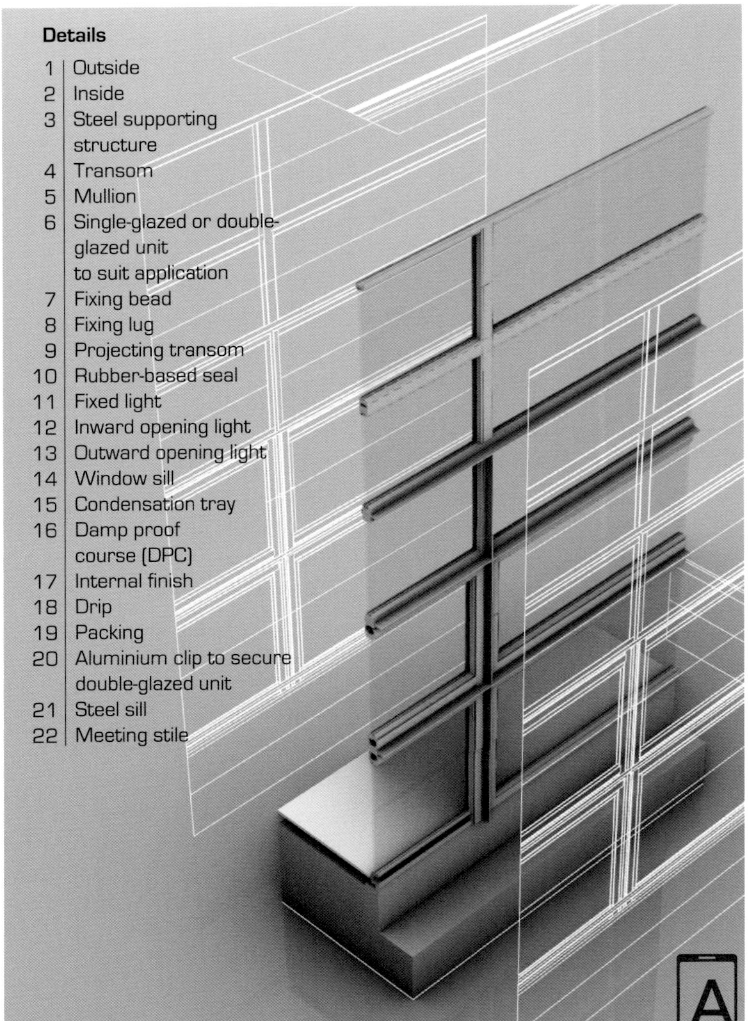

a| Prototype

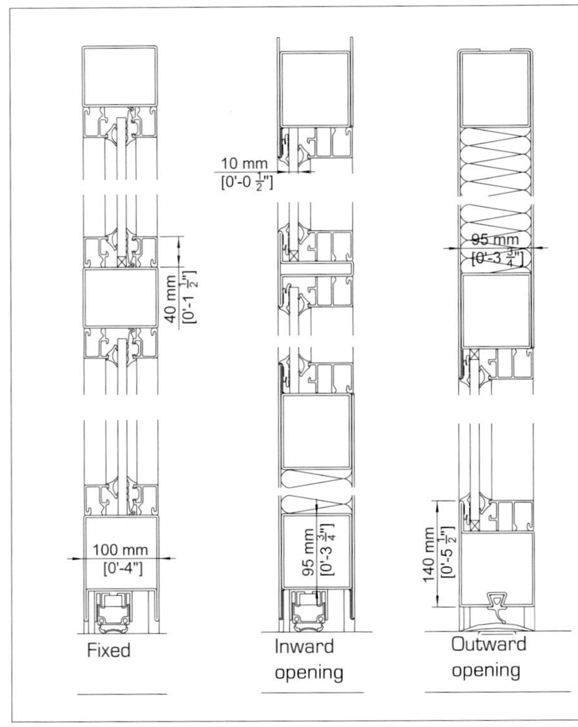

b| Vertical sections. Pressed steel single glazed fire resisting doors without thermal breaks

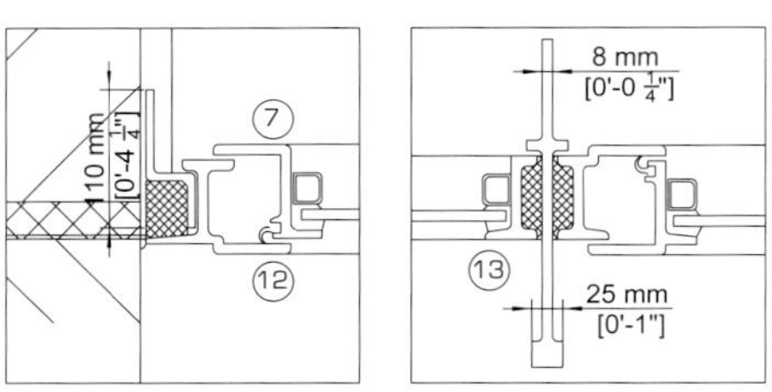

c| Vertical sections. Rolled steel frames with single glazing

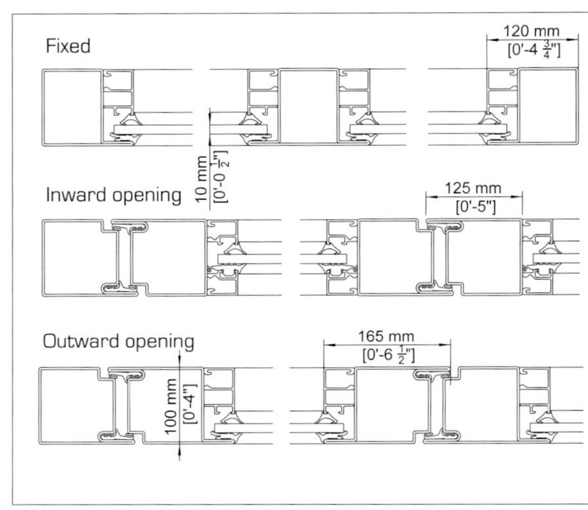

d| Horizontal sections. Pressed steel single glazed fire resisting doors without thermal breaks

When windows, either fixed or opening lights, are joined together to form a glazed wall, mild steel T-section fins are set between the windows to form a stiffening frame (MCE_170, b). The adjacent frames are screwed to the fins in order to contribute to the overall stiffness of the frame. Unlike aluminium, the T-sections used are different from those set vertically in order to provide a projecting drip. Profiles are designed so that they are glazed from either the inside or the outside. The choice of system is usually dictated by the method of glass replacement, which may be from an internal floor or from an external ladder or cleaning and maintenance cradle.

Outward opening lights have profiles with edges that project over the front of the supporting frame to exclude rainwater. A projecting drip is provided at the top to protect the vulnerable top joint. Any rainwater that drips into the inner surface of the frame is drained away around the edge of the frame and out at the bottom of each frame. Rubber-based seals are now used to assist with reducing air infiltration through the window.

Inward opening lights use the same profiles with their projecting edges set against the frame in a similar way. At the window head, a projecting fin is not needed as a drip, since the outer frame is lapped over the inner light. Rubber-based seals are used to reduce air infiltration. Windows and glazed walls are fixed to adjacent masonry walls with steel lugs that allow them to be fixed away from the edge of the concrete or masonry. These lugs are concealed by internal finishes. When fixed into a timber or pressed steel edge frame, steel glazing can be fixed directly through the frame into the supporting material.

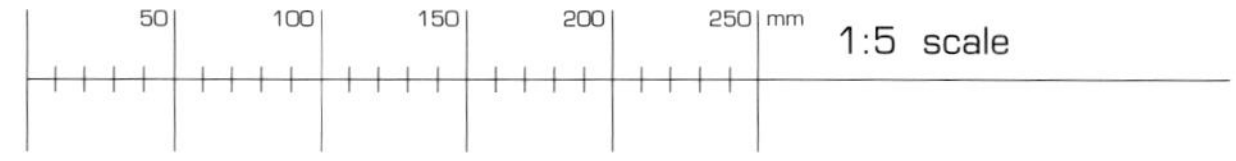

f | Vertical sections. Pressed steel single glazed fire resisting doors without thermal breaks

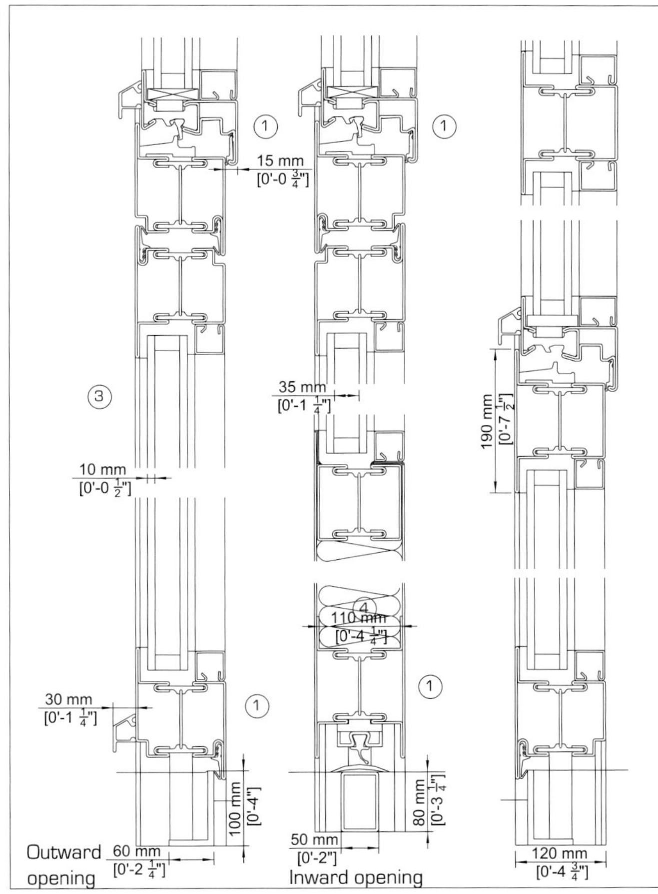

e | Horizontal and vertical sections. Pressed steel doors without thermal breaks

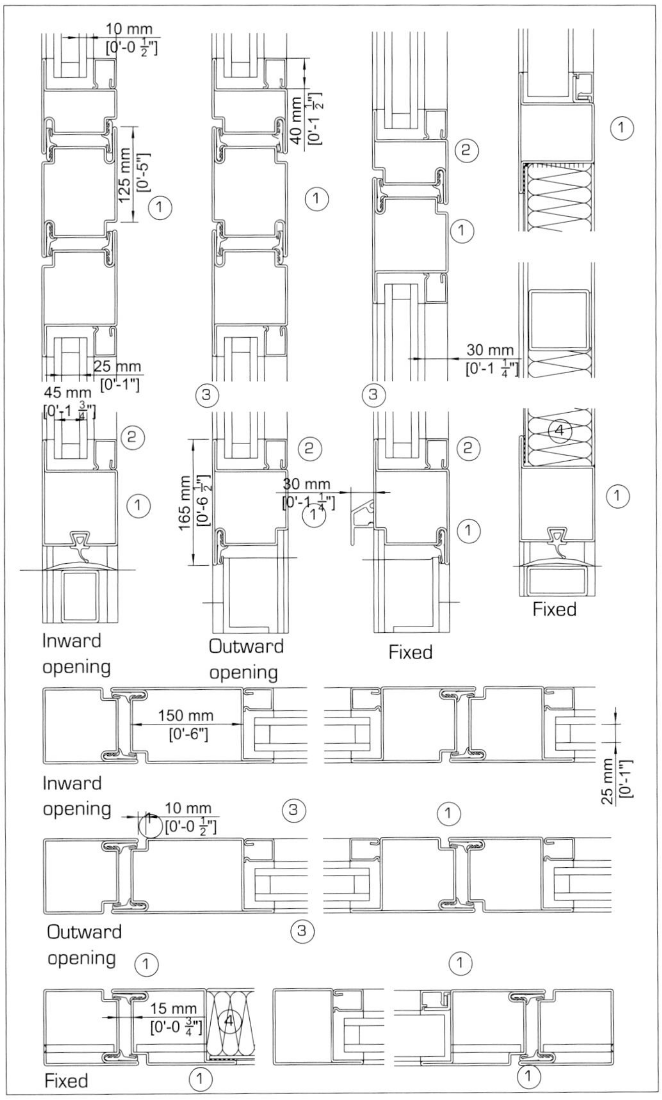

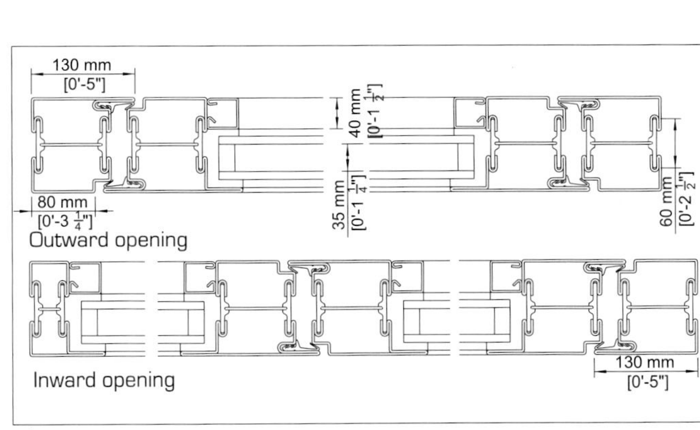

g | Horizontal sections. Pressed steel single glazed fire resisting doors with thermal breaks

Additional stiffening for larger walls is provided by tube or box sections set away from the glazed wall. The glazing is fixed to the support frame with steel cleats and brackets. These provide a method of taking up tolerances and alignment adjustments between the windows and their supporting frame.

Doors are fixed to the supporting frame in the same way as windows, but their construction is more robust. They have thicker sections at the sill and have a horizontal rail at mid-height to provide stiffness. Some also have mild steel plates, glazed in at sill level as kicking plates. The plates have an internal insulation core if used in conjunction with double glazing. The thinness of steel window frames in relation to aluminium allows door sills tend to be visually more discrete than those in aluminium, with a low upstand formed by the sill at floor level. Doors, in common with other sills, have a condensation channel which collects water at the bottom of the glazed wall and directs it to the outside. These channels are needed, since the frame is not thermally broken and condensation can occur on the internal face in temperate climates.

Large-scale glazing

Large-scale glazing uses a carrier frame of mullions and transoms formed in steel box sections similar to that in aluminium. The carrier frame has an indented groove at the front to allow fixing toggles to be inserted into it. Rubber-based air seals are set against the carrier frame onto which the double glazed units are set. The glass is secured with pressure plates and integral rubber-based seals in the same way as aluminium glazed wall systems. The pressure plate is fixed with a bolt that is secured into the toggle. In aluminium construction, a pro-

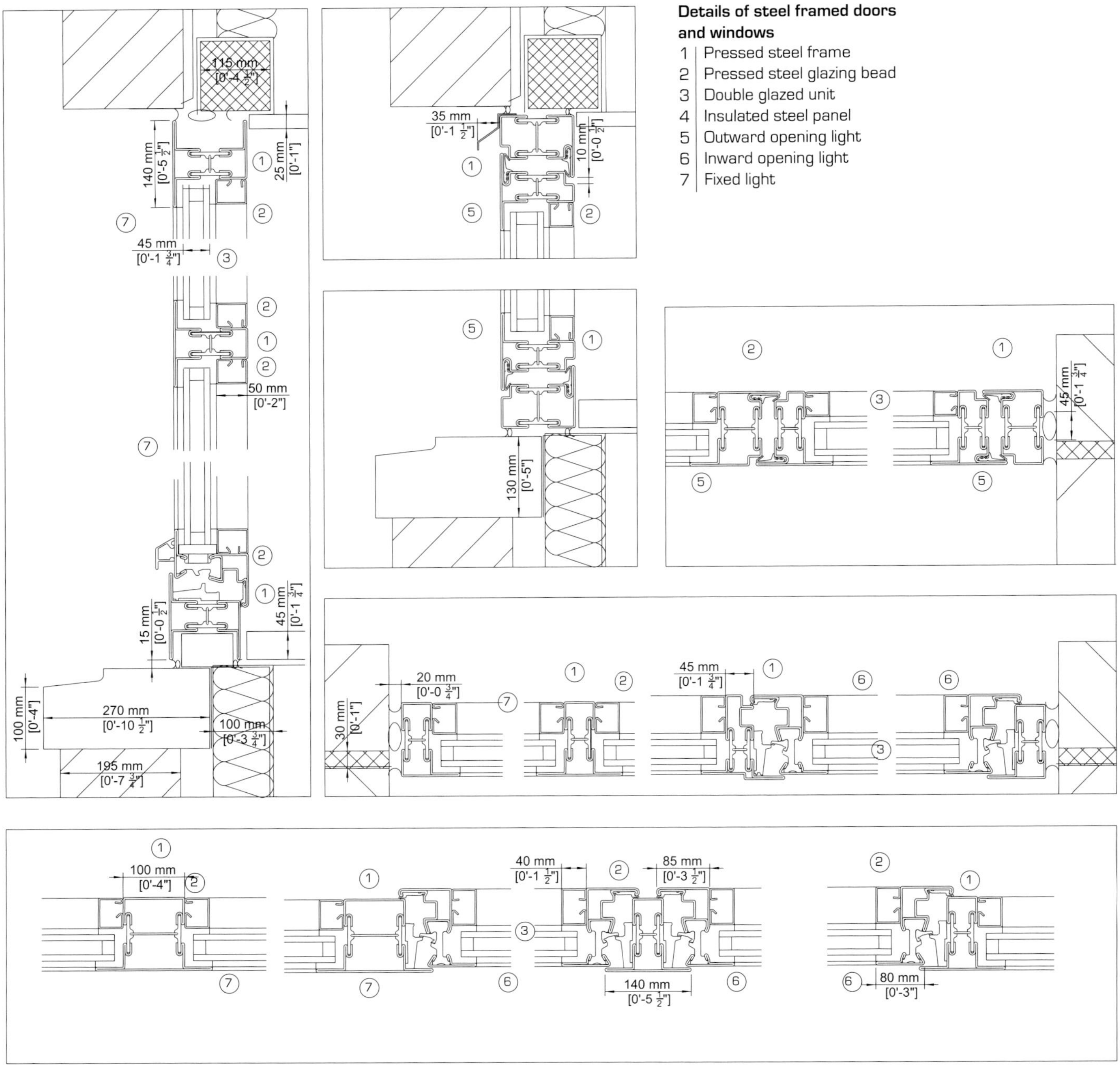

a | Vertical and horizontal sections. Pressed steel windows, thermally broken

jecting groove forms part of the extrusion to which the bolt is fixed. Rolled or pressed steel cannot incorporate such a complicated and rigid profile. The rubber-based air seals fixed to the box sections have an additional lip which tucks down over the top of the double glazed unit below. This provides a sealed inner chamber for drainage and pressure equalisation in the manner of aluminium glazed walls. Manufacturers provide systems to suit different glazing layouts and spans. The shape of the box section can also vary from thin long boxes to a rectangular box to a pointed aerofoil-type section (c). The overall sight lines are thinner than an equivalent in aluminium. Special corner pieces are made within manufacturers' systems which are similar to those in aluminium. These special profiles make corners more elegant than they would be by combining structural sections, which would necessitate the adding of sheet metal in corners.

Parapets are formed by glazing a metal flashing into the transom (horizontally) at the top of the wall and extending the flashing down the parapet wall behind which the roof build-up is retained. An advantage of steel glazed walls in box sections is that they can be combined with a glazed roof. In this detail a coping is glazed into the roof member adjacent to the wall at one end and glazed into the top transom of the wall at its other end. The coping profile is usually formed to project slightly forward of the top face of the pressure plate in order to provide protection from maintenance equipment and ladders that can damage the parapet coping. Sills at ground level have metal flashings glazed into them to project down over adjacent construction, such as an insulated concrete upstand. An aluminium or EPDM foil is also glazed in behind the flashing to provide a watertight seal against adjacent construction such as a floor slab. Slot drains at sill level are increasingly used to conceal the

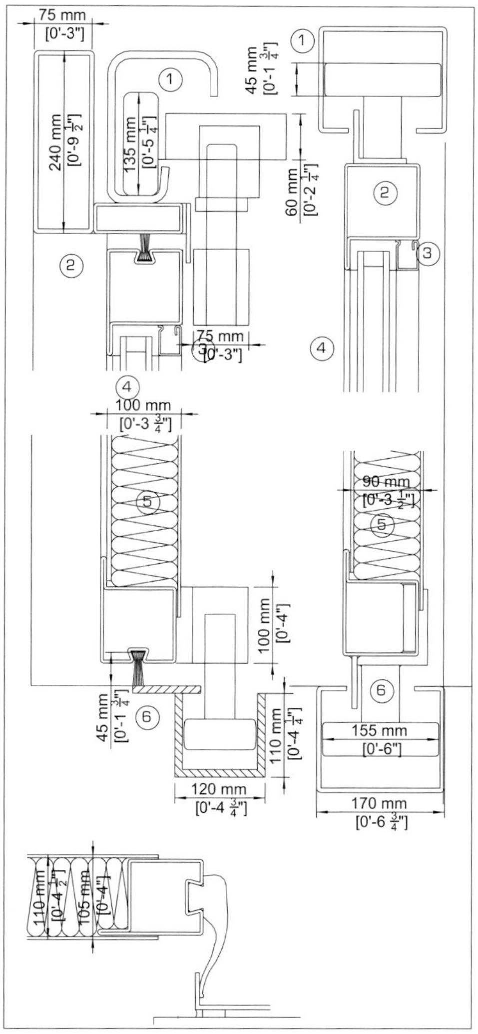

b | Vertical and horizontal section. Top rolling sliding door with brush seals

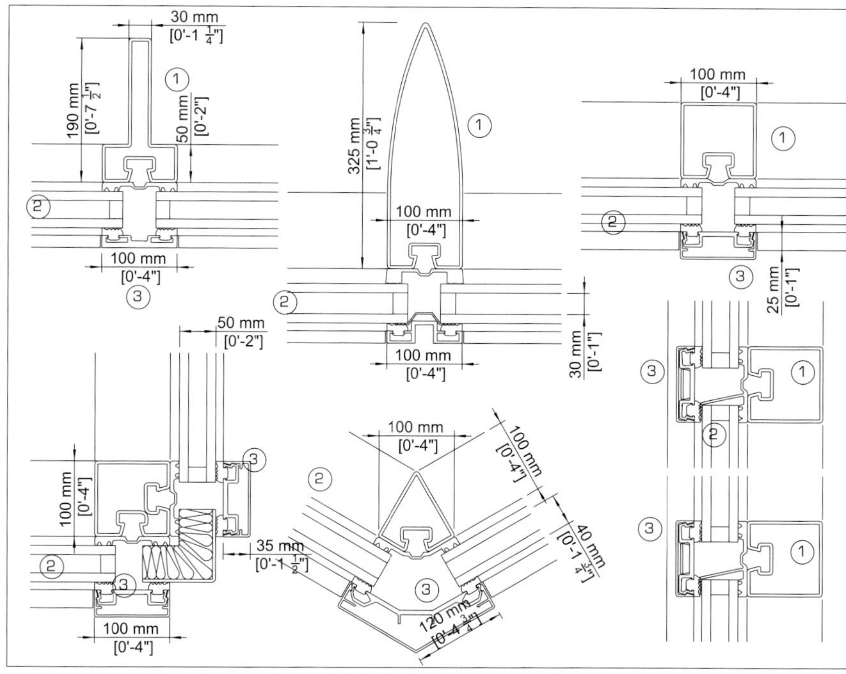

c | Horizontal sections.

Steel framed curtain walling detail

1 | Pressed steel glazing section
2 | Double glazed unit
3 | Pressed steel pressure plate

Steel framed sliding door detail

1 | Rolling wheel
2 | Pressed steel frame
3 | Pressed steel glazing bead
4 | Double glazed unit
5 | Insulated steel panel
6 | Guide at bottom of door
7 | Rubber-based seals between opening doors

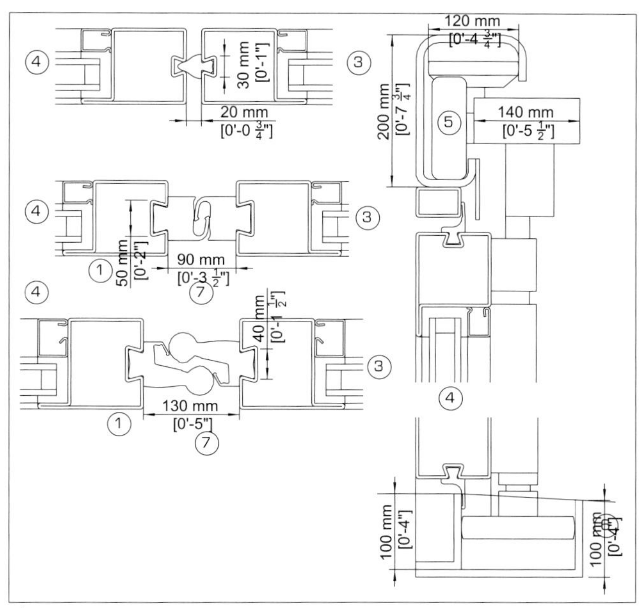

d | Horizontal and vertical sections.
Sliding/folding door with rubber seals

minimum 150mm (6in) upstand required between the lowest transom and the external ground level for waterproofing purposes. The use of an external slot drain also allows the external level to be the same as the internal floor level while providing a semi-concealed upstand.

Metal flashings and an EPDM foil behind it are also used to seal steel glazing against areas of adjacent construction such as masonry and concrete walls (MCE_176, a). Opening lights and doors are set into steel glazing as separate items glazed with an opening. As with aluminium glazed walls, the door or opening light has an additional frame which is visible from the outside. Door and window lights are formed from pressed steel sections which are folded together to form a family of profiles to suit different sizes and glass types (MCE_172, b). Steel-framed windows and doors are also made as separate items for glazing into openings in masonry walls. In this instance they are fixed through the frame into the adjacent structural wall. An EPDM foil or silicone sealant is then used to seal the gap between the steel window/door and the adjacent concrete or masonry wall.

Steel glazing with box sections has a limited thermal break provided by the rubber-based seals. Similarly, doors can have a modest thermal break introduced that avoids condensation forming on the inside face of the door in temperate climates. Doors and opening lights have rubber-based seals between the light and the carrier frame to reduce air infiltration. Steel glazing is usually finished as either galvanised or polyester powder coated. PVDF is not used on steel. Sometimes a combination of galvanising and coating is used, but galvanising can distort small components, making it much less popular than polyester powder coating.

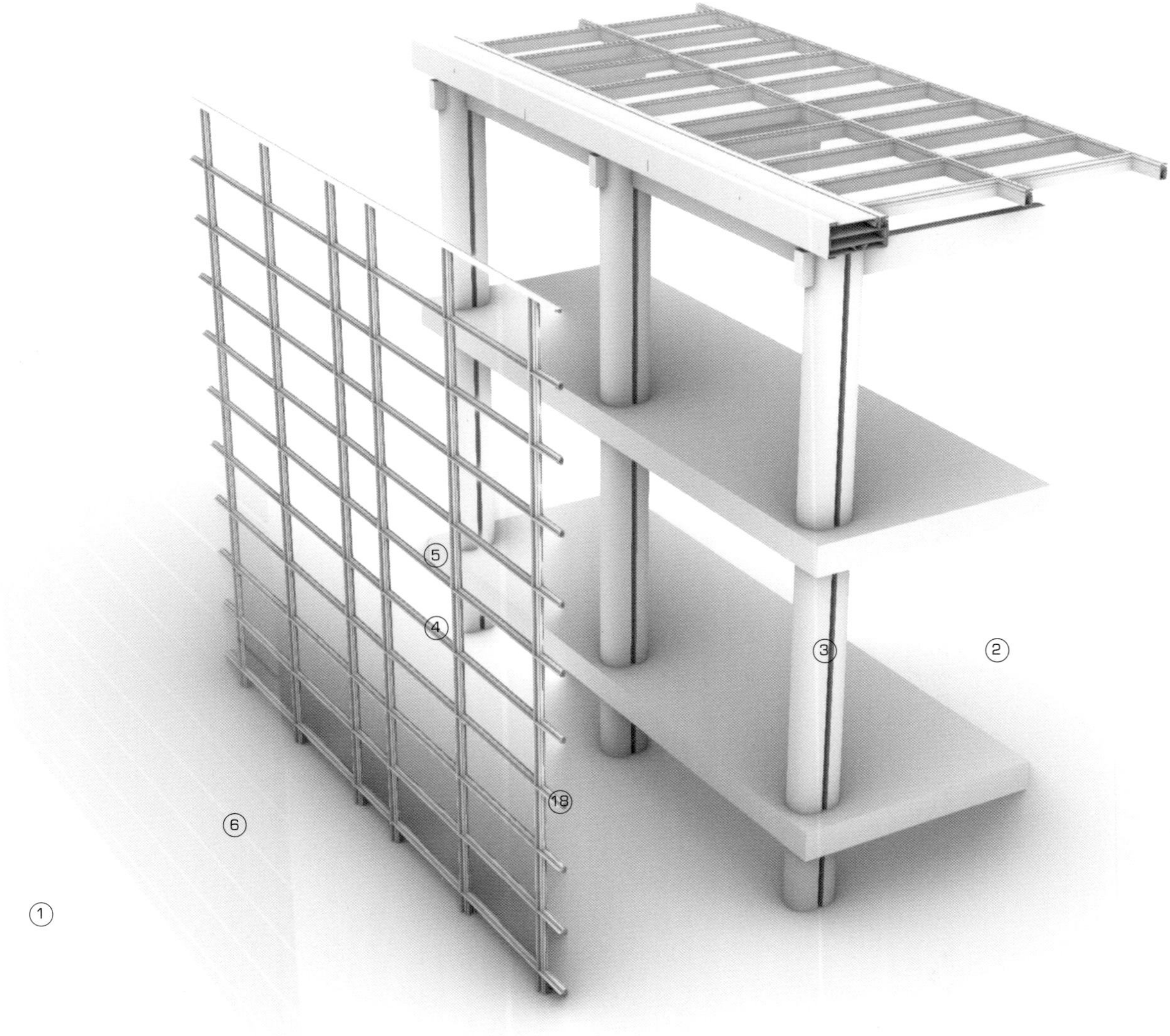

a | 3D exploded view showing steel window facade fixed to concrete frame

b | 3D view showing steel window facade fixed to concrete frame

Details

1 | Outside
2 | Inside
3 | Concrete supporting structure
4 | Transom
5 | Mullion
6 | Single-glazed or double-glazed unit to suit application
7 | Fixing bead
8 | Fixing lug
9 | Projecting transom
10 | Rubber-based seal
11 | Fixed light

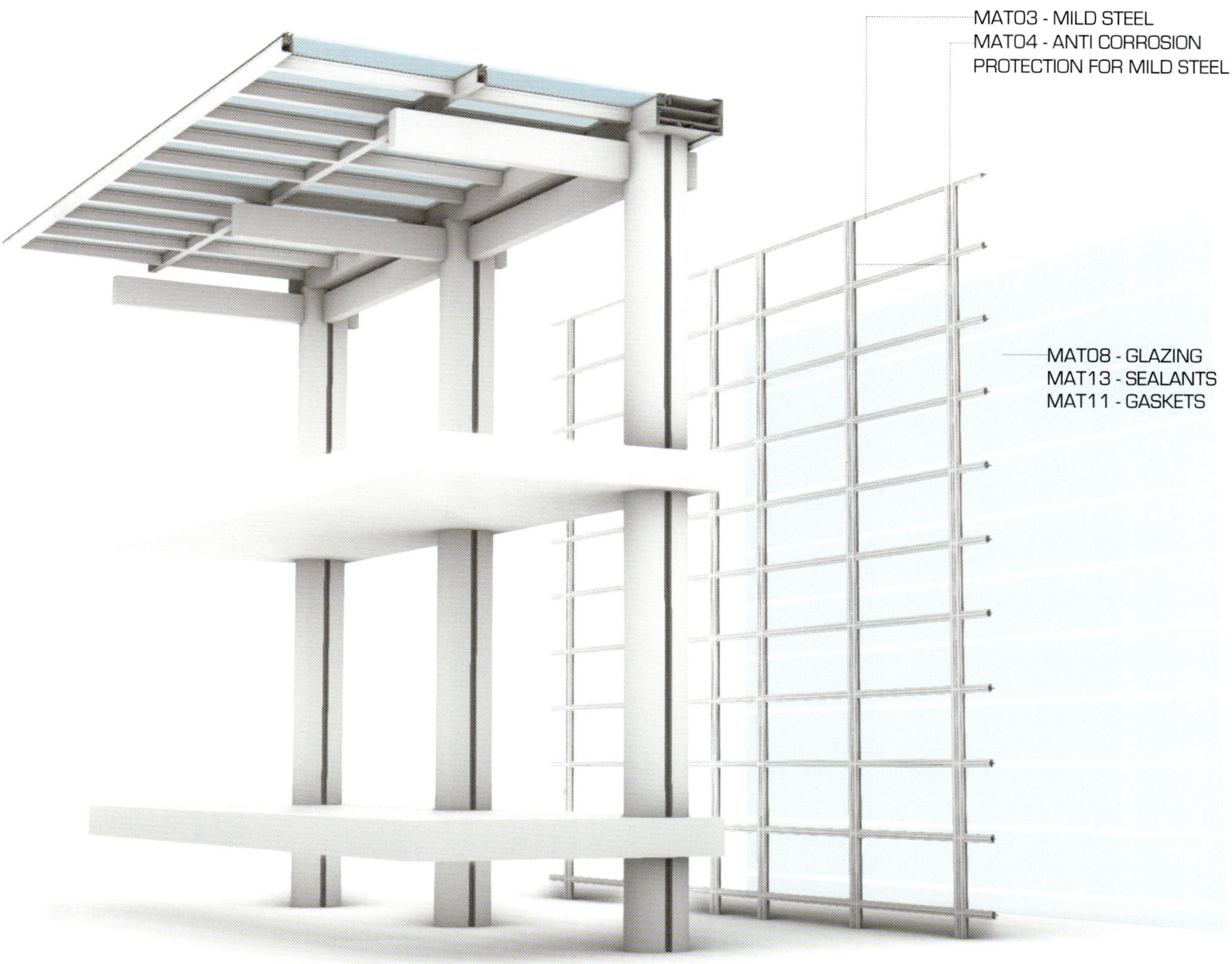

c | 3D exploded view of steel window facade components with material specification references

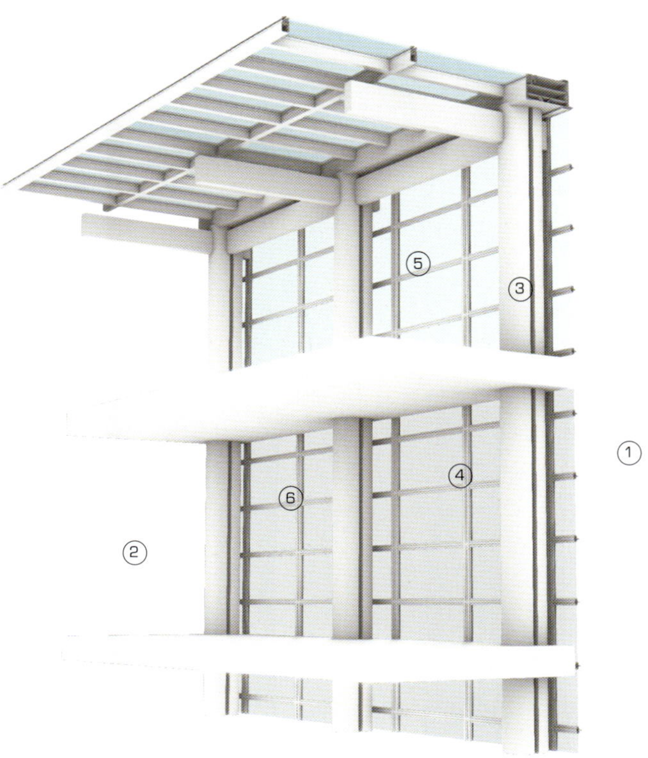

12 | Inward opening light
13 | Outward opening light
14 | Window sill
15 | Condensation tray
16 | Damp proof course (DPC)
17 | Internal finish
18 | Drip
19 | Packing
20 | Aluminium clip to secure double-glazed unit
21 | Steel sill
22 | Meeting stile

d | 3D view showing steel window facade fixed to concrete frame

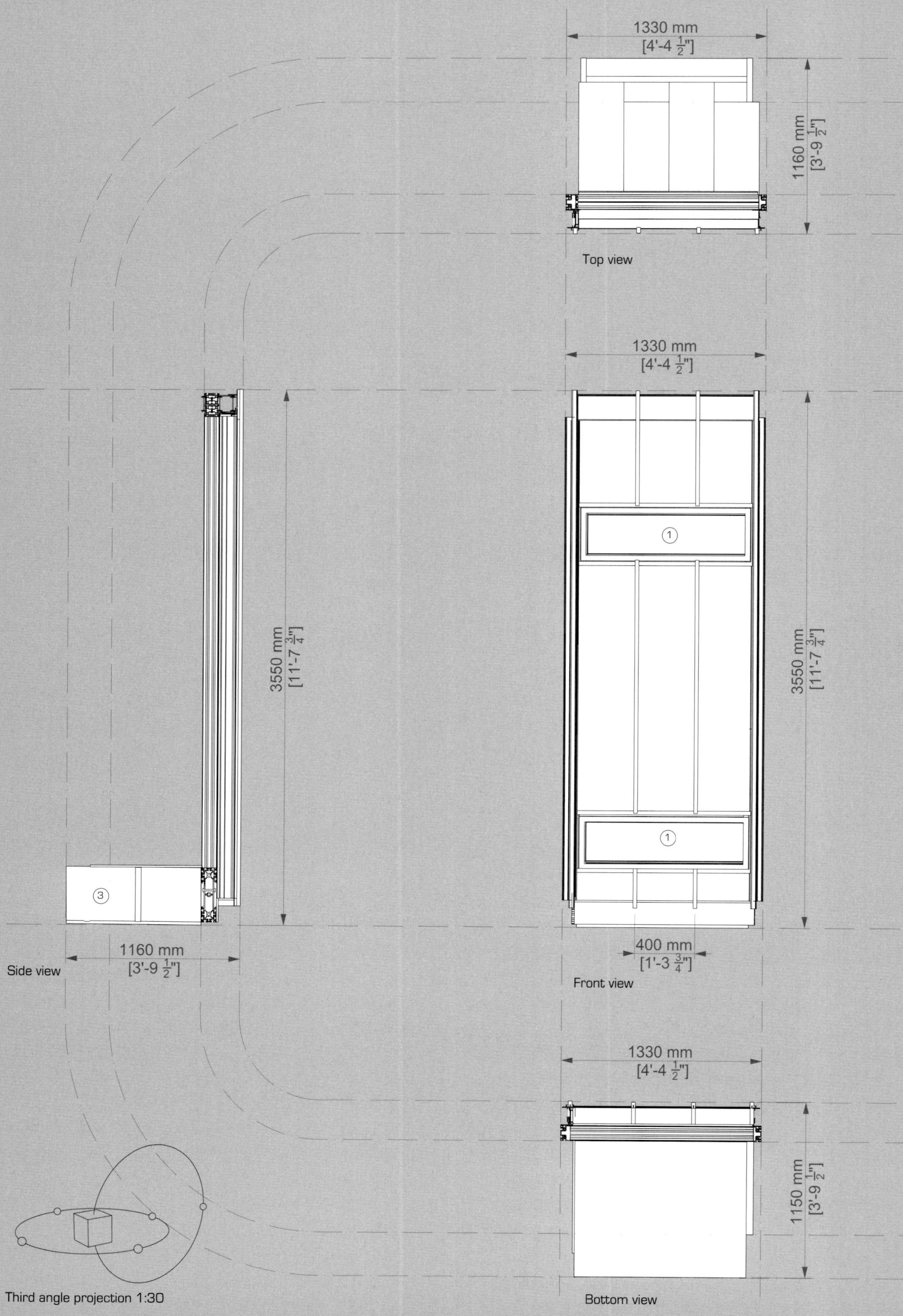

Third angle projection 1:30

Details

1 | Opening door or window
2 | Fixed light
3 | Floor slab
4 | Thermal insulation
5 | Surrounding wall
6 | Internal finishes
7 | Window sill

3D view of the assembly

Side view

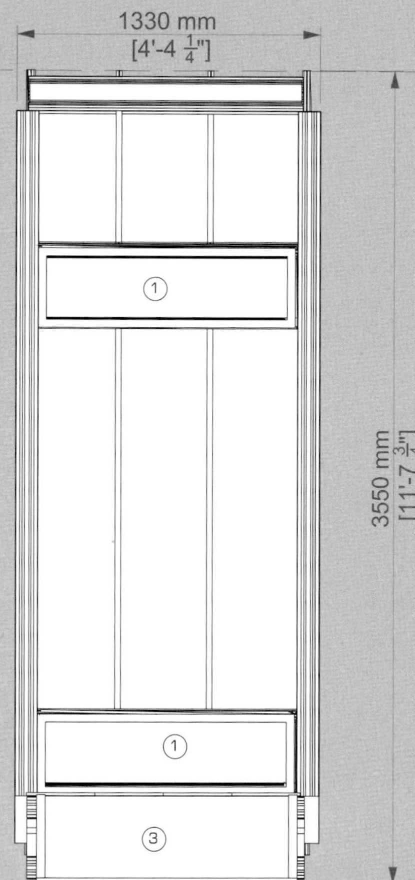

Back view

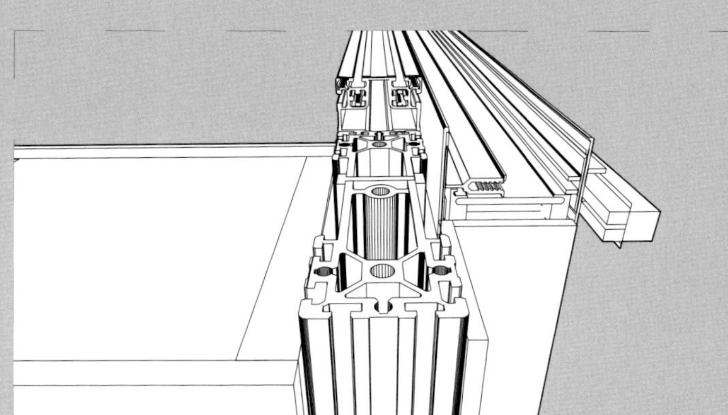

3D view of the assembly

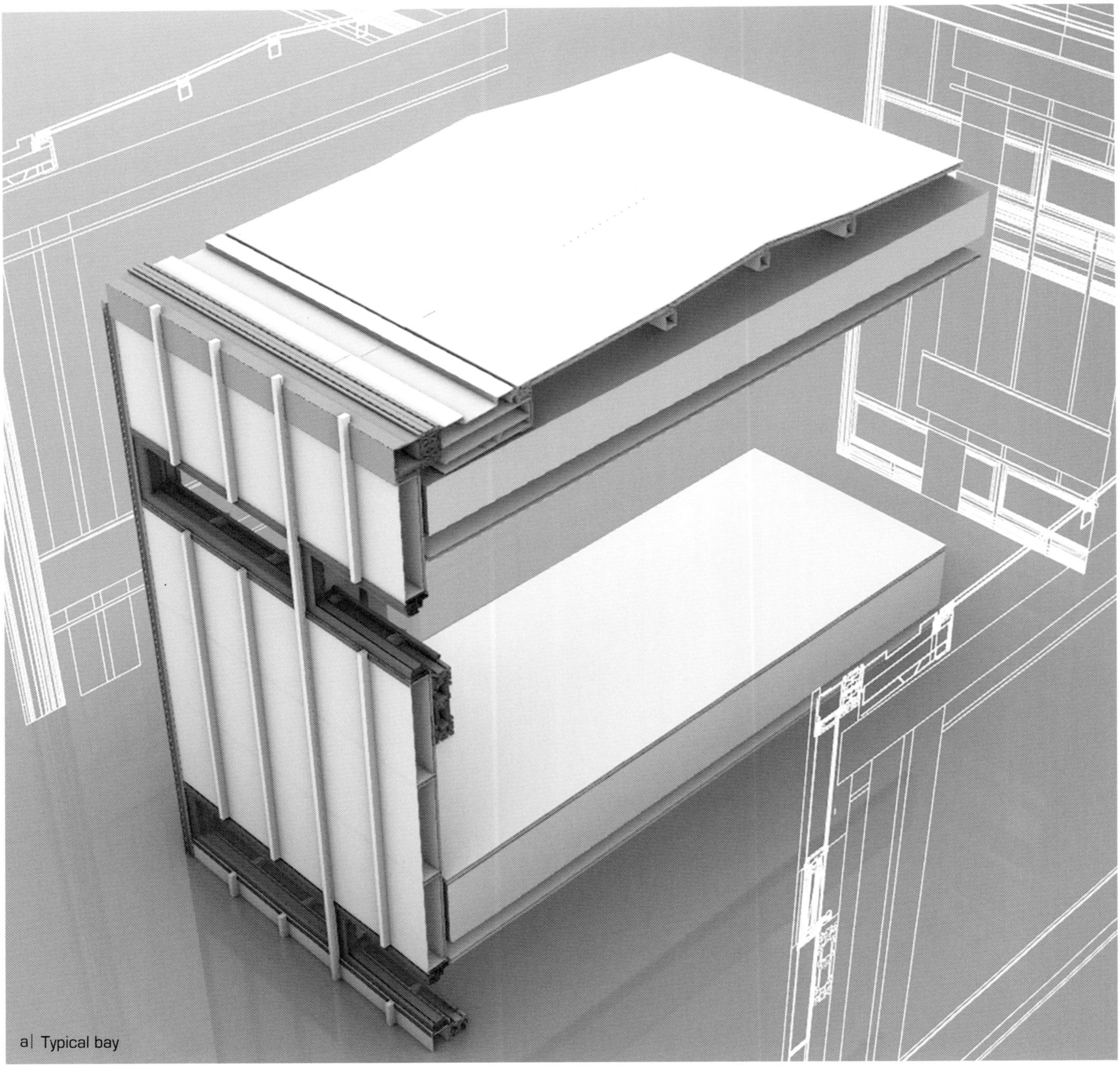

a| Typical bay

Aluminium windows are used either as individual lights set into a structural opening or as 'window walls', where windows are still set into openings but are linked together with metal or opaque glass panels to give the appearance of a continuous glazed wall.

The essential difference between aluminium windows and aluminium glazed walls (or 'curtain walls') is that windows are set individually into structural openings, whereas curtain walling uses a continuous carrier frame fixed forward of the supporting structure. Windows are not technically inferior to curtain walling, but are a different method of glazing. Window systems are well suited to apartments, where the discontinuity of framing is preferred for acoustic reasons, avoiding flanking sound passing from apartment to apartment. 'Window walls' of windows linked by glass or metal panels is often preferred where a fire separation zone is needed between apartments, when combined with thermal and acoustic mass.

Windows in openings

When fixed into a structural opening in a wall formed in another material, aluminium windows have drips and flashings to seal against the adjacent material (MCE_182, a). Specific interface details of windows with adjacent materials are discussed elsewhere in this book, under the heading of each wall type. This section deals specifically with the issues of glazing aluminium windows and doors into any structural opening.

Aluminium windows and doors are often preferred for their thermal and acoustic insulation properties, together with low levels of air infiltration. They have a generally higher performance in these areas than steel or timber frames, and their durable finish gives them a consistent appearance over 20 years, depending on the type of finish used. Their main disadvantage is the wide sight lines required to accommodate smooth movement of the opening light in the frame as well as the seals and thermal breaks required. Manufacturers are ever striving to make the sight

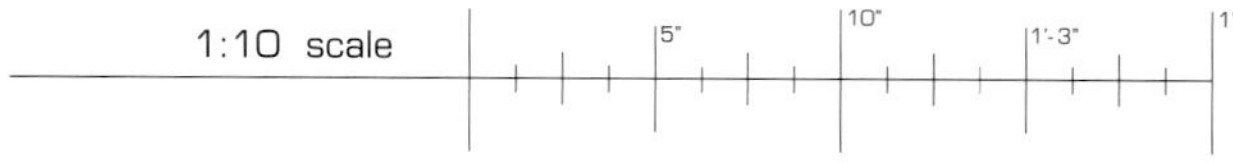

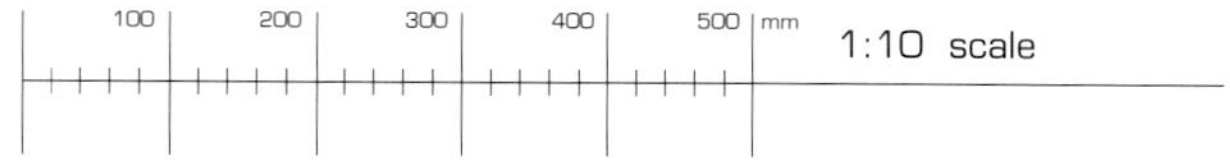

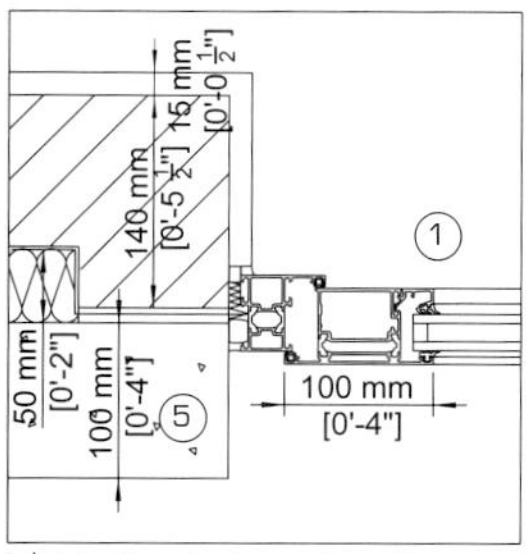

b | Horizontal section. External opening door

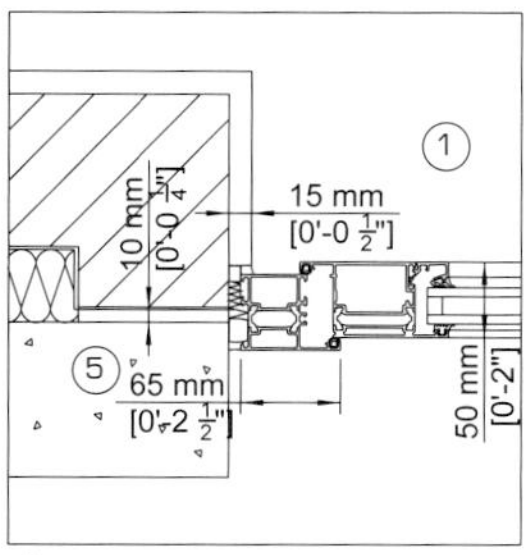

c | Horizontal section. Internal opening door

f | Prototype

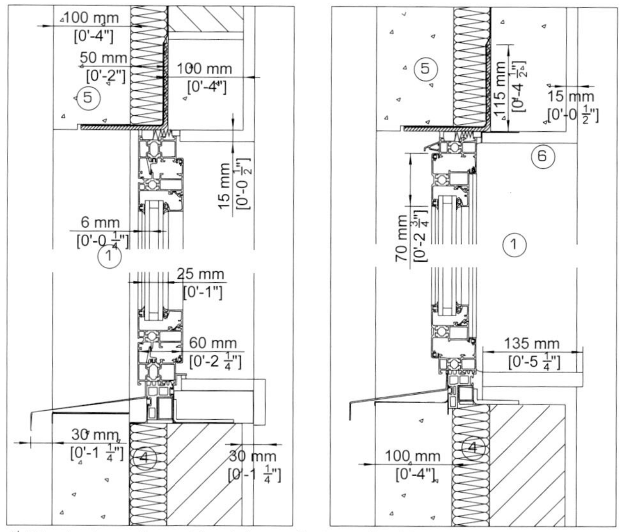

d | Vertical sections. Aluminium windows fixed into openings in masonry/concrete walls

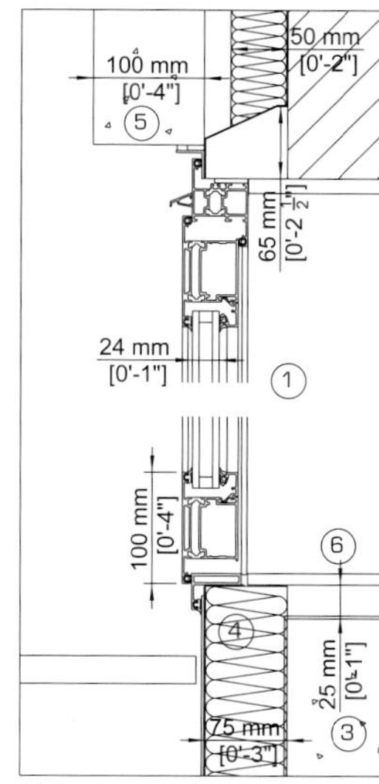

e | Vertical section. External opening door

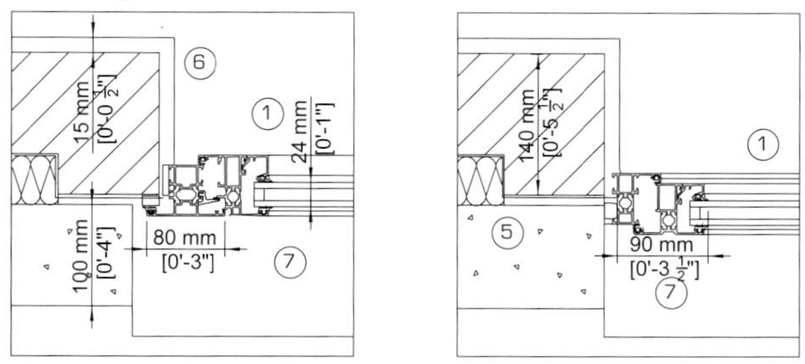

g | Horizontal sections. Aluminium windows fixed into openings in masonry/concrete walls

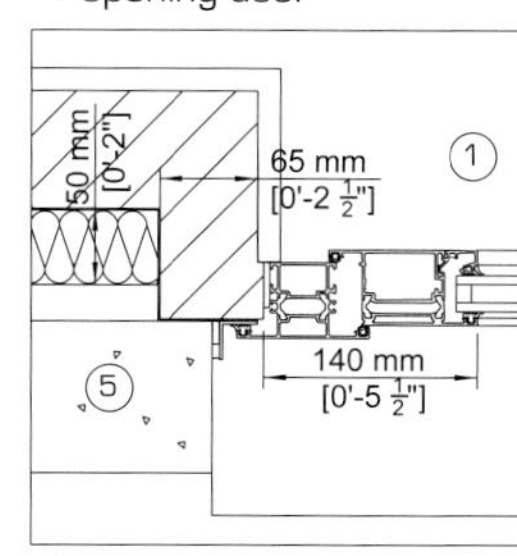

h | Horizontal section. Internal opening door

Details

1 | Opening door or window
2 | Fixed light
3 | Floor slab
4 | Thermal insulation
5 | Surrounding wall
6 | Internal finishes
7 | Window sill

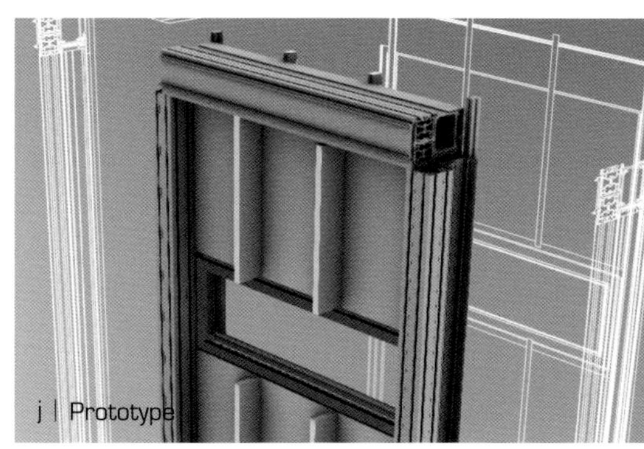

j | Prototype

lines thinner, but this usually results in the depth of the window and door increasing substantially in order to maintain structural stability.

In common with aluminium curtain walls, aluminium windows are pressure equalised and internally drained, providing two lines of defence against rainwater ingress. The outer line of defence is provided by a rubber-based seal fixed to the window frame. Any rainwater penetrating this outer seal is drained away in a ventilated cavity within the window frame and is released back to the outside through the sill profile (d).

There is an increased use of 'flipper' seals within window frames to both keep any rainwater that penetrates the outer seal in a chamber at the front of the section (similar to unitised curtain walling) and to provide a thermal break in the void within the window frame. This prevents air temperature and humidity levels in the outer chamber from passing all the way through the void in the frame and reaching the inner seal. The inner seal is then used only as an air seal rather than a full weathertight seal. This method is similar to that used in unitised curtain walling, where two air chambers are used at the joint between panels. There has been an increased use in recent years of thermal breaks in both the frame and the opening light in aluminium windows. The breaks are positioned to coincide with the rubber-based flipper seal to reduce the thermal bridge across the complete window section.

Aluminium frames for doors and windows differ substantially in that window frames are narrower but deeper than doors (MCE_182, c) in order to provide much lower levels of air and water infiltration. Door frames are generally wider so as to provide stability for the infill panel of glass but are not usually thermally broken in order to provide a rigid frame to support the infill glass (MCE_183, g). Window frames are often used for doors where these lower levels of air and water infiltration are required, as in doors to balconies in apartments, for example.

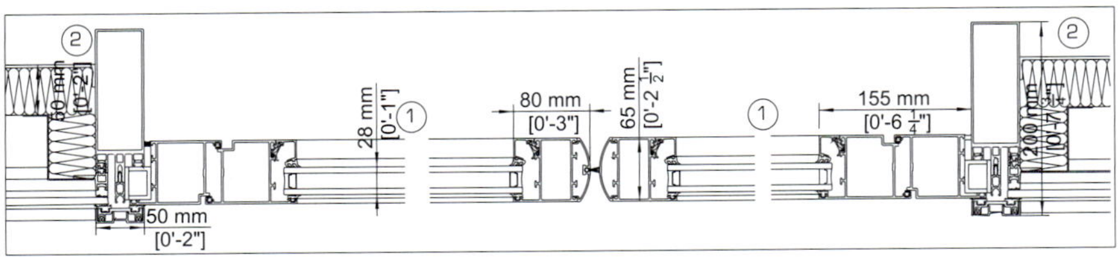

a| Horizontal section.

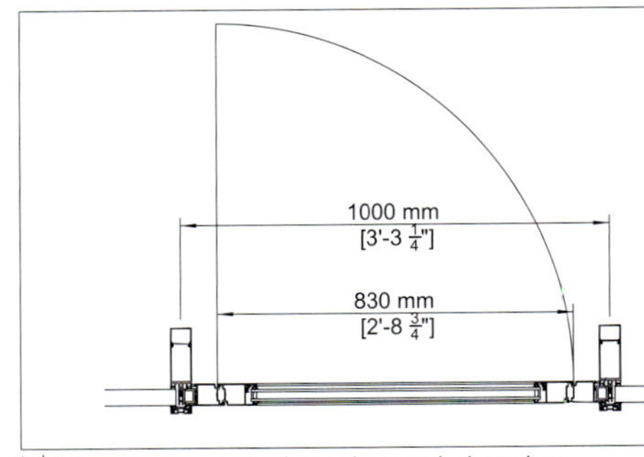

b| Horizontal, vertical section and elevation. Door fitted into a stick curtain wall system

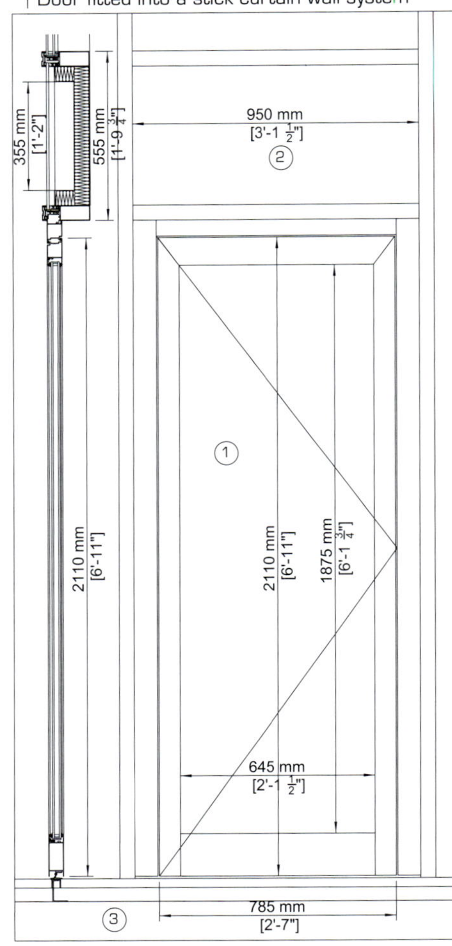

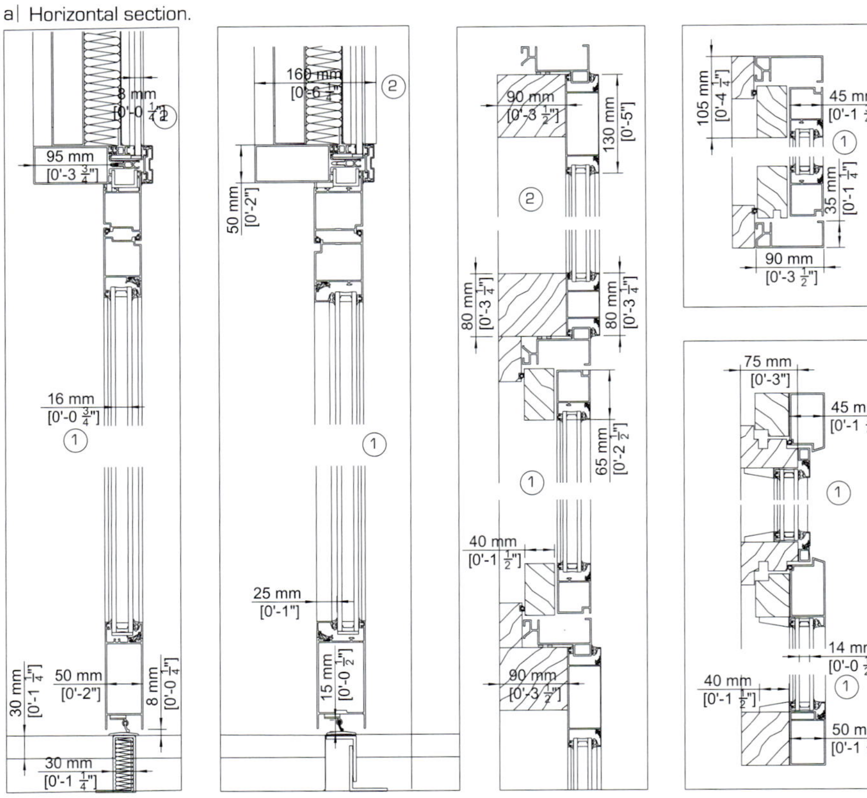

c| Vertical sections. Door fitted into a stick curtain wall system

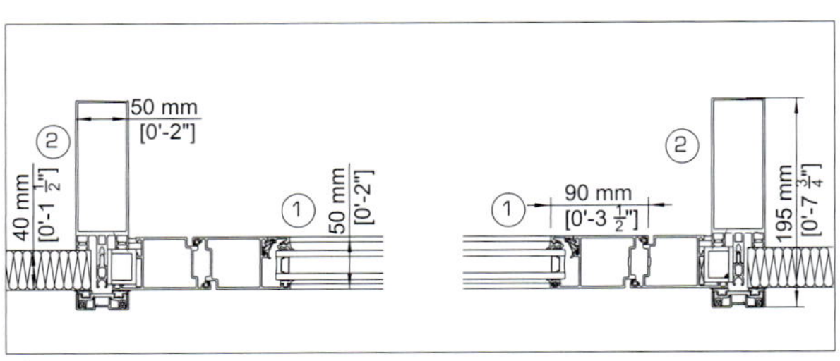

d| Horizontal section. Door fitted into a stick curtain wall system

Doors using door profiles can reach 3.0 to 3.5 metres in height without wide sight lines whereas this is not the case for doors using window frame profiles which can reach a maximum of around 2.4 metres high. Above this height, it is difficult to use the frame to support the glass. For doors exceeding 2.4 metres in height using window sections, it is visually necessary to silicone bond the frame to the glass so that the frame and glass become a single structural component. The glass is then used to stiffen the frame rather than being a non-structural infill panel within the frame. This is more expensive and also difficult to replace the glass if broken without replacing the complete frame. If the frame is replaced, then the colour of the replacement door may not match the surrounding framing. Maximum widths of aluminium glazed doors are a function of the height, in order to restrict the overall weight of the door leaf, but 850mm wide doors (for a 1700mm wide pair of doors) are common. The minimum width of the opening door is about 250mm to 300mm depending on door height and width. The minimum dimension is determined both by the minimum manufacturable size of double glazed unit and by the need to open the light smoothly.

Top-hung lights, often set above fixed aluminium windows and opening doors to provide night-time natural ventilation, have a minimum height of about 800mm to 900mm in order to accommodate the opening mechanism. Fixed lights above doors can be reduced to a height of around 200mm (MCE_185, g). Top-hung, bottom-hung and tilt/turn windows use similar sections to side-hung windows. Tilt/turn windows have the advantage of combining the functions of side-hung windows with their ability to open completely, and the ability of bottom-hung windows to provide high-level natural ventilation has led to their increased use.

Sliding doors are usually top-hung, sliding on rollers set at the door head. The bottom of the door is restrained by guides that both hold the door in place and allow it to move without being obstructed by dirt and grit

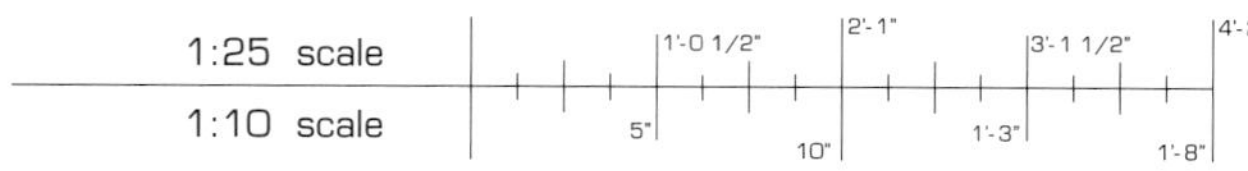

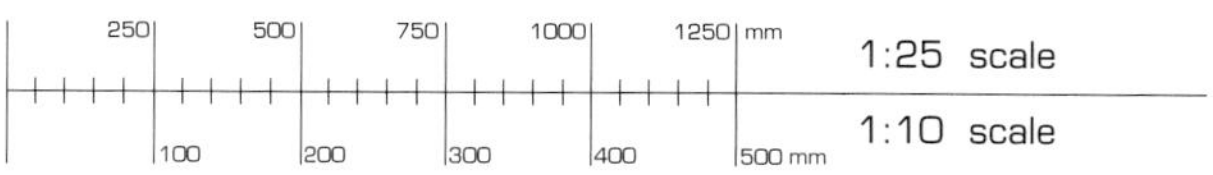

k | Prototype

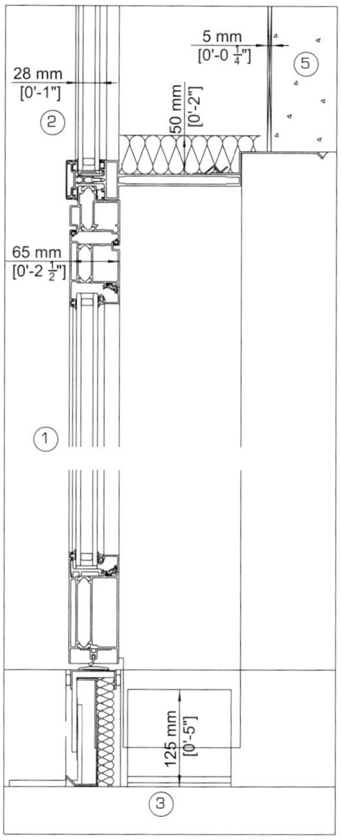

f | Vertical section. Outward opening thermally broken door. Fixed into stick curtain walling

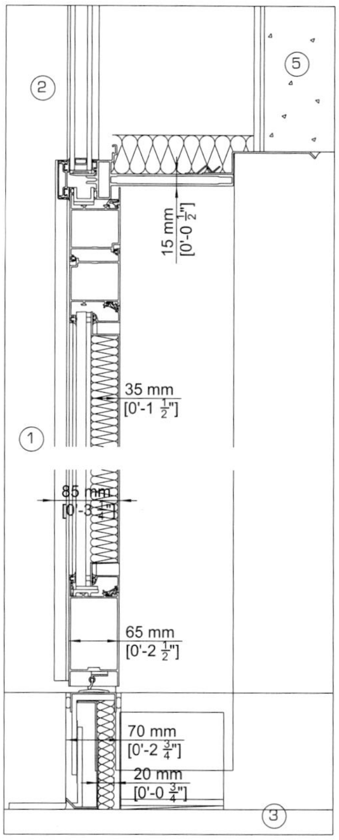

g | Vertical section. Inward opening thermally broken door. Fixed into stick curtain walling

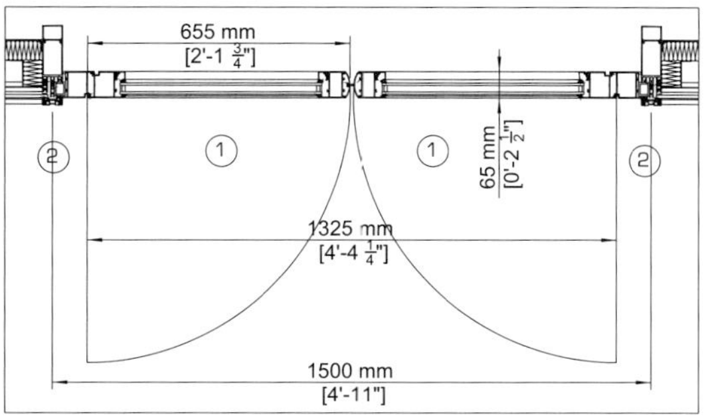

h | Horizontal section. Aluminium doors fixed into a stick curtain wall system

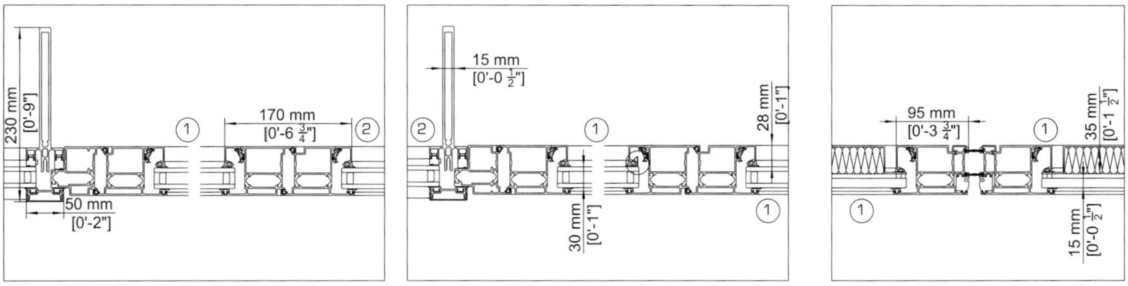

j | Horizontal sections.

Details

1 | Opening door or window
2 | Fixed light
3 | Floor slab
4 | Thermal insulation
5 | Surrounding wall
6 | Internal finishes
7 | Window sill

accumulating in the bottom of the sill profile. Bottom-rolling doors, which are supported on the sill profile and are restrained at their head, are much less commonly used for external doors since dirt and grit can hinder their movement. There are two types of sliding door: those that roll in a constant plane and those which lift out and then slide. The first type has higher levels of air infiltration associated with side-hung door profiles. The second type has much lower levels of air infiltration but has wider sections, creating wider sight lines.

When aluminium windows are glazed into a structural opening, the joint between window and adjacent wall is made either with a sealant, such as silicone, or with an EPDM foil glazed into the window frame and sealed against the face of the surrounding wall (MCE_181, b and c). The choice of seal is dependent upon the material used in the surrounding wall, but silicone seals are generally used where the adjacent wall is of sealed construction, such as an open-jointed wall, sheet metal or render on concrete. EPDM foil is used where the adjacent wall has an open-jointed finish in front of the structural wall, as in rainscreen construction; in metal or stone, for example. The EPDM foil is sealed against the waterproof backing wall used in this type of construction. This method makes it necessary to fix the windows and seal them to the backing wall before the external cladding is fixed. This is often an advantage, where aluminium windows and backing walls provide a weathertight enclosure at an earlier stage than would be achieved with more traditional methods, where the windows are glazed in afterwards. The early date for weathertight enclosure allows construction work internally to commence at the same time as the external cladding is being installed. Forming a seal with sealant, with silicone or a polysulphide for example rather than EPDM, allows the window to be installed afterwards. Since the joints are visible, and are formed with a tool from outside, care should be taken to choose a colour which is compatible with both the window and the adjacent construction.

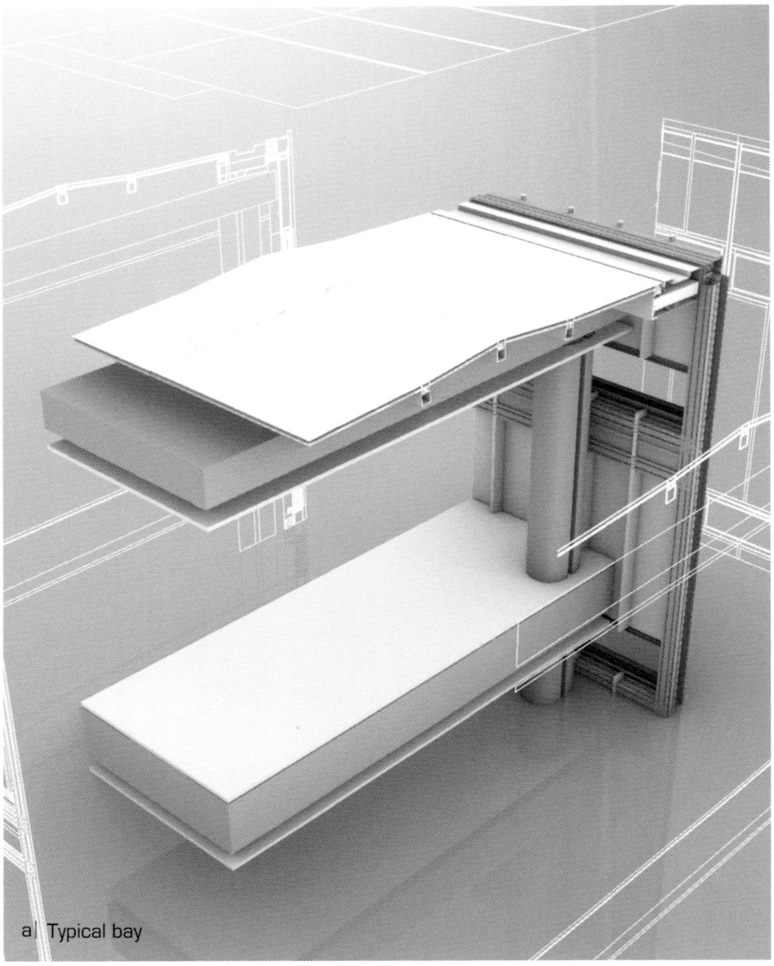

a | Typical bay

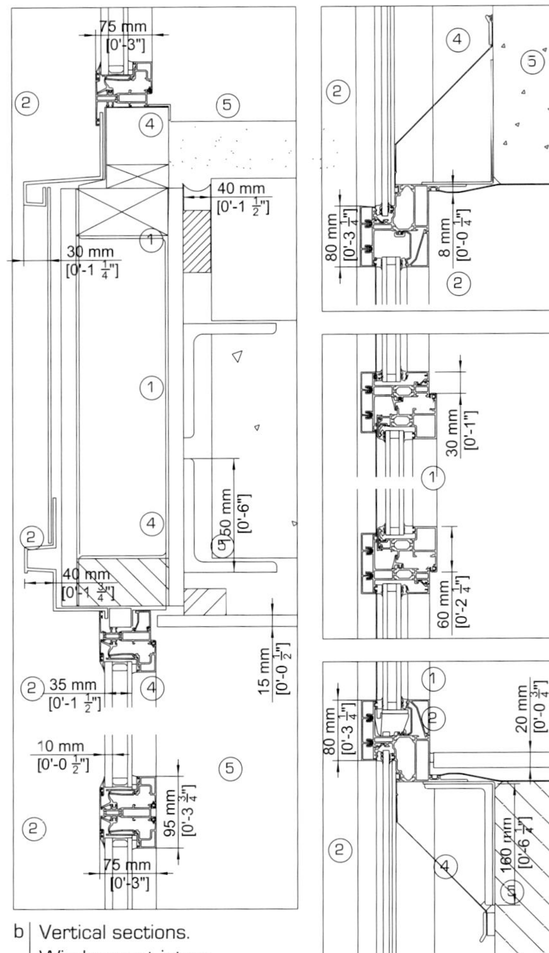

b | Vertical sections. Windows set into a glazed rainscreeen

Details

1 | Opening door or window
2 | Fixed light
3 | Floor slab
4 | Thermal insulation
5 | Surrounding wall
6 | Internal finishes
7 | Window sill

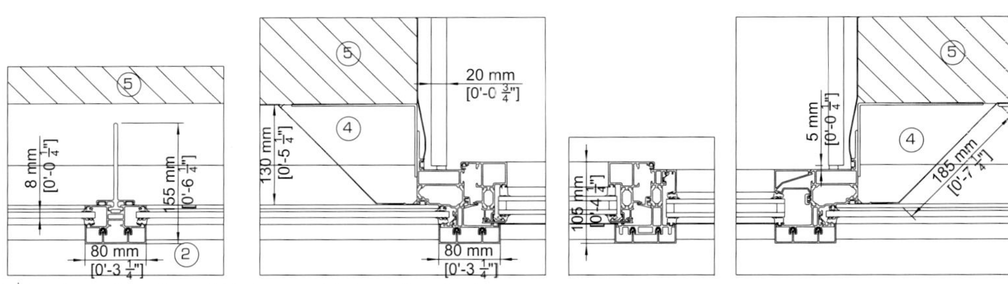

c | Horizontal sections. Windows set into a glazed rainscreeen

Window walls

This system is particularly useful where a discontinuity is required, usually for fire or acoustic reasons, while maintaining an overall modular facade. They are usually less expensive than glazed curtain walling. The opaque areas are usually filled with either an aluminium sheet (coated to match the colour of the adjacent window frames) or with opacified glass. These infill panels are supported by the window frames and do not have a support frame behind, as would be used in curtain walling systems.

In a window wall, the aluminium profile visible from outside is adapted to give the same appearance as the junction between opaque spandrel panels. This gives the external framing an overall consistent appearance associated with other modular glazed wall types (a). Windows are supported either directly onto a floor slab or concrete/block upstand, or alternatively are set forward of them on brackets (g). If the window is sitting on a support, then the opaque panel adjacent to it is usually set forward of the window in order to accommodate its thermal insulation. When the window is set forward of the opening, then the glazing in the aluminium window can be aligned with the opaque glazing. This glazing type is commonly used with both ribbon windows and rows of individual windows.

An advantage of window walling is that the opaque spandrel panels can be made in a different form of construction from the windows. Insulated composite panels can be used for example as metal rainscreen panels, or open-jointed opaque glazing. This can make the system very economic when compared to curtain walling. If the spandrel panels are not set into a framing system that is internally drained and ventilated like the windows, then the void behind the opaque panels must be drained at the head of the window below. In addition, the thermal insulation behind the spandrel panel forms a continuity with the double glazed unit of the window.

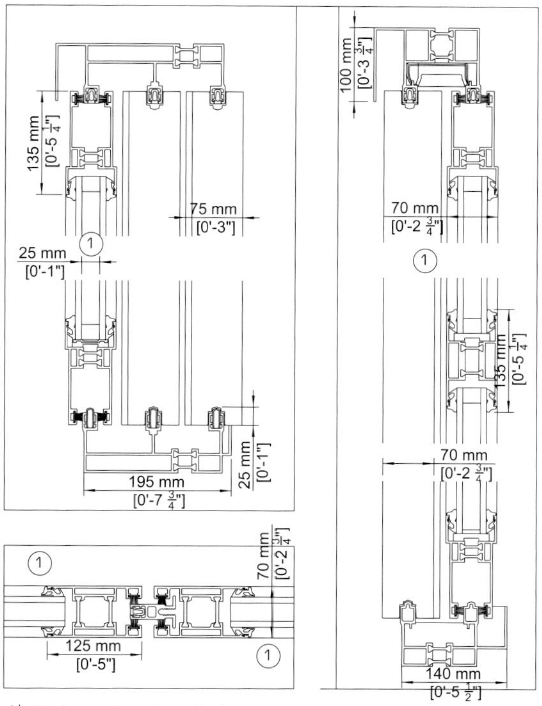

d | Horizontal and vertical section.
Aluminium windows set into masonry openings

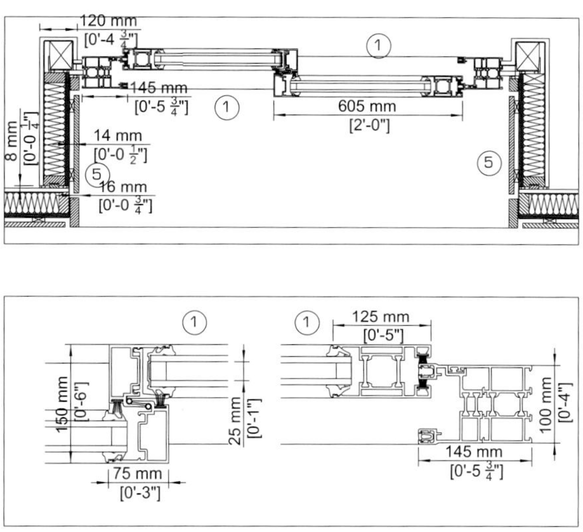

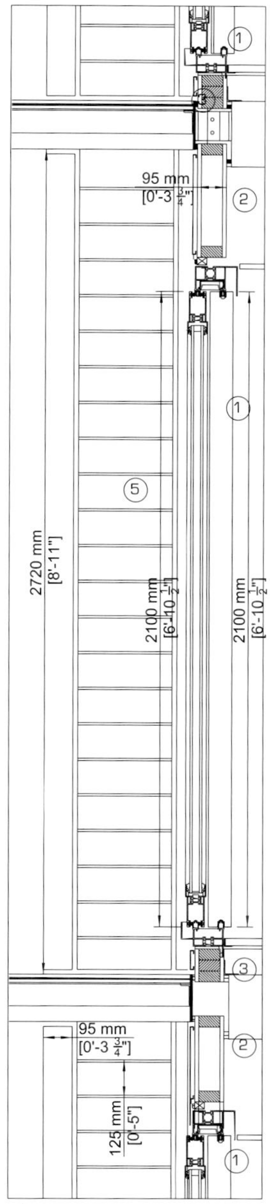

f | Vertical section.
Sliding aluminium doors

e | Prototype

Composite windows

A recent development has been in the use of aluminium windows with an inner frame in timber to support the outer aluminium profile. This is effectively an aluminium window wall with a timber carrier frame. From the interior, the framing has an all-wood appearance. However, the glazing sections used are those from windows rather than curtain walling (MCE_186, a). Some manufacturers make windows with timber frames behind them to form a timber framed unitised system. The outer joint is an open-jointed rainscreen. This protects the inner seals between timber frames from the worst effects of windblown rain. The timber frames are screwed or bolted together. The complete carrier frame is fixed back to floor slabs with steel or aluminium brackets in the manner of glazed curtain walling. These brackets are described in the section on stick glazing and unitised glazing.

g | Prototype

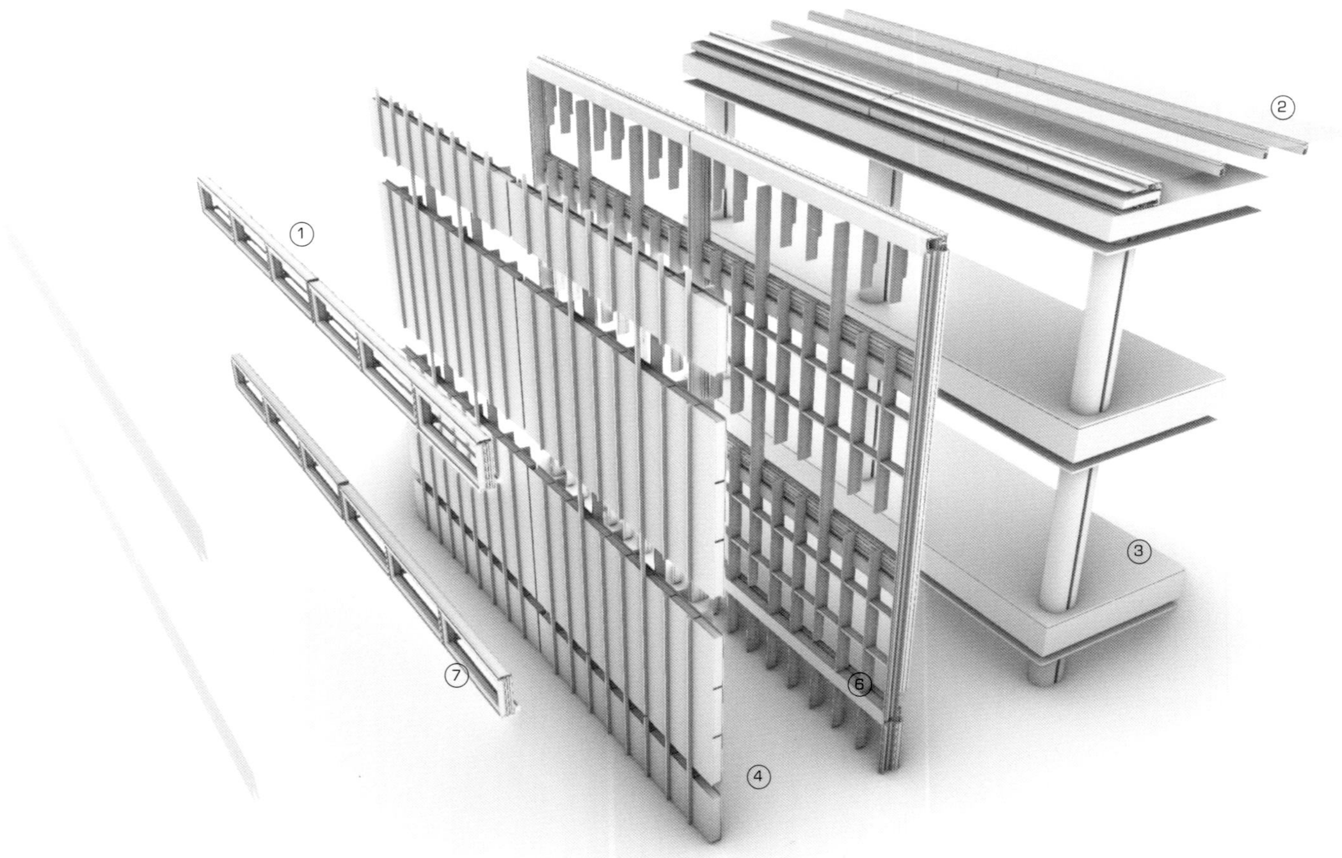

a | 3D exploded view showing aluminium window facade fixed to concrete frame

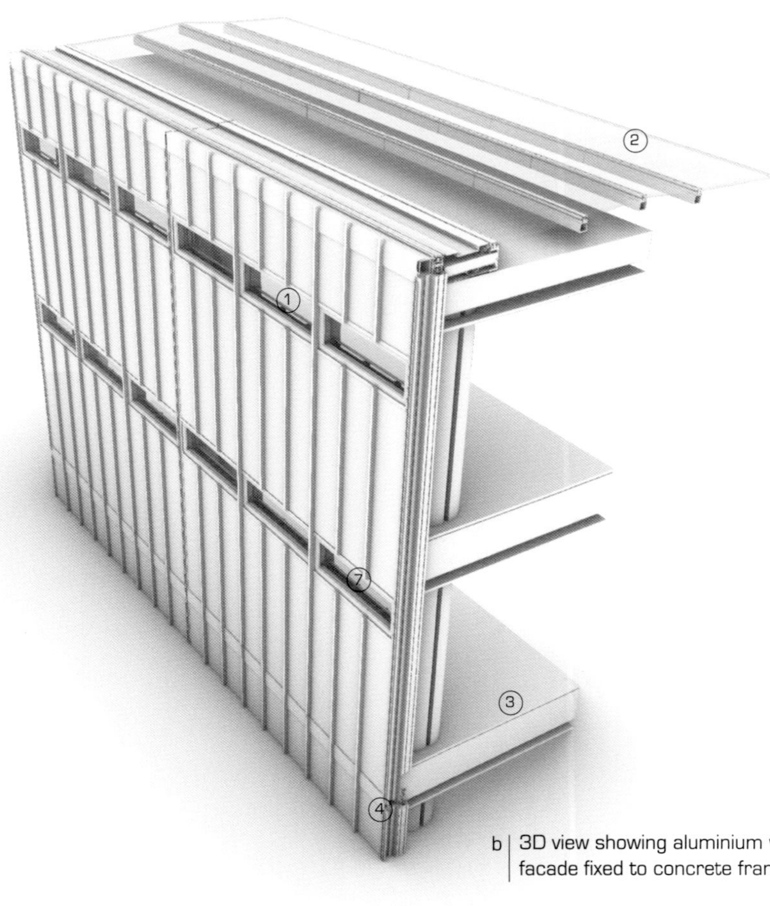

b | 3D view showing aluminium window facade fixed to concrete frame

Details

1 | Opening door or window
2 | Fixed light
3 | Floor slab
4 | Thermal insulation
5 | Surrounding wall
6 | Internal finishes
7 | Window sill

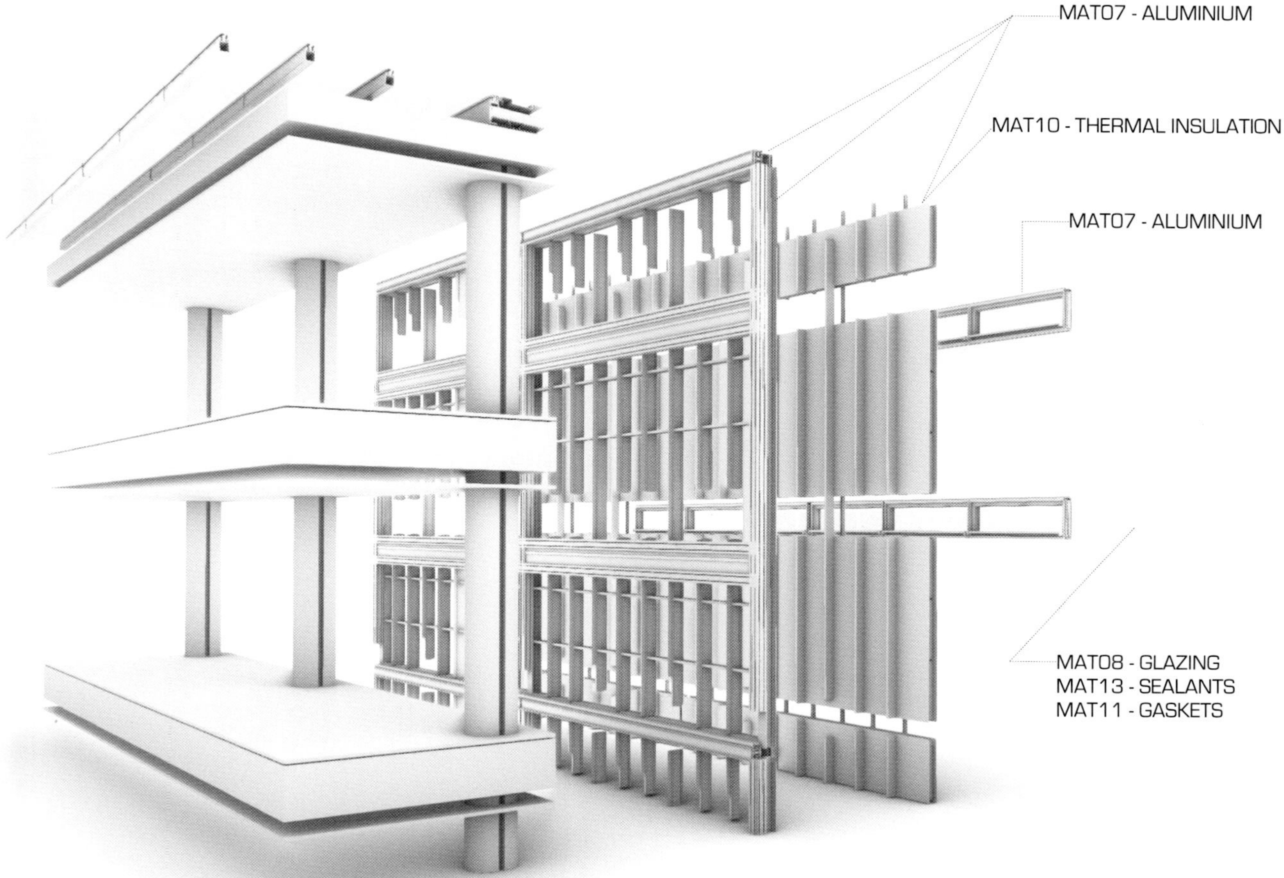

c | 3D exploded view of aluminium window facade components with material specification references

d | 3D view showing aluminium window facade fixed to concrete frame

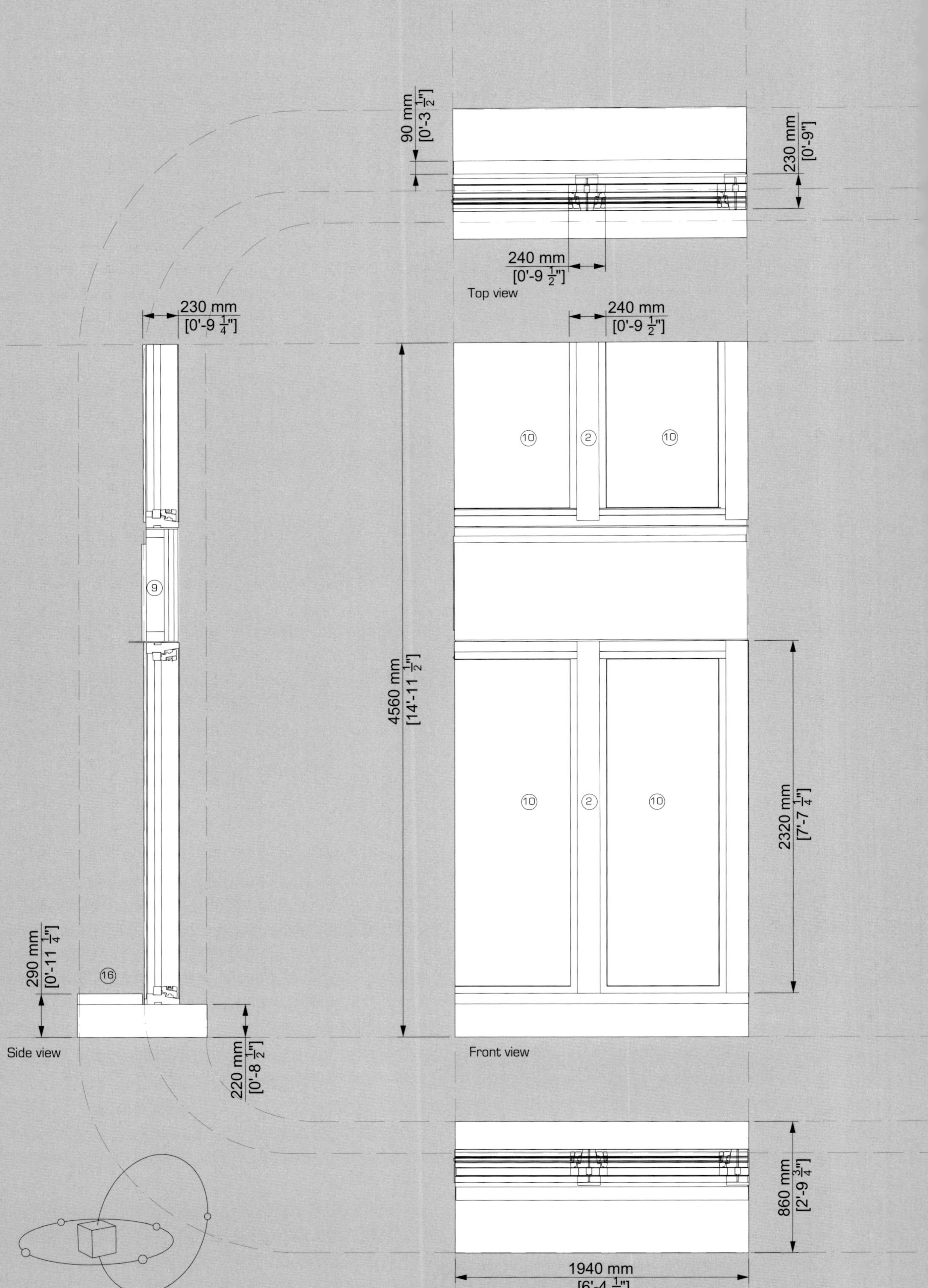

Third angle projection 1:30

Details

1 | Fixed light
2 | Supporting structure
3 | Outside
4 | Inside
5 | Head
6 | Sill
7 | Metal stiffening insert
8 | Timber infill panel
9 | Thermal insulation
10 | Single-glazed or double-glazed unit to suit application

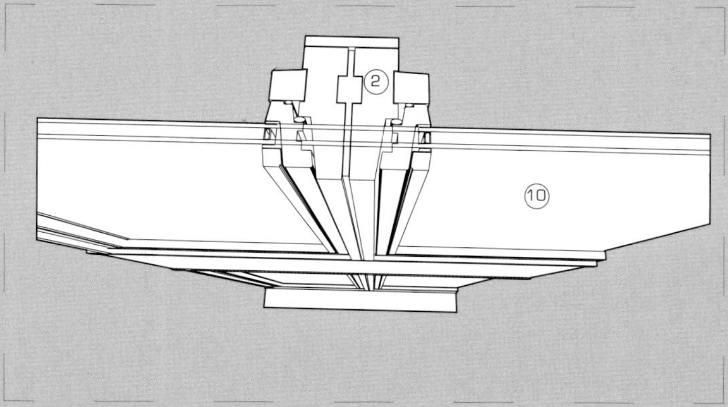

3D view of the assembly

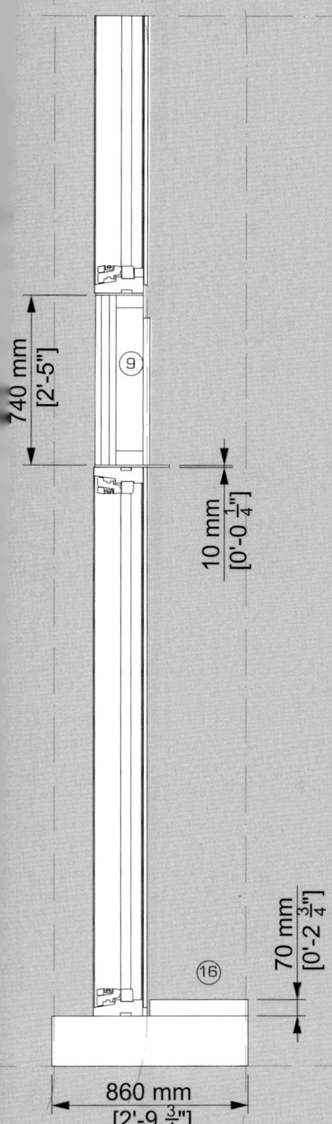

Side view

Back view

3D view of the assembly

11 | Fixing bead
12 | Rubber-based seal
13 | Inward opening light
14 | Outward opening light
15 | Damp proof course (DPC)
16 | Internal finish
17 | Flashing to seal against adjacent wall
18 | Vertically sliding sash
19 | Sliding door frame top-hung or bottom-rolling
20 | Window or door jamb

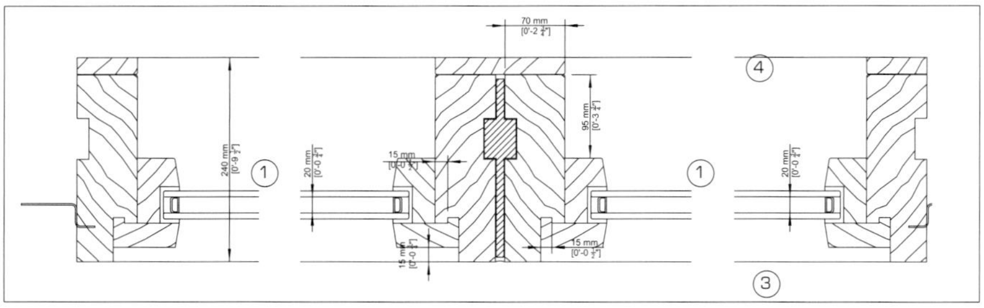

a| Horizontal section. Fixed casement window

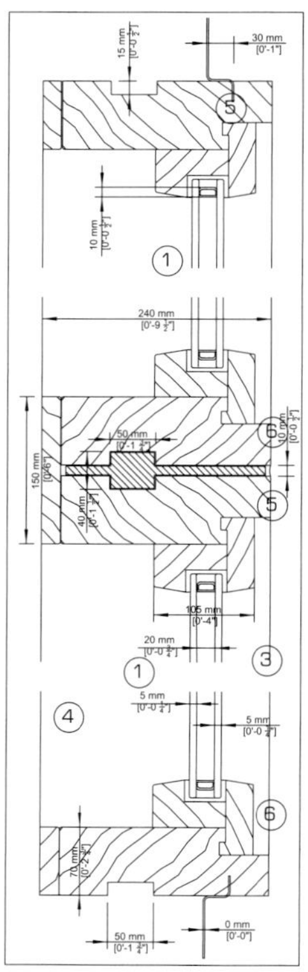

b| Vertical section. Fixed casement window

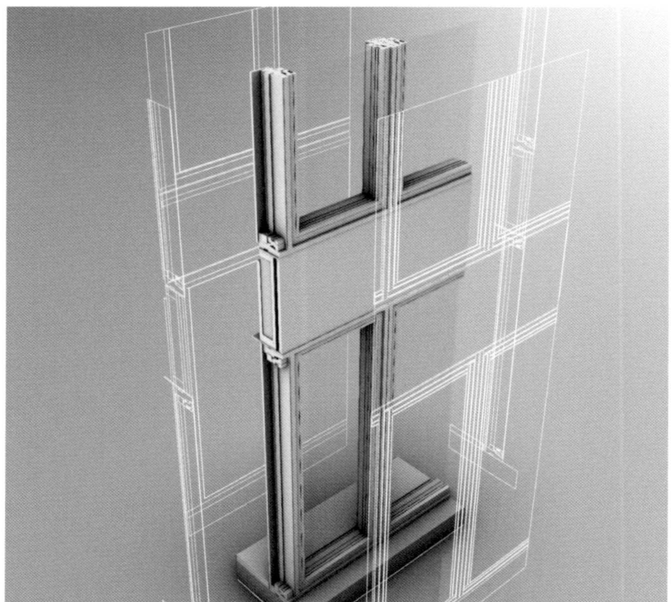

c| Prototype

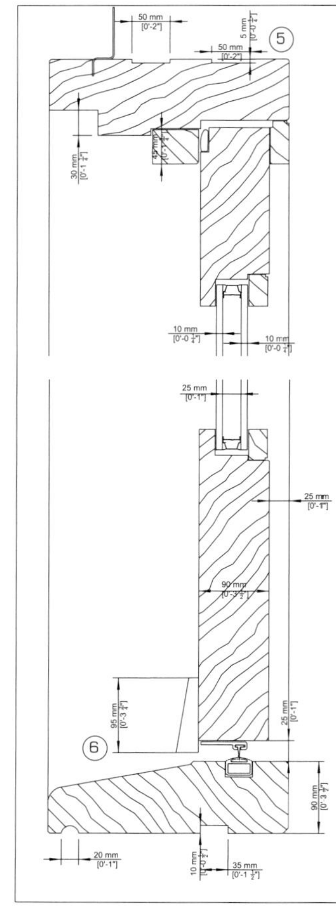

d| Vertical section. Inward opening casement doors

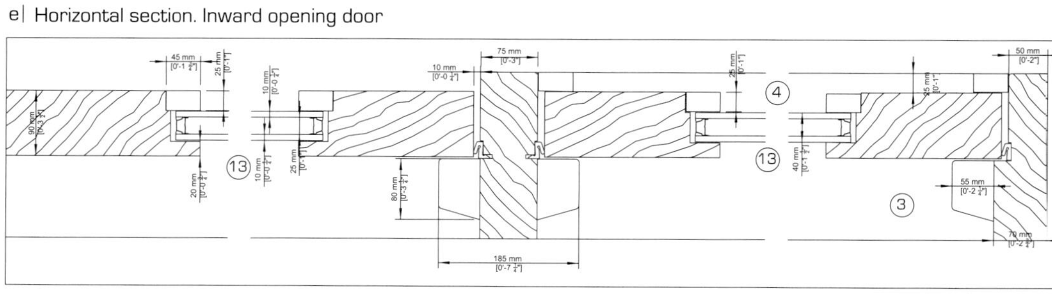

e| Horizontal section. Inward opening door

Like aluminium windows, timber windows are used either as individual components or as assemblies of 'window walls' of opening windows and fixed lights which are linked together.

Window walls

These are formed either as individual windows linked together and reinforced by a secondary frame, usually steel flats, or are formed as full curtain walling systems. Where timber sections form the full structural support for the glazing, the glazing is secured with timber glazing beads. Where individual windows are linked together with steel reinforcement, additional support may be needed from a secondary structure. Steel tubes, T-sections or box sections are typically used to avoid much larger equivalent sections in timber. In common with other framed glazing systems, timber windows are required to be held in a rigid frame, with only small structural deflections allowed under wind load. For this reason, cable-assisted structures are seldom used, since they are associated with higher deflections than framed supporting structures. These reinforcements or additional supports are usually set inside the building to avoid penetrating the exterior seal of the timber frames. Whereas aluminium glazing systems can accommodate support brackets penetrating the outer seals with rubber-based gaskets and silicone seals, the high moisture movement associated with wood makes this much less practical with timber windows.

Where windows are linked together and reinforced with a mild steel flat, a drainage groove is cut into the outside of the frame as part of a standard window frame. Any water that passes through the outer seal is drained away in the groove which is used on all four sides of a typical frame. Water is drained away to the outside at the sill (MCE_192, b). The reinforcement does not extend forward of the drainage groove in order to allow water to be drained away unimpeded. All-timber frames, in the manner of glazed curtain walling, may have mullions and transoms of different depths reflecting their different structural requirements, and each may be in a different

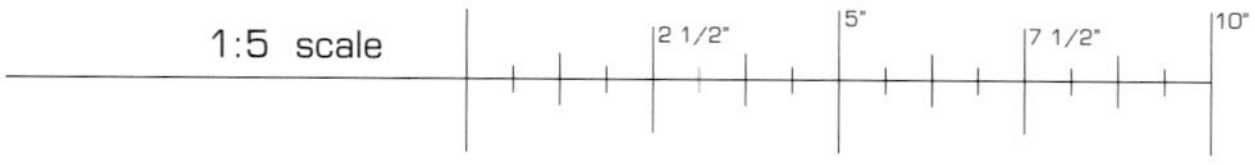

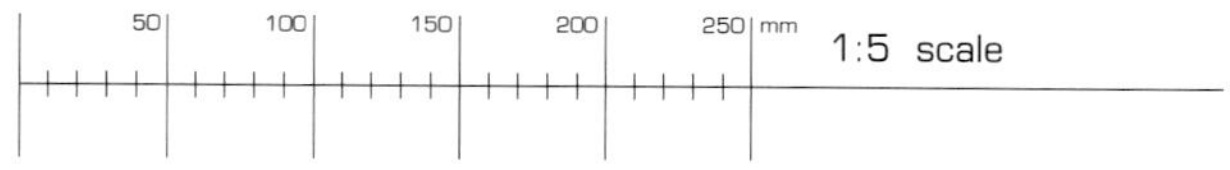

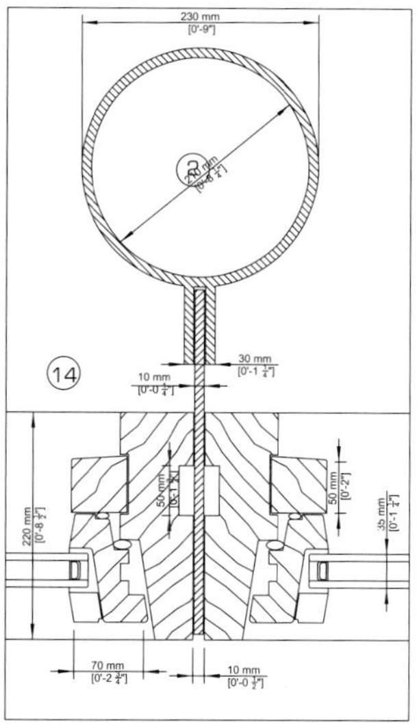

f | Horizontal section. Top hung window in window wall

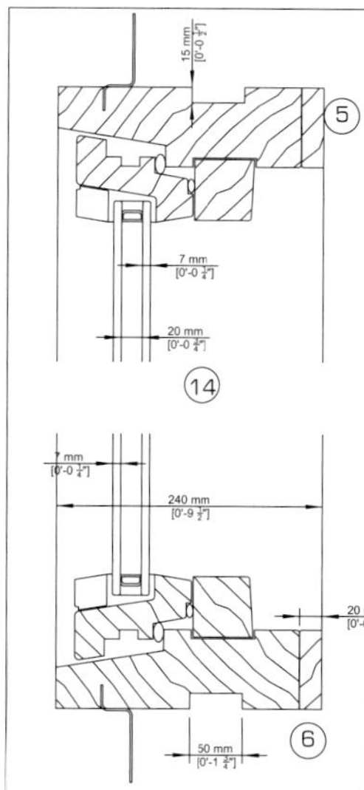

Details

1 | Fixed light
2 | Supporting structure
3 | Outside
4 | Inside
5 | Head
6 | Sill
7 | Metal stiffening insert
8 | Timber infill panel
9 | Thermal insulation
10 | Single-glazed or double-glazed unit to suit application
11 | Fixing bead
12 | Rubber-based seal
13 | Inward opening light
14 | Outward opening light
15 | Damp proof course (DPC)
16 | Internal finish
17 | Flashing to seal against adjacent wall
18 | Vertically sliding sash
19 | Sliding door frame top-hung or bottom-rolling
20 | Window or door jamb

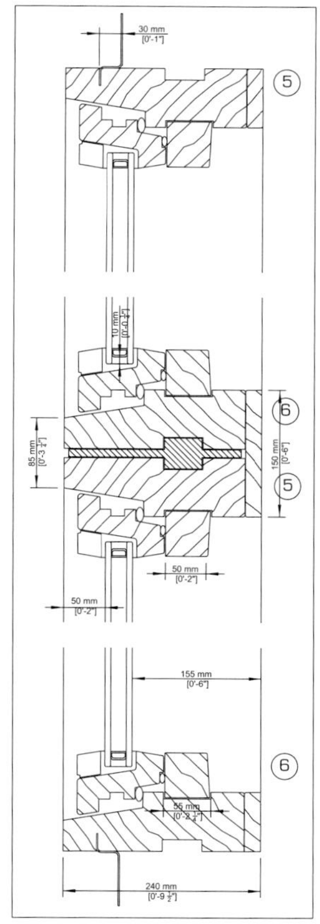

g | Vertical section. Top hung window in window wall

h | Horizontal section. Top hung window

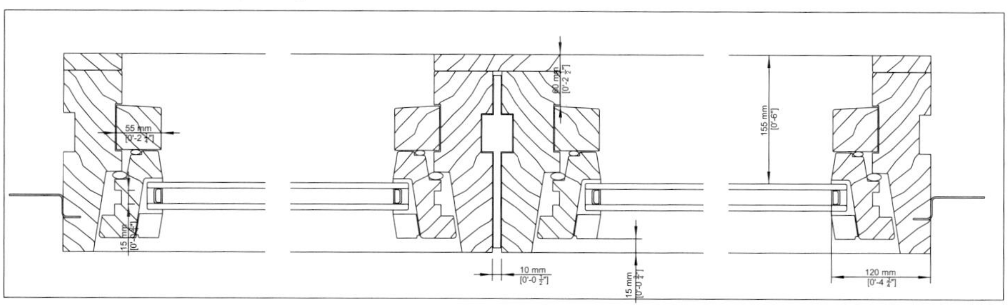

j | Horizontal section. Side hung window wall with different framing methods

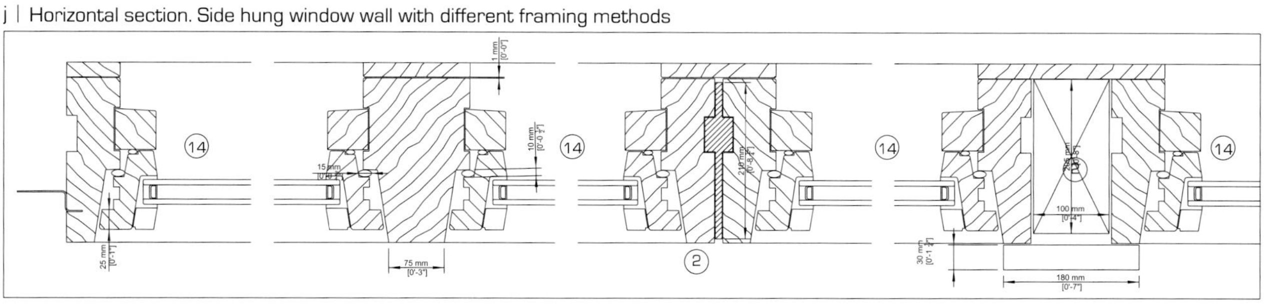

wood type. Hardwoods and softwoods can be mixed, but moisture movement needs to be taken into account. An advantage of solid timber sections over aluminium extrusions is the ability of timber to be routed to form a variety of junctions between mullion (vertical) and transom (horizontal). Slots and continuous grooves can be cut into the sections to give them a visual richness that is often lacking in metal glazed walls. Support brackets fixed to the inside face are formed in aluminium or mild steel.

Timber framed window walls typically span up to two floors, due to the self weight of the panels. This is due to the need to bind the frame into large single units to avoid the effects of thermal movement shrinking of the sections. If timber sections are not tied together, they tend to warp and twist when exposed to outside elements. Even if the surfaces are painted or sealed with varnish, any movement due to moisture will crack the outer finish and allow further movement to occur. Timber sections are jointed with either tongue-and-groove or rebated joints. Where tongue-and-groove joints are used, the linking material can be a durable hardwood or aluminium. Sections can also be bolted together if the sections are of sufficient size, such as around 75 x 50mm (3in x 2in).

Glazed walls can be joined relatively easily to adjacent wall construction with seals and lapped edges associated with traditional windows (MCE_192, a). Corners are straightforward to form as either a structural corner or as a fully glazed corner. Structural corners have two glazing sections brought together, which are reinforced with an additional corner post. This post is set into routed grooves in the edges of the two window frames to provide additional stiffness at the corner, binding the three timber sections into a single structural component.

Window design

Windows in recent years have developed to provide better airtight and better watertight barriers in their design. Air infiltration in opening lights has

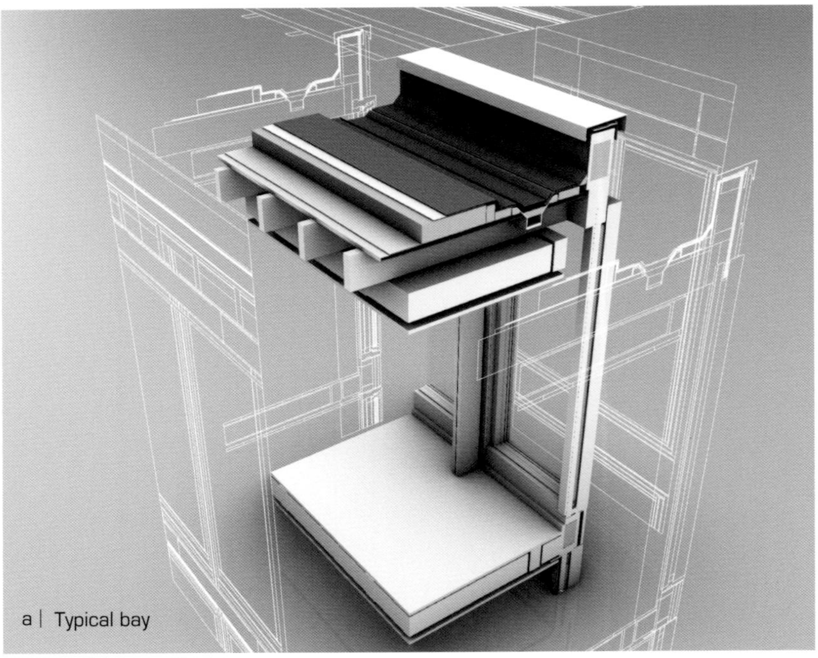

a | Typical bay

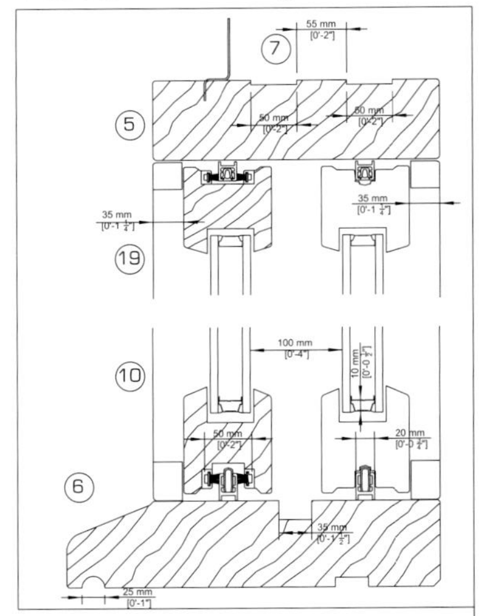

b | Vertical section. Sliding window

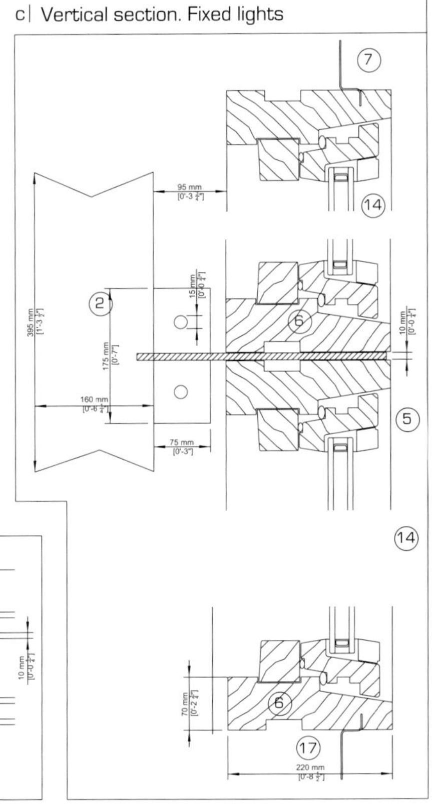

c | Vertical section. Fixed lights

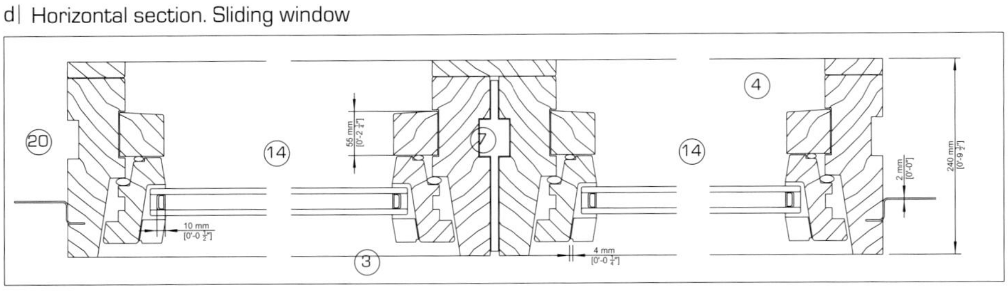

d | Horizontal section. Sliding window

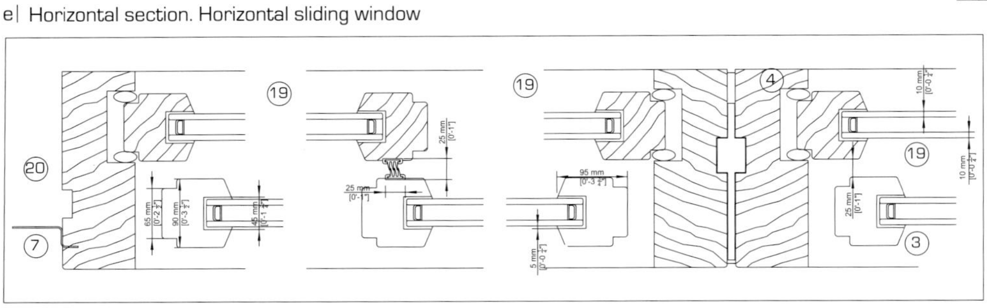

e | Horizontal section. Horizontal sliding window

been reduced when the window is closed by the addition of rubber-based seals, often held in place by aluminium clips so that they can be replaced when worn out (d). Watertightness of opening lights has been improved by the use of pressure-equalised rebates in the window section so that any water entering the profiles at the sides is drained away from an outer cavity without being drawn into the inner seal by capillary action (pressure differential) between the inside and outside of the framing.

An outer seal prevents excessive water infiltration while the inner air seal also serves as an acoustic barrier, providing better sound insulation than traditional windows without those seals. The inner rubber-based air seal is usually 'vulcanised' or fused at the corners of the frame to maintain performance around the complete perimeter of the window. The drainage groove or channel around the frame of opening lights sometimes has an aluminium profile attached to allow the timber profile to be simpler. Holes are drilled into the aluminium profile at sill level to allow water to drain to the outside of the frame. This metal profile is painted to reduce the visibility when the window is open.

Seals between the window and the opening into which the window is fixed have also been improved by the use of folded aluminium and UPVC profiles. These profiles are set fixed on all edges of the window by setting them into a groove in the window frame and projecting them over the face of the surrounding wall. This makes the sealing of the gap between window and surrounding wall much easier by avoiding the need to seal the butt joint created by setting the window directly into the opening. This projecting profile is more reliable than the traditional weather bar used in the sill with sealed butt joints on the jambs and head.

There has been an increase in the control of moisture movement in the timber used. When timber is kiln-dried to reduce its water content prior to manufacture, the process is now better controlled to avoid excessive

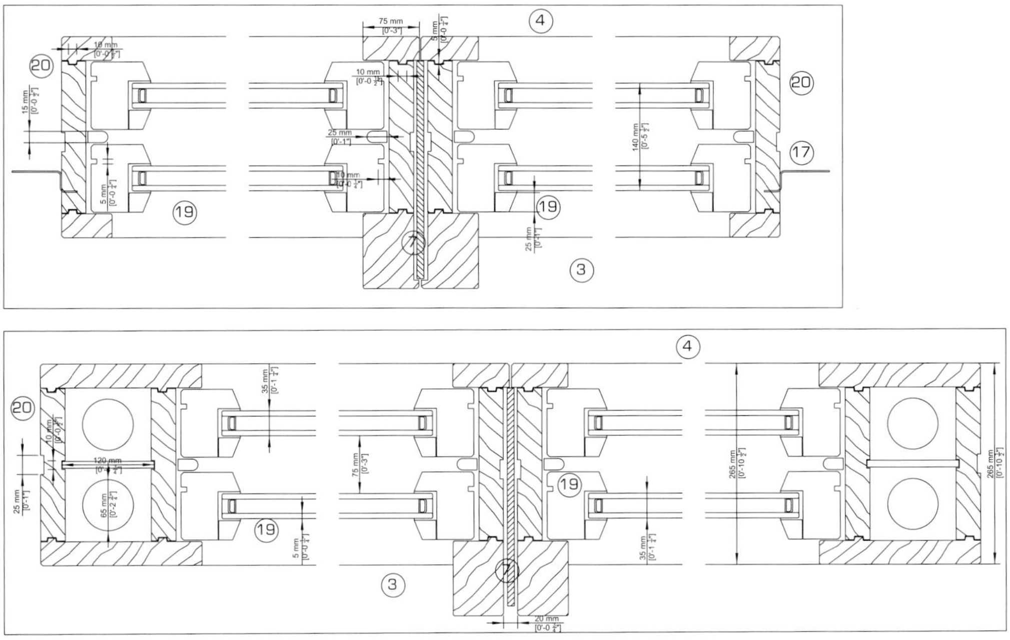

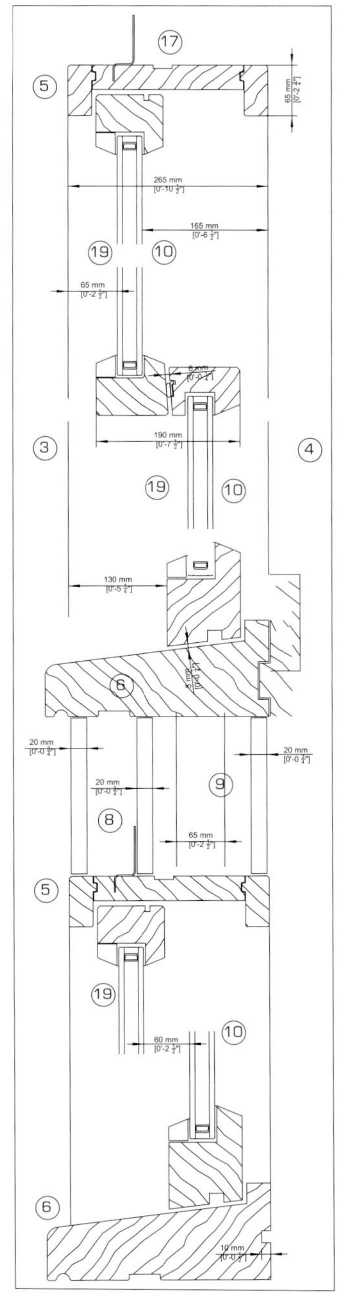

f | Horizontal and vertical sections. Sliding sash window

Details

1 | Fixed light
2 | Supporting structure
3 | Outside
4 | Inside
5 | Head
6 | Sill
7 | Metal stiffening insert
8 | Timber infill panel
9 | Thermal insulation
10 | Single-glazed or double-glazed unit to suit application
11 | Fixing bead
12 | Rubber-based seal
13 | Inward opening light
14 | Outward opening light
15 | Damp proof course (DPC)
16 | Internal finish
17 | Flashing to seal against adjacent wall
18 | Vertically sliding sash
19 | Sliding door frame top-hung or bottom-rolling
20 | Window or door jamb

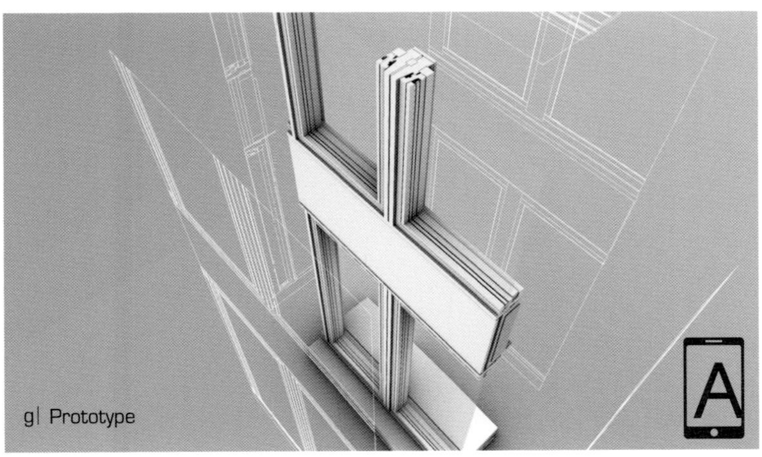

g| Prototype

shrinkage afterwards. Timbers are dried to around 15% moisture content, depending on the species used. Treatments to timbers are also changing to avoid toxic runoff while providing some protection against colour fading from UV radiation from sunlight. Timber still requires re-coating and cleaning to avoid the surface visually fading and causing damage to the timber itself. Some more economic timber windows have traditionally suffered from poor jointing at the corners of the frame. These have been improved across all window types in recent years with the use of double mortice and tenon joints and wood glues with better resistance to heat and moisture.

With the improvement in protective coatings for timber windows, particularly those on the outside face, has come the use of rounded edges to profiles. As is the case with coatings to aluminium profiles, it is very difficult to coat the edge of a sharp corner. A radiused edge allows the protective coating to maintain a better constant thickness at the vulnerable corners. Typically, rounded edges allow an 80% layer thickness of coating (with 100% on adjacent flat surfaces) compared with around 5% coating thickness for a sharp angle. With the increased use of double glazed units in timber windows, rubber-based gaskets are also increasingly being used to allow the moisture movement of the window frame and timber glazing beads to move without reducing the water tightness of the seal between the frame and double glazed unit (MCE_195, f). Soft seals, including those made from wet-applied silicone or silicone extrusions, perform much better in this respect than the harder and more brittle putties and sealants used previously.

Aluminium trims are also used to secure the glass, but the appearance from outside is obviously a mixture of timber and aluminium rather than all-timber, making the window more of a composite design associated with timber window walls. This trend will probably continue due to the greater reliability of aluminium for glazing beads and trims.

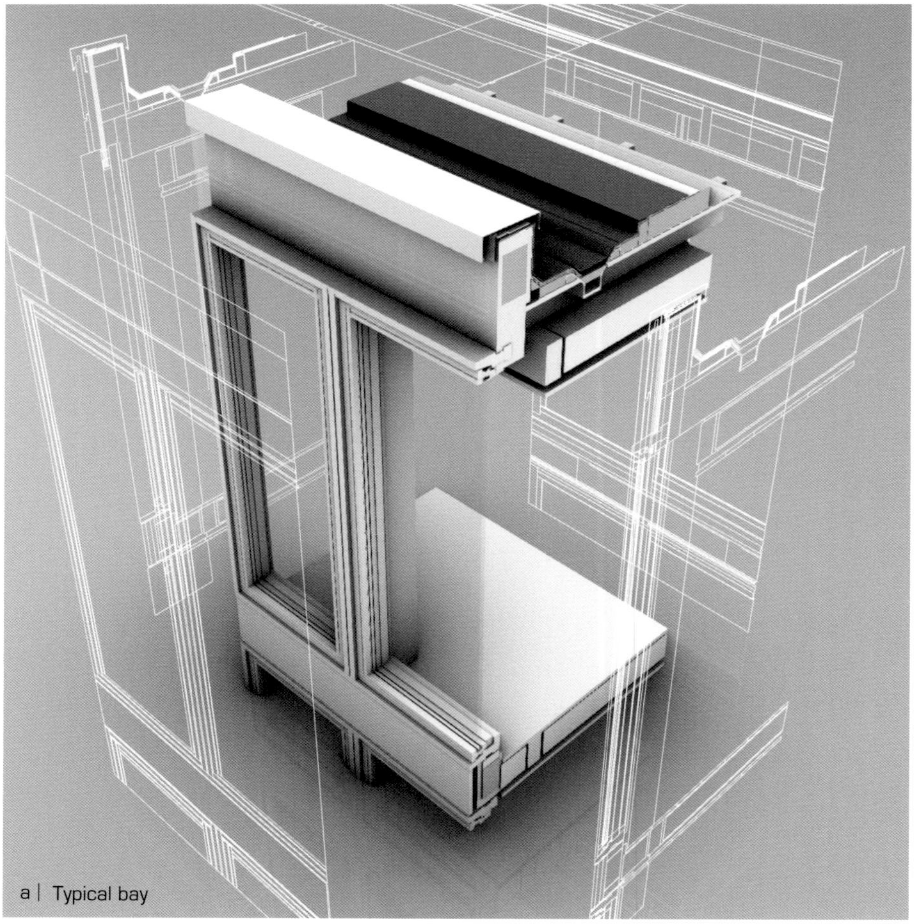

a | Typical bay

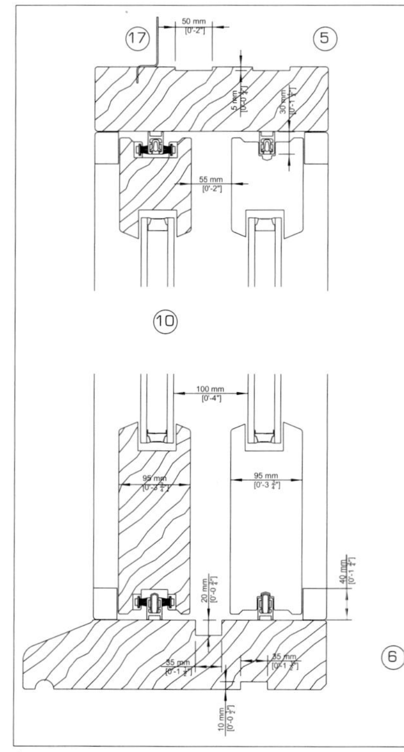

b | Vertical section. Sliding doors

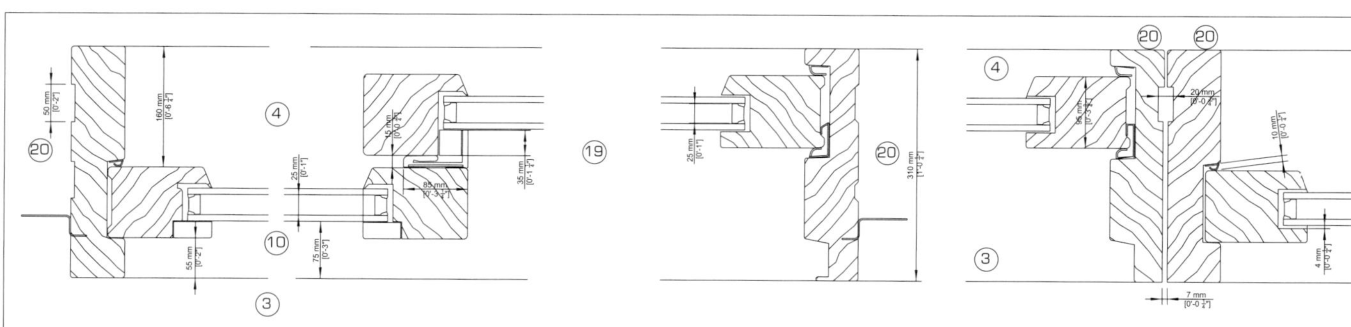

c | Horizontal section. Sliding doors

Windows in openings

Where timber windows are set into an opening rather than being part of a window wall, the most common materials used are masonry block, brick or timber boarding. With all these materials, timber windows are fixed either into the reveals of the opening or are fixed onto the face of the opening (a). The position of the window in the opening has more influence over the junction with the surrounding wall than the choice of material for the wall.

In generic examples, the window is fixed to the outside face of the wall, which would typically be reinforced concrete or concrete block. This would be used where a massive, sound-insulating structure, such as an apartment building, is clad in a different material such as timber rainscreen panels or terracotta, which are often used in a smooth continuous plane without reveals. This construction is ideal for rainscreen construction. It allows the window to be sealed to the structural wall or backing wall with a lapped joint around the opening and fitted independently of that opening. This permits the wall to be set out on a precise grid that would be difficult to achieve by setting the window into the opening. This is because the construction tolerances associated with concrete frame construction are considerably higher than those used for timber windows. Windows would typically be sealed against the concrete wall with projecting profiles that would lap under a waterproof layer on the surrounding wall. This creates a deep reveal on the inside face of the wall which has finishes applied to create a tidy junction with the internal face of the window.

In some cases, the window is set into the reveal of the opening. Because the window has to be smaller than the opening size in order to install it, the joint width has to be accurately formed around the window. Sometimes the window frame is built into the wall as the surrounding wall is constructed, or else a template is used to avoid accidental damage to the window during construction.

1:5 scale 2 1/2" 5" 7 1/2" 10"

50 100 150 200 250 mm 1:5 scale

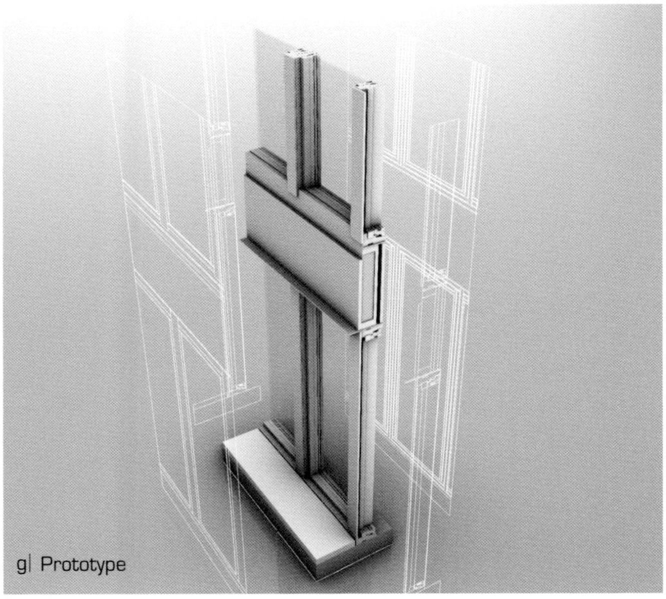
g| Prototype

TYPE 1 TYPE 2 TYPE 3

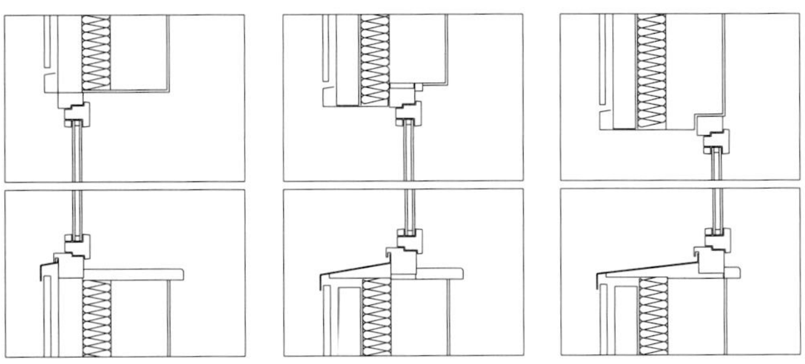

Details

1 | Fixed light
2 | Supporting structure
3 | Outside
4 | Inside
5 | Head
6 | Sill
7 | Metal stiffening insert
8 | Timber infill panel
9 | Thermal insulation
10 | Single-glazed or double-glazed unit to suit application
11 | Fixing bead
12 | Rubber-based seal
13 | Inward opening light
14 | Outward opening light
15 | Damp proof course (DPC)
16 | Internal finish
17 | Flashing to seal against adjacent wall
18 | Vertically sliding sash
19 | Sliding door frame top-hung or bottom-rolling
20 | Window or door jamb

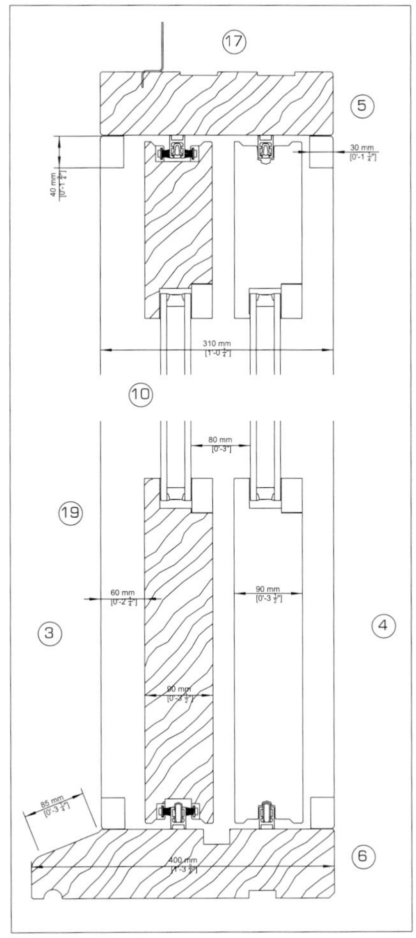

e| Vertical section. Sliding doors

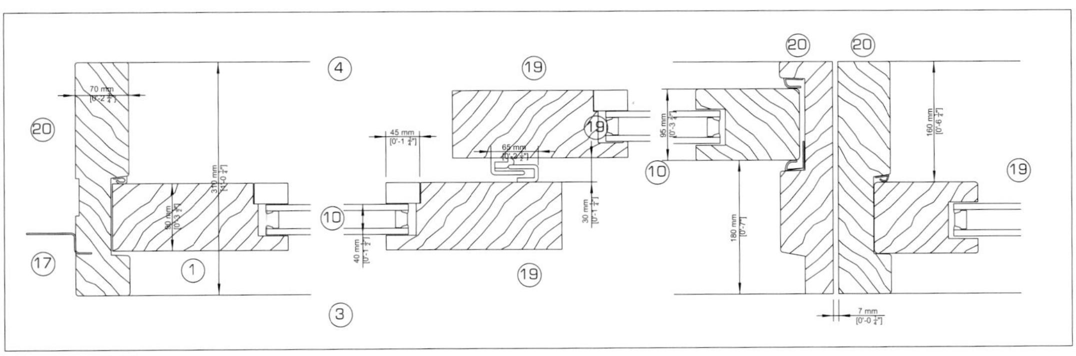

f | Horizontal section. Sliding doors

Alternatively, the window could be set into the outer wall construction and in this case needs to allow water to drain the top of the opening to the sides, as in masonry cavity wall construction. In the previous case, the window does not have this additional requirement, and so is protected from water penetration by the wall itself. However, rainwater needs to be drained at the bottom of the opening, where a sill is provided. The window is lapped over the sill, often with an additional weather bar for protection against water ingress through capillary action (MCE_192, b).

The window can also be set into a stepped reveal. This has the advantage of giving greater construction tolerance when fitting the window as well as creating a lapped joint between the structural opening and the window, which is easier to seal than an equivalent butt joint, where a backing strip is needed to give a surface to which the sealant can adhere.

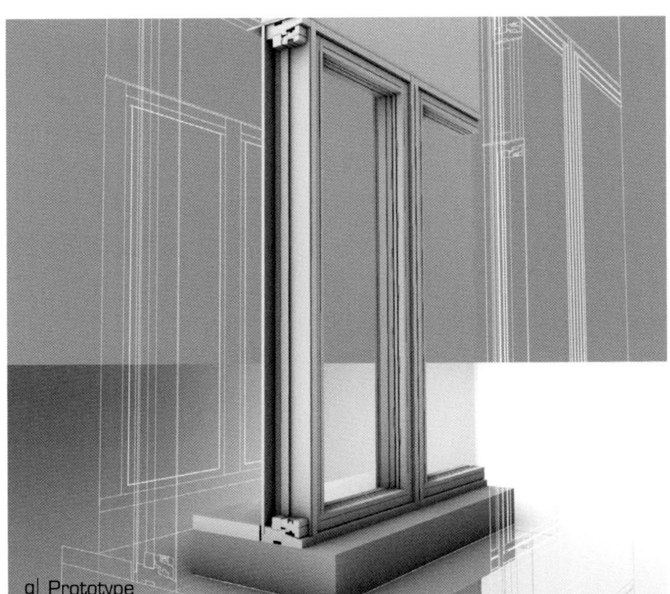
g| Prototype

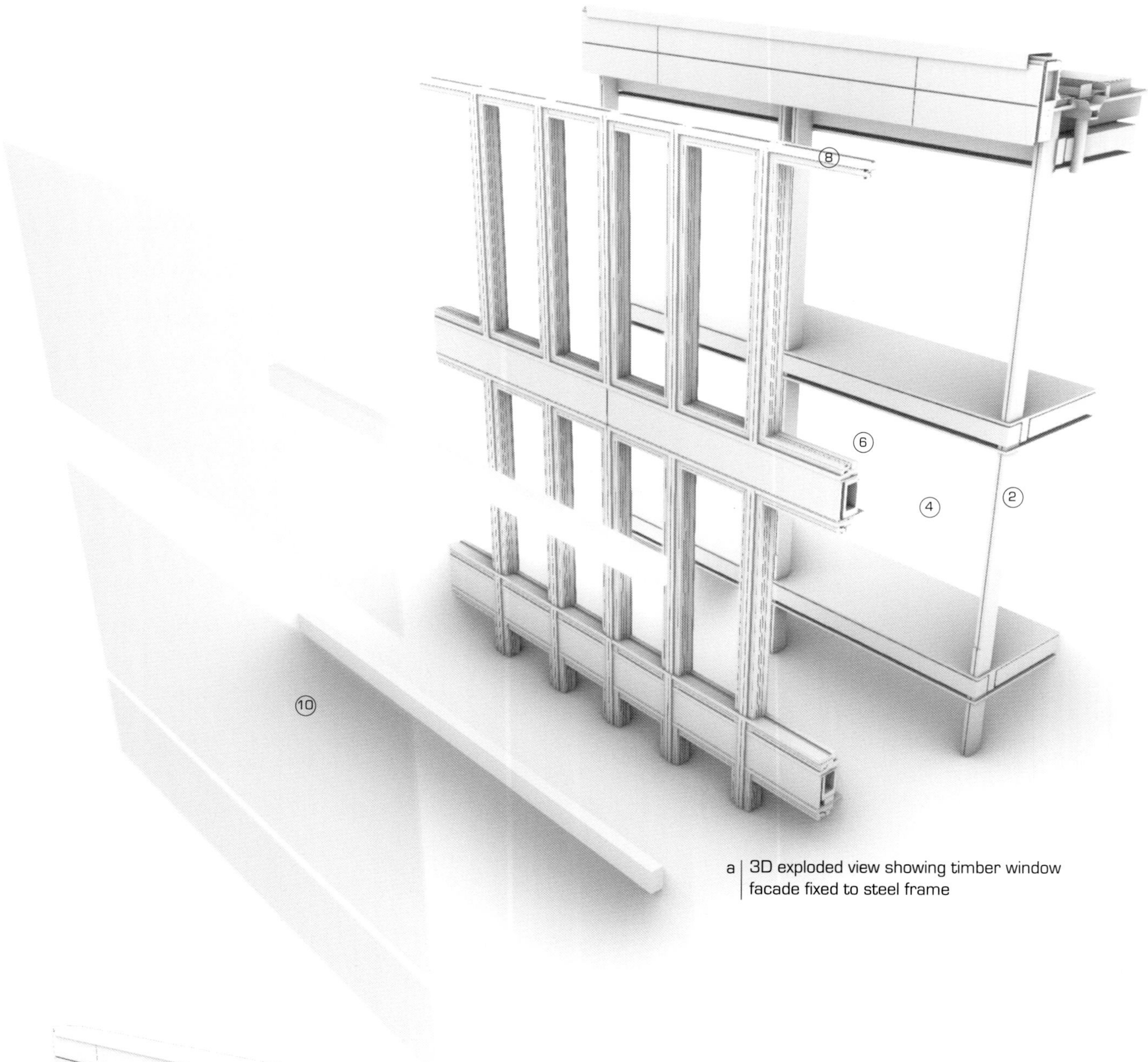

a | 3D exploded view showing timber window facade fixed to steel frame

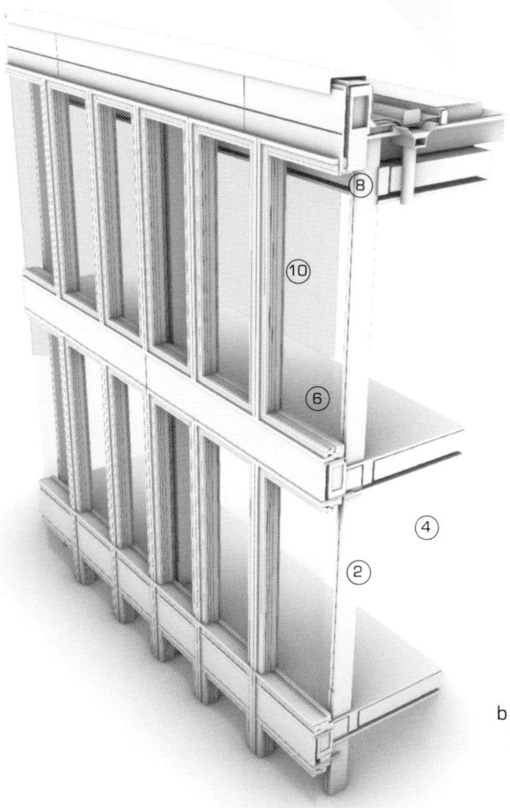

b | 3D view showing timber window facade fixed to steel frame

Details

1 | Fixed light
2 | Supporting structure
3 | Outside
4 | Inside
5 | Head
6 | Sill
7 | Metal stiffening insert
8 | Timber infill panel
9 | Thermal insulation
10 | Single-glazed or double-glazed unit to suit application

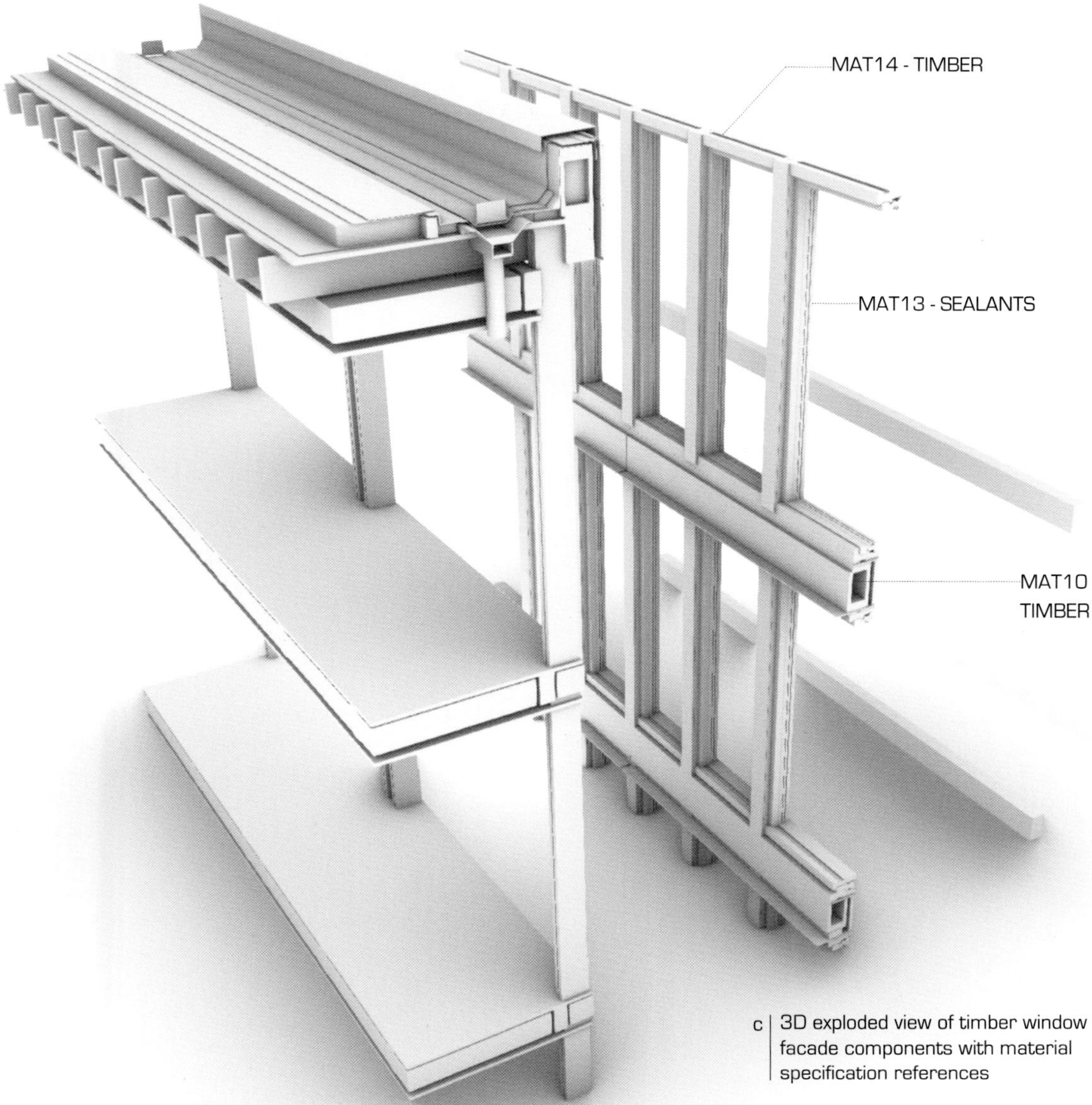

c | 3D exploded view of timber window facade components with material specification references

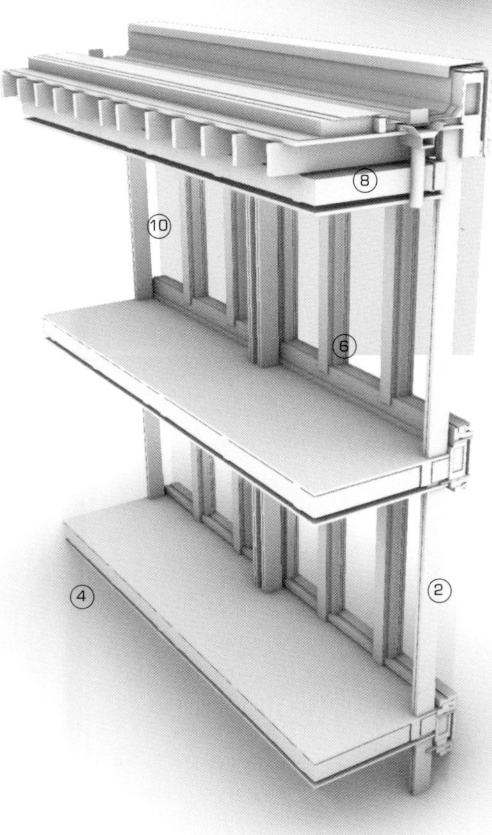

11 | Fixing bead
12 | Rubber-based seal
13 | Inward opening light
14 | Outward opening light
15 | Damp proof course (DPC)
16 | Internal finish
17 | Flashing to seal against adjacent wall
18 | Vertically sliding sash
19 | Sliding door frame top-hung or bottom-rolling
20 | Window or door jamb

d | 3D view showing timber window facade fixed to steel frame

Example project of concrete facades
KAFD Metro, Riyadh, Kingdom of Saudi Arabia
Facade Engineer: Newtecnic
Architect: Zaha Hadid Architects

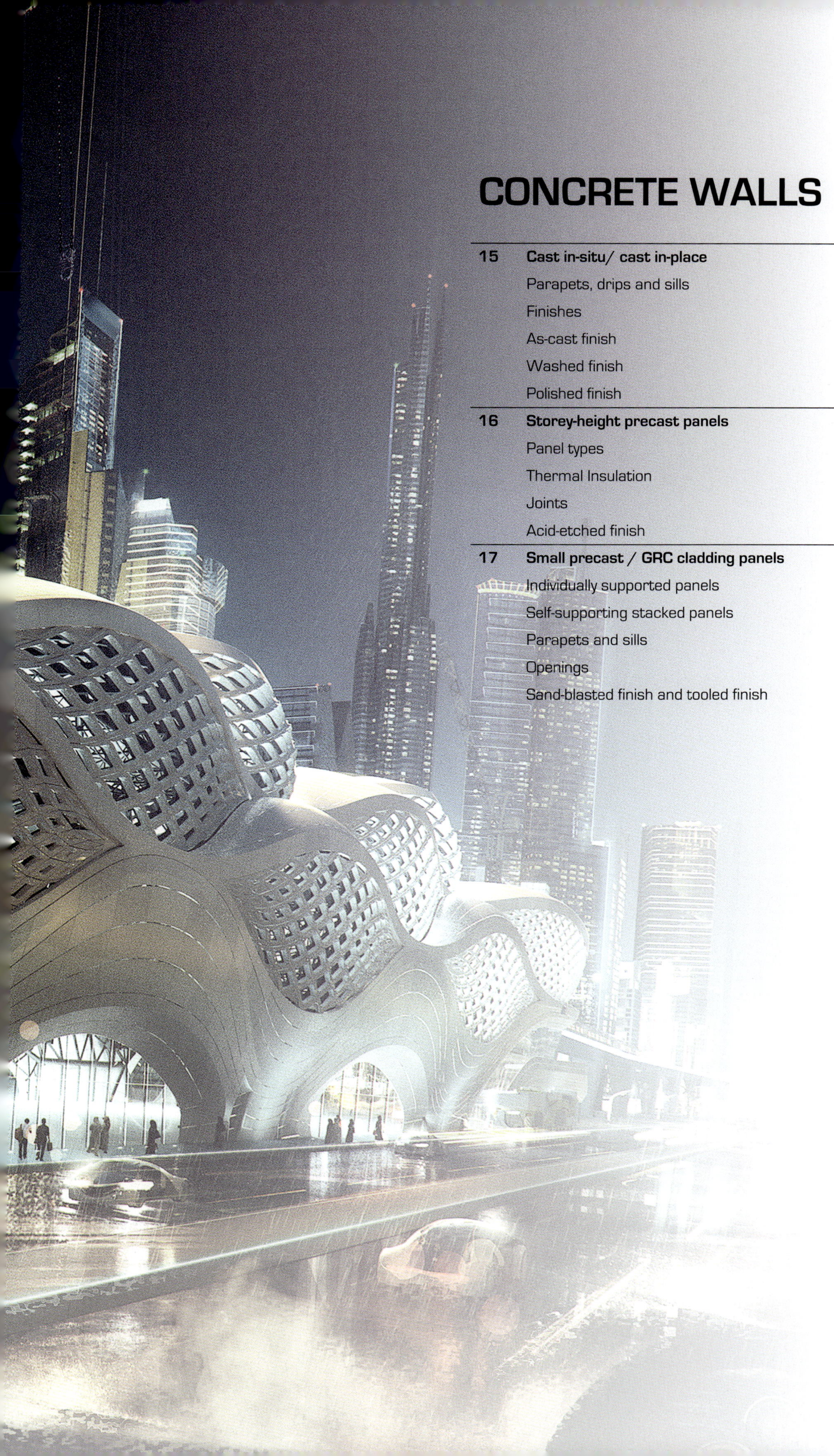

CONCRETE WALLS

15 Cast in-situ/ cast in-place

Parapets, drips and sills

Finishes

As-cast finish

Washed finish

Polished finish

16 Storey-height precast panels

Panel types

Thermal Insulation

Joints

Acid-etched finish

17 Small precast / GRC cladding panels

Individually supported panels

Self-supporting stacked panels

Parapets and sills

Openings

Sand-blasted finish and tooled finish

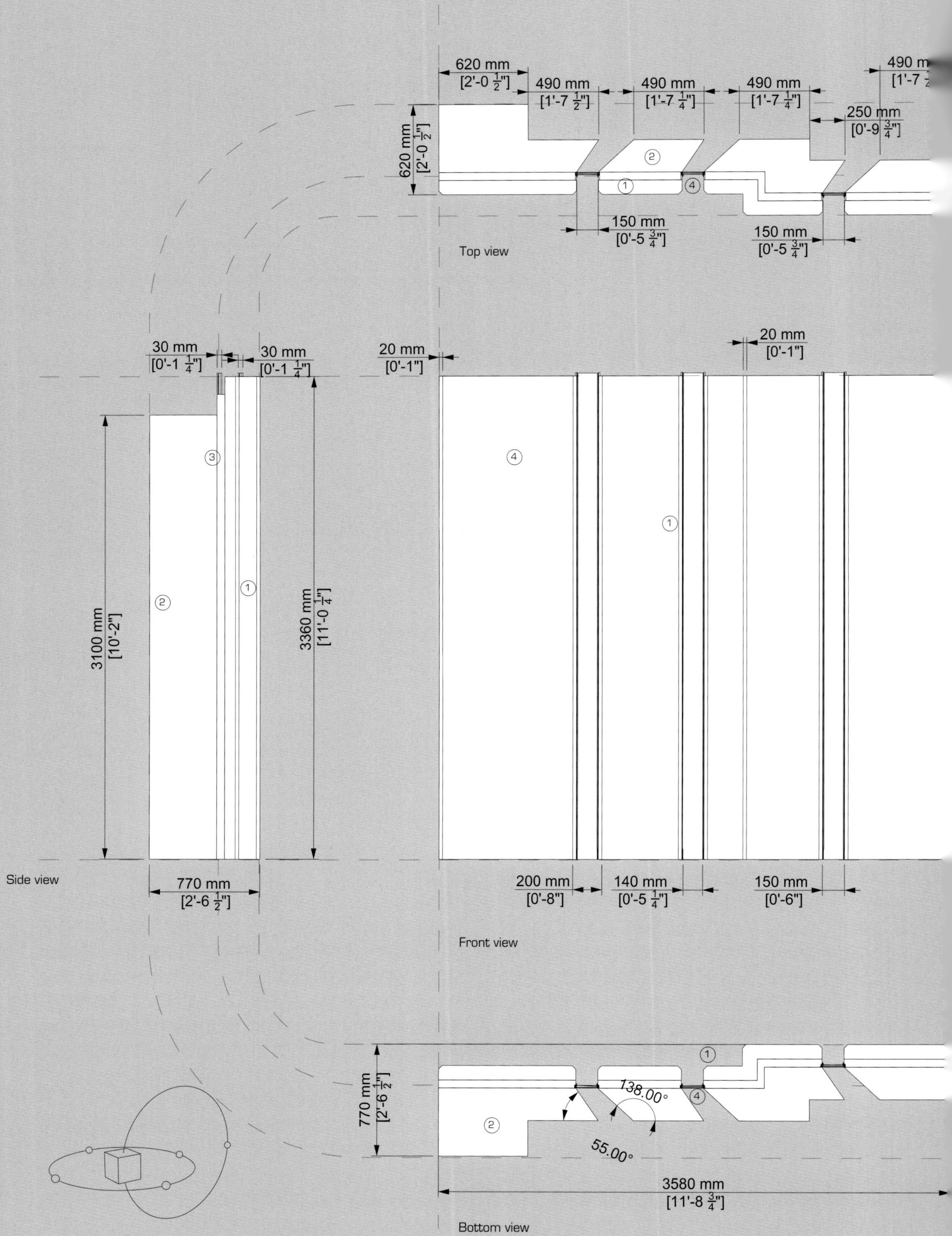

Third angle projection 1:30

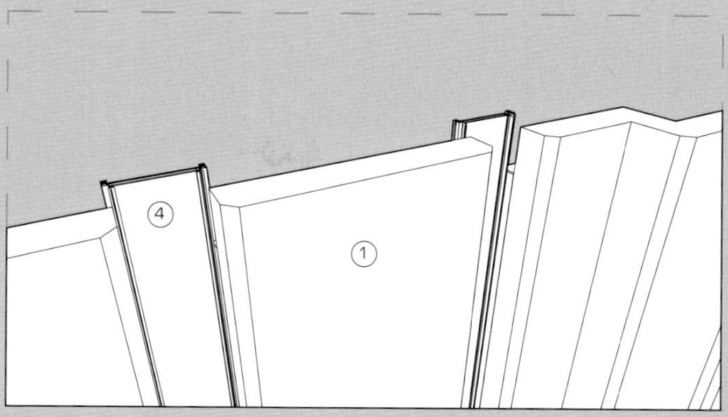

3D view of the assembly

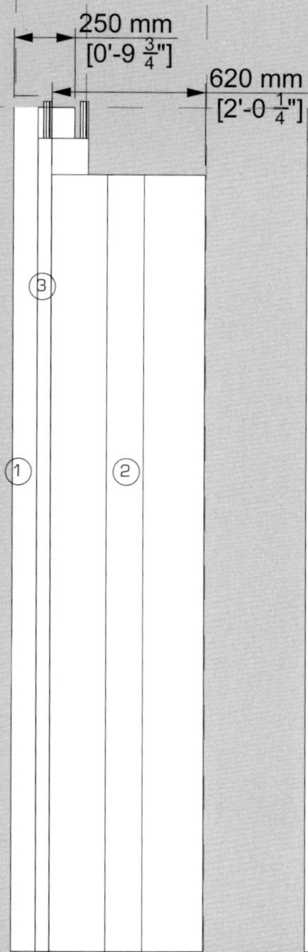

Side view

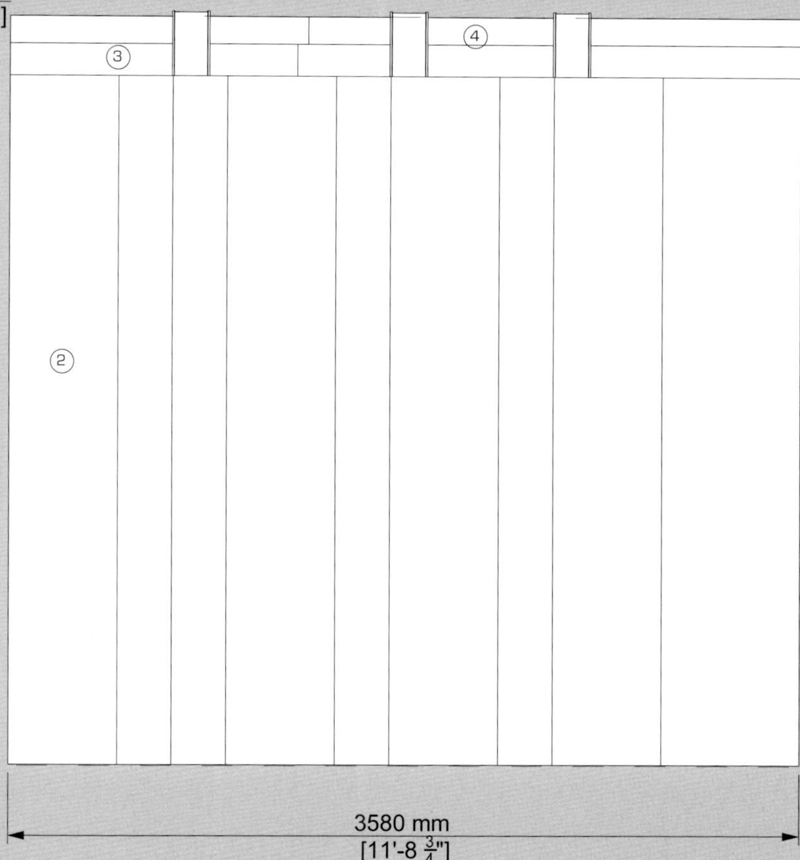

Back view

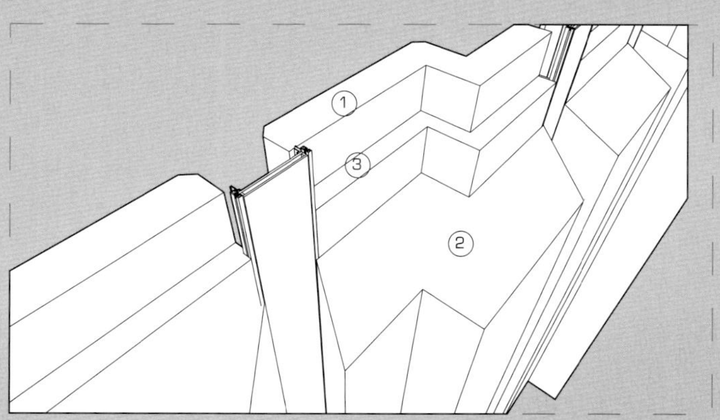

3D view of the assembly

Details

1 Concrete external wall
2 Concrete internal wall
3 Thermal insulation
4 Window frame
5 Waterproof membrane
6 Metal parapet flashing
7 Internal finish
8 Metal sill
9 Drip
10 Metal lined gutter

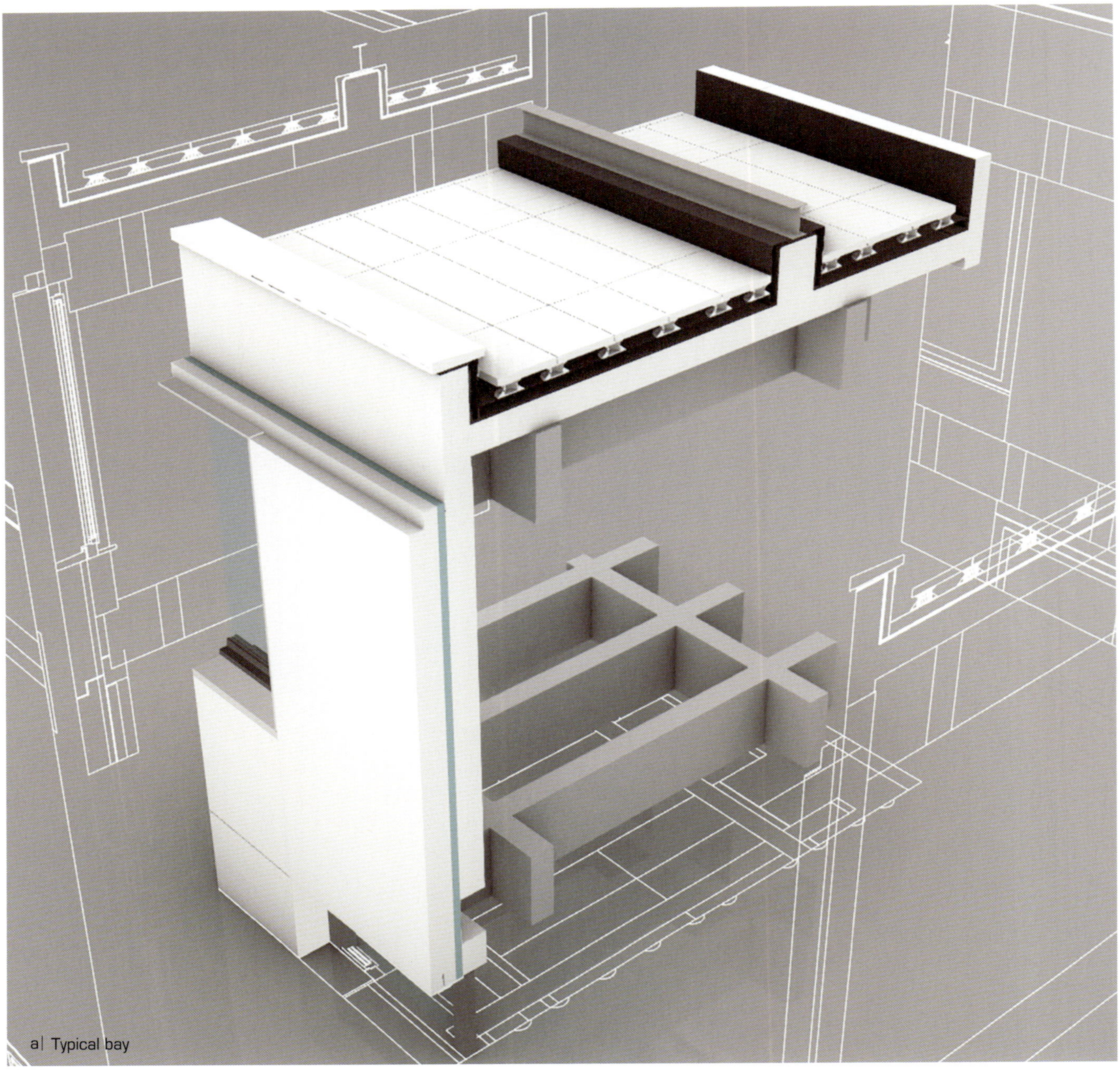

a| Typical bay

An essential difference between concrete and other materials used in facade construction is that concrete is poured in place into moulds, or into formwork, rather than being manufactured as a standard size component in a factory. Whereas metal, glass, masonry, plastics and timber are made to standard dimensions in the form of sheets or sections, concrete is cast, either on site or in a factory as precast panels. Although there are few constraints on the size of a single cast element in concrete, in practice an essential determinant of concrete panel size is the amount of concrete that can be poured at one time. With precast concrete the essential constraint on panel size is the weight that can be lifted by a crane on site.

In-situ concrete is dependent on the formwork in which the concrete is cast. An understanding of formwork is important to appreciate how to control the appearance of joints and of bolt holes through the concrete, since the formwork is the negative impression or mirror image of the final concrete.

The design of in-situ concrete walls has changed in recent years to include thermal insulation, either set within the concrete during pouring or fixed on the internal face or external face after the wall has been cast. The position of the insulation within the construction affects the use of the thermal mass of concrete and its ability to contribute to night-time cooling as part of an overall reinforced concrete structure. Continuity of thermal insulation is increasingly important in the interface with glazed openings and doors to avoid thermal bridging. Continuity of double glazing and thermal insulation in the concrete wall is probably the biggest recent change in this form of construction in recent years. In-situ concrete walls are formed as either a structural concrete wall with thermal insulation on the inside face with an internal dry lining, or as a diaphragm/double wall of concrete with thermal insulation set between the two skins of concrete.

Details

1 | Concrete external wall
2 | Concrete internal wall
3 | Thermal insulation
4 | Window frame
5 | Waterproof membrane
6 | Metal parapet flashing
7 | Internal finish
8 | Metal sill
9 | Drip
10 | Metal lined gutter

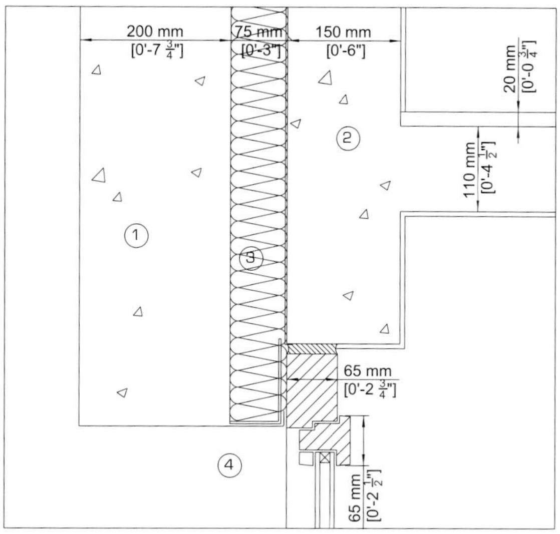

c| Vertical section. Head and sill of window opening

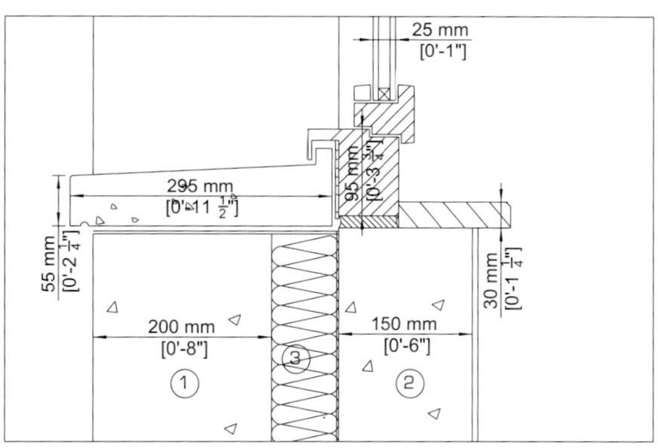

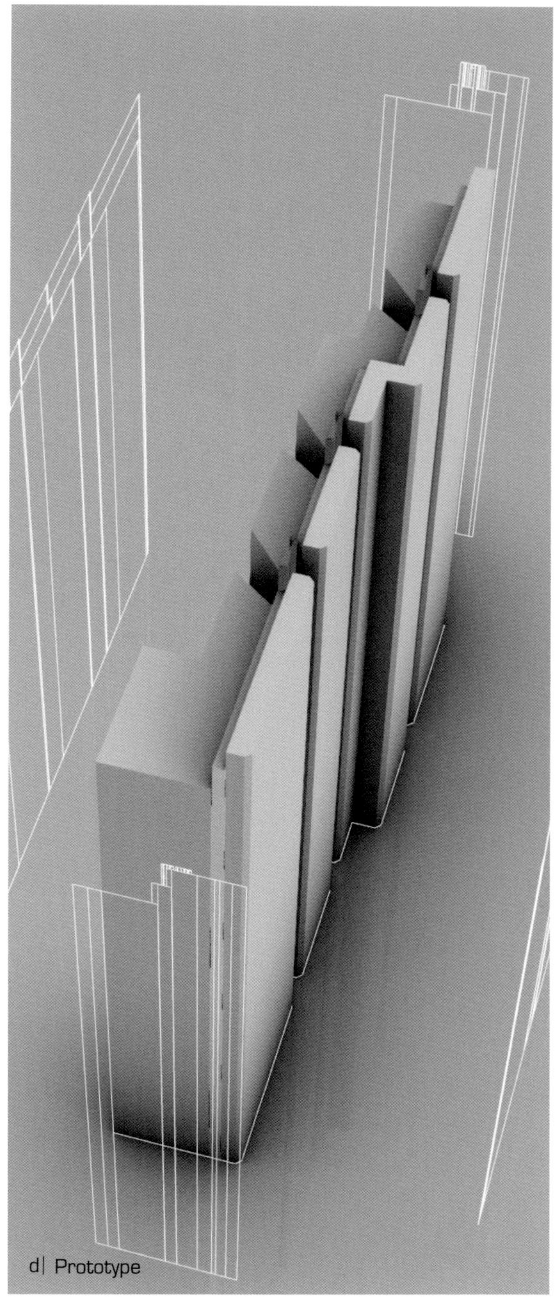

d| Prototype

b| Vertical section. In situ cast concrete wall cast in 2 skins with rigid closed cell insulation between skins

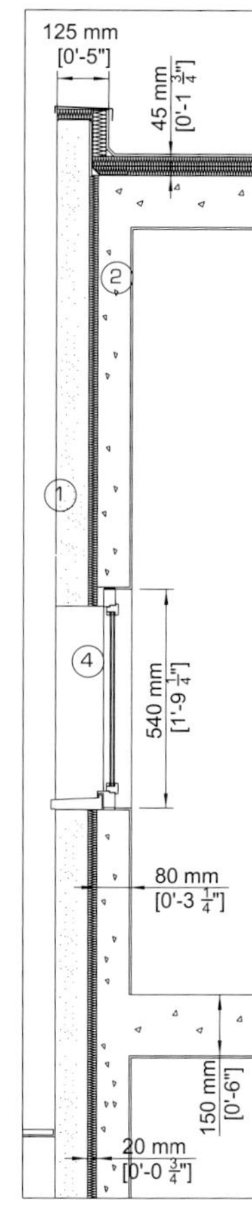

The first method has the benefit of economy, particularly where the thermal mass of the concrete wall is not required for use in night-time cooling. The wall zone for the insulation can also be used to fit windows, so that the opening lights and doors do not have to be set into specially cast recesses in the concrete, which makes it more expensive. Drips are cast into the tops of window reveals to reduce staining to the adjacent wall beneath caused by rainwater runoff from dust that accumulates on flat surfaces (MCE_204, b). Parapets have metal copings that prevent staining on the wall beneath by directing rainwater towards a gutter immediately behind the parapet.

With the second method, in-situ concrete walls are formed as 'sandwich' walls with internal insulation (MCE_204, a). These have either insulation broken by strips of concrete that link across the two skins (forming limited thermal bridges) or are formed as two separate concrete skins linked only by stainless steel brackets and ties. The second type is increasingly popular as it avoids thermal bridging with its associated risk of condensation forming on the inside of the wall in temperate climates.

Parapets, drips and sills

The detailing of openings, parapets and sills follows the same principles for a single wall of structural concrete with internal thermal insulation. When concrete is used as an exposed external finish, sills, parapets and drips are detailed to ensure that rainwater is thrown as clear as possible from the external wall surface. When an additional outer material is used, typically a rainscreen in a wide range of materials, the concrete wall can be formed economically with no regard for visual appearance, since the material is not visible.

Since horizontal or slightly sloping surfaces catch dust, roof overhangs and deep eaves are often incorporated into the design to avoid wash-off

a| Prototype

b | Vertical section. Junction at ground level for typical wall condition and for window/door condition. In situ cast concrete wall cast in 2 skins with rigid closed cell insulation between skins.

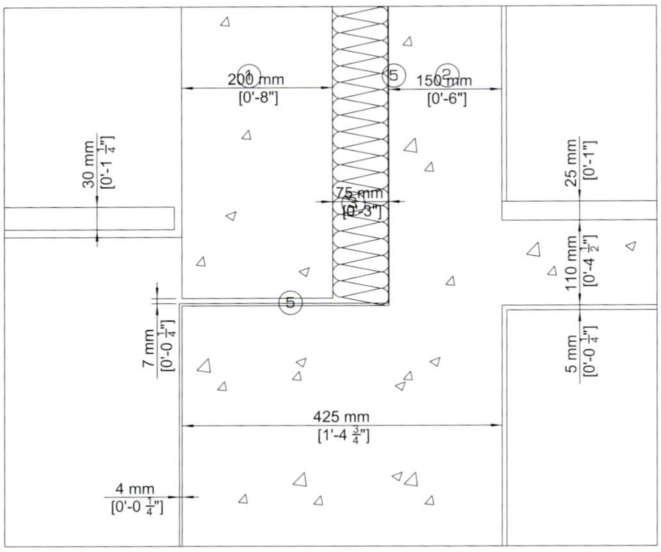

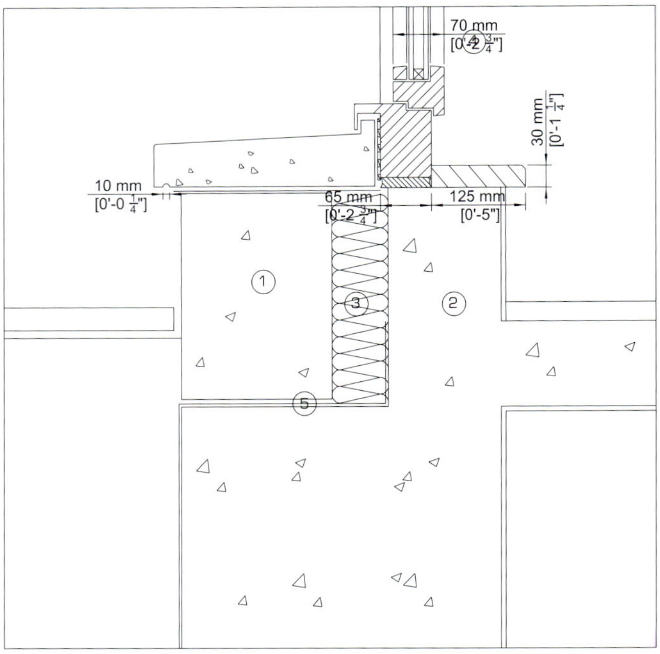

from rain. Dust washed away by rainwater causes the deposit of dirt and dust in adjacent areas of wall. If overhangs are not a part of the design, then water from horizontal or slightly sloping surfaces is thrown clear of the wall by projecting sills and flashings (MCE_205, d). In highly polluted environments, dust-catching wall textures are usually avoided, and smooth finishes are often preferred. However, where protective treatments are applied to the concrete to reduce porosity, this can lead to greater run-off across the facade which can increase staining, and must be taken into account in the design of the external wall.

Finishes

The main influence in the base colour of concrete used for wall construction is in the choice of cement, with fair-faced visible concrete walls using either a grey cement base or a white cement base. The physical properties of these two cement types are very similar.

The colour of finished concrete from a grey cement base is also grey, as expected, but this can vary with the water/cement ratio, the porosity of the shuttering, vibration conditions, formwork stripping time and weather conditions. The grey type can also blacken with rain due to the presence of iron oxide. However, grey cement-based architectural quality concrete can achieve an even colour when pouring methods and conditions are kept consistent. When grey cement is used with a moderate to high level of pigment content, as-cast or treated concretes are much less prone to colour variations due to the strong covering capacity of the pigments. White cement is much more expensive than grey cement but it is not subject to the colour variations of grey cement, nor is it subject to rain darkening.

The most common finishes used for in-situ cast concretes are an as-cast finish, a washed finish and a polished finish. The less common types are described in the following sections on precast panels, but

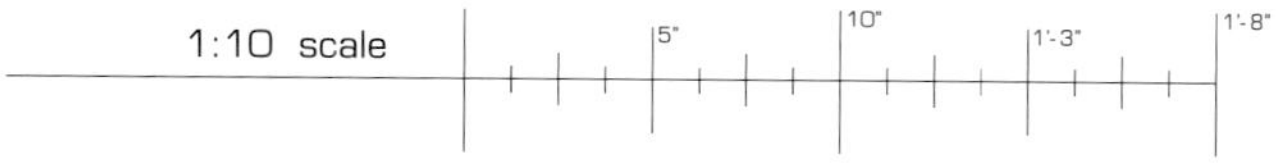

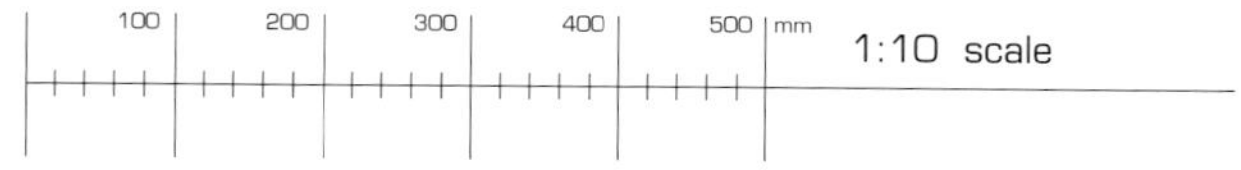

Details

1 | Concrete external wall
2 | Concrete internal wall
3 | Thermal insulation
4 | Window frame
5 | Waterproof membrane
6 | Metal parapet flashing
7 | Internal finish
8 | Metal sill
9 | Drip
10 | Metal lined gutter

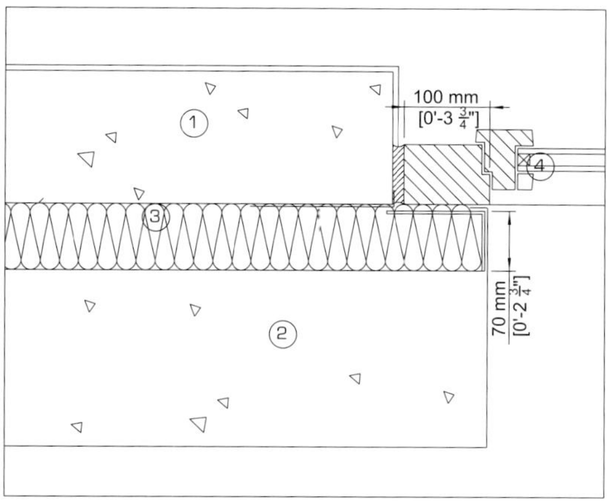

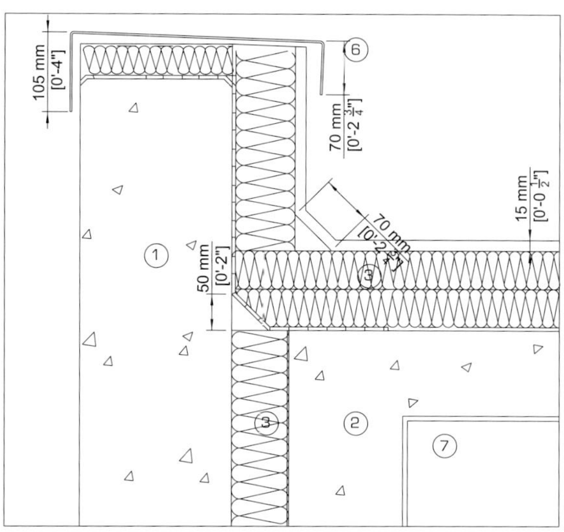

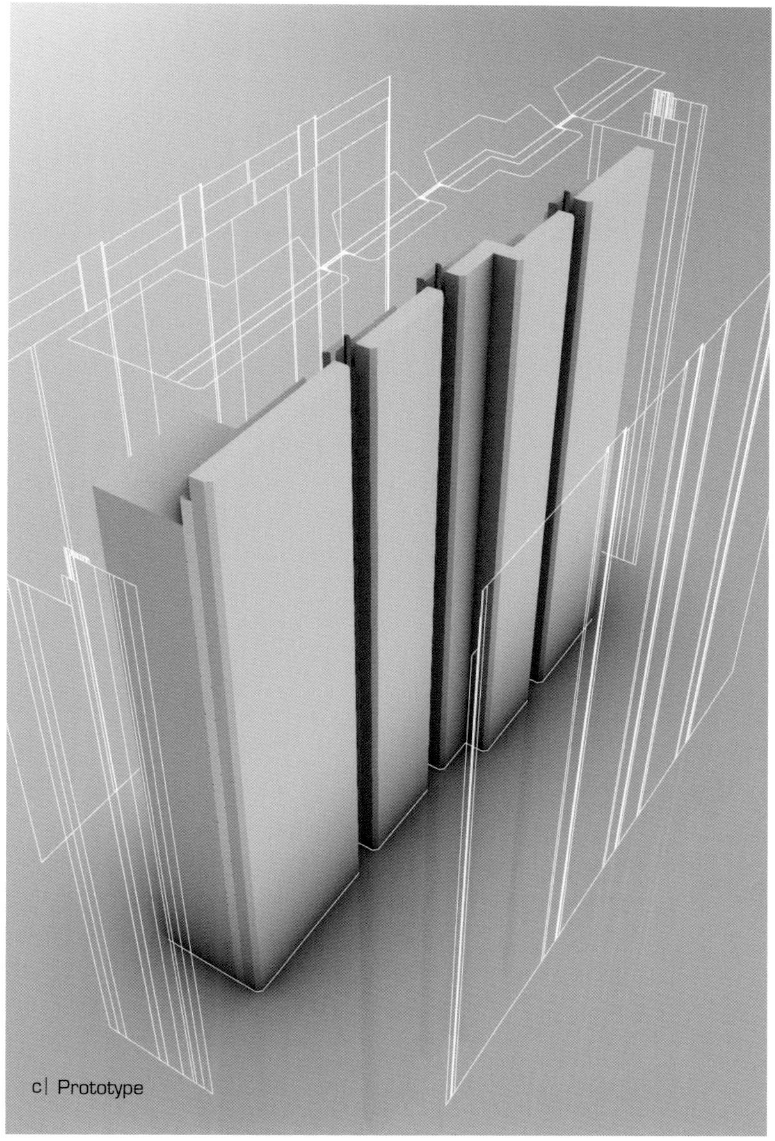

c| Prototype

d | Horizontal and vertical section. Junction at parapet level for typical wall condition and for window jamb. In situ cast concrete wall cast in 2 skins with rigid closed cell insulation between skins.

can also be used for in-situ cast concrete walls, depending on the ability to work on large areas of completed wall surface.

As-cast finish

Smooth concrete finishes can have their surfaces broken by small air bubbles of entrained air, but these hardly affect the surface any more than would be found in natural sandstones and limestones. Colour variations from the use of grey cement result not from its natural colour, but its colour after pouring, and from fine particles of sand becoming segregated in certain areas as a result of vibration. If these fine particles appear in the face of the wall, a marbling effect similar to that of natural stone will occur (MCE_207, d). If the particles are different from the shade of the facing, staining will occur. As a result, the tinting of clear mixtures with fine particles of dark sand is avoided. Smooth concrete is usually self-coloured, at least in large areas, to avoid colour variations associated with using pigment additives. Smooth and visually consistent natural finishes are achieved largely by both the accuracy of the proportions of the materials in the concrete mix, including water, and the care in preparing and setting the formwork.

Textured surfaces with either a shallow or deeper contoured profile can be formed either with specially fabricated shuttering boards or with an additional lining that is set on the inside face of the shuttering boards, against which the concrete is poured. An additional lining is usually flexible and is made of either polystyrene board, which can be used only once, or polyurethane sheet or silicone rubber sheet which can be used several times. Silicone rubber moulds are made by pouring the material into a positive-shaped mould made from a non-cohesive material such as sand, making its use expensive, but capable of forming complex forms in the surface of as-cast concrete. Joints between these specially-made shuttering boards are usually formed as grooves to avoid uneven and blurred lines resulting from attempts to bond their edges.

Details

1 | Concrete external wall
2 | Concrete internal wall
3 | Thermal insulation
4 | Window frame
5 | Waterproof membrane
6 | Metal parapet flashing
7 | Internal finish
8 | Metal sill
9 | Drip
10 | Metal lined gutter

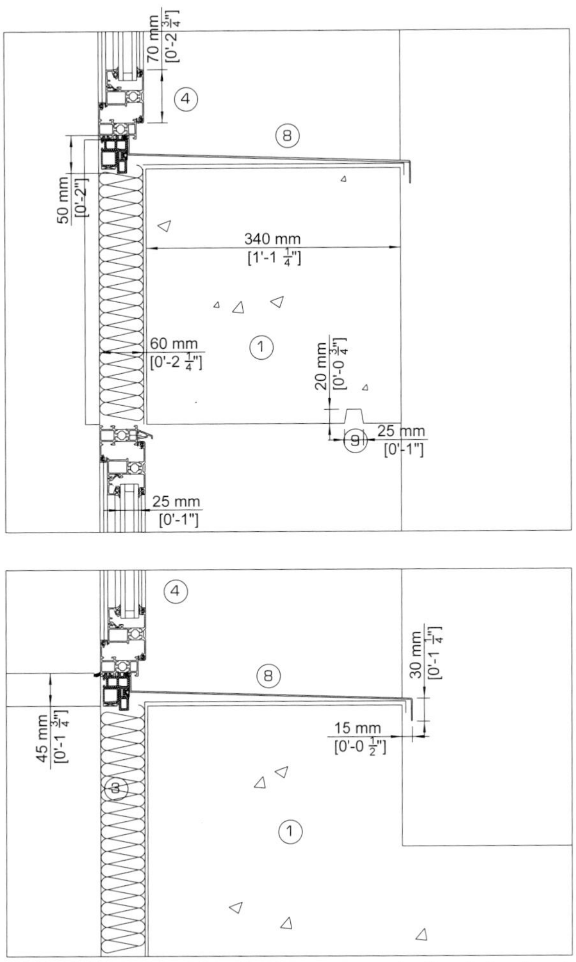

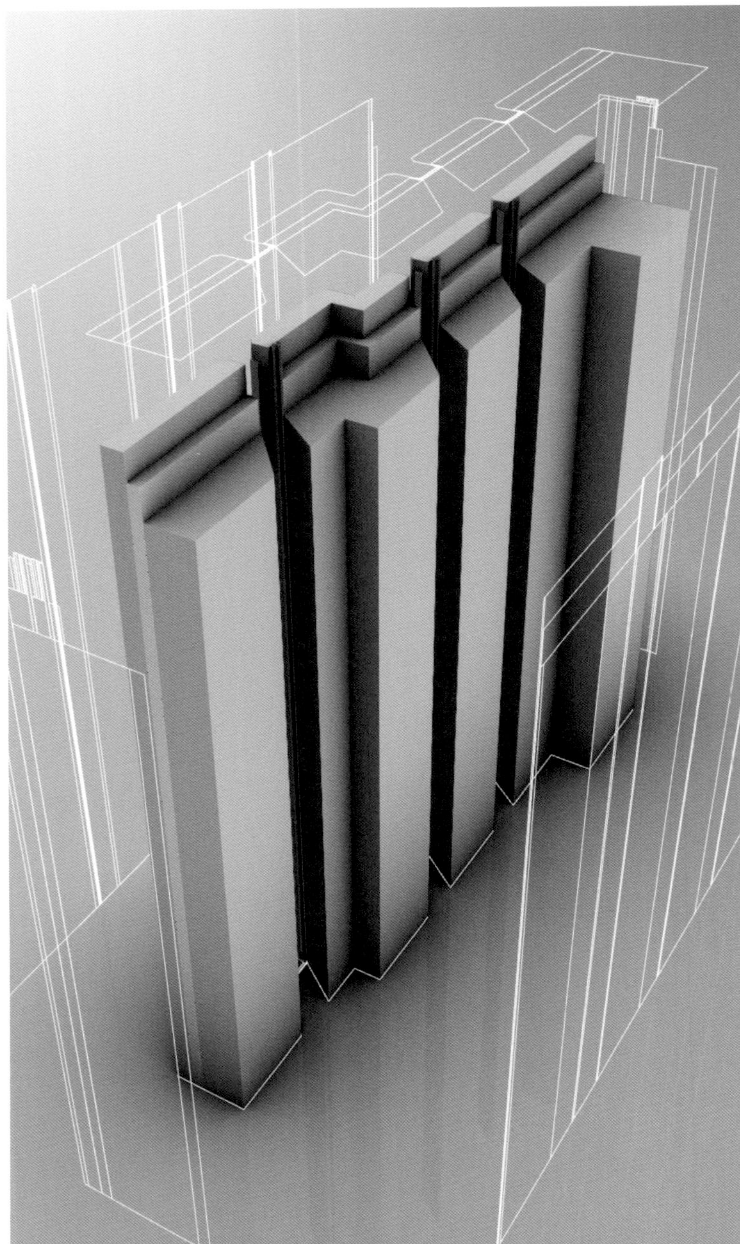

b| Prototype

a | Vertical sections. Junction bewteen panels. Panels insulated on inside face.

Washed finish

The washing out or deactivation of fresh concrete is carried out in two ways, either with 'faces shuttered' or with 'faces not shuttered'. The faces shuttered method involves the application of a product on the shutter panels that deactivates, retards or eliminates hydration of the cement. The product is applied with either a brush or by spraying. After stripping the shuttering, the external faces are washed with a water jet which removes the hydrated skin and, depending on how deep the effect of the deactivator is, reveals either the grains of sand or, more commonly, the coarse aggregate.

The faces not shuttered method is done either by spraying deactivator on the green (fresh) concrete and washing it as just described, or by direct washing before the cement has fully set. Usually the method is completed with a last wash using an aqueous acid solution in order to remove traces of hydrated cement soiling the exposed aggregate, leaving stains on the outside face. When this is completed, the mineralogical nature of the coarse aggregate is exposed, giving a textured appearance. Some aggregates, such as limestone, can become dulled or change colour on contact with the acid. In deep washing, this method is particularly suitable for bringing out the visual characteristics of coarse aggregate. It leads to a very different finish depending on the shape, either crushed or rolled, the mineralogical type (silica or limestone), particle size, but also the mix composition in terms of the density of stones revealed on the surface.

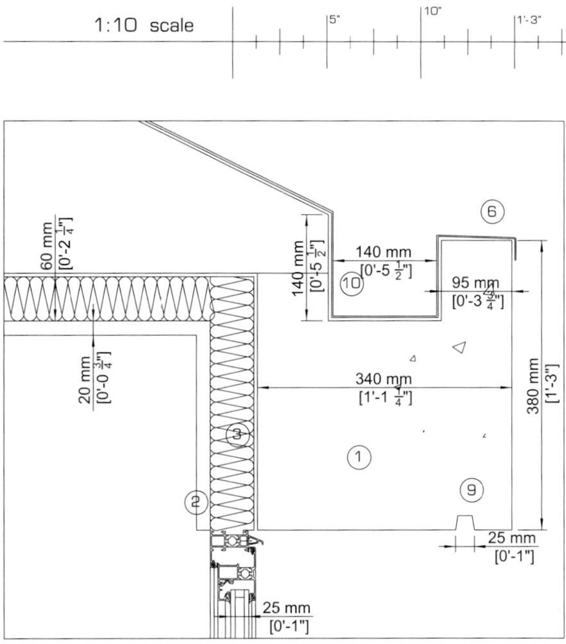

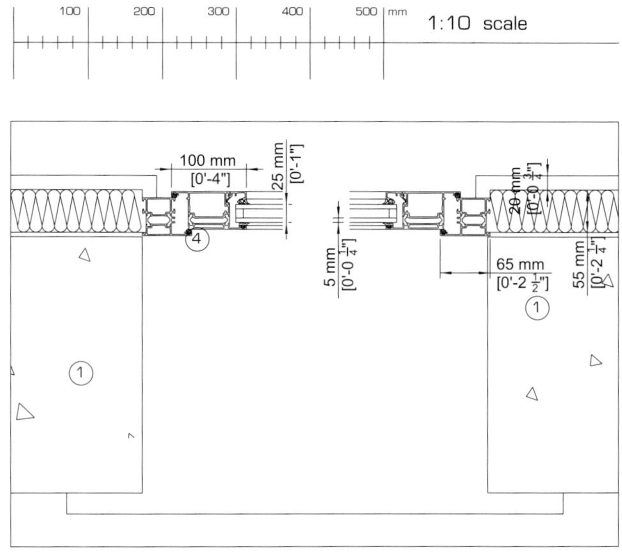

c | Vertical and horizontal section. Junction at parapet level for typical wall condition and for window jamb. In situ cast concrete wall cast as a monolithic construction with internal thermal insulation.

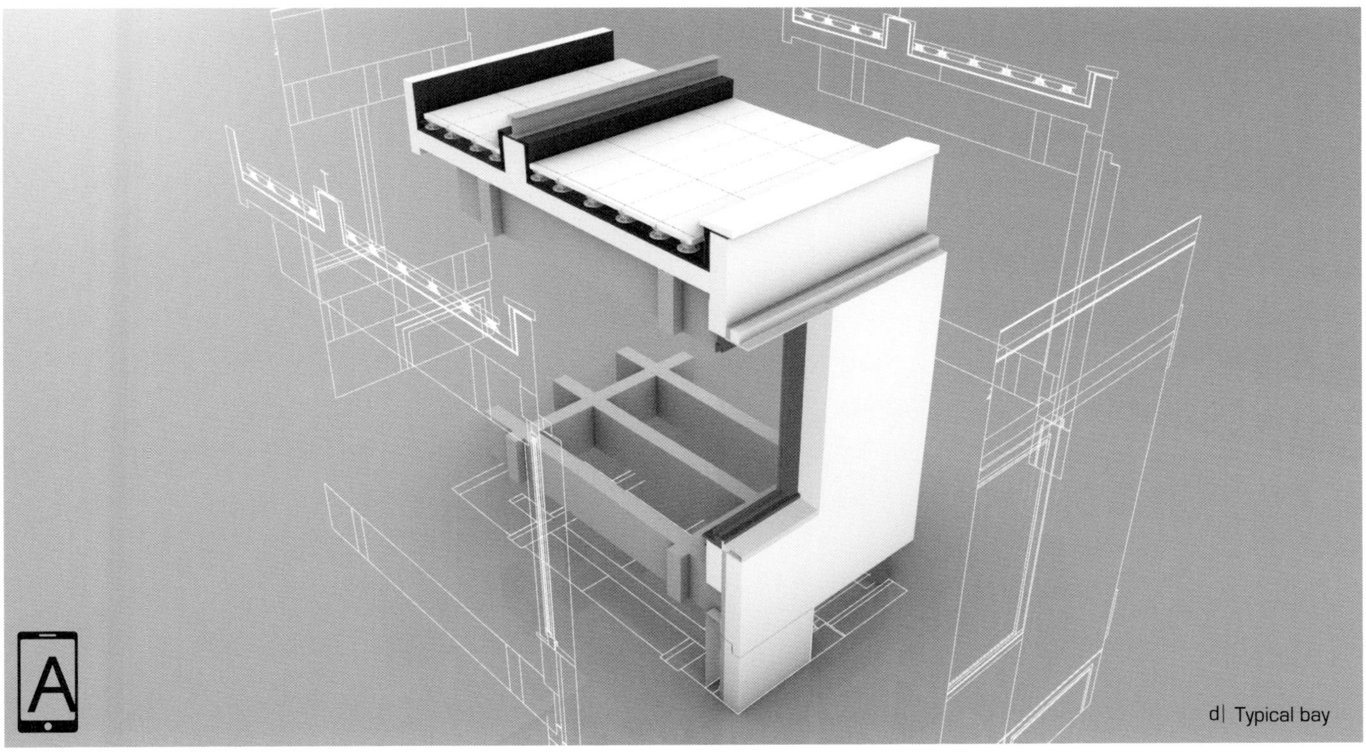

d| Typical bay

Polished finish

Concrete walls can be polished to varying amounts of smoothness by abrasive grinding, using a grinding wheel which is lubricated by water. The outer skin of the concrete is removed by abrasion to a depth of between 1mm and 2mm, with one pass of a grinding wheel which exposes some of the fine and coarse aggregate behind the surface. A second pass with a finer abrasion disc removes the big scratches left by the first pass, which are particularly visible on darker surfaces. With polishing by disc, a filler is applied to the surface to fill the air bubble pockets and small honeycombing. After this filler has hardened, the surface is again polished with successively finer grains. As the scratches are removed, the colours of aggregate reveal themselves, but the surface does not shine naturally. The concrete wall surface can, however, be made shiny with the application of a varnish. Concrete walls can be polished further to create a satin finish, and further still to provide a gloss finish. A protective clear coating can also be applied at this point.

Polishing is easier to apply on flat surfaces than on roughened or curved areas of concrete. Polishing exposes the colour of the minerals within the concrete, giving rich colour effects from the cement, the sand, and the coarse aggregate materials in the mix. This method provides a self-coloured finish with a surface that is less prone to retain dust, which washes off, causing staining, and is easy to maintain by simple washing, undertaken using the same methods of maintenance access used for glazed walls.

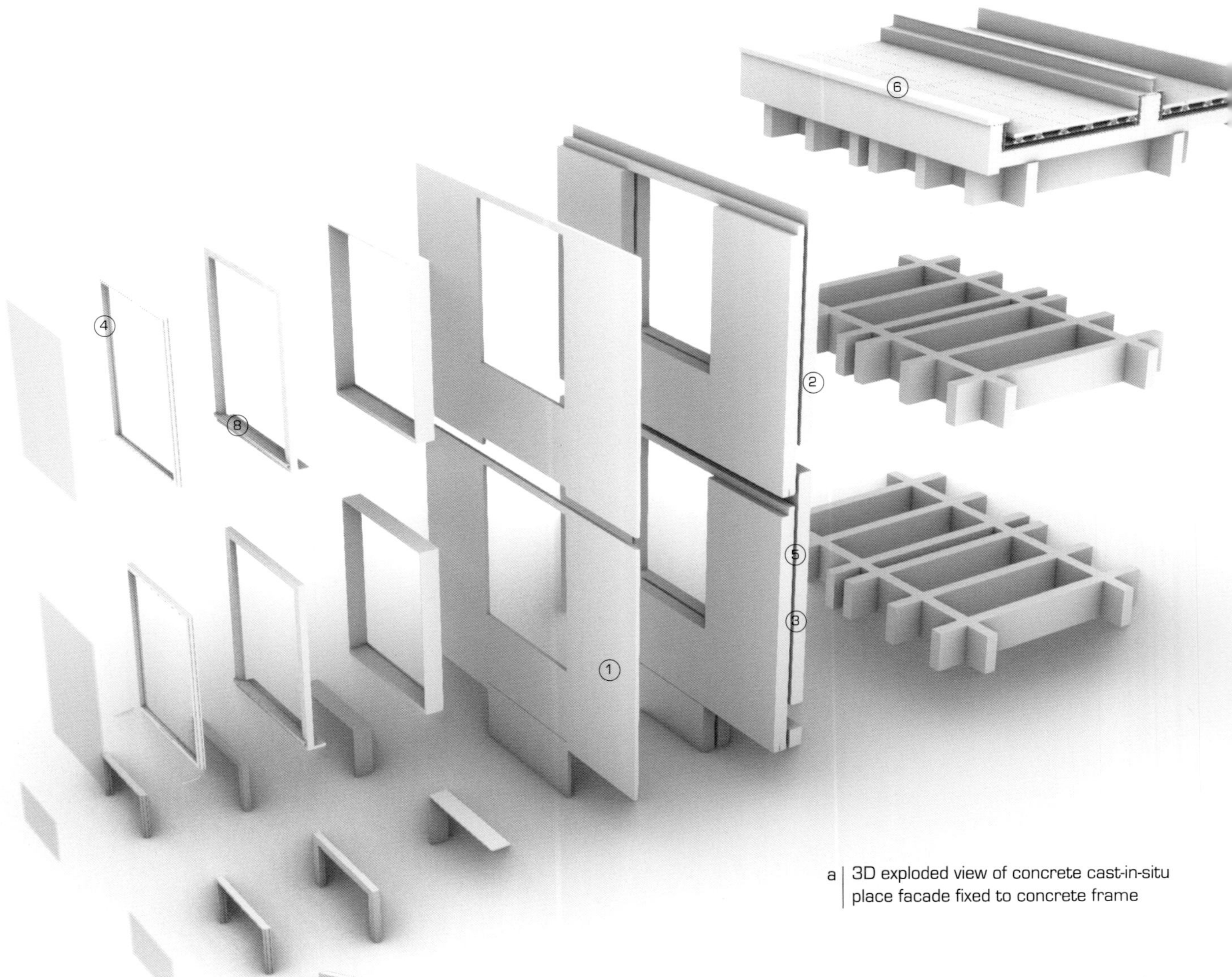

a | 3D exploded view of concrete cast-in-situ place facade fixed to concrete frame

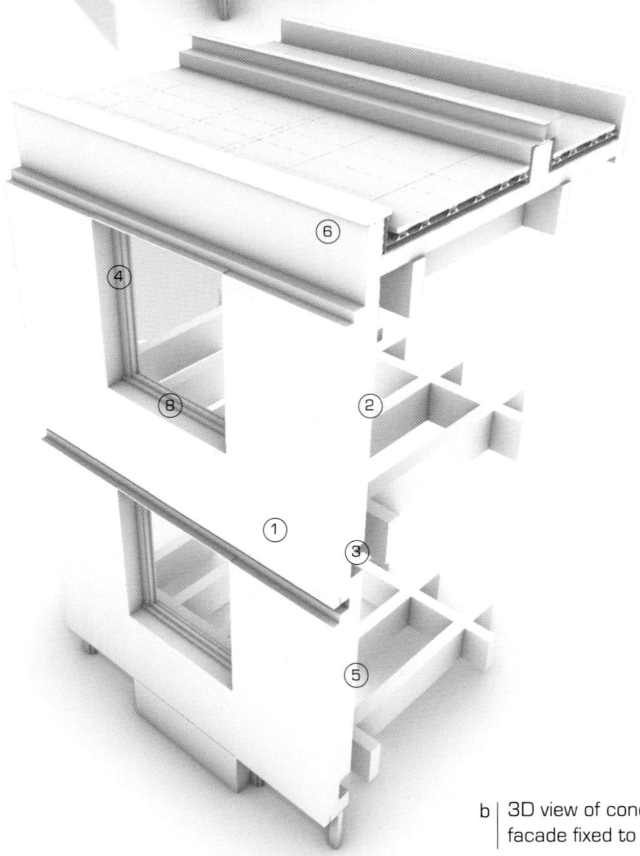

b | 3D view of concrete cast-in-situ facade fixed to concrete frame

Details

1 | Concrete external wall
2 | Concrete internal wall
3 | Thermal insulation
4 | Window frame
5 | Waterproof membrane
6 | Metal parapet flashing
7 | Internal finish
8 | Metal sill
9 | Drip
10 | Metal lined gutter

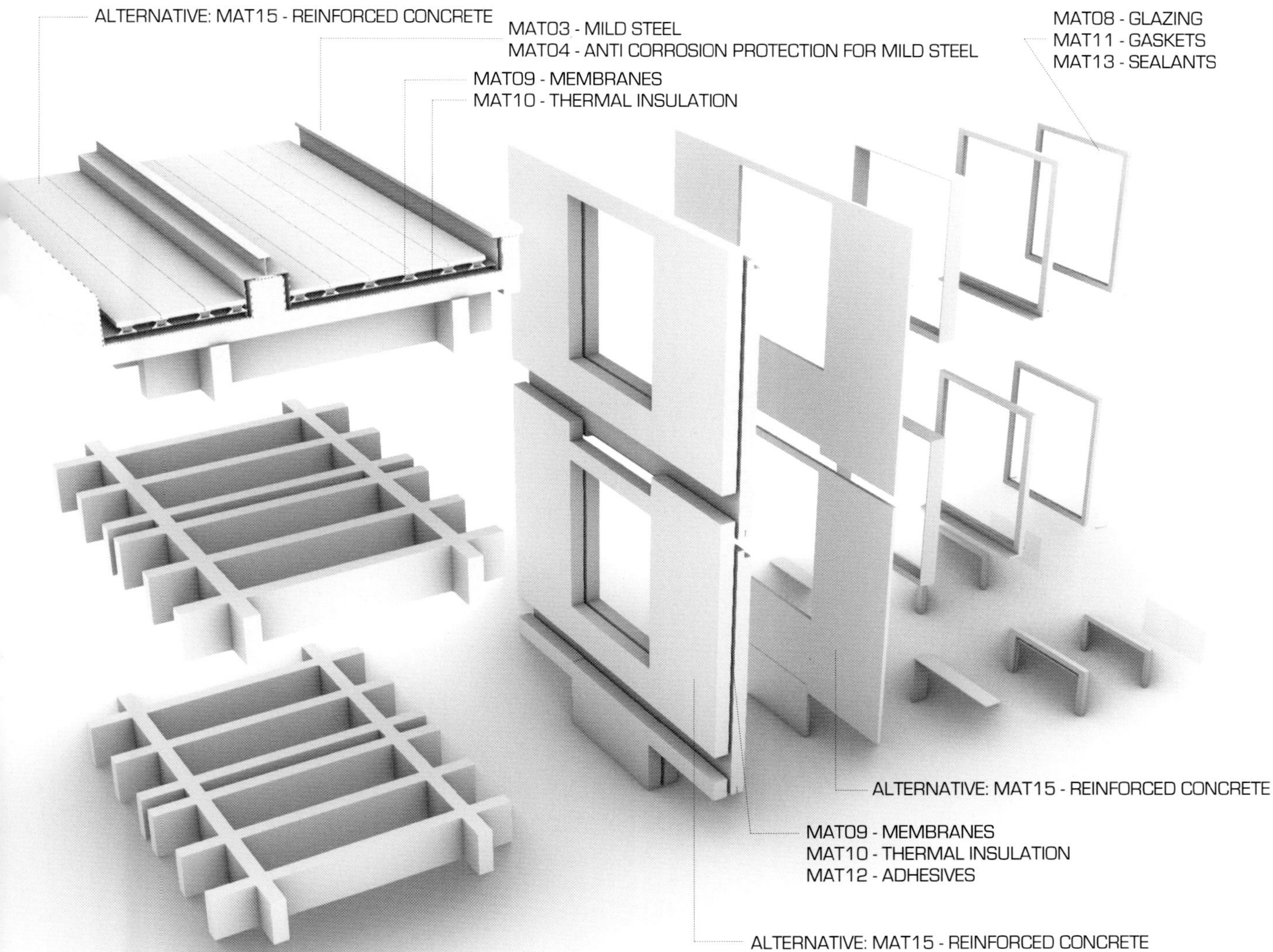

c | 3D exploded view of concrete cast-in-situ facade componenets with material specification references

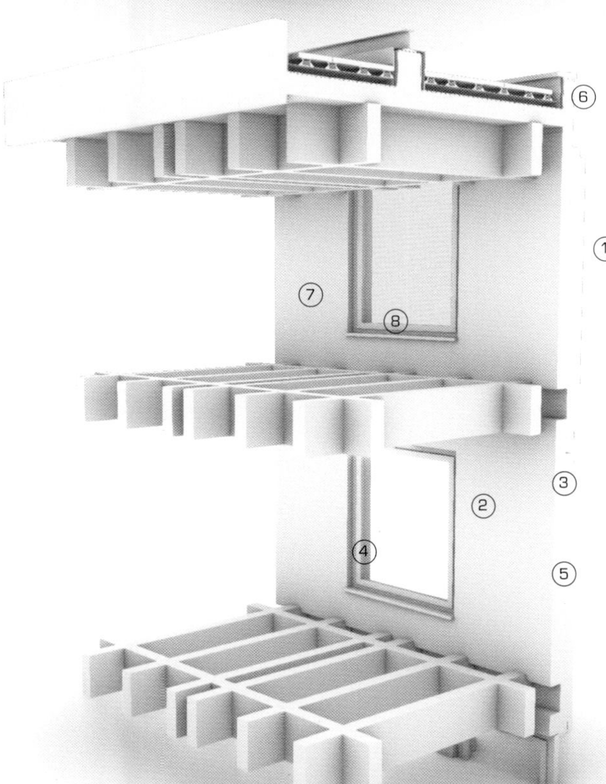

d | 3D view of concrete cast-in-situfacade fixed to concrete frame

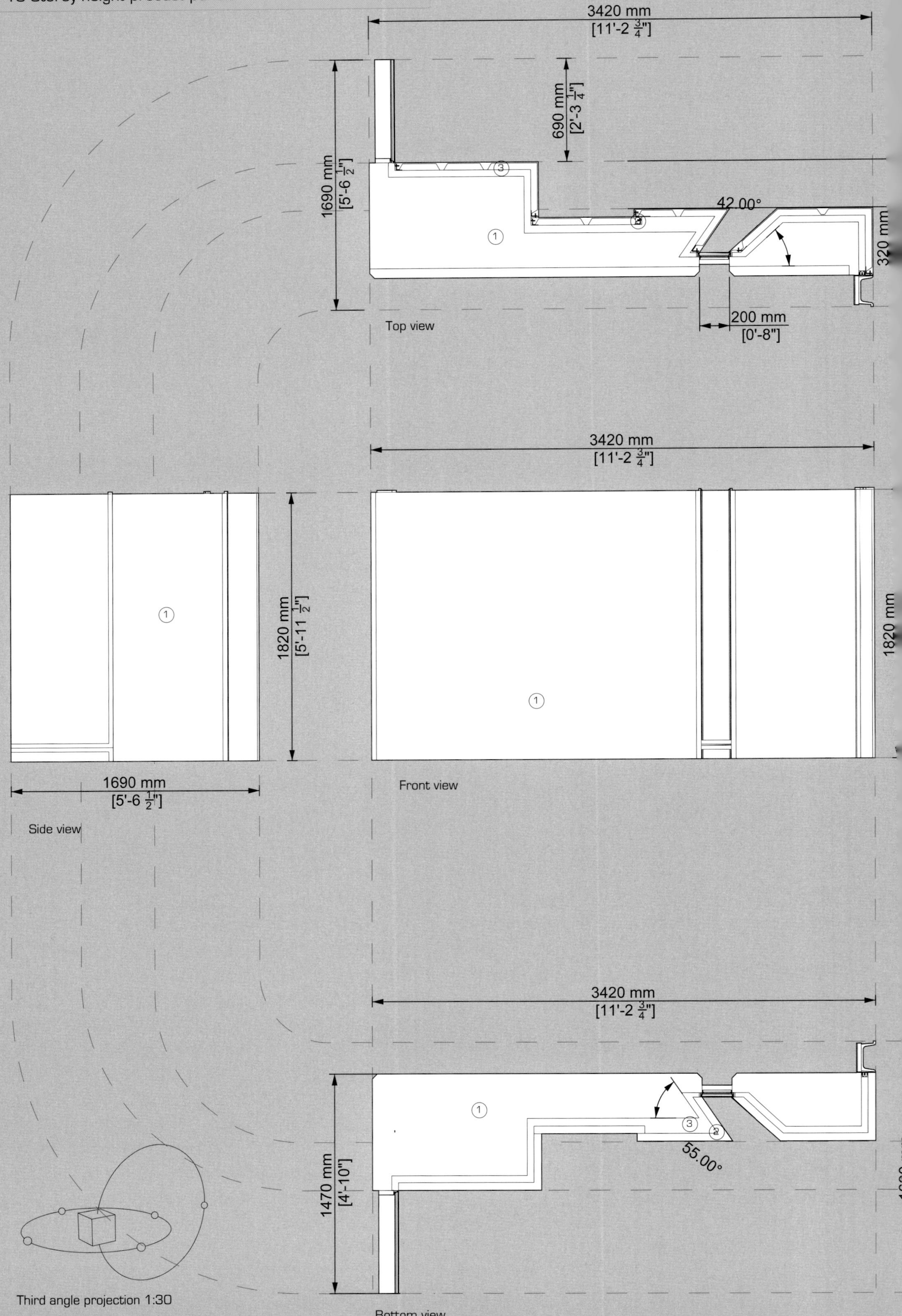

Third angle projection 1:30

Details

1 | Concrete backing wall
2 | Precast concrete panel
3 | Thermal insulation
4 | Vertical baffle joint between panels
5 | Vertical butt joint between panels
6 | Horizontal lap joint
7 | Horizontal butt joint

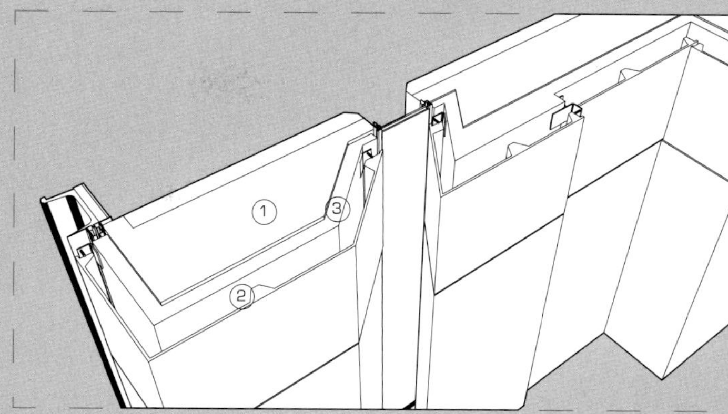

3D view of the assembly

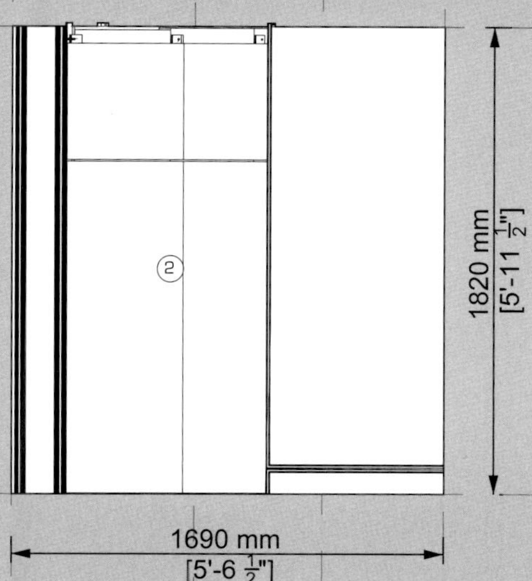

Side view

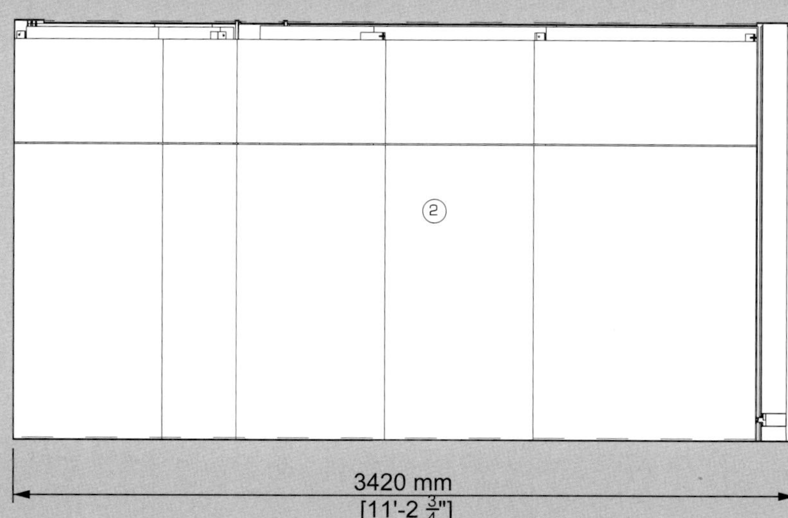

Back view

8 | Window opening
9 | Stainless steel dowel or angle
10 | Rubber-based baffle
11 | EPDM or silicone-based seal
12 | Concrete column
13 | Concrete corner panel
14 | Concrete coping
15 | Roof assembly

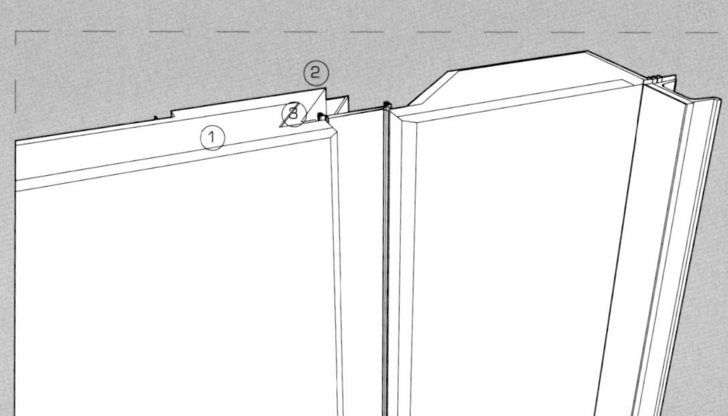

3D view of the assembly

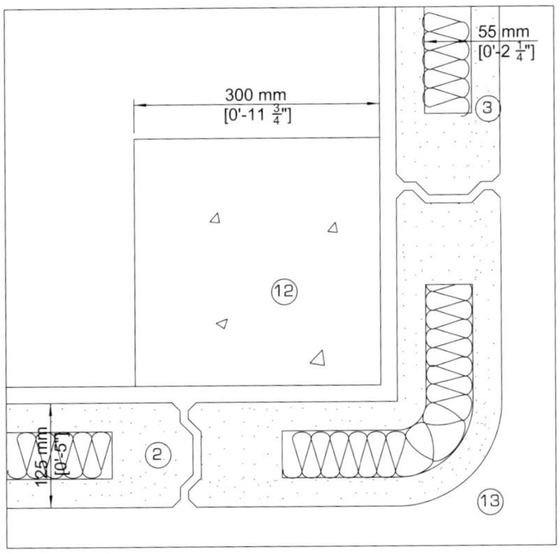

d | Horizontal sections. Panel junctions at corners. In situ cast concrete wall cast in 2 skins with rigid closed cell insulation between skins.

Details

1 | Concrete backing wall
2 | Precast concrete panel
3 | Thermal insulation
4 | Vertical baffle joint between panels
5 | Vertical butt joint between panels
6 | Horizontal lap joint
7 | Horizontal butt joint
8 | Window opening
9 | Stainless steel dowel or angle
10 | Rubber-based baffle
11 | EPDM or silicone-based seal
12 | Concrete column
13 | Concrete corner panel
14 | Concrete coping
15 | Roof assembly

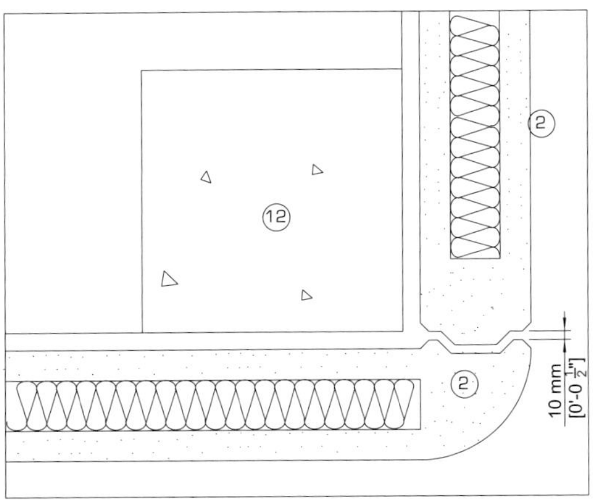

a| Horizontal section. Typical precast concrete wall construction.

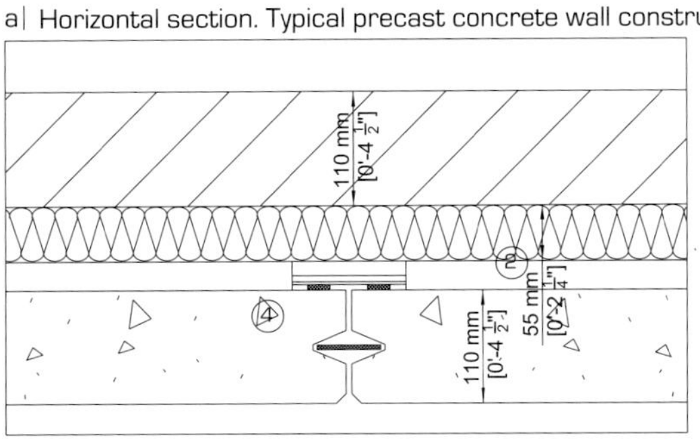

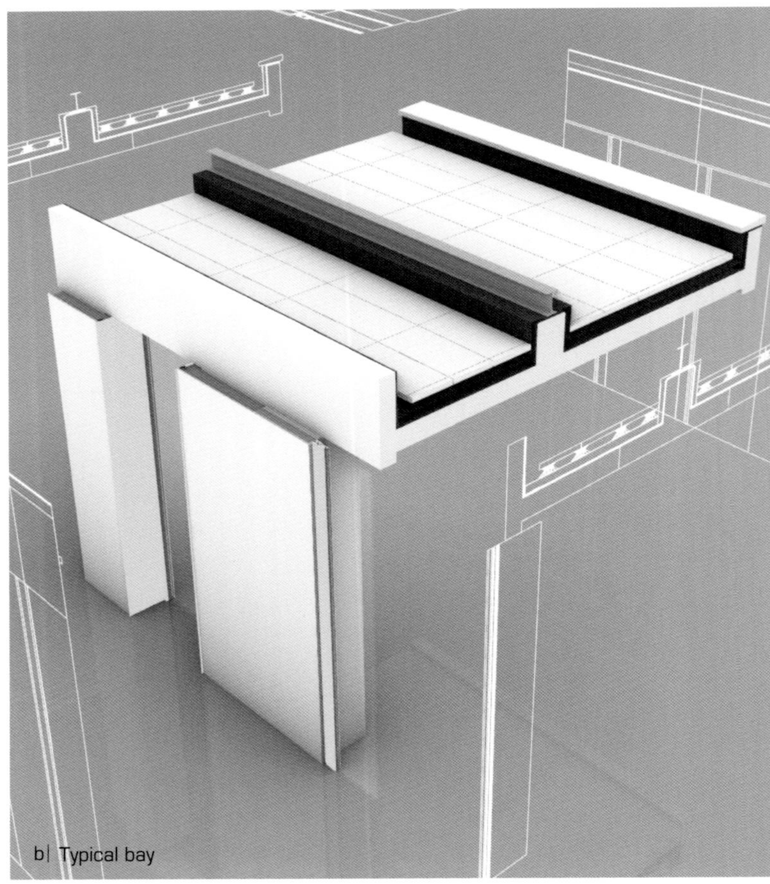

b| Typical bay

While in-situ concrete is a loadbearing form of construction, precast concrete can be formed as either loadbearing walls or as non-loadbearing cladding panels (MCE_213, c). Loadbearing panels are increasingly popular, since they provide a weathertight structural wall with high fire resistance combined with high acoustic insulation and thermal mass. In loadbearing construction, panels are stitched together to form a monolithic structural wall. Non-loadbearing cladding panels have maintained their popularity by providing greater freedom in design, from a visual point of view, than the structural constraints imposed on loadbearing panels.

Loadbearing types consist of units stacked together to support floors by transferring their own weight and a floor loading to a foundation. The unit/floor connection is usually made with a pin joint rather than a moment (rigid) joint since these are difficult to create in precast concrete, as the amount and length of built-in reinforcement is high, as well as the fact that the tensile forces within the unit can be high (MCE_214, b). The horizontal stability, which is not provided by the pin joint, is usually provided by service cores elsewhere in a building. In addition to floor-height loadbearing panels, loadbearing precast spandrel panels are used. These are in effect structural beams spanning between columns. Principles of connections to floor slabs are the same as for full-height panels, but loads from the panels and floors are taken back to structural columns rather than down through the wall panels.

Non-loadbearing cladding panels are fixed back to the primary structure either by concrete brackets forming an integral part of the panel, by stainless steel brackets, or by a combination of both. Typically panels are supported on an edge beam at the base of each panel and are restrained at the top with stainless steel brackets (a). Some cladding panels are top-hung in order to benefit from the tensile qualities of the steel reinforcement or framing (MCE_214, a). Cladding panels are usu-

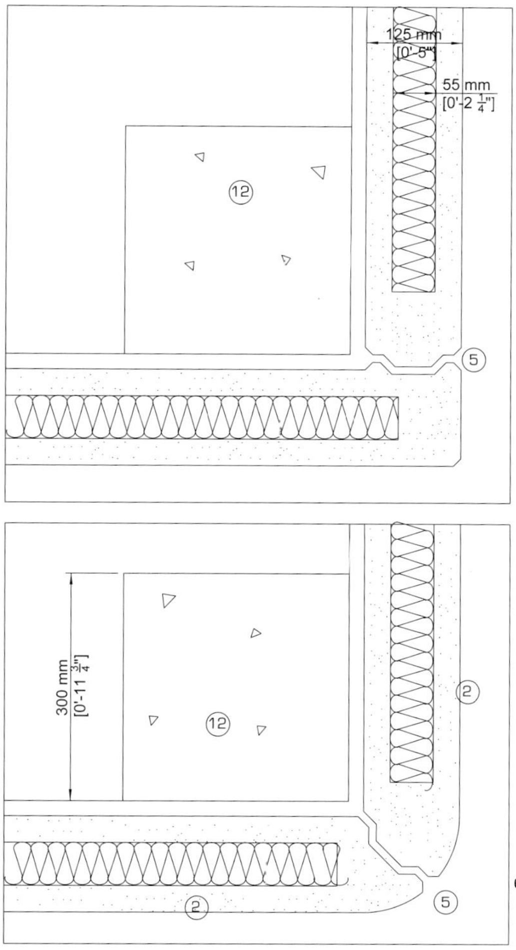

c | Prototype

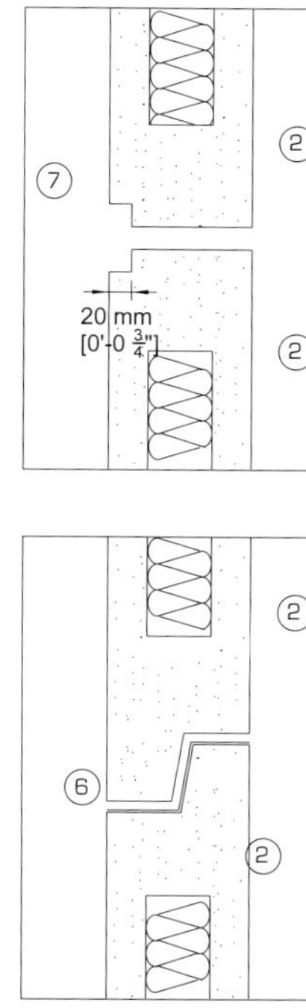

d | Vertical section. Panel to panel junctions. Precast concrete wall cast in 2 skins with rigid closed cell insulation between skins.

e | Horizontal sections. Panel to panel junctions. Precast concrete wall cast in 2 skins with rigid closed cell insulation between skins.

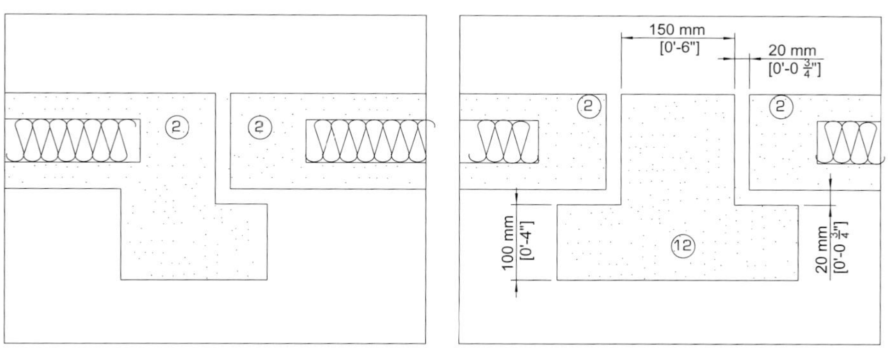

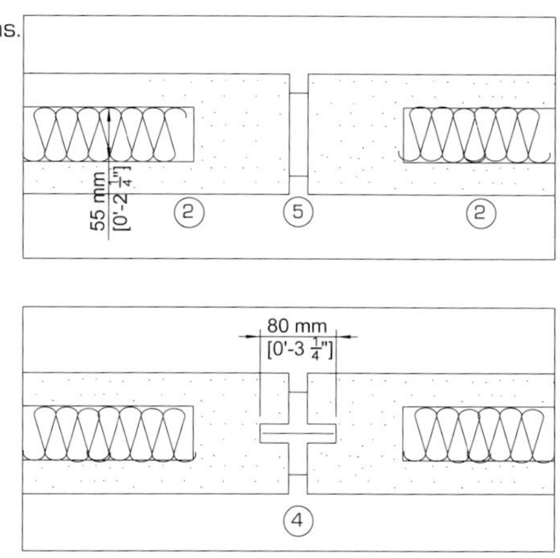

ally made to span one storey high, with a width or height of panel up to a maximum of 3600mm in order that the panel can be transported on a standard flat trailer. In addition, weight usually has a maximum of around 10 tonnes to prevent cracking during lifting and to allow a regular site crane to be used.

Panel types

Precast concrete panels are formed mainly by either forming the finish in the bottom of the mould, the top of the mould with another material to the face of the panel, or as an insulated sandwich panel.

Bottom-formed panels, where the finish is set on the bottom of the mould, are used where additional linings, typically polystyrene board, polyurethane sheet or silicone rubber sheet, as described in the text on in-situ cast concrete, are used to create a textured surface in the face of the panel. Sometimes, ceramic tiles or stones are laid in the bottom of the panel in order to bond to the concrete poured on top. The small size of terracotta or ceramic tiles allows them to be laid at the bottom of the mould and be individually bonded to the concrete.

In top-formed panels, an additional decorative layer is applied to the top of the poured concrete. The thickness of the decorative layer is usually 25mm–30mm. The concrete in the decorative layer is minimised since it is usually much more expensive than grey structural concrete. Alternatively, larger scale facing pieces than those which can be bonded directly to the concrete in the bottom of the panel, such as terracotta, ceramic tiles or natural stone, are mechanically fixed to the top of the concrete in the mould as a decorative finish. Due to its larger size and weight, stone is usually fixed with pins or stainless steel brackets back to the precast panel. The greater size of stone panels over tiles makes them susceptible to relatively significant thermal movement for which the pins or brackets allow.

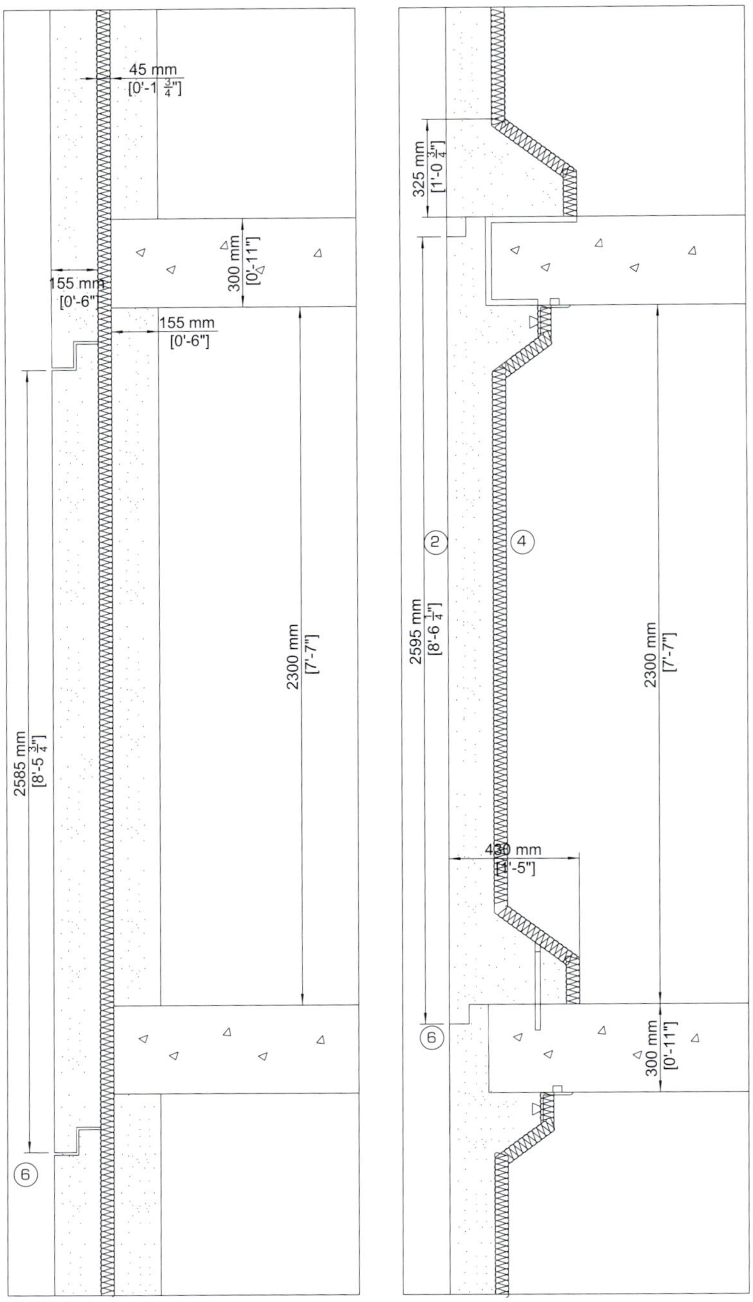

b | Vertical sections. Panel assemblies.
Type 1 sandwich insulation, type 2 insulation on inner face

a | Prototype

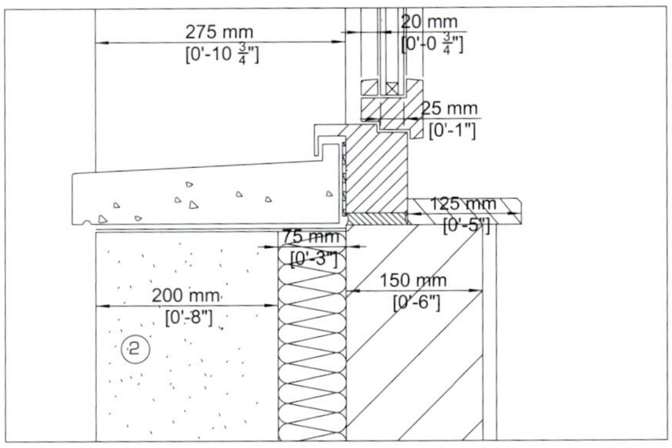

c | Vertical section. Junction at window jamb. In situ cast concrete wall cast in 2 skins with rigid closed cell insulation between skins.

Another method of construction is to form an insulated sandwich panel where the inner part of the wall is loadbearing. Thermal insulation is set between the decorative and structural layers of concrete. This removes the thermal bridge across the panel and, if the interior concrete face is exposed, provides a thermal mass for night-time cooling in the building (MCE_217 d). The temperature of the outer lining can reach about 70°C in summertime, depending on geographical location. The sandwich panel ensures that the outer facing only is susceptible to thermal movement, the continuous insulation is kept dry and the inner massive loadbearing structure is protected from external temperature variations, reducing its own thermal movement. In addition, the thermal capacity of the inner structure provides a regulation of temperature variations within the building from internal heating and heat sources by absorbing surpluses and supplementing deficiencies. External temperature changes in the daily cycle can be absorbed by the outer layer, with the thermal insulation keeping the internal heat in and external solar radiation out.

Thermal Insulation

Both loadbearing and non-loadbearing types without integral insulation have thermal insulation fixed to the internal face of the concrete wall (b). However, the thermal mass of the construction is lost. Where an additional external finish is used, typically a rainscreen panel in another material, the thermal insulation can be set on the outside of the precast panels, which maintains the ability to use the thermal mass of the interior face of the wall.

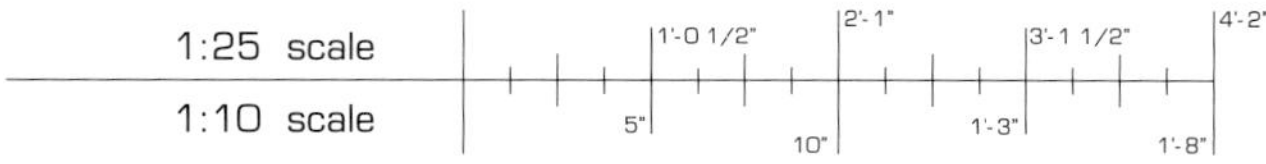

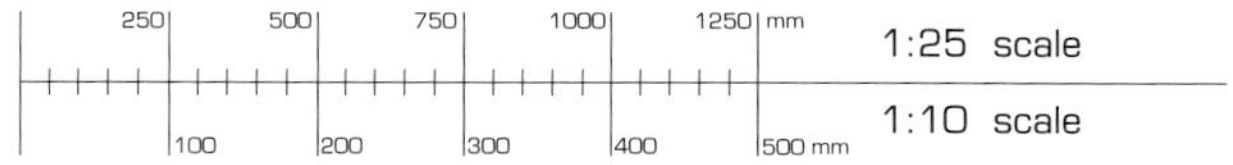

d| Prototype

Details

1 | Concrete backing wall
2 | Precast concrete panel
3 | Thermal insulation
4 | Vertical baffle joint between panels
5 | Vertical butt joint
6 | Horizontal lap joint
7 | Horizontal butt joint
8 | Window opening
9 | Stainless steel dowel or angle
10 | Rubber-based baffle
11 | EPDM or silicone-based seal
12 | Concrete column
13 | Concrete corner panel
14 | Concrete coping
15 | Roof assembly

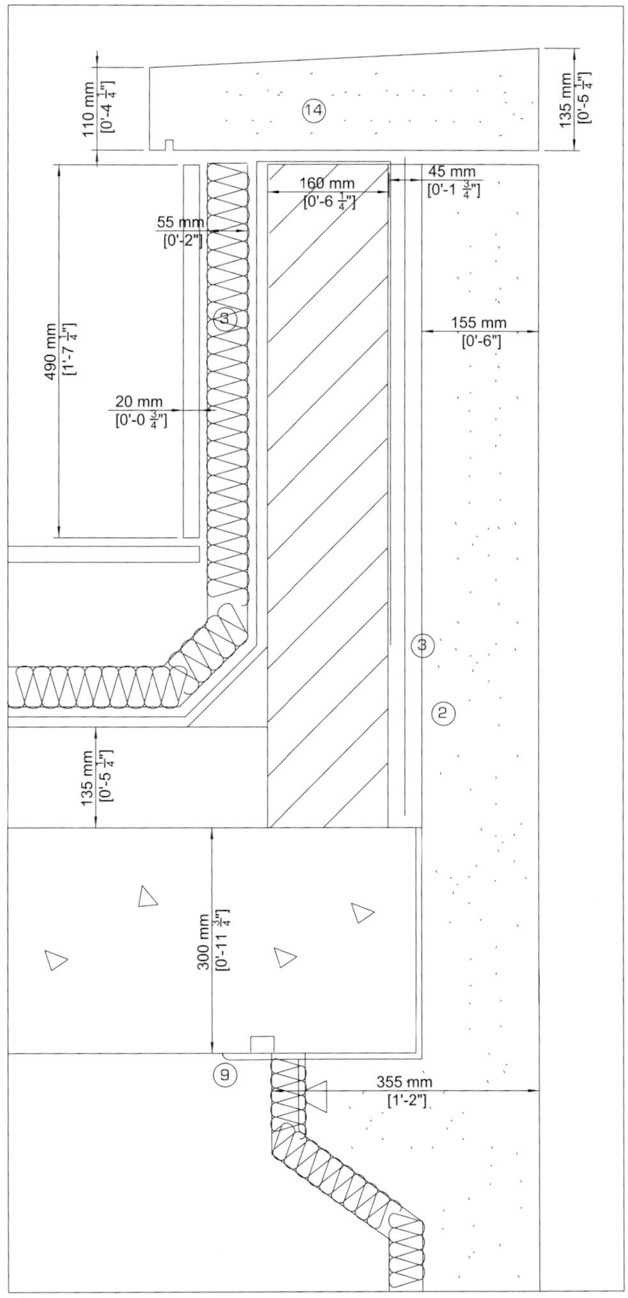

e | Vertical section. Junction at parapet level for typical wall condition and for window jamb. In situ cast concrete wall cast in 2 skins with rigid closed cell insulation between skins.

When thermal insulation is set on the inside face of precast panels, windows can be set directly above the insulation, avoiding the need to fit the window directly into the opening in the concrete. The window, typically aluminium or timber, has metal profiles that form a sill that is continuous with the window, avoiding any source of leaks across the gap between window and precast panel. Where windows are set directly into openings in precast panels, the opening often has a rebate (step) cast into the edge to provide more protection against rainwater penetrating the joint (MCE_216, b).

Joints

Joints between panels are of open or closed type. The open type has an inner membrane seal and an outer flexible mastic seal on the inner face. The closed type has a single outer seal of flexible mastic. These principles apply to both vertical and horizontal joints. In the horizontal joints, a step in the panel may be provided to assist in the weather tightness of panels (MCE_213, a), which are upwards of 150mm thick. Minimum joint widths are typically 10–12mm thick for panels ranging from 1800mm wide to 2400mm wide, increasing to a joint width of 16–18mm for panels 6000mm wide. These widths help visually reduce any variations in the width of the joint as a result of joining panels. Visually, joints are recessed as shadow gaps in order to conceal small differences of alignment between panels.

Open joints are internally drained and ventilated in the manner of unitised curtain walling. In an open joint, windblown rain is allowed to pass through an outer gap between two panels that form a vertical joint. An EPDM baffle strip is set into a continuous slot formed in the edges of

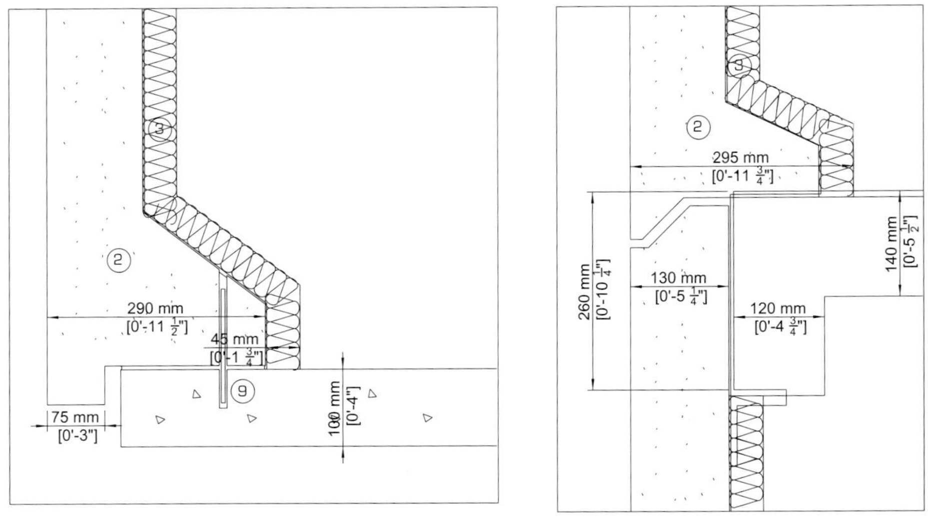

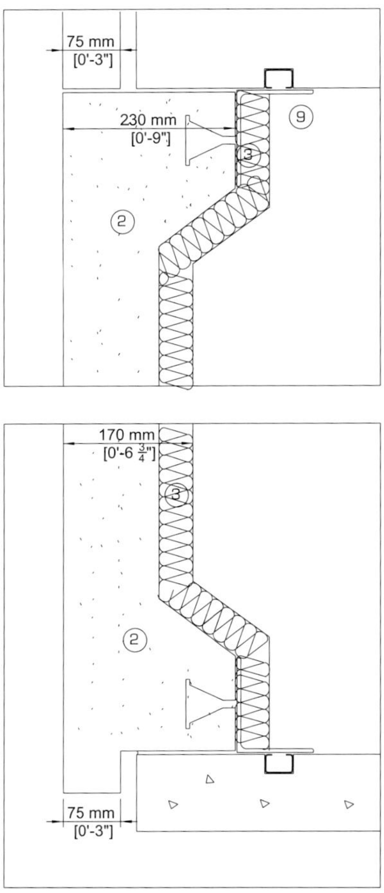

a | Vertical sections. Junction between panels. Panels insulated on inside face.

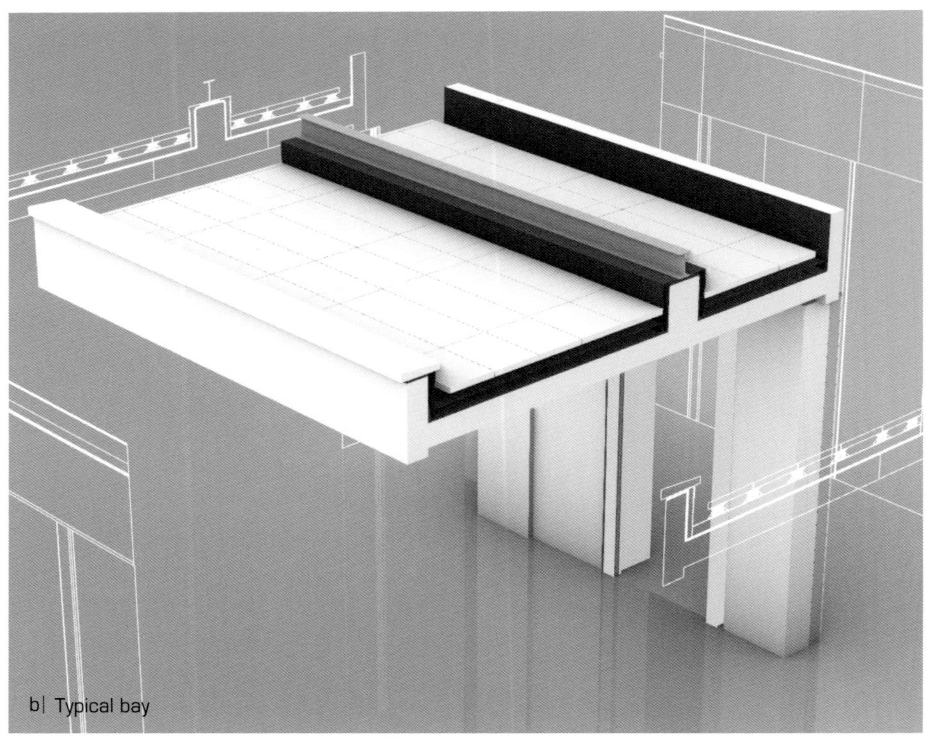

b| Typical bay

two adjacent panels and water entering this joint, which forms an inner chamber, is drained away from the slot and out through the horizontal joint below. The interior face of the joint is either closed with an air seal formed using wet or extruded silicone, in the case of a non-loadbearing cladding panel, or is mechanically fixed to the panel below and grouted with a cement-based material in the case of loadbearing panels.

Horizontal open joints between panels are made by forming a step in the profiles of the top and bottom edges (a) of adjacent panels in order to drain water out of both vertical and horizontal joints, as well as to prevent rainwater from flowing directly through the joint. The horizontal stepped joint is used for both loadbearing and non-loadbearing type panels, but the detail varies with the way the upper panel is either supported on the lower panel in a loadbearing panel or is separated from the lower panel in a non-loadbearing panel. However, the general principle for all panel types remains the same. This ensures that water running down the EPDM baffles in the vertical joints is drained out where it meets the horizontal joints. A continuous horizontally-set EPDM strip is set on the inner face of the joint, in the top of the lower panel. The EPDM strip in the vertical joint is lapped over the face of the EPDM on the horizontal joint. The rear of the joint, adjacent to the interior face of the panel is either closed with an air seal as for vertical joints, or is mechanically fixed to the panel below and grouted in the same way as the vertical joints.

Closed joints are sealed on the outside face with a wet applied silicone or polysulphide sealant. Any water that penetrates this outer seal is drained away in the void behind. Water is released through weep holes (small holes) in the horizontal joints. Closed joints are more commonly used with loadbearing panel construction where panels are bonded together at joints and where most of the joint is required to be filled to provide structural continuity from panel to panel.

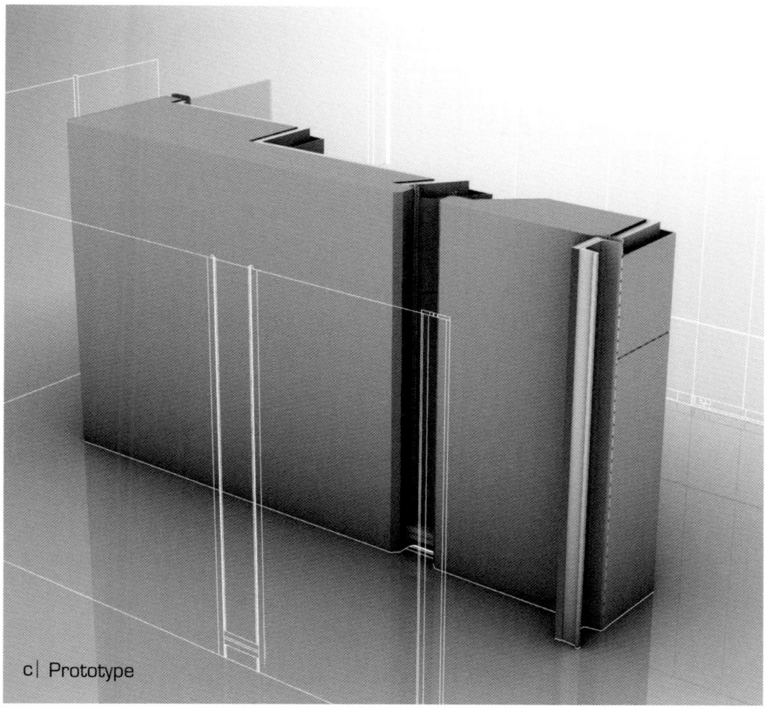
c| Prototype

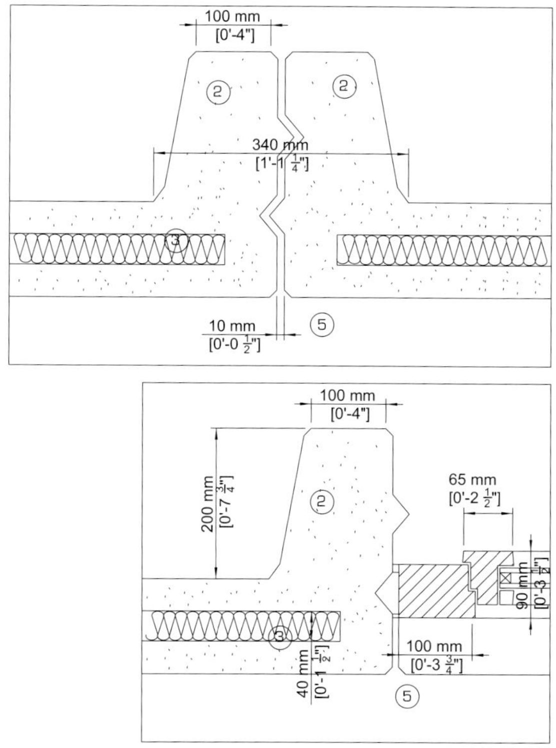

d | Horizontal sections. Panel to panel junctions

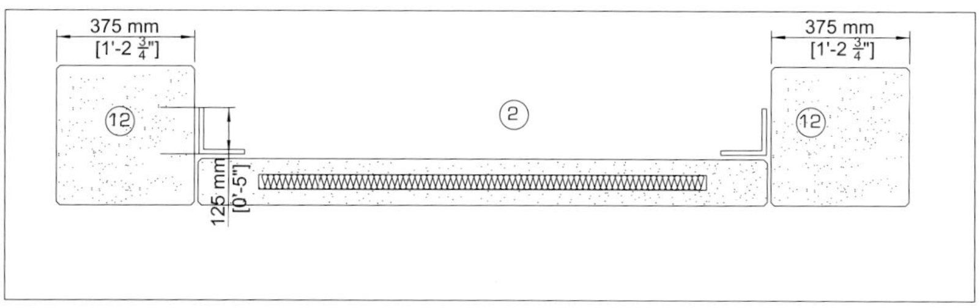

e | Horizontal sections. Panel to panel junctions

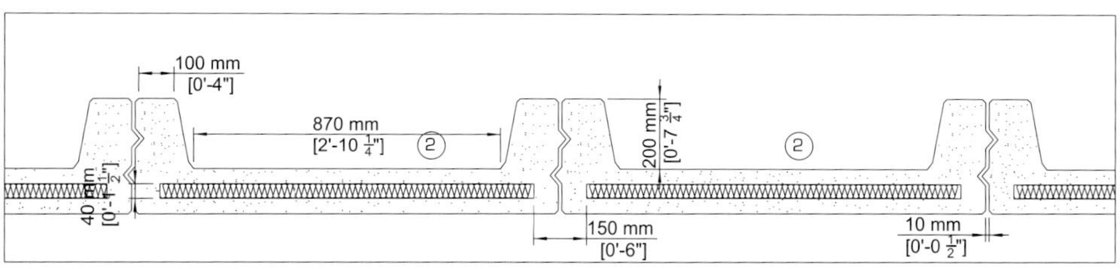

Details

1 | Concrete backing wall
2 | Precast concrete panel
3 | Thermal insulation
4 | Vertical baffle joint between panels
5 | Vertical butt joint
6 | Horizontal lap joint
7 | Horizontal butt joint
8 | Window opening
9 | Stainless steel dowel or angle
10 | Rubber-based baffle
11 | EPDM or silicone-based seal
12 | Concrete column
13 | Concrete corner panel
14 | Concrete coping
15 | Roof assembly

Acid etched finish

In addition to the finishes described in the previous section on in-situ cast finishes, which can also be used in precast panels, acid etching is possible. Acid etching is particularly well adapted to precast panels since they are cast flat, where acid treatments can be carefully controlled.

In cured panels, the surface of the precast concrete is treated with hydrochloric acid, which is then rinsed off. This method first removes the cement skin, then strips away the grains of sand within the concrete beneath, depending on the concentration of the acid solution and length of time for which it is applied. The acid etching method can be used to create either a weathered finish or an acid-washed finish. A weathered concrete effect is produced by submerging a precast concrete panel in an acid bath. This process is most commonly used when all faces of the unit need to be treated. With acid-washed concretes, a gel containing acid is applied to the surface to be treated. This method is well adapted to treatment on specific areas of concrete, to form a surface texture or pattern on the concrete wall.

Acid etching is well suited to exposing small aggregate in the panel surface by exposing only a small amount of the concrete texture beneath its skin after casting has taken place. The precipitation of salts from the acid-cement and aggregates quickly neutralises this, which can only be continued by rinsing and applying more acid. Acid etching always requires rinsing to remove the precipitated salts and neutralise the surface. Acid etching attacks limestone aggregates, and sometimes more quickly than the cement, while the silica-based aggregate remains. The surface texture can be quite different, depending on the fine aggregate material, being more granulated in the case of silica and less coloured in the case of limestone.

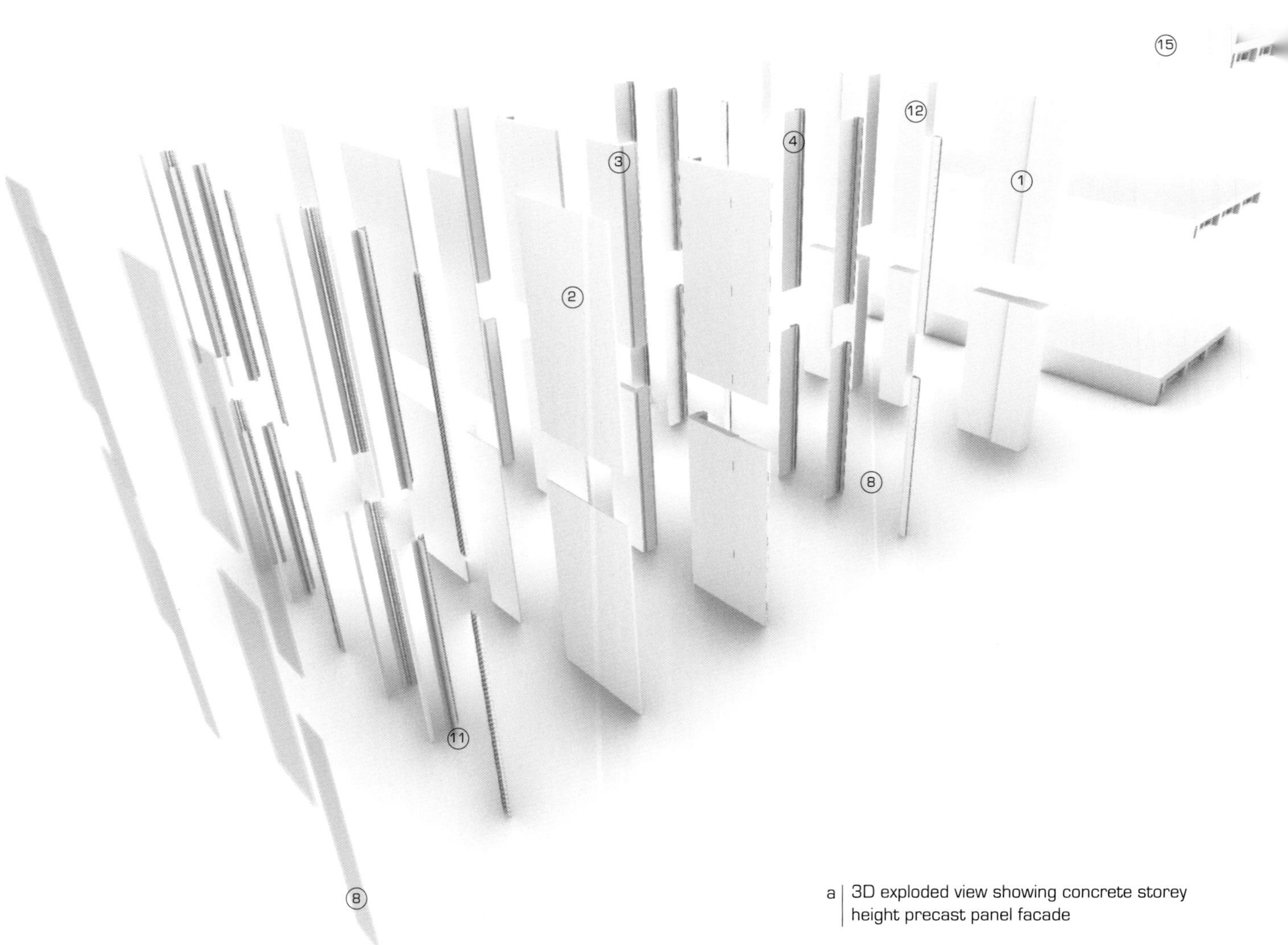

a | 3D exploded view showing concrete storey height precast panel facade

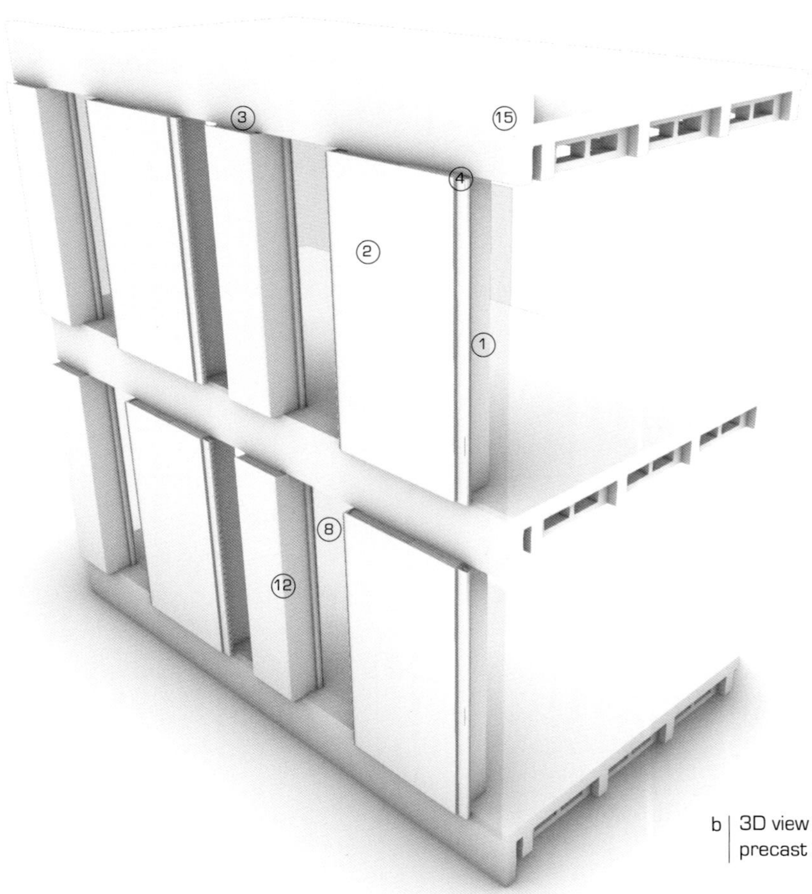

b | 3D view showing concrete storey height precast panel facade

Details

1 | Concrete backing wall
2 | Precast concrete panel
3 | Thermal insulation
4 | Vertical baffle joint between panels
5 | Vertical butt joint between panels
6 | Horizontal lap joint
7 | Horizontal butt joint

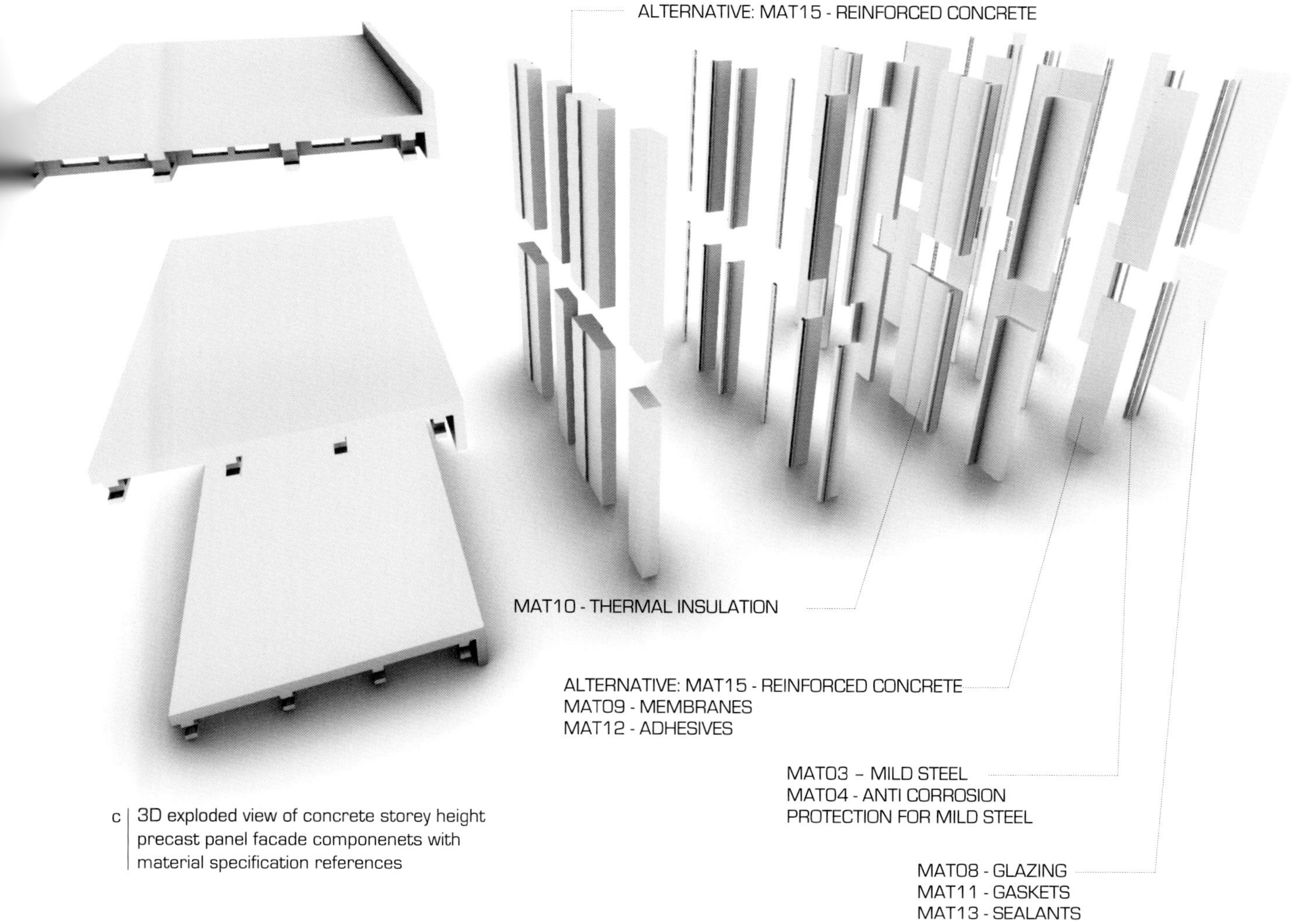

c | 3D exploded view of concrete storey height precast panel facade componenets with material specification references

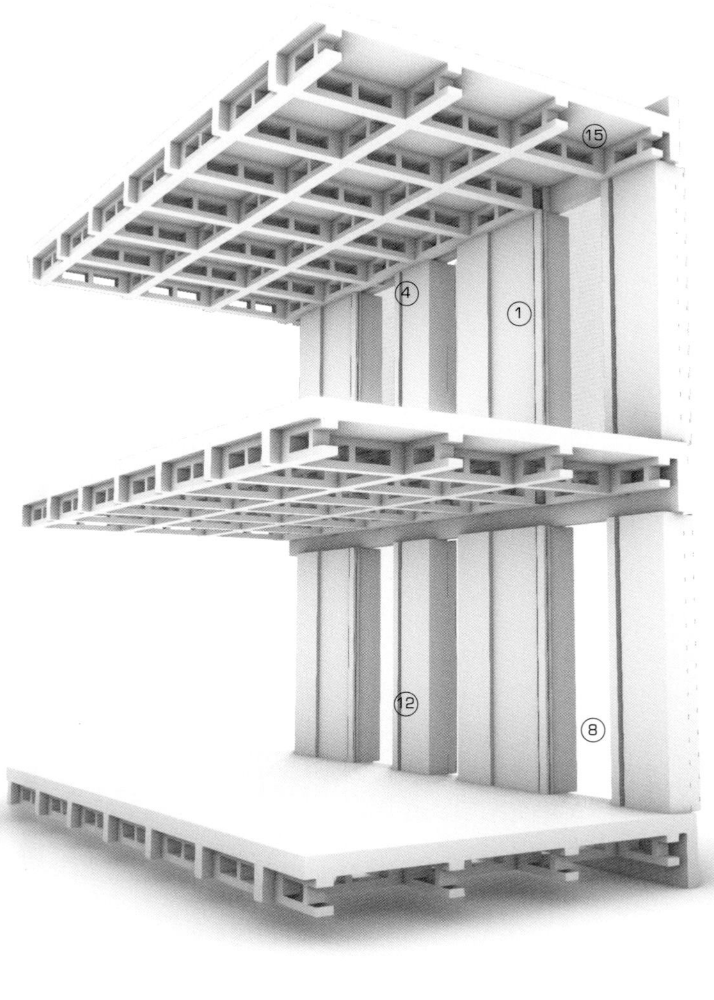

8 | Window opening
9 | Stainless steel dowel or angle
10 | Rubber-based baffle
11 | EPDM or silicone-based seal
12 | Concrete column
13 | Concrete corner panel
14 | Concrete coping
15 | Roof assembly

d | 3D view showing concrete storey height precast panel facade

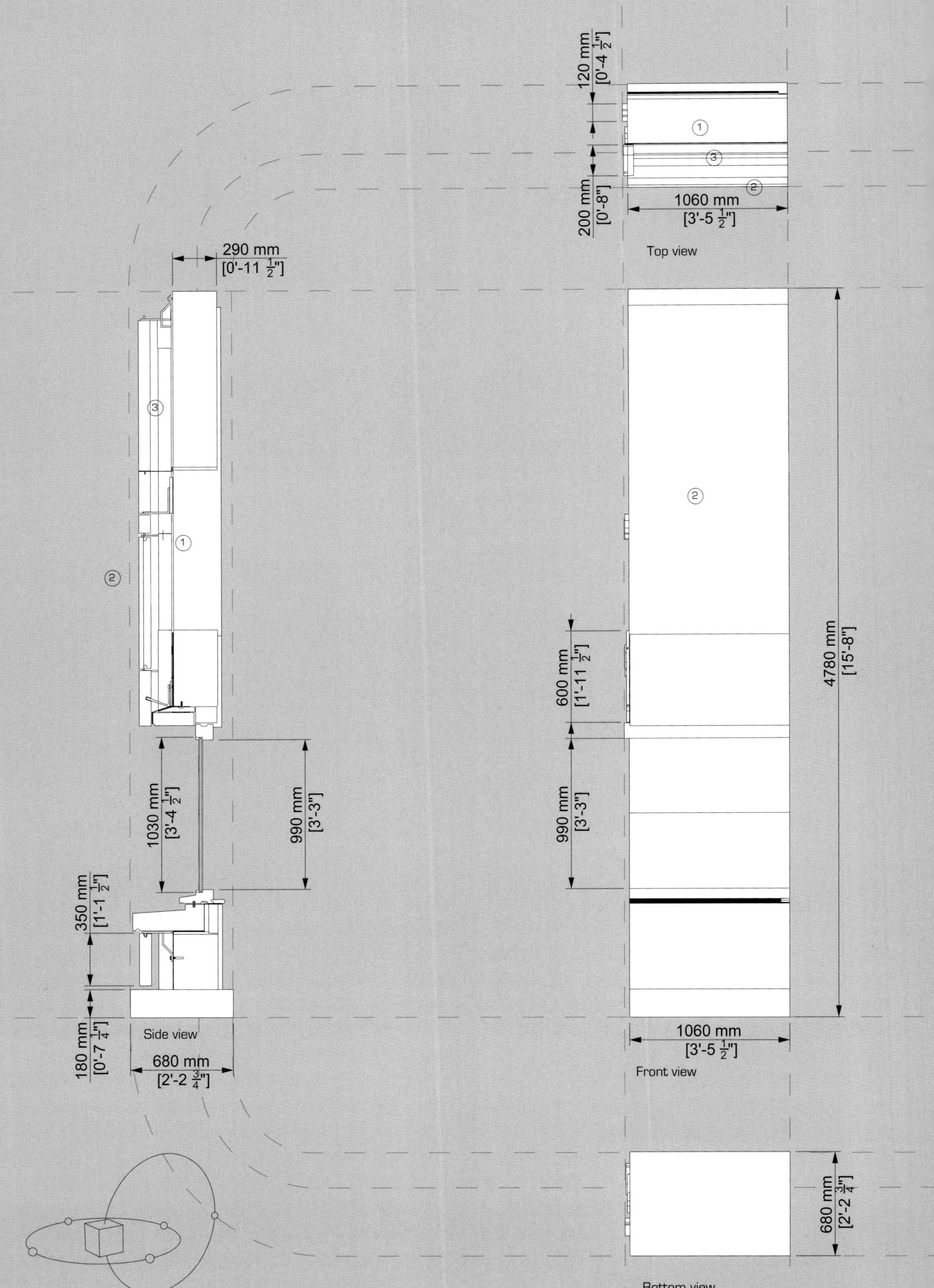

Third angle projection 1:30

Details
1 | Backing wall
2 | Precast concrete panel
3 | Closed cell thermal insulation
4 | Vertical joint open or closed type
5 | Horizontal joint (typically a lap) open or closed type
6 | Internal finish
7 | Window frame

3D view of the assembly

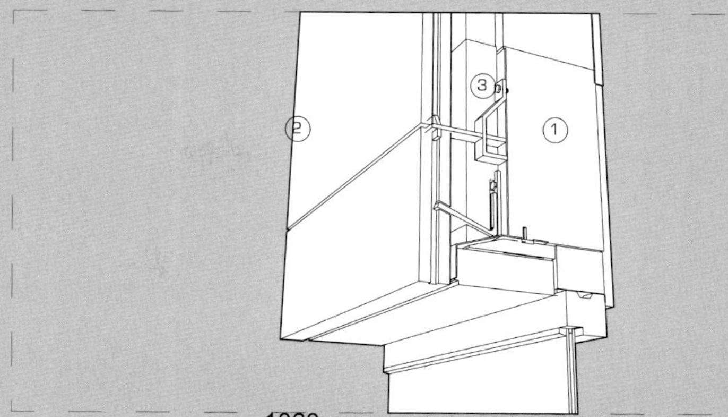

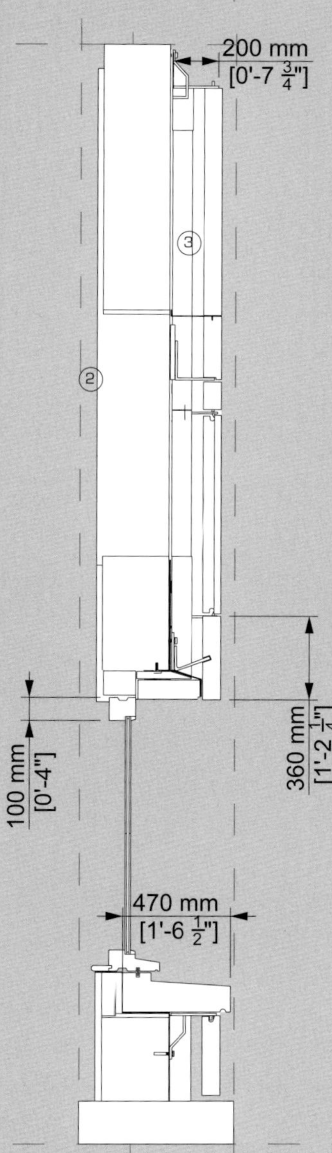

Side view

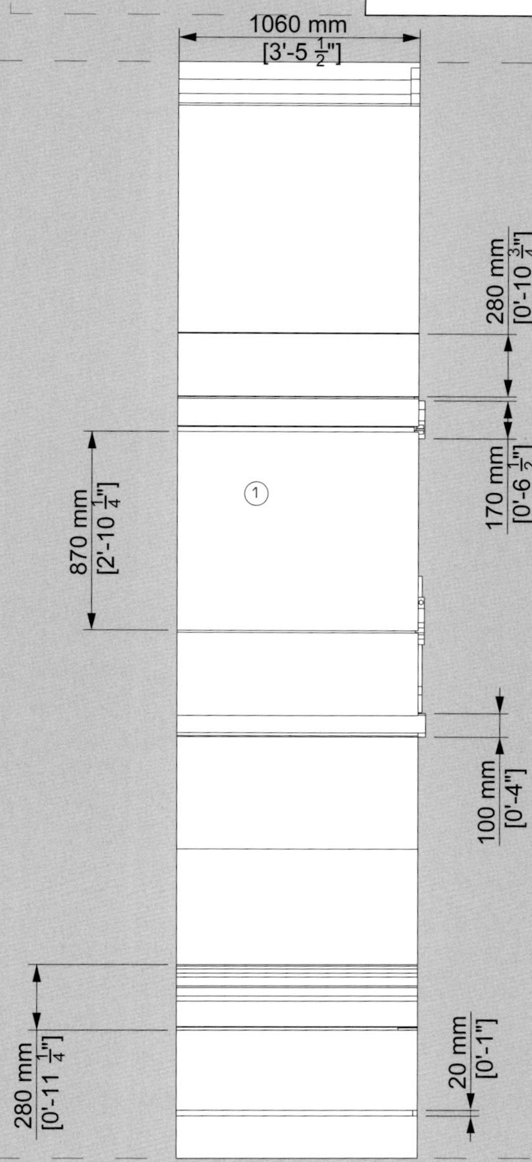

Back view

8 | Reinforced concrete column
9 | Metal corner trim
10 | Metal parapet coping
11 | Concrete floor deck
12 | Adjacent panel in different material
13 | Waterproof membrane
14 | Precast concrete coping

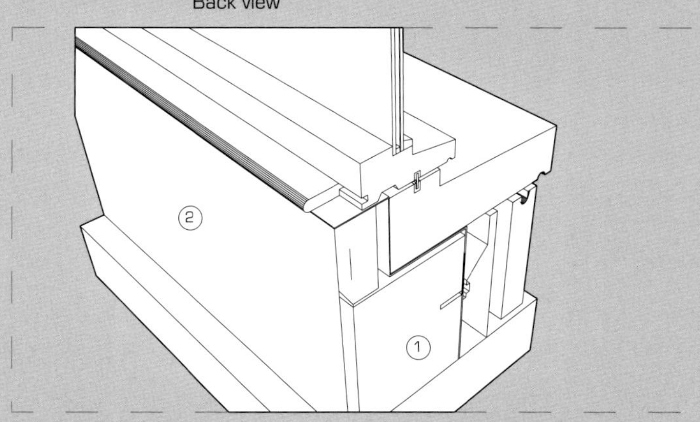

3D view of the assembly

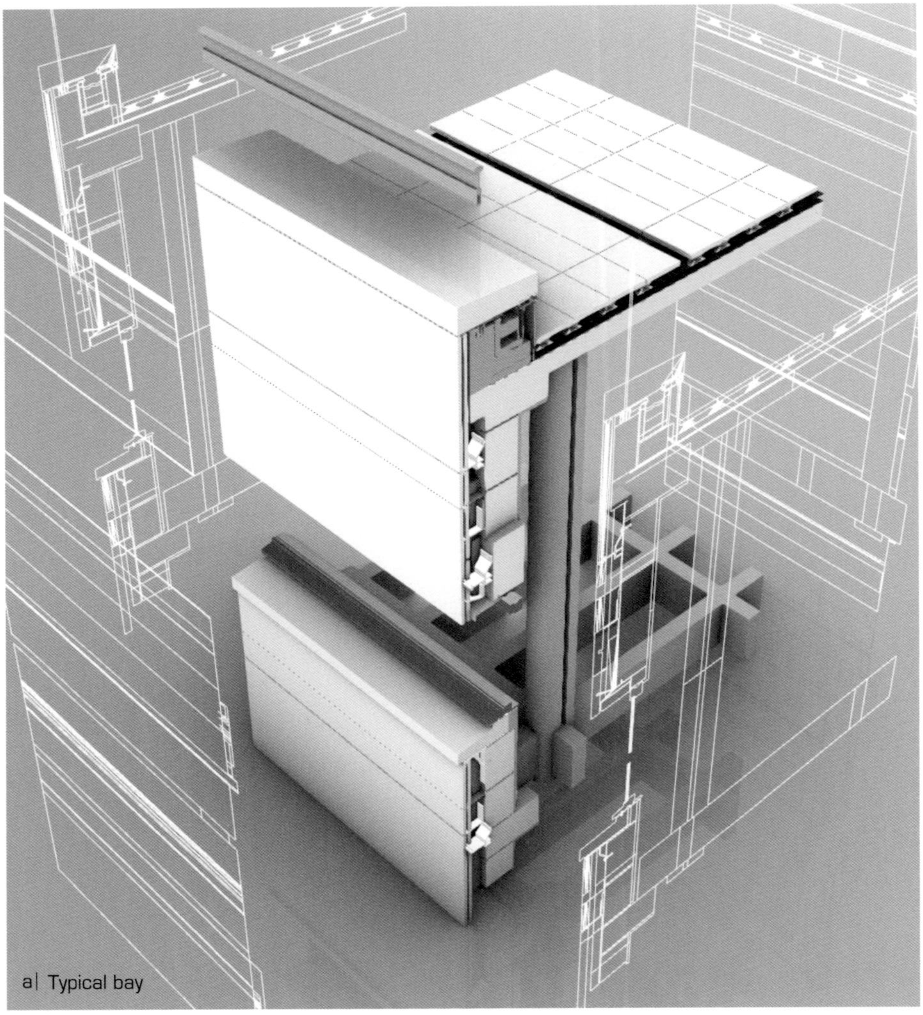

a| Typical bay

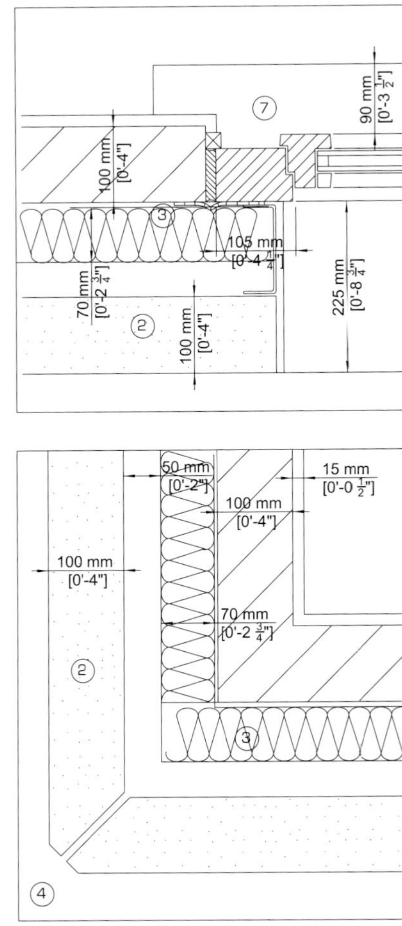

b | Horizontal sections. Panel to panel junction and window jamb.

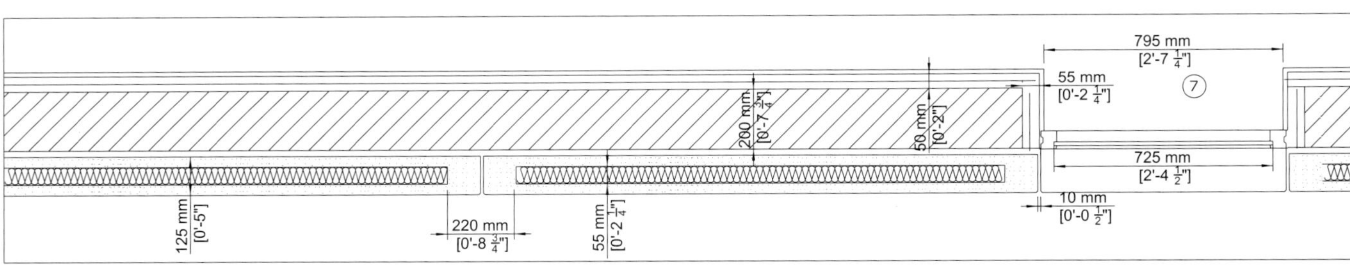

c | Horizontal section.

Smaller precast panels than those used in storey height sizes have been undergoing considerable development over the past 20 years. They are made as either individually supported panels or as self-supporting stacked panels.

Individually supported panels are fixed directly to a background supporting wall, usually formed either in reinforced concrete or in concrete block. Panels are usually open-jointed to allow them to dry easily after wetting in rain and to allow rainwater to drain easily down the backs of the panels or down the face of the backing wall (d). Spandrel panels are usually part of a proprietary precast concrete system of frame, floor slab and spandrel panel, with ribbon windows added as separate elements or full-height glazing in areas where sprinklers are not required. Self-supporting panels are a more recent development and comprise precast panels which are stacked together in the manner of metal composite panels, with a tongue-and-groove connection between panels in the horizontal joints (MCE_224, b). These panels have the advantage of having internal insulation together with external and internal faces that require no further treatment.

Individually supported panels

This system allows narrower joints to be formed between panels than those possible in full-height panels, and allow a varied, non-rectangular layout to be used with a joint pattern independent of the backing wall (MCE_225, e). Individually supported panels use a fixing method which is similar to stone cladding panels, but with the possibility of much bigger panel sizes. Natural stone cladding is limited to sizes that can be cut from a stone block, which is usually around 2000mm x 2000mm x 2000mm, depending on stone type. This yields stones which are a maximum of around 1500mm x 750mm, or 1500mm x 1000mm depending on the strength of the stone. Precast concrete panels can be much larger, typically 1500mm x 3000mm and can be supported

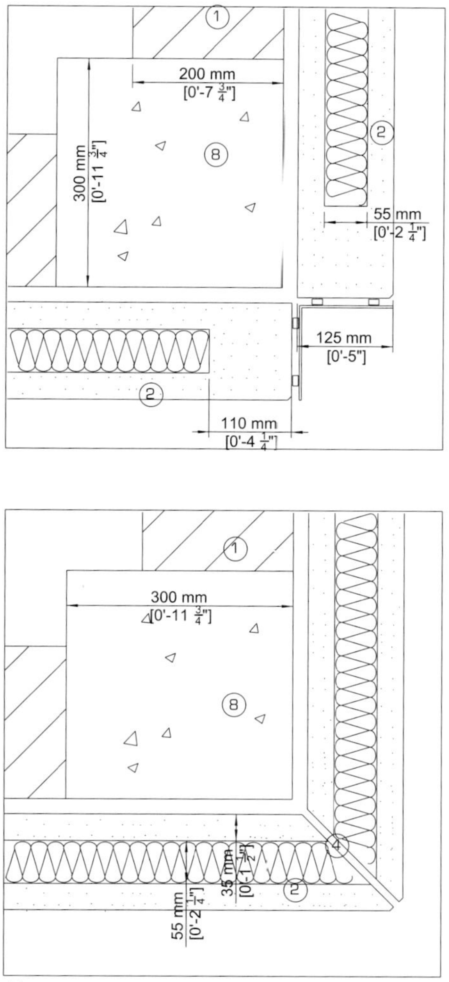

d | Horizontal sections. Panel to panel at corners

Details

1 | Backing wall
2 | Precast concrete panel
3 | Closed cell thermal insulation
4 | Vertical joint open or closed type
5 | Horizontal joint (typically a lap) open or closed type
6 | Internal finish
7 | Window frame
8 | Reinforced concrete column
9 | Metal corner trim
10 | Metal parapet coping
11 | Concrete floor deck
12 | Adjacent panel in different material
13 | Waterproof membrane
14 | Precast concrete coping

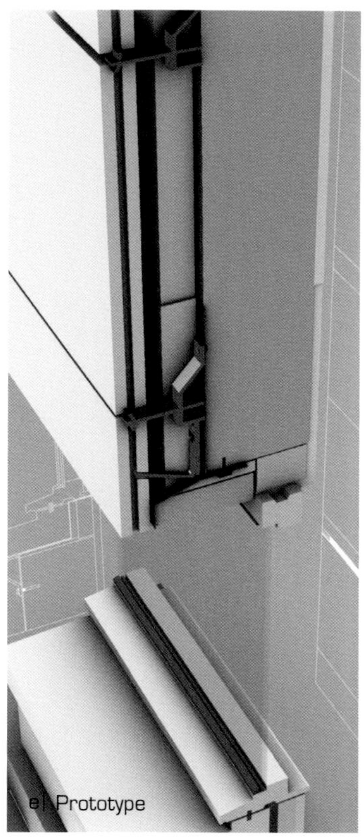

e | Prototype

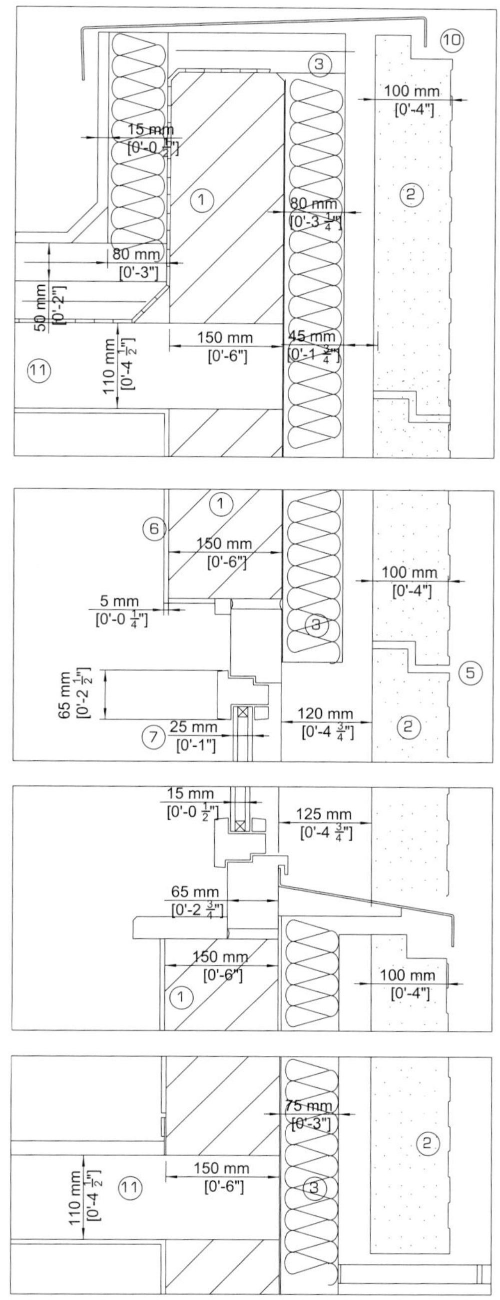

f | Vertical section. Junction at window jamb and at parapet.

on stainless steel fixings back to floor slabs and the primary structure (MCE_225, d). An advantage of precast concrete over natural stone in smaller panels is that corner units and non-rectangular shaped units can be formed more easily and economically than in stone, though the number of panel types is restricted for any project in order to keep the system economic. Small precast concrete rainscreen panels are used increasingly in apartment buildings, where large areas of opaque wall can be given a high degree of finish in the material while benefiting from the casting ability of the material.

Panels are most commonly fixed with stainless steel angles which are either cast-in or are bolted to the concrete. The angles are secured to brackets fixed to a backing wall. Slotted holes provide adjustment vertically, horizontally and laterally (MCE_226, a). These panels have the advantage of being able to have a rich surface texture as a result of the casting process in an individual mould. Slots, grooves and complex profiling can be incorporated in the manner of profiled metal cladding. Backing walls are usually waterproofed in the outer face. Closed cell thermal insulation is set on the outside of this to insulate the building structure. An alternative method is to use metal foil-faced insulation which is semi-rigid. The insulation is fixed directly to the blockwork backing wall, with an outer metal foil face that provides the full weather protection. Joints between strips of insulation are sealed with adhesive foil tape.

Like stone cladding, panels are usually supported on short lengths of stainless steel angle at each floor level, back to the floor slab. This overcomes the risk of progressive collapse of a cladding panel, where a failure in one panel would cause it to drop onto the panel below, causing further collapse down the facade. Panels at floor level are fixed directly to the slab, with fixings designed so that they can take the full load of the panels

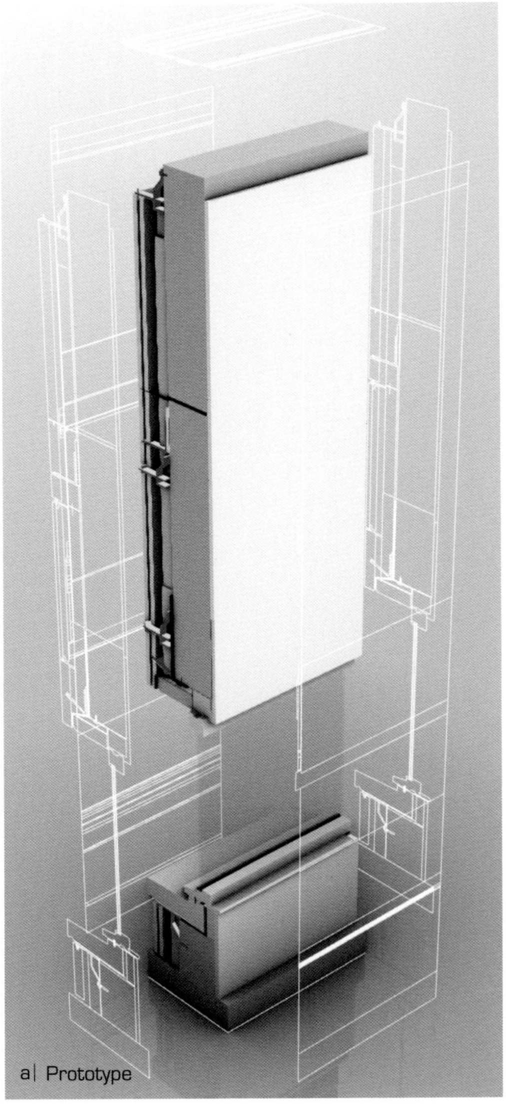

a| Prototype

b | Vertical section. Typical precast concrete wall constructions

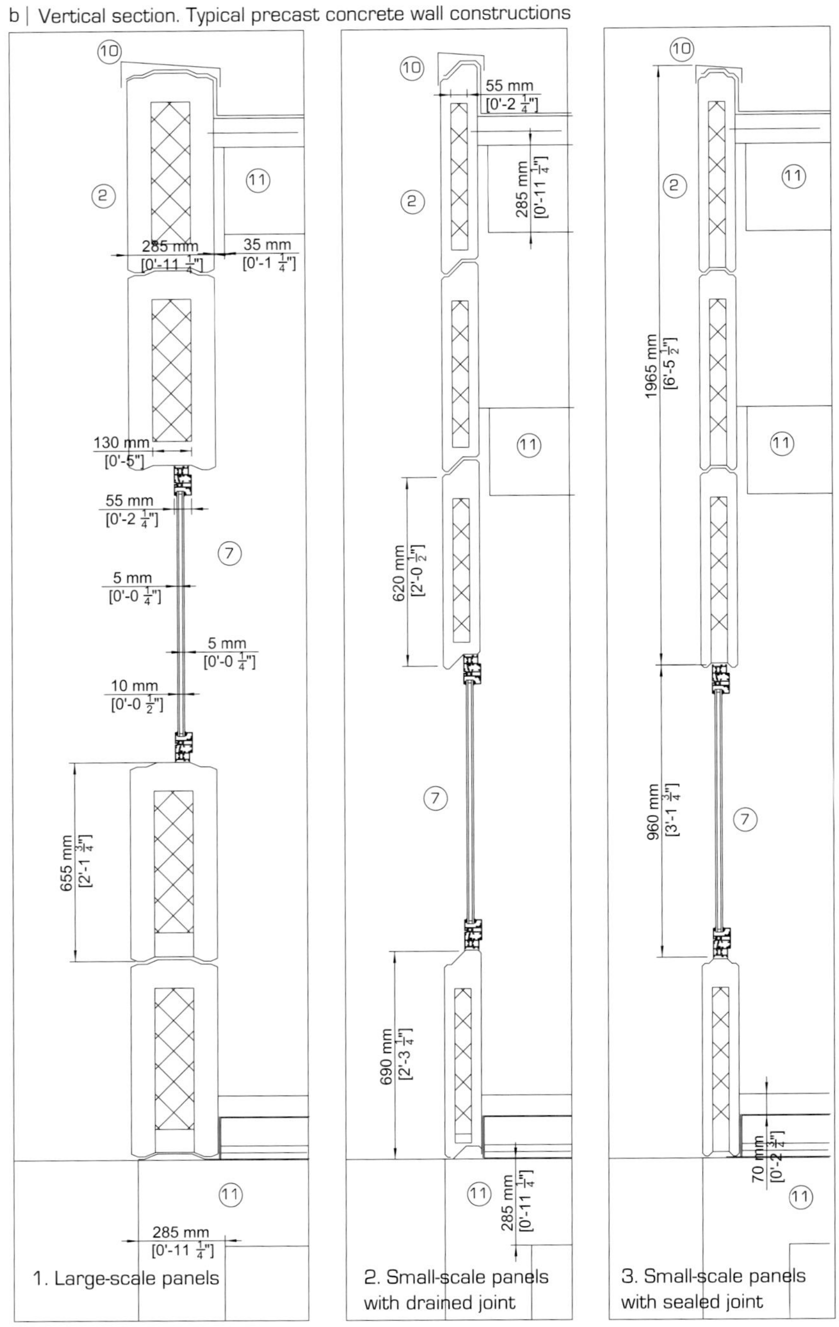

Details

1 | Backing wall
2 | Precast concrete panel
3 | Closed cell thermal insulation
4 | Vertical joint open or closed type
5 | Horizontal joint (typically a lap) open or closed type
6 | Internal finish

immediately above, in the event of fixings to panels above failing either partially or completely.

Self-supporting stacked panels

Stacked panels have the advantage of being supported at the base of the wall in single floor buildings up to around 10 metres, and at every floor level in buildings of more than one storey. When supported at their base, stacked panels are set on a concrete beam or strip foundation forming part of the primary structure (b). Panels are stacked with continuous vertical joints where they are restrained by columns, in either reinforced concrete or steel, but concrete is more common due to the greater ease of connecting the panels to the primary structure. The need to support panels at joints results in their being wide in order to maximise the distance between columns or support posts. As a result they are not very tall in order to reduce their weight to that which can be lifted by a modest sized crane, typically with a lifting capacity of 4 to 6 tonnes.

The thermal insulation in panels does not create a complete thermal break from outside to inside, since the concrete block has concrete faces on its top and bottom edges in order to serve as a method of creating a tongue and groove joint in the horizontal joints (b). If the concrete panel were formed as a sandwich panel then mechanical fixings, such as stainless steel cramps, would be needed both to hold the two sides of the concrete panel together and to connect them together as they are stacked vertically. This would create a diaphragm wall rather than a monolithic loadbearing wall, making it much more expensive as a result of all the mechanical fixings needed. However, the thermal bridge across the vertical and horizontal joints in the panel can result in pattern staining as a result of heat transmission across the panel. This is particularly noticeable on the internal face of the panel.

Stacked panels fabricated in thicknesses of 75mm to 100mm can form part of a complete cavity wall construction of 200mm–300mm

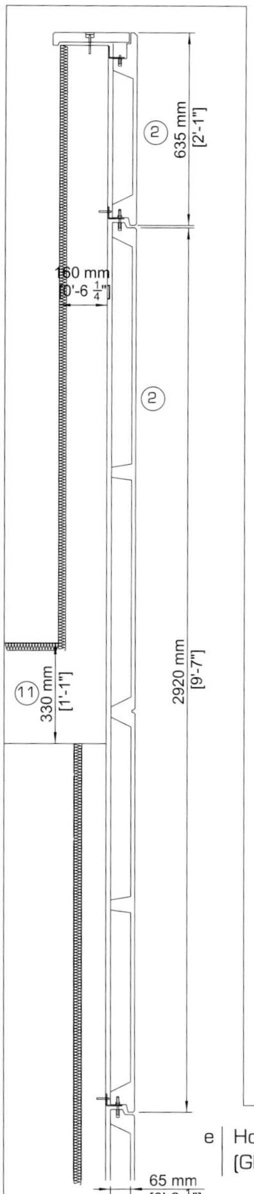

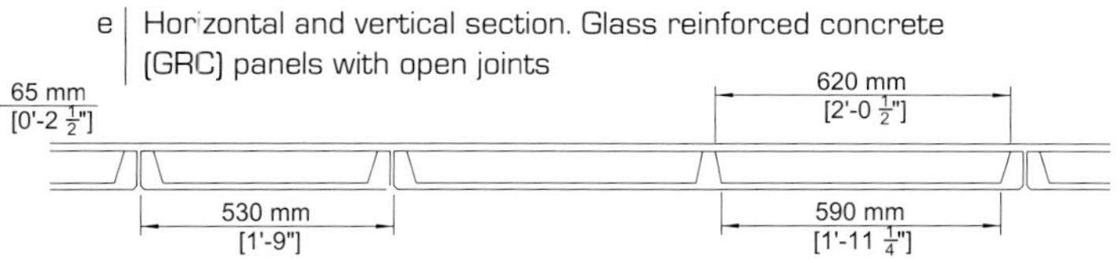

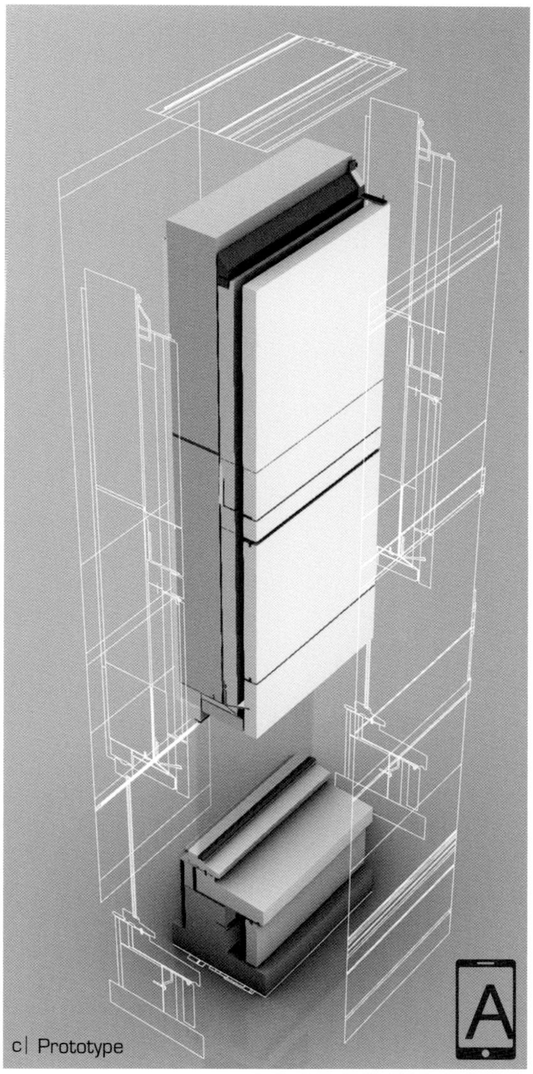

c | Prototype

e | Horizontal and vertical section. Glass reinforced concrete (GRC) panels with open joints

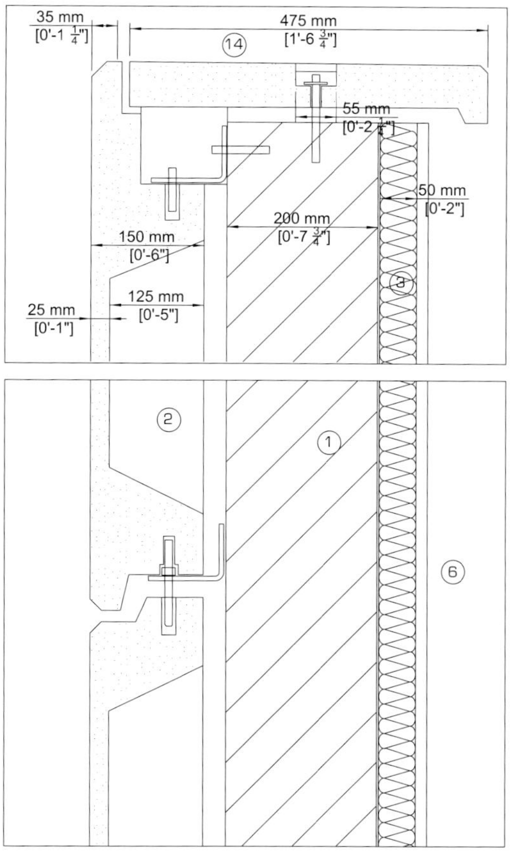

d | Vertical section. Parapet and panel to panel joint. Glass reinforced concrete (GRC) panels with open joints.

7 | Window frame
8 | Reinforced concrete column
9 | Metal corner trim
10 | Metal parapet coping
11 | Concrete floor deck
12 | Adjacent panel in different material
13 | Waterproof membrane
14 | Precast concrete coping

thickness when used as a self-supporting outer skin. The inner wall can vary widely in its construction from concrete block to light gauge metal stud wall with a waterproofed outer facing. The cavity between the outer concrete panel wall and the inner backing wall has a ventilated cavity of 50mm–75mm width. The cavity is ventilated at the top and bottom of the wall.

Parapets and sills

The detailing of parapets and sills for both individually supported panels and for stacked panels follows the same principles. At parapet level, self-supporting concrete panels have the advantage of not needing an additional inner parapet wall but require a pressed aluminium coping over the top of the wall to both keep rain from penetrating the top panel (e), where it is most vulnerable, as well as to throw rainwater off the coping inwards to an inner gutter at the edge of the roof. The visible vertical face of the coping is usually kept to a minimum in order to reduce its visual impact on the facade. A precast coping is also used instead of aluminium, but this is essentially for visual reasons. The advantage of an aluminium coping is that it can be pressed to fold over the inside face of the parapet wall, which protects the roof membrane from the effects of sunlight. It also provides an overcloak to the roofing membrane which wraps up the inside face of the parapet wall and across the top of the precast panel, where it forms a complete weathertight seal to the top of the wall. A coping made from precast concrete is less easy to handle, but provides effective protection to the waterproof membrane beneath.

The base of walls in precast panels are detailed in a similar way to other forms of loadbearing masonry. A continuous damp proof membrane (DPM) extends up from the outside of the structure beneath and forms a damp proof course (DPC) which is positioned so that it extends at least 150mm above external ground level. The DPC is positioned so

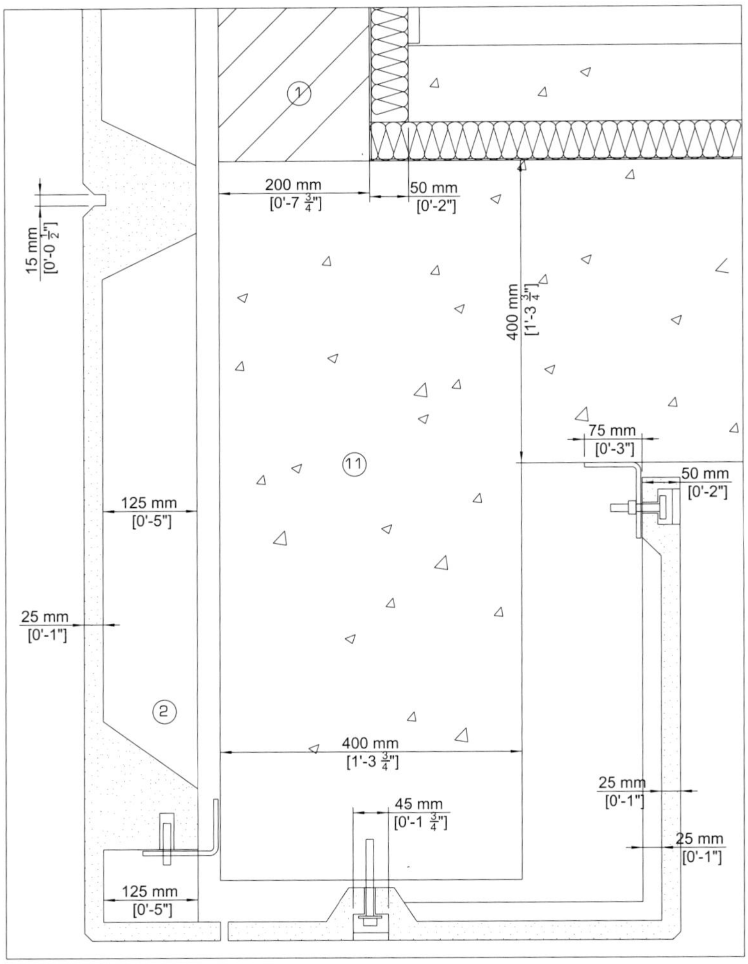

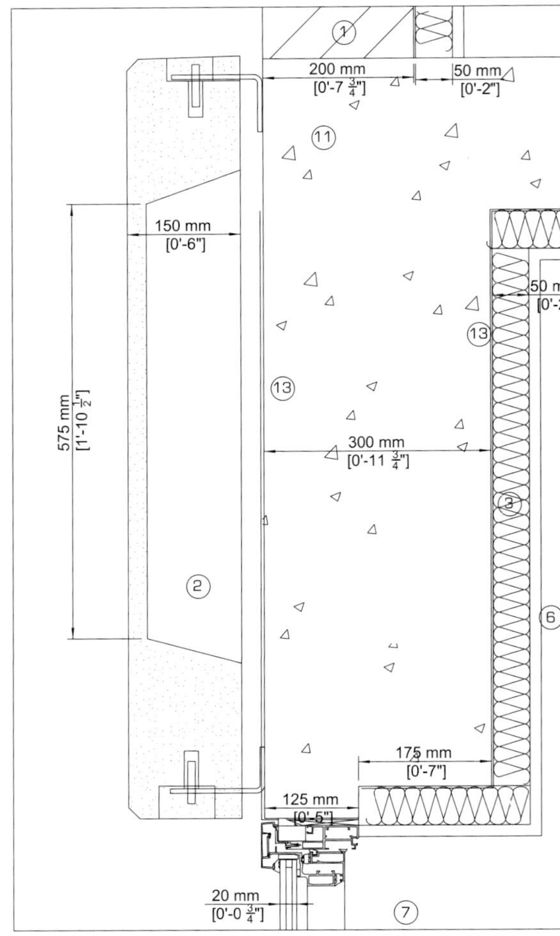

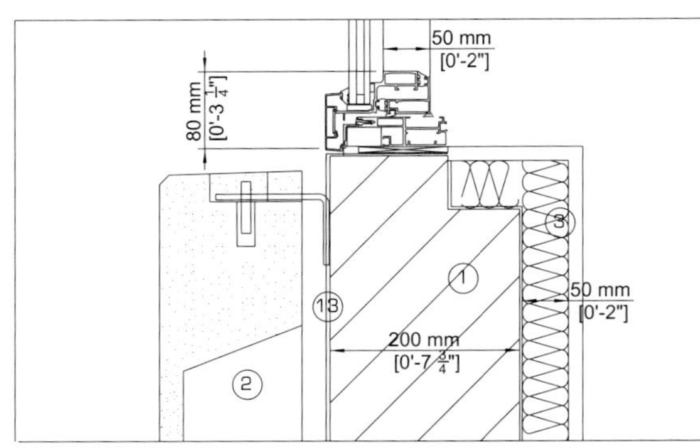

a | Vertical sections. Panel soffit and window junctions. Glass reinforced concrete (GRC) panels with open joints.

Details

1 | Backing wall
2 | Precast concrete panel
3 | Closed cell thermal insulation
4 | Vertical joint open or closed type
5 | Horizontal joint (typically a lap) open or closed type
6 | Internal finish

that the lowest row of panels at ground level appears to sit on the ground with its bottom course level with the adjacent ground or pavement. This avoids the need for a 150mm–200mm plinth around the base of the wall which is popular in loadbearing brick construction in order to avoid rainwater that is splashed off the ground soaking into the base of the wall and causing damp penetration on the inside of the wall at floor level.

Openings

The characteristic tongue-and-groove profile of horizontal joints makes it necessary to provide a complete edging to the openings formed in precast panels. Special precast panels for openings are not usually made, since this adds considerable cost to the system. The most common method of forming an edge to a window or door opening is to use metal sheet on all sides of the reveal to the opening. This usually provides a thin edge as well as the ability to be continuous with a window or door frame, and provides an integral sill to the opening (MCE_222, c).

Sand-blasted finish and tooled finish

In addition to the finishes described in the texts on both in-situ concrete and precast concrete panels, the techniques of sand blasting and tooling are commonly used on small cladding panels.

The sand blasting of concrete panels is usually done with iron filings rather than sand particles. Concrete is blasted with differing amounts of abrasive iron filings at high pressure, depending on the hardness of the concrete wall panel. This slightly erodes the face of the concrete, either back to the surface sand grains or more deeply to the coarse aggregate beneath. The essential visual characteristic of this method is that it abrades all aggregates and dulls them as a result, producing a characteristic matt surface. Sand blasting abrades both in propor-

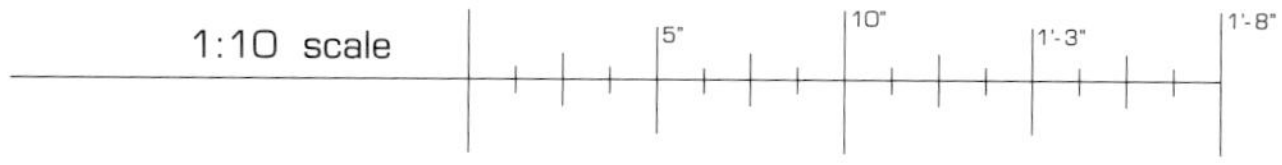

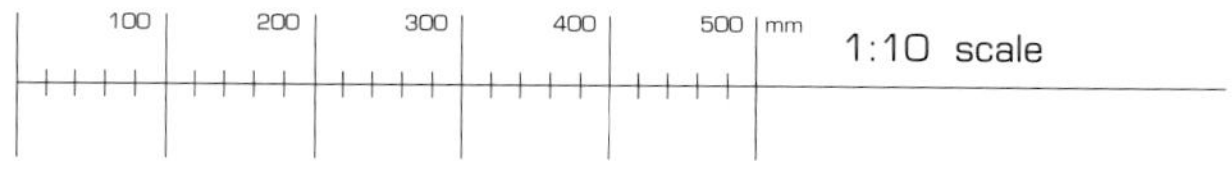

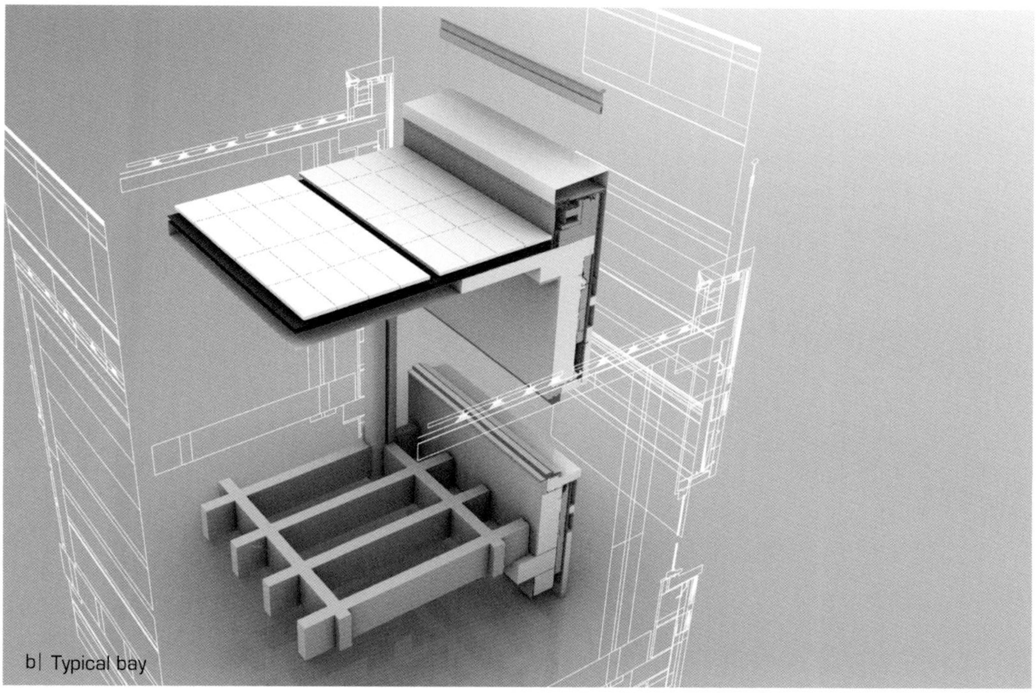
b| Typical bay

7 | Window frame
8 | Reinforced concrete column
9 | Metal corner trim
10 | Metal parapet coping
11 | Concrete floor deck
12 | Adjacent panel in different material
13 | Waterproof membrane
14 | Precast concrete coping

c | Horizontal sections. Glass reinforced concrete (GRC) panels with open joints

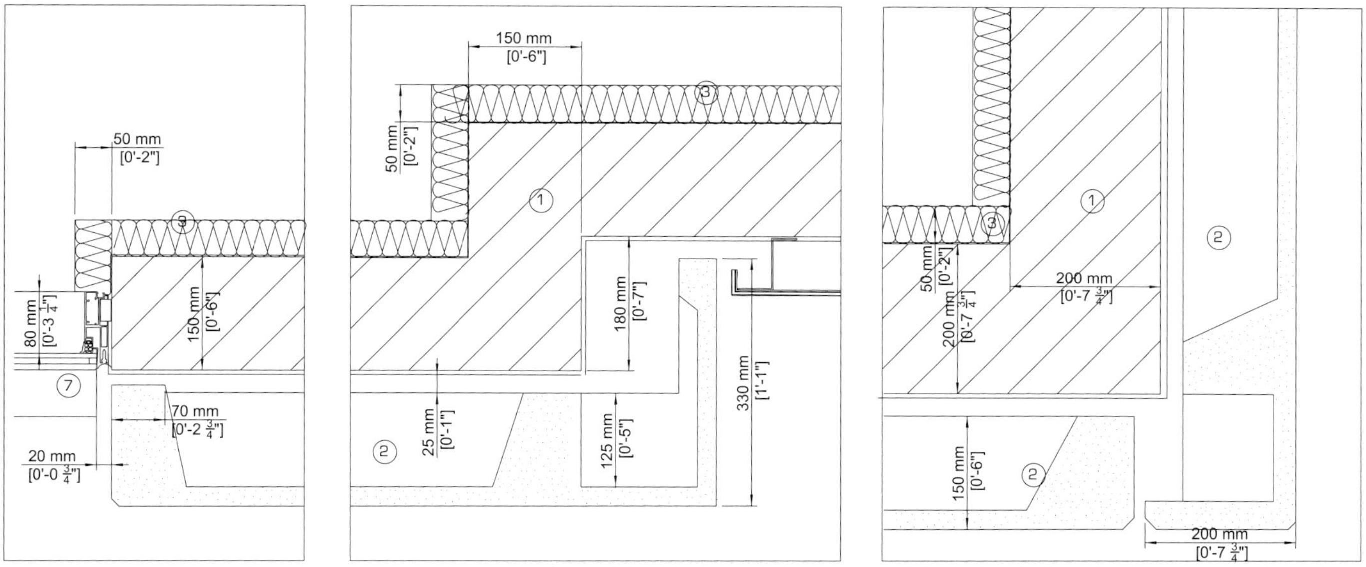

tion to the hardness of the concrete surface and to the proportion of coarse aggregate in the cement mix. This leads to different surface textures, depending on the type of small aggregate and the quality and degree of hydration of the cement mix. Sand blasting allows the finish to be obtained gradually and controlled visually as work progresses, without the need for rinsing which slows down the process. Acid etching also allows very localised working with the use of stencils.

Tooled concrete is a method of texturing concrete by either striking the material with chisel or pick, or by rotation, using diamond-tipped burring tools to make grooves of any shape, depth and distance apart. A claw chisel has a toothed flat head that breaks the surface, while a pick is a chisel with a sharp point that can be used to create a roughened surface. A boasting chisel is a grooving chisel used to make a set of parallel grooves in the face of the concrete. These textures usually stop short of corners of walls to give a crisp edge.

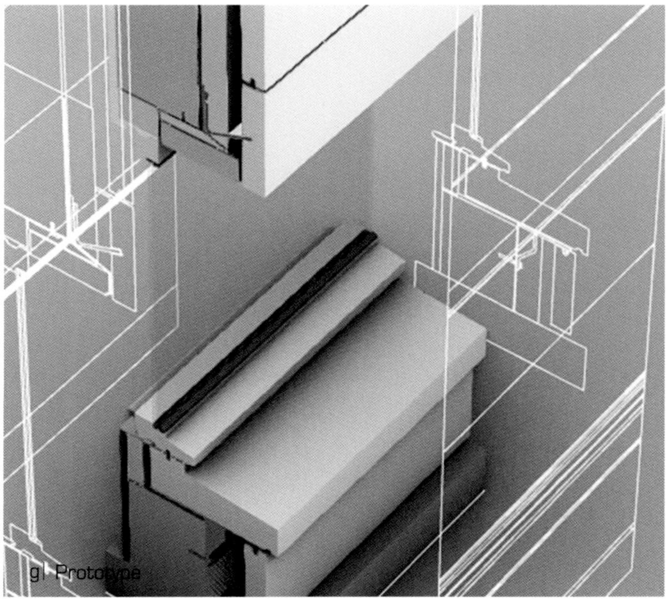
g| Prototype

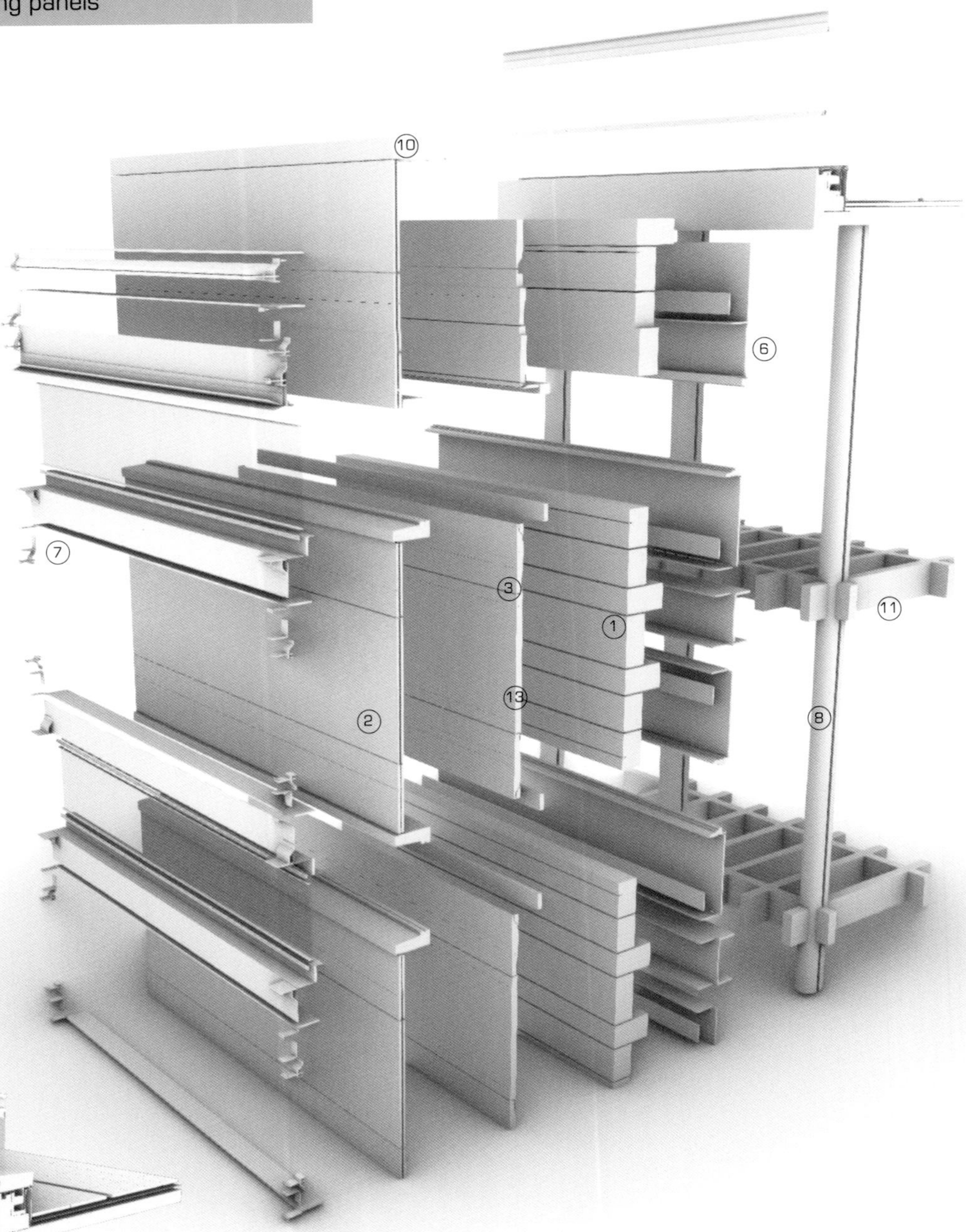

a | 3D exploded view showing concrete small precast panel facade fixed to concrete frame

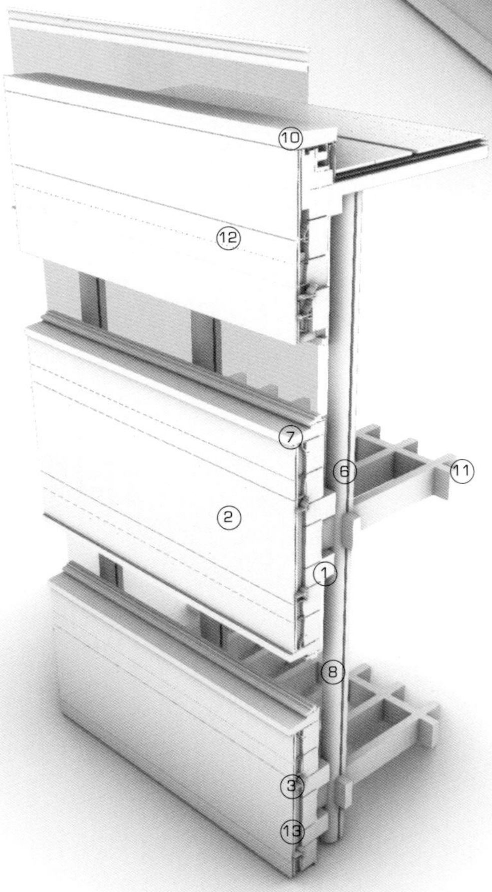

Details

1 | Backing wall
2 | Precast concrete panel
3 | Closed cell thermal insulation
4 | Vertical joint open or closed type
5 | Horizontal joint (typically a lap) open or closed type
6 | Internal finish
7 | Window frame

b | 3D view showing concrete small precast panel facade fixed to concrete frame

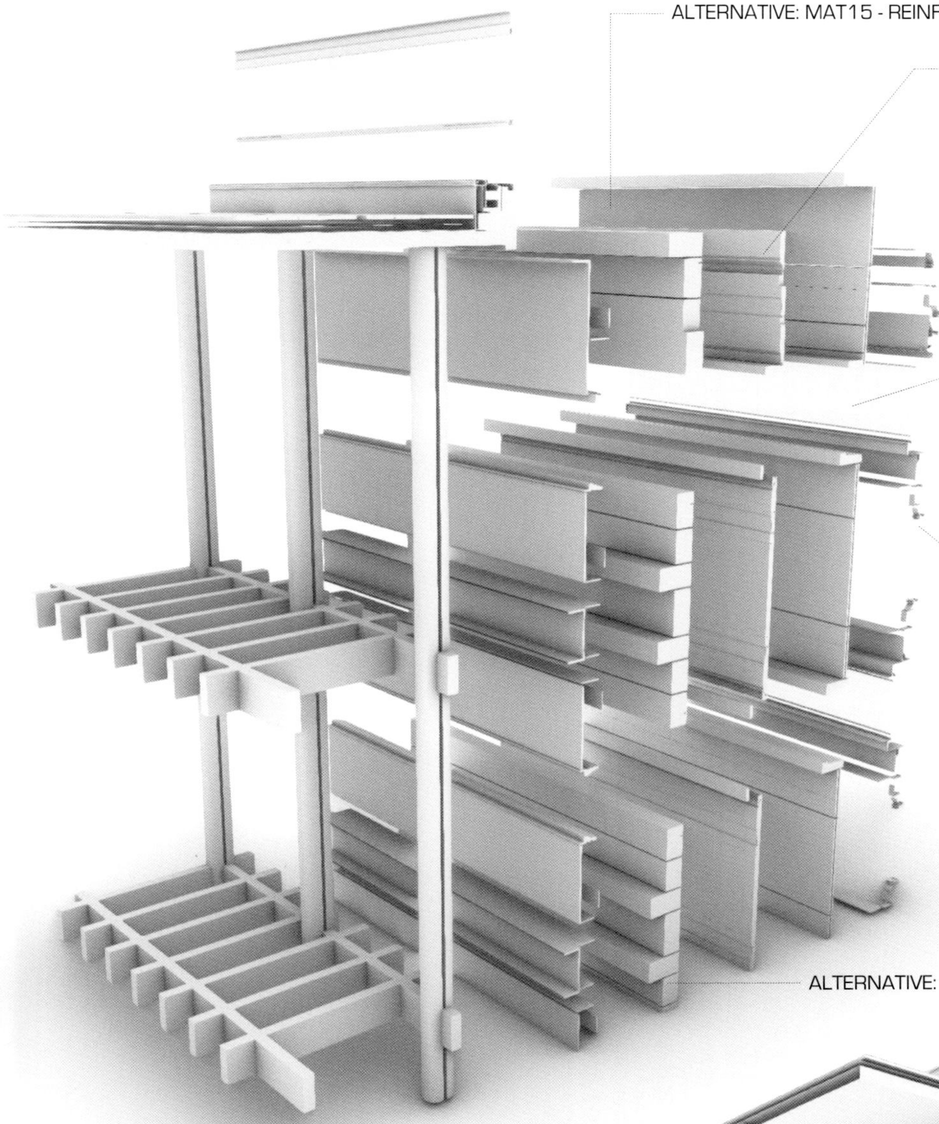

c | 3D exploded view of concrete small precast panel facade componewith material specification references

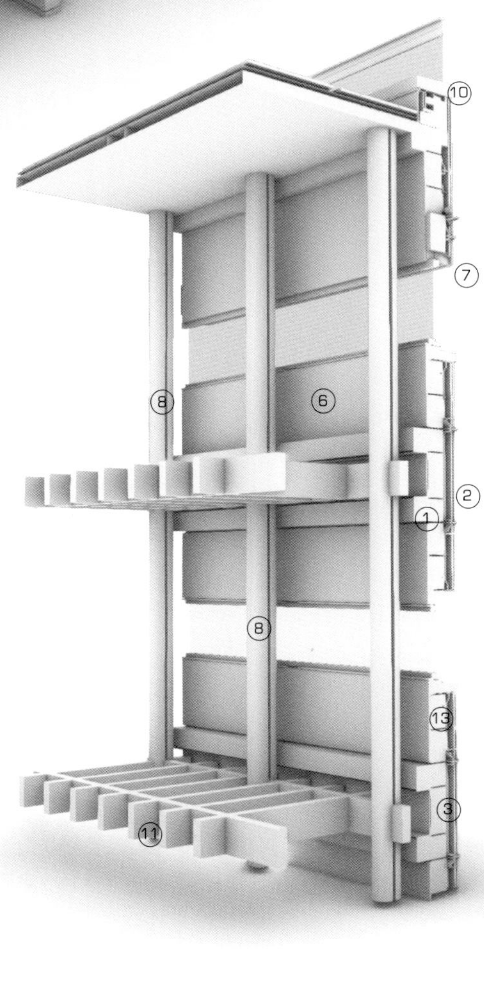

8 | Reinforced concrete column
9 | Metal corner trim
10 | Metal parapet coping
11 | Concrete floor deck
12 | Adjacent panel in different material
13 | Waterproof membrane
14 | Precast concrete coping

d | 3D view showing concrete small precast panel facade fixed to concrete frame

Example project of masonry facades
Federation Square, Melbourne, Australia
Facade Consultant to Atelier One: Andrew Watts of Newtecnic
Architect: Lab Architecture Studio

MASONRY WALLS

18 **Masonry cavity walls**
- Brick
- Ground level
- Window and door openings
- Eaves and parapets

19 **Masonry cavity walls**
- Concrete block
- Wall structures
- Ground level
- Opening in walls
- Eaves and parapets

20 **Stone cladding**
- Fixings
- Cladding to precast concrete
- Panels
- Closed joints
- Movement joints
- Stone finishes

21 **Terracotta rainscreens**
- Manufacture of panels
- Corner pieces
- Fixing systems
- Panel sizes
- Openings

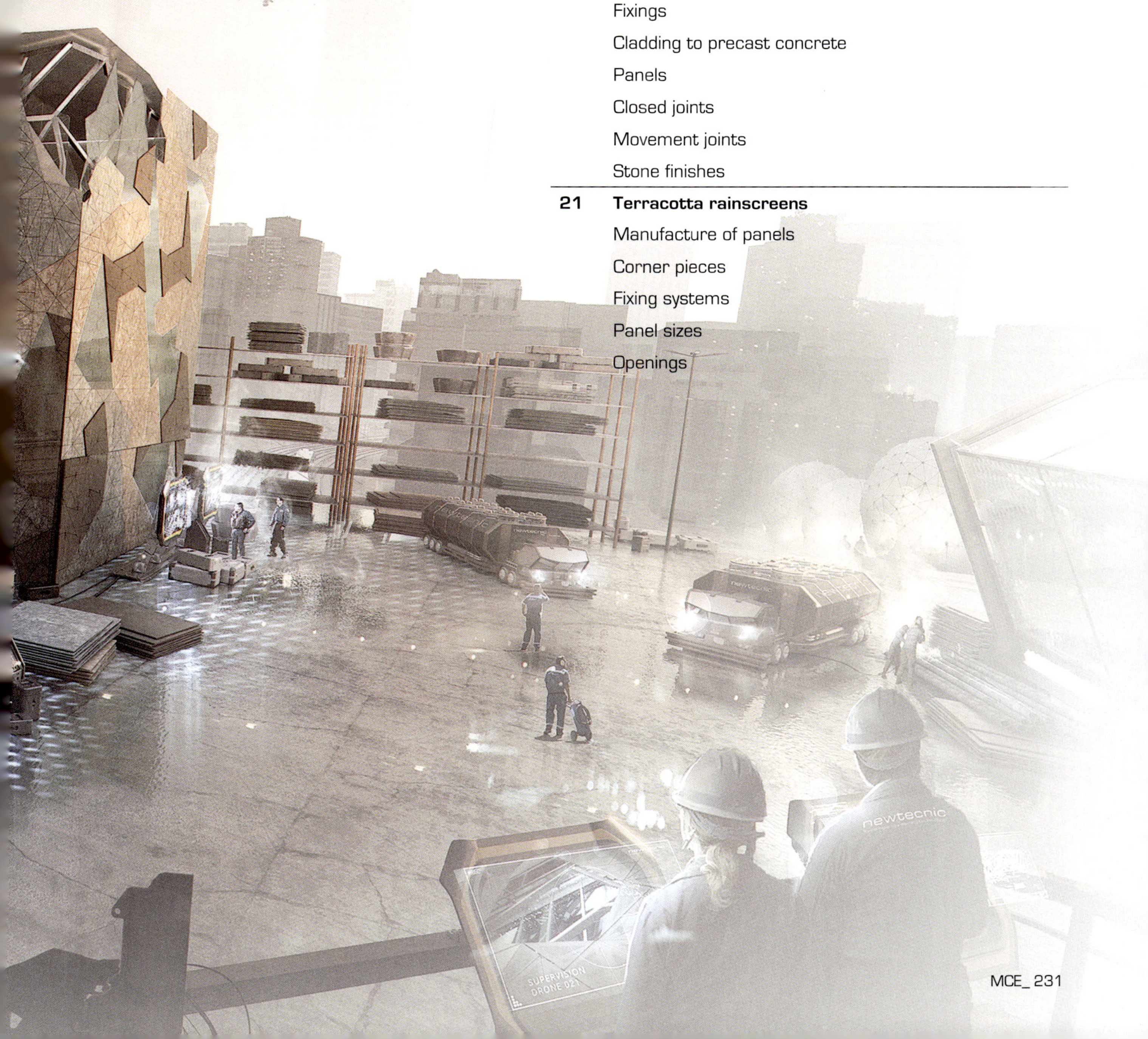

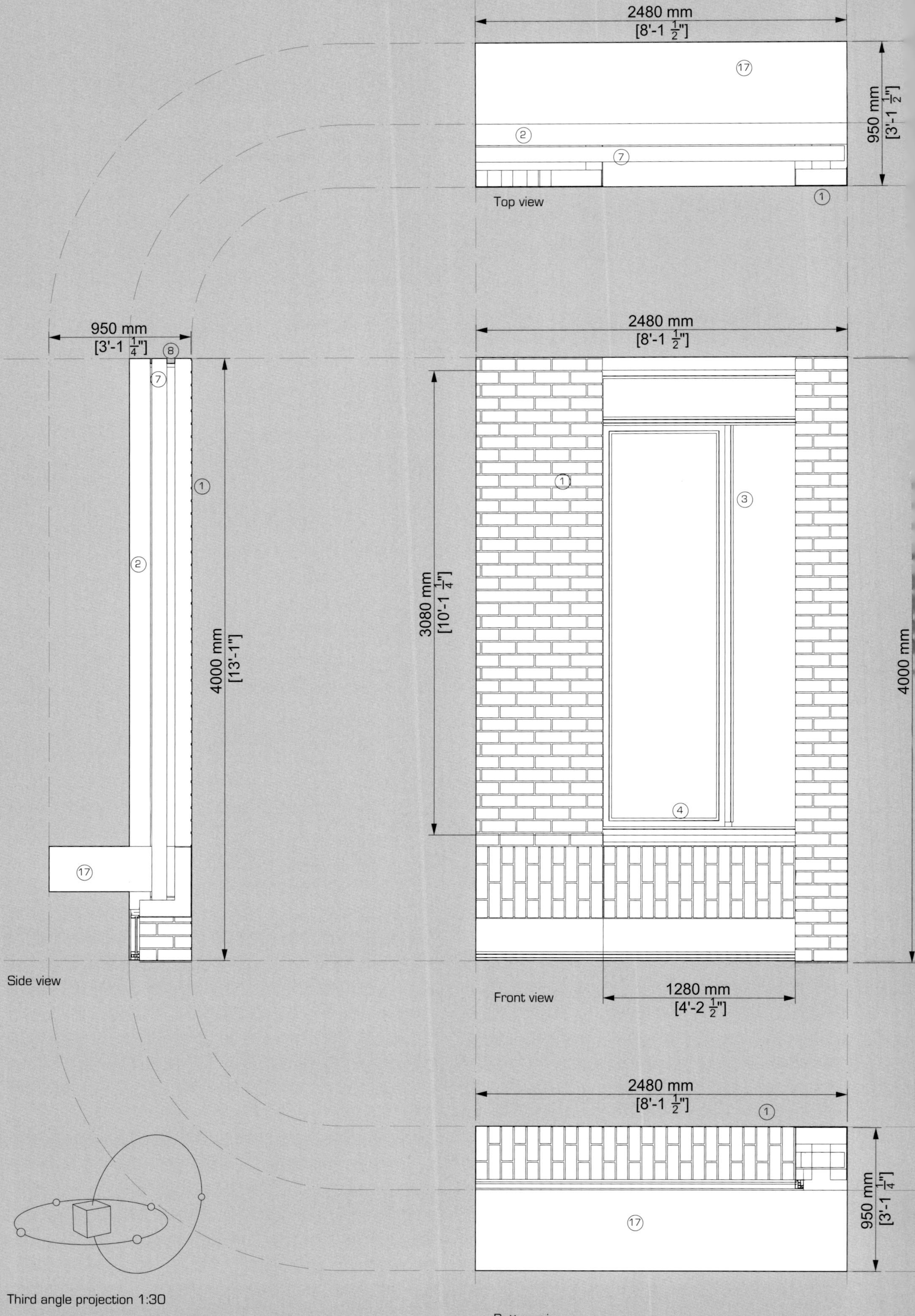

Third angle projection 1:30

Details

1 | Outer brick skin
2 | Inner blockwork or brick skin
3 | Timber framed window
4 | Timber sill
5 | Cavity closer
6 | Internal plaster finish or dry lining/drywall
7 | Thermal insulation in cavity
8 | Air cavity (sometimes omitted where insulation fills cavity)

3D view of the assembly

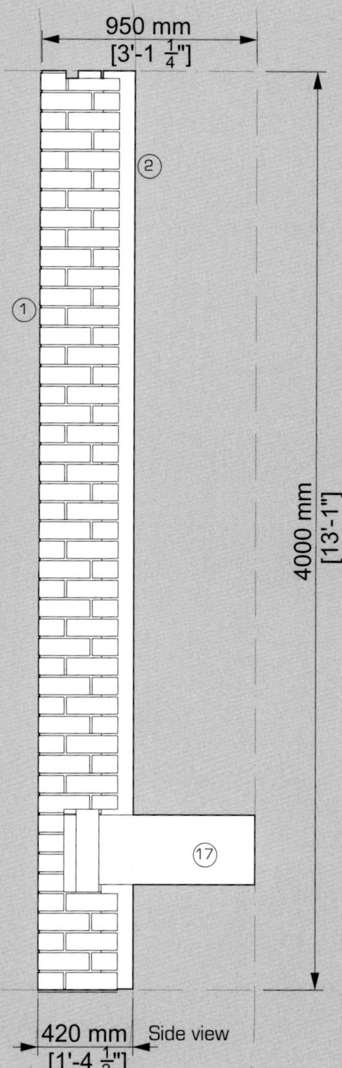

Side view

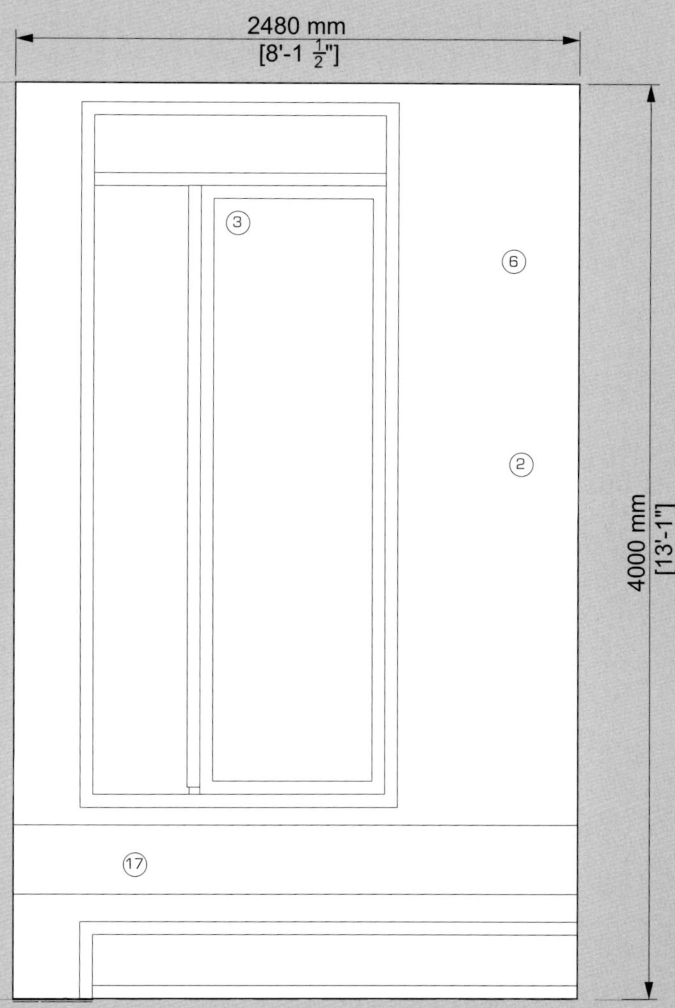

Back view

9 | Inner concrete lintel or beam
10 | Pressed steel lintel
11 | Steel angle
12 | Metal coping
13 | Precast concrete sill
14 | Damp proof course (DPC)
15 | Brick sill
16 | Movement joint
17 | Ground floor slab
18 | Steel cavity wall tie
19 | Foundation

3D view of the assembly

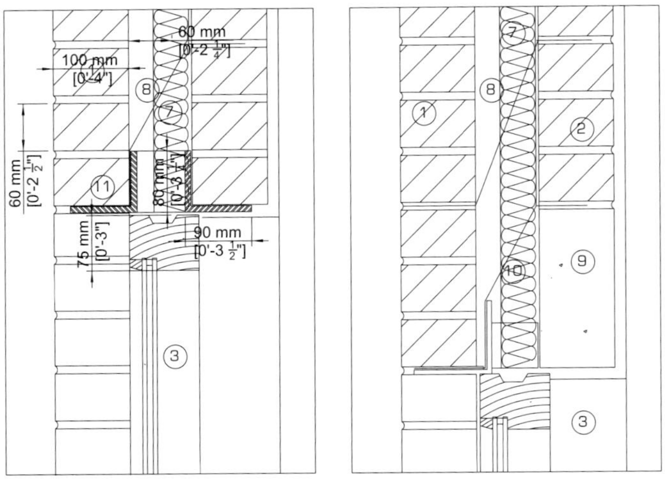

a | Vertical section. Typical connection or restraint to primary structure

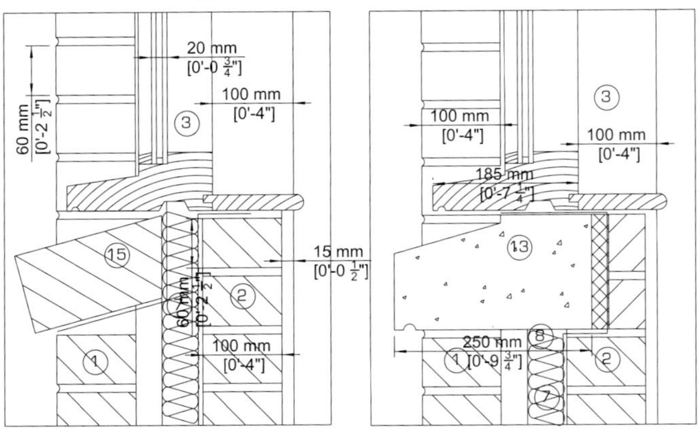

c | Vertical section. Typical connection or restraint to primary structure

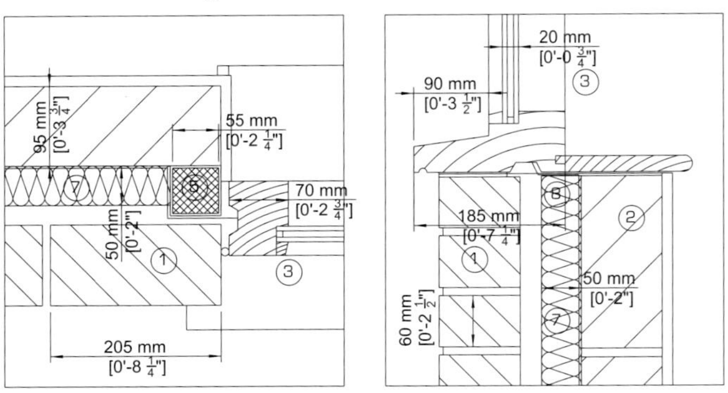

d | Horizontal and vertical section. Typical window at sill and head

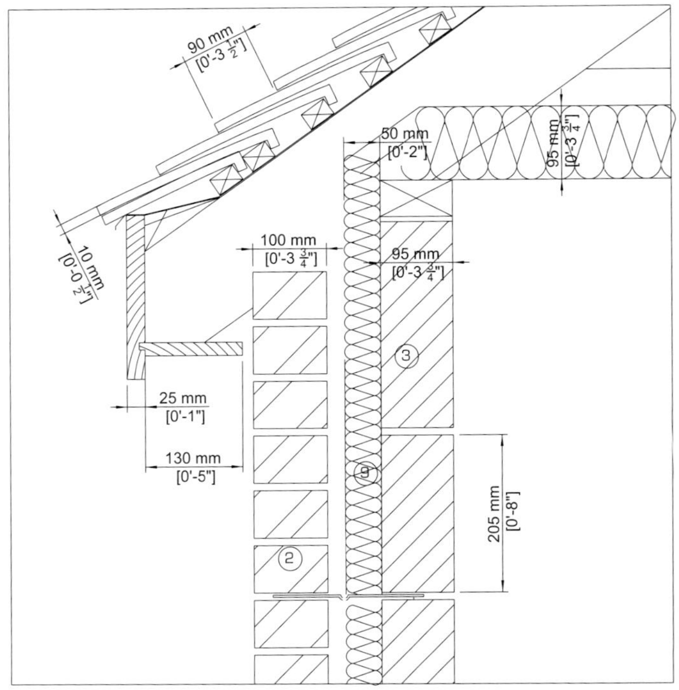

b | Vertical section. Typical eaves to pitched roof

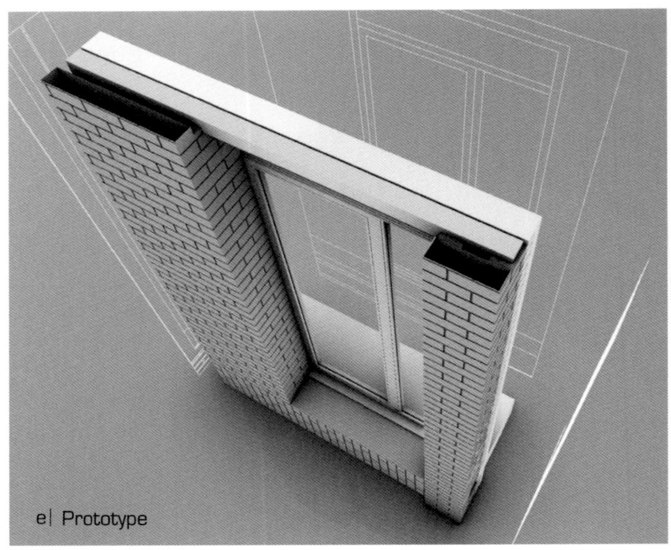

e | Prototype

Brick cavity walls have the advantage of providing a cavity within the depth of a wall for drainage of rainwater before it is allowed to be absorbed too far into the wall construction. Whereas loadbearing brick walls use the overall wall thickness to stop the passage of rainwater from outside to inside, cavity walls use two skins of brickwork separated by a ventilated air gap (MCE_235, j). The inner skin is increasingly formed in concrete block, hollow terracotta block or timber studwork. Thermal insulation is usually set on the external face of the inner skin in order to keep the building structure insulated.

The outer brick skin is usually only 100mm/4in thick (one brick width) with both skins being supported either at ground level or at intermediary floor levels (MCE_235, j). In the case of buildings up to two storeys, the outer skin is built up from the foundation and is restrained back to the floor slab at first floor level and at roof level. The inner skin is built off each floor slab and is restrained at the head by the floor slab or roof structure above (MCE_240, b). Taller buildings also have the inner skin built off floor slabs in the same way, but the outer skin is supported at each floor level on a continuous stainless steel angle, bolted back to the floor slab. A damp proof course (DPC) is set on top of the steel angle to drain the cavity. Slots called weep holes are formed in the vertical joints immediately above the steel angle in order to allow water in the cavity to drain out. In addition to being restrained at floor slabs, the outer skin is also tied at intermediary points back to the inner skin with stainless steel ties (MCE_239, b). These are set typically at 450mm/18in horizontal centres and 900mm/3ft vertical centres.

Historically, inner and outer skins were tied together with floor slabs to form diaphragm walls, but their use has been reduced with concerns over the thermal bridge created by linking the two skins of brick. Current practice uses the outer skin as cladding to a drained and ventilated void behind, with an inner skin that is waterproofed with a high level

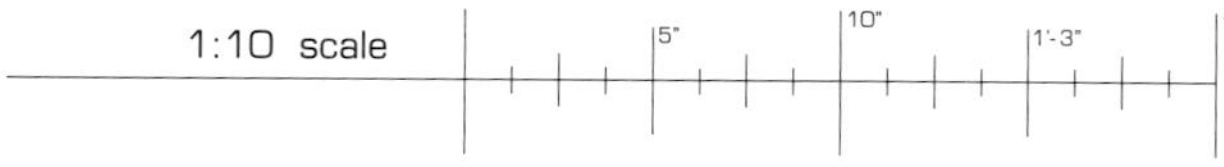

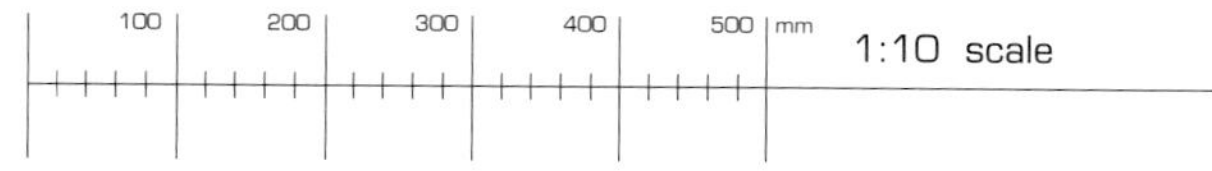

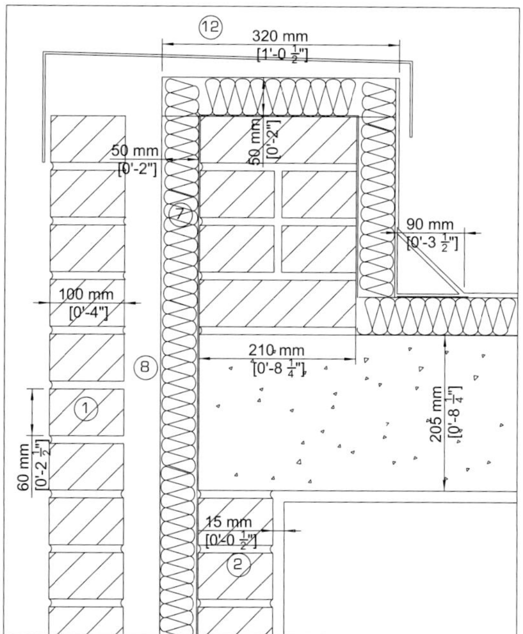

f | Vertical section. Junction at parapet, support of outer leaf at floor slab

g | Prototype

9 | Inner concrete lintel or beam
10 | Pressed steel lintel
11 | Steel angle
12 | Metal coping
13 | Precast concrete sill
14 | Damp proof course (DPC)
15 | Brick sill
16 | Movement joint

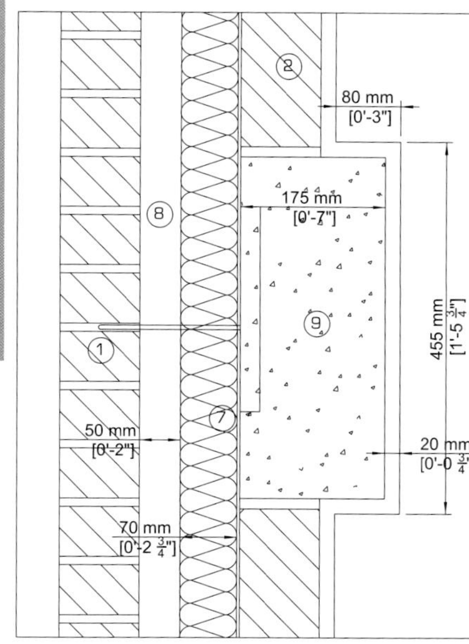

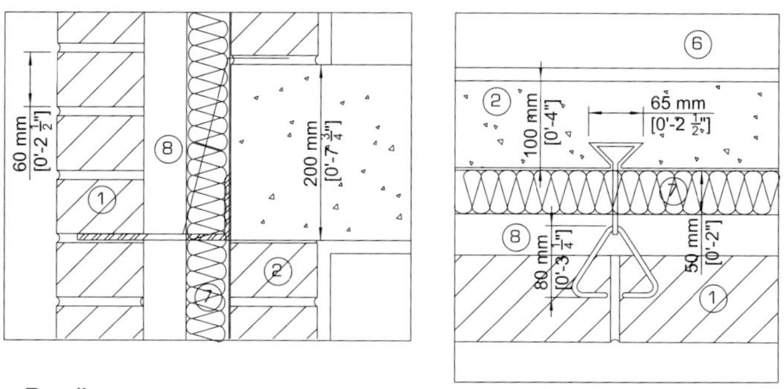

h | Horizontal sections. Typical connection or restraint to primary structure

Details

1 | Outer brick skin
2 | Inner blockwork or brick skin
3 | Timber framed window
4 | Timber sill
5 | Cavity closer
6 | Internal plaster finish or dry lining/drywall
7 | Thermal insulation in cavity
8 | Air cavity (sometimes omitted where insulation fills cavity)

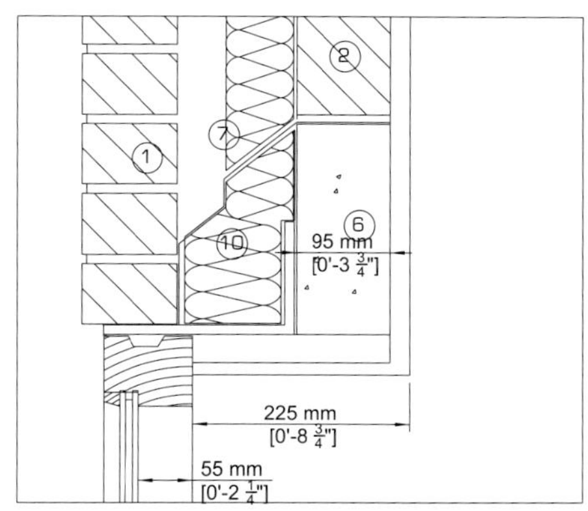

j | Vertical section. Window head

of thermal insulation. Brick types used in cavity construction vary widely, from traditional clay bricks which are either hand-formed or wire-cut to extruded hollow terracotta bricks. The cavity is vented top and bottom to ensure the free passage of air through the cavity. This allows the cavity to dry as well as drying out the outer skin of brick, which can become fully saturated in a rainstorm. The cavity is bridged only by openings for windows and doors.

When an opening is formed, the cavity is closed by a lintel (beam) to support the inner and outer skins at the top of the opening, and a sill piece at the bottom of the opening. The sides are closed by either returning one of the two skins to meet the outer skin or by setting an insulated cavity closer, usually made as an aluminium extrusion filled with thermal insulation. Because the downward passage of water inside the cavity is blocked by the lintel at the head of the opening, a 'cavity tray' is set above the lintel to drain water out of the cavity. A cavity tray is formed from a bituminous felt strip which is set into a horizontal joint in the inner skin and set sloping downwards across the face of the cavity. The bottom of the tray is set on a horizontal course on the outer skin. Weep holes are left in the vertical joints immediately above the DPC to allow water to drain out. The ends of the DPC are tucked down into a vertical DPC set into the jambs (sides) of the opening. The vertical DPC is in turn linked to a DPC set under the sill to form a complete watertight seal to the opening which is drained and ventilated to the outside wall. A similar principle of cavity tray is used to drain water at roof level and at ground level (MCE_236, a). In addition, a DPC is used at ground level to avoid water being drawn up into the wall construction to the inside face of the wall within the building. DPCs are also used beneath parapets and copings as well as at the junction of wall and pitched roof. The use of the cavity tray together with a damp proof course are the two most important principles of cavity wall detailing.

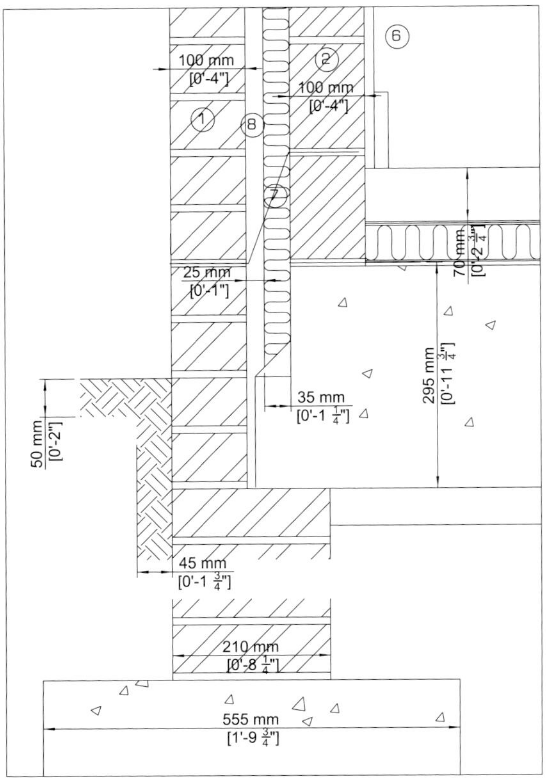

Details

1 | Outer brick skin
2 | Inner blockwork or brick skin
3 | Timber framed window
4 | Timber sill
5 | Cavity closer
6 | Internal plaster finish or dry lining/drywall
7 | Thermal insulation in cavity
8 | Air cavity (sometimes omitted where insulation fills cavity)
9 | Inner concrete lintel or beam
10 | Pressed steel lintel
11 | Steel angle
12 | Metal coping
13 | Precast concrete sill
14 | Damp proof course (DPC)
15 | Brick sill
16 | Movement joint

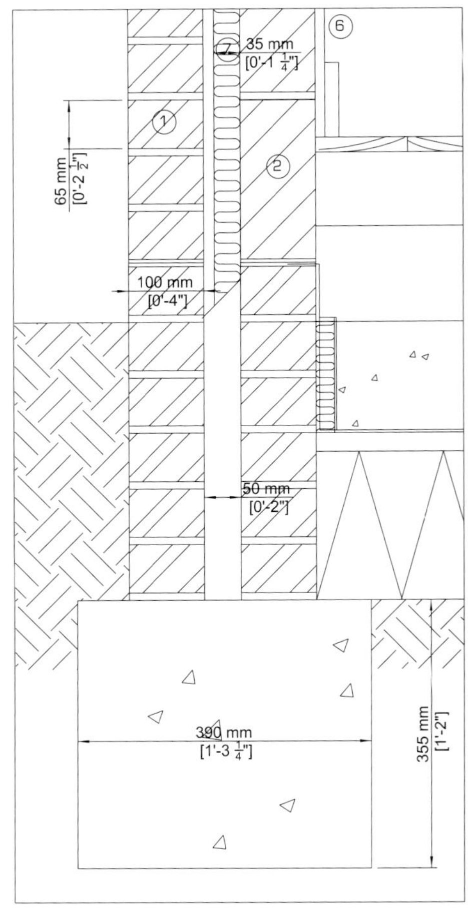

a | Vertical sections.
Junctions at ground floor level

b | Typical bay

Ground level

The detailing of the DPC at ground floor level is dependent upon the difference in height between ground floor level in the building and the adjacent external level. The DPC in the outer skin is set at around 150mm above external ground level (MCE_235, j). The DPC for the internal skin is set at the same level if the step up from outside to inside is around 150mm. If the difference between outside and inside levels is around 300mm/12in then the DPC is stepped up from outer skin to inner skin in the same place, but a separate DPC is added to the inner skin at the same level as the bottom of the cavity tray. The aim of the damp proofing is to provide a continuity of protection from underneath the ground floor or lowest basement slab up into the wall construction.

The cavity is filled below ground floor level to avoid it filling with water which would eventually damage the construction, particularly from freezing in winter in temperate climates. Until recently, thermal insulation was usually terminated at ground floor level. More recently, the thermal insulation continues down below ground where it is continuous with thermal insulation set on top of or beneath the floor slab to provide a completely insulated building enclosure.

Window and door openings

Since a window or door frame has a profile which is the same on all four sides, with an external modification for the sill, the profile of the opening into which it fits must be consistent on all sides. Visual considerations are equally important in the detailing of cavity wall openings. The depths of the reveal can make the outer skin appear more 'massive' if the reveal is made at least one brick deep or more like a thin skin if the window is set forward of the reveal. Whereas a loadbearing wall can incorporate a structural brick arch, this is much more difficult in a cavity wall. This is because the outer and inner skins of a cavity wall are kept separate, except where they are bridged by lintels. Because they are linked only at

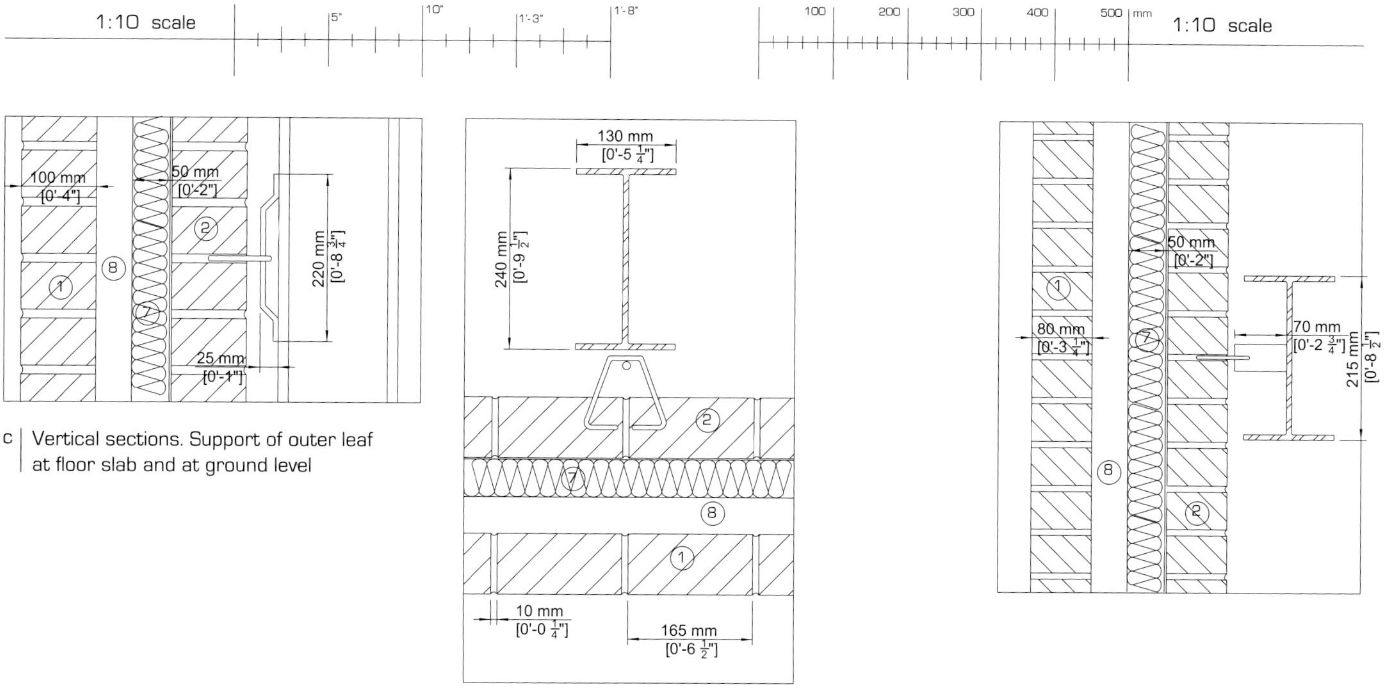

c | Vertical sections. Support of outer leaf at floor slab and at ground level

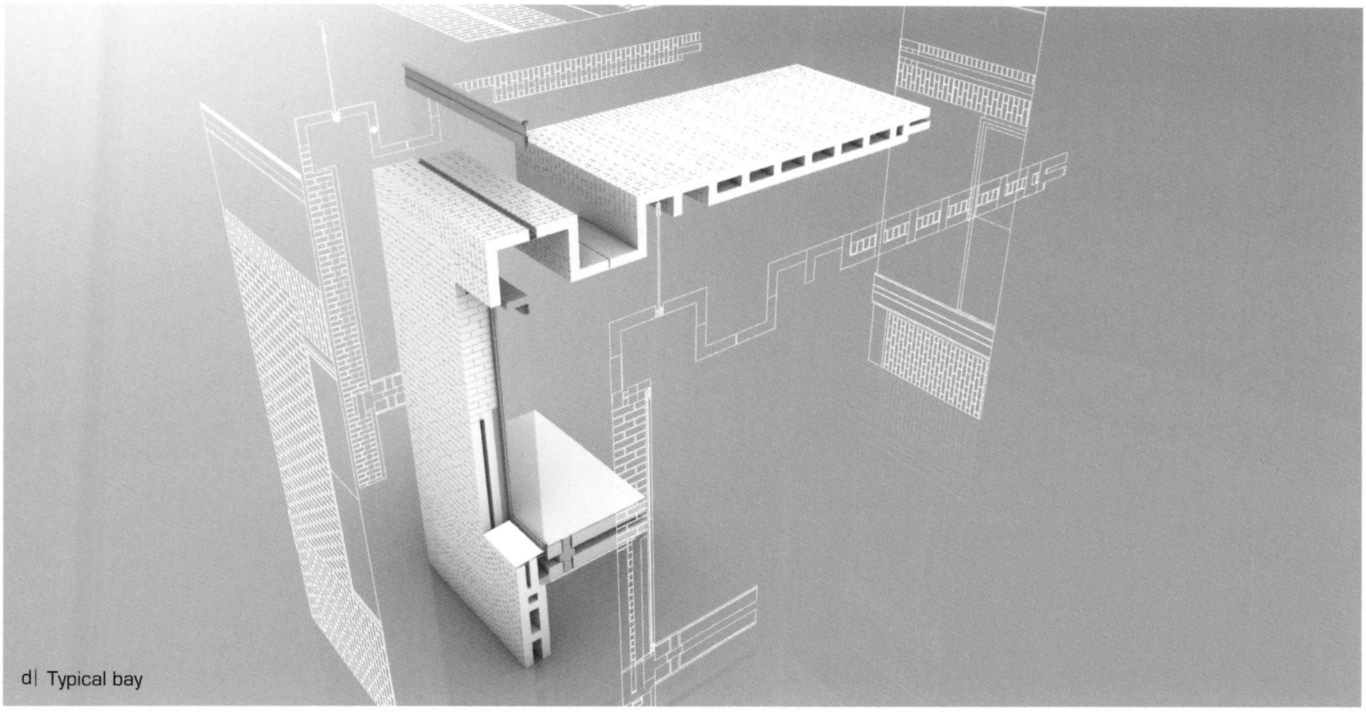

d| Typical bay

these points, the lintel needs to be a separate structural element which supports both skins without exerting lateral forces in the adjacent brickwork (MCE_234, d), where cracking could occur in both skins. For this reason, simply supported lintels are used which exert balanced loads on both skins. Both reinforced concrete and steel types are used, which have the advantage of supporting a course of bricks to give the appearance of a flat arch (d). Reinforced concrete lintels appear as beams visible in both elevation and on the soffit (underside), while steel lintels are visible only on the soffit of the opening (d), making it clearly visible that the bricks across the top of the opening on the outer face are not self-supporting unlike load-bearing wall construction. For this reason brickwork supported by a steel lintel is often not set on edge to form an arch, but is coursed the same as the brickwork above, since any arch supported by a steel lintel is decorative. However, an advantage of the steel lintel is that it can incorporate a cavity tray (MCE_234, d), with rainwater being drained through weep holes in vertical joints in the bottom course.

Lintels are supported on both skins at their bearing points. A reinforced concrete lintel spans across both skins either in a profile that also forms a cavity tray for wide spans or as a flat lintel for short spans. In some examples, the lintel has a beam section supporting the inner skin which provides the spanning element and a toe supporting the outer skin. A DPC is set onto the lintel to drain water. In other cases, the DPC can be set above the lintel depending on the configuration of the wall construction. In both cases thermal insulation is set on the inside face of the wall to avoid a thermal bridge. Steel lintels can be used either as a pressed mild steel lintel which has a profile for supporting both skins, or as reinforced concrete lintel with a stainless steel shelf angle bolted to it (MCE_241, c). In other examples, the lintel has a DPC set on top of the profile formed by the lintel, where the inner block has no structural function but is instead used to infill the gap and provide a background for plaster. Optionally, the DPC can be set onto a stainless steel angle. The angle

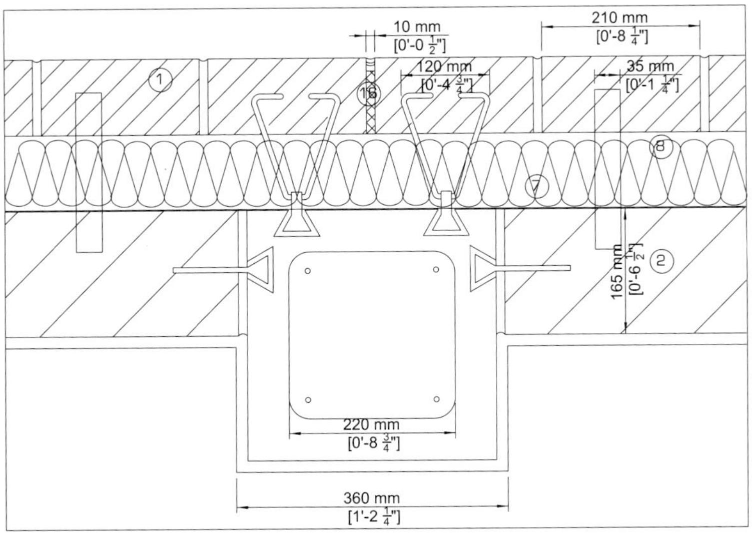

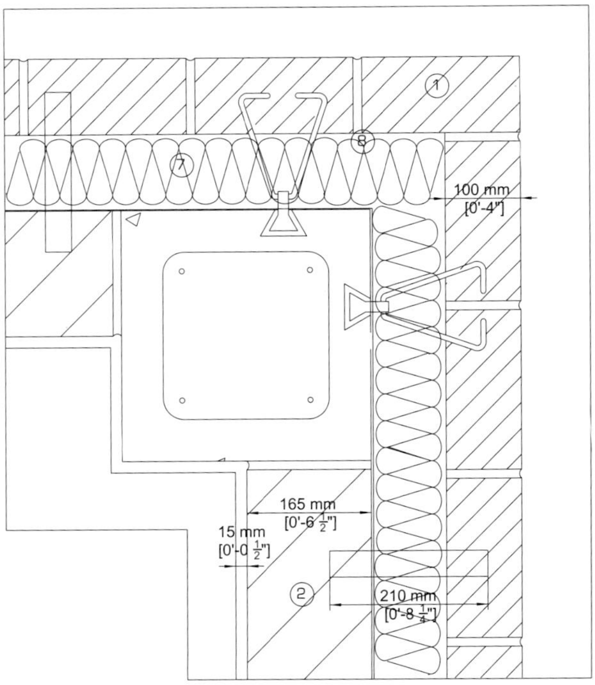

Details

1 | Outer brick skin
2 | Inner blockwork or brick skin
3 | Window
4 | Timber sill
5 | Cavity closer
6 | Internal plaster finish or dry lining/drywall
7 | Thermal insulation in cavity
8 | Air cavity (sometimes omitted where insulation fills cavity)

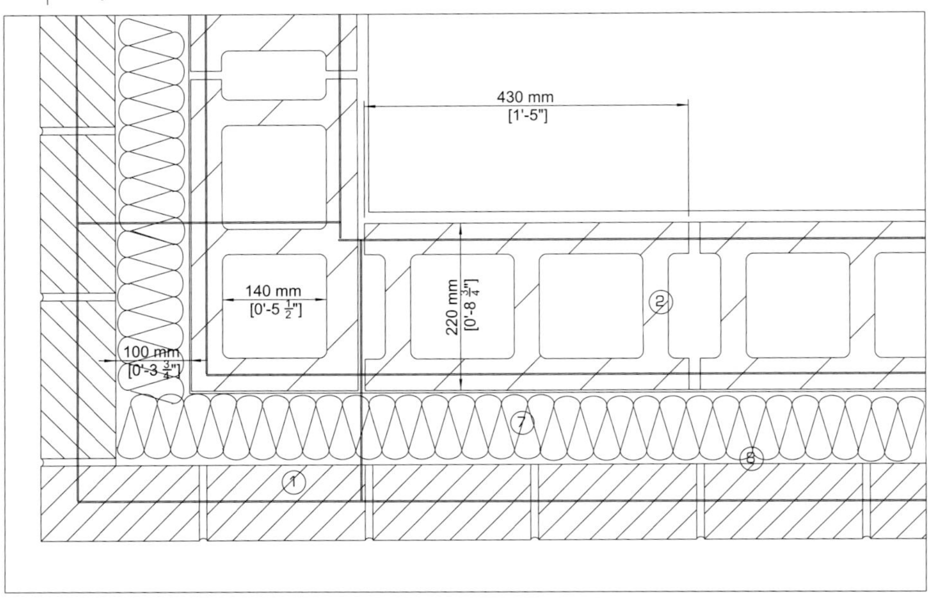

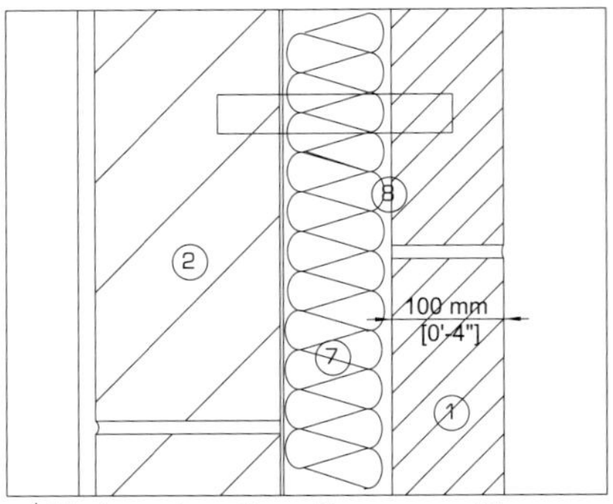

a | Horizontal sections. Corner conditions

9 | Inner concrete lintel or beam
10 | Pressed steel lintel
11 | Steel angle
12 | Metal coping
13 | Precast concrete sill
14 | Damp proof course (DPC)
15 | Brick sill
16 | Movement joint

supports the outer skin, while the reinforced concrete inner beam supports both inner and outer skins.

Eaves and parapets

At the eaves, the bottom of a pitched roof terminates against the top of a brick cavity wall. While there are many variations for the roof itself, the top of the wall maintains a consistent principle of closing the cavity at the top with a brick or block that allows the load from the roof structure to be supported on the inner skin (MCE_239, d). Alternatively, the roof structure may be supported on a column set into the inner skin of the wall or on blockwork piers, also forming part of the inner skin. The closing of the cavity wall at the top allows for continuity of thermal insulation from cavity wall to roof structure while allowing the roof construction to be ventilated where required, and the top of the cavity in the wall to be ventilated (a). The top of the wall usually has plastic spacers set into the horizontal or vertical joints to ensure that air movement can occur within the cavity. A DPC is set on the underside of the brick or block that closes the cavity to ensure continuity between the damp proof membrane on the external face of the inner skin and the waterproofing layer of the roof.

Parapets are closed at the top by a coping, usually in reinforced concrete or stone. A DPC is set beneath the coping to stop the passage of water downwards. Below this, rainwater can enter the cavity from both inner and outer skins and this is prevented by extending the waterproof layer from the roof up the side of the inner skin up to coping level. Thermal insulation is usually continued up the inside of the cavity wall as well as up the external face of the inner skin to avoid a thermal bridge through the inner skin (MCE_235, f and j).

1:10 scale 5" 10" 1'-3" 1'-8"

100 200 300 400 500 mm 1:10 scale

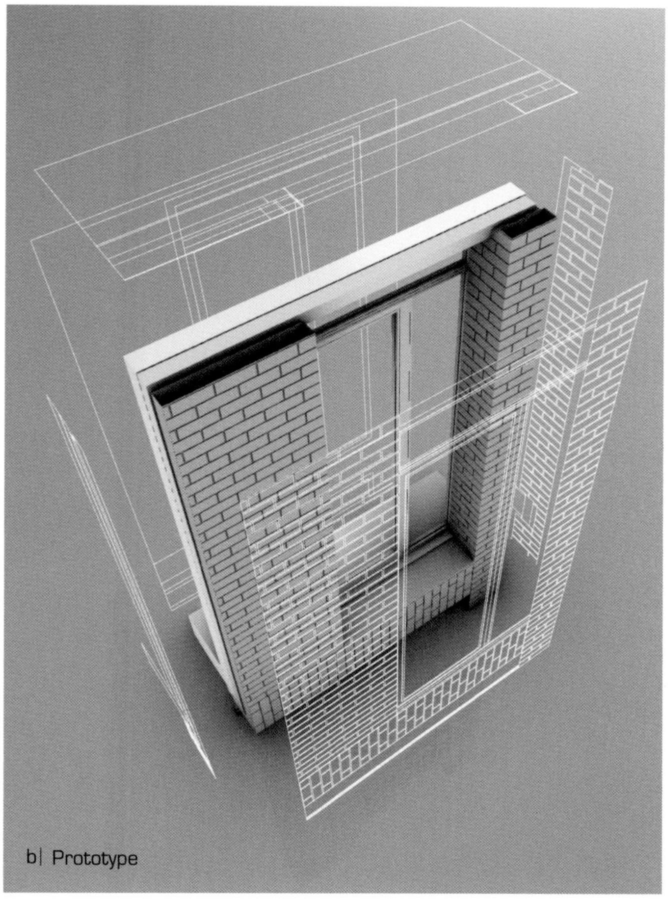
b| Prototype

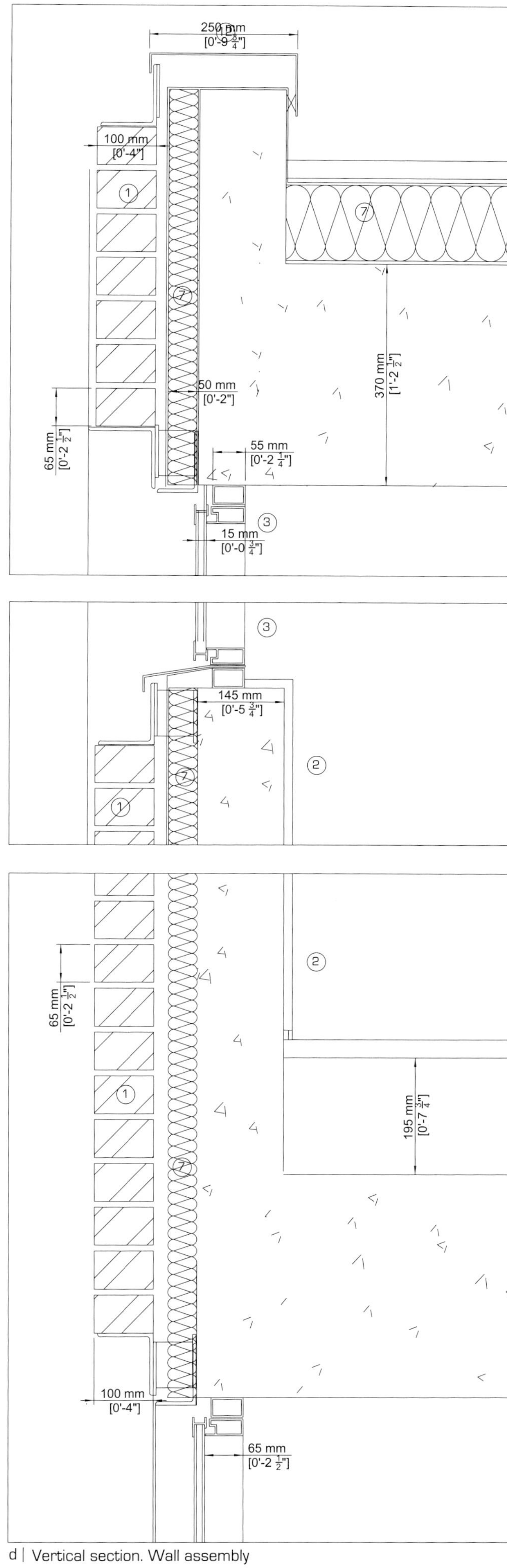

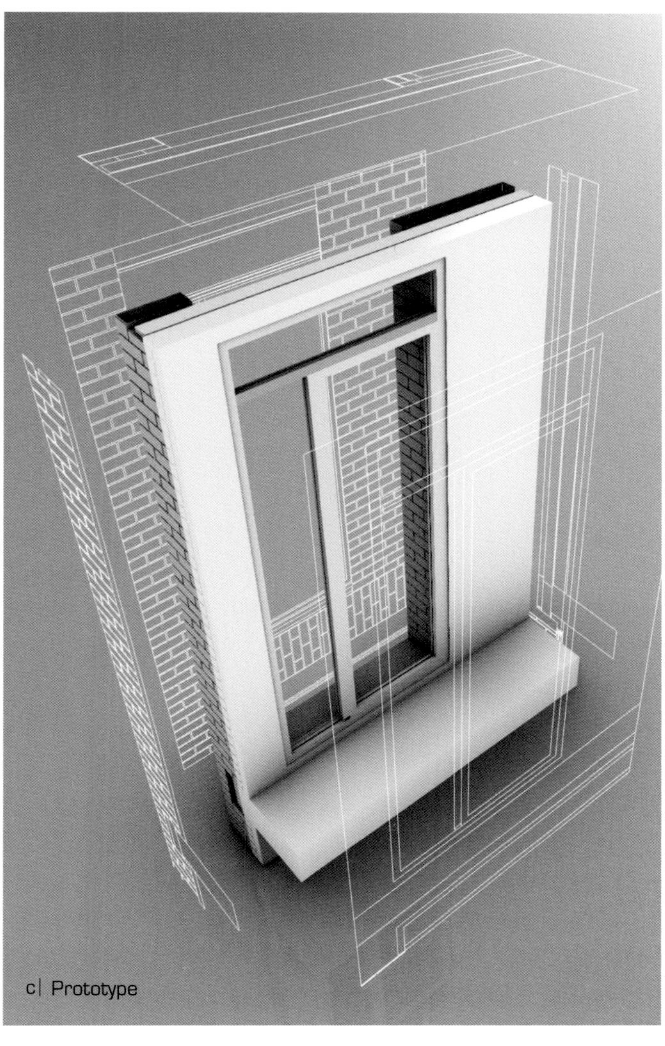
c| Prototype

d | Vertical section. Wall assembly

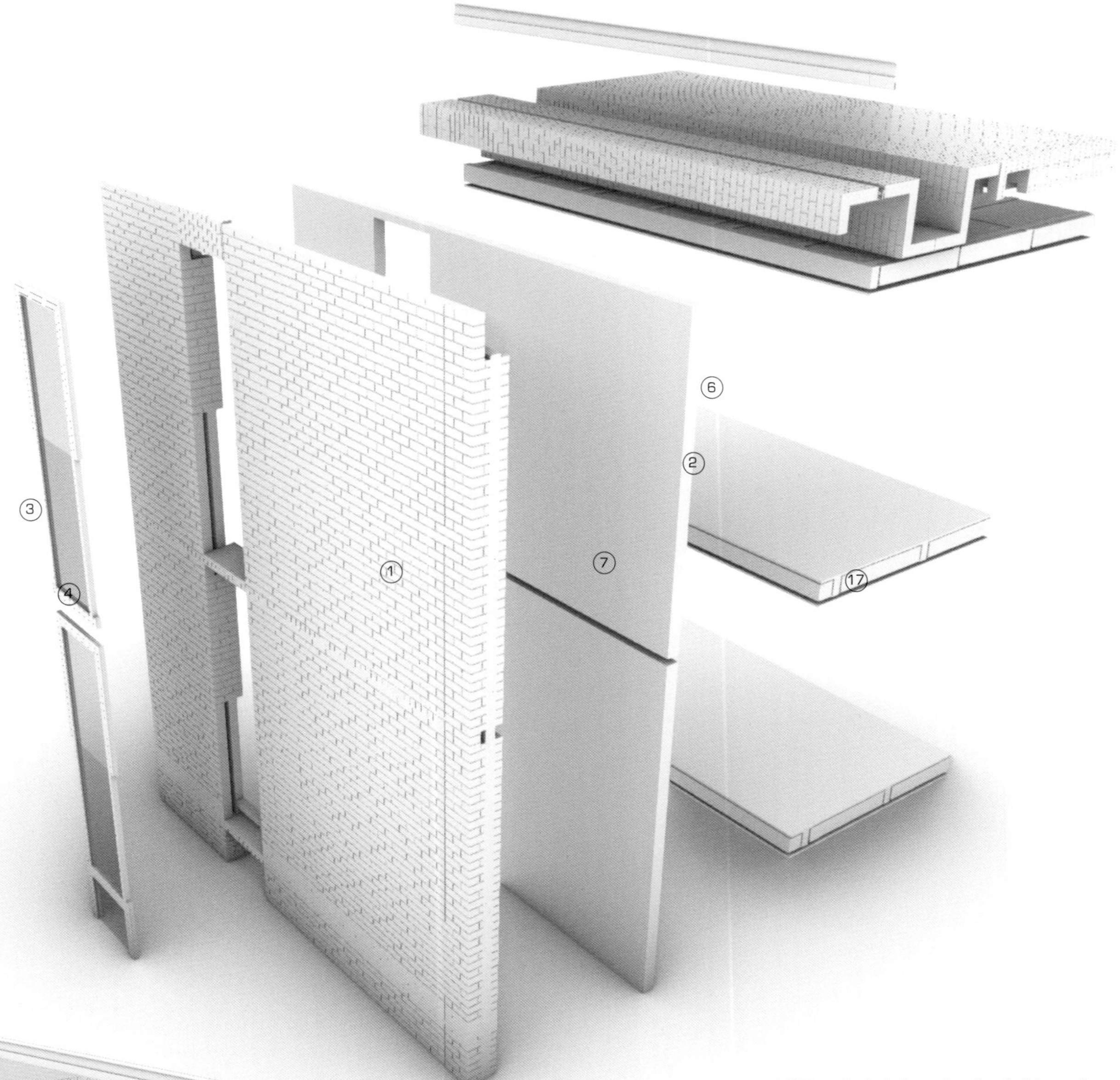

a | 3D exploded view showing brick cavity wall fixed to timber framed structure

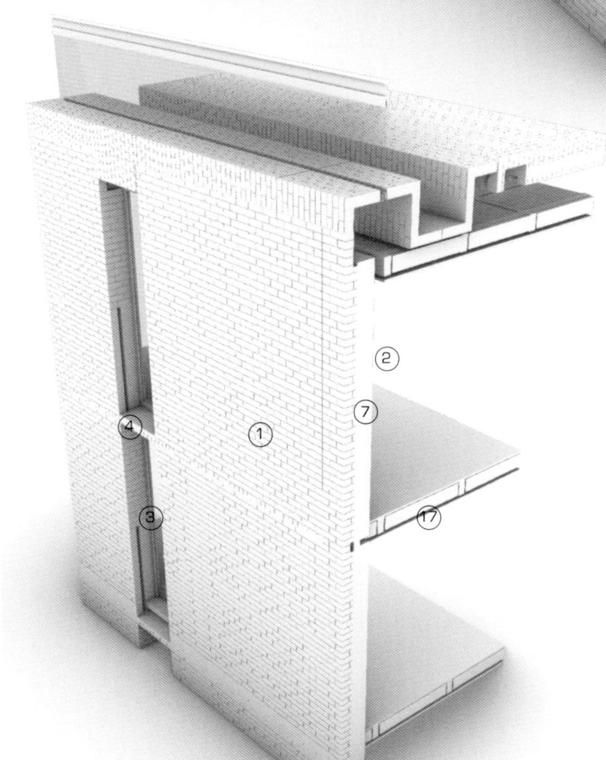

b | 3D view showing brick cavity wall fixed to timber framed structure

Details

1 | Outer brick skin
2 | Inner blockwork or brick skin
3 | Timber framed window
4 | Timber sill
5 | Cavity closer
6 | Internal plaster finish or dry lining/drywall
7 | Thermal insulation in cavity
8 | Air cavity (sometimes omitted where insulation fills cavity)

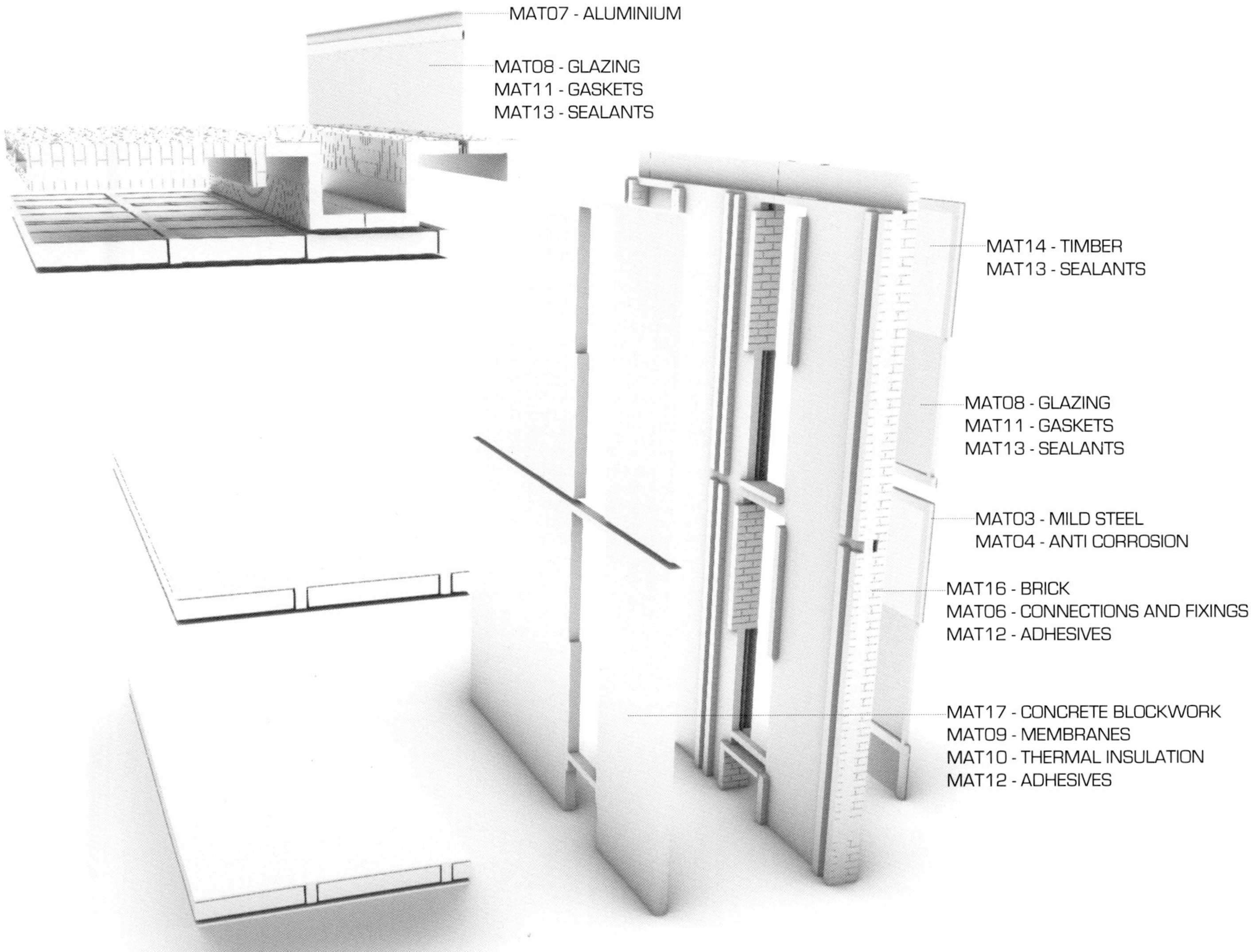

c | 3D exploded view of brick cavity wall facade components with material specification references

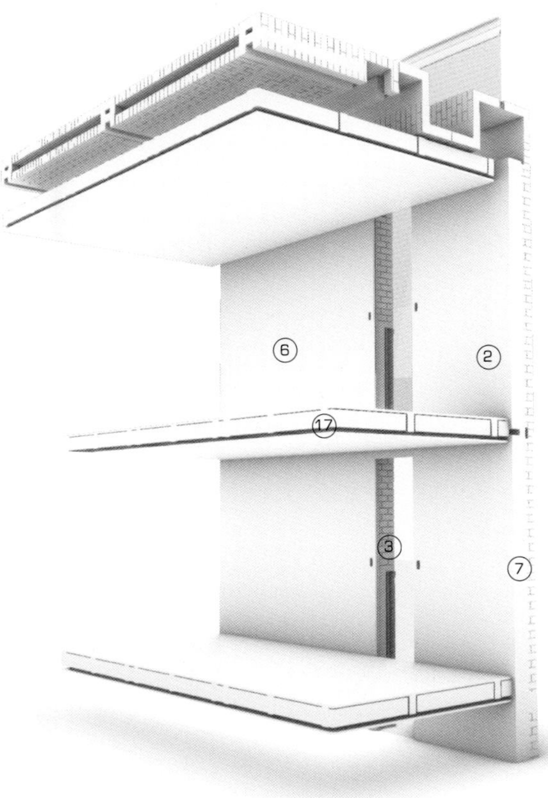

9 | Inner concrete lintel or beam
10 | Pressed steel lintel
11 | Steel angle
12 | Metal coping
13 | Precast concrete sill
14 | Damp proof course (DPC)
15 | Brick sill
16 | Movement joint
17 | Ground floor slab
18 | Steel cavity wall tie
19 | Foundation

d | 3D view showing brick cavity wall facade fixed to timber framed structure

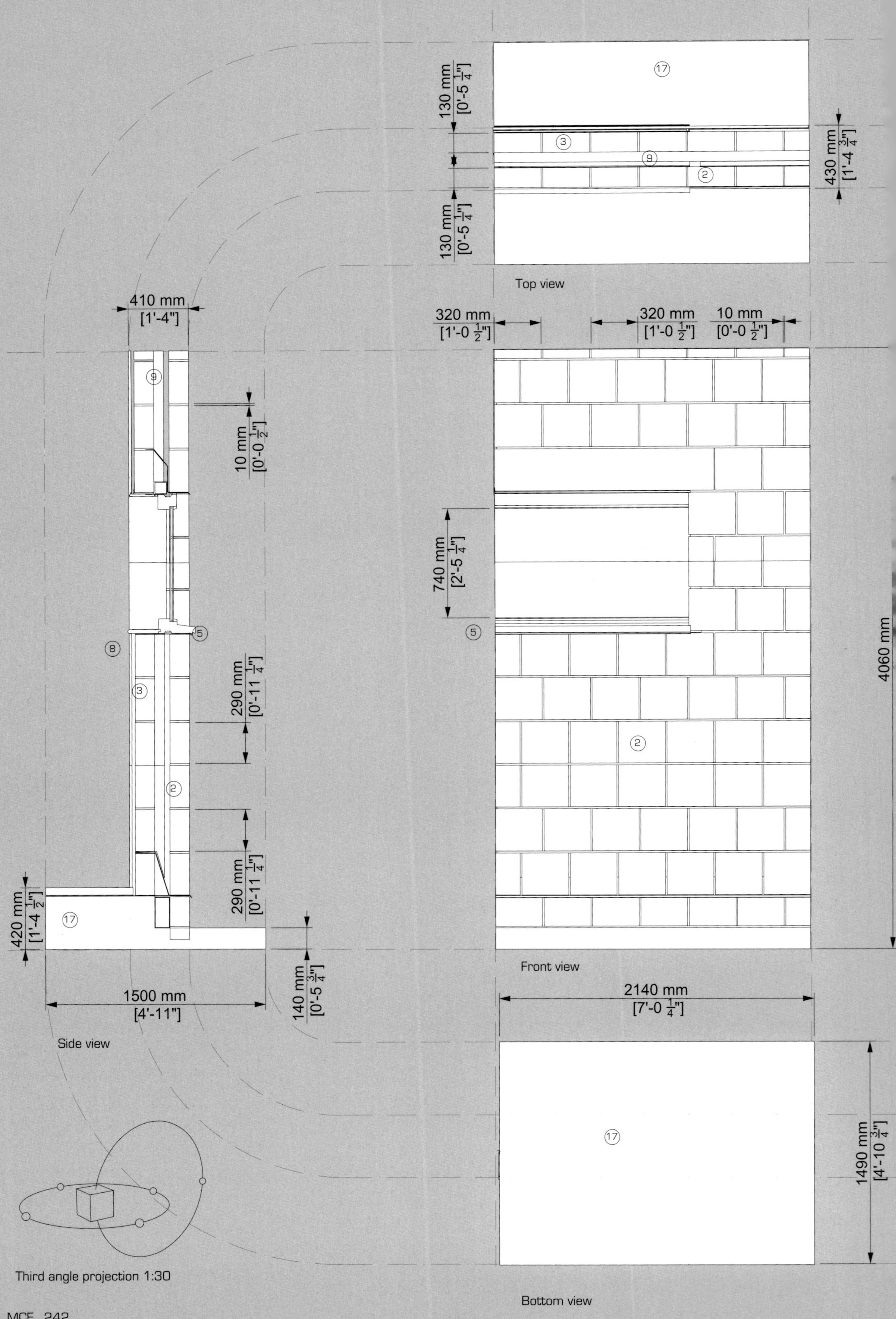

Third angle projection 1:30

Details

1 | Stone or concrete coping
2 | Outer stone skin with brick backing shown to create deep window reveals or where required structurally
3 | Inner block skin
4 | Timber framed window/door
5 | Stone or precast concrete sill
6 | Timber inner sill
7 | Cavity closer
8 | Internal plaster finish or dry lining/dry wall
9 | Thermal insulation in cavity

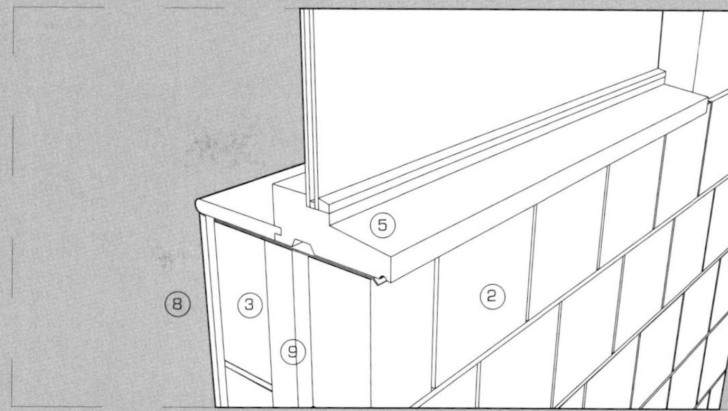

3D view of the assembly

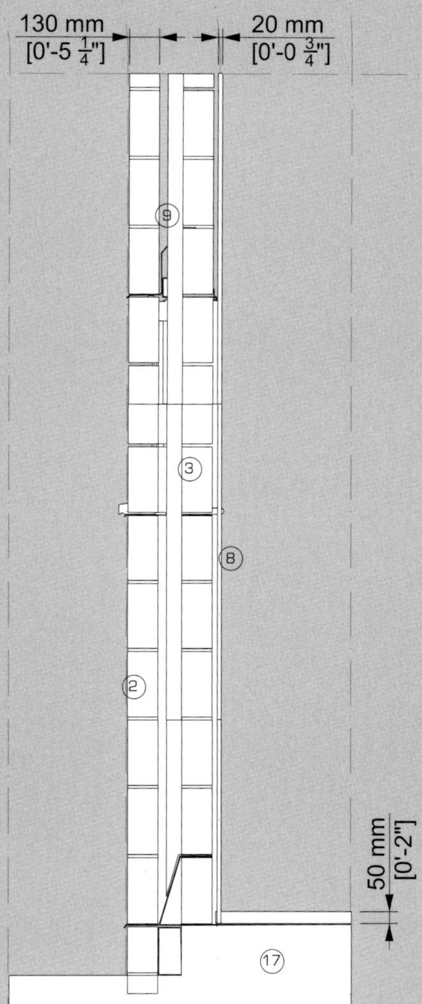

Side view

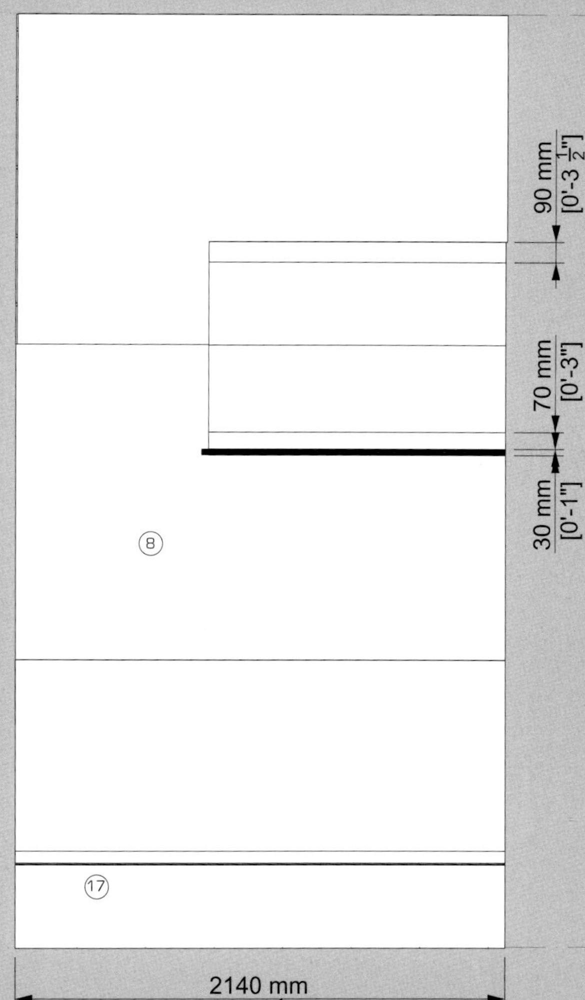

Back view

10 | Air cavity (sometimes omitted where insulation fills cavity)
11 | Inner precast concrete lintel
12 | Outer precast concrete lintel or stone flat arch
13 | Damp proof course (DPC)
14 | Movement joint
15 | Timber framed inner skin with quilt insulation
16 | Cement fill
17 | Foundation
18 | Vertical DPC
19 | Structural column
20 | Floor construction

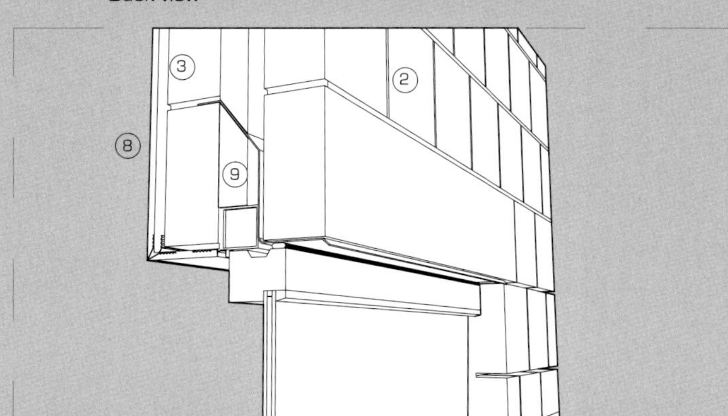

3D view of the assembly

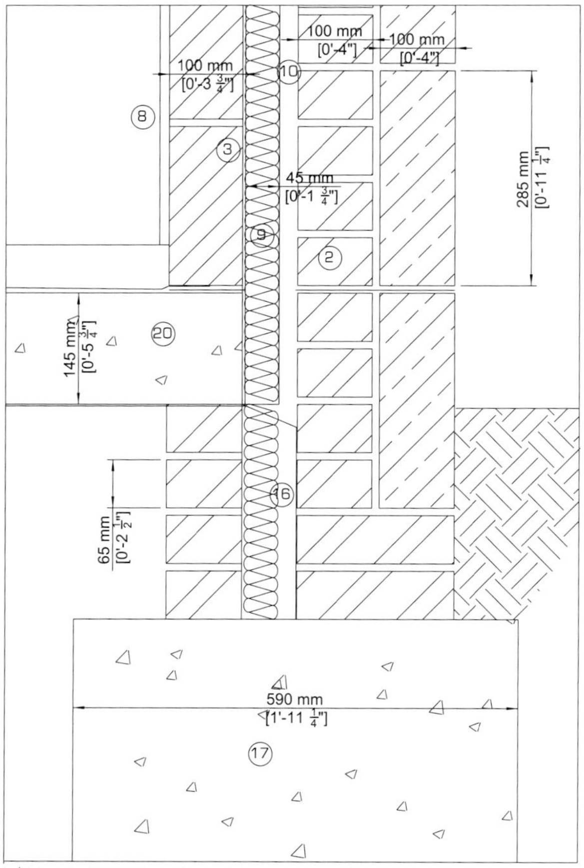

b | Vertical section. Junction at ground floor level

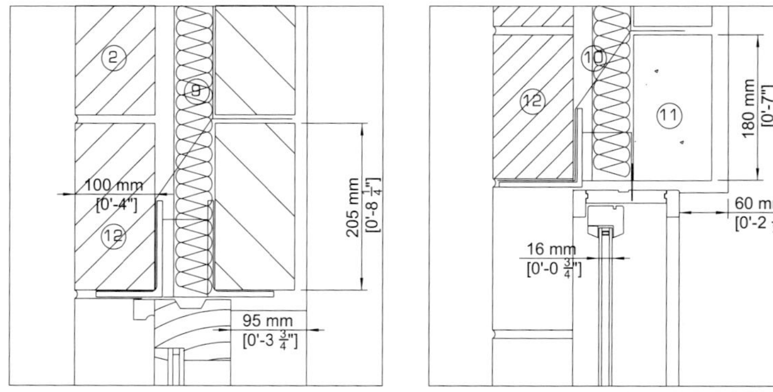

a | Vertical sections. Typical window at sill and head

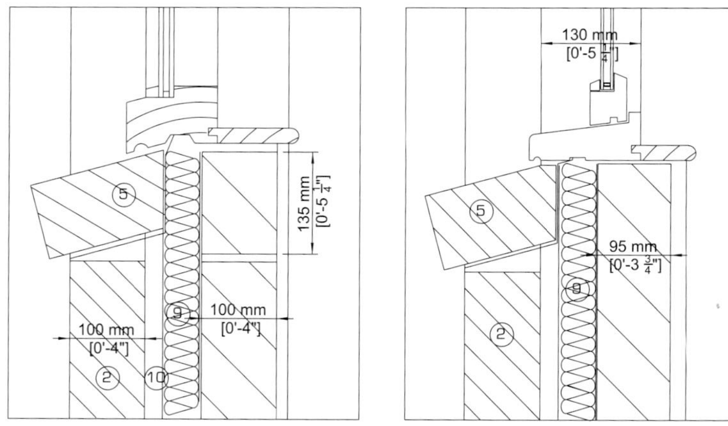

Details

1 | Stone or concrete coping
2 | Outer stone skin with brick backing shown to create deep window reveals or where required structurally
3 | Inner block skin
4 | Timber framed window/door
5 | Stone or precast concrete sill
6 | Timber inner sill
7 | Cavity closer
8 | Internal plaster finish or dry lining/dry wall
9 | Thermal insulation in cavity

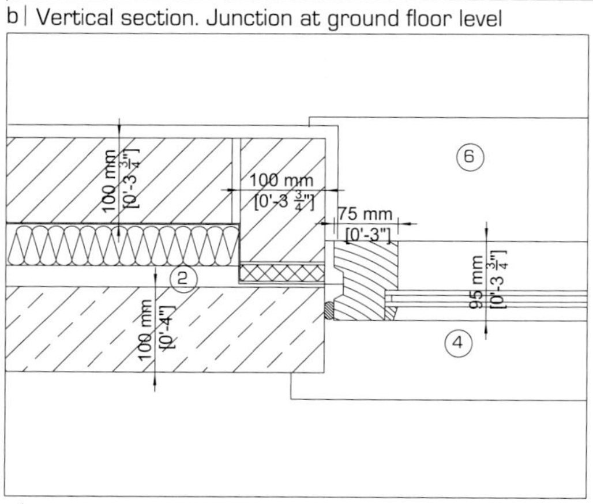

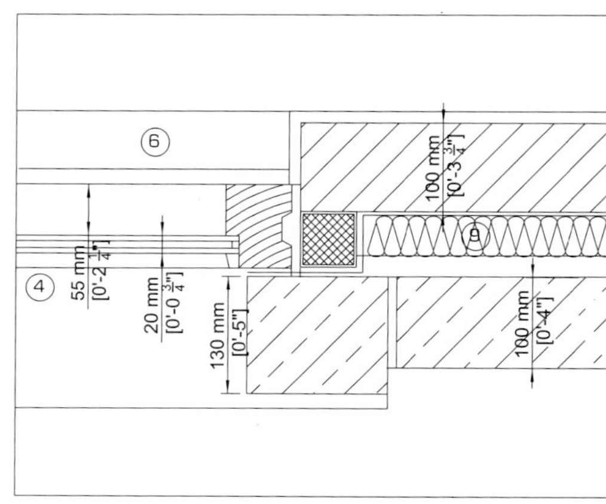

c | Horizontal sections. Typical window jamb

The principles of cavity wall design are set out in the previous text on brick cavity walls. The same principles can be applied for use when stone and concrete blockwork are used to form an outer skin.

When stone is used, it can be used either as an outer skin approximately 100mm/4in thick, or with thInner stone that is bonded to brickwork that together forms a 'composite' outer skin (MCE_250, a). When used as a 100mm/4in thick skin in order to be self-supporting, the stone becomes expensive, so sandstones and limestones are most commonly used. Composite-type outer skins suit stone that is 40mm to 50mm thick (1.5in–2.0in), bonded to a 100mm/4in wide brick skin. This method suits granites and denser limestones.

In single outer skins of stone, the material is ventilated on both sides (MCE_246, a), allowing it to dry out easily, avoiding a situation where stone would dry from only the outer face, which would draw dirt and dust out into the outside face.

The biggest difference in detailing between brick cavity walls and stone / block cavity walls is that there are fewer joints in stone / block due to their large size. This means that there are fewer opportunities to design a damp proof course (DPC) with floor slabs, for example, particularly where shelf angles are used (MCE_246, d). The smaller unit size of brick makes it a very flexible material when detailing; stone / block requires careful co-ordination of stone / block size and floor to floor heights to allow for windows to be suitably placed. For this reason, alternating bands of thick and thin stone are used in coursing. This allows horizontal joint lines to be provided at shelf angles, cavity trays and DPCs at ground level without disturbing the stone pattern with additional horizontal joints. Unlike open-jointed stone cladding, where the mortar is omitted, the mortar and joint profile have a big visual impact.

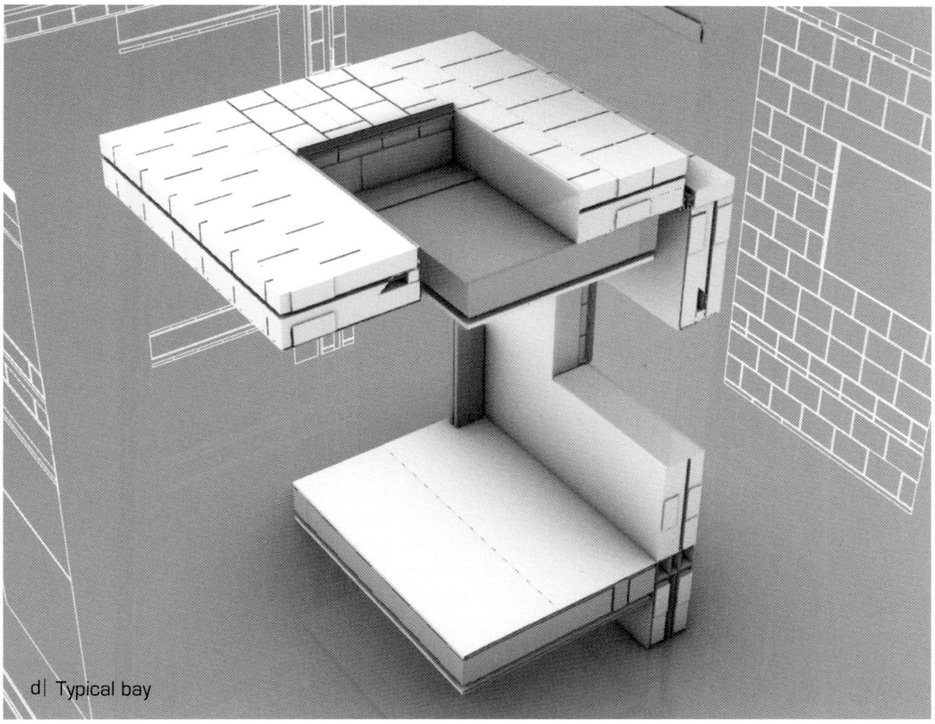

d | Typical bay

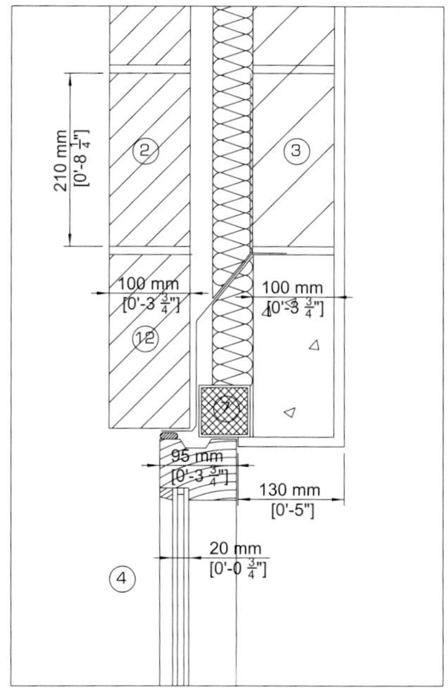

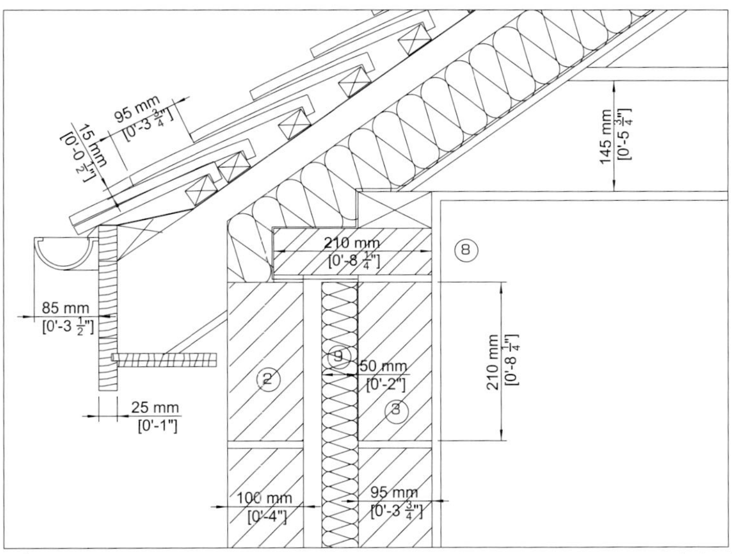

10 | Air cavity (sometimes omitted where insulation fills cavity)
11 | Inner precast concrete lintel
12 | Outer precast concrete lintel or stone flat arch
13 | Damp proof course (DPC)
14 | Movement joint
15 | Timber framed inner skin with quilt insulation
16 | Cement fill
17 | Foundation
18 | Vertical DPC
19 | Structural column
20 | Floor construction

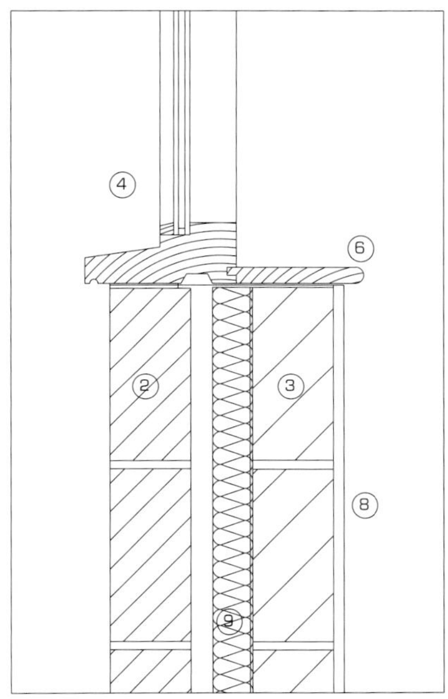

e | Vertical sections. Typical window at cill and head, parapet and eaves to pitched roof

Wall structures

Loadbearing cavity walls used to support two-storey structures on their inner skins are very common in housing construction in Europe and North America. Vertical movement joints are provided at around 7500mm/25ft centres, or else are avoided altogether in the construction by keeping lengths of wall within these dimensions. When cavity walls are used with large-scale building frames in either steel or concrete, the inner leaf is no longer loadbearing and instead the complete wall construction forms a cladding to the frame. When reinforced concrete frames are used, the junction between inner skin, typically concrete block or terracotta block, is straightforward with a gap between the two to allow for structural movement in the frame. Stainless steel sliding anchors are used either in the sides of the inner skin panel where it meets the column, or at the head where it meets the floor slab (MCE_246, d). The outer skin runs continuously in front of it. The situation is more complicated with a steel frame, where the column needs to be protected from corrosion from water vapour in the cavity. Typically the column is painted to form a protective coating and thermal insulation is set across the face of the steelwork to provide a continuity of thermal insulation. Sometimes the outer skin is restrained with cavity ties fixed to the face of the reinforced concrete or steel column (MCE_247, h). This is particularly useful when forming a movement joint in lengths of brickwork, or movement joints that form part of the building structure which typically occur at columns in the building frame. The vertical movement joint is filled with two-part polysulphide sealant that also matches the colour of the mortar as closely as possible and provides a seal that can accommodate the structural movement within the cavity wall.

Ground level

An essential design consideration at ground level is to provide a DPC at a minimum 150mm/6in above ground floor level with a horizontal joint

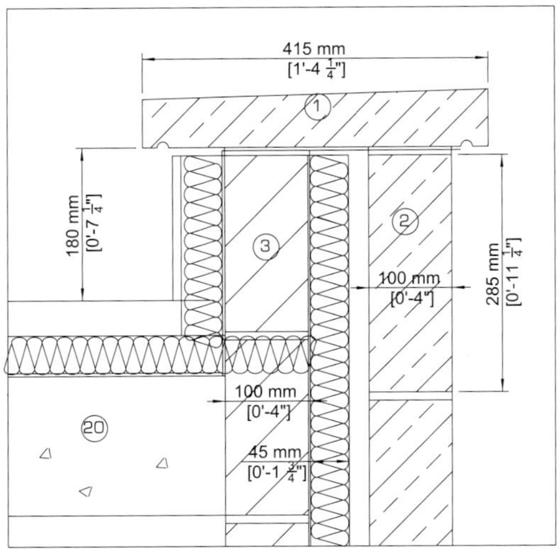

a | Vertical section. Typical parapet

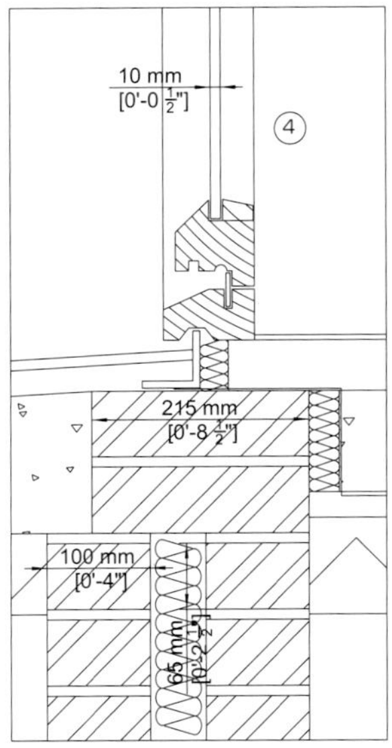

b | Vertical section. Typical threshold

Details

1 | Stone or concrete coping
2 | Outer stone skin with brick backing shown to create deep window reveals or where required structurally
3 | Inner block skin
4 | Timber framed window/door
5 | Stone or precast concrete sill
6 | Timber inner sill
7 | Cavity closer
8 | Internal plaster finish or dry lining/dry wall
9 | Thermal insulation in cavity
10 | Air cavity (sometimes omitted where insulation fills cavity)
11 | Inner precast concrete lintel
12 | Outer precast concrete lintel or stone flat arch
13 | Damp proof course (DPC)
14 | Movement joint
15 | Timber framed inner skin with quilt insulation
16 | Cement fill
17 | Foundation
18 | Vertical DPC
19 | Structural column
20 | Floor construction

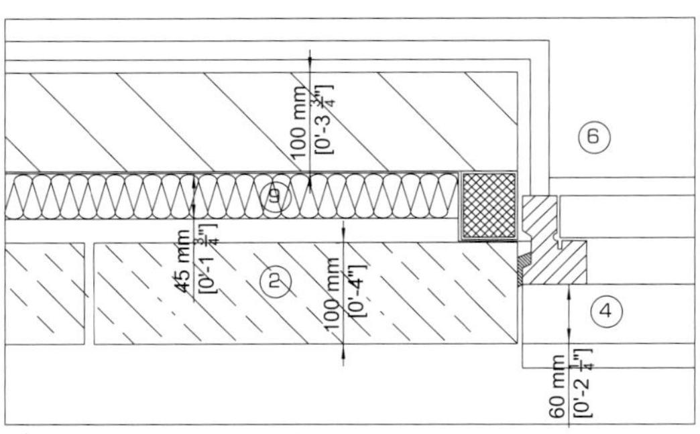

c | Horizontal section. Window jamb

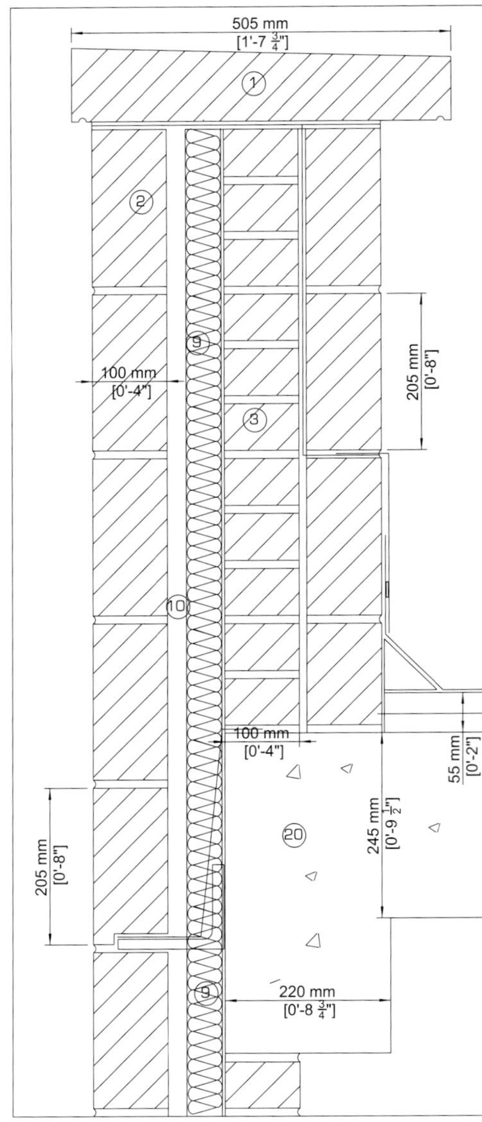

d | Vertical section. Typical parapet

that can be accommodated within the design. Granite is often used below DPC level to avoid staining from rain splashing from the adjacent ground and from dust and damage in urban locations. The granite, usually 30mm to 40mm thick (1.2in to 1.5in), is bonded to brickwork or concrete blockwork to make up the difference in thickness of the outer skin to a minimum of 100mm (4in).

In order to avoid dampness reaching the inside of the building, a mixture of DPCs is used that provides a continuity between DPC and the damp proof membrane for the floor slab. Different combinations of DPCs are used for entry into the building at the same level as outside when the floor level is a minimum of 150mm (6in) above external level. When internal and external levels are almost the same, the DPC is set at ground level and is continuous with the DPM rising from below. An additional stepped DPC is set 150mm (6in) above external level to drain the cavity above (b). When internal and external levels are very different, a first DPC steps down from internal floor level to a horizontal joint in the outer skin, 150mm (6in) above external level, where a second DPC on the inner skin extends down the internal face of the wall down to join the DPM.

The cavity below ground level is filled to avoid water building up in any voids, particularly where foundations for pads for steel or reinforced concrete columns are continuous with the cavity wall. Foundations need to be filled to ground level.

There is always a difficulty in providing a continuity of thermal insulation down the cavity and under the slab due to the need to build the masonry skins directly off the foundation. The thermal insulation in the cavity is taken down as far as possible to overlap with the insulation under the floor slab to reduce the thermal bridge to a minimum.

e| Prototype

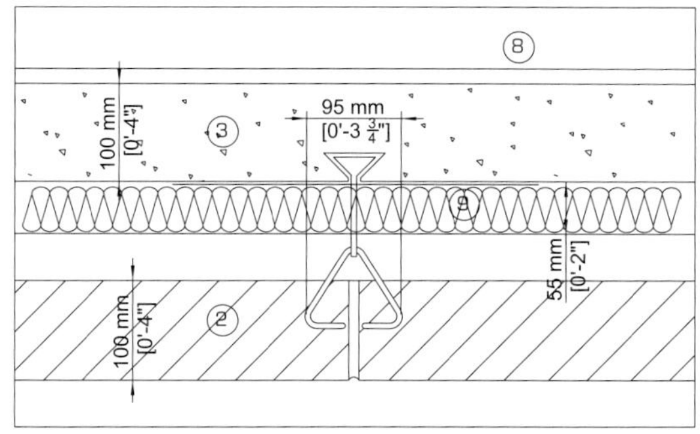

f | Horizontal section.

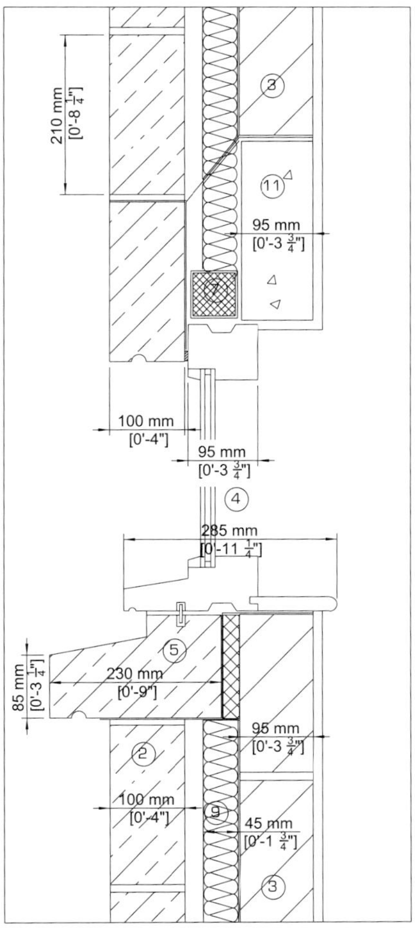

g | Vertical section. Typical window at sill and head

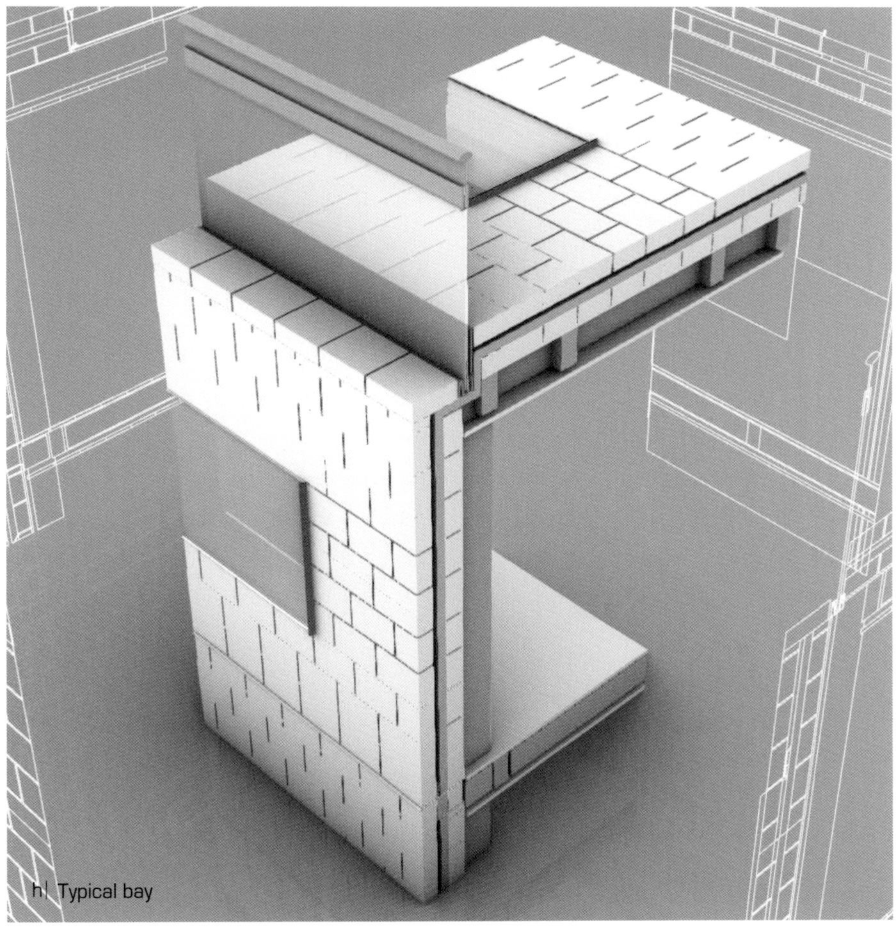
h| Typical bay

Openings in walls

This section discusses variations and details additional to those set out in the previous section on brick cavity walls and focuses on different materials for the inner skin.

When a timber framed inner skin is used, the window is usually an integral part of that inner skin so that a complete enclosure is formed in timber with masonry used as an outer skin providing some lateral stability. The outer masonry skin can be returned to form a reveal up to around 125mm (5in) depending on the thickness of insulation in the cavity. Alternatively the timber window can be set forward close to the line of the outer face with a timber sill projecting forward of the wall (MCE_249, j). This gives the outer wall an appearance of a brick texture, giving a planar look to the material.

With reinforced concrete inner skins, it is difficult to set wall ties into the material to restrain the outer material. Since it is not practical to cast in wall ties in the precise location where they are needed, which does not allow for later adjustment, a set of narrow stainless steel channels are post-fixed to the concrete. Wall ties are then fixed to these channels (MCE_248, b), which provide the adjustment required. In recent years, outer masonry skins have been made as prefabricated panels on some projects when fixed back to a reinforced concrete wall. Panels of stone, terracotta or even brick are bonded together with mortar and then held in a steel edge frame, which is fixed to the inner skin. The concrete inner skin is waterproofed with bituminous paint and thermal insulation set on the outside face of the inner skin. The steel angle on the bottom edge forms a closer to the window below, while the steel on the top edge forms a sill to the window above. The vertical steel angles are usually concealed within the cavity to give a continuous masonry appearance. The masonry is sometimes restrained within the panel

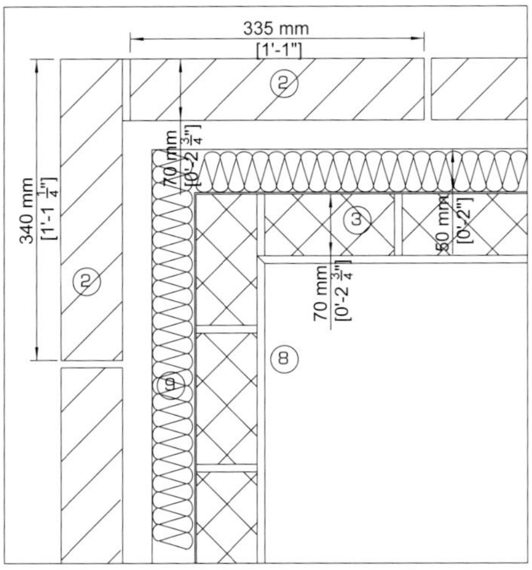

a| Horizontal section. Corner condition

Details

1 | Stone or concrete coping
2 | Outer stone skin with brick backing shown to create deep window reveals or where required structurally
3 | Inner block skin
4 | Timber framed window/door
5 | Stone or precast concrete sill
6 | Timber inner sill
7 | Cavity closer
8 | Internal plaster finish or dry lining/dry wall
9 | Thermal insulation in cavity
10 | Air cavity (sometimes omitted where insulation fills cavity)
11 | Inner precast concrete lintel
12 | Outer precast concrete lintel or stone flat arch
13 | Damp proof course (DPC)
14 | Movement joint
15 | Timber framed inner skin with quilt insulation
16 | Cement fill
17 | Foundation
18 | Vertical DPC
19 | Structural column
20 | Floor construction

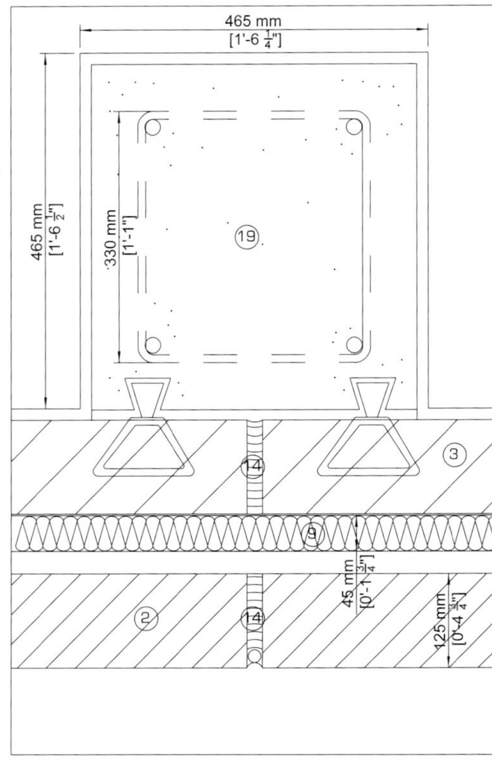

b | Horizontal section. Typical connection or restraint to primary structure

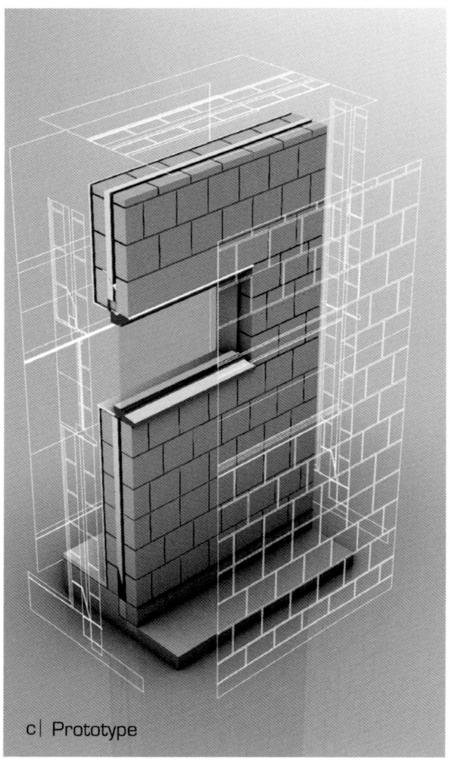

c| Prototype

d| Horizontal section. Typical connection or restraint to primary structure

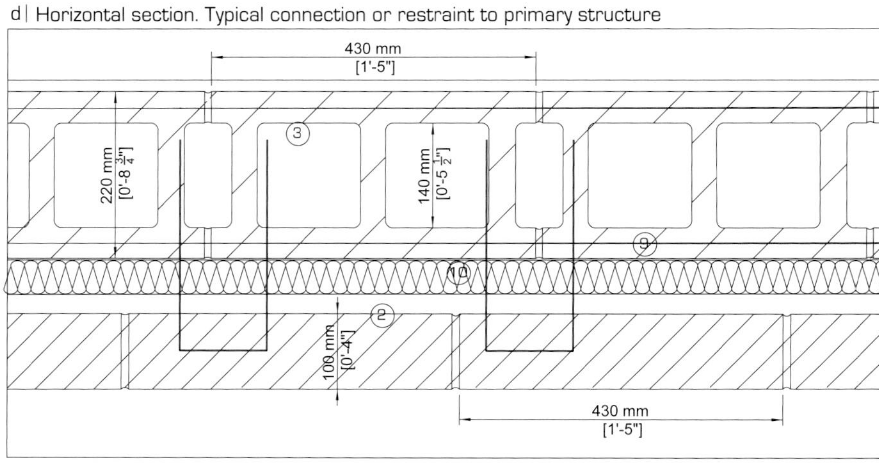

by vertical stainless steel rods which are tensioned against the frame to form lightly prestressed panels. Vertical joints between panels are sealed, typically with a polysulphide sealant.

Eaves and parapets

Eaves to cavity walls in stone and block are generally to either a reinforced concrete slab or to a timber pitched roof. When a reinforced concrete slab is used and is visible from the outside, both inner and outer skins need a joint for expansion at the head. A compressible durable seal is used, recessed from the face of the outer skin to avoid a colour clash with the mortar below. Stainless steel angles with slotted connections are also used to restrain masonry at the head, or even a stainless steel channel if the underside of the concrete is revealed. When the concrete slab soffit is concealed with additional cladding, a horizontal joint in the outer skin is usually placed to align with the cladding to avoid the appearance of the stone disappearing behind the soffit. The outer skin continues with another course to close against the slab soffit. Where a downstand beam is used to support the concrete roof slab above, the beam is usually aligned with the inner skin of the cavity wall. A stepped DPC is set the other way round from its usual position, draining from the outer leaf to the inner leaf, but water running down the underside of the DPC is drained harmlessly down the outer face of the inner skin. If timber is used in a projecting flat roof the DPC is set on top of the wall, on which roof timbers are supported. A continuity of thermal insulation is provided. The timber roof structure can also be ventilated to allow it to be kept dry while maintaining thermal insulation (MCE_245, e). For parapets, the inner skin is thickened up when used as a balustrade. A handrail on top of the coping is fixed by drilling through the top and bolting it to the inner skin beneath. The coping is cut to receive the balustrade or handrail, unless the uprights supporting the handrail pass between the joints. The adjacent area of flat roof or gutter usually has a waterproofing layer returning up the wall,

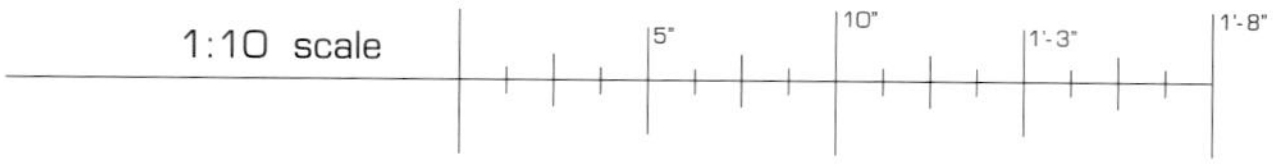

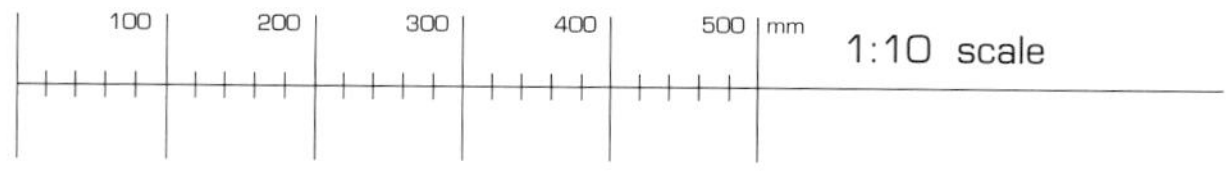

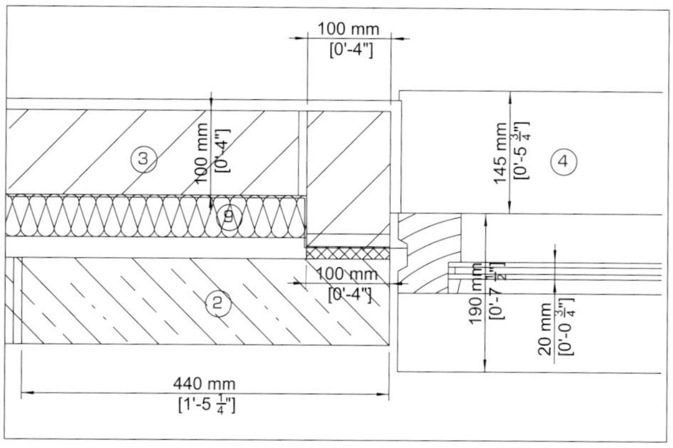

e | Horizontal section. Window jambs

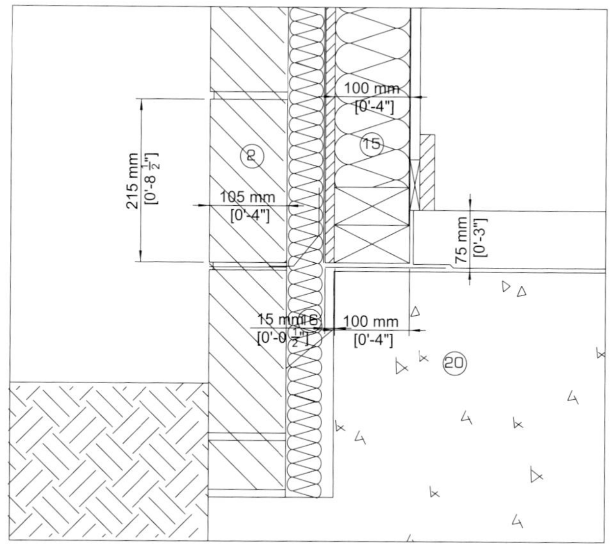

f | Vertical section. Junction at ground floor level

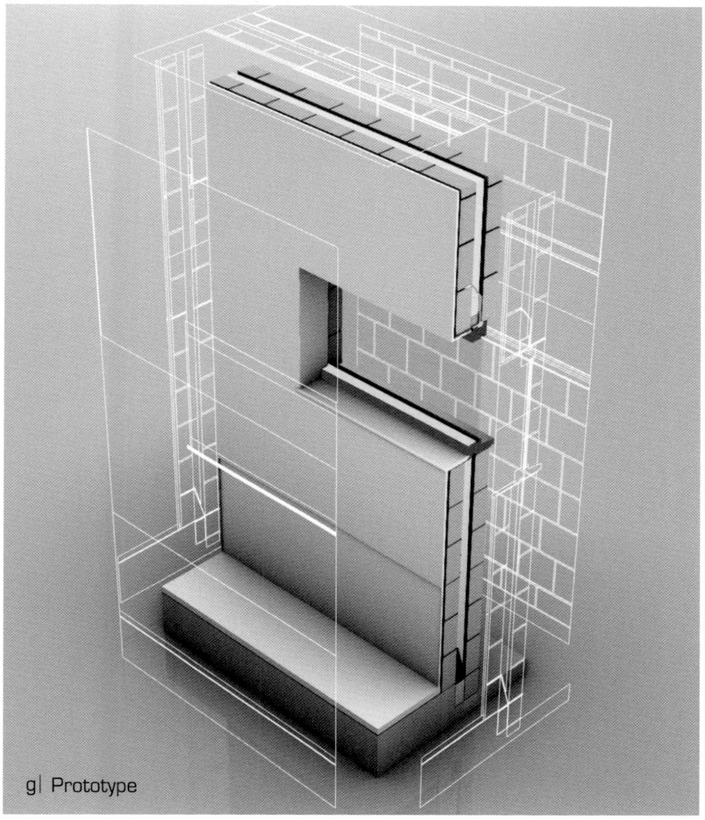

g | Prototype

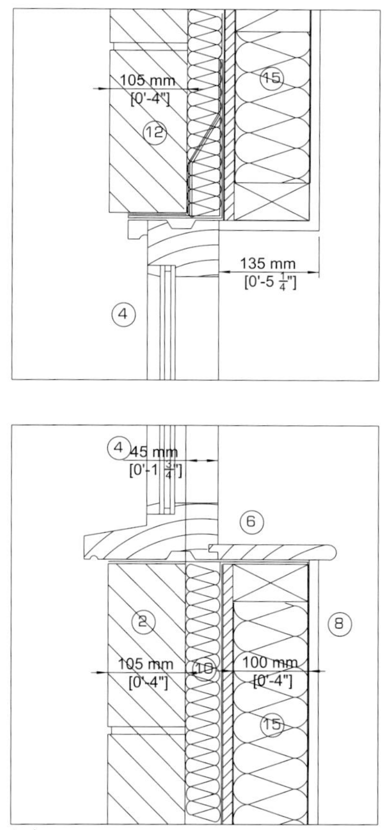

h | Vertical section.
Typical window at sill

regardless of the roof finish. The waterproofing layer is set into horizontal joints in outer masonry skin. A metal flashing is set into the same horizontal joint and is set over the top of the waterproofing to protect it from damage (MCE_246, d). Metal copings are used increasingly on parapets in order to match the appearance of windows and doors, particularly where metal sills are used. The same principles apply as for concrete copings, with a DPC set on top of the masonry wall. Drips are formed on either side of the vertical face to ensure that water is thrown clear of the wall. For all parapets, the waterproofing layer is continued up the wall to become continuous with the DPC in low parapets. For high parapet walls, a stepped DPC is used to drain water back to the inner skin to ensure that water inside this part of the cavity wall is drained immediately, especially in very exposed conditions. A stepped DPC is used instead of a regular flat DPC under the coping.

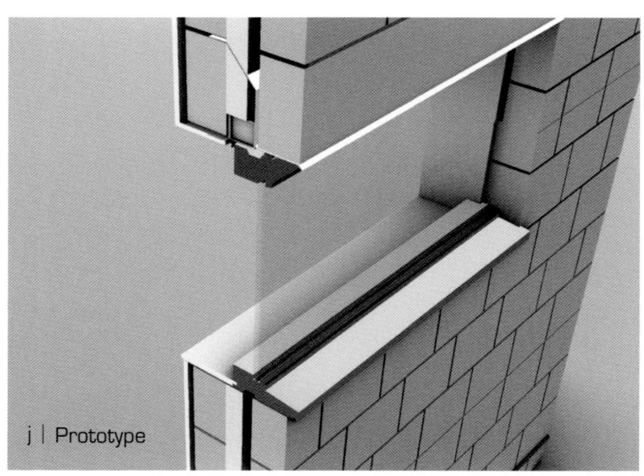

j | Prototype

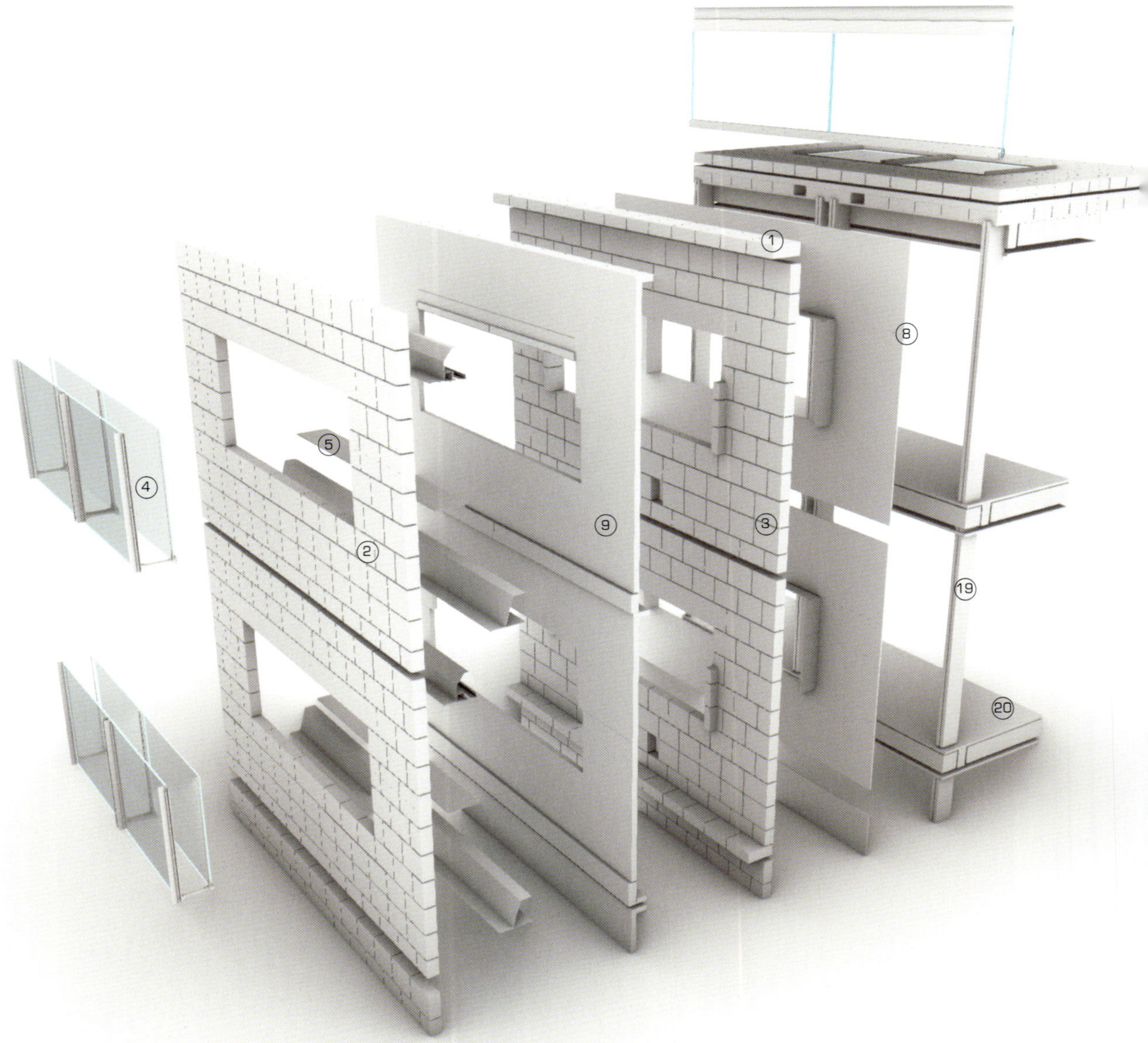

a | 3D exploded view showing concrete block facade fixed to timber frame

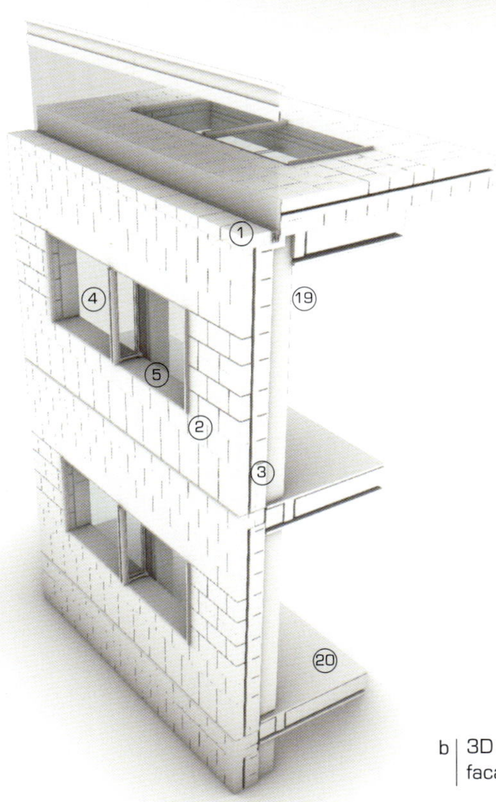

b | 3D view showing concrete block facade fixed to timber frame

Details

1 | Stone or concrete coping
2 | Outer stone skin with brick backing shown to create deep window reveals or where required structurally
3 | Inner block skin
4 | Timber framed window/door
5 | Stone or precast concrete sill
6 | Timber inner sill
7 | Cavity closer
8 | Internal plaster finish or dry lining/dry wall
9 | Thermal insulation in cavity
10 | Air cavity (sometimes omitted where insulation fills cavity)
11 | Inner precast concrete lintel

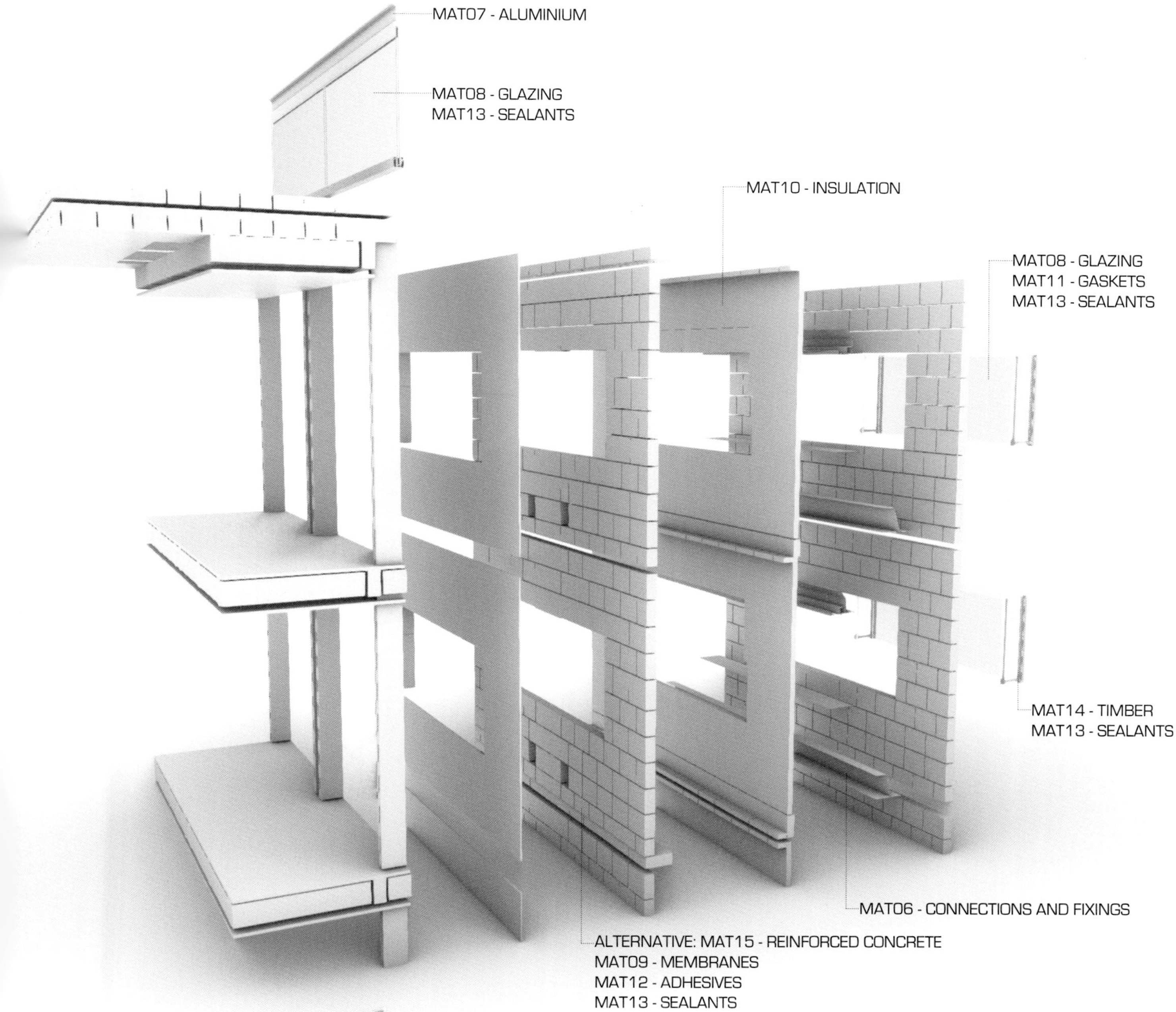

c | 3D exploded view of concrete block facade components with material specification references

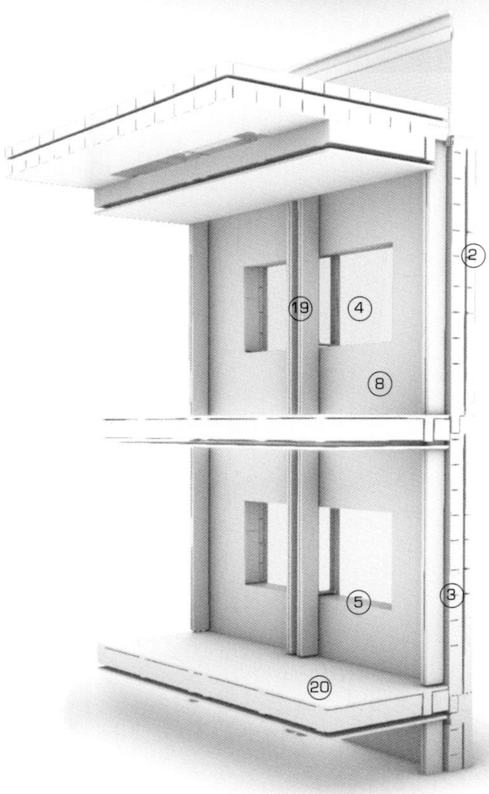

12 | Outer precast concrete lintel or stone flat arch
13 | Damp proof course (DPC)
14 | Movement joint
15 | Timber framed inner skin with quilt insulation
16 | Cement fill
17 | Foundation
18 | Vertical DPC
19 | Structural column
20 | Floor construction

d | 3D view showing concrete block facade fixed to timber frame

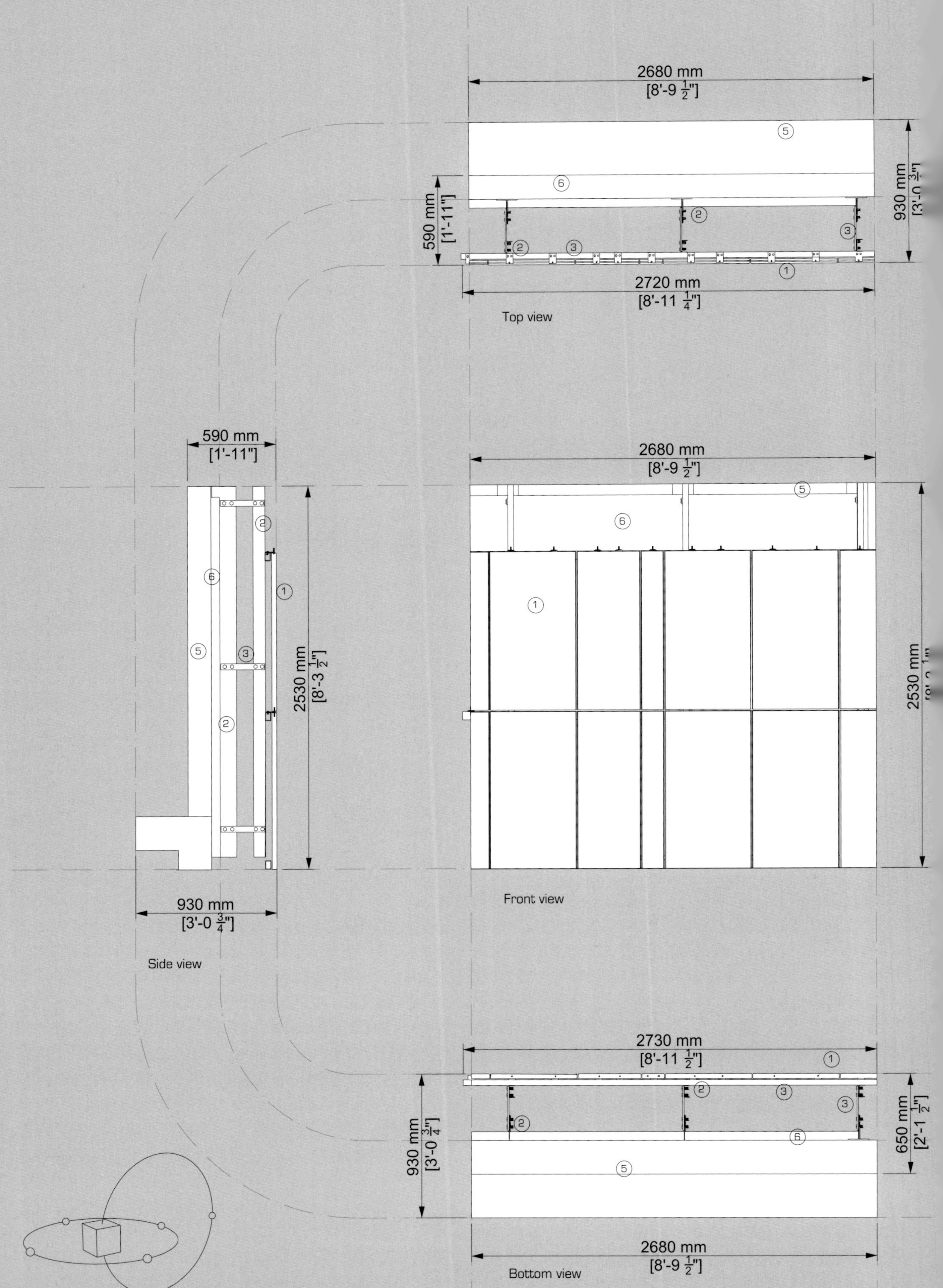

Third angle projection 1:30

Details for stone supported on carrier system

1 | Stone panel
2 | Extruded aluminium carrier system
3 | Stainless steel fixing pins
4 | Stack joint
5 | Backing wall. Concrete shown
6 | Thermal insulation

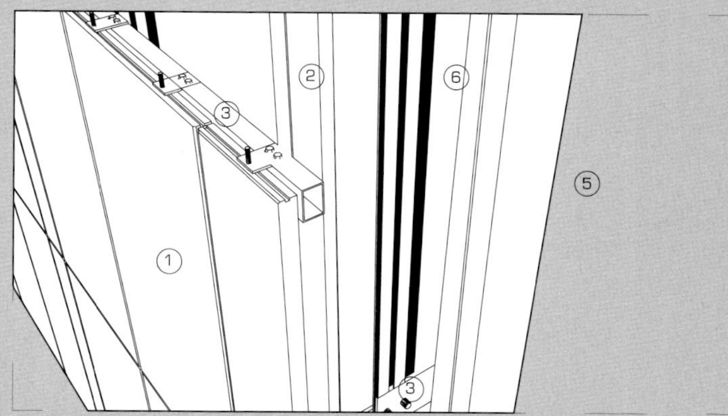

3D view of the assembly

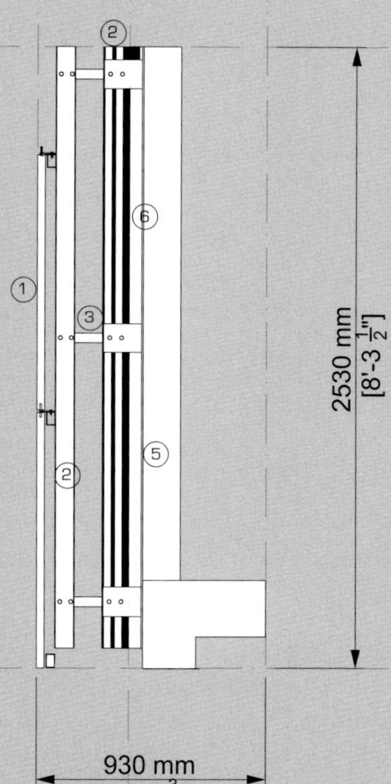

Side view

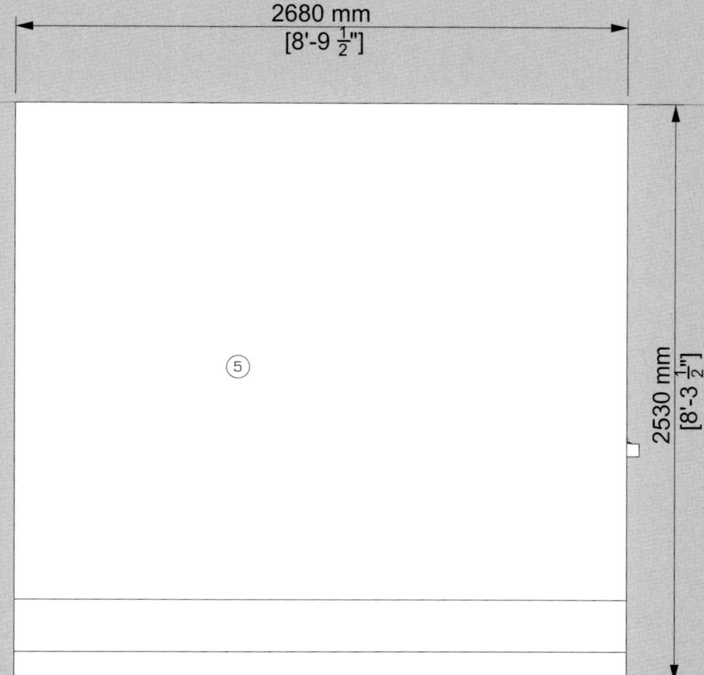

Back view

7 | Adjacent wall. Rainscreen panels shown
8 | Open joint configuration as shown
9 | Support brackets
10 | Restraint brackets
11 | Joint between stone panels
12 | Floor slab
13 | Stone coping

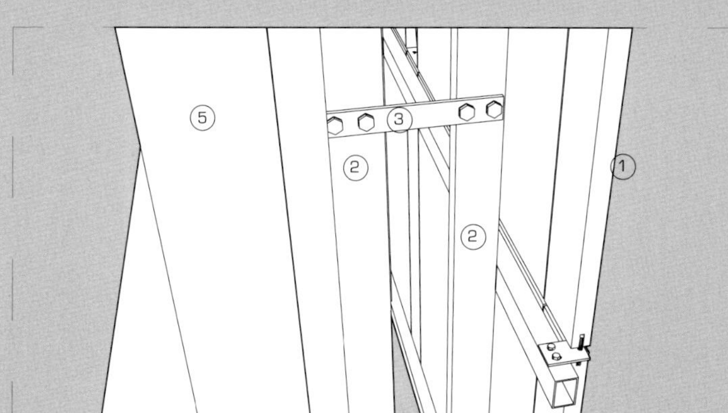

3D view of the assembly

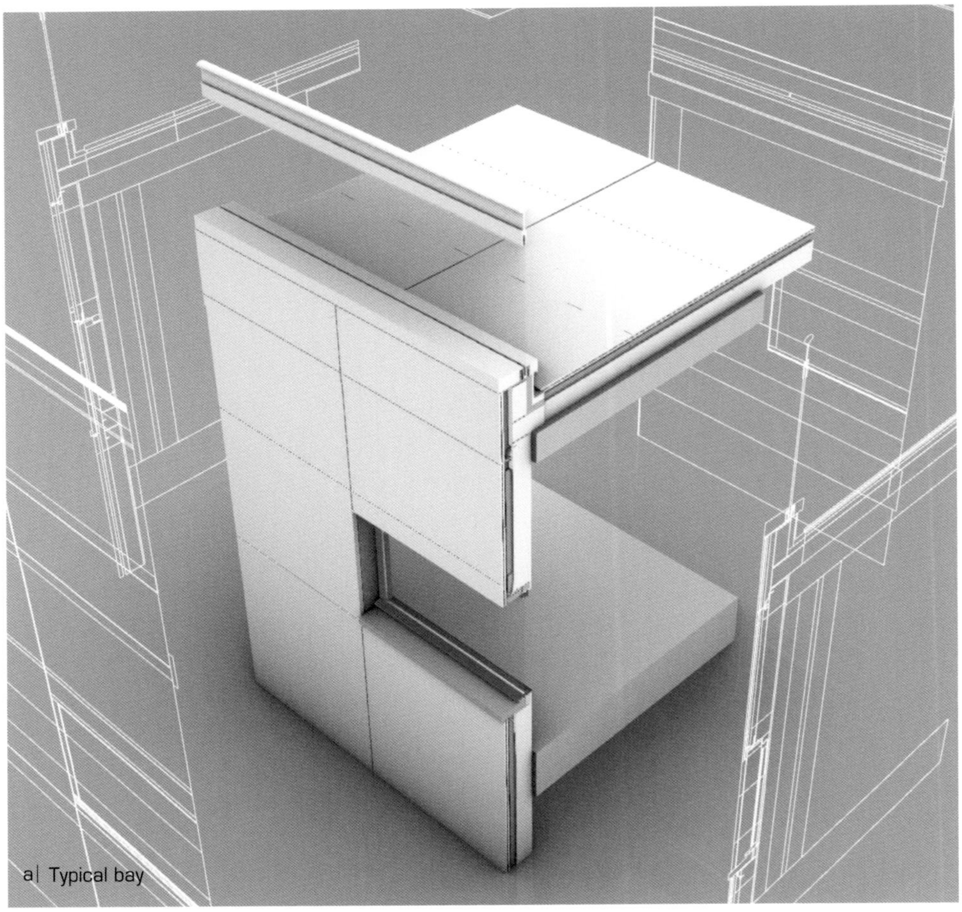

a| Typical bay

b | Horizontal and vertical section. Typical stone carrier system made from extruded aluminium. Stone joints with both straight and staggered joints to show flexibility of carrier systems

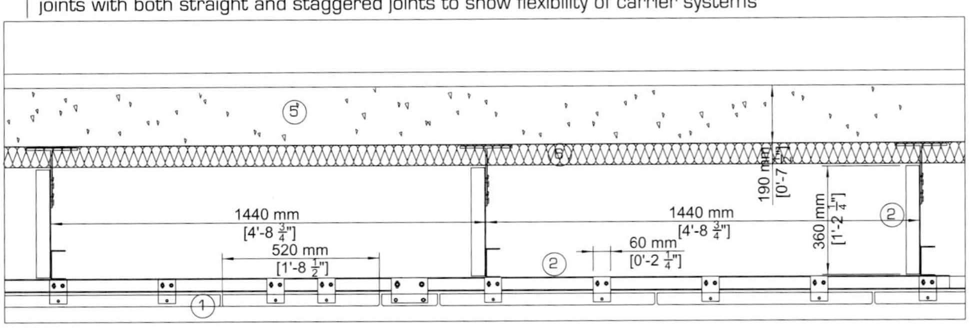

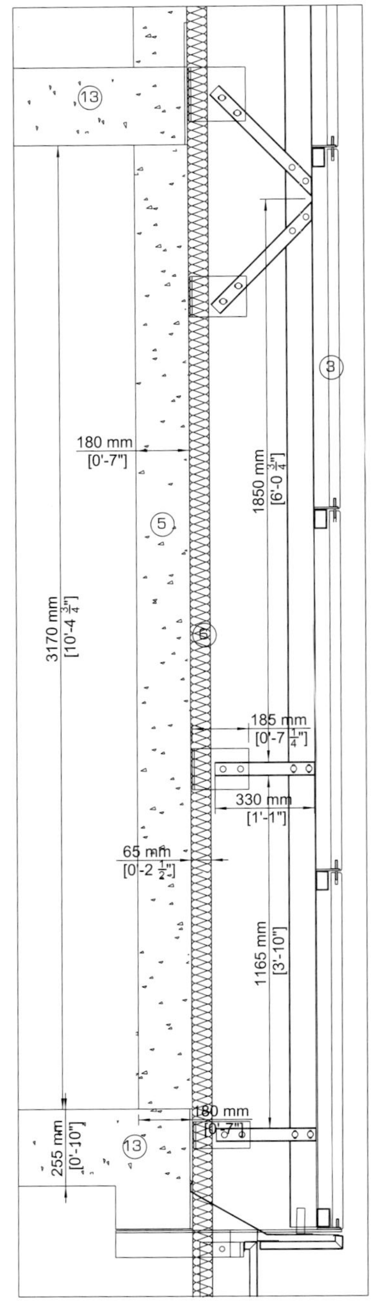

Cut stone used for wall cladding can be either sealed with mortar or sealant or, alternatively, be open jointed following rainscreen principles. In both cases the stones are fixed to an insulated backing wall (MCE_257, c). The stone types used in cladding are granites, sandstones, limestones and slates. Marbles are sometimes used, but are not generally considered to be as durable as the other stone types. Their weights vary from 2500kg/m² for limestone to 2750 kg/m² for granite.

An essential difference between stone and other cladding materials is that the material may still not have been cut from the quarry at the time the material is chosen for a building project, particularly for larger projects. The use of stone for cladding requires considerable planning, from setting the range of colours, tone and surface marks that will be used for the cladding, to establishing the physical properties of the stone from the actual quarry chosen, to using this data to complete a structural design to establish stone panel sizes and thicknesses.

Stone is varied in its durability and strength due to variation in the bed of the stone. Some quarries have consistent stone, while in others its properties may vary considerably. Test data on the physical properties of the stone in each quarry are obtained, particularly its strength, in order to establish the panel size and thickness that will be used. The thickness of stone for a facade application is usually established by structural calculation. Flexural strength, also called the modulus of rupture, is usually the most important structural consideration. Codes of practice often set out minimum thicknesses for different panel sizes in different stones, but this is only a general guide and calculation is usually undertaken for facade applications.

Samples are usually chosen to set the range of colour and texture that are to be used in a facade. Because the stone required for a particular project is not always cut before construction starts, the quarry confirms the amount of time needed to cut and finish the stone panels,

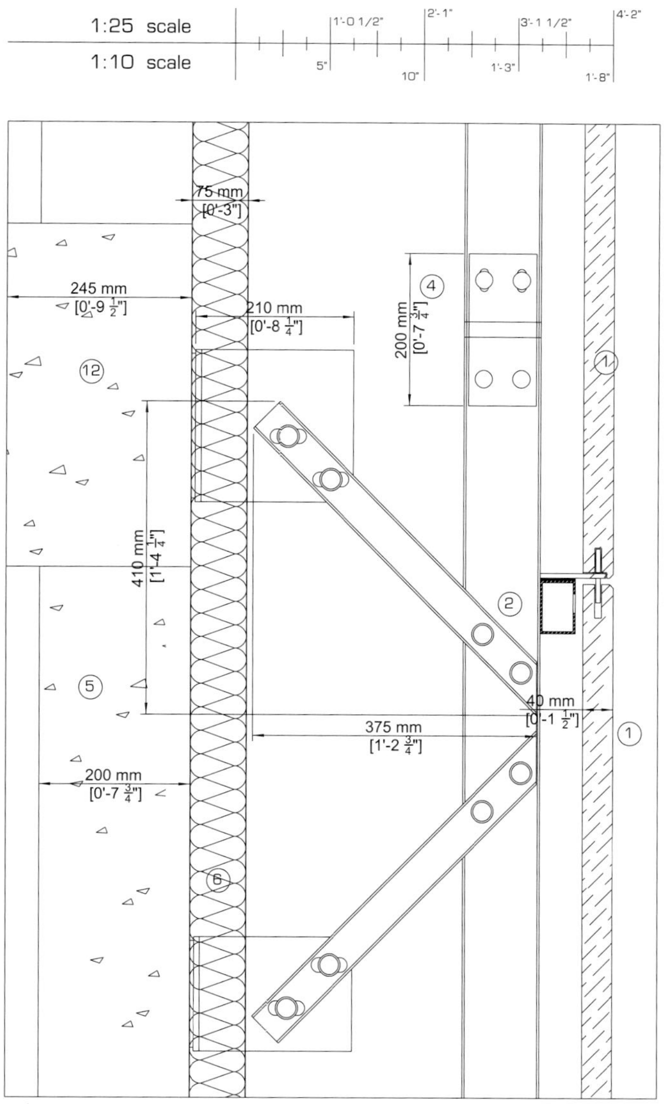

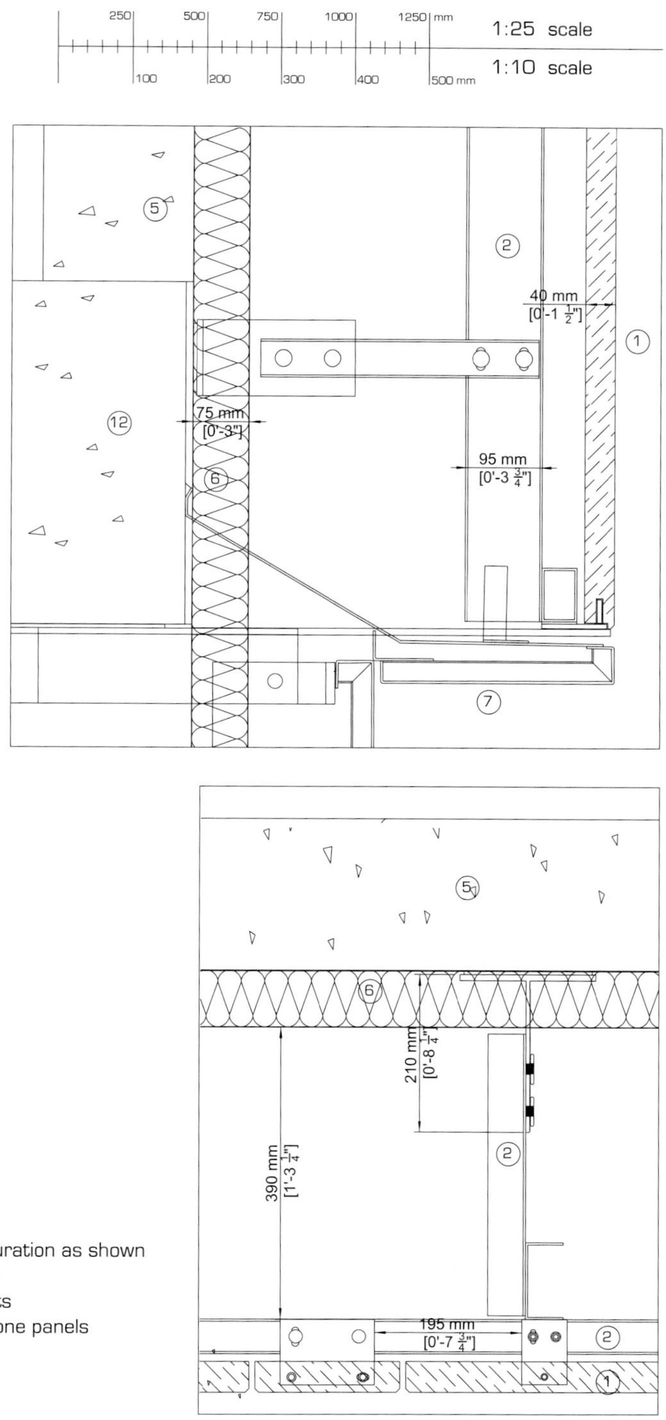

c | Vertical sections. Typical stone carrier system made from extruded aluminium

Details for stone supported on carrier system

1 | Stone panel
2 | Extruded aluminium carrier system
3 | Stainless steel fixing pins
4 | Stack joint
5 | Backing wall. Concrete shown
6 | Thermal insulation
7 | Adjacent wall. Rainscreen panels shown
8 | Open joint configuration as shown
9 | Support brackets
10 | Restraint brackets
11 | Joint between stone panels
12 | Floor slab
13 | Floor slab
14 | Stone coping

which can take considerably longer than cladding panels in other materials. Because the material is natural, care is taken to avoid defects such as cracks and holes in the material that will affect its durability as a cladding material. For slates, natural markings are a part of the nature of the stone and the material is judged for its particular application.

Fixings

Stainless steel is most commonly used for fixings rather than other materials because of its resistance to corrosion combined with high strength and rigidity. Stone fixings allow for a three-way adjustment to ensure a proper fit vertically, horizontally and laterally (in or out from the facade) (MCE_257, c). Where anchor slots or inserts are cast into the floor slab or reinforced concrete wall, these are needed to be accurately positioned even though they provide adjustment for the final position of the stone. The type of fixing used is dependent mainly on the stone thickness.

Loadbearing fixings, supporting the stones, are usually set on the bottom edge of the stone panel, though side fixings are sometimes used (MCE_254, a), depending on the weight and strength of the panels. Two brackets at the bottom of the stone are used, with not more than four brackets in total. Triangular panels usually have one fixing in each corner. At vertical movement joints, adjacent stones are carried on separate supports, each side of the movement joint. The end of the fixing supporting the stone is around 50mm wide.

Restraint fixings are used to resist positive and negative wind loads as well as imposed loads from maintenance equipment. The slot in the stone is either a hole, slot or rebate that is not visible from the stone face (MCE_257, c). These are usually located at 1/5 points for stack bonded stones and ¼ points for ½ bonded stones, and at least 75mm (3in) from the corner. A maximum of four restraint fixings per stone is used. Restraint fixings typically consist of a stainless steel pin, or dowel,

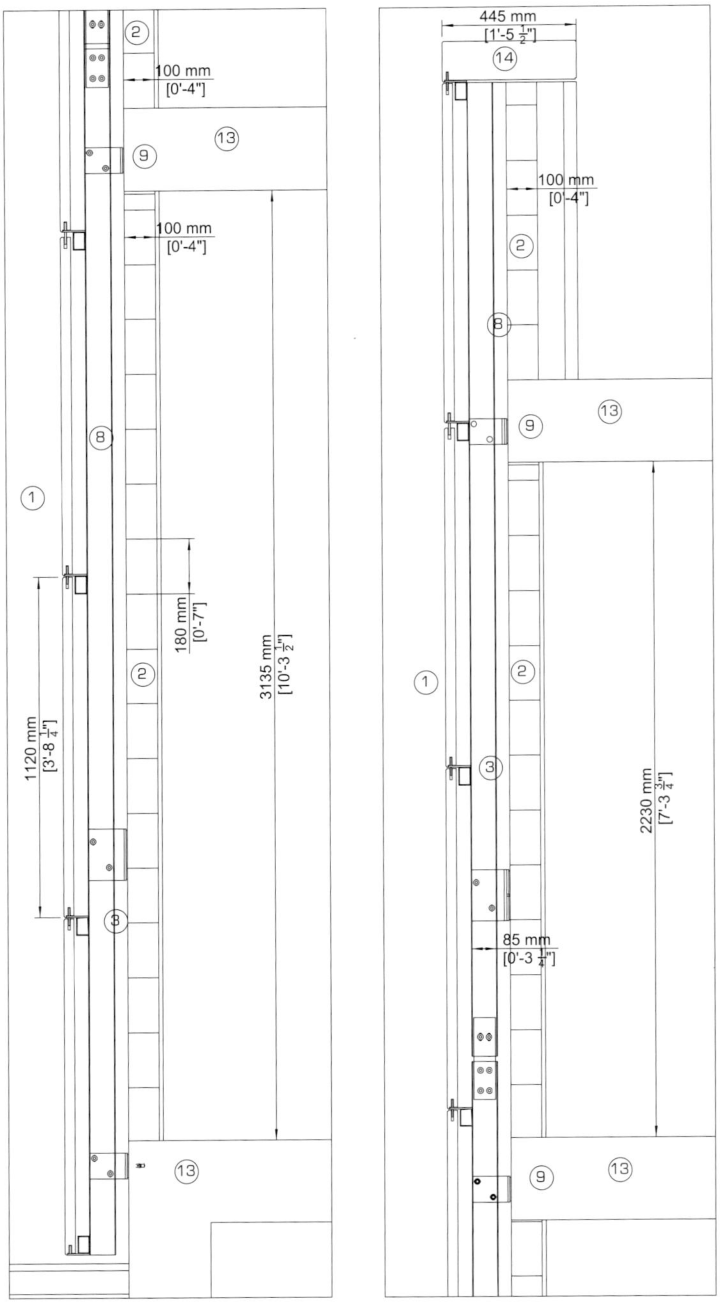

a | Vertical sections. Typical stone carrier

b| Prototype

set into a stainless steel flat section in an 'L' shape (MCE_257, c). The pins are typically 3mm (0.118in) diameter for stone up to 30mm (1.2in) thick and 5mm (0.2in) for stone of greater thicknesses.

Face fixings are normally used with marble and granite. The bolts function as both loadbearing and restraint fixings and are set away from the corners in the manner of bolt fixed glazing. A bolt is fixed in each corner, usually at a distance equal to three times the stone thickness but to a maximum distance of around 75mm (3in) from the corner. Smaller stones are fixed with fewer fixings, and triangular panels usually have a fixing in each corner only.

Fixings used to support stones that clad soffits, such as the underside of a concrete slab, are suspended from bolts or hangers which slide into anchorages cast into the supporting structure.

Cladding to precast concrete panels

Any of the stones listed at the beginning of this section can be used as a facing to precast concrete panels, but granite is most commonly used due to its higher strength, allowing it to be used relatively thinly. Thicker sandstones and limestones are also used, however. Stone is fixed to concrete panels with dowel pin fixings which are inclined at 45º to 60º to suit the size of stone and hold it in place. Pins are usually set at around 200mm (8in) centres both vertically and horizontally with a pin thickness of around 5mm. 50% of dowels are set in each direction to provide a balanced support. Each dowel has a flexible rubber-based washer around 3mm (0.118in) thick to allow for the movement between stone and the concrete background. The dowel penetrates 2/3 the thickness of the stone and around 60mm–75mm into the concrete (2.3in–3in).

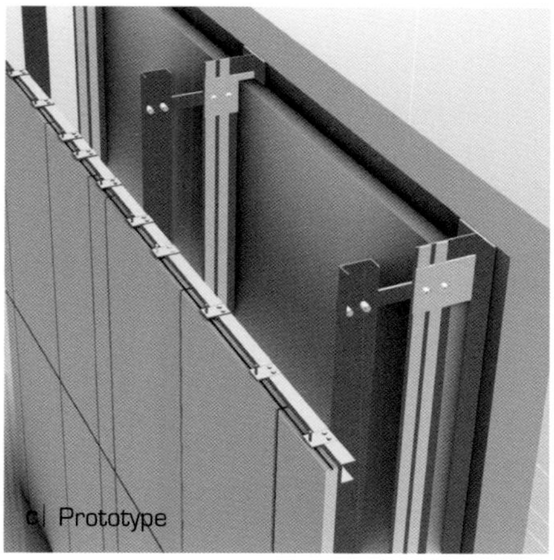

c| Prototype

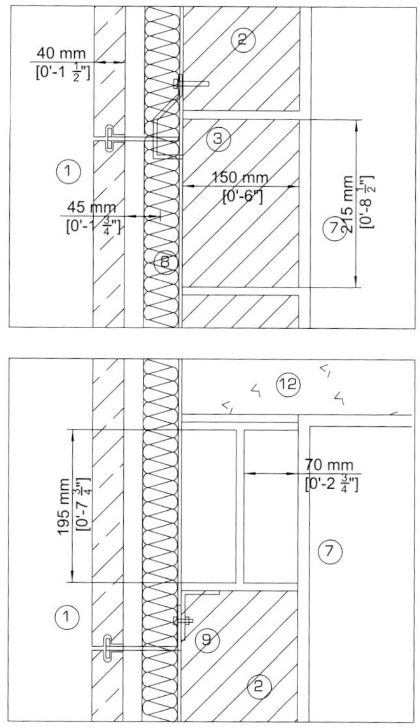

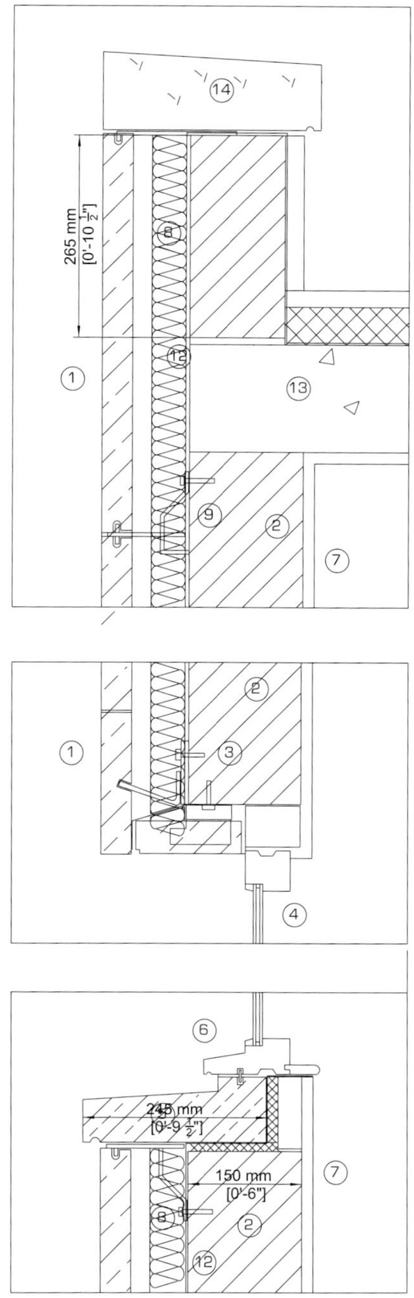

Details for stone on individual fixings

1 | Stone panel
2 | Backing wall, typically concrete block
3 | Stainless steel fixings (a wide range is available)
4 | Timber framed window
5 | Stone sill
6 | Timber inner sill
7 | Internal plaster finish or dry lining/ drywall
8 | Thermal insulation in cavity
9 | Stainless steel restraint at each floor level
10 | Air cavity
11 | Precast concrete lintel
12 | Waterproof membrane
13 | Floor slab
14 | Stone coping

d | Horizontal and vertical sections. Typical stone carrier system made from individual stone fixings

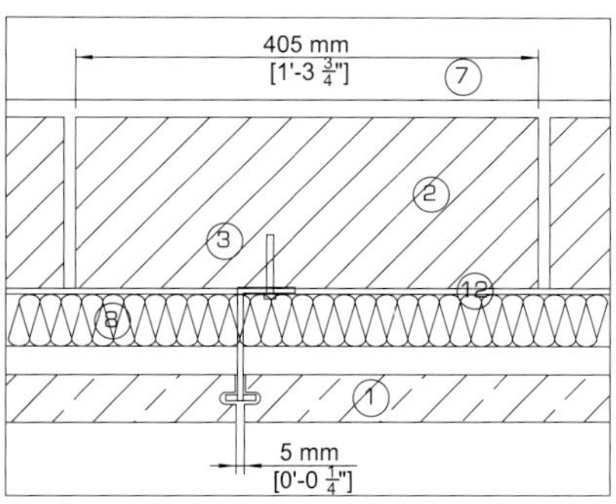

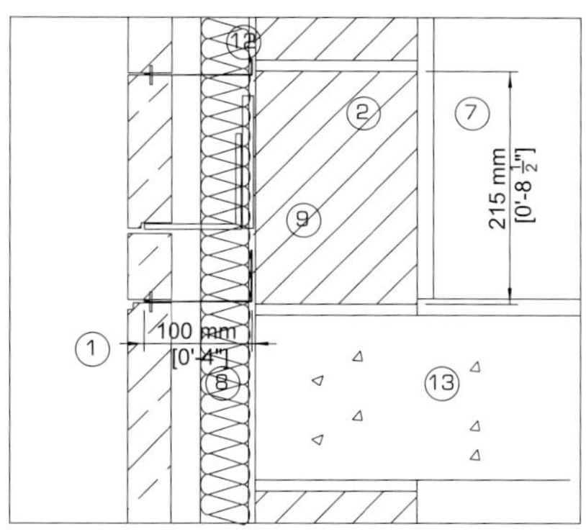

Joints

Joints are either of closed or open type. Closed joints use either mortar or proprietary sealant. Closed joints are used where the cladding is supported at each floor level on stainless steel angles with jointed stones. With open-jointed stones, each panel is individually supported in a rainscreen construction, where rainwater passing through the joints is drained away down either the back of the stones or down the face of the backing wall.

Closed joints

Closed joints need to be loadbearing and watertight and must also accommodate the relative movements of cladding and the supporting building structure. The types of jointing or pointing (the outer finish of mortar or sealant) will depend on the type, size, thickness and surface finish of cladding units. Stones are not usually butted up against one another, since any movement of the unit or of the structure cannot be accommodated, causing damage to the stones.

Joints in sandstone and limestone are usually filled with cement/ sand mortar or cement/sand/lime mortar. Granite and slate typically use proprietary sealant, such as two-part polysulphide. Mortar used for pointing is made frost resistant when used in temperate climates, and of similar strength to the jointing mortar, which is the structural mortar behind. Neither mortar should be stronger than the stone. For limestone and sandstone a mortar of 1/1/5 for cement/lime/sand is typically used, or 1/2/8 for cement/lime/crushed stone particles, where the colour of the mortar is blended with the crushed stone.

For narrow joints used with granite or slate, a high cement-to-sand ratio is used. Joints wider than around 4mm are filled with a weaker mix to reduce shrinkage cracks. The maximum widths of mortar-filled

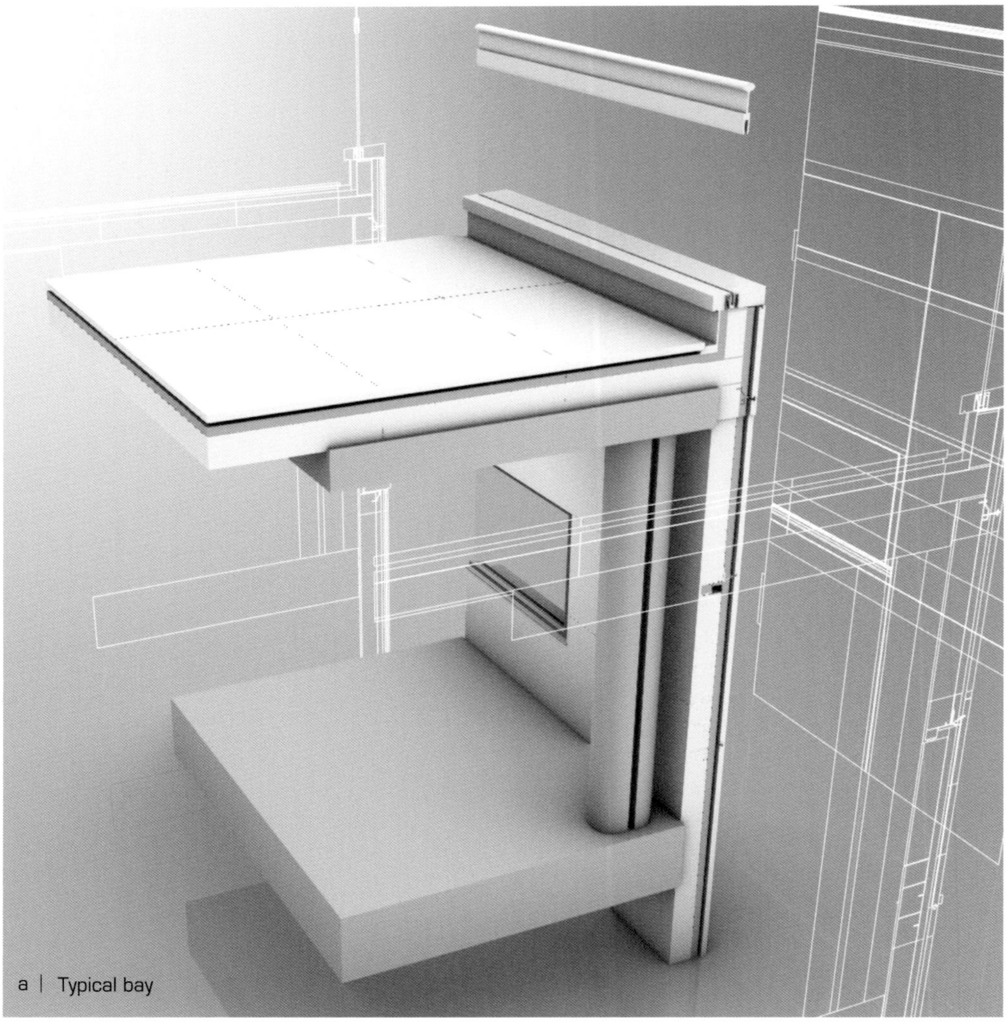

a | Typical bay

b | Horizontal section. Typical stone carrier system made from individual stone fixing

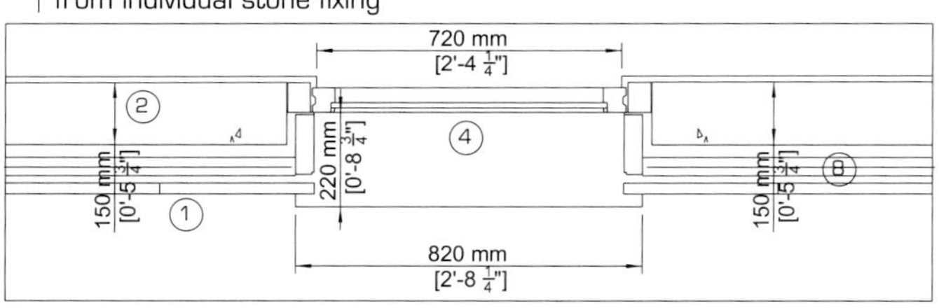

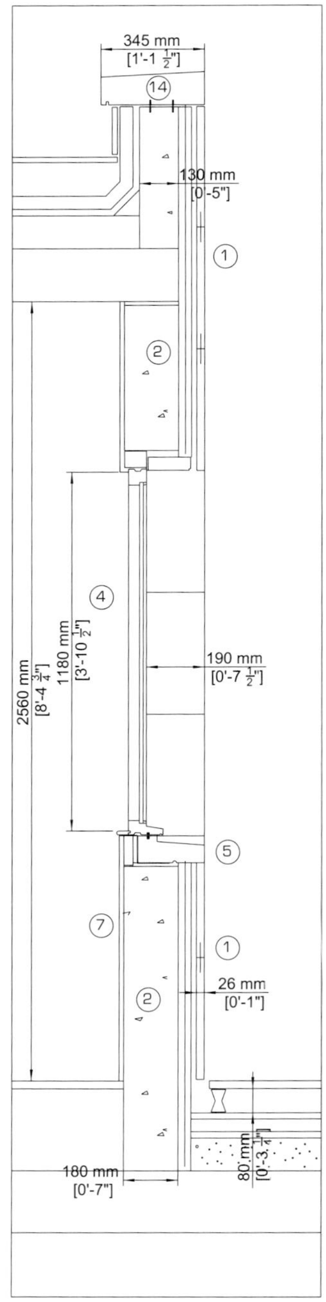

c | Vertical section. Typical stone carrier system made from individual stone fixing

joints are around 12mm (0.5in) but sealant filled joints can be up to around 30mm (1.2in), depending on the proprietary product used. Joint widths are usually a function of the cutting tolerance of the stone, around 2mm difference in the cut line of the stone, depending on stone type and the cutting machine used. Modern machines can cut stones to within 1mm accuracy. Joint widths of 4mm are common, but this can rise to a maximum of 12mm when required for visual reasons (MCE_256, b), particularly when the joint is recessed. Granite, slates and hard limestones and sandstones can have a joint width at a minimum of 3mm, while soft sandstones and limestones can be laid with a minimum 5mm joint. When a proprietary sealant is used, the minimum joint width for all stone types is usually around 5mm.

Movement joints

Horizontal movement joints are used to deal mainly with floor slab deflections, and to a lesser extent vertical shortening, in the structural frame. This horizontal joint is usually provided at slab level, where the stone cladding is supported from either short lengths of stainless steel angle, or a continuous shelf angle (MCE_254, b). The joint occurs immediately below the stainless steel angle, where vertical deflection will occur. Horizontal movement joints can be set at intervals of two storeys if the stone and support brackets or frame can be designed to span the height. The joint width is usually a minimum of 15mm, but 20mm to 25mm (0.8in–1in) are common with reinforced concrete structures. The joint is formed either as a sealant or as a step in the stone, where the upper stone projects forward of the stone below to conceal the wider joint. In both joint types the joint is made watertight, for closed joints and for open joints, where rainscreen cladding principles are used.

Vertical movement joints are provided to deal with racking in the structure as well as movements in the cladding itself. Where movement joints occur in the building structure, usually following a continuous ver-

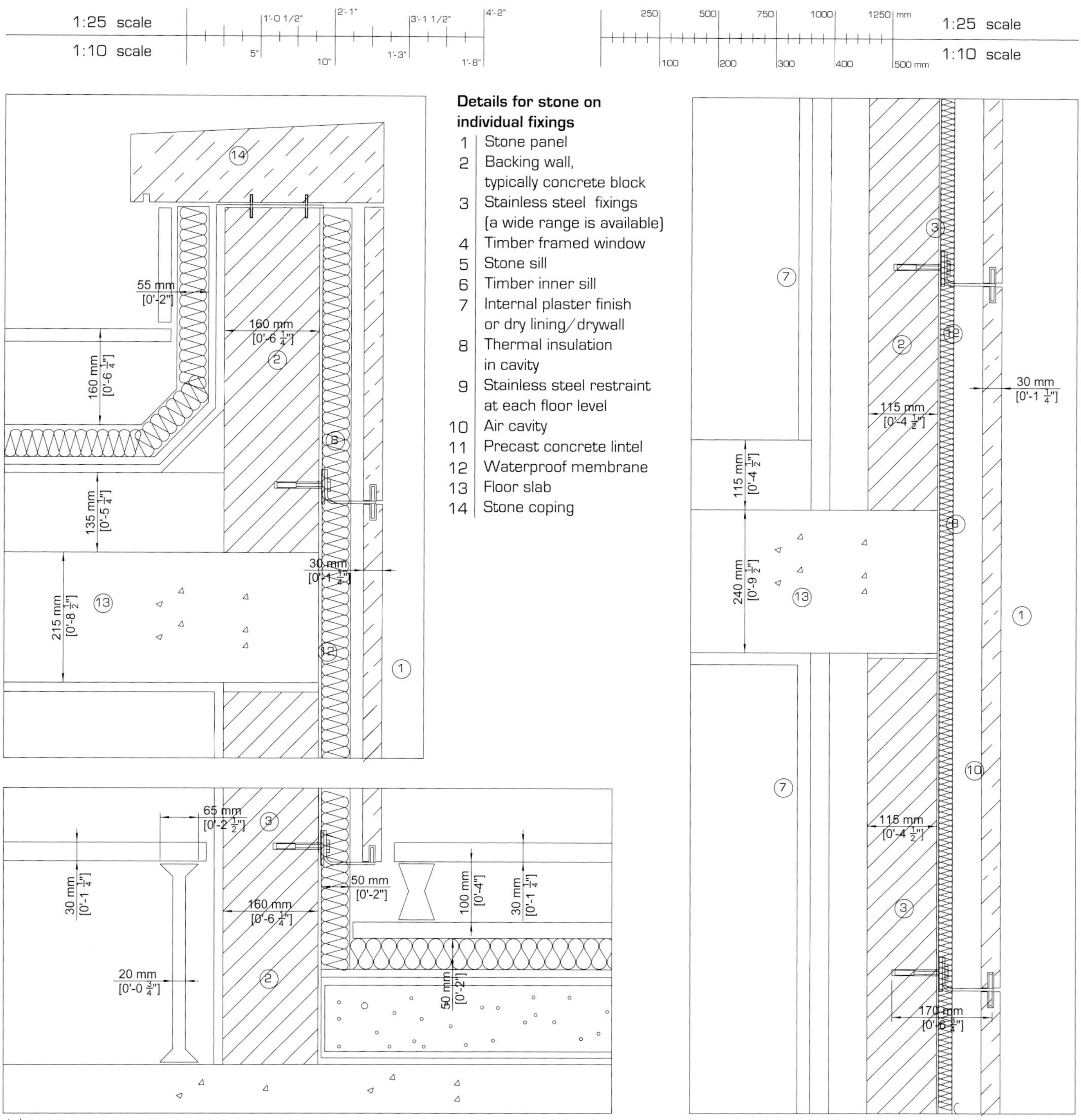

d | Vertical sections. Typical stone carrier system made from individual stone fixing. Parapet detail

tical line where it intersects with the facade, a vertical movement joint is provided in the facade in the same location. The distance between joints is typically around 6 metres in a continuous run of stone cladding with closed joints. The joint width corresponds to the expected movement in the cladding, but where sealed joints are used, the joint width is dependent on the amount of movement that the sealant is required to accommodate. Minimum widths of vertical movement joints are around 10mm (0.5in). Vertical movement joints are extended into parapets and copings.

Stone finishes

Granites, limestones, sandstones and slates are using an increasingly wide range of finishes, with finishes associated with one stone type being used for another. The main finishes are as follows: A rubbed finish is a smooth finish made by rubbing stone with an abrasive material (typically used for limestone and sandstone); a honed finish has a dull polish (used for all types); a polished surface has a high gloss (typically used for granite and hard limestone); a flamed finish is obtained by passing a hot flame over the stone surface (typically used for granite and slate); a riven finish, where the stone is cut on its cleavage plane (typically used for slate, sandstone or limestone); and tooled, where the material is worked, leaving tool marks and is used mainly on sandstone or limestone. In addition, stones can be filled with cements or proprietary fillers to conceal natural voids in the stone. The surface is then coated or polished.

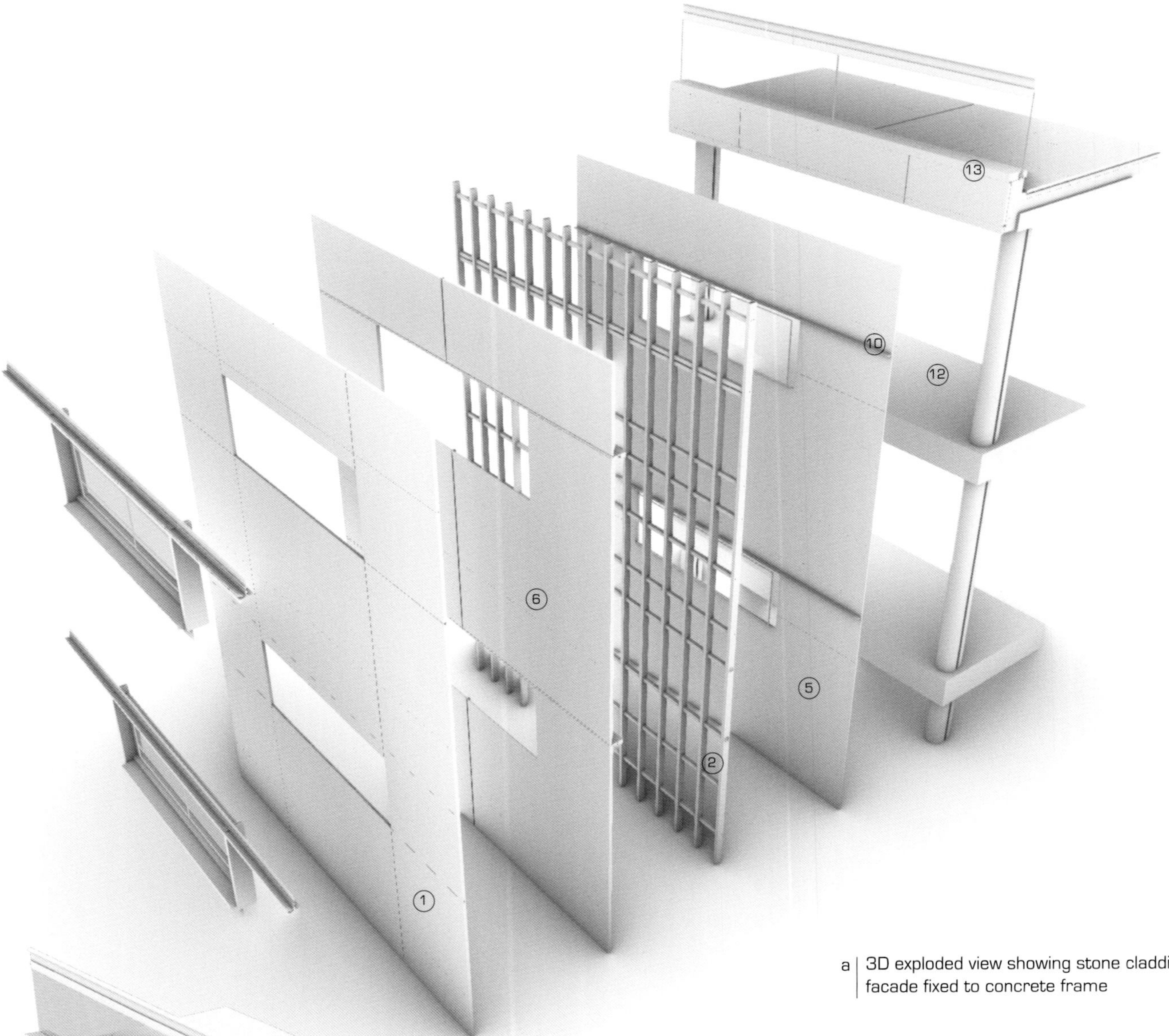

a | 3D exploded view showing stone cladding facade fixed to concrete frame

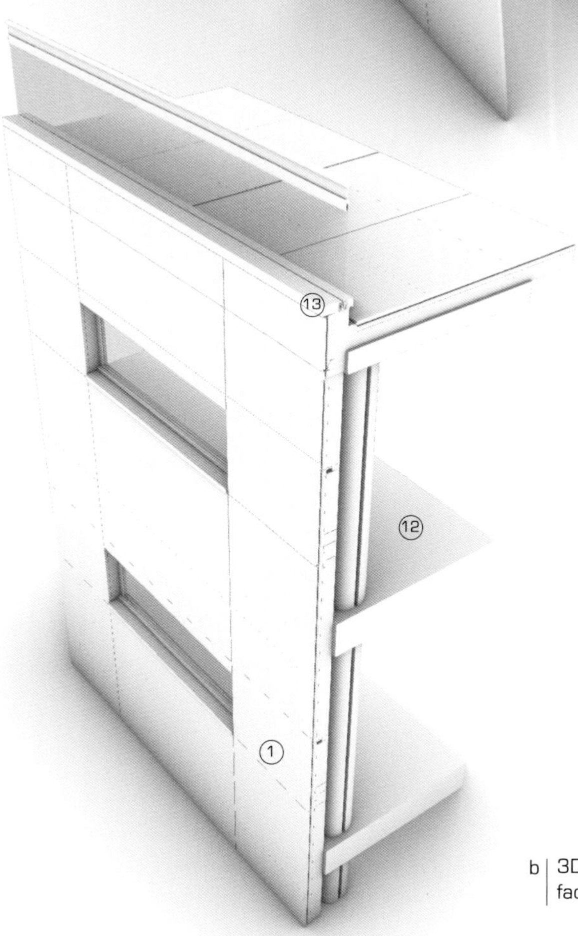

Details for stone supported on carrier system

1 | Stone panel
2 | Extruded aluminium carrier system
3 | Stainless steel fixing pins
4 | Stack joint
5 | Backing wall. Concrete shown
6 | Thermal insulation

b | 3D view showing stone cladding facade fixed to concrete frame

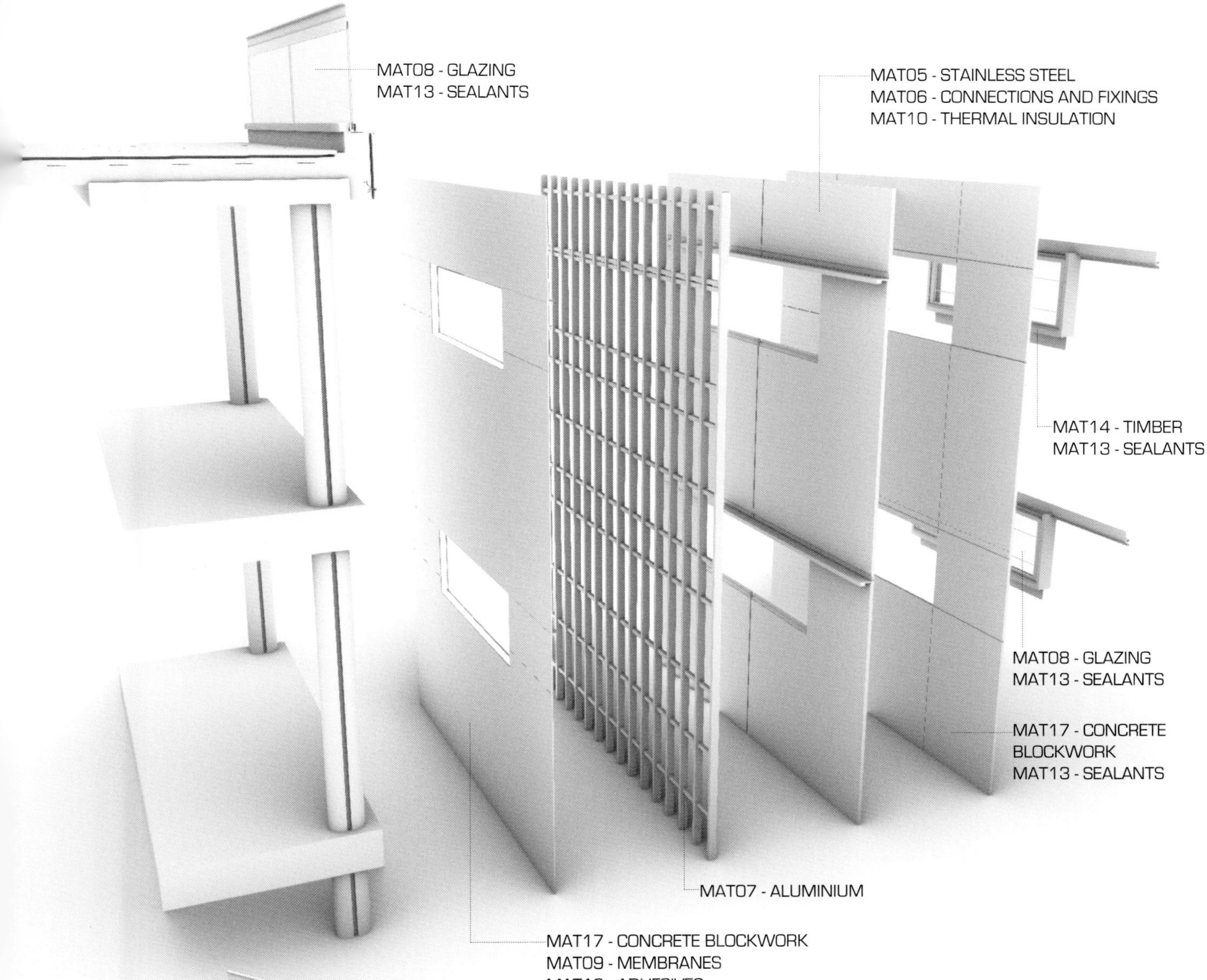

c | 3D exploded view of stone cladding facade components with material specification references

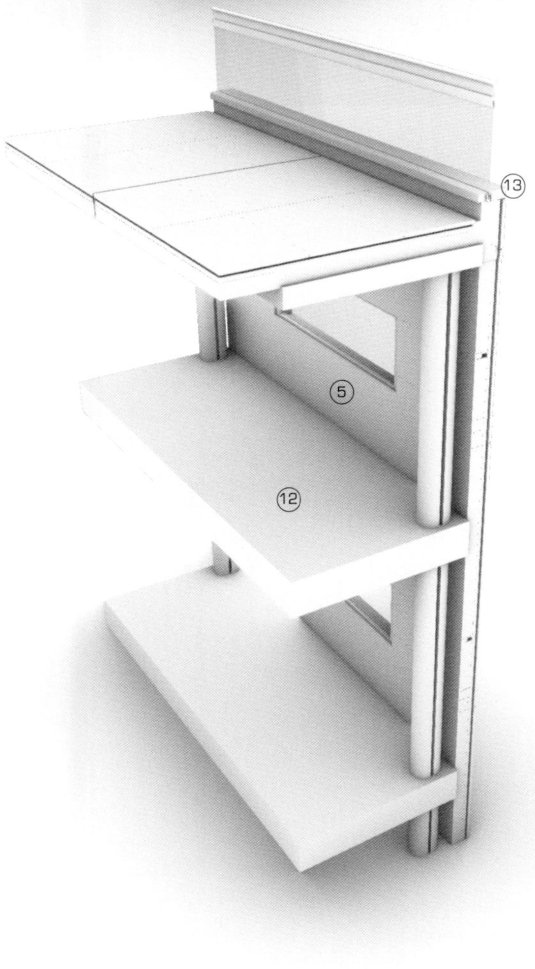

7 | Adjacent wall. Rainscreen panels shown
8 | Open joint configuration as shown
9 | Support brackets
10 | Restraint brackets
11 | Joint between stone panels
12 | Floor slab
13 | Stone coping

d | 3D view showing stone cladding facade fixed to concrete frame

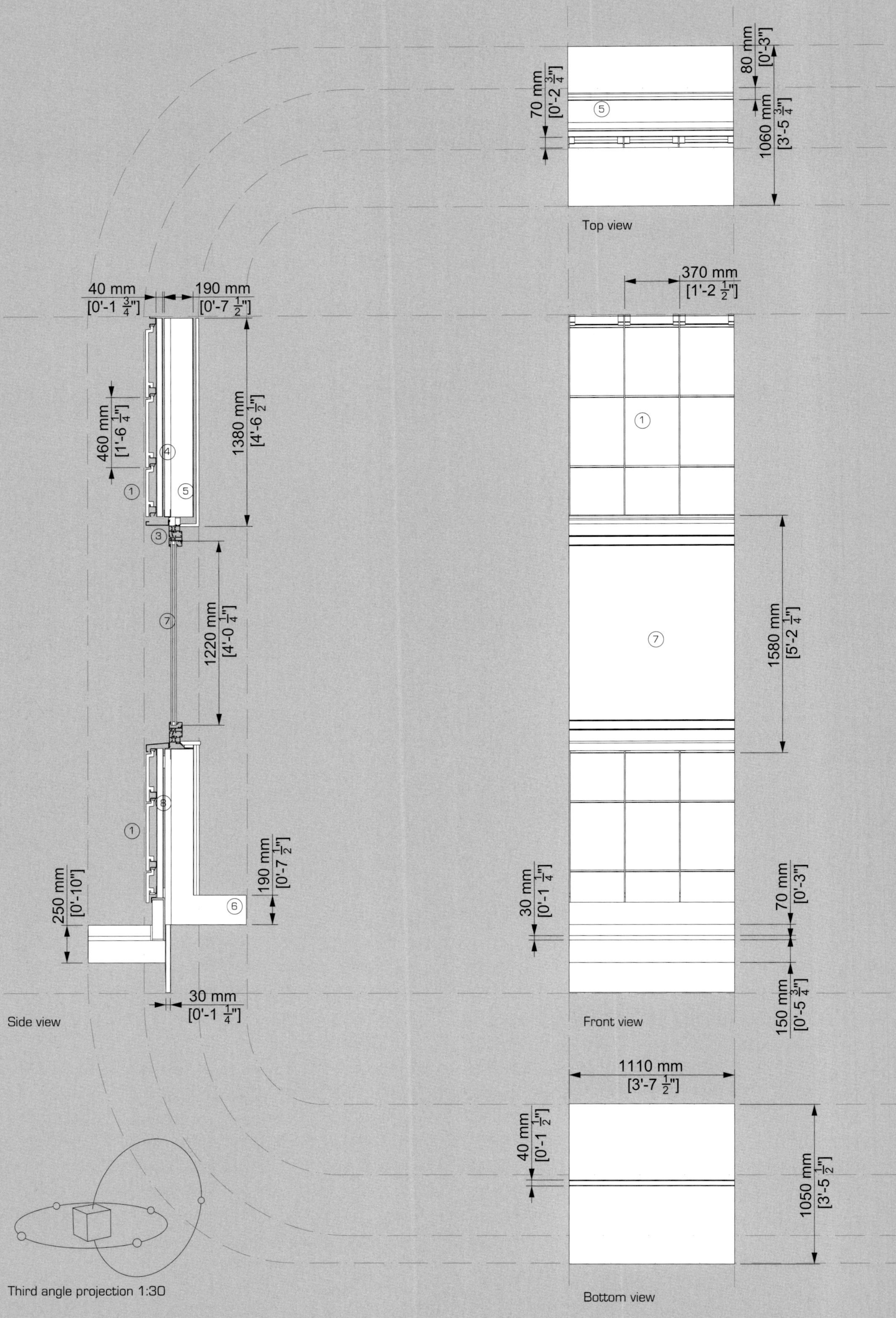
80 mm [0'-3"]
70 mm [0'-2 3/4"]
1060 mm [3'-5 3/4"]
Top view
370 mm [1'-2 1/2"]
40 mm [0'-1 3/4"]
190 mm [0'-7 1/2"]
460 mm [1'-6 1/4"]
1380 mm [4'-6 1/2"]
1220 mm [4'-0 1/4"]
1580 mm [5'-2 1/4"]
190 mm [0'-7 1/2"]
250 mm [0'-10"]
30 mm [0'-1 1/4"]
70 mm [0'-3"]
150 mm [0'-5 3/4"]
30 mm [0'-1 1/4"]
Side view
Front view
1110 mm [3'-7 1/2"]
40 mm [0'-1 1/2"]
1050 mm [3'-5 1/2"]
Third angle projection 1:30
Bottom view

Details

1 | Terracotta rainscreen
2 | Extruded aluminium carrier frame
3 | Support brackets, typically aluminium
4 | Thermal insulation
5 | Backing wall
6 | Floor slab
7 | Metal framed window
8 | Waterproof membrane
9 | Structural column

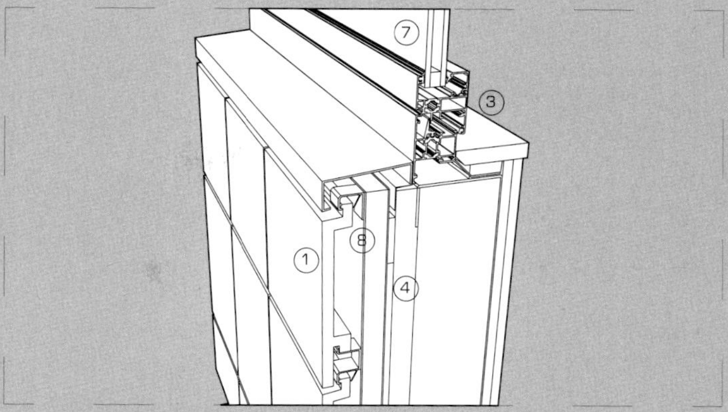

3D view of the assembly

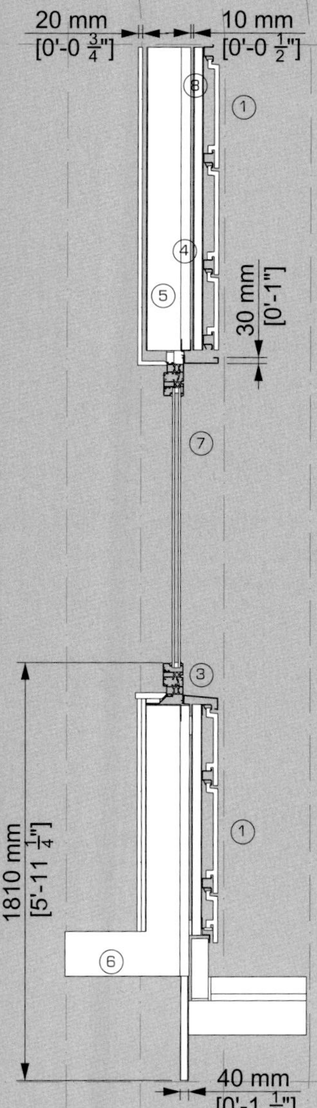

Side view

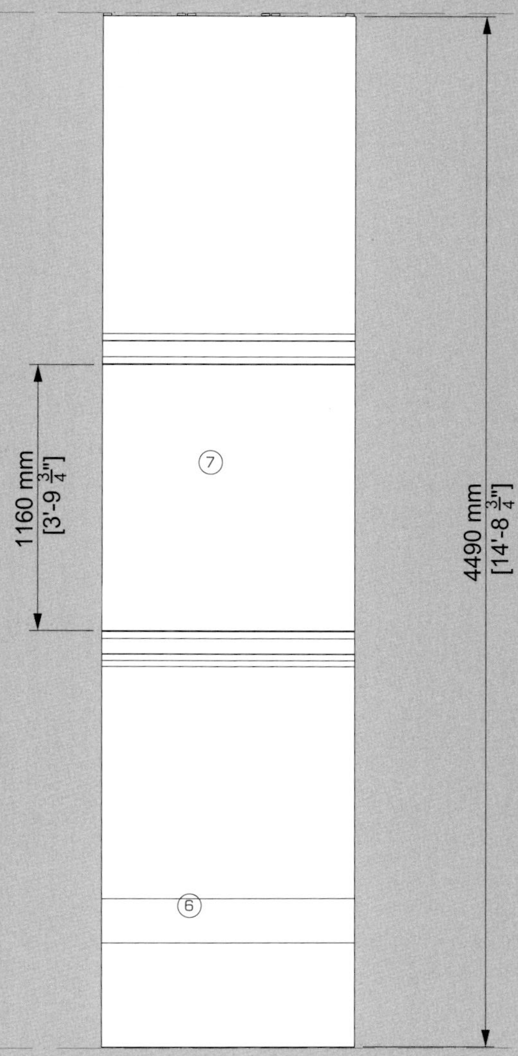

Back view

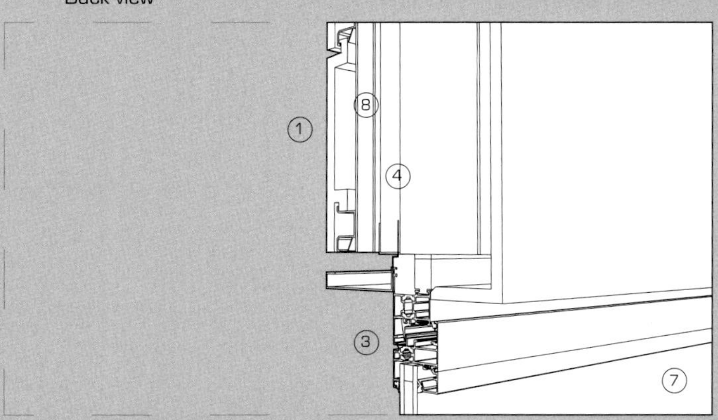

3D view of the assembly

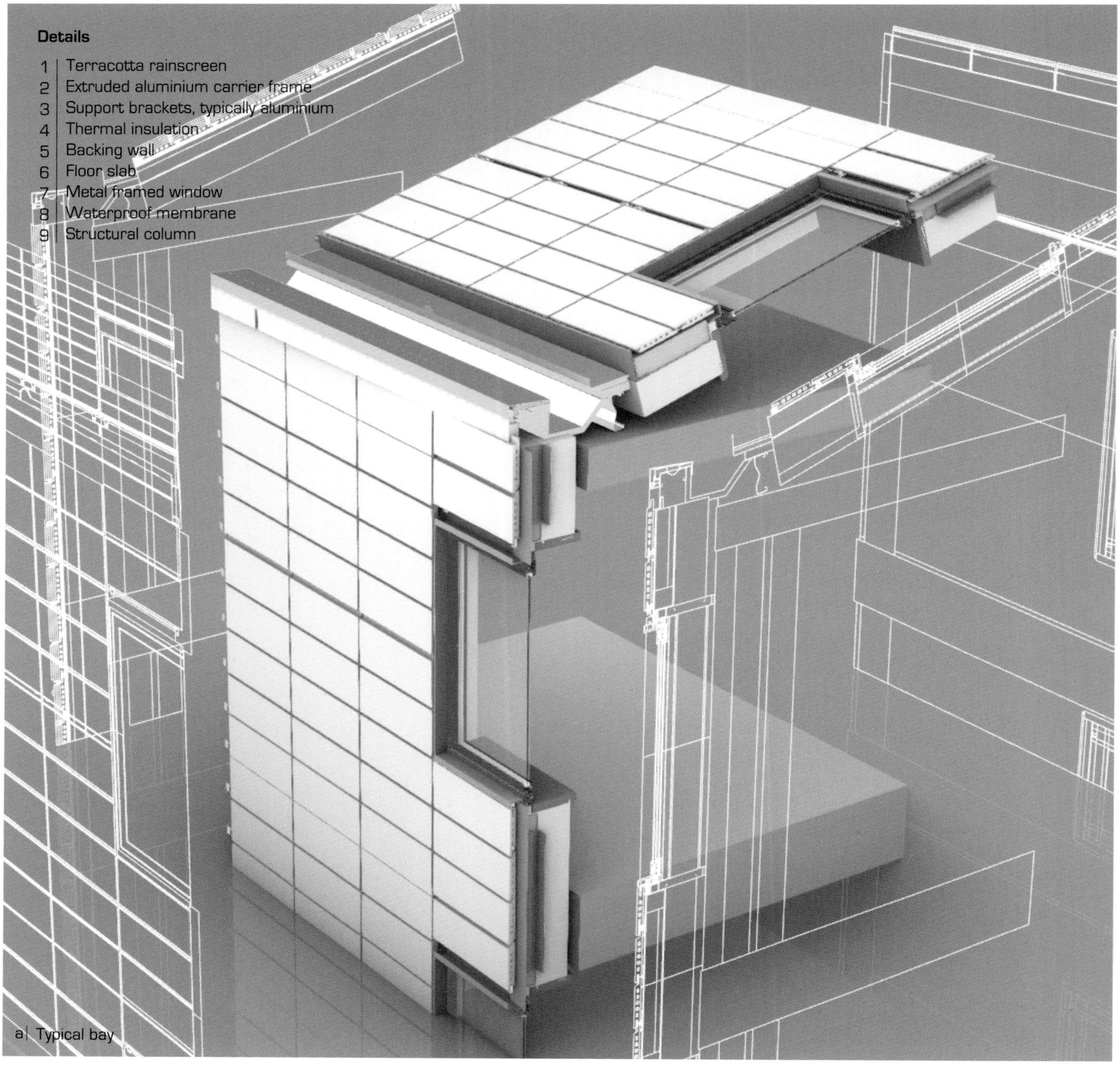

a| Typical bay

The essential principles of rainscreen cladding are discussed in the section on metal rainscreens (MCE_86). Terracotta rainscreens have developed into patented proprietary systems from prototypes within the last ten years. Over that time the sizes of terracotta panels have increased and fixing systems have developed for use as solar shading screens to glazed walls as part of an overall rainscreen system for a building facade. Hollow terracotta sections are reinforced with aluminium sections set into them to form louvred screens that can match with adjacent areas of cladding. Terracotta is fixed either on rails, into aluminium or stainless steel panels, or on individual brackets like masonry cladding with concealed fixings (MCE_267, g). Vertically set or horizontally set rails are used to suit a range of joint arrangements that imitate traditional masonry bonds, or can be stack-bonded in the manner of wall tiling or glass blocks. Terracotta has been developed for use in rainscreens both from building blocks and bricks, as used in loadbearing masonry construction, and from decorative tiles, where many of the glazed finishes are derived. Recently developed systems have interlocking panels, to provide crisp joints, and double wall sections to provide long spanning tiles with high flexural strength, or modulus of rupture, combined with lightness in weight (a). The range of glazed finishes has developed considerably in the past few years to give a very wide range of textures and colour mixes derived from contemporary pottery.

Manufacture of panels

Terracotta is made from natural clay that is extruded and fired in a kiln. Powdered clay is mixed with water in the factory to achieve a controlled level of water content. It is then extruded through dies that draw the material along a conveyor belt where it is wire cut to the required length. The use of dies makes the manufacture of terracotta panels very flexible, giving it the ability to make new shapes and sections for each new project with relative ease. The die creates different heights and depths of block, with hollowing out of the interior to keep the mate-

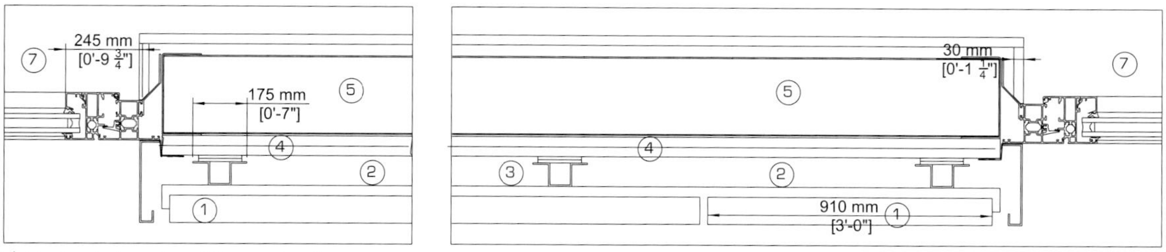

b| Horizontal section. Window jamb

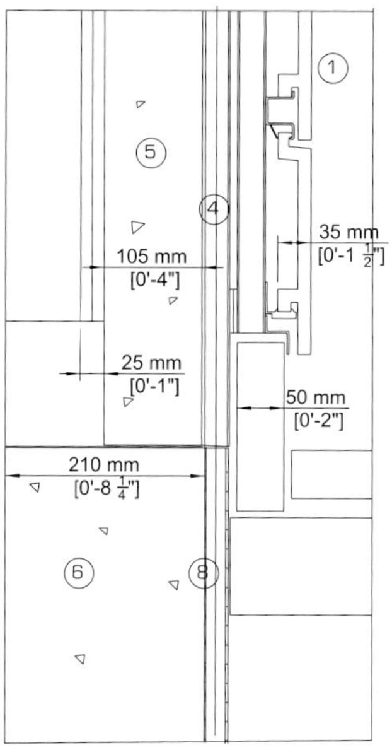

c| Vertical section. Junction with ground

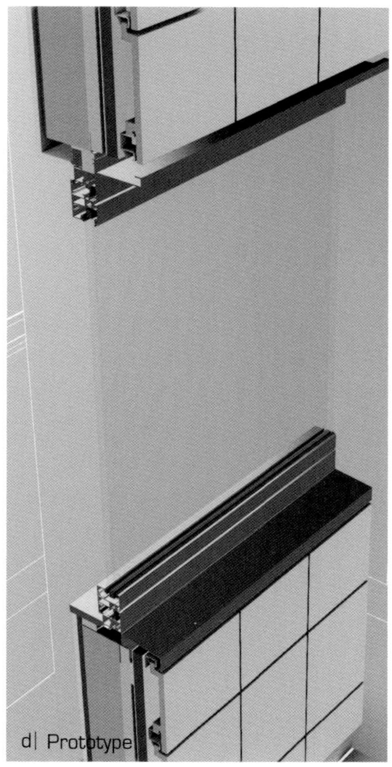

d| Prototype

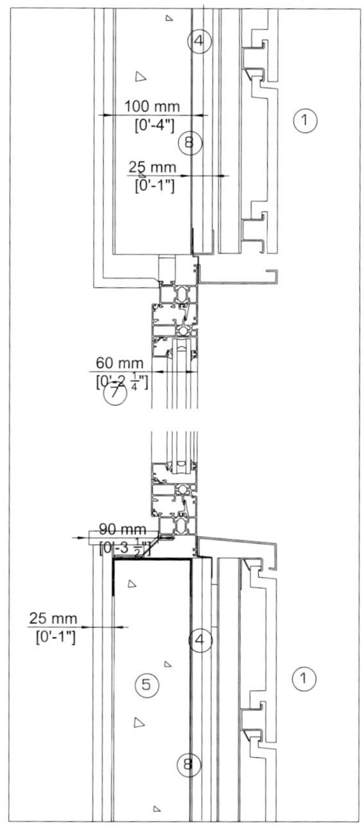

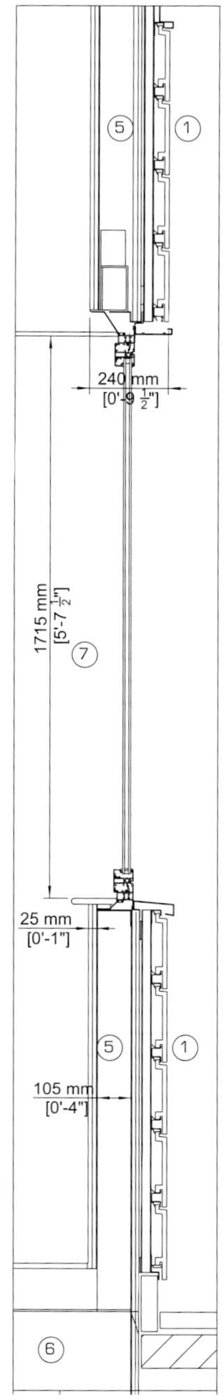

f | Vertical sections. Window at head and sill

e| Horizontal section. Window jamb

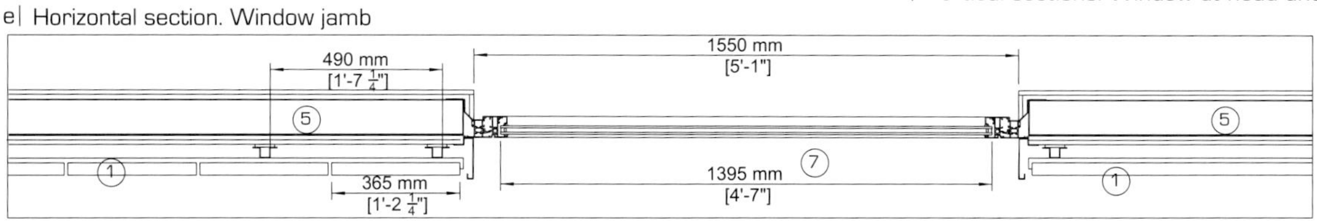

rial relatively light and easy to handle, allowing it to be made in long pieces if required. The extruding process gives flexibility to the material that is similar to the creation of aluminium sections for glazed walls and windows. When the material is extruded and cut, it is dried and fired in kilns of different types depending on the tile size and shape. Some terracotta panels are machined for firing in order to provide the precise profile needed for the fixing system as well as to provide precise smaller joint widths between panels.

Because terracotta panels have two extruded edges and two cut edges, it is important in arranging panels to avoid a cut edge being revealed at a corner. This is because the surface finish and colour of the end face will not match that of the front face. The ends of extruded terracotta differ from terracotta and fired clay bricks in this respect. The ends of panels are usually concealed with aluminium trims, sometimes at the corners but typically around window openings (d).

Corner pieces

Special shapes can be formed by hand to match with the standard extruded tiles, such as corner pieces and decorative elements (MCE_266, c). Corner pieces are made by pressing, usually with a maximum length of 150mm (6in) on one leg and 300mm (12in) on the other leg. Large corner pieces are made by hand by joining two sections together, but these currently produce less reliable results that can lack a straight and crisp edge. Manufacturers often provide extruded sill sections for parapets and window sections to suit wall constructions of 300mm to 500mm (12in–20in) in width. These have slopes to either one side or to both sides from the middle of the extrusion. The fired terracotta is either left in its natural colour or is glazed with a wide range of glazes. A glazed finish can give the material a more visual sparkle by making the material more reflective, which also provides better durability from staining. However, water absorption of regular terracotta panels is between 3% and 6%, with a density of around 2000kg/m³,

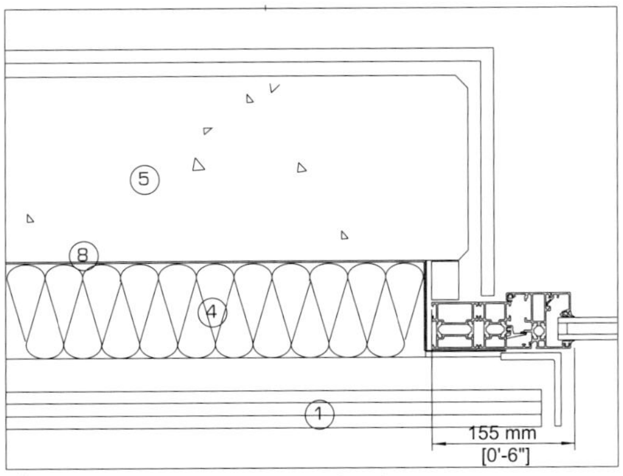

a| Horizontal section. Window jamb

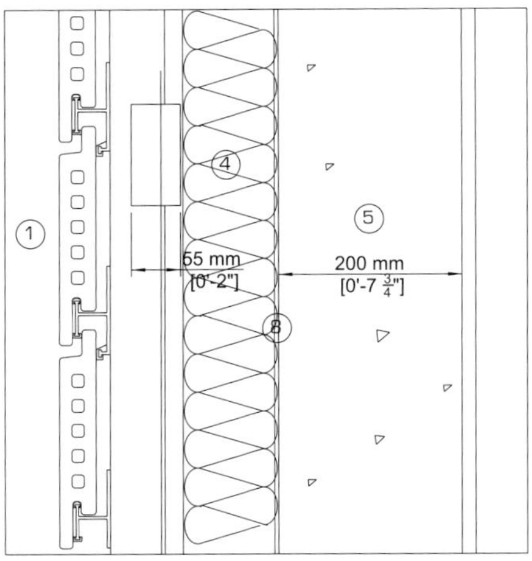

b| Vertical section. Panel-to-panel junctions

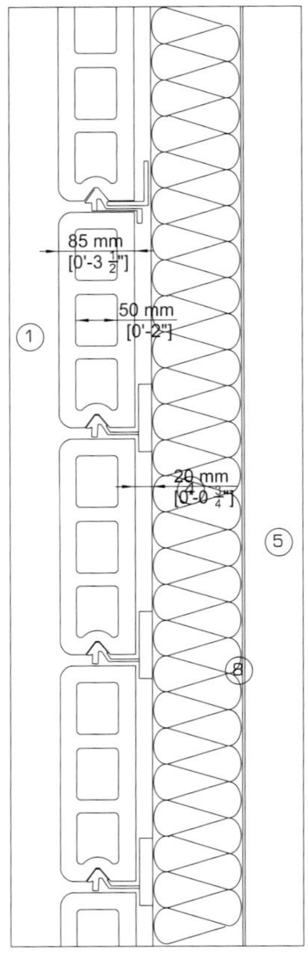

d| Vertical section. Panel-to-panel junctions

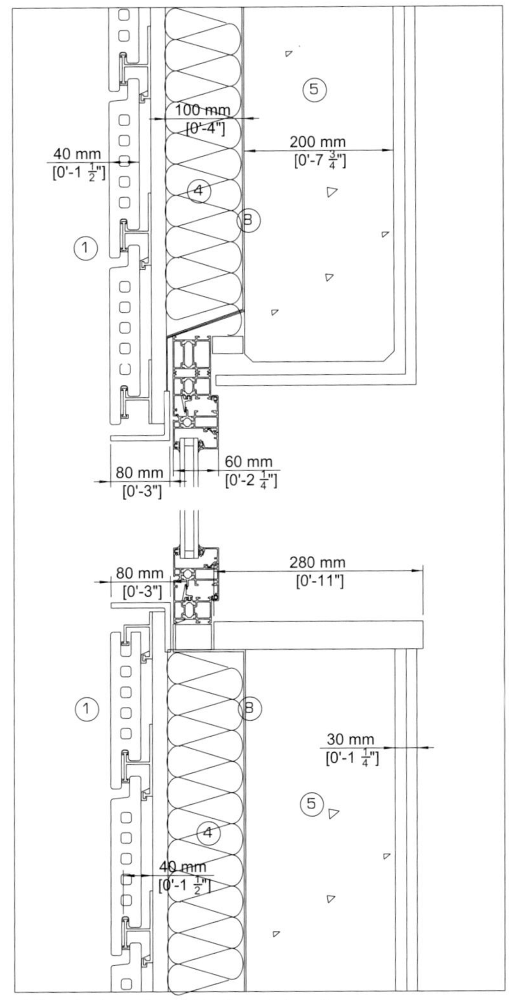

e| Vertical section. Window at head and sill, junction with ground

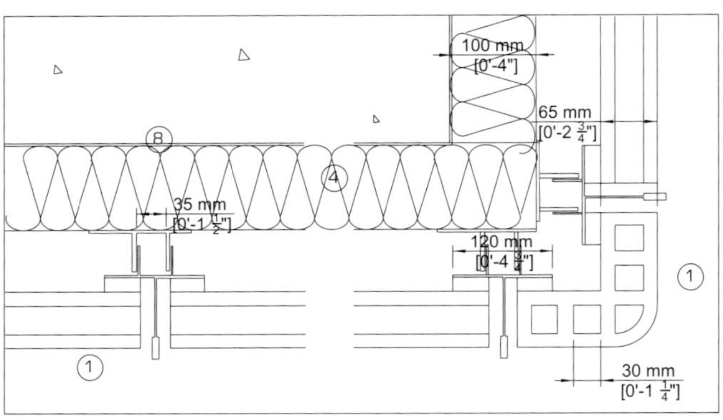

c| Horizontal section. Corner conditions

Details

1 | Terracotta rainscreen
2 | Extruded aluminium carrier frame
3 | Support brackets, typically aluminium
4 | Thermal insulation
5 | Backing wall
6 | Floor slab
7 | Metal framed window
8 | Waterproof membrane
9 | Structural column

making the use of glazes not very important in excluding rainwater but important more for visual reasons.

Fixing systems

Terracotta tiles have been used in traditional construction for a long time, but their use as extruded panels in open-jointed rainscreen construction is relatively new and has been undergoing considerable development over the past 10 years. The extruded nature of the material used for cladding allows it to be supported on clips from both behind and inside the material. Depending on panel size and thickness, the material can be made solid or hollow to suit a range of fixing systems. Smaller, solid panels have continuous support clip profiles on the top and bottom at the back of the panel which forms part of the material when extruded (MCE_265, c). Some manufacturers use the hollows of the extrusion to hold fixing pegs at their ends to provide a fully concealed fixing system.

The panel is then fixed back to support rails which are set either vertically or horizontally, made from aluminium for their ability to be formed precisely as extrusions for ease of fixing. Vertical rails are well suited to 'stack bonded' terracotta, where joints form a rectilinear grid of continuous vertical and horizontal joints. Horizontal rails are well suited to staggered bonds of panels that imitate the stretcher bond used in masonry cavity wall construction. Because vertical joints are not continuous, at least twice as many vertical rails would be needed as those needed for a stack-bond arrangement. Since horizontal joints are continuous, horizontal rails are used to fix courses of terracotta.

Vertical rails are continuous, and form a setting-out grid for stack bonded tiles, which are fixed back to rails with extruded aluminium clips (MCE_269, g). Each manufacturer has a proprietary fixing system which makes easier the process of fixing panels to form even joints around the terracotta panel edges.

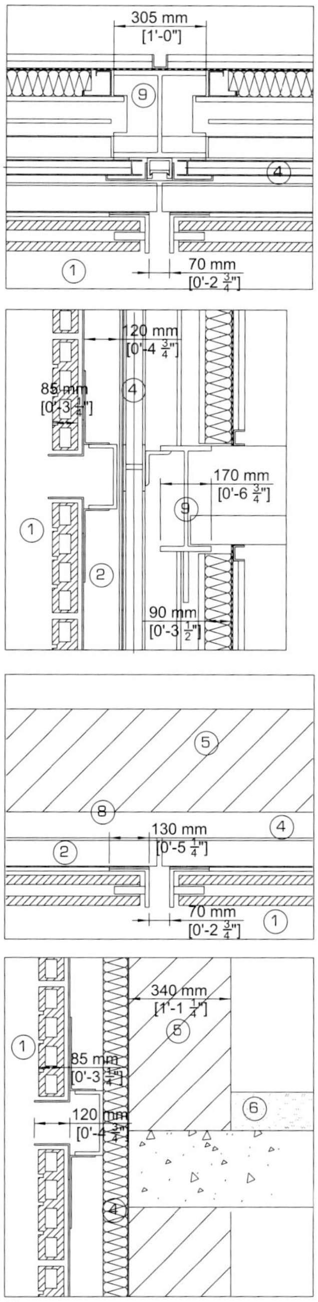

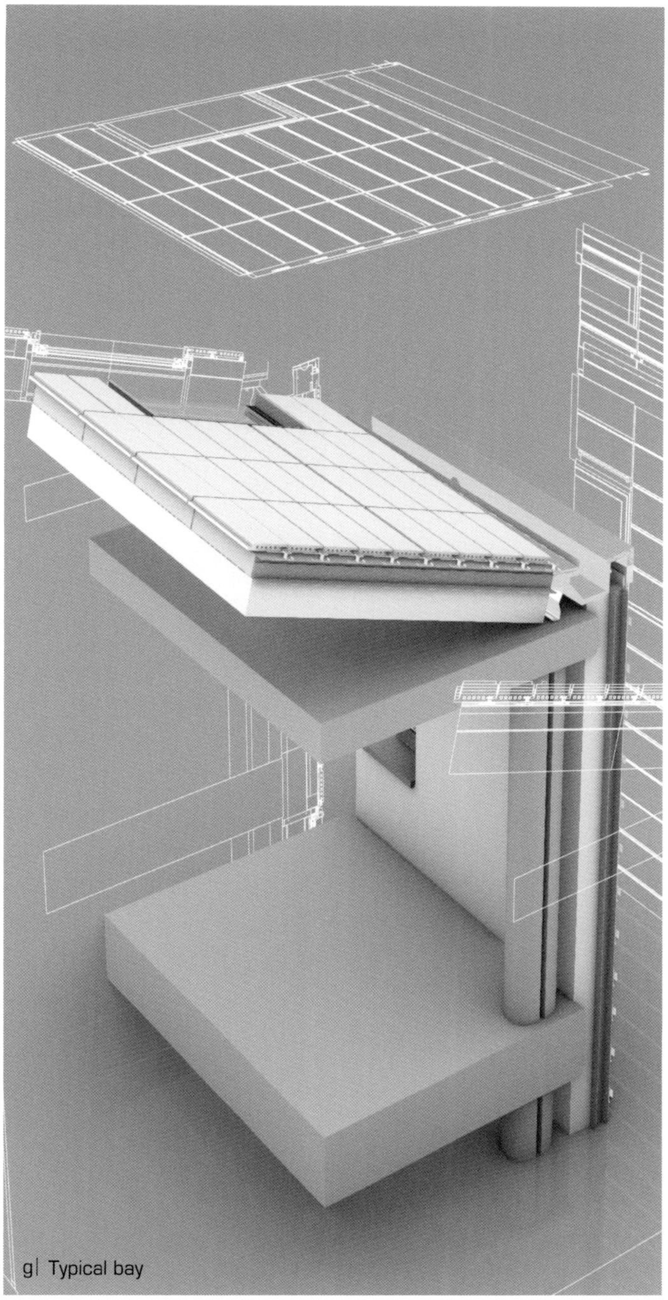

g| Typical bay

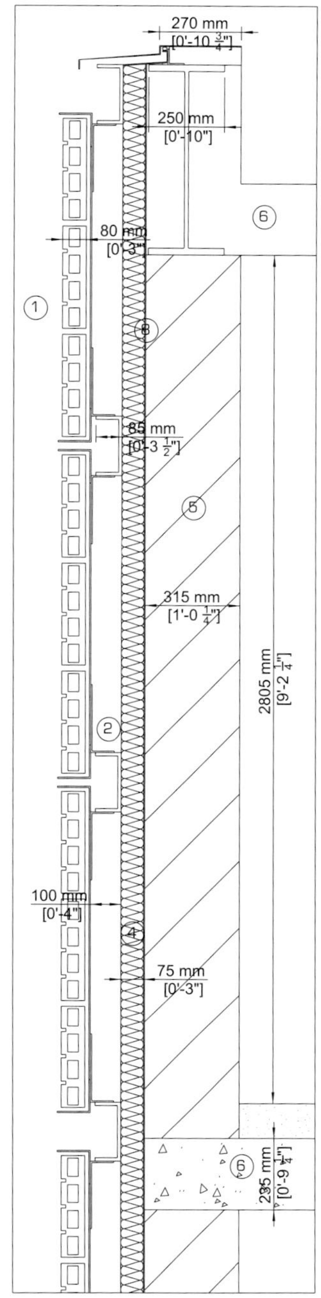

f | Horizontal section. Panel-to-panel junctions

h| Vertical section. Panel-to-panel junctions

Horizontal rails are not continuous in order to allow water to run down the backing wall without being impeded by the brackets. An alternative method of fixing horizontal rails is to set them forward of the backing wall on brackets to allow water to pass between the rails and the backing wall. Some manufacturers use stainless steel components mixed with aluminium components since the former material is considered to be more durable.

Horizontally supported systems usually have extruded aluminium sections which, like vertically supported systems, are fixed back to a continuous backing wall such as concrete blockwork (g). Aluminium brackets are fixed to the backing wall from around 1000mm to 2000mm (3ft 3in to 6ft 6in) horizontal centres, depending on rail size. The use of a horizontal clip ensures that most of the rainwater is excluded at horizontal joints by clipping the panels together in a way that ensures most of the water runs back out of the joint. Terracotta tiles used in horizontally supported systems have a stepped edge that projects into the cavity that clips into the aluminium profile. The bottom edge of the terracotta panel above laps over the front of the horizontal support profile in order to conceal it. A continuous lip is formed in the back of the panel, as part of the terracotta extruding process, which serves as a support bracket to take the load of the panel onto the supporting rail (g). Vertical joints are either left open or have a plastic or black coated aluminium strip as a baffle to prevent most of the rainwater from entering the joint (MCE_268, a). The baffle also serves as a visual screen to the cavity behind where there is a risk of daylight reflecting back out of the cavity to reveal the backing wall behind. An advantage of keeping vertical joints open is that the gaps allow the void behind to be better ventilated, instead of relying only on ventilation points at the top and bottom of the wall as is the case in masonry cavity walls. Joint widths of terracotta panel rainscreens vary from 2mm to around 10mm (up to 0.5in), depending on the size of panel and type of fixing system chosen.

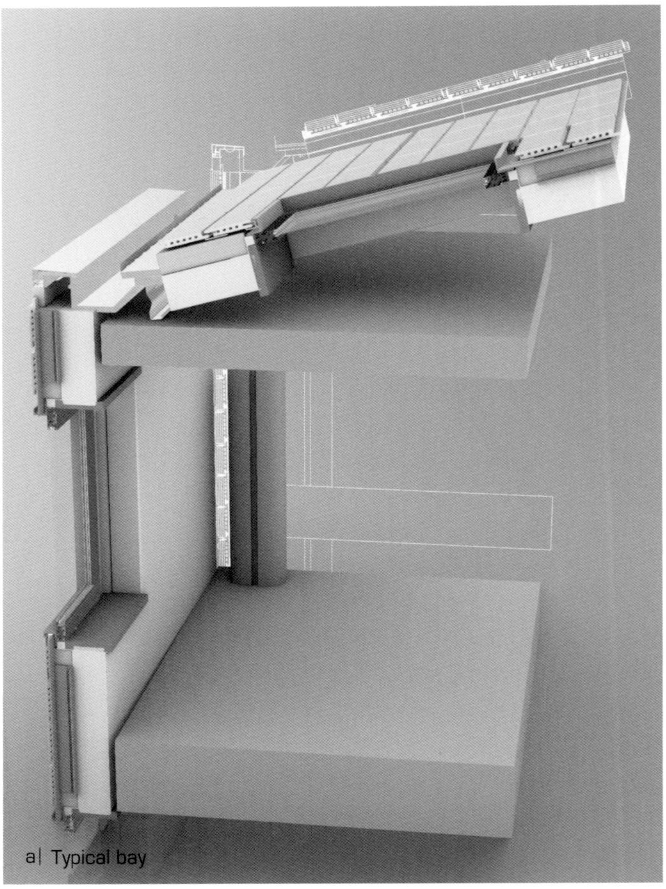

a| Typical bay

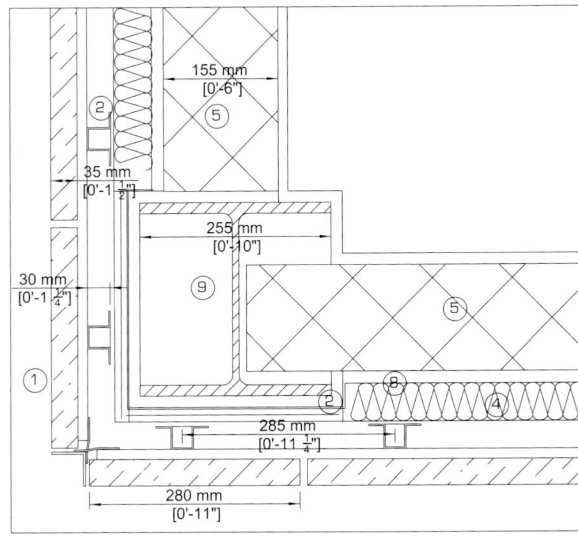

b| Horizontal section. Corner condition

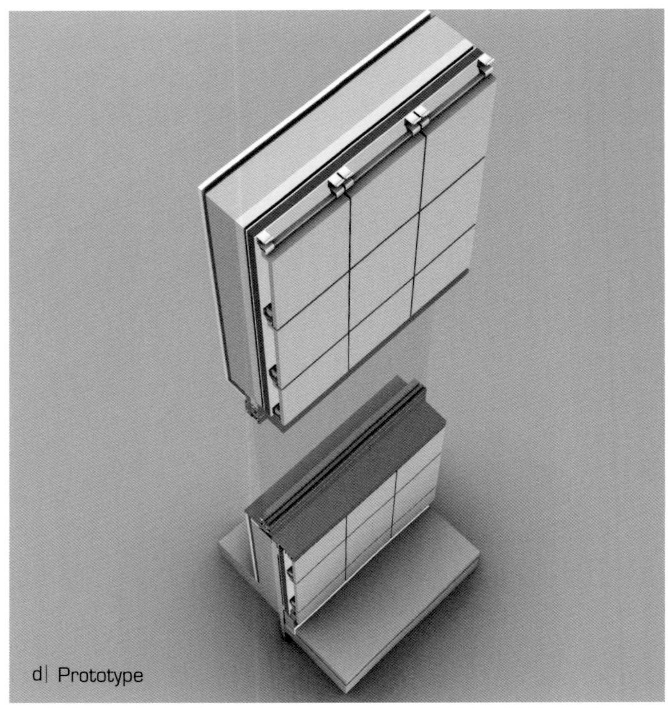

d| Prototype

c| Horizontal section. Corner conditions, window jamb

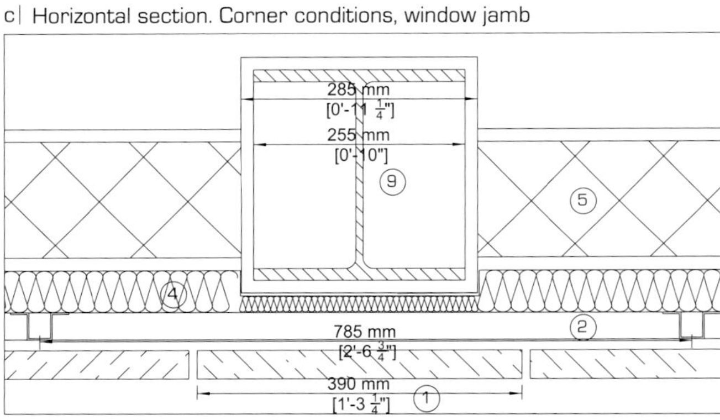

Panel sizes

For the largest tiles, formed as 'planks' up to around 1500mm long x 600mm wide x 40mm thick (5ft x 2ft x 1.5in), a substantial aluminium support section is needed behind the panels. The extrusions, set at the ends of each panel, sometimes project forward of the terracotta in order to provide enough stability in the section. This gives the facade a characteristic appearance of vertical bays of panels, where only vertical backing supports are used, divided by the visible edge of the aluminium support. Corner pieces can be made in sizes of 250mm x 300mm (10in x 12in) in height, which often do not match with the maximum size that can be manufactured for the planks, but this constraint will no doubt be overcome in the next few years. Thinner terracotta panels of 30mm (1.2in) thickness are used, in sizes with a maximum length of around 800mm (2ft 8in) and corresponding maximum height of 300mm (12in). These thinner panels have maximum corner panels of 150mm (6in) on one leg and 300mm (12in) on the other leg. The minimum sizes that can be accommodated with the vertical rail system are terracotta panels around 200mm long x 200mm high (8in x 8in), with a thickness of 30mm to 40mm (1.2in–1.6in). All panel dimensions vary with a tolerance of around ±1.0mm in their length, since terracotta is wire-cut as it comes out of the extrusion machine, and is subject to a tolerance of around ±1.5mm in its width due to shrinkage during firing.

Openings

Windows and doors can be set very easily into terracotta rainscreen cladding. Extruded aluminium trims which project from window openings can be used to form a crisp edge to a window when it is set back behind the face of the terracotta (MCE_269, a), as is typically the case since the window is usually fixed into the backing wall to which the rainscreen cladding is fixed. Alternatively, the window frame can form part of a continuous horizontal or vertical metal trim that visually divides

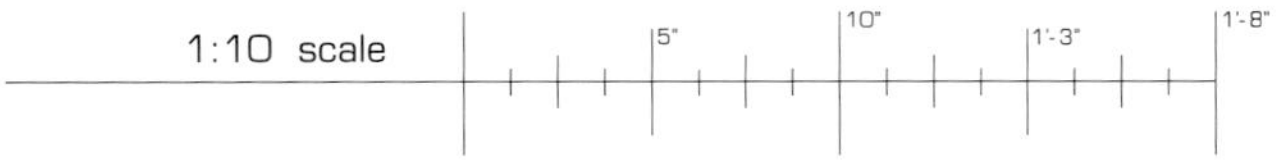

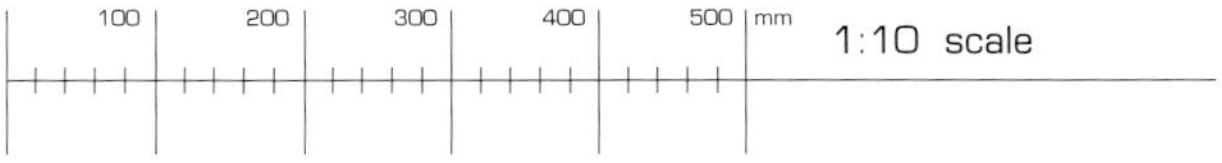

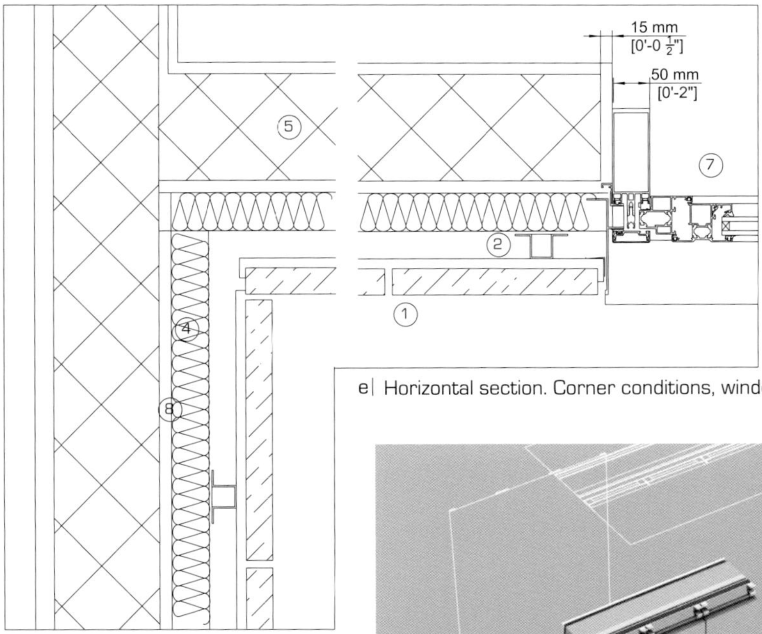

e| Horizontal section. Corner conditions, window jamb

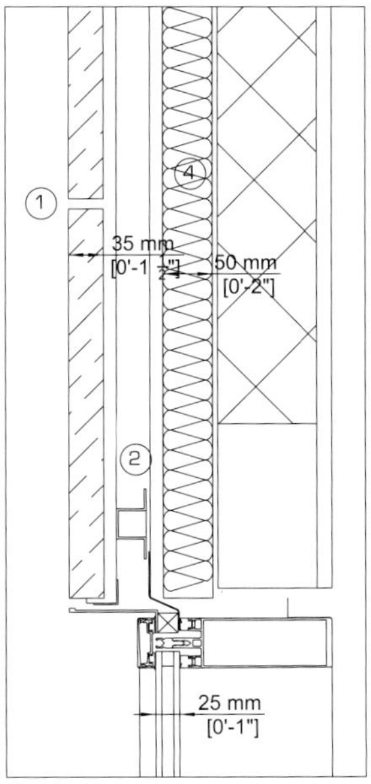

f| Vertical section. Window at head and sill

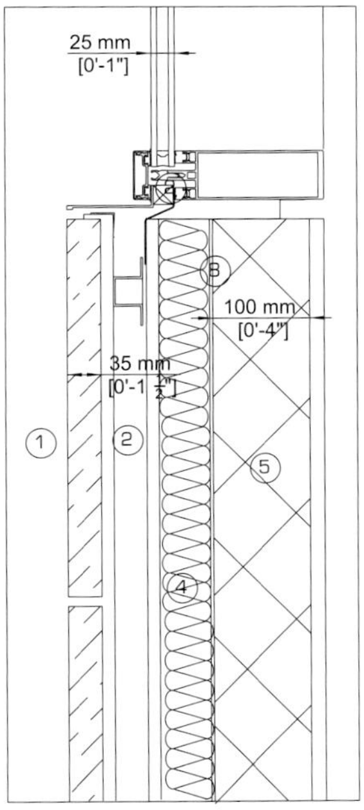

Details

1 | Terracotta rainscreen
2 | Extruded aluminium carrier frame
3 | Support brackets, typically aluminium
4 | Thermal insulation
5 | Backing wall
6 | Floor slab
7 | Metal framed window
8 | Waterproof membrane
9 | Structural column

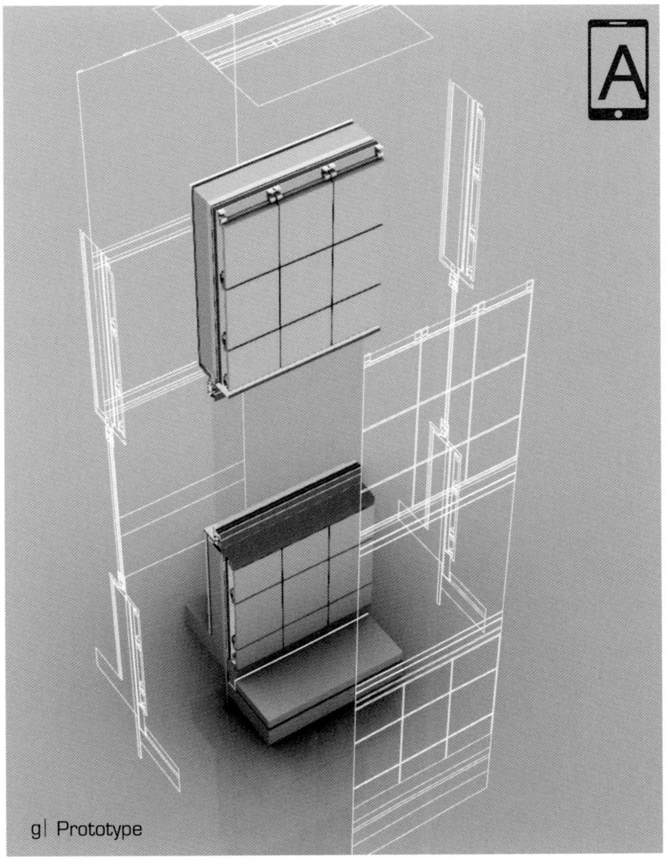

g| Prototype

the terracotta into panels, typically storey height. The trims are made from either folded aluminium or rolled steel channels (g). Trims are not continuous with the structure behind in order to avoid thermal bridging. The flat, planar nature of terracotta has led to the increased omission of reveals around window openings by pushing the window frame forward to align with the face of the terracotta panels. Bays of terracotta are also increasingly mixed with other materials such as metal rainscreen panels and metal louvre panels where materials change without the need for any special detailing at the junction, since all panels are in a continuous rainscreen arrangement. Where terracotta reveals are introduced into window openings, corners are mitred with open joints if special corner panels are not used. This principle is also used at internal and external corners in the facade (MCE_268, b).

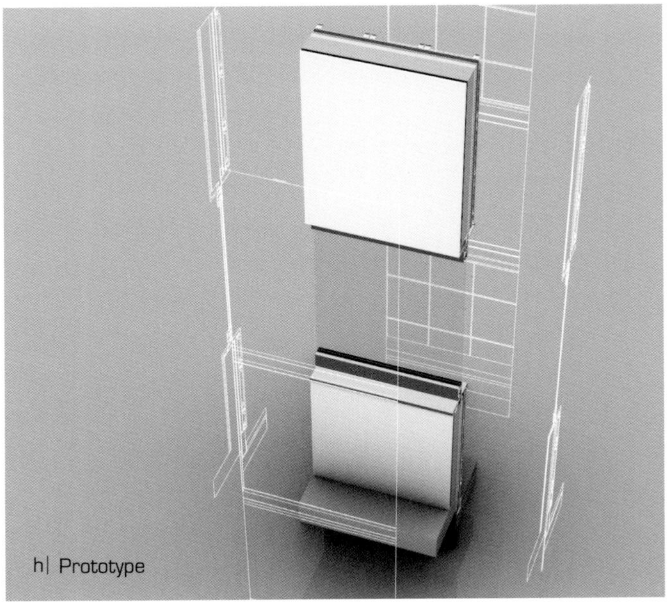

h| Prototype

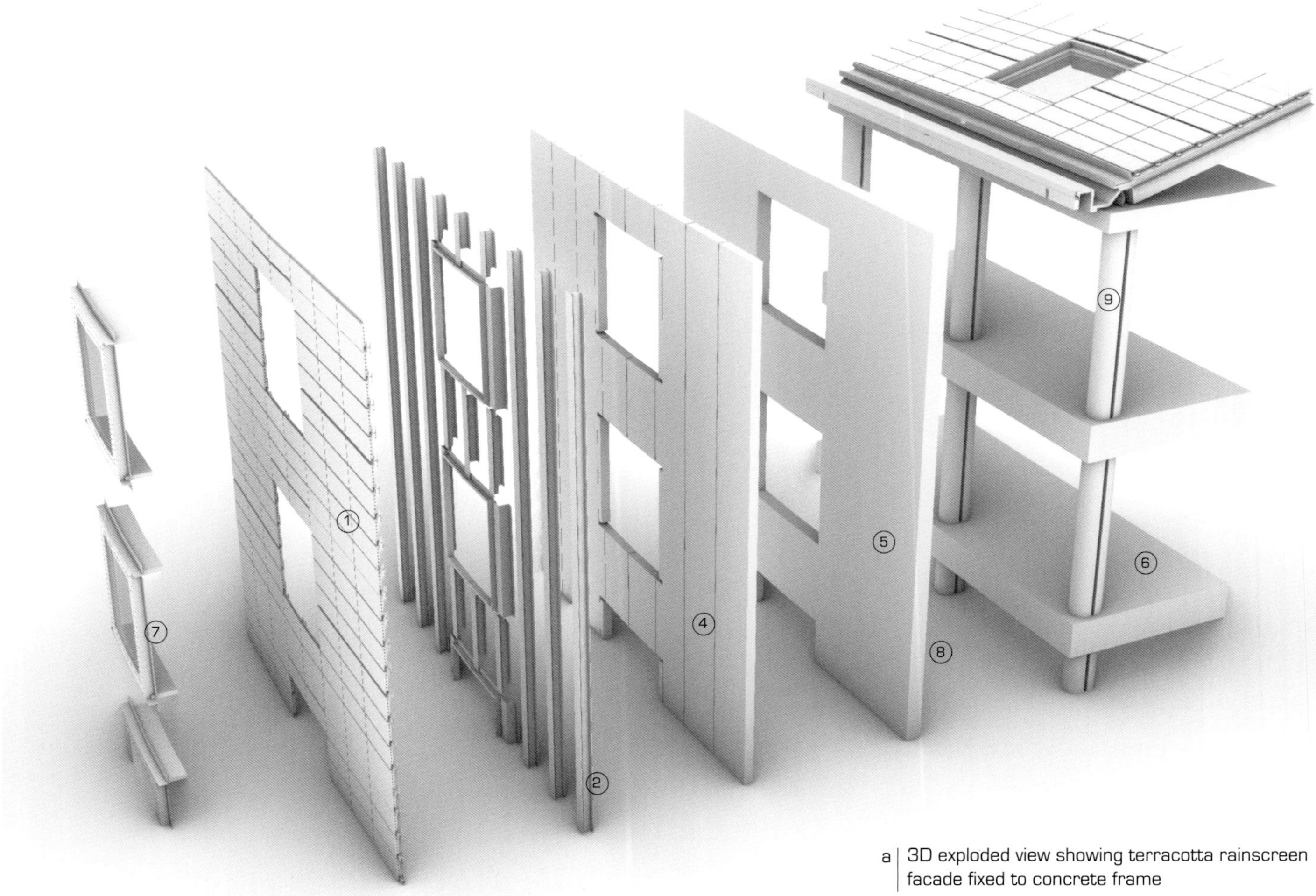

a | 3D exploded view showing terracotta rainscreen facade fixed to concrete frame

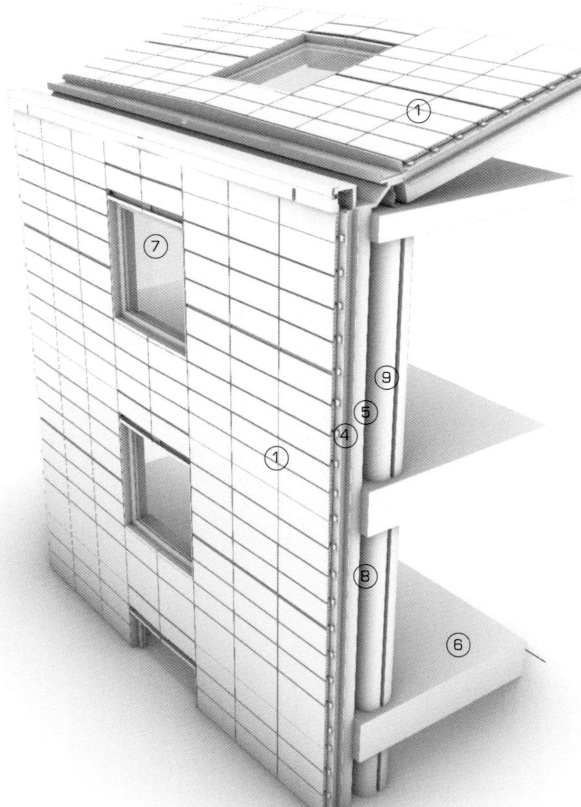

b | 3D view showing terracotta rainscreen facade fixed to concrete frame

Details

1 | Terracotta rainscreen
2 | Extruded aluminium carrier frame
3 | Support brackets, typically aluminium
4 | Thermal insulation
5 | Backing wall
6 | Floor slab
7 | Metal framed window
8 | Waterproof membrane
9 | Structural column

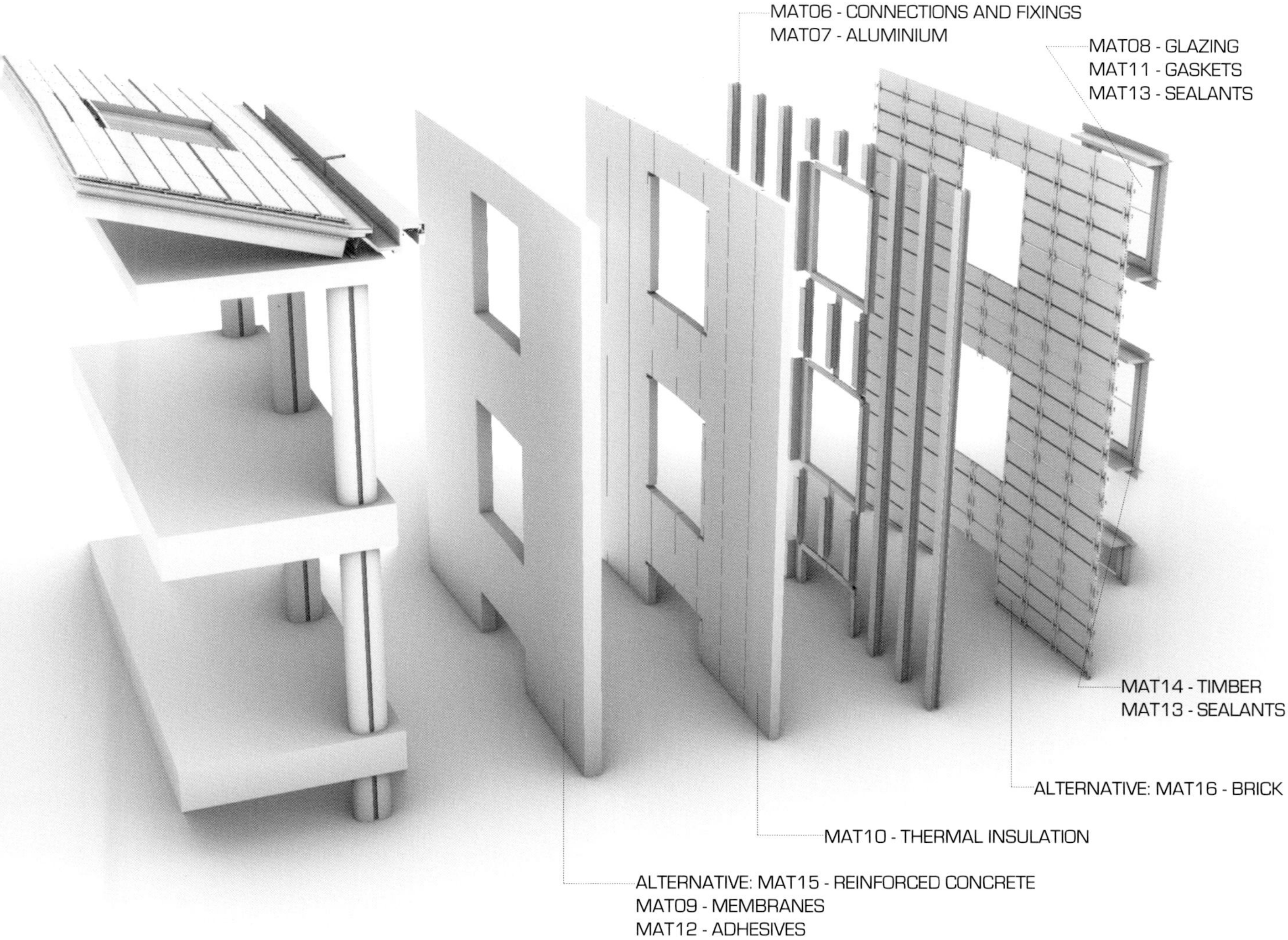

c | 3D exploded view of terracotta rainscreen facade components with material specification references

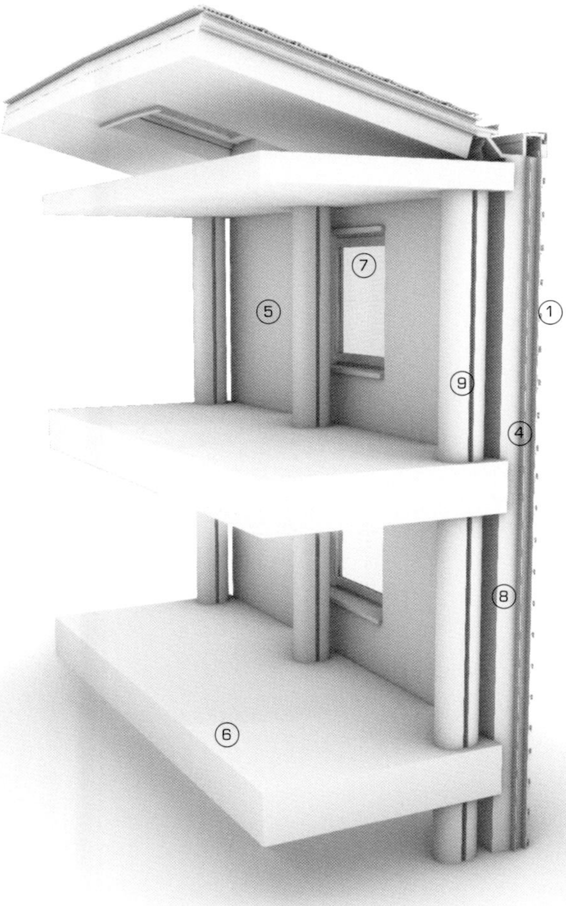

d | 3D view showing terracotta rainscreen facade fixed to concrete frame

Example project of polymer facades
Galaxy Soho, Beijing, China
Facade Consultant to ZHA: Newtecnic
Architect: Zaha Hadid Architects

POLYMER WALLS

22 **Sealed panels**

GRP panels

Polycarbonate cladding

23 **Rainscreens**

Flat polycarbonate sheet

Multi-wall polycarbonate sheet

Profiled polycarbonate sheet

Plastic-composite flat panels

UPVC board cladding

UPVC windows

GRP panels

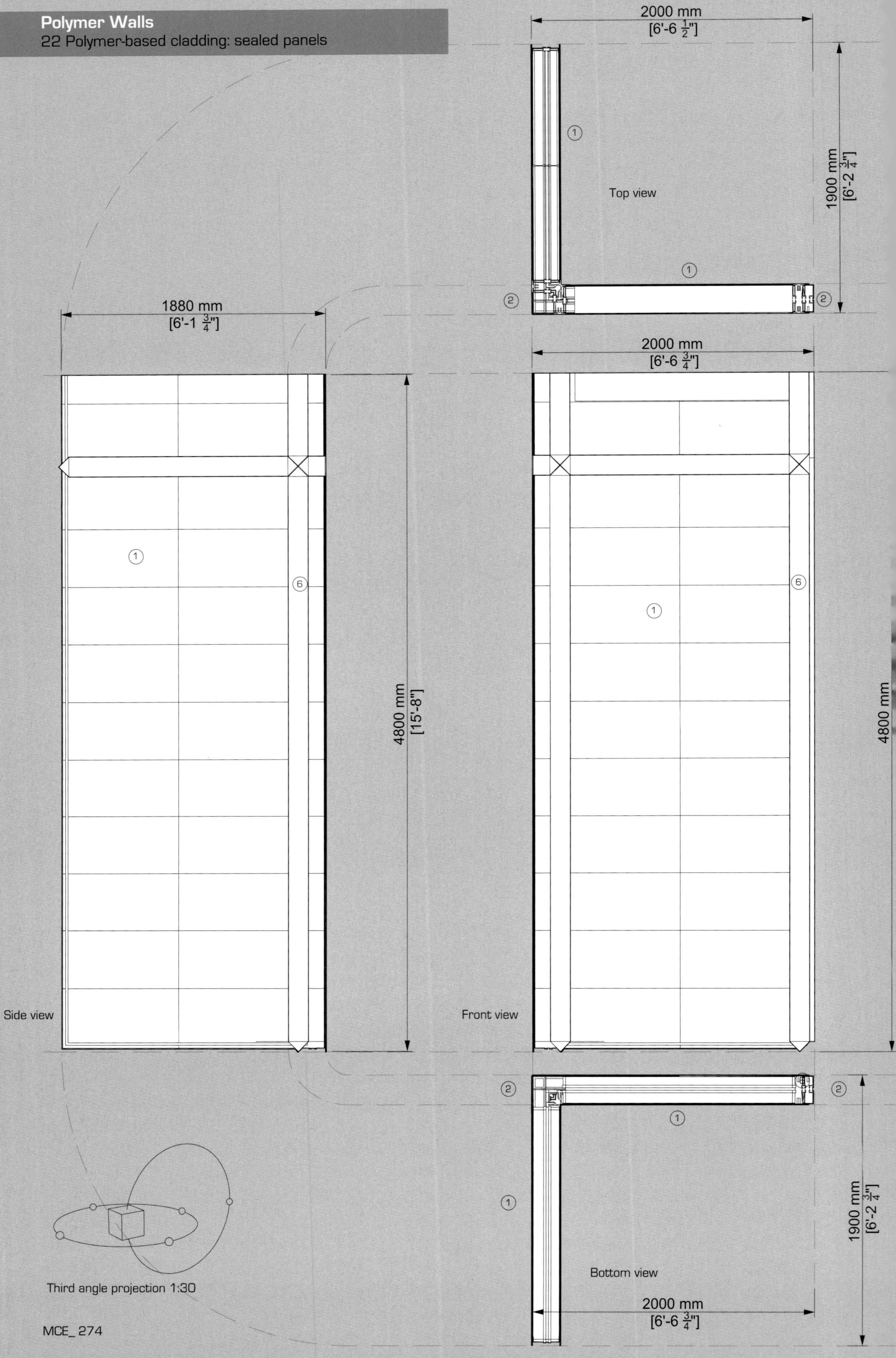
2000 mm
[6'-6 1/2"]
1
Top view
1900 mm
[6'-2 3/4"]
1
2
2
1880 mm
[6'-1 3/4"]
2000 mm
[6'-6 3/4"]
1
6
4800 mm
[15'-8"]
6
1
4800 mm
Side view
Front view
2
2
1
1
1900 mm
[6'-2 3/4"]
Bottom view
2000 mm
[6'-6 3/4"]
Third angle projection 1:30

Details

1 | Translucent GRP cladding panel, insulated
2 | Thermally broken extruded aluminium framing
3 | Opaque GRP cladding panel, insulated
4 | Inside
5 | Outside
6 | Recessed cover cap
7 | Window inserted into framing
8 | Adjacent wall. Metal rainscreen shown
9 | Insulated corner panel
10 | Metal cover strip
11 | Supporting structure

3D view of the assembly

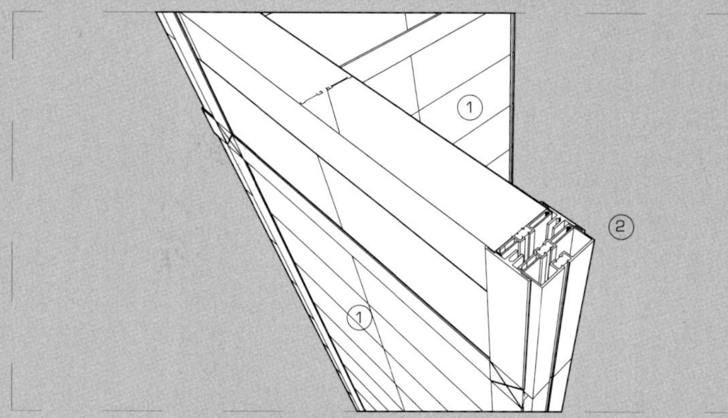

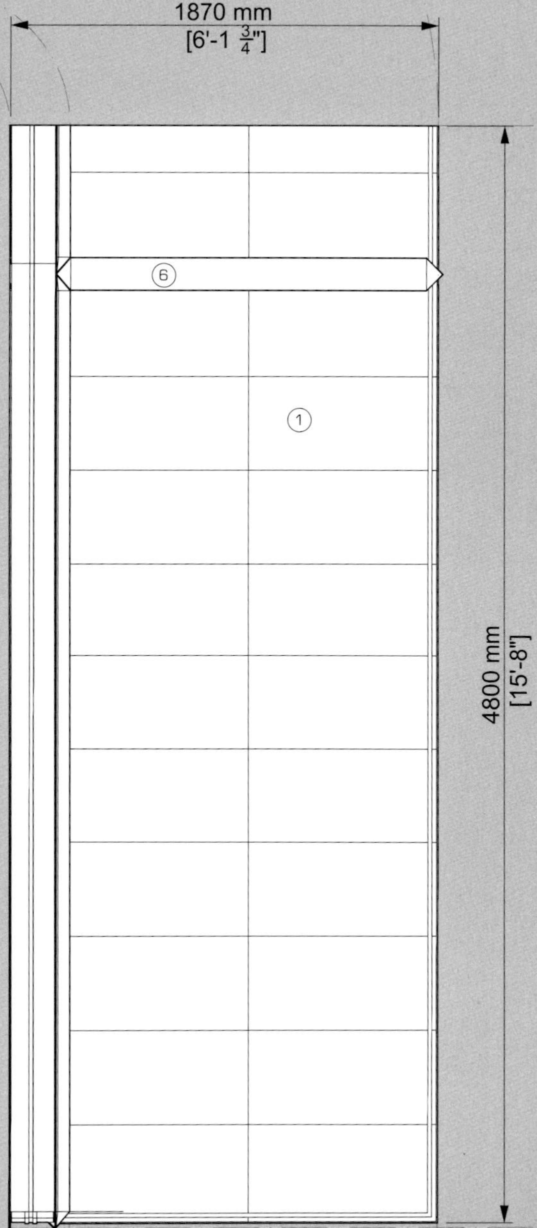

Side view

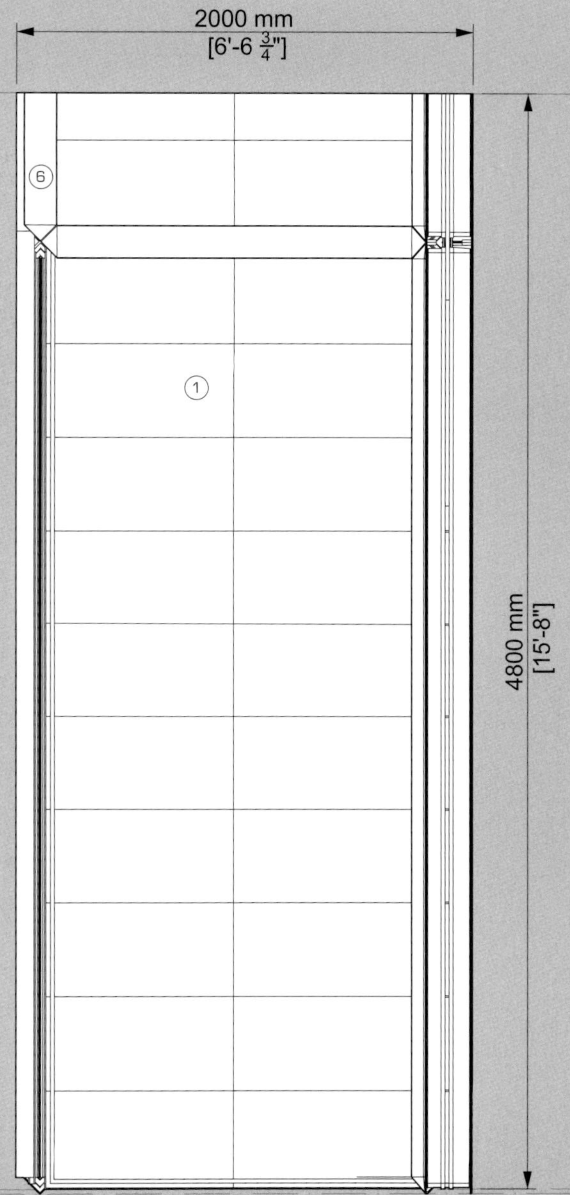

Back view

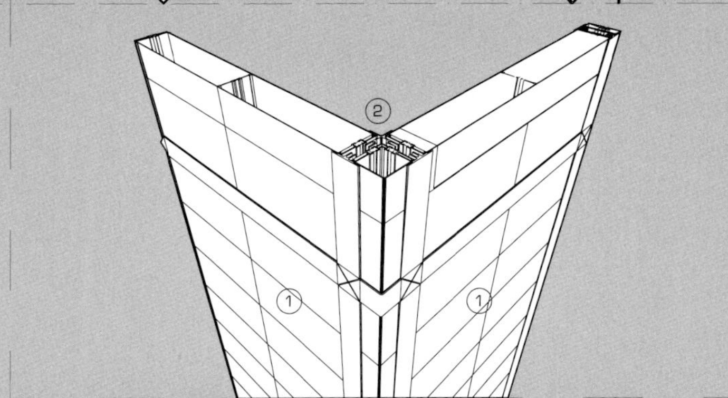

3D view of the assembly

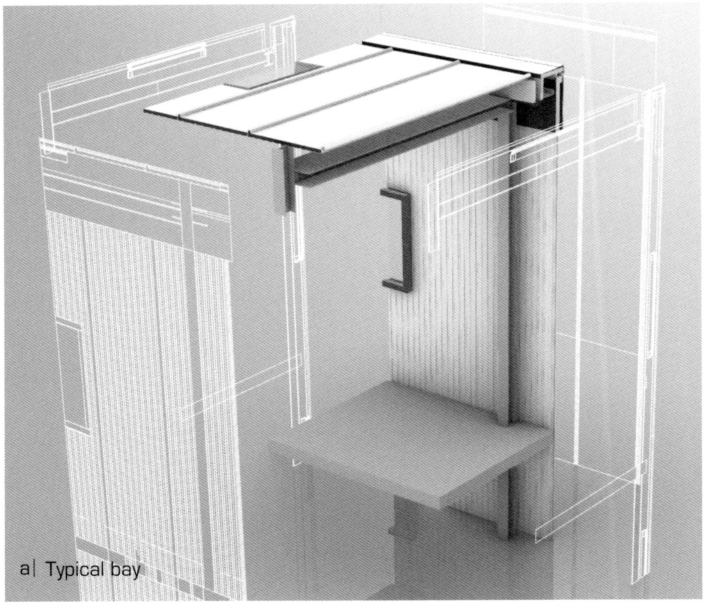

a| Typical bay

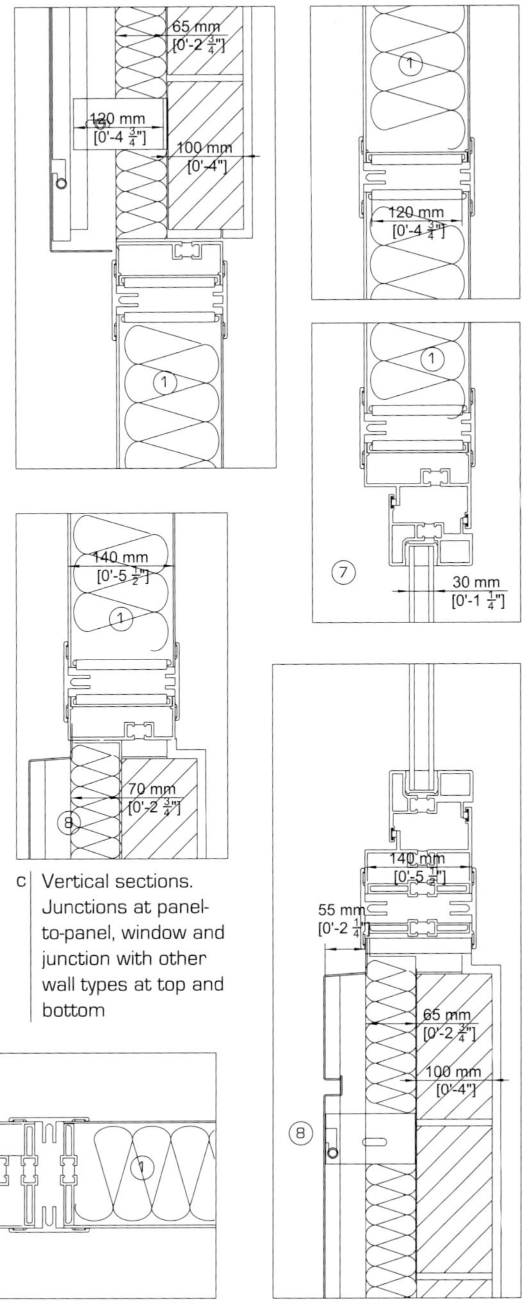

c| Vertical sections. Junctions at panel-to-panel, window and junction with other wall types at top and bottom

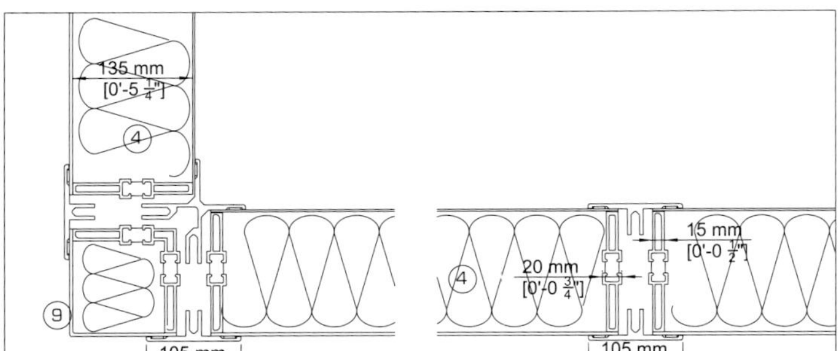

b| Horizontal sections. Junctions at panel-to-panel, window and junction with other wall types at top and bottom, corner

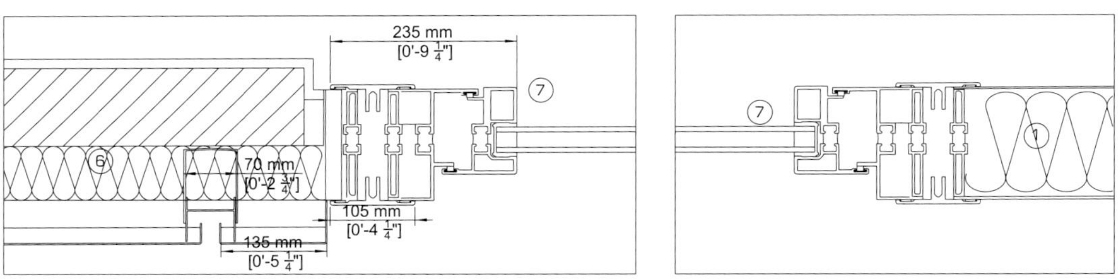

Plastics are resinous, polymer-based materials, and they are used for both sealed cladding and rainscreens, which are discussed in the following two sections. The materials used are principally glass fibre-reinforced polyester (GRP), polycarbonate and UPVC. Since the properties of these materials are generally not as familiar to readers as other materials, a brief description of each of these materials is given here.

GRP is a composite material made from thermosetting polyester resins (that set hard and do not melt when reheated) which are mixed with glass fibre mat. This composite material has high tensile, shear and compressive strength combined with lightness and resistance to corrosion. However, like aluminium it deflects considerably under high loads and requires stiffening; but the material is stiffer than other plastics. GRP is not combustible and can reach one hour fire resistance in some cladding applications. Glass fibre mat is a flexible sheet material made from fibres drawn from molten glass. Its tensile strength is much greater than that of steel. Polyester resin, with which the fibre is combined, forms a solid material when a chemical catalyst is added. GRP panels are formed in a mould where glass fibre cloth is laid into a mould and coated with resin and catalyst. An alternative method is to spray a mixture of glass fibre and resin into a mould. The face of the mould is coated with a releasing agent to allow the GRP to be removed when it has set hard. GRP sections are made by pultrusion, where fibres are pulled through a die drawing strands of the material to form continuous sections in the manner of aluminium extrusions. Pultrusions are beginning to be used as structural components in footbridges, where their durability is considered to be better than painted aluminium or steel structures. Whereas sections are made by an expensive pultrusion method, panels are made by hand as a craft-based activity in a workshop. The production of GRP panels is economic, requiring neither high temperatures for manufacture nor expensive equipment.

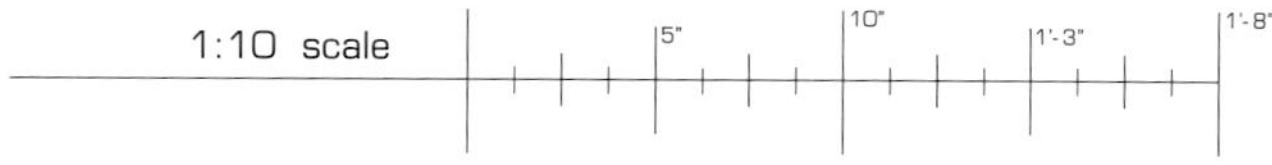

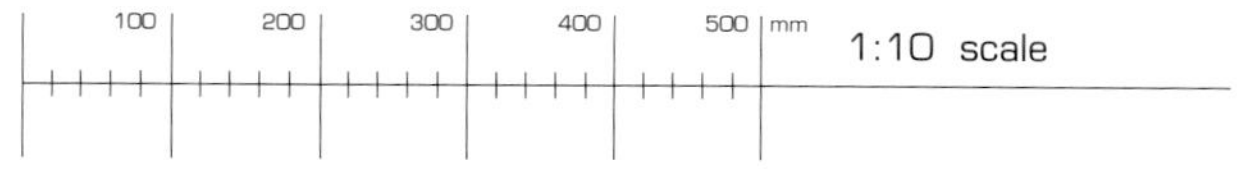

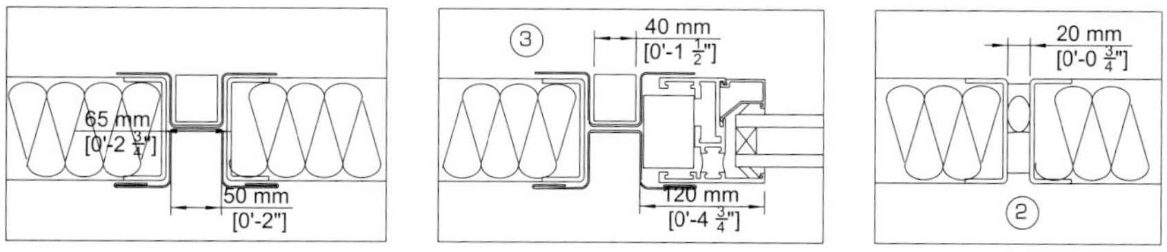

d| Horizontal sections. Junctions at panel-to-panel, corners

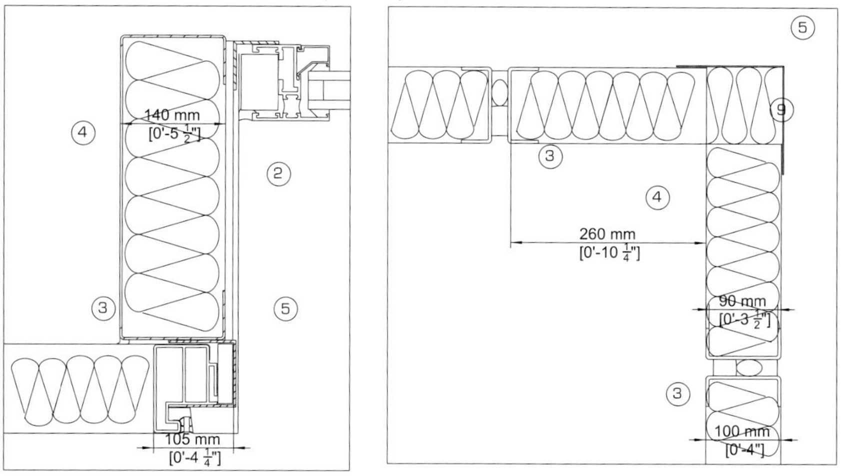

e| Prototype

Details

1 | Translucent GRP cladding panel, insulated
2 | Thermally broken extruded aluminium framing
3 | Opaque GRP cladding panel, insulated
4 | Inside
5 | Outside
6 | Recessed cover cap
7 | Window inserted into framing
8 | Adjacent wall. Metal rainscreen shown
9 | Insulated corner panel
10 | Metal cover strip
11 | Supporting structure

f | Horizontal sections. Corner conditions

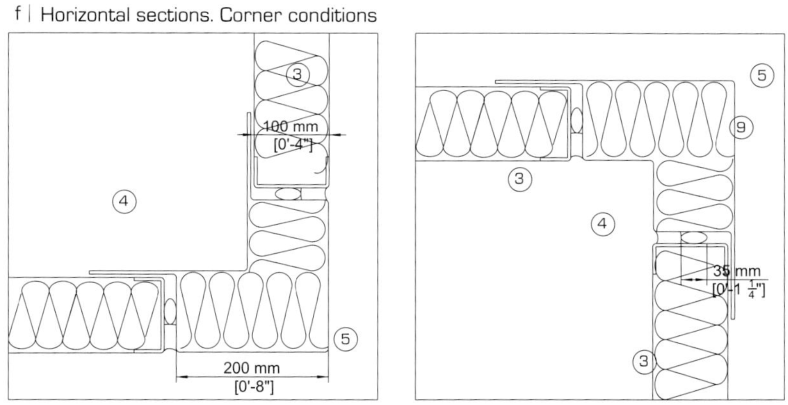

Polycarbonate is a thermoplastic; that is, it melts at high temperatures. It is used in cladding for its translucency and transparency, especially where its high thermal insulation values are needed. Polycarbonate is made by melting a polymer and extruding it into strands which are chopped to produce polycarbonate granules. The granules are then extruded or moulded to form sheet materials. Polycarbonate is extruded in single wall, double wall or triple wall materials. Twin wall sheet is an extrusion of two layers separated by fins, giving the material greater rigidity, combined with an air gap that provides additional thermal insulation (MCE_281, e). The maximum sheet size is approximately 2000mm x 6000mm. The material has a tendency to yellow with age, which can be overcome with an acrylic coating. The material can also be moulded into complex shapes. It is also used for its high strength, ductility and lightness in weight. Because the material is combustible, its use in facades is limited. An advantage of the material over glass is its impact resistance, which is higher than that of toughened glass or laminated glass. The main disadvantage of polycarbonate over glass is that it is less durable and scratches easily, which makes the surface dull with time. In addition, polycarbonate has high thermal expansion; it expands up to 20% more than glass.

UPVC, or PVC-u is an unplasticised or rigid PVC, used in plastic-based cladding mainly for window frames and weatherboarding. The material can be easily extruded as complex sections to provide an economic material with low thermal conductivity, making it ideal for window frames where a plastic thermal break would be provided in an equivalent aluminium window. Like polycarbonate, it is combustible but is not easily ignited, burning slowly if exposed to a flame source. If the flame source is removed, then the flame will self-extinguish. The material will soften if exposed to a direct heat source. UPVC is available in a range of colours, weathering well, but is susceptible to fading, particularly with brighter colours. It is a tough material, but also flexible.

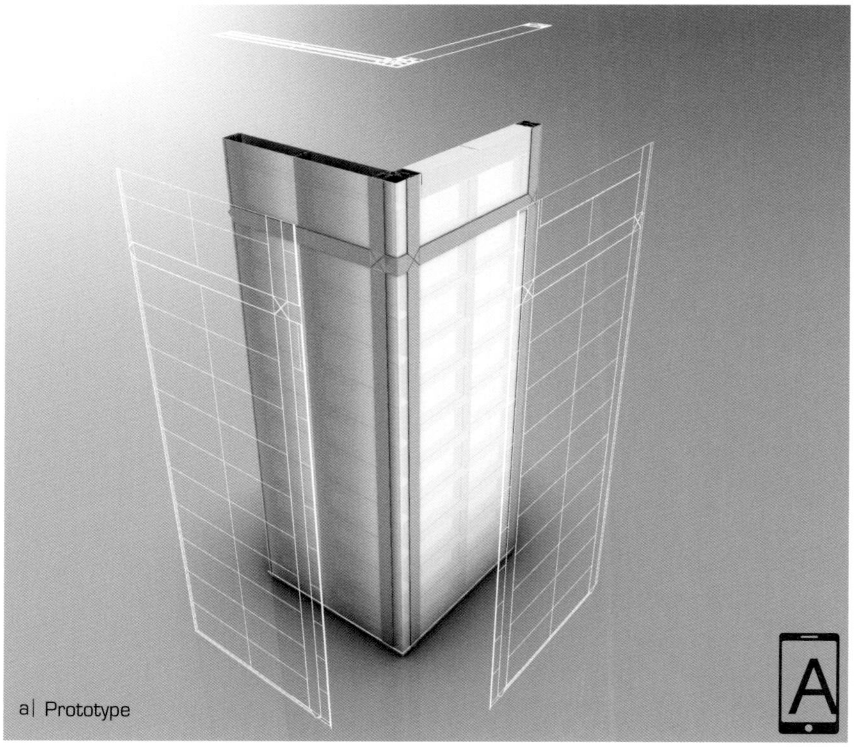

a| Prototype

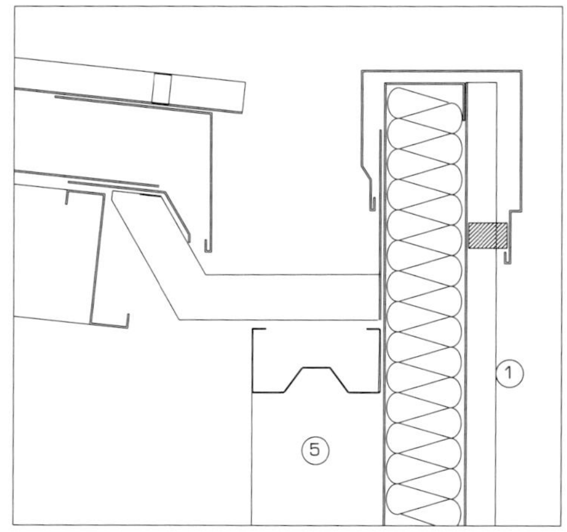

b | Vertical section. Junction at parapet

c | Horizontal section. Corner condition

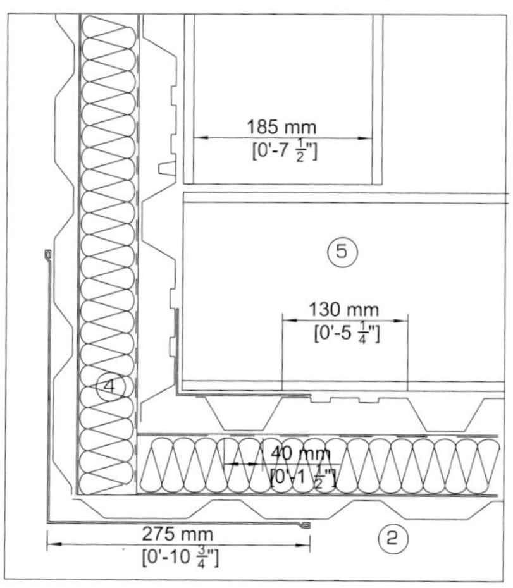

Details

1 | GRP or UPVC rainscreen panels
2 | Twin wall polycarbonate sheet
3 | Support rail
4 | Thermal insulation
5 | Supporting structure
6 | Metal drip
7 | Floor slab
8 | Metal parapet coping
9 | Backing wall

d | Vertical section. Junction at sill

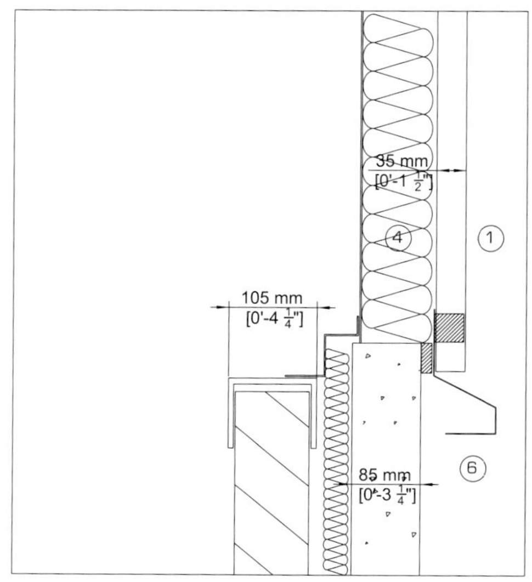

GRP panels

GRP cladding panels can be made either as separate panels glazed into an aluminium pressure plate system with a secondary supporting structure behind, or by forming aluminium extrusions within the panels (a). An advantage of GRP panels is their lightness in weight, combined with being moulded, allowing them to be made in large panel sizes, up to 6000mm x 1500mm (20ft x 5ft). They are more economic than an equivalent insulated glazed wall. Panel thicknesses are usually 70–75mm to provide structural stability and thermal insulation.

Where panels are fixed into aluminium pressure plates, the cover capping over the face of the plate is also aluminium. This can present a difficulty in colour matching between the capping and the adjacent panel. Where the aluminium is powder coated or PVDF coated, the GRP is coloured with a resin applied to the face of the mould during manufacture. This can lead to colour variation between the cappings and the panels, which can either form part of the design, or else different colours are used for each. GRP panels are glazed in a similar way to glass, with a thin panel edge (MCE_279, f) which can be compressed, held between EPDM gaskets by an extruded aluminium plate. Panels are made from two moulded GRP skins which are bonded either side of the rigid insulation. In common with metal composite panels, GRP panels have undergone much development in the use of glues to avoid delamination between the outer skins and the insulation core. Edges of the GRP panels are bonded together to form a sealed panel (MCE_281, e). Windows are glazed into the pressure plates as separate panels (MCE_280, c). Windows are rarely glazed directly into GRP panels since it is difficult to provide a drained and ventilated void within a GRP panel. Earlier versions of GRP panels, used in the 1960s and 1970s, used a rubber gasket to seal the joint between window and panel, as was used in car windscreens at that time. But the lack of a second line of defence behind this seal led to the leaks. The same prob-

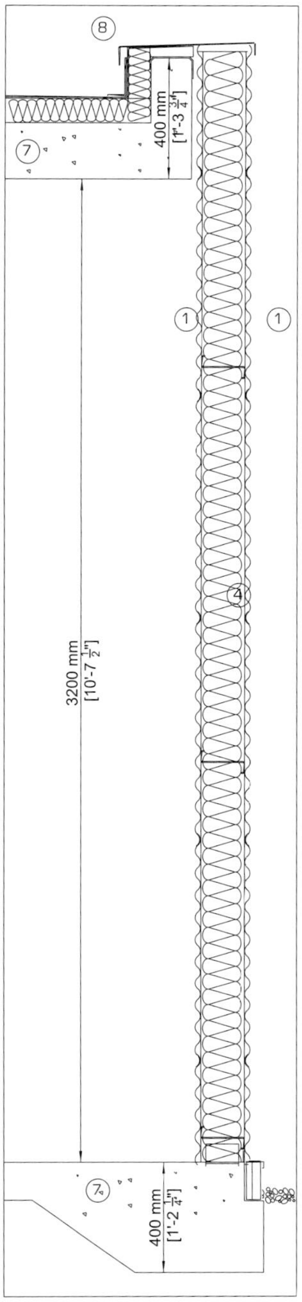

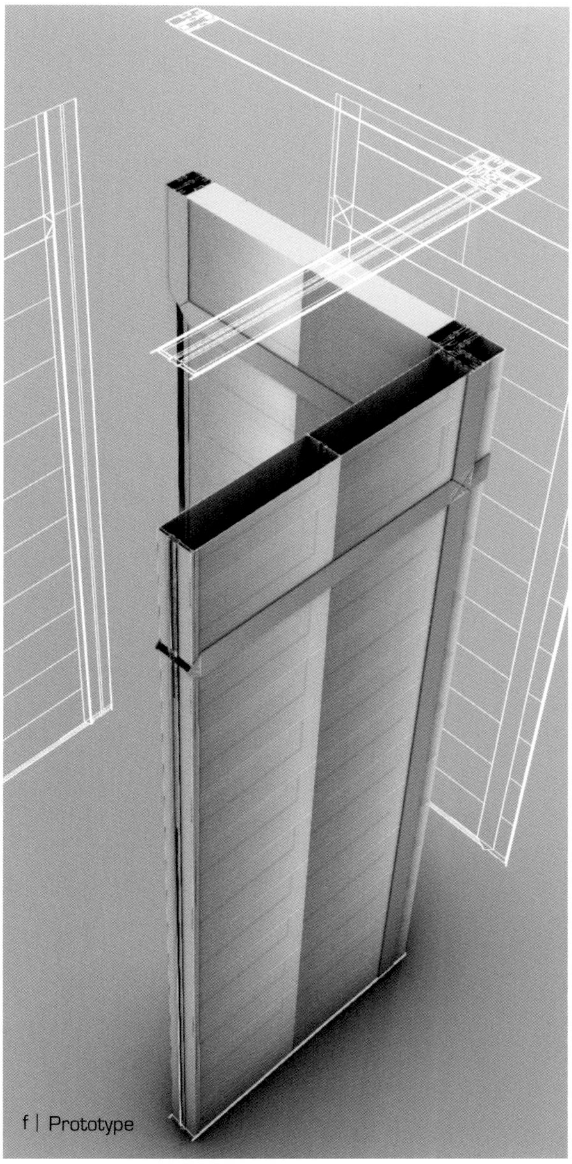
f | Prototype

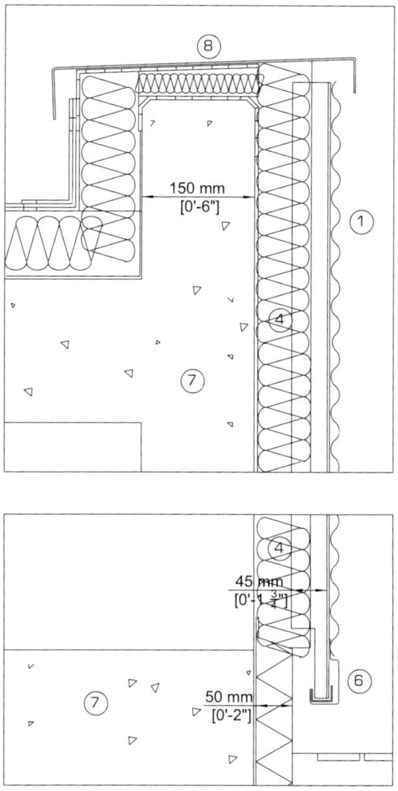

e | Vertical section. Junctions at parapet and sill

g | Horizontal section. Corner condition

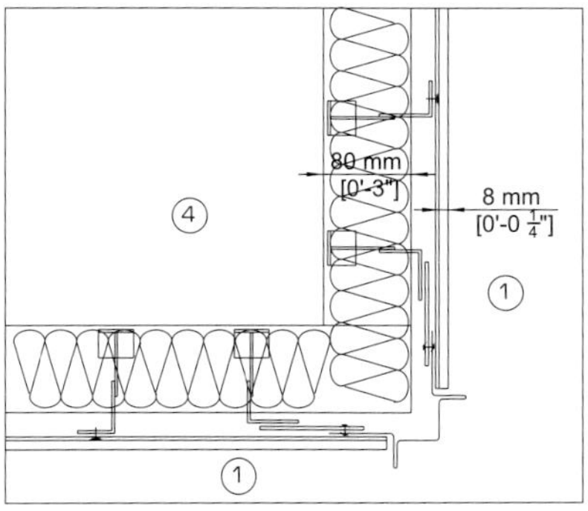

h | Vertical section. Two skins of polycarbonate fixed to aluminium support frame

lem was encountered in car windows, which moved on to using silicone bonds. Sills and parapets are formed with GRP or pressed aluminium copings in the same way as stick glazing systems (f), discussed in the Glass chapter. Copings are glazed into the pressure plate framing in the same way as a panel.

GRP panels can also be fixed together with extruded aluminium sections set within the depth of the panel (MCE_281, e). This is a more economic solution which is closer to metal composite panels in its approach than the aluminium pressure plate system, which is based on stick-built glazed walls. GRP panels are clipped into aluminium extrusions which have a thermal break set into the section. Panels are also stiffened internally with aluminium I-sections or T-sections where large-scale panels are used. Panels can be opaque or translucent, like the previous pressure- plate-based systems described. Where panels are transparent, the internal aluminium framing within the panel forms a visible grid, resembling traditional Japanese Shoji screens. These internal ribs are typically set from 300mm x 300mm centres to 300mm x 600mm centres. In these translucent panels the void between the two skins can be filled with translucent insulation quilt to increase thermal insulation, while still allowing a diffused light to pass through the panel. Mineral wool quilt is typically used, but fixing the insulation requires care to avoid the material from later sagging, which is clearly visible through the GRP panel. Light transmission without additional thermal insulation is typically around 15% with a U-value of 1.5W/m² ºC, which is similar to an argon-filled double glazed unit, and a shading factor of 20%, which provides a high level of shading for a 'glazed' wall. Unlike the pressure plate system, windows can be glazed into the panels, giving the possibility of a rich mix of windows, doors and translucent panels without the need for complex framing (MCE_276, c). Window frames can form part of the T-section extrusion around a window. Integrating the window frame into the extrusion supporting the GRP panels avoids potential

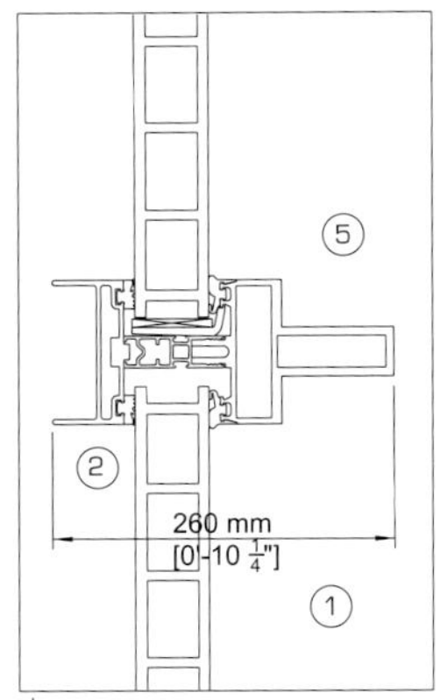

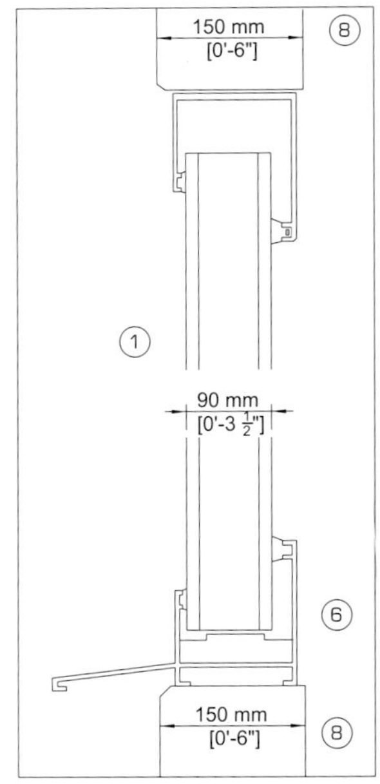

a | Vertical sections. Panel-to-panel junction and head and sill condition

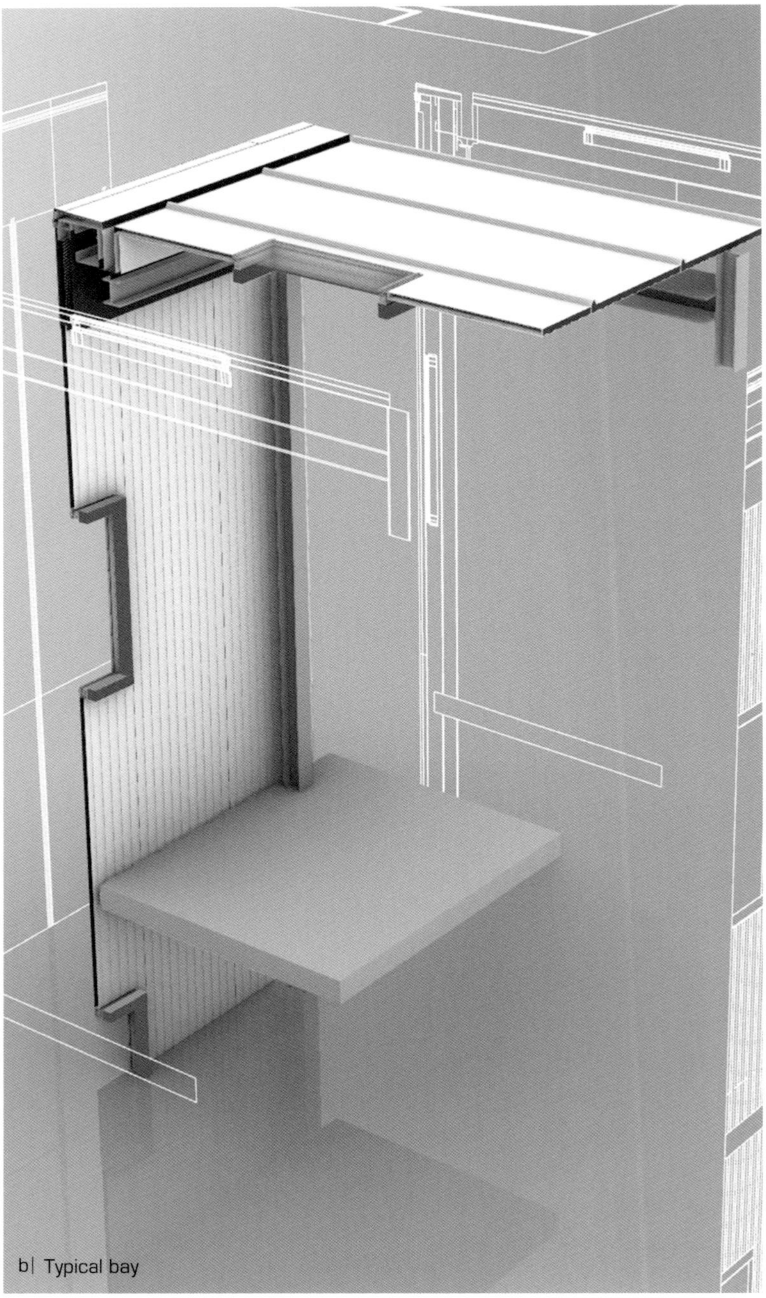

b | Typical bay

c | Prototype

leaks associated with silicone-sealed butt joints when a separate window frame and panel frame are fixed together. The integrated window frame allows water to be drained from the frame. GRP panels can be glazed into large structural openings, from floor to ceiling for example, or can form a complete glazed wall, restrained by a secondary steel frame. When glazed into an opening, the edge T-section aluminium profiles are sealed against the adjacent concrete floor slab with silicone, at both top and bottom. When fixed to a secondary support frame, panels are supported at each floor level on metal brackets in either aluminium, mild steel (if internal) or stainless steel (if exposed to the weather). Sills and copings are formed with the methods described in the section on metal composite panels.

Polycarbonate cladding

Polycarbonate is used for wall cladding as either profiled sheet (m) or as twin wall / triple wall sheet (MCE_283, d). Where profiled sheet is used, it is typically used in long lengths to avoid horizontal joints formed by lapping sheets over one another. The sheet is orientated typically with the profile lines running vertically to allow rainwater to run off with a minimum of visible staining to this translucent material, but horizontally orientated sheet is increasingly being used (MCE_283, c). When set vertically, sheets are joined by either a straightforward lap of the material, which results in visible staining through the translucent material unless an opaque sealant is used. Alternatively, an aluminium top hat section is used, set between two adjacent sheets used as described in the section on profiled metal sheet. The manufacture of polycarbonate cladding is restricted only to the sheet material, and accessories such as sill profiles, coping profiles and window trims are not usually available. Aluminium trims are used instead, and are fixed so as to reduce their visibility to a minimum. Trims are fixed in the same way as for profiled wall cladding described earlier (MCE_281, e).

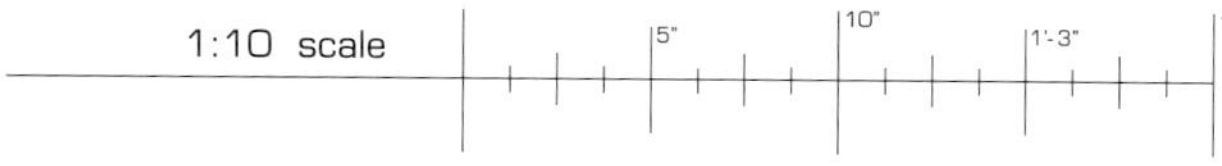

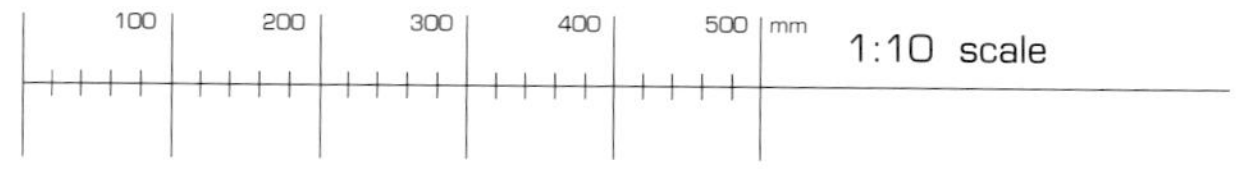

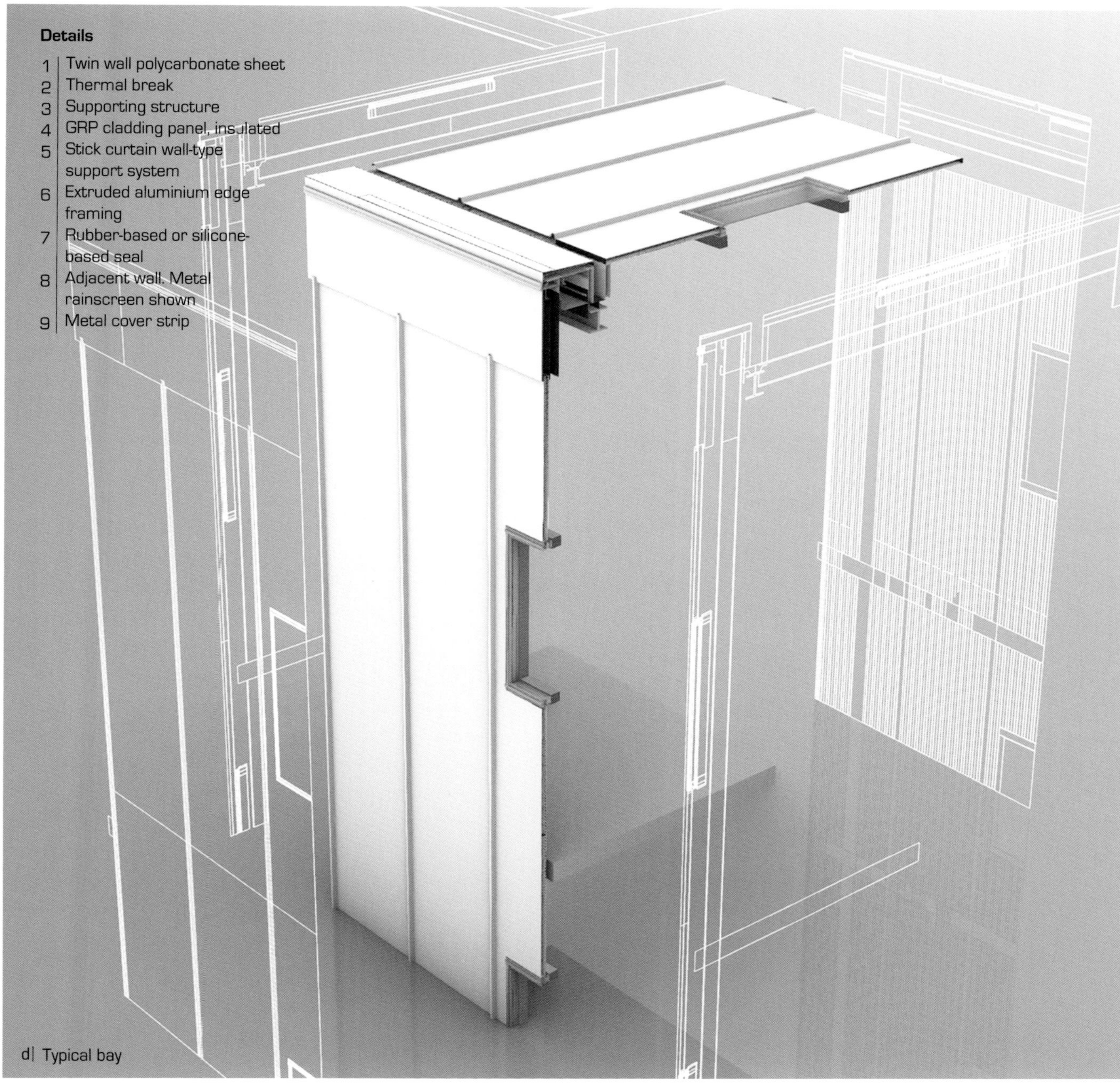

d| Typical bay

Twin wall polycarbonate is fixed with either conventional aluminium framing for windows, or framing for stick glazing curtain walling (MCE_283, d). Some manufacturers provide extruded I-sections, similar to those used in GRP cladding, to clip the twin wall sheets to provide a completely lightweight system (MCE_282, a). This is a very economic form of cladding which can be screen printed to create visually dramatic translucent facades. Its lack of fire resistance makes it a less attractive material for warehouses and factory buildings, where its translucency and thermal insulation, combined with economy, would make it a much more popular material. As with profiled polycarbonate sheet, other standard components are not usually manufactured, and folded aluminium sections are used for drips and parapet copings instead of polycarbonate sections, which are expensive to produce as new profiles.

e| Prototype

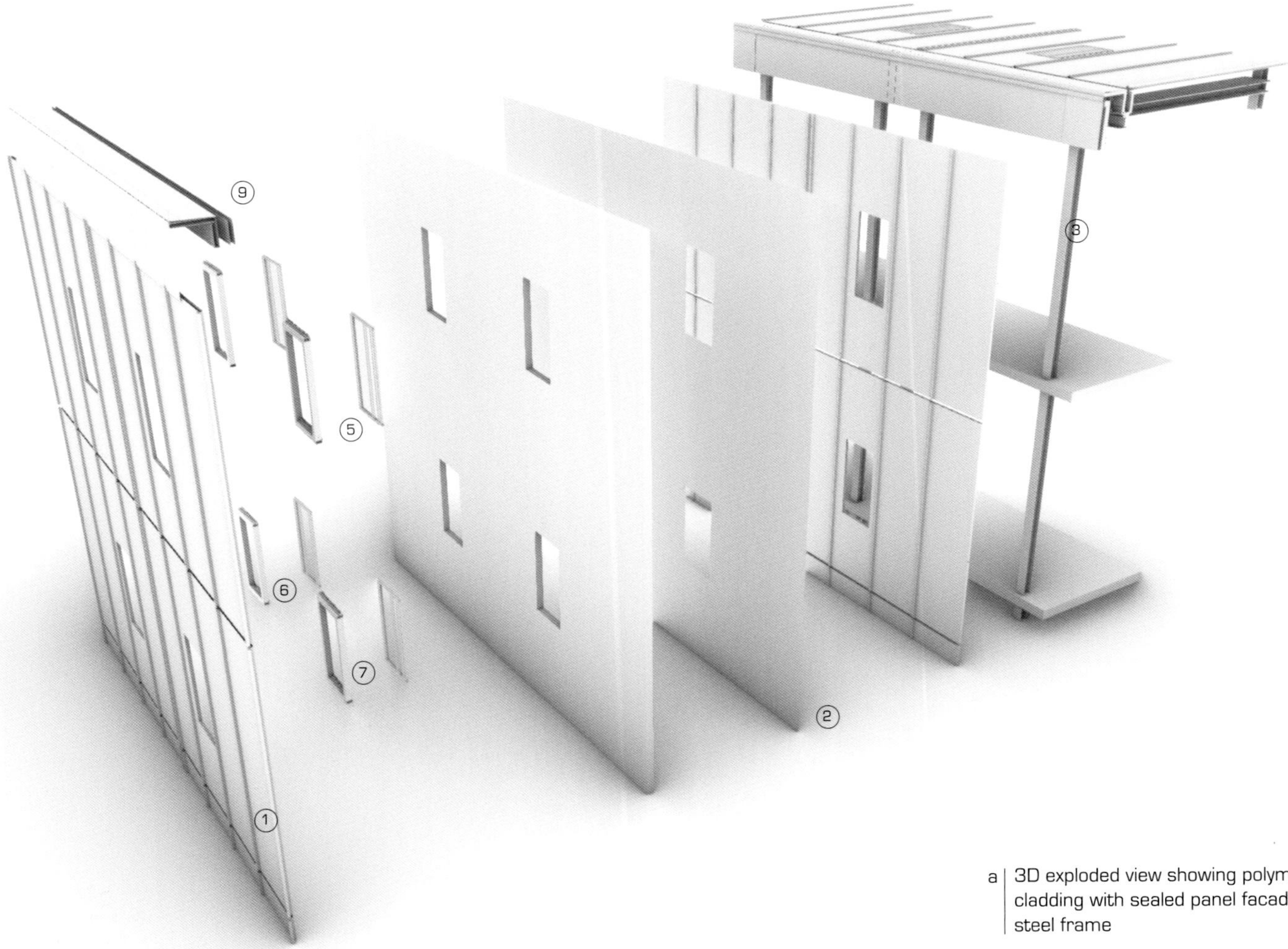

a | 3D exploded view showing polymer-based cladding with sealed panel facade fixed to steel frame

b | 3D view of polymer-based cladding with sealed panel facade fixed to steel frame

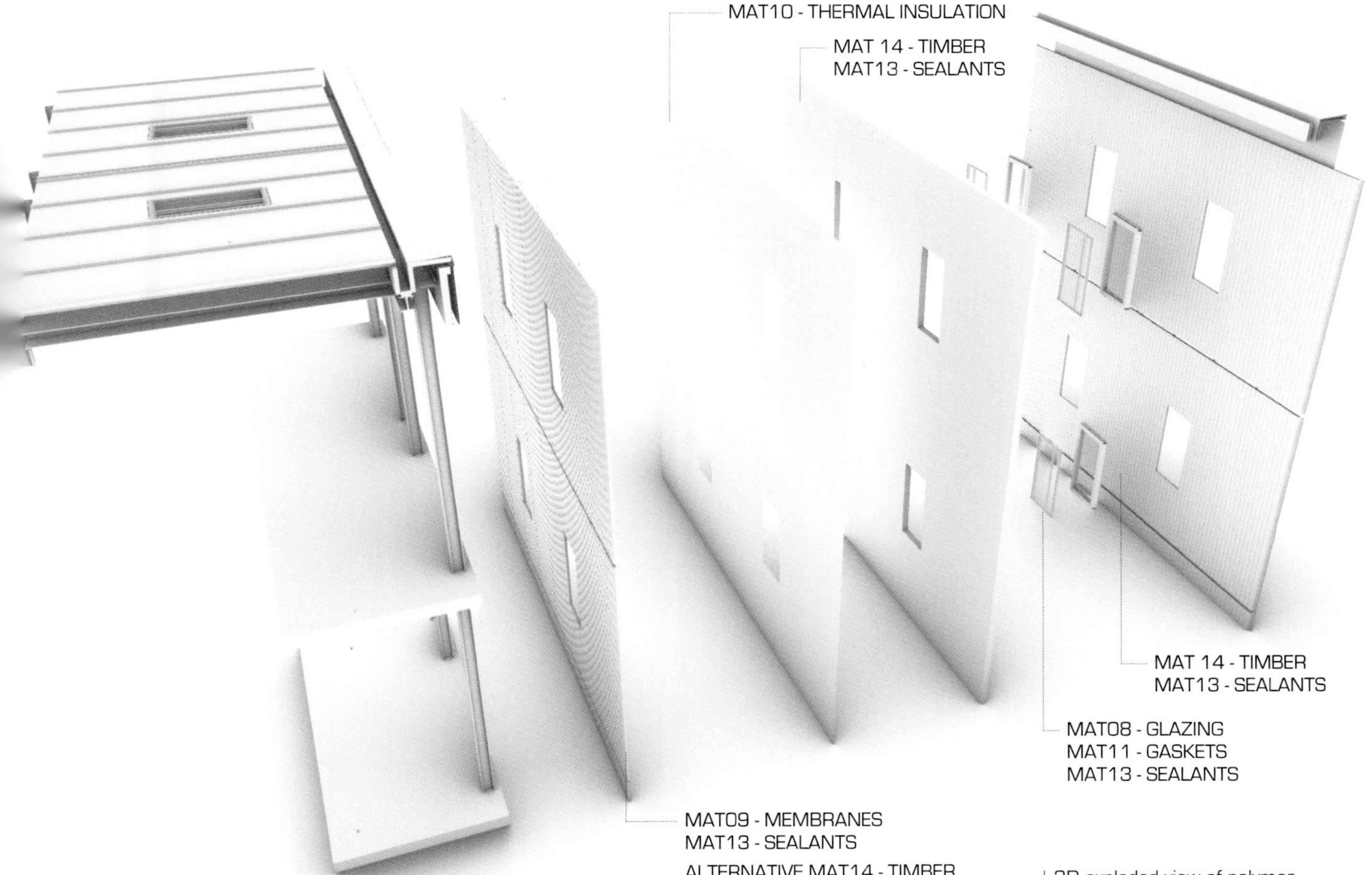

c | 3D exploded view of polymer-based cladding sealed panel facade components with material specification references

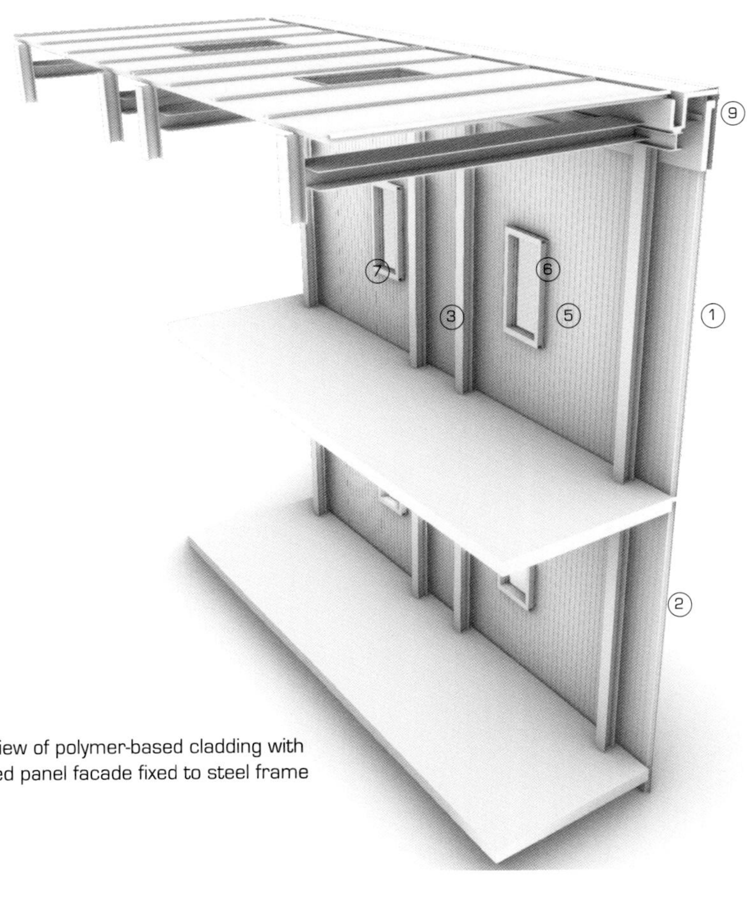

Details

1 | Twin wall polycarbonate sheet
2 | Thermal break
3 | Supporting structure
4 | GRP cladding panel, insulated
5 | Stick curtain wall-type support system
6 | Extruded aluminium edge framing
7 | Rubber-based or silicone-based seal
8 | Adjacent wall. Metal rainscreen shown
9 | Metal cover strip

d | 3D view of polymer-based cladding with sealed panel facade fixed to steel frame

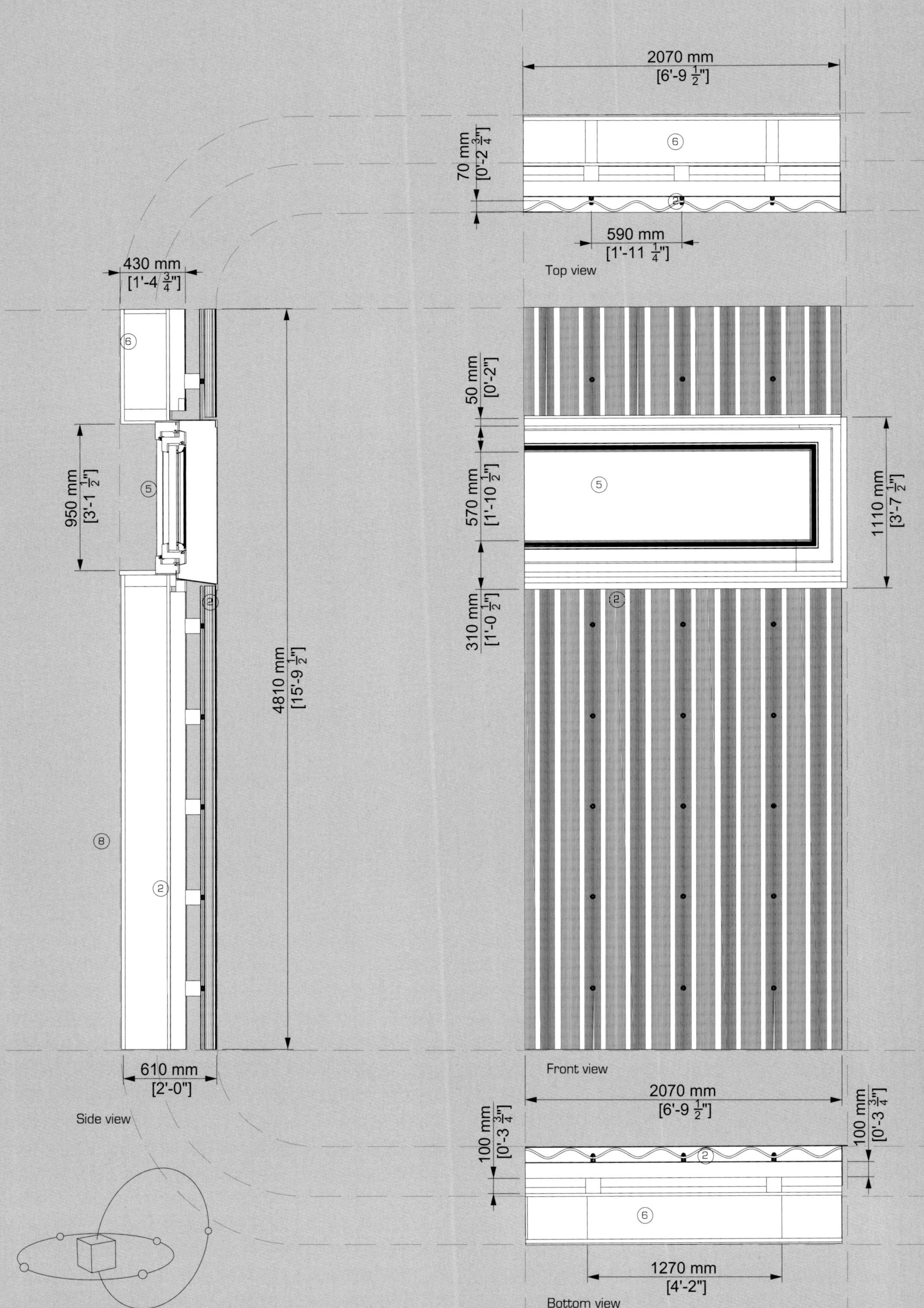

Third angle projection 1:30

Details

1 | UPVC section
2 | Plastic-composite flat panels
3 | UPVC board cladding
4 | GRP cladding panel
5 | Glazed unit
6 | Thermal insulation
7 | Backing wall
8 | Support rails
9 | UPVC window
10 | Plastic-composite coping

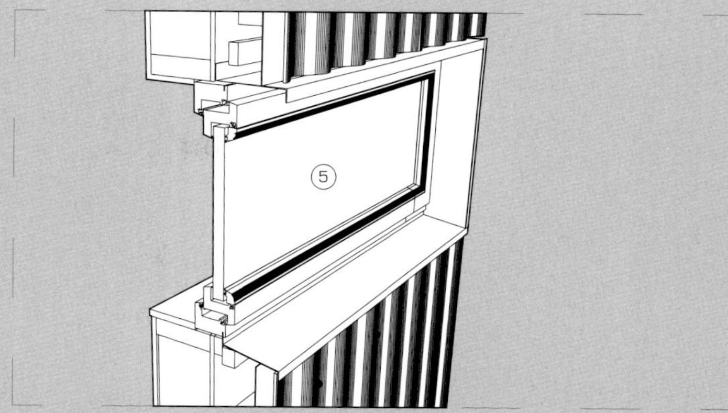

3D view of the assembly

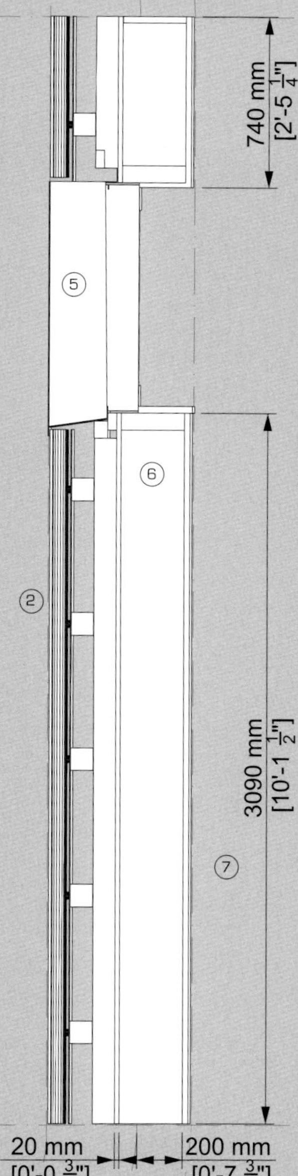

Side view

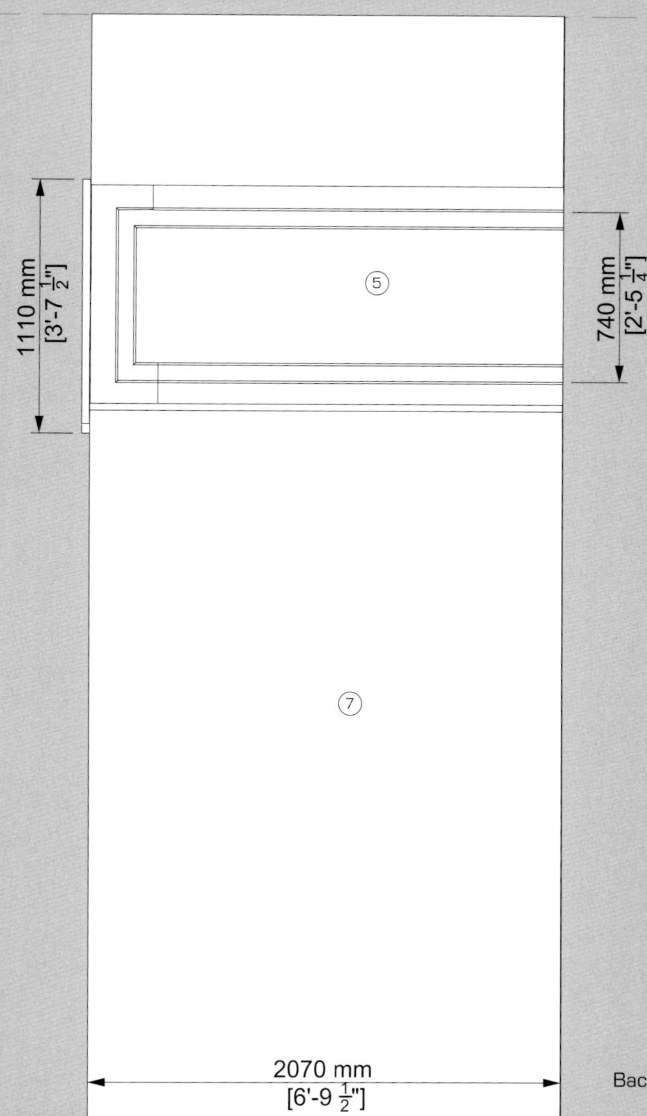

Back view

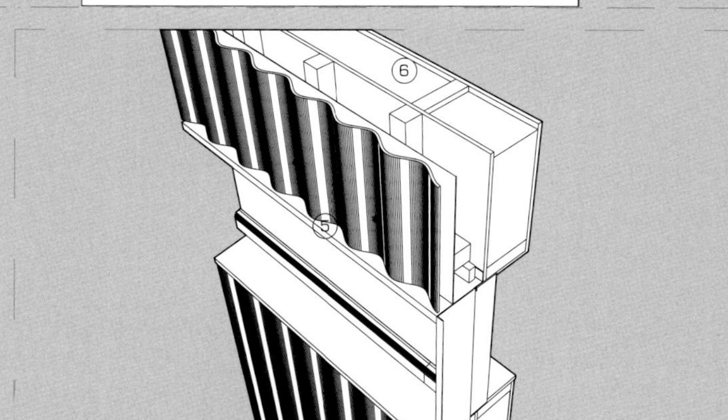

3D view of the assembly

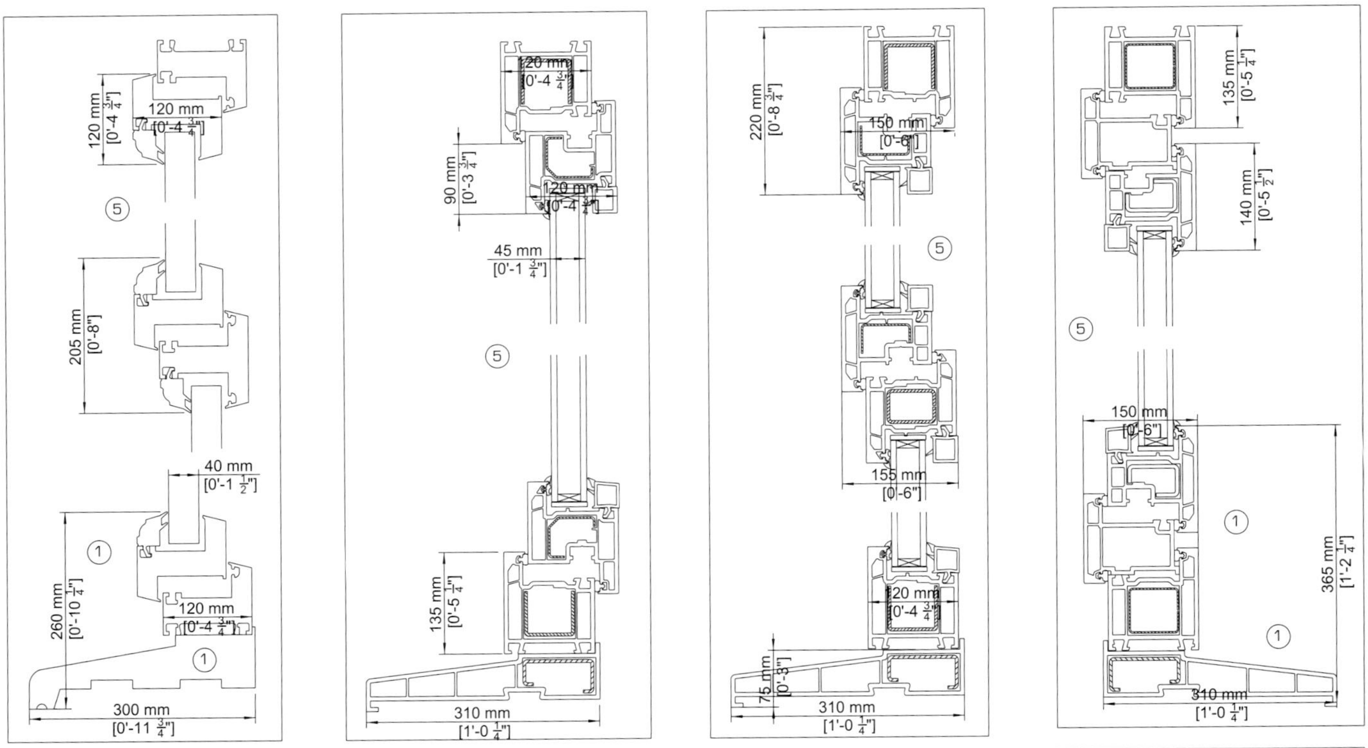

a | Vertical sections. Window at head and sill, louvre

b| Prototype

Details

1 | UPVC section
2 | Plastic-composite flat panels
3 | UPVC board cladding
4 | GRP cladding panel
5 | Glazed unit
6 | Thermal insulation
7 | Backing wall
8 | Support rails
9 | UPVC window
10 | Plastic-composite coping

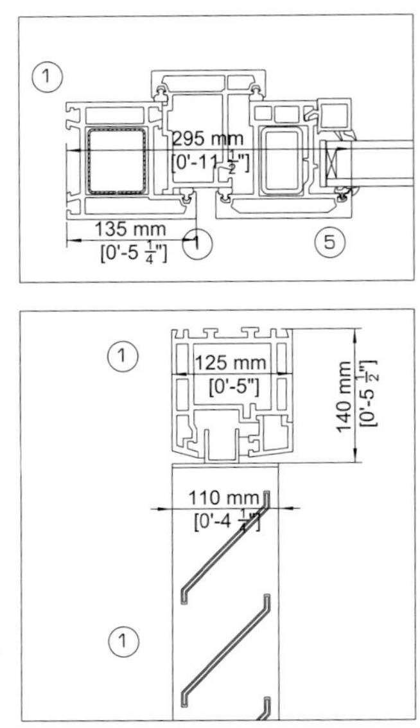

c | Horizontal section. Window jamb, internal corner

The main types of plastic-based rainscreen are flat panels, cassette panels, profiled sheet and overlapping tiles. They are used as either outer screens to glazed walls, typically as solar shading, or as rainscreen panels to an opaque wall, typically in-situ concrete or concrete blockwork. The materials used are either polycarbonate or glass fibre-reinforced polyester (GRP). Acrylic and UPVC, while softer than both these materials, are used for window frames and special moulded elements. In addition, composites other than GRP (described in the previous section on plastic-based cladding) are used for rainscreens. Thermosetting polymer resins can be mixed with cellulose fibres to provide sheet materials with high durability which fade little in sunlight. In common with rainscreens in other materials, panels or sheets are fixed with either visible point fixings, vertical/horizontal rails with partially concealed framing members, or partially interlocking panels where there is no view through the joints.

Flat polycarbonate sheet

Opaque flat sheet is fixed as rainscreen cladding panels using similar techniques to bolt fixed glazing and clamped glazing, with open joints between panels (b). Flat sheet is made in a wide range of colours. Sheets are held captive at their corners with an aluminium clamp on both sides, with a set of bolts squeezing the clamps together without the need to drill through the polycarbonate. Since the cost of drilling polycarbonate sheet is much lower than glass (glass has to be heat-strengthened after drilling), point fixings are more common when this technique is used. Bolts are much simpler than those used for glass, since the material is much lighter, though polycarbonate has higher thermal expansion than glass. Polycarbonate sheets usually have UV protection on both sides to avoid yellowing with age. The problem of yellowing has now been largely overcome with higher quality sheet materials. Sheet sizes are around 2000mm x 3000mm (6ft6in x 10ft) and 2000mm x 6000mm (6ft6in x 20ft), varying between manufacturers,

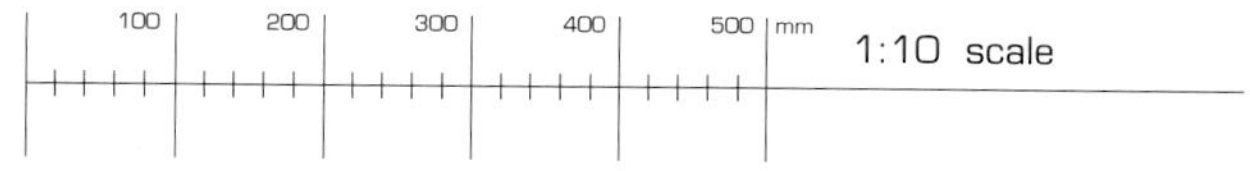

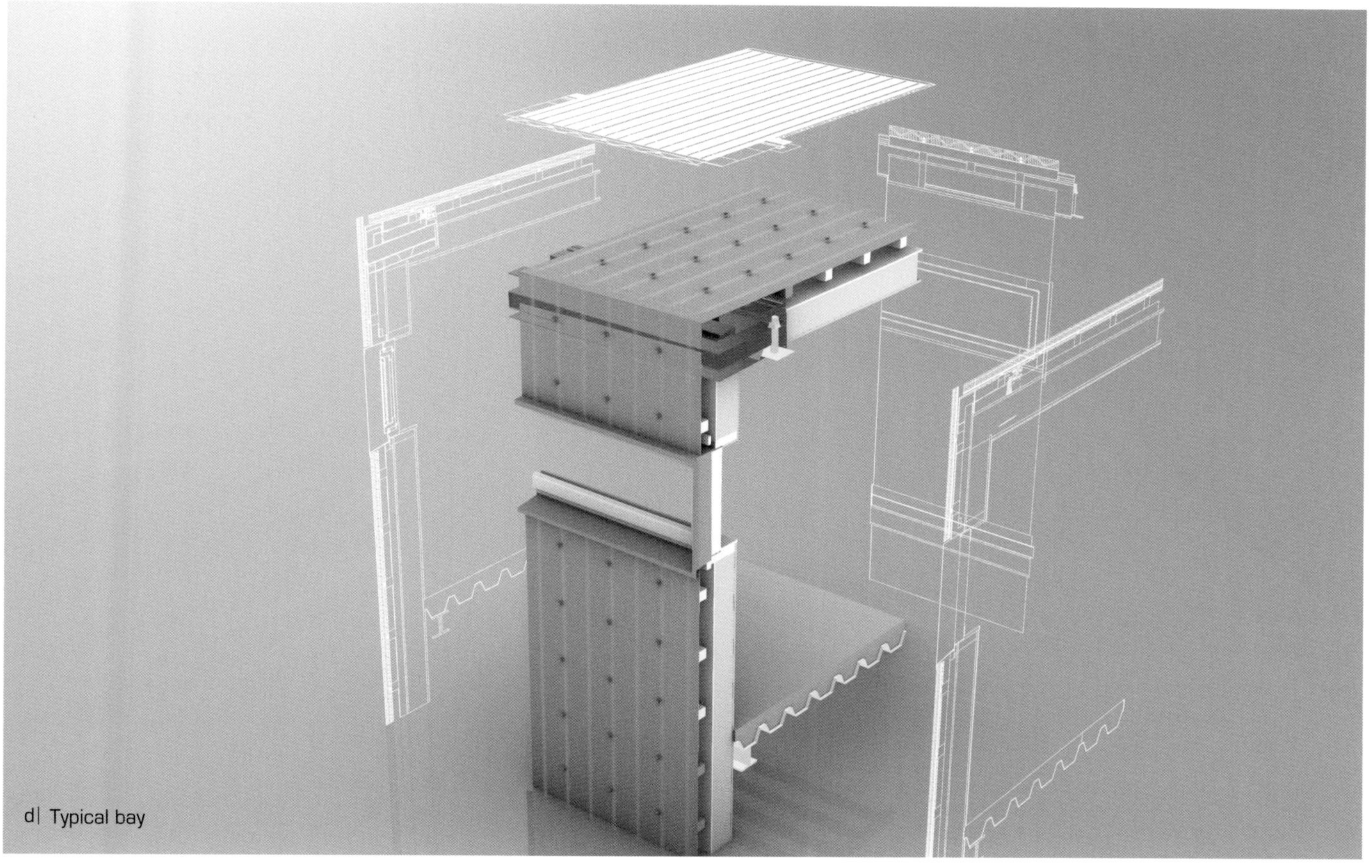

d| Typical bay

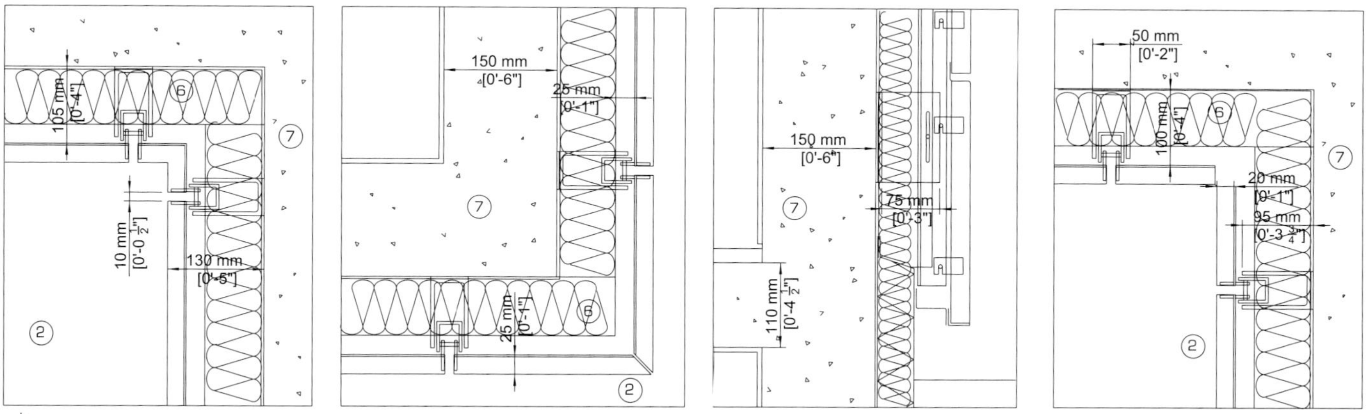

e | Horizontal section. Internal corner, external corner

in thicknesses from 3mm to 8mm (0.118–0.3mm). When opaque colours are used rather than the translucent or clear types, hook-on fixings can be bonded to the rear face of the panel with structural silicones. The panel can then be attached to vertically or horizontally set rails without the fixings being visible (d). An alternative fixing method is to fix the panels through the front face as visible fixings. Panels are set onto continuous rails, running vertically or horizontally, which are positioned behind the panels. A layer of EPDM or silicone is set between the panels and framing to provide a smooth consistent surface for the panels. The panels are fixed with screws from the front through holes pre-drilled in the panel. These screws then have a decorative capping applied, such as a dome-head screw that is secured into the head of the fixing screw that is already in place.

Multi-wall polycarbonate sheet

Like flat sheet, multi-wall sheet can be used for rainscreens, where its main advantage is the ability to provide large, flat panels rather than its high level of thermal insulation. Thicknesses are from 4mm to 32mm (0.15–1.2in) in sheet sizes from 1000mm x 6000mm (3ft3in x 20ft) to 2000mm x 7000mm (6ft6in x 23ft). The material can be screen- printed or coated to provide a wide variety of colours on a large format sheet. As with flat sheet, the material can be coated to give UV protection on one side or both sides to avoid yellowing. The sheet material is used in increasing numbers of walls, from twin wall to five or six walls thick. Panels are fixed using the same methods as flat sheet, with thicker sheets being supported up to around 1800mm centres (6ft). Panels can be curved by setting them into a pre-formed aluminium edge frame, when the material can be curved to a minimum radius of 1500mm (5ft) for twin wall sheets and around 4000mm (13ft) for the thickest sheets.

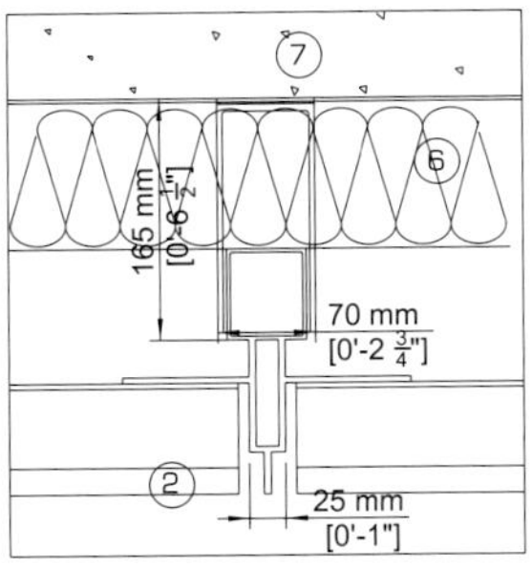

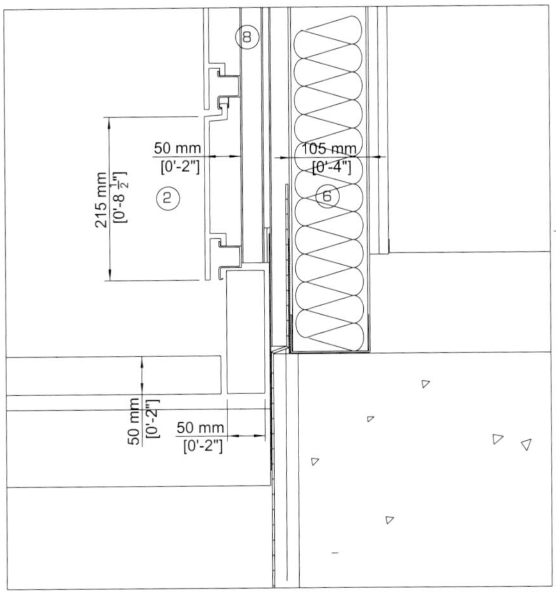

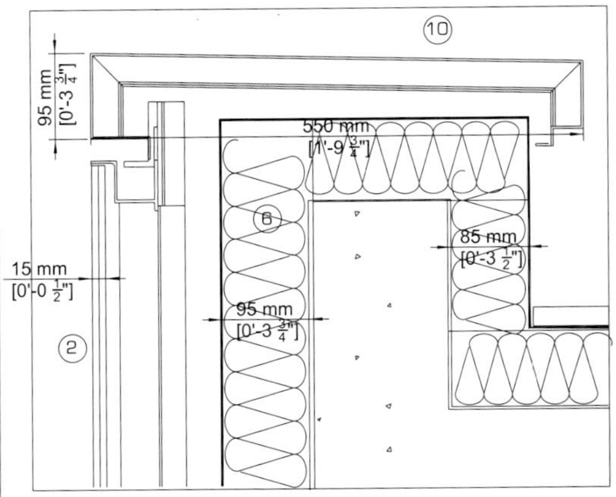

b | Vertical section. Panel-to-panel junction at top of wall assembly

Details

1 | UPVC section
2 | Plastic-composite flat panels
3 | UPVC board cladding
4 | GRP cladding panel
5 | Glazed unit
6 | Thermal insulation
7 | Backing wall
8 | Support rails
9 | UPVC window
10 | Plastic-composite coping

a | Horizontal sections. Internal corner, external corner, panel-to-panel junction

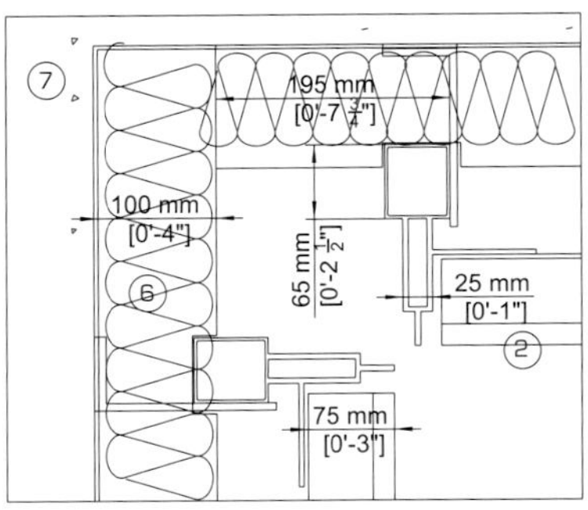

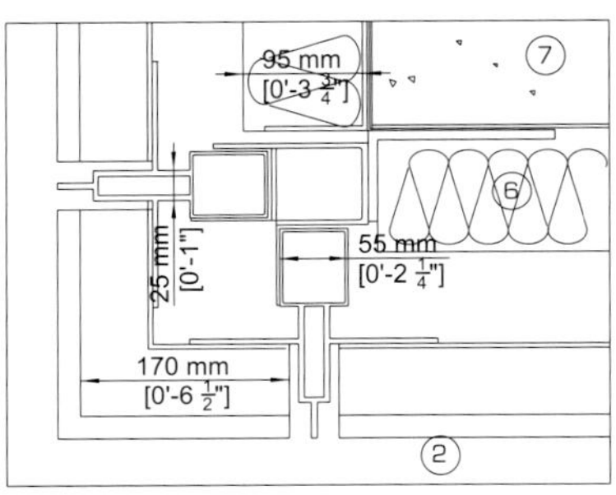

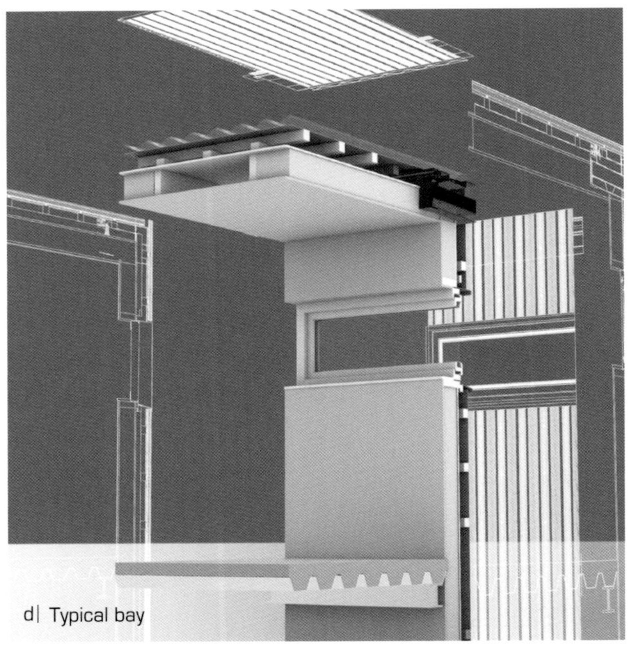

d | Typical bay

c | Vertical sections. Panel-to-panel junction, at ground level

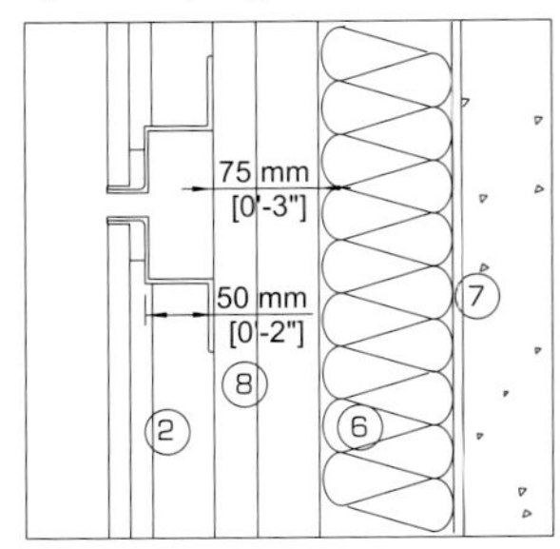

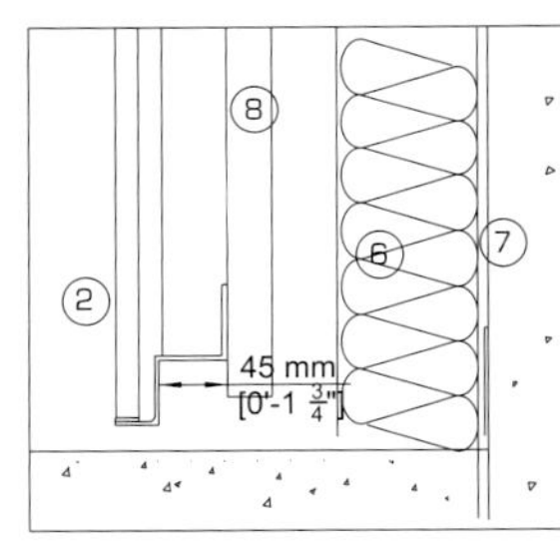

Profiled polycarbonate sheet

Profiled sheets are made from extruded polycarbonate resin, manufactured to provide transparent panels with around 90% light transmission, made to match steel and aluminium sheeting from other companies. They are used mainly to provide daylight in roofs and in wall cladding in industrial buildings without the need for special rooflight panels. Sheets are made in profiles to suit sheets by a particular manufacturer with whom they have an agreement. They are produced in both sinusoidal (curved) profiles and trapezoidal (zigzag) profiles in lengths of up to around 10 metres (32ft) with thicknesses of around 1mm. The width of most sheets is around 1200mm (4ft). They can also be used for rainscreen wall cladding, as materials in their own right. Vertically set sheeting is being used in city centre buildings, far from its use as an economic roofing material. Its high impact resistance, lightness in weight and long term high weather resistance make polycarbonate very suitable for dramatic cladding structures where weight is a critical issue. Profiled sheets are fixed with self-tapping sealed screws, with a watertight washer on the outside (MCE_255, c). More elaborate point fixings are difficult to use due to the profiled nature of the sheet, where a screw or bolt is attached through the peak or trough of the profile. Sheets are lapped on all sides, creating a shadow where they lap, which can be concealed by the support structure behind. Alternatively, sheets can be lapped in the horizontal direction, regardless of the sheet orientation, and can be joined by aluminium top-hat profiles in the vertical direction in the manner of profiled metal sheeting. Sill, drip and coping profiles are made from either extruded UPVC, GRP or extruded aluminium (MCE_289, g). The material can be curved to a minimum radius of around 4000mm (13ft) for a sheet of 50mm (2in) thickness. Profiled polycarbonate sheet is also made in a translucent white colour to provide a light transmission of around 45% and in grey colours with light transmission of around 35%.

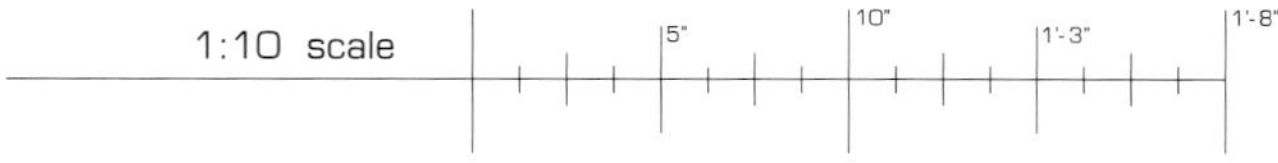

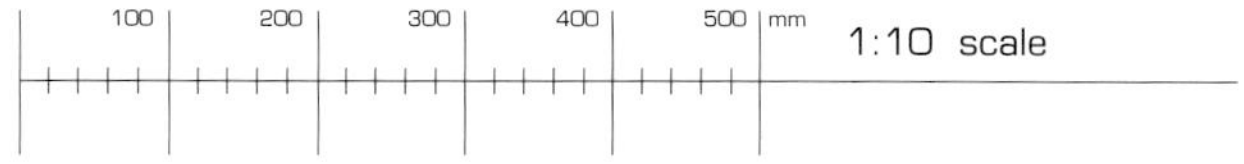

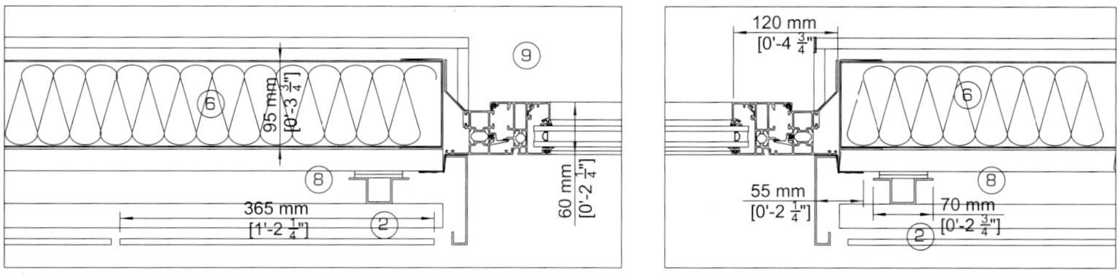

e | Horizontal section. Window opening

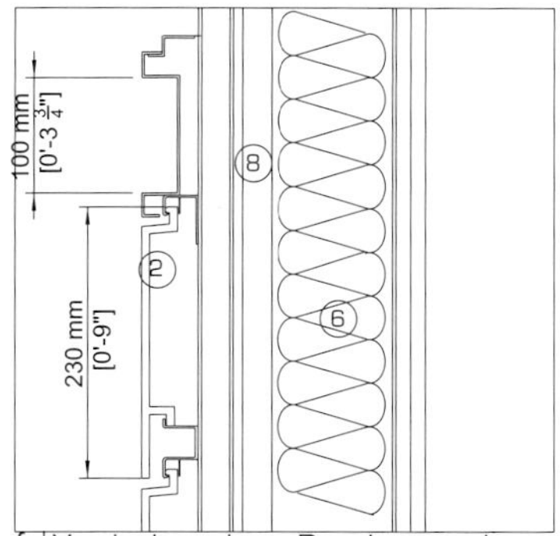

f | Vertical sections. Panel-to-panel junctions of wall assembly

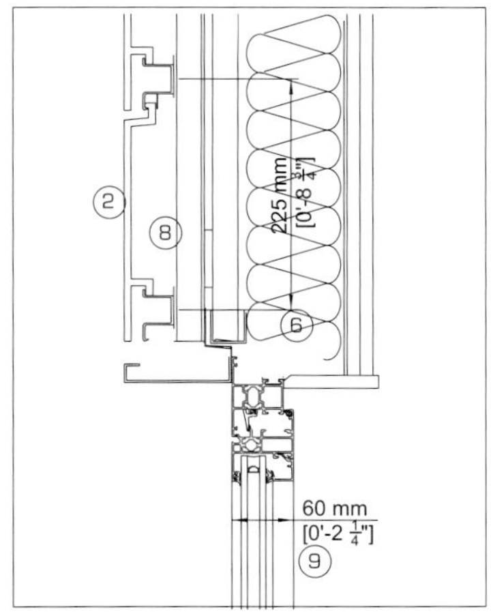

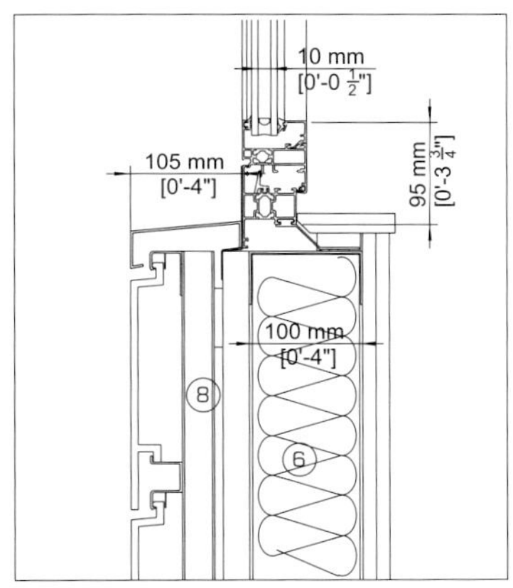

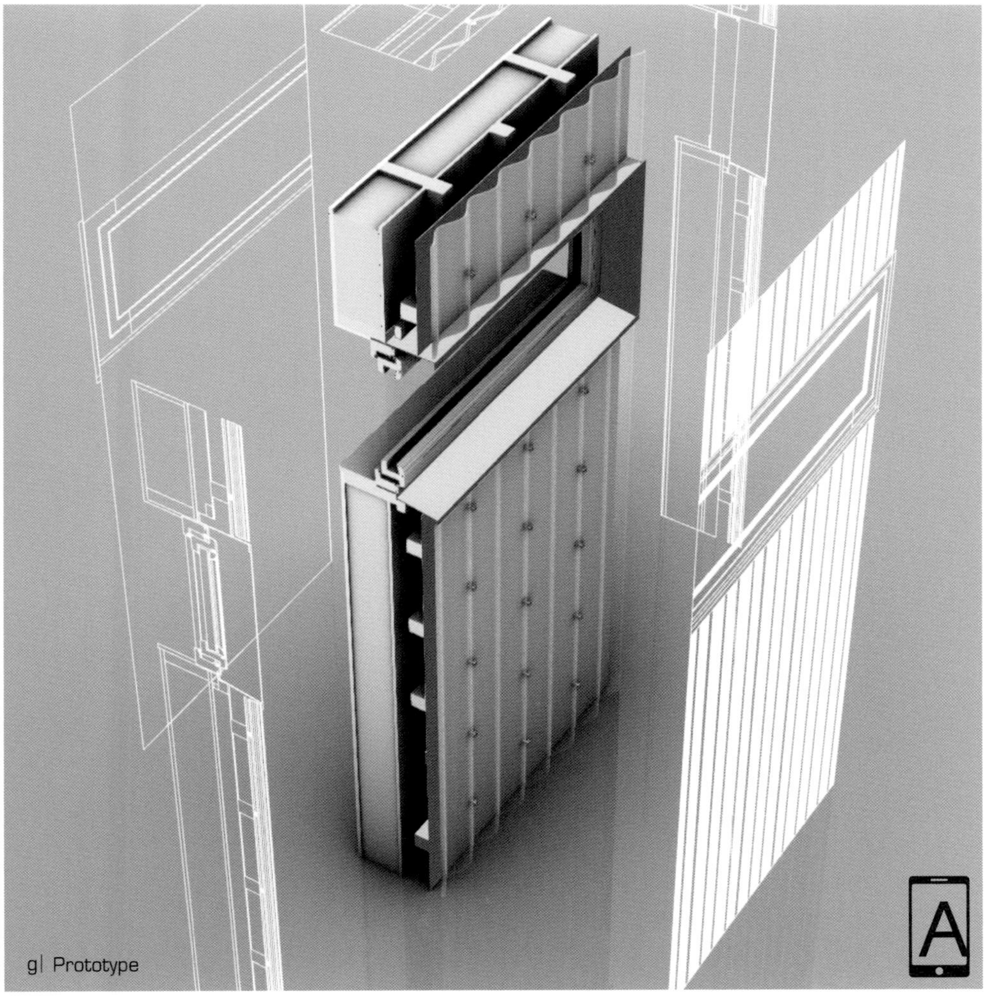

g | Prototype

Polymer-composite flat panels

Flat panels can also be made from thermosetting resins mixed with cellulose fibres. This is usually a mixture of 70% softwood fibre and 30% resin, manufactured at high temperature and pressure. Panels can be coloured on one side, both sides or have different colours on each side. Panels are smooth with an almost impervious sealed surface. Although the finish colour is formed by using pigmented resins as a top coat in the mould, sawn edges do not require repainting or protective cover since the colour extends all the way through the material. Rainscreen panels formed in this material are very resistant to the long-term effects of weather, with high UV resistance and colour stability and high fire resistance. Plastic-composite flat panels can be cut, drilled and routed on site, making them more like timber in terms of thier flexibility during construction. Components do not all have to be cut off-site in a workshop. Because the sawn edges are smooth, they can be laid as overlapping tiles in the manner of timber shingles, a technique used in sheet metal cladding (MCE_291, g). Their high impact resistance and impervious surface make them well suited to demanding conditions where damage can occur easily. The material is made in sheet sizes from 3600mm x 1800mm (12ft x 6ft), 3000mm x 1500mm (10ft x 5ft) and 2500mm x 1800mm (8ft x 6ft) in thicknesses from around 5mm to 12mm (0.25–0.5in). Corner panels and parapets are also manufactured in the same material. Plastic-composite flat panels can be fixed with visible fixings or concealed fixings used for flat sheet and multi-wall polycarbonate sheet (MCE_290, b).

UPVC board cladding

Extruded UPVC boards are used as a substitute for timber cladding, and are often fixed back to a timber frame in the same way as timber boarding. Although the main advantage of this material is its low maintenance when compared to painted timber boards, the material is beginning to be used in its own right, without the imitation of the lan-

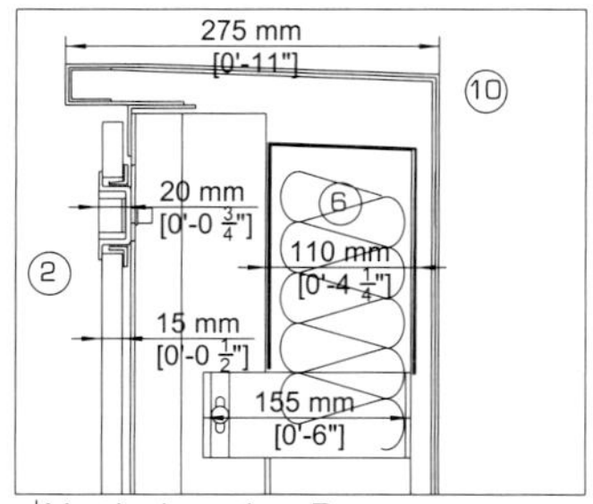

a | Vertical section. Parapet

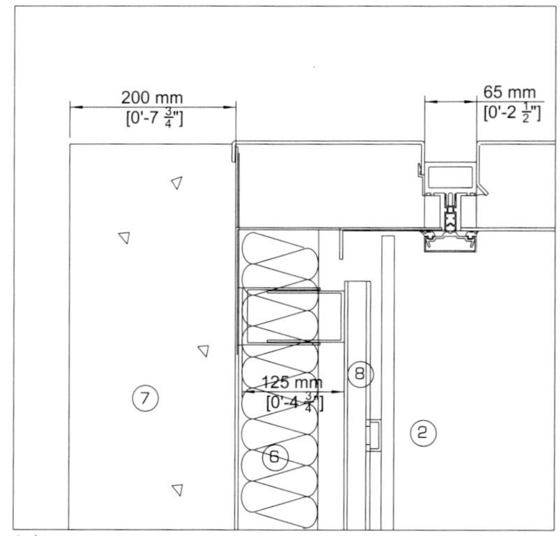

b | Horizontal section. Internal corner

Details

1 | UPVC section
2 | Plastic-composite flat panels
3 | UPVC board cladding
4 | GRP cladding panel
5 | Glazed unit
6 | Thermal insulation
7 | Backing wall
8 | Support rails
9 | UPVC window
10 | Plastic-composite coping

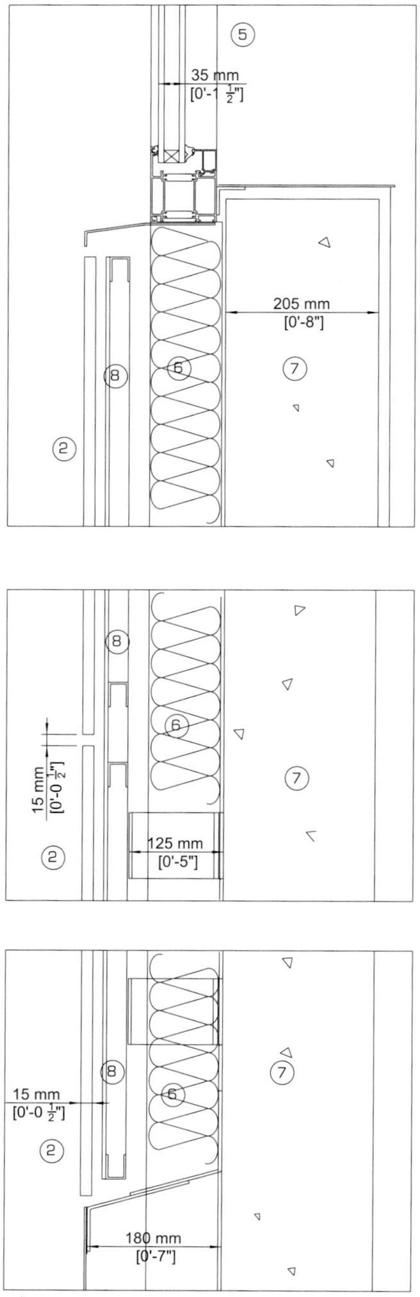

c | Vertical sections. Window opening, ground level

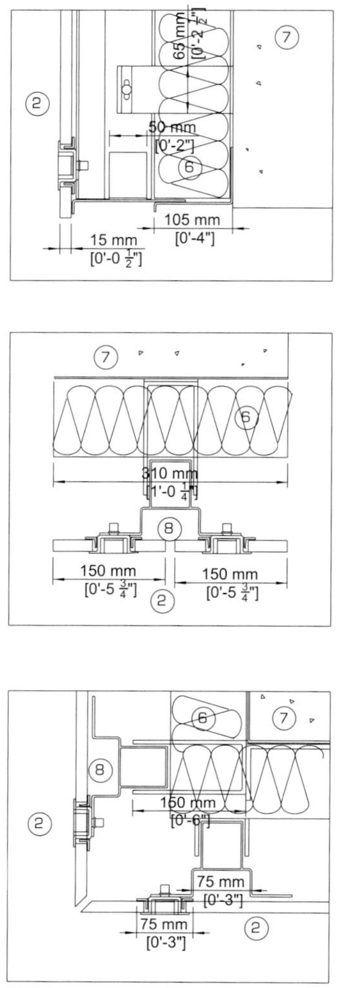

d | Horizontal sections. Panel-to-panel junctions

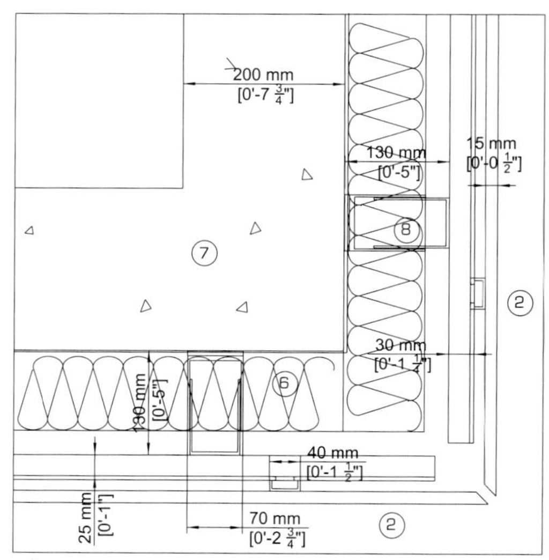

e | Horizontal section. External corner

guage of timber wall cladding. Boards are made as extrusions in lengths of up to around 5000mm (16ft), in widths from 250mm to 300mm (10–12in). Thicknesses match those of timber boarding at 18–20mm (around 0.7in). Boards are nailed back to timber frames or screwed back to aluminium frames and are fixed at around 600mm (2ft) along the length of the panels. Manufacturers have specially formed corner pieces in a range of angles to suit typical situations of 45º angles both internally and externally. Corners can also be glued with solvent adhesives or sealed with silicone sealants. Other components specially made for UPVC cladding include end covers for the exposed ends of extrusions, flashings for sills, copings and window surrounds. Boards are fixed back to battens that provide a ventilated void behind to allow the boards to function as a rainscreen. Where the supporting wall is timber framed, as is typically the case, this wall is faced in a waterproof membrane with plywood sheathing behind that forms part of a timber stud wall. UPVC board cladding provides a high level of thermal insulation, with U-values at around 0.15 W/m² ºC. UPVC boards are subject to relatively high levels of thermal expansion when compared to GRP, at 7x10–5 per ºC for UPVC (like polycarbonate) and 2.5x10–5per ºC for GRP. A length of UPVC at 3000mm (10ft) long will expand about 6mm (0.25in) with a 25ºC increase in its temperature.

UPVC windows

Extruded UPVC frames are used for windows, typically in rainscreen panels in plastic-based materials. UPVC can be extruded to provide complex and accurate sections that can be appreciably more economic than those in aluminium. The material has a much lower thermal conductivity than aluminium, and so requires no additional thermal break.

1:10 scale | 5" | 10" | 1'-3" | 1'-8"

100 | 200 | 300 | 400 | 500 mm 1:10 scale

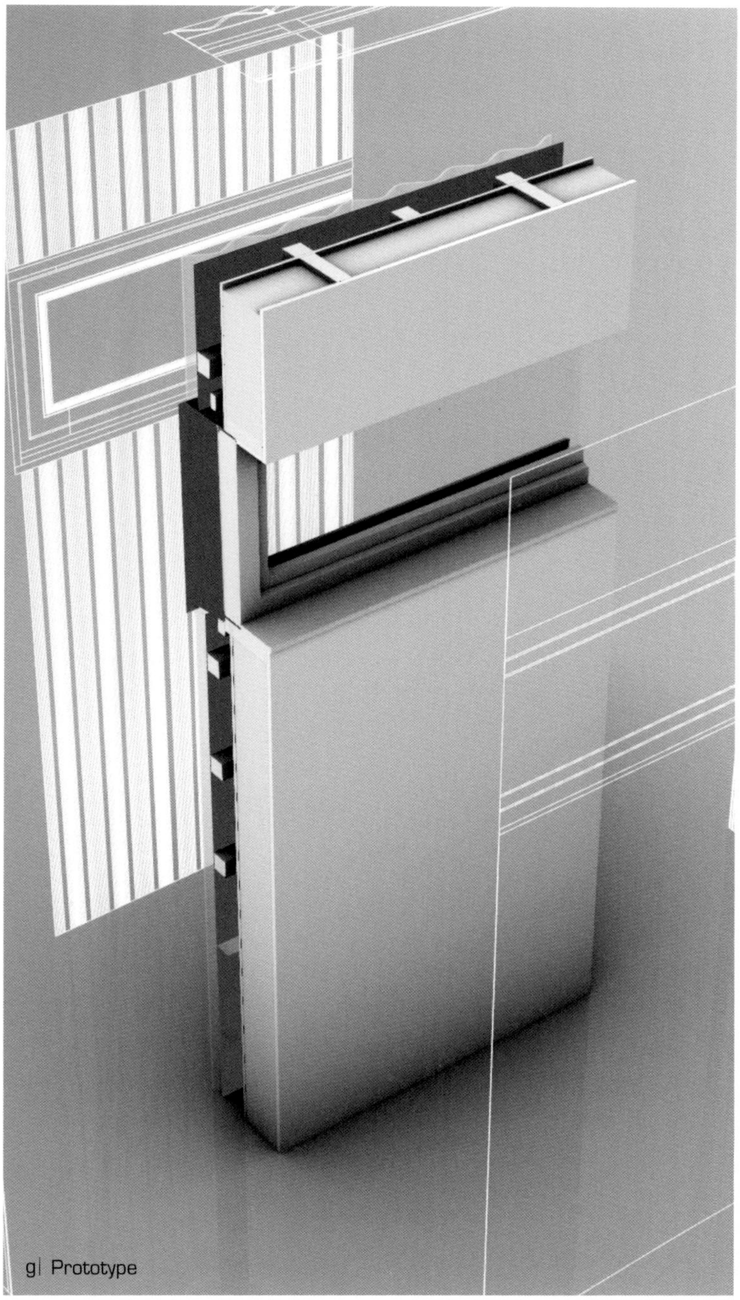

g| Prototype

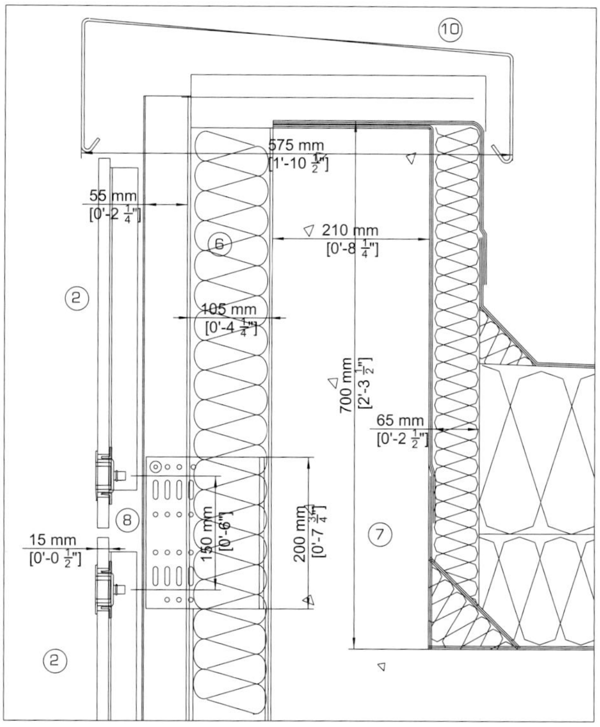

f | Vertical section. Parapet condition

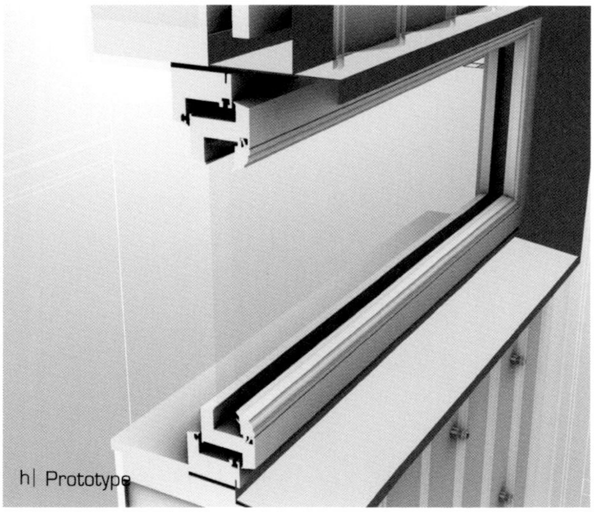

h| Prototype

Windows have an inner chamber which is drained and ventilated in the manner of aluminium windows, with rubber-based seals providing the necessary air seals (MCE_286, a). Windows are typically sealed with silicone around their edges. U-values for window frames can reach as low as 1.4 W/m² ºC, which is comparable with highly insulated double glazed units. Some UPVC window sections have a galvanised steel internal core reinforcement to stiffen the frames (MCE_286, a), usually as folded sections that are held by the walls of the extrusion. Corners of frames are bonded together to avoid corrosion of the steel sections. UPVC windows have developed considerably in recent years with higher impact resistance and higher crack resistance, particularly in the low winter temperatures where the material is more vulnerable.

GRP panels

Polycarbonate is generally a more expensive material than GRP, making GRP more suitable for lower-cost applications. However, GRP has one advantage over all the other plastic-based materials, which is its ability to be moulded easily and economically. When used as rainscreen panels the material needs a top gel coat to avoid its fibres being seen. The ability to see the fibres through the material make it very unsuitable for transparent or translucent panels, but for moulded rainscreen panels it is possible to introduce some 3D modelling into the facade panels. GRP can be bonded to honeycomb panels formed in the same material, to produce large panels with high fire resistance. The face of the panel can be screen printed to any design, with the use of photographic images being increasingly popular.

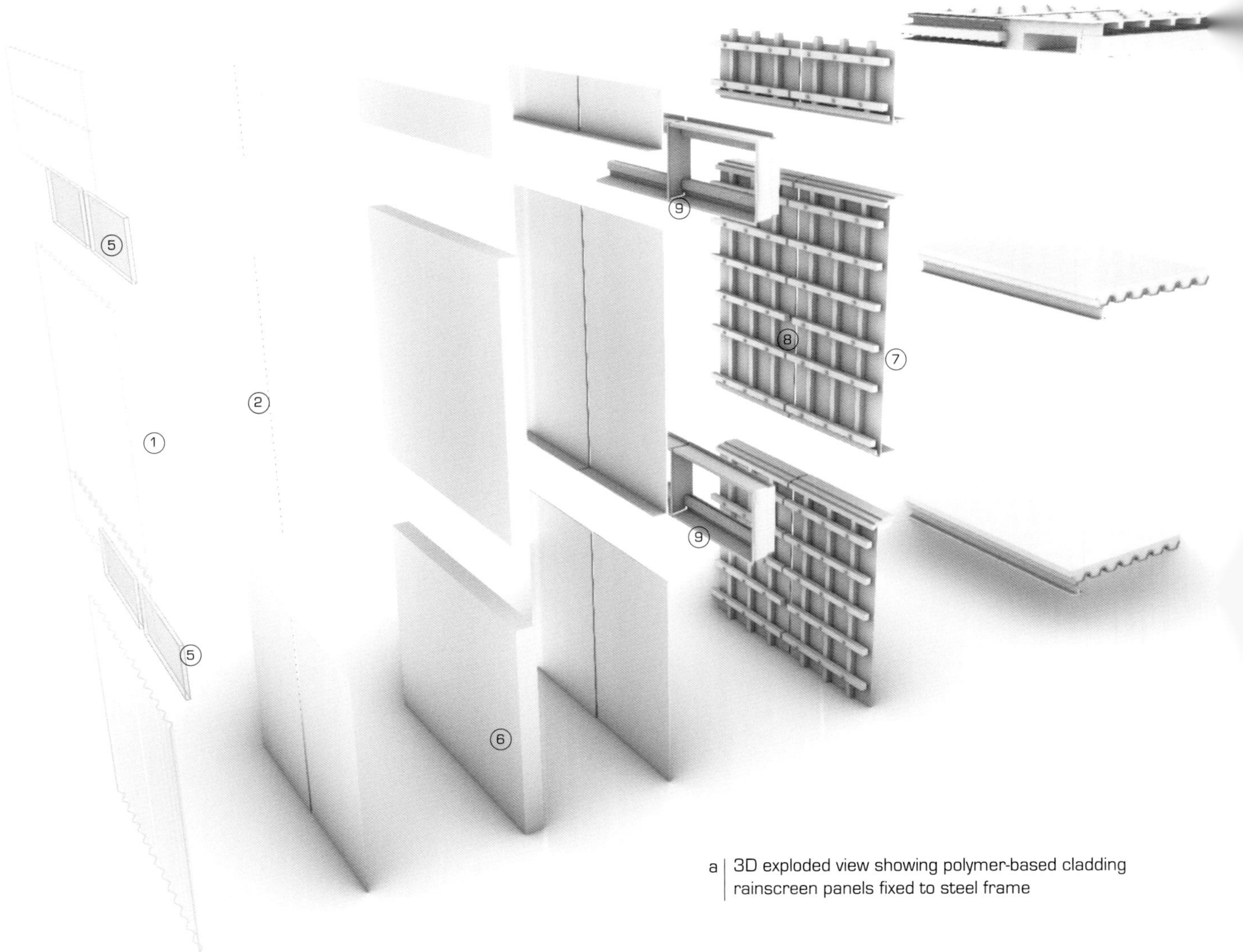

a | 3D exploded view showing polymer-based cladding rainscreen panels fixed to steel frame

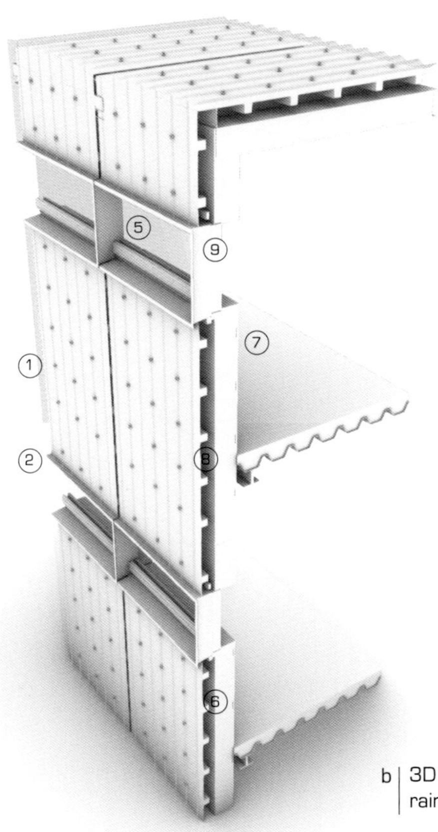

b | 3D view showing polymer-based cladding rainscreen panels fixed to steel frame

Details

1 | UPVC section
2 | Plastic-composite flat panels
3 | UPVC board cladding
4 | GRP cladding panel
5 | Glazed unit
6 | Thermal insulation
7 | Backing wall
8 | Support rails
9 | UPVC window
10 | Plastic-composite coping

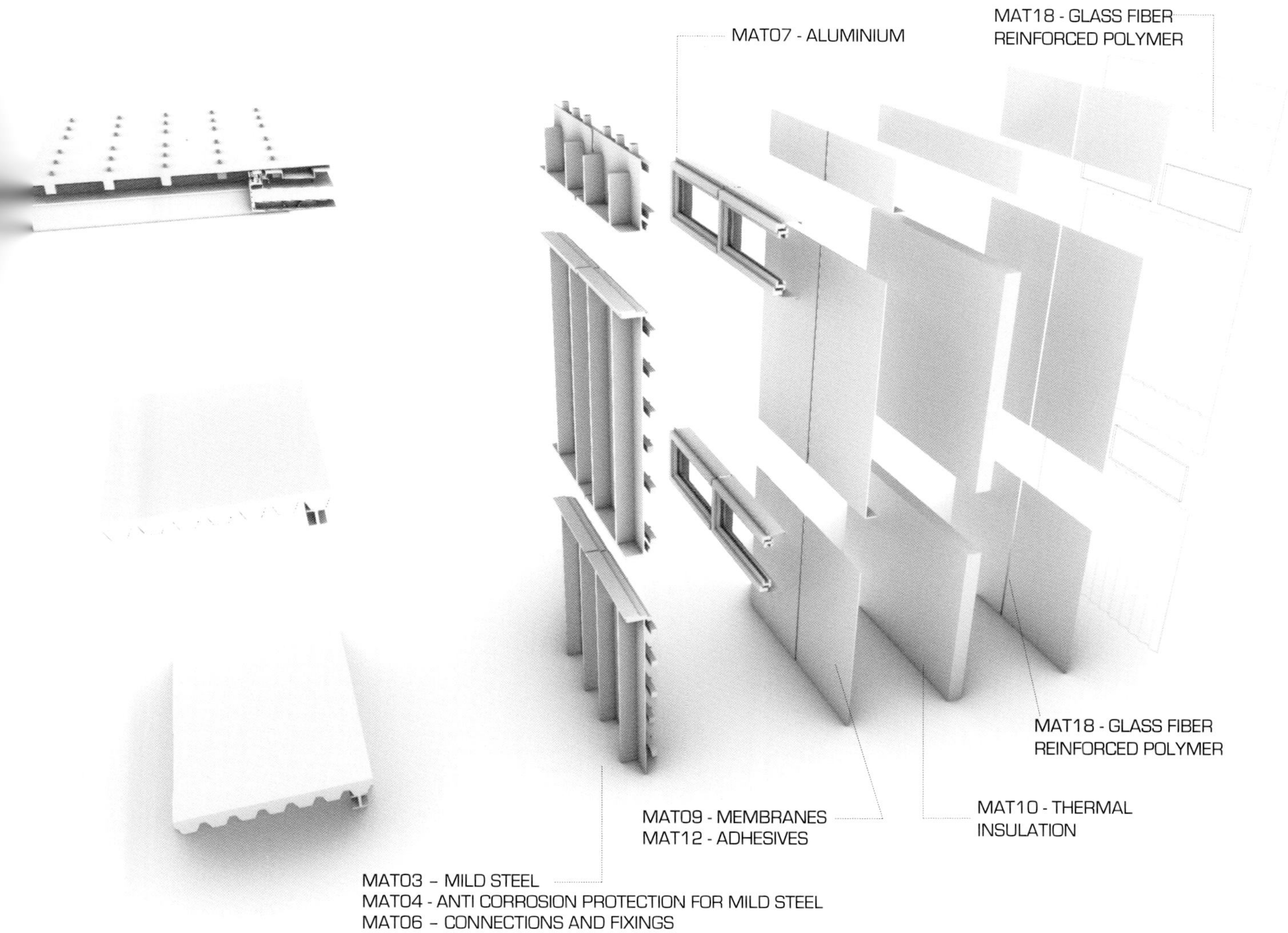

c | 3D exploded view of polymer-based cladding rainscreen panel components with material specification references

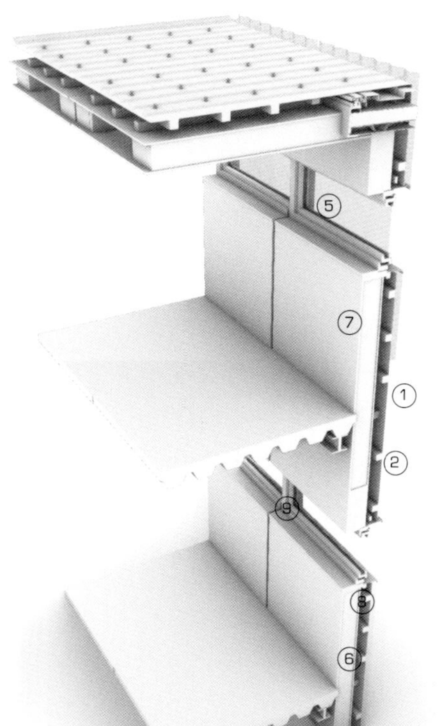

d | 3D view showing polymer-based cladding rainscreen panels fixed to steel frame

Example project of timber facades
Harbour Housing, Hong Kong
Facade Engineer: Newtecnic
Architect: Newtecnic

TIMBER WALLS

24 Cladding the timber frame

Timber frames

Ground level

Upper floors

Corners

Roof eaves and parapets

25 Cladding panels and rainscreens

Timber boards

Finishes

Cladding panels and rainscreens

Plywood sheets

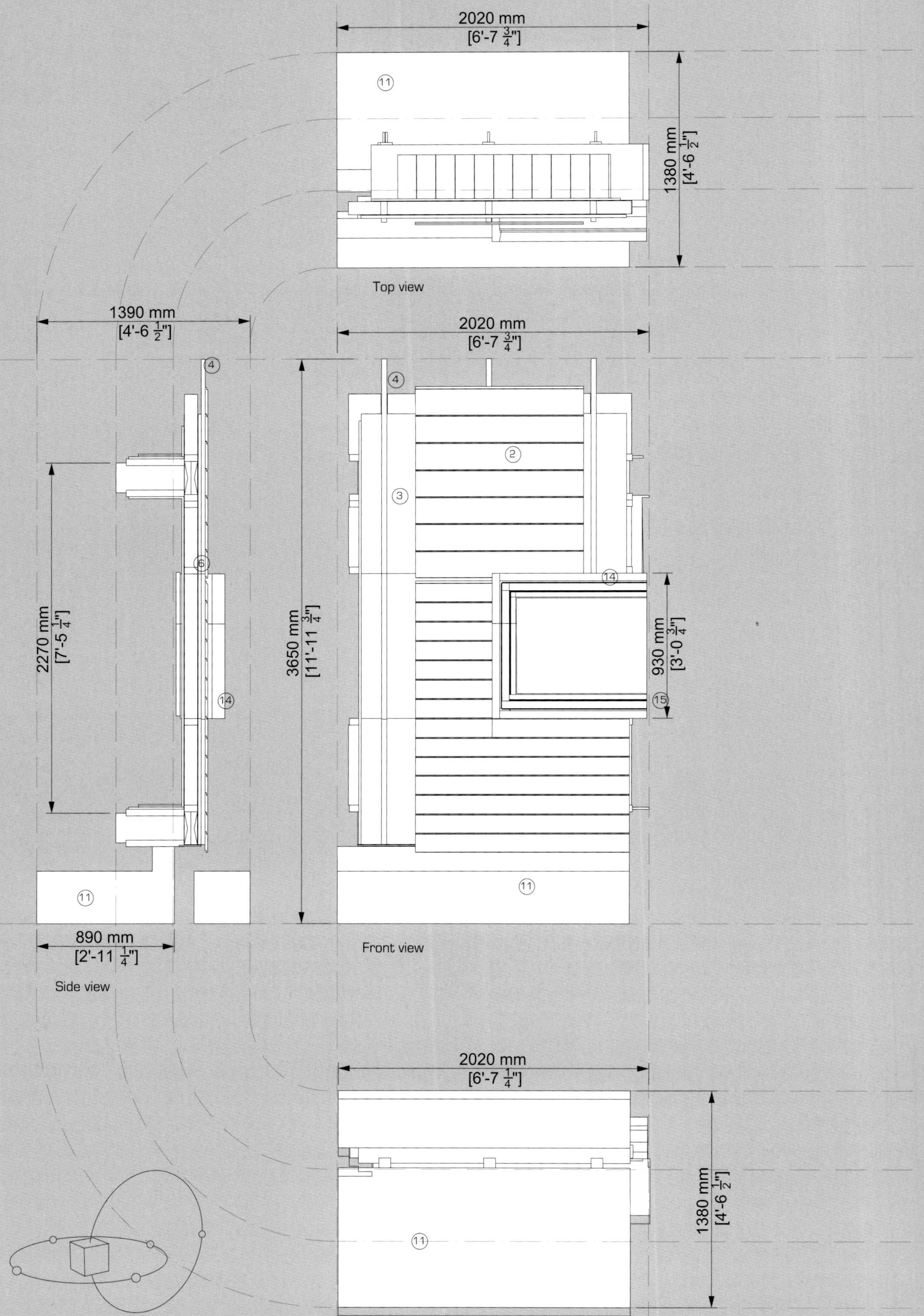

Third angle projection 1:30

Bottom view

Details

1 | Metal parapet flashing
2 | Timber boards
3 | Plywood sheathing
4 | Timber studs
5 | Timber rail
6 | Breather membrane
7 | Window flashing
8 | Damp proof course

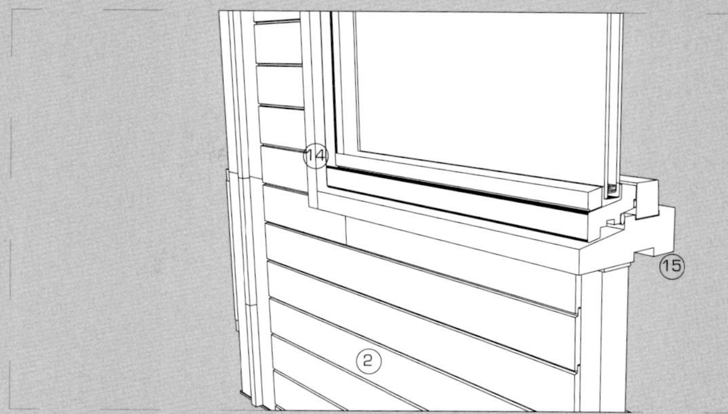

3D view of the assembly

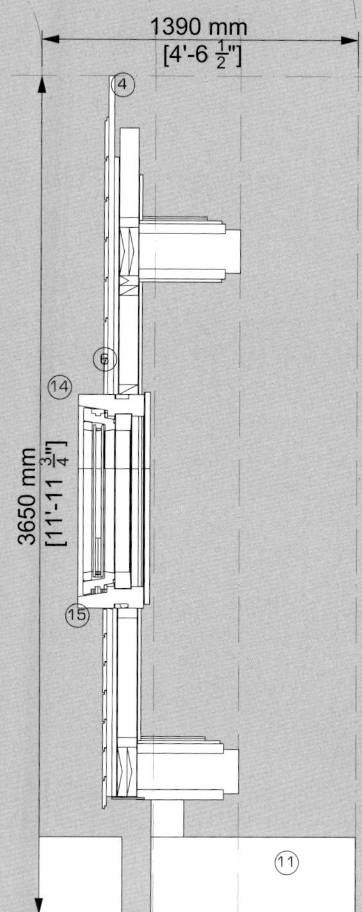

Side view

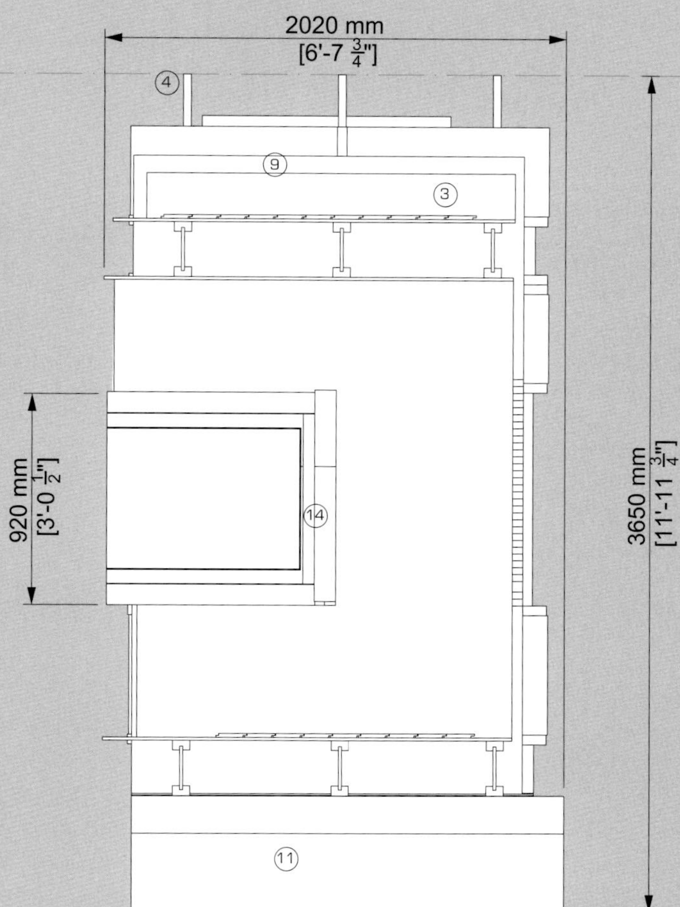

Back view

9 | Vapour barrier
10 | Timber roof
11 | Concrete ground slab
12 | Internal plaster finish or dry lining/dry wall
13 | Thermal insulation quilt set within timber frame
14 | Timber framed window/door
15 | Timber sill
16 | Air gap

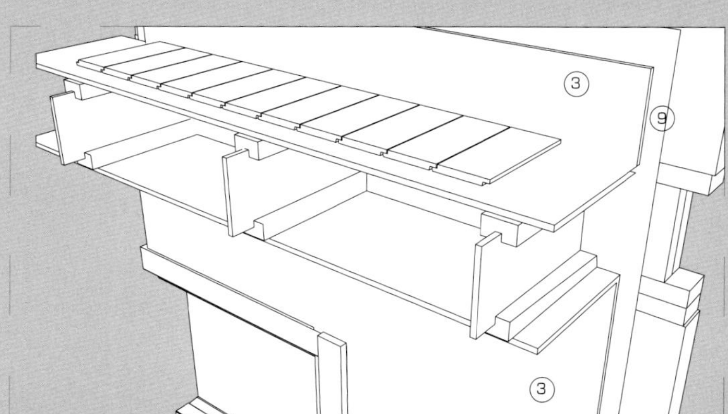

3D view of the assembly

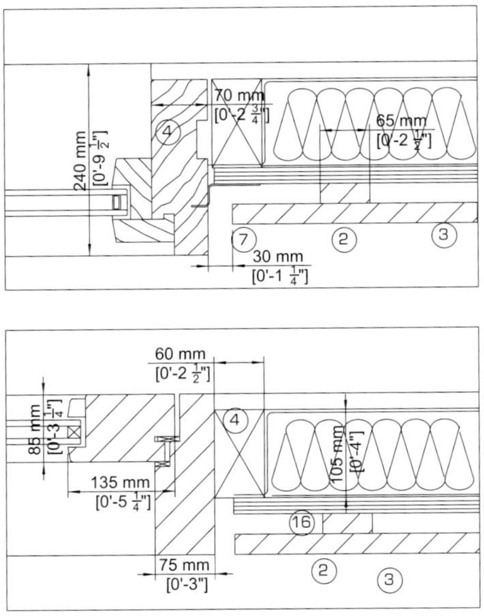

a| Horizontal sections. Window jambs

Details

1 | Metal parapet flashing
2 | Timber boards
3 | Plywood sheathing
4 | Timber studs
5 | Timber rail
6 | Breather membrane
7 | Window flashing
8 | Damp proof course
9 | Vapour barrier
10 | Timber roof
11 | Concrete ground slab
12 | Internal plaster finish or dry lining/dry wall
13 | Thermal insulation quilt set within timber frame
14 | Timber framed window/door
15 | Timber sill
16 | Air gap

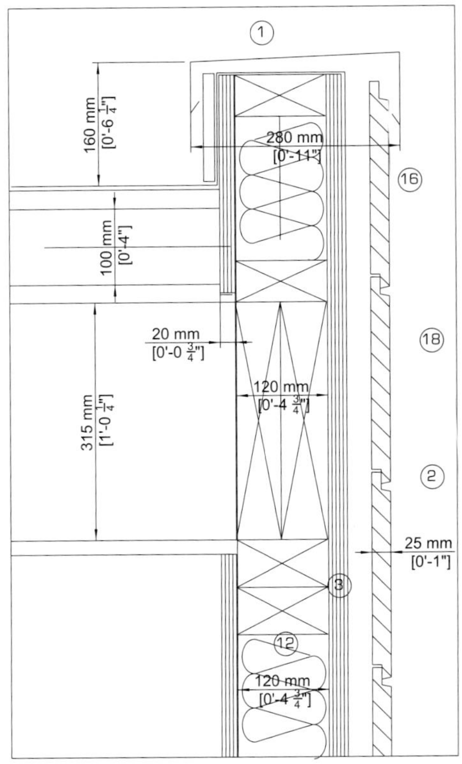

b| Vertical section. Parapet

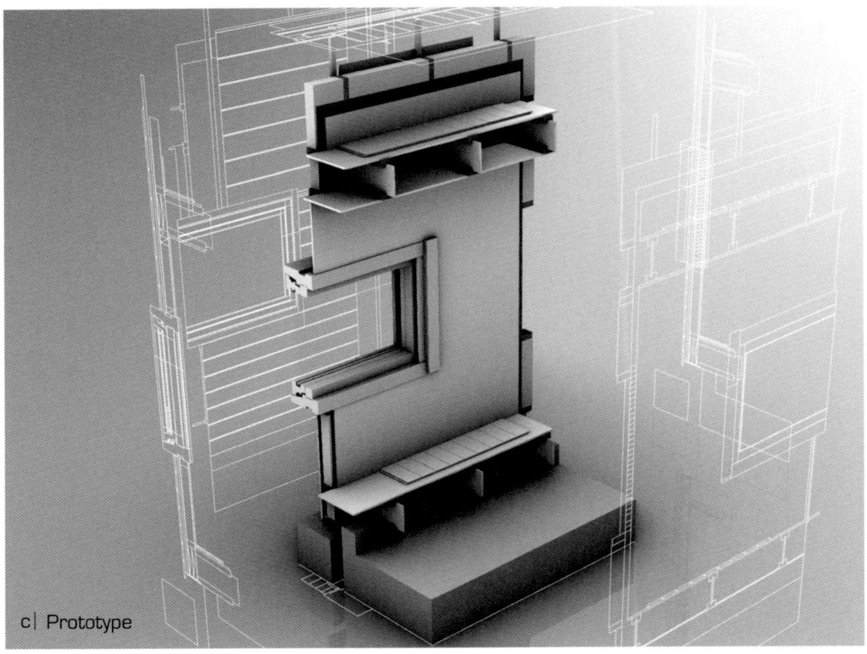

c| Prototype

Timber cladding has traditionally been used on loadbearing timber framed walls, as discussed in this section. Its more recent application as cladding panels and as rainscreens is discussed in the second section of this chapter. There are two traditional generic forms of timber frame that use small section timbers to form framed loadbearing walls: the platform frame and the balloon frame. More historical types that use large timber sections with an infill in timber or another material are not discussed here, as their application is currently very limited. Both the platform frame and the balloon frame are based on softwood sawn timber sections around 100mm x 50mm (4in x 2in). These are economic, and set at close centres of around 400mm (16in) in order to be a multiple dimension of plywood sheets and related board materials used to form the sheathing for the walls and floor decking. Since the floor joists are set at the same centres as the vertical studs, board materials are used for the floor decking in the same size. The platform frame comprises studs spanning from floor to floor (MCE_300, b), with the timber floor structure being supported at each storey height set of timber frames. The balloon frame, which is now used less, is enjoying a revival in light gauge steel sections. This method has vertical framing members which are continuous, with the intermediary floors being supported by the wall running continuously past it.

Timber frames

Timber frames comprise vertical sections called 'studs' fixed to horizontal members called 'rails'. The studs run vertically continuously, with discontinuous horizontal members which are called 'noggins'. The outer face of the timber frame is clad with plywood sheathing to provide lateral bracing (MCE_304, a), typically 12mm–18mm thick (0.5in-0.75in), depending on the structural requirement of stiffening the frame. Timber boards can also be used as sheathing; this is an expensive solution. Framing members are typically formed from 100 x 50mm (4in x 2in) softwood sections at 400mm (16in) vertical centres which are nailed

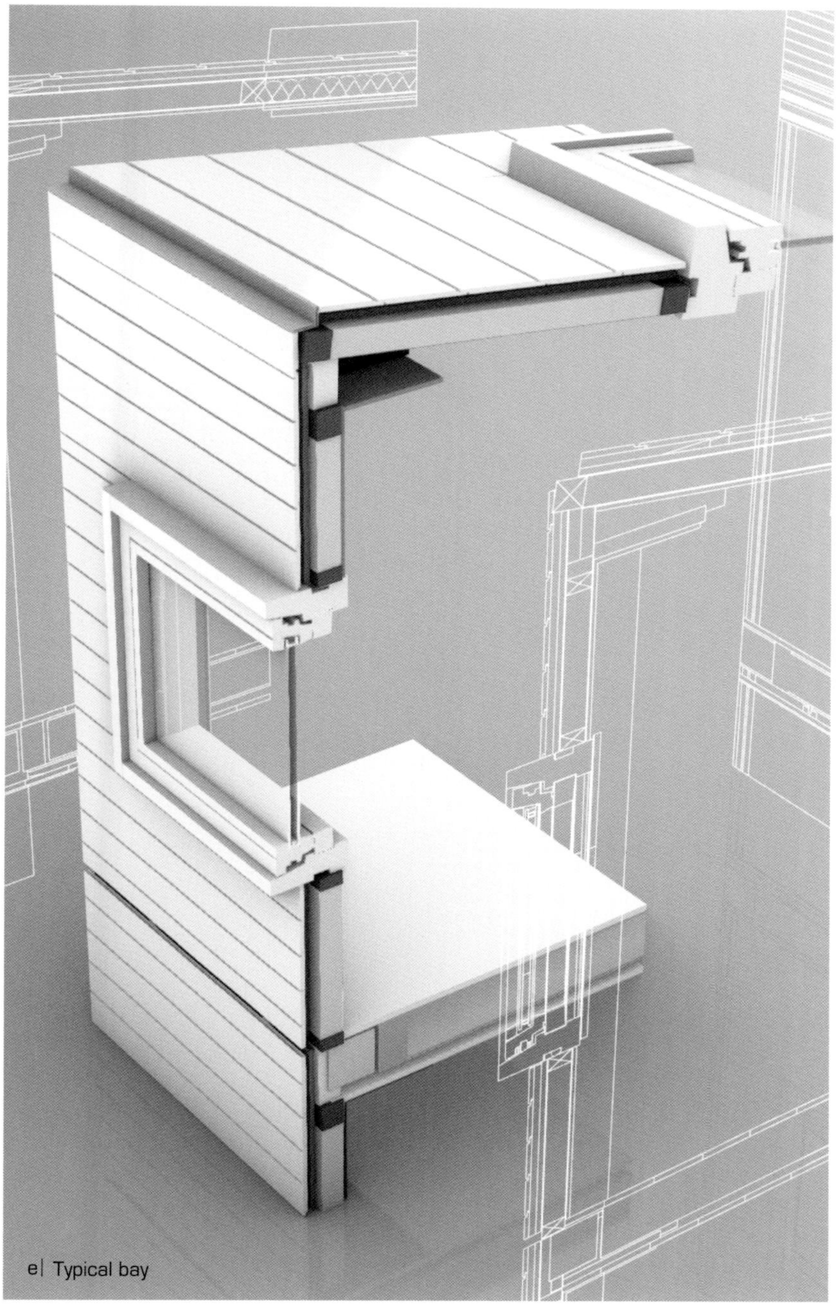

e| Typical bay

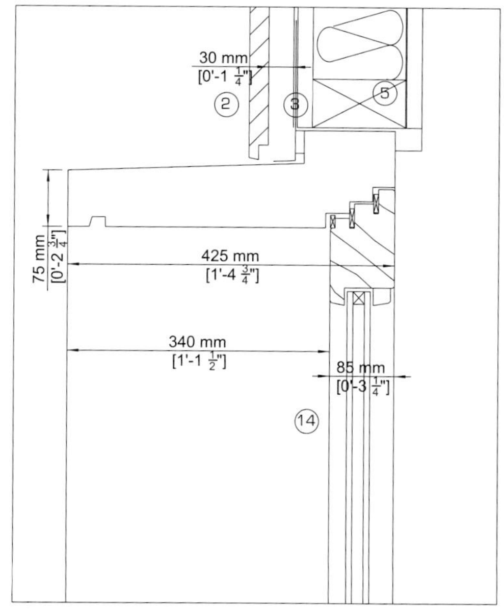

d| Vertical section. Window openings

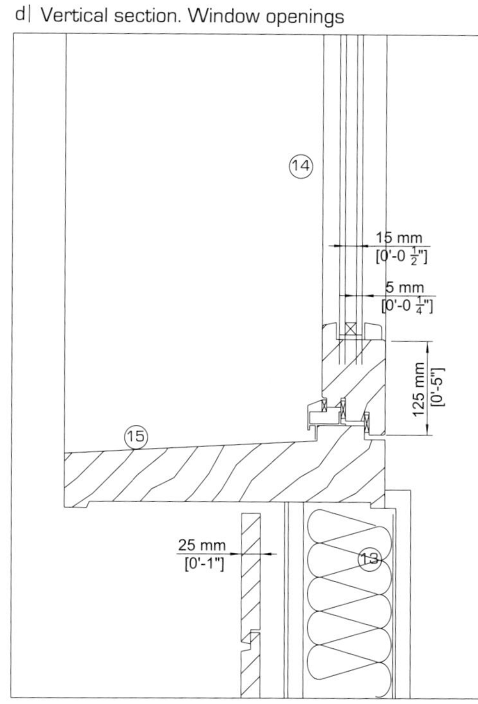

together. Mild steel corner brackets and cleats are commonly used to make the connections more reliable and easy to form, either on site or in a workshop. Voids between the framing members are filled with thermal insulation. A breather membrane is then fixed to the face of the sheathing layer. This provides a waterproof barrier which also allows the vapour to escape to allow the timber wall to release and absorb moisture with changes in the weather. Outer timber cladding boards are then fixed on the outside of the breather membrane (MCE_304, b).

When timber cladding is used on timber frames, the softwood stud walls that form the wall structure can use the timber cladding to stiffen the wall. Traditionally, this has been done by fixing the boarding directly to the timber frame (d), with a breather membrane set between the frame and the timber boards to allow the outer cladding to 'breathe', or dry out and absorb moisture, in response to the changes in weather conditions. Alternatively, the timber boards are fixed to battens which are set forward of the breather membrane or waterproofing layer to ensure that the timber is ventilated on all sides. When battens are used, the timber boards are less able to stiffen the frame if the battens are not set in alignment with every stud forming the frame of the wall behind.

The inside face of the timber framed wall has a continuous vapour barrier, typically thick polythene sheet, set onto it, in temperate climates, on the 'warm in winter' side of the wall. The inner face of the wall is then finished with plasterboard (drywall) to provide an internal finish.

Ground level

This text focuses on different conditions at ground level, upper floor levels and at roof level. The lightweight nature of timber cladding has led to its use with a raised timber ground floor, often with no concrete ground slab. This makes the junction with the ground quite different from those

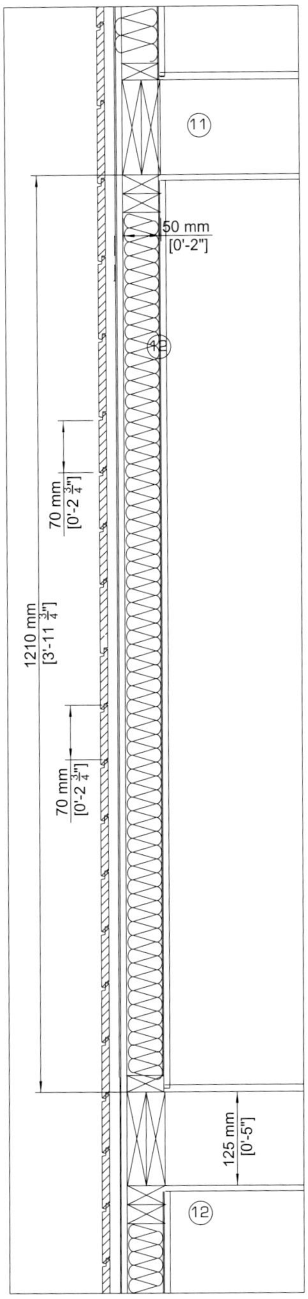

b| Vertical section. Wall assembly

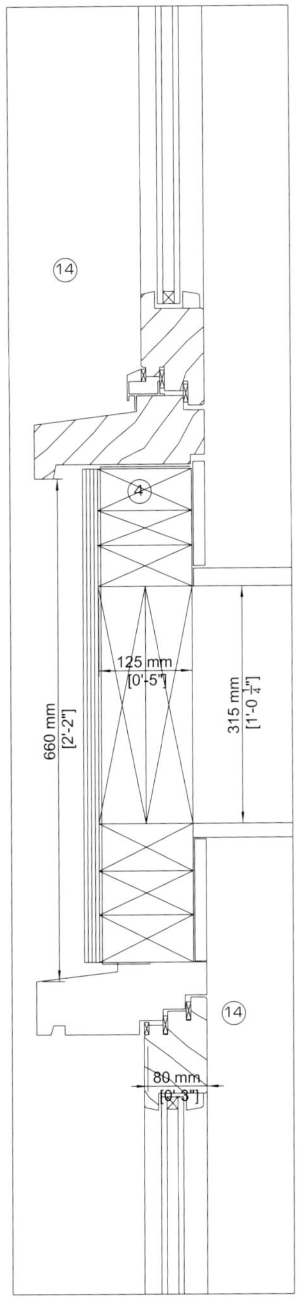

c| Vertical section. Door opening

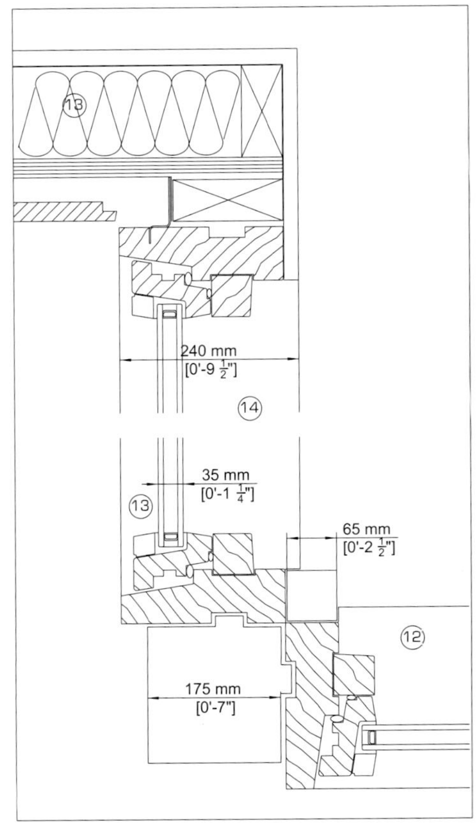

a| Horizontal section. Window jamb

Details

1 | Metal parapet flashing
2 | Timber boards
3 | Plywood sheathing
4 | Timber studs
5 | Timber rail
6 | Breather membrane
7 | Window flashing
8 | Damp proof course
9 | Vapour barrier
10 | Timber roof

of the other materials described in this book. Details around windows and doors are as those described in the section on timber window walls in the Glass chapter.

Timber cladding at ground level stops a minimum of 150mm (6in) above external ground level to avoid rainwater splashing up the timber (b), which would cause staining and deterioration of the material. Cladding is usually supported at ground level on a concrete slab or edge beam that forms part of the concrete wall. Alternatively, the wall can span between concrete pads at 3000mm–5000mm centres (10ft–16ft), with timber beams at the base of the wall to provide support between pads. Where a concrete slab is used, the edge of the slab has traditionally been exposed as a base to the wall. With growing concerns of increased thermal insulation at the ground level of a building, thermal insulation has been introduced on the visible edge of the slab between ground level and the base of the timber cladding. The insulation requires an outer protection, typically thin concrete slab or brickwork.

The timber wall frame is not fixed directly to the floor slab, but instead is usually set on a continuous timber section which is first fixed to the concrete slab to provide both a level surface to set the timber in place and to be an easier fixing to use than simply nailing (a). A damp proof course (DPC) is set beneath the continuous timber base plate to protect the timber, the DPC usually extending down the vertical face of the concrete slab where it connects with the damp proof membrane (DPM) beneath the concrete slab or the vertical face of the basement wall. The DPC will also be continuous with a DPM set on top of the concrete slab. Floor finishes are then applied in the same depth as the continuous baseplate, allowing internal skirting boards to be fixed to the bottom rail of the timber frame, which sits at the finished floor level. Where concrete pads are used, the timber beam is set into stainless

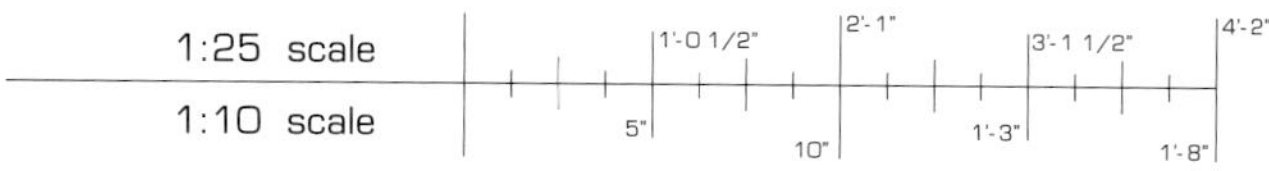

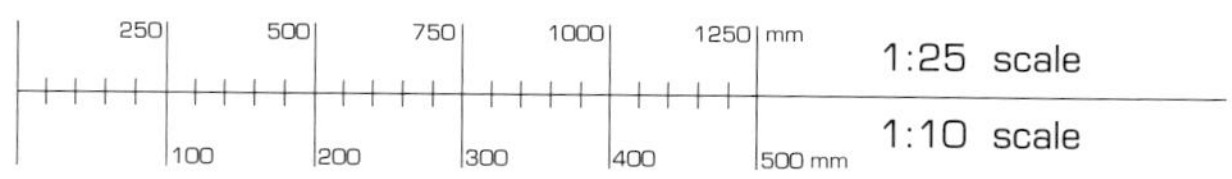

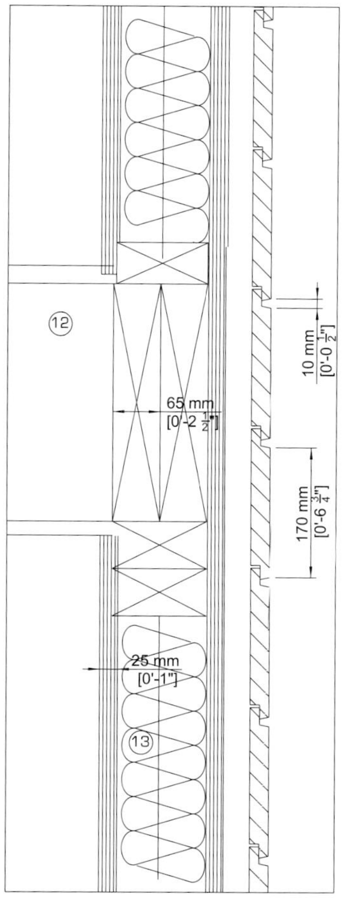

d| Vertical section. Floor junction

11 | Concrete ground slab
12 | Internal plaster finish or dry lining/dry wall
13 | Thermal insulation quilt set within timber frame
14 | Timber framed window/door
15 | Timber sill
16 | Air gap

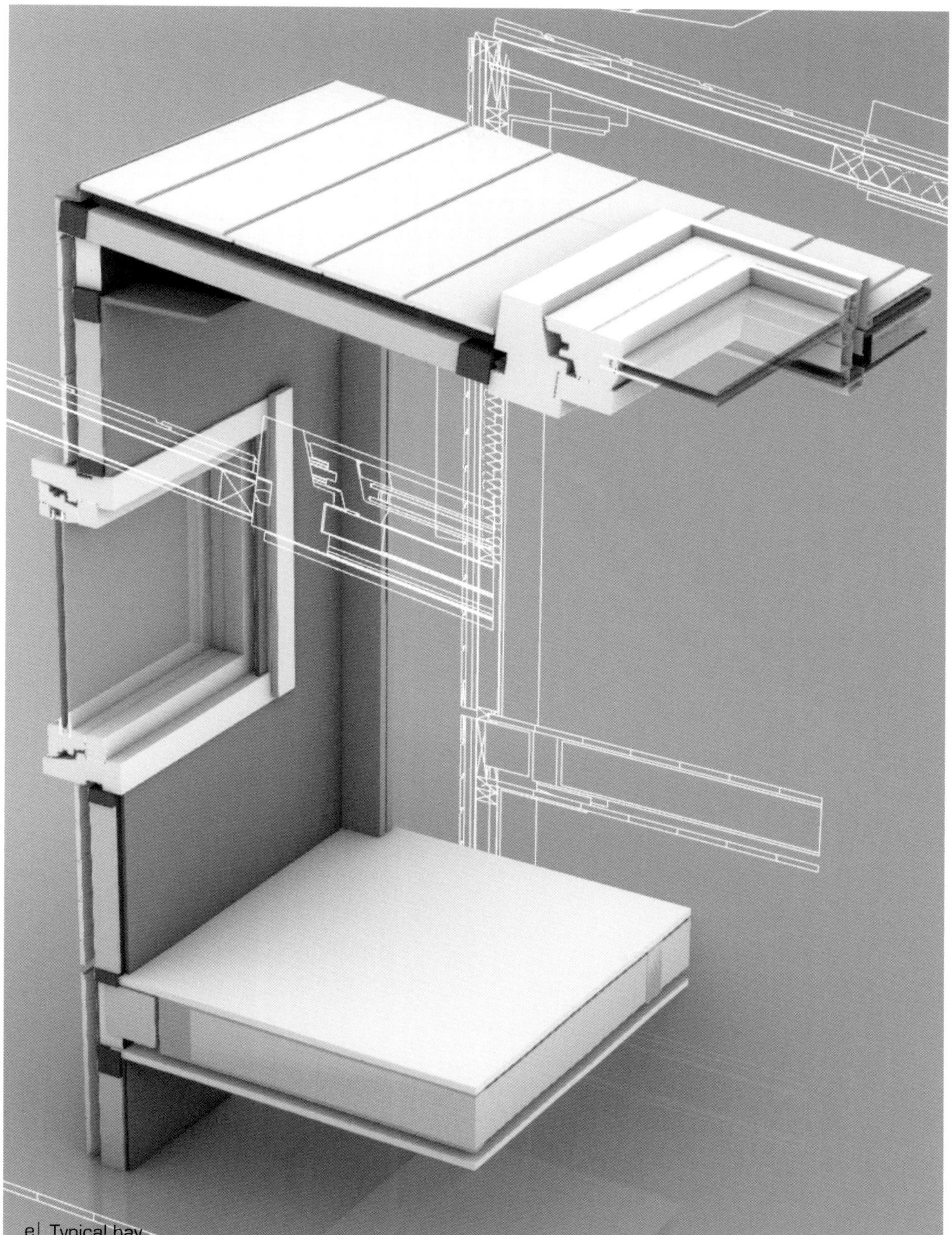

e| Typical bay

steel shoes which are fixed to the concrete pads. The pads may extend below ground level to form foundations where the ground is allowed to continue underneath the building. Alternatively, the panels may sit on a concrete floor slab set below the timber ground floor to prevent the growth of any vegetation beneath the raised ground floor. Gravel is often set at ground level in the void below a raised timber floor. Door thresholds at ground level usually have a raised profile to prevent water from penetrating through the opening, and these are also fixed directly onto the continuous baseplate. An additional DPC is set on top of the baseplate to avoid moisture penetration through the door threshold.

Timber can also be supported on brick walls set at a minimum of 150mm (6in) above external ground level and be supported on a concrete strip foundation or ground beam. As bricks go below ground they usually change to either a dense concrete block or a dense engineering quality brick. A raised floor is then set into this brick wall (MCE_302, a). The void beneath the timber floor is ventilated with air bricks that encourage cross ventilation. This avoids stagnant air in the void from damaging and eventually rotting the timber floor. Instead of supporting the floor on the brick wall, floors can be supported separately as pads, typically in mild galvanised or stainless steel posts. This allows the ground floor slabs to be built before the timber wall is started, and avoids any risk of long-term damage to the floor joists from contact with a damp wall below DPC level. An alternative method of supporting the low-level brickwork is on stainless steel lintels spanning between the concrete pads that support the timber floor. The brick base can be avoided completely by the use of beams spanning between concrete panels as described previously.

Upper floors

Timber framed wall panels are used to support upper floors also in timber. Floor joists are set directly into the timber frame, where they are

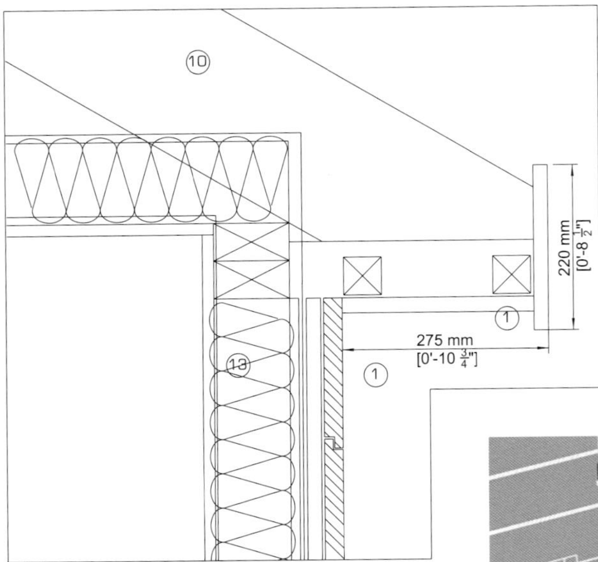

a | Vertical sections. Floor junction, eaves to pitched roof, 2 ground conditions

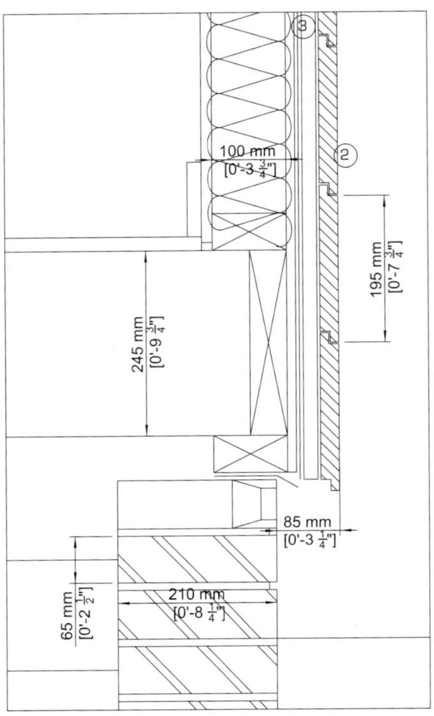

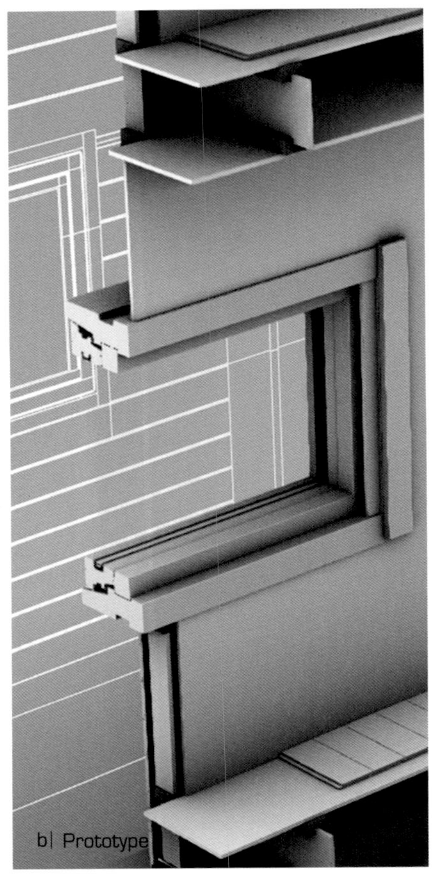

b| Prototype

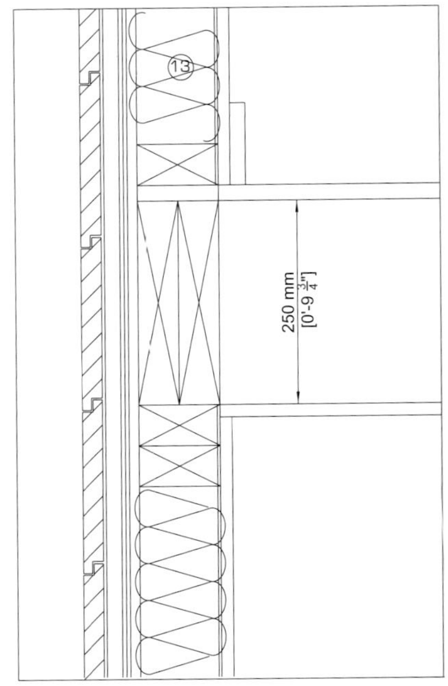

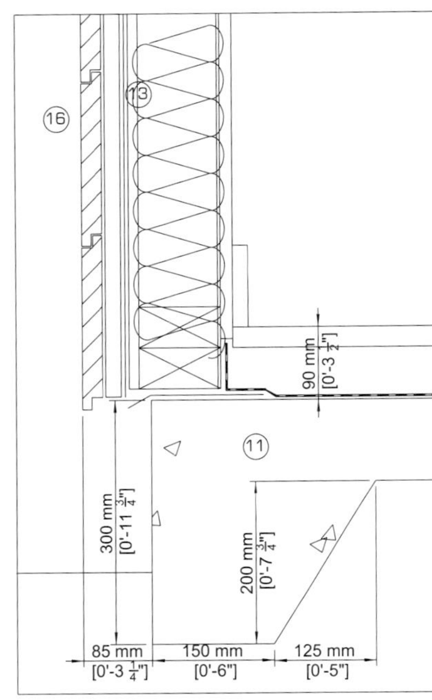

supported, either on the top of the floor-height panel below (MCE_300, b) or to the sides of the timber studs if the floor extends over two or more floors in the balloon frame (a). Where timber wall frames span only from floor to floor, as is more commonly the case, floor joists are fixed to the face of timber beams that form part of the wall frame instead of penetrating the frame itself. This allows the gap between wall and floor to be filled with timber, which drastically reduces sound transmission between floors. Because the timber floor and wall form a continuous structure, there is no need to form a horizontal movement in the timber cladding outside. This gives the continuous surface of timber cladding across two or three floors that is characteristic of platform frame construction.

Corners

The most common corner formed in timber cladding is a single timber bead set so that the timber boards on both sides butt into the corner bead (MCE_304, b). If a breather membrane is used behind the cladding, then an additional waterproof flashing is added to the corner. This is formed in a durable polymer-based sheet or metal sheet. Alternatively, the boards can be allowed to make a corner with a butt joint, and an additional L-shaped timber trim formed from two separate timber sections is added on the face of the corner to protect the exposed end grain of one of the sides forming the corner. These trims also provide a visual tidiness to the corners. Boards can be joined with a mitred joint (45º) without any cover strip but the timber used must be of the highest quality to avoid the joint opening up with moisture movement, and very high levels of workmanship are required. A waterproof layer or flashing behind the mitred joint is essential to the waterproofing performance of this detail. Internal corners are formed using the same sets of timber beads or mitred joints.

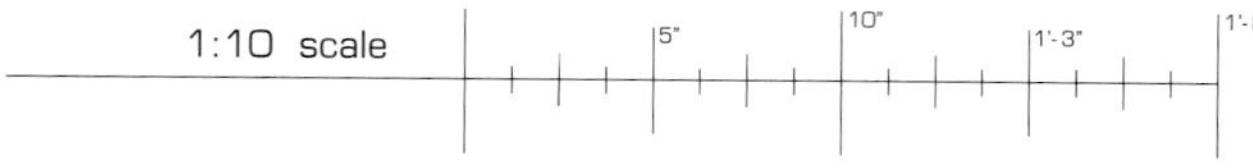

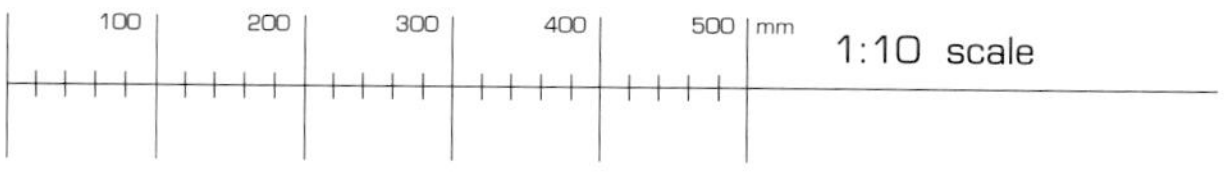

Details

1	Metal parapet flashing	10	Timber roof
2	Timber boards	11	Concrete ground slab
3	Plywood sheathing	12	Internal plaster finish or dry lining/dry wall
4	Timber studs	13	Thermal insulation quilt set within timber frame
5	Timber rail	14	Timber framed window/door
6	Breather membrane	15	Timber sill
7	Window flashing	16	Air gap
8	Damp proof course		
9	Vapour barrier		

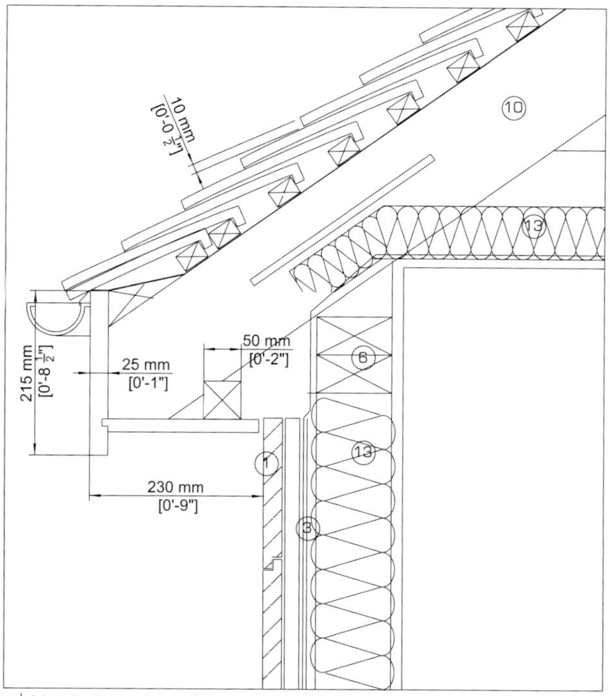

c| Vertical sections. 2 eaves conditions to pitched roofs

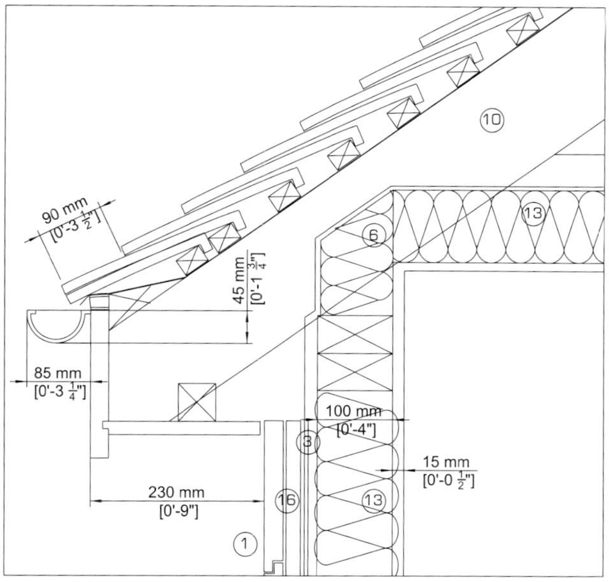

Roof eaves and parapets

Overhanging eaves from pitched roofs or flat roofs are very common with timber cladding, as they provide protection to the wall beneath from the worst effects of weathering from water running directly down the facade. A continuous timber section, or wallplate, is set on top of the timber frame, creating a double thickness of timber at the top of the wall. This serves as a base for sloping rafters or for flat joists (MCE_305, d). For large- scale roof structures, deeper timber sections are used, set on end, as used for connections to upper floor structures. The waterproof layer on the outside of the timber framing, or breather membrane, is made continuous with that of the roof structure. In the case of overhanging eaves, the roof may be ventilated or sealed, but in both cases the membranes must be continuous to avoid water penetrating the joint. On the inside face of the wall, the vapour barrier must be continuous from wall to roof, or wall to ceiling level, beneath the roof (c).

d| Prototype

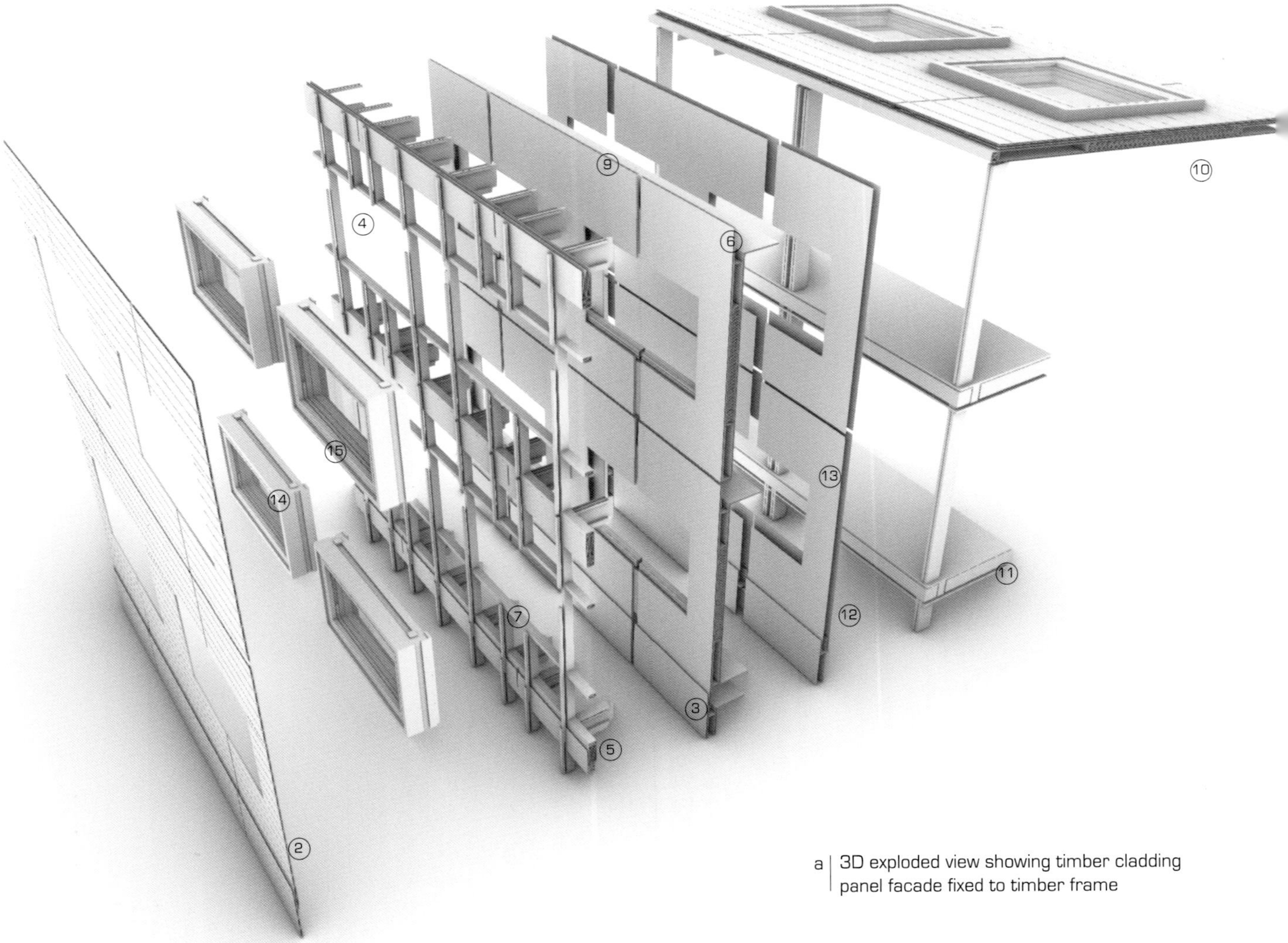

a | 3D exploded view showing timber cladding panel facade fixed to timber frame

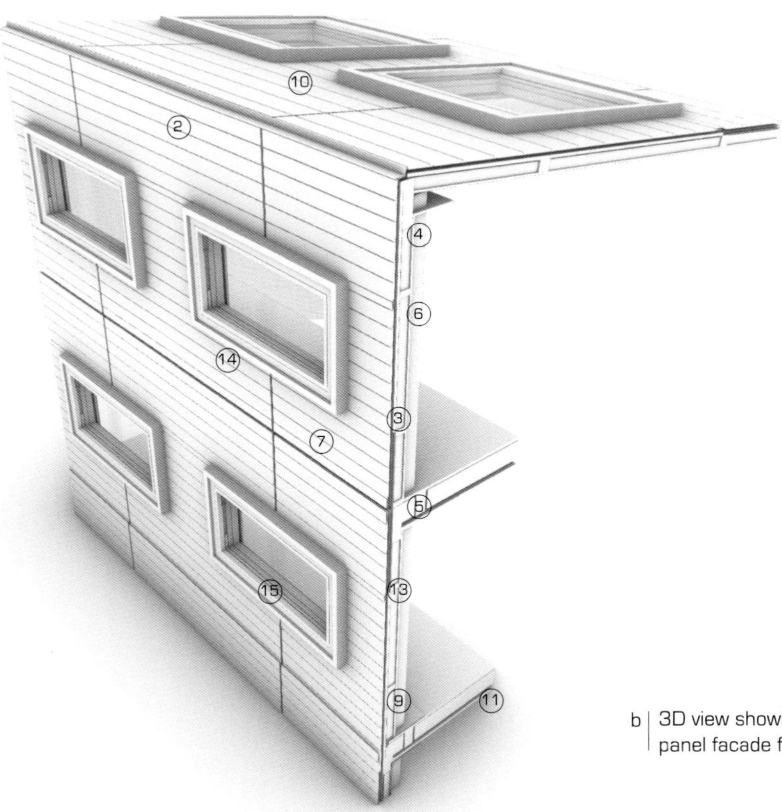

b | 3D view showing timber cladding panel facade fixed to timber frame

Details

1 | Metal parapet flashing
2 | Timber boards
3 | Plywood sheathing
4 | Timber studs
5 | Timber rail
6 | Breather membrane
7 | Window flashing
8 | Damp proof course
9 | Vapour barrier

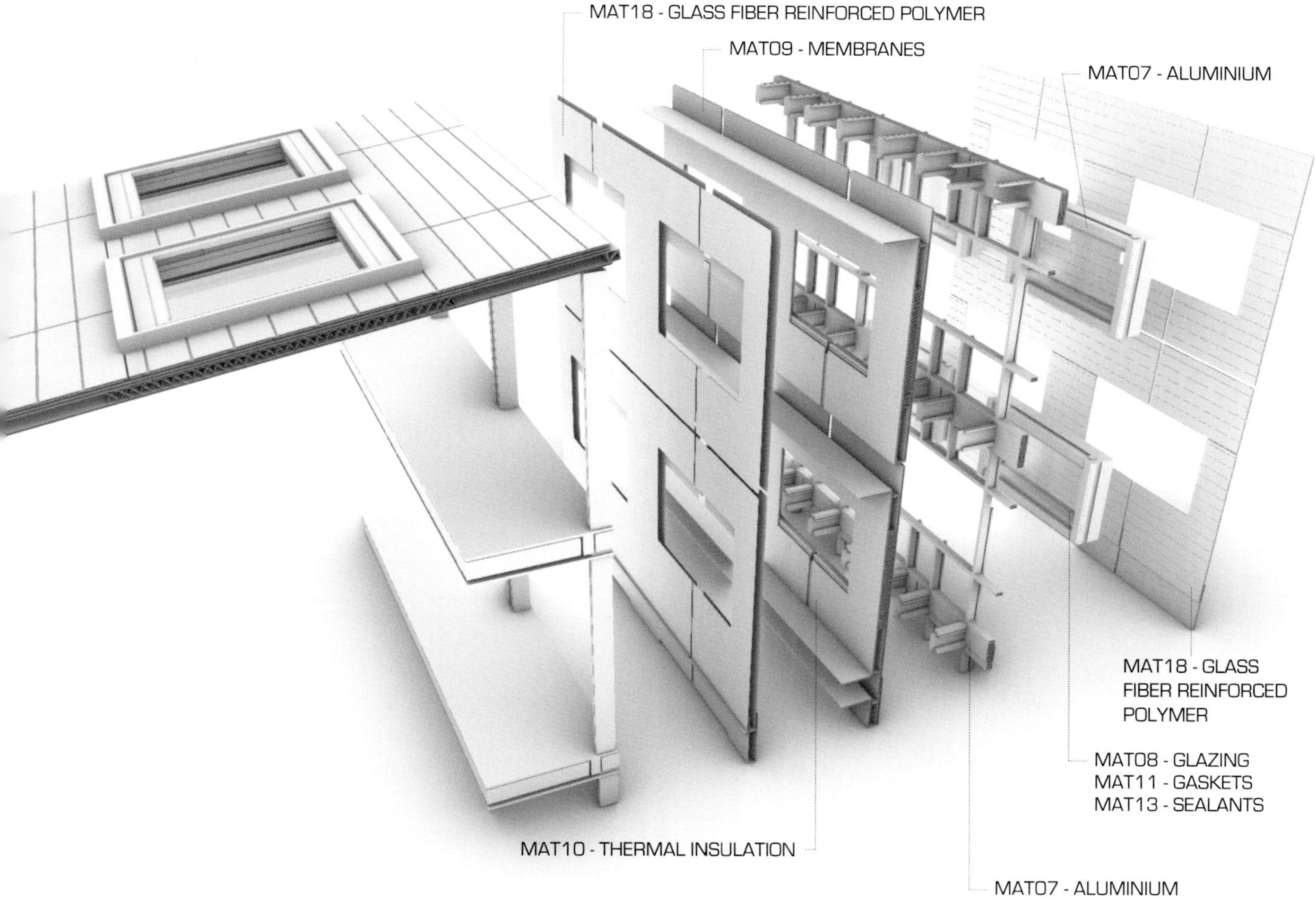

c | 3D exploded view of timber cladding panel facade components with material specification references

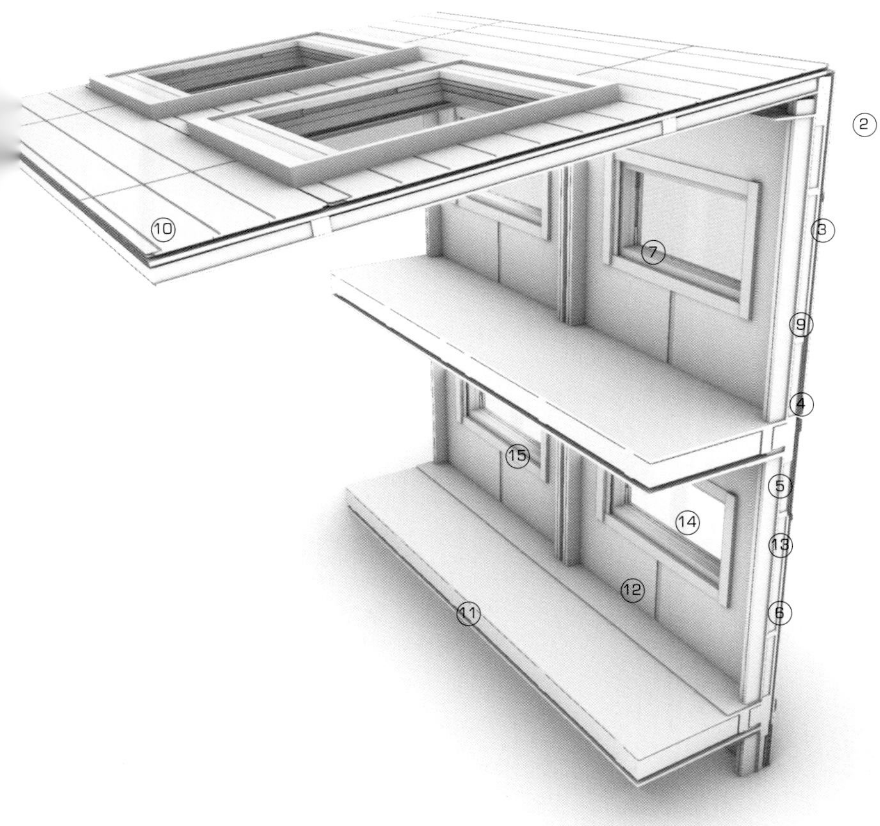

10 | Timber roof
11 | Concrete ground slab
12 | Internal plaster finish or dry lining/dry wall
13 | Thermal insulation quilt set within timber frame
14 | Timber framed window/door
15 | Timber sill
16 | Air gap

d | 3D view showing timber cladding panel facade fixed to timber frame

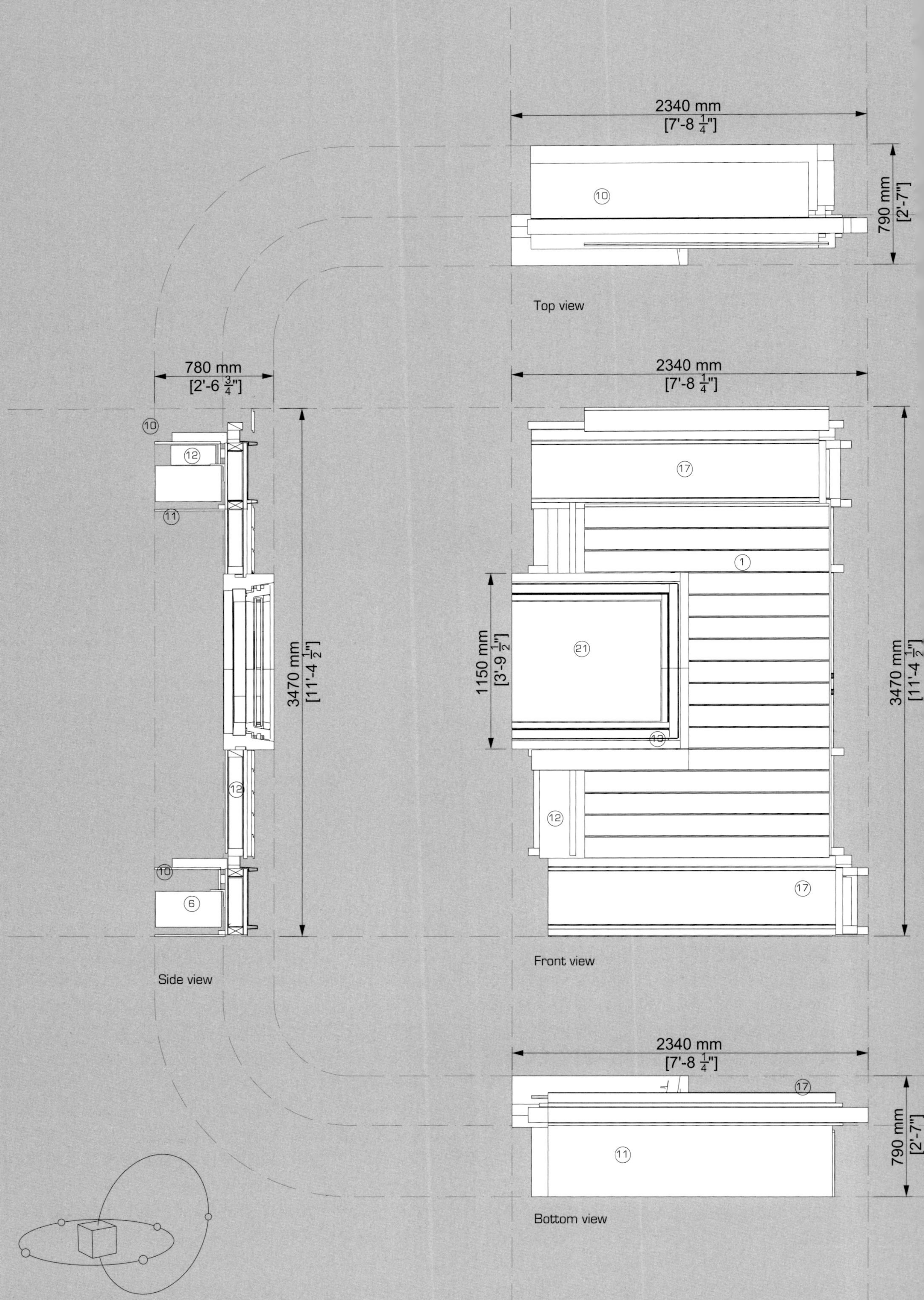

Third angle projection 1:30

Details

1 | Timber boards
2 | Plywood sheathing
3 | Timber studs
4 | Timber rail
5 | Breather membrane
6 | Foundation
7 | Damp proof course
8 | Vapour barrier
9 | Timber floor
10 | Concrete floor
11 | Internal finish

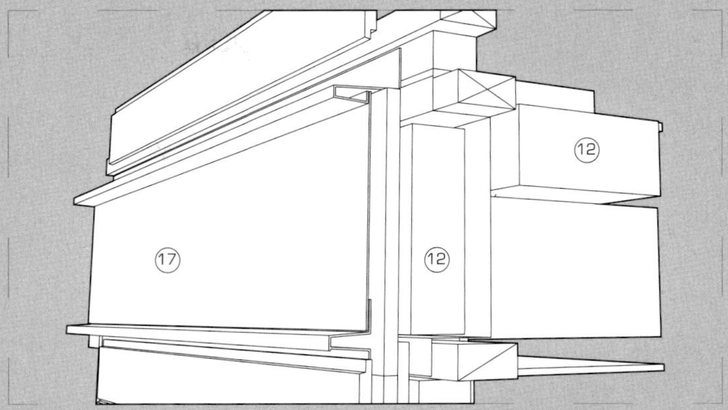

3D view of the assembly

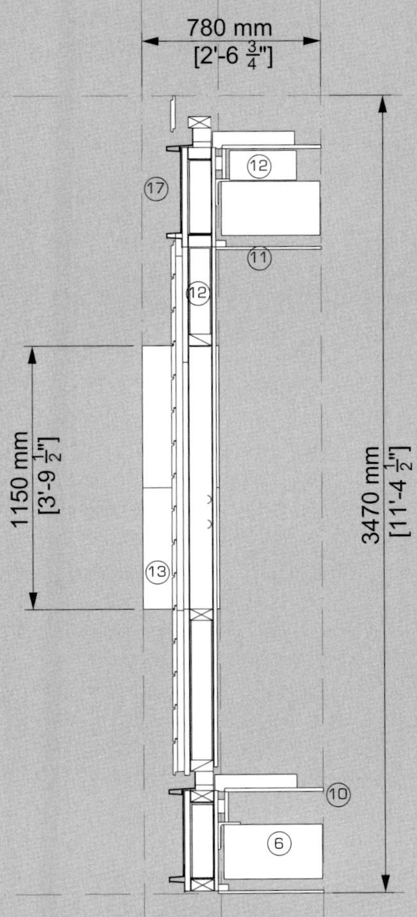

Side view

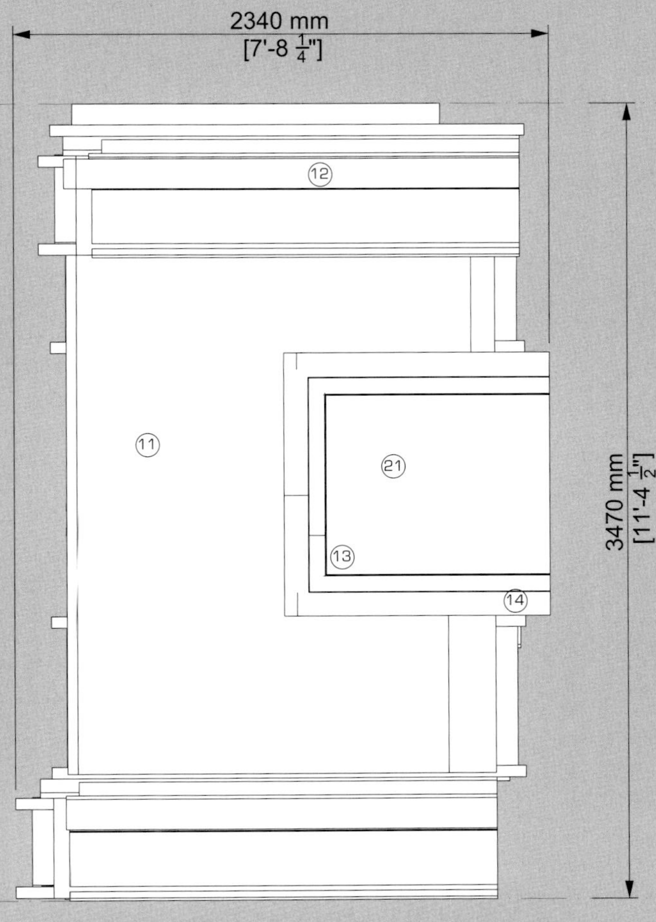

Back view

12 | Thermal insulation quilt set within timber frame
13 | Timber framed window/door
14 | Timber sill
15 | Air gap
16 | Sliding timber louvre panel
17 | Metal facing
18 | External plywood facing
19 | Cover strip
20 | External floor deck
21 | External glass wall in twin wall
22 | Structural timber frame

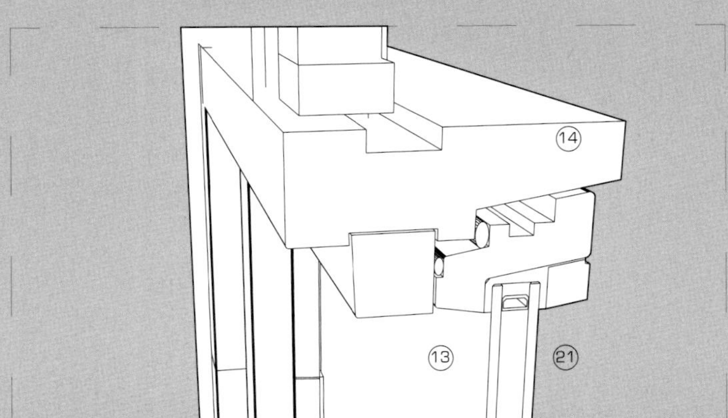

3D view of the assembly

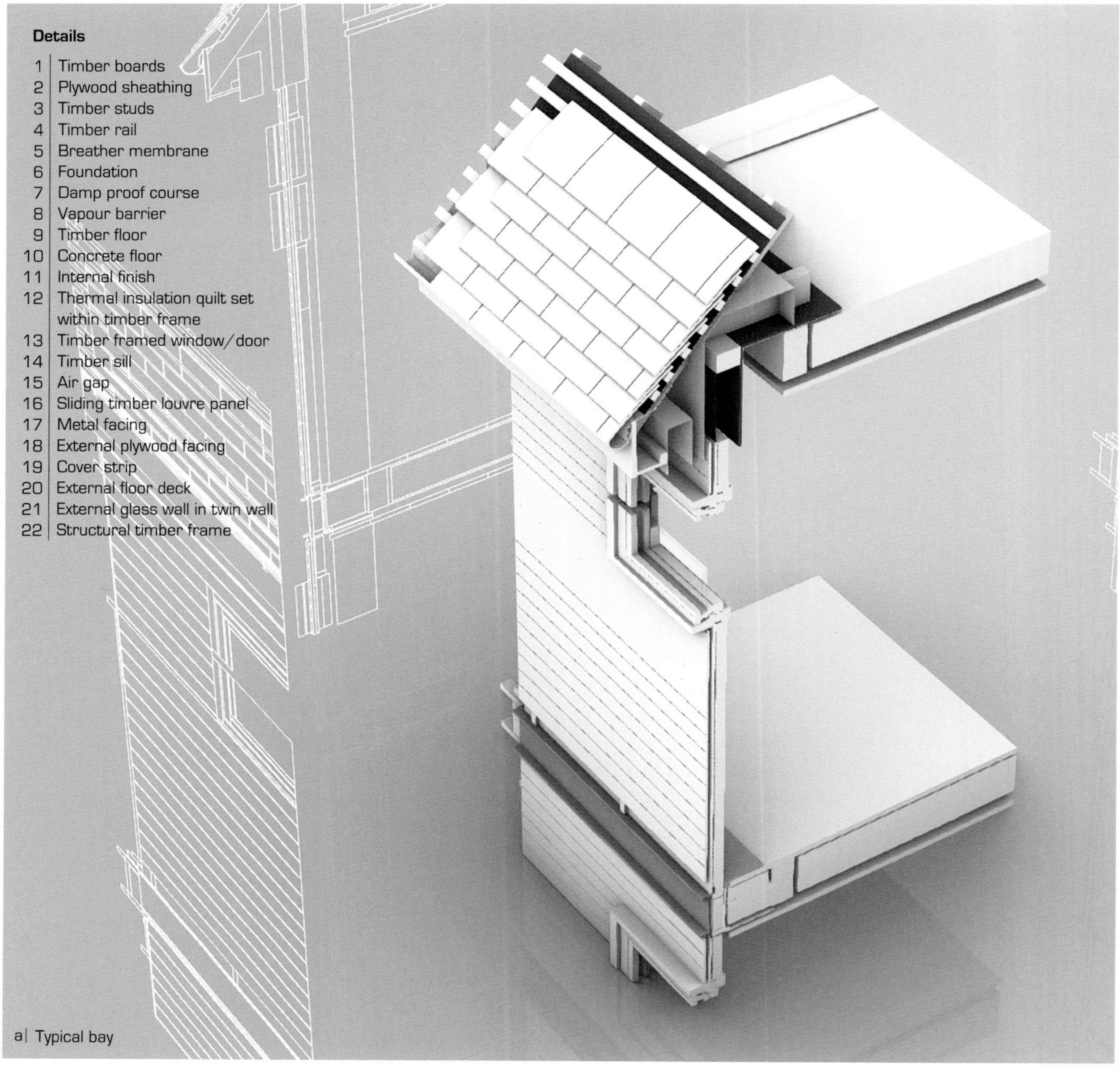

a| Typical bay

Timber boards

Timber boards for cladding panels and rainscreens are sawn and are available in many timber species, though environmental concerns have led to an increasing preference for locally-grown timber for large-scale applications. This is because timber has a negative effect on CO_2 levels, reducing the levels in the atmosphere. The CO_2 emissions created as a result of cutting and working timber before it reaches a local building site are still less than the amount of CO_2 consumed by the timber during its growth. The highest timber grades are used for wall cladding boards, since exposure to the weather will involve considerable temperature and humidity variations, as well as fading in sunlight. Softwoods are generally used for the cladding of timber frames, with hardwoods being more commonly used for cladding panel and rainscreen applications. Where hardwoods are used, those species that are naturally durable are used. Where less expensive, less durable timbers are used, these require higher levels of finishing and maintenance. For softwoods, preservative treatment is used, but with an increasing scrutiny of the methods, since some preservatives can damage ground around the site as a result of water run-off from the timber into the ground around the timber wall. Softwood boards are made usually in 250mm (10in) widths, with trimmed boards with profiles routed into them usually trimmed down to 150mm–200mm widths (6in–8in). Wider boards cannot generally be used due to the risk of the material curving in section.

The increased use of hardwoods in rainscreens, for their durability, has led to different thicknesses and sections of timber being used. This has led to the increased use of louvred timber screens and panels in front of glazed windows and doors to provide solar protection while contributing to the texture of the timber facade (MCE_309, b).

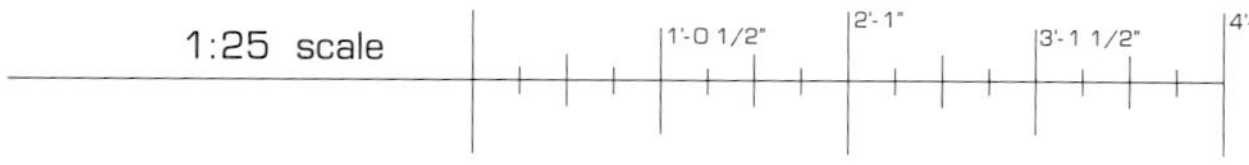

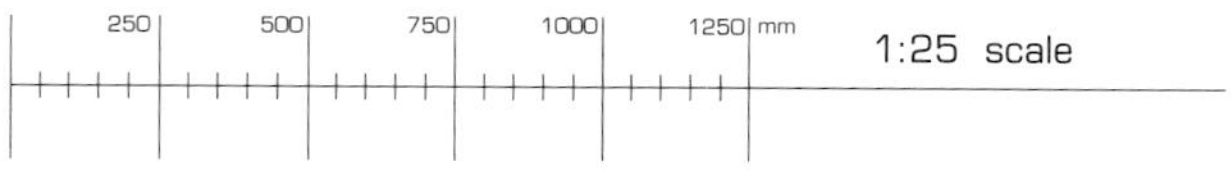

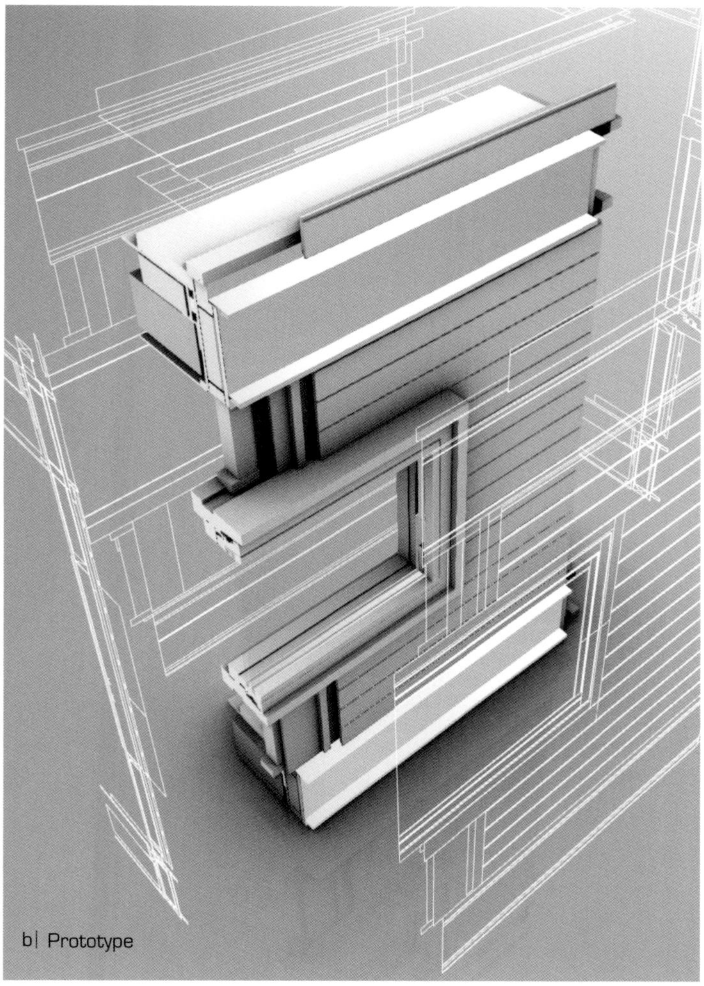

b| Prototype

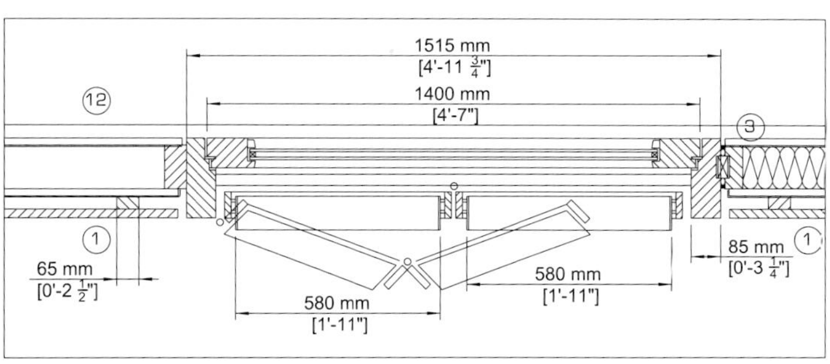

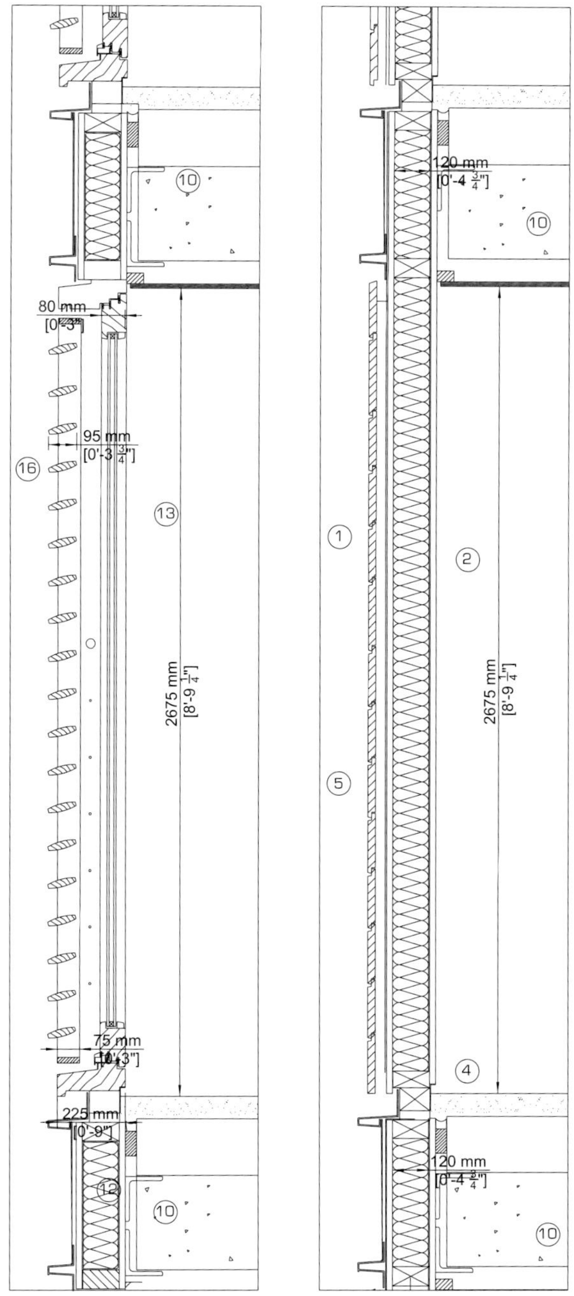

c| Horizontal and vertical sections. Cladding panel in twin wall configuration for taller buildings

All timbers vary in moisture content with changes in temperature and air humidity; this being one of the essential aspects to be considered in timber detailing. Most timbers used in cladding will have a moisture content of around 5% to 20% when in use. Similar levels are found in timbers from timber suppliers, and are classified as 'dry', 'kiln dried' or 'seasoned'. However, some softwoods are supplied unseasoned, depending on the supplier, so the amount of timber drying and seasoning is critical to the way that timber is cut on site to fit a junction. Unseasoned timber will shrink as it dries, and consequently an allowance is made for later shrinkage by increasing the overlap between timbers where applicable, or making tighter butt joints where these are used. Unseasoned timbers are fixed in place soon after delivery to site to avoid any twisting or warping of the timber. As a result, lapped boards are not nailed together where they overlap, and nails and screws are secured so that the timber can move without it splitting or being damaged.

Finishes

Timber cladding is finished with the timber being left either as supplied, with preservative applied or injected by the supplier; or alternatively is given coats of preservative in clear, stained, or opaque finish on site with preservatives that repel rainwater, wood stains and paint. Paints can be oil-based or acrylic, while preservatives are clear and can be used as a finish that penetrates the depth of the wood without appreciably changing its appearance. It can also be used before staining or painting the timber. Preservatives help to prevent moisture absorption as well as reduce fungal growth. This is because they enhance the life of the timber but do not prevent the material changing colour and fading to a silver-grey appearance. Site-applied finishes and regular maintenance help overcome the effects of the weathering of timber.

In addition to the use of preservatives and coatings, the orientation of timber boards is critical to the long-term performance of timber clad-

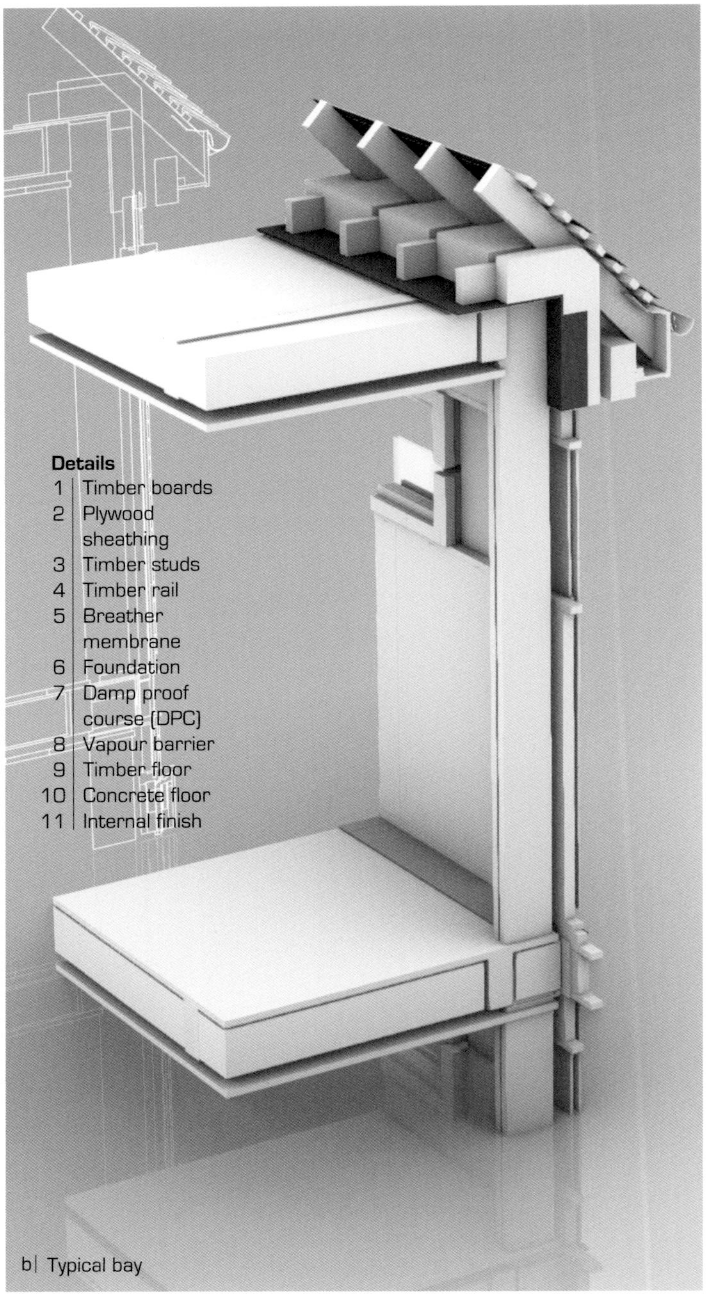

b | Typical bay

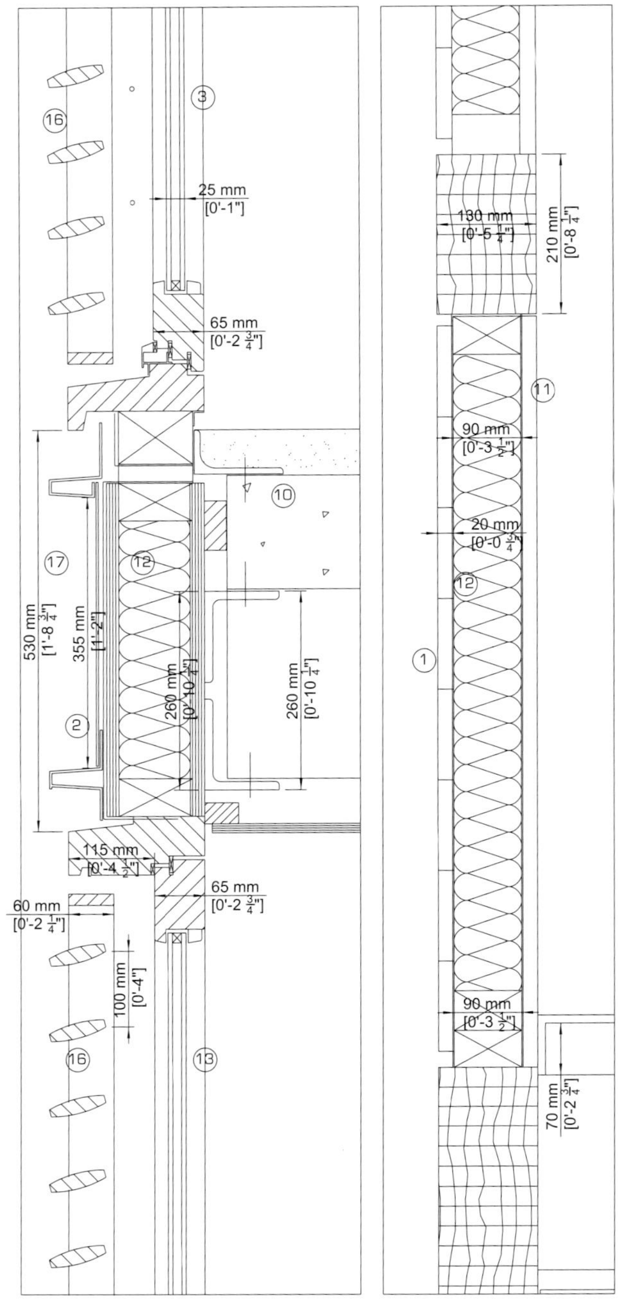

a | Vertical sections. Junction with floor slab

ding. The most common types of jointing of boards is 'shiplapping', where timber boards are set horizontally and lapped over one another with the upper board lapped over the top of the board below to protect it from rainwater ingress. Shiplapping can be assisted by the use of 'feathered' or wedge-shaped boards to give the lapping a more elegant appearance. Tongue-and-groove boards are used to give a continuous flat appearance, while having the advantage of locking boards together into a continuous plate-like structure (MCE_309, c). Boards are typically around 20mm (0.75in) thick, made as long as possible at around 3000mm–3500mm (10ft–11ft6in), to avoid vertical joints which are a potential source of rainwater penetration except in rainscreen configuration. Where tongue-and-groove boards are used, the groove is set on the underside to avoid water accumulating when the boards are set horizontally or diagonally. Where tongue-and-groove boards are set vertically, the groove is set away from the prevailing wind direction to avoid windblown rain being blown into the joint. Joints between boards are never sealed along their long edges with silicone or mastic sealants as this prevents the timber from drying properly, which would cause the material to deteriorate and eventually rot. However, sealants are used on the ends of timber boards when the complete external surface is sealed with paint.

Cladding panels and rainscreens

Timber cladding to a platform frame or a balloon frame is continuous, forming an integral part of the wall structure (a). In contrast, timber cladding panels are fixed to building frames in reinforced concrete, steel and timber. In this application, cladding panels follow the principle of other forms of cladding, requiring the pre-fabrication of panels and an allowance for structural movement in the supporting frame associated with larger-scale structures. Cladding panels can also be faced with plywood rather than timber boards, in rainscreen applications. Because of the higher moisture movement associated with timber than with other

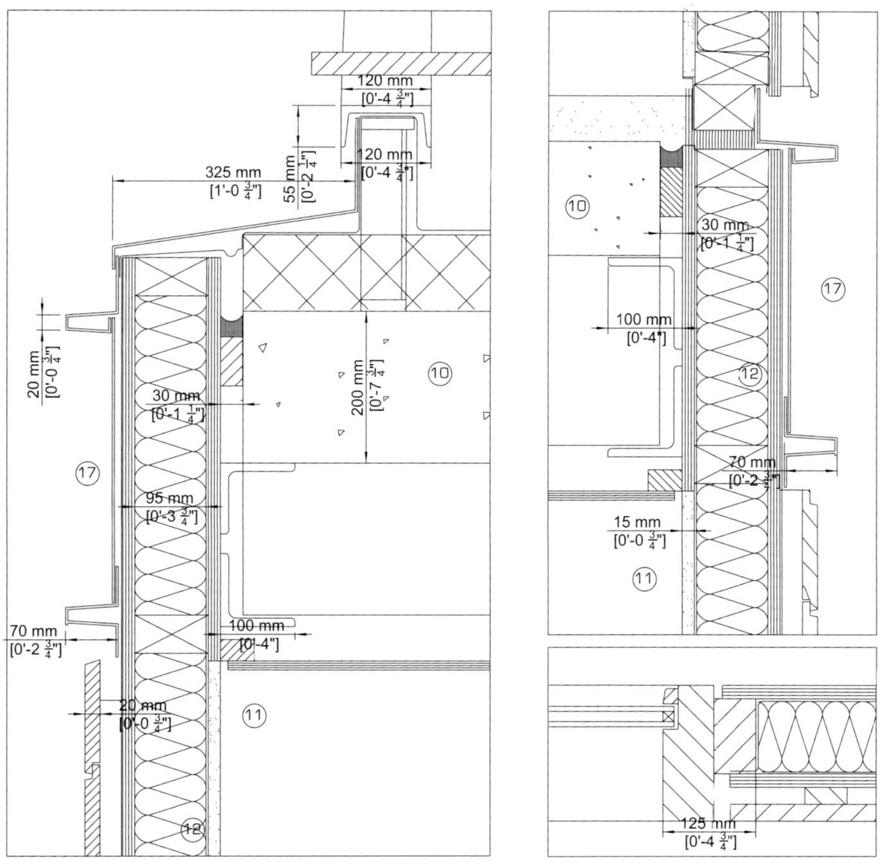

12 | Thermal insulation quilt set within timber frame
13 | Timber framed window/door
14 | Timber sill
15 | Air gap
16 | Sliding timber louvre panel
17 | Metal facing
18 | External plywood facing
19 | Cover strip
20 | External floor deck
21 | External glass wall in twin wall
22 | Structural timber frame

c| Vertical and horizontal sections. Typical build-up of cladding panel

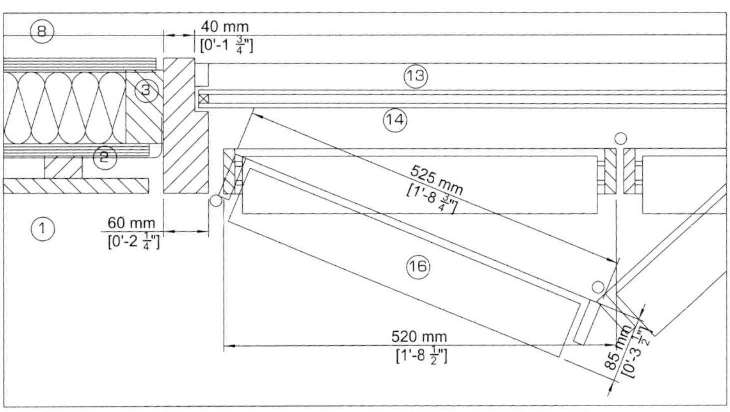
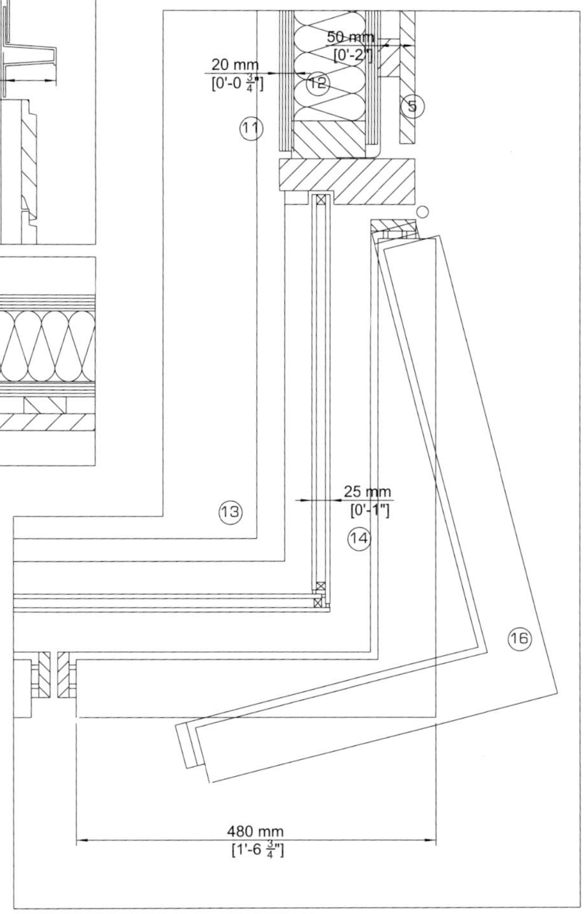

d| Horizontal section. Window jambs

materials, junctions between cladding panels require an allowance for movement as a result of changing moisture levels in the material.

When reinforced concrete or steel frames are used, the timber cladding panels are set forward of the floor slabs in the manner of glazed curtain walling (MCE_308, a). Panels extend from floor to floor, being either hung from a floor slab or else sitting on a bracket on the floor below (MCE_312, b). In taller buildings, timber cladding is used in conjunction with a twin wall construction with an inner timber wall and an outer open-jointed glazed wall (MCE_314, b). The outer glazed wall provides protection against wind, dust and noise, allowing the windows to be opened for natural ventilation. The outer wall provides a 'thermal buffer' to reduce the effects of heat and cold at different times of the year. The two walls are set 700mm-1000mm (2ft 3in-3ft 3in) apart to allow access from the inner timber wall to the zone between the two walls for maintenance and cleaning access. The outer glass screen also provides protection to the timber cladding from the effects of rain and wind, allowing the material to maintain its appearance without full exposure to outside conditions. Vertical joints between panels have a stepped joint to allow for deflections in floor slabs between panels, following principles of glazed curtain walling. This stepped joint is covered on the outside with timber boards, set forward of the face of panels on battens in rainscreen configuration.

The construction of panels follows the same principle of timber cladding described in the previous section. Horizontal joints have an inner chamber formed between two adjacent panels. Any rainwater that penetrates the outer seal, which is also kept open in some designs, is drained down an inner chamber where the water is discharged through the horizontal joint at floor level. The outer timber cladding has continuous vertical joints to allow panels to be fixed sequentially (MCE_309 b), unless the outer timber boards are

a| Vertical section. Cladding panel in twin wall configuration for taller buildings

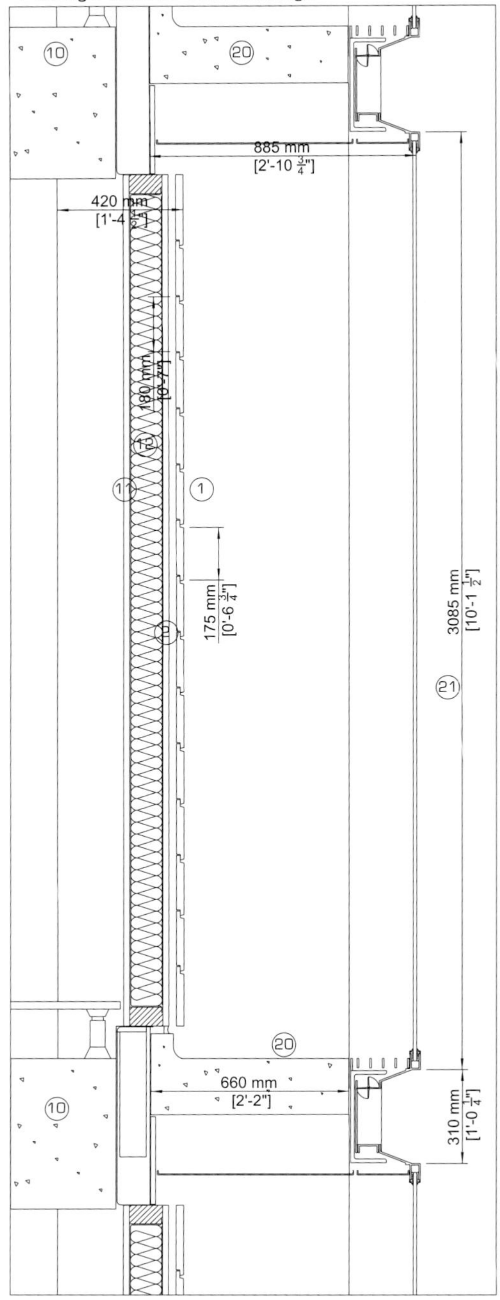

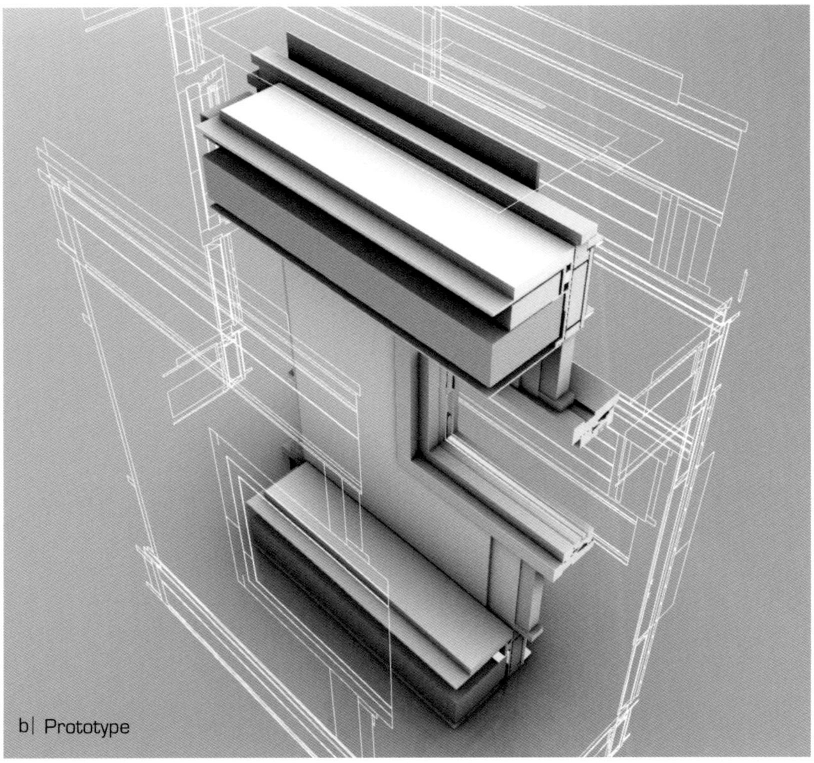

b| Prototype

c| Horizontal sections. Typical caldding panel with additional solar shading or guarding, allowing door to be opened above ground level if panel is fixed closed

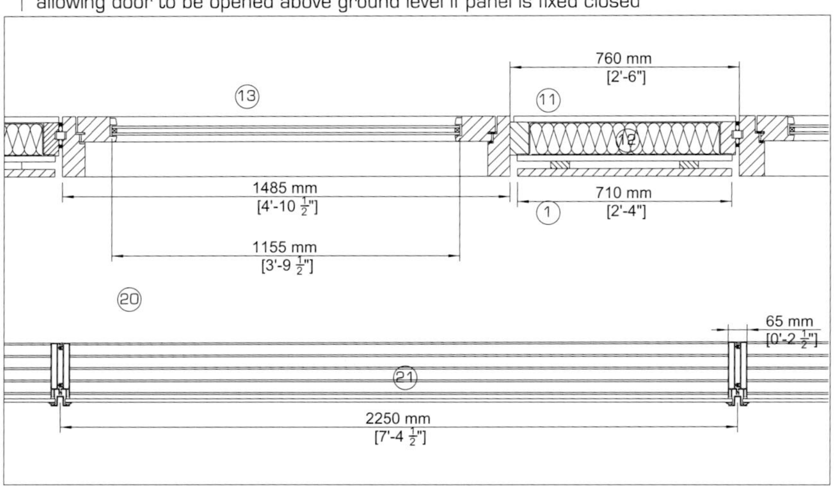

applied after the panels have been installed, but this is contrary to the panel-based form of construction.

Where timber frames are used to support timber cladding panels, cladding is set between floor structures rather than forward of them, since there is no significant thermal bridge from outside to inside, allowing the structural frame to be exposed on the outside. Where timber frames are exposed, laminated timber is often used, since it can be formed in beams of significant loadbearing capacity, and is able to form a frame rather than the continuous loadbearing wall of the platform frame. Cladding panels are set into openings in the laminated timber frame, with panels supported at their base on the tops of beams. Floor decks in timber are then fixed to the side of the laminated timber beams. Timber panels are fixed at their base to the beam beneath, but have a sliding restraint at the top to allow the slab and panel above to deflect without damaging the panel below. A metal flashing at the base of the panel drains water at its base and throws it clear of the beam beneath in order to avoid staining of the timber beam. The outer timber rainscreen cladding is set flush with the outer face of the laminated timber frame to avoid any views into the waterproofing layer behind.

The use of computer numerically controlled (CNC) machines ensures the accurate cutting of components to make timber wall panels with the close tolerances of high quality, large-scale construction, particularly housing, where timber has undergone a huge revival, partly due to its lower levels of embodied energy in its construction. The use of composite aluminium/timber windows, with their high performance, low U-values and acoustic attenuation, has given timber cladding panels greater impetus for large-scale projects where traditionally the thermal and weather resisting performance of timber windows and frames has been much poorer than that of aluminium or UPVC windows and doors.

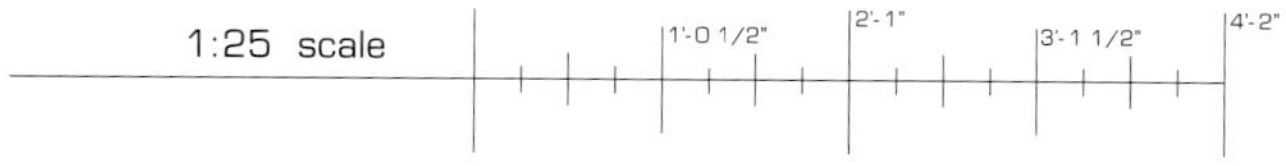

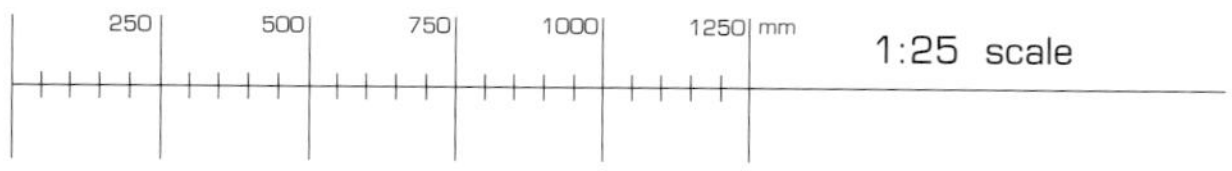

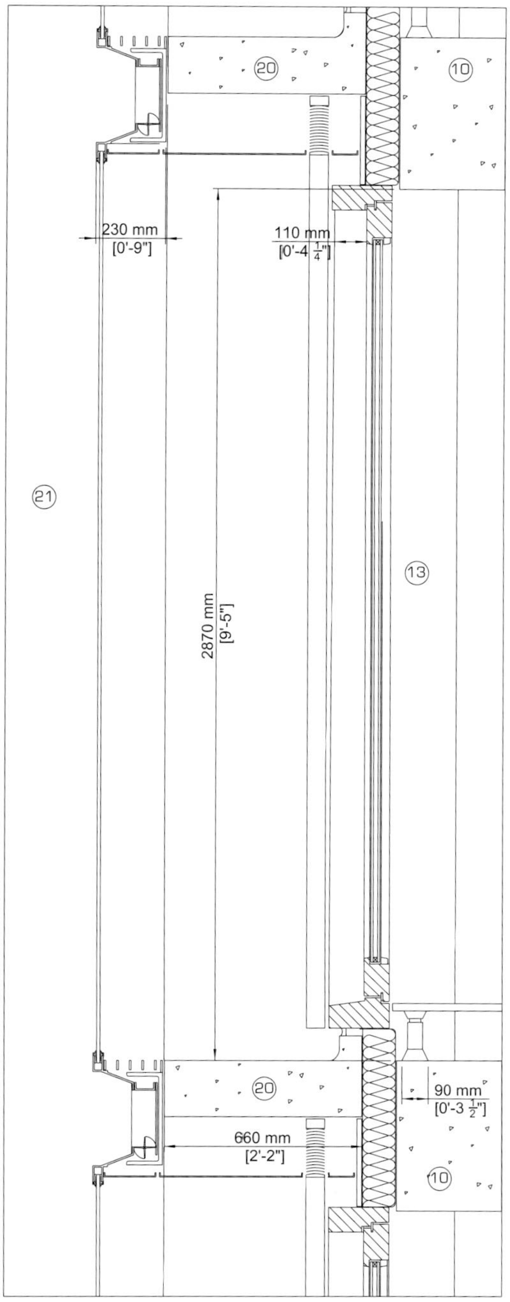

Details

1 | Timber boards
2 | Plywood sheathing
3 | Timber studs
4 | Timber rail
5 | Breather membrane
6 | Foundation
7 | Damp proof course
8 | Vapour barrier
9 | Timber floor
10 | Concrete floor
11 | Internal finish
12 | Thermal insulation quilt set within timber frame
13 | Timber framed window/door
14 | Timber sill
15 | Air gap
16 | Sliding timber louvre panel
17 | Metal facing
18 | External plywood facing
19 | Cover strip
20 | External floor deck
21 | External glass wall in twin wall
22 | Structural timber frame

d | Vertical section. Cladding panel in twin wall configuration for taller buildings

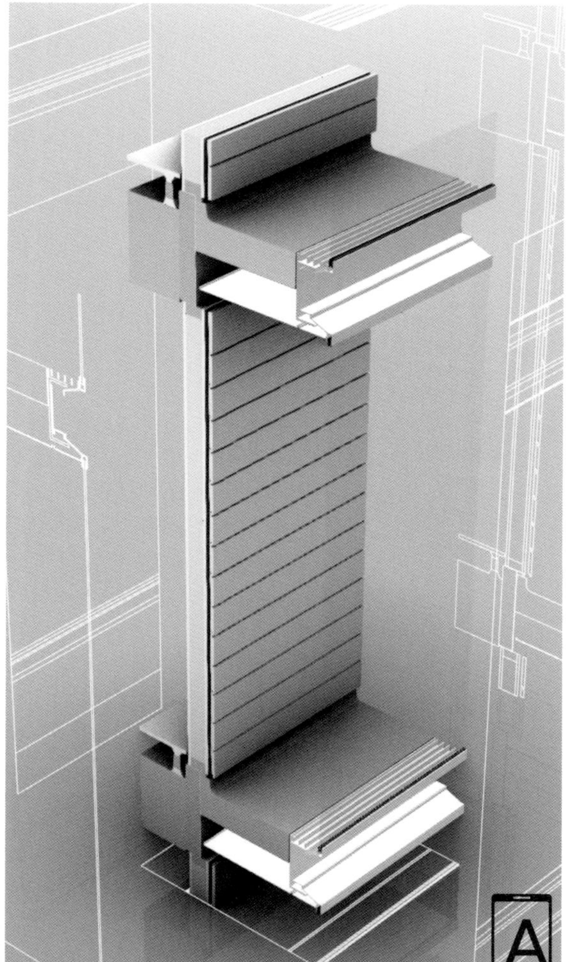

e | Prototype

f | Horizontal section. Cladding panel in twin wall configuration for taller buildings

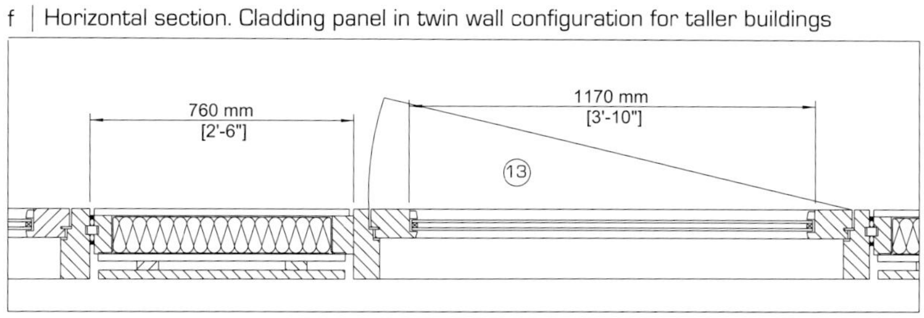

Plywood sheets

When plywood sheets are used with timber cladding panels, the edges of the sheets are protected, since the edges of plywood are susceptible to damage and rot if left exposed in an external environment. Edges can be protected with metal trims and flashings, where an aluminium Z-shaped profile is set on the back of one sheet, projecting out through a vertical or horizontal joint, and is set over the face of the adjacent panel. Timber strips can also be set over joints to protect the outside face of the plywood but also allow the rear face to be ventilated into a cavity to allow the material to be kept dry. Since plywood sheets can now be supplied in very large sizes, bays of plywood can be made from a single sheet (MCE_315, c), reducing the need for edge protection of the plywood. Plywood sheets are also lapped in the horizontal joints to create a visual effect like shiplapped timber boards, or are shingled in the manner of sheet metal cladding. The increased use of rainscreen timber cladding with a waterproof backing wall has increased the freedom for timber panel design.

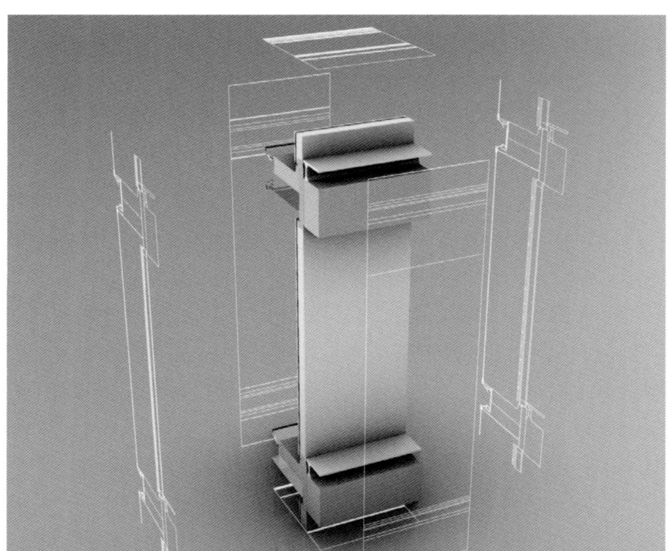

g | Prototype

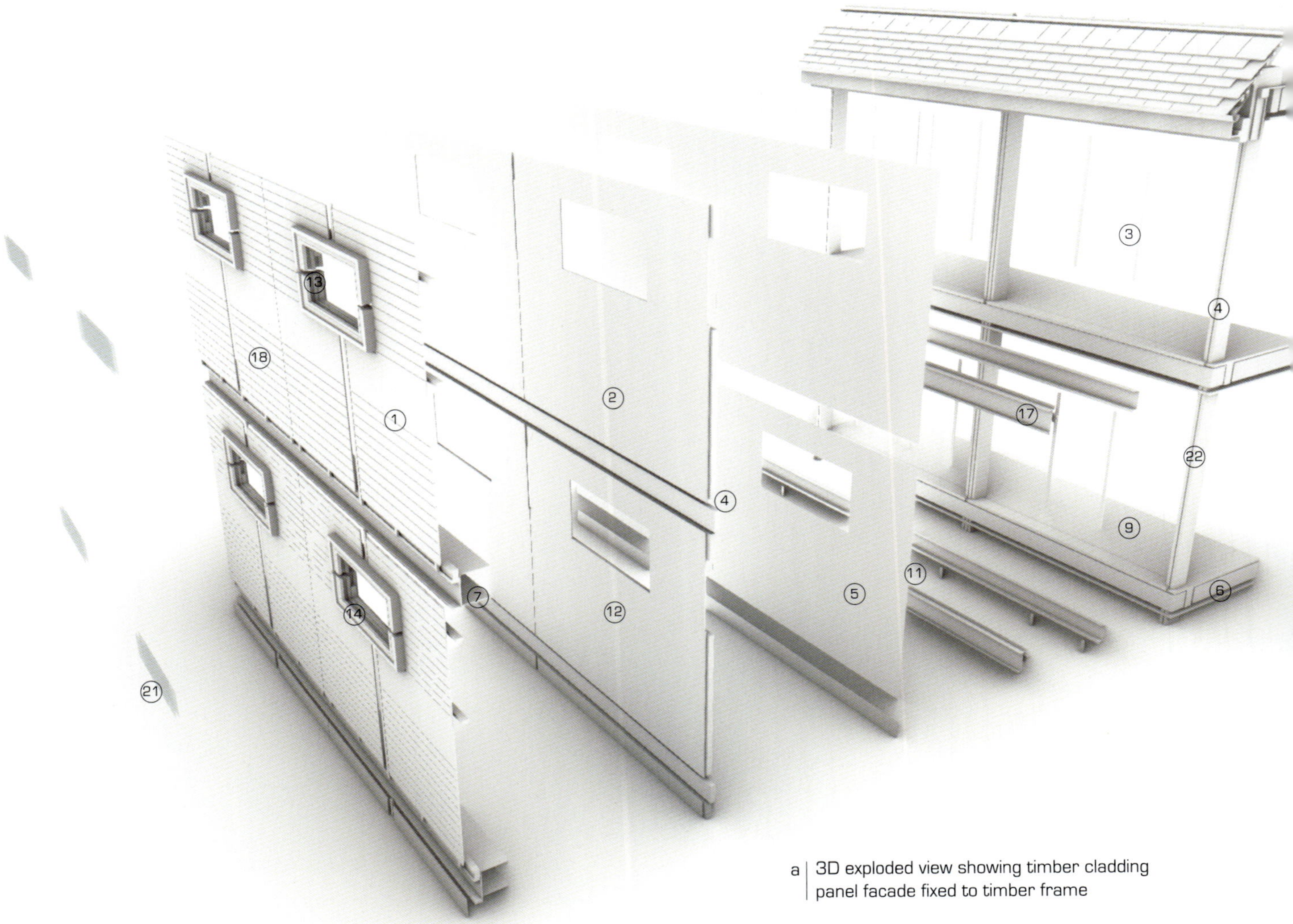

a | 3D exploded view showing timber cladding panel facade fixed to timber frame

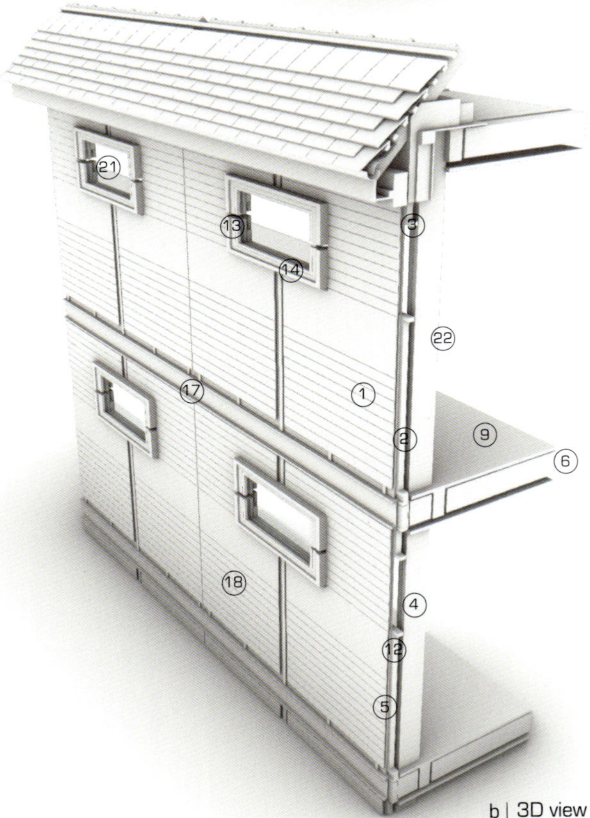

b | 3D view showing timber cladding panel facade fixed to timber frame

Details

1 | Timber boards
2 | Plywood sheathing
3 | Timber studs
4 | Timber rail
5 | Breather membrane
6 | Foundation
7 | Damp proof course
8 | Vapour barrier
9 | Timber floor
10 | Concrete floor
11 | Internal finish

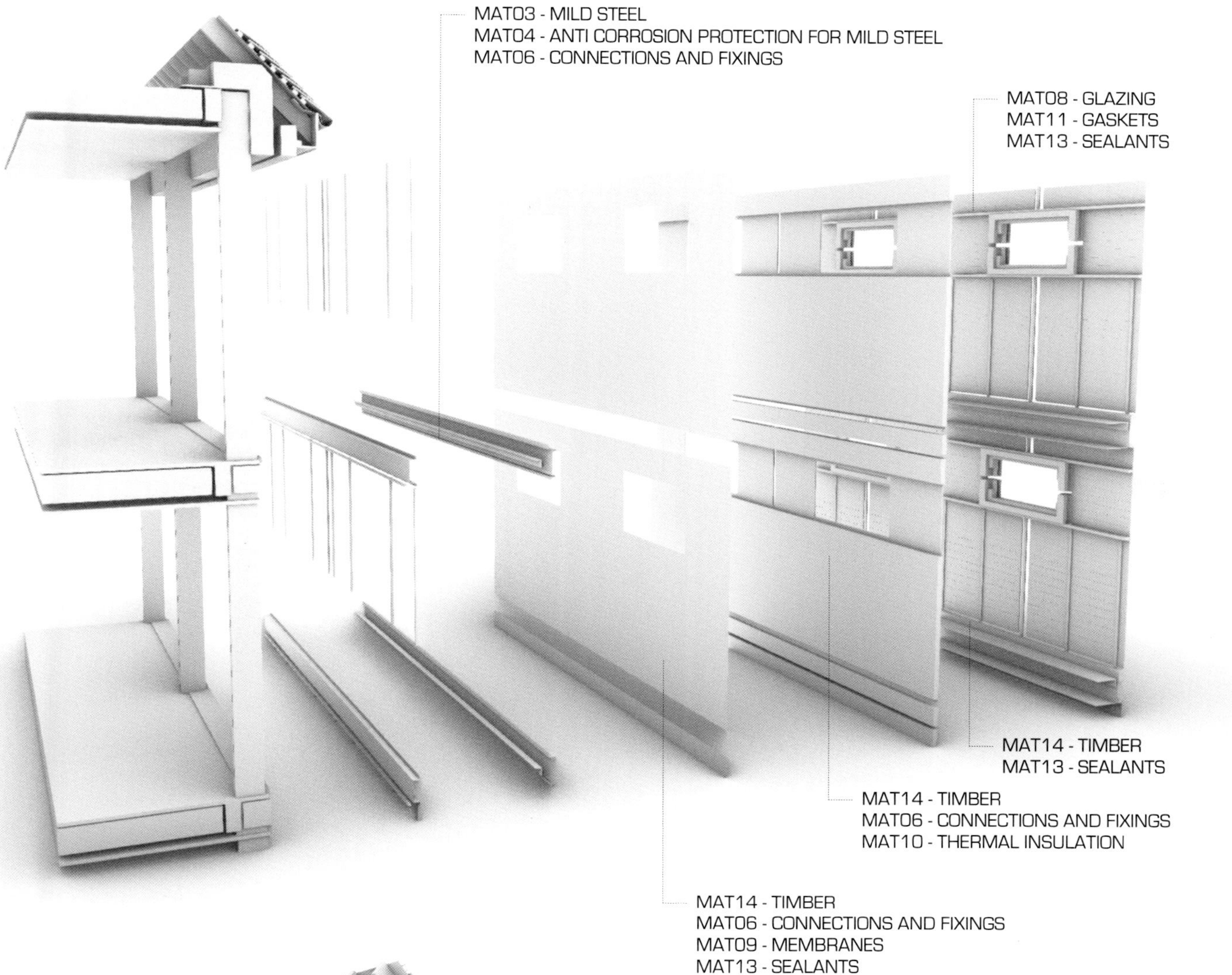

c | 3D exploded view of timber cladding panel facade components with material specification references

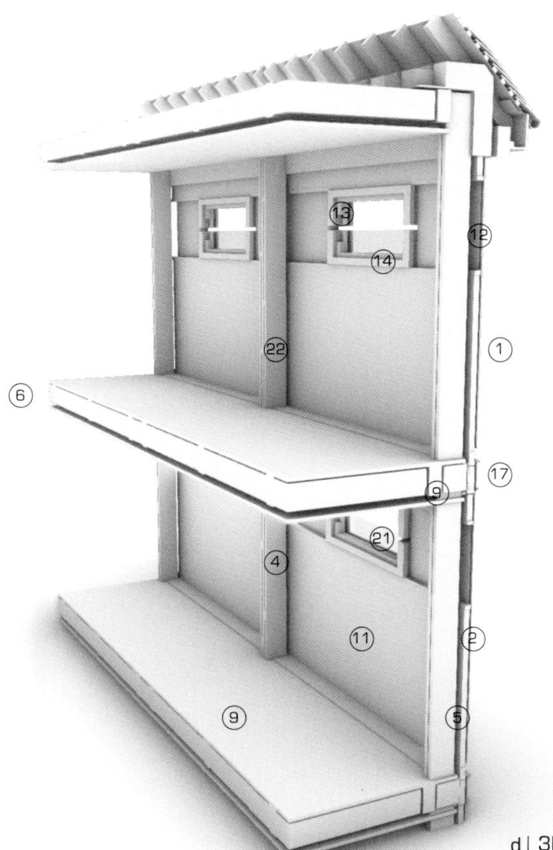

12 | Thermal insulation quilt set within timber frame
13 | Timber framed window/door
14 | Timber sill
15 | Air gap
16 | Sliding timber louvre panel
17 | Metal facing
18 | External plywood facing
19 | Cover strip
20 | External floor deck
21 | External glass wall in twin wall
22 | Structural timber frame

d | 3D view showing timber cladding panel facade fixed to timber frame

Example project of metal facades
Meixihu IC&A Centre, Changsha, China
Facade Engineer: Newtecnic
Architect: Zaha Hadid Architects

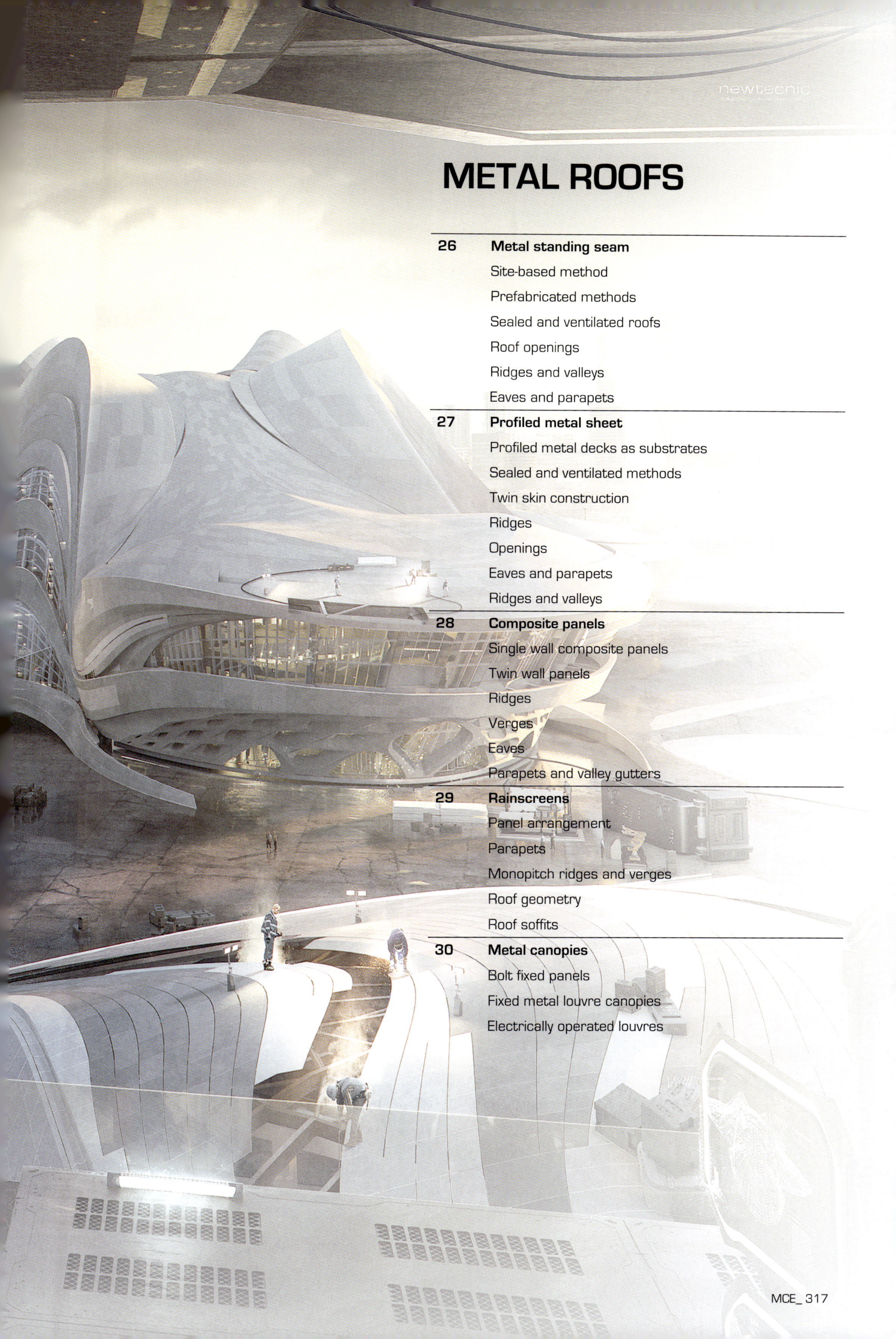

METAL ROOFS

26 **Metal standing seam**
- Site-based method
- Prefabricated methods
- Sealed and ventilated roofs
- Roof openings
- Ridges and valleys
- Eaves and parapets

27 **Profiled metal sheet**
- Profiled metal decks as substrates
- Sealed and ventilated methods
- Twin skin construction
- Ridges
- Openings
- Eaves and parapets
- Ridges and valleys

28 **Composite panels**
- Single wall composite panels
- Twin wall panels
- Ridges
- Verges
- Eaves
- Parapets and valley gutters

29 **Rainscreens**
- Panel arrangement
- Parapets
- Monopitch ridges and verges
- Roof geometry
- Roof soffits

30 **Metal canopies**
- Bolt fixed panels
- Fixed metal louvre canopies
- Electrically operated louvres

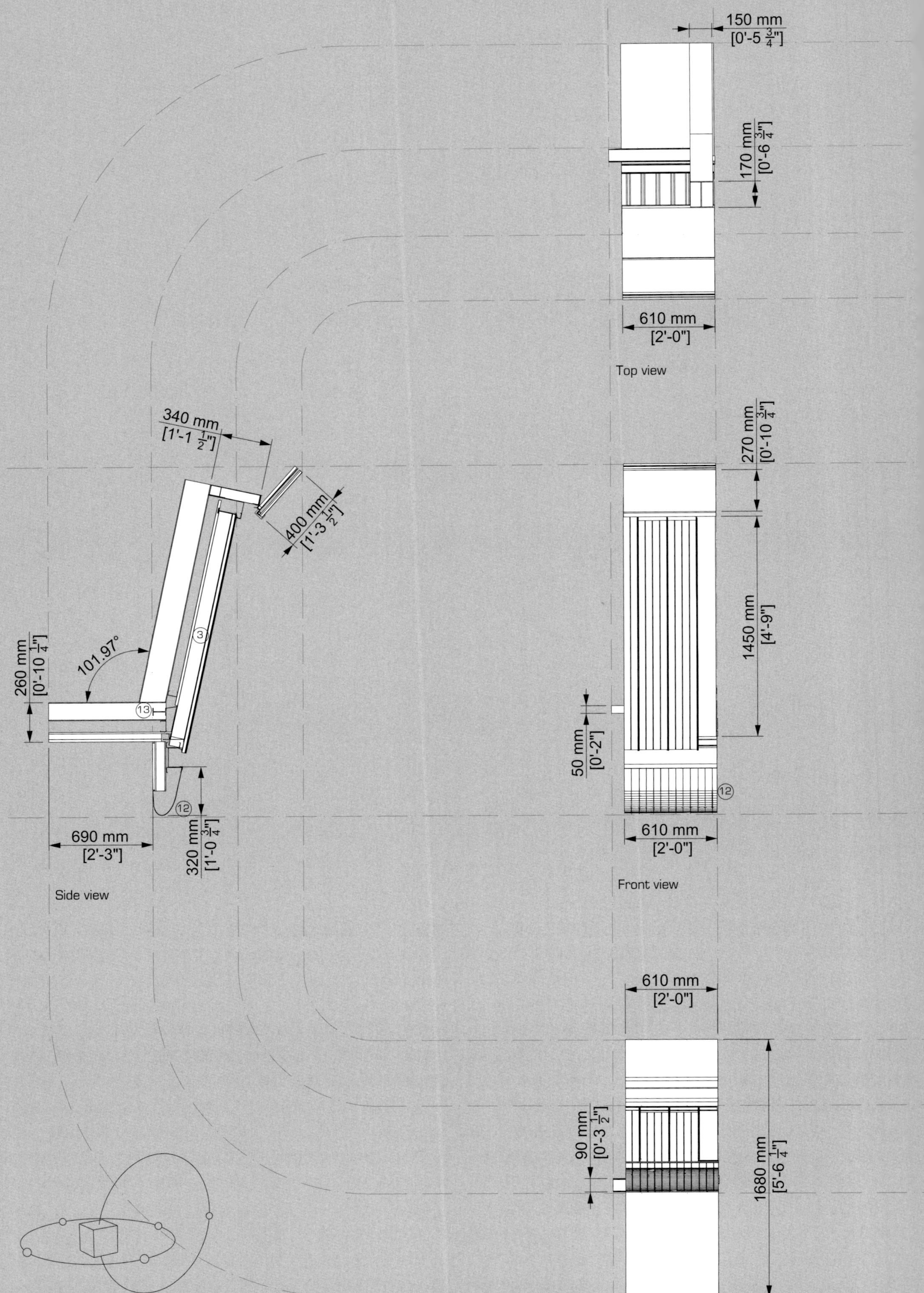

Third angle projection 1:30

Details

1 | Outer standing seam sheet
2 | Inner lining sheet
3 | Fibre quilt thermal insulation
4 | Vapour barrier
5 | Purlin or structural beam
Secondary purlin
6 | Folded metal gutter
7 | Folded metal drip
8 | Metal fascia panel
9 | External wall

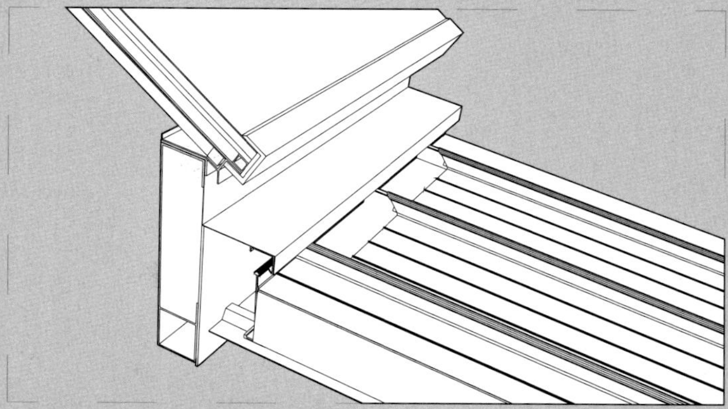

3D view of the assembly

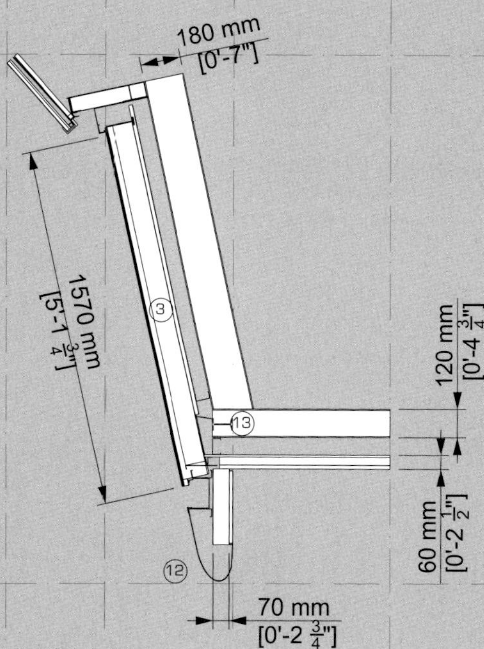

Side view

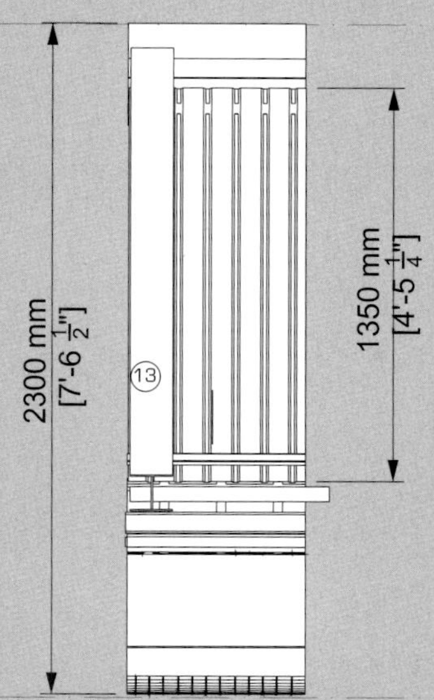

Back view

10 | Outer sheet fixing
11 | Bracket
12 | Curved eaves sheet
13 | Structural frame
14 | Ridge piece
15 | Metal flashing
16 | Rooflight
17 | Pipe or duct penetration
18 | Parapet flashing

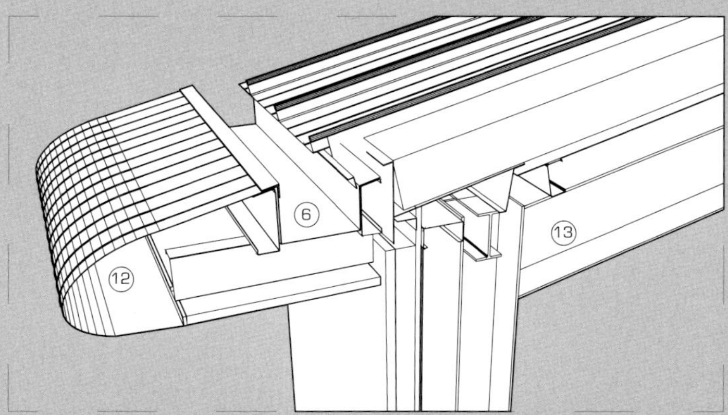

3D view of the assembly

b| Typical bay

Details

1 | Outer standing seam sheet
2 | Inner lining sheet
3 | Fibre quilt thermal insulation
4 | Vapour barrier
5 | Purlin or structural beam
Secondary purlin
6 | Folded metal gutter

a| Vertical section. Typical roof assembly with acoustic layer

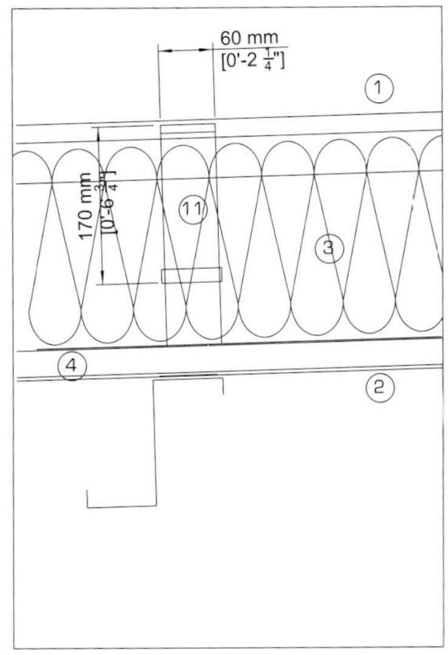

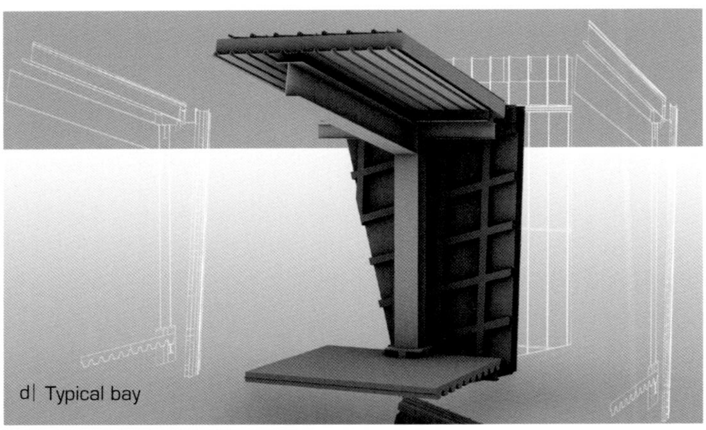

d| Typical bay

c | Vertical section. Typical roof assembly with acoustic layer

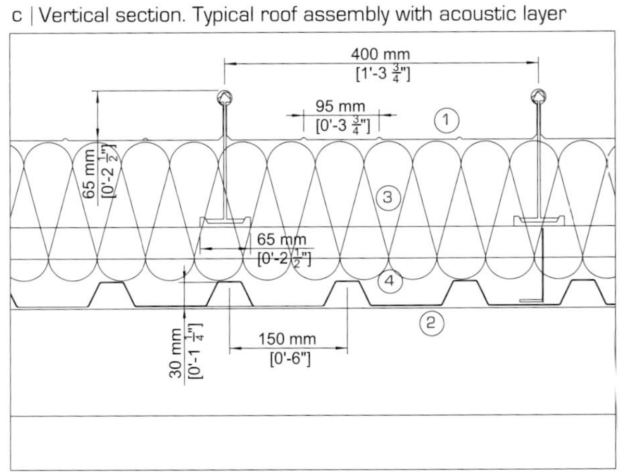

Standing seam roofs are increasingly being used for industrial and commercial buildings in preference to profiled metal sheet, where concealed fixings and low roof pitches are required for visual reasons. This is because standing seam roofing is both economic and has crisp, uninterrupted joint lines that allow it to be made a visible part of the building design, often with as much architectural presence as the facade beneath. The main advantage of standing seam roofs over profiled metal roofs is that almost no fixings pass through from outside to inside the construction. This gives the roof surface a visually crisp appearance with very few visible fixings. The standing seams allow the technique to be used on very low pitch roofs (b).

The traditional method of forming a standing seam roof is to set the sheet onto a timber substrate, and to fold the long edges of the metal upwards to form a standing seam joint. However, this method is increasingly giving way to prefabricated systems where the sheet metal is folded to a specific profile either in a factory or on site with a rolling machine. The folded metal is then secured with a clip-based fixing system rather than onto a continuous substrate. Both types are discussed in this section.

Site-based method

This method of fixing sheet is well suited to small-scale applications, or where complex geometries are used. These applications make the use of prefabrication both unnecessary and uneconomic, due to the time needed to make special junctions and edges on site. The use of a single sheet metal profile and angle support clips used in prefabricated methods is usually too inflexible for such conditions.

In this traditional method of forming standing seam roofs, timber boards or plywood sheet are used to form a continuous substrate, or supporting surface (MCE_322, a). Standing seams are formed by timber strips of rectilinear or curved section which are set at 450–600mm centres down the slope of the roof, corresponding to the width of the sheet

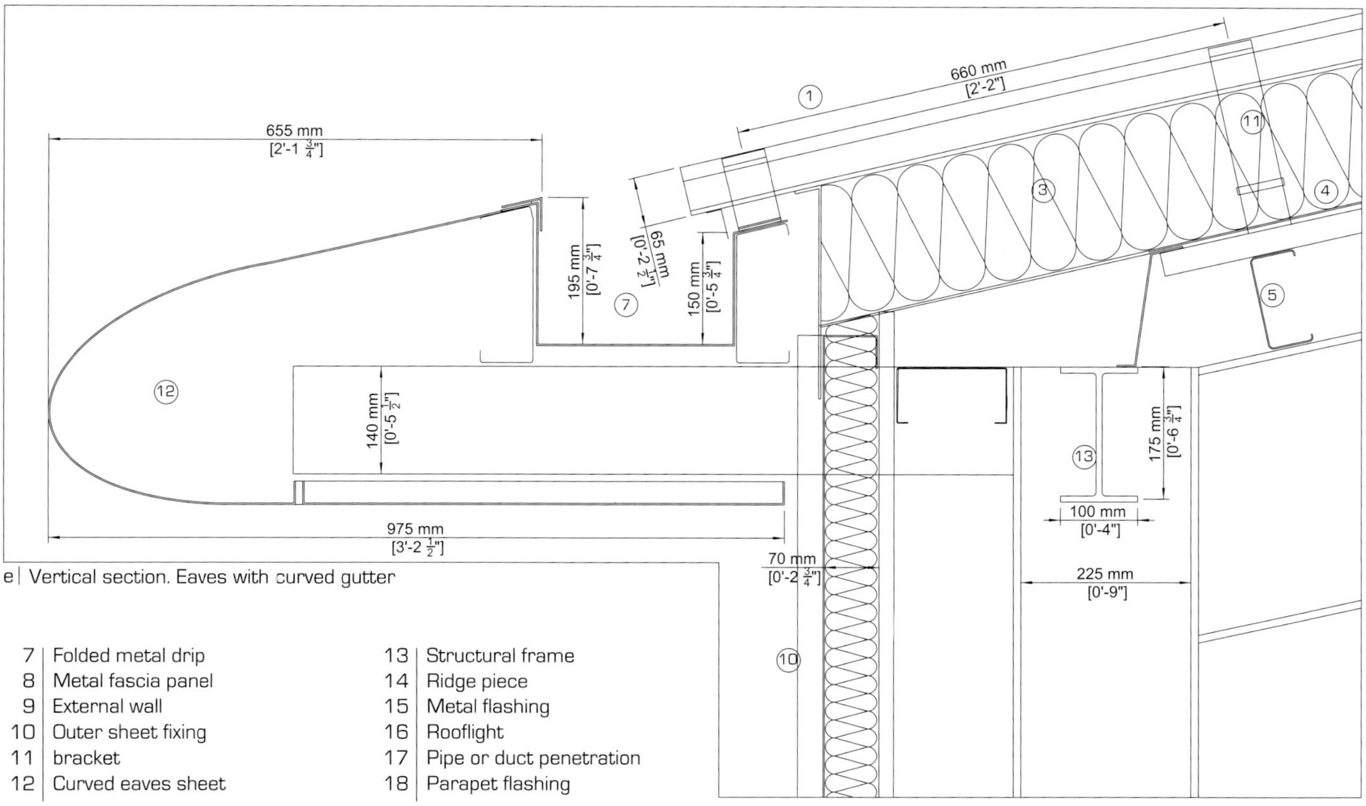

e | Vertical section. Eaves with curved gutter

7 | Folded metal drip
8 | Metal fascia panel
9 | External wall
10 | Outer sheet fixing
11 | bracket
12 | Curved eaves sheet
13 | Structural frame
14 | Ridge piece
15 | Metal flashing
16 | Rooflight
17 | Pipe or duct penetration
18 | Parapet flashing

g | Vertical section. Typical roof assembly without acoustic layer

f | Vertical section. Typical roof assembly without acoustic layer

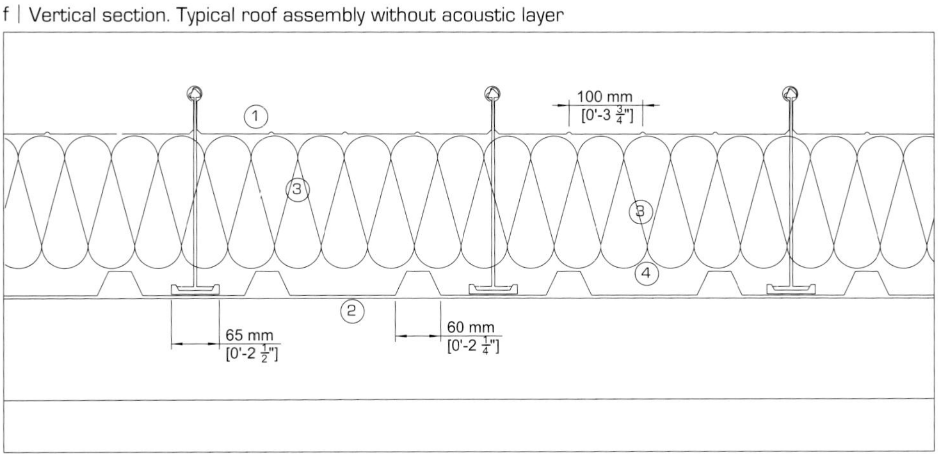

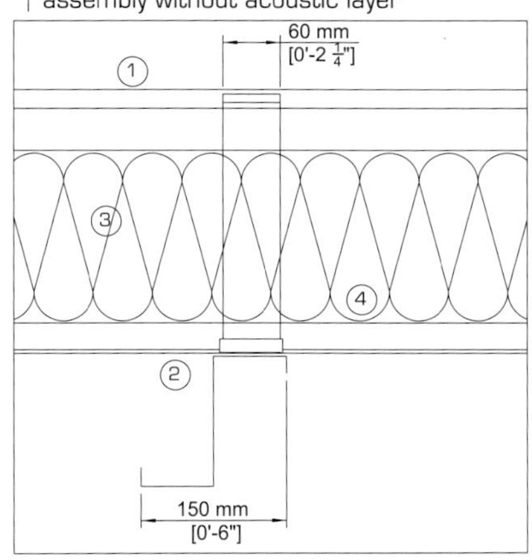

metal used. Sheet metal is laid along the length of the roof from top to bottom, with the sides of the sheet folded up and over the timber battens. Successive strips of metal sheet are lapped over the next to form a continuous sealed surface. The standing seam joint is formed by folding the metal together to form a seal. Because the roof is formed, effectively, as a series of linked 'gutters', the standing seam between each gutter is above the level of the water draining down it. Rainwater is avoided being drawn through the joint by capillary action by one of two methods, where the joint is either sealed or ventilated. In a sealed joint the seam is pressed tight, as in a traditional lead or copper roof, either by folding the metal over itself to form a thin seam, or by forming the metal over a timber roll or section. In a ventilated joint, a small gap is left between the folded sheets to allow air to pass through but not rainwater.

Sheet metal is fixed to the timber upstand strips either by clips, which avoid penetration of the sheet metal, or by a mechanical fixing through one side of the sheet. The fixing is applied to the side which has the adjacent sheet lapped over it, in order to avoid rainwater passing through the fixing penetration. Timber-based substrates are increasingly being replaced by profiled metal sheet, which has a much greater spanning capability than plywood sheet, helping to reduce the cost of the supporting structure. Timber-based substrates, typically plywood or timber boards, require support at 400–600mm centres. Profiled metal sheet is increasingly being used as a substrate, as it provides a self-finished soffit (underside) to the space below the roof. This is particularly useful where acoustic ceilings are used, with perforated sheet that has a paint coating on its underside.

Prefabricated methods

The most common configuration of prefabricated standing seam roofs is a structural deck, typically reinforced concrete or profiled metal sheet, with insulation set on top and an outer (upper) sheet supported on brackets set

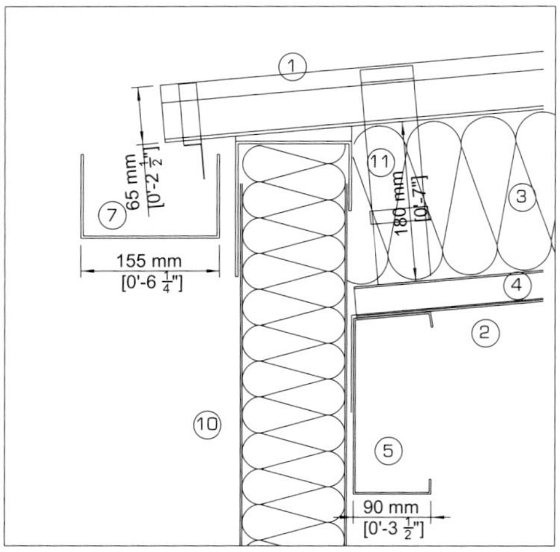

a | Vertical section. Eaves with exposed gutter

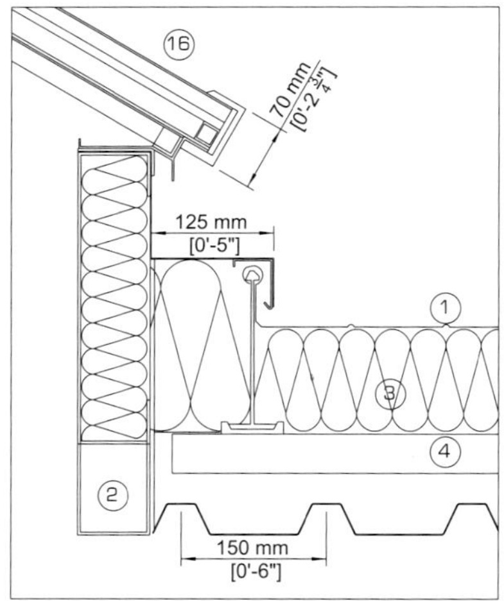

b | Vertical section. Junction with rooflight

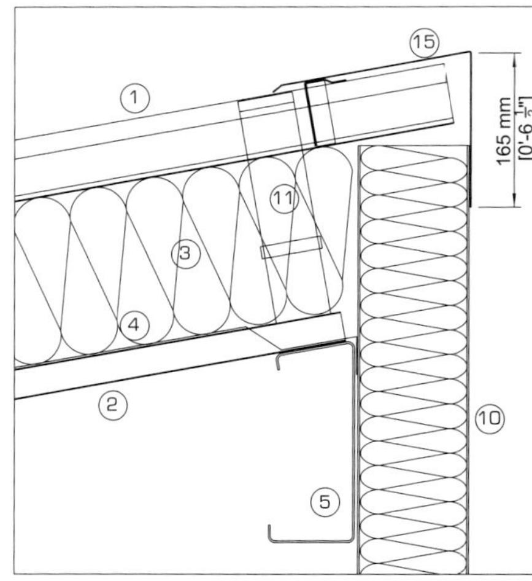

c | Vertical section. Monopitch ridge

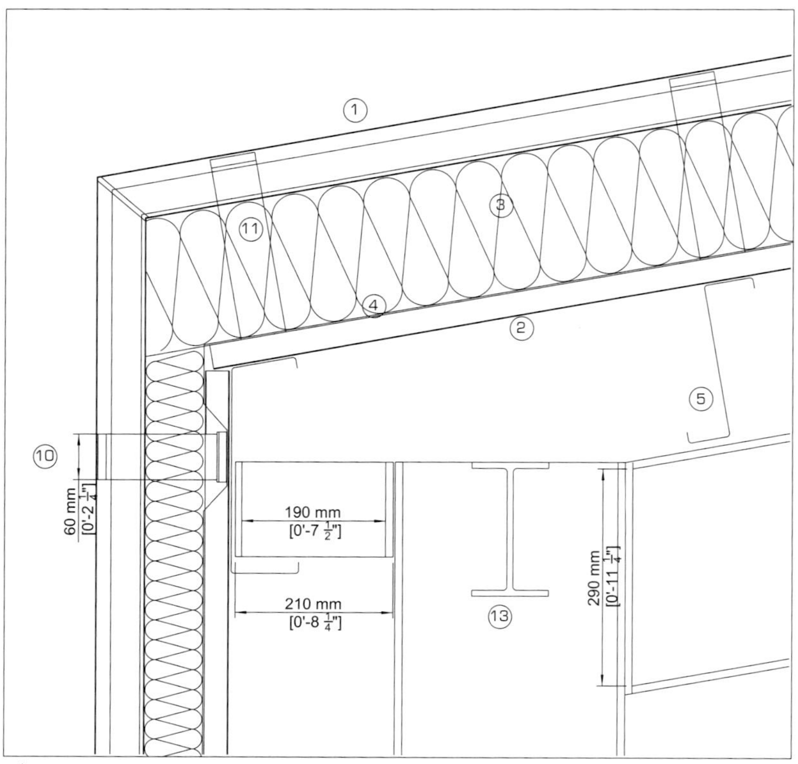

d | Vertical section. Folded eaves

Details

1 | Outer standing seam sheet
2 | Inner lining sheet
3 | Fibre quilt thermal insulation
4 | Vapour barrier
5 | Purlin or structural beam
6 | Secondary purlin
7 | Folded metal gutter
8 | Folded metal drip
9 | Metal fascia panel
10 | External wall
11 | Outer sheet fixing bracket
12 | Curved eaves sheet
13 | Structural frame
14 | Ridge piece
15 | Metal flashing
16 | Rooflight
17 | Pipe or duct penetration
18 | Parapet flashing

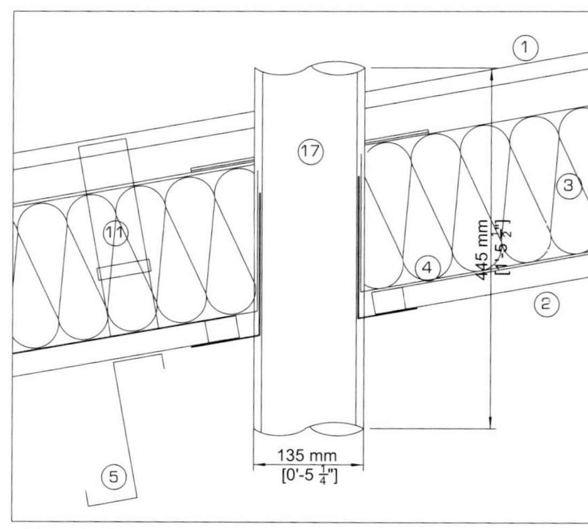

e | Vertical section. Roof penetration

onto the structural deck. An alternative configuration is to fix the brackets supporting the outer sheet to a set of metal purlins. A metal liner sheet is set below the purlins to support the thermal insulation quilt, set between the purlins. A vapour barrier is set between the insulation quilt and the liner tray on the warm (in winter) side of the thermal insulation. In hot, humid countries an additional vapour barrier is set on top of the insulation where the risk of interstitial condensation is from the outside as well as the inside.

For both construction configurations the roof pitch can go down to 1°, after taking into account any structural deflections that would further reduce this angle. Metal sheets can be made up to 40 metres in length, but road transport is difficult, with long sheets (longer than a trailer length) requiring special arrangements for road transportation in most countries. For larger projects, long sheets are formed on site with a rolling machine that can form the profile of the standing seam sheet to any length required, the profile being formed from metal coil.

Support brackets are usually T-shaped and are fixed either to the structural deck or to purlins with self-tapping screws. The brackets are usually made from extruded aluminium in order to provide a profile that is both thick enough to form a rigid connection and sufficiently precise in section to retain a given standing seam profile in place. Metal sheet is formed in long lengths of folded trays which are then fixed onto the support brackets. Finally, the standing seam joints are crimped to form a seal, usually with a 'zip up' tool that travels along the joint and across the roof, sealing the joint as it moves along. This fixing method gives very smooth and straight joints, but the long lengths of sheet metal forming the roof surface can result in 'oil canning', where part of the metal surface appears to be crinkled as a result of uneven thermal expansion. Generally, thermal expansion is accommodated by allowing the long lengths of metal sheet to slide over the support clips, with the sheet itself fixed rigidly in only a few places along its length.

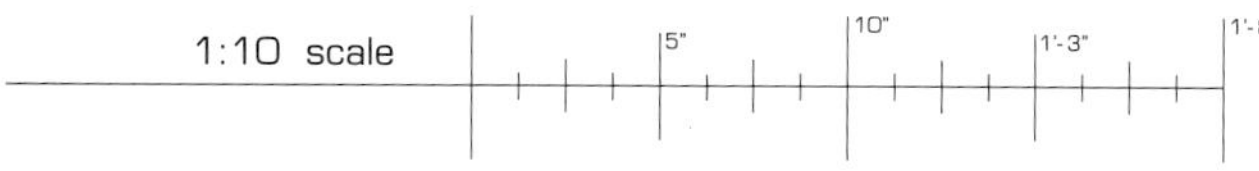

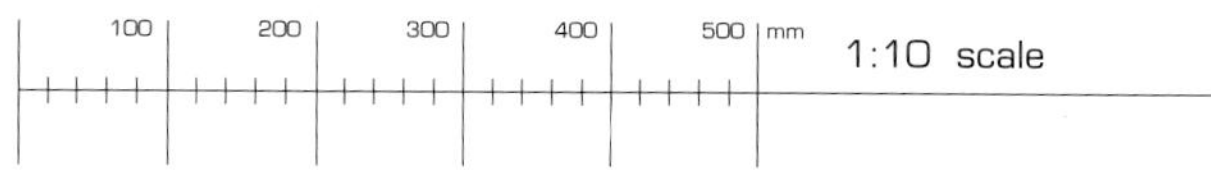

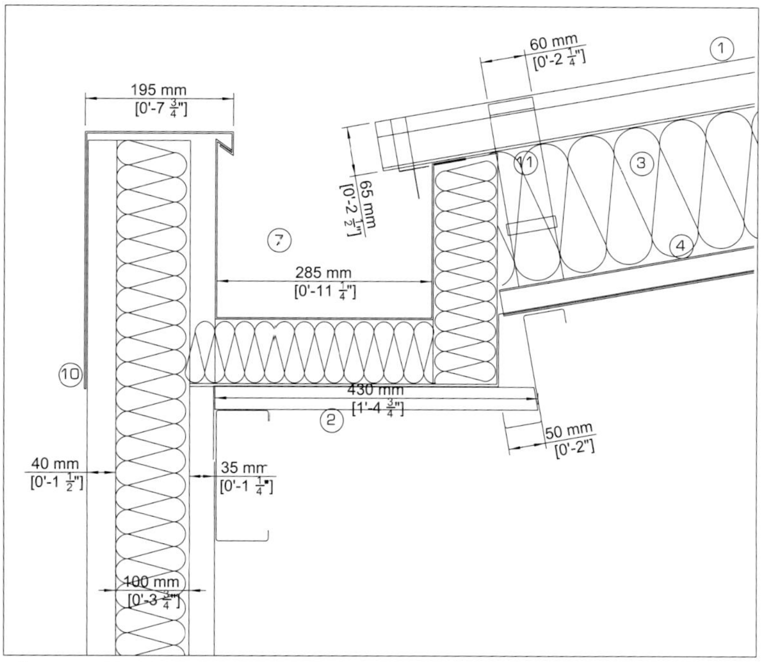

f | Vertical section. Concealed eaves gutter

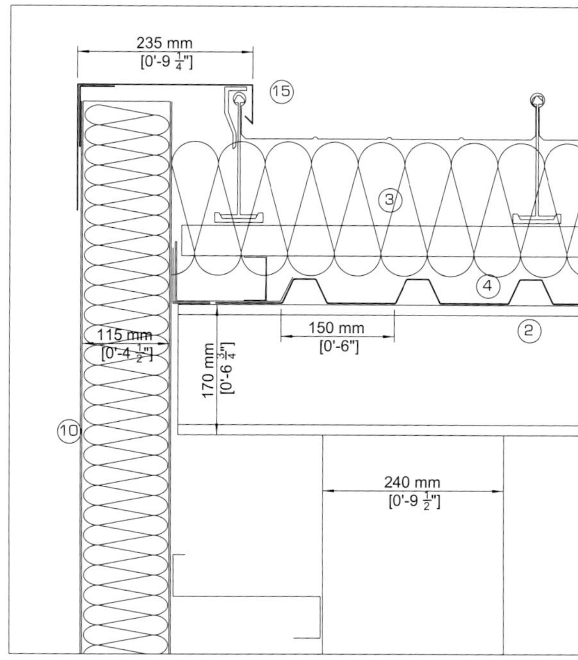

g | Vertical section. Verge

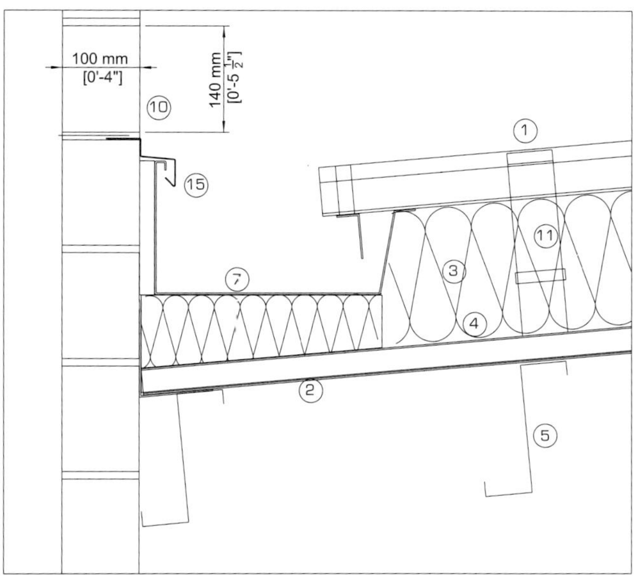

h | Vertical section. Parapet gutter at abutment

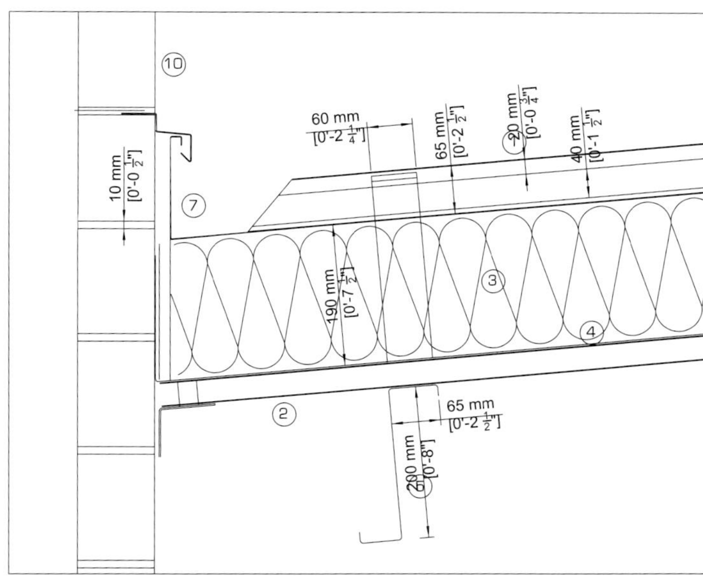

j | Vertical section. Parapet gutter at abutment

Standing seam sheets can form shallow curves by gently bending the metal on site, or alternatively by curving the sheets in the factory to give a smooth appearance (MCE_320, b). Small radius curves are formed by crimping the sheet in the factory, where the material is mechanically formed with small local folds. Sharp-folded corners are made by welding two sheets together along the fold line.

Sealed and ventilated roofs

The choice of a sealed or ventilated roof configuration depends upon both the metal used and the nature of the construction beneath. Ventilation is essential if the thermal insulation is set some distance below the metal sheet, as in a pitched roof where the insulation is set horizontally, above a closed ceiling (a). Ventilated roofs are also suitable where the external roof form has a complex geometry. This allows the thermal insulation to be set horizontally beneath, while the roof form can be free of the constraints of making the entire roof construction follow the same geometry. Ventilated roofs have slots at ridges, valleys and perimeter gutters to allow air to pass through the construction to ensure that the roof build-up remains dry. Sealed roofs use sealing strips in the standing seam joints to make them both airtight and watertight. Moisture vapour that builds up inside the construction is released by slots for passive ventilation at the ridge and eaves.

Unlike other metals, zinc sheet requires ventilation on the underside to avoid corrosion from water vapour trapped inside the construction. Ventilation for zinc roofs has traditionally been provided with open-jointed timber boards, but this method is being replaced on larger applications by a plastic-based woven mat set onto a substrate, which allows air to pass across the inside face of the zinc while using a continuous substrate material beneath.

Details

1 | Outer standing seam sheet
2 | Inner lining sheet
3 | Fibre quilt thermal insulation
4 | Vapour barrier
5 | Purlin or structural beam
Secondary purlin
6 | Folded metal gutter
7 | Folded metal drip
8 | Metal fascia panel
9 | External wall
10 | Outer sheet fixing
11 | bracket
12 | Curved eaves sheet
13 | Structural frame
14 | Ridge piece
15 | Metal flashing
16 | Rooflight
17 | Pipe or duct penetration
18 | Parapet flashing

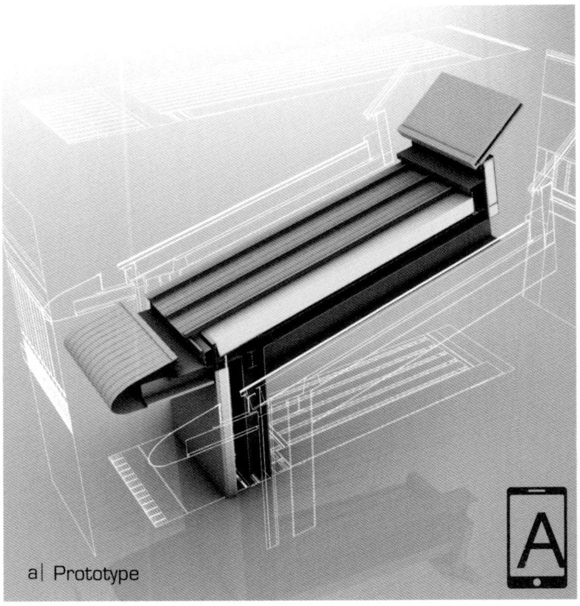

a| Prototype

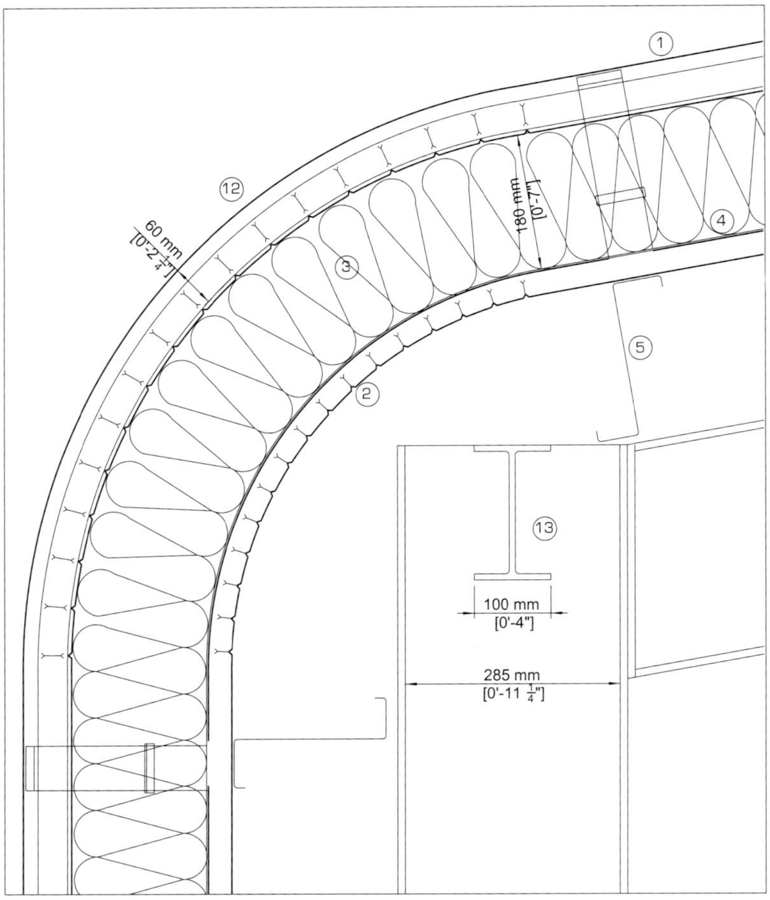

b| Vertical section. Curved ridge

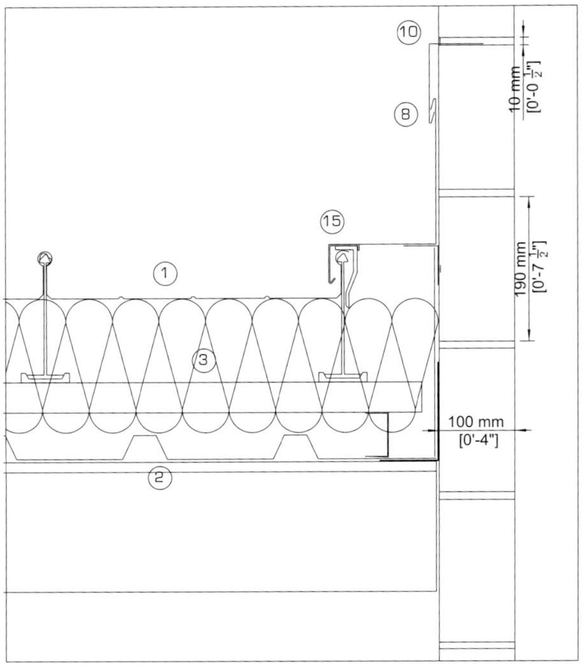

c| Vertical section. Abutment with masonry wall

Roof openings

Rooflights are set into standing seam roofs in one of two ways: either by forming a pressed metal upstand around the opening so that the rooflight projects up around 150mm above the level of the roof, or by setting it level with the roof finish and forming a gutter around the edges of the rooflight. This second method avoids the need for rooflights to appear as projecting box-like forms in an otherwise smooth, continuous roof plane. If the rooflights are set into the opening, then a metal flashing is fixed around its edge, being lapped under the standing seam roof on its top edge. At the sides the flashing forms a standing seam with the adjacent joints running down the roof, and its bottom edge lapped over the top of the roof sheet immediately below it. Penetrations for small ducts and pipes through the standing seam roofs use simplified flashings. An upstand is formed as a continuous skirt around the projecting pipe and a flashing, usually welded to the pipe, folds over the top of the upstand to avoid rainwater from penetrating the joint. The base of the 'skirt' flashing is bonded to the metal sheet roof covering and sealed, typically with a silicone-based bond and seal. Standing seams that clash with the base of the skirt flashing are stopped and closed above and below the penetration.

Ridges and valleys

Ridges are formed by a variety of methods. In some cases a folded or curved metal sheet is set over the gap between the two sides of the roof on the level of the top of the standing seam. The gap between the top and bottom of the standing seam is closed with a formed metal filler piece, or strip. Elsewhere, a sharp ridge line can be formed as a standing seam joint, with the seams meeting the ridge being terminated to avoid the need for visually bulky filler that could be seen from below. Alternatively, the ridge can be treated as a soft fold, without any break in the material. Although this may appear to be the most straightforward of the ridges to form, the alignment of the ridge piece is as critical as in previously discussed versions, in order to create a

1:10 scale 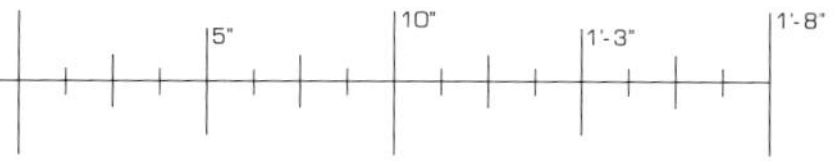

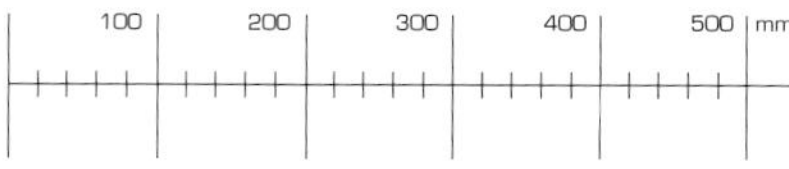 1:10 scale

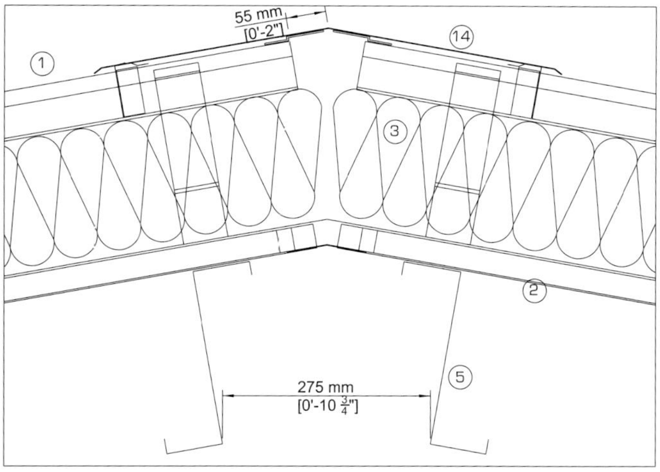

d | Vertical section. Ridge with flashing.

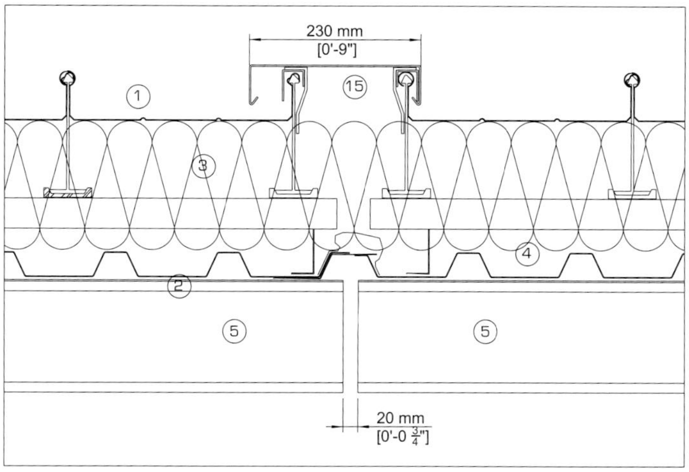

e | Vertical section. Expansion joint

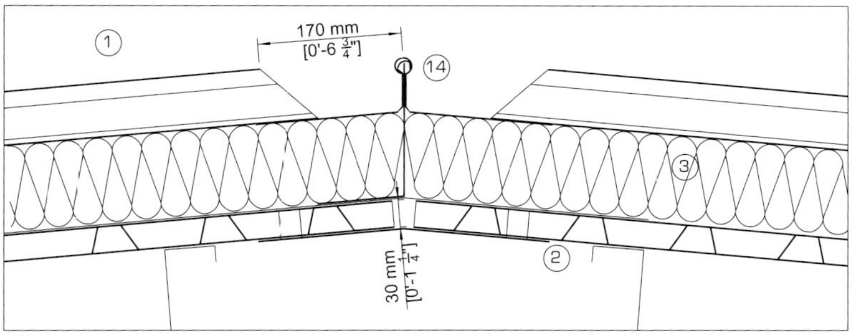

f | Vertical section. Ridge with flashing

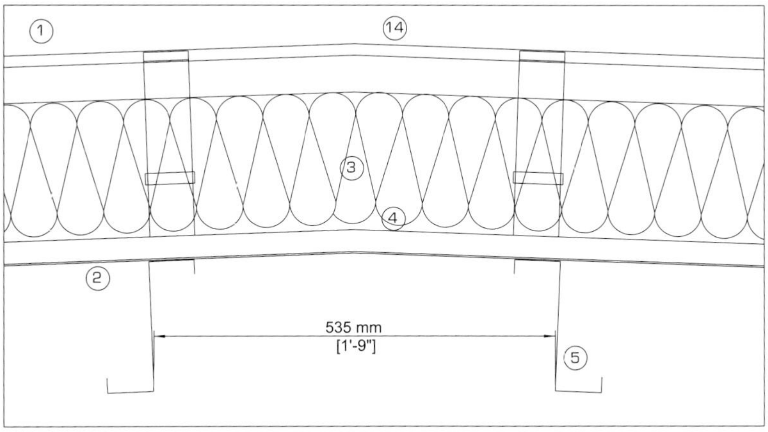

g | Vertical section. Curved ridge

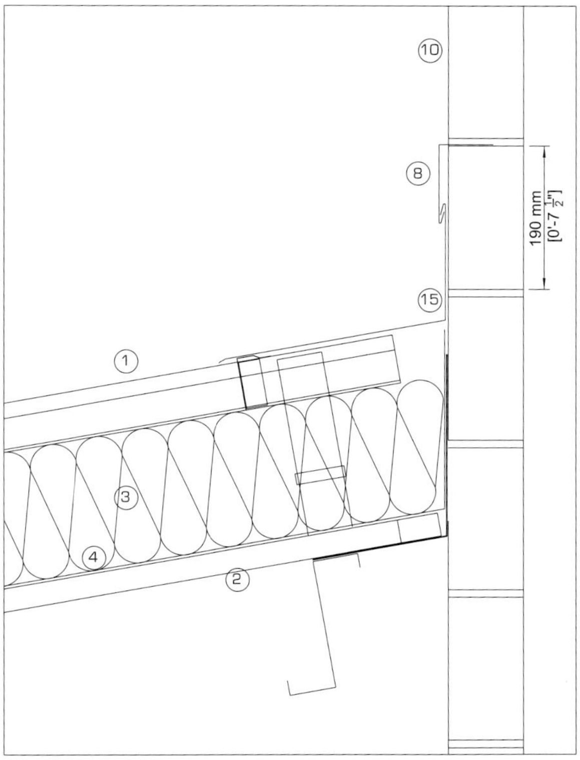

h | Vertical section. Abutment with masonry wall

straight ridge line. Shadows from the sun cast across the ridge will reveal any waviness in the line of the top of the roof.

Valleys are formed by lapping the ends of the standing seam roof into a folded metal tray forming a continuous gutter. The gutter is often welded at the joint between one length of folded sheet and the adjacent sheet in order to avoid the possibility of a water leak in the gutter. Because of the reduced roof depth at gutters, thermal insulation set below it is often thinner than the adjacent areas of roof (MCE_323, f). The reduced thermal insulation can be improved upon by using either higher performance insulation in that area, or by deepening the structure of the roof beneath where this is possible, to allow the depth of thermal insulation to be increased. If the roof is ventilated, the gap formed between the top of the gutter upstand and the underside of the standing seam roofing allows the passage of air into the roof void without the need for ventilation slots (MCE_326, a), visible from below, being set into the outer face of the roof.

Eaves and parapets

Eaves are formed in a similar way to a valley, with a gutter set at the edge of the roof. Increasingly, gutters are being integrated into roof forms in order to avoid a weak visual line formed by a gutter which is not continuous with the smooth lines of the roof. When additional closer pieces are used, such as bull-nose profiles, the metal panels are usually designed to be drained and ventilated to the exterior, and the line of waterproofing continues up the external wall to the underside of the standing seam roof.

Parapets are formed by taking the side of the gutter that is adjacent to the external wall up to the parapet coping, where it is terminated with a rubber-based seal bonded to the top of the external wall. A parapet coping is set into this seal as an overcloak flashing and second line of defence against rainwater penetration (MCE_323, g).

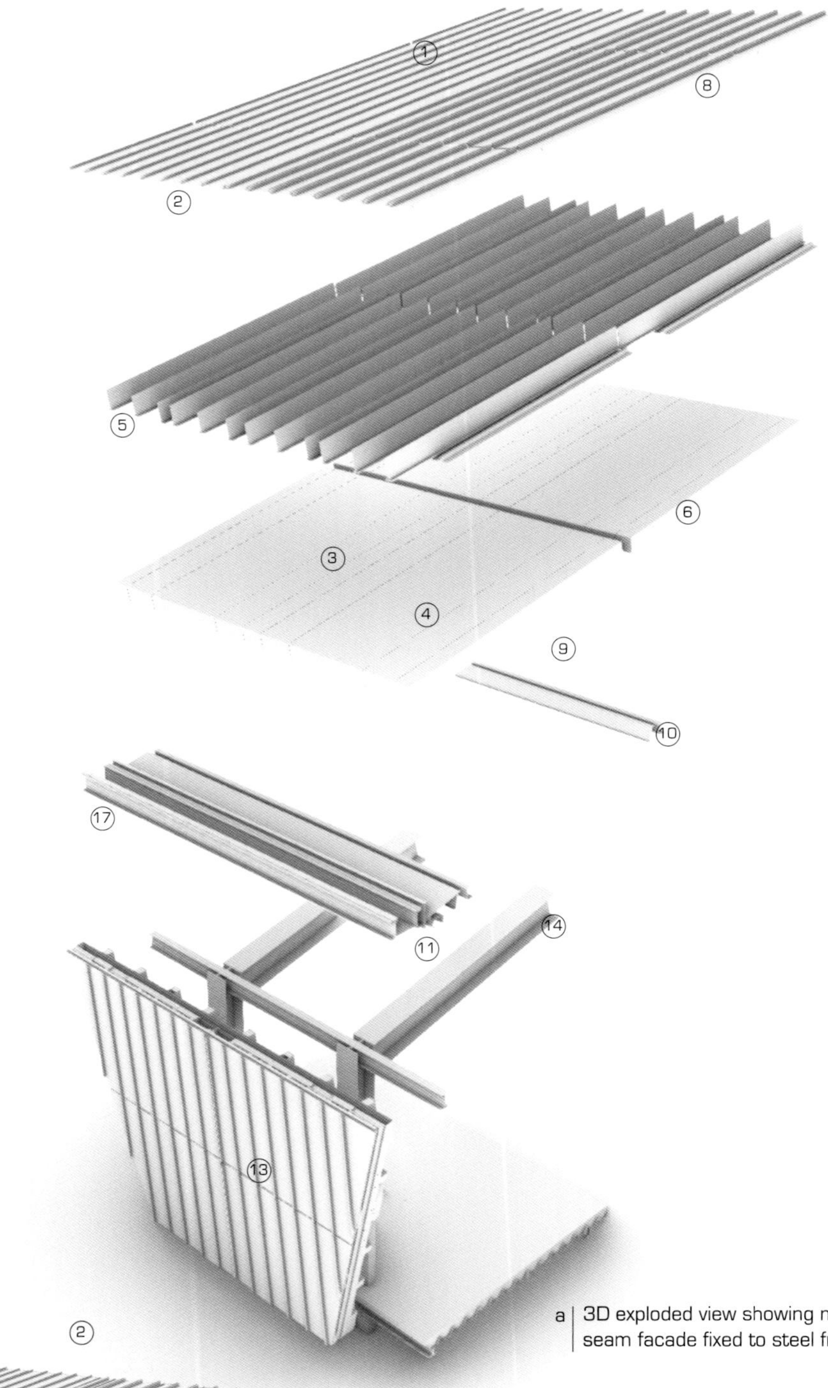

a | 3D exploded view showing metal stand seam facade fixed to steel frame

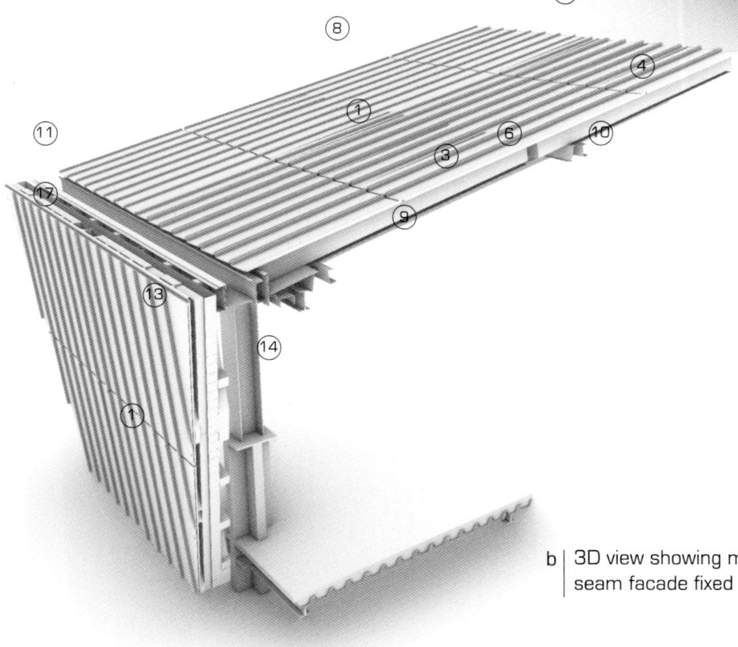

b | 3D view showing metal standing seam facade fixed to steel frame

Details

1 | Metal sheet
2 | Standing seam joint
3 | Breather membrane
4 | Thermal insulation
5 | Substrate, typically timber/metal rafters with plywood facing
6 | Vapour barrier
7 | Drywall/dry lining if required
8 | Outer standing seam sheet
9 | Inner lining sheet

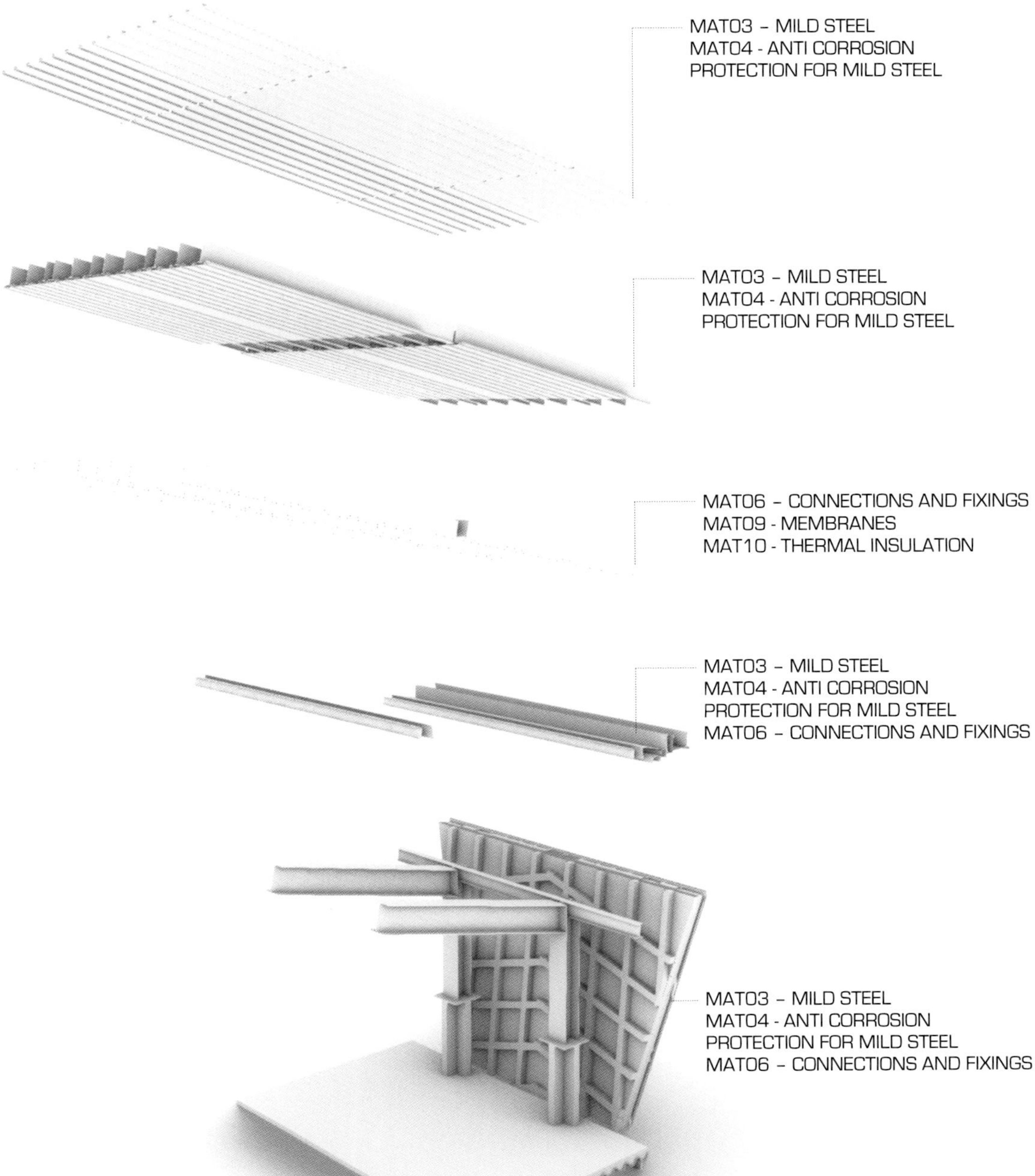

c | 3D exploded view of metal standing seam facade components with material specification references

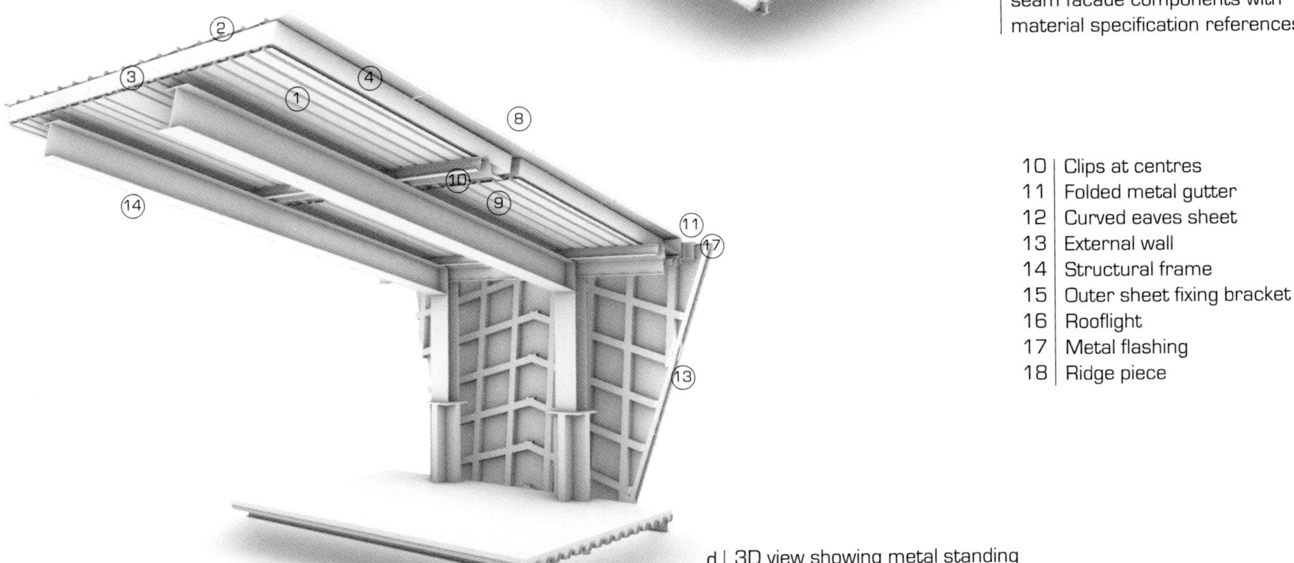

d | 3D view showing metal standing seam facade fixed to steel frame

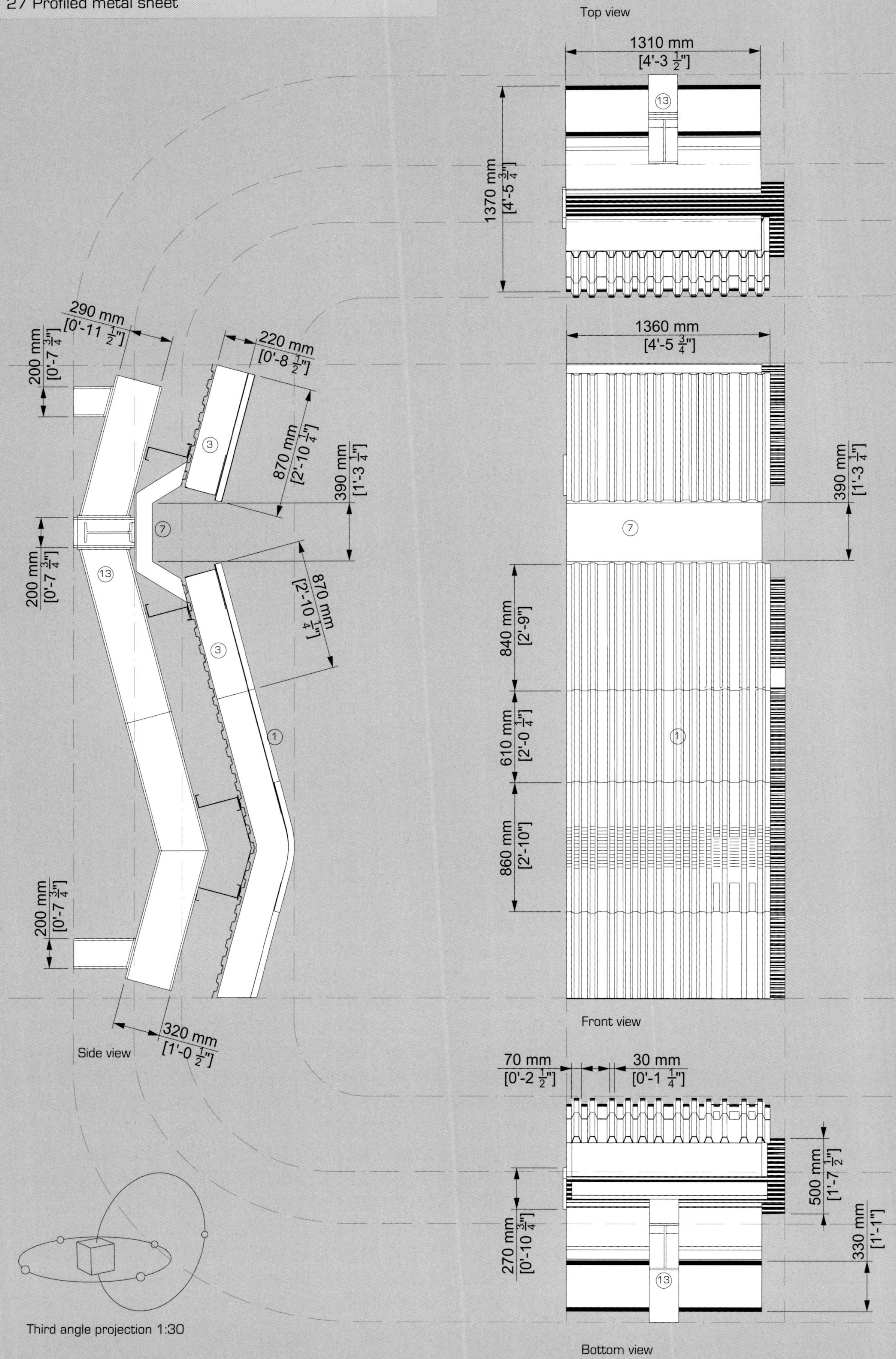
Top view
1310 mm
[4'-3 1/2"]
1370 mm
[4'-5 3/4"]
13
1360 mm
[4'-5 3/4"]
390 mm
[1'-3 1/4"]
7
840 mm
[2'-9"]
610 mm
[2'-0 1/4"]
1
860 mm
[2'-10"]
Front view
290 mm
[0'-11 1/2"]
220 mm
[0'-8 1/2"]
200 mm
[0'-7 3/4"]
3
870 mm
[2'-10 1/4"]
390 mm
[1'-3 1/4"]
7
200 mm
[0'-7 3/4"]
13
870 mm
[2'-10 1/4"]
3
1
200 mm
[0'-7 3/4"]
320 mm
[1'-0 1/2"]
Side view
70 mm
[0'-2 1/2"]
30 mm
[0'-1 1/4"]
500 mm
[1'-7 1/2"]
270 mm
[0'-10 3/4"]
330 mm
[1'-1"]
13
Bottom view
Third angle projection 1:30

Details

1 | Outer profiled metal sheet
2 | Inner lining sheet
3 | Fibre quilt thermal insulation
4 | Vapour barrier
5 | Purlin or structural beam
6 | Profiled metal structural deck
7 | Folded metal gutter
8 | Folded metal drip
9 | Metal fascia panel
10 | External wall

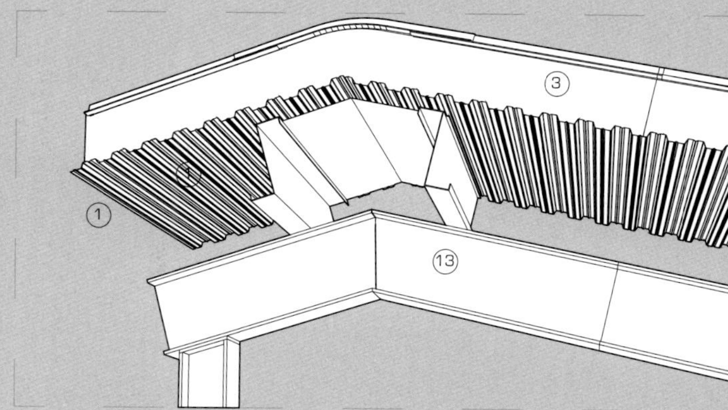

3D view of the assembly

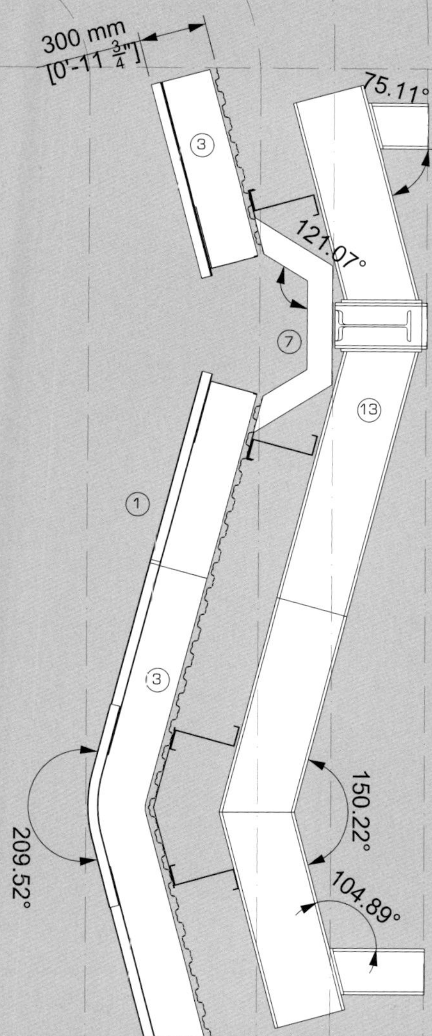

Side view

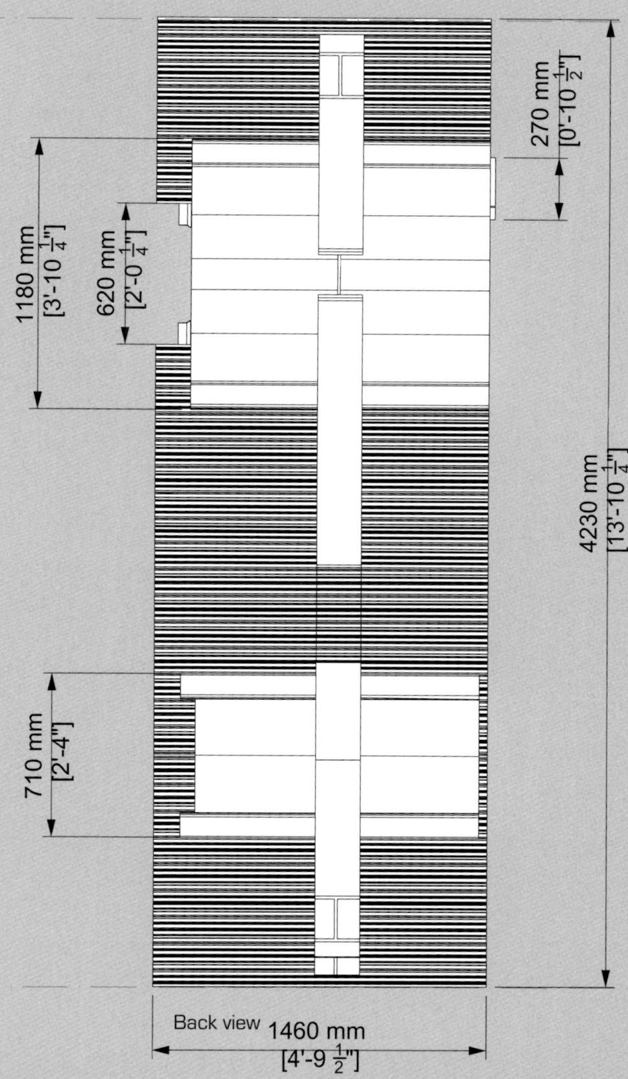

Back view

11 | Outer sheet fixing bracket
12 | Curved eaves sheet
13 | Structural frame
14 | Ridge piece
15 | Metal flashing
16 | Rooflight
17 | Pipe or duct penetration
18 | Parapet flashing
19 | Vented filler piece

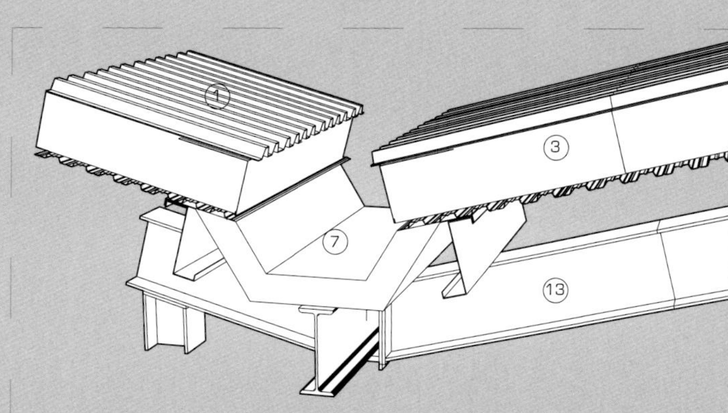

3D view of the assembly

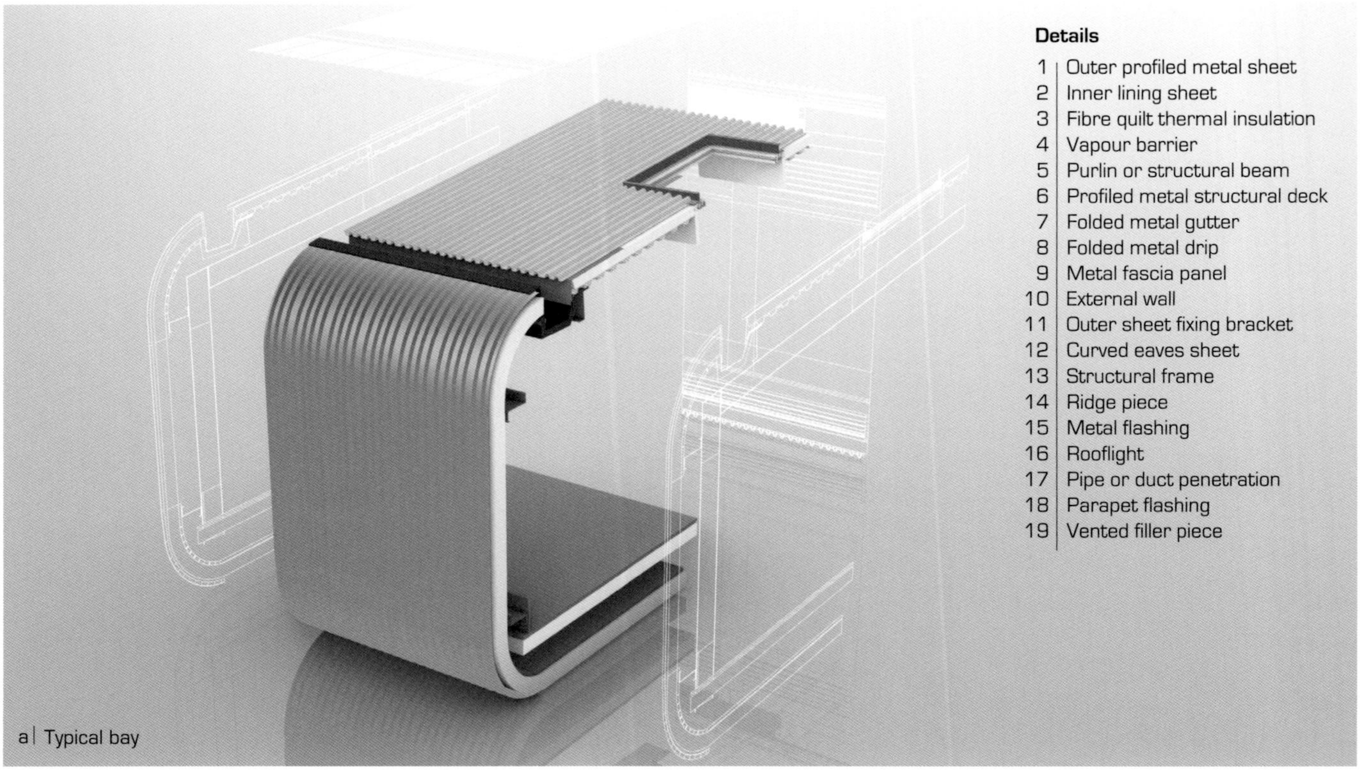

a| Typical bay

b| Vertical sections. Typical roof assembly

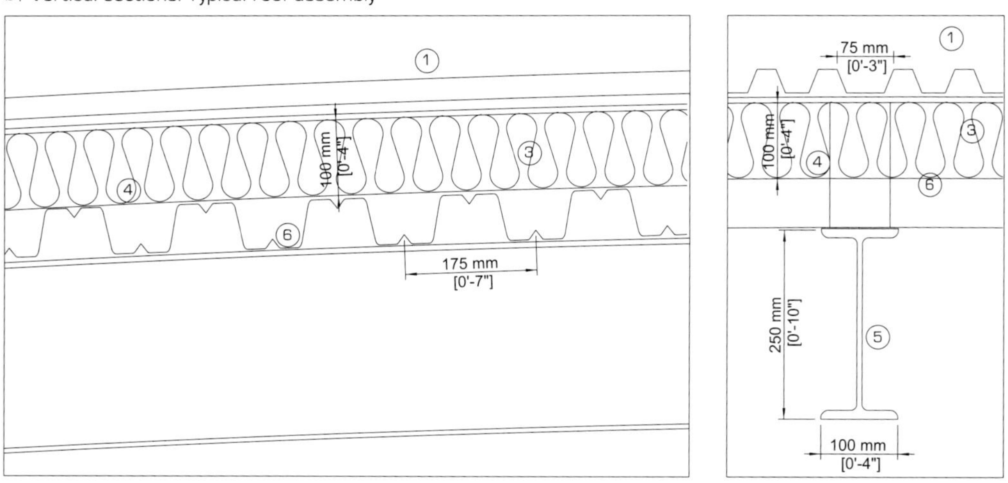

c| Typical bay

The main advantage of profiled metal sheet over other metal roof types is the ability of the material to span economically up to around 3.5 metres between primary structural supports. This self-supporting ability of the material, combined with its weather-resistant, painted coating applied during manufacture, allows for it to be used as both a substrate material for a finish in a different material set onto it, or as a single-layer structural and weatherproof material. Where standing seam roofing, with its high projecting folds, is suited to long, straight, or gently curving spans, profiled metal sheet can both span between supports and form complex geometries. It is this flexibility of being both structural deck and waterproofing layer that has an advantage where the interior finish is designed to be in a different material, such as dry lining or decorative boarding. In recent years roof pitches have greatly reduced to make the roof as flat as possible, usually for visual reasons. Most profiled sheet is laid to a minimum pitch of around 4°. Standing seam roofs can go down to a 1° pitch, depending on the geometry of the roof.

When used as a substrate, profiled metal sheet can be cut to form complex geometries, typically supported by a steel frame to create a three-dimensional form. Profiled sheet with an overall depth of 50mm (2in) is used typically, but much deeper sections are used for spans above 3.5 metres (10ft) to around 6.0 metres (20ft), with a depth up to around 200mm (8in). The deep sections are also used in composite roof decks when filled with concrete. For a steel profile, sheets of 0.7mm thick are used for the outer skin; for aluminium a 0.9mm thick sheet is used. Steel is galvanised and coated, while aluminium is mill-finished or coated.

Profiled metal decks as substrates

Where profiled metal is used as a deck rather than as a roofing material, a lightweight build-up is usually applied, since the metal decking is chosen where a lightweight roof is required. The lightweight finishes used are typically an additional layer of profiled sheet, metal standing

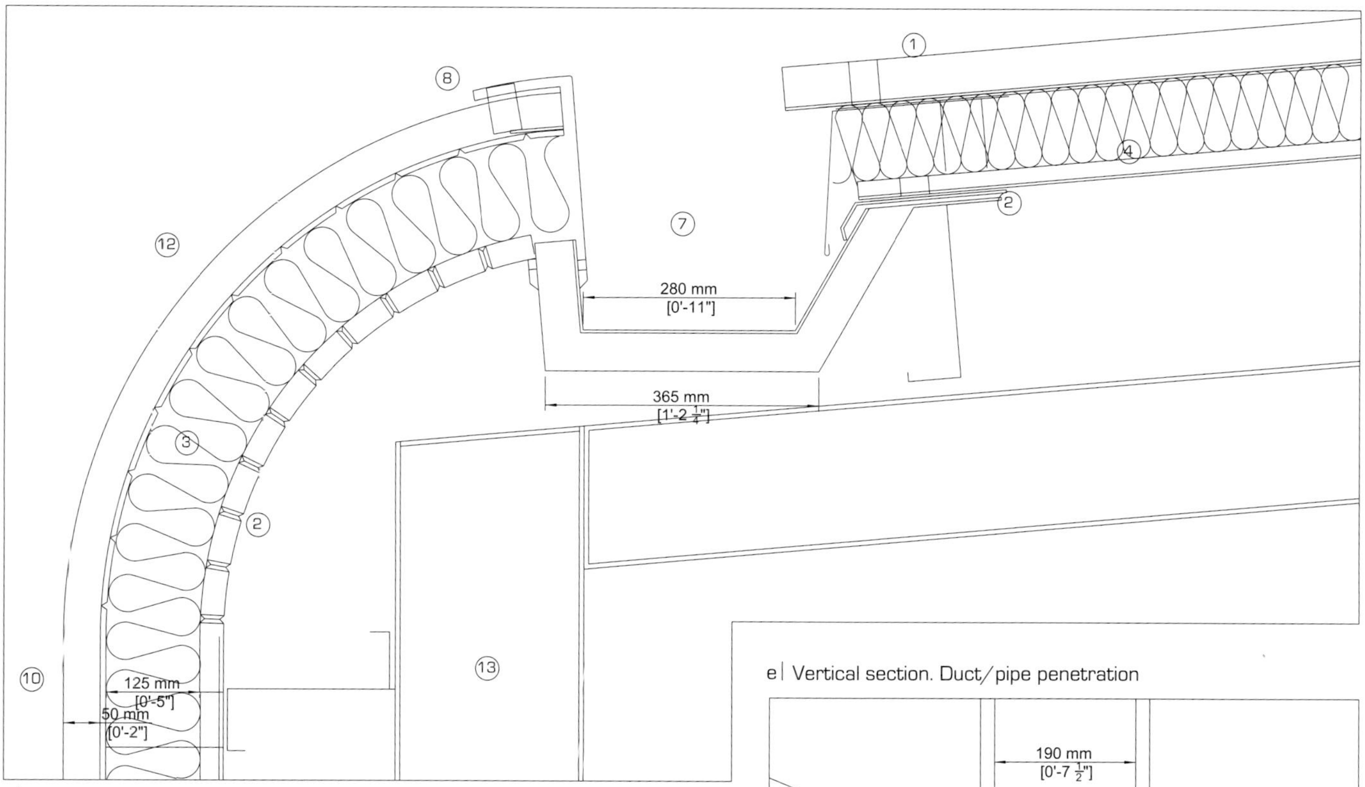

d | Vertical section. Curved eaves with concealed gutter

e | Vertical section. Duct/pipe penetration

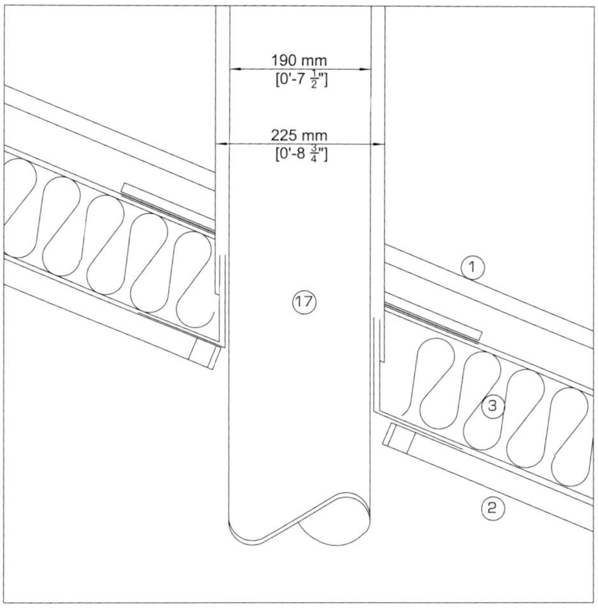

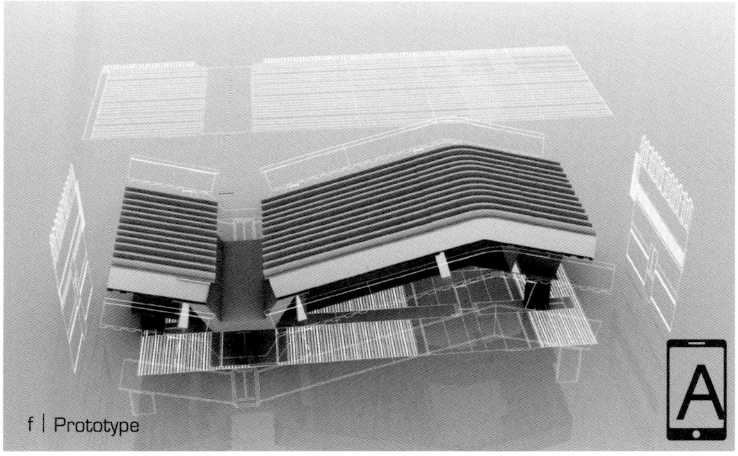

f | Prototype

seam (discussed in the previous section), membranes (mainly elastomers) and light-planted roofs. A typical build-up is of closed cell insulation set onto the profiled metal deck, rigid enough to span across the peaks of the profile without deflecting significantly when it is walked upon for maintenance access, which would otherwise stretch the joints in the membrane. A single-layer membrane is then set onto this insulation, usually an elastomeric membrane that can be left exposed to the effects of the sun without damage. Sometimes a thin layer of smooth pebbles is laid on top to keep the sun off the membrane and allow maintenance access without risk of puncturing the membrane. The closed cells of the material ensure that any water vapour trapped in the construction is not absorbed by the insulation which would otherwise cause its deterioration.

Profiled metal roof sheeting

When used as a finish material, called 'roof sheeting', profiled metal sheet provides a continuous weatherproof skin with the ability to be curved in one direction. A limitation of the material is that openings for rooflights, edges, and junctions with other materials are not easily integrated into the profile of the sheet (f). Even simple rectilinear openings have few standard profiles to close off the gap between the flat flashing and the gaps between the peak and trough of the profile. However, one of the main advantages of profiled metal sheet is its ability to be curved easily, where the supporting structure beneath requires only a few structural members to be curved, and most framing can be straight (c). Used as roof sheeting, the material is lapped on all four sides like metal standing seam roofs. The laps are made long enough to avoid capillary action through the joint. This simple jointing system provides large areas of reliable, weathertight roofing that can be installed quickly.

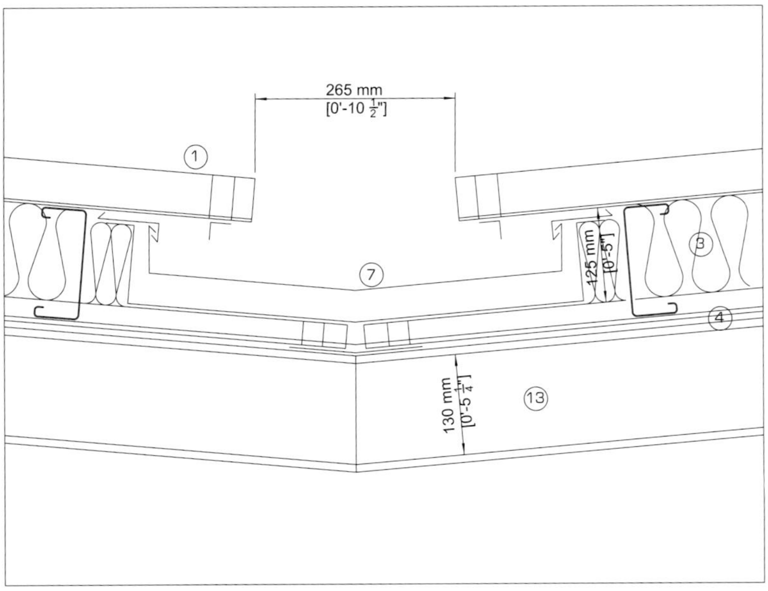

a | Vertical section. Folded metal valley gutter

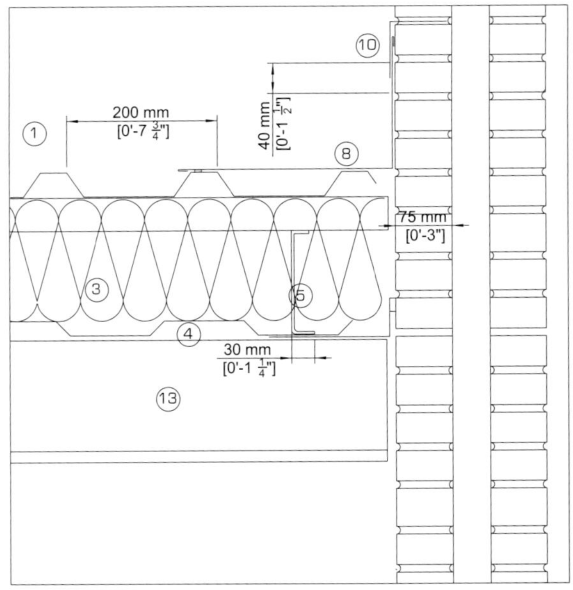

b | Vertical section. Abutment with masonry wall

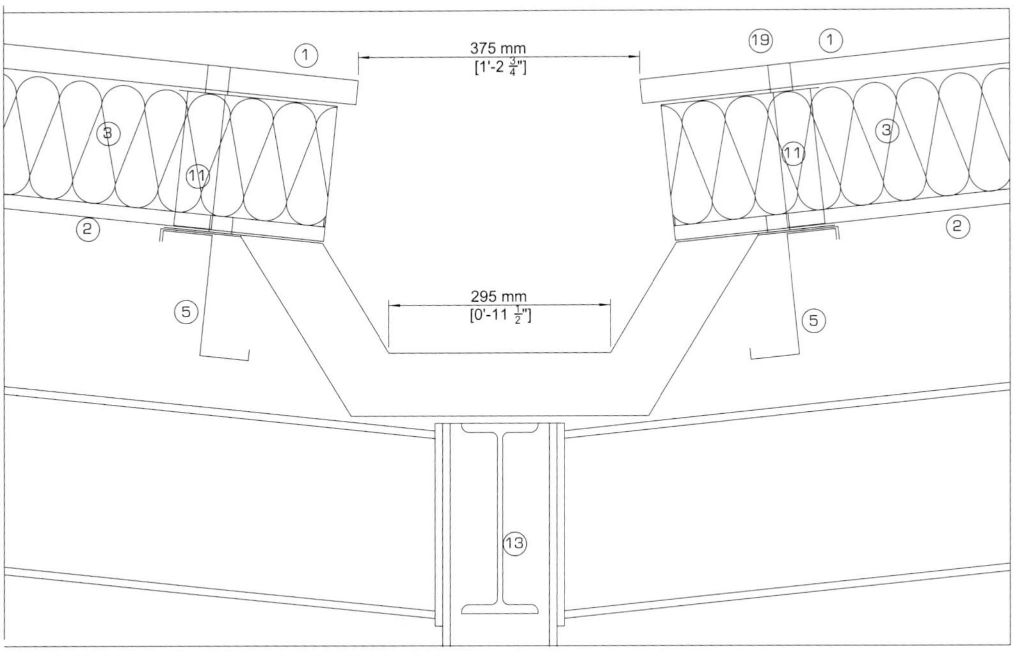

c | Vertical section. Prefabricated valley gutter

Details

1 | Outer profiled metal sheet
2 | Inner lining sheet
3 | Fibre quilt thermal insulation
4 | Vapour barrier
5 | Purlin or structural beam
6 | Profiled metal structural deck
7 | Folded metal gutter
8 | Folded metal drip
9 | Metal fascia panel
10 | External wall
11 | Outer sheet fixing bracket
12 | Curved eaves sheet
13 | Structural frame
14 | Ridge piece
15 | Metal flashing
16 | Rooflight
17 | Pipe or duct penetration
18 | Parapet flashing
19 | Vented filler piece

Sealed and ventilated methods

In common with standing seam roofing, profiled metal roofs can be used as a roof covering in either sealed or ventilated construction. Ventilation is used mainly where a timber supporting structure is used, with the timber being ventilated to avoid rot in the material from moist air in the void that would otherwise be trapped within the construction. This is discussed further in the section on timber pitched roofs. The following topics in this section deal with the use of profiled sheet as a sealed roof covering.

In sealed roofs, the thermal insulation usually fills the voids in between the inner and outer skins, but ventilators are often provided at the ridge and eaves to allow some breathing through the ribs of the profiled sheet (MCE_335, f). This helps to keep the insulation completely dry. Because the inner lining sheet presents a hard surface under the roof, perforated sheets are used to improve sound absorption. Sound is allowed to be absorbed partly by the insulation quilt. This helps to reduce reverberation, particularly in noisy internal environments. A vapour barrier is set between the thermal insulation and the thin layer of acoustic insulation beneath.

Twin skin construction

Profiled metal sheet roofs have the ability to conceal the supporting structure within the depth of the roof construction. This gives a smooth finished appearance to the inside face of the roof. An outer metal sheet is supported on metal roof purlins, and an inner lining tray which supports the thermal insulation is fixed to their underside. The purlins are usually Z-shaped galvanised steel types, typically 1.5mm thick, with nylon washers or sleeves between the purlins and the outer and inner sheets to provide a thermal break as well as a pad to seal the screw fixings on the outer roof sheet. In recent years the Z-shape section has developed into a wide range of section types. Sheets are fixed with self-

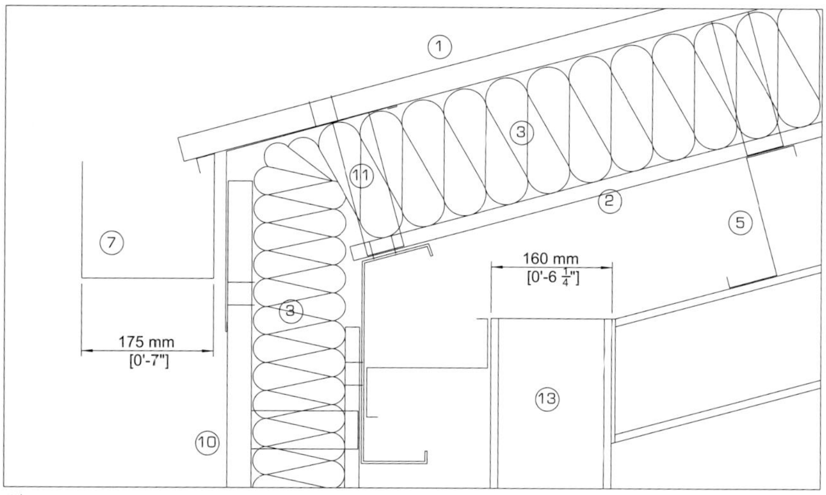

d | Vertical section. Eaves with exposed gutter

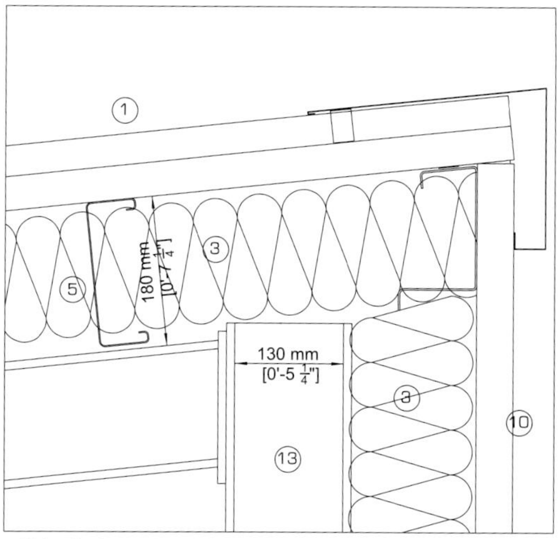

e | Vertical section. Monopitch ridge

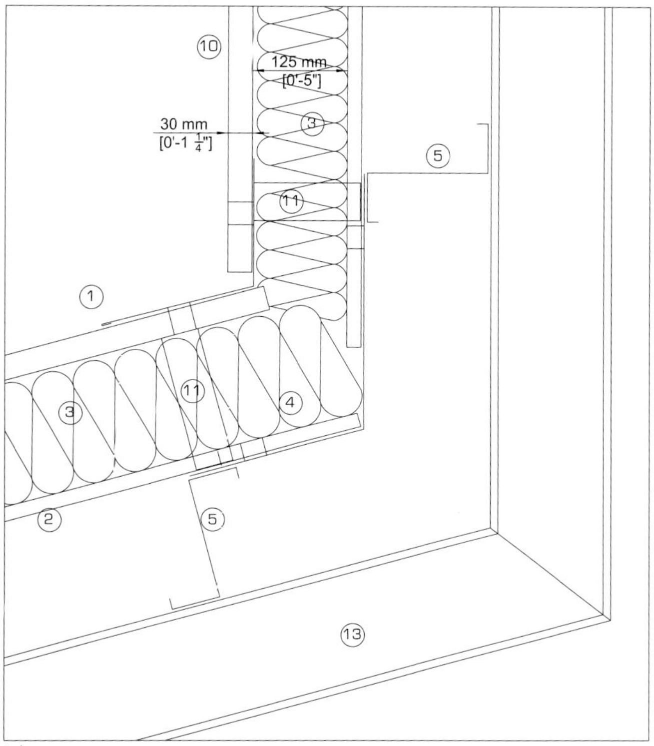

f | Vertical section. Abutment with vertical roof panel

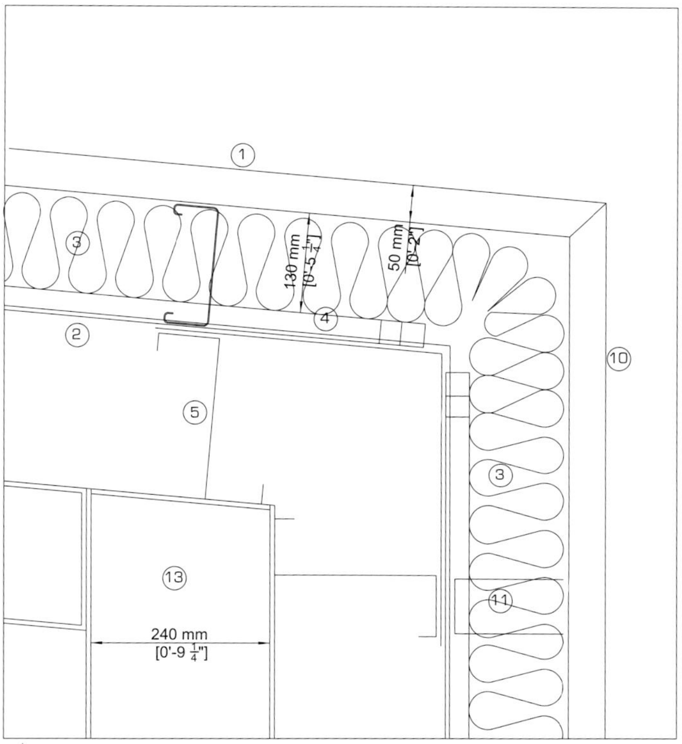

g | Vertical section. Folded eaves

tapping screws which, in addition to fixing the sheets to the supporting structure, are also required to be weathertight. A vapour barrier is provided on the warm (in winter) side of the insulation, between the liner tray and the thermal insulation.

This construction method contrasts with composite panels, where the outer skin, mineral fibre thermal insulation quilt and inner lining are combined into a single panel that is fixed onto a supporting structure which remains visible. The supporting roof structure, visible from below the roof, is either left exposed on its underside or is concealed with a layer of dry lining.

Laps between profiled sheets along their top and bottom edges are sealed with butyl sealant strip. Two strips are normally used, one at the end of the external lap and the other at the top end of the internal lap. The outer seal provides protection against capillary action of rainwater being drawn up into the lap between sheets, while the other provides a vapour barrier that avoids moisture, generated inside the building, from condensing in the joint. Self-tapping screws that hold the sheets in place clamp the two sealed surfaces together. Sheets are normally lapped 150mm over one another, while laps between sheets on their side edges are made with a single lap of profile, with a single seal of butyl tape set at the centre of the lap. In common with standing seam roofs, the thermal insulation quilt is usually 150-200mm thick in order to achieve a U-value of 0.25 W/m^{2}K.

Ridges

Ridges at the junction of a double-pitched roof use a folded metal strip to form a continuous ridge sheet. These sheets can be folded on a straight line or formed to a curved line. The void beneath is filled with thermal insulation in addition to that used beneath the profiled sheet. Closer strips are used to seal the gaps in the profiled sheet where it meets the ridge cover strip set on top of the profiled sheet.

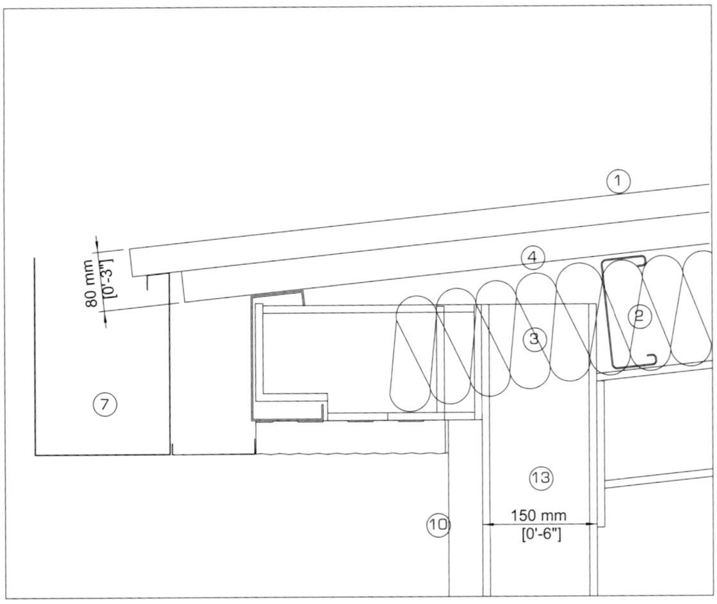

a | Vertical section. Eaves with semi concealed gutter

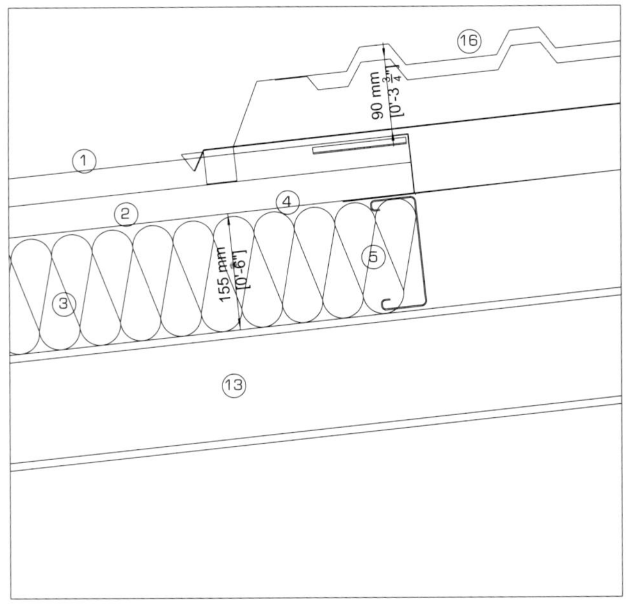

b | Vertical section. Rooflight formed as profiled sheet

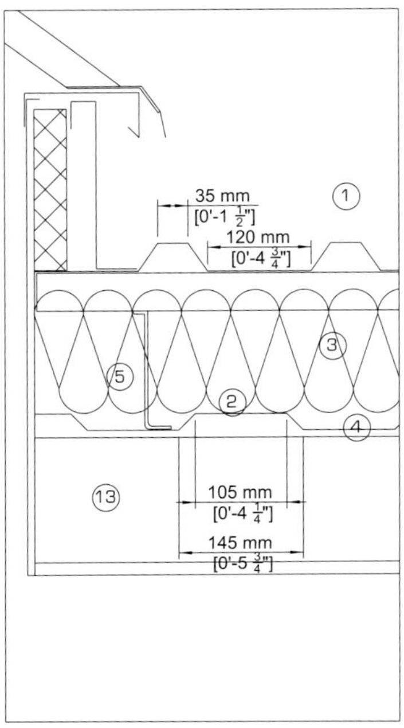

d | Vertical section. Rooflight formed with upstand

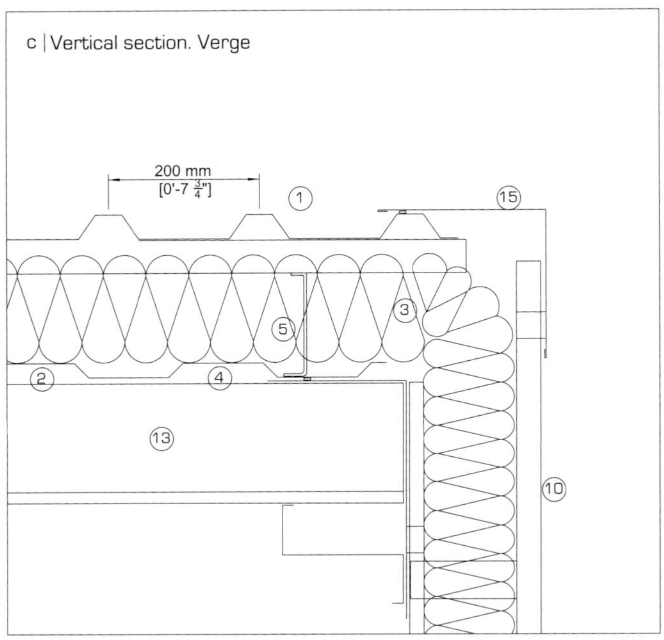

c | Vertical section. Verge

Details

1 | Outer profiled metal sheet
2 | Inner lining sheet
3 | Fibre quilt thermal insulation
4 | Vapour barrier
5 | Purlin or structural beam
6 | Profiled metal structural deck
7 | Folded metal gutter
8 | Folded metal drip
9 | Metal fascia panel
10 | External wall
11 | Outer sheet fixing bracket
12 | Curved eaves sheet
13 | Structural frame
14 | Ridge piece
15 | Metal flashing
16 | Rooflight
17 | Pipe or duct penetration
18 | Parapet flashing
19 | Vented filler piece

Where a pitched roof meets an abutment with an adjacent wall, the cover strip between wall and roof is folded up the wall and is fixed to it. A flashing projecting from the wall is lapped over the outside of the ridge strip in order to direct rainwater over it and down onto the roof (MCE_332, b).

Openings

In common with rooflights fixed into standing seam roofing, a gutter is required along the top edge of rooflights (b), along which water running down the roof is directed to the sides. The gutter can be formed in front of the metal sheet as in a parapet detail, or be concealed by setting the profiled sheet close to the rooflight and concealing the gutter, leaving only a 50mm gap between roof sheet and rooflight. The gutter can be concealed for small rooflights which carry only little rainwater and so do not need to be very big. The sides of rooflights parallel to the slope, and along the bottom edge, have flashings lapped down from the rooflights and onto the adjacent roof sheets.

Penetrations for small ducts and pipes through the roof are sealed with a flashing, usually welded to form a single upstand collar around the pipe and fixed to the roof deck (MCE_331, e). Also in common with standing seam roofs, a counter flashing is welded or bonded to the upper part of the pipe, above the roof, which laps down over the flashing to protect it from rainwater running down the pipe above the roof level.

Eaves and parapets

Parapets are formed by creating a gutter at the base of the profiled sheet, then continuing the line of the gutter up to a parapet coping with laps over the top (MCE_333, d). The gutter is formed from a single folded sheet to avoid the possibility of leaks, and is lapped under the vertical sheet that forms the seal between gutter and coping. Parapets in profiled metal sheet are often used where the external wall is formed in a different material, so that the parapet forms a visual break between them. However, many profiled metal roofs are used in conjunction with

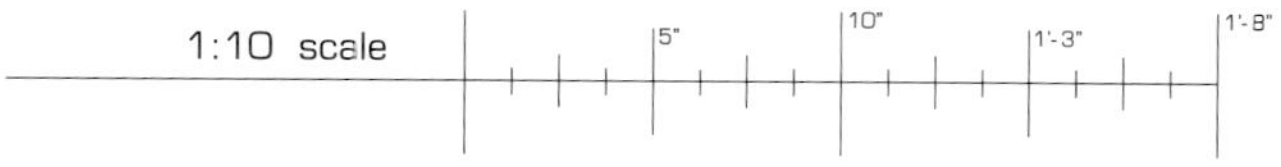

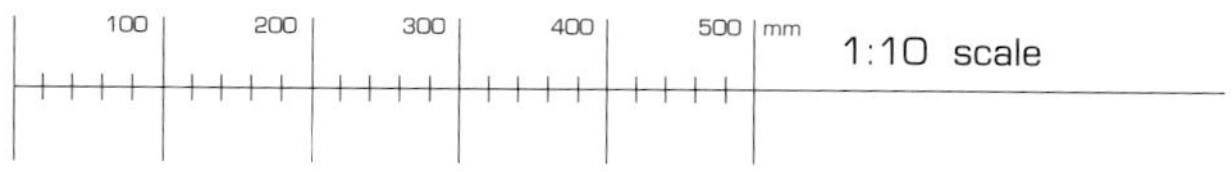

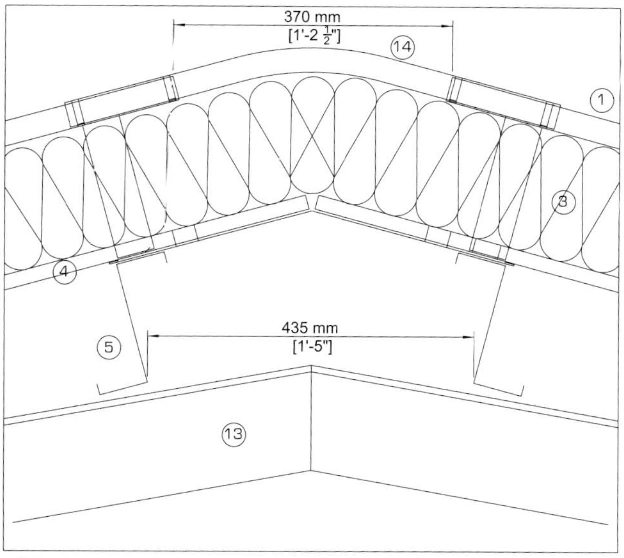

e | Vertical section. Ridge with flashing

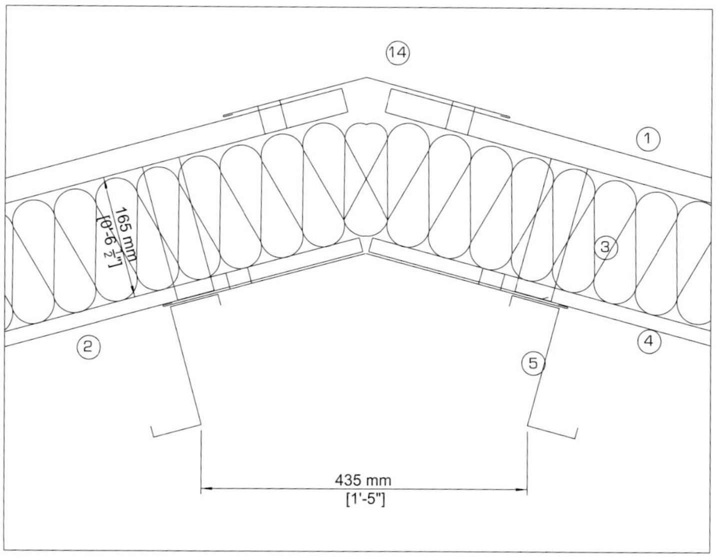

f | Vertical section. Ridge with flashing

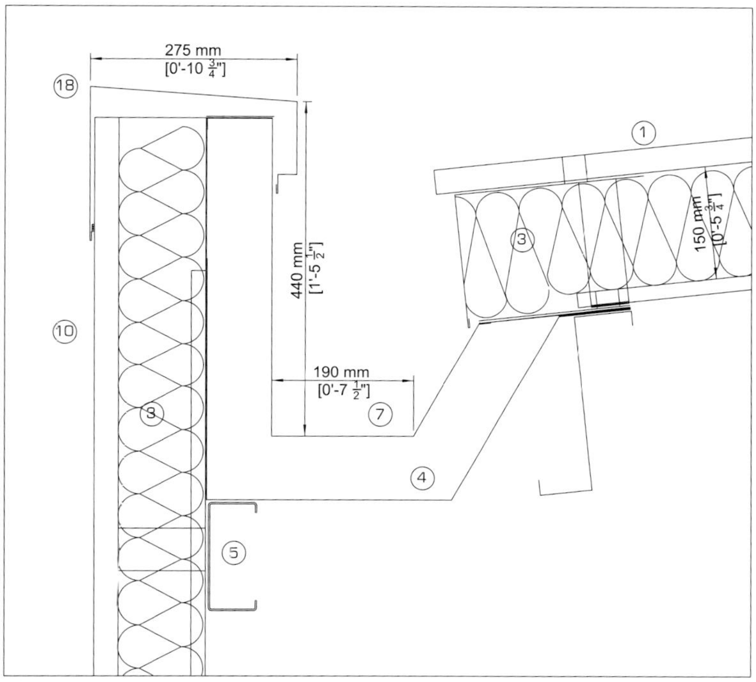

g | Vertical section. Parapet gutter

h | Prototype

walls in the same material using a concealed gutter that allows wall and roof to be continuous. Manufacturers provide standard curved pieces to different radii in profiled metal sheet that allow the same material to form a gently curved edge (MCE_331, f) or, alternatively, a sharply curved termination to the roof which can be lapped into the profiled sheet forming the wall beneath. Eaves pieces with sharp edges are also manufactured as part of proprietary systems. These are formed to different angles by welding two profiled sheets together to form a continuous smooth fold in the roof. Curved pieces are usually made by crimping the material along the vertical edges of the profiled sheet, to give a characteristic appearance, but these pieces are increasingly being made with a continuous smooth appearance.

Ridges and valleys

These folds in metal roofs are formed using the same methods discussed in the previous section on standing seam roofs. However, whereas standing seams can be cut down to form a flat ridge without a projecting ridge piece, this is not possible in profiled sheet, and instead a folded ridge piece is fixed to the upper surface of the profiled sheet. The gaps between the ridges and troughs of the profiled sheet are filled with a proprietary metal filler piece (f), usually forming part of the manufacturer's system. Ventilated roofs do not require this filler piece, with the resulting gap between the folded ridge sheet and the profiled sheet being usually sufficient to provide ventilation into the construction.

Valleys are also formed in a similar way to that discussed in standing seam roofs, with a ventilation gap provided in the gap between the gutter and the underside of the roofing sheet (MCE_332, a).

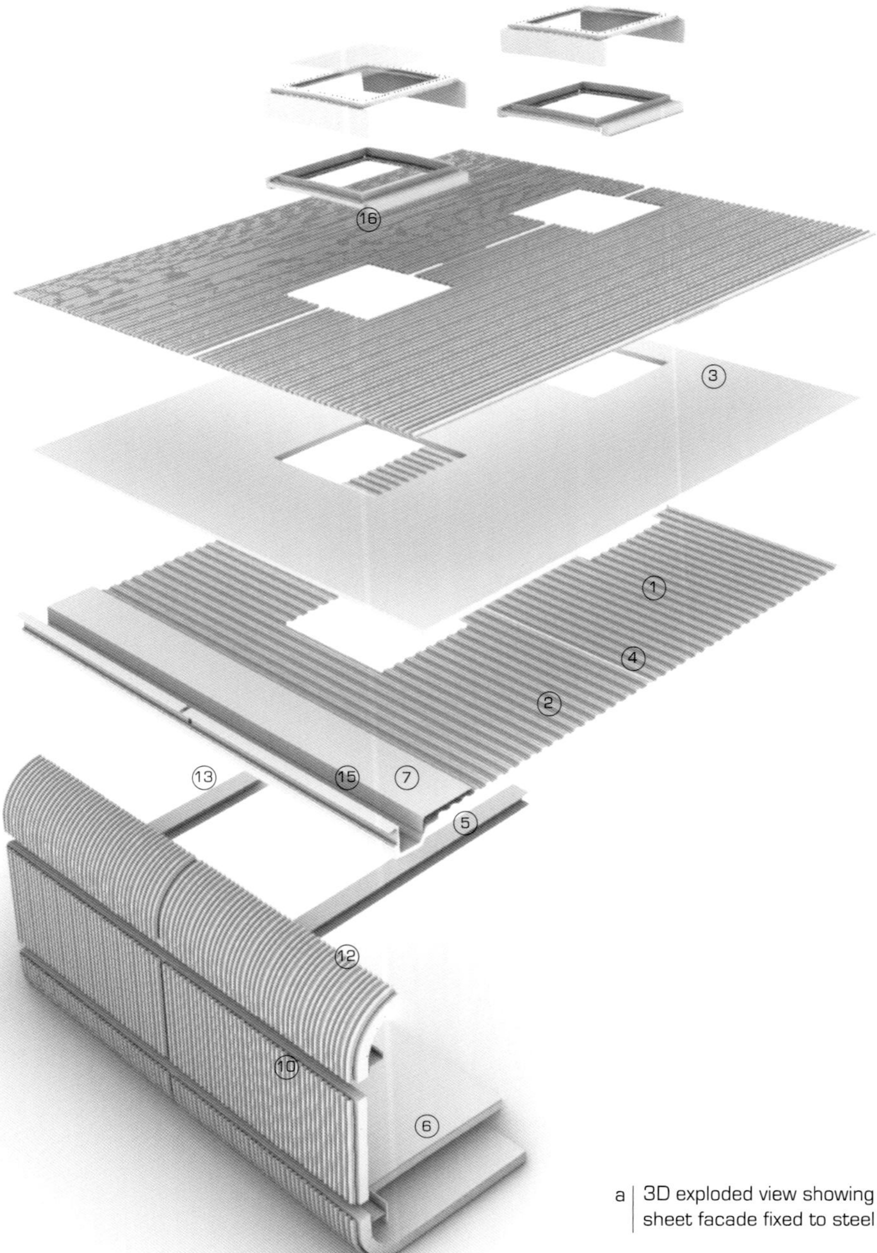

a | 3D exploded view showing profiled metal sheet facade fixed to steel frame

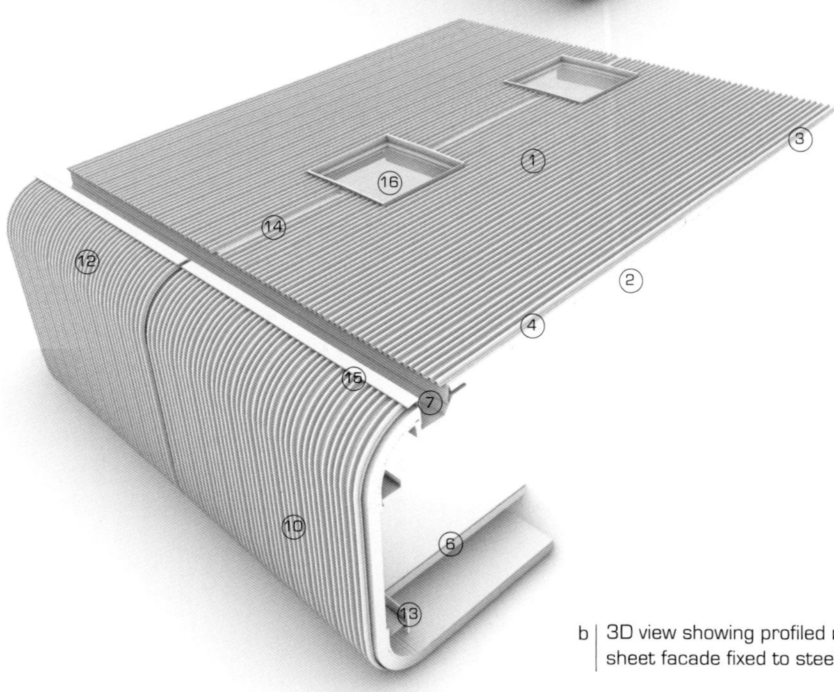

b | 3D view showing profiled metal sheet facade fixed to steel frame

Details

1 | Outer profiled metal sheet
2 | Inner lining sheet
3 | Fibre quilt thermal insulation
4 | Vapour barrier
5 | Purlin or structural beam
6 | Profiled metal structural deck
7 | Folded metal gutter
8 | Folded metal drip
9 | Metal fascia panel
10 | External wall

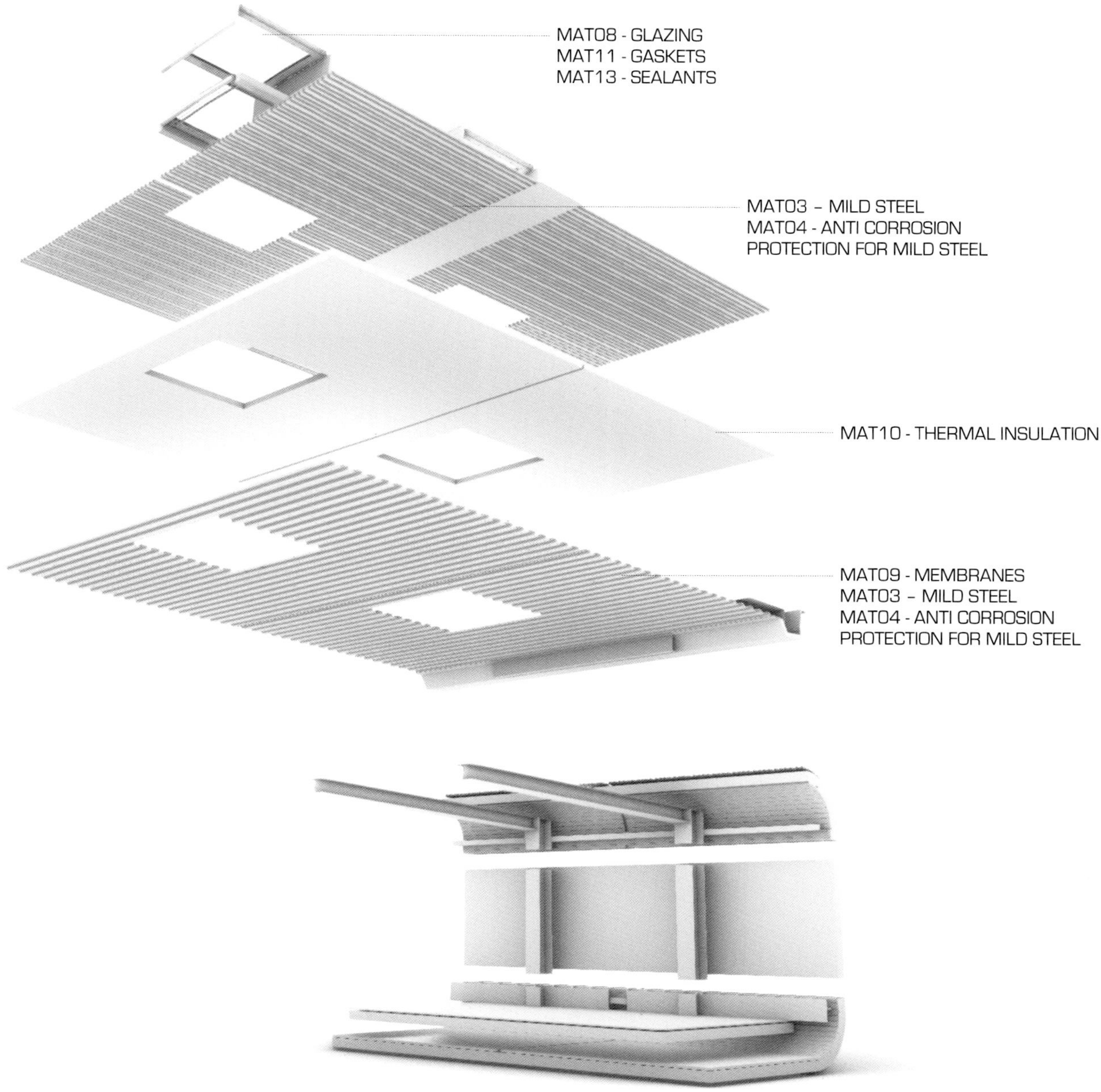

c | 3D exploded view of profiled metal sheet facade components with material specification references

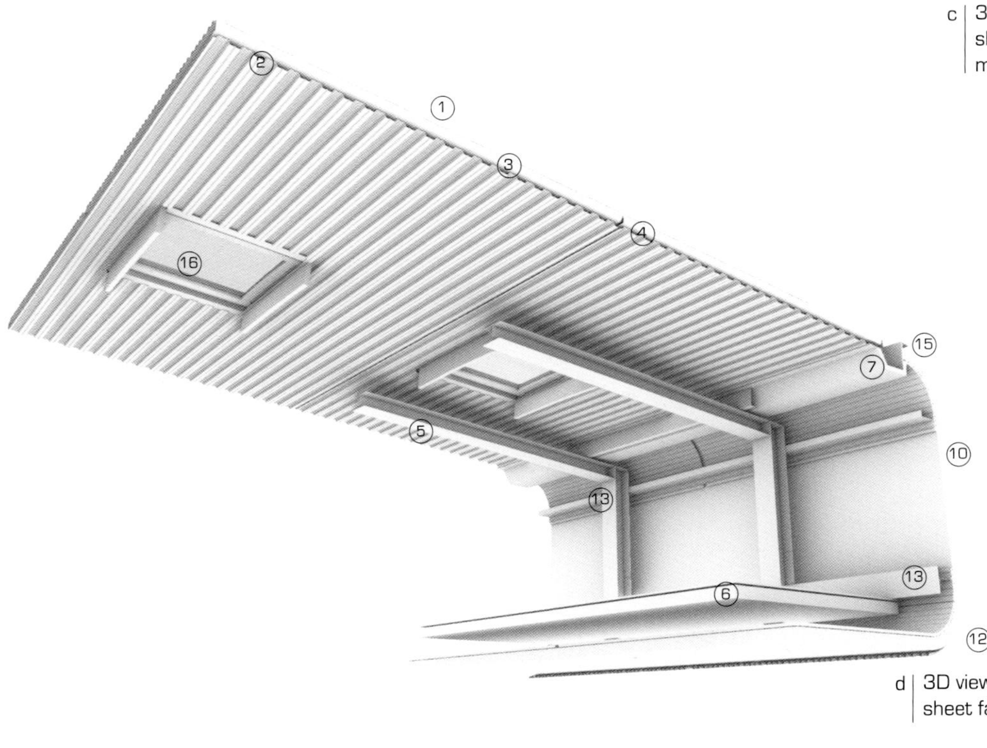

11 | Outer sheet fixing bracket
12 | Curved eaves sheet
13 | Structural frame
14 | Ridge piece
15 | Metal flashing
16 | Rooflight
17 | Pipe or duct penetration
18 | Parapet flashing
19 | Vented filler piece

d | 3D view showing profiled metal sheet facade fixed to steel frame

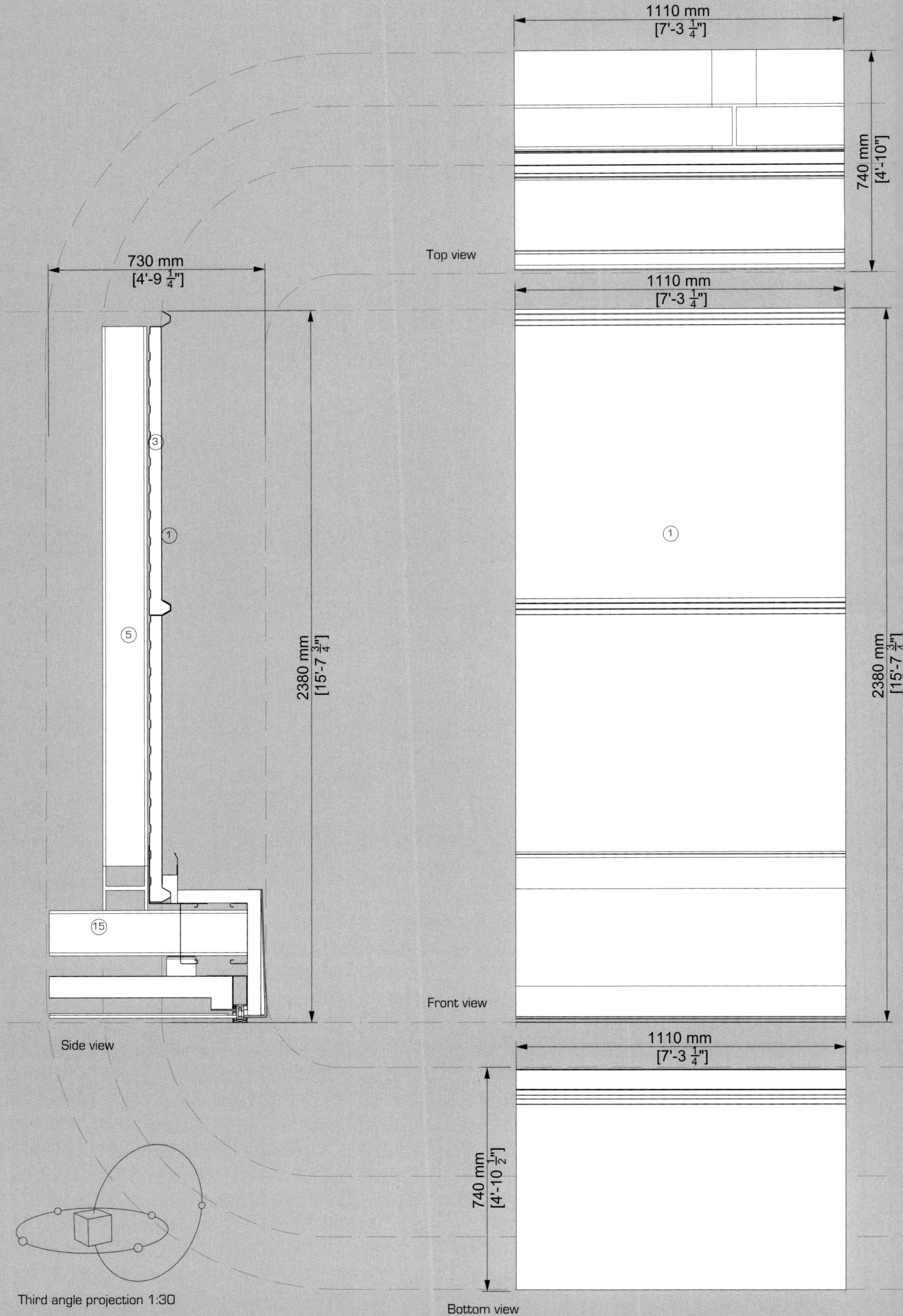
1110 mm
[7'-3 1/4"]
740 mm
[4'-10"]
Top view
730 mm
[4'-9 1/4"]
1110 mm
[7'-3 1/4"]
3
1
5
15
2380 mm
[15'-7 3/4"]
1
2380 mm
[15'-7 3/4"]
Front view
Side view
1110 mm
[7'-3 1/4"]
740 mm
[4'-10 1/2"]
Third angle projection 1:30
Bottom view

3D view of the assembly

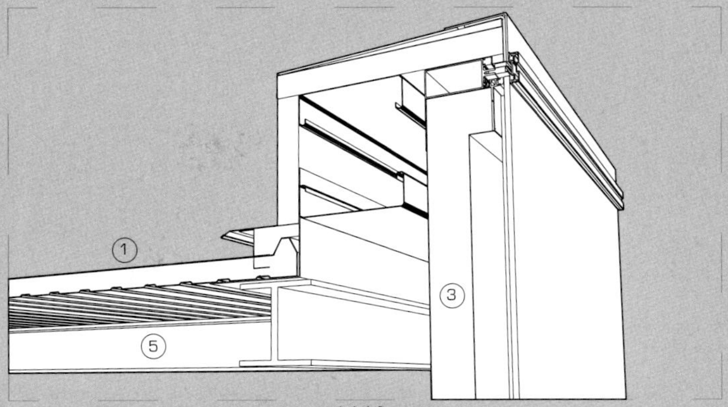

Details

1 | Metal rainscreen panel
2 | Single-layer membrane
3 | Composite panel
4 | Folded metal coping
5 | Purlin or structural beam
6 | Secondary purlin
7 | Folded metal gutter
8 | Folded metal drip

Side view

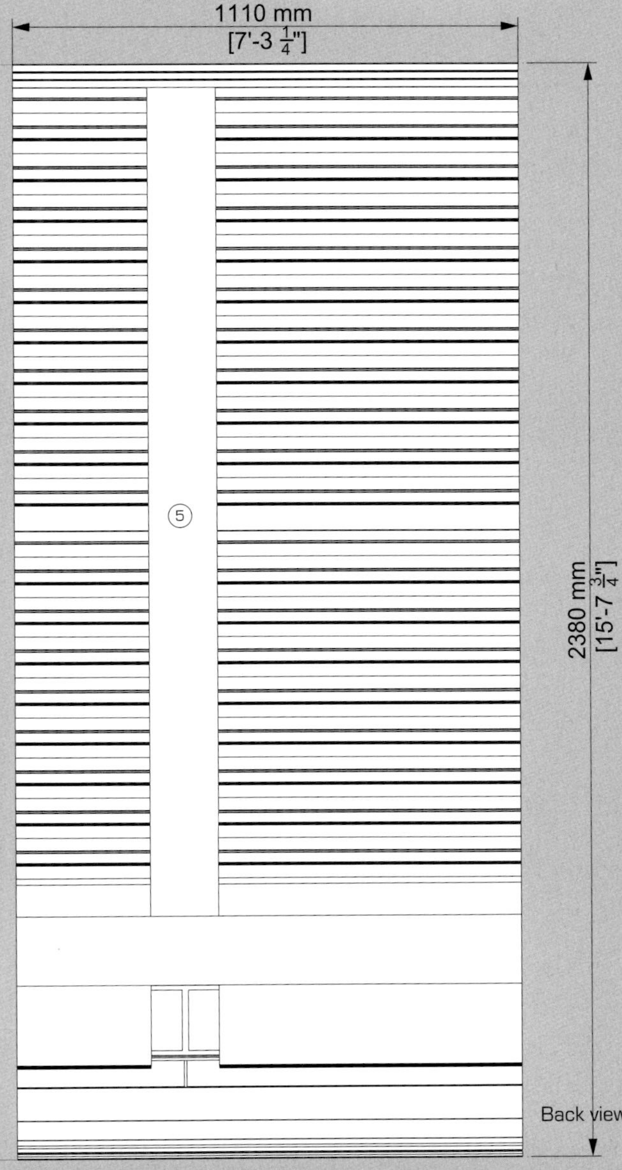

Back view

9 | Metal fascia panel
10 | External wall
11 | Outer sheet fixing bracket
12 | Panel 1
13 | Panel 2
14 | Ridge piece
15 | Structural frame
16 | Pipe or duct penetration

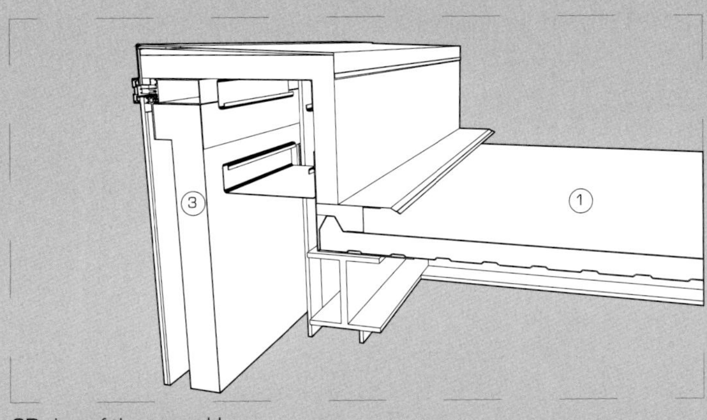

3D view of the assembly

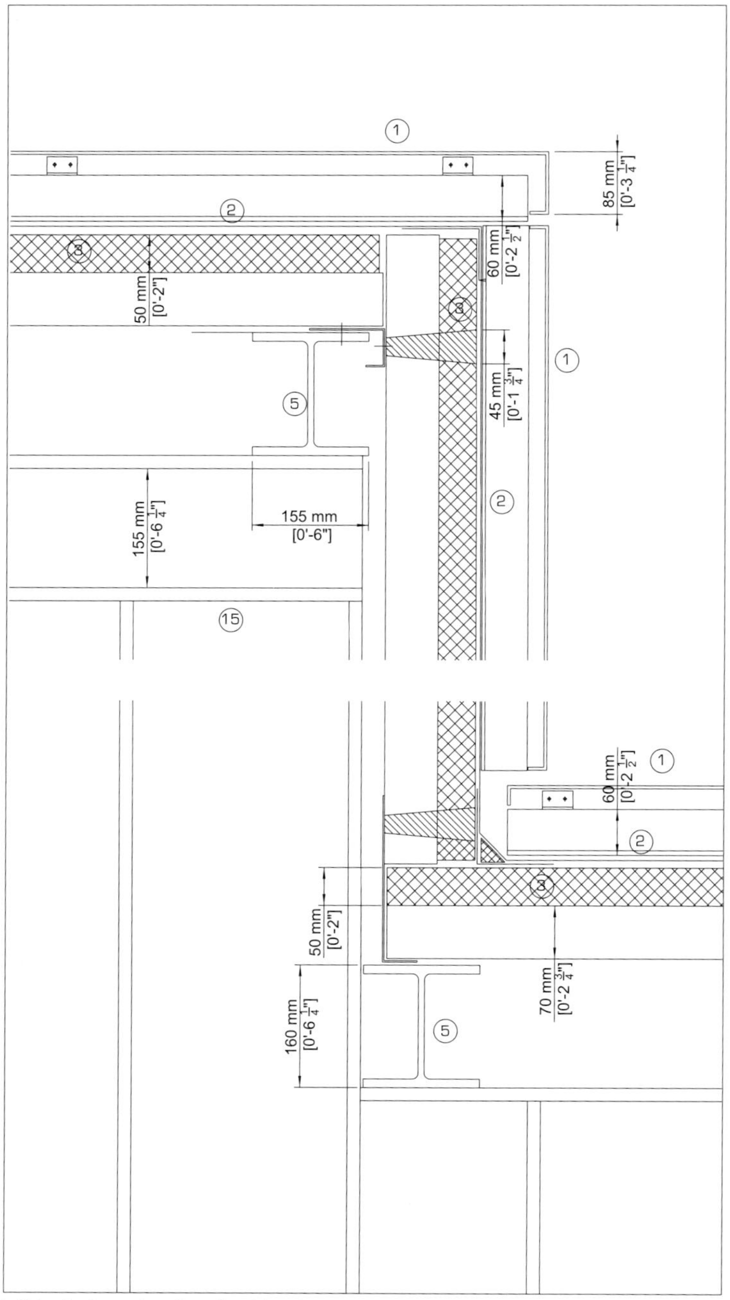

b | Vertical section. Panels with rainscreen covering at step in roof

Details

1 | Metal rainscreen panel
2 | Single-layer membrane
3 | Composite panel
4 | Folded metal coping
5 | Purlin or structural beam
6 | Secondary purlin
7 | Folded metal gutter
8 | Folded metal drip
9 | Metal fascia panel
10 | External wall
11 | Outer sheet fixing bracket
12 | Panel 1
13 | Panel 2
14 | Ridge piece
15 | Structural frame
16 | Pipe or duct penetration

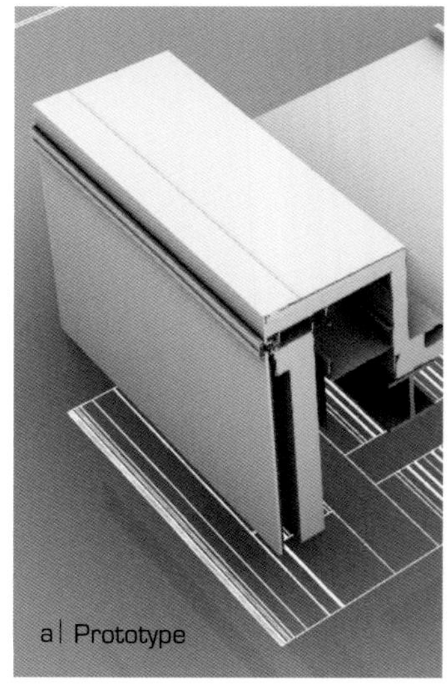

a | Prototype

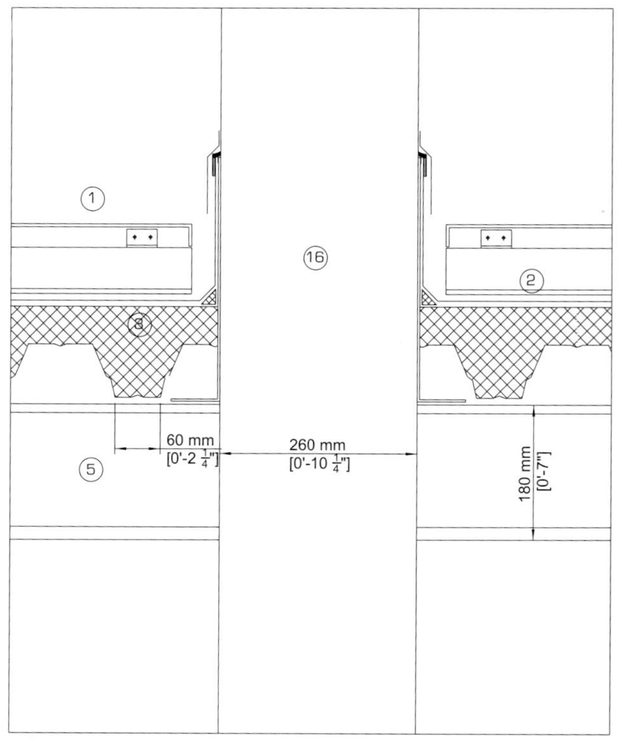

c | Vertical section. Pipe / duct penetration

There are two types of composite panel used for roofs: twin wall and single wall panels. The first is a development of profiled metal sheet, where outer sheet, thermal insulation and inner sheet are combined into a single panel (MCE_345, e). These are used in pitched roofs and have an appearance very similar to that of profiled metal roofs. Their main advantage over profiled metal roofs is the speed of erection on site, but they are usually a little more expensive than an equivalent profiled metal sheet roof. The second composite roof panel type, single wall panels, consists of profiled metal sheet bonded to a layer of foam-type thermal insulation. The insulation is laid face-up to receive a separate waterproofing layer, typically a single-layer membrane (a). The membrane is then typically finished in rainscreen panels or smooth pebbles, depending on the geometry and required appearance of the roof.

A development in composite roof panel construction which has been slow to appear commercially is the truly interlocking panel with integral gutter, where the principles of composite wall panels would be applied to roof panels, making for use in (nominally) flat roofs. These panels would have the smooth face and edges used in wall panels to provide a construction that would be both economic and very elegant. The joints between panels would form gutters that would create a connected grid of drainage channels that could form a crisp joint line between panels. This development of composite panel design has yet to be commercially available in a technically reliable system.

Single wall composite panels

This panel type has a single sheet of profiled metal on the lower load-bearing face of the panel which is bonded to a foam-based insulation that fills all the voids in the profiled sheet, providing a smooth, flat upper surface. The depth of the thermal insulation is determined by the U-value

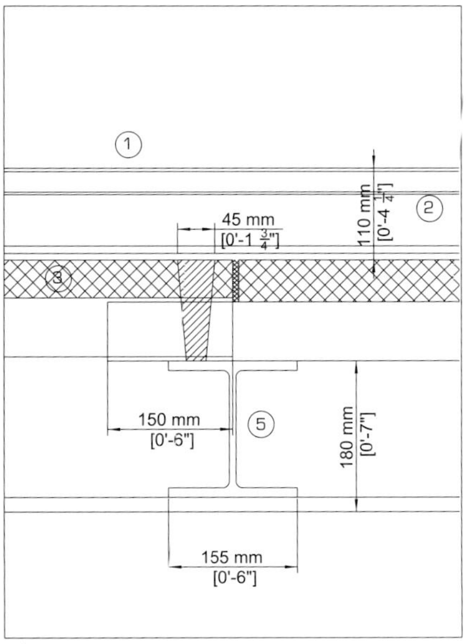

d | Vertical section. Typical panel to panel junction

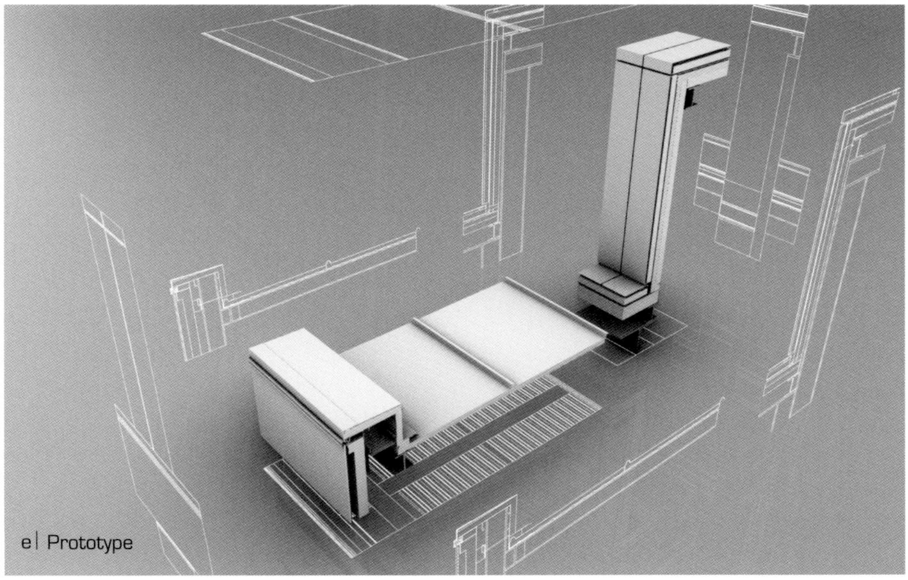

e | Prototype

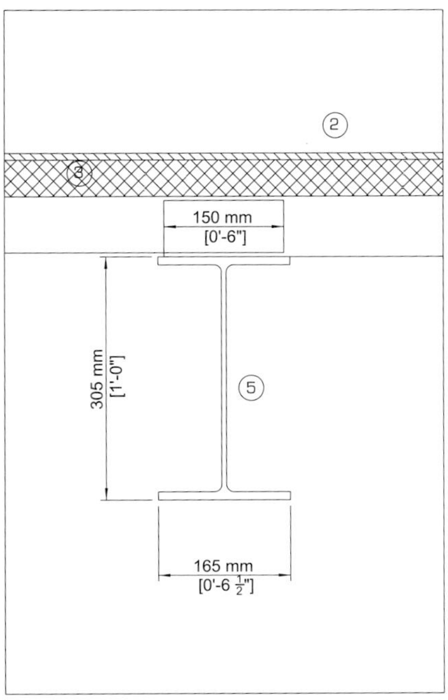

f | Vertical section. Typical panel to panel junction

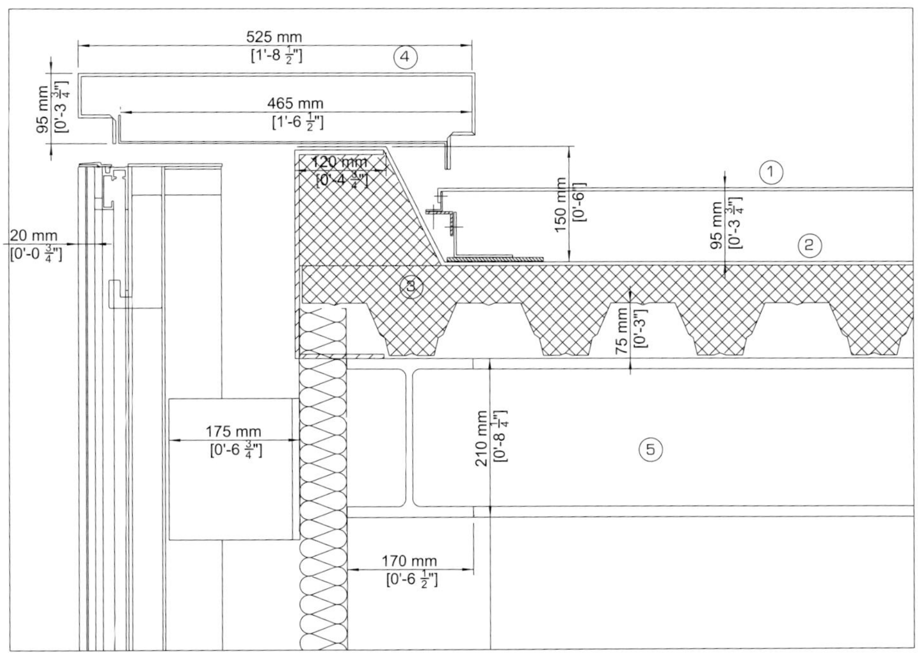

g | Vertical section. Parapet edge

required, and manufacturers are often flexible in this regard. The upper face of the panel is waterproofed with an independent membrane, typically an elastomeric type that requires no upstands or special joints between sheets. Panels are usually set butted up to one another, with the gap between panels filled with foam-based thermal insulation. A separating layer is usually set between the waterproof membrane and the insulated panel to allow movement to occur freely in both the membrane and the composite panel substrate. The sheets of waterproof membrane are bonded or torch-welded together by lapping one sheet over the other, or by using bonding strips in the same material that form part of the proprietary system. The membrane is often protected with a lightweight covering of smooth pebbles that can be walked upon for maintenance access without puncturing the surface. Metal rainscreen panels are also used to protect the membrane from the effects of direct sunlight. At ridges and folds in the roof geometry, panels are joined in the same way, with the gaps between panels filled typically with a foam insulation applied by injection on site. The membrane sheets are usually joined at the fold in the roof, and a strip of the same material is bonded along the joint to provide a weathertight seal.

At parapet gutters, an upstand is formed in the same insulation material as that used in the composite panels, the upstand being bonded to the composite panel beneath. The outer edge of the upstand is sealed and stiffened with a metal strip fixed to the underside of the composite roof panel or the supporting structure beneath (a). An additional length of membrane is then bonded to the top of the upstand and is mechanically fixed or bonded to the top of the adjacent external wall, typically formed in composite wall panels or glazed curtain walling. This parapet flashing is then protected by a folded metal coping set onto it.

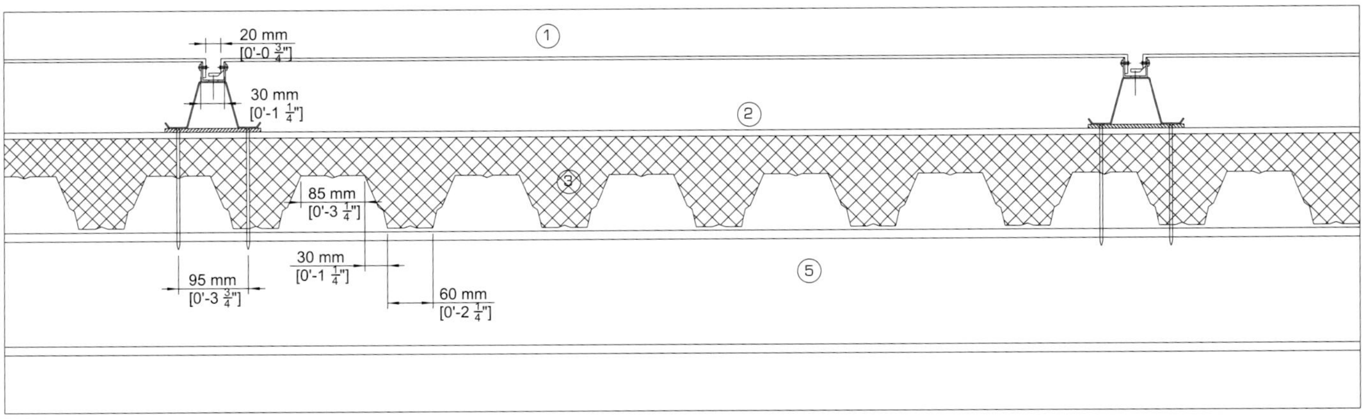

a | Vertical section. Panel to panel junction at fold

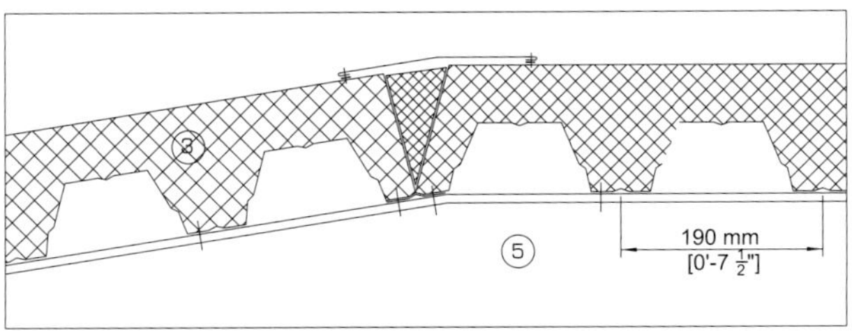

b | Vertical section. Panels with rainscreen covering

Details

1 Metal rainscreen panel
2 Single-layer membrane
3 Composite panel
4 Folded metal coping
5 Purlin or structural beam
6 Secondary purlin
7 Folded metal gutter
8 Folded metal drip
9 Metal fascia panel
10 External wall
11 Outer sheet fixing bracket
12 Panel 1
13 Panel 2
14 Ridge piece
15 Structural frame
16 Pipe or duct penetration

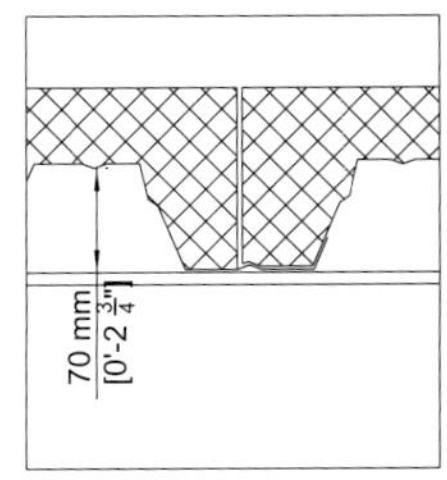

c | Vertical section. Panel-to-panel junction

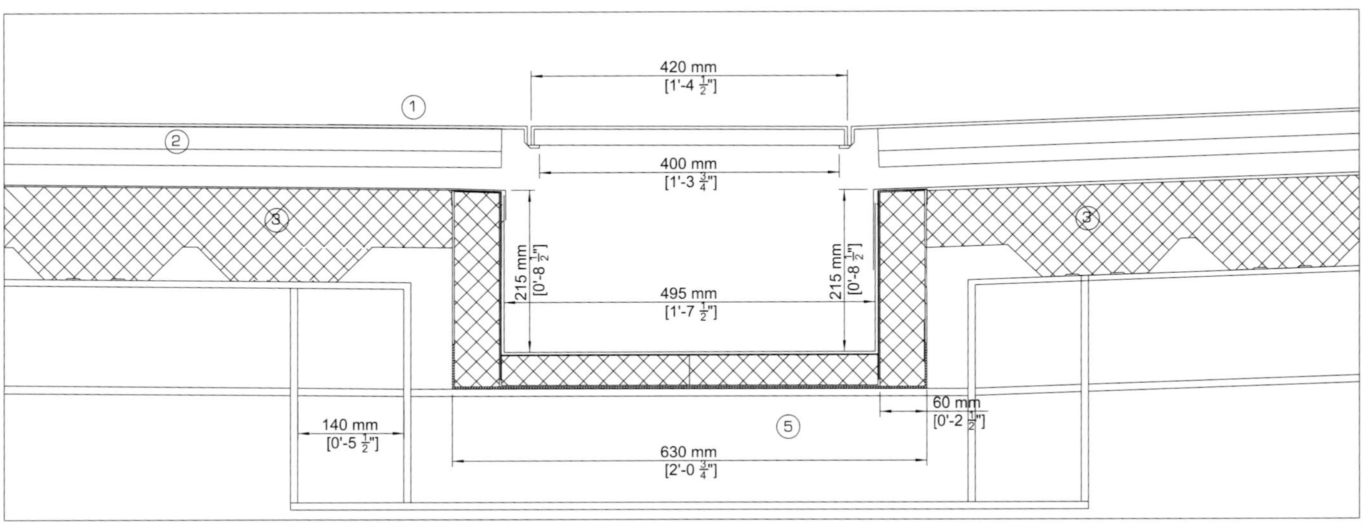

d | Vertical section. Folded metal gutter set between panels

Single-faced composite roof panels are well adapted to complex roof forms, where a lightweight, well-insulated roof can be provided without the need for complex junctions that would be needed with twin wall panels. The use of a separate waterproof membrane on a substrate that is easy to form into a smooth continuous surface allows junctions to be formed easily, particularly around roof penetrations such as duct openings and pipe penetrations, where waterproof membranes are well suited due to the ease with which they can be cut, formed and sealed on site. Gutters can be formed by using the rapid site assembly method of setting lengths of composite panel together, while a membrane bonded to the upper surface gives a watertight finish (MCE_344, c). Composite panels forming a valley gutter or parapet gutter are fixed rigidly to reduce the amount of structural movement that would otherwise damage the membrane, which is typically bonded to the base and edges of the gutter in order to closely follow its shape.

Twin wall panels

This panel type, which combines the separate components of profiled metal sheet, has two joint types: a double seam with a cap on top in the manner of standing seam roofing, or a single lap of profile in the manner of profiled metal decking. With the first method, panels have raised edges on their long sides running down the slope. The raised edges are butted together and sealed with butyl tape. A metal capping is fixed over this joint to provide a weathertight seal which sheds water onto the panels either side of the joint. This method gives a distinct visual appearance of wider joints. The second method has an uninsulated rib of the outer sheet projecting from the panel on one long side, which laps over the adjacent panel. This gives a continuous ribbed appearance to the roof that is visually no different on its outside face to profiled metal roof cladding. Both methods have lapped joints on their short edges, where an uninsulated edge projects down to form a lap joint very similar to that used in profiled metal roofs. These horizontal joints are also sealed with

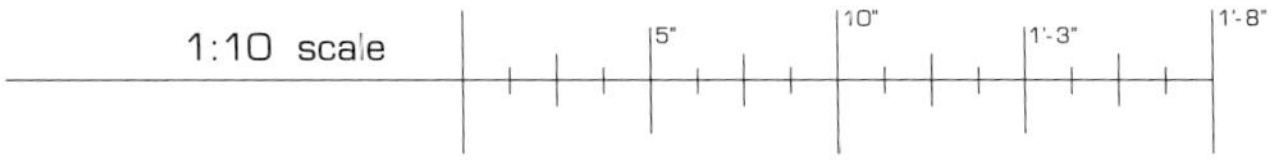

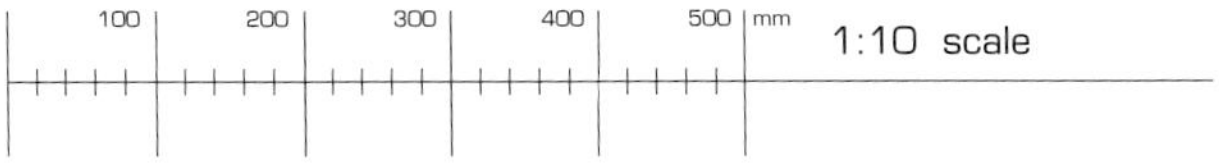

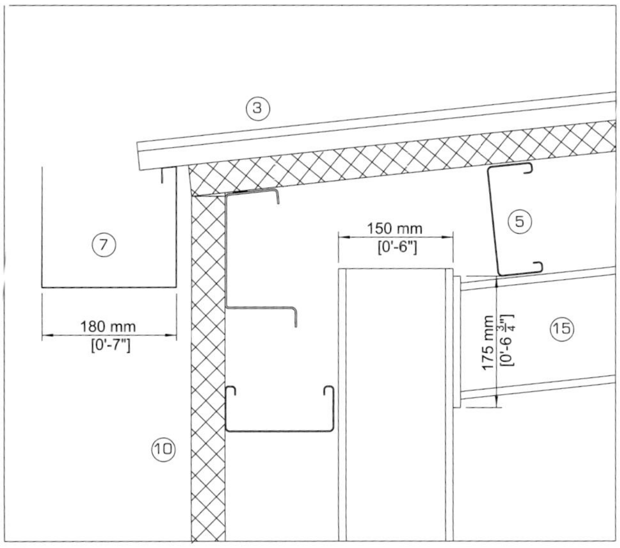

e | Vertical section. Eaves with exposed gutter

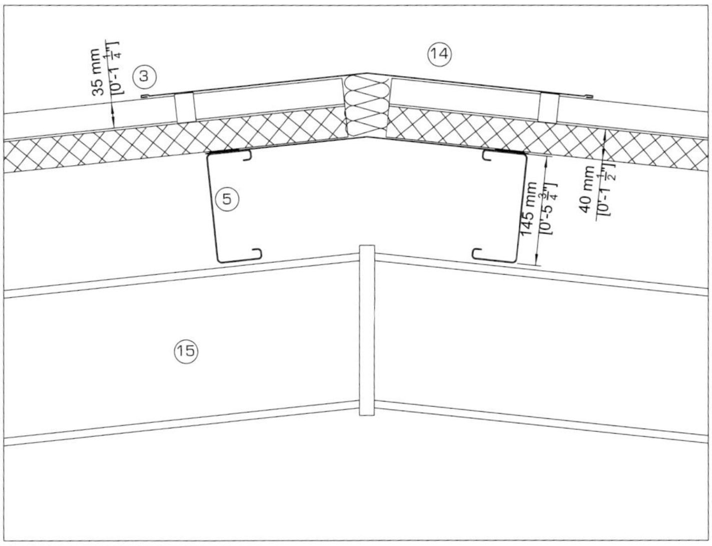

f | Vertical section. Ridge with flashing

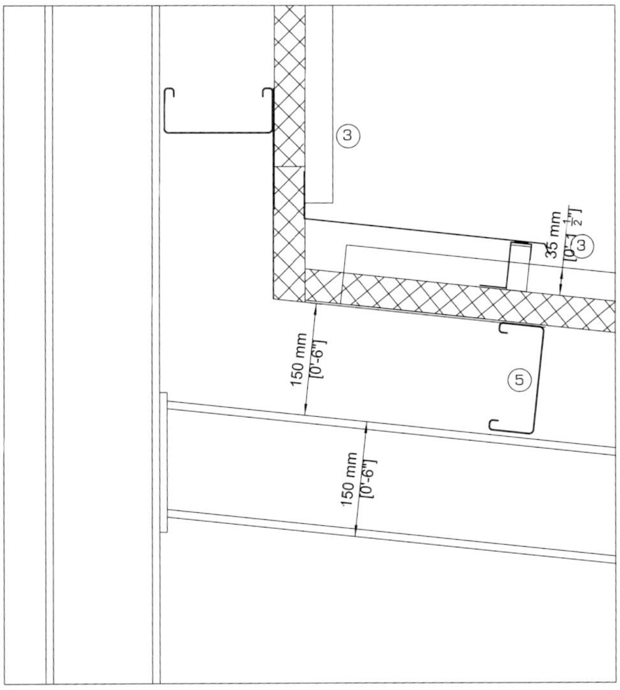

g | Vertical section. Abutment with vertical roof panel

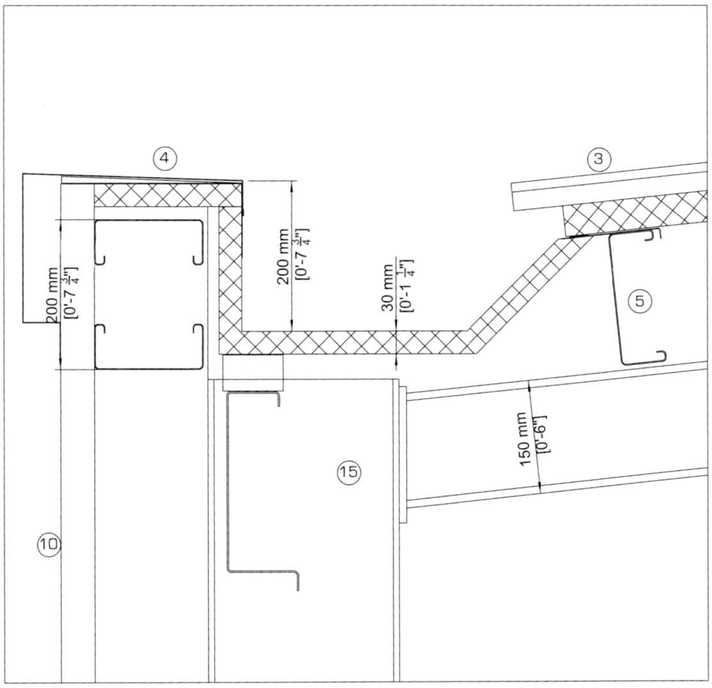

h | Vertical section. Parapet gutter

butyl tape to avoid capillary action from rainwater outside, and to prevent the passage of water vapour into the joint from inside the building.

Ridges

Ridges for twin wall composite panels are formed by fixing a metal flashing over the junction between the panels. The flashing is profiled to match the profile of the panels onto which it sits (MCE_341, e). Manufacturers often make these profiles as part of their proprietary system. Alternatively, a flat ridge flashing is used which sits on top of the profiled upper sheet. The gaps are then closed with a profiled filler piece as is used in profiled metal roof construction. The angle between the meeting panels is closed by adjusting the angle of the fold of the ridge flashing on site, or by folding a flat flashing over the joint to give a smooth curve across the ridge line. The ridge flashing has visible fixings which are difficult to conceal, so that precise positioning is essential to the overall visual appearance of the ridge. The gap between the panels is filled with thermal insulation on site, with either mineral fibre quilt or, more frequently, with the same foam-based insulation used to manufacture the panels (MCE_341, e). Insulation is injected into the gap to provide a U-value to match that of the adjacent panels.

The inner face of the panels forming the ridge is sealed with a folded metal sheet, typically fixed to adjacent roof purlins and sealed against the inner face of the composite panel to provide a continuous vapour barrier. This inner trim is made either flat or profiled to suit the composite panels used.

Verges

Verges between composite panels and the adjacent wall construction are formed with a folded metal closer. Where composite panels form the wall construction, a folded metal strip is sealed to the top of the last 'peak' on the roof panel and is then sealed against the wall panel (MCE_341, e). Alternatively, a Z-section closer piece is fixed and sealed to profiled sheet and the flashing is fixed to the hip of the closer piece.

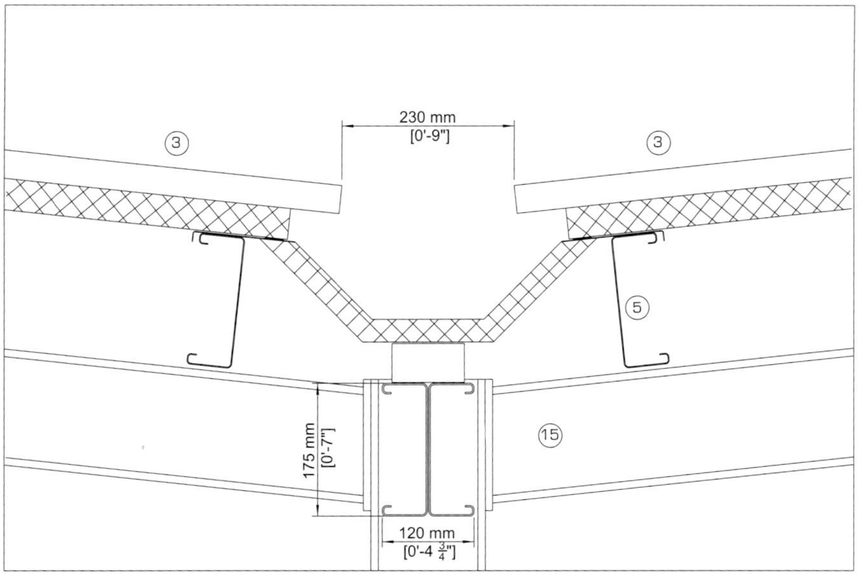

a | Vertical section. Prefabricated valley gutter

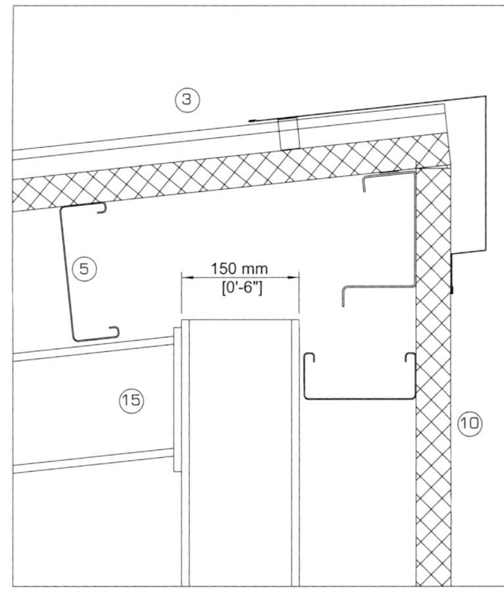

b | Vertical section. Monopitch ridge

Details

1	Metal rainscreen panel	7	Folded metal gutter	13	Panel 2
2	Single-layer membrane	8	Folded metal drip	14	Ridge piece
3	Composite panel	9	Metal fascia panel	15	Structural frame
4	Folded metal coping	10	External wall	16	Pipe or duct penetration
5	Purlin or structural beam	11	Outer sheet fixing bracket		
6	Secondary purlin	12	Panel 1		

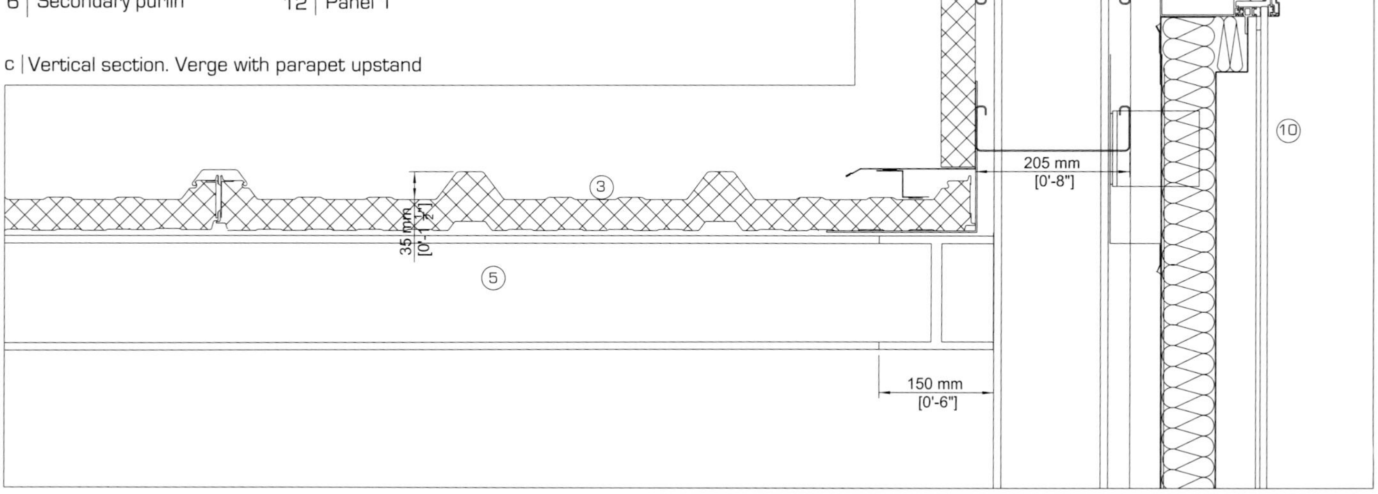

c | Vertical section. Verge with parapet upstand

Where glazed curtain walling is used, the metal flashing is glazed into the top transom of the glazed wall. As with ridges, the gap between the roof panel and the wall construction is filled with thermal insulation, and a folded closer strip is applied to the internal face of the joint to provide a vapour barrier between the insulation and the inside of the building.

If the verge overhangs the external wall instead of meeting it directly, then the void created by the overhang can be either ventilated or sealed. If sealed, then the overhanging verge follows the same detailing principles. If the eaves void is ventilated, then the edge of the verge is sealed to the edge of the roof, while the wall beneath is sealed to the underside of the composite metal roof in order to provide a continuous weather seal without any break in the thermal insulation.

Verges that terminate in a parapet use a flashing that is set behind the vertical panel, or onto it to avoid a thermal bridge, and is folded to be sealed onto the composite roof panel in the same way as for the eaves detail. The inner face of the joint is sealed with a folded metal closer strip to provide a vapour barrier.

Eaves

These are formed in a similar way to profiled metal panels, by lapping the edge roof panel over a gutter. The gutter is closed against the underside of the composite panel either by folding it outwards and sealing it or by folding the top edge inwards, up the underside of the panel (MCE_343, g and h). The gutter is supported by either a bracket beneath the gutter that is fixed back to the primary structure, or by a metal support arm set onto the top of the composite panels 'peaks' and cantilevering out to support the top of the gutter. An essential advantage of the metal support arm on the roof is that it does not require any penetration through the wall construction to the supporting structure, but can be fixed through to the roof panels which do not usually require

1:10 scale

1:10 scale

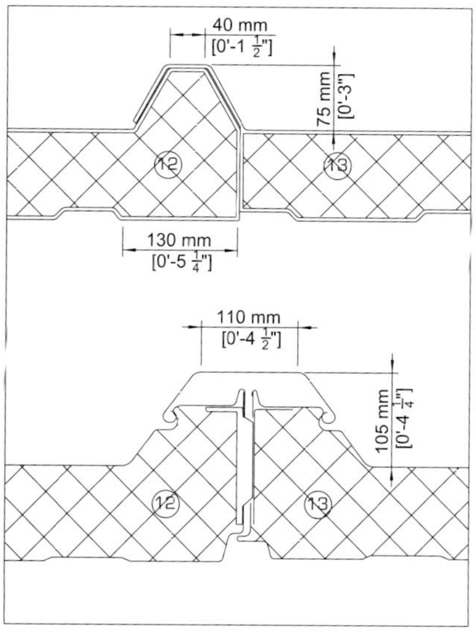

d | Vertical section. Panel to panel junctions

e | Vertical section. Verge

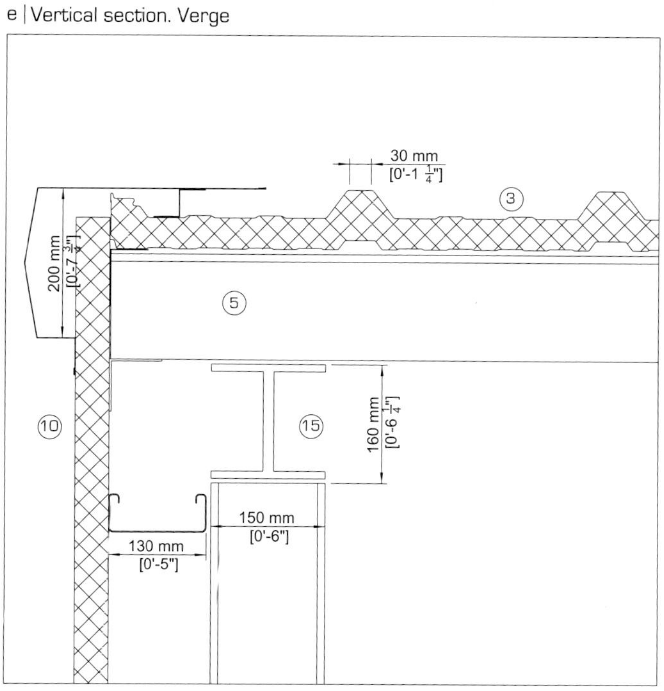

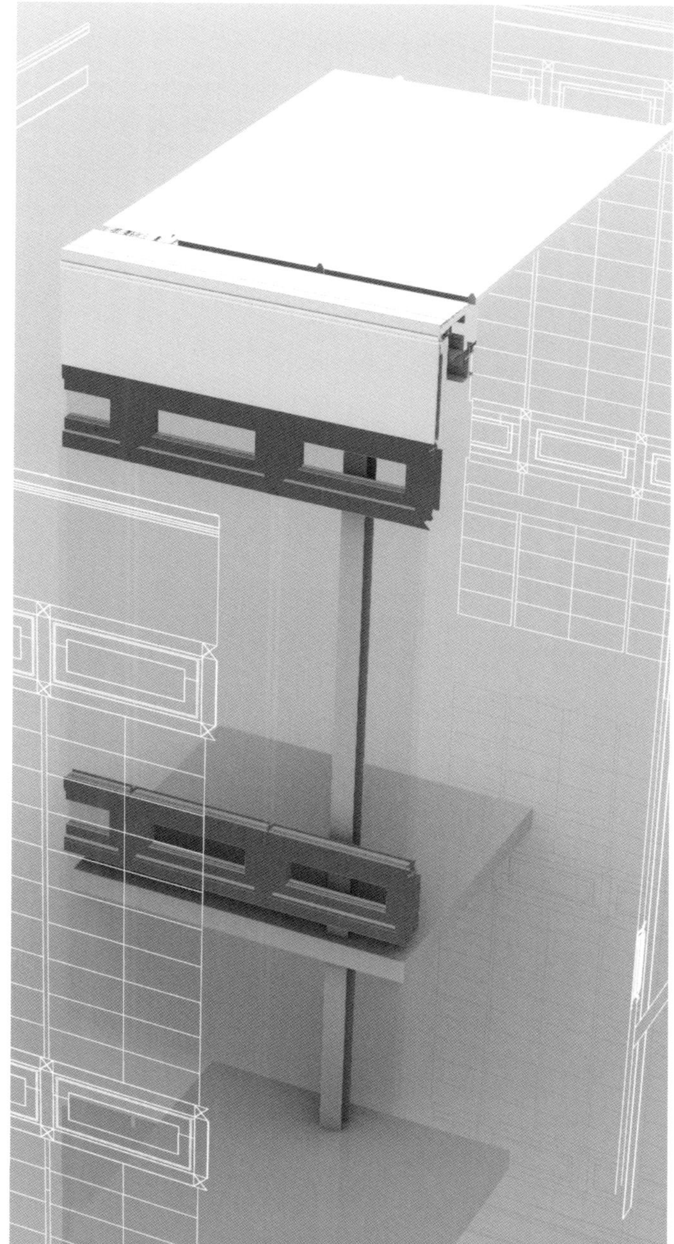

f | Typical bay

a fixing to pass all the way through the wall panel, thus reducing the risk of any rainwater penetration through the building envelope.

Parapets and valley gutters

Unlike eaves gutters, parapet gutters are thermally insulated, since they form part of the external envelope (MCE_343, g and h). The gutter is usually prefabricated to form part of the overall composite panel system. Since an advantage of composite panel construction is the increased speed of construction on site over other metal roofing methods, this advantage would be lost if the gutters took much longer to fix than the panels themselves. The gutter shape, in cross section, is formed to provide continuity in the thermal insulation from roof panel through to the adjacent parapet wall. Seals are provided to avoid any water from penetrating the seals and passing into the building, which might occur in the event that the rainwater outlets become blocked and the entire gutter becomes filled with water up to the level of the outer (top) face of the composite roof panels.

Valley gutters are formed in a similar way, with the seal between roof panel and prefabricated gutter being critical to avoid water penetration in the event of the valley gutter being filled to capacity in the event of a blocked rainwater outlet.

The tops of parapets are closed with a pressed metal coping that is folded down over the face of the external wall and the inside face of the parapet to provide a complete weathertight seal. The top of the coping is usually inclined towards the inside face of the wall (into the gutter) to avoid dust that settles on horizontal surfaces from being washed down the face of the external wall during rain. The coping is usually made from a minimum 0.7mm thick steel sheet or 3mm thick aluminium sheet. The void immediately beneath the coping is faced with thermal insulation to avoid a thermal bridge occurring through the coping into the air space behind.

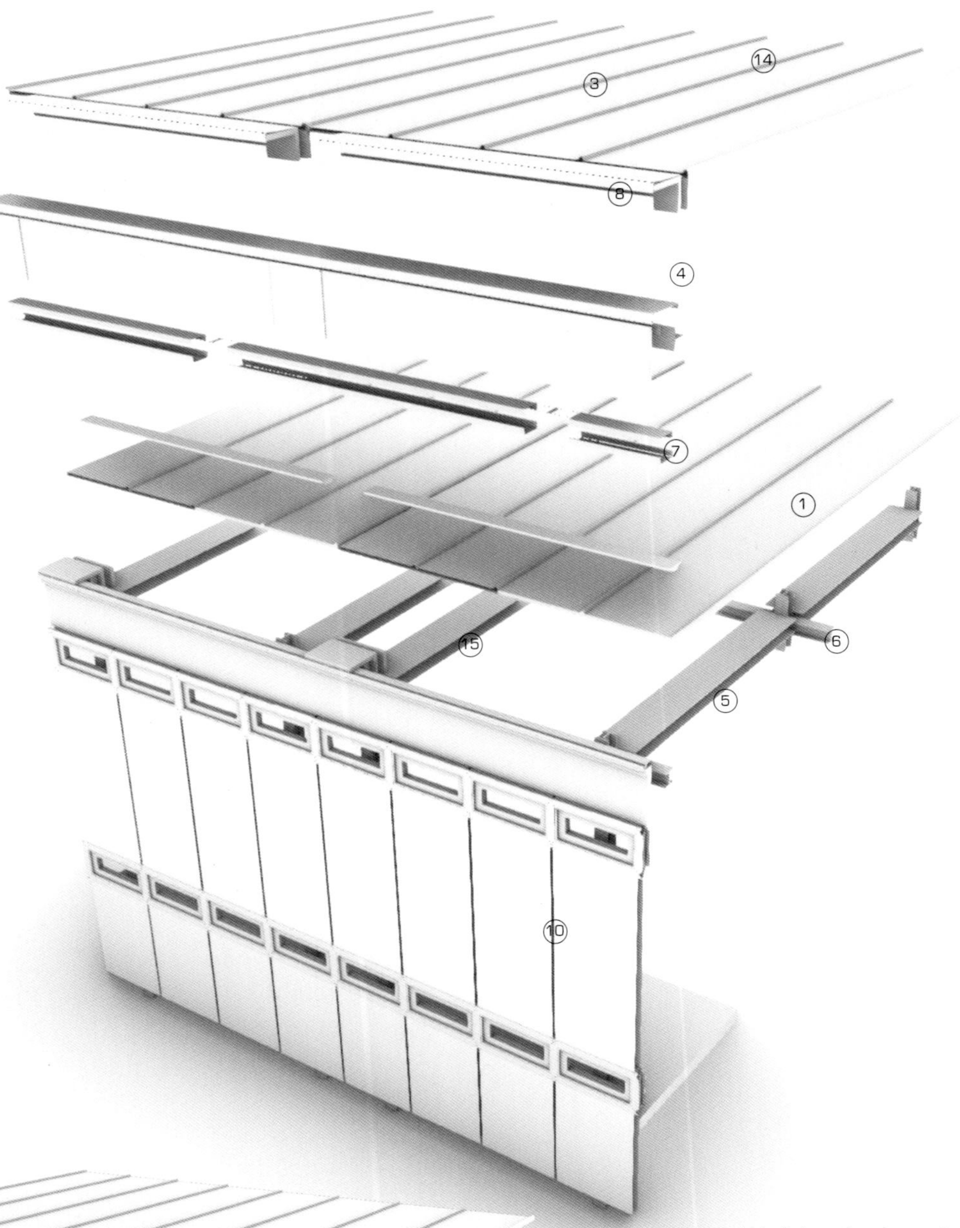

a | 3D exploded view showing metal composite panel facade fixed to steel frame

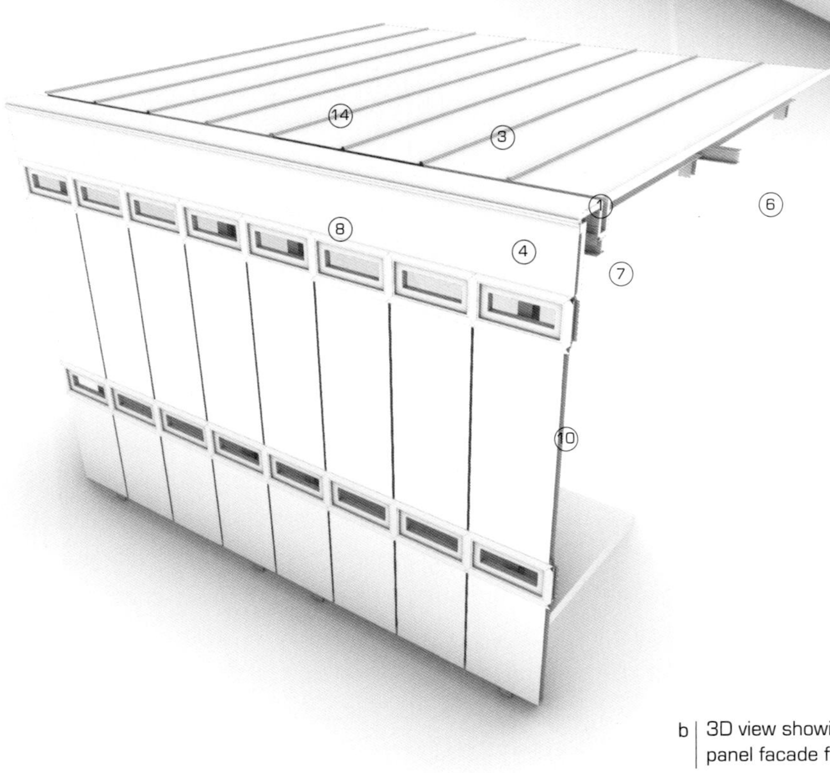

Details

1 | Metal rainscreen panel
2 | Single-layer membrane
3 | Composite panel
4 | Folded metal coping
5 | Purlin or structural beam
6 | Secondary purlin
7 | Folded metal gutter
8 | Folded metal drip

b | 3D view showing metal composite panel facade fixed to steel frame

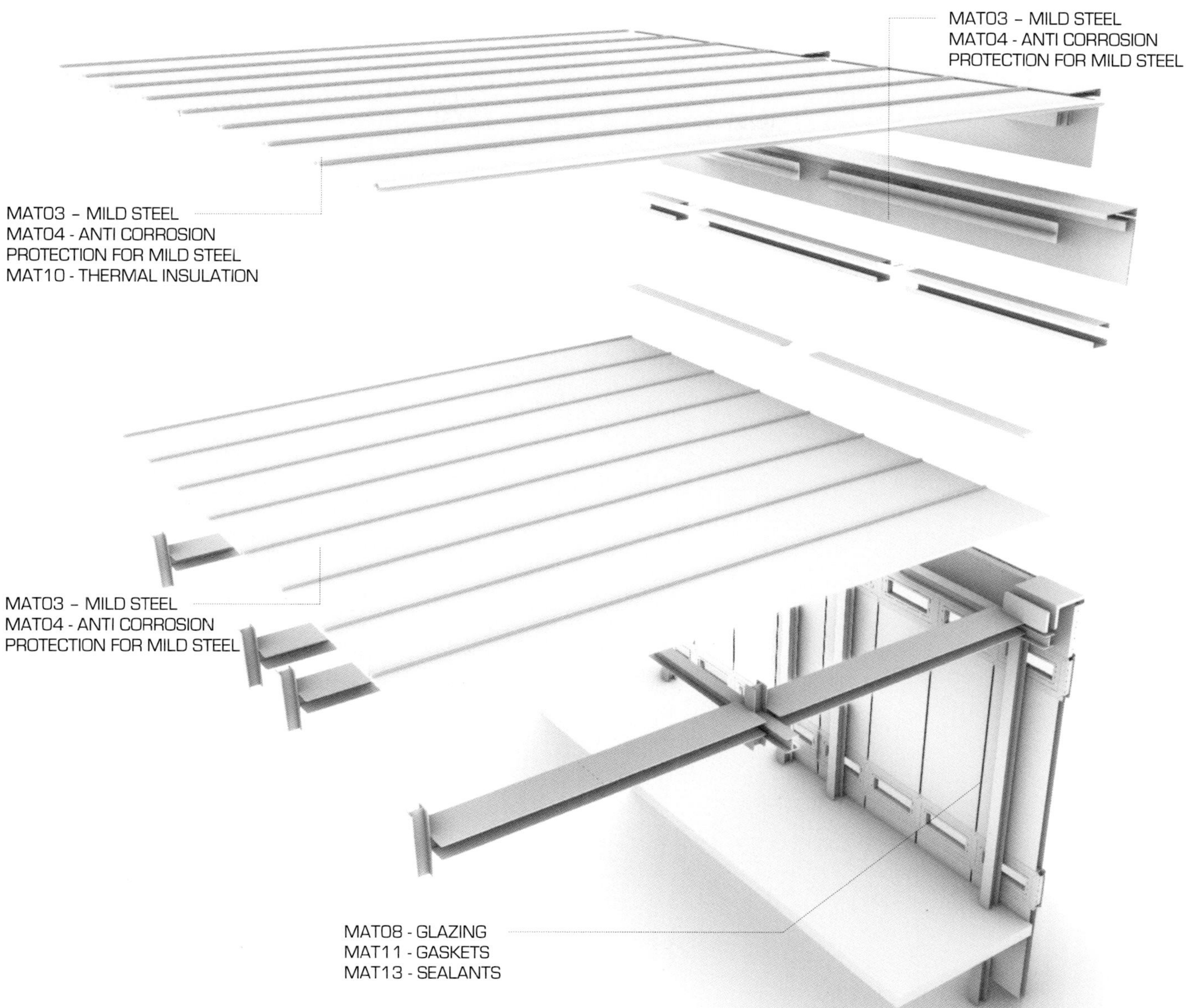

c | 3D exploded view of metal composite panel facade components with material specification references

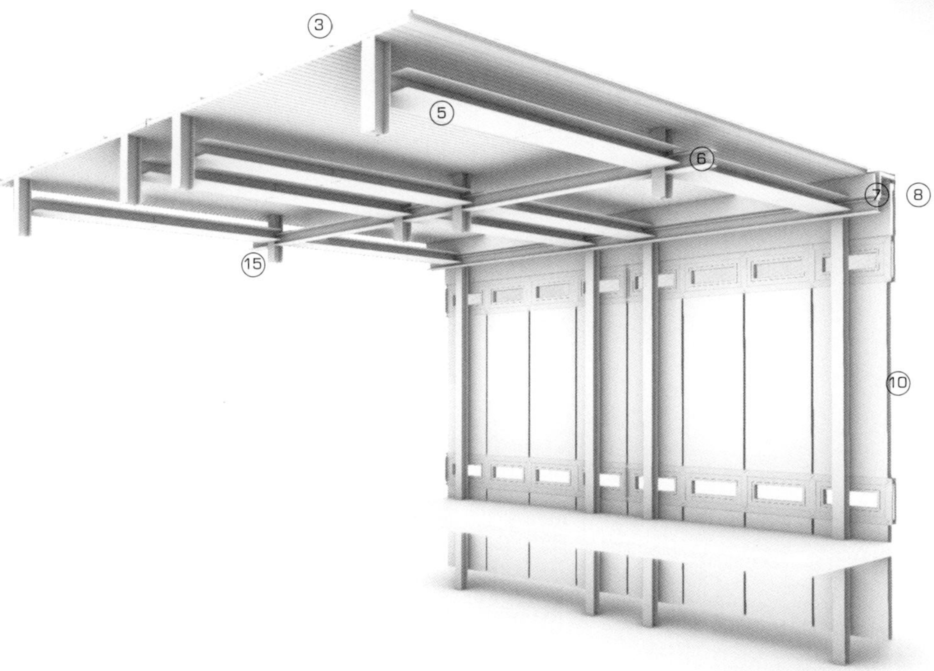

9 | Metal fascia panel
10 | External wall
11 | Outer sheet fixing bracket
12 | Panel 1
13 | Panel 2
14 | Ridge piece
15 | Structural frame
16 | Pipe or duct penetration

d | 3D view showing metal composite panel facade fixed to steel frame

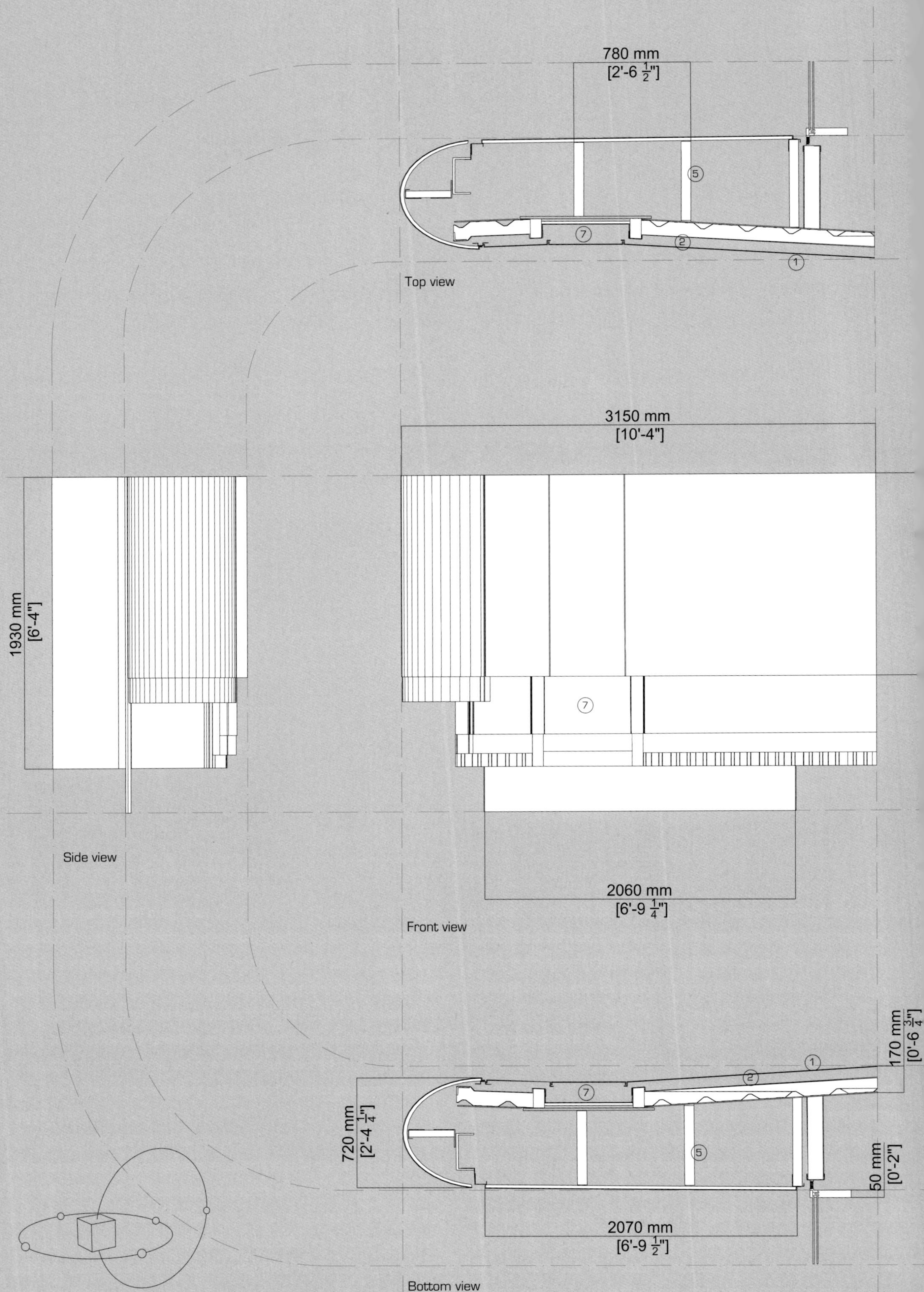

Third angle projection 1:30

Details

1 | Metal rainscreen panel
2 | Single-layer membrane
3 | Composite panel
4 | Folded metal coping
5 | Purlin or structural beam
6 | Secondary purlin
7 | Folded metal gutter
8 | Folded metal drip

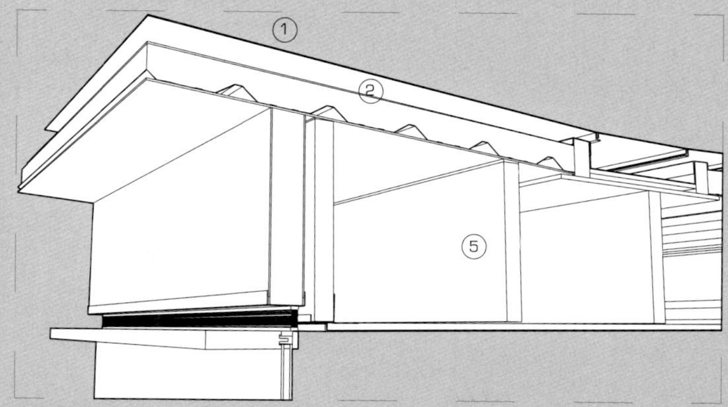

3D view of the assembly

Side view

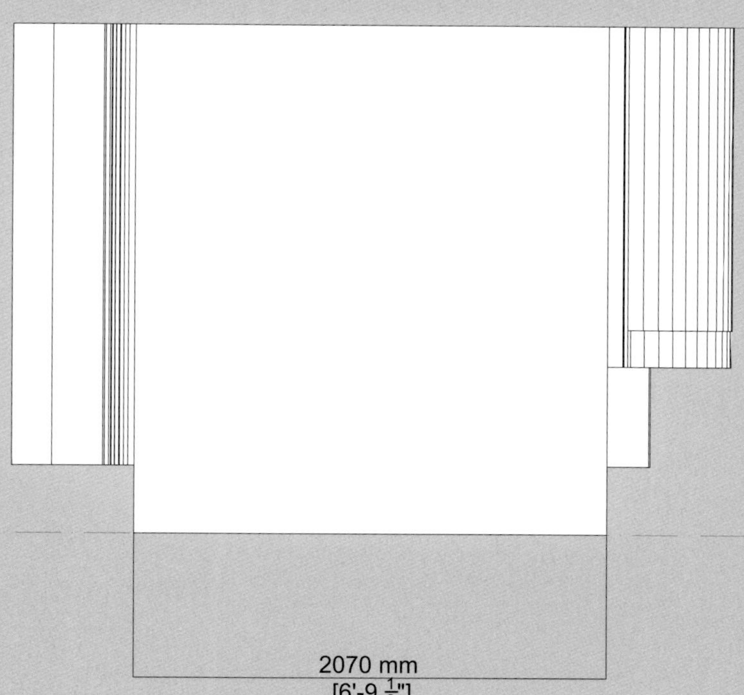

Back view

9 | Metal fascia panel
10 | External wall
11 | Outer sheet fixing bracket
12 | Panel 1
13 | Panel 2
14 | Ridge piece
15 | Structural frame
16 | Pipe or duct penetration

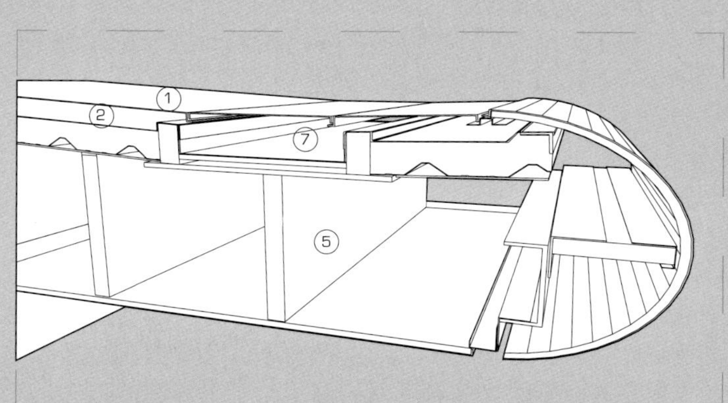

3D view of the assembly

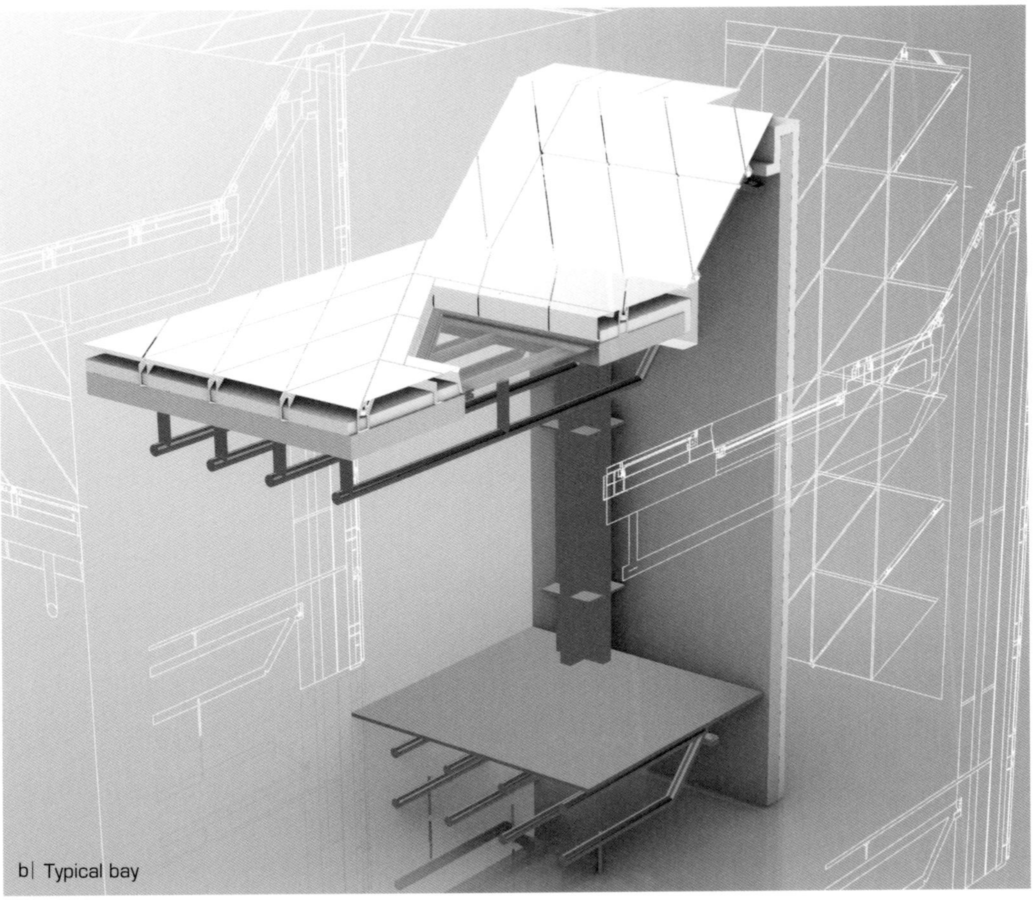
b| Typical bay

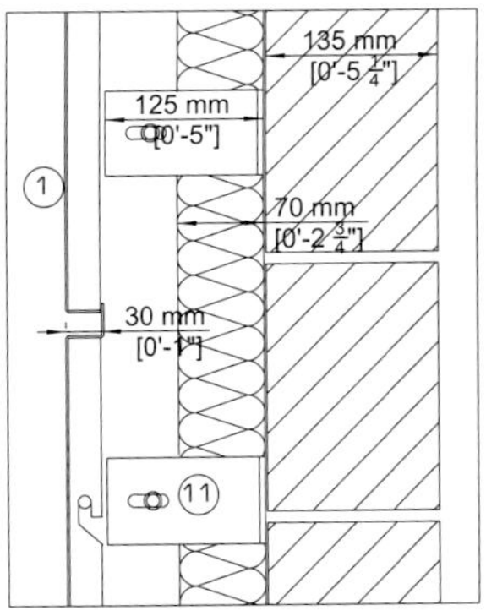

a| Vertical section.

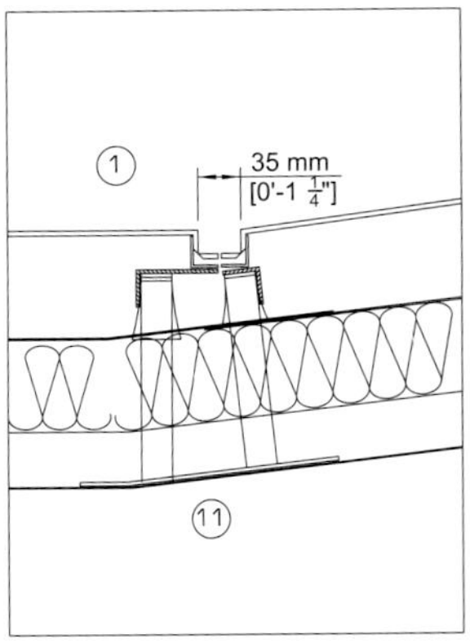

c| Horizontal section.

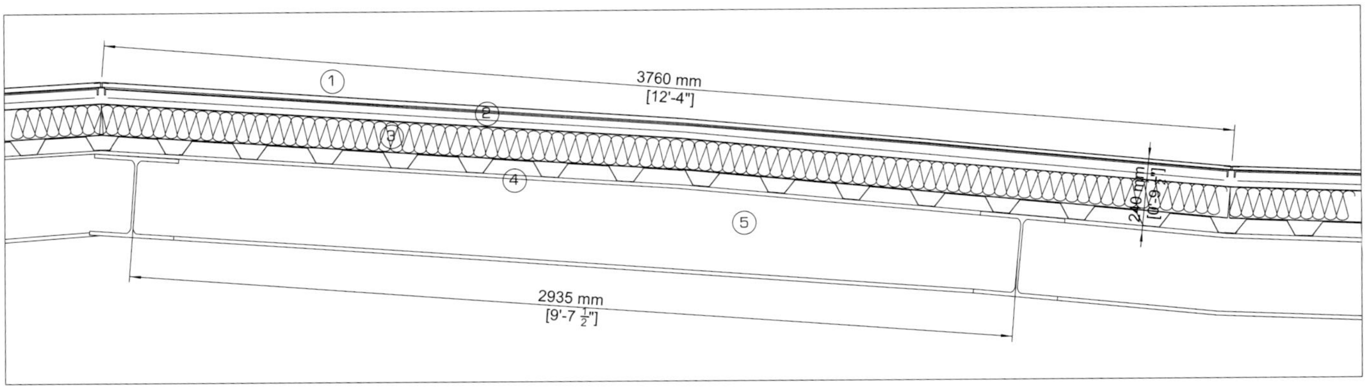

d| Vertical section. Typical cross section

The use of metal rainscreen panels is relatively new in roof construction, and has developed from its use in external walls. In facades, rainscreen panels are used in a configuration that allows most of the wind-blown rain that reaches a metal panel facade to drain down its face. Joints between metal panels are left open-jointed, so that only a small amount of rainwater passes through them. Rainwater is drained away in the cavity behind the rear face of the cavity being sealed, typically with a single layer membrane on an insulated, lightweight, backing wall. A commonly used alternative is of a bitumen-based paint on a reinforced concrete or concrete block backing wall. Closed cell thermal insulation is usually set on the outside face of the backing wall, with a cavity between the thermal insulation and the inner face of the metal rainscreen panel (b).

In contrast, metal rainscreens for roofs vary considerably from the configuration used in external walls. Firstly, most of the rain falling onto a metal rainscreen roof is not usually drained away on the outer layer of panels, unless the roof has a relatively steep pitch or curved section. Rainwater is still expected to drain onto the waterproofing layer beneath as if the panels were not in place. The main function of the rainscreen panels on roofs is to protect the membrane from the effects of the sun (heat and UV radiation) as well as from the worst effects of windblown rain. Rainscreen panels provide a lightweight covering that forms part of the visual language of the external walls. Although smooth pebbles are also used to protect waterproofing membranes on roofs, pebbles and gravel are obviously not suited to sloping or curved roofs. The use of metal rainscreen panels is well suited to these roofs which form a visible part of the design. These panels allow traditional roof elements, which are usually visually dominant such as gutters, parapets and ridges, to be accommodated within a smooth, continuous finish, allowing roofs to take on the visual characteristics of external walls and become a 'facade' in their own right.

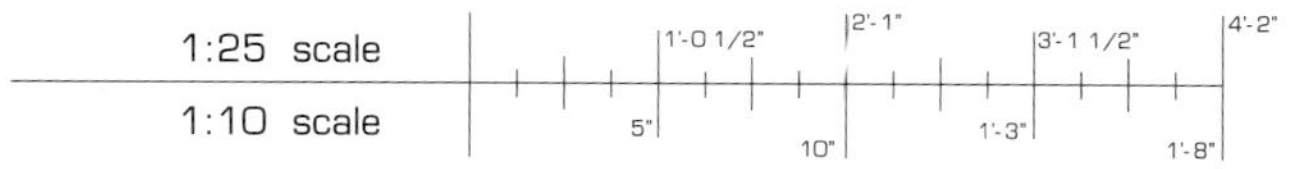

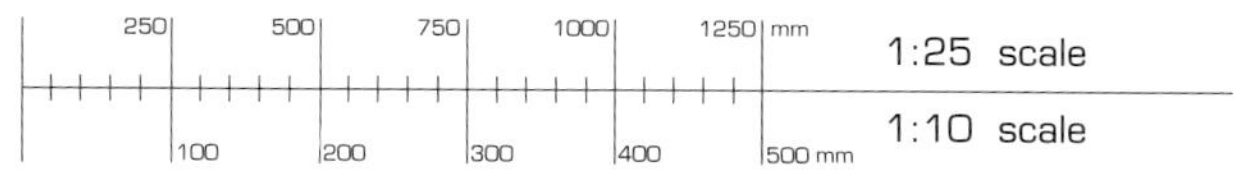

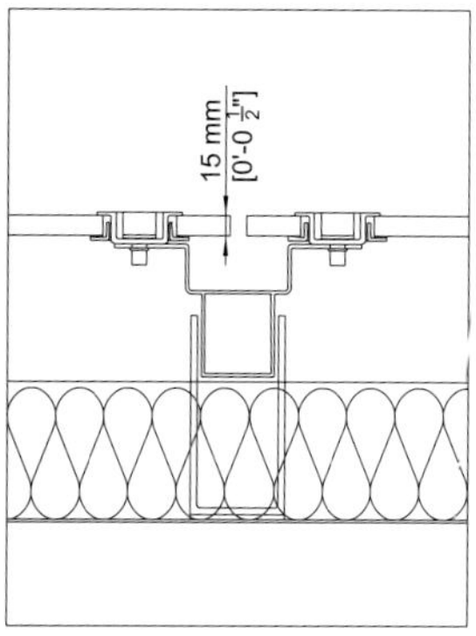

c | Horizontal section.

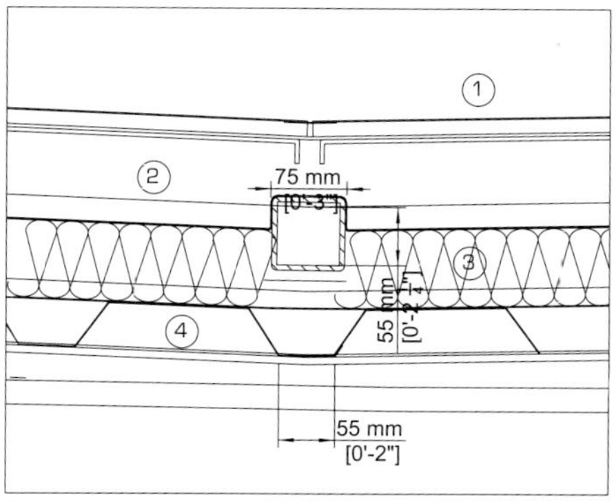

f | Vertical section. Panel-to-panel junction

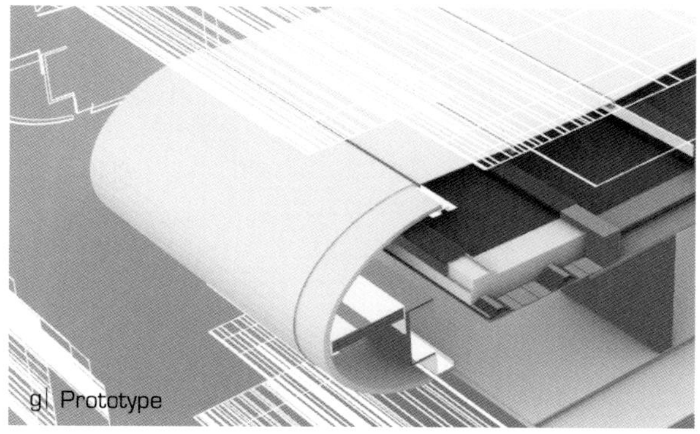

g | Prototype

Details

1 | Metal rainscreen panel
2 | Single-layer membrane
3 | Composite panel
4 | Folded metal coping
5 | Purlin or structural beam
6 | Secondary purlin
7 | Folded metal gutter
8 | Folded metal drip
9 | Metal fascia panel
10 | External wall
11 | Outer sheet fixing bracket
12 | Panel 1
13 | Panel 2
14 | Ridge piece
15 | Structural frame
16 | Pipe or duct penetration

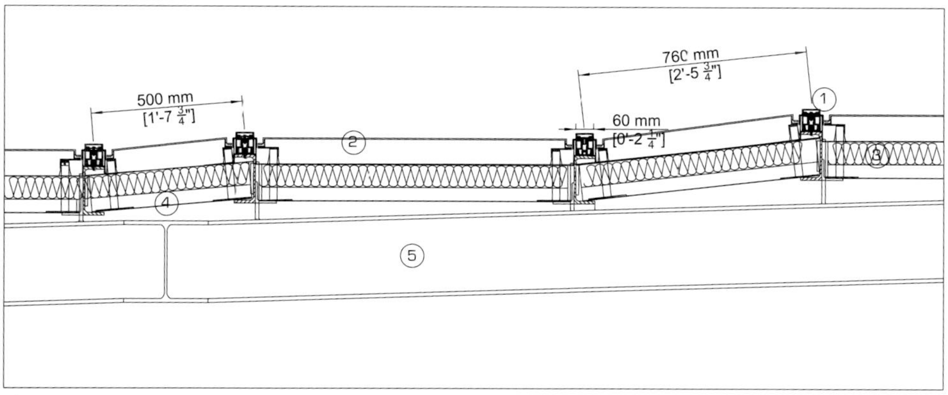

h | Vertical section. Typical long section

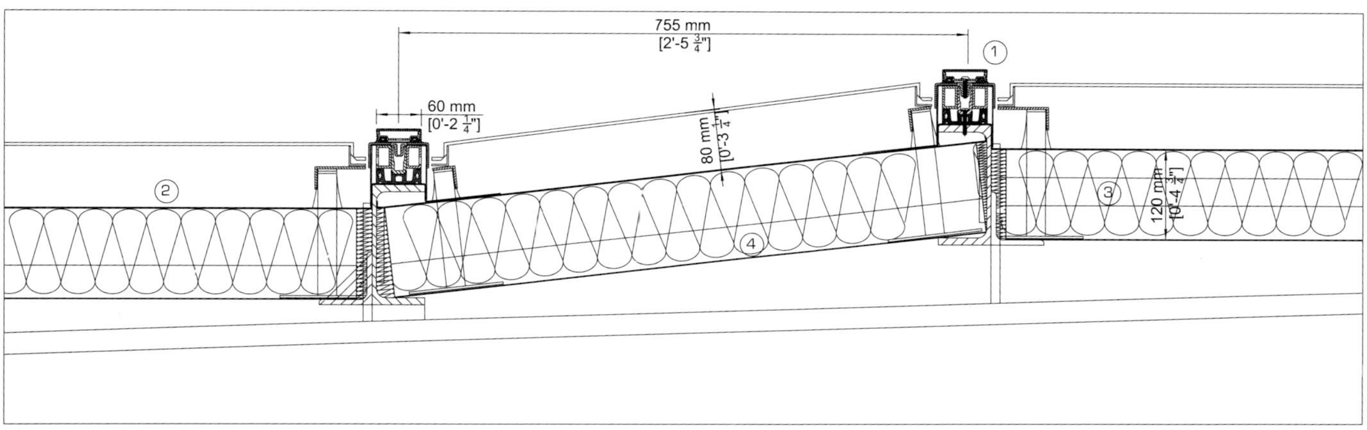

j | Vertical section. Panel-to-panel junction

Although metal sheet is used, metal and plastic composites are becoming increasingly popular as they are less likely to become dented from foot traffic during maintenance work, which helps to maintain the flatness of metal surface associated with composite metal materials. The 'oil canning' effect of depressed or dented panels, resulting from regular maintenance access, is usually avoided by using these composite sheet materials. In such materials a thin layer of plastic is faced with two thin sheets of aluminium which are bonded to the plastic sheet core. The size of metal rainscreen panels is restricted more by the panel width that can be walked upon for maintenance access than by the sheet size available. The maximum size of metal sheet is usually in a width of 1200mm or 1500mm metal coil (MCE_356, a). Composite sheets are typically 1000mm to 1200mm wide, in lengths from 2400mm to 3000mm, depending upon the manufacturer. In practice, panels may only be around 600mm wide if they do not have additional supporting framing beneath to stiffen them. Framed panels can reach the maximum sizes already mentioned, but care must be taken to avoid the pattern staining or denting that can reveal the frame behind during the lifecycle of the roof.

Panel arrangement

Rainscreen roof panels are typically arranged either as panels laid in a flat grid, or as lapped panels, where the bottom edge is lapped over the top of the panel beneath. Side joints remain open-jointed and are set in the same plane in this configuration. Panels in a flat grid are set onto metal Z-sections, which are either bonded to the top surface of the waterproof membrane to avoid any risk of water penetration through mechanical fixings, or are set above the membrane on support pads. These pads are usually covered with a waterproofing membrane to reduce the number of fixing penetrations through this layer. The rainscreen panels are then screw-fixed to the Z-section with brackets that avoid the screw fixings being seen at a distance where this is a visual

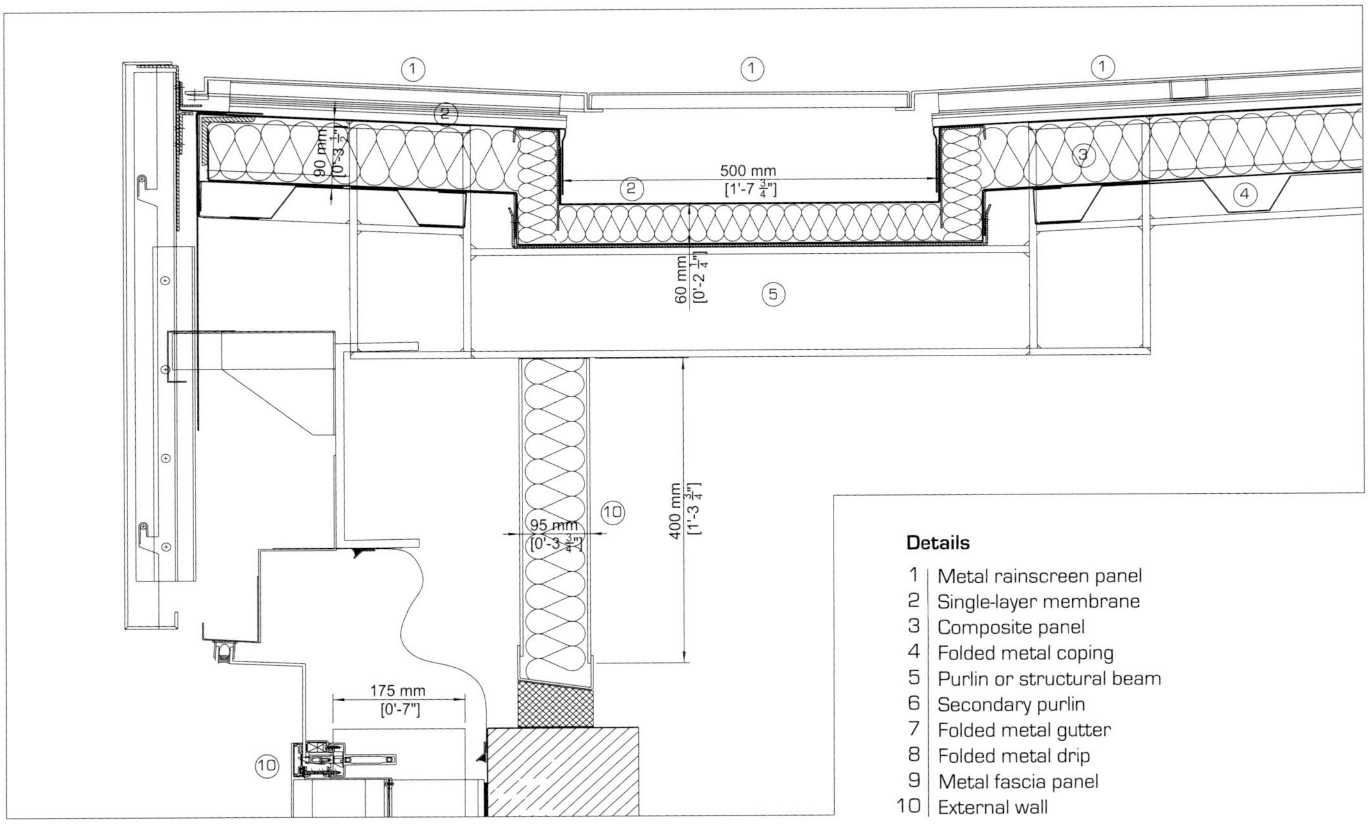

a| Vertical section. Concealed parapet gutter

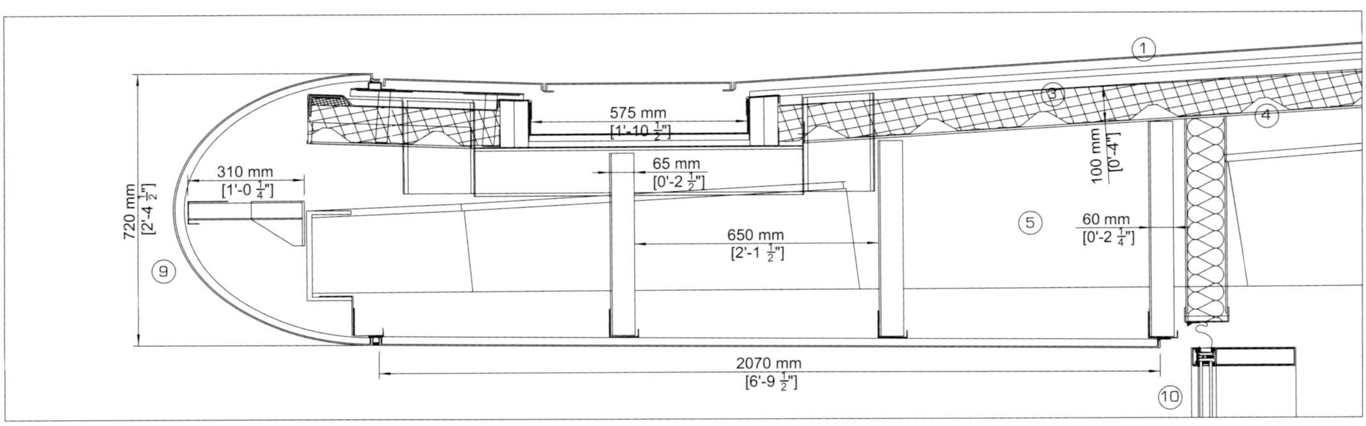

b| Vertical section. Concealed parapet gutter

requirement. Unlike rainscreen panels for walls, roof panels cannot be secured easily on a hook-on-type support system while being set near a horizontal plane. Since panels must have the ability to be removed easily and regularly for maintenance access, screw fixings in all panel corners are most commonly used. Concealed fixings are more difficult to accommodate, though such systems are likely to appear over the next ten years as demand for this roof system increases.

Panels are made usually by folding the edges down to form a tray, then outwards to form a rim around the panel. Holes are drilled at the corners to allow access to screw fixings beneath. Short lengths of bracket are bonded or riveted to the sides of the tray beneath the projecting panel rim. The panels are then screw-fixed at the brackets to the supporting rails or Z-sections by passing the screw through the hole in the rim. This allows the screw fixings to be concealed from view, while allowing each roof panel to be removed without affecting adjacent panels.

An alternative fixing method for panels is to set screws in each corner of the face of the panel onto a support rail system. Screw heads are difficult to coat in a colour that matches the panel, and even if this is achieved, scratching of the panel is likely as a result of removing panels for maintenance access. Screws are usually left as a self-finish, and have a particular appearance when viewed as a complete set of roof panels. Countersunk screws have the least visually obtrusive appearance.

Parapets

An advantage of rainscreen panels for parapets is that the gutter and upstand can be made without either element being visible. Consequently, eaves, monopitch ridges and verges can have a similar outward appearance of an uninterrupted panel layout extending from the roof down to the external wall. Parapets are formed only by a gutter that also provides the necessary upstand height for the parapet

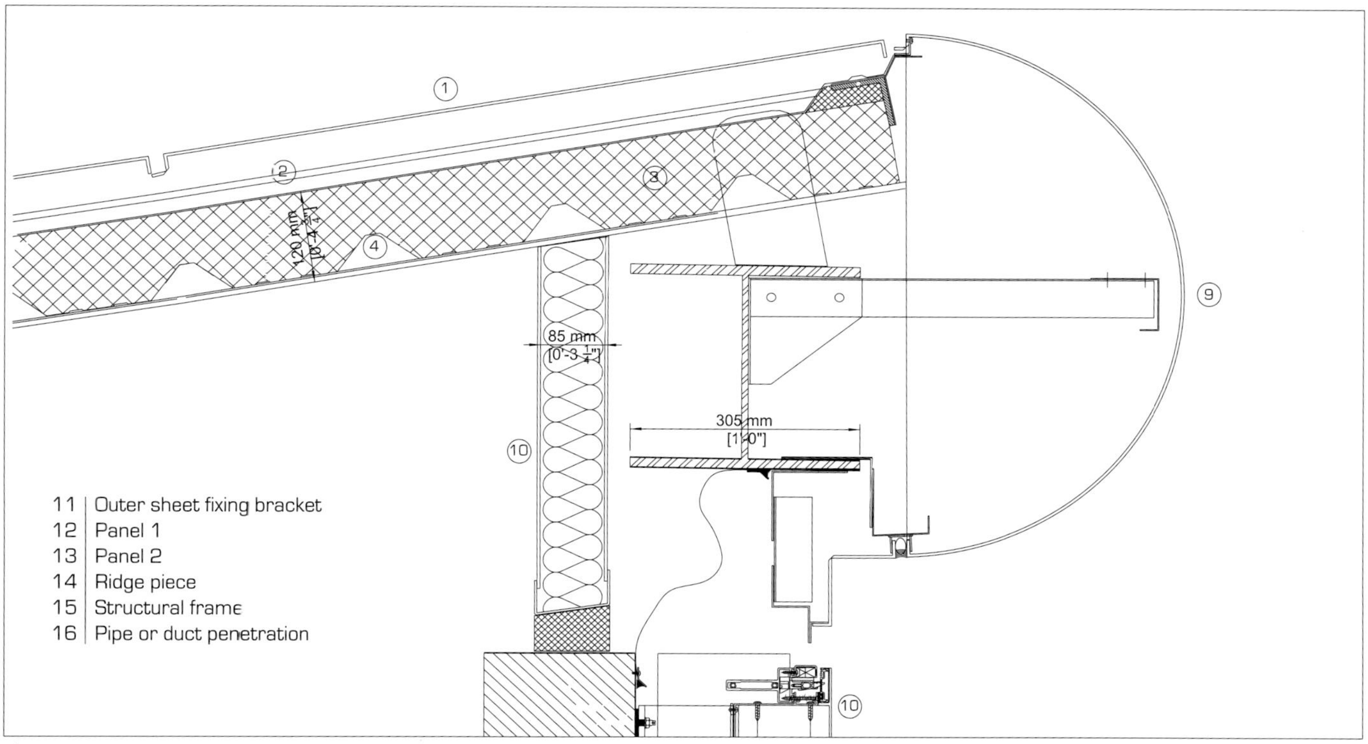

c| Vertical section. Monoridge with curved fascia panel

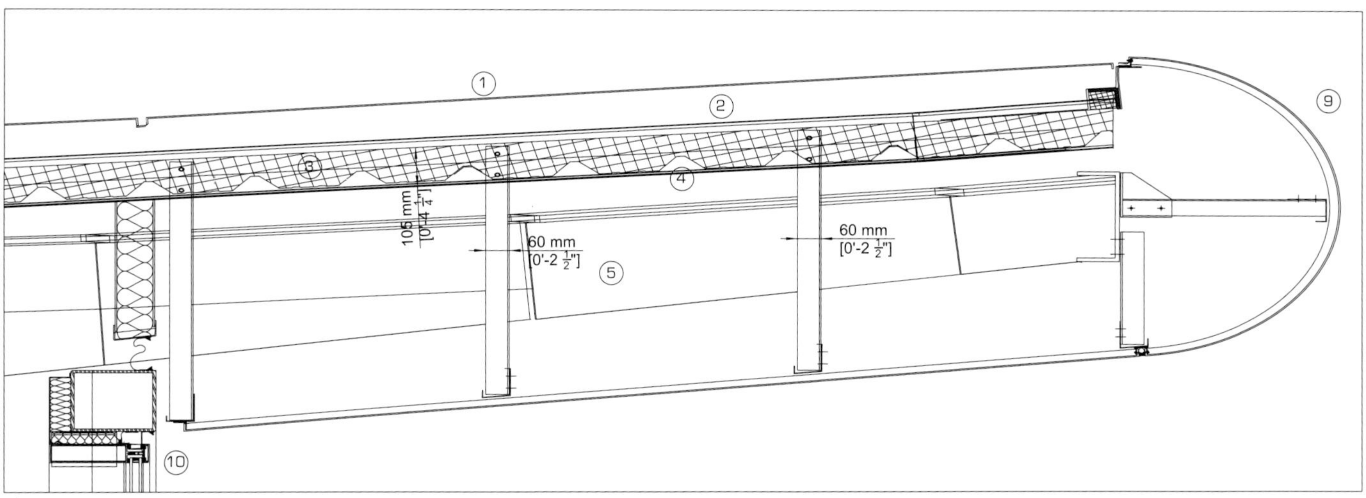

d| Vertical section. Monoridge with cantilevered panel

itself. A coping is then formed by using the same rainscreen panels as elsewhere on the roof in order to provide a continuity of appearance. The waterproof membrane beneath is then sealed against the wall construction or against the side of the sealed roof deck, which is closed off with a folded metal strip in the case of profiled metal sheet or composite metal panels being used (a).

The depth of the roof construction is set vertically, which is fixed in front of the parapet. Unlike parapets in profiled metal or composite panels, the vertical fascia panel can extend up to the top of the wall in order to conceal the coping flashing (a). In other types of metal roofing, the coping extends over the top of the wall, resulting in a thin visual edge to the top of the wall. This coping line can be concealed from view in rainscreen roof panel construction. The external wall beneath is typically sealed up to the underside of the roof deck.

The parapet gutter can be covered with a perforated or slotted metal cover in the same material and finish as the adjacent rainscreen panels. Since rainwater runs off each panel at its edges onto the membrane below, there is no need to leave the gutter uncovered, as is the case with other metal roof systems. Water running down the membrane is drained directly into the gutter, and water underneath the gutter cover drains through slots or perforations.

Monopitch ridges and verges

These ridge types are formed by sealing the waterproof membrane against the side of the roof deck to provide a watertight enclosure that is continuous with the external wall beneath. The rainscreen panel at the edge of the roof is linked directly to a fascia panel. Where a curved panel is structurally independent of the external wall beneath, the roof deck is allowed to deflect and move under normal structural loading without affecting the wall beneath (c). If wall and roof were rigidly fixed

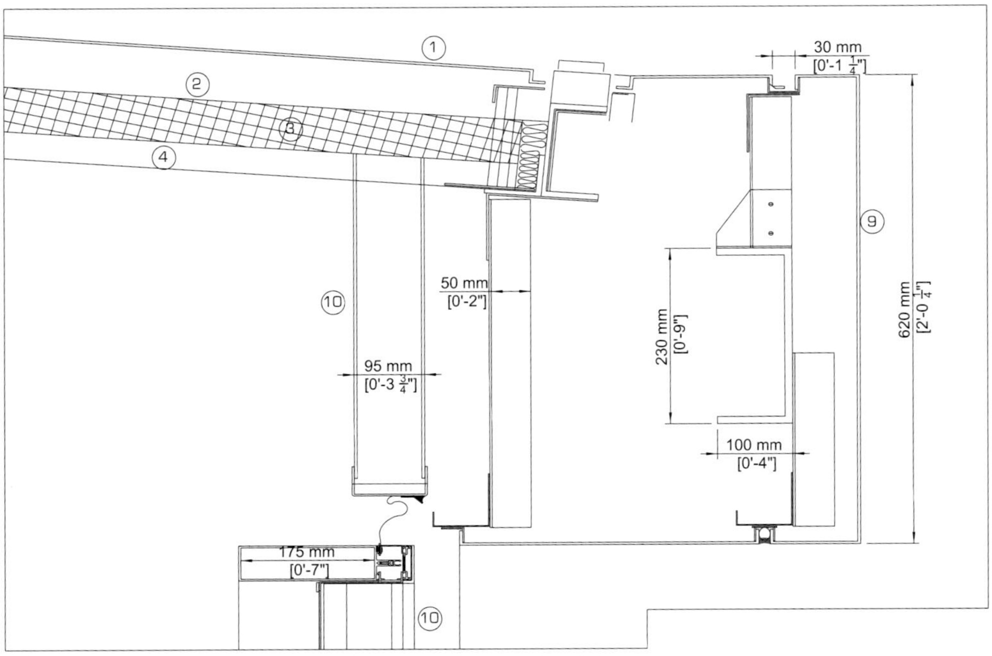

a| Vertical section. Verge with projecting panel

Details

1 Metal rainscreen panel
2 Single-layer membrane
3 Composite panel
4 Folded metal coping
5 Purlin or structural beam
6 Secondary purlin
7 Folded metal gutter
8 Folded metal drip
9 Metal fascia panel
10 External wall
11 Outer sheet fixing bracket
12 Panel 1
13 Panel 2
14 Ridge piece
15 Structural frame
16 Pipe or duct penetration

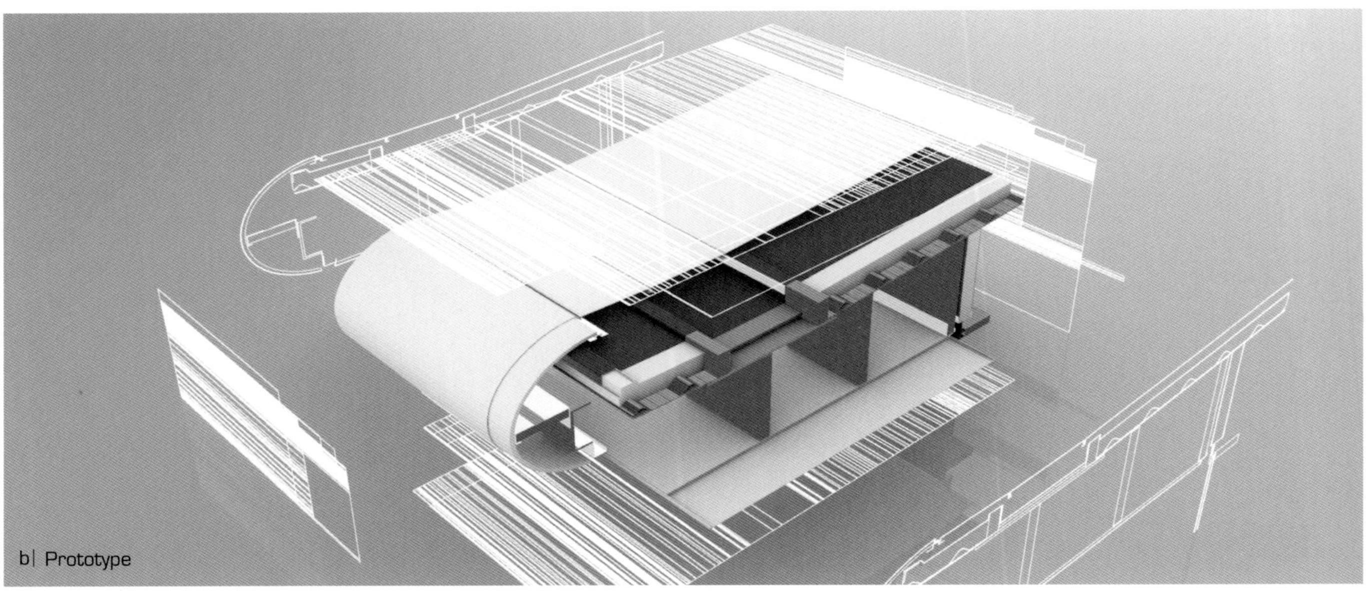

b| Prototype

together, the wall would deflect with the roof to an extent that would be too much for curtain wall systems, whose movements are very restricted. This is a common detail for roofs where the expected structural movements are higher than the modest movements allowed with curtain wall facades.

A seal between the external wall and the underside of the roof is provided by two layers of EPDM sheet with flexible insulation quilt set between them. Folded metal sheet can also be used, but is less flexible in the longitudinal direction (along the length of the roof) than EPDM sheet.

Verges are similar to monopitch ridges, but differ in that they require an upstand at the roof edge to prevent rainwater from spilling off the roof. The upstand can be low if the verge is flat, and if relatively small amounts of rainwater are flowing along this edge of the roof. If large amounts of water flow are expected then the roof can be angled upwards slightly to avoid an upstand that extends above the line of the rainscreen roof panels. An alternative solution is to introduce a gutter at the edge of the verge to give greater height to the verge upstand without it becoming visible, and without breaking the continuous line of rainscreen panels.

Roof geometry

An advantage of rainscreen panels is their ability to form complex geometries from flat panel components. Since the panels are not required to be waterproofed, they do not require any joints between panels that would become difficult and expensive for roofs with complex geometries. Flat panels can be fixed then finished with another rainscreen panel to create a set of gently curved panels that are turned in either one or two directions. Panels are increasingly being twisted in two directions in order to create genuinely curved roof finishes. Alternatively, panels can be set flat but with each panel set at different angles

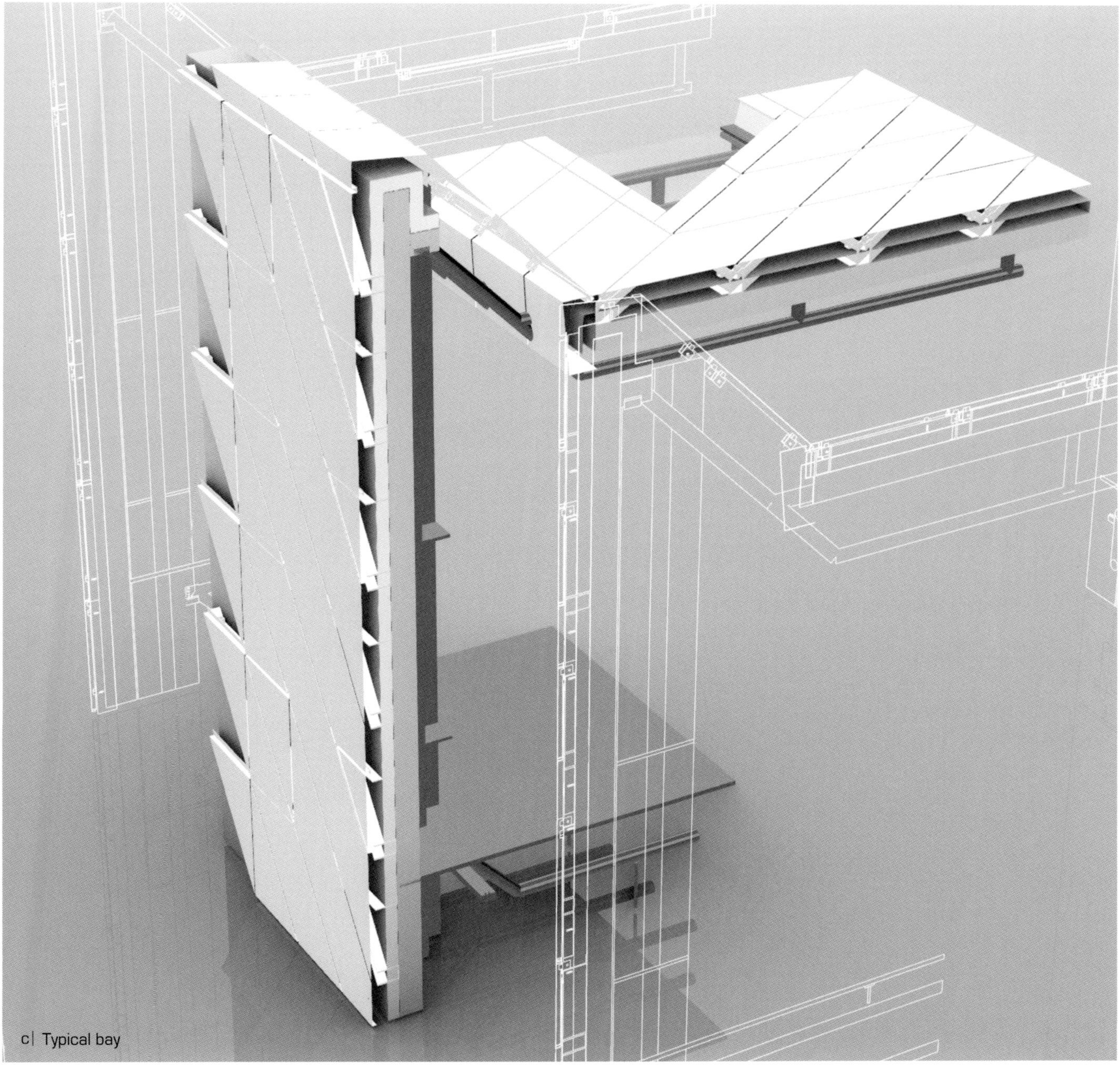

c| Typical bay

to create a facetted roof section. Although the drawn examples here show lightweight roof configurations to suit the lightweight nature of the metal rainscreen panels, any compatible roof substrate can be used, from reinforced concrete slabs to timber shell structures.

Although most rainscreens are set at around 100mm above the membrane (b), some can be up to 1000mm above the roof deck in order to accommodate mechanical plant equipment and duct outlets for mechanical ventilation within the building. This allows rainscreen panels to provide weather protection for equipment as well as a visual screen to these items, which can have a considerable visual impact on a roof. Rainscreen panels set 1000mm above the roof are fixed to a secondary support framework, typically of cold-formed steel sections or aluminium extrusions. The framework is fixed at its base to pads or Z-section profiles in the same way as the rainscreens set close to the roof membrane.

Roof soffits

Metal rainscreen roof panels can also be used as soffit panels to either an overhanging eaves or parapet, or alternatively to form a complete soffit to the underside of a roof. When mixed with perforated or slotted panels and louvres, a roof can become a single visual entity, set above its external walls as a separate and distinct element (b). This can be achieved without the need for an expensive structure, or for an expensive waterproofing layer, since the membrane is concealed from view and is chosen for its performance rather than for its visual qualities. The easy removal of panels for maintenance access both on the roof, and into the ceiling void from the soffit panels beneath, make it an economic and easy-to-use system for roofs of complex geometry. The range of colours for composite sheet and for coatings to sheet metal has increased considerably in the past ten years, making this a roofing method set for new developments.

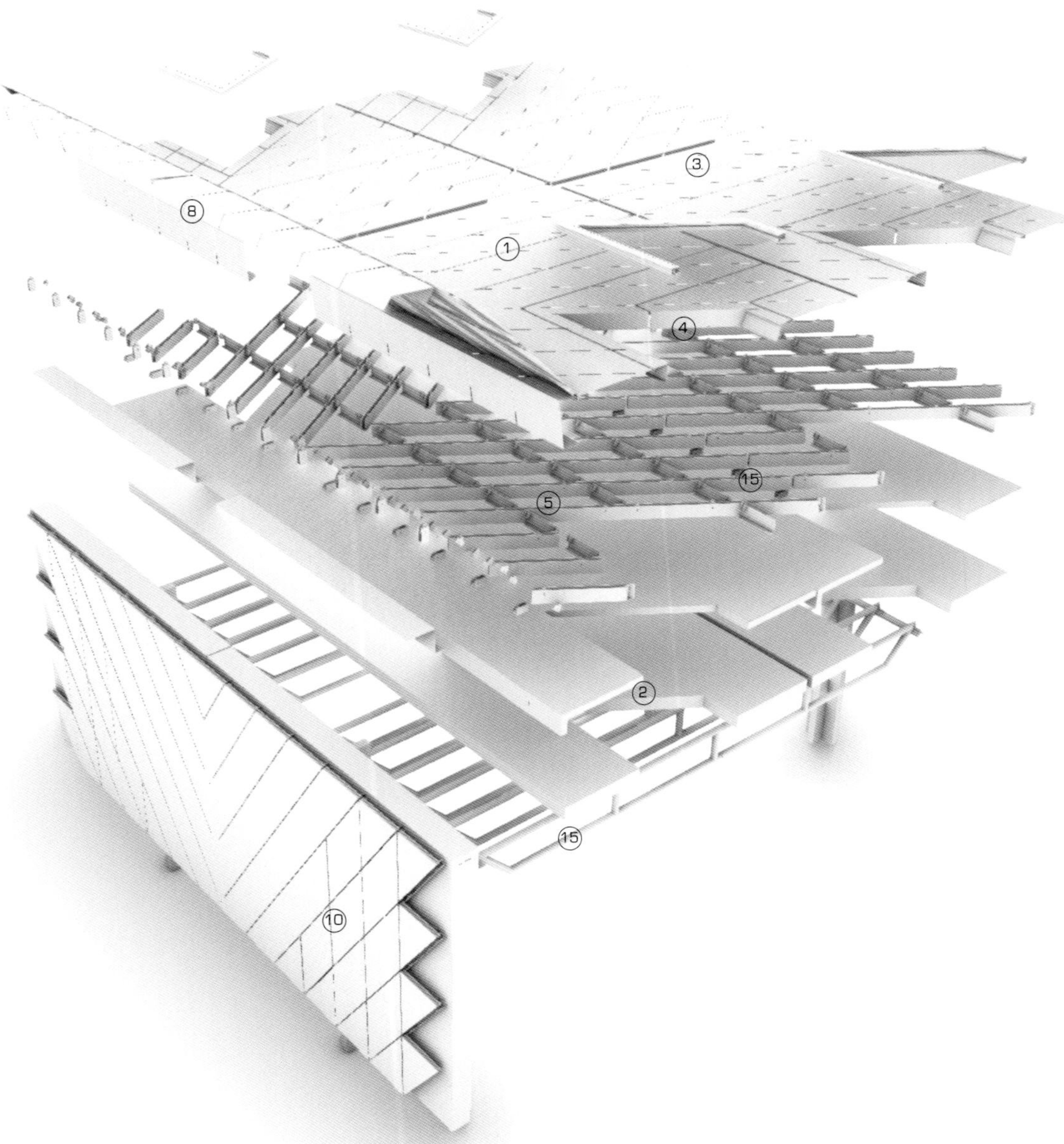

a | 3D exploded view showing metal rainscreen facade fixed to steel frame

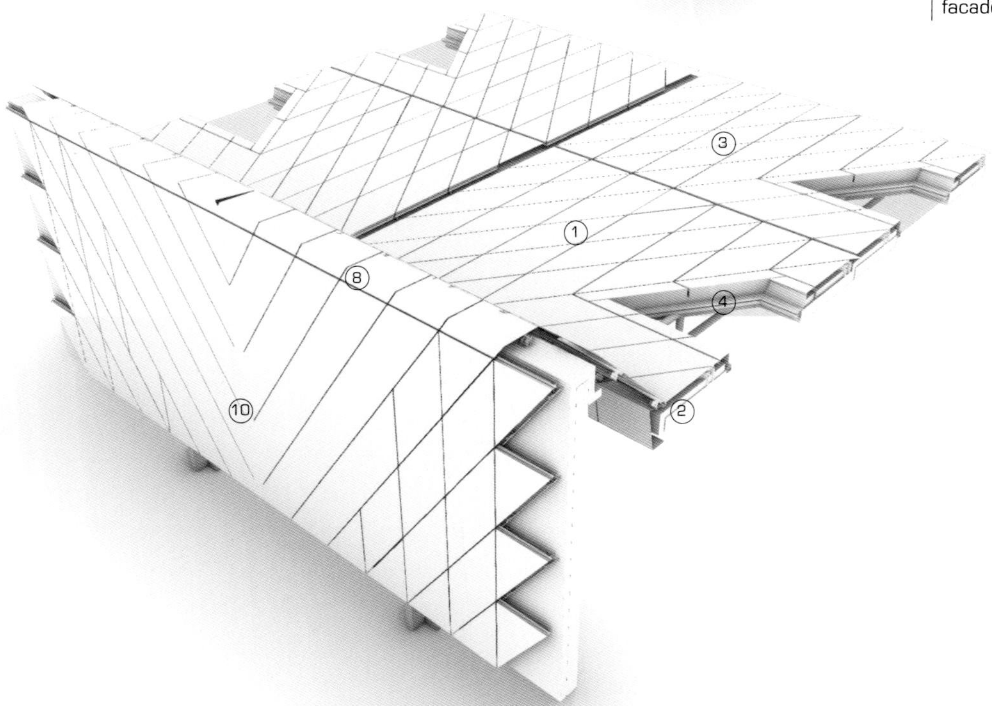

b | 3D view showing metal rainscreen facade fixed to steel frame

Details

1 | Metal rainscreen panel
2 | Single-layer membrane
3 | Composite panel
4 | Folded metal coping
5 | Purlin or structural beam
6 | Secondary purlin
7 | Folded metal gutter
8 | Folded metal drip

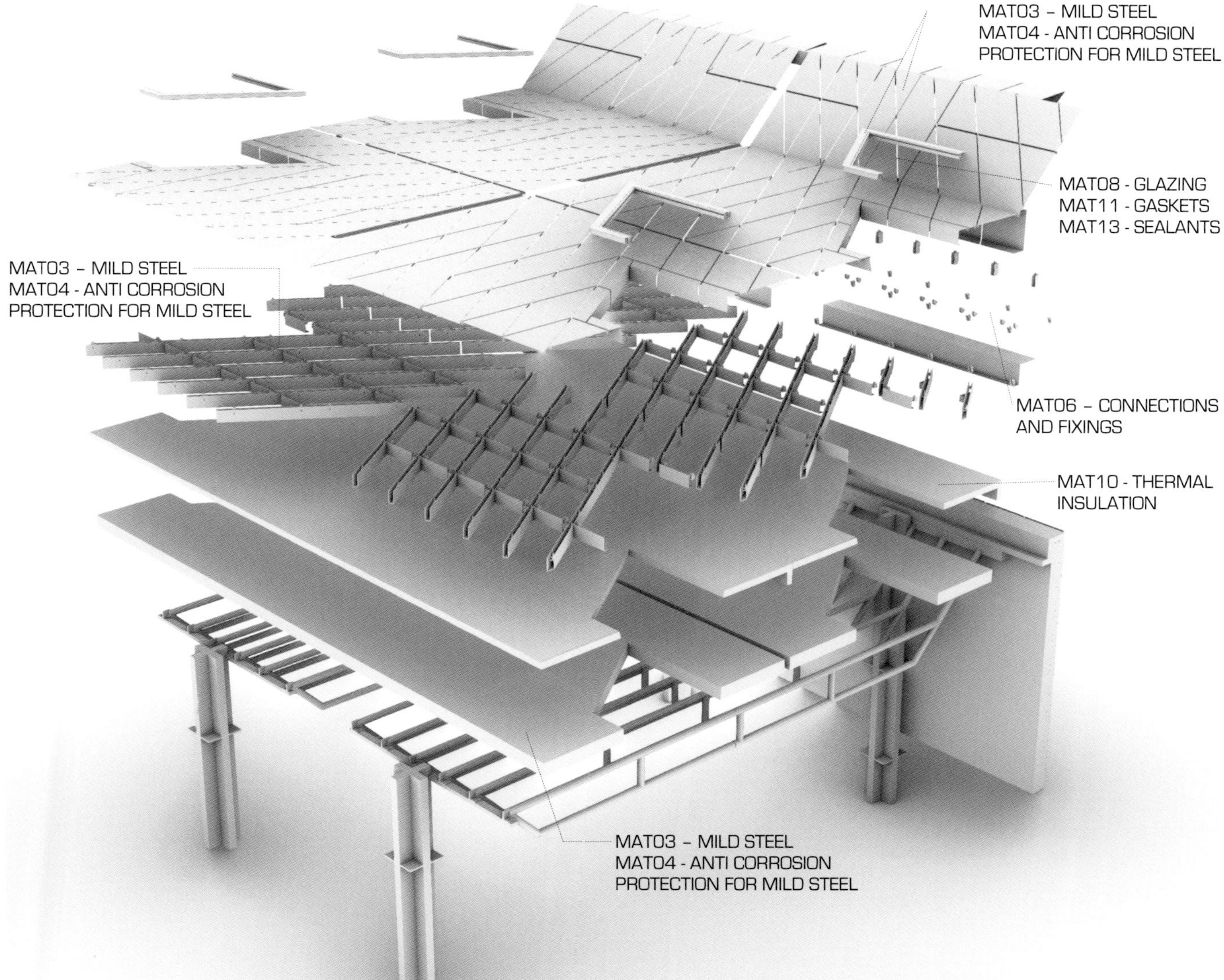

c | 3D exploded view of metal rainscreen facade components with material specifications references

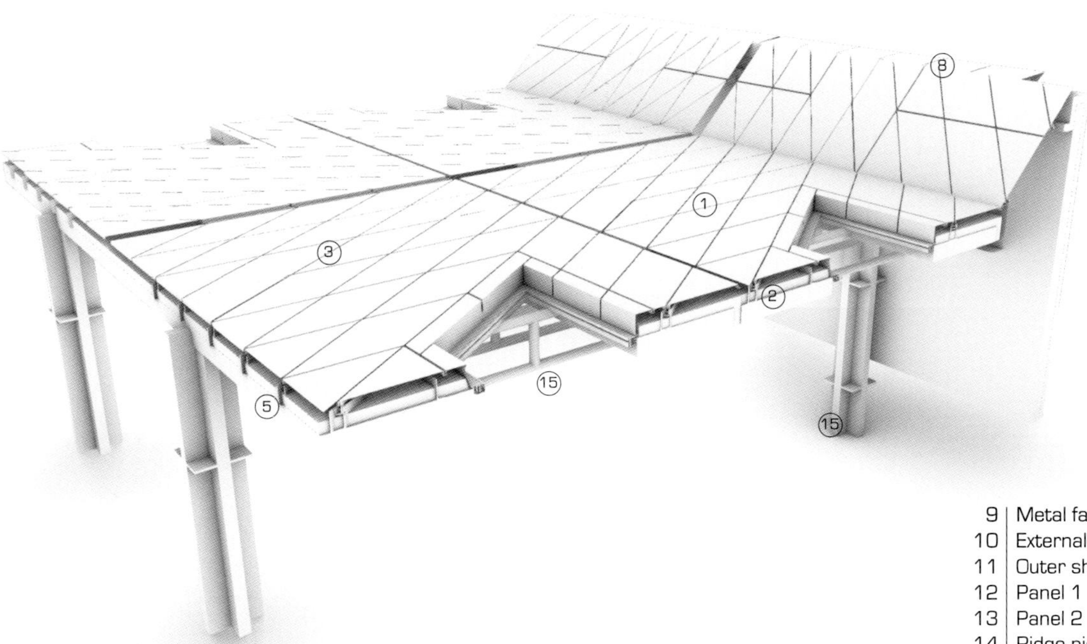

9 | Metal fascia panel
10 | External wall
11 | Outer sheet fixing bracket
12 | Panel 1
13 | Panel 2
14 | Ridge piece
15 | Structural frame
16 | Pipe or duct penetration

d | 3D view showing metal rainscreen facade fixed to steel frame

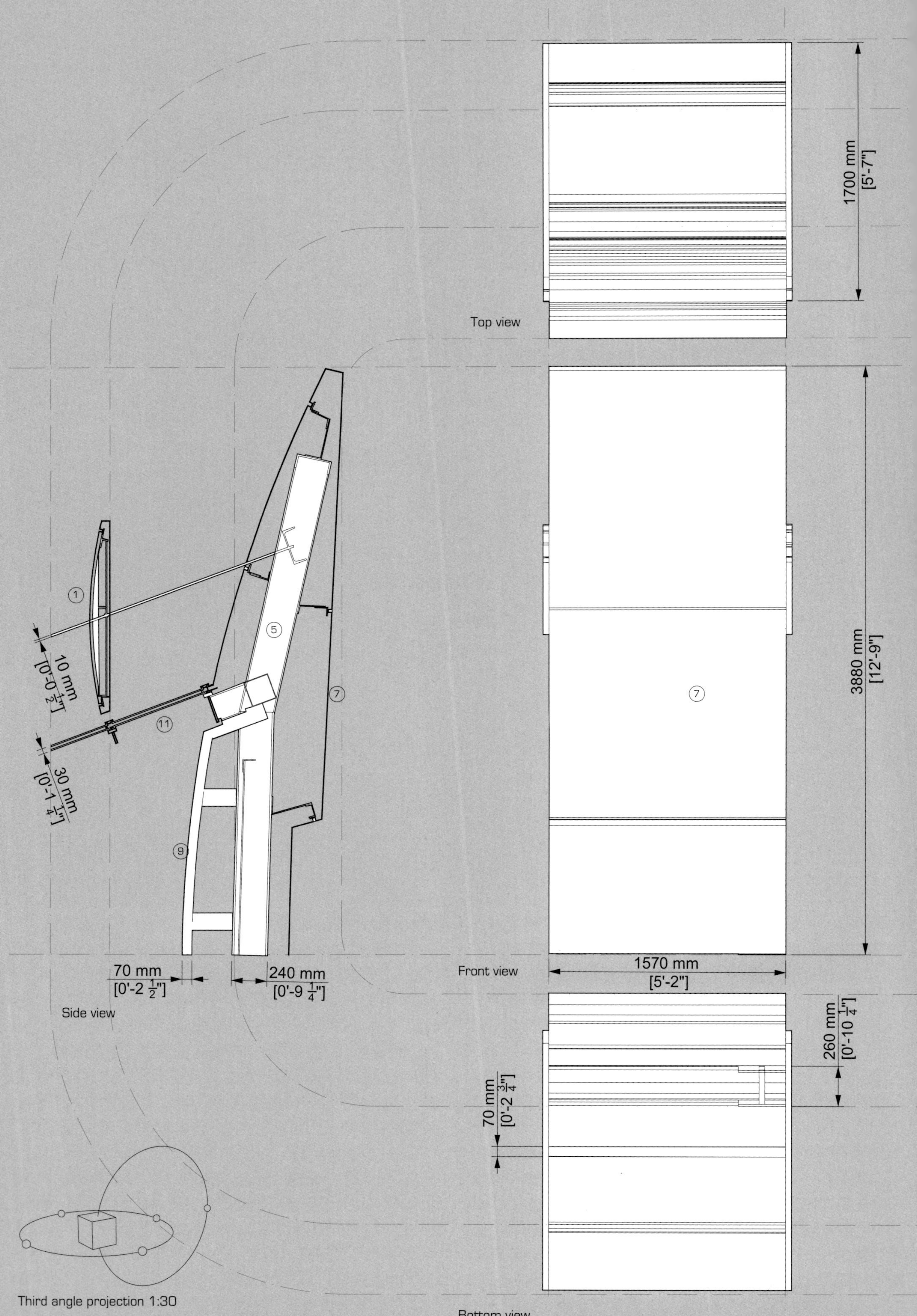
Top view
1700 mm
[5'-7"]
Front view
3880 mm
[12'-9"]
1570 mm
[5'-2"]
Bottom view
260 mm
[0'-10 1/4"]
70 mm
[0'-2 3/4"]
Side view
10 mm
[0'-0 1/2"]
30 mm
[0'-1 1/4"]
70 mm
[0'-2 1/2"]
240 mm
[0'-9 1/4"]
1
5
7
9
11
Third angle projection 1:30

Details

1 | Extruded aluminium louvre blade
2 | Mild steel box section
3 | Mild steel tube
4 | Structural pin connection
5 | Mild steel I-section
6 | Bolt fixed metal panel
7 | Aluminium sheet
8 | Mild steel or aluminium support frame
9 | Thermal insulation
10 | Silicone seal
11 | Glazed wall

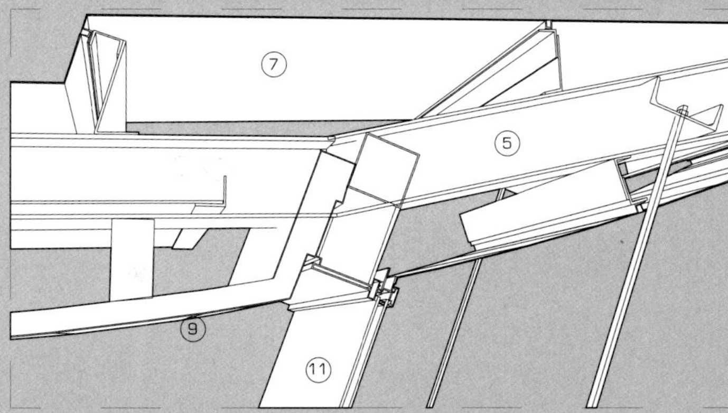

3D view of the assembly

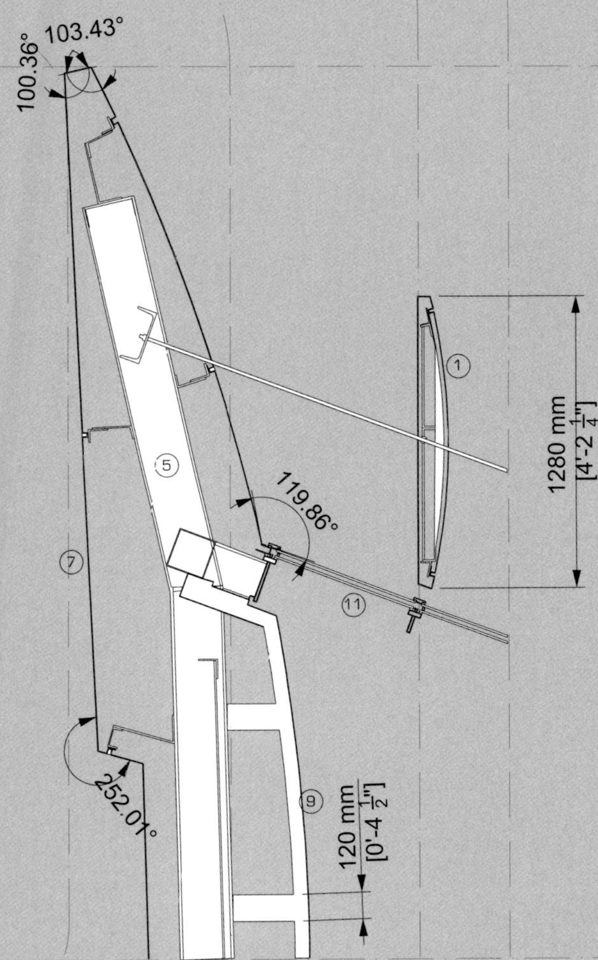

Side view

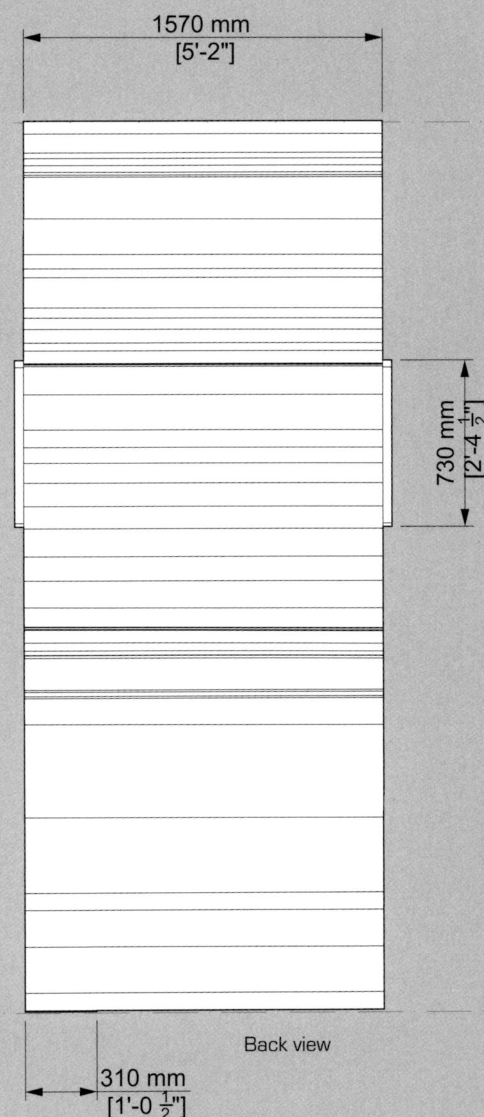

Back view

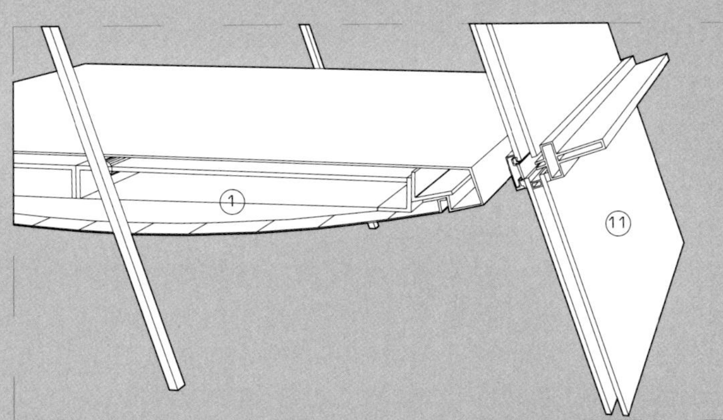

3D view of the assembly

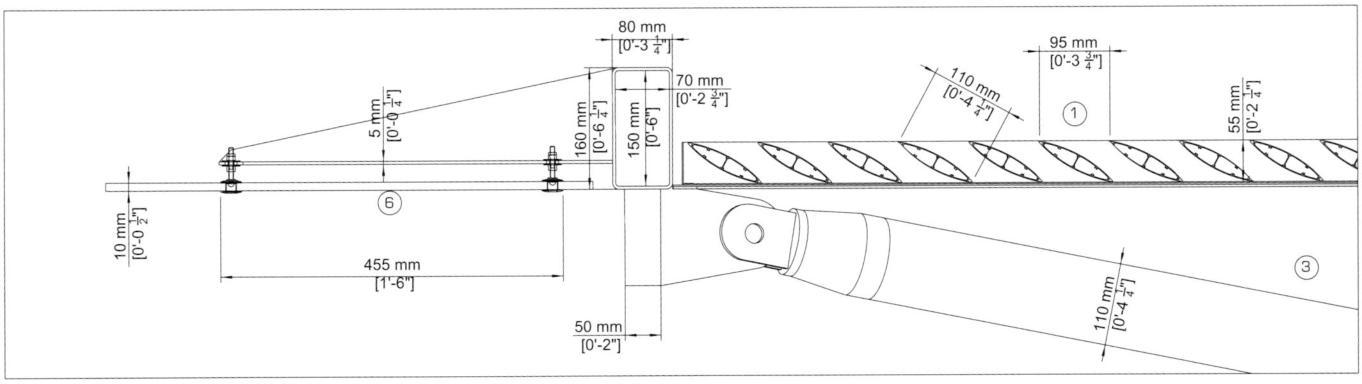

a| Vertical section. Canopy edge

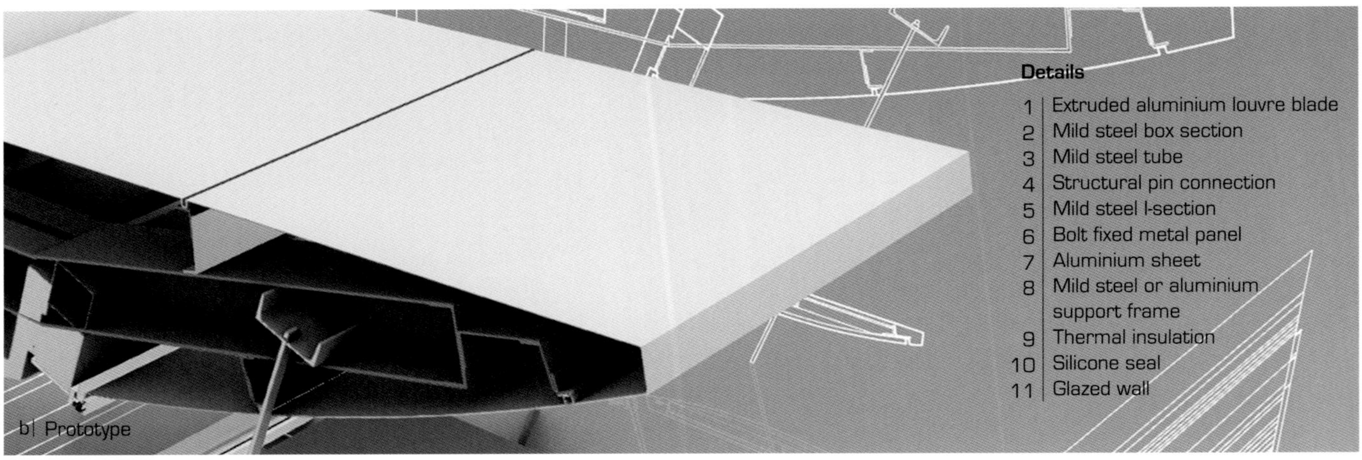

b| Prototype

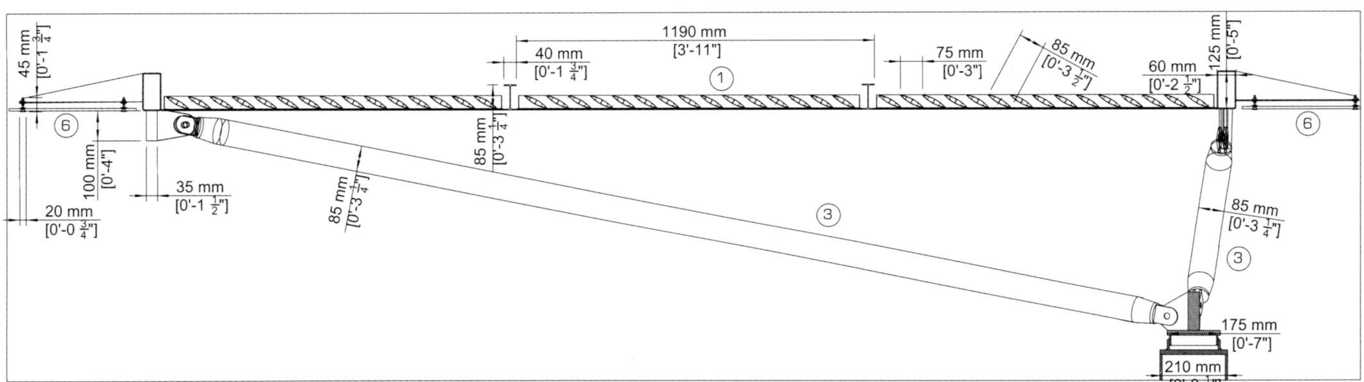

c| Vertical section 1:25. Canopy assembly

Bolt fixed panels

Metal panels can be bolt fixed onto the canopy edges. Flat panels, such as minimum 3mm thick aluminium sheet and metal / composite sheet, can be bolt fixed using proprietary systems destined for use in point fixed glazing (d). The use of oversized holes or slotted holes in the metal panel ensures that the metal panels can move with thermal expansion without bowing or bending.

Gutters and rainwater outlets can be formed in sheet metal. The rainwater outlet can be finished to match the adjacent metal finishes so as not to detract from the overall visual form of the canopy.

The canopy can form part of a glazed wall, and is thus thermally insulated. The insulation usually follows the form of the outside of the canopy to keep the voids within the structure at a similar temperature to the internal temperature of the building. An alternative is to form the insulation with the shortest distance between the glazed walls above and below, so that the structure and void are at the external temperature. The warm roof solution is usually preferred.

Fixed metal louvre canopies

Arrangements of metal louvres are used as canopies to provide solar shading while still allowing daylight to pass through the canopy. Louvre blavdes are set typically at 45° to the vertical in order to block the passage of direct sunlight but allow the light to be reflected off its surfaces down to the space beneath the canopy. Louvre sections are created from folded strips of aluminium or mild steel sheet, but these have limited stiffness and stability, requiring restraint along their length to hold their straightness (c). Greater stiffness is provided by extruded aluminium sections, where the elliptical section is most commonly used, mainly for its ability to reflect daylight in a way that reveals its 3-dimensional form, enhancing its appearance. Sections are either a

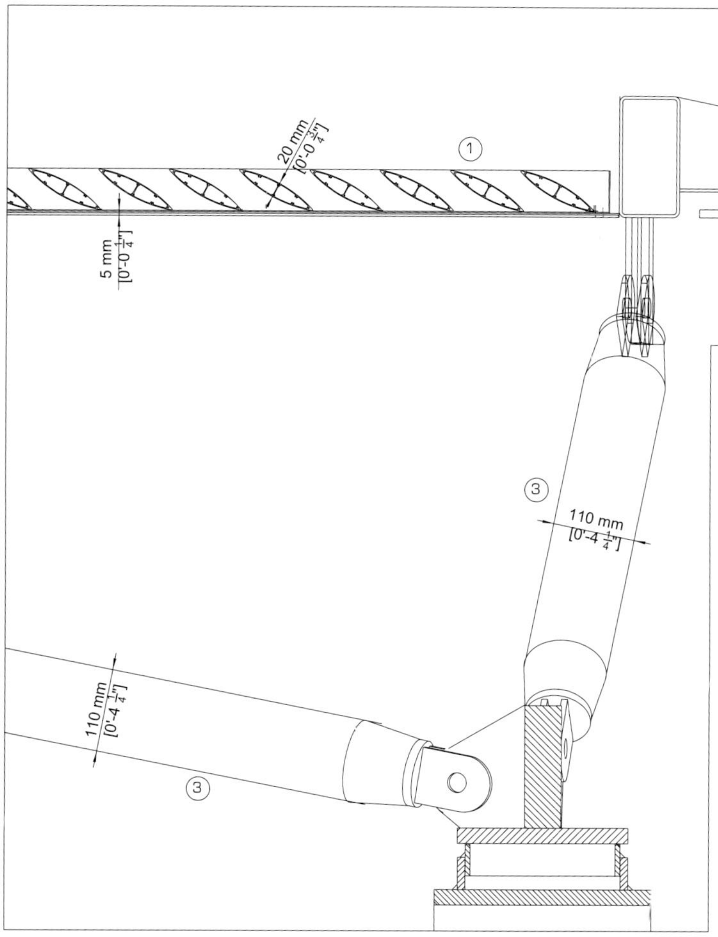

d| Vertical section. Canopy edge and support

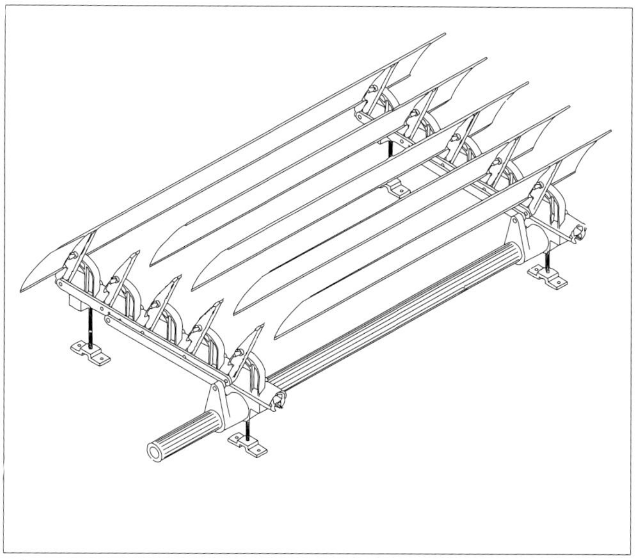

e| Isometric view of assembly. Electrically operated louvres

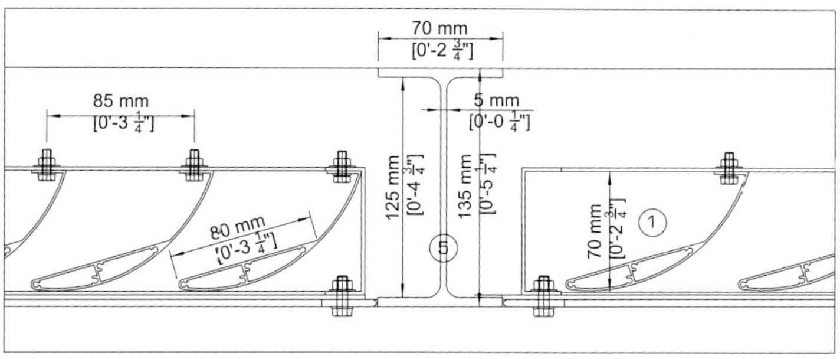

f | Vertical sections. Typical louvre sections

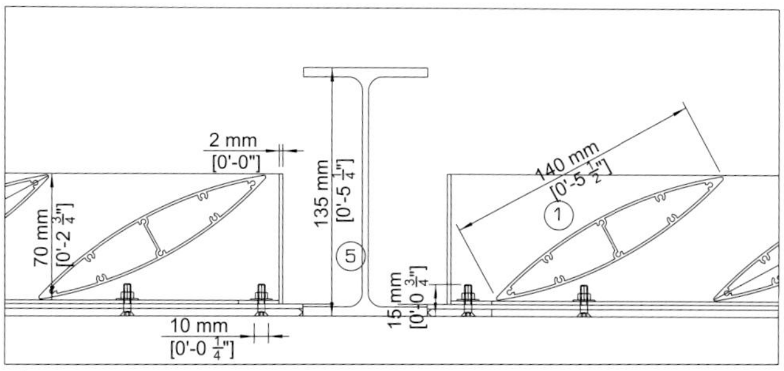

g| Vertical sections. Typical louvre sections

half ellipse or a full ellipse. Flat louvre arrangements provide much less visual vibrancy when viewed from below. Extruded aluminium sections require end caps usually for visual reasons, and these are either fixed with countersunk screws into the wall of the section or are welded and ground smooth. Where end caps are screwed to the ends, the aluminium profile has screw ports that form part of the extrusion into which the screws are fixed. Aluminium extrusions can be made in lengths of up to around 6000mm (20ft), and are supported at centres to suit their structural depth. An elliptical section will span typically 1500mm for a 75mm to 100mm deep section while a 250mm deep section will span 2500mm, depending upon design wind speed and relative loads. When fixed at their ends, a fixed louvre assembly can be made without visible fixings.

In some cases, the supporting structure comprises a mild steel frame fabricated from box sections, supported by tube sections that spring from points below the roof. The box sections are shown bolted together to form a flat frame structure, into which are set louvre panels, prefabricated and finished in a factory, then fixed to the supporting mild steel frame on site. The steel tubes are fixed to the flat frame with pin connections. Two flat plates are welded to the ends of the tubes, and a single plate forming a cleat is fixed to the underside of the flat frame and the base support below the roof. The fork ends of the tubular supports and the support cleats to which it connects are fixed together with face-mounted or countersunk bolts to form a visually crisp connection. The tubes have tapered ends which is typical of this type of steel construction. The aluminium louvre panels are fixed to the supporting flat frame with brackets that are welded to the sides of the bottom flange of the steel I-section. The aluminium louvre panel is supported on these brackets, with a nylon spacer between them to allow for thermal movement. The louvre panel is fixed to the cleat bracket with a countersunk bolt.

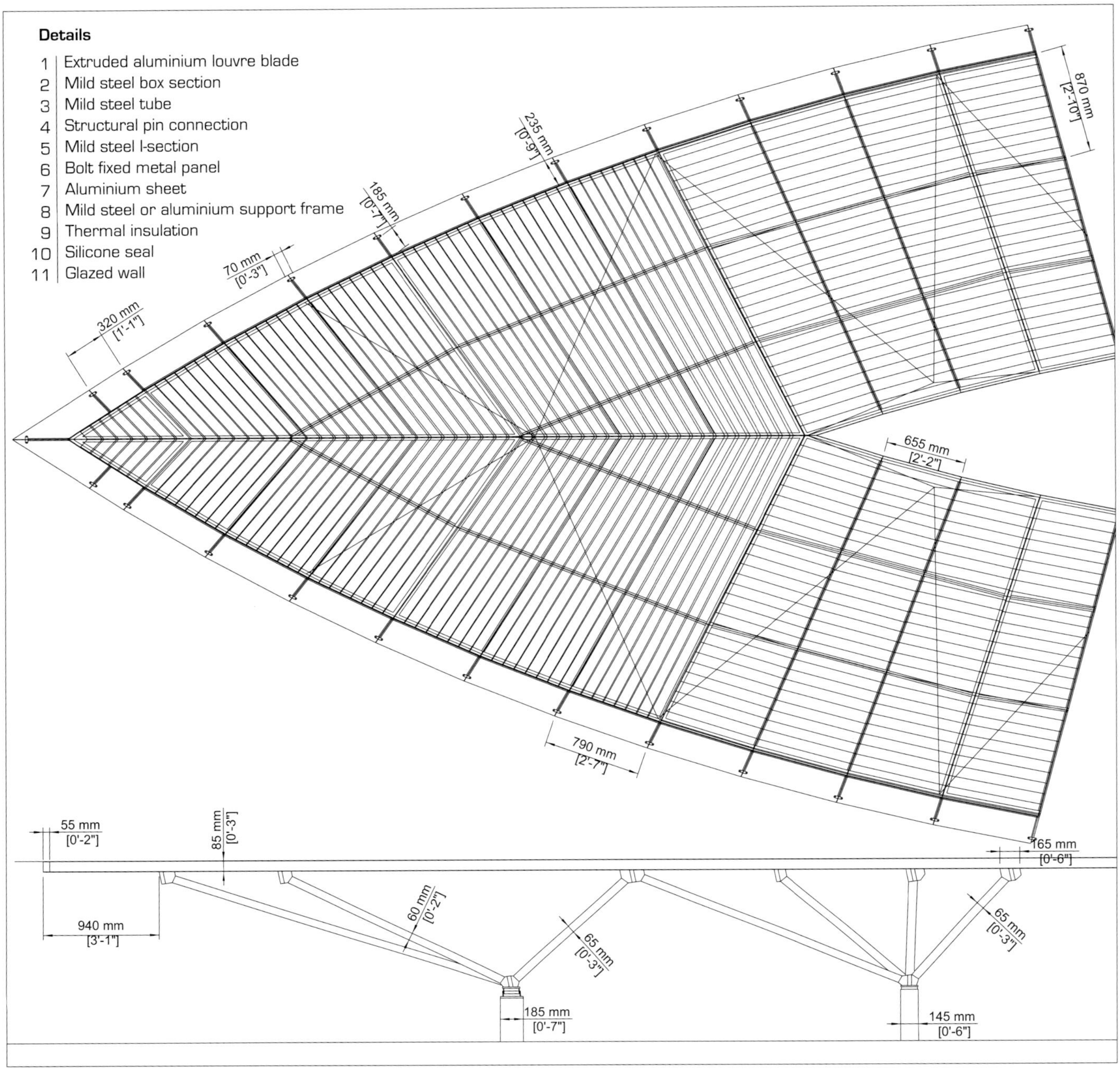

a| Horizontal and vertical section 1:50. Typical canopy assembly

Metal canopies

Metal canopies use all the techniques available to metal: standing seam, profiled sheet, composite panel and rainscreen panels with a membrane beneath. However, the preferred technique for metal canopies is usually none of these, but instead a method suited to the small-scale construction of complex geometry. This method uses metal panels with folded edges, sometimes called cassettes; the joints between panels being sealed with silicone sealant. Panels can be individually folded to different shapes to form a canopy of complex geometry. There has been a gradual move towards larger metal panel sizes, even a complete form made from metal coil, which imitates the homogeneous forms found in GRP membranes and tent structures, but with the more durable finishes associated with metal panels. In common with other canopy types, metal canopies are expected to be fully watertight. The generic example here follows this principle.

In generic examples, the metal canopy also serves as a sealed roof (MCE_365, d) to part of a glazed wall. An inclined glazed wall meets a metal canopy at mid-height which might form part of an entrance, or may serve as solar shading to a facade. The sealed outer layer of the canopy comprises metal panels which are folded to form an overall curved profile, each panel being made from metal sheet which is folded at its edges to form a tray. The folded edges form the surface to which the silicone sealant adheres. An alternative method of joining metal panels together is by riveting them together, leaving a hairline joint between the sheets. This is done either by butting flat sheet together and riveting the two sides together, or by folding the sheets down and riveting the underside. In some situations, the sheets are fixed with countersunk rivets fixing the edges of each sheet in place. A compressible polymer sheet is set beneath the sheets to provide a weathertight seal. Alternatively, a similar sheet material can be set between the riveted panels to provide a seal. In practice, both seals

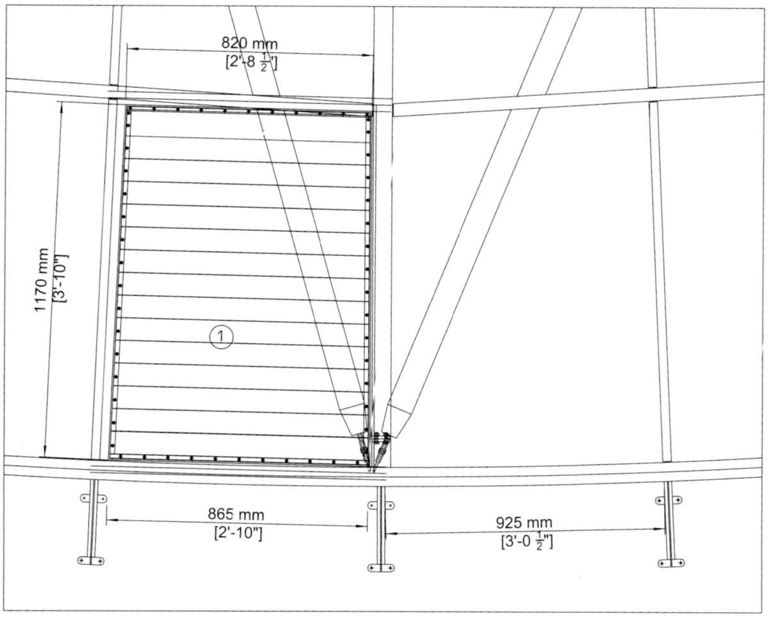

b| Vertical section. Typical louvre panel

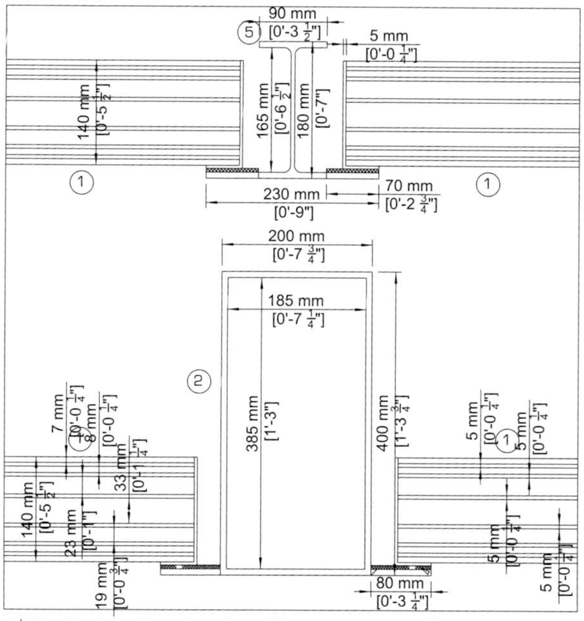

c| Vertical section. Junction of louvre panel and frame

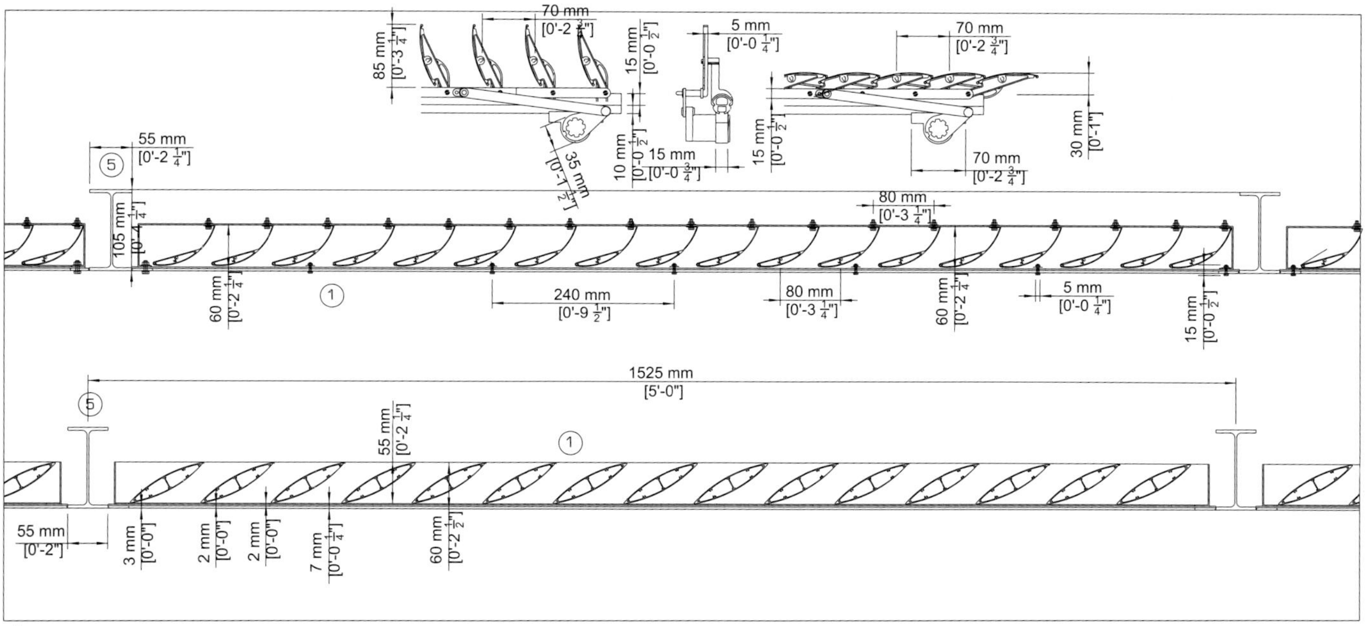

d| Vertical section. Typical louvre panel

discussed are difficult to achieve where a high weathertight performance is required. Instead, a silicone seal of around 10mm wide is used between panels which are fixed with brackets back to a supporting structure (MCE_361, d). This provides a reliable seal that maintains its weather tightness in the long term.

The supporting structure is made from either mild steel or aluminium sections. Aluminium sections are usually preferred for their durability but mild steel is often used for its greater rigidity (MCE_361, d). Mild steel is galvanised, painted, or both; while aluminium, with its greater durability, can be natural, anodised or cromated (similar to anodising) depending on the individual application. Metal panels of large size, made from 1200mm x 2400mm (4ft x 8ft) sheet, can result in a gentle oil-canning effect around their edges. This gives panels their characteristic soft edge, but they look smooth and consistent, and this is usually accepted visually as part of their appearance. If much thicker sheet is used, there is a disadvantage in increased cost as well as a greater difficulty in working the sheet to form smooth-shaped pressed panels. Aluminium panels are typically either PVDF coated or polyester powder coated, while steel sheet is usually polyester powder coated only. The use of anodised aluminium as a finish on sheet has increased in recent years as a result of the greater reliability of the finish, which until recently has suffered from uneven colour consistency. In many cases, the silicone used to seal the joints can be applied in a variety of monotone tints, ranging from white, to greys, to black. Some silver-grey colours match well with silver-coloured panels to give an overall appearance of a single colour for the complete canopy, particularly when recessed joints are used between panels. Flush joints in silicone tend to give the canopy surface an homogenous appearance, giving it more the appearance of concrete than of metal. This can detract from the crisply fitted assembly of panels which is characteristic of metal and is also one of the advantages of the material.

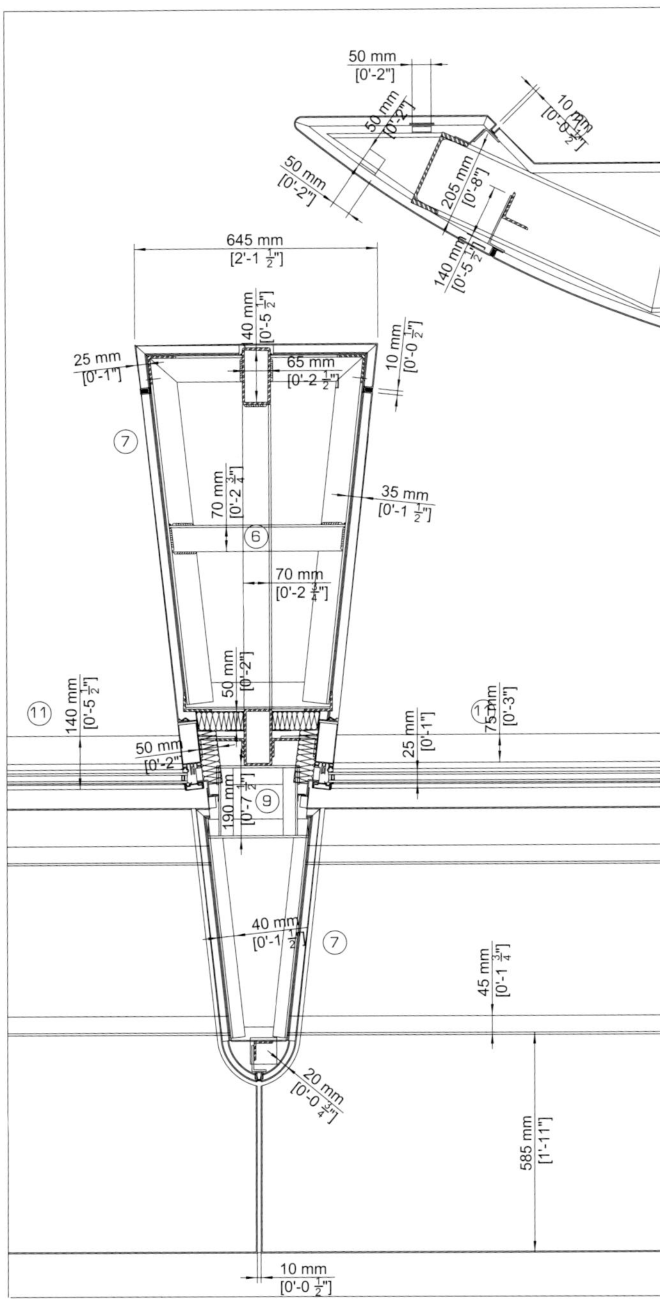
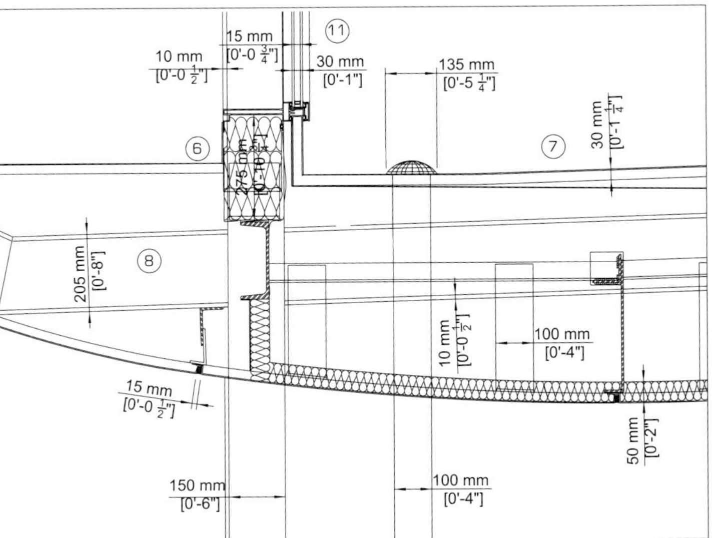

a| Horizontal and vertical section. Typical metal canopy

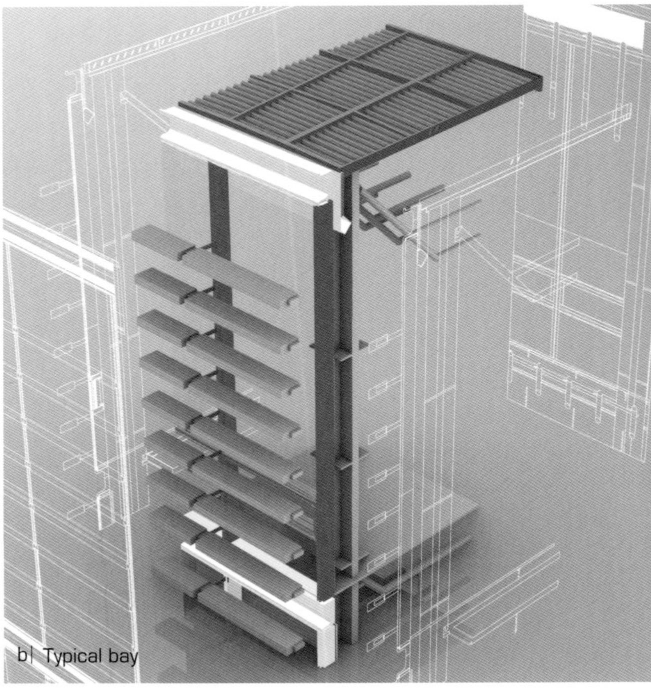

b| Typical bay

Metal panels can bring together the full set of assembly techniques available to the material within a single assembly. With folded panels, the edges are welded and ground smooth, and the complete panel is coated to a single consistent colour. The use of countersunk screws, painted on site in a colour matching that of the panels, enhances the overall appearance.

In generic examples, the glazed wall above the canopy is sealed to it by extending the metal sheet forming the external gutter up into the bottom transom (horizontal glazing bar) of the glazing system to form a continuous seal with the glazed wall above. The thermal insulation within the canopy is continued in order to meet the thermally broken transom where it forms a continuity with the double glazed units above. The metal sheet forming the top of the canopy on the internal side of the wall is lapped up the internal face of the glazed wall framing where it is folded under the bottom transom. The metal sheet forming the top of the canopy becomes a vapour barrier on the internal side of the wall, rather than the full weather barrier used on the external face of the canopy.

The glazed wall on the underside of the metal coping is joined in a way that disturbs the line of the canopy as little as possible as it passes it from outside to inside. The metal sheet on the exterior face of the wall is glazed into the top transom of the wall below (c). Any excess rainwater that is blown into the joint drips off the top of the transom pressure plate and capping. Small amounts of rainwater that find their way into the transom are taken away through the ventilated internal drainage system of the glazed wall framing. The metal sheet forming the underside of the canopy on its internal face is taken as close to the glass as possible before being folded back to align with the transom.

Metal canopies are increasingly using a single seal between panels to form a weathertight joint. This method uses the application method of

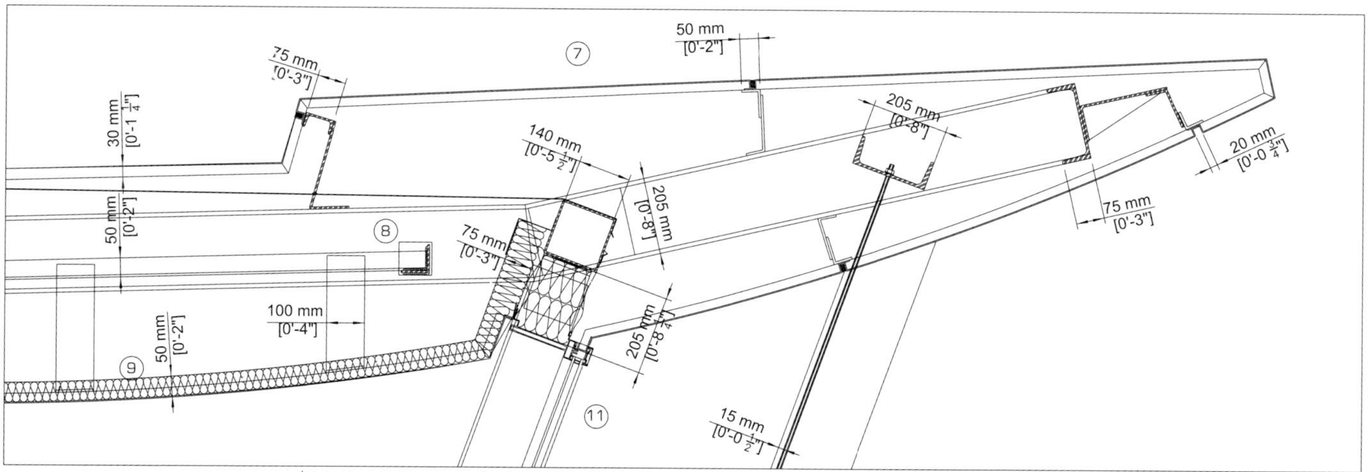

c| Horizontal and vertical section. Typical metal canopy

Details

1 | Extruded aluminium louvre blade
2 | Mild steel box section
3 | Mild steel tube
4 | Structural pin connection
5 | Mild steel I-section

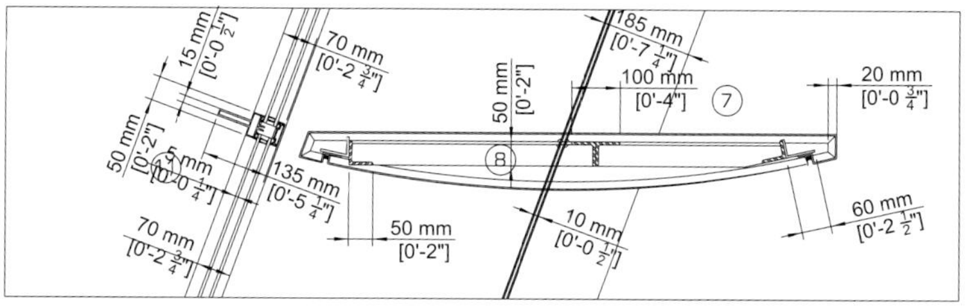

6 | Bolt fixed metal panel
7 | Aluminium sheet
8 | Mild steel or aluminium support frame
9 | Thermal insulation
10 | Silicone seal
11 | Glazed wall

d| Prototype

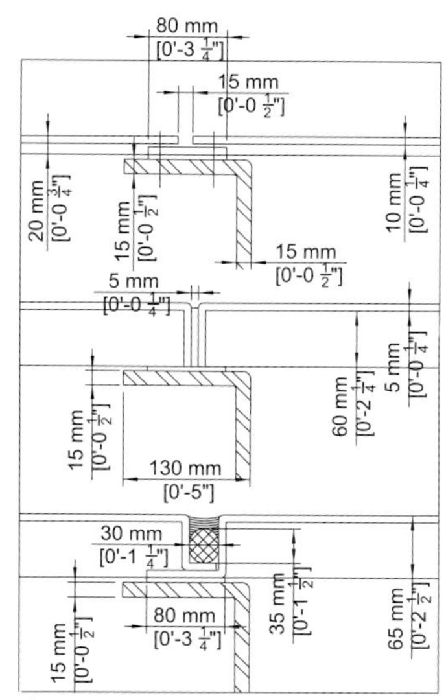

e| Vertical sections. Joint details

bolt fixed glazing, where a reliable silicone seal between double glazed units is the norm. However, where a foam-based backing rod is used to support the silicone applied from the outside, workmanship on site must be of the highest quality to ensure a durable weathertight seal. If this is difficult to achieve, as where the edge of the panel has a small edge return, or where panels are not very rigid due to their geometry or manufacturing method, a second line of defence for weather tightness is provided by an elastomeric or thermoplastic waterproof membrane. The ends of the waterproofing sheet at their top and bottom ends are fixed into the glazing system above and below, where small amounts of water that penetrate the outer silicone seal are allowed to drain to the outside.

Electrically operated louvres

Louvre canopies are also used for horizontal and inclined glazed roofs, excluding up to 90% of solar heat gain when set at a 45° angle. Louvre blades are 75mm–100mm (3in-4in) in typical proprietary systems but blades up to 300mm (12in) wide can be made as a single extrusion. Nylon sleeves and washers are used at the connection of moving parts, rather than metal, to avoid the need for regular lubrication. Louvres can be solid or perforated to different percentages of solid to void, from around 10% void to a maximum of 50%, though the latter is difficult to fabricate.

Louvres are fixed by steel pins into a sliding aluminium section at each end of the profile. The louvres are also fixed at their centre in section. As the sliding aluminium rod moves, the aluminium louvres move together, opening and closing together. The sliding rods are connected at each end of the louvres that are fixed to a supporting frame of aluminium I-sections. The single tube is powered by an electric motor, and as it turns, the sliding arms move through the arrangement of gears.

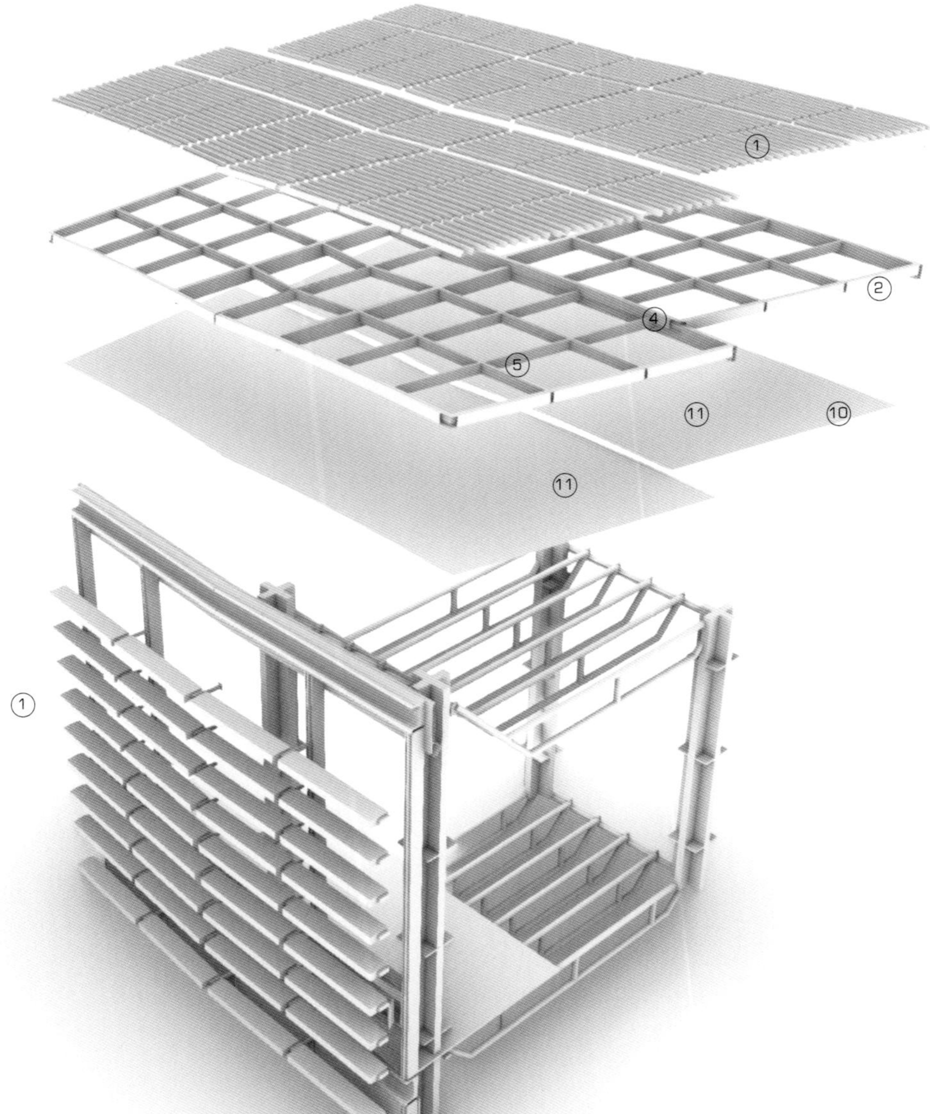

a | 3D exploded view showing metal canopy fixed to steel frame

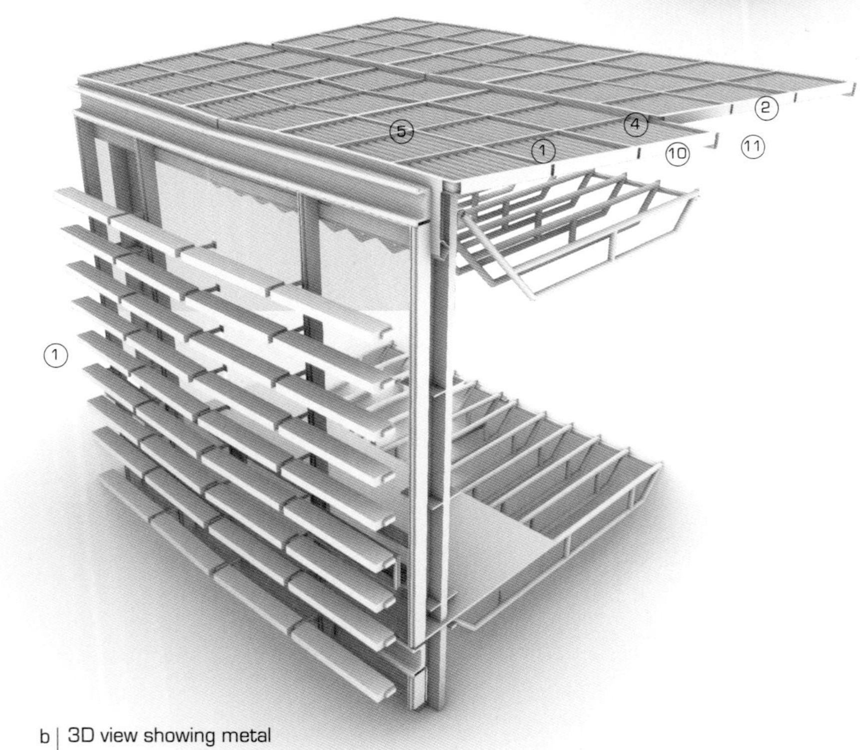

b | 3D view showing metal canopy fixed to steel frame

Louvres are typically a maximum of around 6000mm long for those 75mm to 100mm deep, requiring support at 1000mm to 1500mm. Sliding arms typically support up to a 6000mm length of louvre blades, giving an overall square shape (in plan) to each set of controlled louvres of 6000mm x 6000mm size which are fixed into the supporting I-section frame. A 100mm deep I-section will span typically 1000mm–1500mm between supports depending on the design of the glazed roof below. The distance between the moveable louvre panels and the glazed roof below is made sufficient to allow for access for cleaning the glass below and the louvre assembly itself.

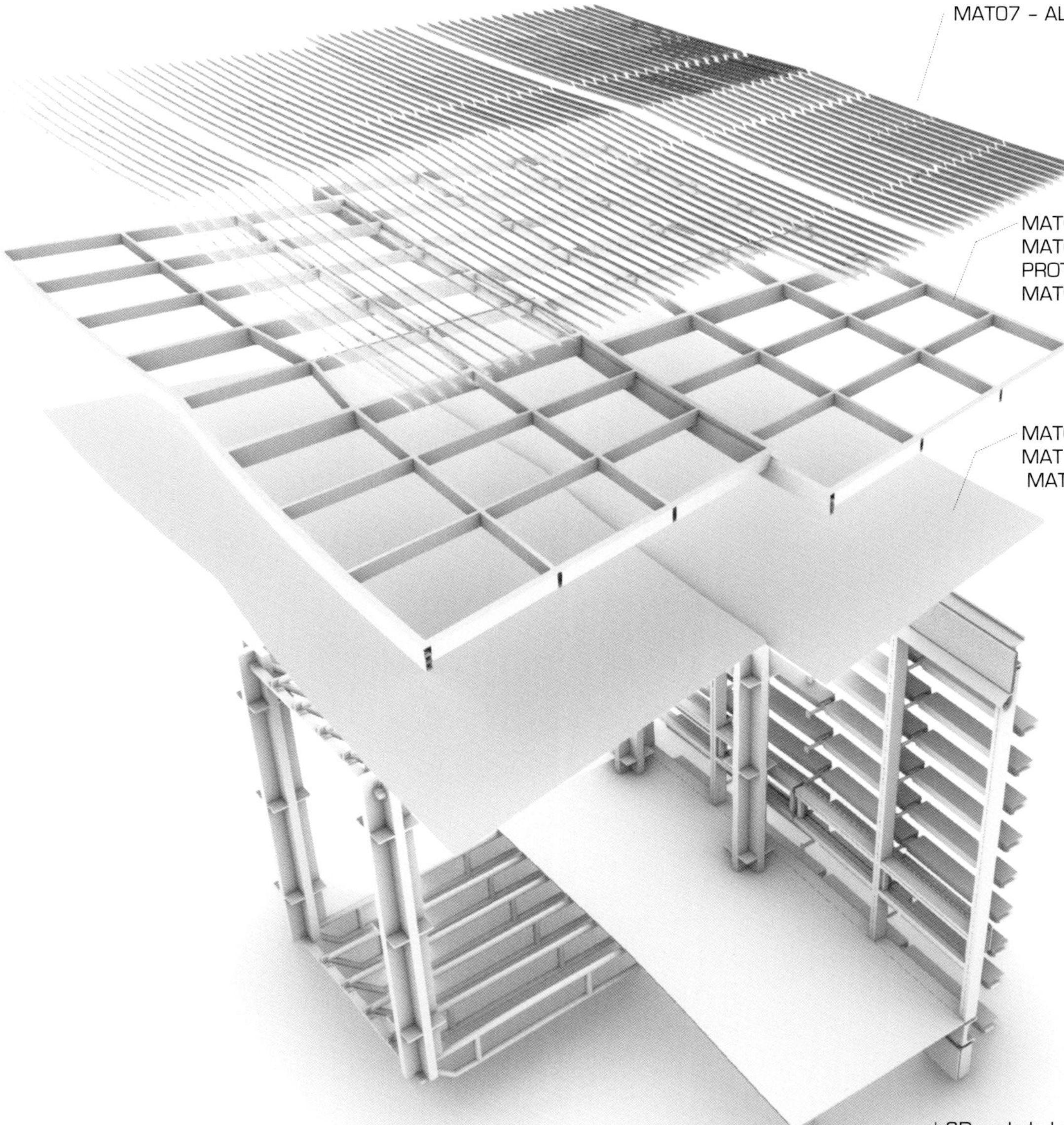

c | 3D exploded view of metal canopy components with material specification references

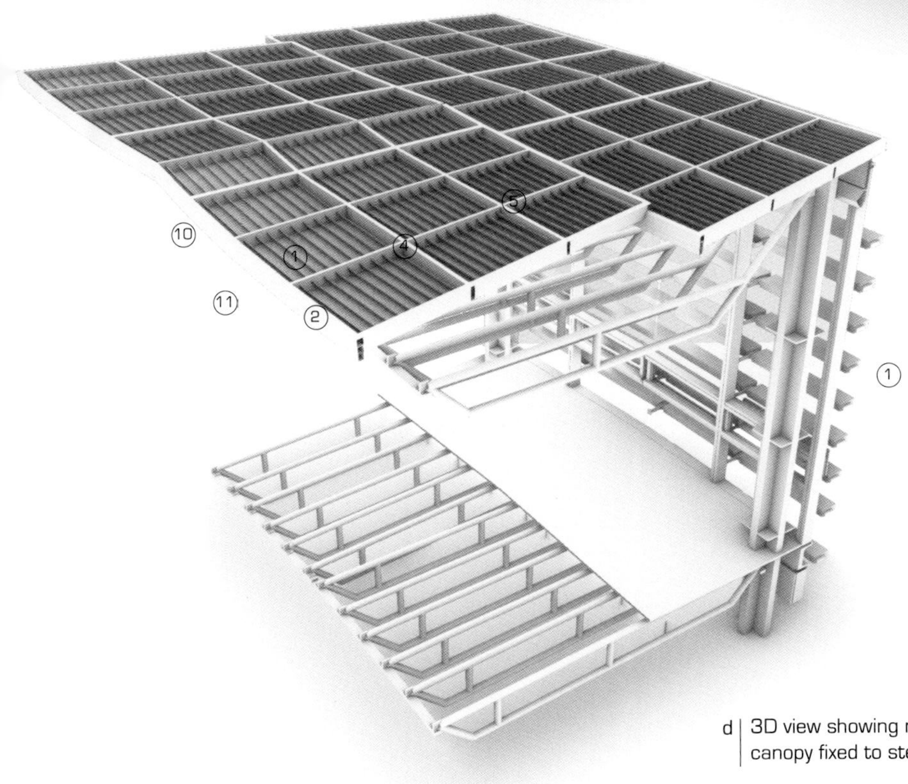

Details

1 | Extruded aluminium louvre blade
2 | Mild steel box section
3 | Mild steel tube
4 | Structural pin connection
5 | Mild steel I-section
6 | Bolt fixed metal panel
7 | Aluminium sheet
8 | Mild steel or aluminium support frame
9 | Thermal insulation
10 | Silicone seal
11 | Glazed units

d | 3D view showing metal canopy fixed to steel frame

Example project of glazed facades
Airport terminal, Egypt
Facade Engineer: Newtecnic
Architect: Newtecnic

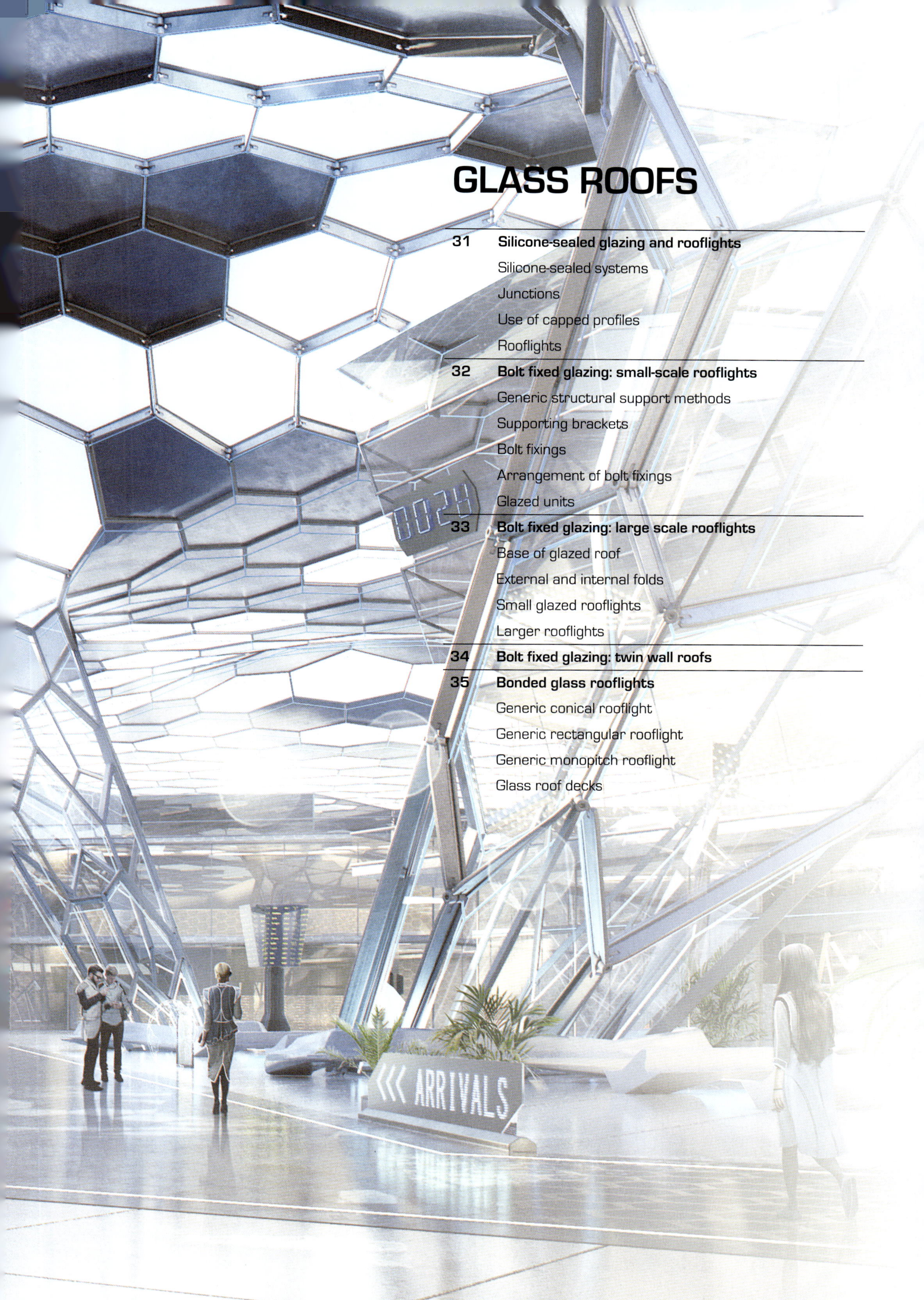

GLASS ROOFS

31 Silicone-sealed glazing and rooflights

- Silicone-sealed systems
- Junctions
- Use of capped profiles
- Rooflights

32 Bolt fixed glazing: small-scale rooflights

- Generic structural support methods
- Supporting brackets
- Bolt fixings
- Arrangement of bolt fixings
- Glazed units

33 Bolt fixed glazing: large scale rooflights

- Base of glazed roof
- External and internal folds
- Small glazed rooflights
- Larger rooflights

34 Bolt fixed glazing: twin wall roofs

35 Bonded glass rooflights

- Generic conical rooflight
- Generic rectangular rooflight
- Generic monopitch rooflight
- Glass roof decks

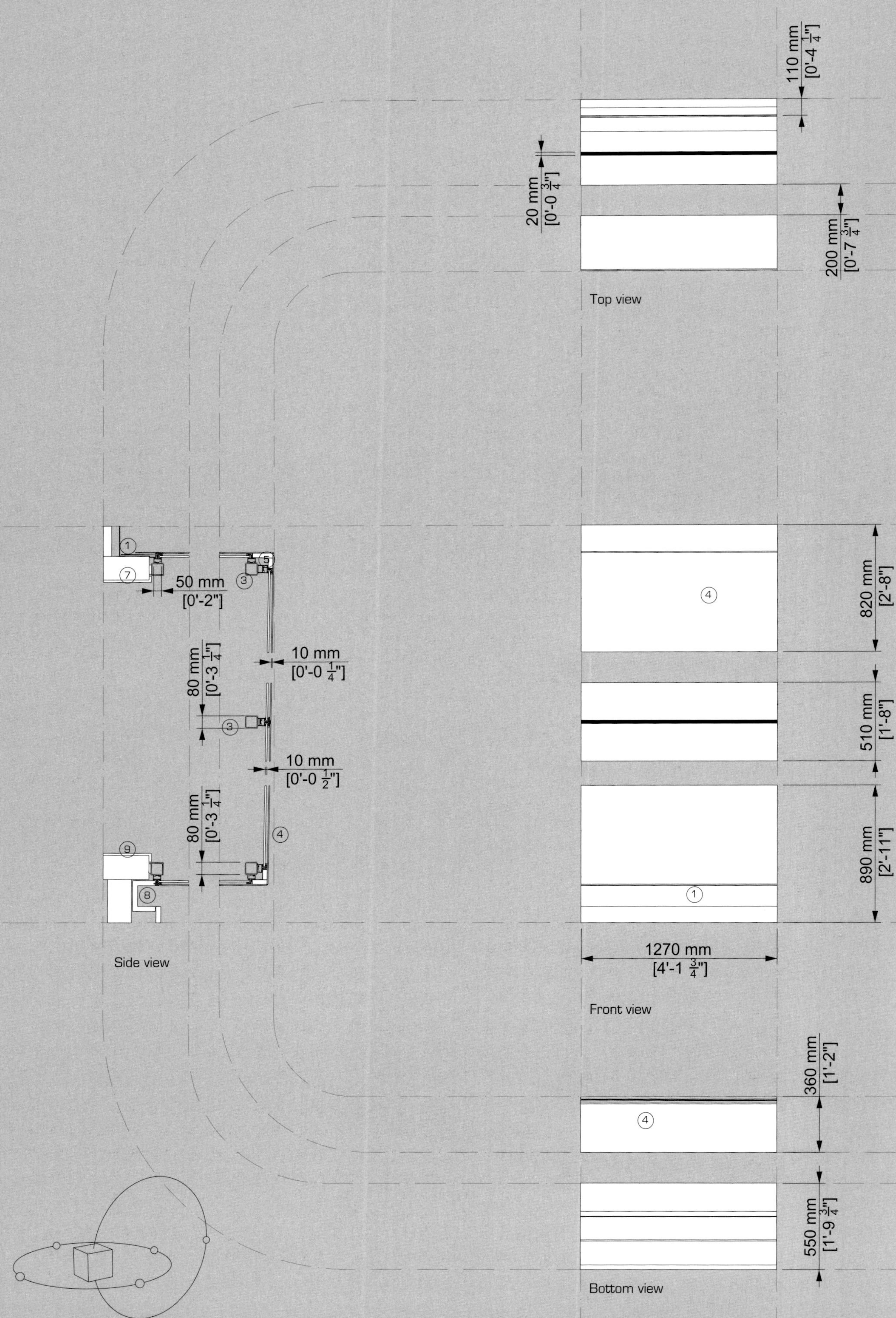

Third angle projection 1:30

1:10 scale

1:10 scale

Details

1 Extruded aluminium glazing profile
2 Pressure plate and capping
3 Mild steel support frame
4 Double glazed unit with recessed edge
5 Thermal insulation
6 Silicone seal
7 Concrete base
8 Gutter
9 Internal finish

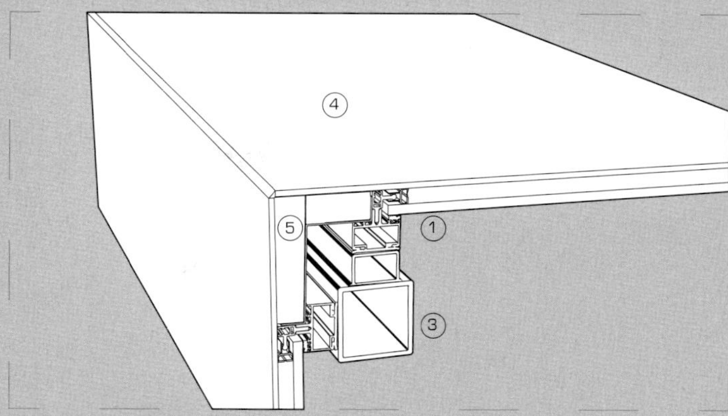

3D view of the assembly

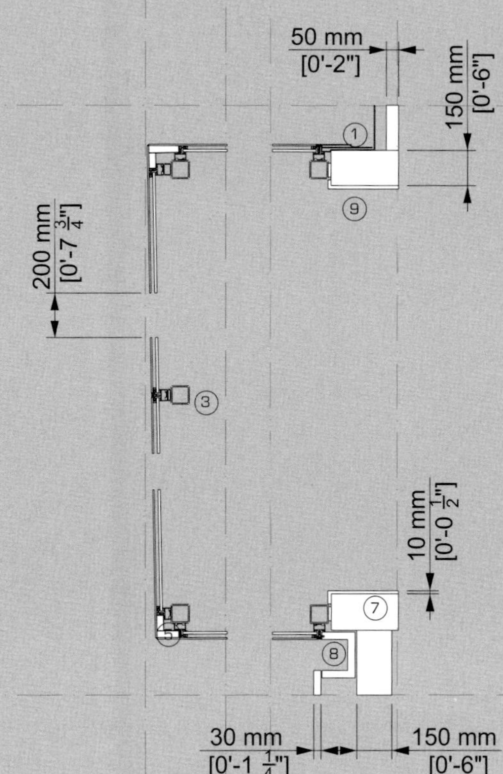

Side view

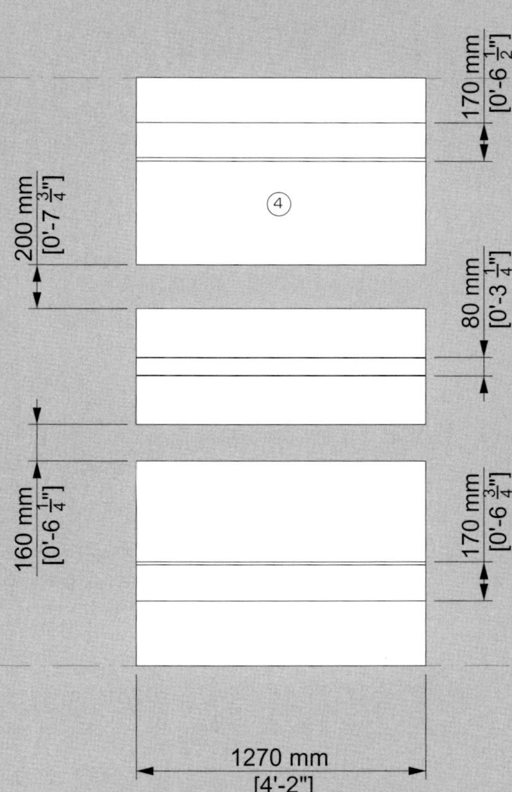

Back view

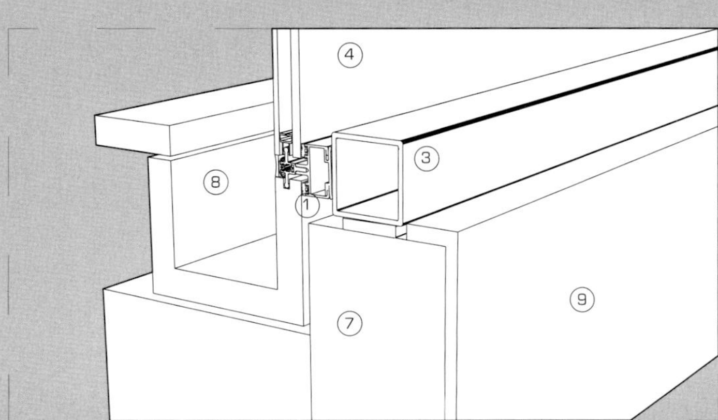

3D view of the assembly

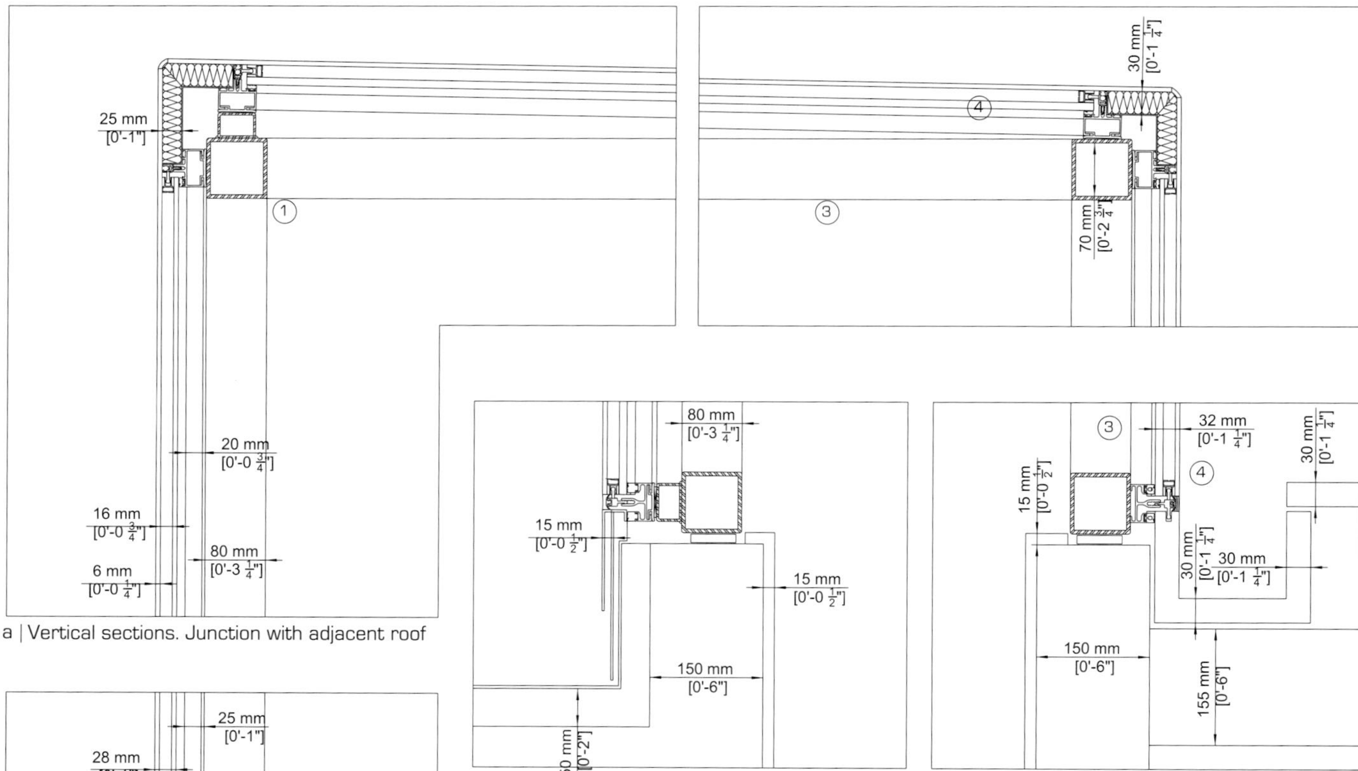

a | Vertical sections. Junction with adjacent roof

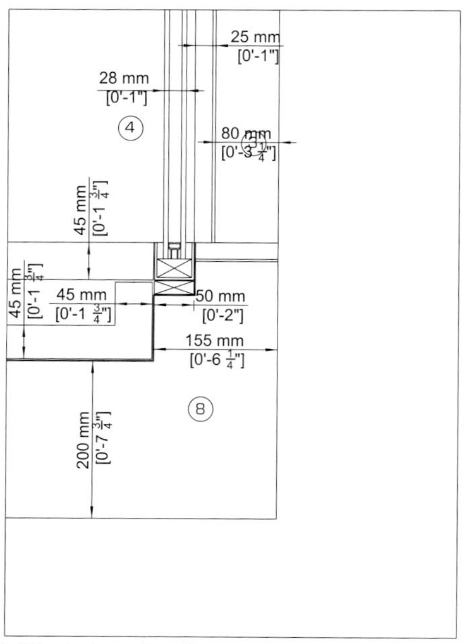

b | Vertical section. Corner

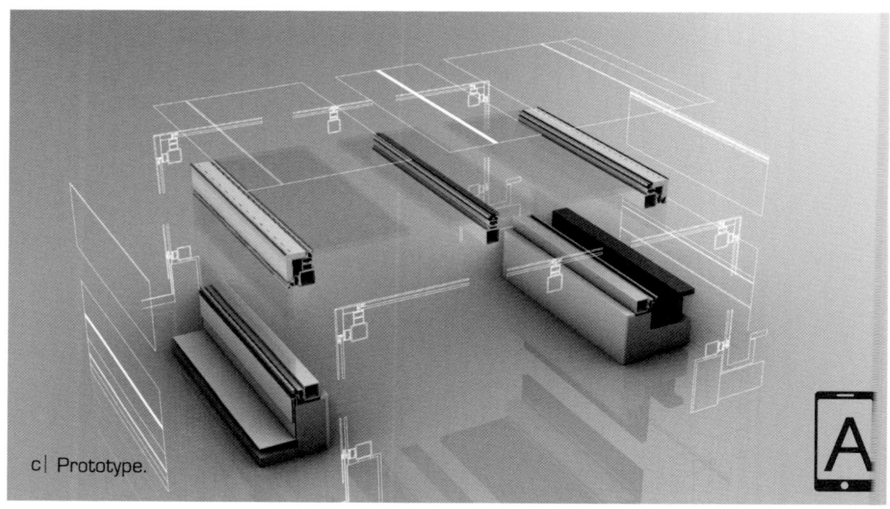

c| Prototype.

Silicone-sealed systems

While capped systems, described in the previous section, suit pitched roofs, they cannot reliably be used on (nominally) flat roofs, where the roof pitch is usually 3° to 5°. This is mainly because the rainwater running down the roof cannot pass the horizontal glazing bars which project above the surface of the glass. The smooth, continuous finish required for flat glazed roofs is achieved with a silicone seal between glass panels that is set flush with the surface of the glass. The glass is clamped in place with short lengths of pressure plate that are recessed below the outer seal. The recessed plates are secured to an aluminium channel which forms an integral part of the double glazed unit and which is continuous around its perimeter. The extruded aluminium channel is recessed into the gap at the edge of the unit usually used to bond the unit together and to seal the edges behind the spacer. The adjacent spacer in the double glazed unit both keeps the glass at a fixed distance apart as well as having desiccant within it to absorb any residual moisture within the sealed cavity between the glass sheets. The recessed aluminium channel is bonded to each glass sheet and also provides the edge seal to the completed unit, as in a regular double glazed unit (MCE_374, d). The short lengths of pressure plate are then set into the gap formed by the recessed channels of abutting glazed units, and are clamped to the glazing bar with self-tapping screws, typically at 300mm (12in) centres. The gap between the glazed unit is sealed with silicone, typically 15mm to 20mm wide, and with a backing strip or 'backing rod' behind it to form a back edge to the silicone seal.

In some cases, the glazing bar has its structural box or fin removed in order to fix it directly to a steel support frame. Square hollow sections are shown here (MCE_374, d). Alternatively, an all-aluminium glazing bar can be used. As with capped systems, small amounts of rainwater that pass through the outer silicone seal are drained away in the condensation channels set below the glass, within the glazing bar. In

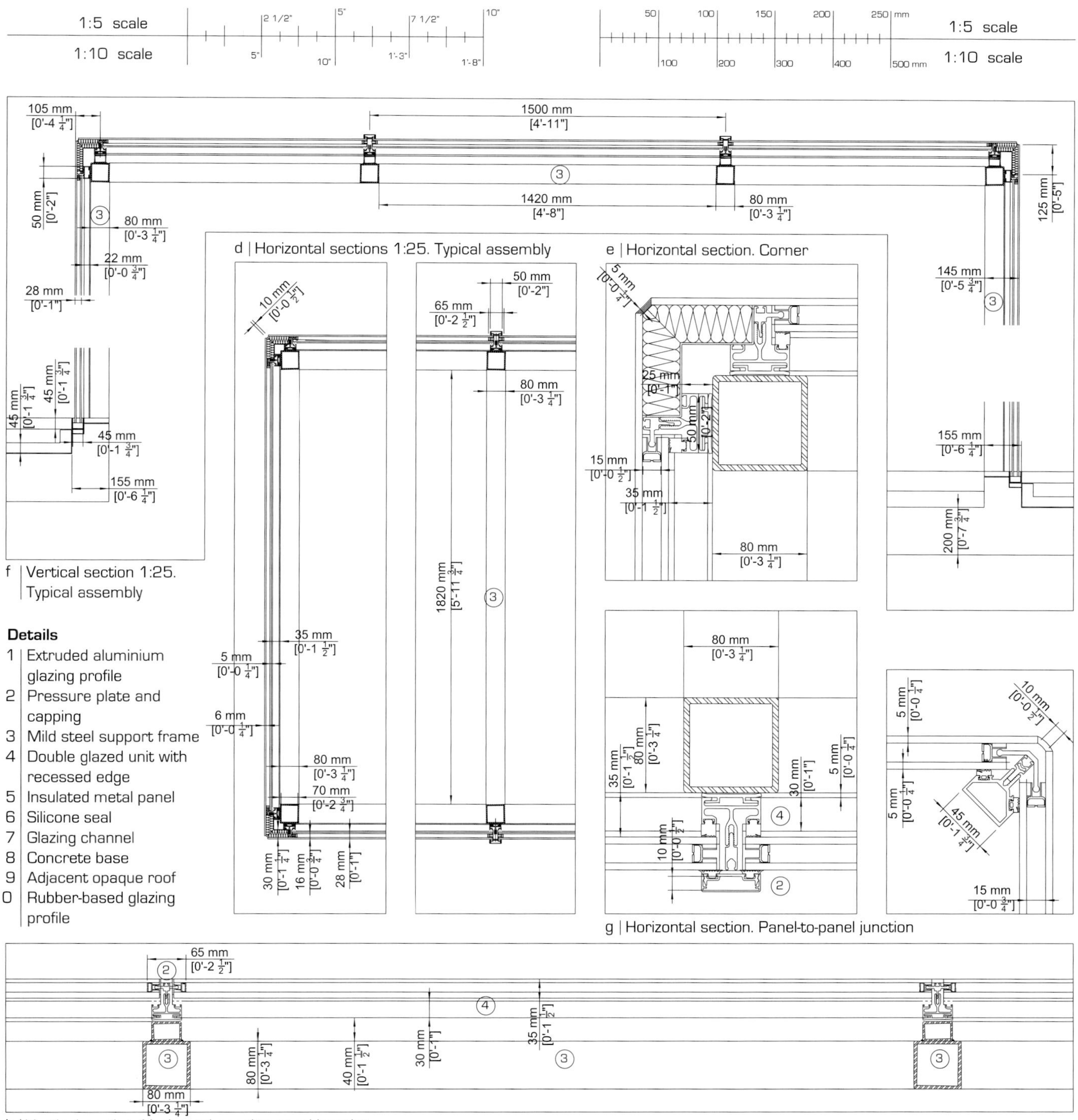

d | Horizontal sections 1:25. Typical assembly

e | Horizontal section. Corner

f | Vertical section 1:25. Typical assembly

g | Horizontal section. Panel-to-panel junction

h | Vertical section. Horizontal panel-to-panel junction

Details

1 | Extruded aluminium glazing profile
2 | Pressure plate and capping
3 | Mild steel support frame
4 | Double glazed unit with recessed edge
5 | Insulated metal panel
6 | Silicone seal
7 | Glazing channel
8 | Concrete base
9 | Adjacent opaque roof
10 | Rubber-based glazing profile

practice, silicone seals are very reliable but are dependent upon good workmanship on site, so the condensation channel is often not used in practice but serves as a secondary chamber to support the inner air seals. Silicone-sealed glazing bars can be used in all directions across a roof, unlike capped systems, since the glazing bars present no barrier to the passage of water.

Junctions

Since the advantage of silicone-sealed systems is of continuous glazed surfaces uninterrupted by visible glazing bars, ridges and valleys are treated as simple folds in the surface of the glazing, since rainwater runs across the complete sealed surface of the glass rather than being directed into gutters across its surface. The short lengths of pressure plate can be folded in the factory to the required angle, while the glazing bar forming the ridge is the same as that used elsewhere on the roof, with some modifications to the angle of the clips that hold the inner EPDM seal in place (g). The edges of roofs are also treated as folds, with rainwater usually allowed to run off the edge into a gutter, either just below the roof, or down to the base of the glazed wall below the glazed roof. An advantage of this system is the ability of the roof to be continuous with a glazed wall in the same system with a simple 'fold', without reducing its weather tightness. Typically the wall is not very high, forming part of a larger glazed roof. Dust that is carried down off the roof during rain is washed down the vertical glazing, rather than being carried away in a gutter, but in practice glazed rooflights require regular cleaning to maintain their crisp appearance.

The folded corner is formed with either a single specially formed glazing bar, or with two glazing bars meeting (g). The recessed lengths of pressure plate are folded to form the required angle, and the silicone is chamfered to form a flat surface between the two meeting glass panels. Silicone is rarely used to make a sharp angle between the two

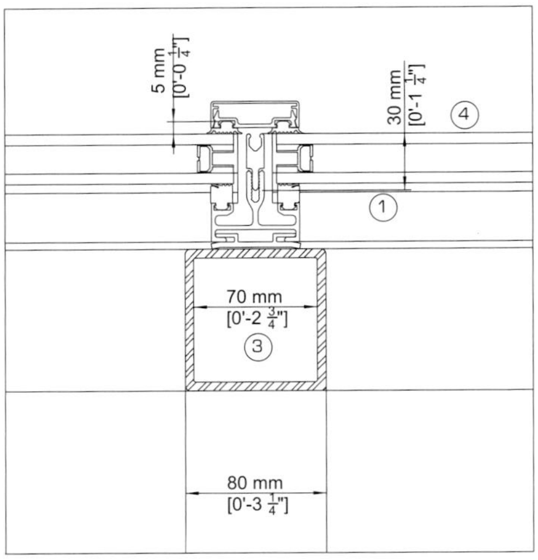

a | Vertical section. Horizontal panel-to-panel junction with capping

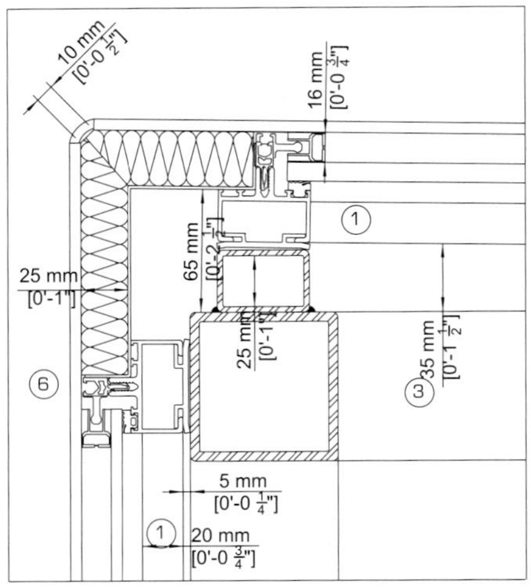

b | Vertical section. Corner

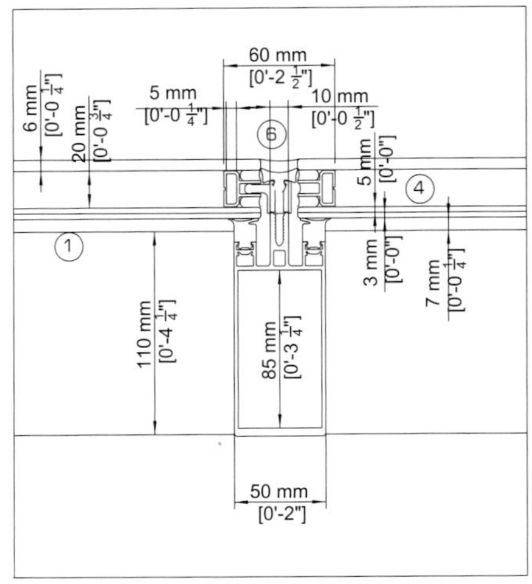

c | Vertical section. Horizontal panel-to-panel junction without capping

Details

1 | Extruded aluminium glazing profile
2 | Pressure plate and capping
3 | Mild steel support frame
4 | Double glazed unit with recessed edge
5 | Thermal insulation
6 | Silicone seal
7 | Concrete base
8 | Gutter
9 | Internal finish

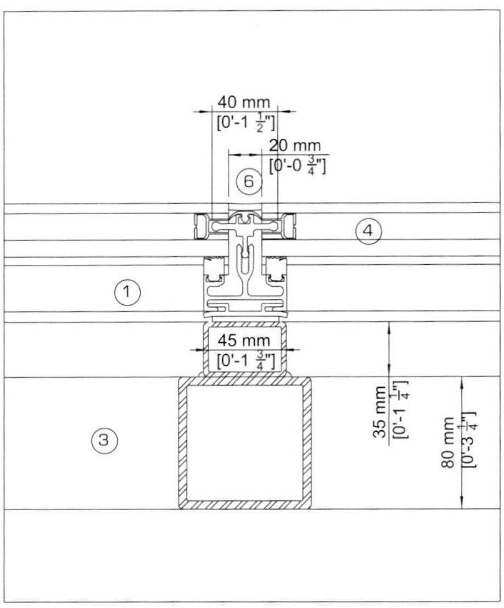

d | Vertical section. Horizontal panel-to-panel junction without capping

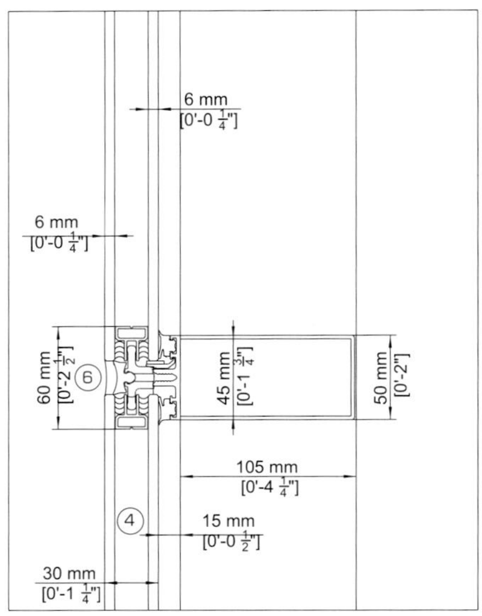

e | Vertical section. Vertical panel-to-panel junction with capping

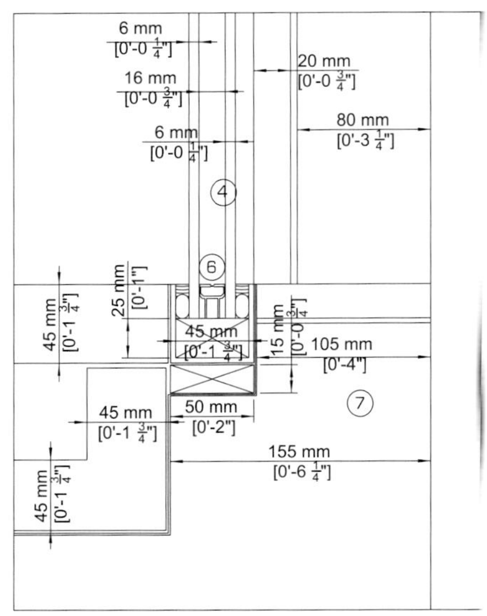

f | Vertical section. Junction with adjacent roof

double glazed units as it is very difficult to achieve a straight line without the assistance of an additional metal angle bedded into the silicone. The alternative method of forming a roof edge requires the edge of the double glazed unit to be coated or 'opacified' to avoid the frame behind being visible through the glass. This is often achieved by stepping the glazed unit, with the outer glass extending to meet the corner, while the inner glass stops at the glazing bar to allow the recessed aluminium angle to be bonded in its usual position adjacent to the glazing bar. The glass unit is secured in the same way, with a recessed length of pressure plate, while the outer glass is cantilevered to meet the adjacent glazed unit at the corner. Thermal insulation is bonded to the outer glass along the cantilevered edge to form a continuity of thermal insulation and avoiding a thermal bridge that will result in condensation occurring in temperate climates. The inner face of the thermal insulation is lined with a vapour barrier, typically a 3mm thick folded aluminium sheet. The opacifying of the glass is done by screen printing on the internal face of the glass. While black is often the preferred colour in order to match with the silicone seals, other colours and patterns are increasingly being introduced in glass manufacture.

Where a gutter is required, an insulated gutter is fitted to the glazing bar. Prefabricated gutters are used where they are seen from the underside, inside the building. They can be glazed into the system following the line of the internal face of the double glazed units. Where the gutter is required to be separated from the glazing structures, typically for structural reasons, a separate flashing can be used, fixed to the glazing bar but independent of the gutter. The gutter is still sealed to the glazing bar with a flexible EPDM gasket to provide a continuous seal.

The base of the glazed roof that terminates in a glazed wall can meet an adjacent concrete roof slab with either a metal glazing channel, or in an upstand and flashing. Where a glazing channel is used, the double

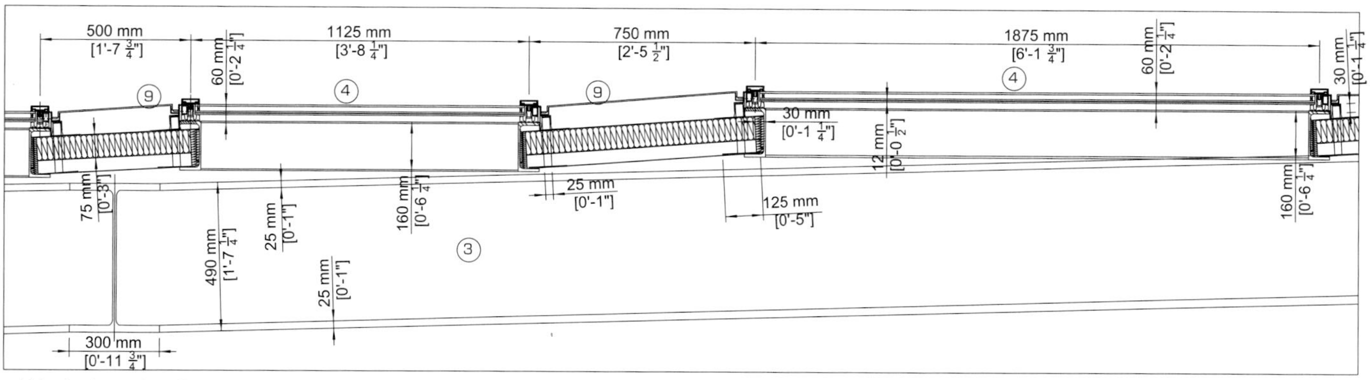

g | Vertical section. Capped glazing. Typical profiles

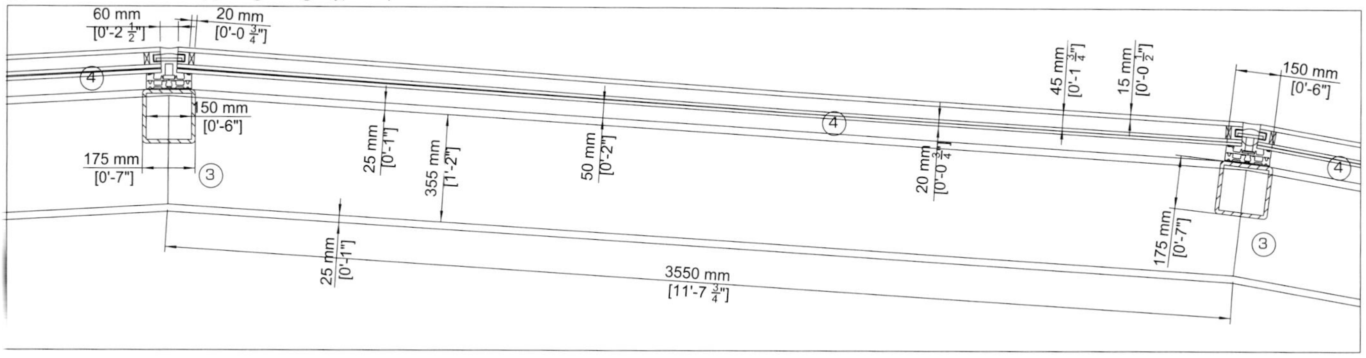

h | Vertical section. Capped glazing

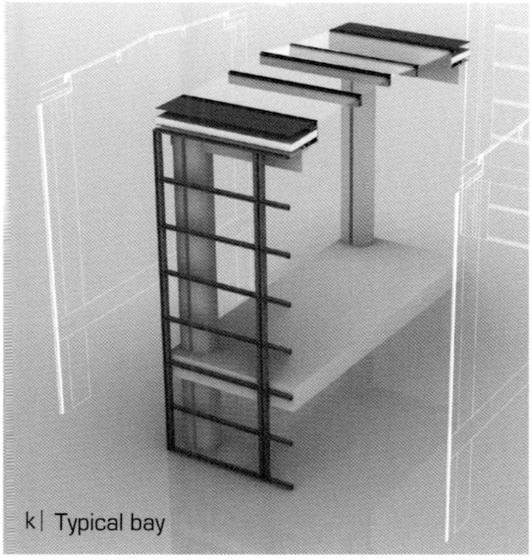

k | Typical bay

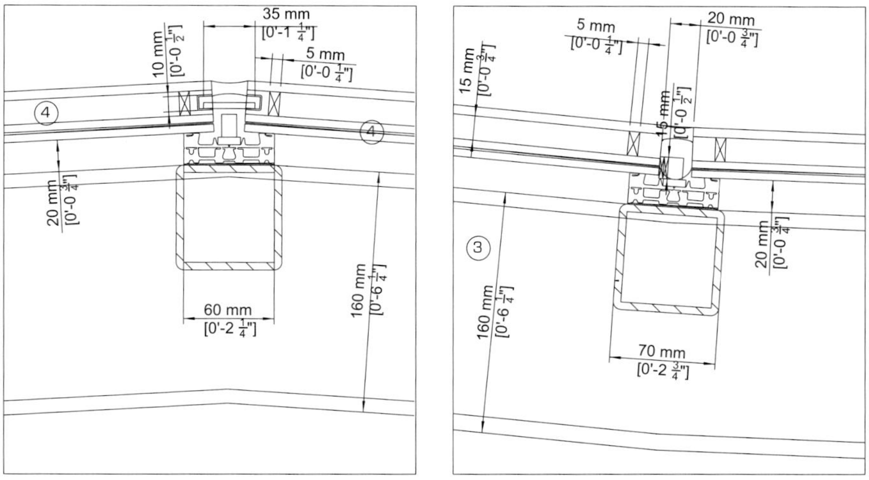

j | Vertical section. Horizontal panel-to-panel junction without capping

glazed unit sits in a profile formed from extruded aluminium, stainless steel or painted mild steel. The glass is levelled on metal shims (short lengths of metal strip), then sealed with silicone. The advantage of the glazing channel is that it can be set flush with the finished internal level to provide a junction with no visible horizontal glazing bar. Alternatively, the glazing can terminate in a horizontal glazing bar, to which an aluminium flashing can be fixed (MCE_376, b). A downstand flashing is more easily fixed to a horizontal glazing bar than a glazing channel. The vertical glazing bars (forming the mullions) are fixed to brackets that are secured to the upstand. A concrete upstand can be used with external insulation and a single-layer membrane. The metal flashing extends down the face of the upstand to protect the joint. As with gutters, an EPDM membrane extends down from the horizontal glazing bar where it is bonded to the membrane. The roofing membrane and the EPDM are always checked for compatibility, but in practice this rarely presents any difficulty.

Use of capped profiles

An advantage of silicone-sealed glazing over fully bonded glazing (discussed in the next section) is its ability to be mixed with capped glazing (MCE_377, f). Since both systems are drained and ventilated (pressure equalised), the same glazing bar can be used in a mixed roof system of flush silicone joints and capped profiles (MCE_377, g). Although this mix is often done for visual reasons, it does allow for easily formed junctions with adjacent areas of roof in different materials, and for a mix of actual panels and glazed panels in a single roof using a reliable drained and ventilated system.

The most common application of this method is where capped profiles are used for the vertically-set glazing bars running down the slope of a roof, while silicone-sealed glazing is used on horizontal joints to allow rainwater to pass down it unimpeded by any projecting glazing bars. Junctions in silicone sealed glazing, such as edges of roofs and

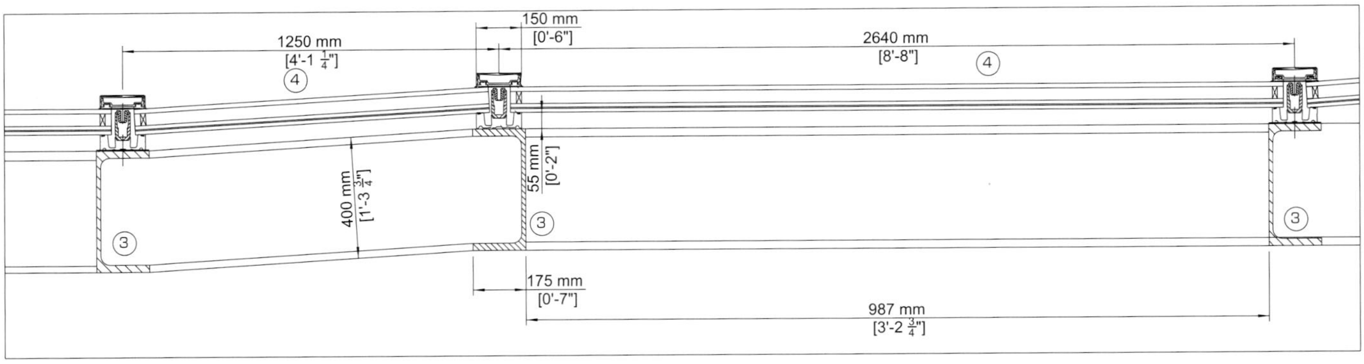

a | Vertical section. Edge of rooflight

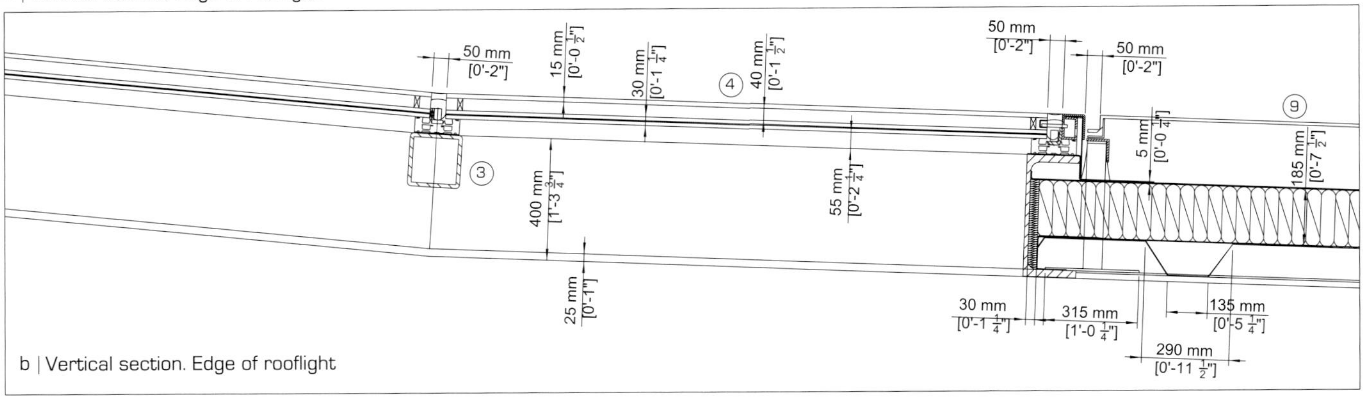

b | Vertical section. Edge of rooflight

Details

1 | Extruded aluminium glazing profile
2 | Pressure plate and capping
3 | Mild steel support frame
4 | Double glazed unit with recessed edge
5 | Insulated metal panel
6 | Silicone seal
7 | Glazing channel
8 | Concrete base
9 | Adjacent opaque roof
10 | Rubber-based glazing profile

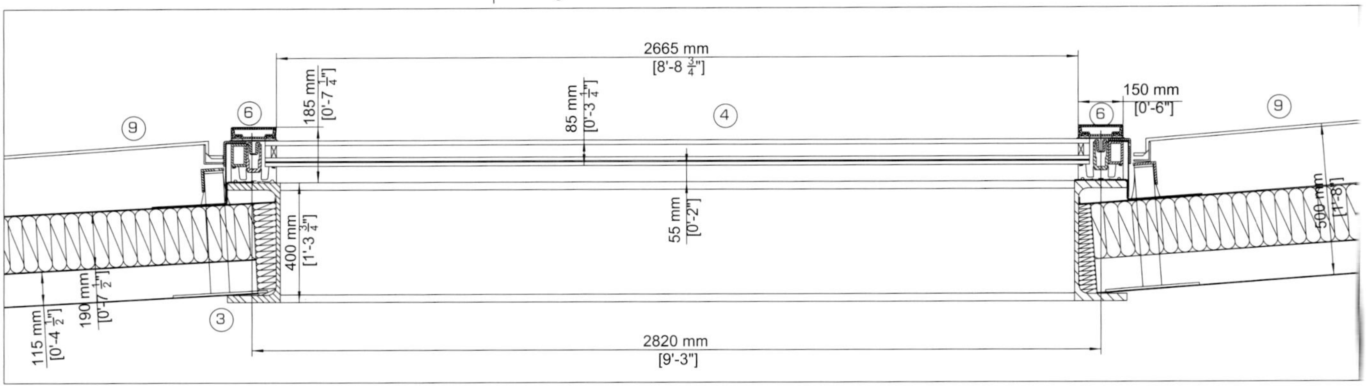

c | Vertical section. Typical profiles

ridges, are formed in the same way, while the capped system follows the folds with continuous pressure plates that are mitred and sealed at the folds. Butyl tape is used as an extra seal at folds, set between the pressure plate and the outer EPDM gaskets. Cover caps are also mitred to give a crisp appearance.

The internal drainage of the mixed glazing system works in the same way, with any water that penetrates the silicone-sealed glazing being drained along the internal condensation channels before draining into the profiles of the capped system and onwards down to the base of the roof. The condensation channel and inner air seal can also be formed in a single EPDM extrusion without the use of any aluminium extrusion, but with a supporting structure behind. In this case, the short lengths of pressure plate holding the glass in place are secured by self-tapping screws that are secured to a supporting structure behind. A hollow bo section can also be used. The EPDM gasket, combined with a steel sup- porting structure, is used as an alternative to the extruded aluminiu glazing bar, usually for larger glass panels where the supporting fi or box profile to the aluminium extrusion would be visually too dee or too wide.

Rooflights

A useful application of combining capped and silicone-sealed glazing i in sealed rooflights, where strips or bands of glazing are combined wit a roof in a different material (MCE_374, f). Traditionally, individual roo lights are formed in upstands that sit high above the level of the adja cent roof. Large numbers of these traditional individual rooflights la the visual elegance of continuously glazed roofs. In recent years th

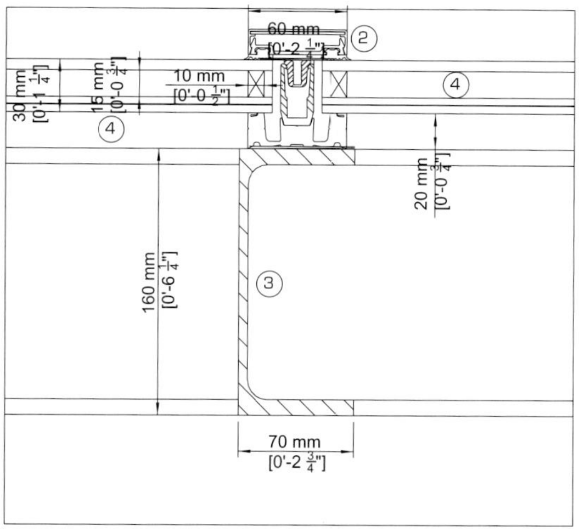

d | Vertical section. Horizontal panel-to-panel junction with capping

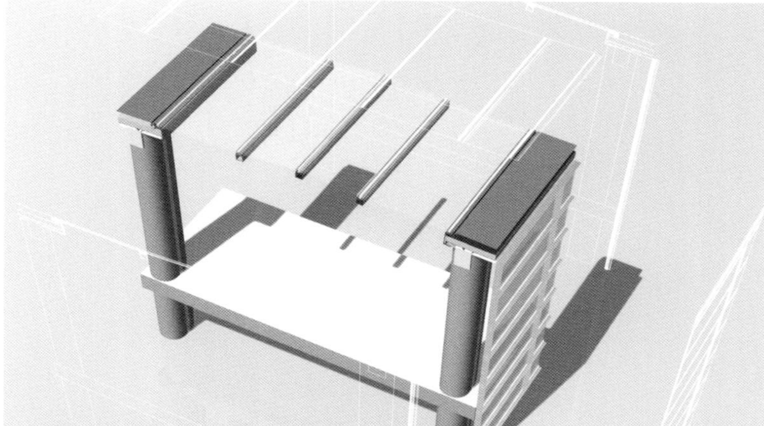

e | Typical bay

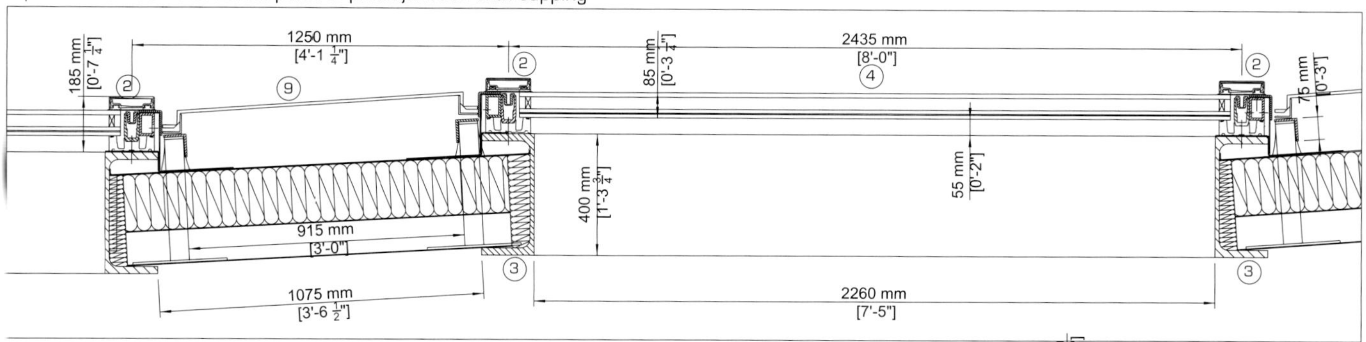

f | Vertical section. Junction of 2 rooflights with opaque roof

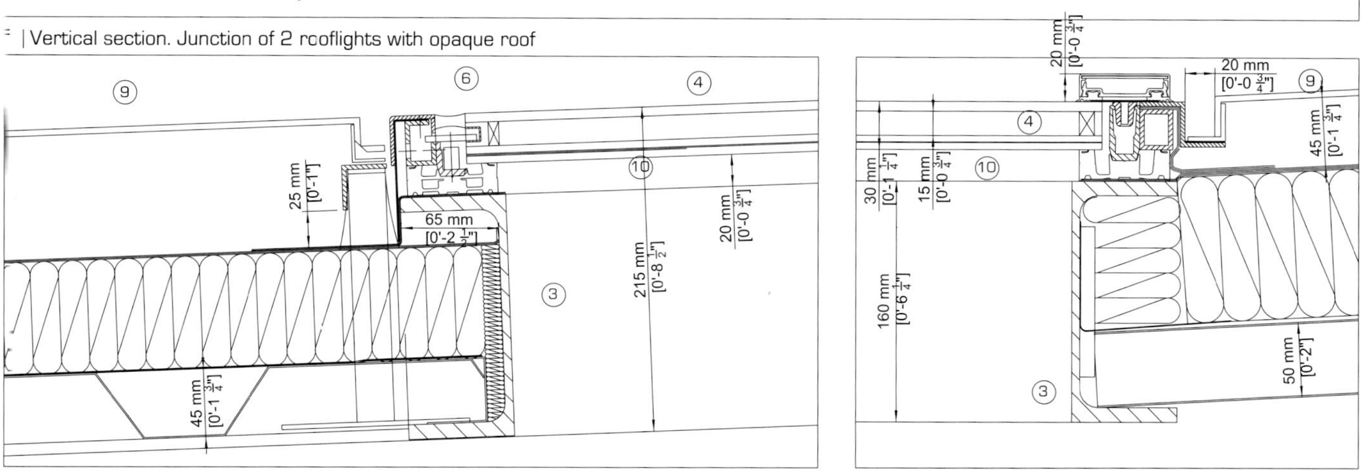

g | Vertical section. Edge of rooflight

has changed, with the increased use of the combination of single-layer membranes and rainscreen panels with capped and silicone sealed glazing to provide individual rooflights that are visually integrated into adjacent areas of opaque roof (MCE_375, g). A single-layer membrane roof can be sealed into the edge of a capped glazed rooflight by clamping the membrane into the glazing system. A metal rainscreen can also be incorporated, sitting completely independent of the rooflight, but set onto the membrane to both protect it and give a visual continuity to the rooflights across the roof.

The edge of the rooflight is formed with a capped glazing profile, while the joints within the rooflight, running across it (at 90°), use a silicone-sealed profile to allow water to run down to the bottom edge of the rooflight. This lowest edge of the rooflight has a silicone-sealed edge with a drip flashing glazed into the edge of the profile to drain water off the edge. The single-layer membrane is tucked under this flashing and is clamped down with the pressure plate that extends the full width of the rooflight. The top edge of the rooflight (running parallel with the roof slope) has a capped profile to allow water to drain around its sides. The condensation channels in all profiles are set at the same level to ensure that any water that passes through the outer seal is drained through a set of linked channels to the bottom edge of the rooflight where it is released above the roof level of the adjacent membrane. Openable rooflights, as shown in the previous section on capped glazing systems, can be incorporated easily, with the use of an additional sub-frame, into which the openable light is set.

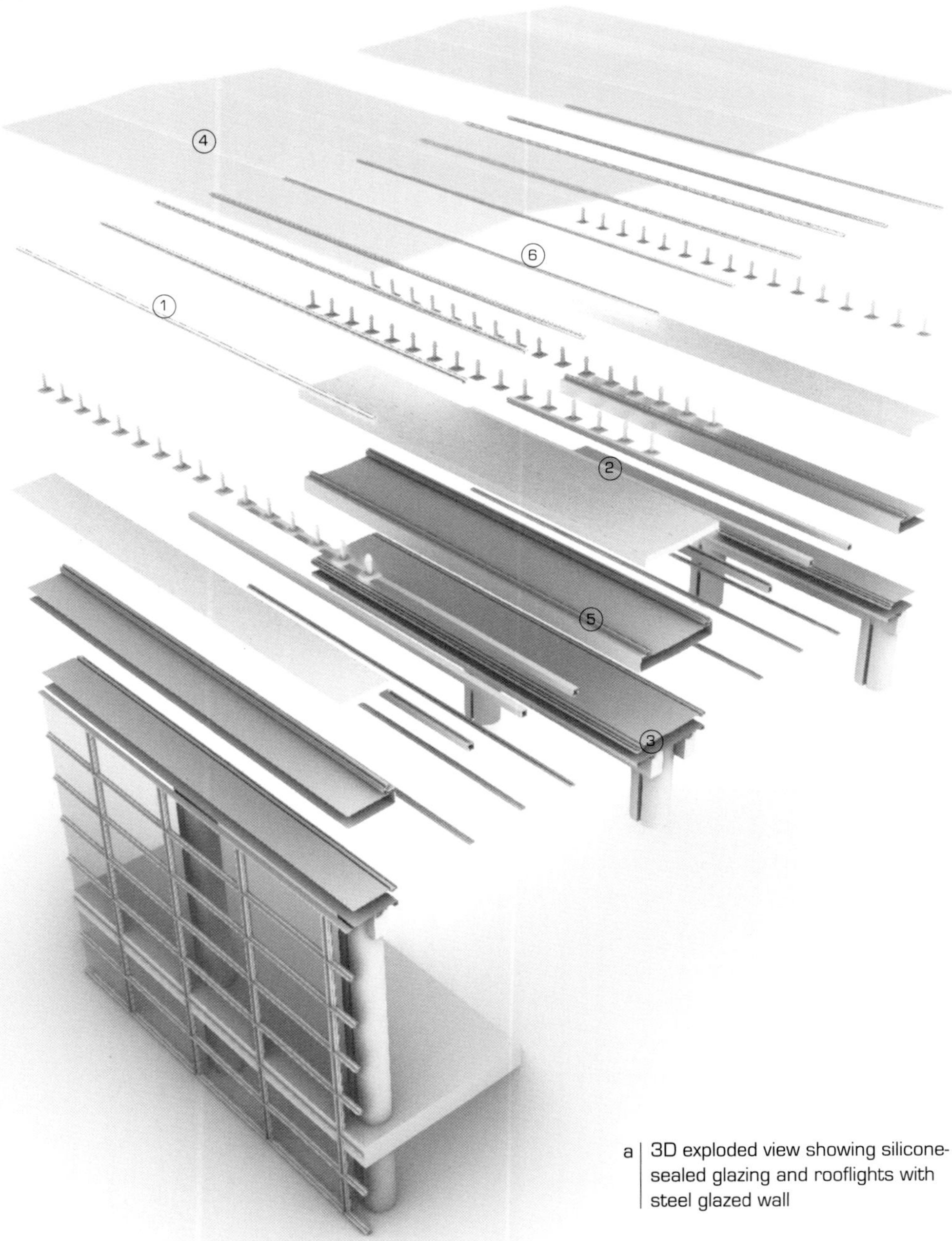

a | 3D exploded view showing silicone-sealed glazing and rooflights with steel glazed wall

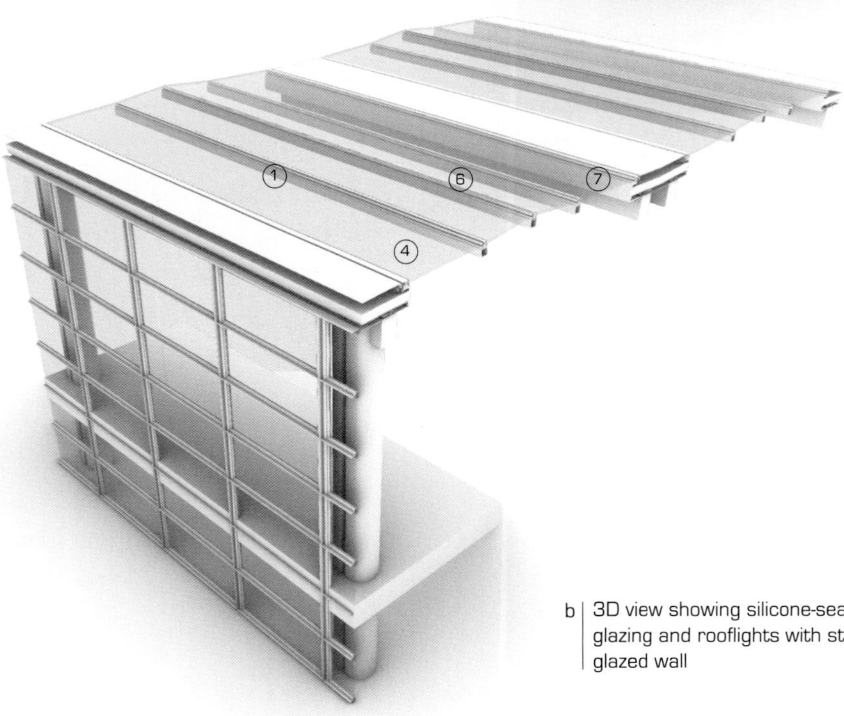

b | 3D view showing silicone-sealed glazing and rooflights with steel glazed wall

Details

1 | Extruded aluminium glazing profile
2 | Pressure plate and capping
3 | Mild steel support frame
4 | Double glazed unit with recessed edge
5 | Insulated metal panel
6 | Silicone seal
7 | Glazing channel
8 | Concrete base
9 | Adjacent opaque roof
10 | Rubber-based glazing profile

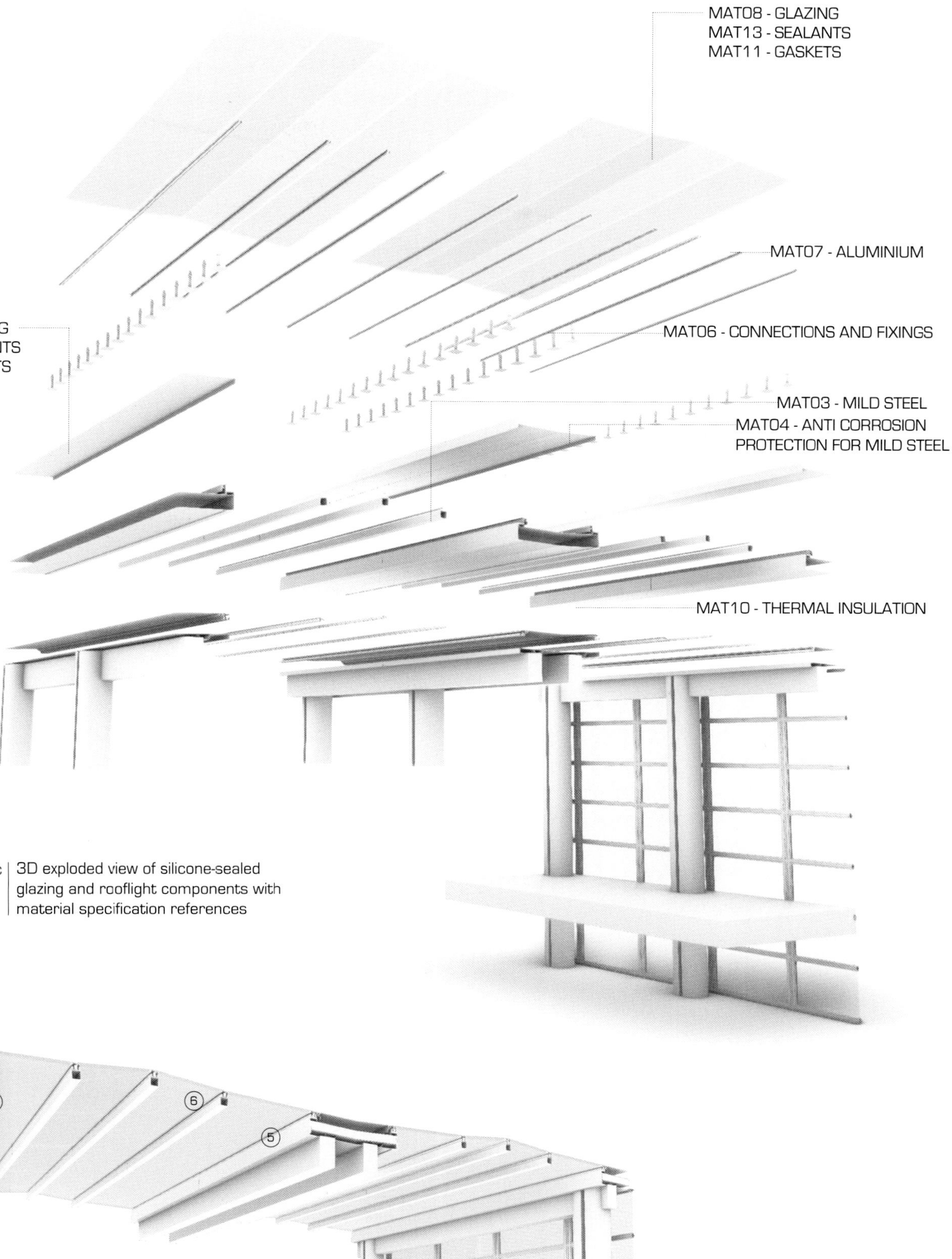

c | 3D exploded view of silicone-sealed glazing and rooflight components with material specification references

d | 3D view showing silicone-sealed glazing and rooflights with steel glazed wall

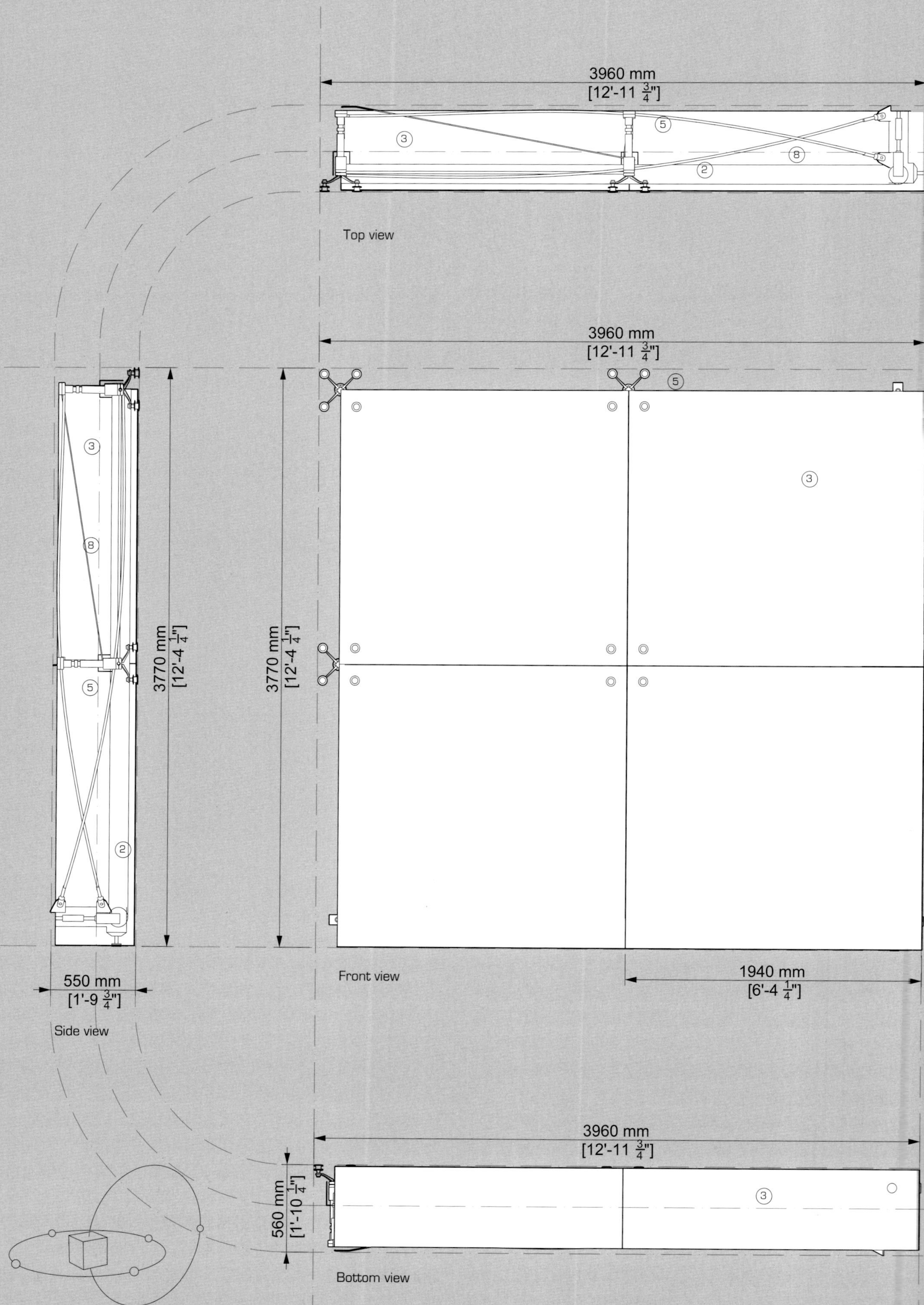

Third angle projection 1:30

1:10 scale

1:10 scale

Details

1 | Mild steel connector
2 | Mild steel support frame
3 | Double glazed unit
4 | Silicone seal
5 | Bolt fixing
6 | Support bracket
7 | Concrete base
8 | Stainless steel cable
9 | Mild steel tension rod
10 | Mild steel plate
11 | Adjacent external wall

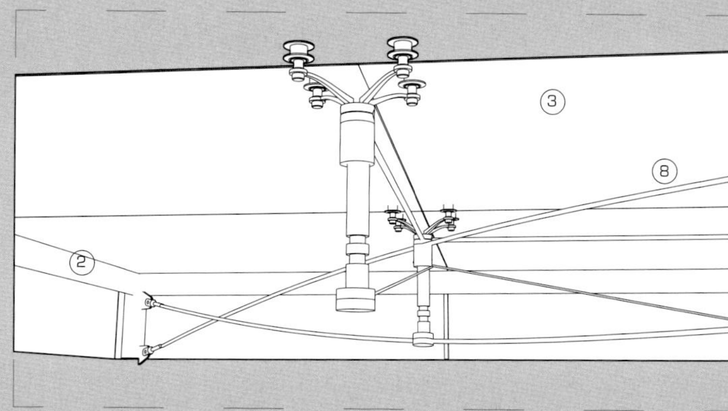

3D view of the assembly

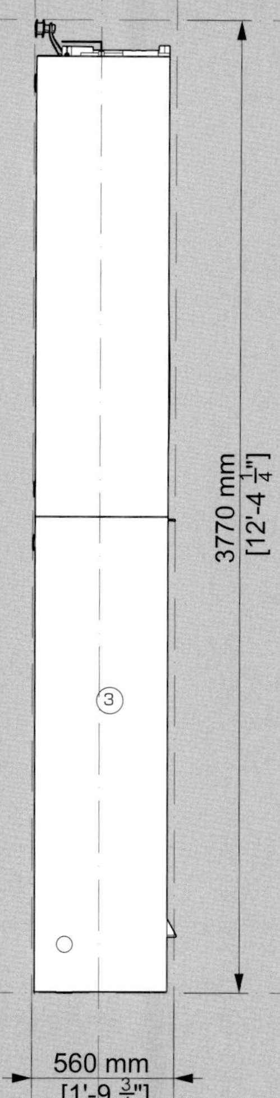

Side view

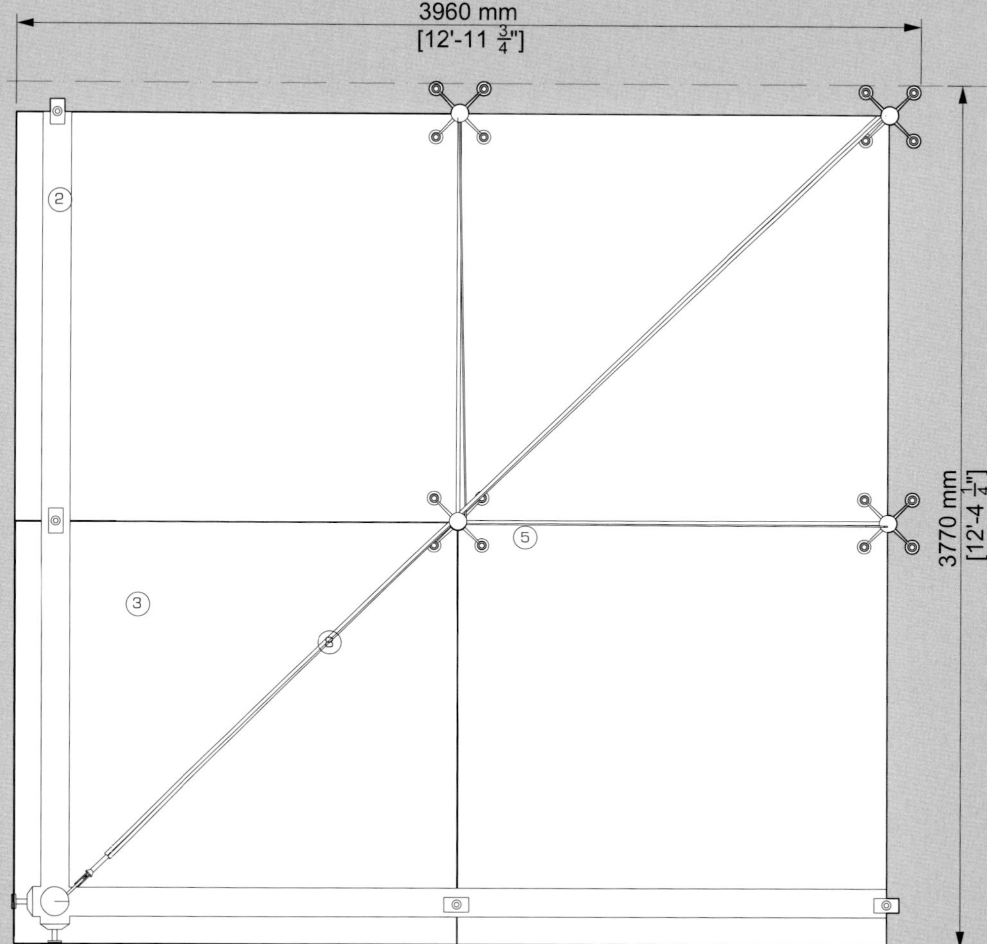

Back view

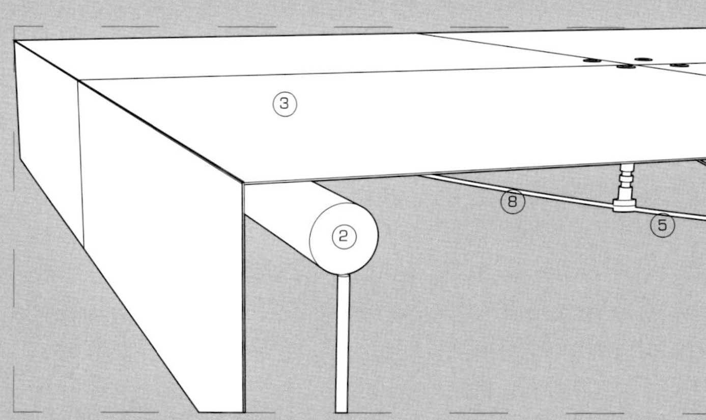

3D view of the assembly

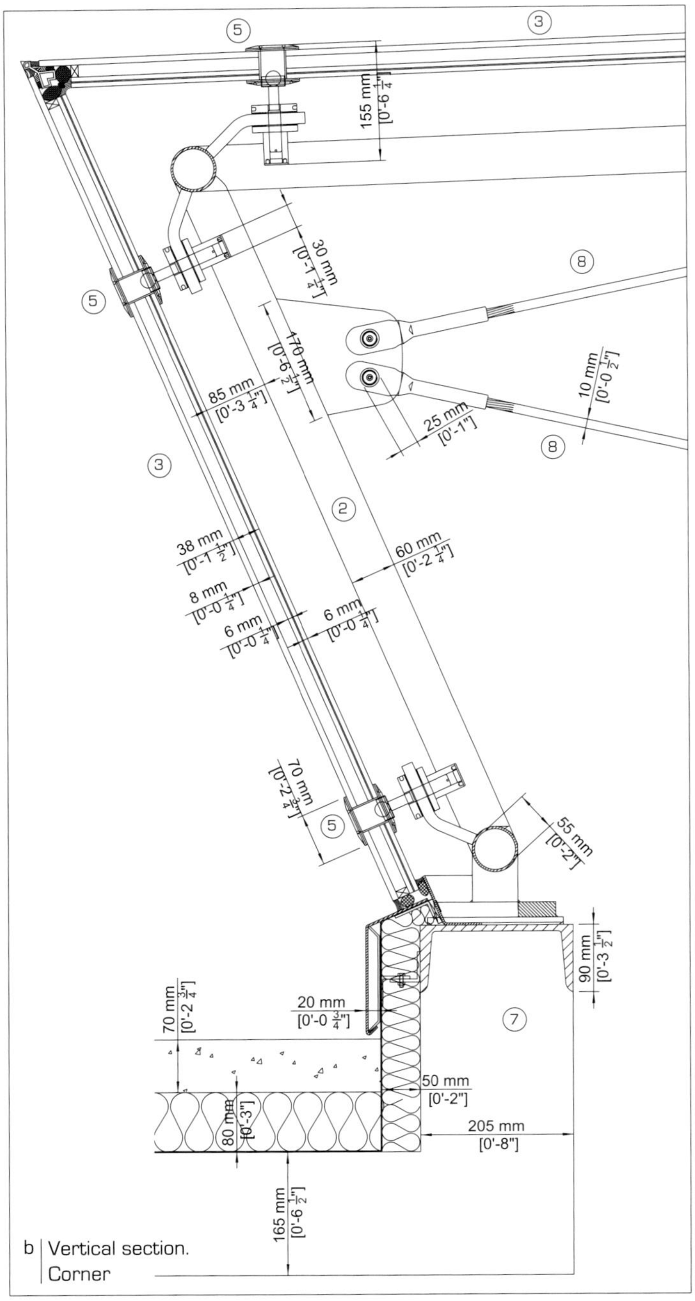

b | Vertical section. Corner

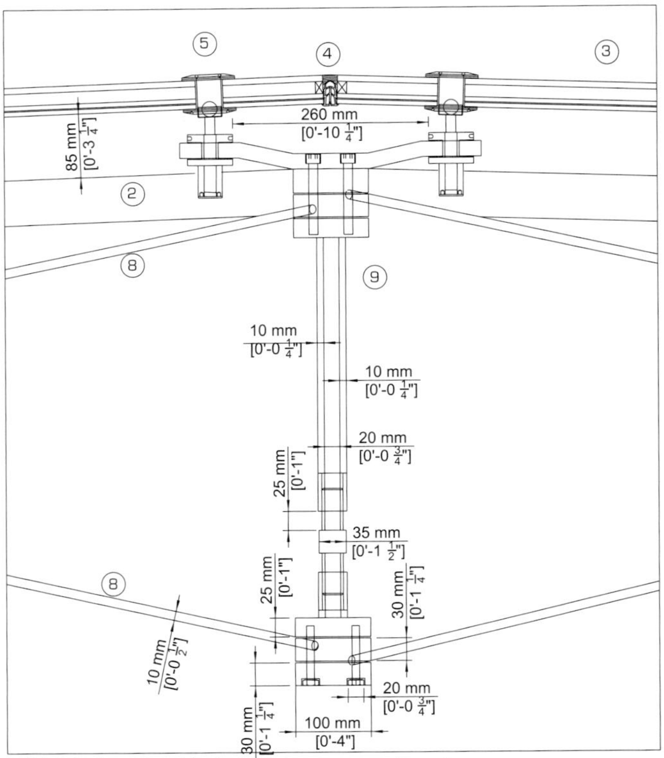

a | Vertical section. Horizontal panel-to-panel junction

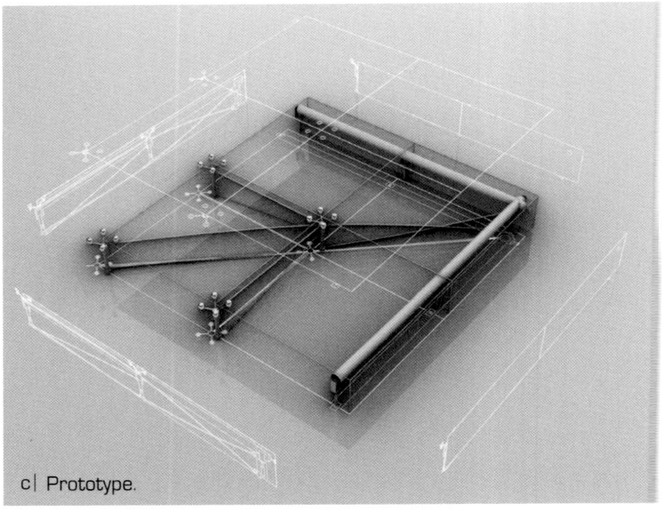

c| Prototype.

This method of glazing for roofs has been adapted from the technique used for glazed walls, where glass is fixed at points with specially designed bolts rather than with a frame supporting the perimeter of the glass. Bolt fixed glazing for facade construction developed from patch-plate glazing in the 1960s, where single glazed sheets of glass are bolted together with mild steel bracket. In the contempoary version, glass fins are used to stiffen the glazed walls to replace the aluminium mullions. The L-shaped patch fittings bolt the fins and glass together, as well as bolting the glass to the supporting structure at the top and bottom of the wall. While this glazing method has been developed and is still in use, the idea of frameless glazing has evolved further into double glazed units being bolted directly to a supporting structure without patch fittings at all (MCE_384, a). This further enhances the essential concept of frameless glazing, which is to provide greater visual transparency than an equivalent framed system.

The patch-plate method of frameless glazing is not used very often in roofs, since the glass fins become glass beams in such applications. While glass beams have been used in modest rooflight applications, there is uncertainty surrounding the difficulty of replacing cracked or damaged beams once the roof is completed. This has limited the use of glass beams to modest applications in glass roofs. In contrast to this, bolt fixed glazing has become increasingly popular for glazed roofs in mainly commercial and public buildings.

During the early 1990s doubts were raised over the reliability of the waterproof silicone seals between glazed units, but these were soon overcome with a mixture of laboratory testing and a better understanding of the workmanship required with silicone jointing. While manufacturers offer proprietary systems for walls, which can include a supporting structural system such as cable trusses (MCE_386, a), roof glazing systems are usually designed for individual applications.

1:10 scale

1:10 scale

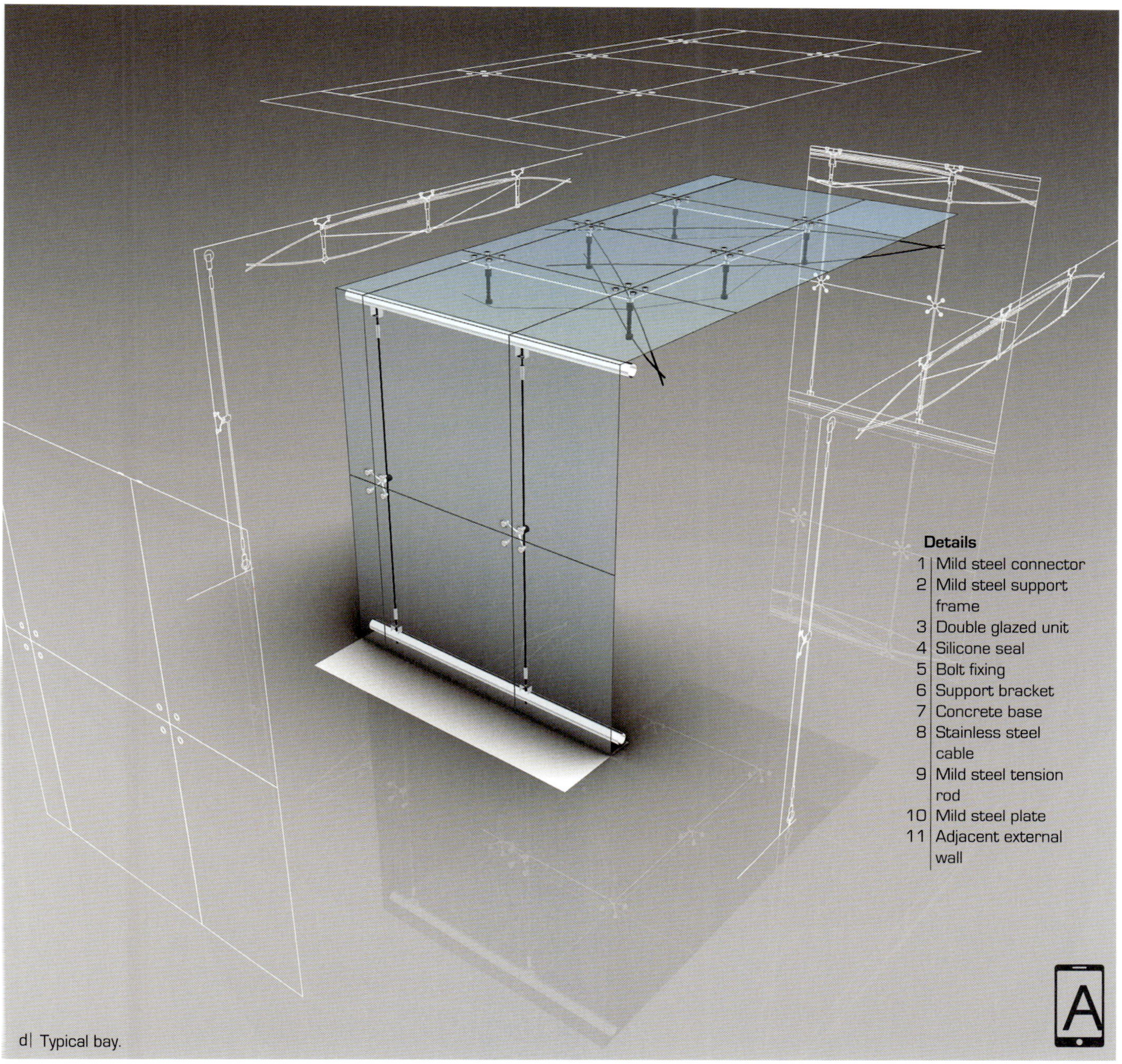

d| Typical bay.

Generic structural support methods

Whereas glazed walls are usually structurally supported by either top-hung or bottom-supported methods, the support of glazed roofs is by trusses, steel sections or purlins that span across the roof opening. For (nominally) flat roofs, there are two common arrangements of supporting beams. A supporting beam can be set under each glass joint, so that bolts are supported on each side of the beam by a short bracket (b). Alternatively, by setting the beam in the middle of alternate glass panels, only half the number of beams are required to support the same three panels of glass. Brackets are cantilevered from the beam to support the edge of the panel above as well as one side of the panel next to it. This method provides greater visual transparency than the first option, but requires larger brackets, which in practice detract little from the increased effect of transparency.

The single tube section shown in the diagrams would suit only a short span (d), as in a rooflight, but large span roofs require deeper beams, usually formed as open trusses in order to maintain the sense of transparency at oblique viewing angles. Triangulated trusses provide both structure and support for the glass, but tend to be visually heavy. Cable trusses are often preferred, but they require a ring beam around the edge of the glazed roof to form a tensile supporting structure like a tennis racquet. The cable trusses, always in tension, require an equivalent surrounding structure in compression to transfer the loads to the main building structure.

Supporting brackets

All these supporting structures require brackets to which the bolt fixings are attached. In the case of a short bracket attached to a beam, a mild steel bracket can be welded to each side of the tube (MCE_386, a). Because the bracket is welded, the adjustment for tolerance between

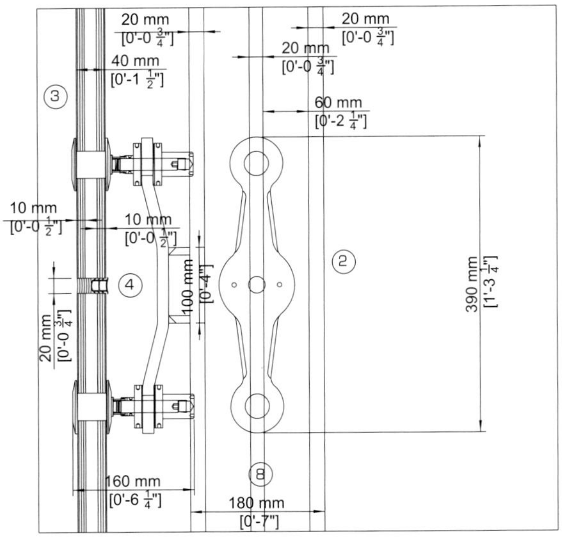

a | Vertical section. Vertical panel-to-panel junction

Details

1 | Mild steel connector
2 | Mild steel support frame
3 | Double glazed unit
4 | Silicone seal
5 | Bolt fixing
6 | Support bracket
7 | Concrete base
8 | Stainless steel cable
9 | Mild steel tension rod
10 | Mild steel plate
11 | Adjacent external wall

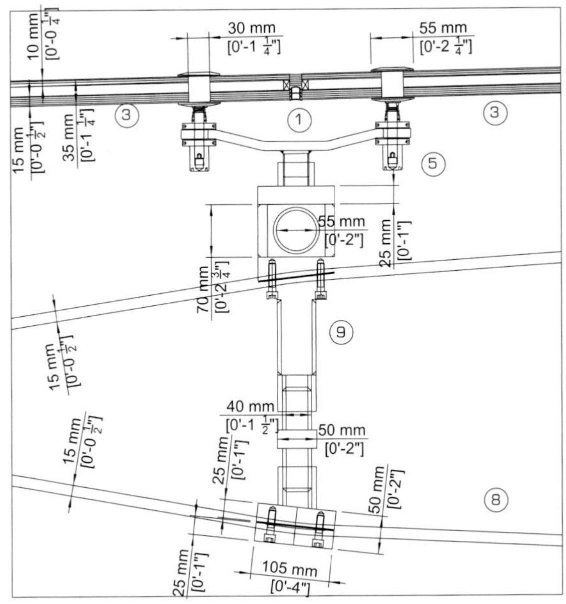

b | Vertical section. Horizontal panel-to-panel junction

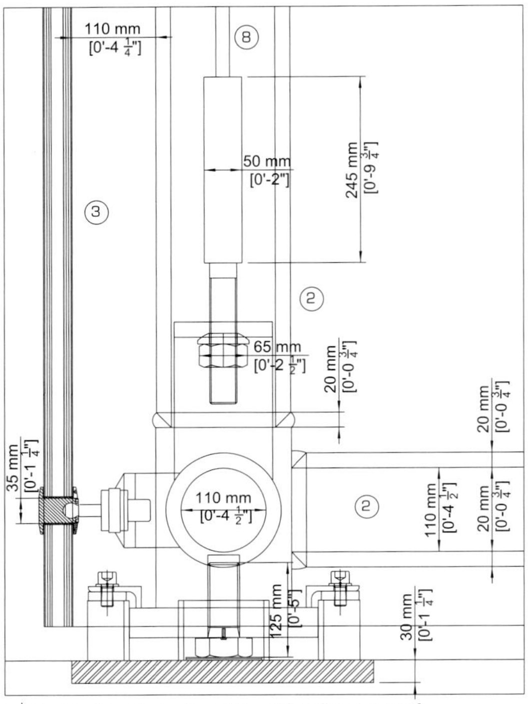

a | Vertical section. Junction with adjacent roof

d| Typical bay.

fixing of the supporting structure and the glass panels is taken out in the position at the connection between the bolt and the bracket. A slotted or oversized hole is cut into the supporting bracket and the bolt supporting the glass is fixed to it. The bolt may be off-centre from the bracket, and each bolt may be in a different position in relation to its neighbour. This can produce an awkward appearance when seen from below the roof, but this is certainly one of the most economic solutions. Where a channel-shaped bracket is welded to the top of the beam, a single bracket, typically mild steel plate, is bolted to the channel bracket. An oversize hole in the bracket is used to make the adjustment for tolerance. The bolts supporting the glass are set at each end of the bracket. The flat bracket can be replaced by a casting, usually in either mild steel or aluminium. The casting for this application would be made from a steel mould which is expensive to manufacture, so a large quantity of cast brackets are needed to make this method economic. Adjustment is made in the same way as the channel bracket example. The beam can become a cable truss, where a supporting bracket in either mild steel plate or a casting is clamped to the cable (g). Here the adjustment for tolerance is made in two places; at the junction of the bracket and the bolt fixing, as well as at the junction of the bracket and the cable clamp. There are, of course, many variations of these types, but these are the most commonly used bracket support methods.

Bolt fixings

The essential component in this glazing method, the bolt fixing, is made typically in stainless steel, and consists of several components that form the complete assembly. The part that passes through the glass has either a disc on each side of the glass to clamp the glass or double glazed unit together, or alternatively is angled to form a countersunk device within the depth of the double glazed unit. The countersunk fitting is set flush with the outer face of the glass, and the face-fixed disc type is set forward of the face of the glass. A polished stainless

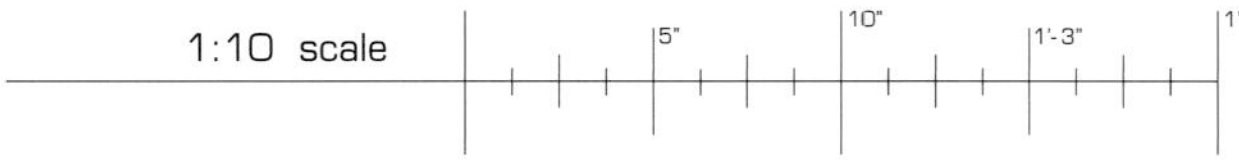

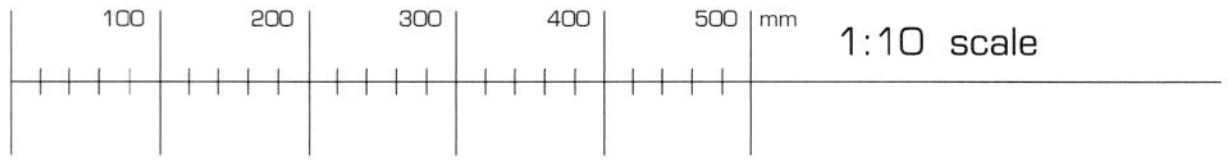

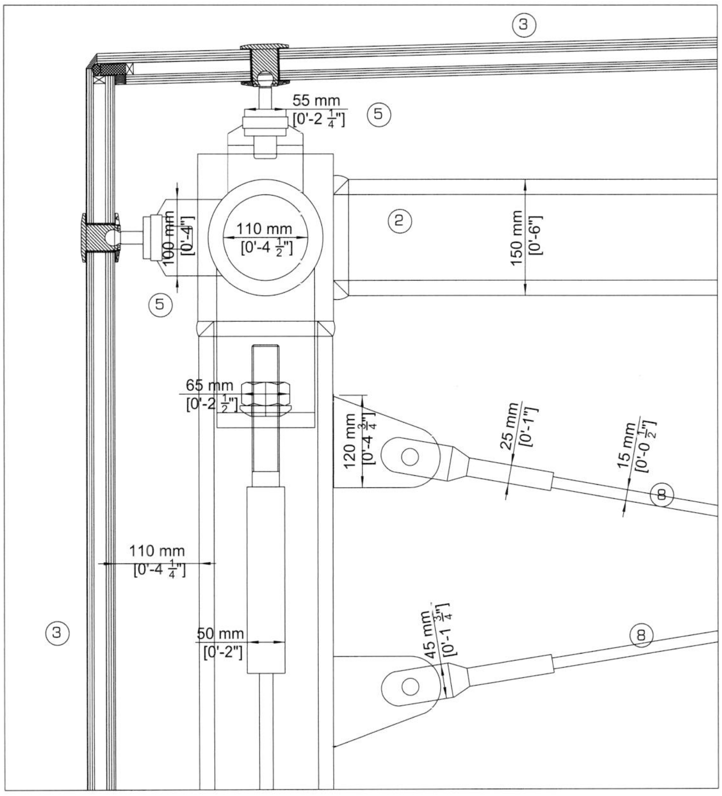

e | Vertical section. Corner

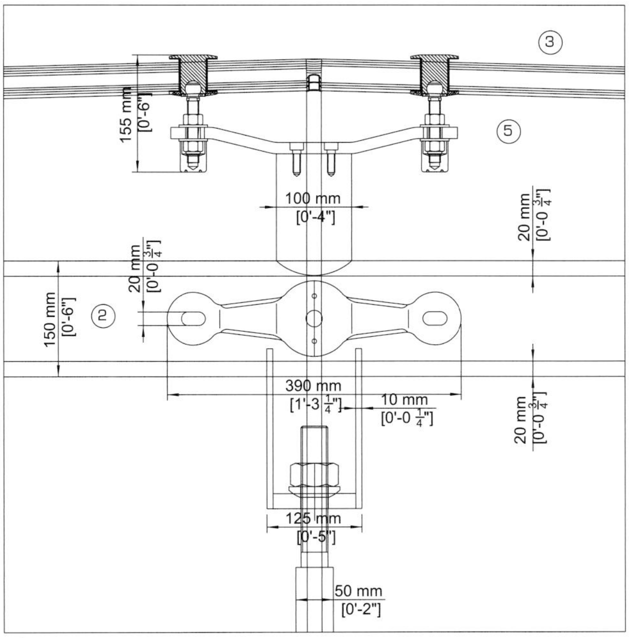

f | Vertical section. Horizontal panel-to-panel junction

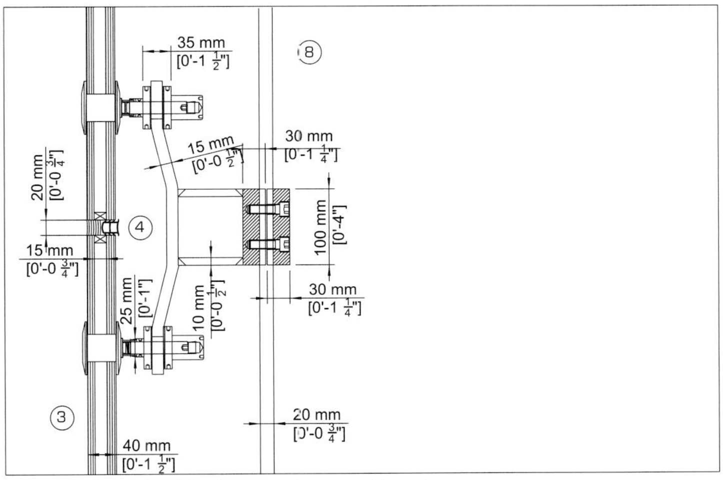

g | Vertical section. Vertical panel-to-panel junction

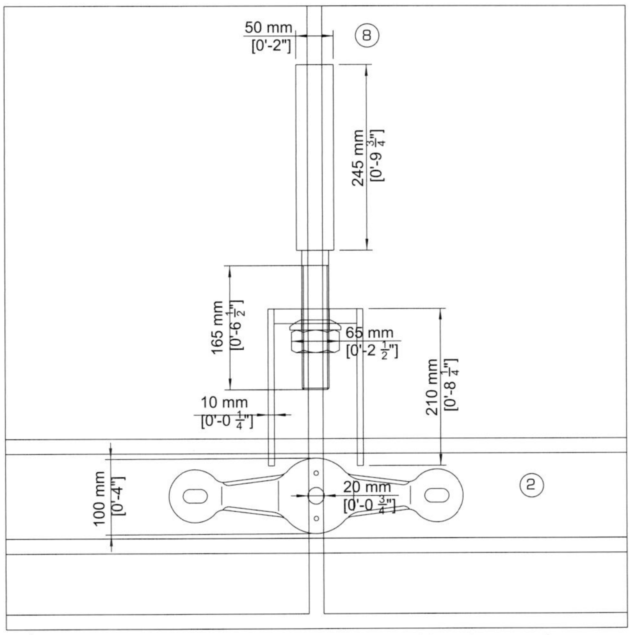

h | Horizontal section. Horizontal panel-to-panel junction at edge

steel finish is mostly used on the outside in order to make it easy to clean and maintain. In the face-fixed type, which is currently the most commonly used fixing, the inner disc screws over the threaded shank that forms part of the outer disc until it is tight up to the inner face of the glass. The threaded shank projects into the building and is able to rotate about a ball bearing where it meets the inner face of the double glazed unit. This allows the double glazed unit to rotate up to around 12° under wind load and associated structural deflections. This swivel joint is essential in avoiding the over-stressing of the glass under full wind load that would otherwise result in breakage of the glass unit. The threaded shank is then used to clamp the complete bolt fixing to a support bracket with either threaded discs or nuts. The visible thread in the shank can be either left exposed, or be covered with threaded sleeves and stop ends (g). This bolt type is used regardless of the orientation of the roof, whether flat or pitched.

Arrangement of bolt fixings

The arrangement of brackets for bolt fixed roofs is dependent upon the position of the fixing bolts. Rectangular-shaped double glazed units can be arranged with supporting beams set in the direction of the long side of the panel. Bolts are positioned in a way that reduces the span of the glass by pushing them away from the edge. The reduced glass span allows the glass thickness to be thinner, making it more economic, particularly given that the material cost of glass increases dramatically with the increase in thickness (the relationship between glass thickness and cost is not linear). Flat plates can be used, subject to structural requirements.

Where the glass fixing bolts are equidistant from the edges of the glass, a cross-shaped bracket allows four glass connections to be made with a single bolt connection between this support bracket and the beam beneath. The cross-shaped bracket is carrying a high load of

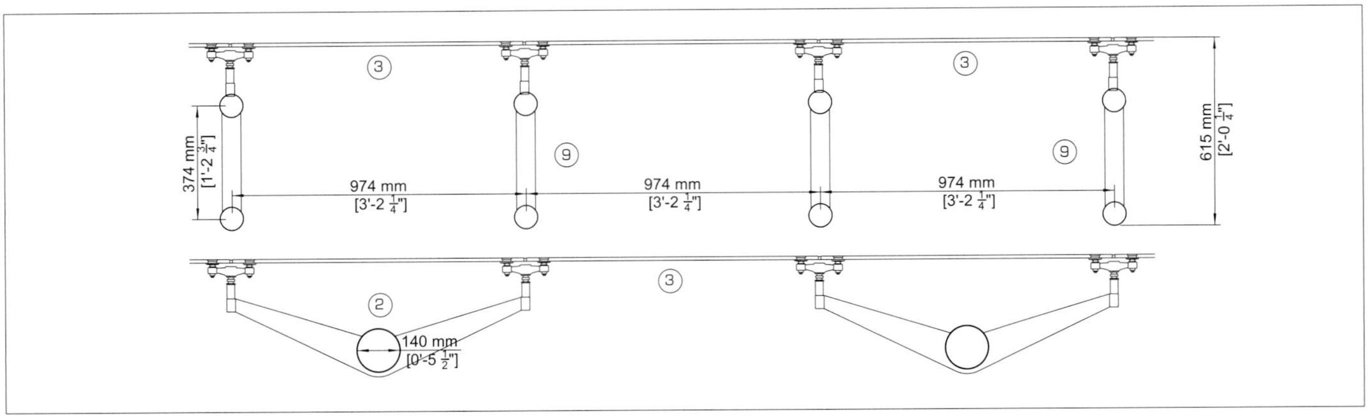

a | Vertical section 1:25. Typical assembly

Details

1 | Mild steel connector
2 | Mild steel support frame
3 | Double glazed unit
4 | Silicone seal
5 | Bolt fixing
6 | Support bracket
7 | Concrete base
8 | Stainless steel cable
9 | Mild steel tension rod
10 | Mild steel plate
11 | Adjacent external wall

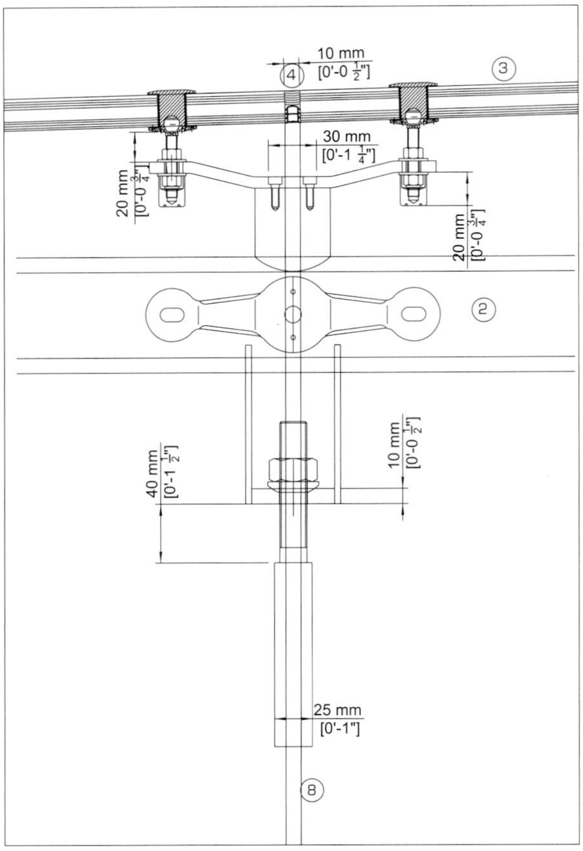

b | Vertical section. Vertical panel-to-panel junction

c | Prototype.

glass to a single point on the beam. Stiffening fins are usually needed on the underside of the cross-shaped bracket. These can be individually welded and ground, but it is often economic to make them as castings, which have a more reliably refined appearance. Mild steel brackets require painting; stainless steel brackets can be buffed or polished to the preferred visual finish.

Glazed units

In common with other glass roof types, the inner glass of a double glazed unit is usually made from laminated glass. In the event of a double glazed unit being broken, the inner laminated sheet remains intact, while the broken pieces of the heat-strengthened or fully toughened outer sheet come to rest on top of the damaged, but intact, inner sheet. The double glazed units are first fixed and adjusted to form even joint widths between all the units. Joints of 20-28mm are used, though around 20mm is the most common joint width (in elevation) that allows for both structural movement and the slight variations in the size of the glass panels (MCE_385, h). Unlike capped roof glazing systems, the entire double glazed unit is visible from both outside and inside; the edges are not set behind pressure plates that conceal any variations in glass panel size. Joint widths up to around 28mm, which is deemed close to the maximum practical joint width for the adhesion of silicone sealant in a regular double glazed unit, are used where brackets penetrate the outer seal from inside the roof to outside. If required, these brackets are used to support external sun shading and maintenance equipment. These brackets are usually in the form of flat plates that are welded to the internal supporting structure, and project through the joint. Although an additional lip around the projecting plate may provide additional protection to water penetration between the silicone and the bracket, in practice it has been found that this detail performs well if the seal is applied to a good level of workmanship.

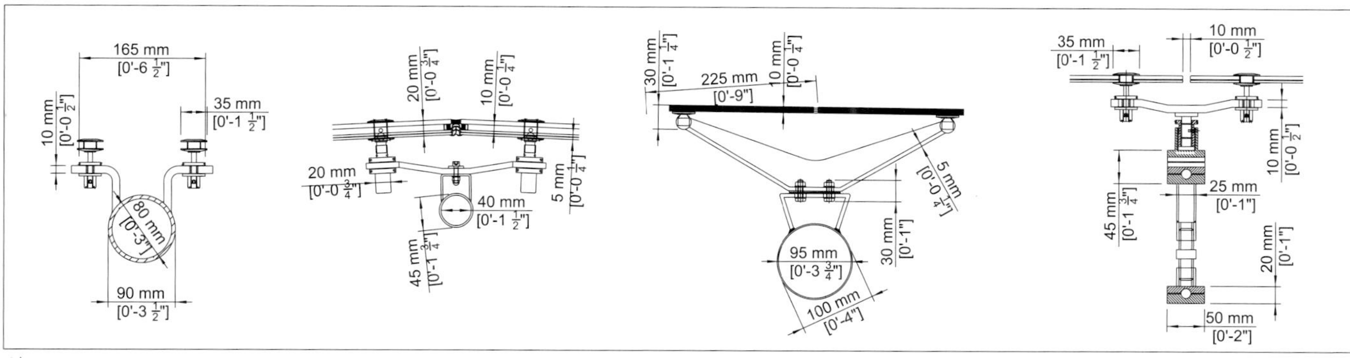

d | Vertical sections. Varying glass fixings

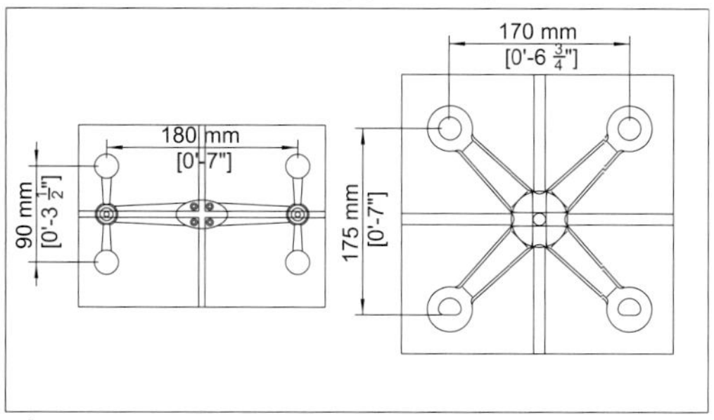

e | Horizontal sections. Bolt fixed arrangements

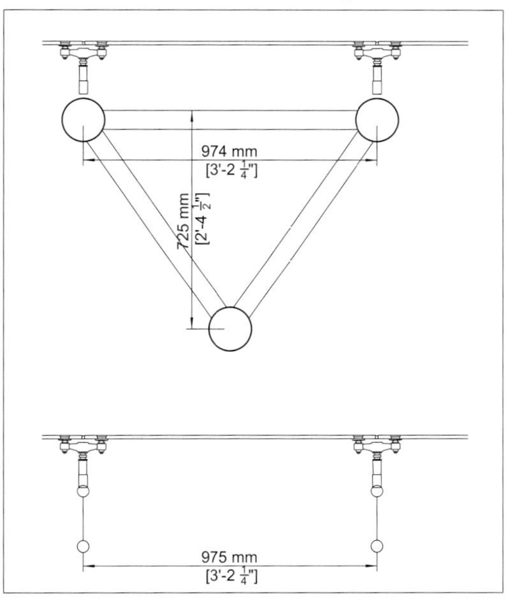

f | Vertical sections 1:25. Bolt fixed arrangements

g| Prototype.

Seals between double glazed units are made as an outer silicone seal with an inner backing rod of extruded EPDM. The gasket has projecting flaps on each side to form a 'fir tree' section which prevents any water that penetrates the external seal from reaching the inner face of the seal. This EPDM gasket also serves as an inner air seal, and provides a crisp appearance of sharp lines in the interior face of the glazed roof.

In the manufacturing of double glazed units, the butyl seal between the spacer and the glass creates a slightly wavy line, visible to the eye when it spreads beyond the face of the spacer. This slightly uneven appearance of the edge of the glazed unit can be concealed by the use of edge 'fritting', or a baked-on screen printed edge in black, which ensures that the edge of the unit has a crisp black edge. This is mainly where the rooflight glass can be seen at close proximity.

The holes in double glazed units, to which the bolt fixings are attached, are made by cutting holes in the glass around 10mm larger than the metal circular sleeve that fits between it. The circular sleeve bonded to the glass is sealed around it to maintain the sealed cavity of the double glazed unit. This is particularly important in the case of argon filled cavities, where the gas improves thermal insulation.

The drilling of the glass itself is now well established in glass manufacturing, with both float glass and laminated glass being drilled before any heat treatment to make them either heat strengthened or fully toughened. Where coated glasses are used, such as solar control coatings or low emissivity (low e) coatings, this is done after heat treatment.

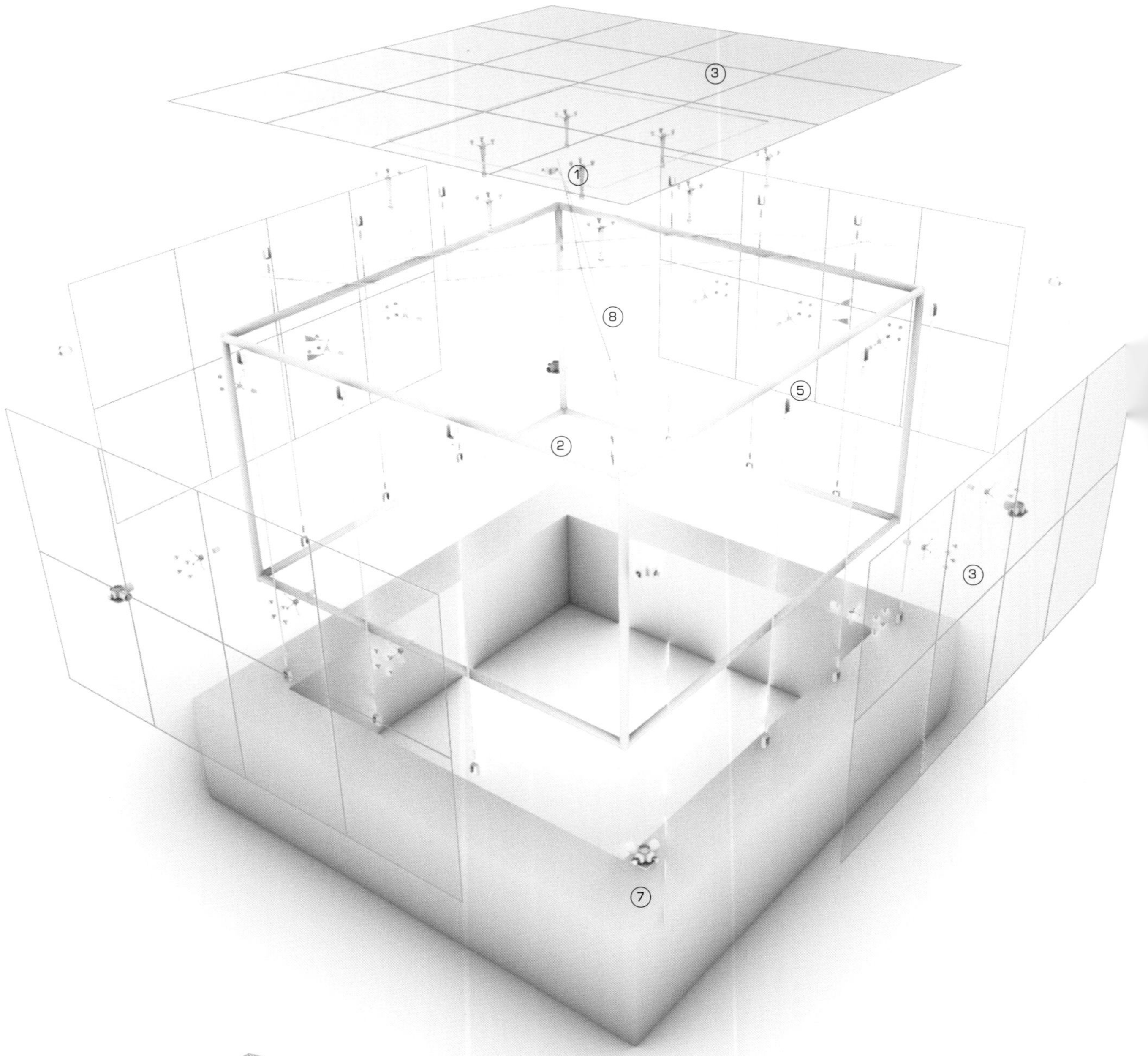

a | 3D exploded view showing small scale bolt fixed glazed rooflights

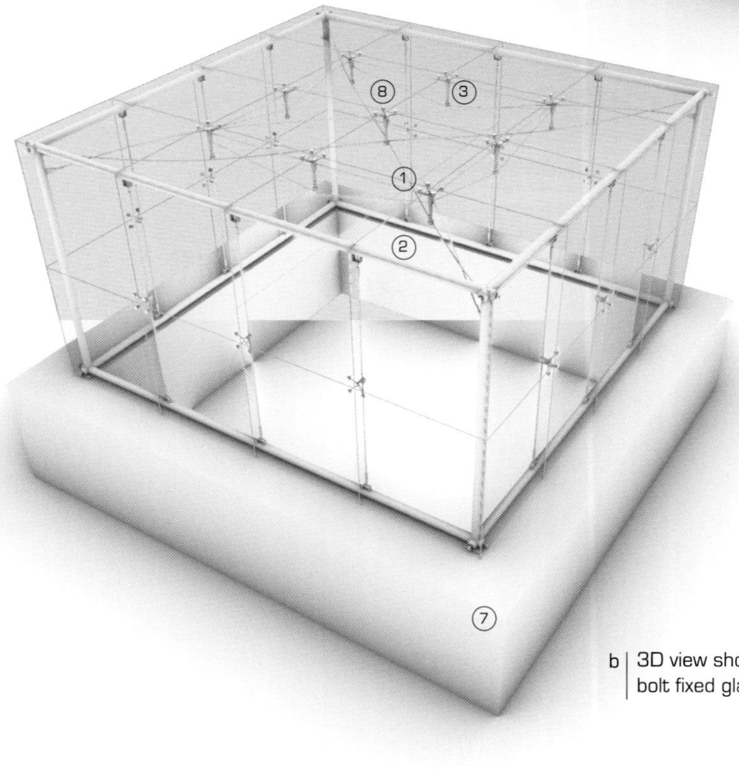

b | 3D view showing small scale bolt fixed glazed rooflights

Details

1 | Mild steel connector
2 | Mild steel support frame
3 | Double glazed unit
4 | Silicone seal
5 | Bolt fixing
6 | Support bracket
7 | Concrete base
8 | Stainless steel cable
9 | Mild steel tension rod
10 | Mild steel plate
11 | Adjacent external wall

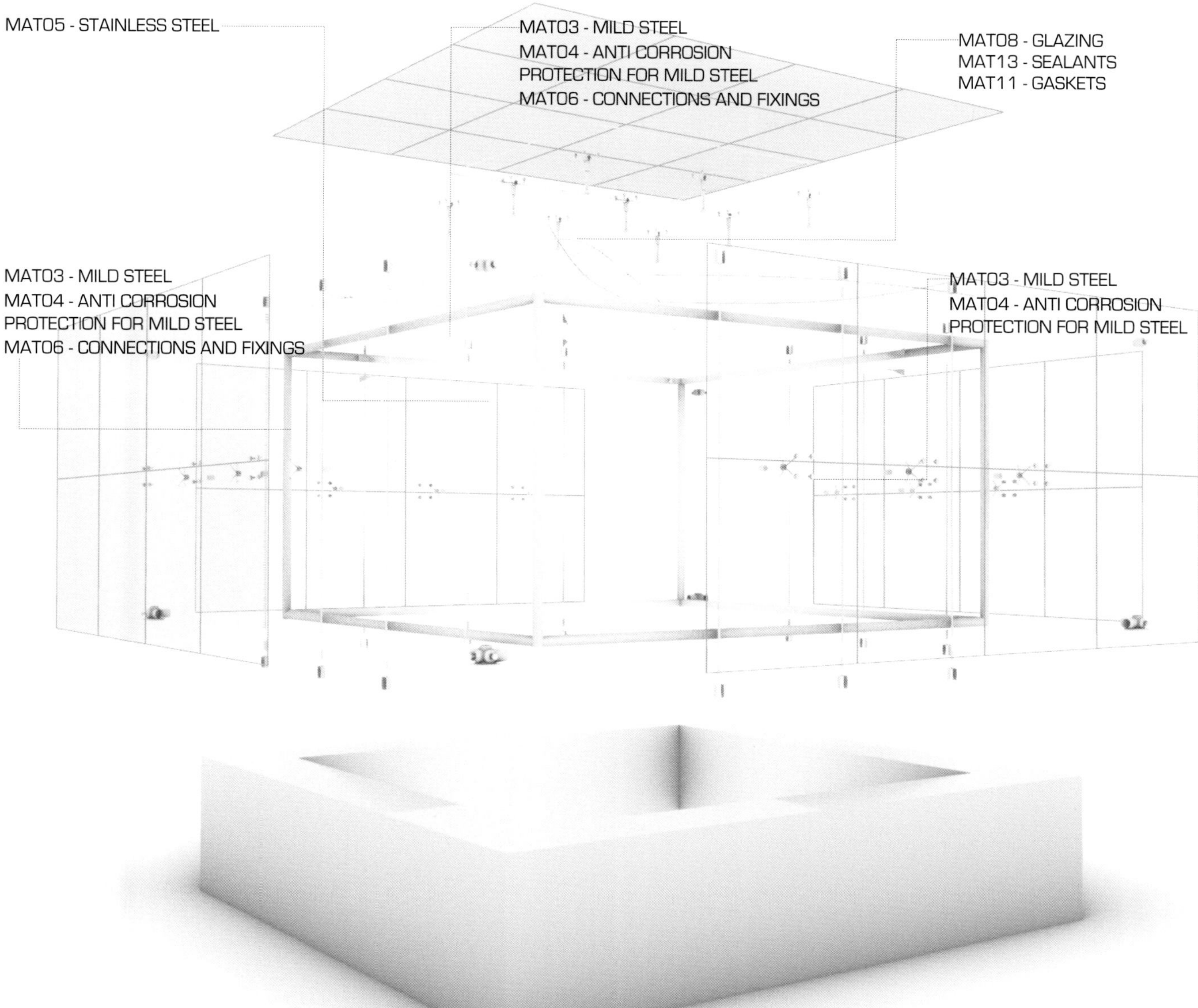

c | 3D exploded view of small scale bolt fixed glazed rooflight components with material specification references

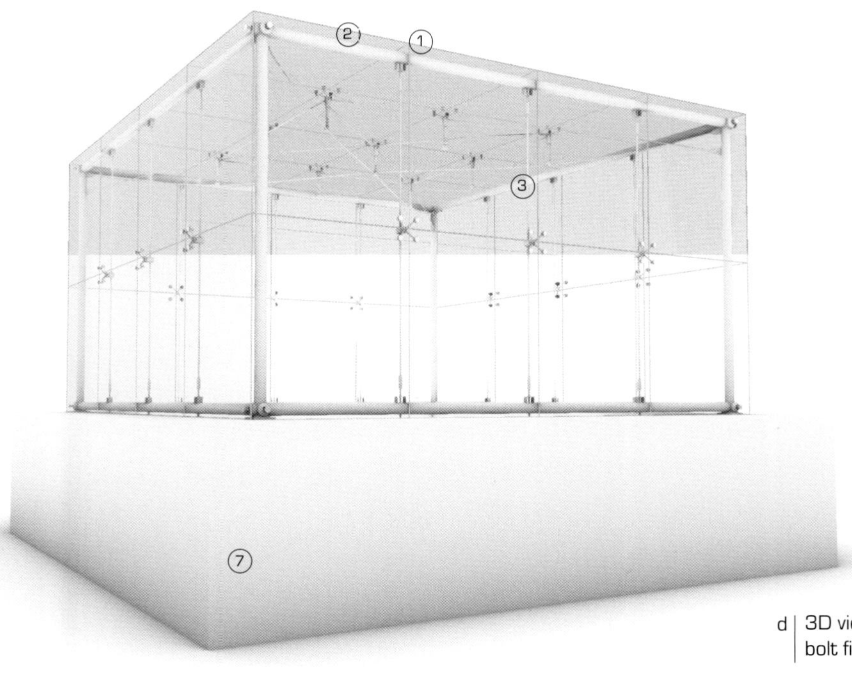

d | 3D view showing small scale bolt fixed glazed rooflights

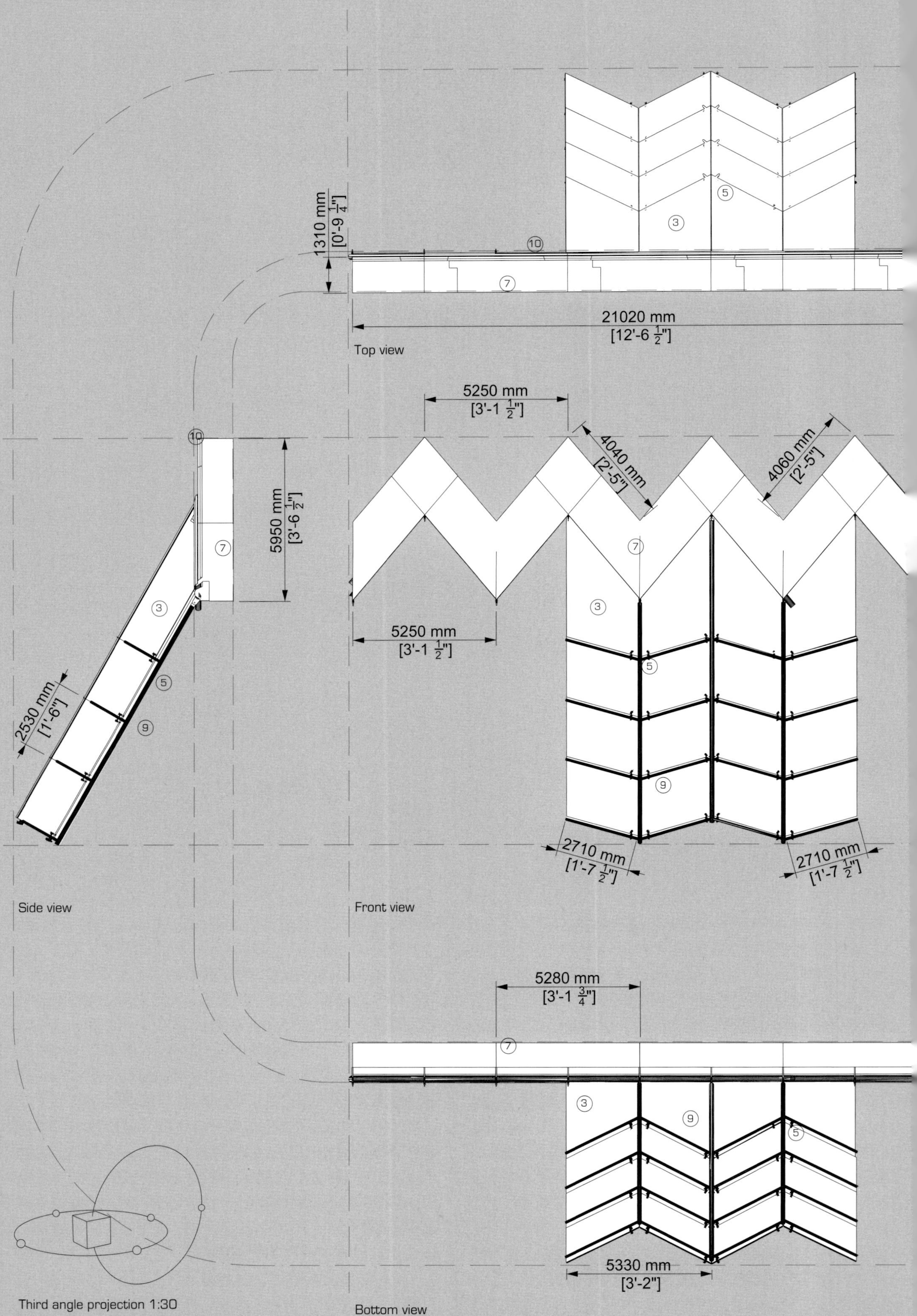
1310 mm
[0'-9 1/4"]
21020 mm
[12'-6 1/2"]
Top view
5250 mm
[3'-1 1/2"]
4040 mm
[2'-5"]
4060 mm
[2'-5"]
5950 mm
[3'-6 1/2"]
5250 mm
[3'-1 1/2"]
2530 mm
[1'-6"]
2710 mm
[1'-7 1/2"]
2710 mm
[1'-7 1/2"]
Side view
Front view
5280 mm
[3'-1 3/4"]
5330 mm
[3'-2"]
Third angle projection 1:30
Bottom view

Details

1 | Mild steel connector
2 | Mild steel support frame
3 | Double glazed unit
4 | Silicone seal
5 | Bolt fixing
6 | Support bracket
7 | Concrete base
8 | Stainless steel cable
9 | Mild steel tension rod
10 | Mild steel plate
11 | Adjacent external wall
12 | Single glazed solar shading glass

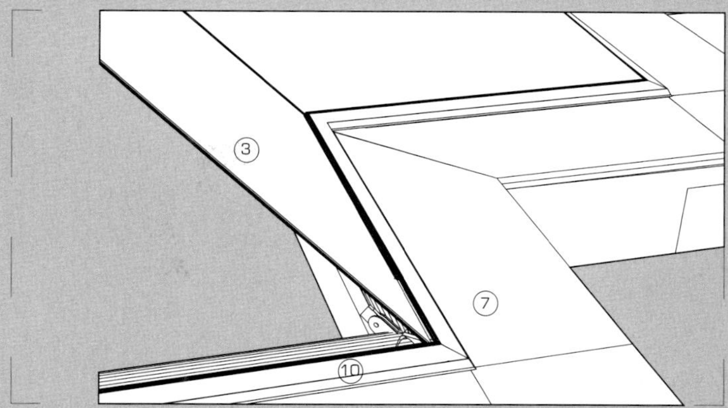

3D view of the assembly

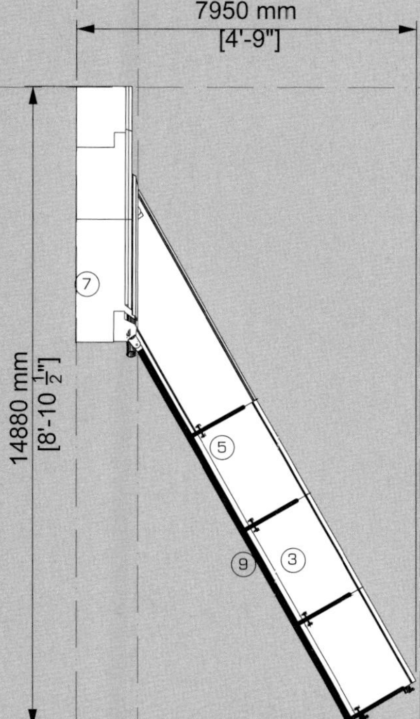

Side view

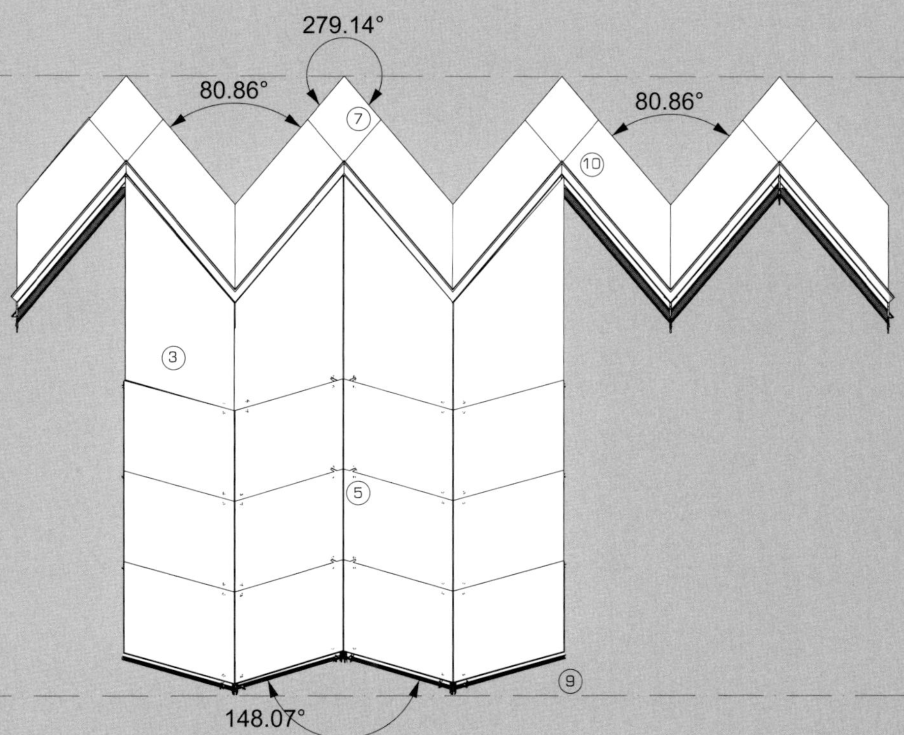

Back view

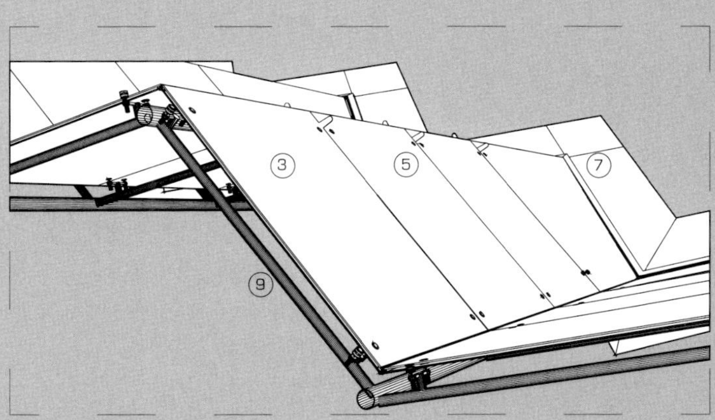

3D view of the assembly

a | Vertical section. Panel-to-panel junction

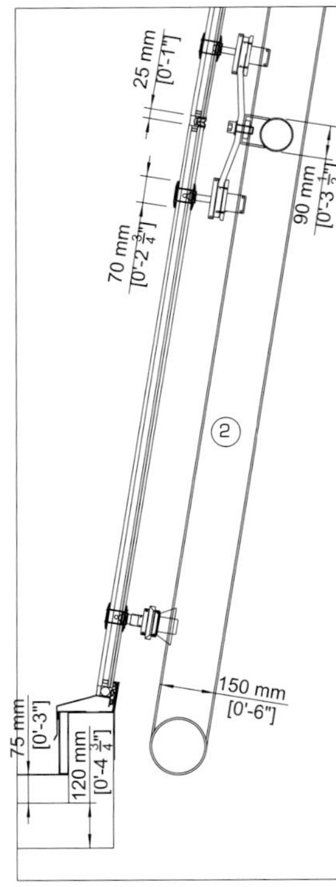

b | Horizontal section. Panel-to-panel junction

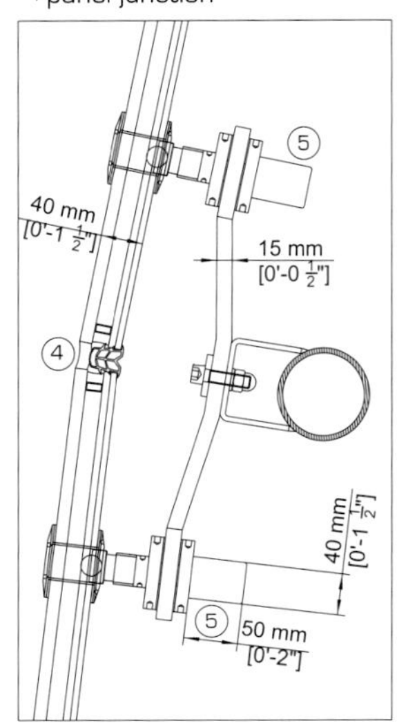

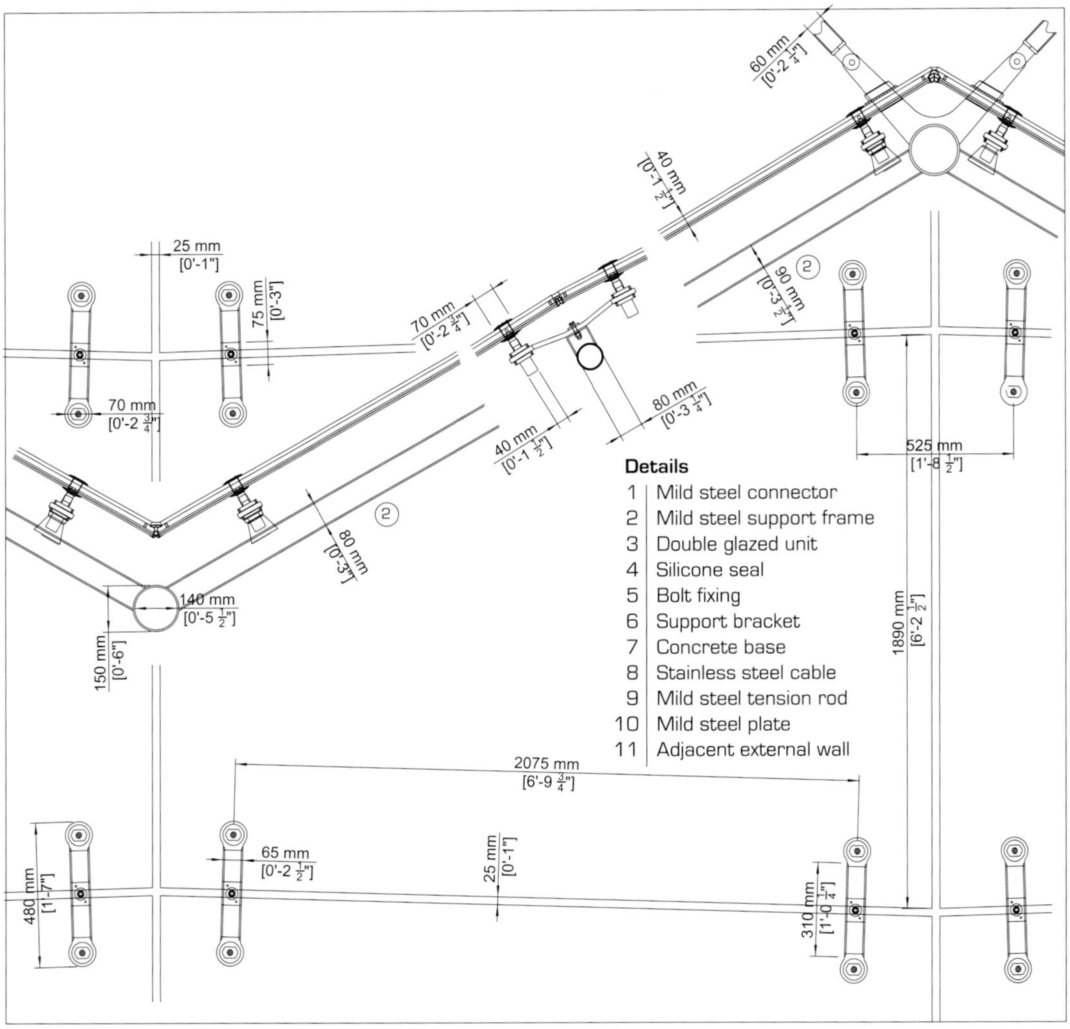

c | Vertical section. Typical assembly

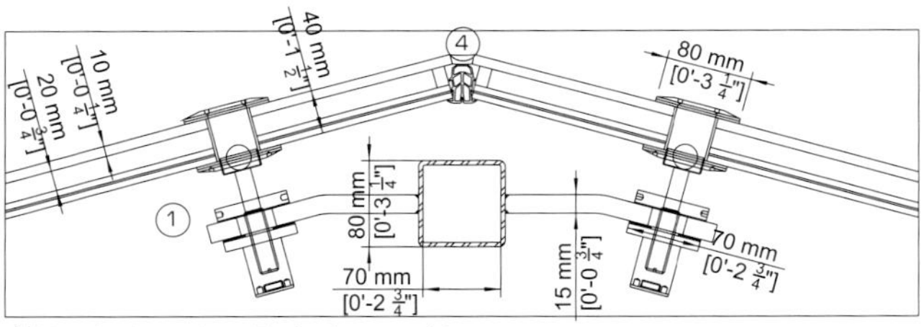

d | Vertical section. Typical assembly

The generic support structures described in the previous section suit a range of roofs in a single plane, whether pitched or (nominally) flat. Supporting structures which are facetted or curved to form complex geometries require a slightly different approach. These supporting structures are dictated more by the demands of the form that is being created rather than by optimising the position of supporting members to maximise visual transparency through the structure. Structures for arches or curved surfaces are usually made with as little structural depth as possible, with the visually successful types using a single 'plane' of structure formed in 'ladder' forms. Circular hollow sections and box sections are preferred for their visual consistency regardless of where the glazed roof is viewed below the roof. The ladder principle is actually two vierendeel trusses linked together, with the short cross members welded to the long members. If additional stiffness is required in part or all of the structure, then cross bracing is typically added. Additional structural stability is provided by folding, curving or ribbing the surface of the structure. This avoids the need to deepen the structure with trusses that significantly reduce the visual transparency offered by bolt fixed roof glazing.

The 'ladders' are bolted or welded together to form the complete structure. If welded, then an inner sleeve is inserted between the sections being jointed and the edges of joints are welded together and painted. An alternative fixing method is to bolt the sections together. This method avoids the bolts being visible, leaving only a hairline joint visible at the junction, giving the joint a minimal appearance and avoiding welds which form a visible ring around the joint. This joint method also allows a faster installation than if welding were used, with the additional benefit that steelwork can be installed with a fully painted finish from the factory if required. In this method, a mild steel plate is welded to each end of the tube being jointed. A rectangular hole is cut into a wall of one of the hollow steel sections, big enough to pass a hand through. Access-

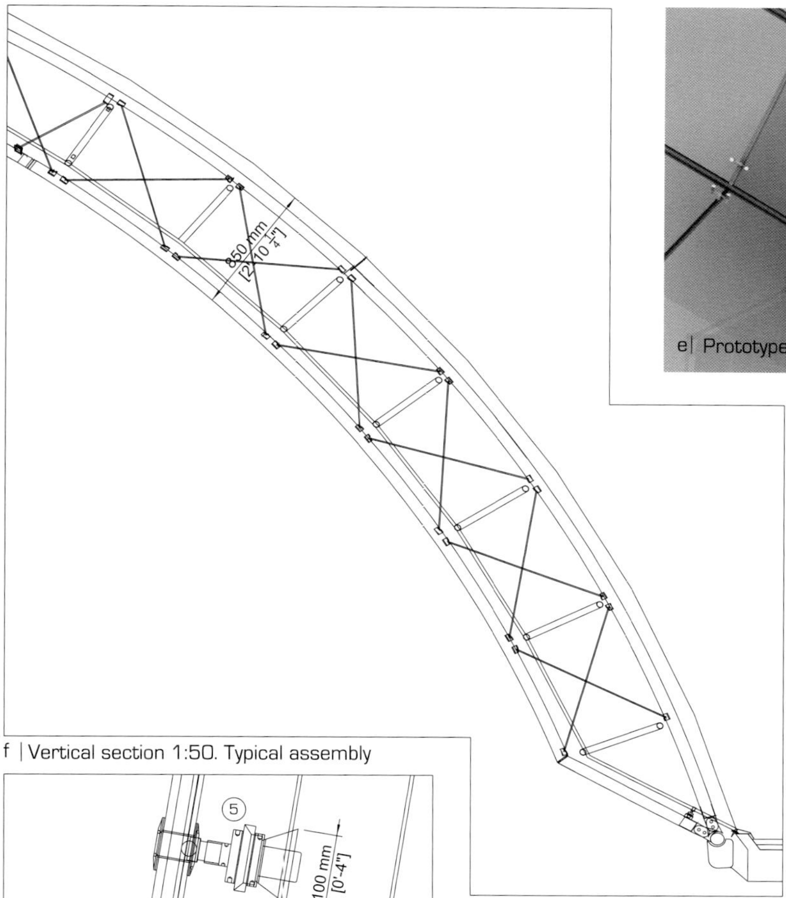

e| Prototype.

f | Vertical section 1:50. Typical assembly

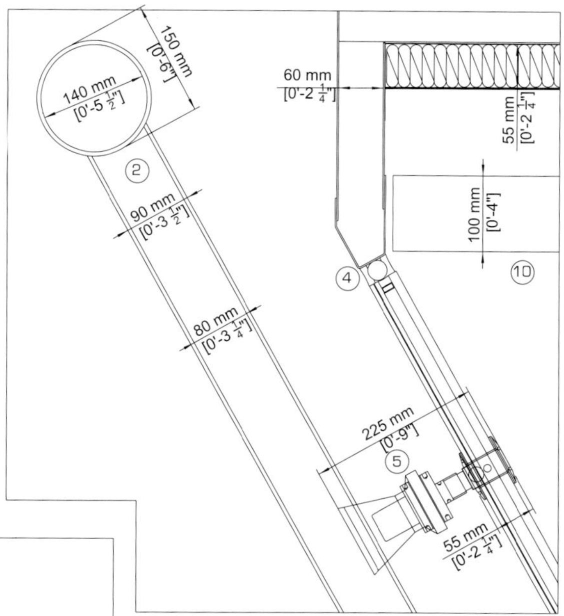

g | Horizontal section. Junction with adjacent wall

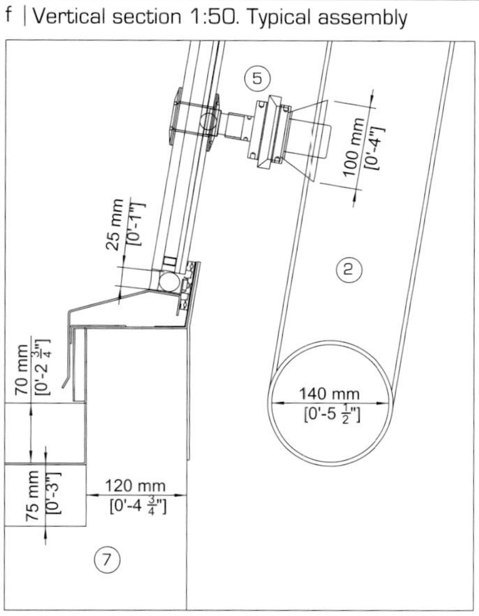

h | Vertical section. Base upstand

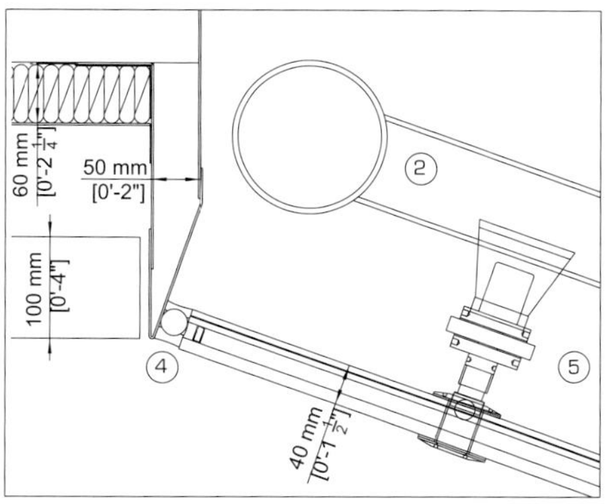

j | Horizontal section. Junction with adjacent wall

ing the inside of the hollow section from the rectangular hole, bolts are fixed through the end plates to fix the two steel sections together. The rectangular hole is then covered with a thin metal plate which is either bonded to the surrounding metal or is screwed into the surrounding wall of the hollow section.

The structural form of the 'ladder' usually corresponds to the layout of the glass panels, since the corners of panels are fixed back to the supporting structure, which is preferably as close to the glass joints (in elevation) as possible (MCE_395, f). In other examples the structure is set in line with the joints in the glass. In some cases, the bolt fixings for the glass are fixed to a tubular supporting structure with mild steel brackets fixed to the main tube (e) (as described in the previous section). The positions of the bolt fixings in relation to the edge of the glass are optimised to reduce the glass to an economic thickness. Bolt fixings can be set mid-distance between two horizontal ladders. This might correspond to a joint between two double glazed units or an additional bolt fixing to secure a large double glazed unit. A v-shaped mild steel bracket is set at 90° to the brackets. The stainless steel bolts are aligned in elevation with adjacent brackets.

Base of glazed roof

Bolt fixed glazed roofs are often fixed to a surrounding reinforced concrete slab (c). An essential feature of junctions in bolt fixed glazed roofs at their perimeter is that there is no mechanical connection between the double glazed unit and the adjacent roof deck, or parapet, since the glass is bolted some distance away from the corner and the edge of the glass is cantilevered from the bolt fixing (MCE_394, c). The adjacent roof or external wall (in a different material) meets it with two sets of silicone seals only, or alternatively with EPDM seals which are bonded to the edge of the glass unit. In some cases, a low-pitched roof meets a reinforced concrete upstand forming part of a concrete roof

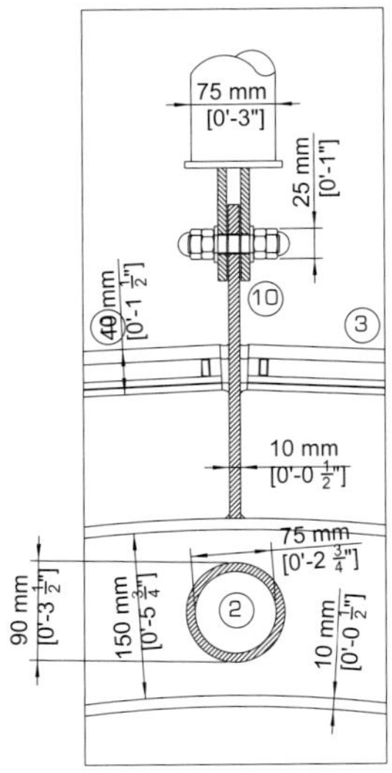

Details

1 | Mild steel connector
2 | Mild steel support frame
3 | Double glazed unit
4 | Silicone seal
5 | Bolt fixing
6 | Support bracket
7 | Concrete base
8 | Stainless steel cable
9 | Mild steel tension rod
10 | Mild steel plate
11 | Adjacent external wall

a | Horizontal section. Penetration of plate through glass joint

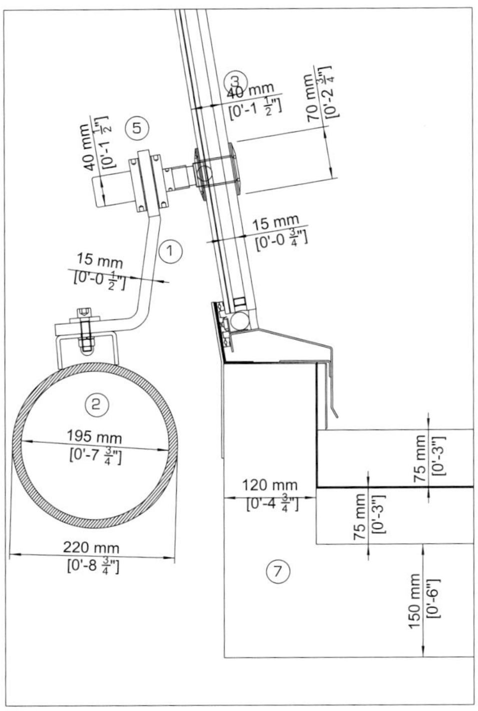

b | Vertical section. Internal fold

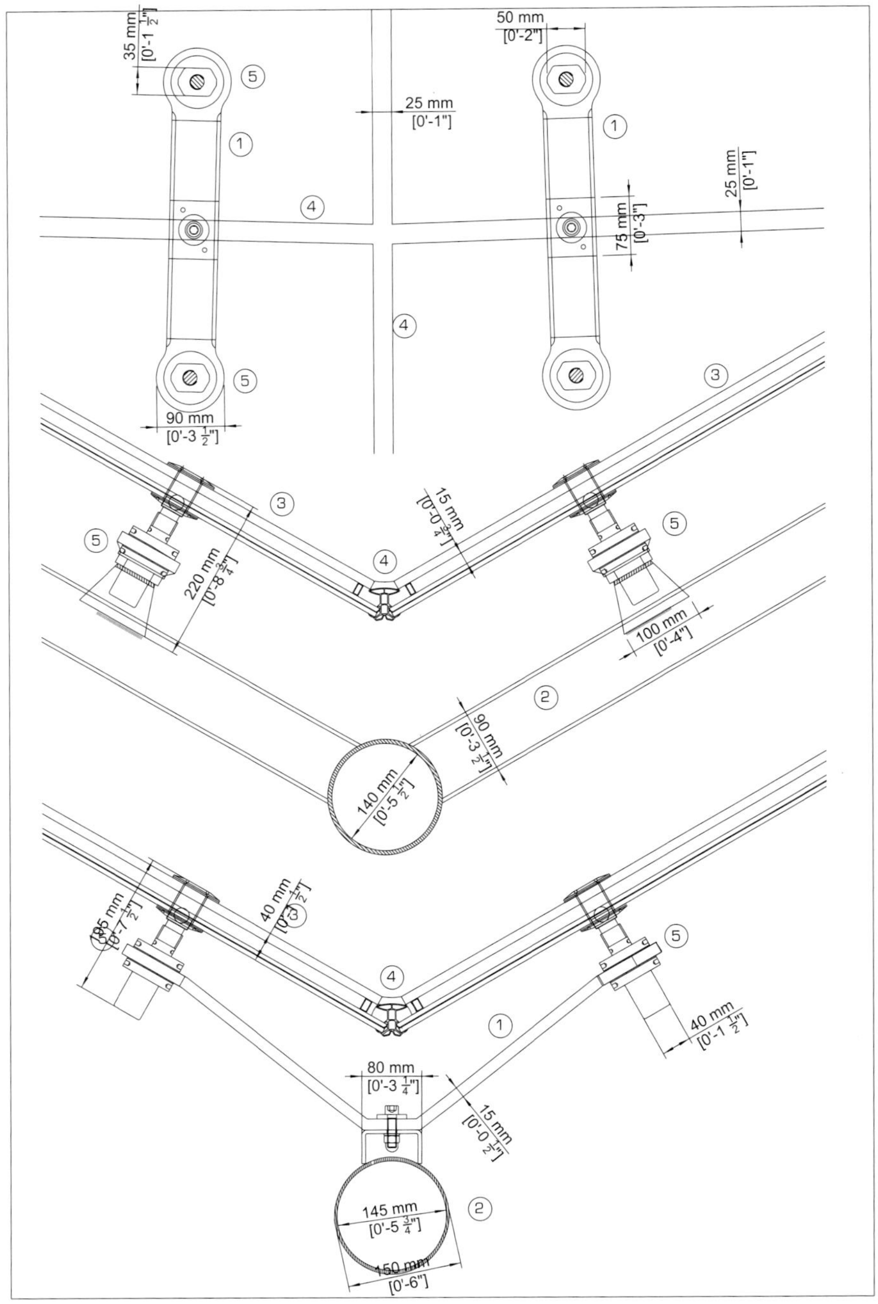

c | Elevation and vertical section. Internal fold

deck. The supporting steel structure is fixed to the edge of the concrete slab, which typically forms a continuous ring beam around the opening to carry the load of the glazed roof. The base of the glazing has a cantilevered edge of the double glazed unit which is sealed to an aluminium flashing. An additional flashing, which is continuous with the waterproofing membrane, is set on the inside face of the glass. This folded aluminium strip is bonded with silicone to the bottom of the double glazed unit. It can be either concealed by the interior finishes or be coated, usually in a PVDF or polyester powder-coated finish. This aluminium strip is bonded to the edge of the waterproofing membrane that extends up the face of the upstand across the top. This combination provides an inner, second line of defence to the flashing above (MCE_396, b). The outer flashing is set on the outer face of the glass, and has a folded edge in order to bond it with a silicone seal, typically 20mm wide to match visually with the other seals between the glass units. Any rainwater which penetrates the outer seal is drained away on the inner metal flashing onto the waterproof membrane. The void between the two flashings is filled with closed cell thermal insulation, sometimes as injected foam to fill the cavity completely. In some instances, an all-metal upstand flashing can be used, while in others the upstand is concealed by an accessible timber deck.

External and internal folds

Changes of direction in bolt fixed roofs that form ridges and valleys are formed as external folds and internal folds respectively. While the joint is formed in the same way as other joints, with an outer silicone seal and an inner extruded EPDM baffle, or air seal, the double glazed units are stepped on their edges to ensure that the joint is not wider than adjacent joints. With an external fold the outer glass is stepped beyond the edge of the unit to maintain a constant width through the depth of the joint (f). The internal fold, on the other hand, can have the inner laminated glass extending beyond the outer glass to perform the same

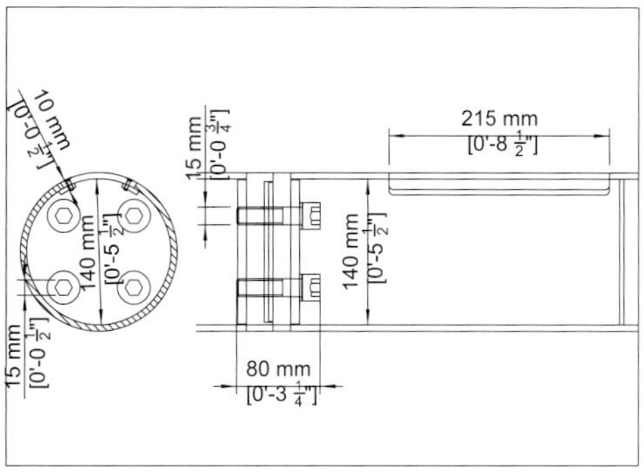

d | Horizontal section. Concealed framing connection

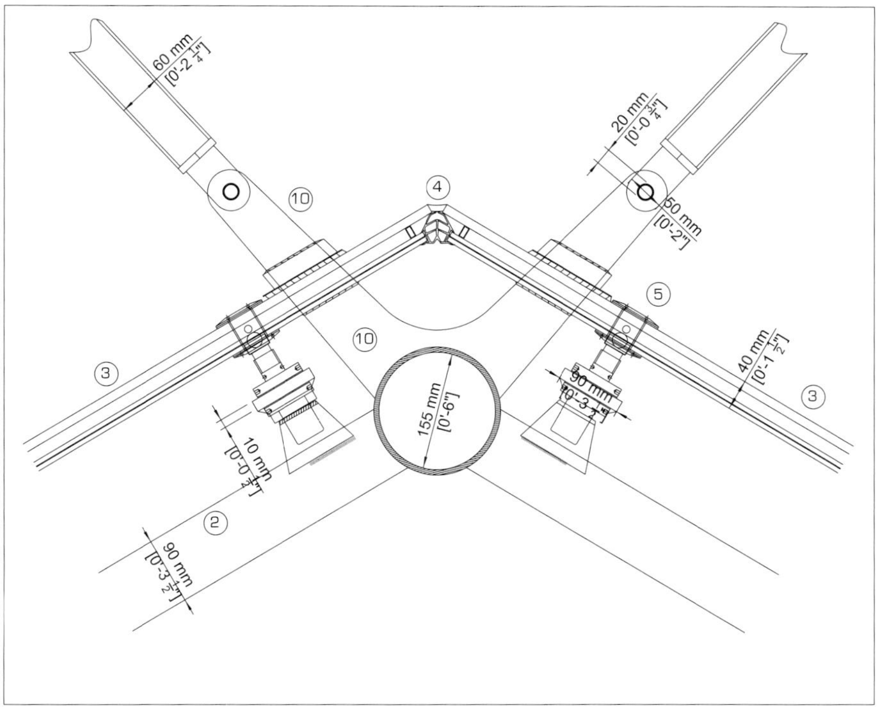

e | Vertical section. Concealed framing connection

f | Prototype.

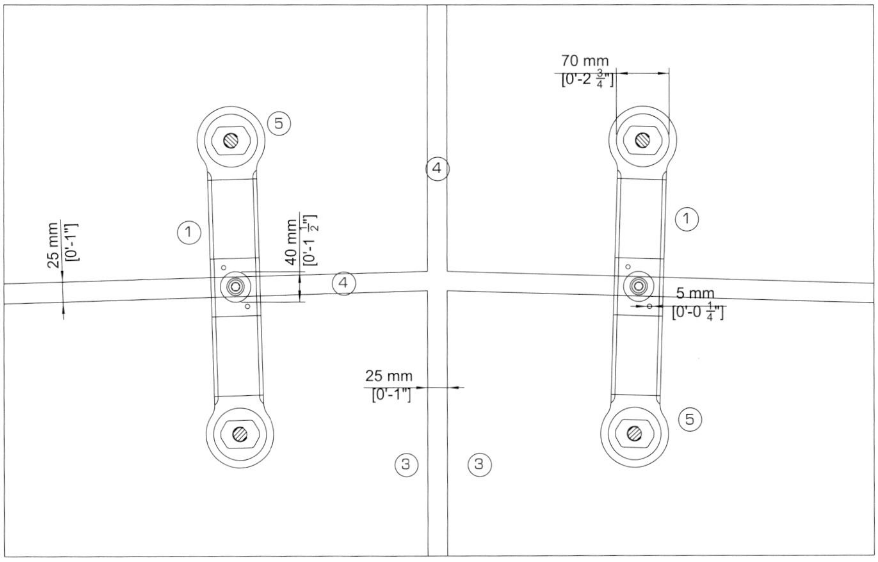

g | Horizontal section. External fold

function. The joint width both externally and internally is made to match the width of adjacent joints in order to allow the same extruded EPDM gasket to be used as an inner seal.

Junctions of the edge of the roof with external walls in other materials are similar to flashings at the base of the roofs, with which these joints can be continuous as the roof perimeter turns from roof to external wall. In some cases, an insulated aluminium closer is bonded to the edge of the double glazed unit and is sealed against the adjacent wall construction. The bolt fixed roof could meet, for example, a metal rain-screen wall, or an external wall clad in a range of panels.

Support brackets for external shading and for rope fixing points for abseil-based cleaning are sometimes designed to penetrate through the joints between the glass joints. This method is described in the previous section. Where these brackets occur at external and internal folds, the brackets are usually designed so as to avoid penetrating the joint at the corner itself. This is done to avoid a complicated junction of the bracket together with the four corners of a joint meeting at the same place. This is both difficult to seal and difficult to give a smooth, continuous appearance as seen from inside the building.

Small glazed rooflights

An essential aspect of small rooflights is the greater number of interfaces with surrounding construction and other materials than is usually the case with large glazed roofs. Where small rooflights have a bolt fixed glazing assembly mixed with gutter elements in a typical single rooflight, the glazing is sealed against the adjacent construction. In some instances, an outer seal is formed against an external roof panel in a different panel. A second inner seal is made with an inner metal panel. Glazing channels can also be used (as described in the previous section) in order to seal the gap, but the glazing channel must

Details
1 | Mild steel connector
2 | Mild steel support frame
3 | Double glazed unit
4 | Silicone seal
5 | Bolt fixing
6 | Support bracket
7 | Concrete base
8 | Stainless steel cable
9 | Mild steel tension rod
10 | Mild steel plate
11 | Adjacent external wall

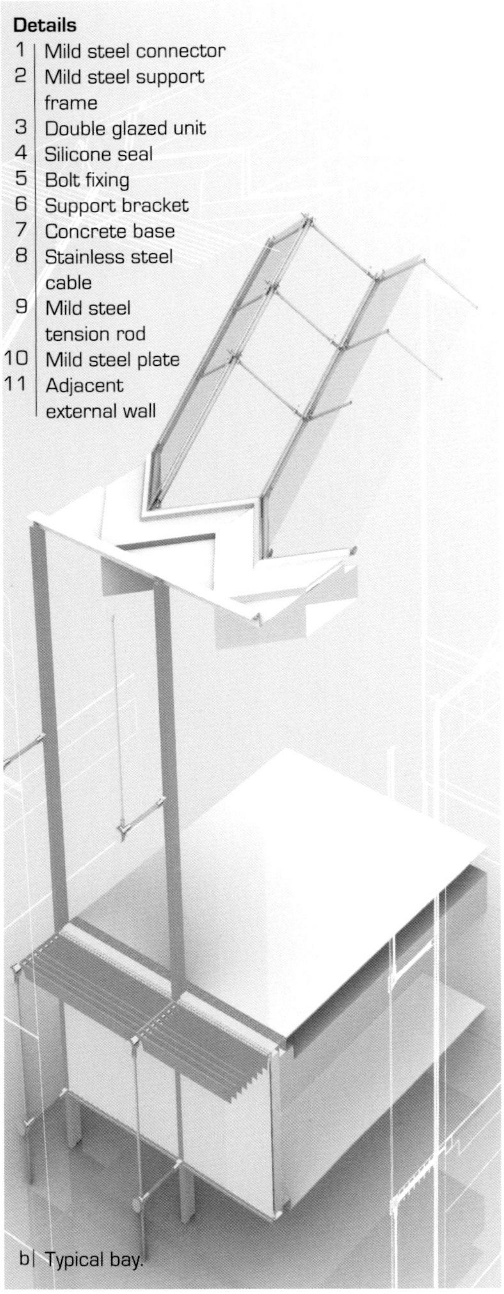

b| Typical bay.

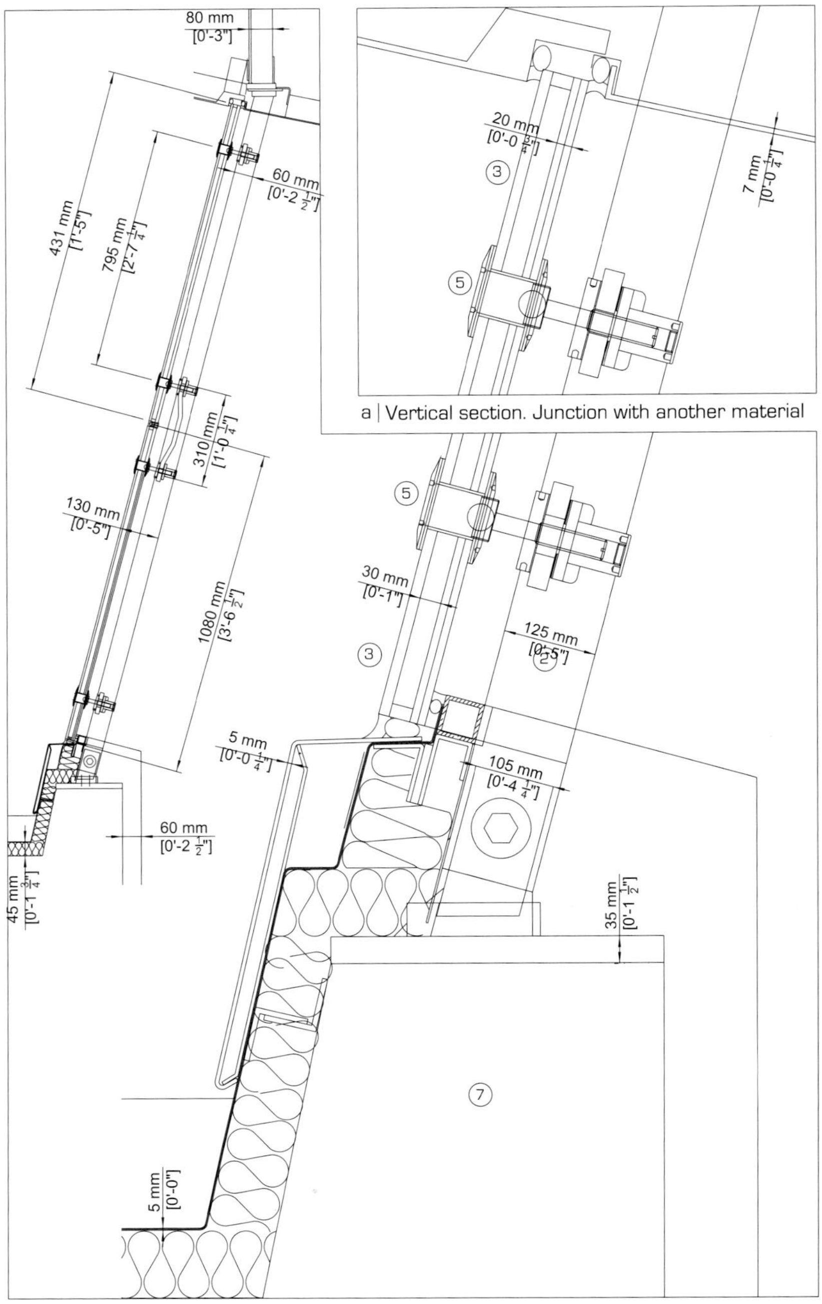

a | Vertical section. Junction with another material

c | Vertical section. Base upstand junction

be connected to a flexible seal such as an EPDM strip to allow the bolt fixed glazing to move independently of the other materials making the junction. In this case it is assumed that these structural movements are small.

In small rooflights the supporting structure is made visually lightweight in order to maximise the effect of the bolt fixed glazing. Stainless steel cables are often used to increase transparency (b). In generic examples, a small rooflight of 3500mm x 3500mm (10ft x 10ft) in plan is made from a mild steel tube supporting structure. The top of the rooflight requires four glass panels in order to span from side to side. The bolt fixings in the centre of the top of the rooflight are supported by two cable trusses spanning diagonally from the corners, intersecting in the centre. The four bolt fixings are secured to a cross-shaped bracket which is in turn bolted to two halves of a clamp bolted to the supporting cables. A vertical mild steel rod forms the central vertical element in the truss. The absence of steel tubes spanning across the centre of the top of the rooflight increases its visual transparency. Junctions at the corner and at the base of this typical small rooflight do not rquire bolt fixings, increasing the sense of transparency. In plan, external corners can be formed by stepping the edges of the double glazed unit in order to maintain a constant joint width for all rooflight joints. The visible area of silicone behind the glass can be concealed with silk screen printing, or 'fritting', usually in a black colour. Although external and internal corners of meeting double glazed units in bolt fixed roofs have a wide sight line, the continuity of the glass and its reflections obscure the effect of the fritting and silicone seals behind.

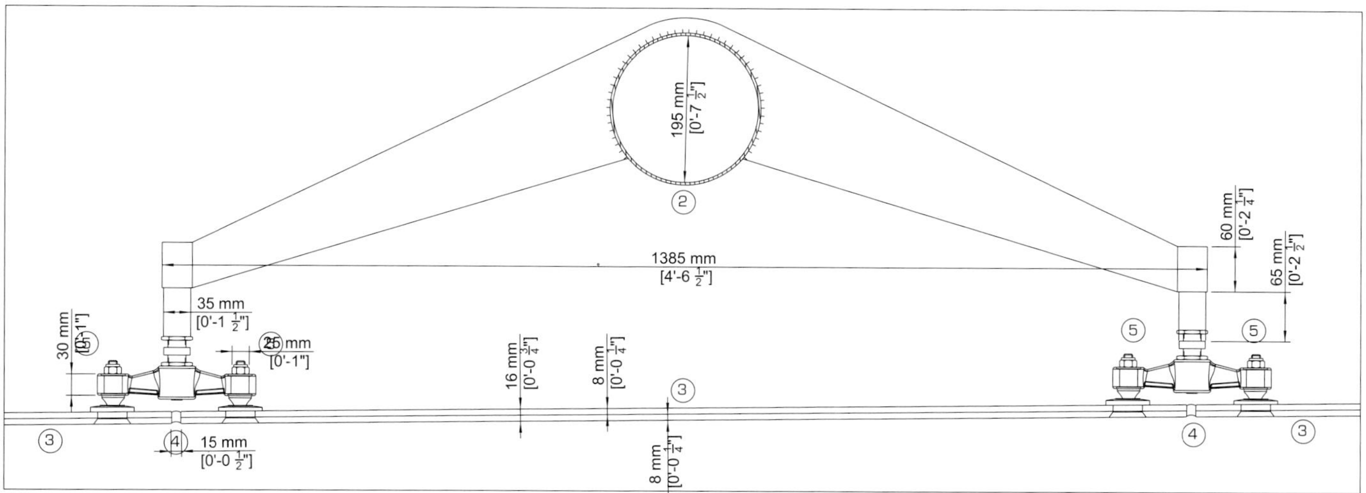

d | Vertical section. Base upstand

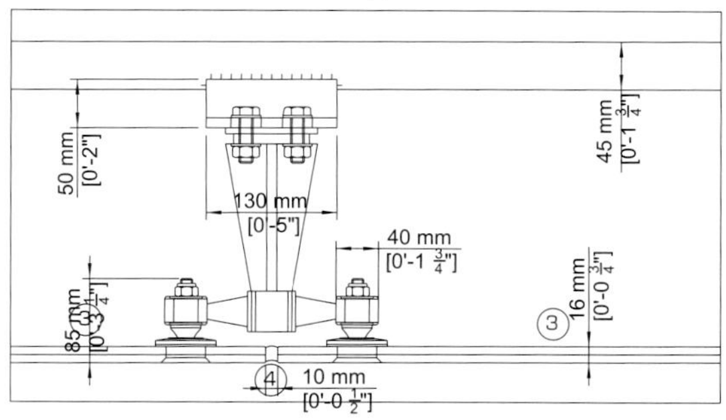

e | Vertical section. Bracket support

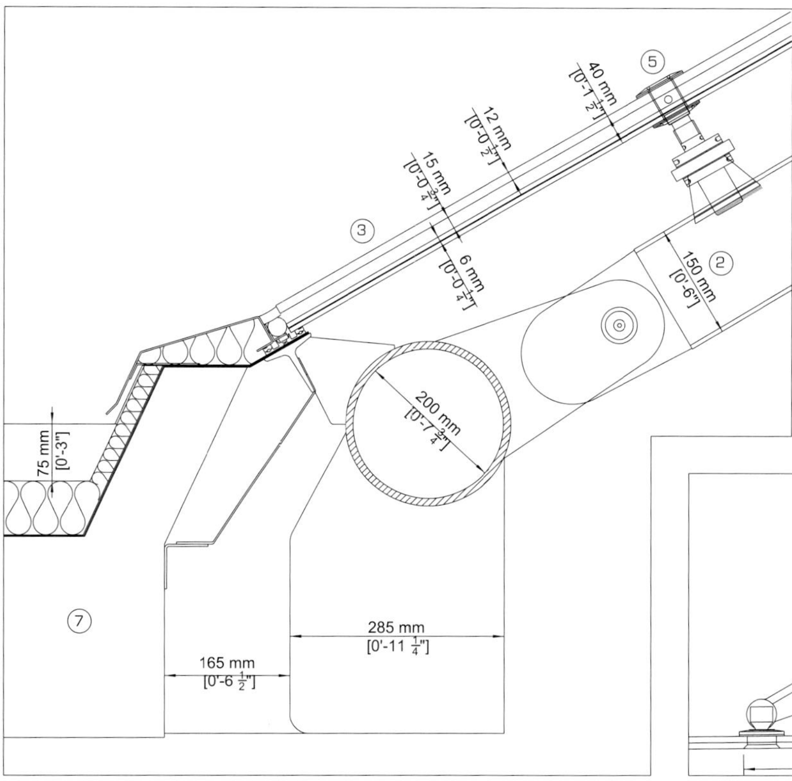

f | Vertical section. Internal fold

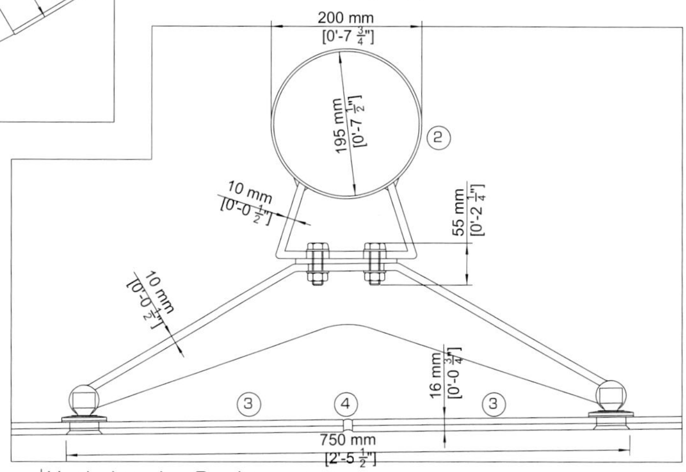

g | Vertical section. Bracket support

Larger rooflights

For bigger rooflights up to around 5.0 metres high and 8.0 metres wide, lightweight hollow mild steel sections, together with stainless steel cables, are commonly used. This rooflight size suits glass sizes which are around 2000mm x 2000mm (6ft6in x 6ft6in) for horizontally-set units and around 2000mm x 2500mm (6ft6in x 8ft) high for vertically-set units. A modest rectangular steel frame forming the edges of the generic rooflight can support stainless steel cables spanning both vertically and horizontally to which bolt fixings and cross-shaped brackets can be fixed. The cross-shaped bracket is fixed to a clamp which is bolted to the cable. For glass joints set directly in front of the tubular steel structure, the bolt fixings are secured directly to a channel-shaped bracket welded to the main supporting steel tube. All bolt fixings in these modest sized rooflights have their adjustment for fixing tolerance made at the junction of the bolt fixing and the support bracket. The glass units set horizontally can be supported at their corner fixings by either a grid of steel tubes immediately below the joint, or with a cable truss (MCE_395, f), which provides greater visual transparency than the all-tube solution. The main stainless steel cables of the truss span across the diagonal corners, meeting at a central vertical tubular post, as described in the previous paragraph for small rooflights. The increased span is assisted by a set of secondary steel cable trusses set at 45° to the diagonal geometry (orthogonal with the glass). The secondary trusses both stiffen the main trusses and provide a fixing point for all bolt fixings (MCE_395, f). The principle of this medium-sized generic rooflight can be adapted to suit a range of individual designs of similar overall dimensions.

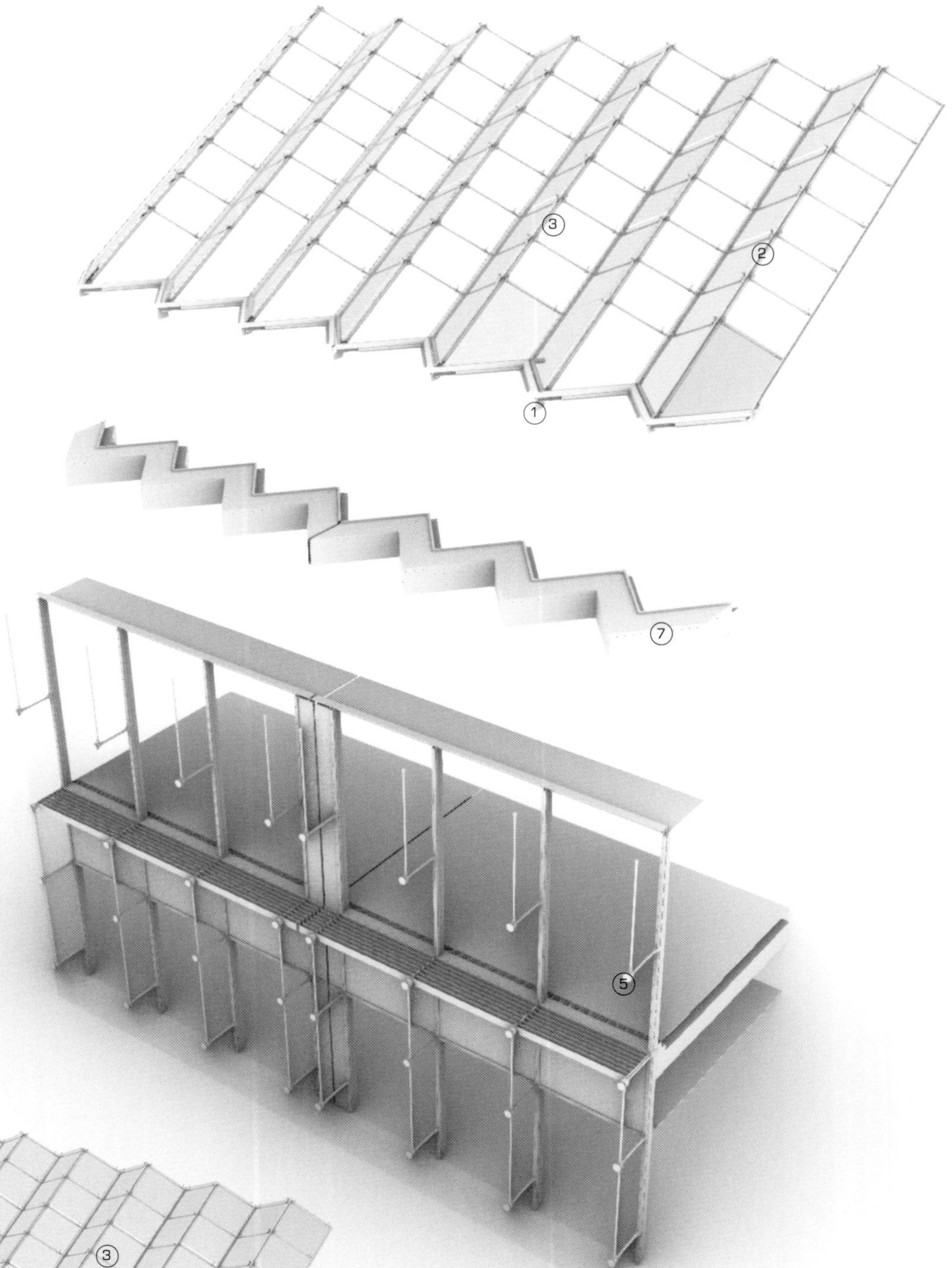

a | 3D exploded view showing large scale bolt fixed glazed rooflights

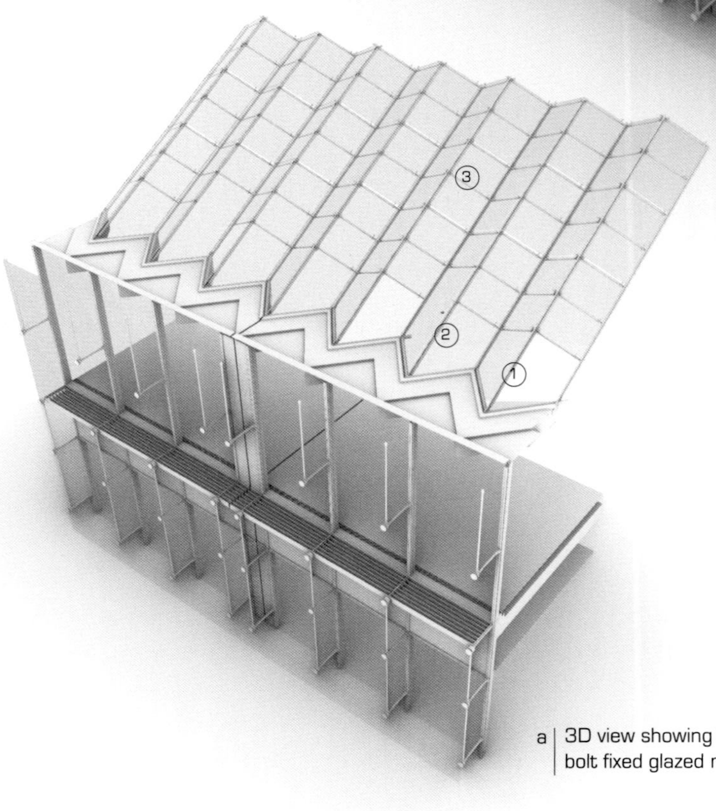

a | 3D view showing large scale bolt fixed glazed rooflights

Details

1 | Mild steel connector
2 | Mild steel support frame
3 | Double glazed unit
4 | Silicone seal
5 | Bolt fixing
6 | Support bracket
7 | Concrete base
8 | Stainless steel cable
9 | Mild steel tension rod
10 | Mild steel plate
11 | Adjacent external wall
12 | Single glazed solar shading glass

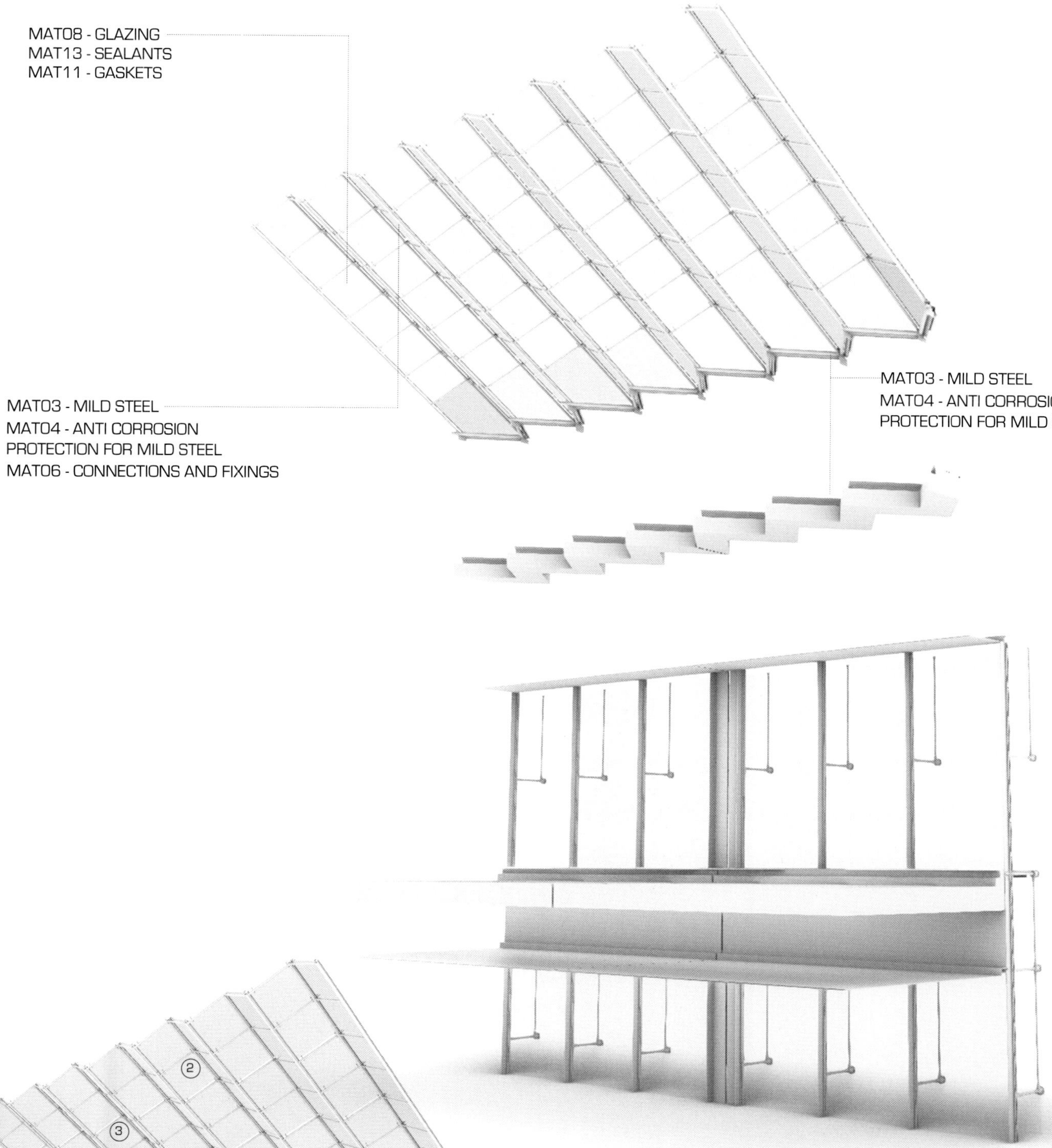

c | 3D exploded view of large scale bolt fixed glazed rooflight components with material specification references

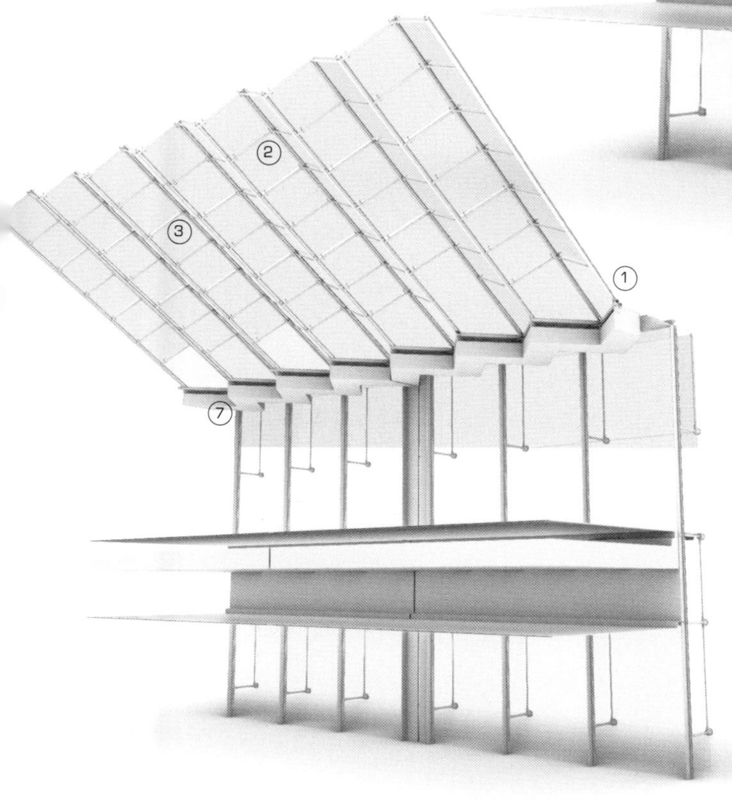

d | 3D view showing large scale bolt fixed glazed rooflights

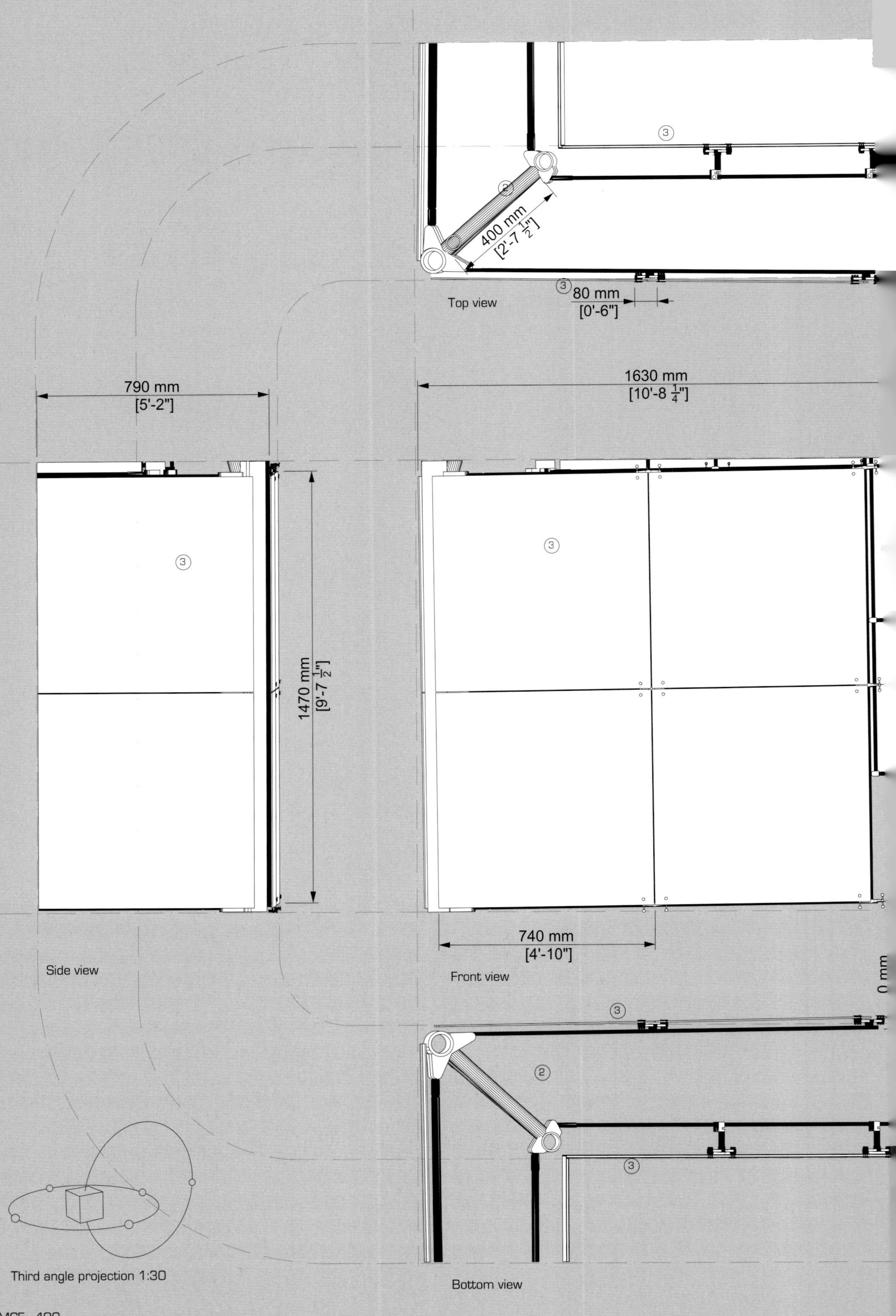

Top view

Side view

Front view

Bottom view

Third angle projection 1:30

Details

1 | Mild steel connector
2 | Mild steel support frame
3 | Double glazed unit
4 | Silicone seal
5 | Bolt fixing
6 | Support bracket
7 | Concrete base
8 | Stainless steel cable
9 | Mild steel tension rod
10 | Mild steel plate
11 | Adjacent external wall

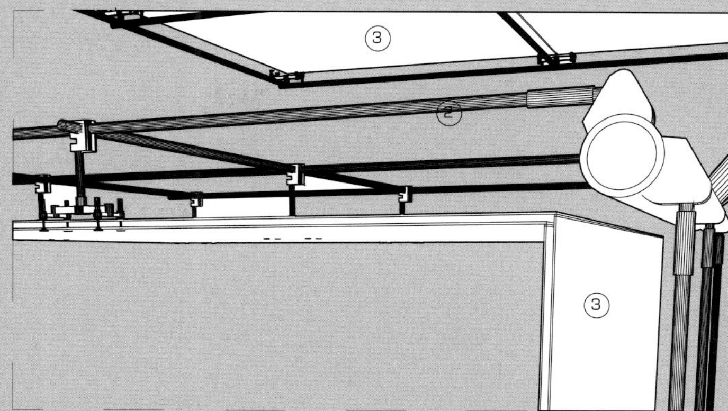

3D view of the assembly

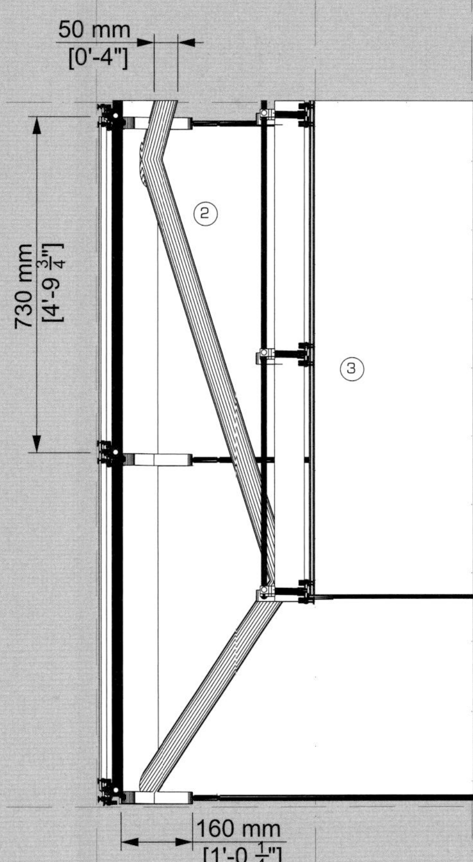

Side view

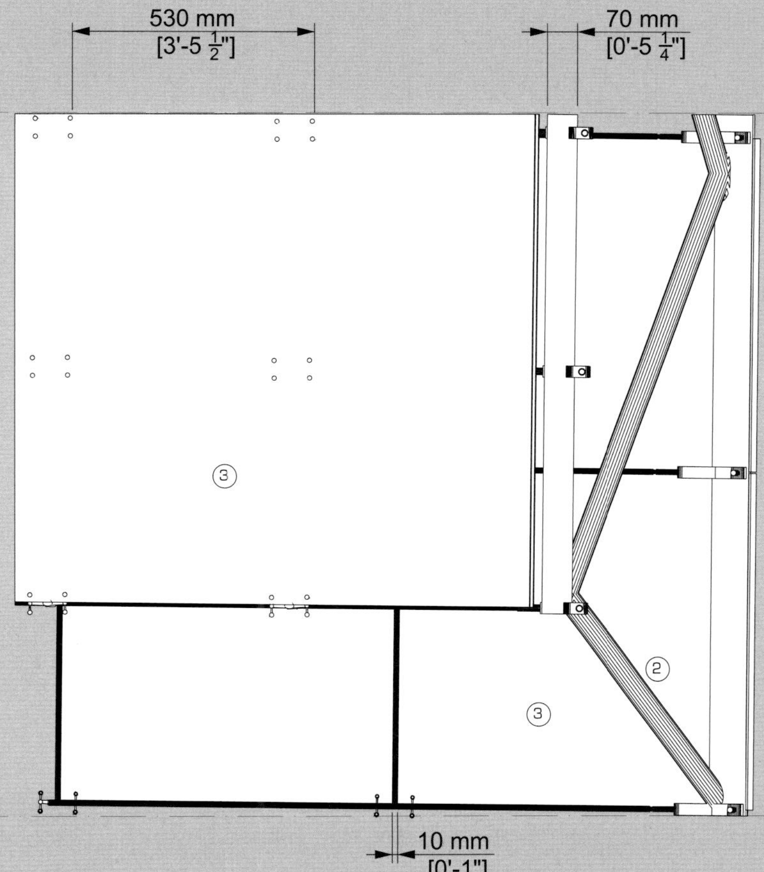

Back view

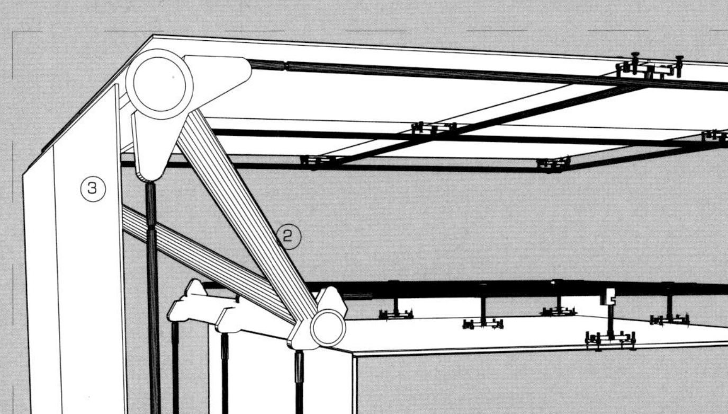

3D view of the assembly

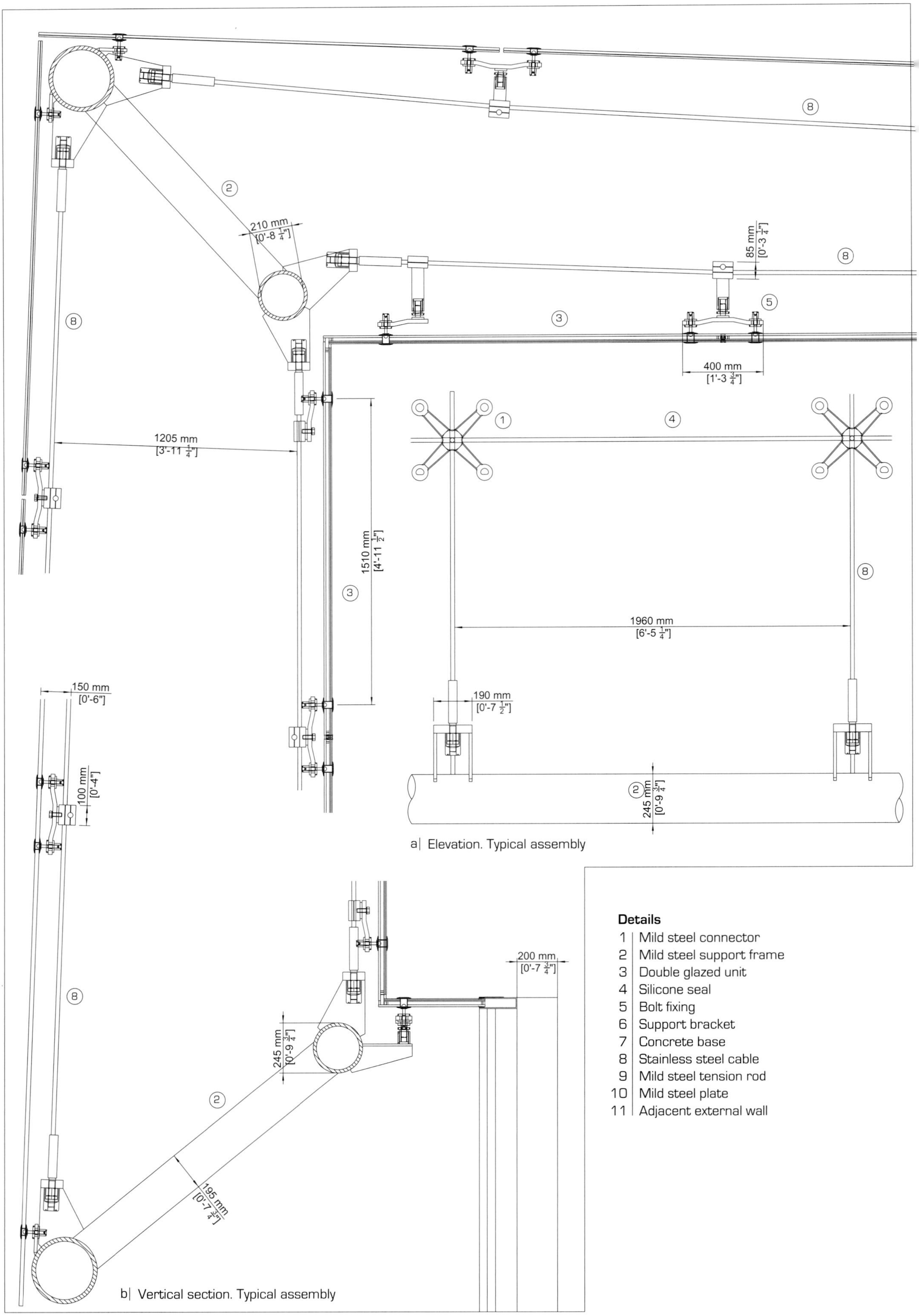

a| Elevation. Typical assembly

b| Vertical section. Typical assembly

Details

1 | Mild steel connector
2 | Mild steel support frame
3 | Double glazed unit
4 | Silicone seal
5 | Bolt fixing
6 | Support bracket
7 | Concrete base
8 | Stainless steel cable
9 | Mild steel tension rod
10 | Mild steel plate
11 | Adjacent external wall

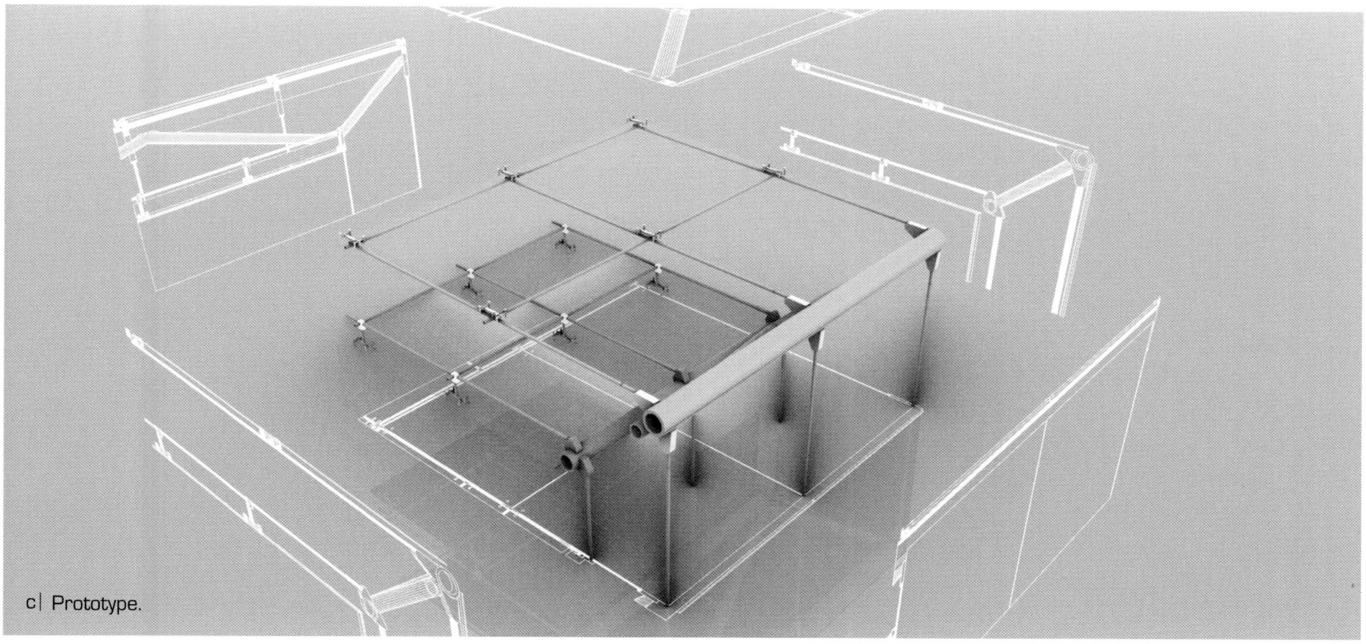

c| Prototype.

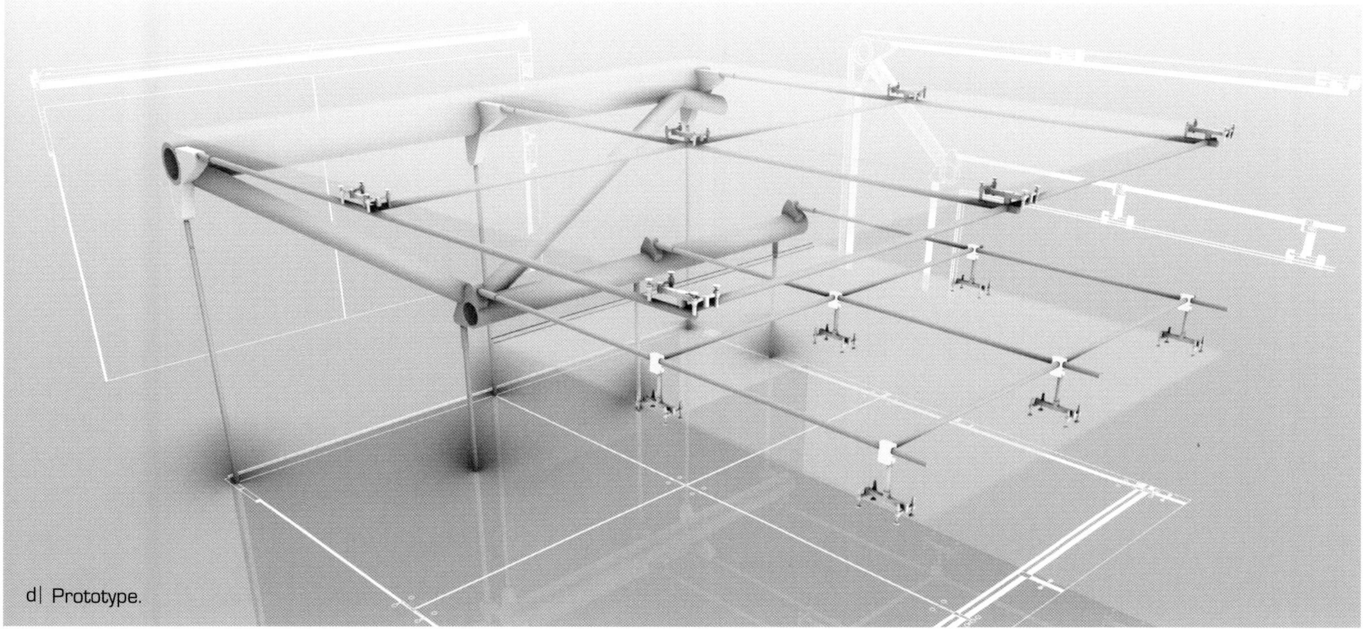

d| Prototype.

Bolt fixed glazing can be used as solar shading in twin wall rooflights. Two generic examples are shown here for a square-shaped rooflight and a circular-shaped rooflight, both of a medium size. The square example is 8.0 metres x 8.0 metres for the inner glass roof, and the circular rooflight has a radius of 9.0 metres across its inner glazing. Both use the same structural principle of a tubular steel frame with stainless steel cables stretched between them, but the position of the structure in relation to the bolt fixed glazing is different for each rooflight. These examples aim to show a visually lightweight solution that suits the quality of visually transparency that can be achieved with bolt fixed glazed roofs rather than more conventional framed solutions that detract from the thin joint lines of this roof glazing method. Both examples have an inner sealed rooflight and an outer layer of open jointed glass solar shading panels. The outer glass can be either screen printed with dots or patterns to provide shading, or have a solar control film set next to the PVB interlayer within the laminated glass. The inner glass in each example comprises sealed double glazed units. A cross-shaped support bracket has been used in each rooflight, and the glazed units are square shaped to keep the examples as generic as possible.

The square rooflight has two rectangular frames, one set inside the other, linked by diagonal members. The inner roof is hung from steel cables which span from side to side in both directions, forming a rectilinear grid of cables. Cross-shaped bolt fixing brackets are set at the intersection of these cables. The vertical parts of the inner sealed rooflights are set out in the same way, so that the supporting structure is entirely outside the building. The external sun shading is fixed in the same way, but with the outer glass fixed above the supporting cables. The rooflight has solar shading on its roof and on three vertical sides, so that from outside the building the supporting structure is seen only

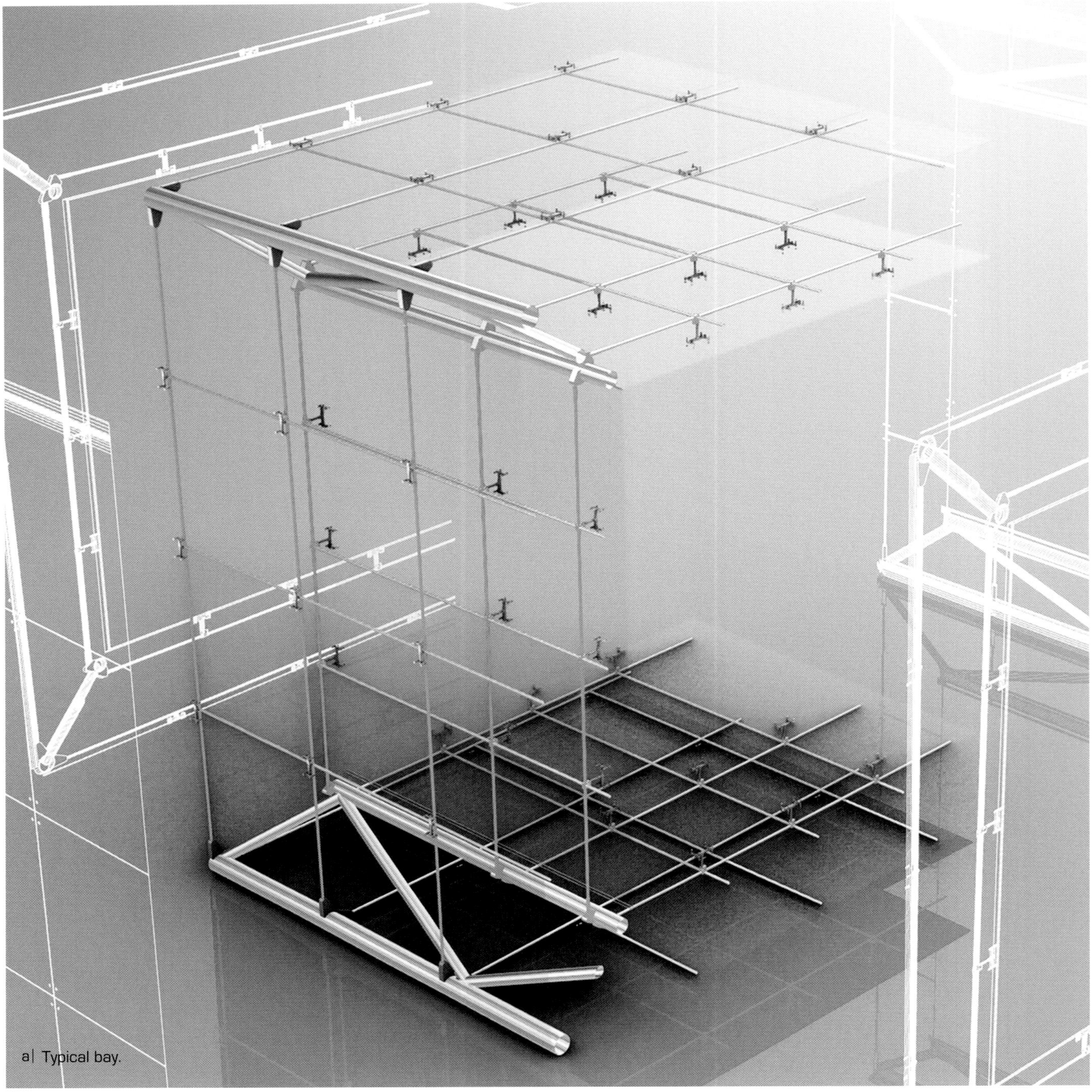

a| Typical bay.

through the shading. The outer shading is tilted out of plane in relation to the inner cube. With this construction method the inner and outer cable structures do not need to follow the same geometry.

The circular rooflight has an outer shading set on top of the inner sealed rooflight to provide partial solar shading. Where the square rooflight has a gap of approximately 1.0 metres between the two skins of glass to provide maintenance access for cleaning, the circular rooflight has a gap of only 600mm around its edge. Where there is not sufficient space for a 600mm to 1000mm gap for maintenance access, a structural tube is removed when access is required.

Cleaning and maintenance is undertaken on both rooflights by walking on the glass itself, with individuals wearing harnesses which are secured to the upper cables, which also serve as a fall arrest cable. The top of the glass is cleaned by passing the rope of the person's harness through the open joints between the glass shading panels. The gap is 20mm-25mm which is sufficient for a rope and securing hook.

The use of bolt fixings instead of a framed rooflight system allows the glass panels to experience deflections under wind load that are higher than those permissible in a rigid, framed system. This allows a single line of cables to be used, where they would have to be formed into (more rigid) trusses if a framed system were used.

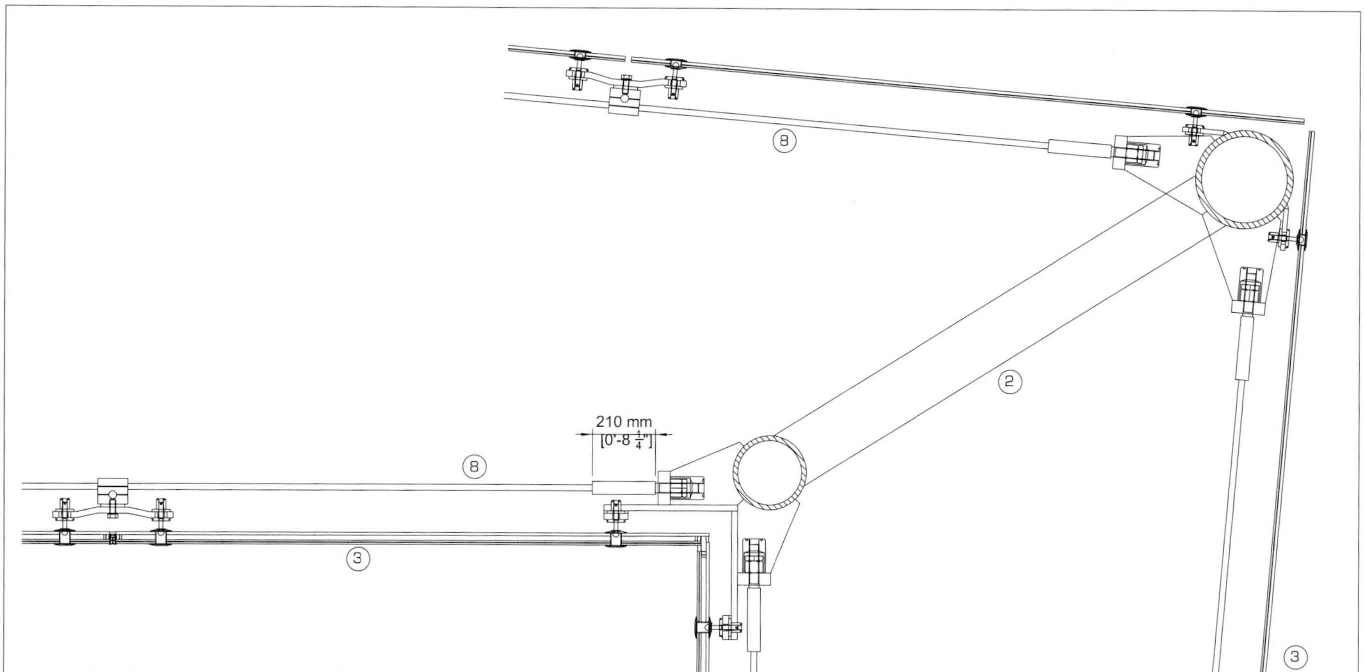

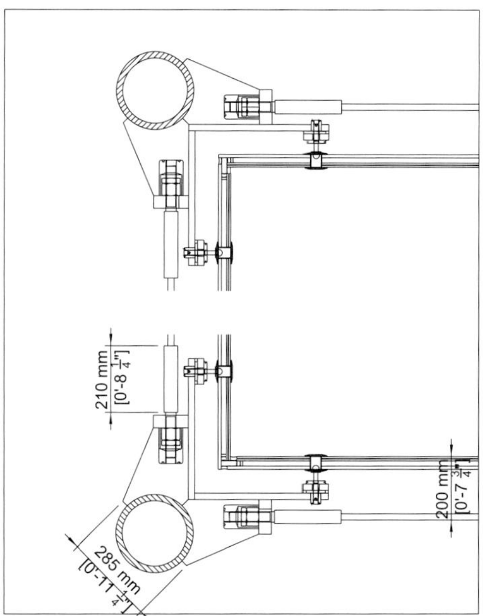

b| Horizontal section. Typical assembly

Details

1 | Mild steel connector
2 | Mild steel support frame
3 | Double glazed unit
4 | Silicone seal
5 | Bolt fixing
6 | Support bracket
7 | Concrete base
8 | Stainless steel cable
9 | Mild steel tension rod
10 | Mild steel plate
11 | Adjacent external wall
12 | Single glazed solar shading glass

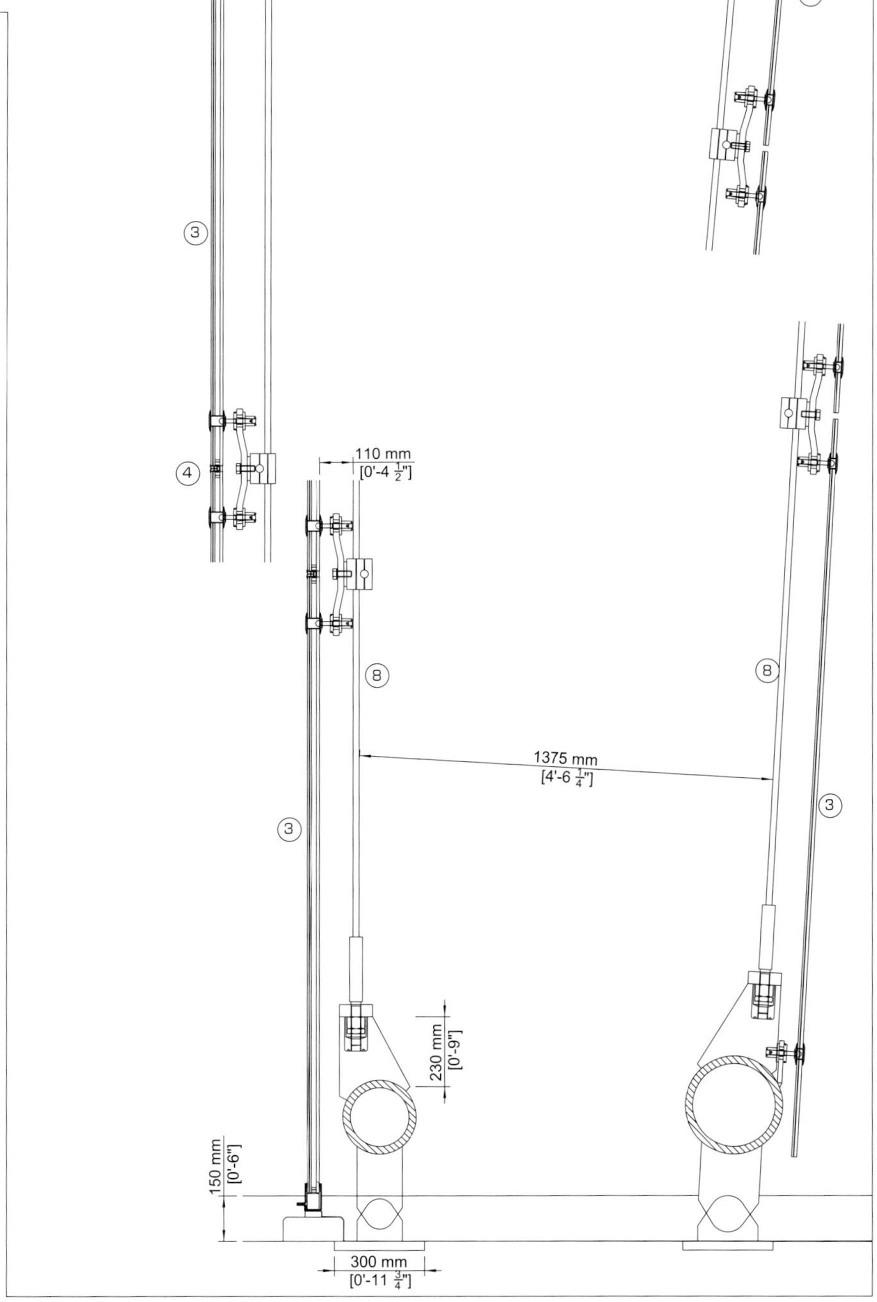

c| Vertical sections. Typical assembly

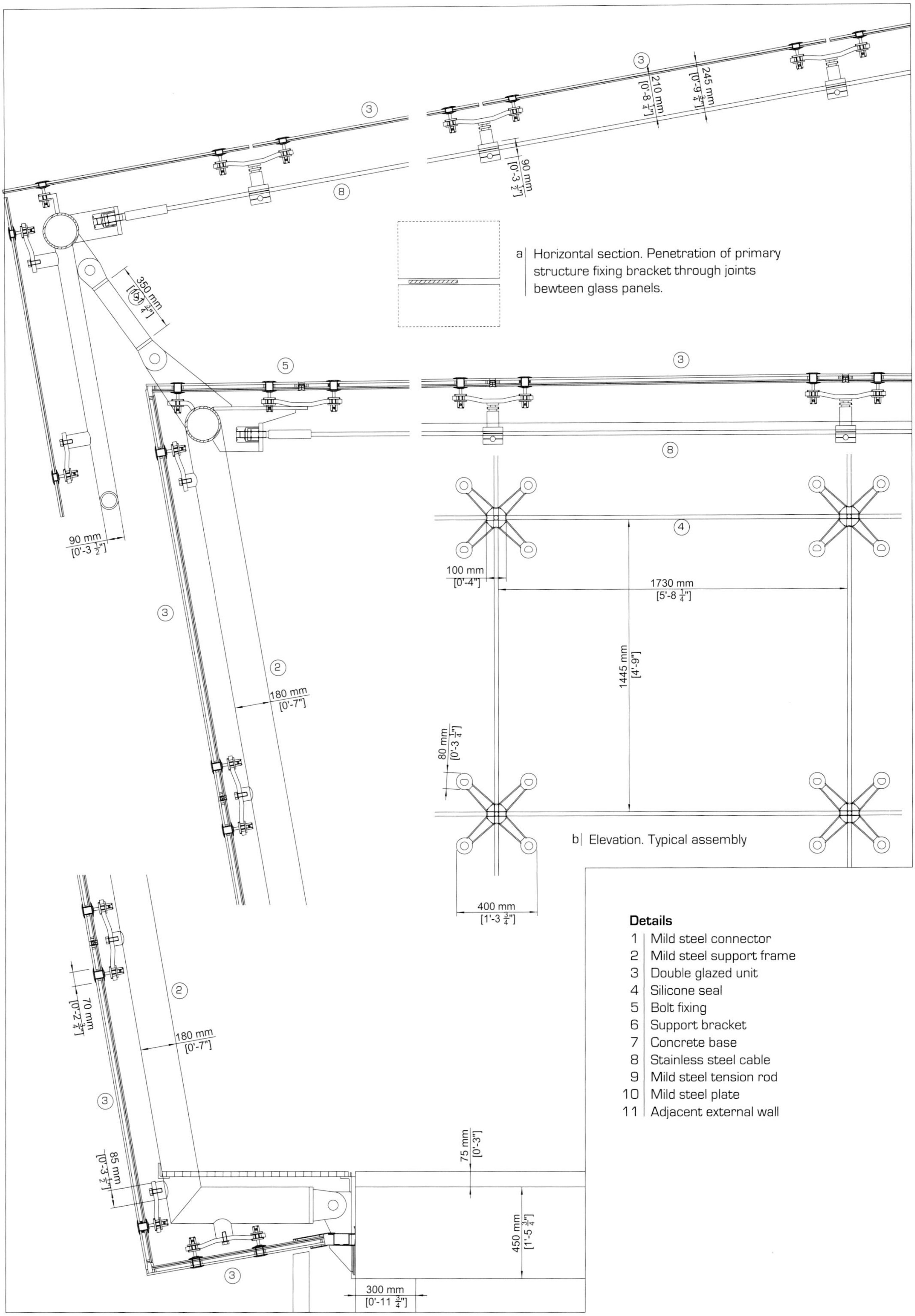

a| Horizontal section. Penetration of primary structure fixing bracket through joints bewteen glass panels.

b| Elevation. Typical assembly

Details

1 | Mild steel connector
2 | Mild steel support frame
3 | Double glazed unit
4 | Silicone seal
5 | Bolt fixing
6 | Support bracket
7 | Concrete base
8 | Stainless steel cable
9 | Mild steel tension rod
10 | Mild steel plate
11 | Adjacent external wall

c| Vertical sections. Typical assembly

1:25 scale 1'-0 1/2" 2'-1" 3'-1 1/2" 4'-2"

250 500 750 1000 1250 mm 1:25 scale

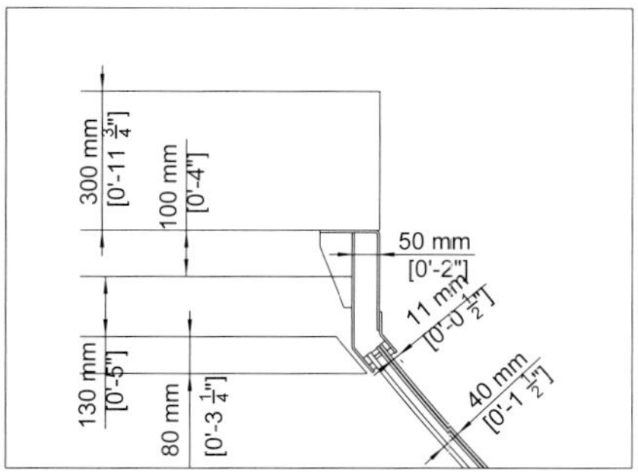

d| Horizontal section. Typical assembly

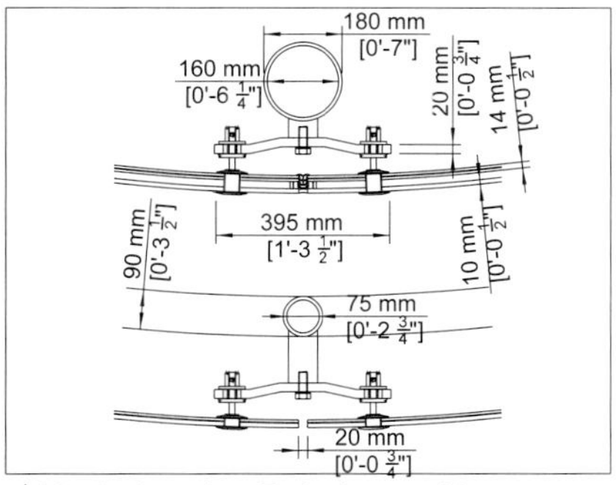

e| Vertical section. Typical assembly

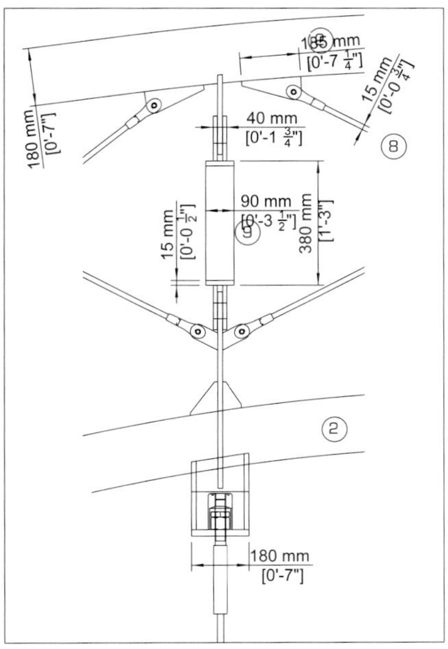

f| Elevation. Typical assembly

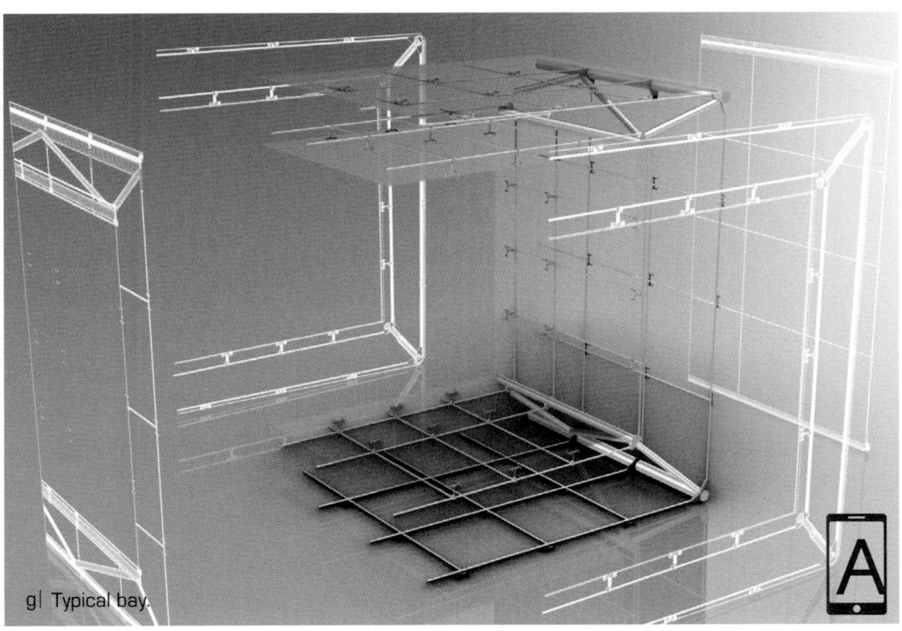

g| Typical bay

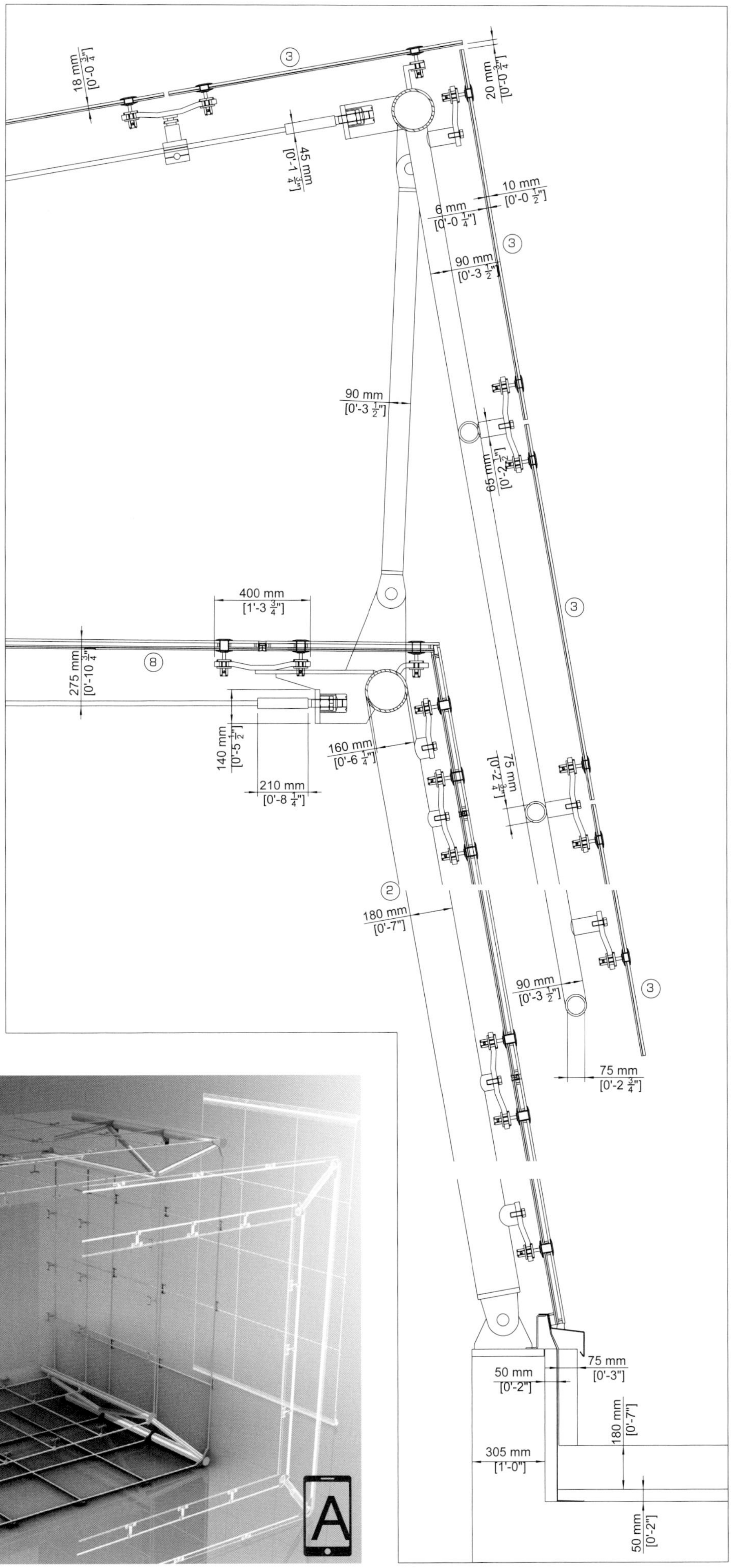

h| Vertical section. Typical assembly

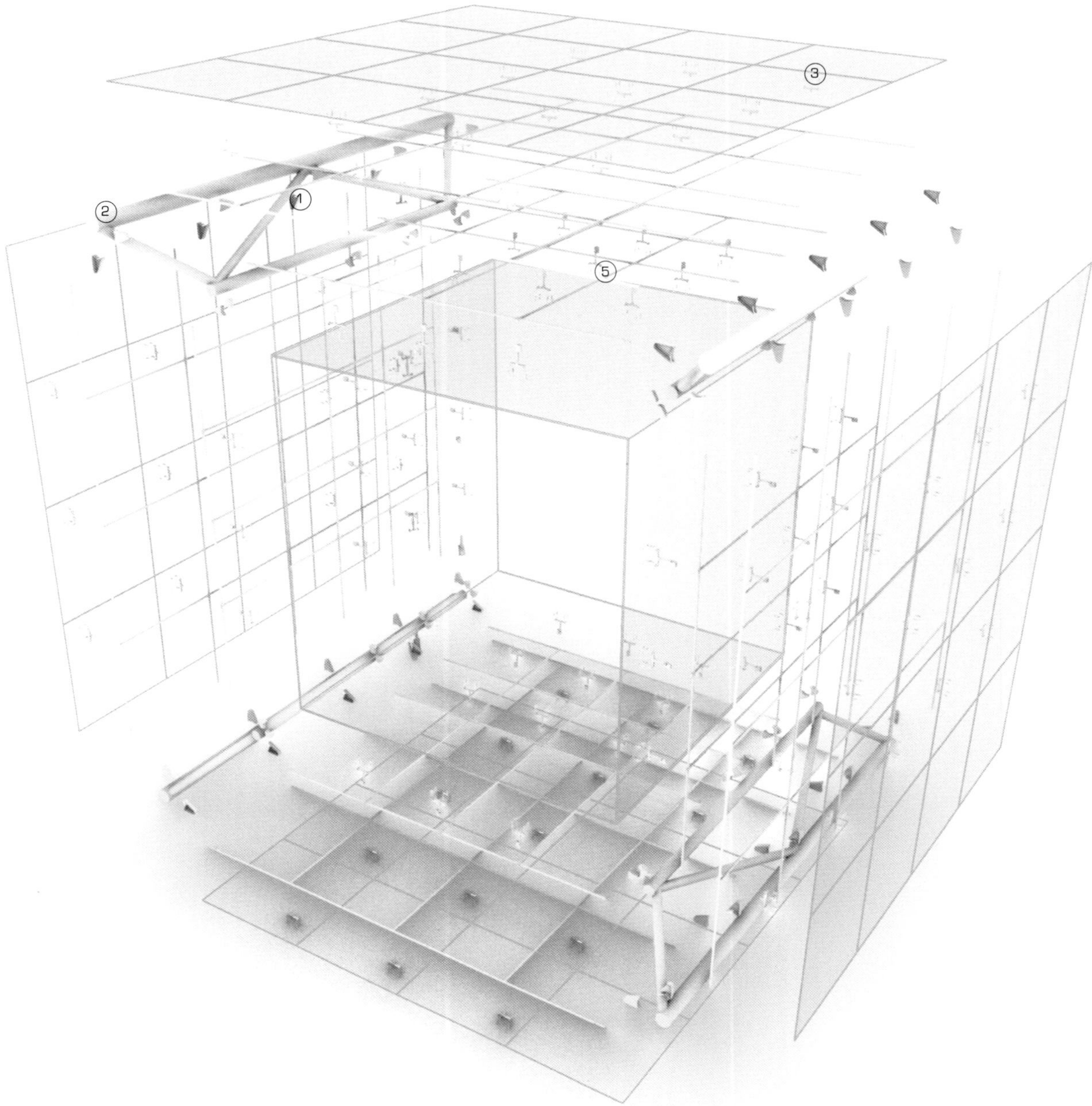

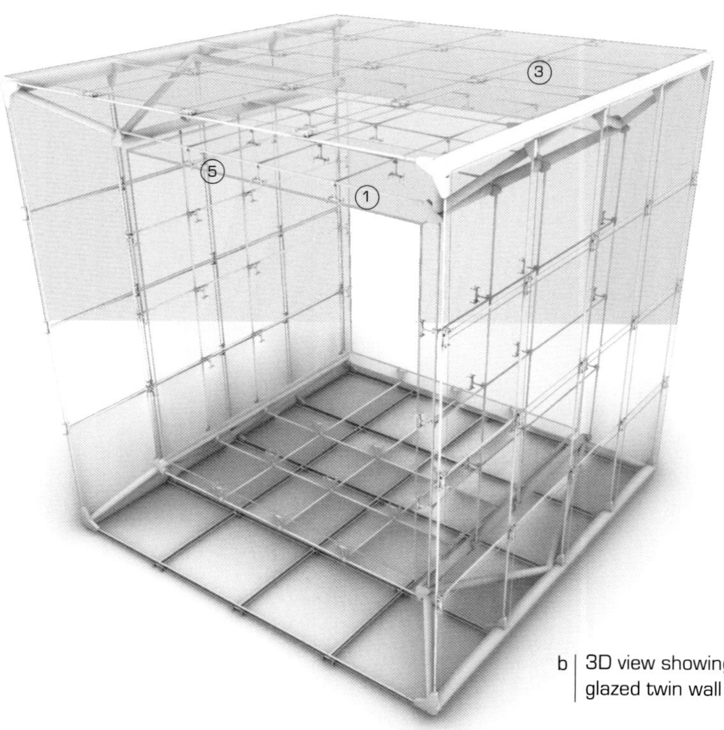

a | 3D exploded view showing bolt fixed glazed twin wall roof

Details
1 | Mild steel connector
2 | Mild steel support frame
3 | Double glazed unit
4 | Silicone seal
5 | Bolt fixing
6 | Support bracket
7 | Stainless steel cable
8 | Mild steel tension rod
9 | Mild steel plate
10 | Adjacent external wall

b | 3D view showing bolt fixed glazed twin wall roof

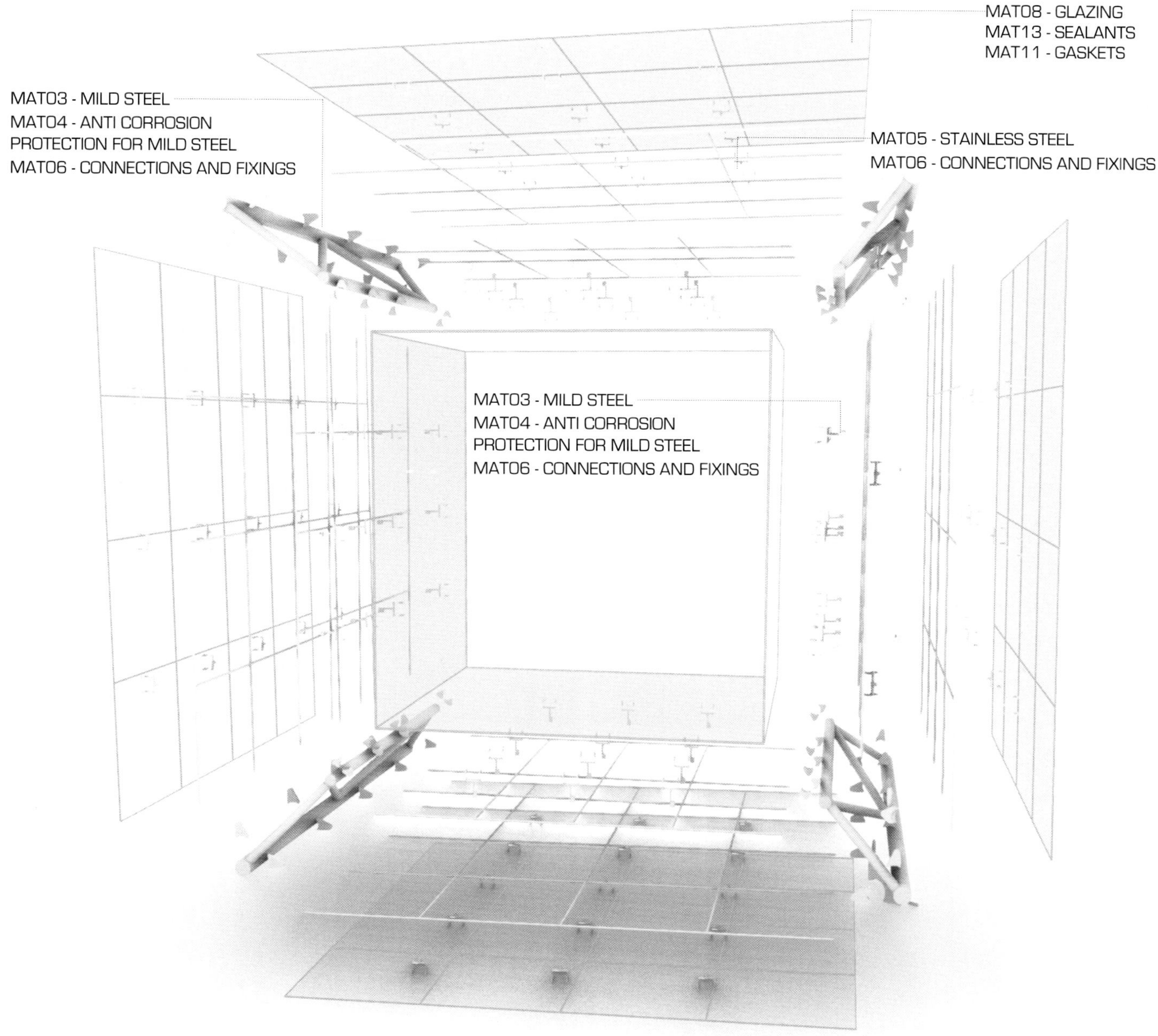

c | 3D exploded view of bolt fixed glazed twin wall roof components with material specification references

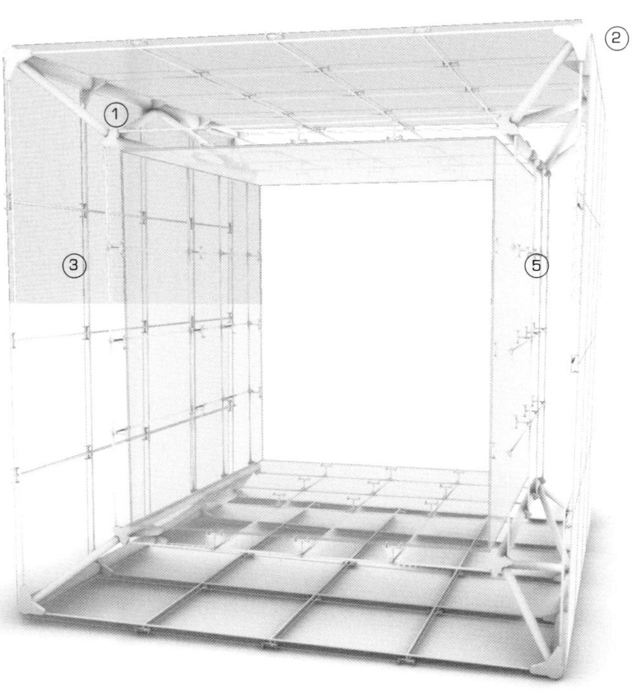

d | 3D view showing bolt fixed glazed twin wall roof

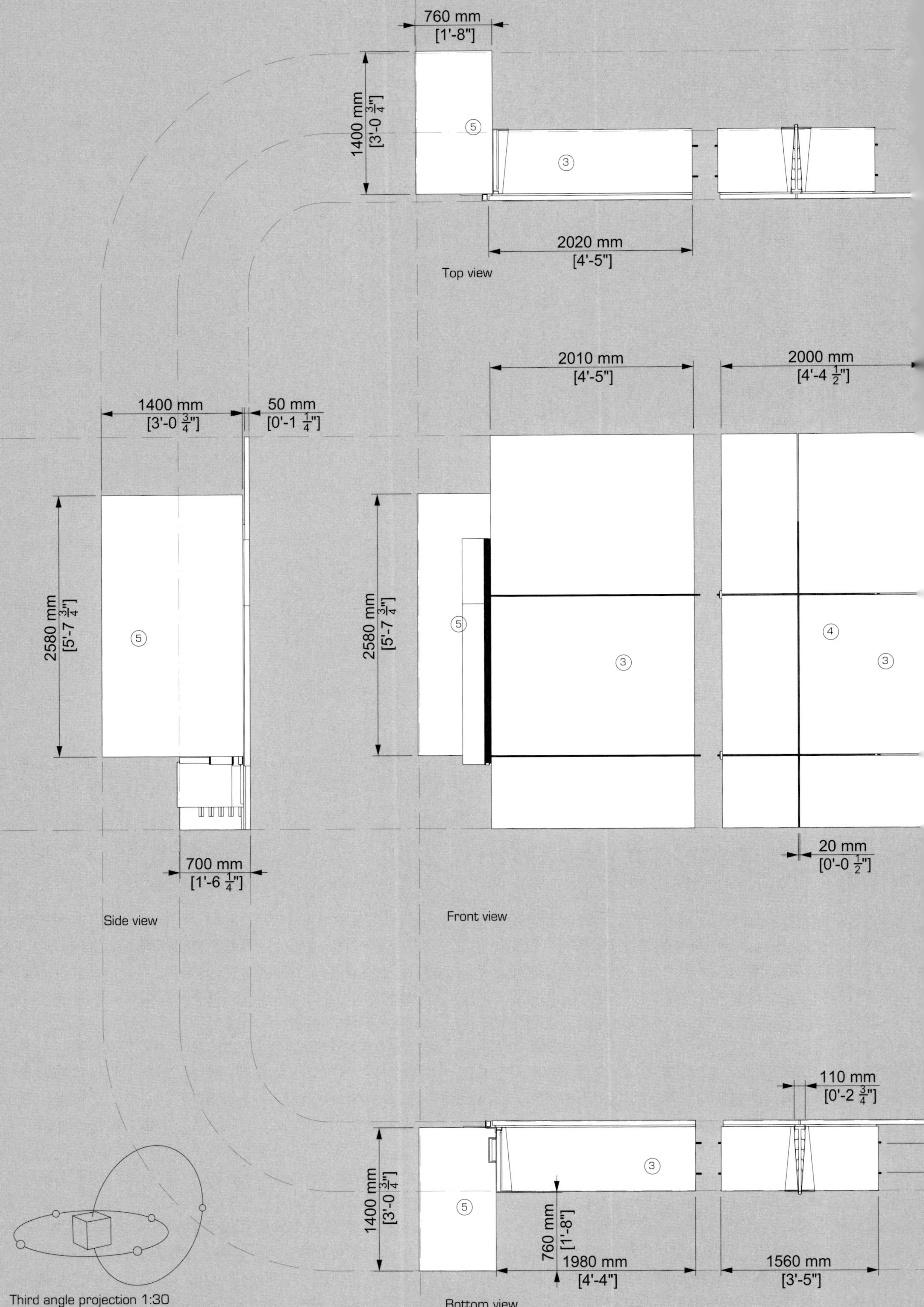

Third angle projection 1:30

Details

1 | Silicone bond
2 | Mild steel support frame
3 | Single glazed laminated glass panel
4 | Silicone seal
5 | Concrete base
6 | Insulated metal panel
7 | Folded metal flashing
8 | Reinforced concrete support frame

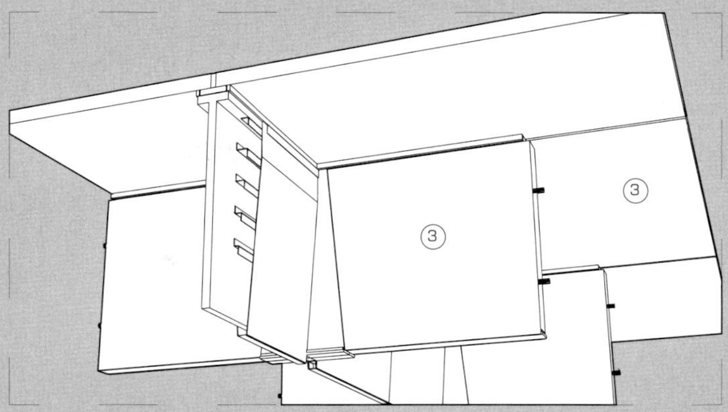

3D view of the assembly

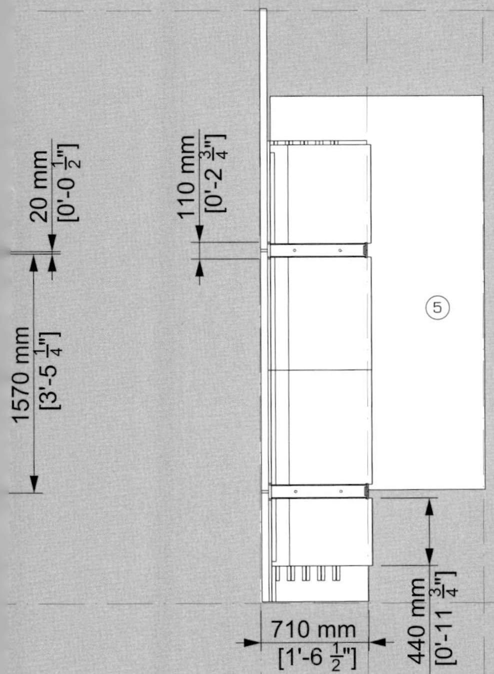

Side view

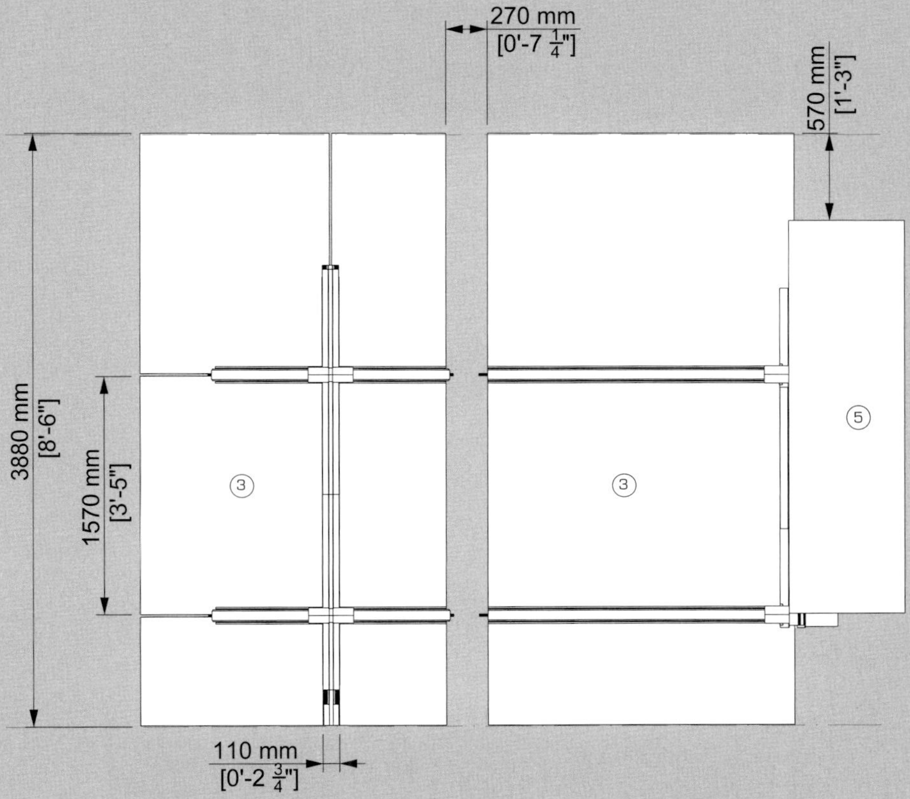

Back view

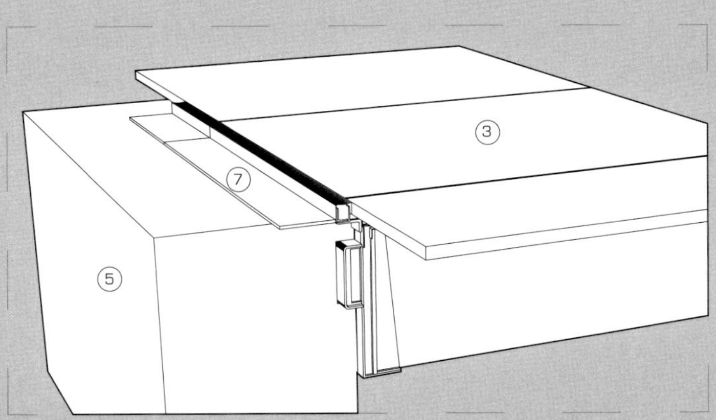

3D view of the assembly

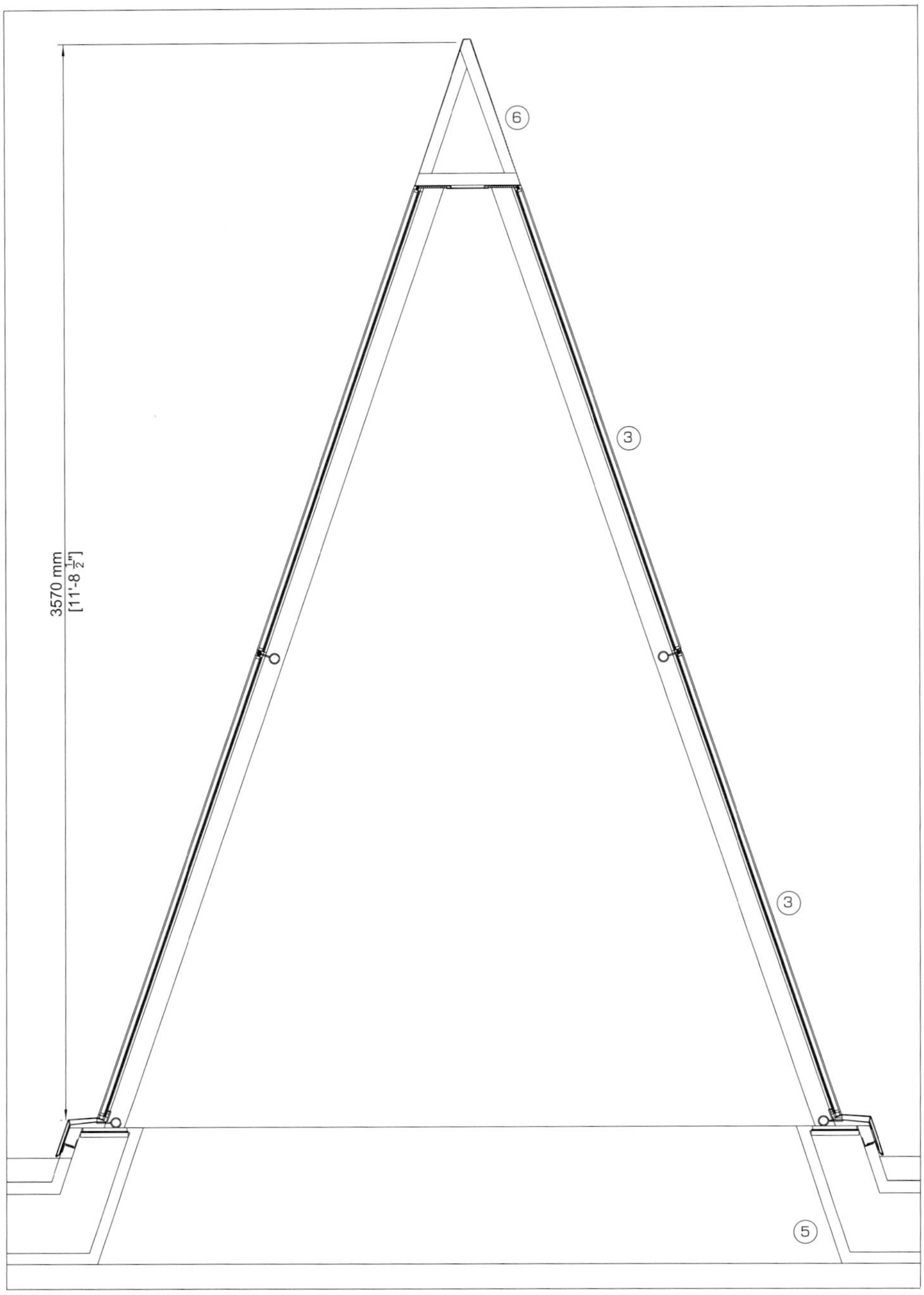

d | Vertical section 1:25. Typical assembly

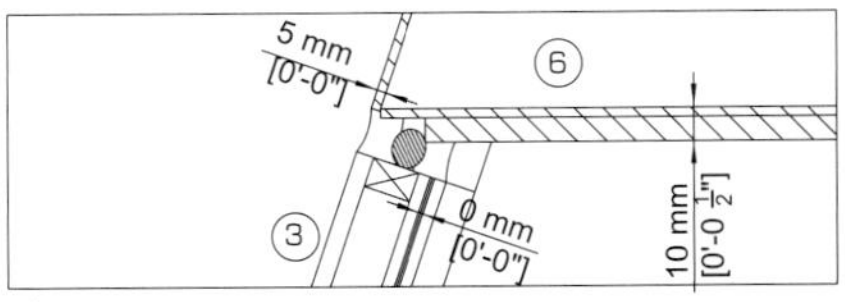

a| Vertical section. Junction with adjacent mater

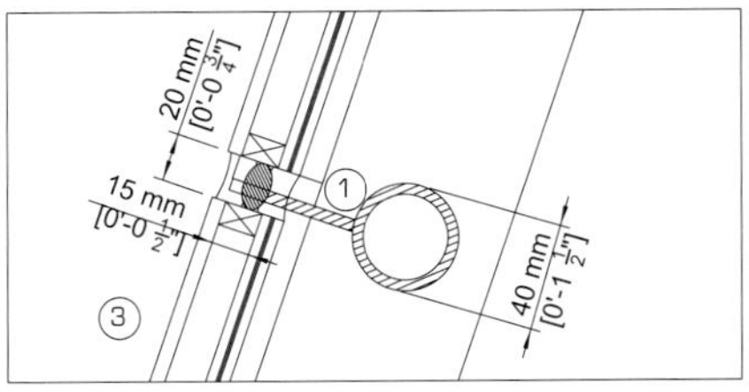

b| Vertical section. Glass to glass junction

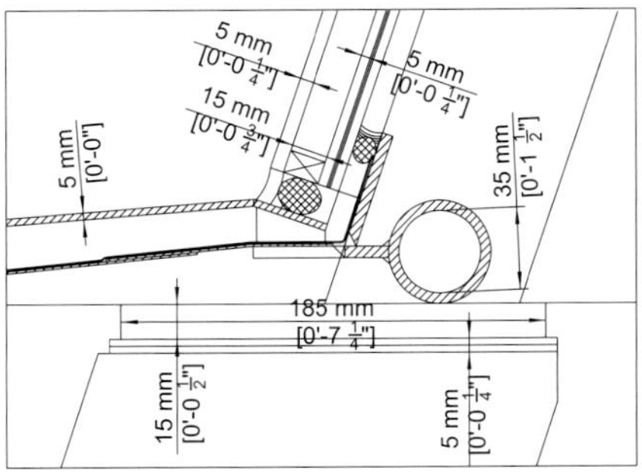

c| Vertical section. Junction with upstand

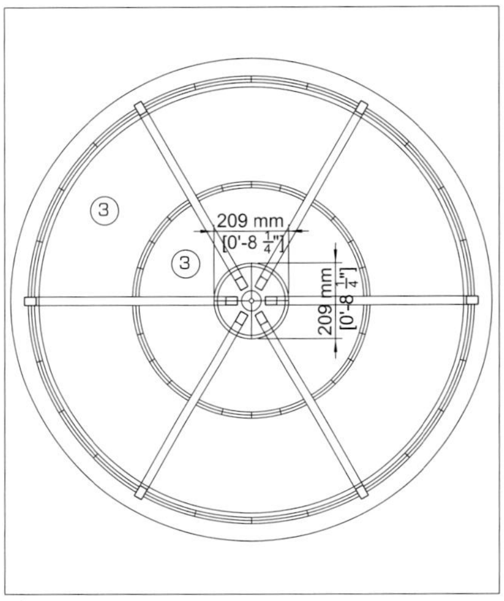

e| Horizontal section 1:25.

The method of silicone bonding glass to aluminium framing is well developed for use in glazed curtain walling to provide visually smooth glass facades with no visible cappings (MCE_414, a). The use of silicone sealed rooflights, discussed in an earlier section, can be taken a step further to become a full bond without the need for the mechanical restraint of pressure plates. In silicone bonded rooflights, the glass is glued to a supporting frame. The glue is also the external seal. This technique is useful for small rooflights, where cappings would be very difficult to fabricate, and in rooflights which are walked upon, where the rooflight is an external glass floor.

Four examples are discussed in this section. The generic conical rooflight (h) has curved double glazed units bonded to an aluminium frame. Silicone bonding avoids cover caps which would have to be curved both vertically and horizontally, and which would be extremely difficult to fabricate. The generic rectangular rooflight can be bonded together without a supporting structure, with the glass providing its own support (MCE_414, c). A flat monopitch rooflight can be bonded to a frame to provide a small rooflight from one double glazed unit. The use of laminated glass sheets can make a glass deck strong enough to walk on. Similar to a glass floor used inside a building, it must also take heavier traffic loadings and be weathertight.

Generic conical rooflight

In a generic conical rooflight, a lightweight steel frame is used to support double glazed units that form a rooflight. The structural frame comprises box sections set vertically, held in place by thin tube sections running horizontally to form a circle (g). The glass panels are supported on steel flat sections which are welded to the horizontally set tube section. The glass is levelled on blocks set onto the horizontal flat section, and the silicone is applied to the joint. At the base, the horizontal metal section projects out to form a flashing over the upstand in which the rooflight is set (f). An additional inner metal upstand can be provided

Details

1 | Silicone bond
2 | Mild steel support frame
3 | Single glazed laminated glass panel
4 | Silicone seal
5 | Concrete base
6 | Insulated metall panel
7 | Folded metal flashing
8 | Reinforced concrete support frame

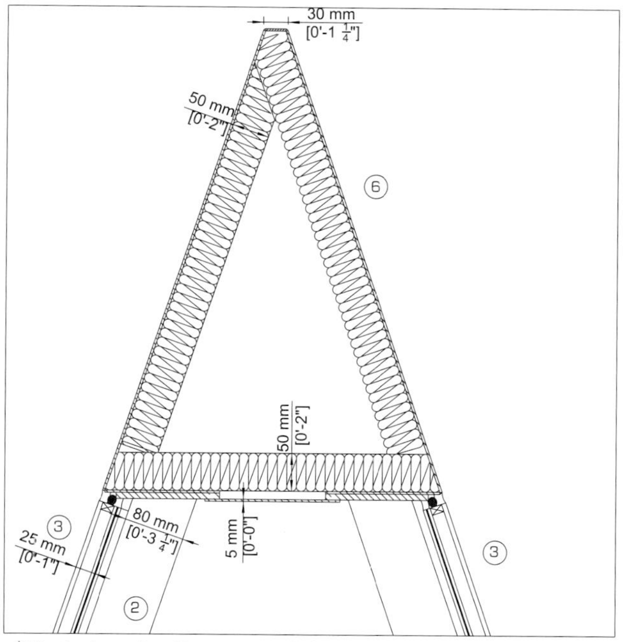

g | Vertical section. Base upstand. Junction with adjacent material

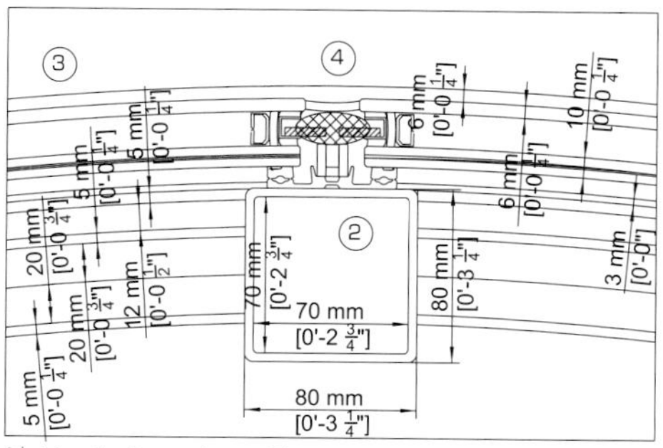

f | Vertical sections. Glass to glass junction

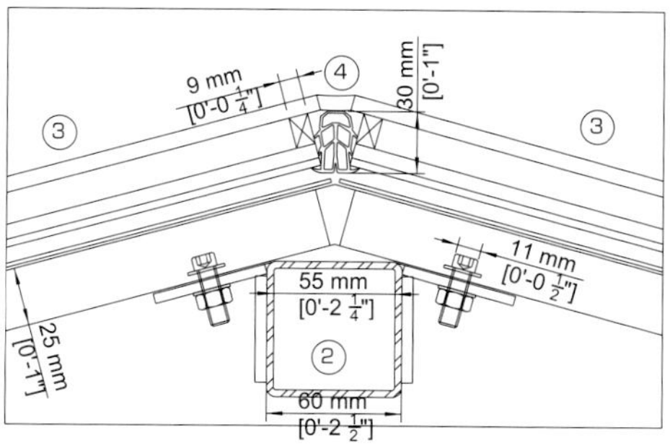

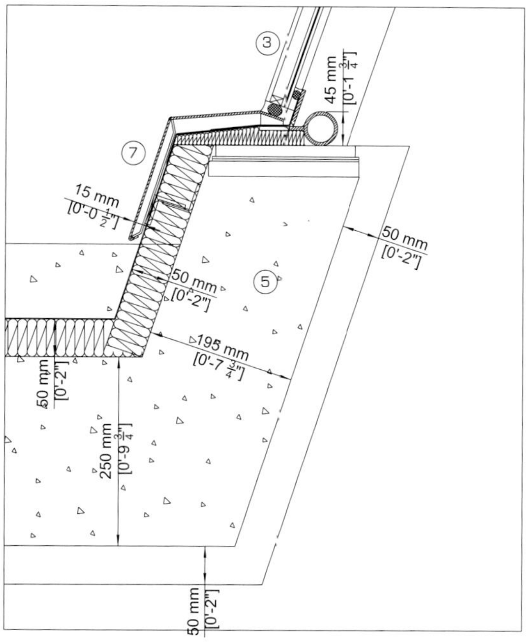

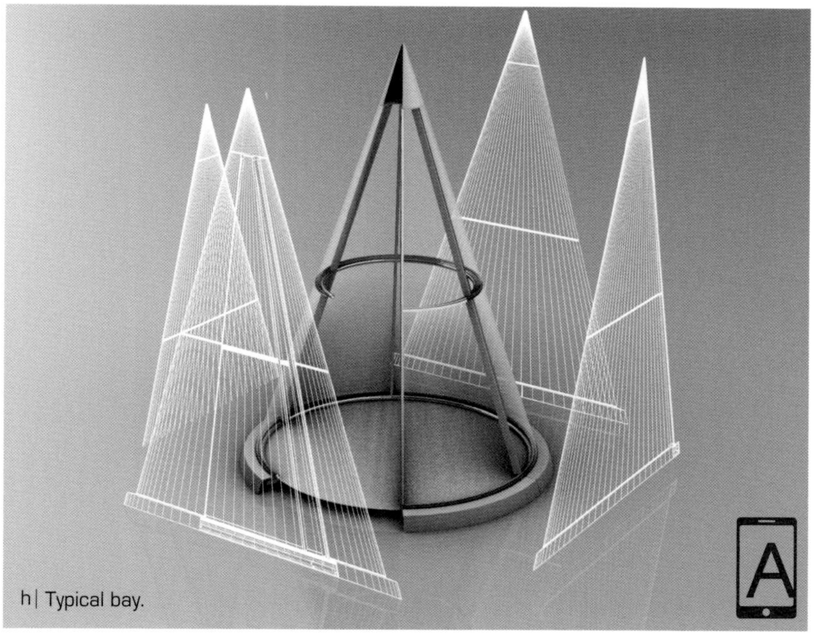

h | Typical bay.

with another silicone seal if there is risk of future flooding from blocked rainwater outlets, for example. The waterproof membrane for the roof slab is continued up the upstand and is bonded to the base of the horizontal section that supports the glass. This provides a complete seal from the glass to the roof membrane, with the metal flashing providing both a protection to this seal and a means of concealing the closed cell thermal insulation set on top of the waterproof membrane.

At the top of the rooflight, a metal panel is used to seal the pointed form. The glass can be continued almost to the top, with only a small pointed metal cover, but this example aims to show how the glass is sealed to another mater al set above it. The metal cover is folded inwards at its junction with the double glazed unit. The fold forms an edge to make a silicone seal. The glass is bonded to another steel flat set below the metal cover. This provides lateral support to the glass unit as well as making an additional inner seal. The metal cover is typically formed from a single piece of aluminium or stainless steel that is welded and ground smooth, and is insulated with injected foam or mineral fibre quilt to provide a continuity of thermal insulation.

The use of silicone bonding in this rooflight avoids the need to use visually obtrusive bolt fixings, allowing the form of the rooflight to be seen more clearly. In some countries, additional restraining clips are required on the outside of the glass, at the corners, but this additional safety feature is dependent upon the individual rooflight design and the local building codes.

Generic rectangular rooflight

The generic rectangular rooflight can be constructed without a supporting frame. Mechanical restraint is provided at the corners in the form of pressure plate clips (MCE_414, b). The double glazed units at the corners are fabricated with a recessed groove on the two sides of

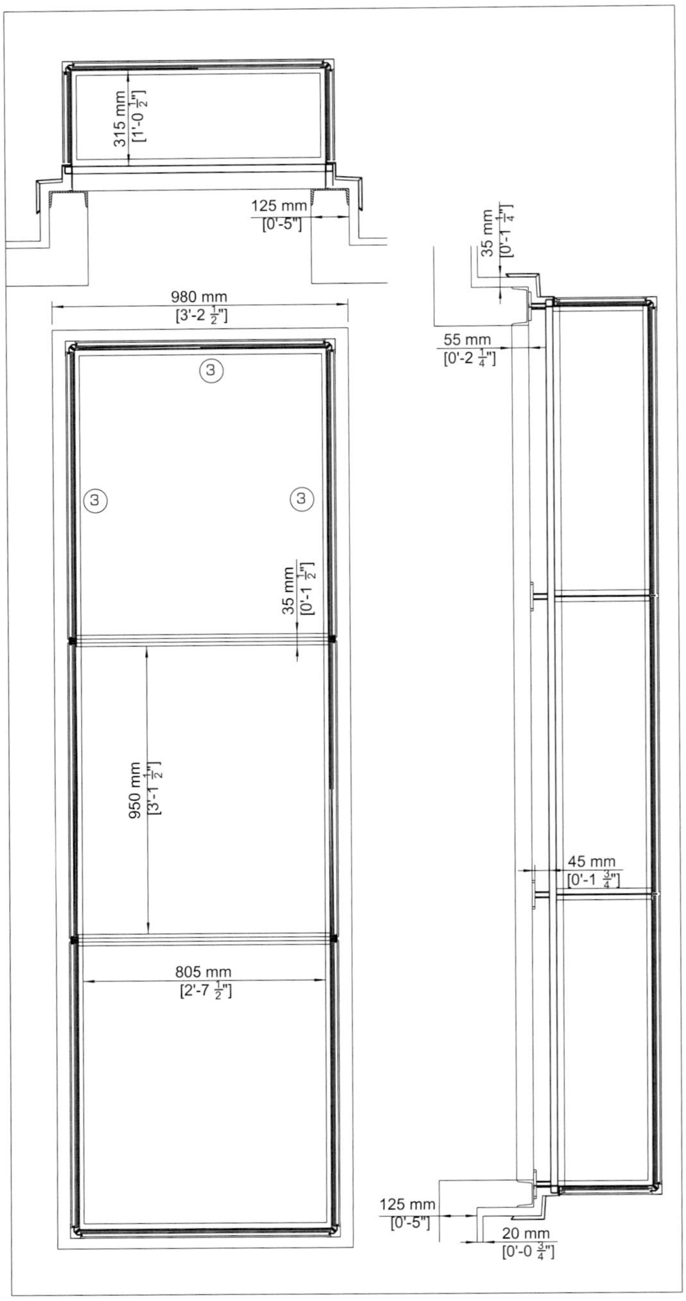

c | Elevation, horizontal and vertical section. Typical assembly

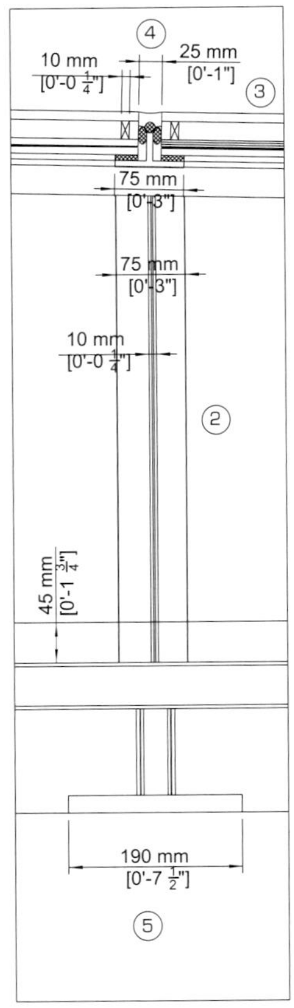

a | Vertical section. Typical assembly

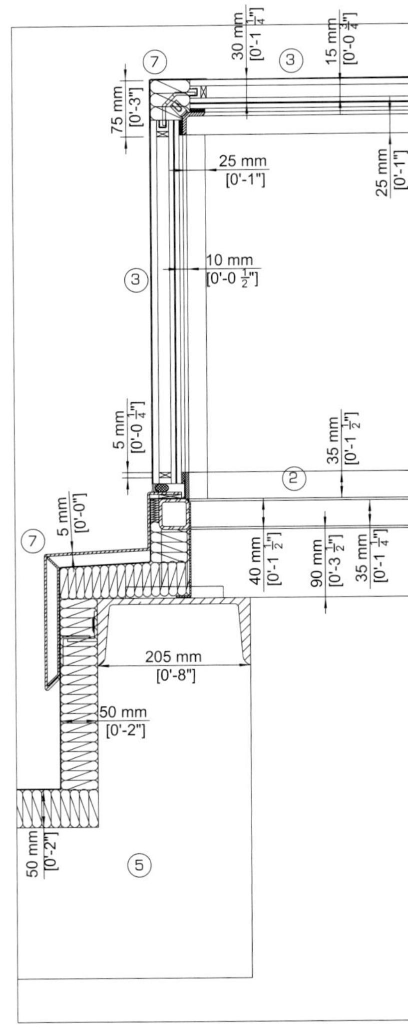

b | Vertical section. Typical assembly

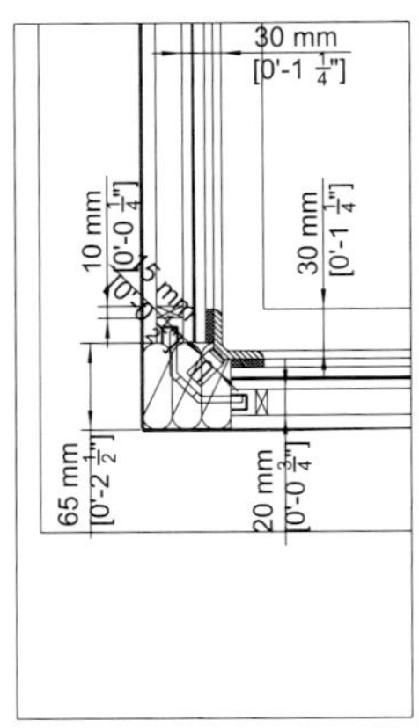

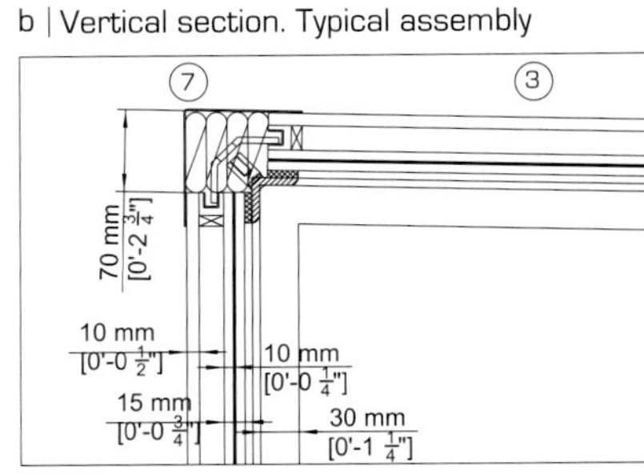

d | Horizontal sections. Typical assembly

the panel forming the corner in order to receive the clip. An alternative method is to fix the metal clip to the outside of the glass at the corner. This avoids the need for specially made corner pieces, but it does form a visible fixing. The modest size of this rooflight allows the horizontally-set glass to span from side to side with no additional support. The corners of the rooflight are stiffened by short lengths of pressure plate which hold the glass in place. The glass has a specially shaped groove in the depth of the double glazed unit to which the pressure plates are fixed. Corner joints have an outer corner piece of folded aluminium which is silicone bonded either to the face of the adjacent glass units, or is folded at 90° to bond it to the side of the unit (d). Glass-to-glass joints between horizontally-set units have a silicone seal with an aluminium angle set on the inside face to provide a second seal (a).

At the base of the rooflight the glass units are seated on aluminium or mild steel sections, which are in turn supported on brackets to the required height of the flashing. These metal brackets are fixed to the adjacent roof deck. The waterproof membrane for the adjacent roof extends up the concrete upstand and up the face of a folded aluminium sheet that forms a complete weathertight seal. Closed cell thermal insulation is applied to the outside of this membrane, and an outer pressed aluminium flashing is fixed to protect both the membrane and the thermal insulation. The insulation forms a continuous layer from the junction with the roof insulation up to the silicone seal in order to avoid any thermal bridges that would cause condensation to form on the inside face of the framing in temperate climates (b). Although thermal insulation can be difficult to install in such situations, the continuity of insulation is essential to avoid the effects of thermal bridging.

The inside face of the double glazed unit at its base is bonded to an inner metal angle. The outside face of the glass is sealed to the metal

e | Horizontal section. Typical assembly

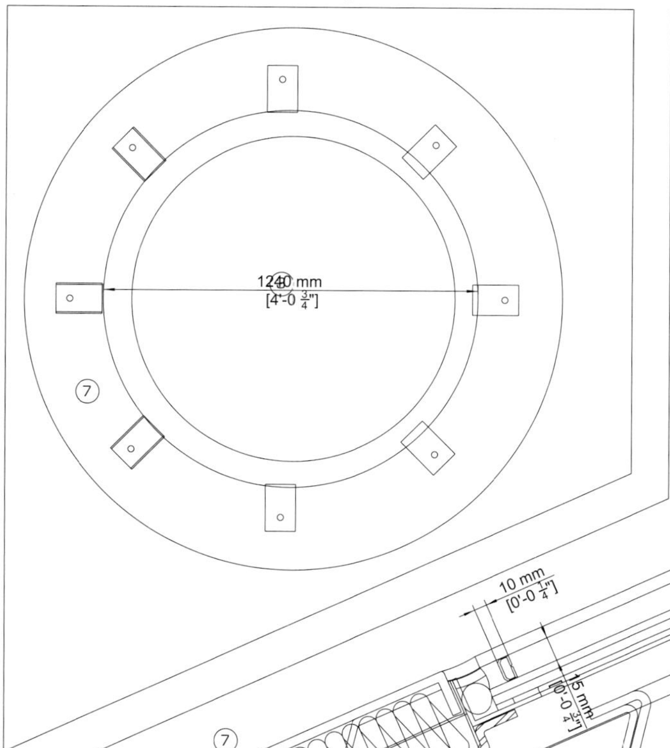

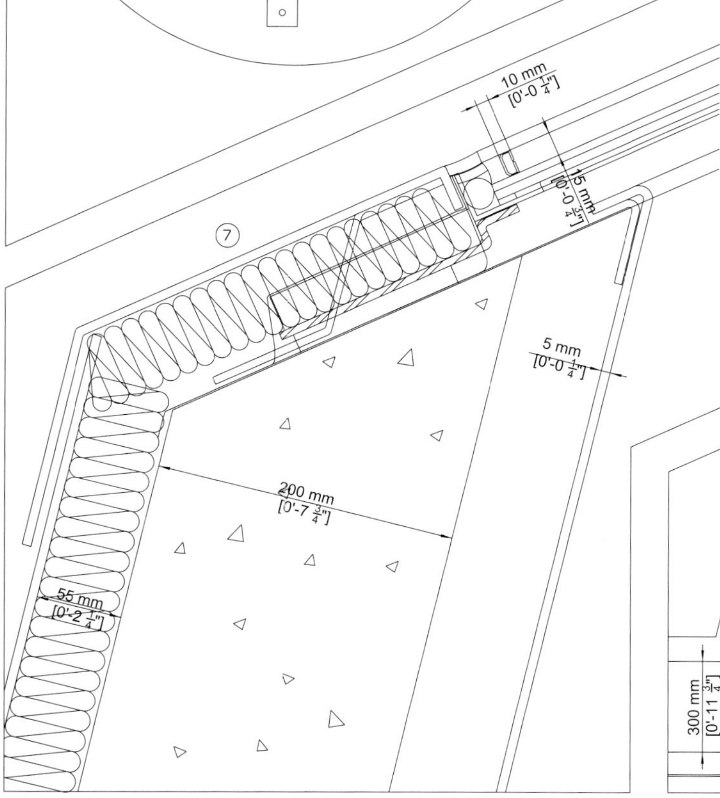

f | Vertical section 1:5. Typical assembly

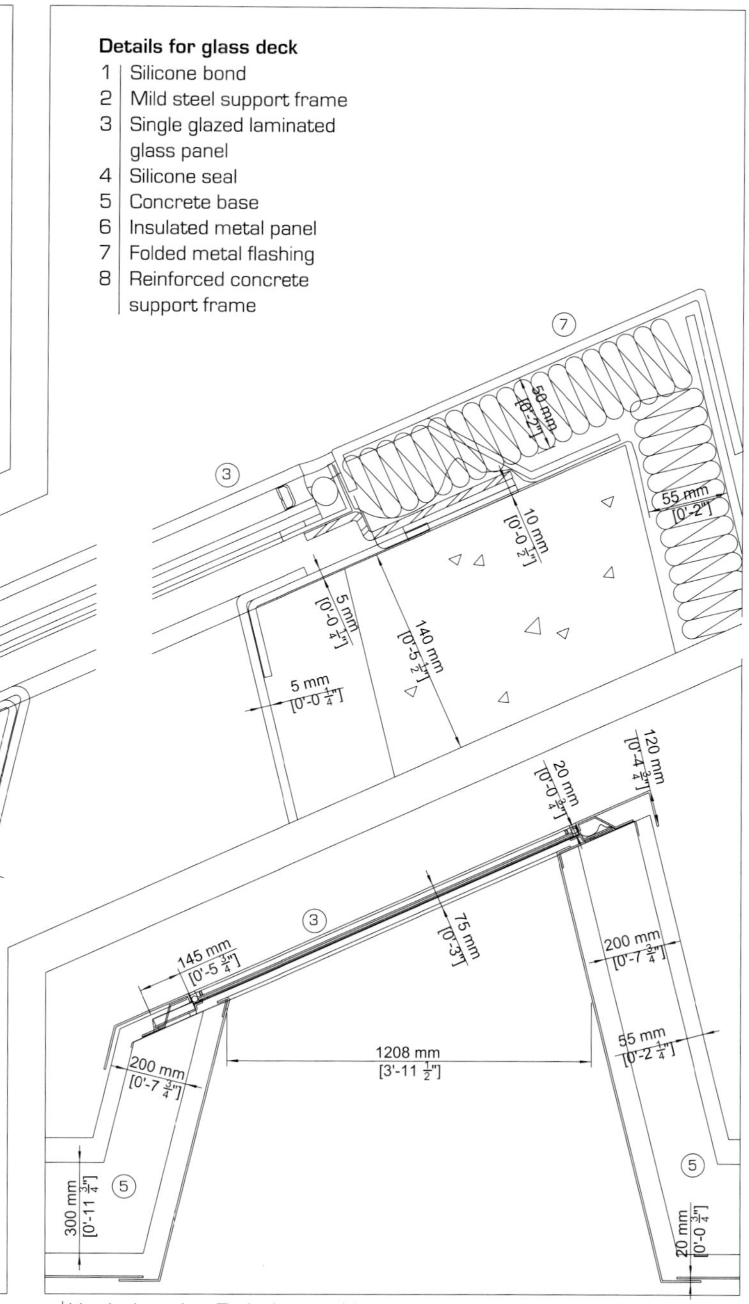

g | Vertical section. Typical assembly

flashing below with silicone. Any water that passes through the outer seal is drained to the external face of the roof membrane beneath.

Small rooflights can have upstands as flashings which can appear to be large in relation to the area of glass. In highly visible or accessible roofs, this can be avoided either by setting the rooflight on the outer edge of the upstand to create a flat appeareance to the flashing, or by adding a layer of decking to conceal the upstand. Decking is usually in the form of open-jointed concrete slabs or timber boards.

Generic monopitch rooflight

The monopitch rooflight provides a visually simple method of closing a roof opening formed in a different material. The circular rooflight is formed by a circular ring of aluminium fixed on Z-shaped brackets which are fixed to the concrete upstand beneath (f). A vertical metal flat section is screwed to the Z-sections and the junction with the flat ring is sealed with silicone. The circular double glazed unit is set into this ring frame in a bed of silicone with spacer blocks to locate it evenly on the ring. The joint between glass and frame is then sealed with silicone using a polymer backing rod. The surrounding waterproof membrane is lapped up the edge of the vertical ring to provide a continuous seal around the rooflight. A metal panel forms the edge of the silicone seal around the glass (g). The continuity of thermal insulation is provided by closed cell insulation set onto the roof membrane that extends up to the silicone seal.

Glass roof decks

Rooflights formed as (nominally) flat, accessible roof decks have been in use over the past ten years. Previously used only inside buildings for walkways and stairs, they are now being used as fully waterproofed external decks, manufactured as proprietary systems. Single glazing is used, since double glazed units are difficult to use as a result of solar

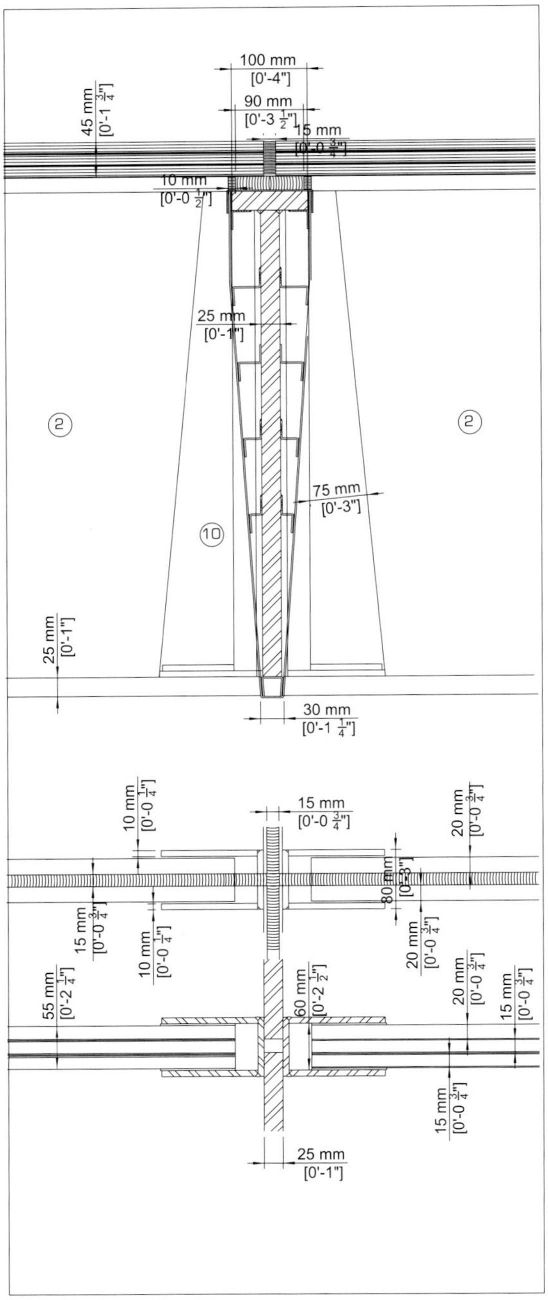

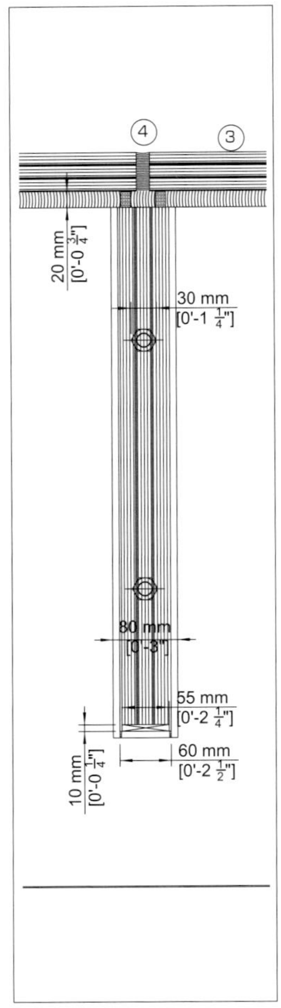

a | Vertical sections. Typical assembly

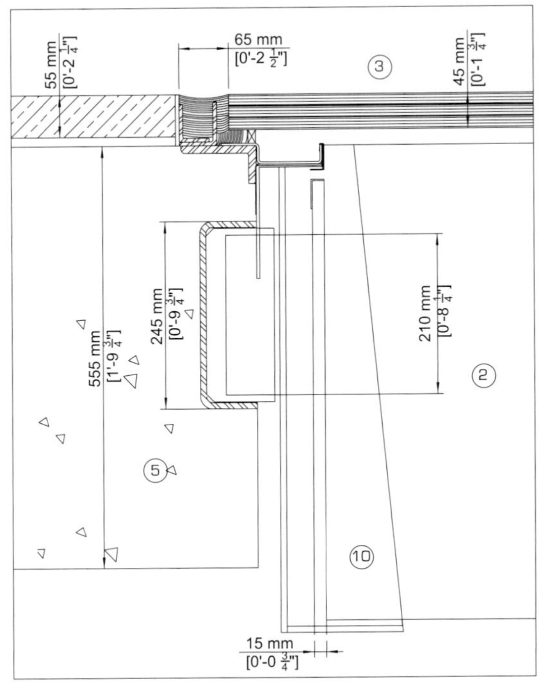

b | Horizontal section 1:100. Typical assembly

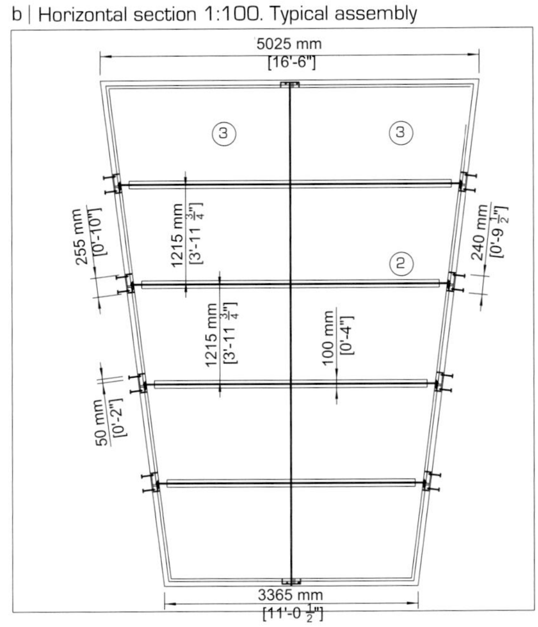

Details for glass deck

1 | Silicone bond
2 | Mild steel support frame
3 | Single glazed laminated glass panel
4 | Silicone seal
5 | Concrete base
6 | Insulated metal panel
7 | Folded metal flashing

gain around the edge of the unit, where the glass is exposed to the outside, but is supported on its underside, allowing heat to enter but not to escape. Where black coloured edge 'fritting' is used, the situation is made difficult. However, double glazed roof decking is in development and will no doubt become much more common over the next ten years.

The glass used is laminated, in common with other rooflights, both to avoid the possibility of any falling objects from penetrating the glass on impact as well as preventing damaged glass from falling immediately into the space below. The glass is set into a series of extruded aluminium angles to which it is bonded on its underside. A gap between the frame and the top of the glass is sealed with silicone of a different type. An additional condensation channel is set below the glass to catch any water that penetrates the silicone seal, or any water that passes through a damaged joint (a).

The details here show alternatives for both steel beam and glass beam supports to the glass panels. The supporting structure has a glass beam set in the centre, spanning the full length of 6000mm (20ft). Its depth is approximately 600mm (2ft), but the beam depth will vary depending upon the individual design. The glass beams here comprise three layers of glass, each 19mm thick, which are laminated together. In the event of one glass being broken, the remaining two glasses will take the full load, avoiding collapse of the beam. The overall thickness of the beam, of approximately 60mm, also provides sufficient bearing for the two glass sheets that meet on top. Each glass has 20–25mm bearing, with a 15–20mm joint width between the glass decking sheets, to suit the individual design. The ends of the glass beams are supported by a metal shoe support, made from either mild steel or aluminium. Stainless steel is used where corrosion is an essential consideration of the design. The metal shoe is bolted back to the supporting structure or reinforced concrete floor slab. Where steel plate is used to provide a

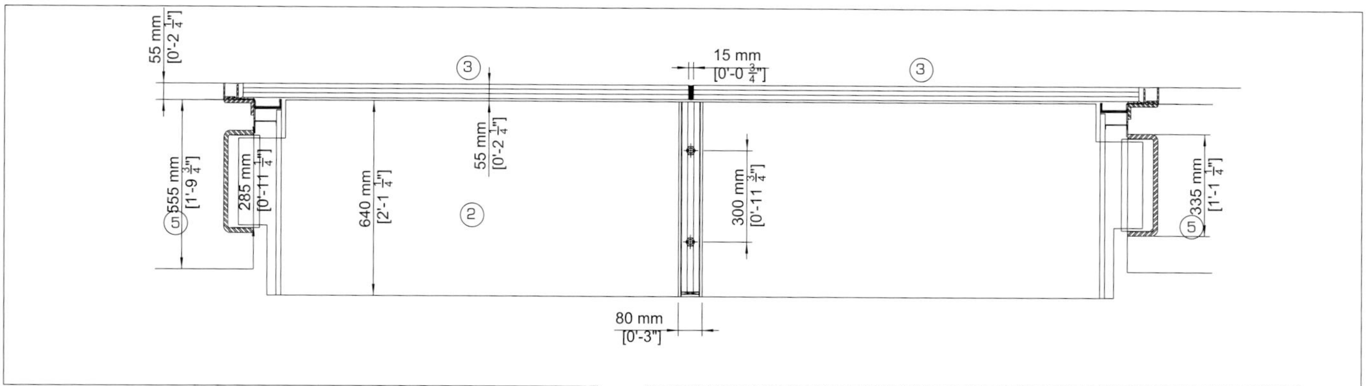

c | Vertical and horizontal sections. Typical assembly

8 | Mild steel plate beam as alternative support
9 | Folded aluminium decorative cover
10 | Steel aluminium support 'shoe'

d | Prototype.

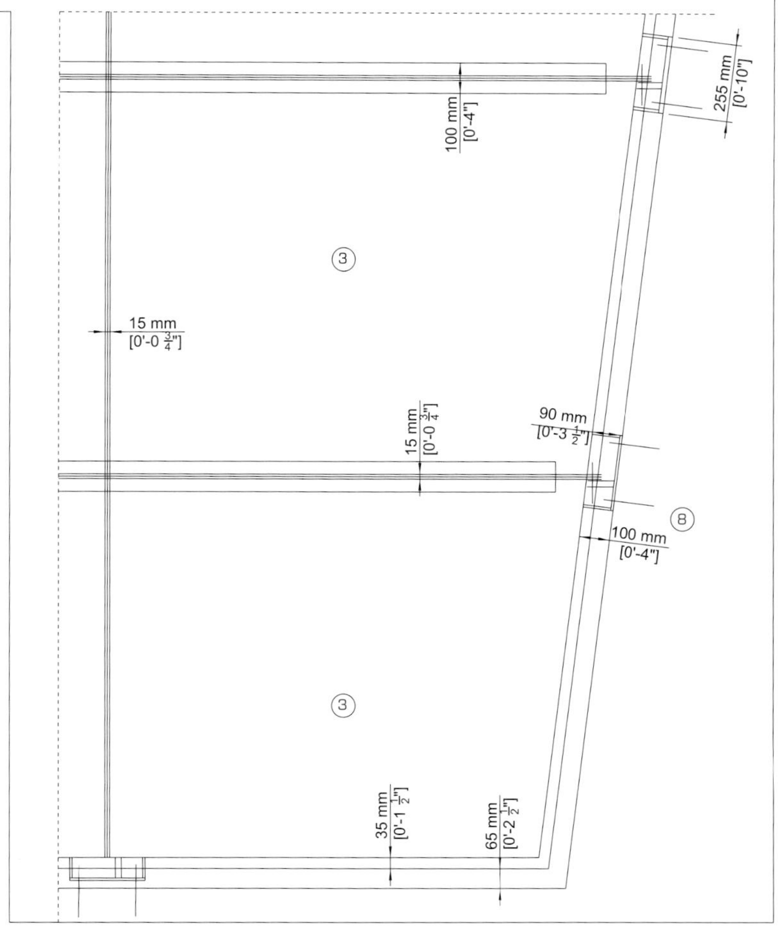

beam instead of laminated glass, a steel flat is welded to the top of the beam to form a T-section that gives enough bearing for the glass deck. The gap between the edge of the glass deck and the adjacent roof finish material is made with a silicone seal.

Typical glass deck panel sizes range from 1000mm x 1000mm (3ft3in x 3ft3in) to 1500mm x 1500mm (5ft x 5ft). Both panel size and overall glass thickness are determinants in the overall rigidity of the glass deck and its associated vertical deflections. Typical vertical deflections for the glass build-ups, given below, range from 0.2mm to 2.0mm, though the acceptable amount of vertical deflection under full design load is dependent upon the specific design application. Typical laminated glass build-ups for the range of panel sizes just mentioned are as follows. The thicknesses of interlayers have been ignored for clarity:

15mm+12mm+12mm = 36mm thick
15mm+15mm+12mm = 42mm thick
15mm+15mm+15mm = 45mm thick

At the junction of glass beams, the secondary glass beam is fixed to the primary beam by either a mild steel shoe, in the manner of timber floor construction, or mild steel cleats in the manner of steel framed floor construction. Both methods require at least one of the glass beams to be drilled in order to bolt the metal component to the beam. When bolting a glass beam to a steel beam, the same principles apply of using a metal shoe. The drilling of glass is a well established technique developed mainly for bolt fixed glazing.

Glass decks usually have an additional layer applied to the top surface of the glass to increase its friction which reduces the possibility of a building user from slipping while walking on the glass. Carborundum is sometimes added to the top surface of the glass for this reason. An alternative method is to 'frit' or screen print the top surface of the glass, which has the added benefit of reducing visibility through the glass from below.

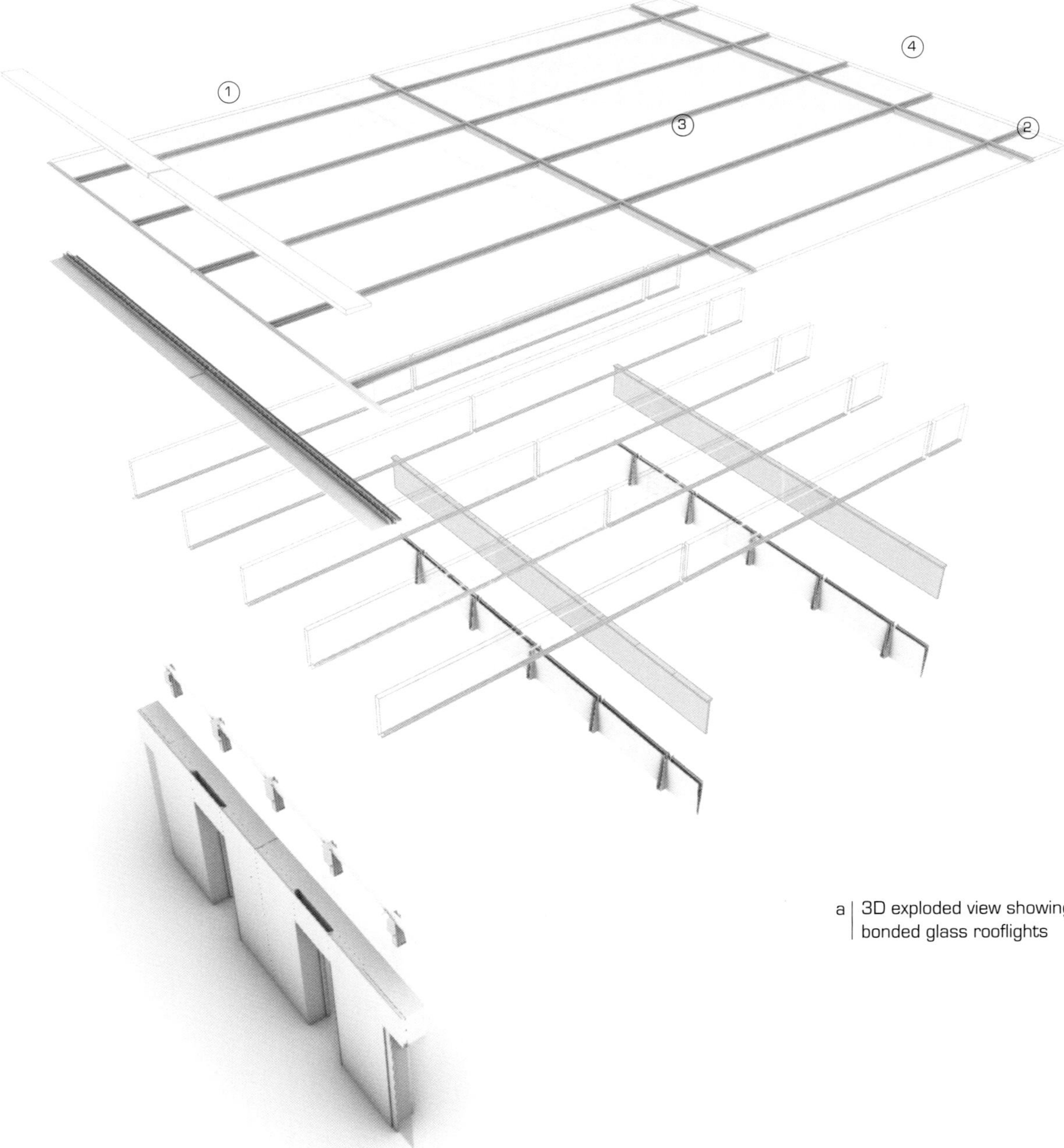

a | 3D exploded view showing bonded glass rooflights

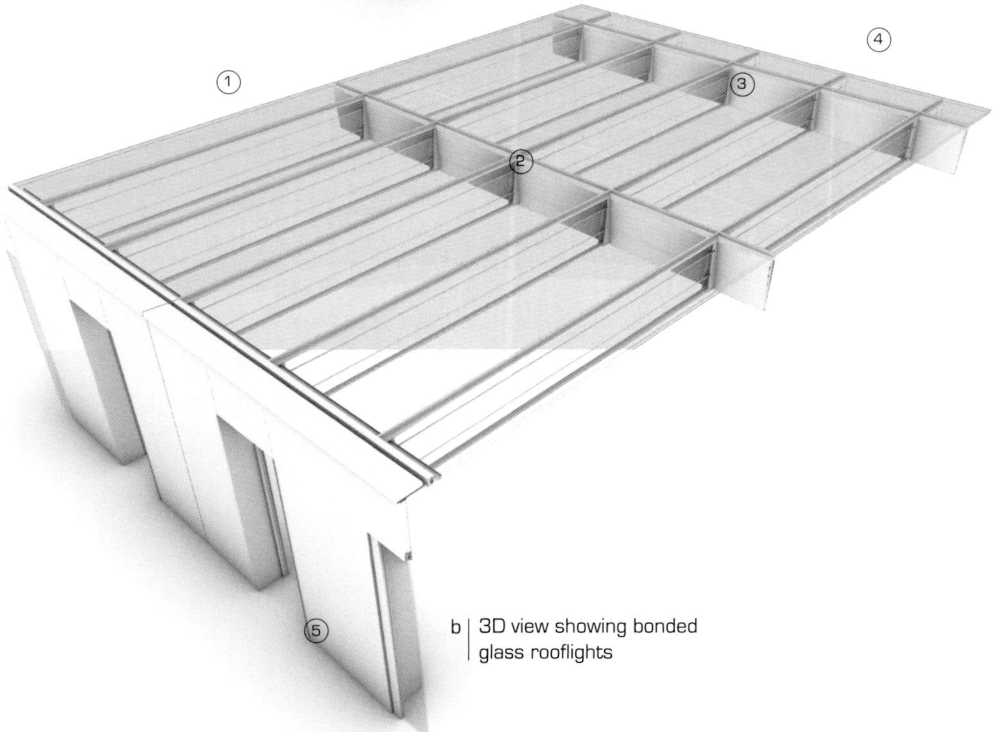

Details

1 | Silicone bond
2 | Mild steel support frame
3 | Single glazed laminated glass panel
4 | Silicone seal
5 | Concrete base
6 | Insulated metall panel
7 | Folded metal flashing
8 | Reinforced concrete support frame

b | 3D view showing bonded glass rooflights

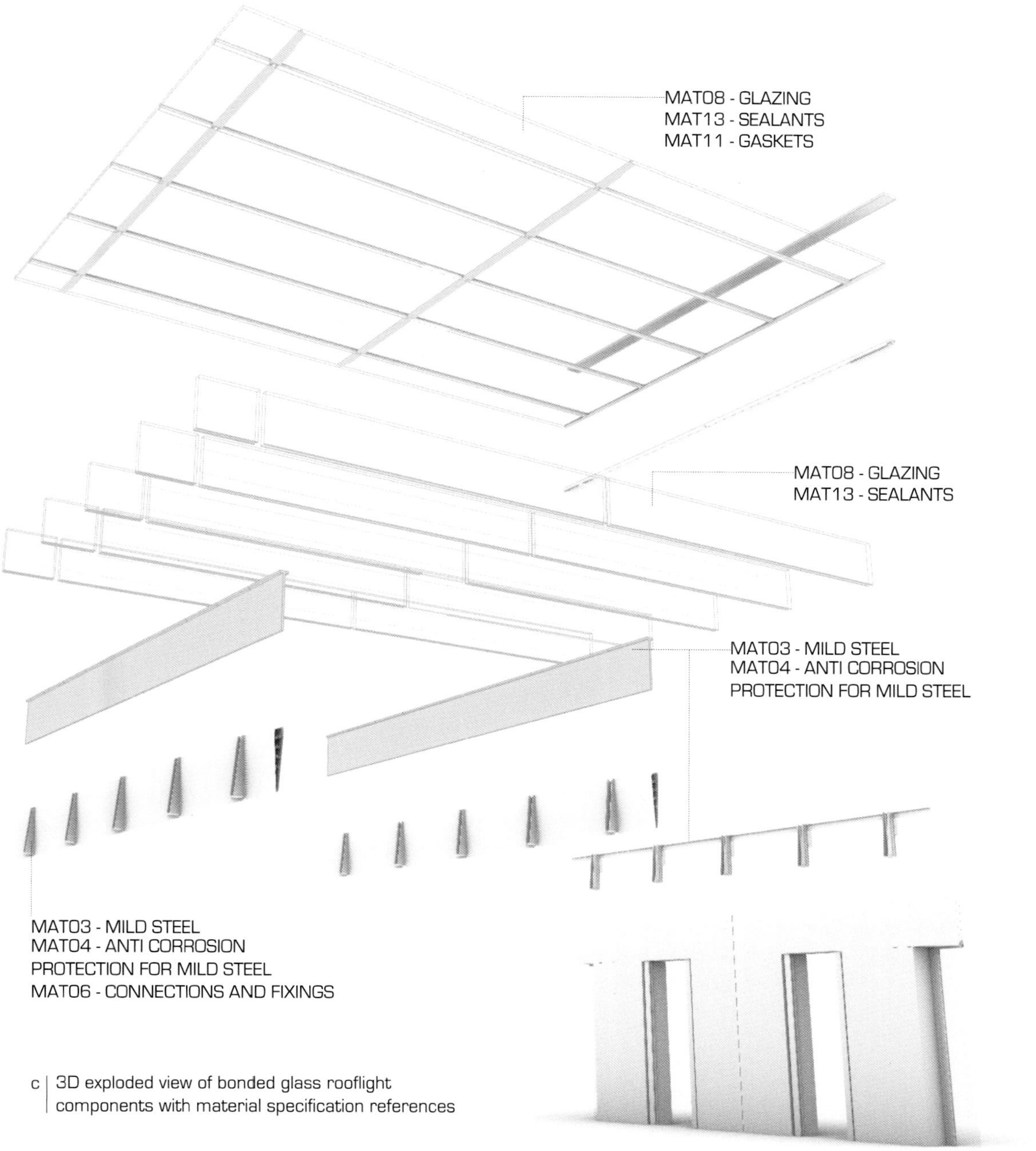

c | 3D exploded view of bonded glass rooflight components with material specification references

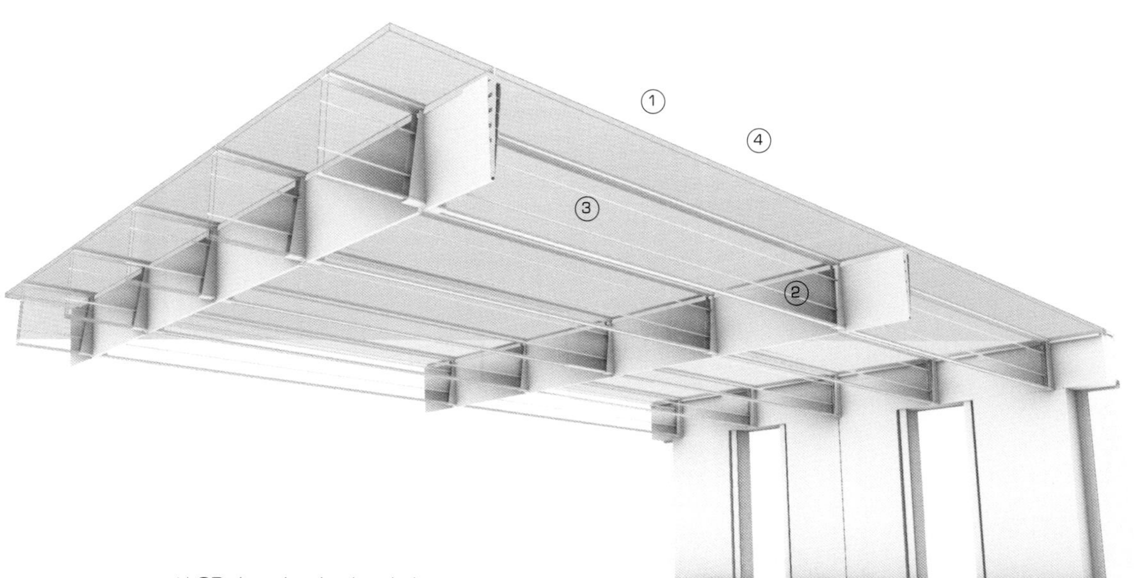

d | 3D view showing bonded glass rooflights

Example project of concrete facades
Airport terminal, Asia
Facade Engineer: Newtecnic
Architect: Newtecnic

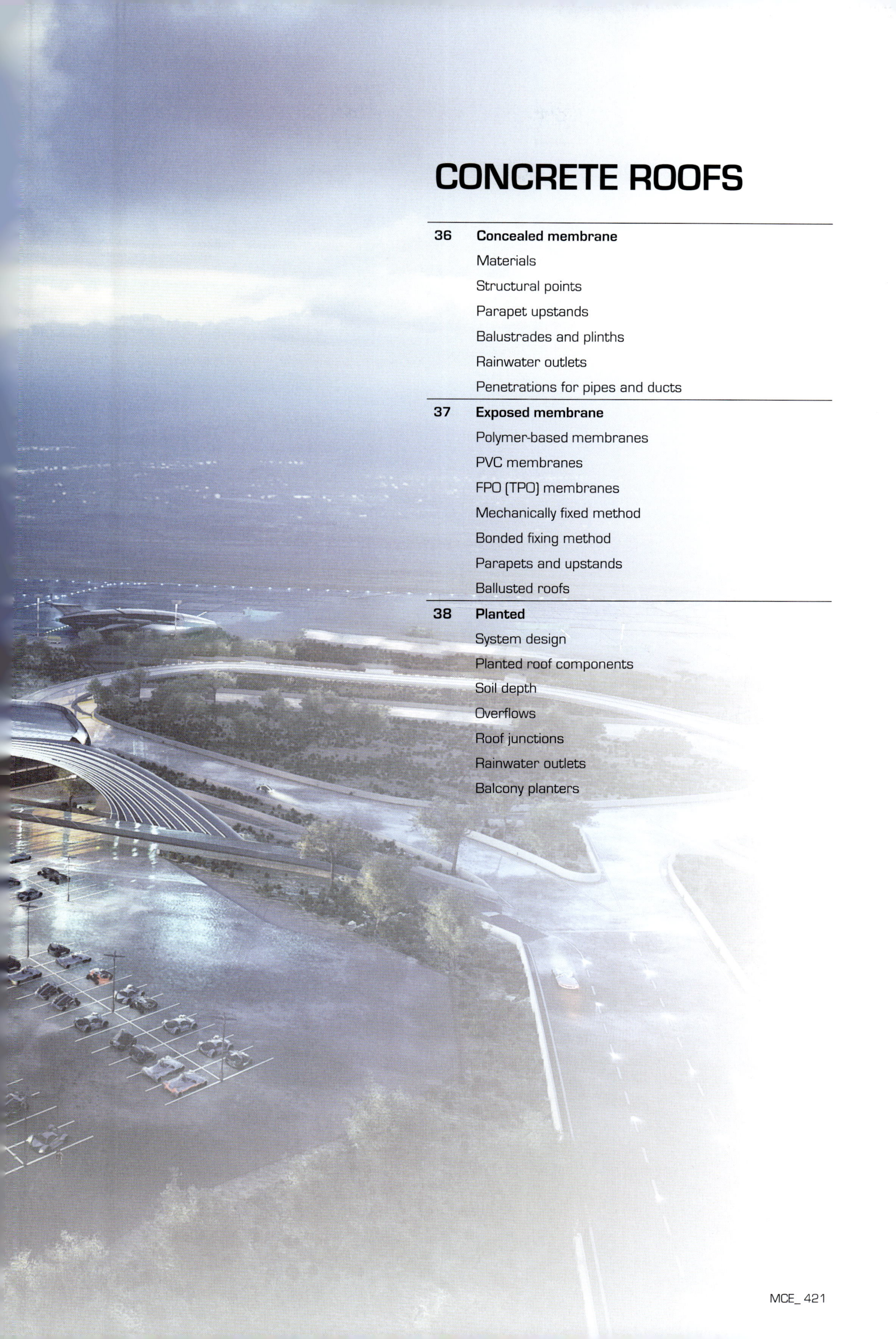

CONCRETE ROOFS

36 Concealed membrane

- Materials
- Structural points
- Parapet upstands
- Balustrades and plinths
- Rainwater outlets
- Penetrations for pipes and ducts

37 Exposed membrane

- Polymer-based membranes
- PVC membranes
- FPO (TPO) membranes
- Mechanically fixed method
- Bonded fixing method
- Parapets and upstands
- Ballusted roofs

38 Planted

- System design
- Planted roof components
- Soil depth
- Overflows
- Roof junctions
- Rainwater outlets
- Balcony planters

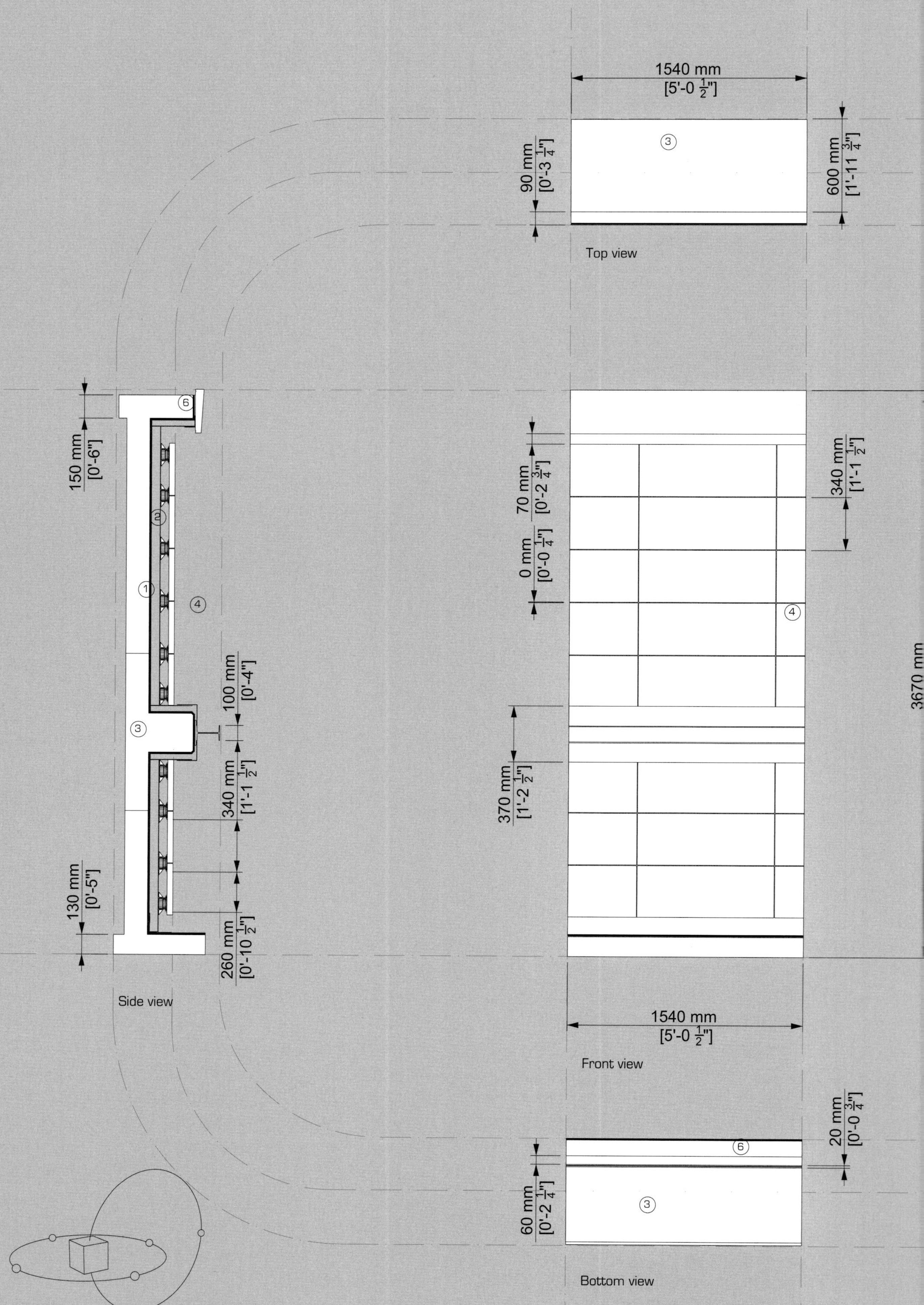

Third angle projection 1:30

Details

1 | Waterproof membrane
2 | Thermal insulation
3 | Concrete deck
4 | Paving slabs
5 | Smooth pebbles
6 | Parapet coping
7 | Rainwater outlet
8 | Slot drain
9 | Opening for overflow

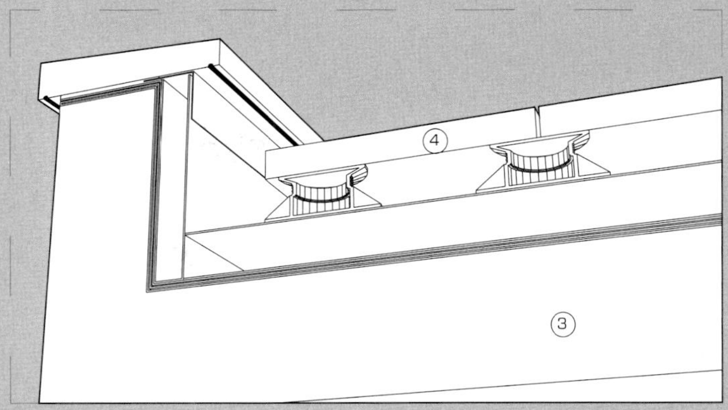

3D view of the assembly

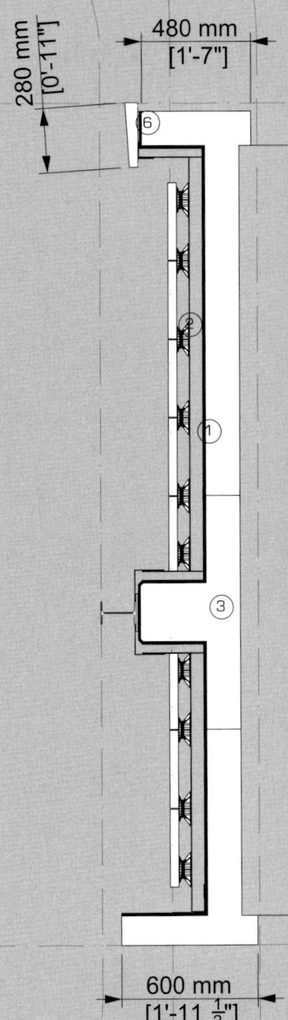

Side view

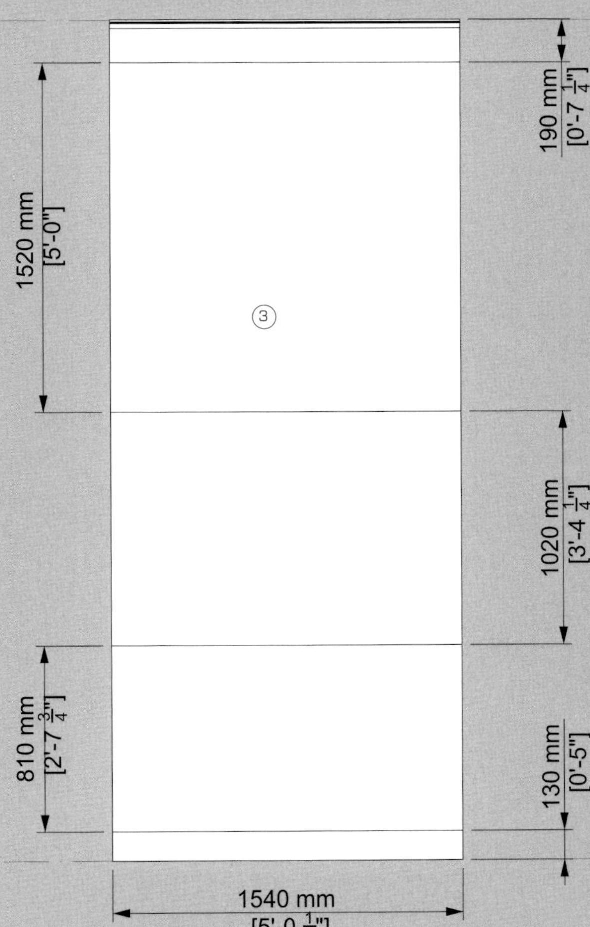

Back view

10 | Reinforcement at fold if required
11 | Proprietary movement joint
12 | Balustrade
13 | Pipe or duct
14 | Filter sheet

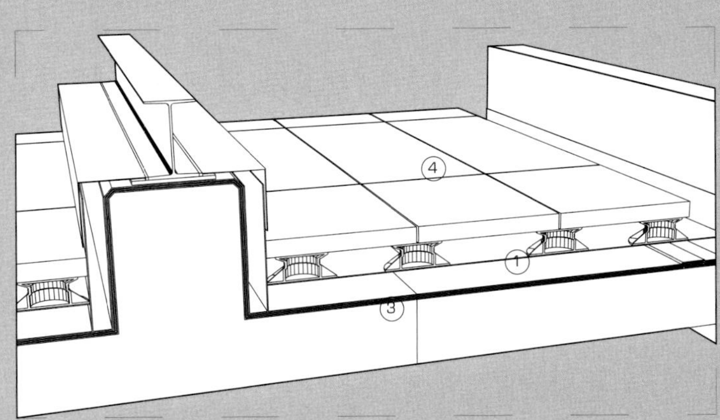

3D view of the assembly

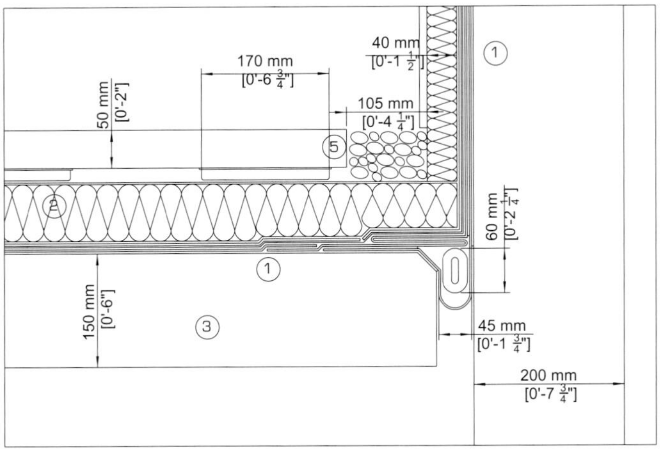

a | Vertical section. Junction of external wall and roof slab

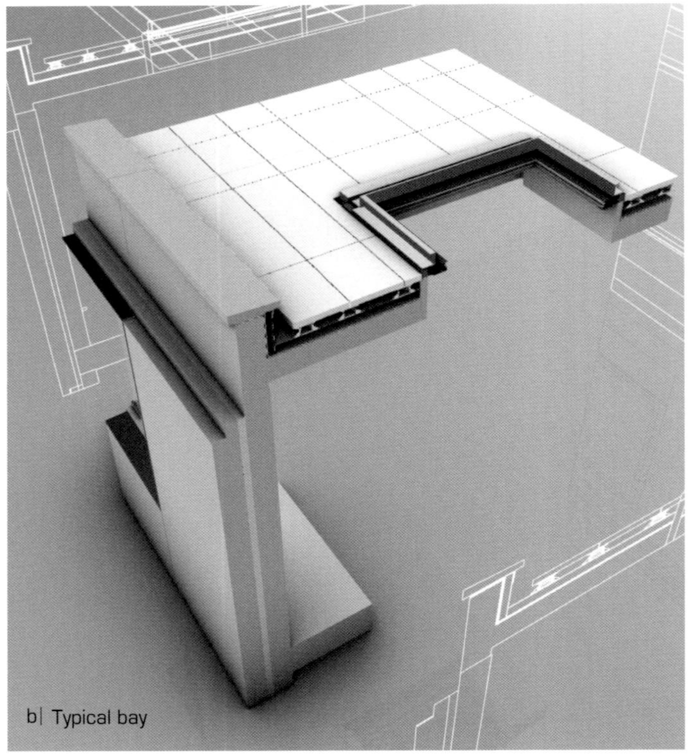

b| Typical bay

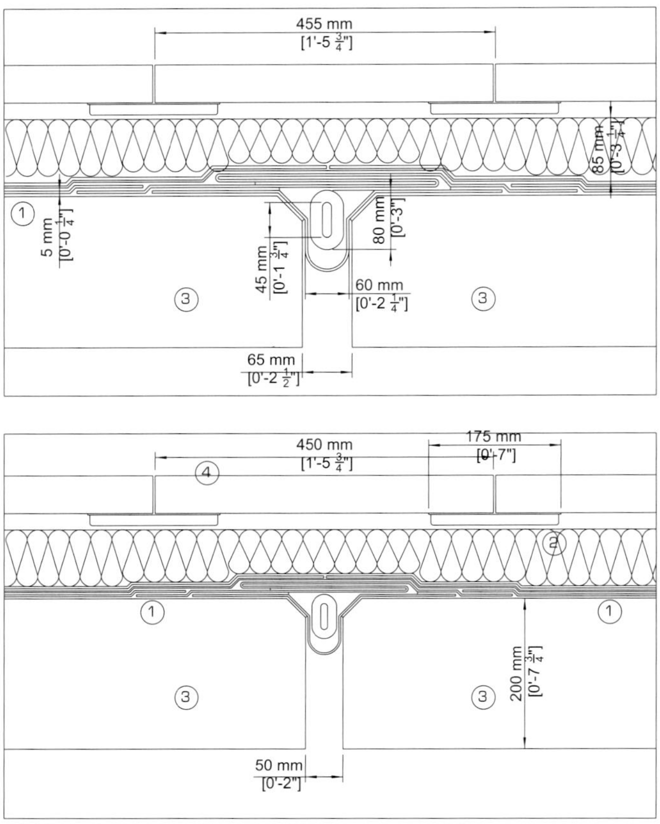

c | Vertical sections. Expansion joint in concrete slab

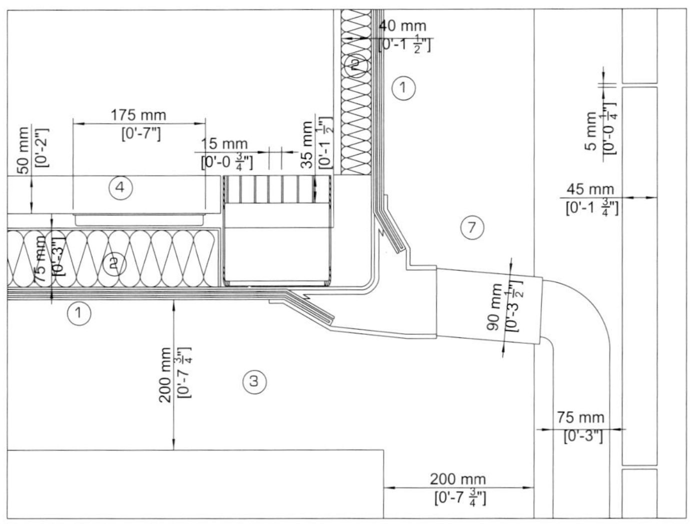

d | Vertical section. 2-way drain outlet

Materials

Bitumen has traditionally been used as a waterproof layer, applied while hot in liquid form onto a concrete roof slab. As it cools it hardens, forming an impervious membrane, but will soften again if heated by the effects of solar radiation. For this reason, in order to keep the material cool, bitumen membranes are concealed by smooth pebbles, paving slabs, usually with thermal insulation set between the bitumen and the pebbles/paving (h). Traditional bitumen roofs are usually laid in two layers, with an overall thickness of around 25mm. One of the limiting factors with bitumen is folding the material through an angle. When the material turns through a right angle from the horizontal roof to a vertical parapet wall, it can pass through a maximum of 45° in a single fold. For this reason 45°angle fillets are used to make a 90° turn from roof to wall.

Modern bitumen-based membranes that are concealed beneath roof finishes are typically a combination of bitumen-based sheet mixed with synthetic rubber to give flexibility combined with a reinforcement to give dimensional stability and tensile strength. This reinforcement often allows the material to be folded through 90°, making its use considerably easier, where angle fillets are not required.

With the development of much thinner membranes in thermoplastics and elastomers, together with their competitive costs, there have been considerable efforts made by manufacturers over the past 20 years to make the bitumen layers thinner, to reduce the material required while enhancing its properties of strength and flexibility. This has been achieved by replacing the thick two-layer method with a mixture of thin layers, still applied in hot liquid form on site, but reinforced with an elastomeric sheet, usually bedded between the layers. This is typically two layers, each 3mm thick with reinforcing layers bedded into the material. This allows the bitumen to accommodate both small amounts of movement at these junctions, as well as the sharp fold in the material,

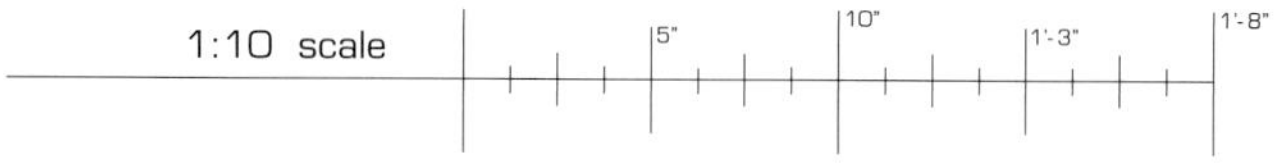

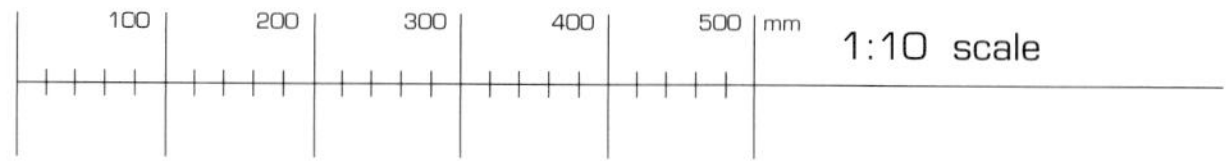

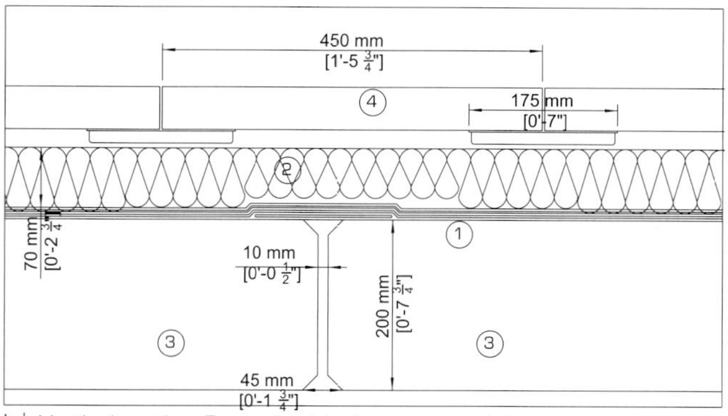

b | Vertical section. Expansion joint in concrete slab

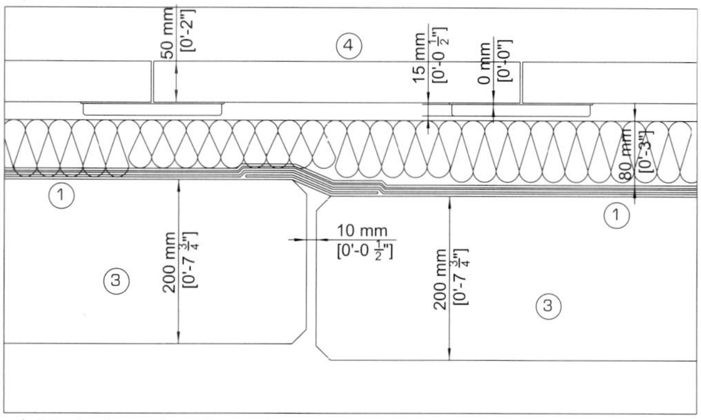

g | Vertical section. Expansion joint in concrete slab

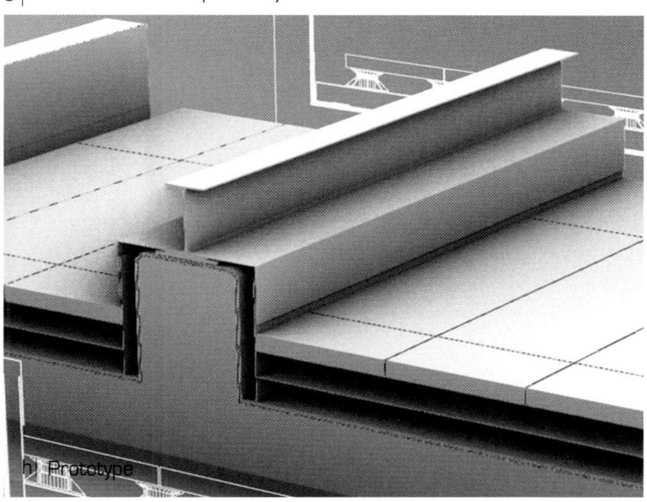

h | Prototype

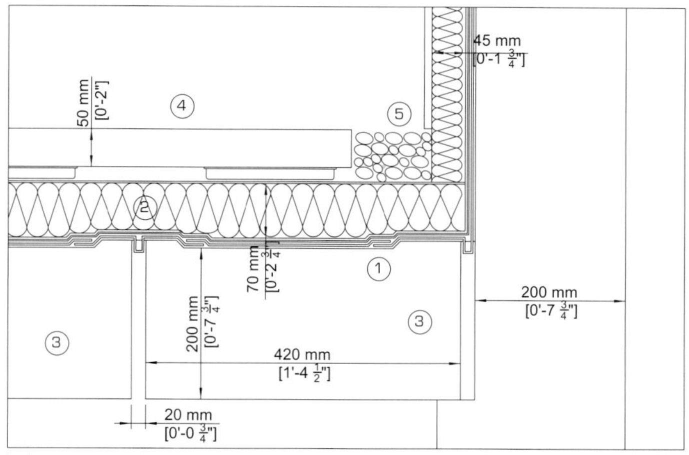

f | Vertical section. Junction of external wall and roof with drain slot

Details

1 | Waterproof membrane
2 | Thermal insulation
3 | Concrete deck
4 | Paving slabs
5 | Smooth pebbles
6 | Parapet coping
7 | Rainwater outlet
8 | Slot drain
9 | Opening for overflow
10 | Reinforcement at fold if required
11 | Proprietary movement joint
12 | Balustrade
13 | Pipe or duct
14 | Filter sheet

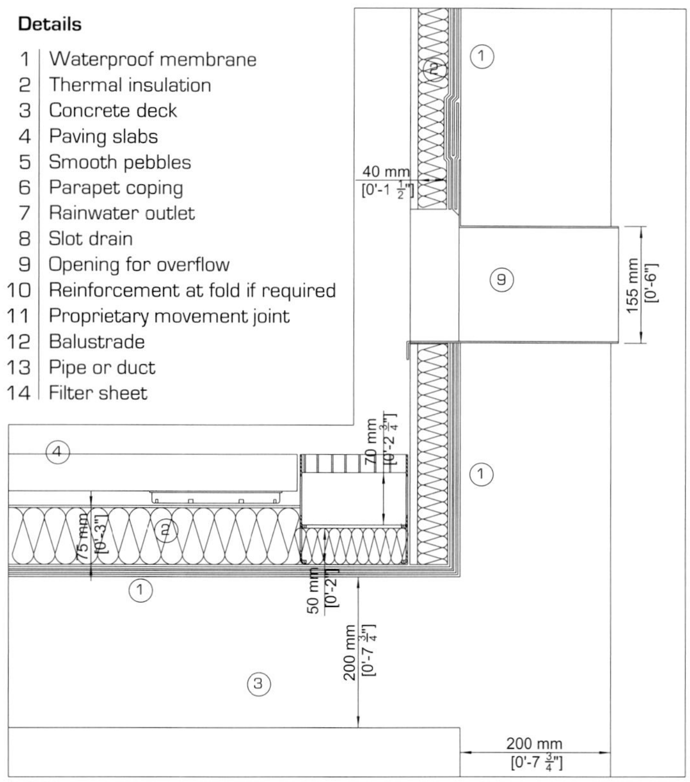

j | Vertical section. Junction of external wall and roof with overflow

which creates a weakness in the membrane which might otherwise be damaged during the life of the building. An outer protective layer is added for vulnerable locations such as at gutters and at upstands.

Concealed membrane roofs are increasingly being laid flat with no fall, unlike exposed membranes where a slight fall is always required. This is partly because bitumen bonds more reliably to a concrete slab than to a thin screed, and partly because the slopes required to create falls in large areas of flat roofs create difficulties in level changes across the roof which can be difficult to drain. The more traditional methods of screeds laid to falls and thick layers of bitumen is used where the roof finish is sealed, such as paving with sealed joints.

Typically, upstands are formed before the main areas of roof are waterproofed in order to allow external walls to be completed. When an area is completed, the roof deck is waterproofed and finished, with areas of flat roof being bonded continuously to the already completed upstands (MCE_428, c). This avoids the need to protect the membrane, which has attendant risks of being damaged before the building is finished.

Concealed membrane roofs are typically in 'inverted' roof configuration with either open joints or sealed joints in the top layer that covers the thermal insulation, usually paving. In the open joint version the membrane, bonded to the concrete slab, is covered by a protection layer with closed cell, rigid thermal insulation set on top. A polyester filter sheet is set on top, with paving or smooth pebble ballast on top to hold the insulation in place as well as to walk on. Pebbles are 20mm–40mm diameter, while paving slabs are around 600 x 600mm in size, 30mm–40mm deep (h). In the sealed joint configuration, the bitumen membrane with its protection layer has a drainage layer on top, onto which is laid a minimum 65mm sand/cement screed, usually reinforced or made sufficiently thick to avoid cracking both in the screed

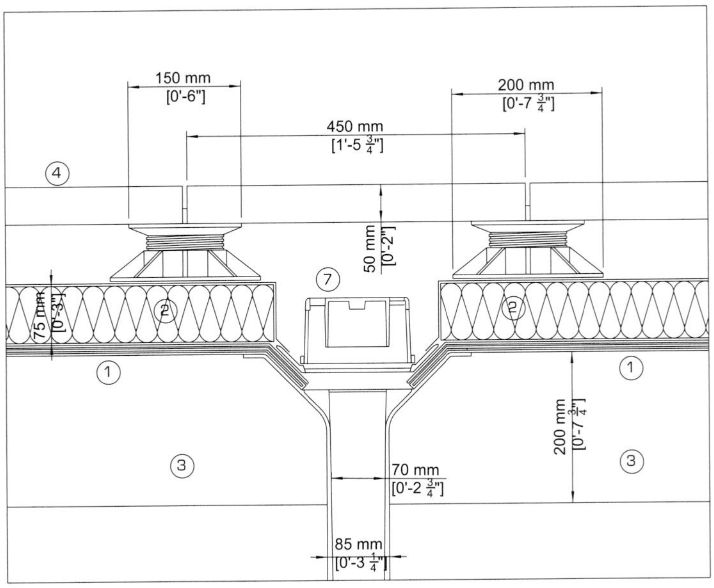

a | Vertical section. Rainwater outlet

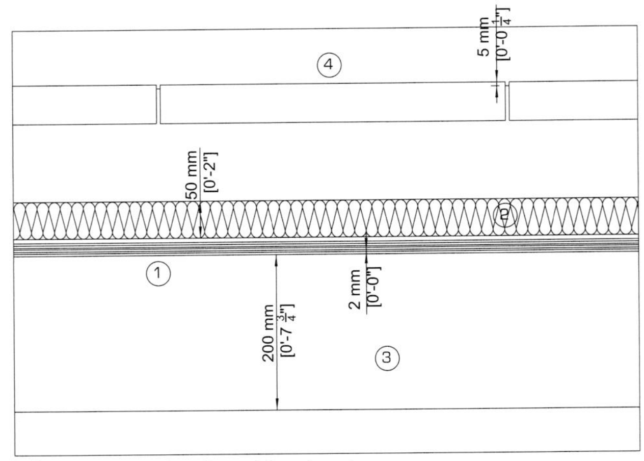

b | Vertical section. Open jointed paving

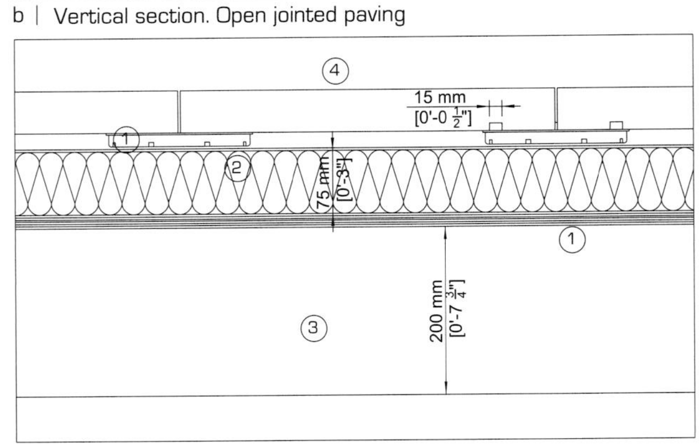

c | Vertical section. Sealed paving

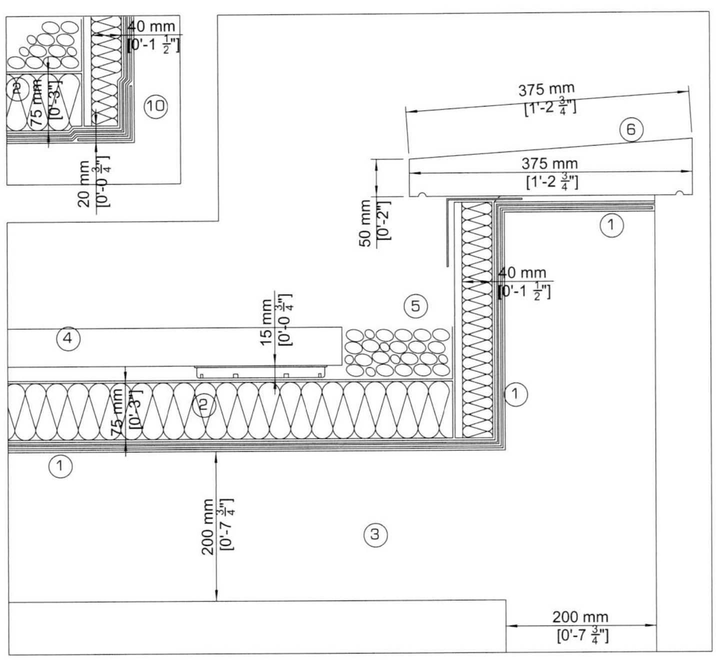

d | Vertical section. Low parapet

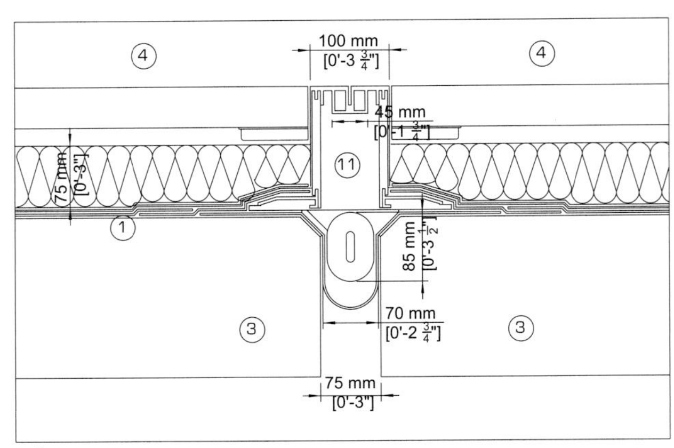

e | Vertical section. Proprietary expansion joint

and the sealed paving above. Paving slabs or blocks are bonded to the screed with mortar and grouted.

Structural joints

The main advantage of concealed membranes is their ability to span movement joints and expansion joints in reinforced concrete slabs with simple, reliable details that require no upstands to form the junction. Expansion joints between concrete slabs of widths between around 10mm to 50mm are formed by stopping the material each side of the joint and setting a rubber-based strip that dips into the gap between the slabs, linking the membranes into a continuous seal. As the gap between the slabs varies with structural movements, the rubber-based strip is allowed to move without being stretched significantly. The joint is protected and reinforced with an additional layer, either flat and bonded on one side only, or formed as a folded, S-shaped cover that folds back over itself, held in place by an additional protection sheet on top (e). The gap between the membrane, dipped into the joint, and the reinforcement cover is filled with a foam backing rod or tube, as used in the glass joints of bolt fixed glazing. The material used for the reinforcement is either the same bitumen-based material, or increasingly a rubber-based strip.

The top of the joint is finished as level as possible with the adjacent areas of roof to allow water to drain freely from the roof. The reinforcing membrane is sometimes folded down into the gap, separated from the membrane below with a foam backing rod. It can be difficult to drain water from this groove at the edge of the slab unless water at this lower level can discharge into a rainwater outlet.

Junctions between concrete roof slabs and walls are treated in a similar way, with the membrane dipping down into the gap between wall and roof and the membrane continuing up the wall (g). Reinforcement is similarly applied, with a rubber-based strip folded through the 90°

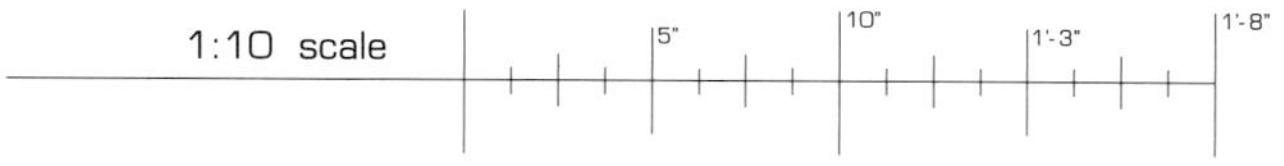

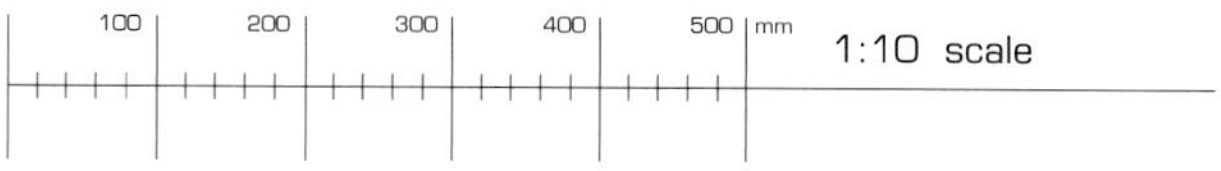

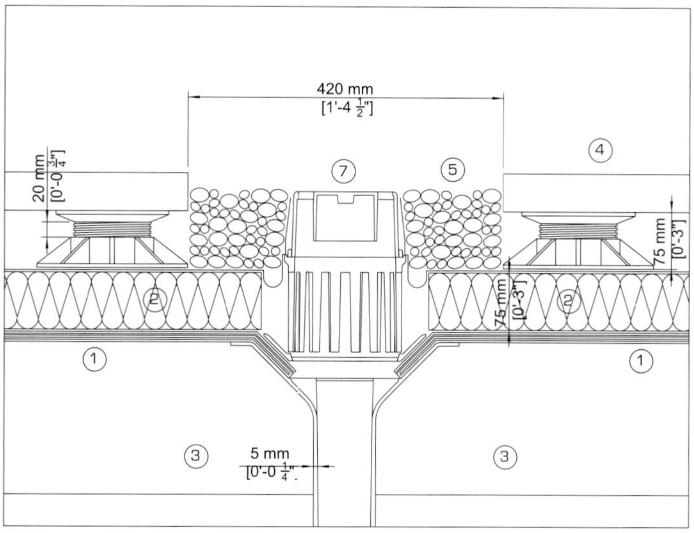

f | Vertical section. Rainwater outlet

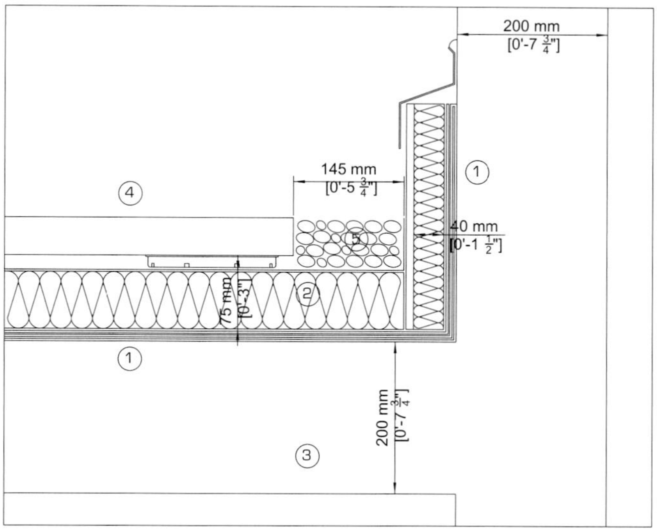

g | Vertical section. Junction of external wall and roof slab

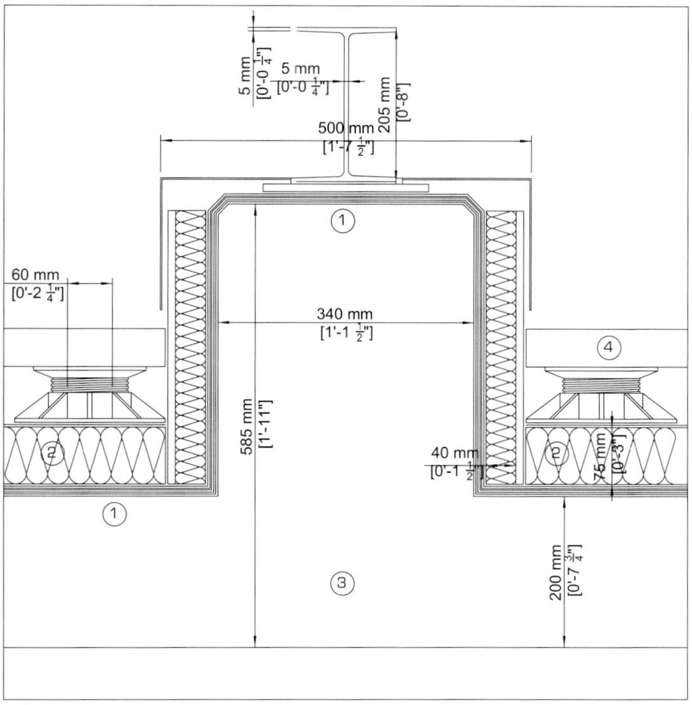

h | Vertical section. Plinth

Details

1 | Waterproof membrane
2 | Thermal insulation
3 | Concrete deck
4 | Paving slabs
5 | Smooth pebbles
6 | Parapet coping
7 | Rainwater outlet
8 | Slot drain
9 | Opening for overflow
10 | Reinforcement at fold if required
11 | Proprietary movement joint
12 | Balustrade
13 | Pipe or duct
14 | Filter sheet

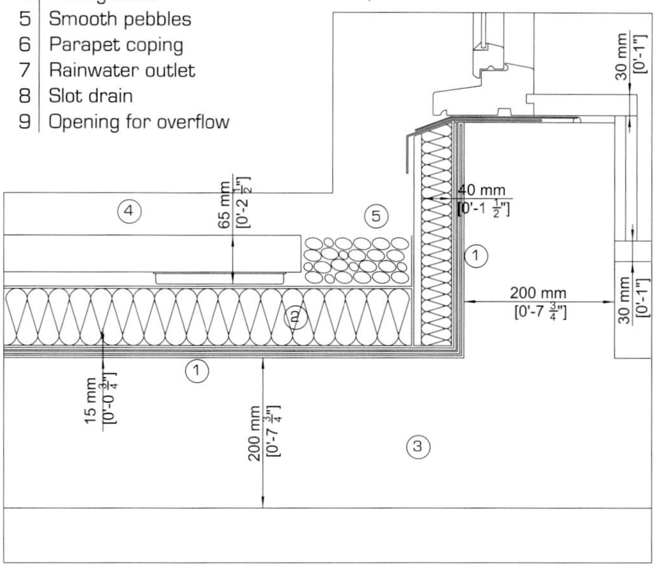

j | Vertical section. Door sill

corner rather than the maximum 45° folds that are usually allowed in traditional bitumen-based membranes.

Proprietary metal-based expansion joints are used to form part of the visible finishes, typically with sealed roof finishes. In this case the seal is formed by bonding a strip of membrane to the metal assembly. The metal expansion joint assembly is fixed on top of the membrane, which still dips down into the joint. A foam backing rod is set on top of the dipped membrane and is covered with a bed of compatible sealant to ensure that any water that penetrates the movement joint is drained to the sides of the movement joint. The waterproof membrane is then formed up out the base of the movement joint assembly to provide a complete seal across the joint.

Joints between concrete panels, such as precast concrete slabs, where only negligible structural movement is expected, are also formed with rubber-based strips. The membrane forms a continuous lapped joint across the rubber-based strip (MCE_430, a), with a foam backing rod being provided where there is a risk of damage during construction.

Parapet upstands

In forming parapet upstands, an essential requirement is to keep the waterproof membrane as well protected from the effects of the sun as elsewhere on the roof. For this reason, thermal insulation is applied to the inside face of the parapet even if this has no direct benefit to the passage of heat through the building. The membrane is turned through a full 90° as shown in the drawings (MCE_430, a), but an angle fillet is required by some manufacturers to limit the angle of any fold to 45°. A reinforcing strip is usually added where a 90° fold is made.

In generic examples, a low parapet wall is terminated by a concrete or stone coping. The waterproof membrane extends up the height of

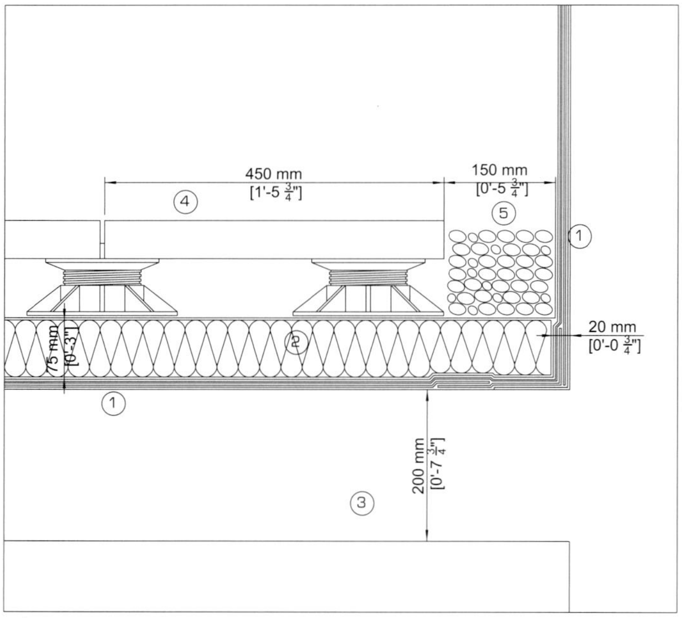

a | Vertical section. Junction of external wall and roof slab

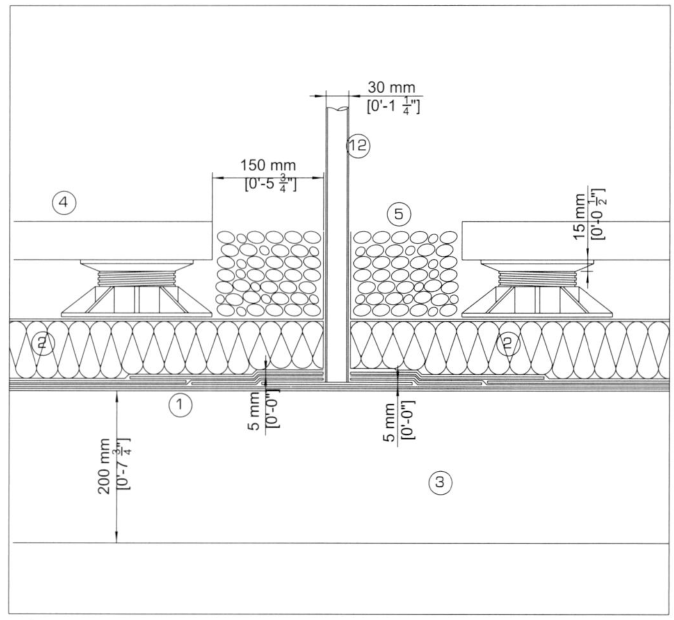

b | Vertical section. Base of balustrade

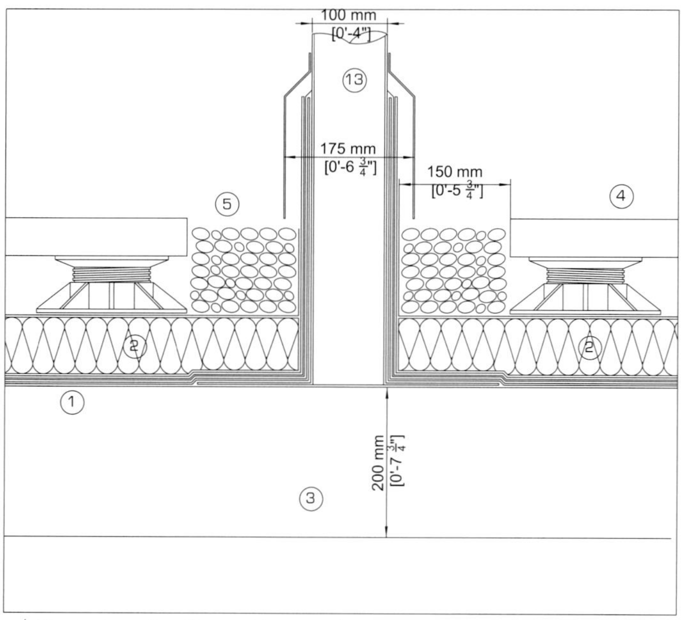

c | Vertical section. Base of balustrade

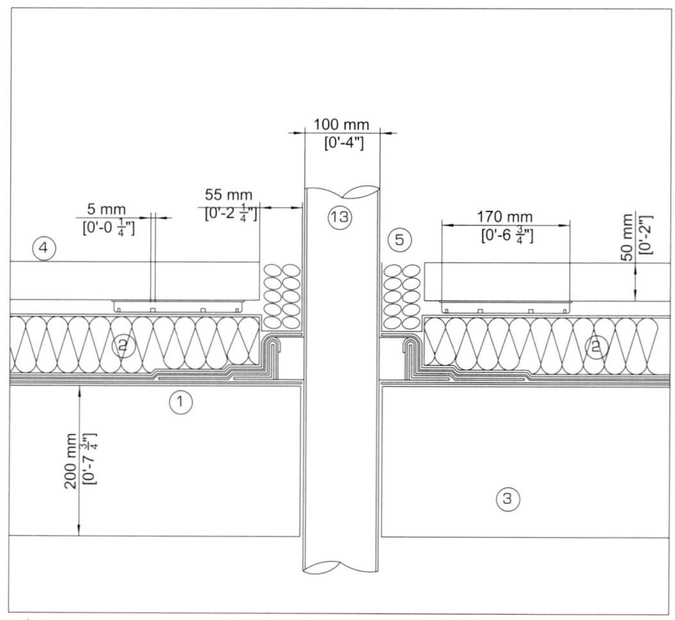

d | Vertical section. Pipe penetrations

the upstand, which runs horizontally to form a full damp proof course underneath the coping. This ensures that the membrane provides a continuous waterproof layer at the junction with the external wall. An additional metal flashing covers the top of the metal faced insulation to protect the junction between the top of the metal facing and the insulation itself. In this case, an open jointed arrangement of paving slabs is adopted, with rainwater draining at the level of the membrane. To assist with the drainage of the rainwater from the parapet coping, a strip of pebbles is shown. This avoids staining of the roof level paving where it meets the parapet upstand by allowing rainwater running off the inside face of the parapet coping to drain through the pebbles to the membrane beneath. In other instances the paving is continued up the parapet upstand. In an example where the sill to the external door covers the opening, and is supported from the base of the opening, the waterproofing stops at the edge of the opening (MCE_427, j). The termination of the membrane varies with each situation.

Balustrades and plinths

Balustrade posts can be fixed to a base plate which is set onto the finished waterproof membrane. The balustrade base plate is bolted through the membrane to the concrete slab below. If the membrane has an additional protection layer, then this is usually omitted around the base plate to give a more reliable seal. An additional membrane is then laid on top of the base plate with the protection layer. Alternatively, a rubber-based seal is bonded to the top of the base plate to provide a secondary seal to the penetrations in the membrane formed by the bolts securing the base plate. The polyester filter sheet, set loose laid on top of the insulation in the detail, is wrapped around the balustrade post below the level of the paving.

Plinths which are formed as short columns for the support of roof-mounted mechanical equipment are waterproofed in a similar way to a parapet upstand. The membrane is folded up through 90° from the

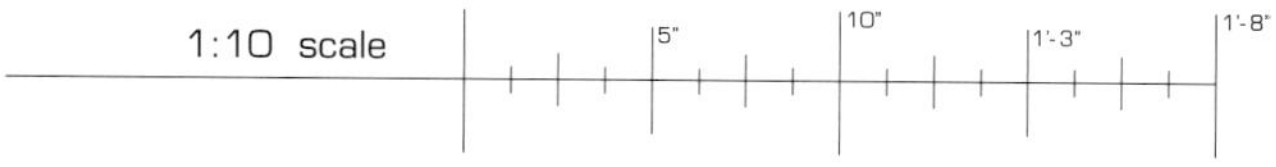

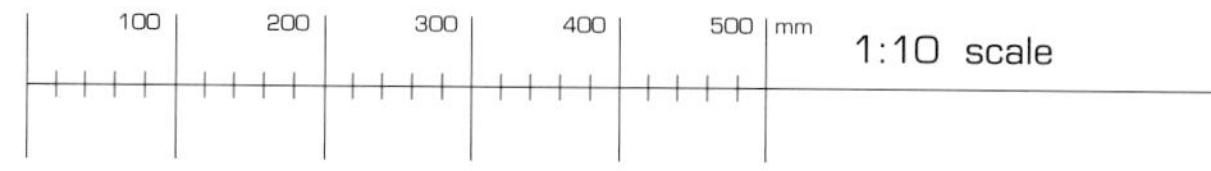

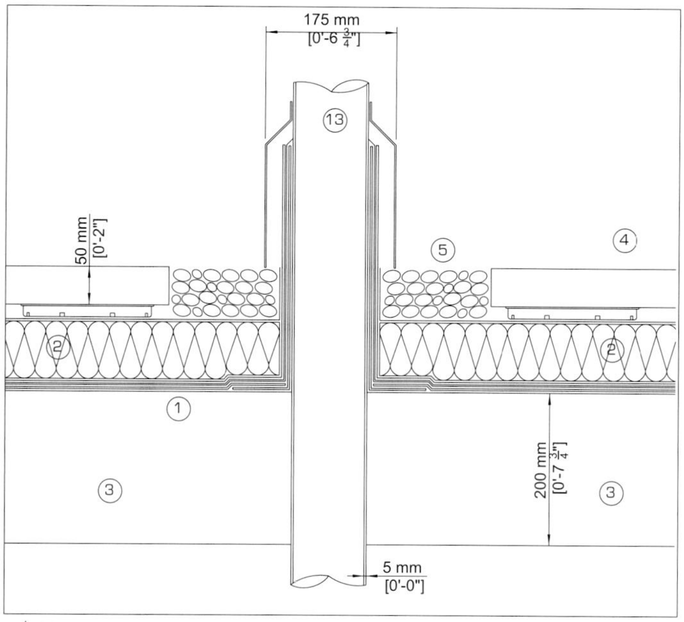

e | Vertical section. Pipe penetration

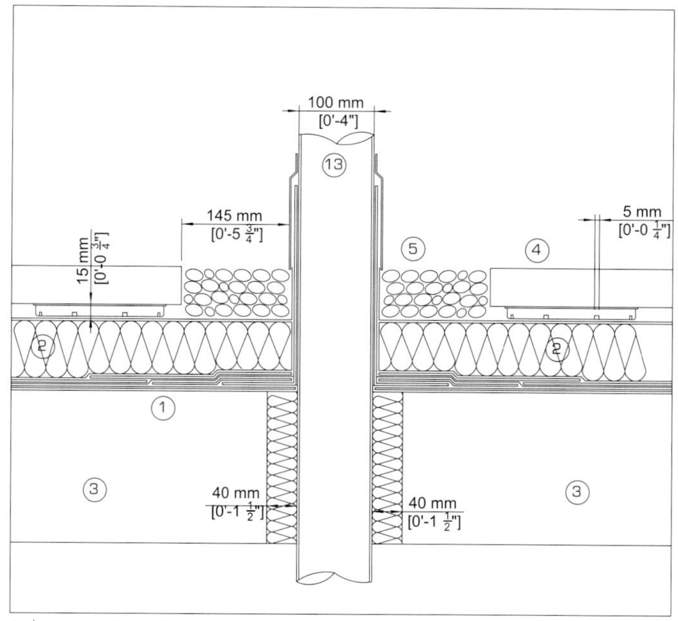

f | Vertical section. Pipe penetrations

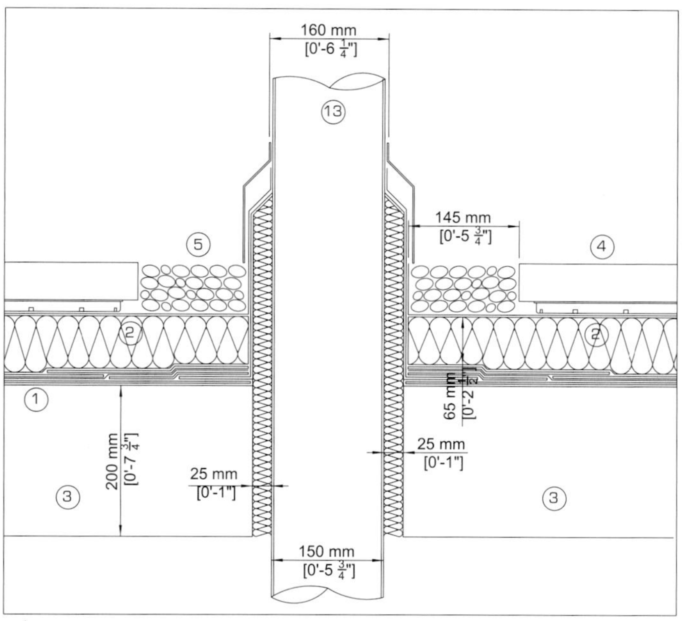

g | Vertical section. Pipe penetrations

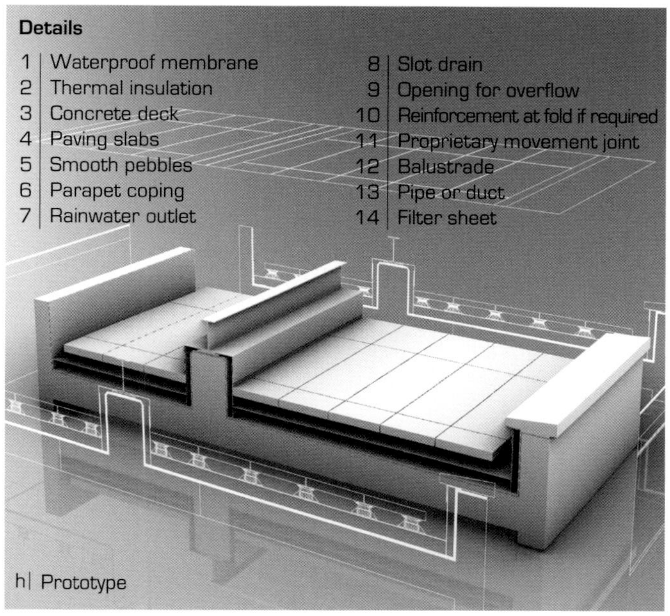

h| Prototype

roof level and is formed to cover the complete plinth. Some configurations incorporate a rail supporting an I-section that would support an air handling unit or a rail for a cleaning cradle trolley. The thermal insulation extends across the complete plinth to prevent a thermal bridge through the roof construction.

Rainwater outlets

Some rainwater outlets are set at the level of the waterproof membrane, whereas other roof configurations are drained at both the level of the sealed paving and the level of the waterproofing layer. The base of the rainwater outlet can alternatively be fixed to the concrete slab. The waterproof membrane is dressed down into the top of the rainwater outlet and the upper part of the rainwater outlet is bolted down onto the part already fixed and sealed. The geotextile sheet is wrapped around the outlet to avoid dirt and debris being washed into the rainwater drainage system. The rainwater outlet could be fixed in the same way, but in this case the cover would be much lower, since rainwater is drained only at the edges of the cover, and not through its full height. The filter sheet could also be tucked down into the edges of the rainwater outlet. Because of its concealed position, the rainwater outlet requires regular visual inspections by lifting up the paving slab immediately above to remove any debris that might collect in the gap between the rainwater outlet and the thermal insulation adjacent to it.

In other examples, the rainwater could be used increasingly to drain roof terraces into external rainwater pipes fixed to the facade, or to a void immediately behind the facade, without the need for a visible opening in the parapet wall that is visually unappealing. The two-way outlet is fixed in two parts. A 25mm gap is shown between the parapet wall and the edge of the paving slab to allow rainwater to drain into the outlet [MCE_425, j].

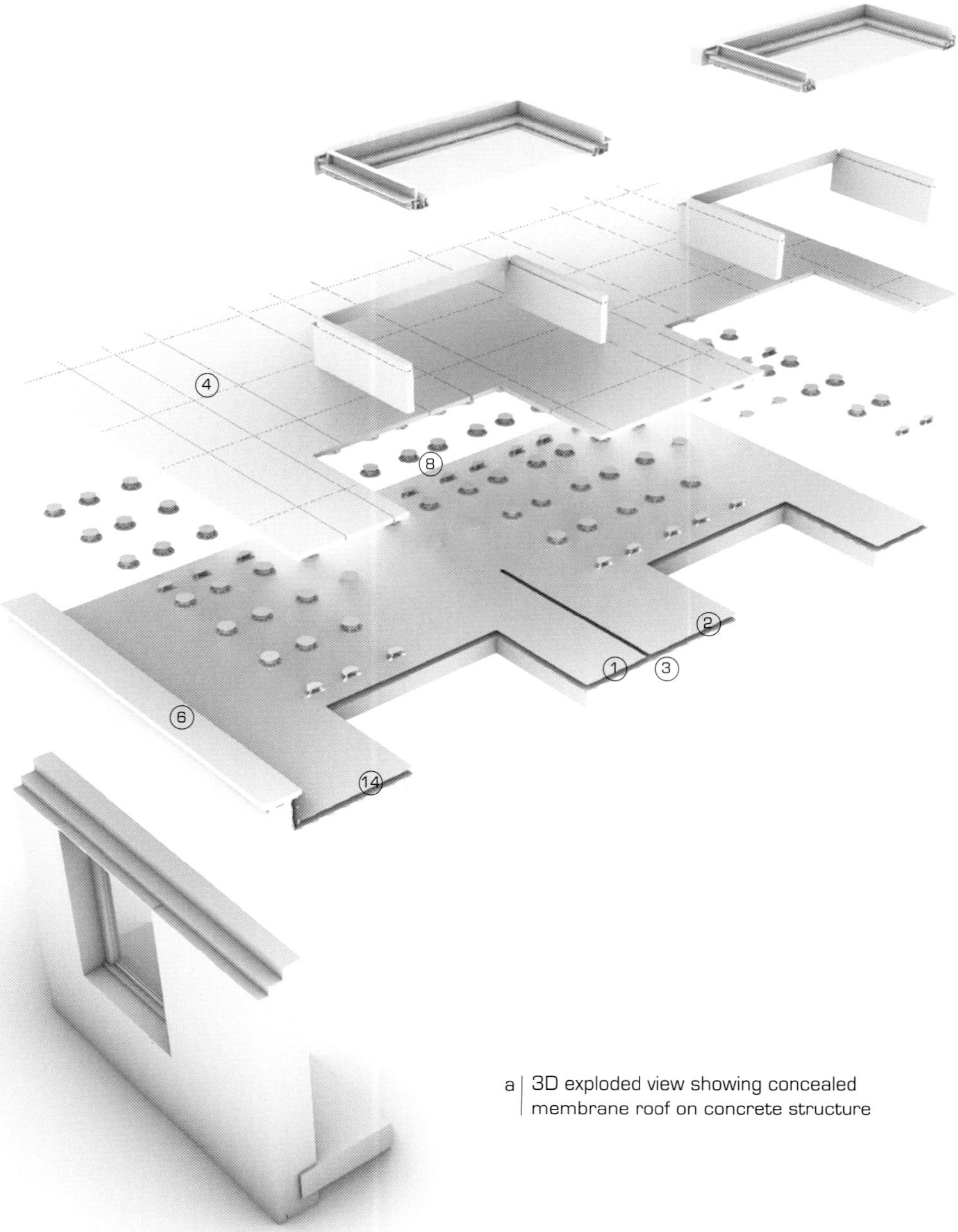

a | 3D exploded view showing concealed membrane roof on concrete structure

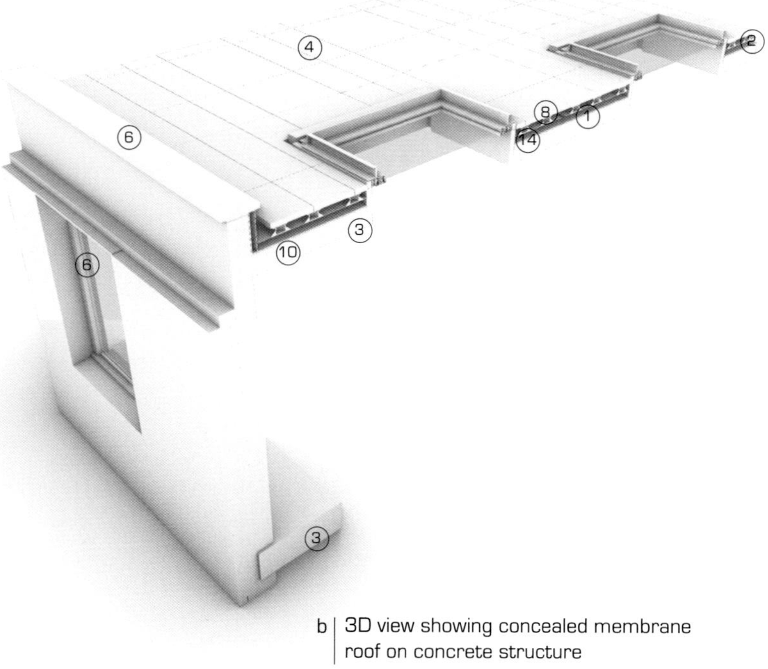

b | 3D view showing concealed membrane roof on concrete structure

Penetrations for pipes and ducts

Penetrations are sealed by either forming an upstand around the opening, or by forming a metal collar around the pipe or duct, similar to the balustrade detail mentioned earlier. Where a concrete upstand is formed, a metal flashing is welded or mechanically fixed and sealed to the penetrating pipe or duct. In some siutations, a metal sleeve is bolted through the membrane and a reinforcing rubber-based disc is bonded to the top of the base plate. The pipe is set into this sleeve and is sealed with a tension clip at the top of the sleeve. The pipe flashing detail can allow the pipe to be both thermally insulated and independent of the enclosing sealed sleeve.

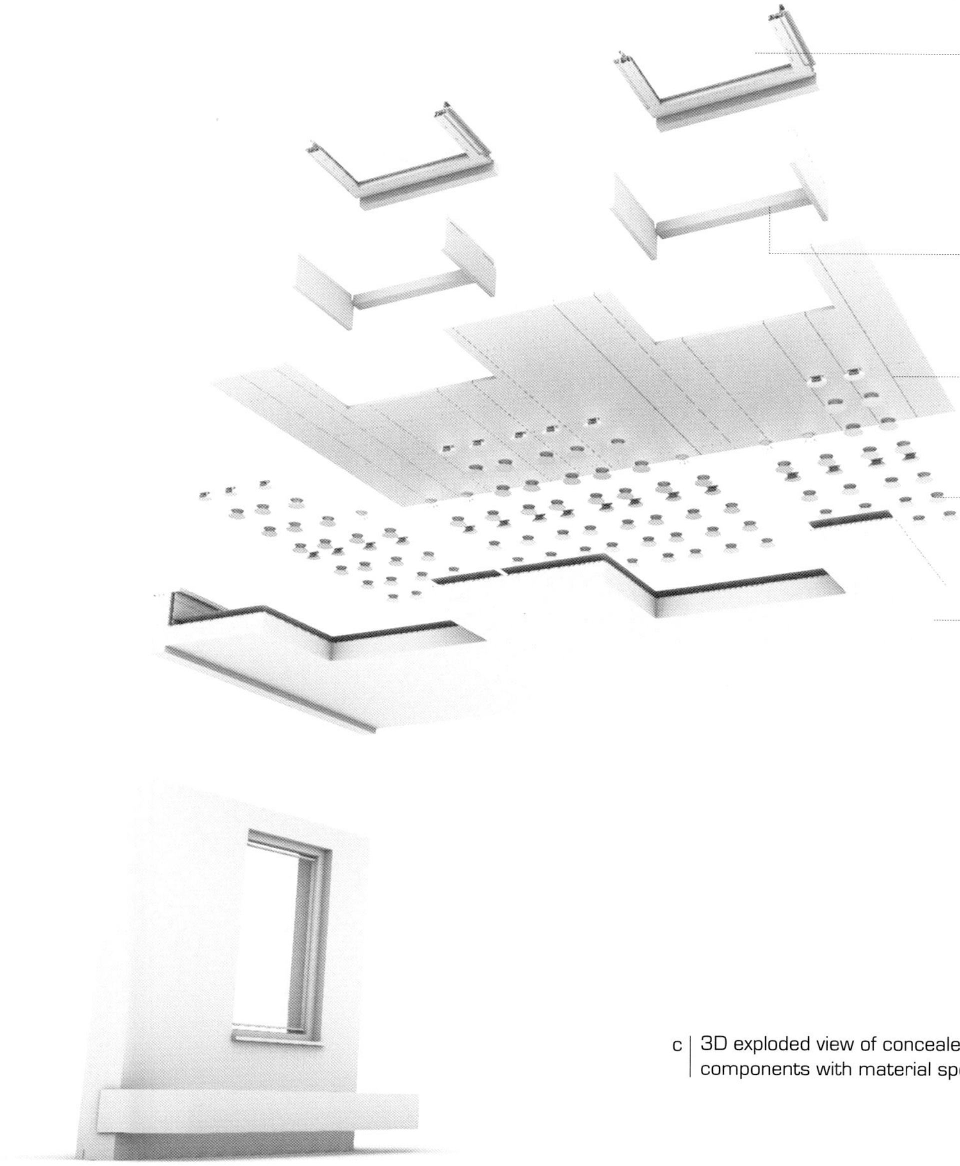

c | 3D exploded view of concealed membrane roof components with material specification references

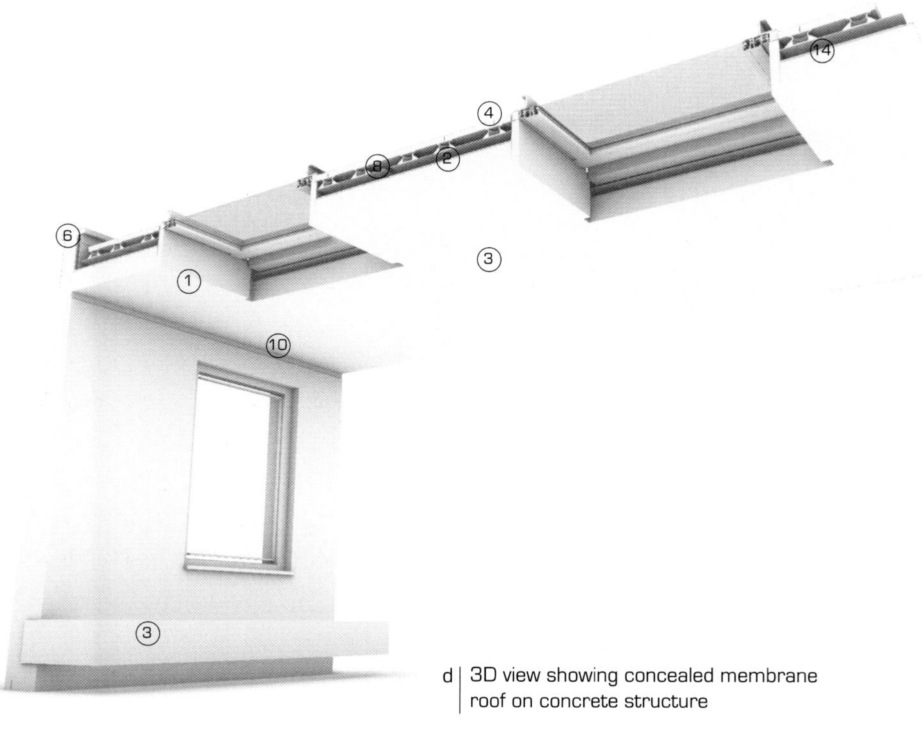

d | 3D view showing concealed membrane roof on concrete structure

Details

1 | Waterproof membrane
2 | Thermal insulation
3 | Concrete deck
4 | Paving slabs
5 | Smooth pebbles
6 | Parapet coping
7 | Rainwater outlet
8 | Slot drain
9 | Opening for overflow
10 | Reinforcement at fold if required
11 | Proprietary movement joint
12 | Balustrade
13 | Pipe or duct
14 | Filter sheet

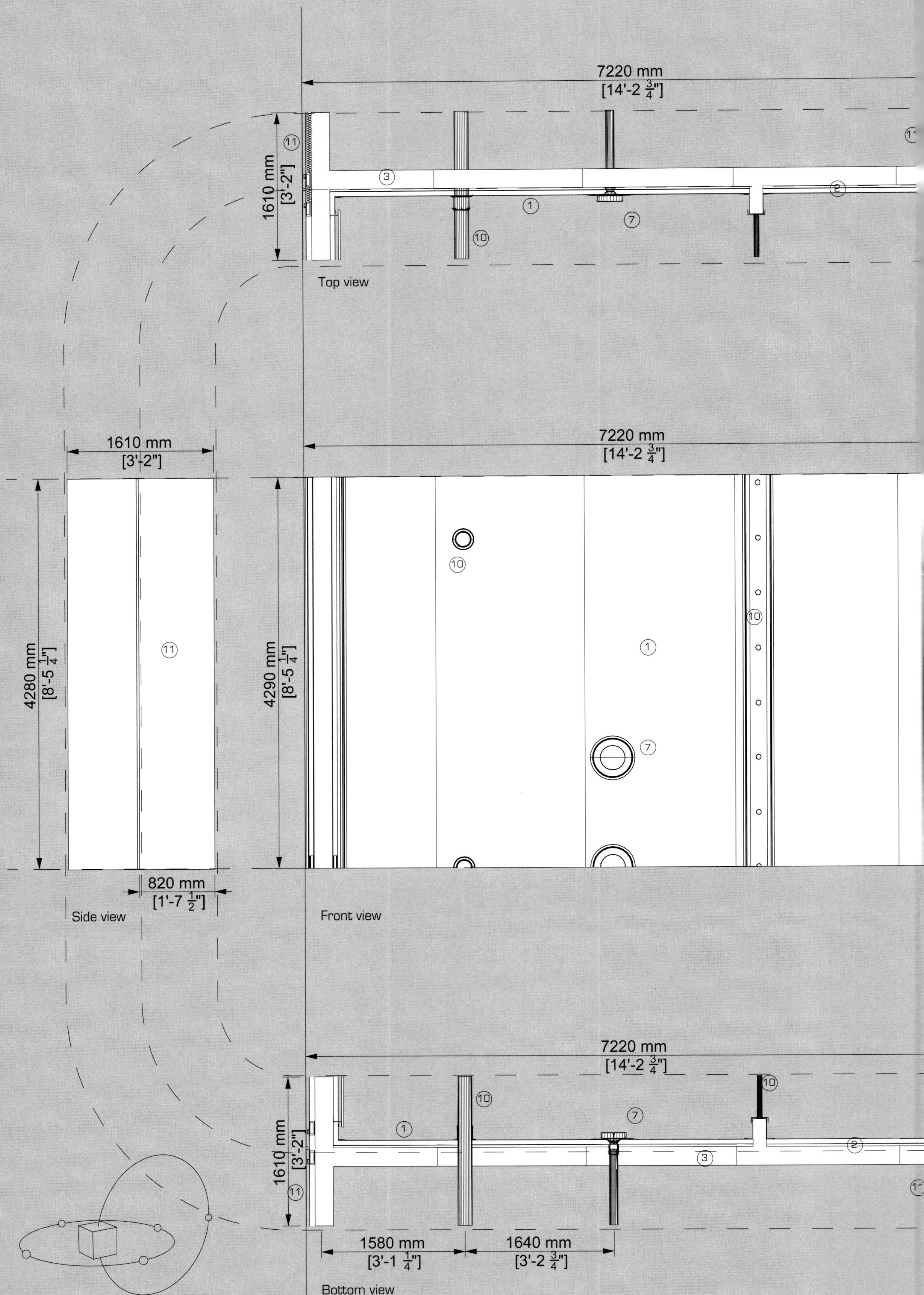

Third angle projection 1:30

Details

1 | Waterproof membrane
2 | Thermal insulation
3 | Concrete deck
4 | Paving slabs
5 | Smooth pebbles
6 | Parapet coping

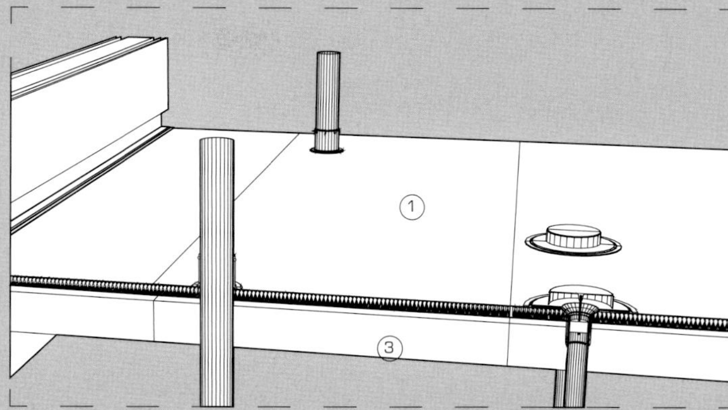

3D view of the assembly

Side view

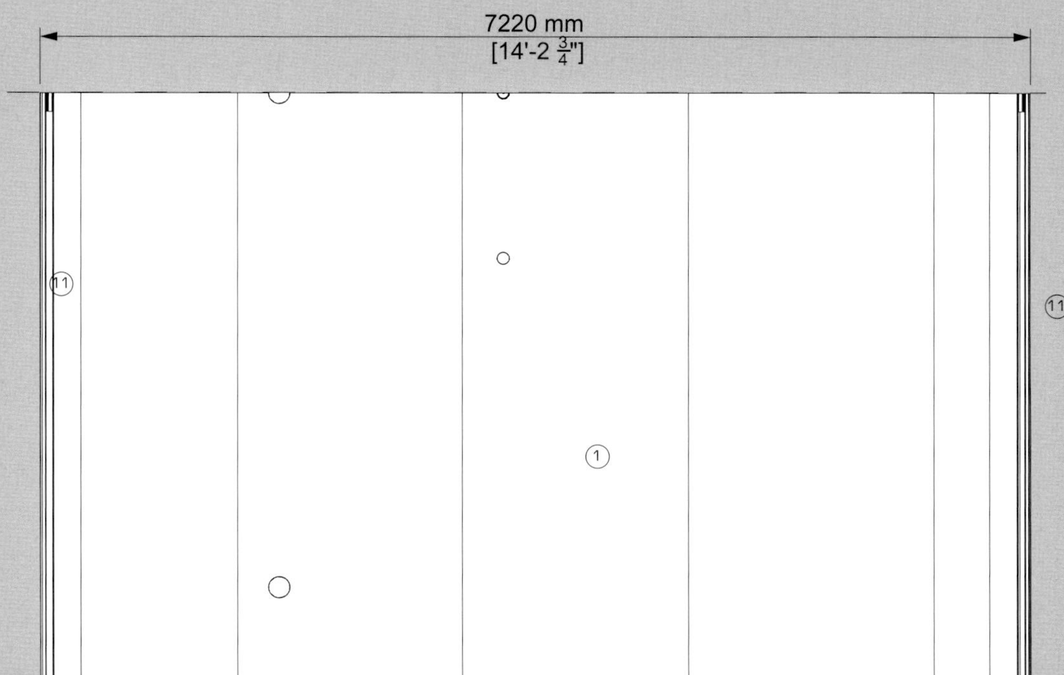

Back view

7 | Rainwater outlet
8 | Opening for overflow
9 | Balustrade
10 | Pipe or duct
11 | External wall
12 | Rooflight

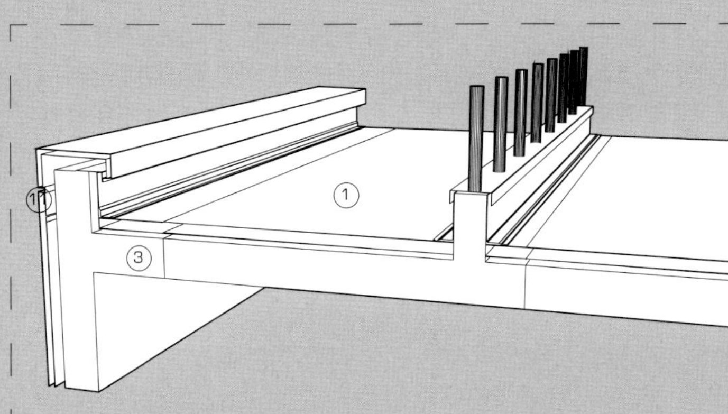

3D view of the assembly

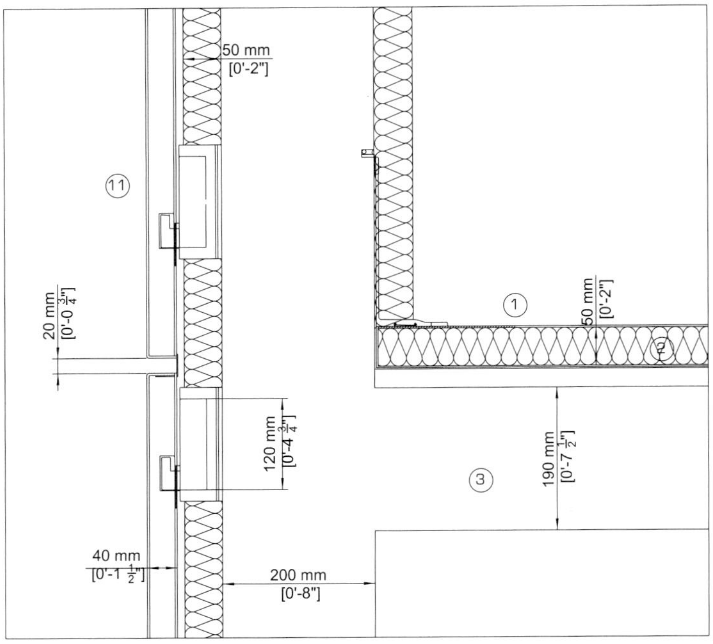

a | Vertical section. Bonded method. Junction of external wall and roof slab

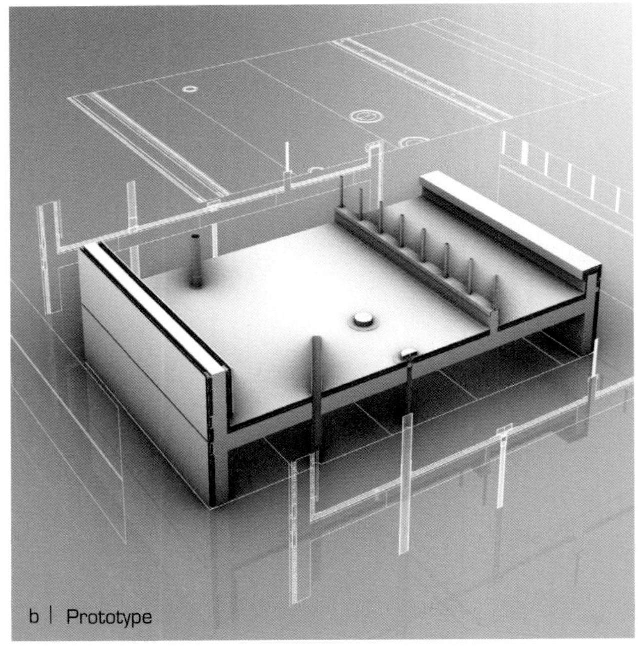

b | Prototype

c | Vertical section. Bonded method. Upstand for balustrade

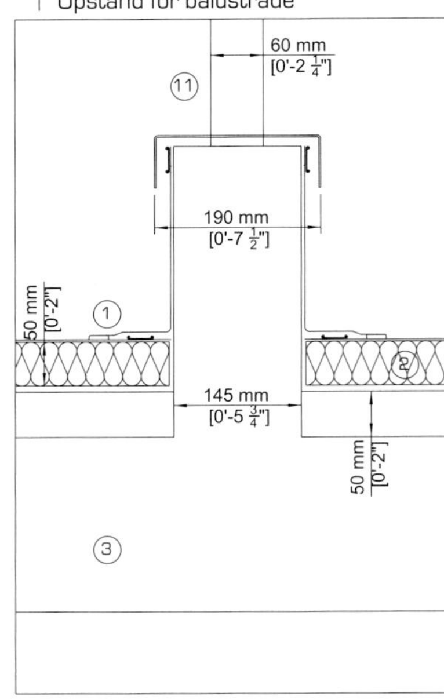

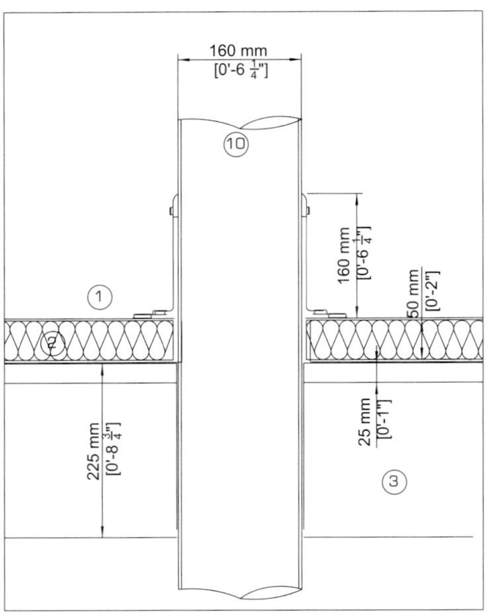

d | Vertical section. Bonded method. Pipe penetration

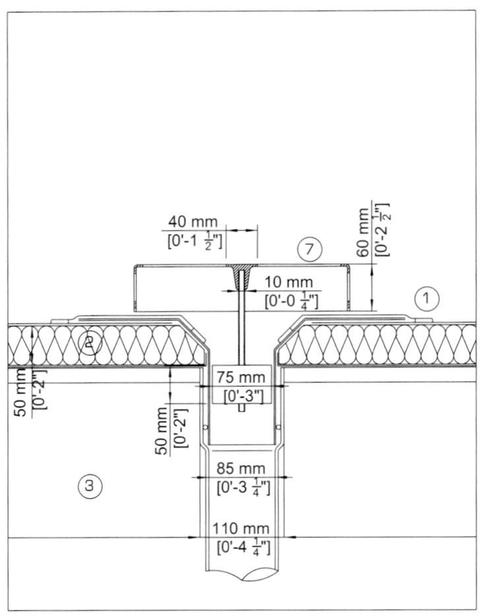

e | Vertical section. Bonded method. Rainwater outlet

Exposed membranes have been used for flat roofs which are not visible from below, but this has changed in recent years as membranes are produced in increasingly smooth and regular finishes. Because of their lightweight nature, they are often used in conjunction with lightweight roofs such as profiled metal deck and timber. This section considers their use in concrete construction, though the same principles of waterproofing can be applied to these other materials.

It was still commonly the case 30 years ago for exposed membranes on concrete roofs to be made from bitumen or bitumen-based sheet materials. This material was expected to last around 10–15 years, after which time the roof covering should have been replaced. In practice these roofs were patched up where leaks occurred since it was considered to be a major undertaking to install a completely new roof covering. Damage to the bitumen-based roof materials was typically the result of a lack of flexibility of the membrane that could not easily accommodate thermal and structural movements as well as deflections in the building structure. Although concrete roofs slabs are less susceptible to thermal movement than some other materials, interfaces between wall and roof, or roof and rooflights, for example, would often result in damage to the roof membrane where the roofing sheet continued across the joint at the junction of the two materials. The structural movement between the roof deck and the adjoining elements would sometimes result in the roofing sheet splitting or tearing, allowing rainwater to penetrate the roof construction. The weakness of these materials was partially overcome by introducing added sacrificial layers. Although this had the advantage of reducing the risk of the membrane being punctured by foot traffic during maintenance, this method did not add any substantial strength to the membrane and damage would occur as before.

The introduction of polymer-based membranes provided economic waterproofing materials that were more flexible than their bitumen-

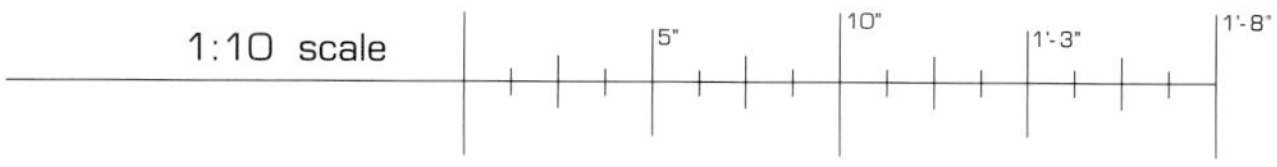

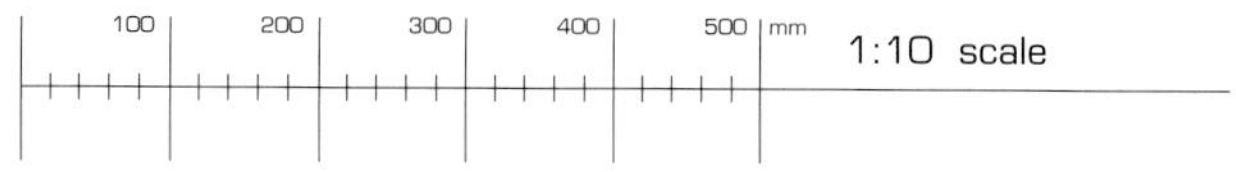

Details

1 | Waterproof membrane
2 | Thermal insulation
3 | Concrete deck
4 | Paving slabs
5 | Smooth pebbles
6 | Parapet coping
7 | Rainwater outlet
8 | Opening for overflow
9 | Balustrade
10 | Pipe or duct
11 | External wall
12 | Rooflight

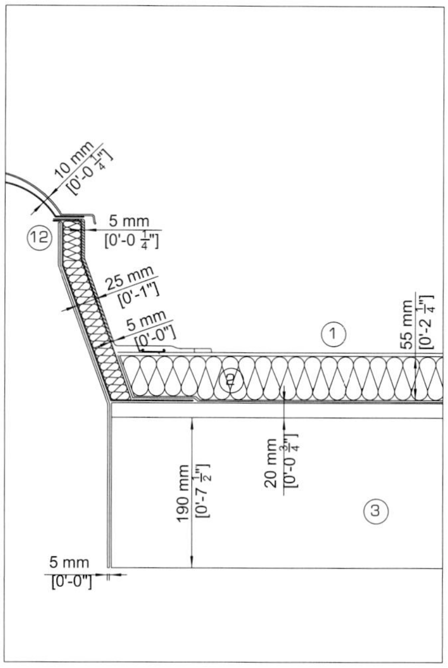

f | Vertical section. Bonded method. Junction with rooflight

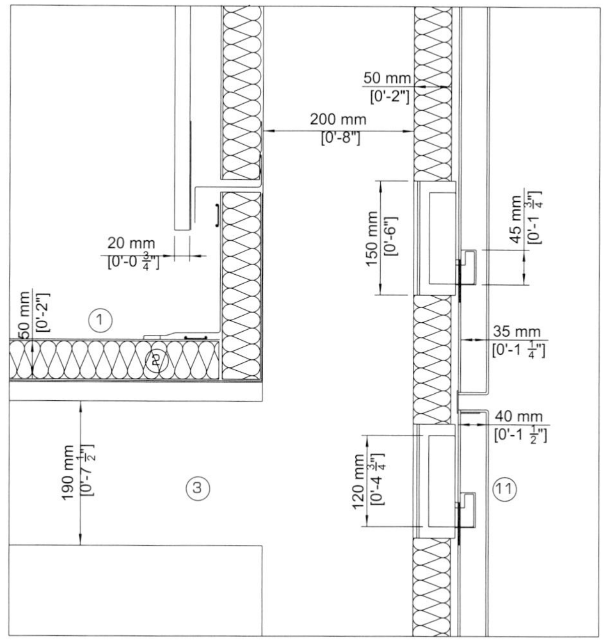

g | Vertical section. Bonded method. Junction of external wall and roof slab

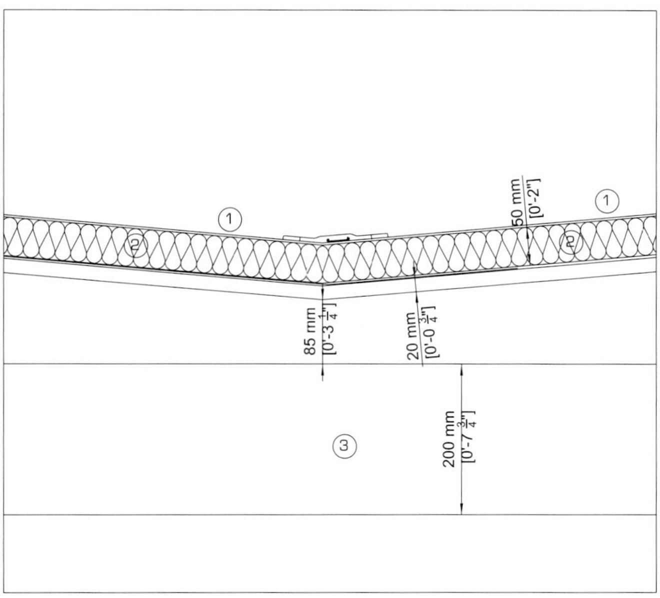

h | Vertical section. Bonded method. Roof fold

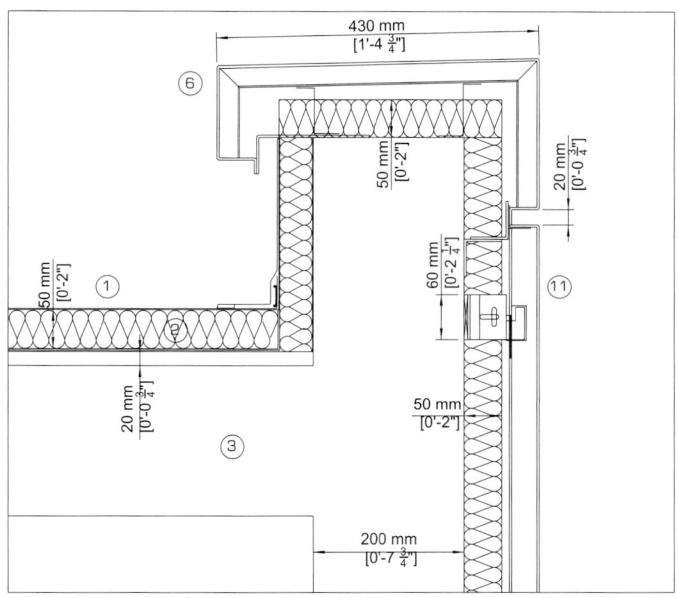

j | Vertical section. Bonded method. Low parapet

based predecessors. First introduced in the 1960s and 1970s, they became much more widely used in the 1980s and 1990s. The increased flexibility of the new sheet materials allowed for greater amounts of movement between adjoining components and assemblies, allowing the detailing of junctions to be relatively straightforward in forming reliable weathertight roof membrane (a). As a result of these developments, polymer modifications were also made in the older-generation bitumen-based materials to make them more flexible, in order to compete with the polymer-based sheet materials. As a result there is now a wide range of exposed membrane materials available to suit different budgets and individual roof designs.

Polymer-based membranes

The main advantage of the polymer-based sheet materials is their ability to be cut and formed to complex shapes, allowing them to take up shapes precisely, sometimes pre-formed in the factory before being delivered to site. Single-layer membranes are very practical on roofs with a large number of penetrations, typically in commercial buildings where mechanical ventilation equipment is regularly being modified or replaced during the lifetime of a building.

Single-layer membranes are made from either elastomeric materials, typically EPDM, or from thermoplastic materials, typically plasticised PVC (PVC-P). Elastomeric materials are very popular in the US, while thermoplastics are preferred in Europe. EPDM (ethylene propylene diene monomer) is a flexible and elastic material that has the appearance of a synthetic rubber. EPDM is manufactured in the limited colours of black, grey and white.

Both elastomers and thermoplastics can be mechanically fixed, bonded or secured with ballast to the concrete deck beneath. Developments in these materials have led to them being used on timber decks and

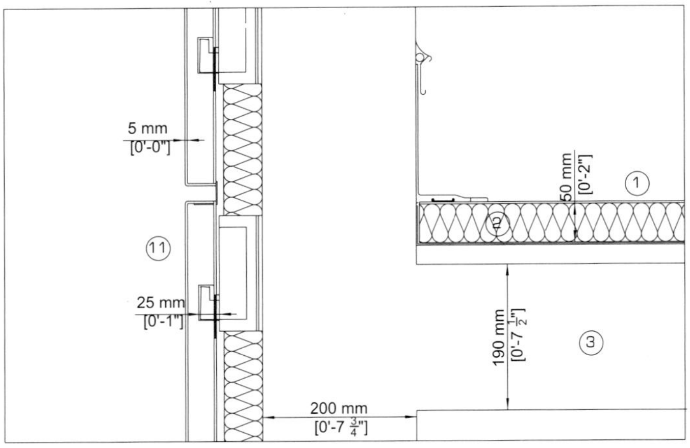

a | Vertical sections. Bonded method. Junction of external wall and roof slab

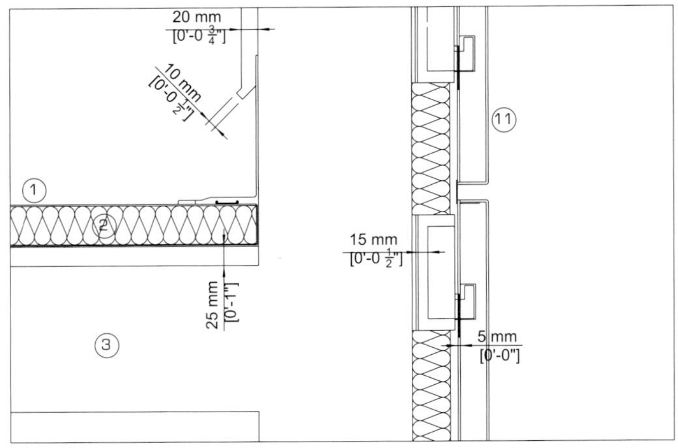

b | Vertical section. Bonded method. Gutter

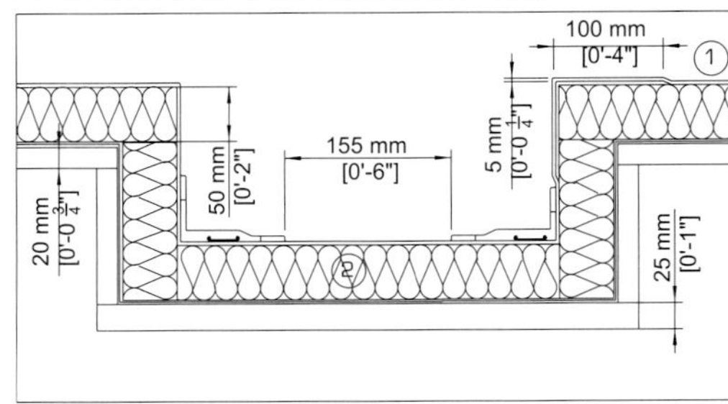

c | Vertical section. Bonded method. Pipe penetrat

Details

1 | Waterproof membrane
2 | Thermal insulation
3 | Concrete deck
4 | Paving slabs
5 | Smooth pebbles
6 | Parapet coping
7 | Rainwater outlet
8 | Opening for overflow
9 | Balustrade
10 | Pipe or duct
11 | External wall
12 | Rooflight

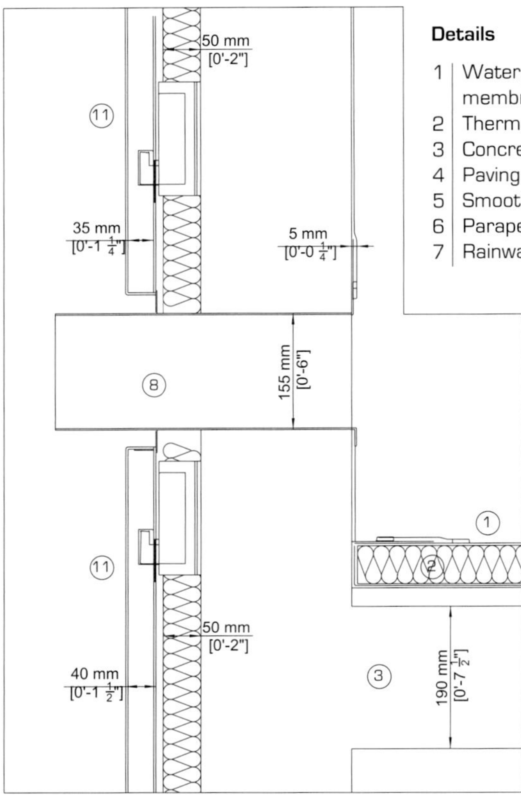

e | Vertical section. Bonded method. Roof overflow

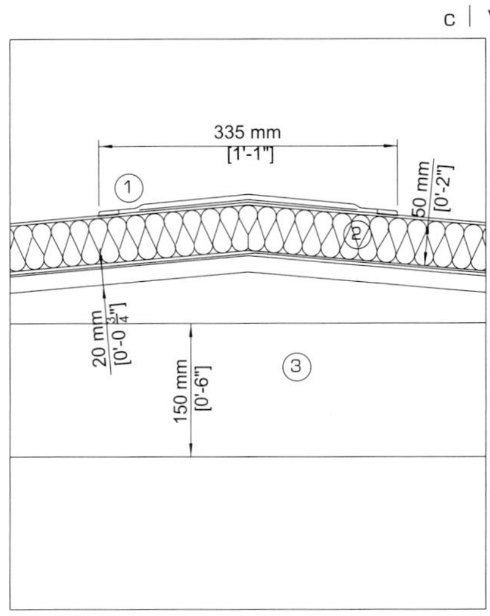

f | Vertical section. Bonded method. Roof fold

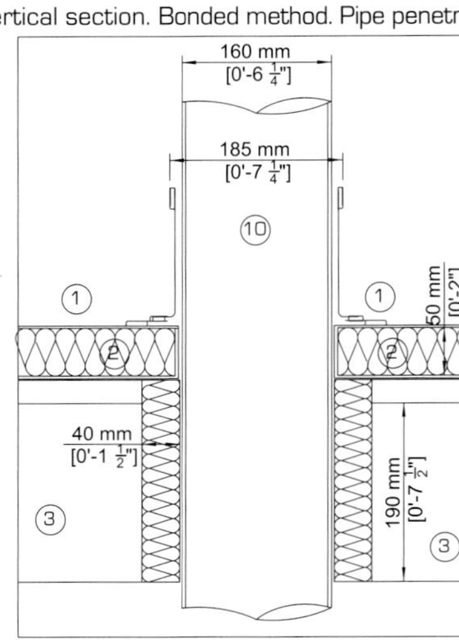

profiled metal decks in addition to the concrete decks discussed here. Both thermoplastic and EPDM membranes can be welded together to form a continuous waterproof sheet. While both material types were glued, there is an increasing use of hot air welding methods, which avoid the need for flame techniques or adhesive bonding methods that can be both slow and can damage adjacent work during their application. In hot air welding, a jet of heated air is used to soften the materials and weld together, applied from a range of tools that are either hand held or fully automated, depending on the application.

PVC membranes

PVC (polyvinyl chloride) roof membranes have been in use since the 1960s as a very lightweight and relatively economic roofing material, and have become widely used in recent years. Membranes in this material were first developed in Europe in the late 1960s and were used in the US from the 1970s onwards. PVC sheet material is usually reinforced with glass fibre to give it increased rigidity that is easier to bond to the substrate.

The PVC used in membranes is plasticised (PVC-P), unlike the unplasticised PVC (PVC-U, or uPVC) used to make window frames and rainwater drainage components. PVC-P is rigid at normal external temperatures, but softens when heated, making it flexible and allowing strips or sheets to be welded together to form a continuous membrane without the need for standing seams or visible joints. Plasticisers and filler material in PVC-P is added to give the material greater flexibility. The material has very low levels of shrinkage, is dimensionally stable and does not creep visibly with age. The material experiences only very small amounts of movement under full wind load.

Membranes are reinforced with glass fibre sheet or polyester fabric. These layers are bonded into the material. The glass fibre provides

1:10 scale

g | Vertical section. Ballusted method. Rainwater outlet

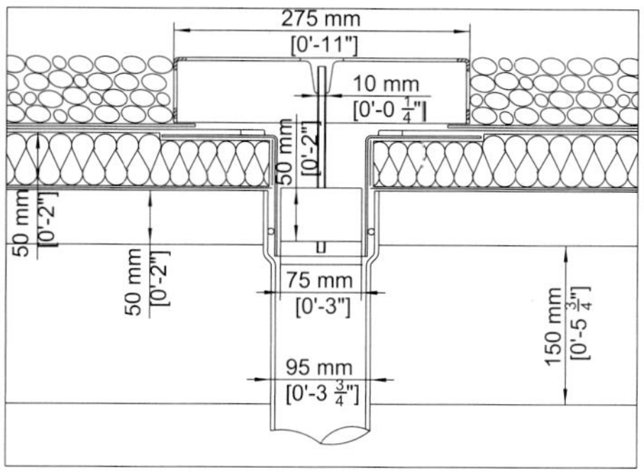

h | Vertical sections. Ballusted method. Junction with rooflight

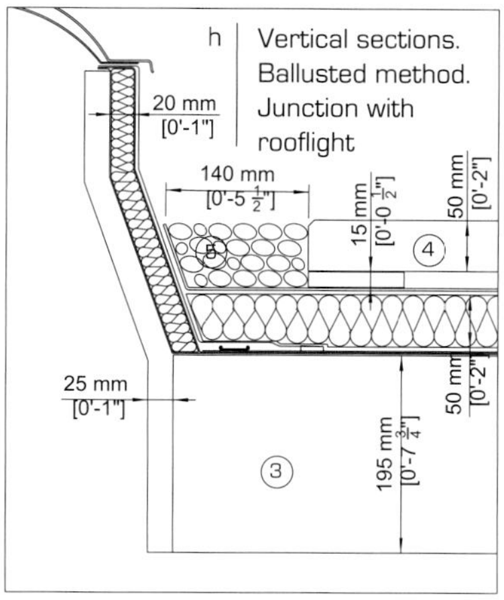

j | Vertical section. Ballusted method. Eaves

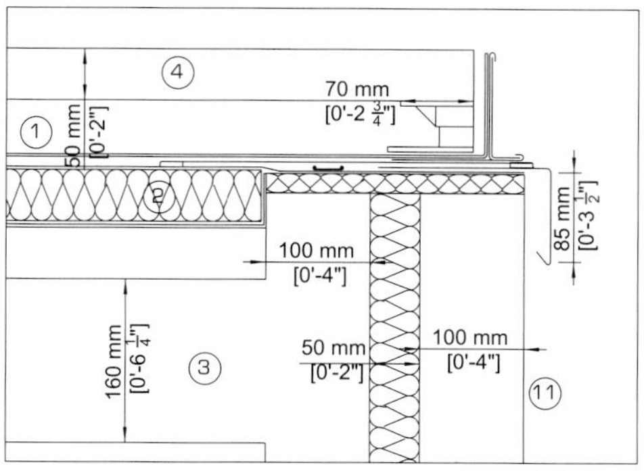

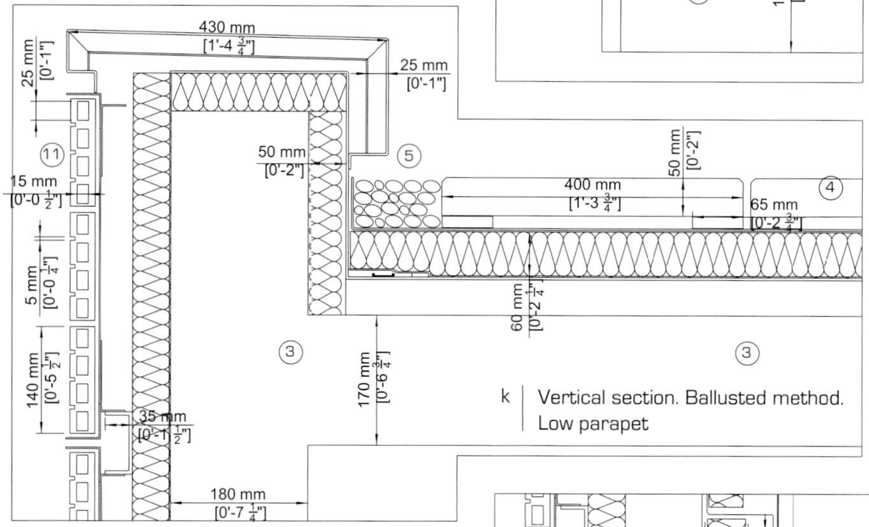

k | Vertical section. Ballusted method. Low parapet

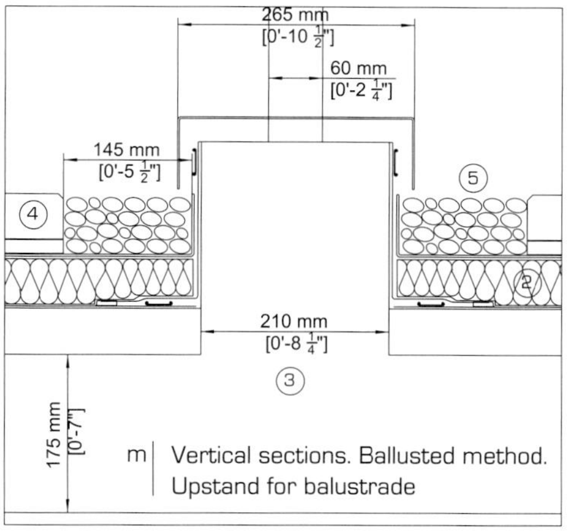

m | Vertical sections. Ballusted method. Upstand for balustrade

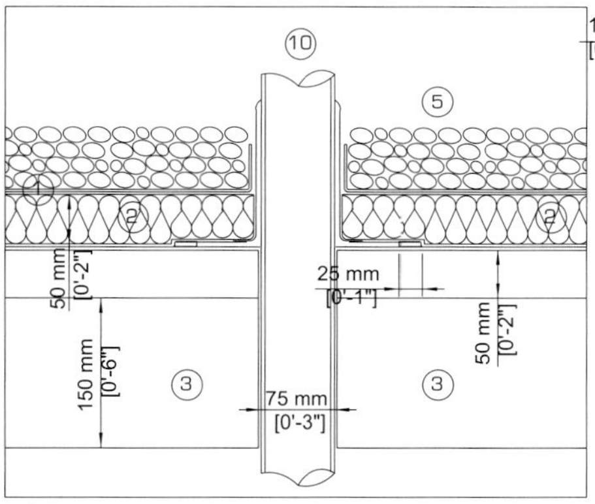

n | Vertical section. Ballusted method. Pipe penetration

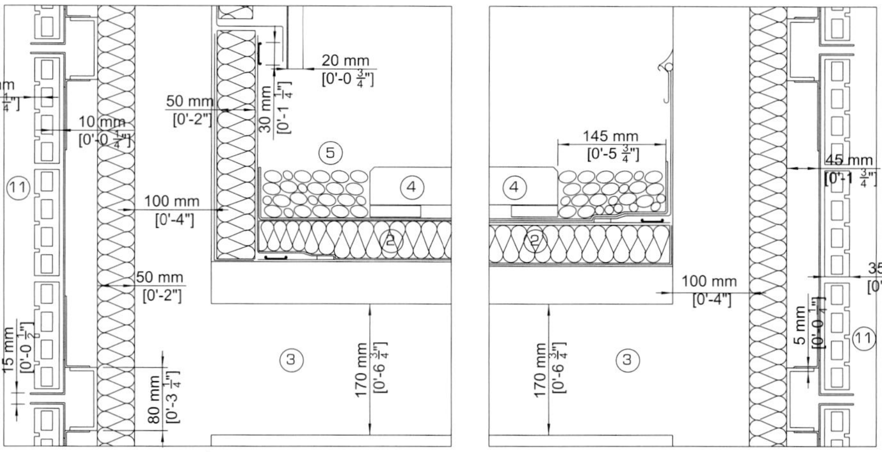

p | Vertical sections. Ballusted method. Junction of external wall and roof slab

dimensional stability, making it more stable for bonding to the substrate. The woven polyester fabric, used in tent membrane structures, has high tensile strength to resist wind loads resulting in mechanical fixing methods being used. A typical build-up for a single layer membrane is a concrete deck with a vapour barrier set on top, with thermal insulation above that, sealed on top with a single-layer membrane (c). PVC-P membranes are typically 1.5mm–3.0mm thick, while EPDM membranes are typically 1.0mm–1.5mm thick.

FPO (TPO) membranes

A recent development in thermoplastic membrane types are polypropylene- and polyethylene-based materials. They have greater flexibility than PVC-P membranes, but still require reinforcement in glass fibre sheet for increased dimensional stability and polyester fabric to give greater tensile strength. Fire retardant is added to provide fire resistance, unless it is PVC-P which is self-extinguishing when flame is applied.

Mechanically fixed method

This fixing method is suited to applications with high wind uplift forces, as bonded systems tend to be limited by the bonding strength of the vapour barrier to which the membrane is itself bonded through the thermal insulation layer, which is typically made from expanded polystyrene board. The vapour barrier is loose laid on the concrete deck and thermal insulation is then mechanically fixed through this barrier to the deck beneath. The spacing of the fasteners varies with the design wind loads. A separating layer of glass fibre sheet is usually laid onto the insulation with an outer single-layer membrane. The membrane is mechanically fixed with pressure-plate bars, similar to those used in glazed curtain walling systems to hold the glass in place. Bars form strips of pressure plate to hold the roof build-up in place. The pressure plates are fixed by bolts at centres along their length to the substrate below.

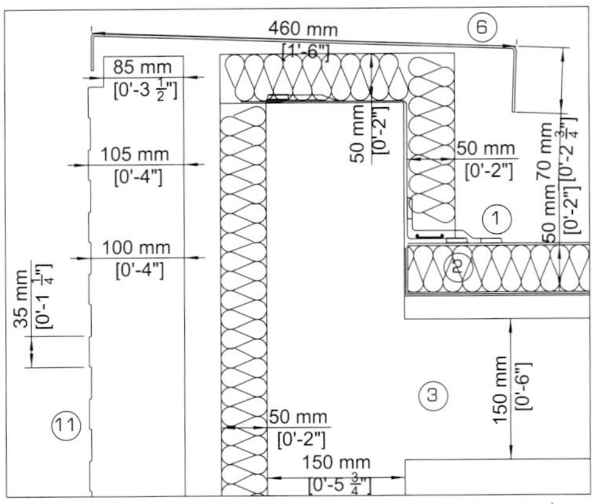

a | Vertical section. Mechanical fastening method. Low parapet

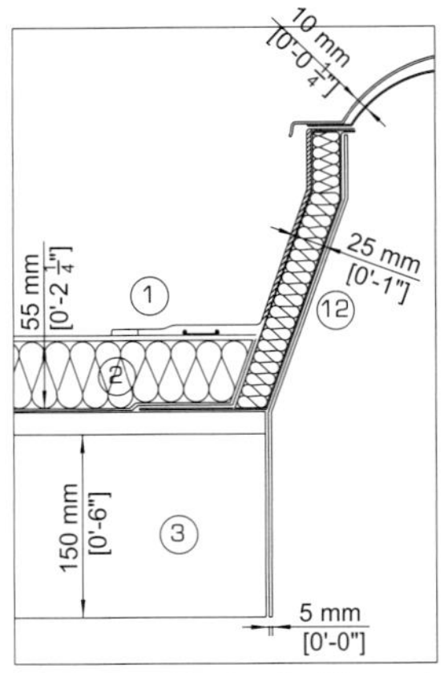

b | Vertical section. Mechanical fastening method. Junction with rooflight

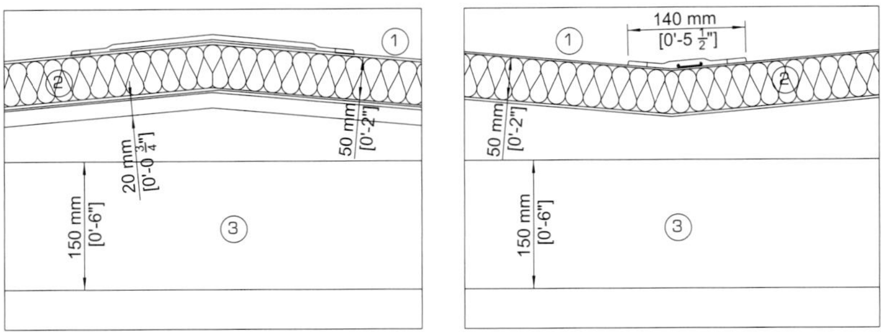

c | Vertical sections. Mechanical fastening method. Roof fold

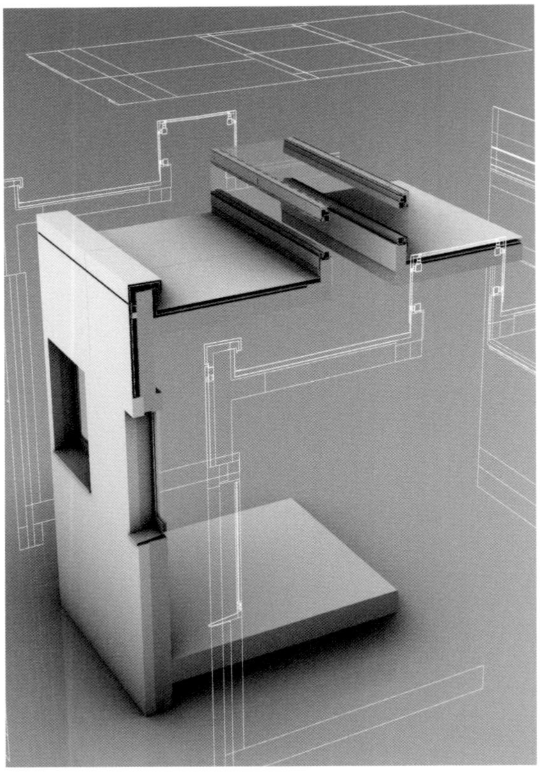

d | Typical bay

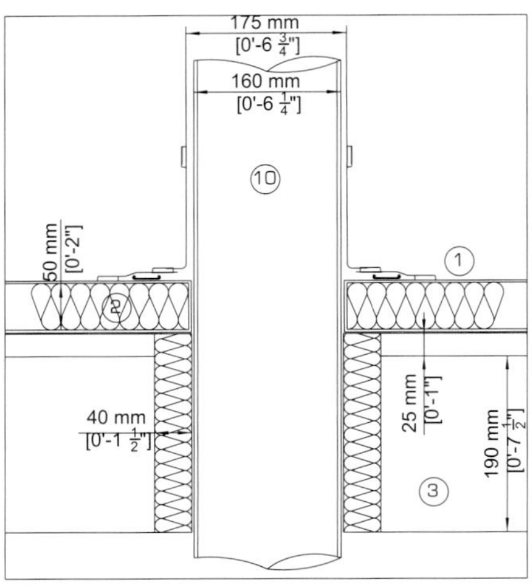

e | Vertical section. Mechanical fastening method. Pipe penetration

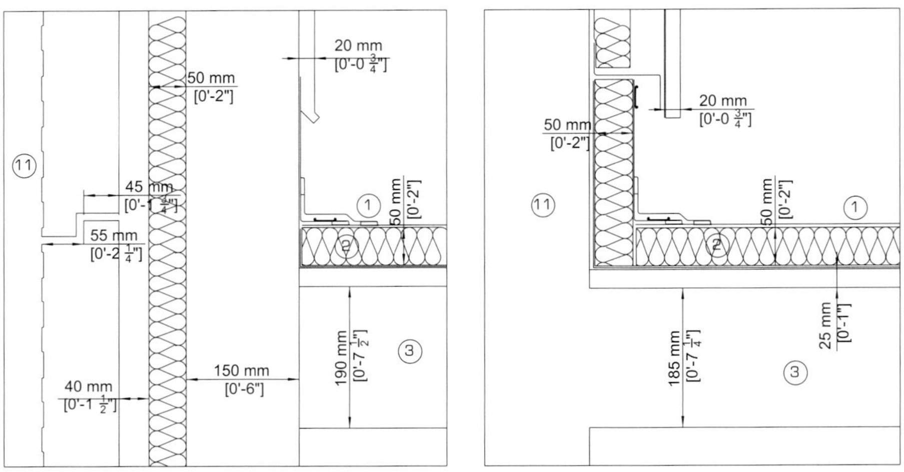

f | Vertical sections. Mechanical fastening method. Junctions of external wall and roof slab

Membranes can also be secured by point fixings rather than by pressure plates. 50–75mm diameter rigid plastic discs are used to hold the build-up in place. These are set at centres to suit the design wind loads. The closed cell rigid insulation is typically made in panel sizes of 1200mm x 2400mm in thicknesses from 25mm to 100mm.

Bonded fixing method

The build-up of materials is the same as that for the mechanically fixed system. The membrane can either be bonded directly to the concrete deck to form a concealed membrane, or be in the exposed configuration discussed here. When bonded directly to the concrete deck, a felt backing layer is usually used to overcome any roughness in the substrate that would puncture the material. In the exposed membrane configuration the vapour barrier is usually bitumen-based and is bonded to the deck. Joints between the vapour barrier sheets are lapped to avoid any risk of vapour passing through the roof structure from inside the building. The thermal insulation is then bonded to the vapour barrier (c). Insulation can also be mechanically fixed with pressure plates to the concrete deck beneath. The membrane is then bonded to the insulation with a continuous layer of bonding adhesive on its underside. Some systems still bond the membrane at points only rather than across the entire surface of the membrane, but this is dependent upon the wind load and the proprietary system used.

Bonded membranes have a visually smooth appearance, making it suitable where the roof surface is seen from points around the building. It can be more difficult to achieve the bonding required for high wind uplift conditions, but this is a matter of individual design. This fixing method

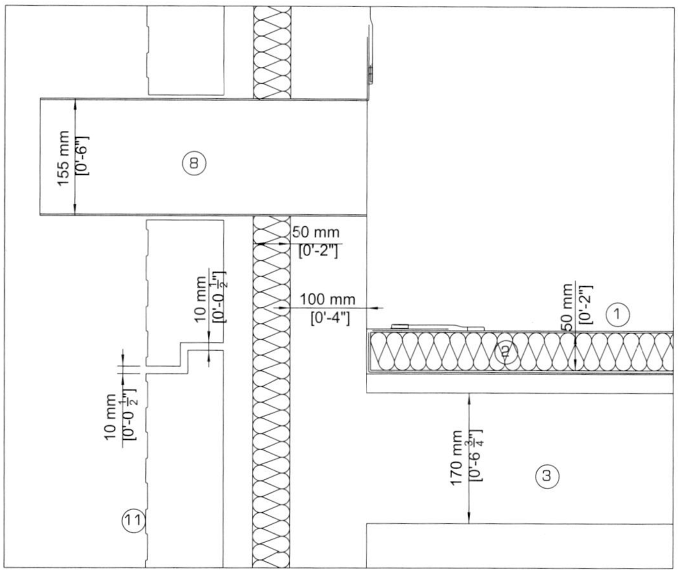

g | Vertical section. Mechanical fastening method. Roof overflow

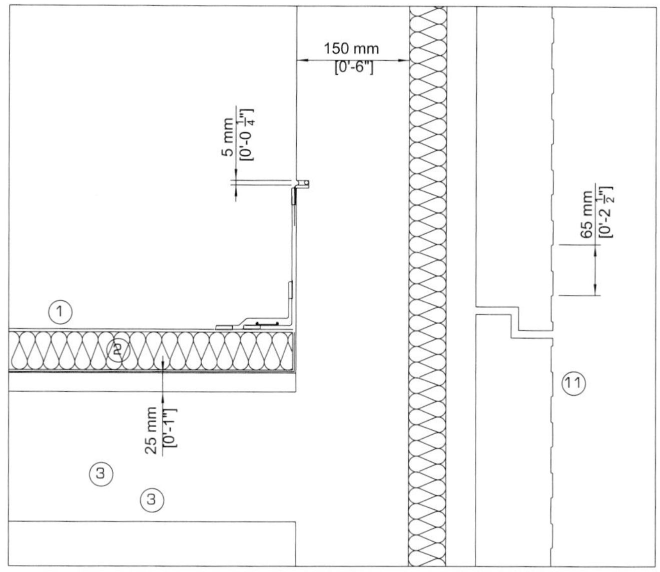

h | Vertical section. Mechanical fastening method. Junctions of external wall and roof slab

Details

1	Waterproof membrane	7	Rainwater outlet
2	Thermal insulation	8	Opening for overflow
3	Concrete deck	9	Balustrade
4	Paving slabs	10	Pipe or duct
5	Smooth pebbles	11	External wall
6	Parapet coping	12	Rooflight

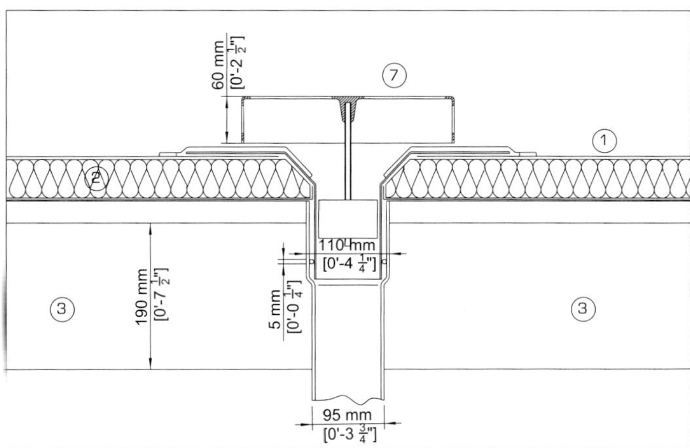

i | Vertical section. Mechanical fastening method. Rainwater outlet

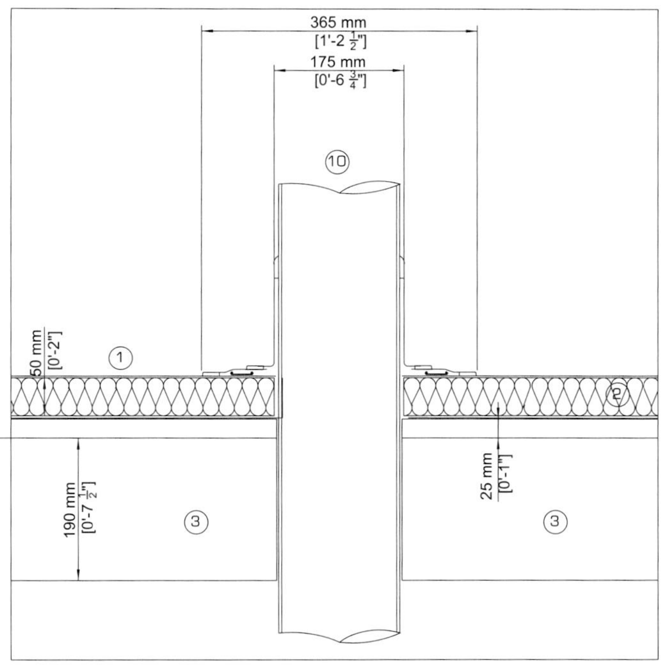

k | Vertical section. Mechanical fastening method. Pipe penetration

still requires mechanical fixing at the edges, and around openings such as rooflights.

Parapets and upstands

Membranes can be bonded or mechanically fixed to parapet upstands. The fixing method that is used on the main area of roof is usually continued on these vertical areas. With mechanical fixing the pressure plate can be fixed either to the upstand or to the flat roof area. The pressure plate forms a junction between the membrane sheet forming the upstand and the membrane sheet of the roof. Intermediate pressure plates are applied horizontally on the upstand when its height exceeds around 500mm, depending on the specific material used.

Ballasted roofs

Concealed membranes which are ballasted use a top layer to weigh down the membrane and insulation rather than use mechanical fixing or bonding methods. A typical build-up consists of a single-layer membrane welded together to form a continuous sealed sheet, which is loose laid onto a concrete deck. A backing felt is used where the concrete is considered to be too roughly finished for the membrane to be laid directly on top. Thermal insulation is loose laid onto the membrane with a filter layer set onto the insulation. Smooth pebbles are spread on top to a depth dependent upon both the weight required, to avoid wind uplift as well as to satisfy visual requirements.

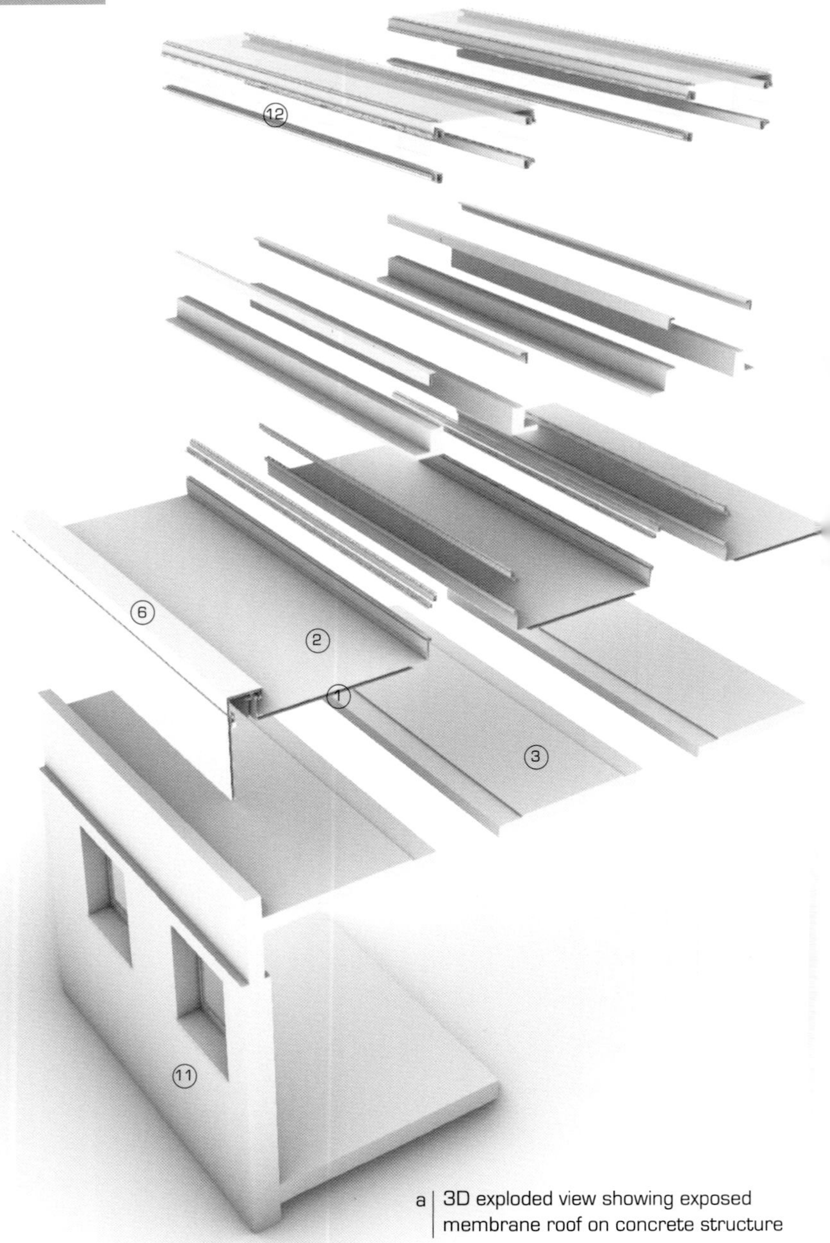

a | 3D exploded view showing exposed membrane roof on concrete structure

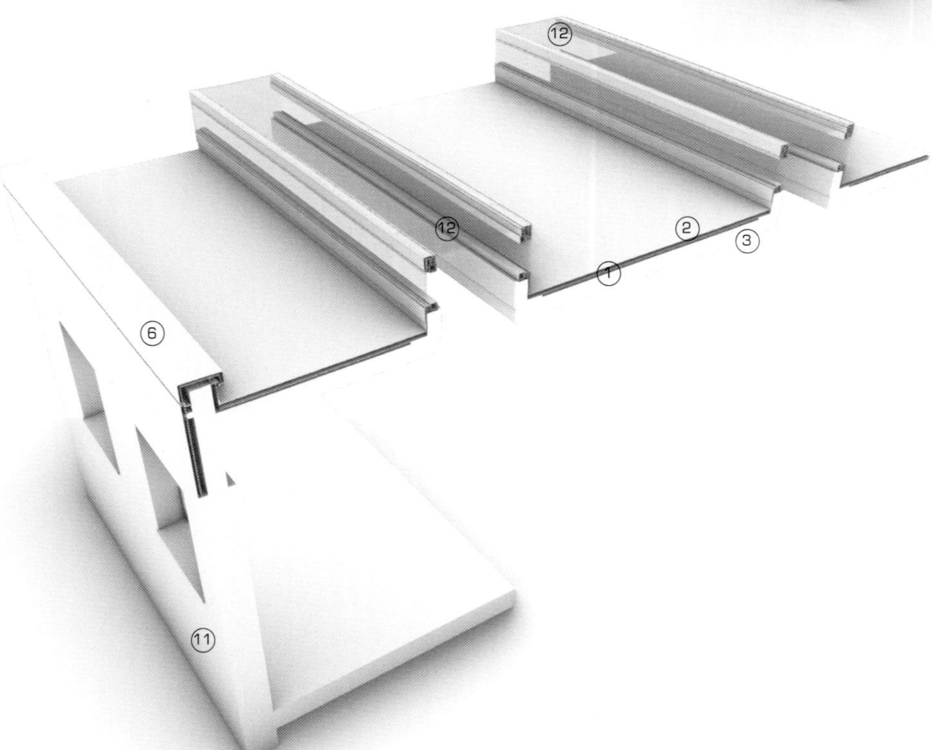

b | 3D view showing exposed membrane roof on concrete structure

Details

1 | Waterproof membrane
2 | Thermal insulation
3 | Concrete deck
4 | Paving slabs
5 | Smooth pebbles
6 | Parapet coping
7 | Rainwater outlet
8 | Opening for overflow
9 | Balustrade
10 | Pipe or duct
11 | External wall
12 | Rooflight

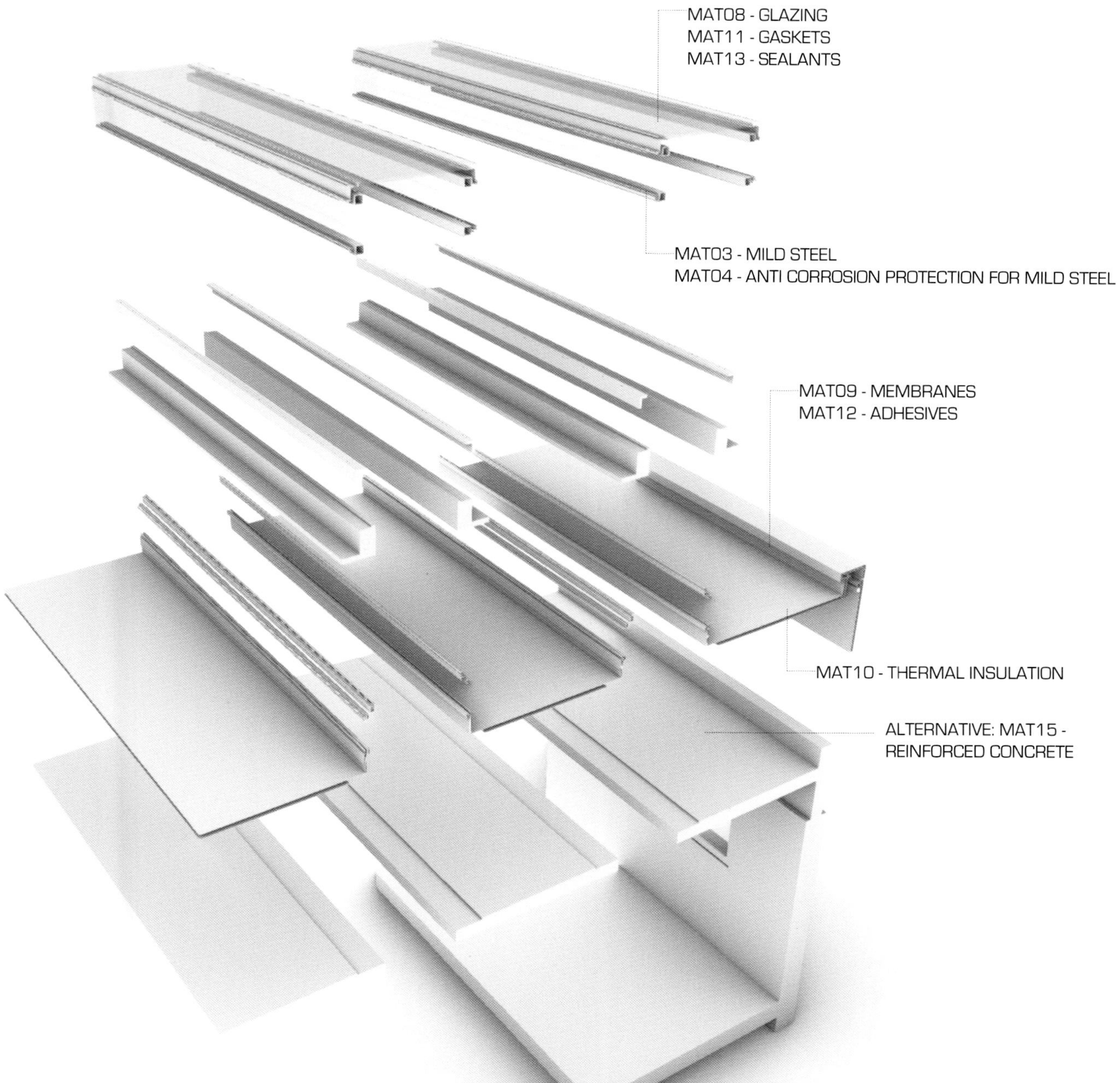

c | 3D exploded view of exposed membrane roof components with material specification references

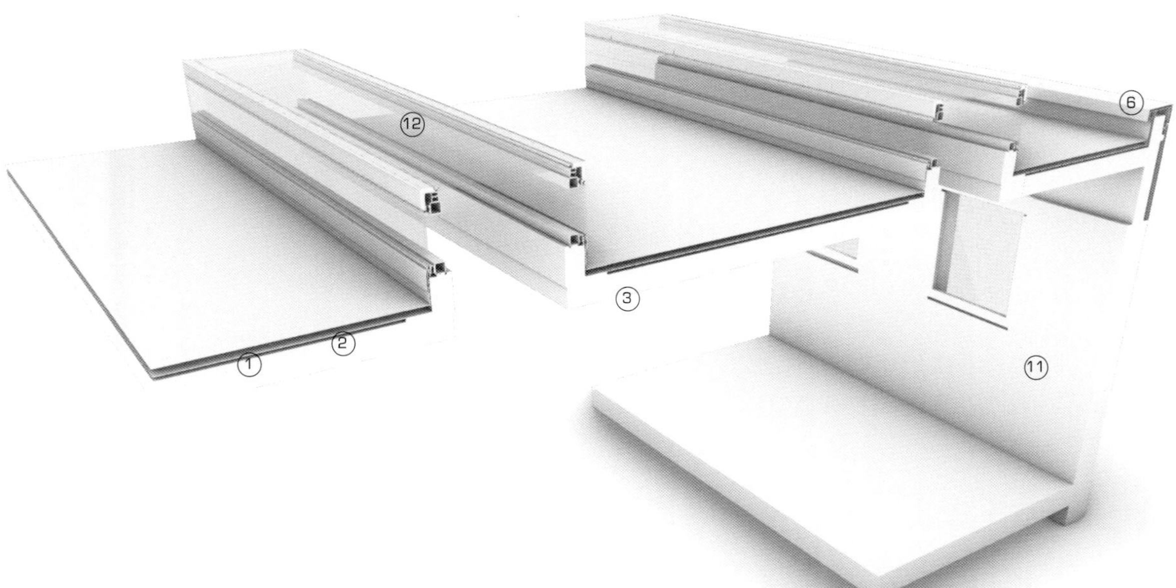

d | 3D view showing exposed membrane roof on concrete structure

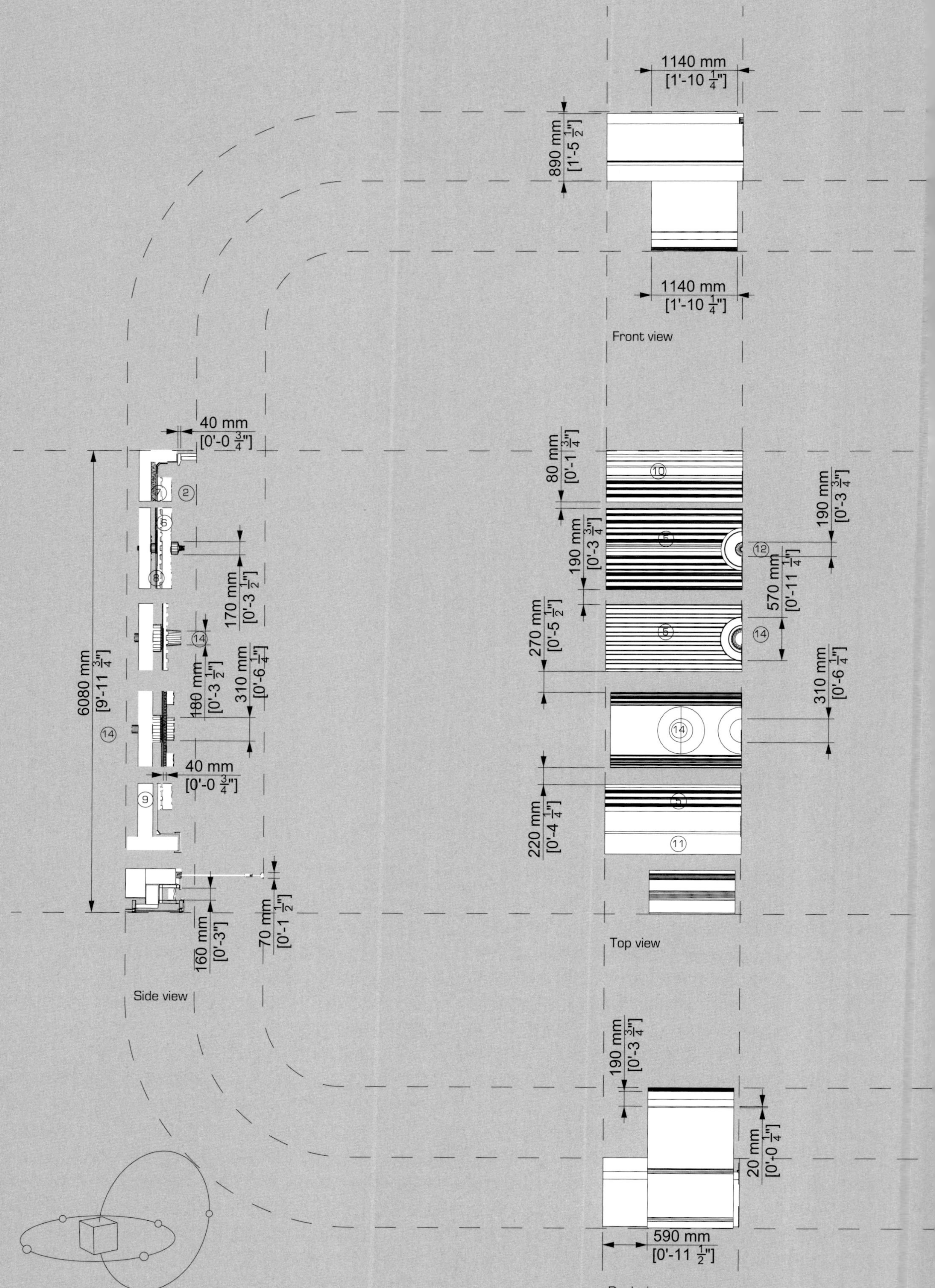

Third angle projection 1:30

Details

1 Light vegetation
2 Heavy vegetation
3 Soil/growing medium
4 Filter sheet
5 Drainage layer
6 Waterproofing layer
7 Thermal insulation

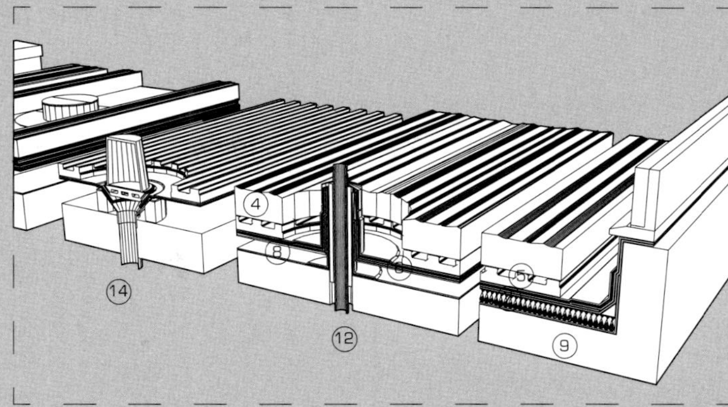

3D view of the assembly

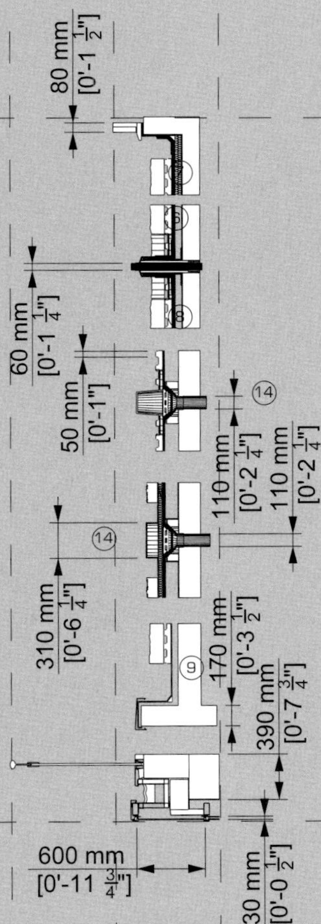

Side view

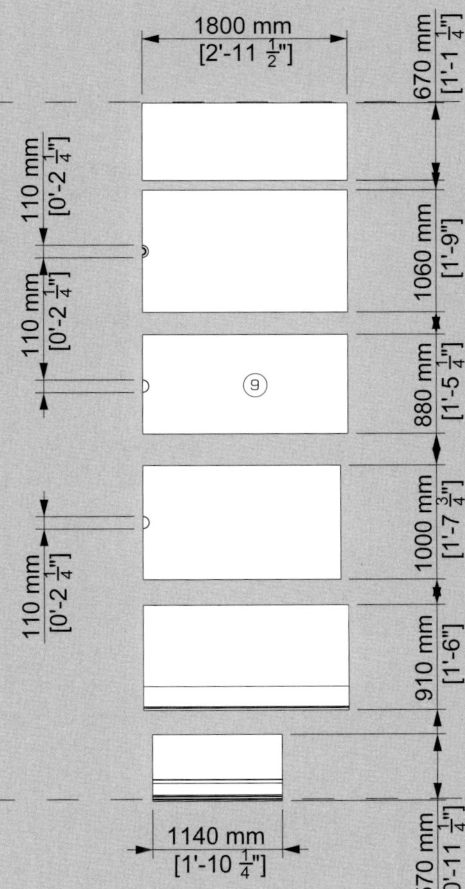

Bottom view

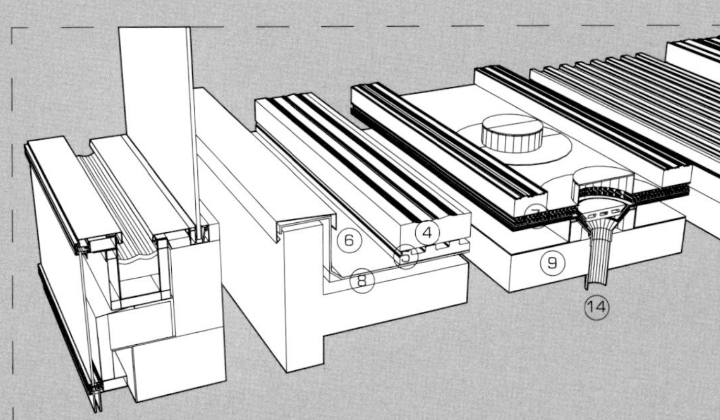

3D view of the assembly

8 Vapour barrier
9 Concrete deck
10 Smooth gravel
11 Coping
12 Pipe / duct
13 Supply pipe
14 Rainwater outlet
15 Wall cladding

Details

1 | Light vegetation
2 | Heavy vegetation
3 | Soil / growing medium
4 | Filter sheet
5 | Drainage layer
6 | Waterproofing layer
7 | Thermal insulation
8 | Vapour barrier
9 | Concrete deck
10 | Smooth gravel
11 | Coping
12 | Pipe / duct
13 | Supply pipe
14 | Rainwater outlet
15 | Wall cladding

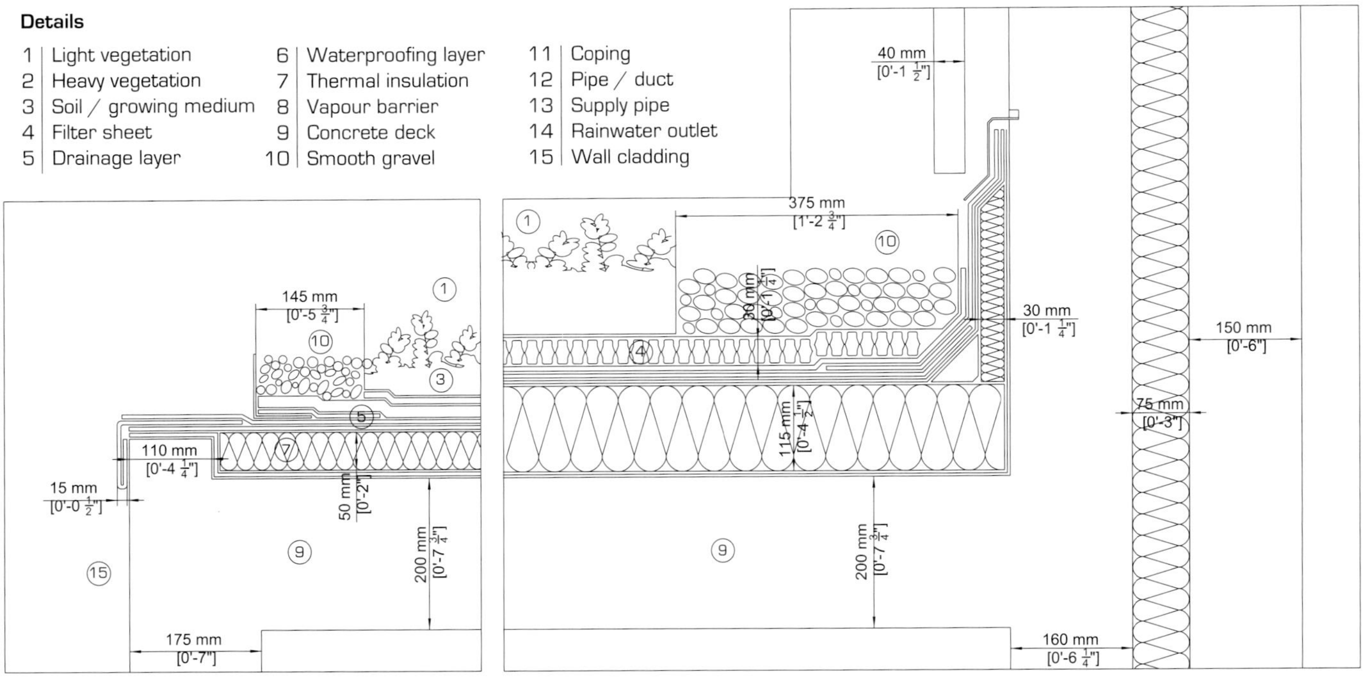

a | Vertical section. Light planting. Eaves and parapet junctions

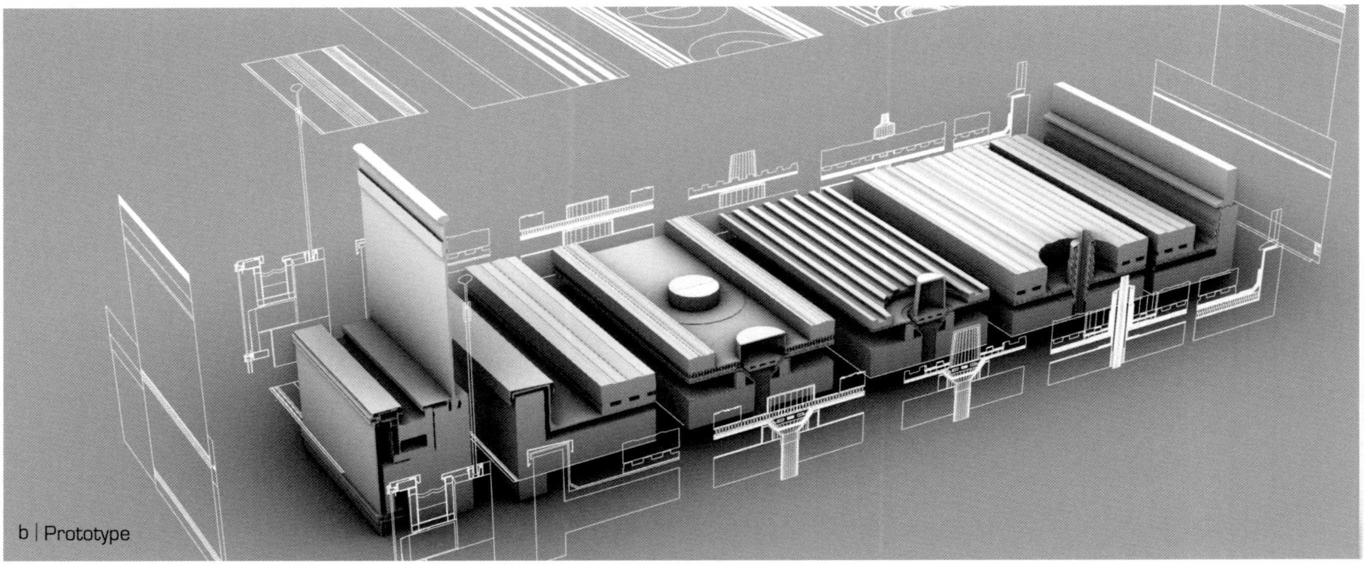

b | Prototype

System design

Concrete decks used for planted areas can be waterproofed with either a concealed membrane or an exposed membrane, as discussed in the previous sections. Planted roofs are of two types: light planted and heavy planted. Unlike other concrete roof types, planted roofs are not always insulated as they often form the roof of underground structures such as car parks, providing a planted roof at ground level.

Light planted roofs have resilient plants that require little or no irrigation, and that will grow in a thin layer of soil or organic growing medium (b). They are not usually used on a roof accessible to building users, but are seen from vantage points around the building. Light planted roofs have plants and flowers that require little maintenance and do not usually have an irrigation system to supply water at controlled times, relying on rainwater and modest amounts of watering during maintenance at specific times of the year. These lightweight planted roofs suit a lightweight deck, such as a thin concrete shell, although profiled metal decks are commonly used as substrates. Maintenance access is provided by the pebble strips at the roof edges or by individual paving slabs that avoid the need to walk across the planting.

Heavy planted roofs permit a wide variety of plants, shrubs and trees to grow on a concrete roof deck. Due to the size and intensity of the planting they require an automated irrigation system, usually from pipes set into the soil that provide a trickle water feed to the soil at specific times which may vary during the course of the year. Heavy planted roofs require regular maintenance, provided by paved paths or by areas of grass.

Both light planted and heavy planted roofs have drainage layers beneath the growing medium that hold water and release it back to the plants when required. This allows the soil depth to be much less than that which would be required for older landscaping methods, where the soil was

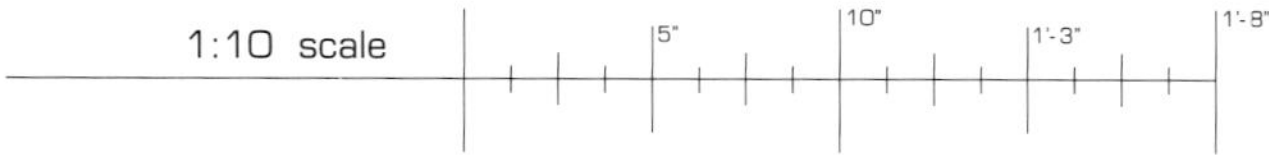

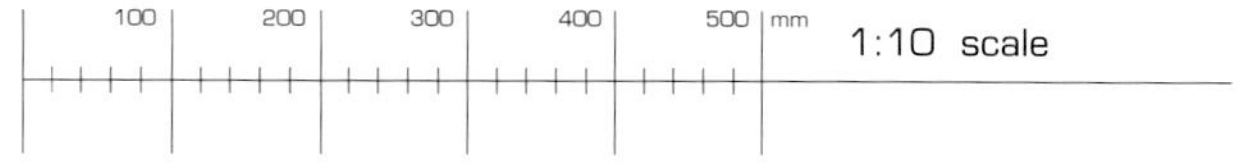

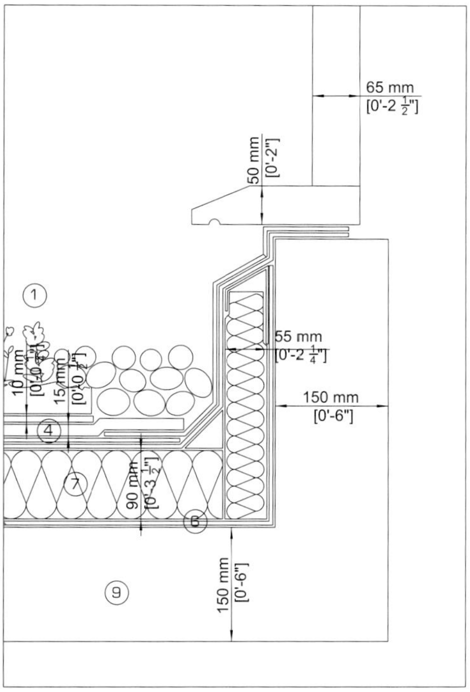

c | Vertical section. Light planting. Sill junction

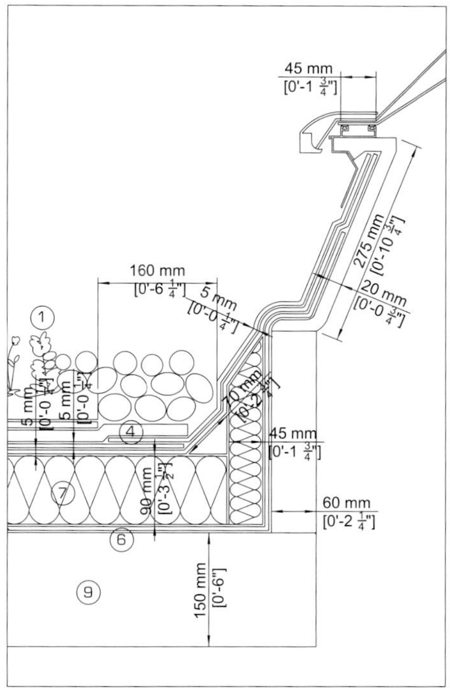

d | Vertical section. Light planting. Rooflight upstand

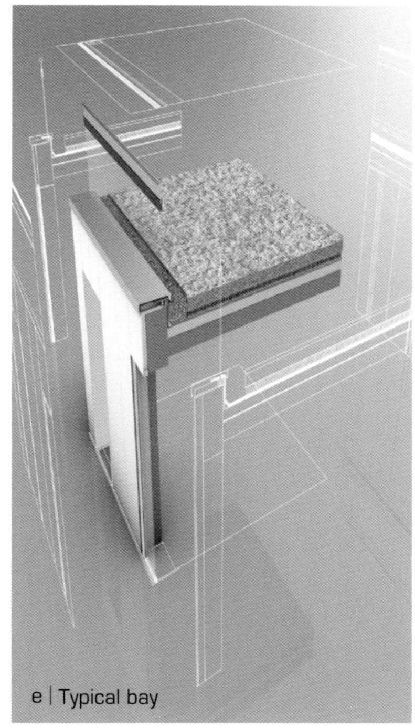

e | Typical bay

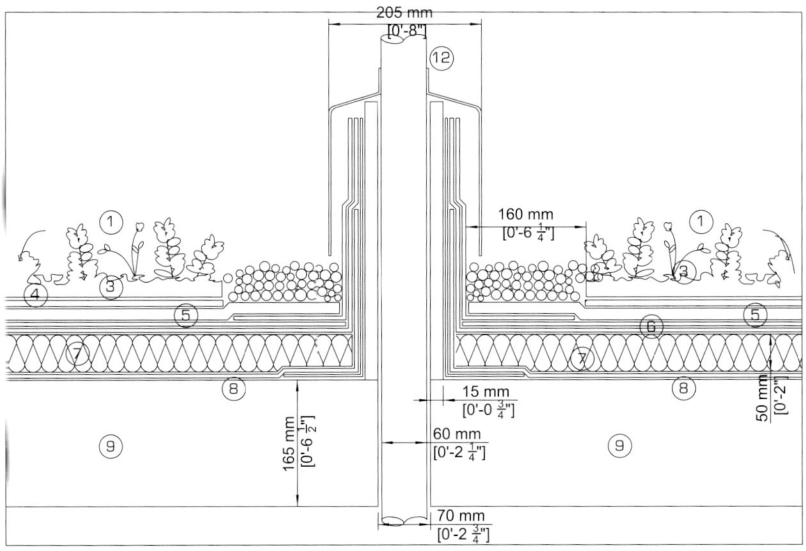

f | Vertical section. Light planting. Pipe penetration

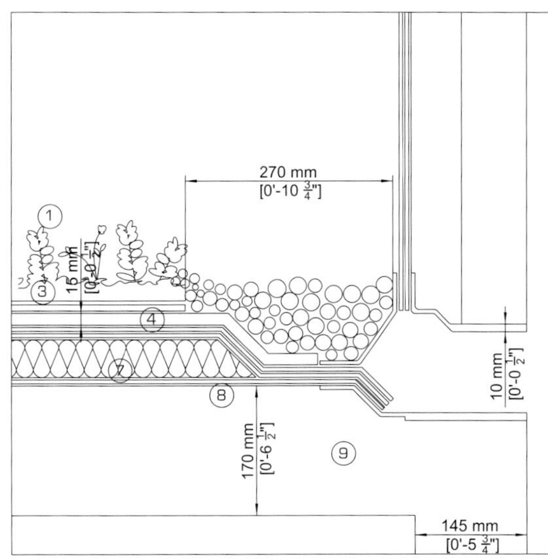

g | Vertical section. Light planting. 2-way outlet

expected to hold all the water. The reduced depth of soil allows planting to be considered for concrete roof structures that would require no significant strengthening to receive the added weight of soil. In terms of drainage it is estimated by manufacturers of proprietary systems that 50% to 90% of rainfall is retained in planted roofs, but this varies considerably with local climate conditions and rainwater drainage provision.

Planted roof components

Both light planted and heavy planted roofs have a similar build-up, comprising typically a top layer of planting, with a growing medium or soil beneath. A filter layer is set underneath, and below this, a drainage layer and moisture mat. Beneath this lowest layer is set thermal insulation if required. Although planted roofs provide a limited amount of thermal insulation from the soil, in practice this is reduced due to the varying amounts of water held within the soil. A root barrier is set beneath the insulation to protect the waterproof membrane, which forms the bottom layer, which is bonded to the concrete roof deck. The root barrier is sometimes bonded to the waterproofing layer, usually when the complete build-up is a single proprietary system.

The essential requirements for stabilising and maintaining plant growth in planted roofs are the provision of nutrients in the growing medium, water retention, soil aeration and drainage. The soils used are aimed to be relatively low in weight but are balanced to suit the nutrients, soil porosity, vapour permeability (from the drainage layer below) and pH values required by the plants chosen. Both the soil mix and the soil depth determine the amount of plant growth that can be expected on a planted roof.

To prevent the passage of organic matter and fine particles into the water drainage system, a filter sheet is set underneath the planting. This sheet is lapped up the sides of the planting, where it meets an upstand, to the level of the planting.

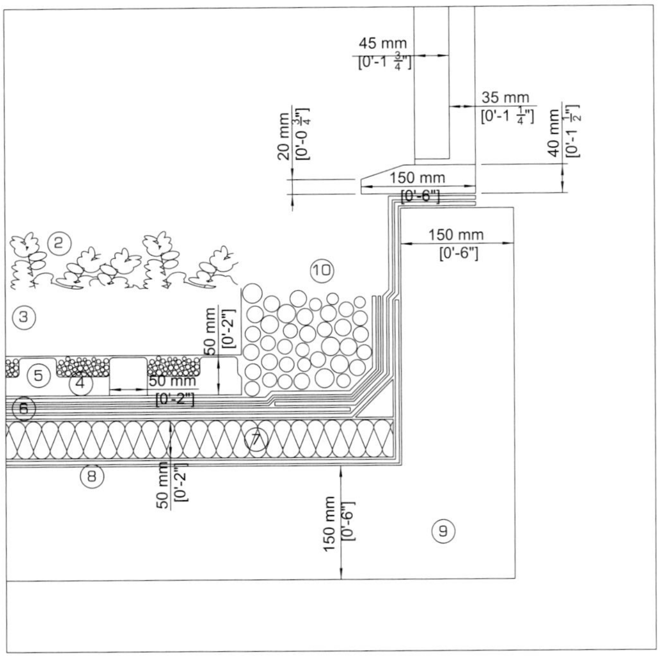

a | Vertical section. Heavy planting. Sill junction

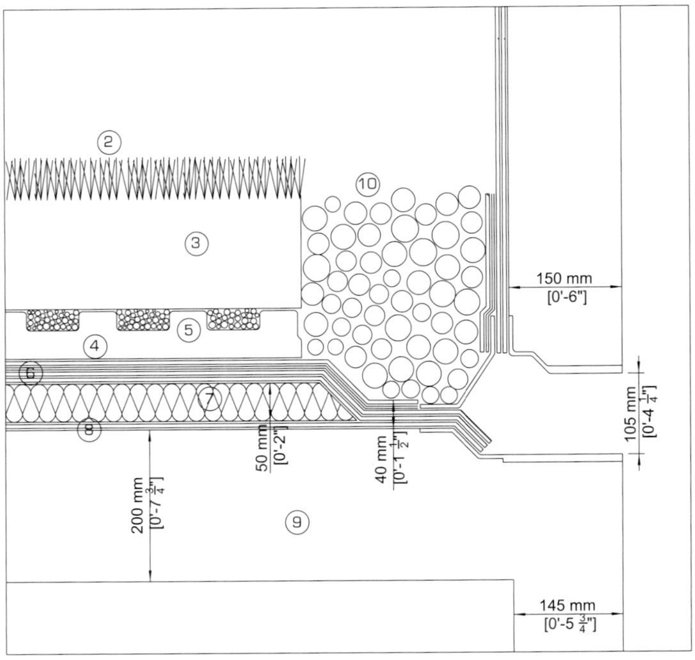

b | Vertical section. Heavy planting. 2-way outlet

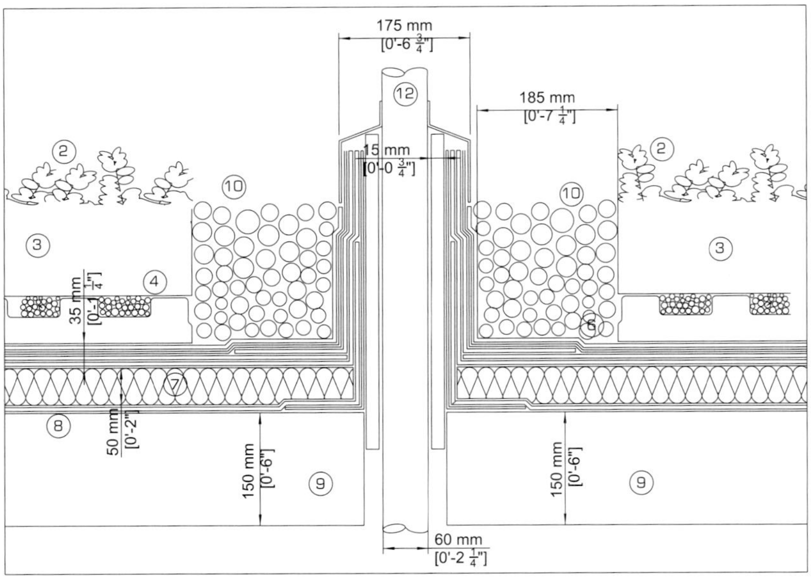

c | Vertical section. Heavy planting. Pipe penetration

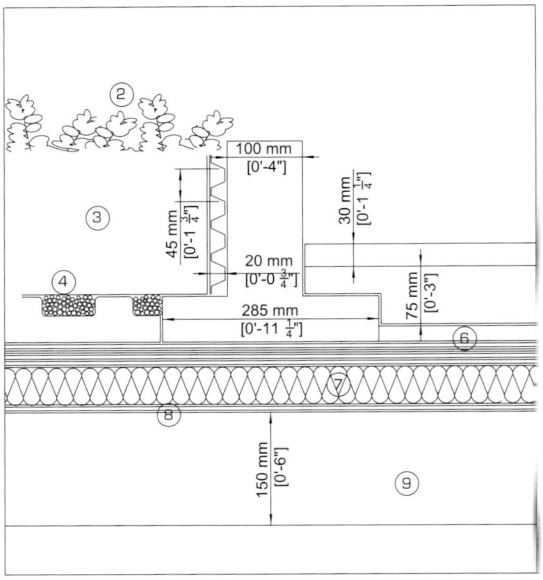

d | Vertical section. Heavy planting. Sill junction. Planter edge

The drainage layer beneath the filter sheet retains water that drains through the planting. Water is retained in profiled troughs in a typically polystyrene egg-crate shaped tray that releases water back to the planting. This method also performs satisfactorily on sloping concrete roofs. Excess water is drained away through gaps between the drainage trays. The egg-crate form allows aeration, permitting the soil to absorb the water stored here. In drier months, water diffuses up through the soil to the plant roots. A moisture mat is often set under this layer to catch water that runs off the drainage layer. The mat is made from a durable fibre that retains moisture and nutrients as well as serving as protection to the root barrier beneath (MCE_450, a). It is not used in inverted roof configurations. In inverted roofs, a root barrier is set immediately below the insulation to protect the waterproof membrane forming the lowest layer. This layer prevents planting roots from damaging the waterproofing. In warm roof construction, the waterproof membrane is set on top of the thermal insulation, positioning the insulation within the building envelope. A vapour barrier is set between the thermal insulation and the concrete deck. In this configuration a moisture mat is set between the waterproof membrane and the drainage layer above.

Soil depth

The soil depth in light planted roofs ranges from 50mm to around 150mm, weighing a minimum of around 70kg/m² of roof area. Water is stored in the growing medium and drainage layer, making it efficient in mild, temperate climates. Light planted roofs can be grown on both nominally flat roofs and on sloping roofs with a pitch of up to 25° to 30°. Heavy planted roofs have a deeper drainage layer to provide greater water storage. The soil depth, in excess of 150mm, requires an automatic irrigation system to provide a reliable water supply coverage of the complete roof.

In inverted roof configurations, the weight of soil and vegetation is made sufficient to avoid wind uplift and the possibility of the insulation floating

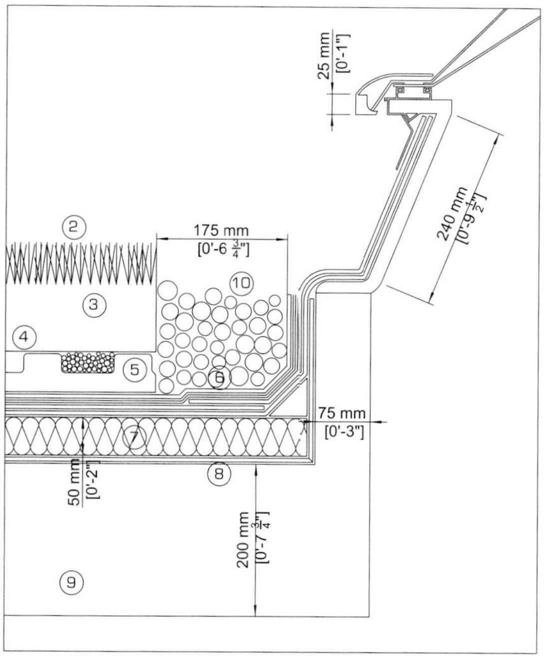

e | Vertical section. Heavy planting. Rooflight upstand

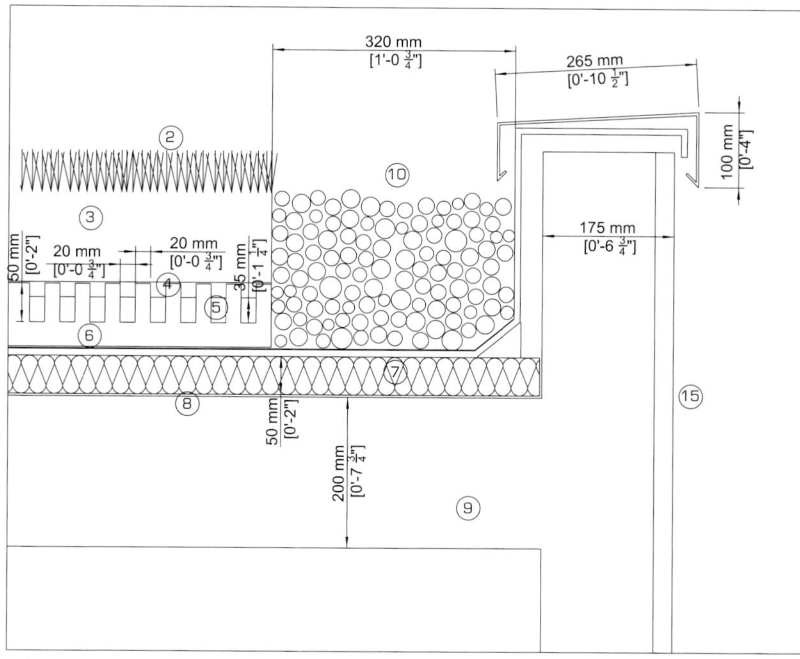

f | Vertical section. Heavy planting. Low parapet

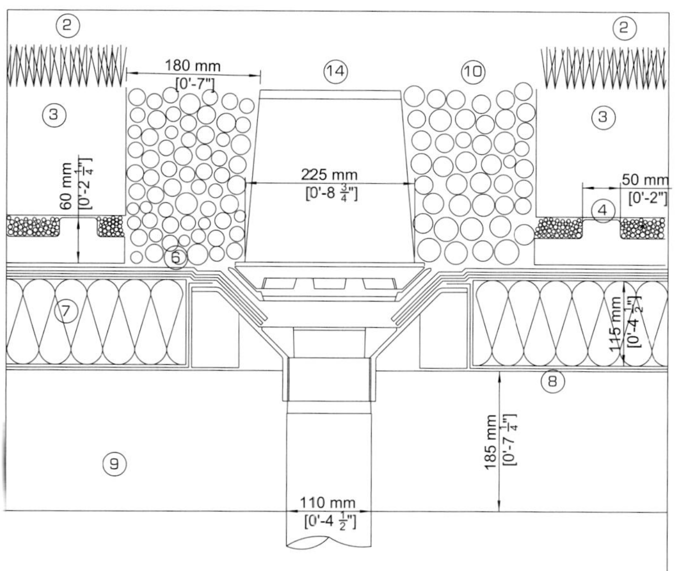

g | Vertical section. Heavy planting. Rainwater outlet

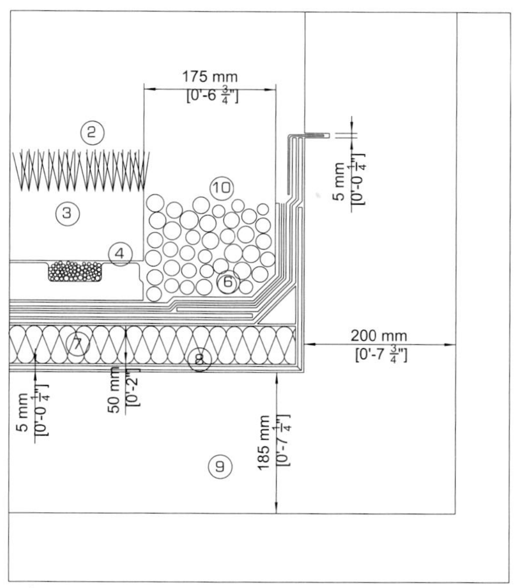

h | Vertical section. Heavy planting. Upstand

Details

1 | Light vegetation
2 | Heavy vegetation
3 | Soil/growing medium
4 | Filter sheet
5 | Drainage layer
6 | Waterproofing layer
7 | Thermal insulation
8 | Vapour barrier
9 | Concrete deck
10 | Smooth gravel
11 | Coping
12 | Pipe / duct
13 | Supply pipe
14 | Rainwater outlet
15 | Wall cladding

on the water during rainfall. Although the ponding is often considered by proprietary waterproofing manufacturers not to affect the waterproofing layer, it can cause lightweight planted roofs to 'float' during rainstorms if this layer is not properly secured.

Overflows

Irrigated heavy planted roofs are usually provided with overflow outlets so that, in the event of the rainwater outlets being blocked, a high rainfall or failure of the irrigation control equipment does not cause the roof to flood with water. Overflows are set typically at a height of between 50mm and 150mm above the planting level to avoid flood damage to both the planting and to the interior of the building. When the roof is laid to falls, some overflows are set at the level of the highest finished roof level in order to avoid planting being damaged should flooding result in landscaping being temporarily submerged in water.

Roof junctions

At upstands and eaves the same principles apply to planted roofs, as discussed in the previous sections on concealed membranes and exposed membranes. The waterproofing extends a minimum of 150mm above the level of the planting, providing a continuity from the roof membrane to the flashing at the top of the upstand or to the adjacent wall construction. Upstands for parapets and door sills, high walls and rooflights are formed by extending the waterproof filter sheet and root barrier up to a minimum of 150mm above the level of the soil or growing medium. The visible membranes and sheets are concealed with thermal insulation, and typically either paving turned on edge (the same paving used for adjacent access paving) or a metal sheet to match that of the parapet coping where a metal coping is used.

Eaves can be formed by metal edge trims, usually from a minimum 3mm thick folded aluminium sheet or stainless steel angle. The filter

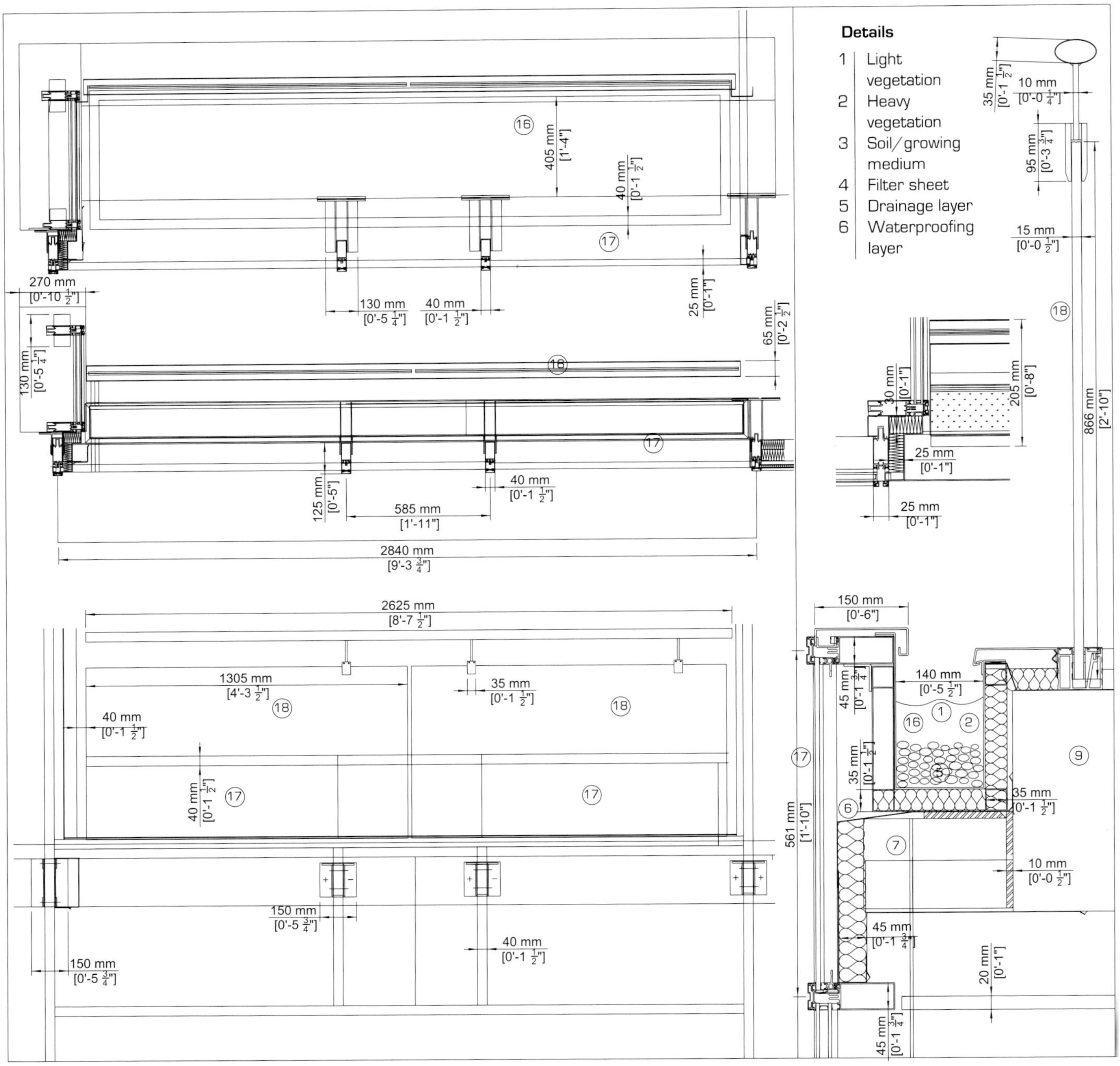

a | Elevation, horizontal and vertical sections. Small balcony planter.

sheet is folded up the inside face of the angle to avoid organic matter and fine particles from being washed down into the drainage layer. The waterproof membrane is bonded to the base of the metal angle where the edge is terminated by a paving slab. Smooth pebbles can be used, provided there is no risk of them being pushed over the edge, particularly during maintenance work. Many planted roofs have low parapet or eaves edges so that the roof can be experienced visually from the outside of the building. A fall arrest system is provided, such as harnesses worn by maintenance personnel, which is attached by a safety line to an anchor point or a cable. Balustrades are provided for building users in conditions with low parapets.

Vegetation barriers are provided at roof perimeters, upstands, duct penetrations and rainwater outlet points to avoid damage to the adjacent construction that would be caused by plants. Pebbled strips with a minimum width of 300mm are used, with river-washed pebbles of 16mm to 32mm diameter (MCE_445, f).

Rainwater outlets

The filter layer is dressed up around the vertical edges of the rainwater outlets. An access cover is provided for maintenance purposes. Water drains at the level of the waterproof membrane, which is dressed into the base of the outlet (MCE_445, g).

Balcony planters

Planting troughs with automatic irrigation and a drainage system can be incorporated at roof level into curtain wall facades. In some cases, a small planter can be integrated into a balcony. The planter is sealed, but should any water leak either from the planter (as a result of damage) or if water penetrates the seals around its top edges, then the water is drained by an internal waterproof tray into the transom below, where it drains to the outside. The curtain wall system, which is drained and ventilated internally allows any water that penetrated these planter seals to drain through its pressure equalisation chambers (a). A glazed balustrade is shown

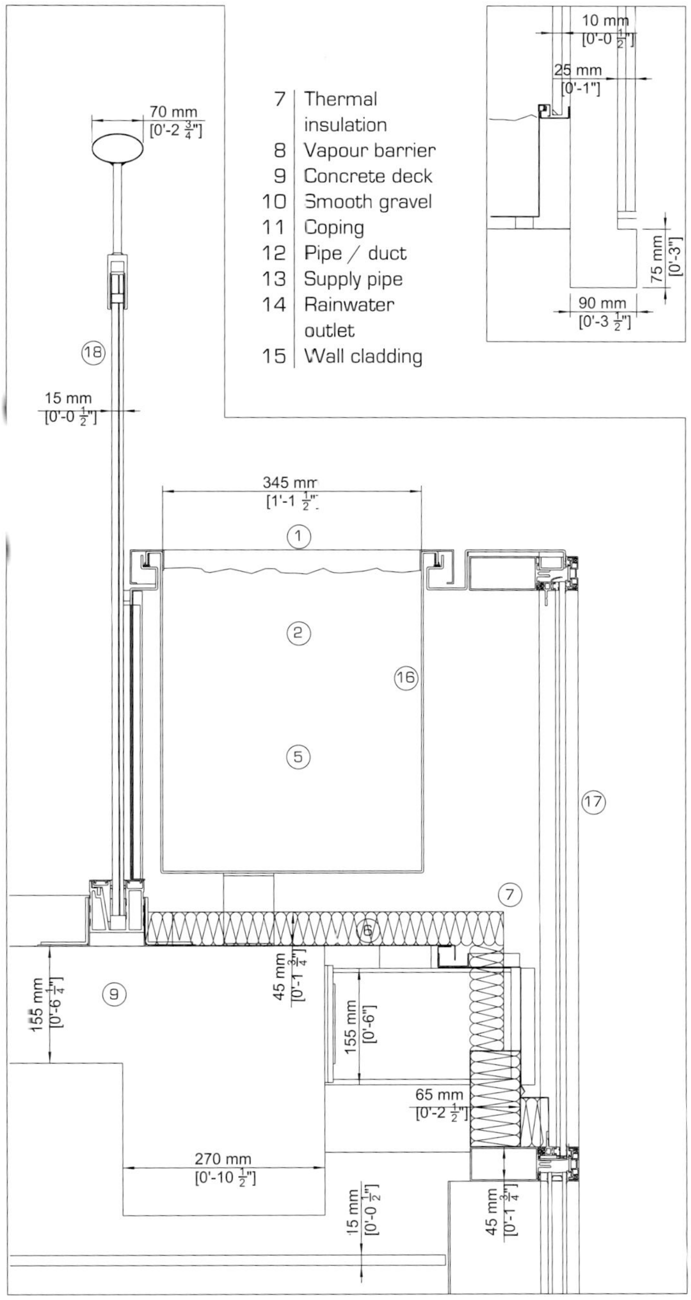

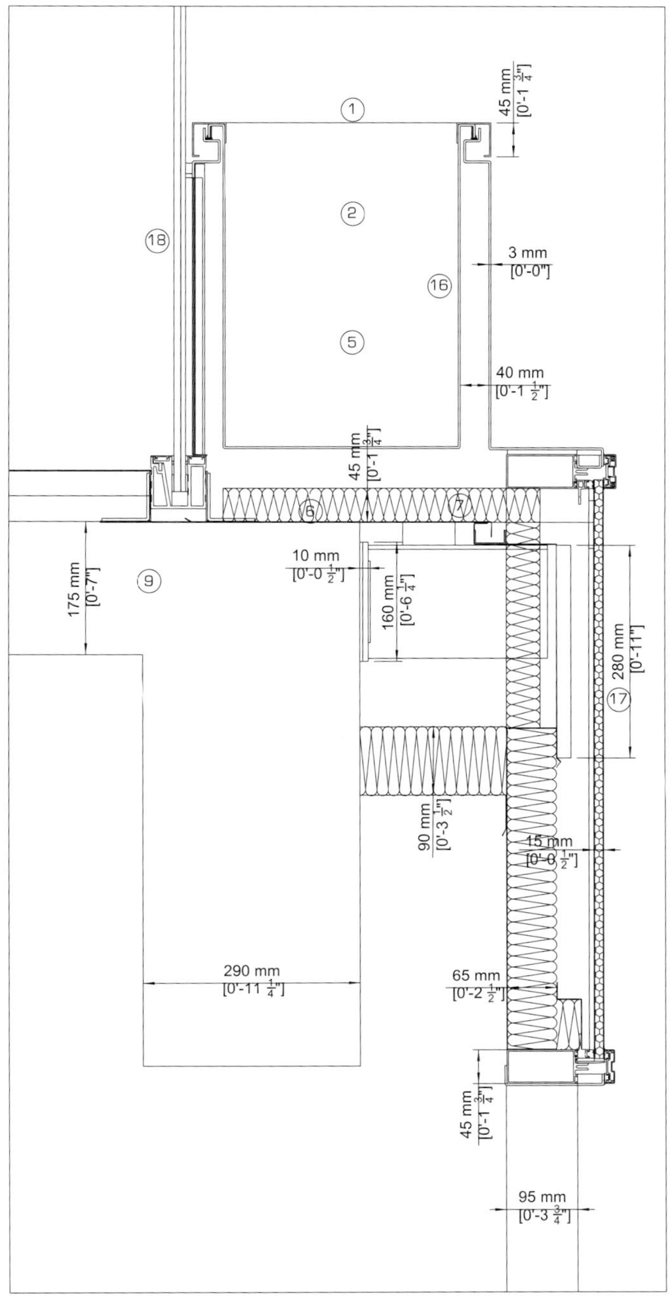

b | Vertical section. Large balcony planters

behind the planters, indicating that the planter would be maintained from the facade cleaning system, typically cleaning cradle. Alternatively, larger versions of this planter can be used that are more suitable for much larger plants. The planters are drained internally in the same way. In all planters water is supplied at one end, typically by a water supply pipe of small diameter set into the roof finishes. The drainage pipe is typically of 50mm diameter for such installations, and is set either within the facade panes or directly in front of the facade. Planter boxes are made from glass reinforced polyester (GRP) which is moulded to form a single, sealed shell. This material is very resilient and can be moulded by hand to suit individual project requirements. Planter boxes made from thermoplastics are much more expensive to manufacture, requiring large numbers to be fabricated at the same size to make them economic. The boxes are set into a metal frame provided within the curtain walling system and are sealed with silicone around their edges. In this example, a metal strip is set on top of the planter to conceal the GRP from view.

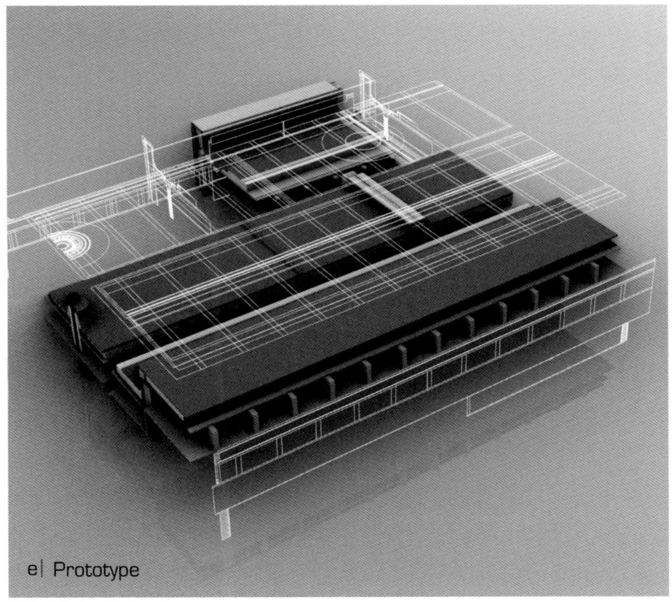

e | Prototype

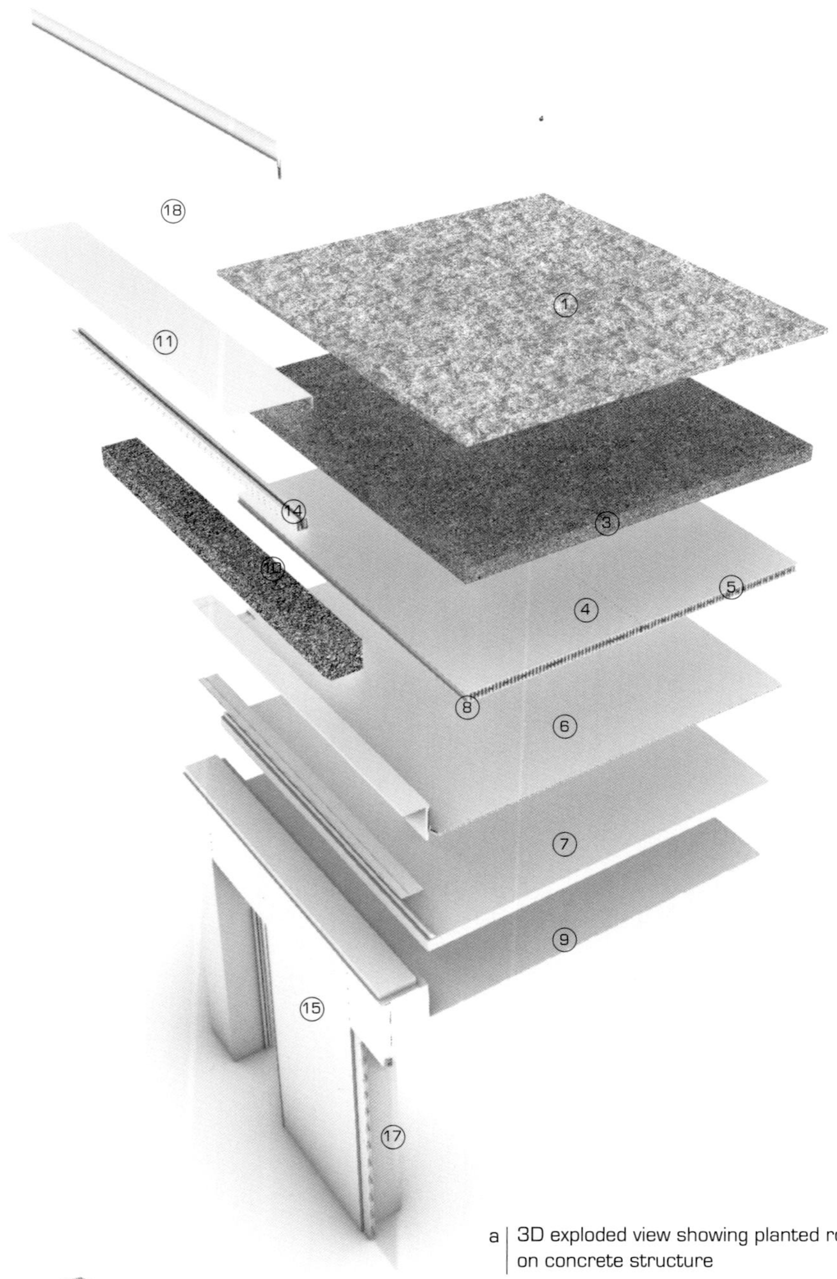

a | 3D exploded view showing planted roof on concrete structure

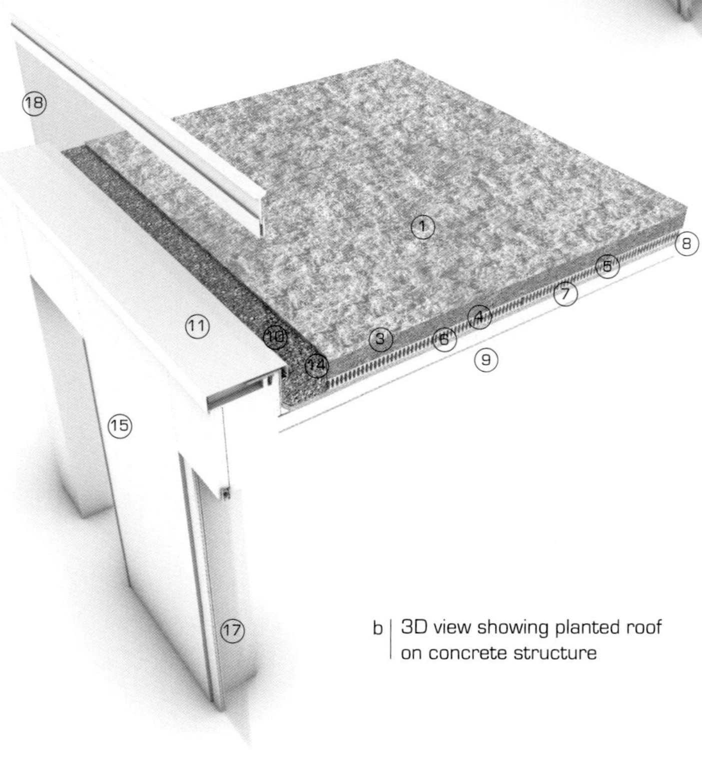

b | 3D view showing planted roof on concrete structure

Details

1 | Light vegetation
2 | Heavy vegetation
3 | Soil / growing medium
4 | Filter sheet
5 | Drainage layer
6 | Waterproofing layer
7 | Thermal insulation
8 | Vapour barrier
9 | Concrete deck

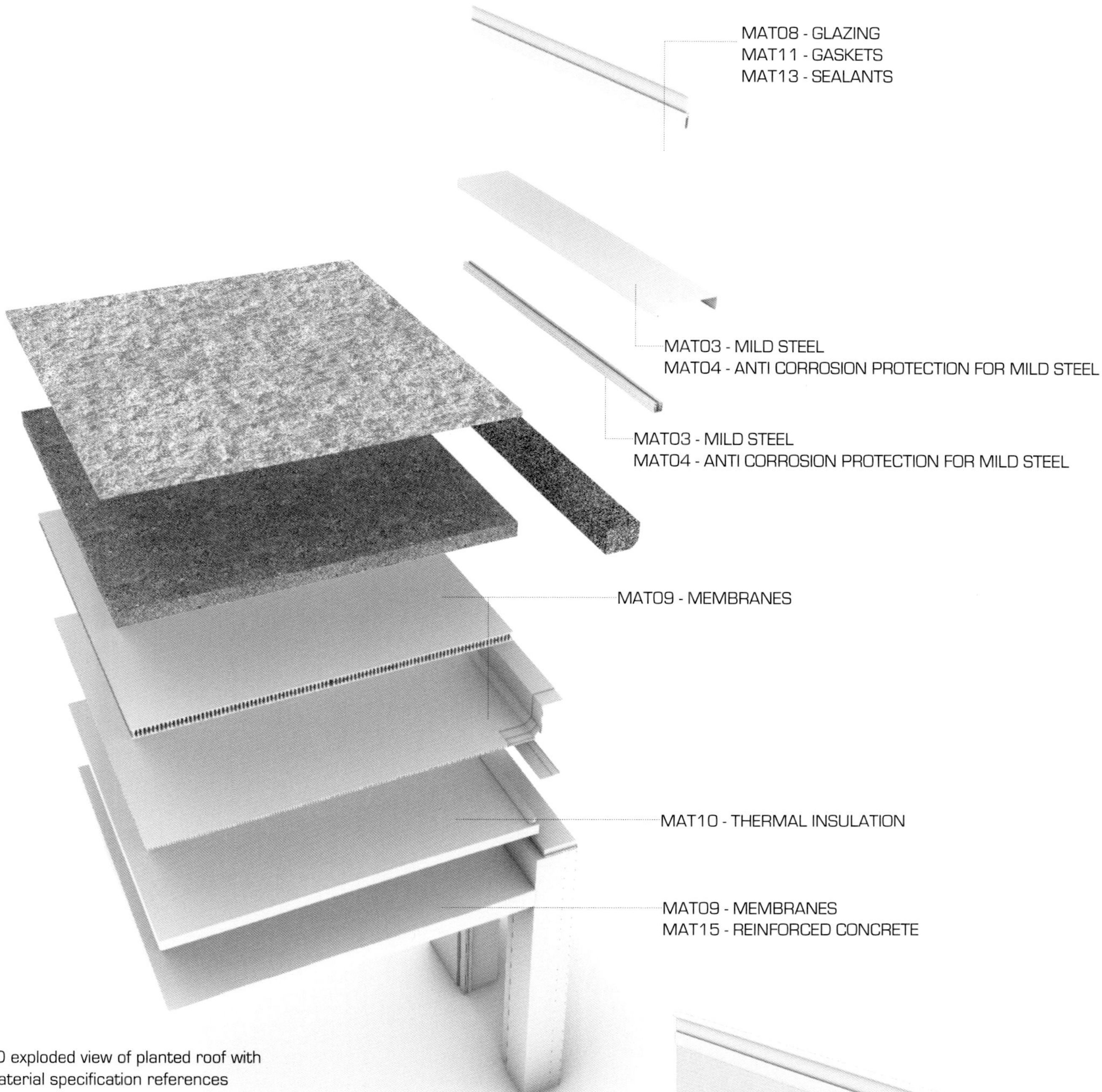

c | 3D exploded view of planted roof with material specification references

10 | Smooth gravel
11 | Coping
12 | Pipe / duct
13 | Supply pipe
14 | Rainwater outlet
15 | Wall cladding
16 | GRP planter
17 | Glazed external wall
18 | Glass balustrade

d | 3D view showing planted roof on concrete structure

Example project of timber roofs
Harbour Housing, Hong Kong
Facade Engineer: Newtecnic
Architect: Newtecnic

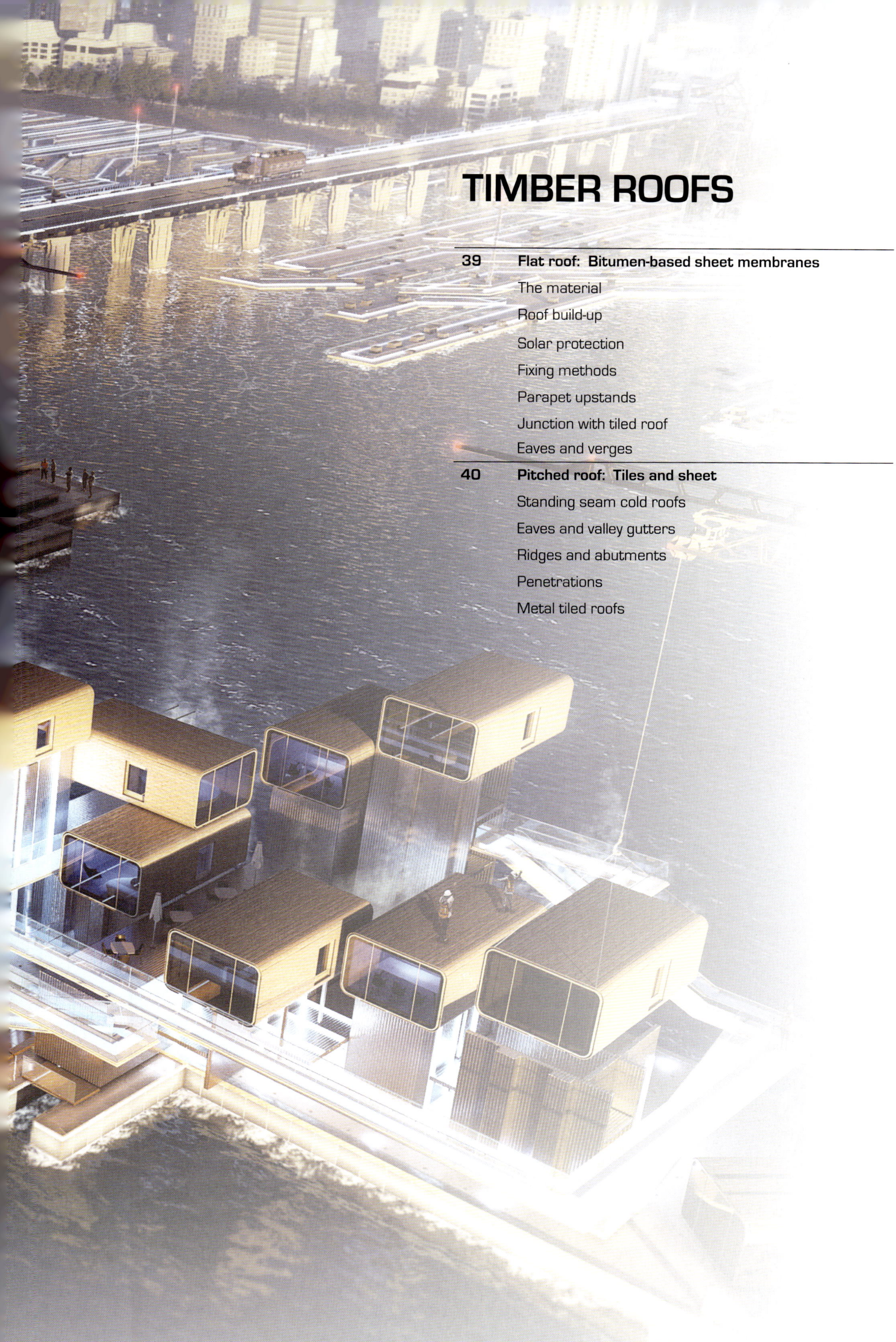

TIMBER ROOFS

39 Flat roof: Bitumen-based sheet membranes

The material

Roof build-up

Solar protection

Fixing methods

Parapet upstands

Junction with tiled roof

Eaves and verges

40 Pitched roof: Tiles and sheet

Standing seam cold roofs

Eaves and valley gutters

Ridges and abutments

Penetrations

Metal tiled roofs

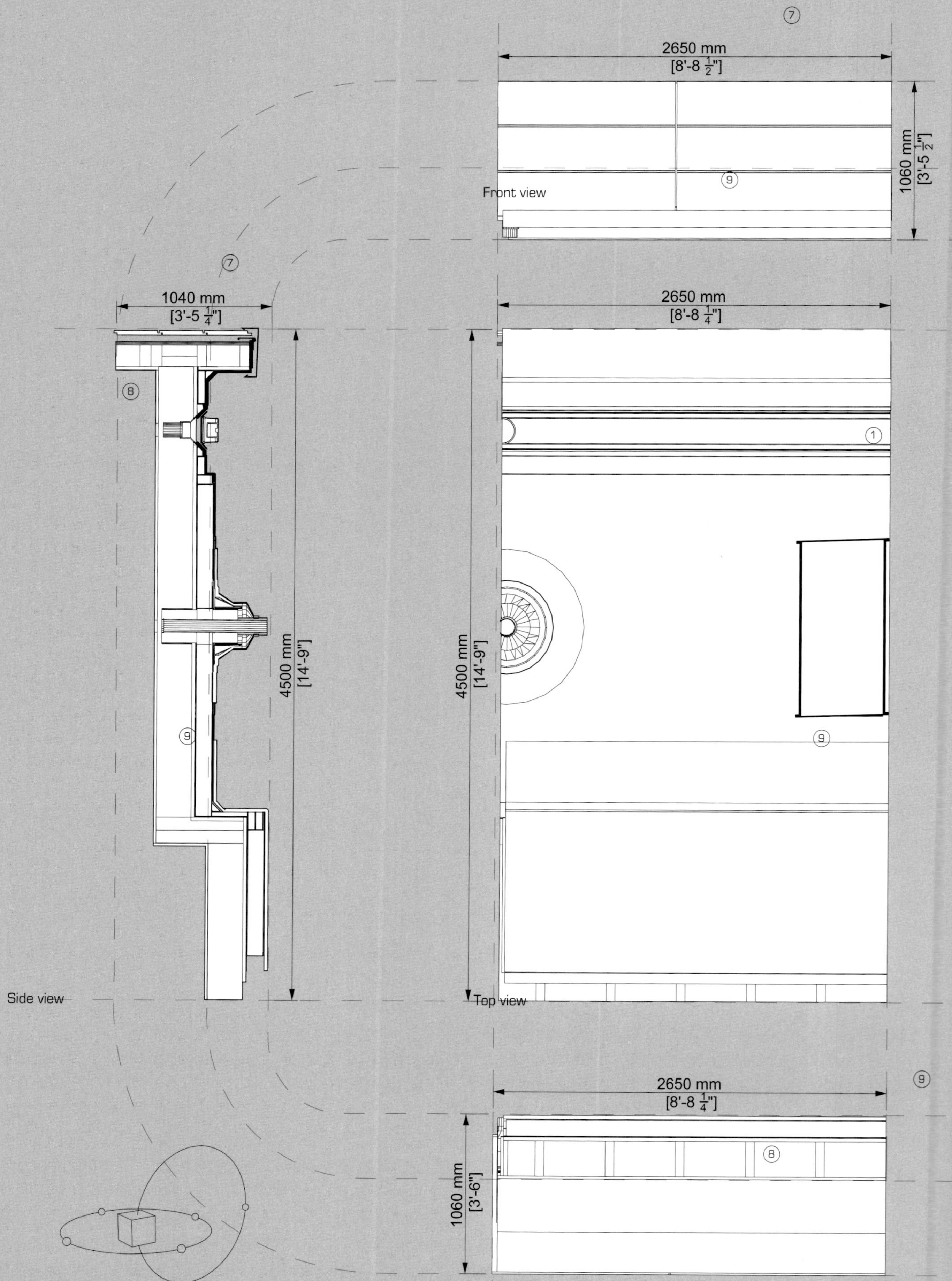

Third angle projection 1:30

3D view of the assembly

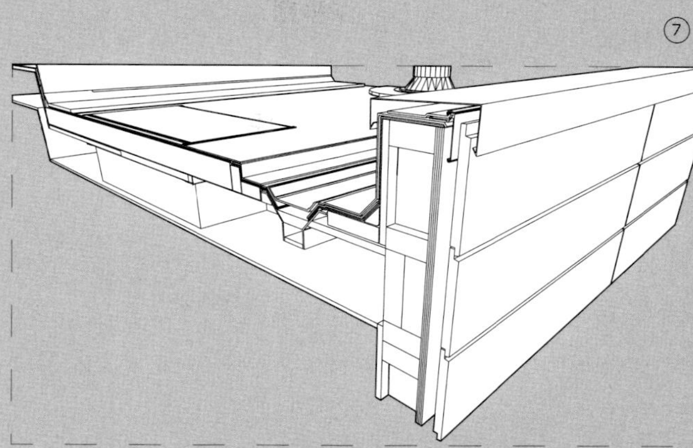

Details

1 | Bitumen-based sheet
2 | Plywood sheet
3 | Rigid thermal insulation
4 | Vapour barrier
5 | Softwood joists
6 | Dry lining/drywall internal finish

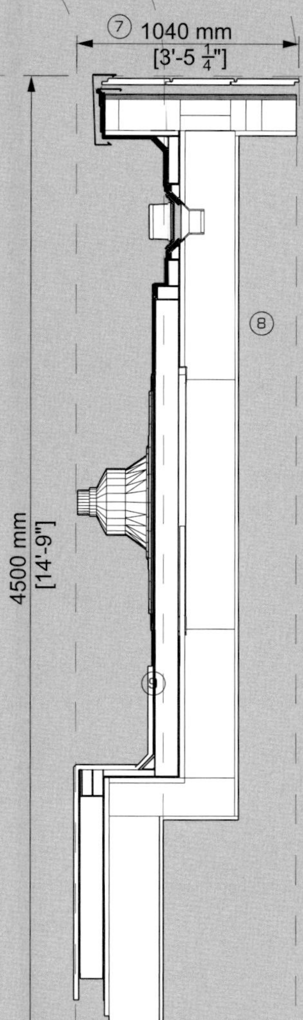

Side view

Bottom view

7 | Metal flashing
8 | Timber upstand
9 | External wall
10 | Angle fillet
11 | Proprietary skirt flashing
12 | Rainwater outlet
13 | Paving bonded to bitumen-based sheet

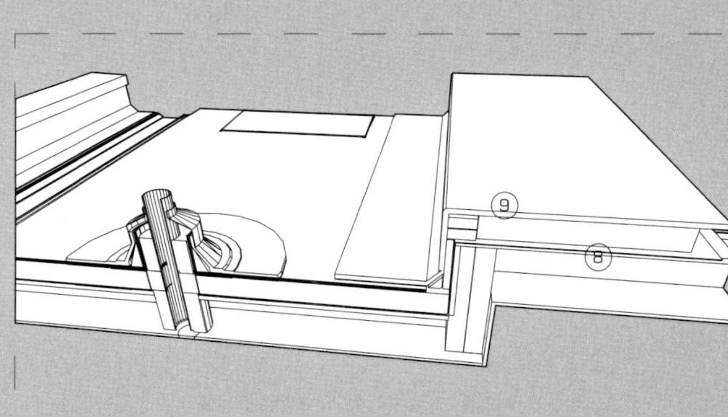

3D view of the assembly

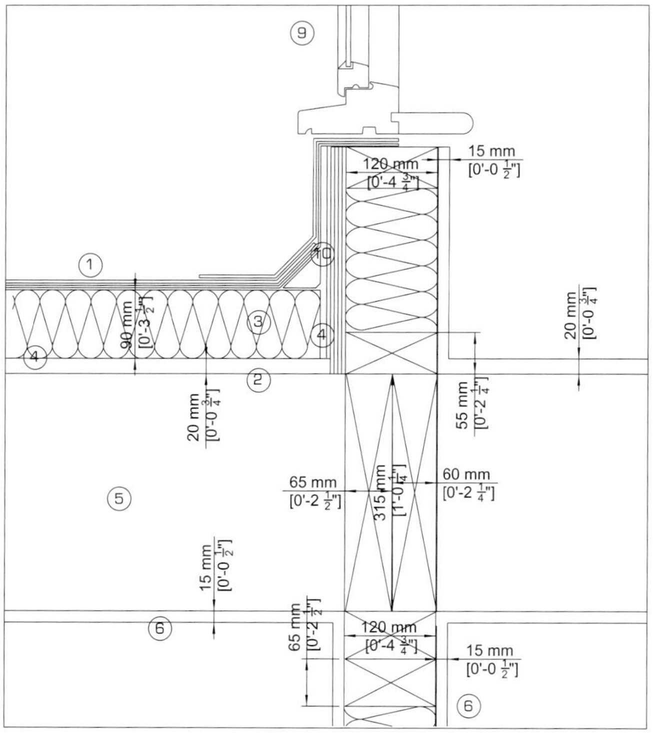

a | Vertical section. Low parapet wall

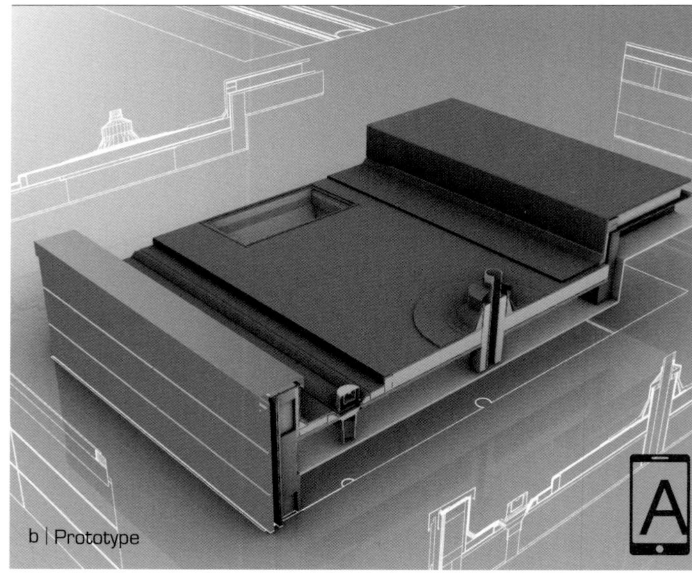

b | Prototype

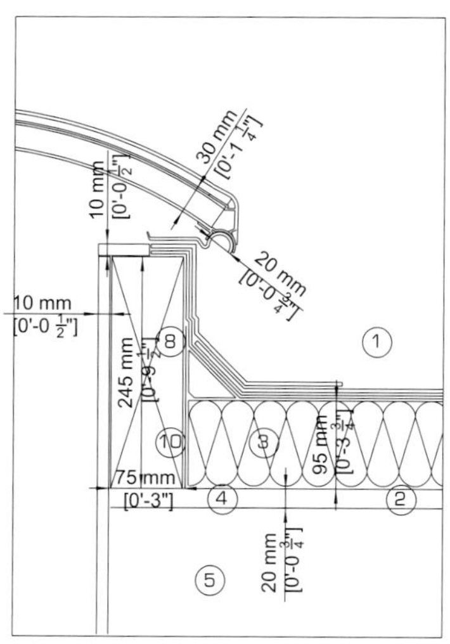

c | Vertical section. Rooflight

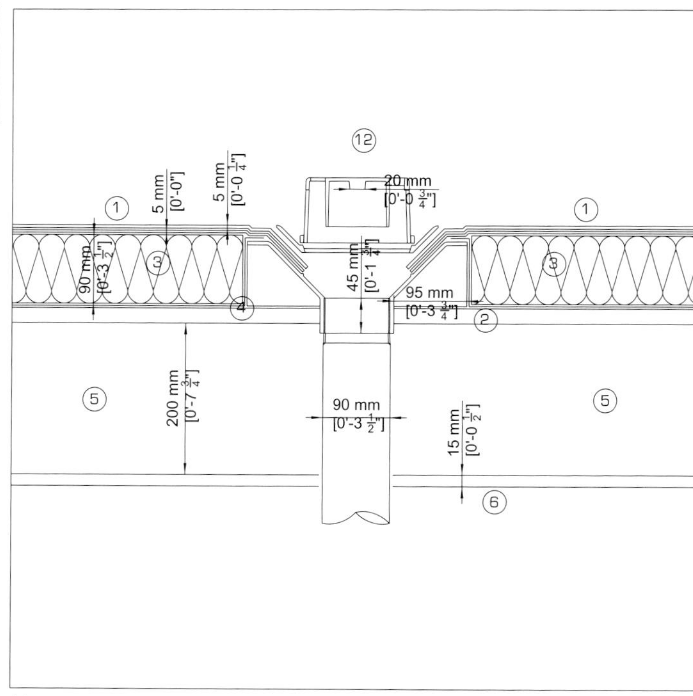

d | Vertical section. Rainwater outlet

This section discusses flat timber roofs that use a bitumen sheet-based membrane as a waterproofing layer in a warm roof, which is a common combination. Although other membrane materials are used on flat timber roofs as both warm and cold roofs, typically formed in elastomeric and thermoplastic membranes, their application is discussed as Exposed Membranes in the Concrete chapter. The principles of detailing in that section can be similarly applied to timber roofs. Bitumen-based sheet can also be used in 'inverted roof', or concealed membrane configuration as described in the Concrete chapter where the detailing is similar, but bitumen-based sheet is generally less robust than the membranes described in that section. Membranes applied to concrete decks are usually laid in hot liquid form and are reinforced to suit the specific conditions of folds and joints occurring within the structural deck. In this section the material is considered as an exposed and visible material on a relatively lightweight deck (b).

Bitumen sheet membranes are economic, and are often used with timber roof decks, which together provide an economic roof design for relatively small-scale applications, or designs with a complex geometry of low-pitched roofs, as is often used in housing and in school buildings. Bitumen-based membranes have developed over the past 25 years to compete with the newer elastomeric and thermoplastic materials by increasing their flexibility and making them thinner, thus requiring less material, which helps to reduce their overall cost. Bitumen-based membranes can also be used with concrete and metal decks, and the principles here can be applied in a similar way to those roof deck types

The material

Bitumen-based sheet is manufactured in roll form in widths of around 1000mm, is black in colour, and is typically mixed with SBS (styrene-butadiene-styrene) polymers or with TPO (thermoplastic polyolefin) polymers. The addition of these polymers raises the melting point which ensures

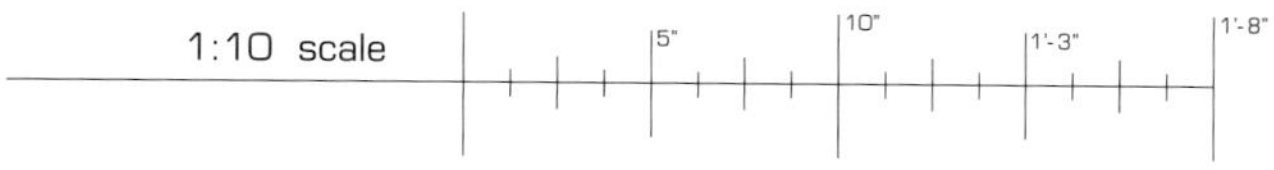

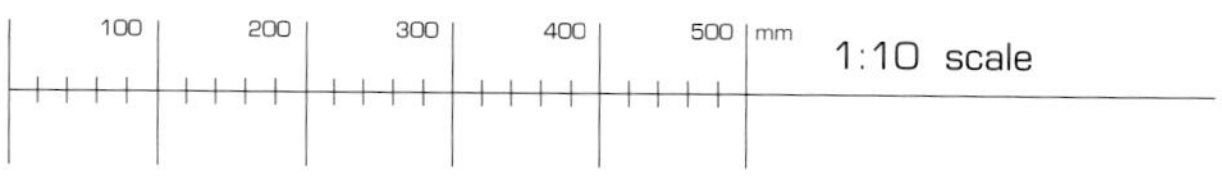

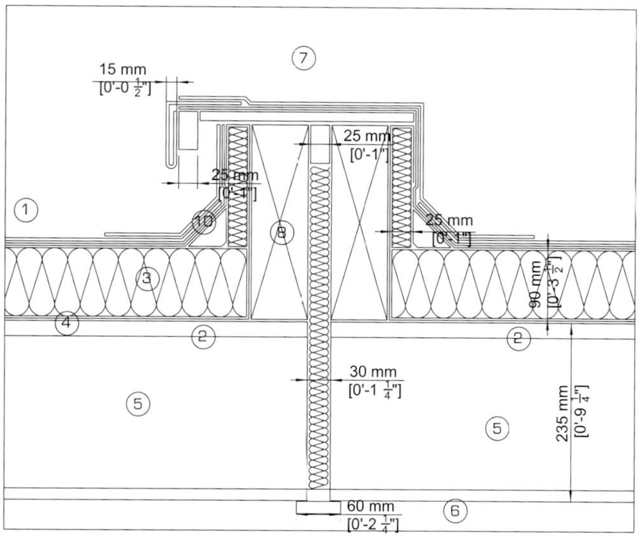

e | Vertical section. Expansion joint

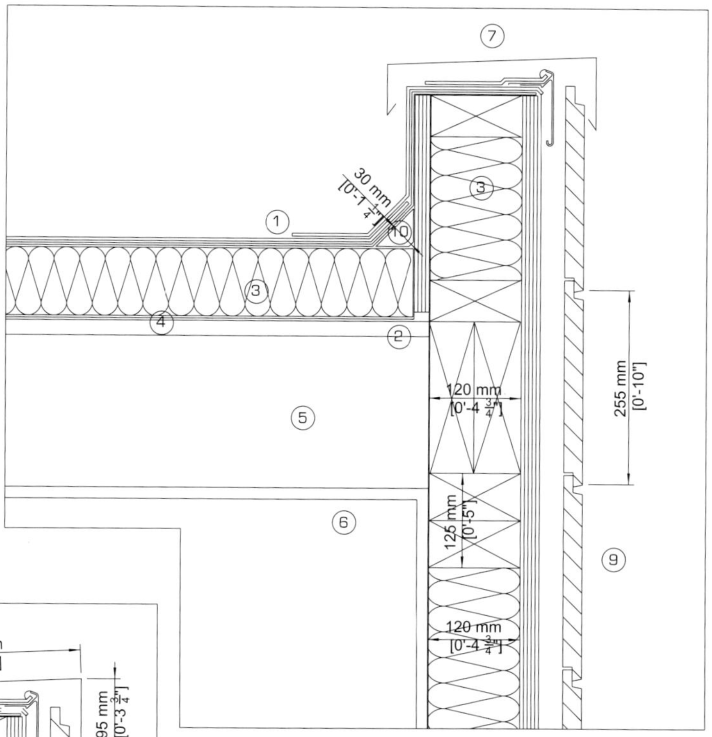

f | Vertical section. Low parapet wall

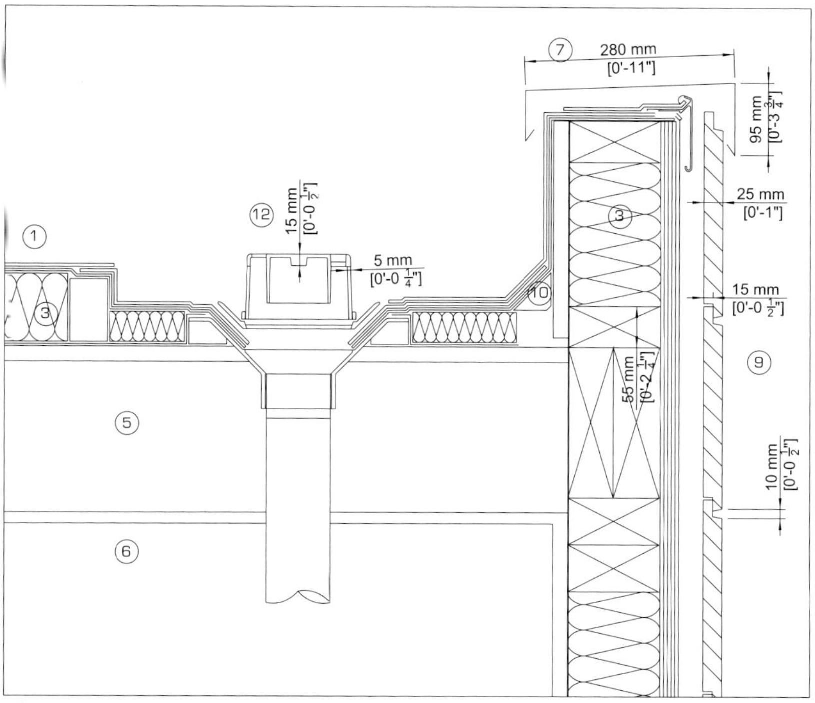

g | Vertical section. Low parapet wall with drain outlet

Details

1 | Bitumen-based sheet
2 | Plywood sheet
3 | Rigid thermal insulation
4 | Vapour barrier
5 | Softwood joists
6 | Dry lining/drywall internal finish
7 | Metal flashing
8 | Timber upstand
9 | External wall
10 | Angle fillet
11 | Proprietary skirt flashing
12 | Rainwater outlet
13 | Paving bonded to bitumen-based sheet

stability in hot weather as well as increasing the flexibility of the material at low temperatures (usually in winter in temperate climates) and enhancing the fire resistance of the material. Bitumen-based sheet often has a glass fibre-reinforced upper face to provide greater dimensional stability and resistance to accidental damage, as well as a polyester-reinforced core to increase tensile strength. These sheet materials are typically around 4mm thick, depending on the proprietary system used. Even with these additives, bitumen-based sheet is slowly oxidised by heat, making the material gradually more brittle, which eventually results in cracks. The polymer additives reduce this effect, particularly the TPO additives which help to increase the life of the material, which can now be up to around 25 years. TPO-modified sheet can be exposed to the effects of the sun, requiring no additional solar protection, since the material provides better UV resistance than older-type bitumen-based membranes. SBS-modified sheet is usually covered with stone chippings or solar reflective paint to protect them from the effects of the sun.

Roof build-up

Where plywood is used to form the structural deck of a timber roof, a thin layer of bitumen is laid onto the deck, applied typically in thick liquid form to seal the joints between the boards, with the bitumen setting to form a smooth substrate. A vapour barrier is set onto the prepared timber deck, the barrier being often bitumen-based as part of a proprietary system. Rigid closed cell insulation such as polyurethane is bedded in hot bitumen onto the vapour barrier to hold the insulation securely in place. A loose-laid perforated isolating layer is set onto the thermal insulation which is used to allow the membrane and thermal insulation to release gases into the isolating layer which are formed as a result of bonding the bitumen to the insulation. The bitumen-based membrane is then bonded to the thermal insulation through the holes in the isolating layer (g).

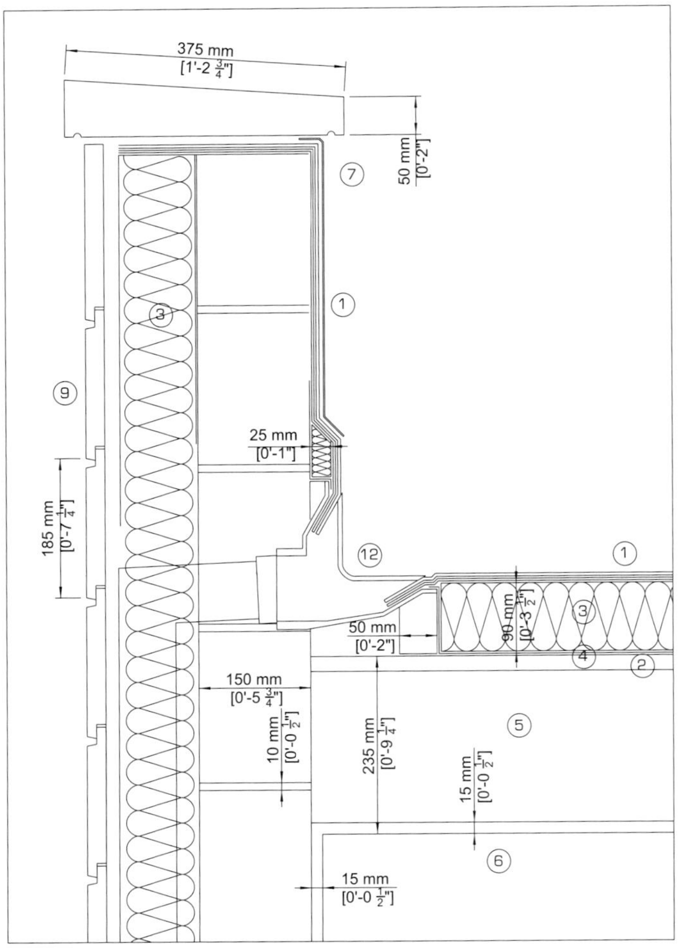

a | Vertical section. 2-way drain outlet

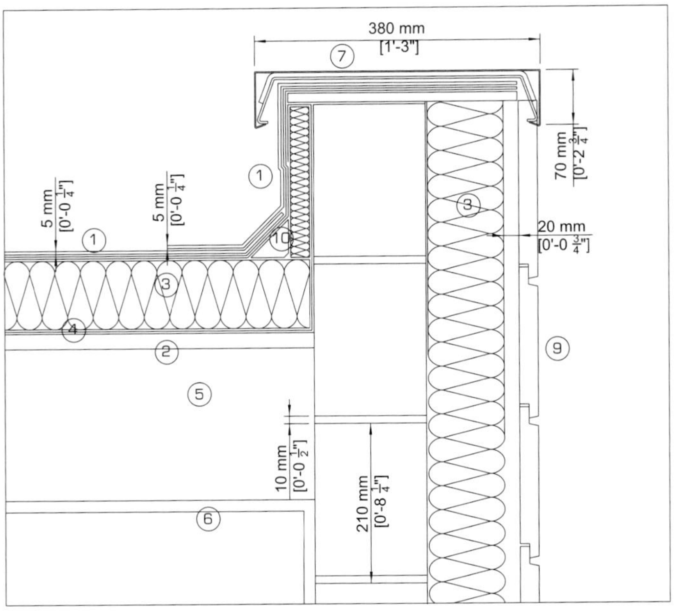

b | Vertical section. Low parapet wall

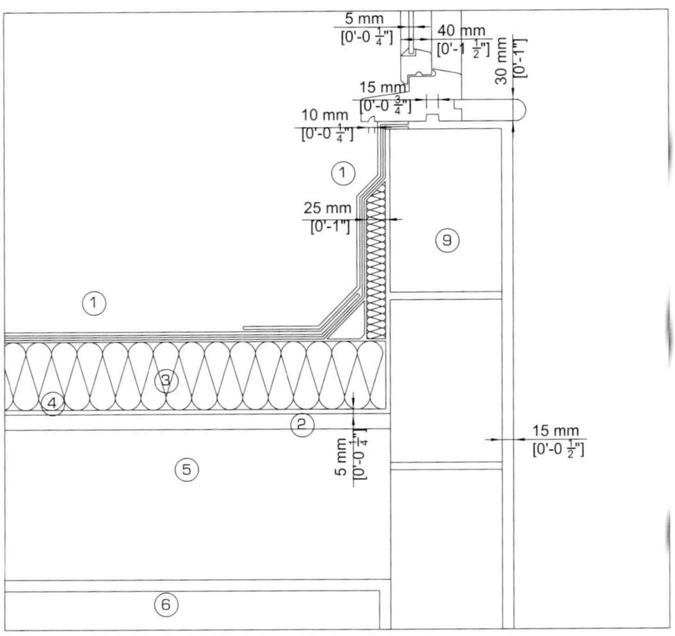

c | Vertical section. Door threshold

Details

1	Bitumen-based sheet	8	Timber upstand
2	Plywood sheet	9	External wall
3	Rigid thermal insulation	10	Angle fillet
4	Vapour barrier	11	Proprietary skirt flashing
5	Softwood joists	12	Rainwater outlet
6	Dry lining/drywall internal finish	13	Paving bonded to bitumen-based sheet
7	Metal flashing		

Solar protection

Resistance to UV radiation is provided by either a coating of fine stone chippings or by aluminium solar reflective paint applied to the visible surface of the bitumen-based sheet. As a result of providing this additional UV protection, these coatings have the additional benefit of reflecting heat, which has the effect of reducing the surface temperature of the roof below that which would otherwise be the case. Solar reflective paint gives the roof a metal appearance, which provides a visual alternative to the characteristic black colour of bitumen-based sheet. Membranes can also be provided with a solar protection layer during manufacture as part of a proprietary bitumen-based membrane system.

Fixing methods

Bitumen-based membranes are fixed typically by either torching, bonding or mechanical fixing methods. With torching, a flame is used to melt an adhesive layer on the underside of the sheet so that the membrane adheres to the substrate. Sheets are lapped by around 100mm to ensure a watertight seal (MCE_460, d). Torches are usually gas fuelled, supplied from a small canister as part of a hand-held tool, or are supplied from a large gas cylinder set onto the roof to a variety of tools, either hand-held or wheeled, for larger-scale applications.

In the bonded method, the sheet is applied cold onto the substrate, the adhesive being either poured and spread in place or spray-applied to the substrate / structural deck. Bitumen-based sheet is laid on top and is rolled into place. Laps are sealed with adhesive, but torching is sometimes used to seal the laps of bonded membranes to allow them to be installed more quickly (MCE_460, d). Like elastomeric and thermoplastic membranes, the hot air welding of laps and joints is being introduced. A tool that provides hot air to the edge of the material is used that allows the material to melt locally and be sealed together. The use of bonding avoids the need for a flame that can damage adjacent finished work.

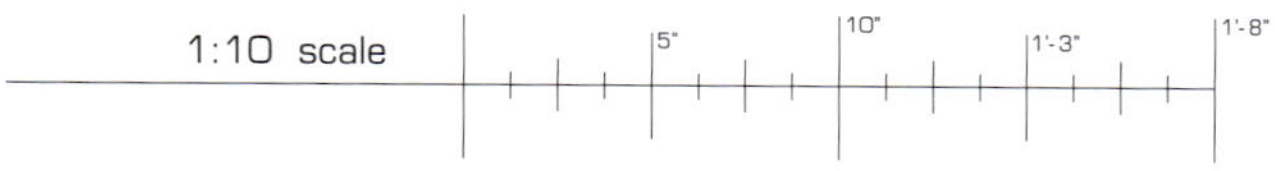

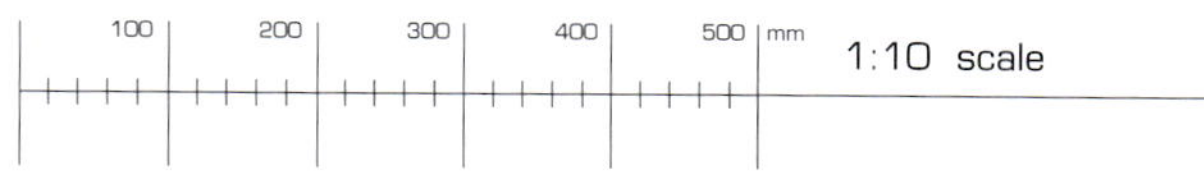

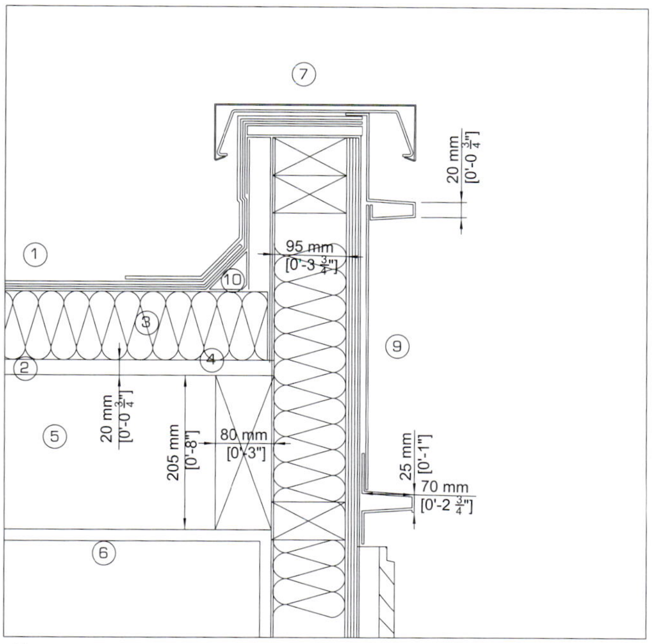

d | Vertical section. Low parapet wall

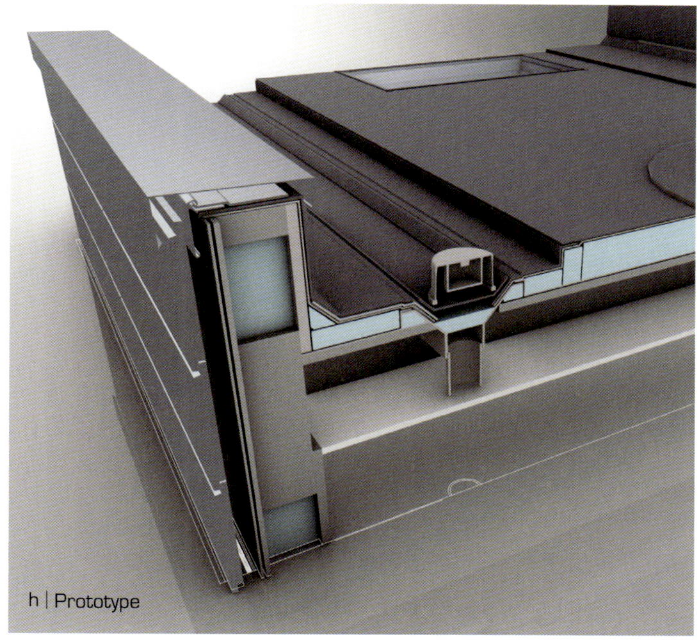

h | Prototype

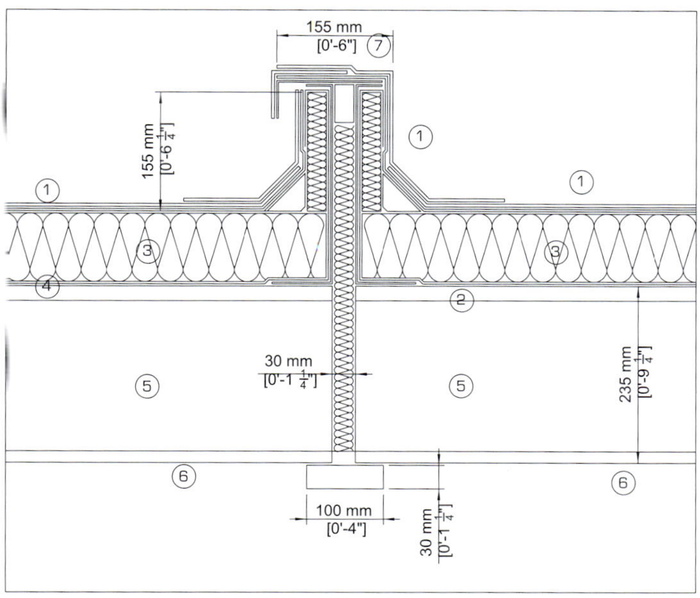

f | Vertical section. Expansion joint

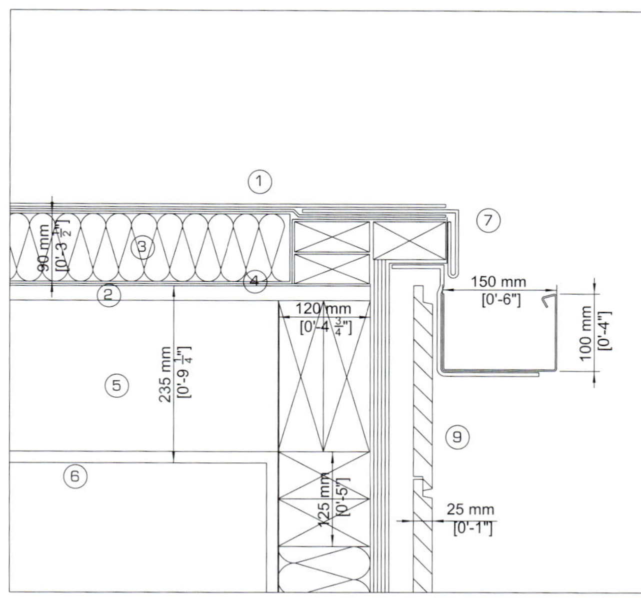

g | Vertical section. Eaves

With the mechanical fixing method no adhesive is required, the membrane being fastened through the insulation layer into the timber deck with disc-type fasteners. Laps between sheets are usually torch-sealed to a width of around 150mm. A vapour barrier, with sealed laps, is laid onto the structural deck or substrate with the insulation set on top. The insulation is mechanically fixed to the deck, with the bitumen-based sheet laid onto the insulation. The mechanical fasteners are fixed within the area of the lap in the sheet through the insulation into the deck beneath. Mechanical fixings are covered by strips or are lapped over the top of the fixings, the laps being sealed by torching, then pressed in place with a roller tool.

Parapet upstands

Upstands in bitumen-based sheets are formed by either fixing the sheet to the plywood face of a timber-framed upstand, or to the face of the thermal insulation, depending on the configuration of the external wall. Where a timber roof deck meets a masonry wall, and a concrete block wall is clad in timber rainscreen panels, the bitumen-based sheet is shown fixed to the face of the upstand. With a low upstand, the membrane continues up the full height and extends across the top of the wall underneath the coping (b). The roof membrane is made continuous with the waterproof seal of the external wall, with the membrane terminating against the bitumen paint finish of the external face of the blockwork wall. The coping can be made from any impervious and durable material. A pressed metal coping overhanging on both sides provides added protection to the membrane as it folds over the top of the wall. Upstands to high parapet walls can be formed by terminating the membrane 150mm above the finished roof level. Above this level the wall is waterproofed with a different method. Typically, concrete walls are waterproofed with bitumen paint and an external panel-based finish, or render. The top of the membrane is protected with a folded metal flashing such as aluminium, which is either bedded into a groove formed in the joint between courses of blockwork, or the metal strip is sealed against the wall with silicone if a concrete wall is used

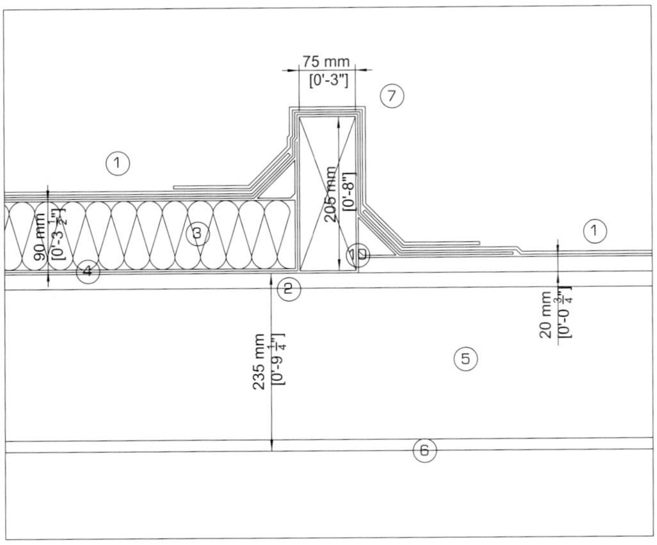

a | Vertical section. Step in roof

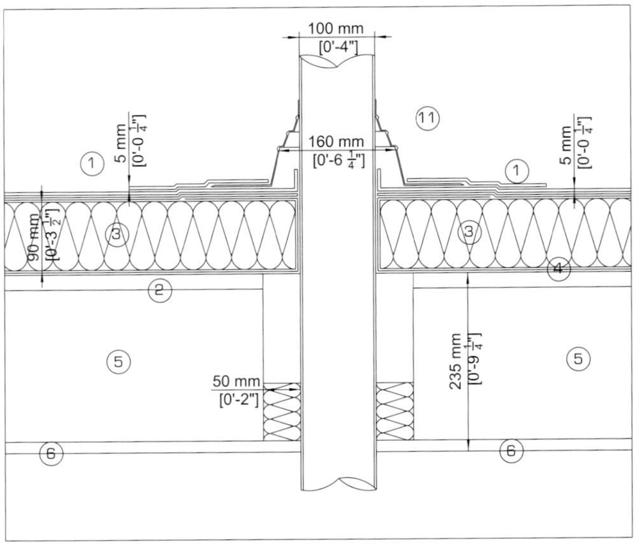

b | Vertical section. Pipe penetration

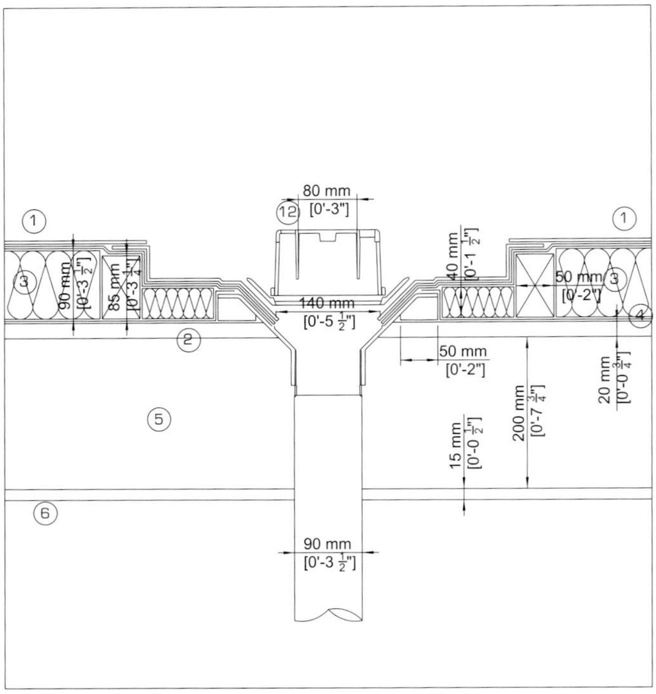

c | Vertical section. Rainwater outlet

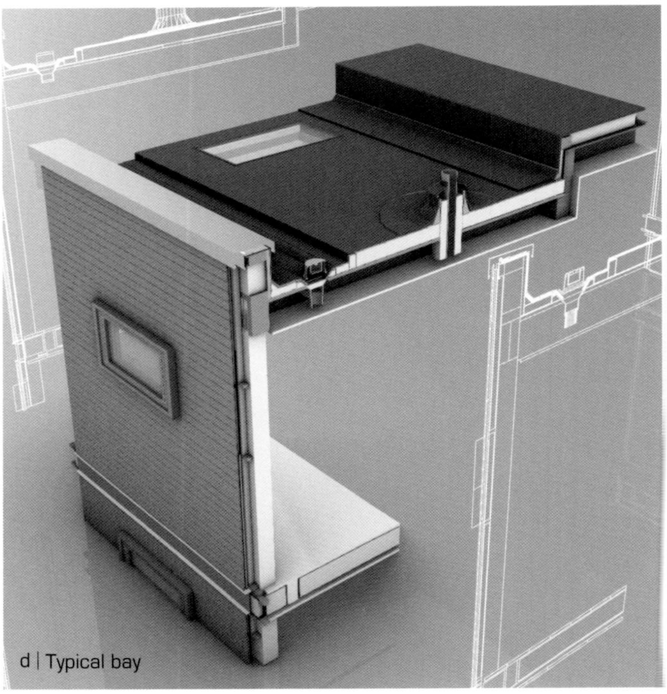

d | Typical bay

(MCE_462, a). The vapour barrier continues up the full height of the membrane. A 45° angle fillet is used to avoid turning the bitumen-based sheet through a full 90° fold in a junction where significant structural movement can be expected.

Upstands for perforations and movement joints are formed in the same way as other upstands (b), but the membrane continues over the top of the upstand. Where an expansion joint is formed, the membrane is separated on one side of the movement joint to open and close as a result of structural movement (MCE_459, f).

Junction with tiled roof

The junction of a flat roof and a tiled pitched roof is formed by extending the base layer and membrane up 150mm above the finished roof level, with an additional reinforcing layer to strengthen the junction. The roofing felt, or breather membrane, of the tiled pitched roof extends down over the top of this membrane to provide a complete weathertight seal across the roof. The bottom top row of tiles is kept clear of this junction in order to avoid the possibility of damage and to ensure that rainwater running down the tiles cannot run back up the gap between the tile and the roof membrane by capillary action. Where a bitumen-based membrane forms a waterproof covering to an area of flat roof at the top of the pitched roof the membrane is folded at the edge to form a drip, so that rainwater cannot track up into the roof construction. A metal drip is sometimes used to give a visually sharper edge to the roof. The vapour barrier beneath the thermal insulation is turned up so that it terminates against the edge of the membrane drip to provide a continuous barrier. A metal flashing extends down from the vapour barrier and is lapped over the top of the tiling to provide a complete seal from the bitumen-based membrane of the roof tiles beneath. The roofing felt or breather membrane under the roof tiling is terminated against the underside of the upper timber deck. Where the membrane meets a rooflight upstand, the bitumen-based

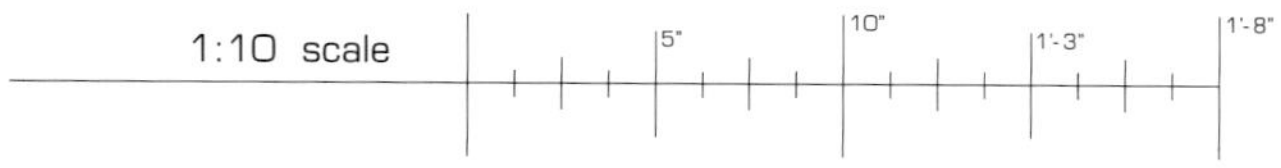

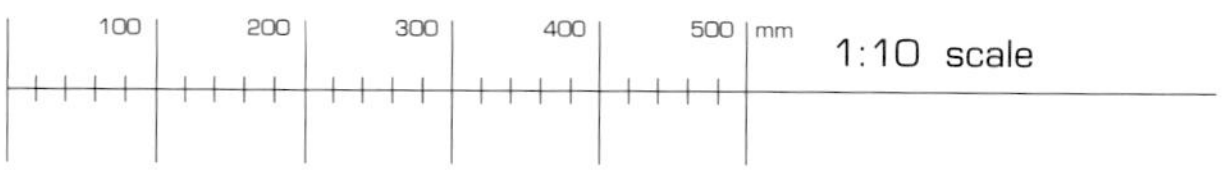

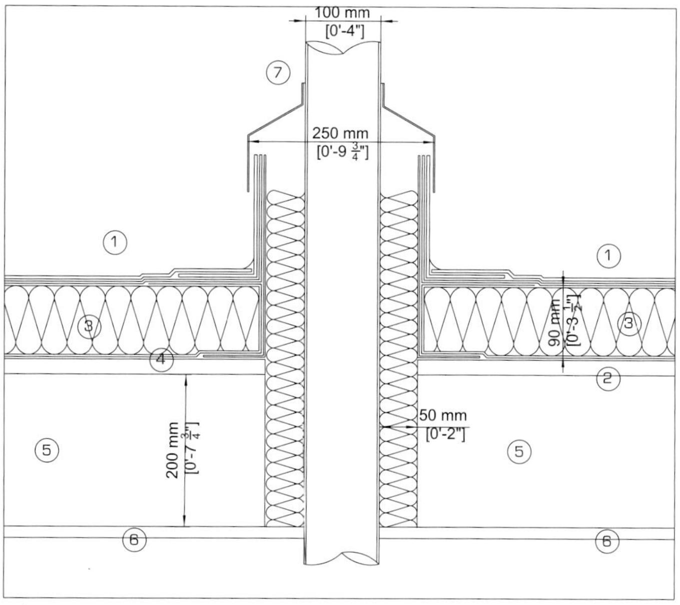

e | Vertical section. Pipe penetration

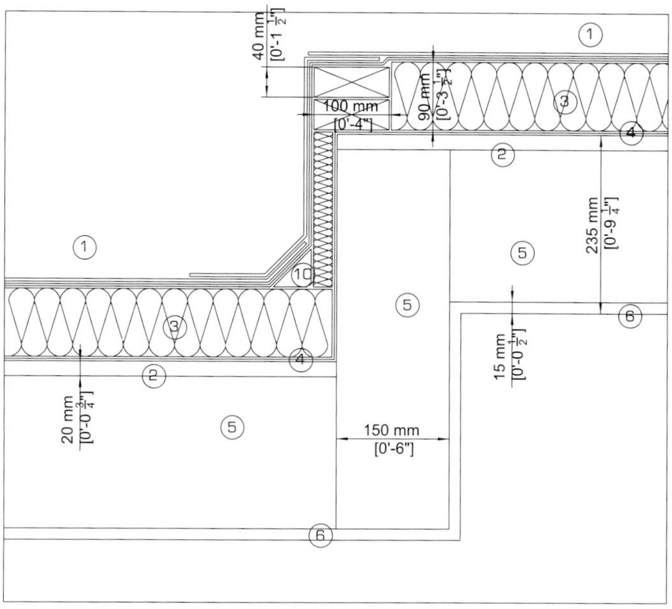

f | Vertical section. Step in roof

Details

1 | Bitumen-based sheet
2 | Plywood sheet
3 | Rigid thermal insulation
4 | Vapour barrier
5 | Softwood joists
6 | Dry lining/drywall internal finish
7 | Metal flashing
8 | Timber upstand
9 | External wall
10 | Angle fillet
11 | Proprietary skirt flashing
12 | Rainwater outlet
13 | Paving bonded to bitumen-based sheet

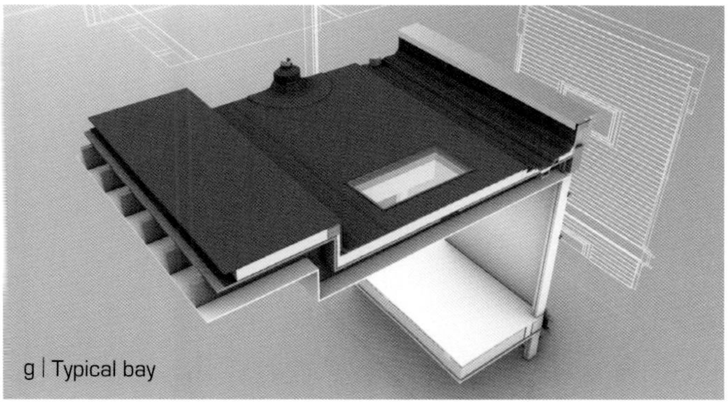

g | Typical bay

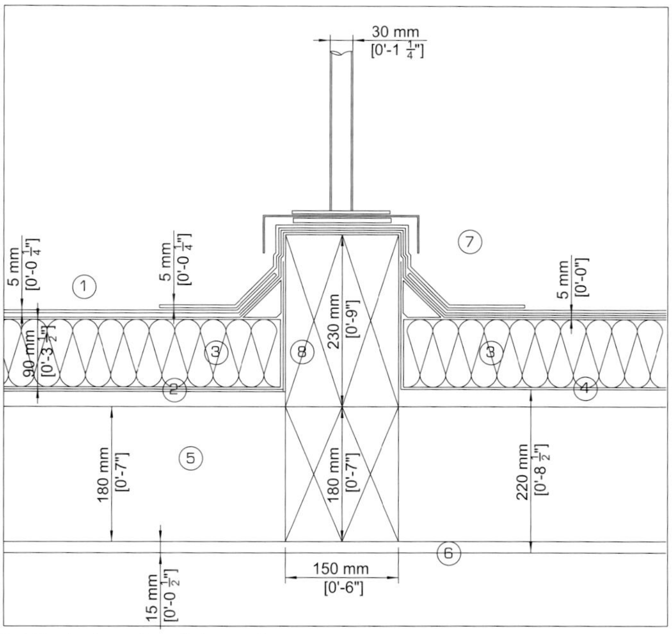

h | Vertical section. Balustrade

sheet is taken over the top of the timber upstand to form a continuous seal with the vapour barrier on the inside of the building (MCE_456, c) where required, as well as the vapour barrier beneath the thermal insulation. The rooflight is fixed to the top of the sealed upstand, typically with a continuous timber glazing bar or metal strip that seals the gap between glazing and upstand. The edge of the rooflight typically includes a drip to avoid water running back up into the joint between upstand and rooflight.

Eaves and verges

The eaves detail is formed with a folded drip formed in the same way as described for the junction with the top of a pitched roof. The gutter is tucked up behind the drip to ensure that all rainwater running off the edge of the eaves is collected by the gutter, and that no water is allowed to run behind the gutter and down the face of the wall below (MCE_459, d) where staining can occur from the dust washed off the flat roof. The vapour barrier under the thermal insulation is continued to the edge of the roof to ensure that the timber sections forming the edge of the eaves are kept dry and ventilated within the roof void beneath (g). The thermal insulation is kept continuous between wall and roof to avoid thermal bridging that reduces the thermal insulation value of the external envelope.

Verges typically extend along the edge of shallow slopes of bitumen-based roofs, and are formed with a low upstand that prevents rainwater from running over the edge of the verge. The membrane is continued up over the top of the upstand. A GRP or metal edge trim is set at the edge to terminate the membrane as well as forming a drip at the top of the external wall (MCE_459, d). An additional sealing strip of membrane is lapped over the edge trim to provide a complete water tight barrier, as well as providing reinforcement to strengthen the joint. The vapour barrier extends up the side of the timber upstand, but it is not necessary to extend it to the edge of the roof due to the extra layer of bitumen-based membrane that provides a full seal.

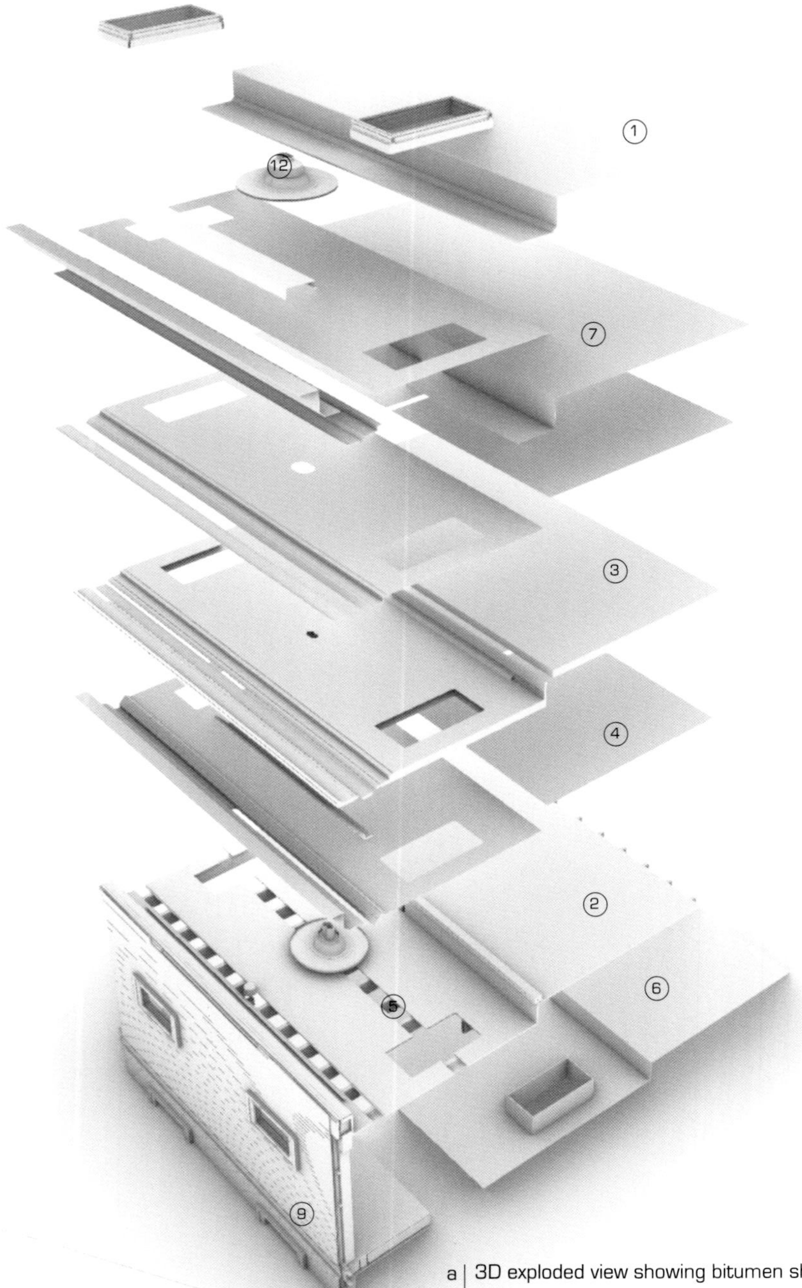

a | 3D exploded view showing bitumen sheet-based membrane on concrete roof

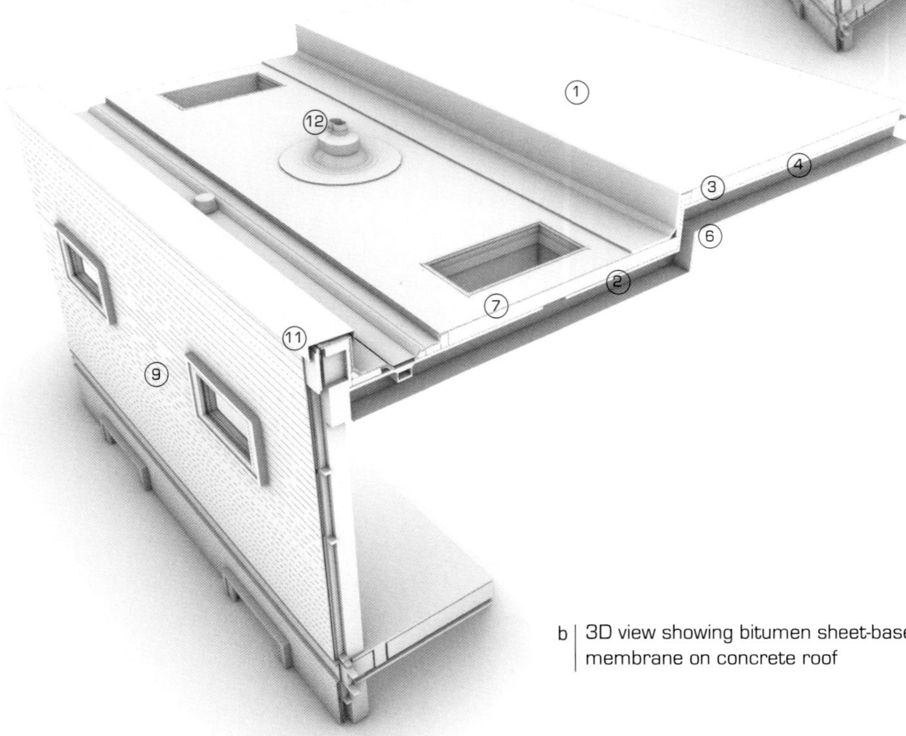

b | 3D view showing bitumen sheet-based membrane on concrete roof

Details

1 Bitumen-based sheet
2 Plywood sheet
3 Rigid thermal insulation
4 Vapour barrier
5 Softwood joists
6 Dry lining/drywall internal finish
7 Metal flashing
8 Timber upstand
9 External wall
10 Angle fillet
11 Proprietary skirt flashing
12 Rainwater outlet
13 Paving bonded to bitumen-based sheet

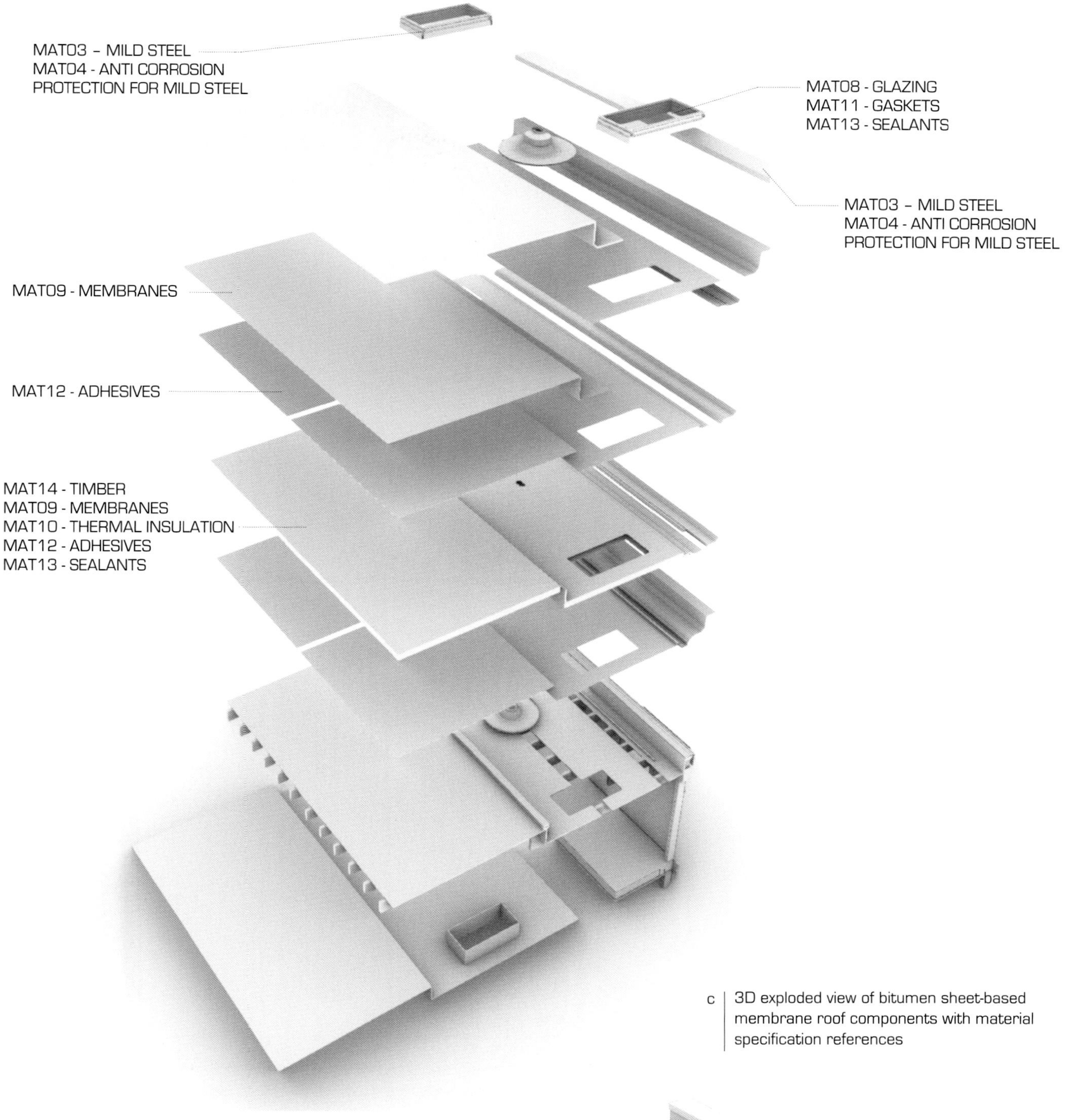

c | 3D exploded view of bitumen sheet-based membrane roof components with material specification references

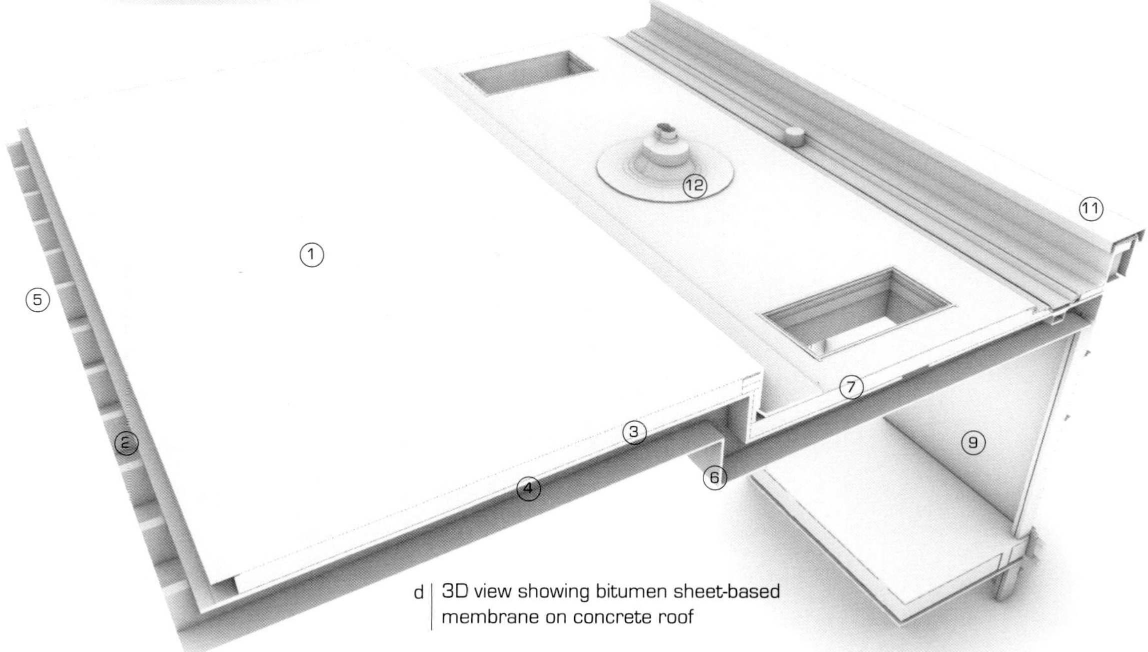

d | 3D view showing bitumen sheet-based membrane on concrete roof

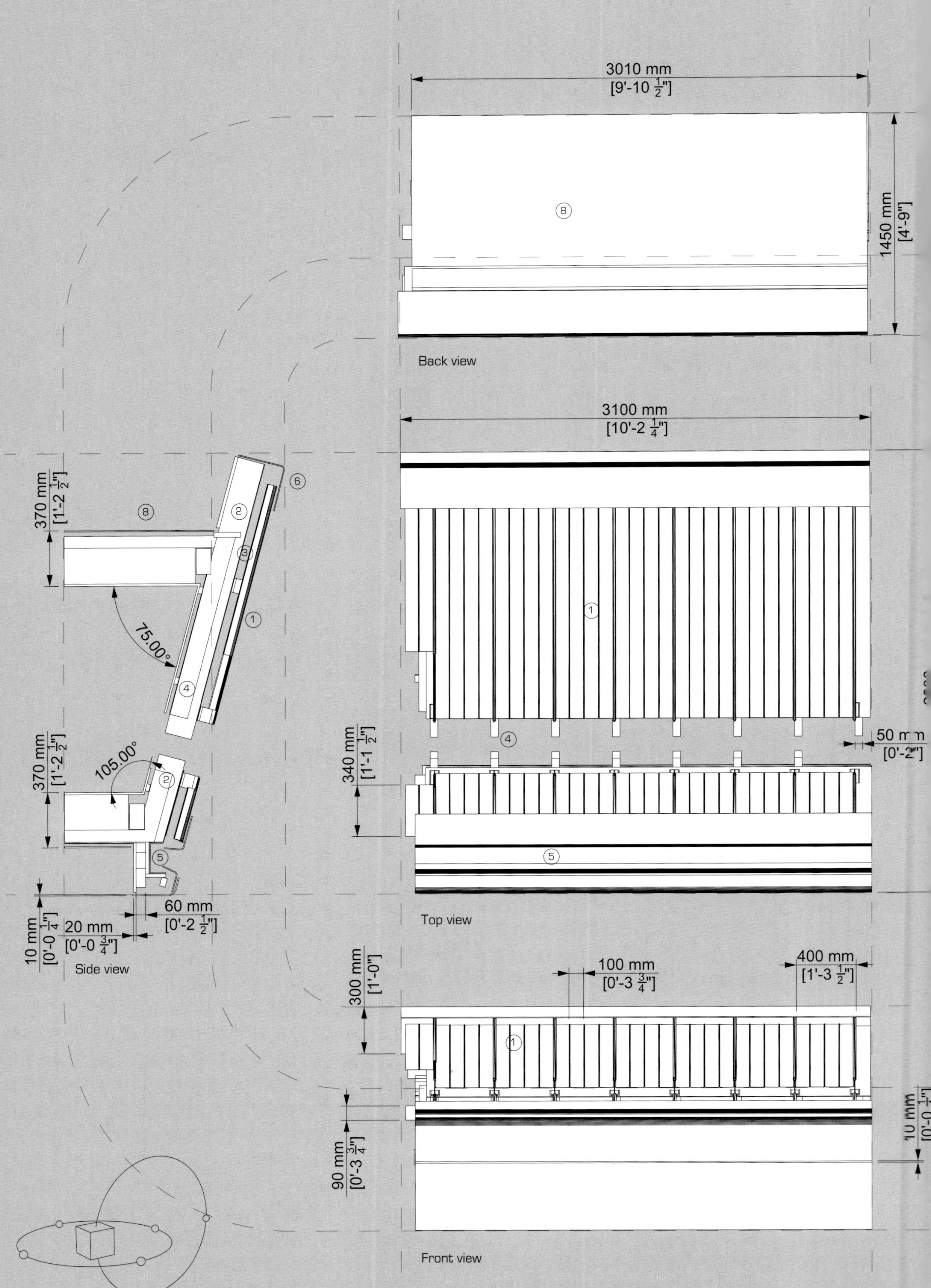

Third angle projection 1:30

Details

1 | Standing seam sheet
2 | Fibre quilt thermal insulation
3 | Vapour barrier
4 | Timber beam/joist
5 | Folded metal gutter
6 | Folded metal drip
7 | Fascia board

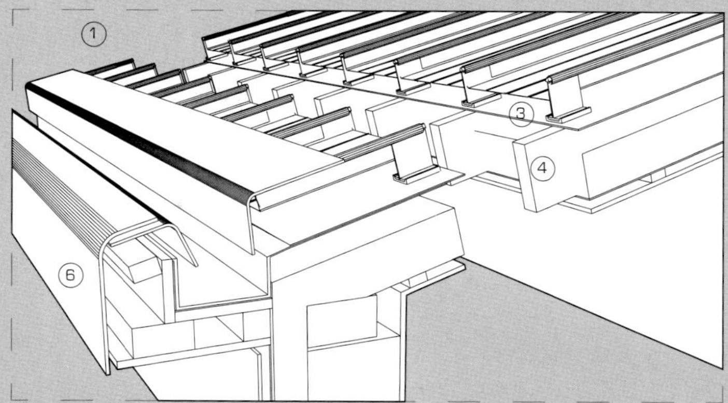

3D view of the assembly

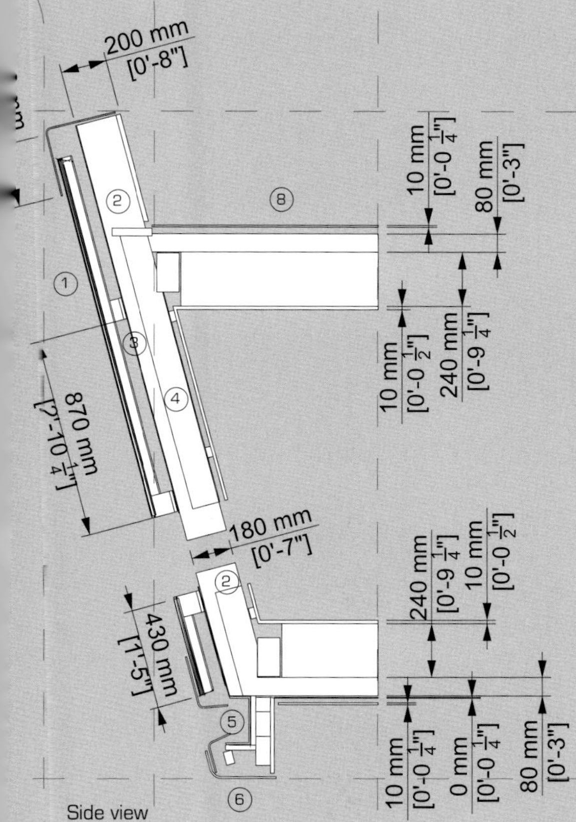

Side view

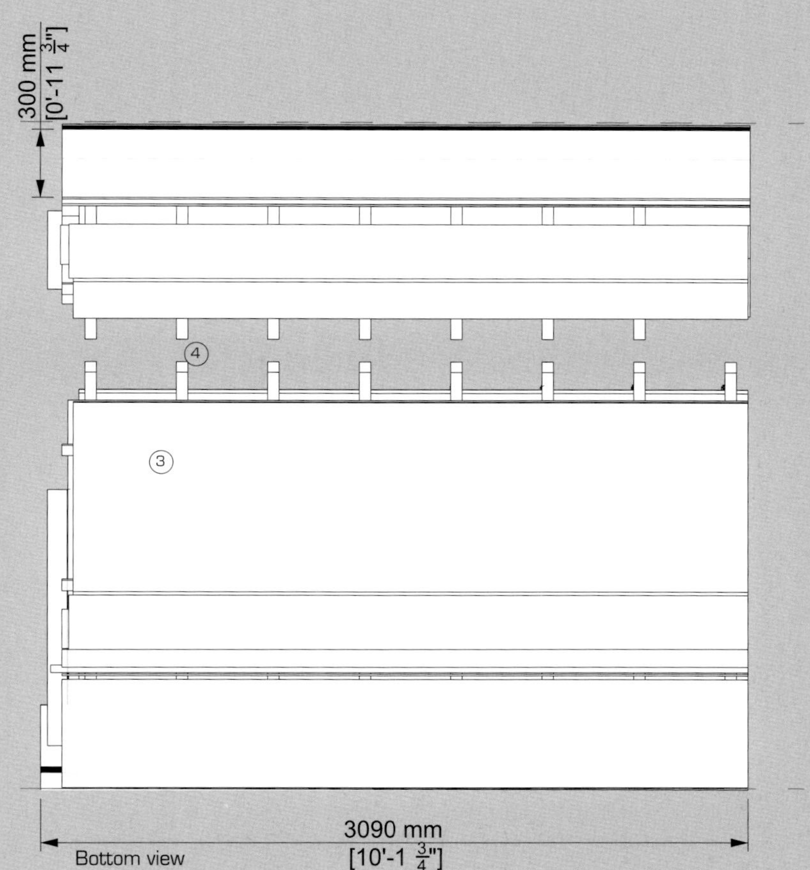

Bottom view

8 | External wall
9 | Metal sheet fixing bracket
10 | Soffit board
11 | Metal supports
12 | Ridge piece
13 | Metal flashing
14 | Parapet flashing

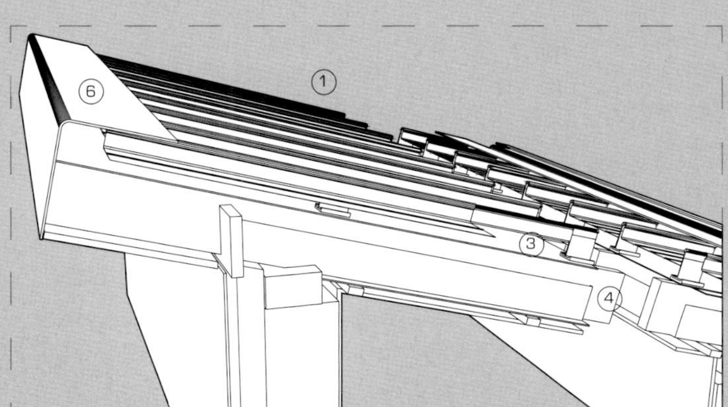

3D view of the assembly

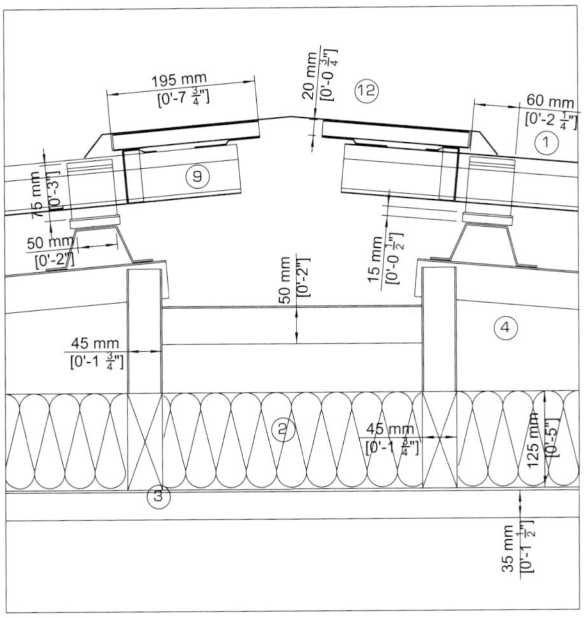

a| Vertical section. Ridge

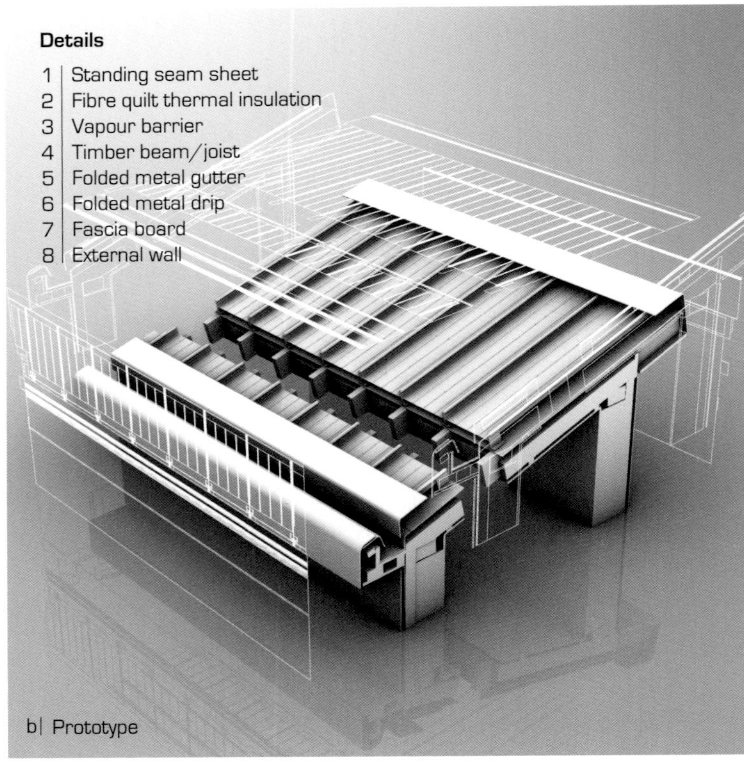

b| Prototype

9 | Metal sheet fixing bracket
10 | Soffit board
11 | Metal supports
12 | Ridge piece
13 | Metal flashing
14 | Parapet flashing

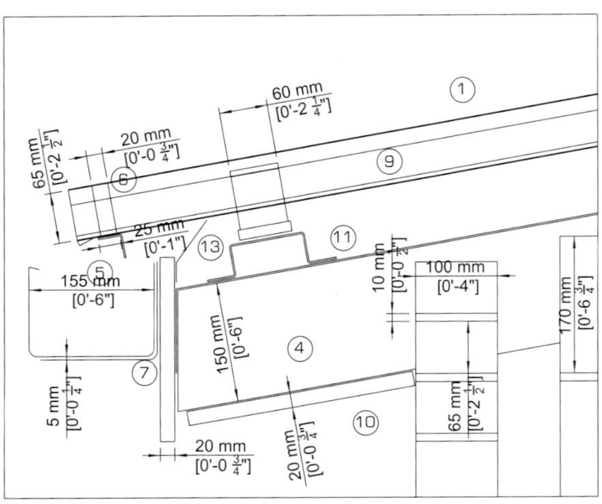

c| Vertical section. Eaves

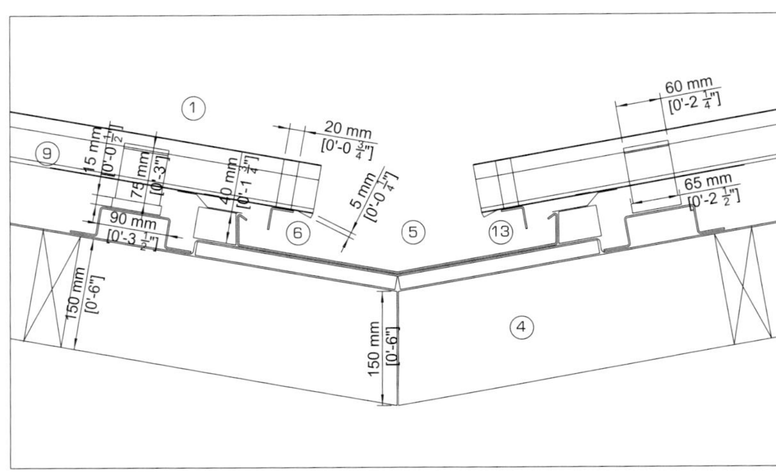

d| Vertical section. Ridge

Standing seam cold roofs

Standing seam roofs are discussed in the Metal chapter as sealed, insulated roof coverings in a warm roof configuration. When used in the cold roof configuration with a pitched timber roof, the standing seam roof is ventilated, with thermal insulation provided at ceiling level. In this build-up a ventilated standing seam metal deck is set onto a timber roof of trusses and purlins. At ceiling level, thermal insulation quilt is set between the ceiling joists (horizontal members). A vapour barrier is set beneath the insulation, positioned on the top surface of a dry lined wall beneath. The roof void is ventilated at the lowest point (eaves or parapet gutter) and at the highest point (ridge or abutment), with verges and parapets at the sides of the roof remaining sealed.

Eaves and valley gutters

Eaves are ventilated by leaving a gap between the top of the fascia board supporting the gutter and the underside of the standing seam roof sheet. A metal angle is fixed in front of the opening to avoid rainwater being blown through the opening into the roof void. Valley gutters are formed in a similar way, with a gap formed between the gutter tray and the metal sheet to provide ventilation to the roof void. The metal tray has an upstand at the edge which is lapped up to the underside of the metal sheet and sealed against the sheet. This ensures that rainwater running down the gutter cannot enter the roof void, while air can still pass into the roof void.

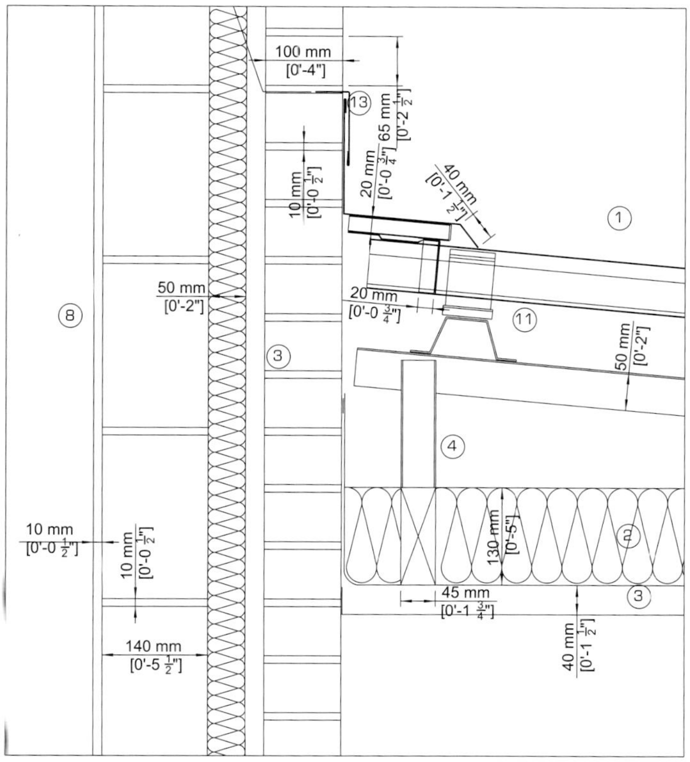

e| Vertical section. Abutment

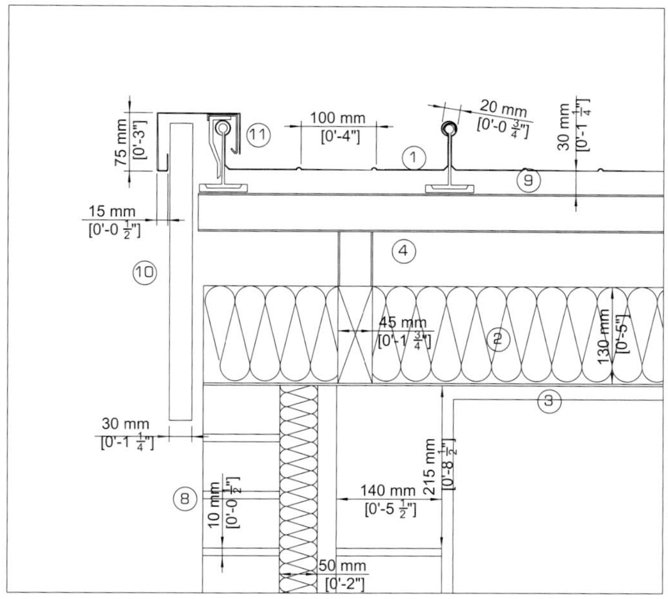

f | Vertical section. Verge

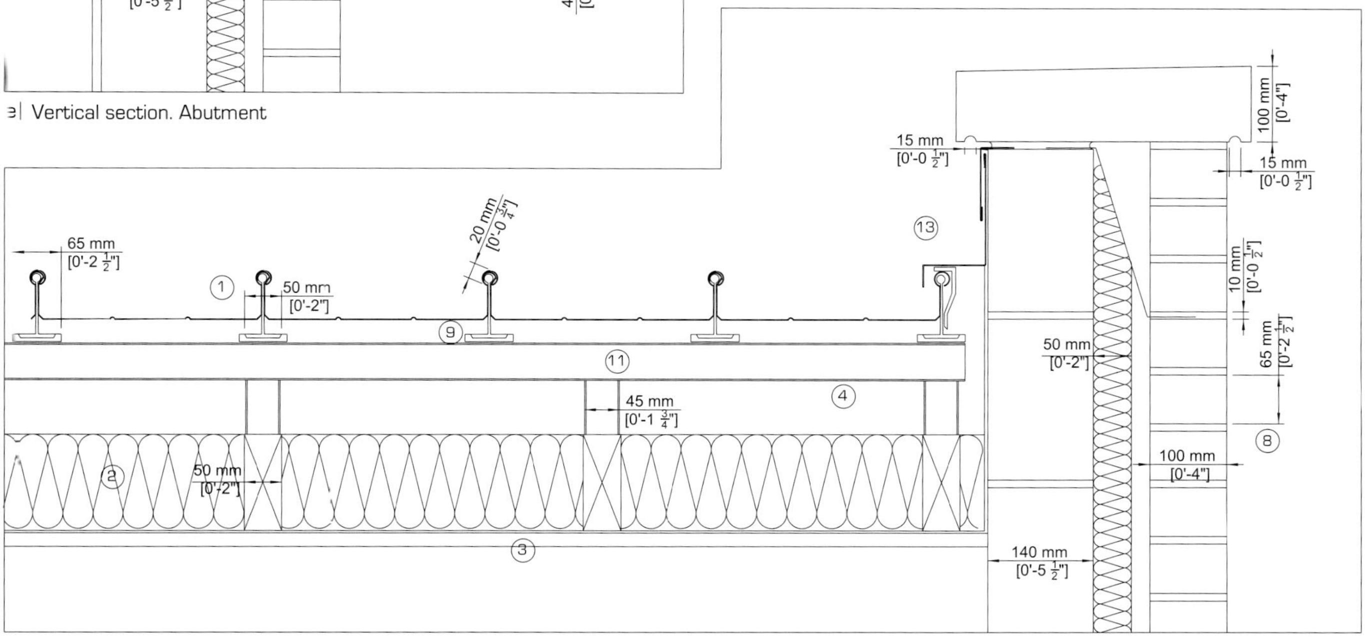

g| Vertical section. Parapet verge

Ridges and abutments

Ridges are formed by creating a gap, typically 100mm between the metal sheets. The gap at the ridge is covered with a folded metal ridge capping which is fixed to the sheets beneath. A narrow gap is maintained between the ridge sheet and the standing seam sheet in order to provide constant ventilation along the ridge. Alternatively, PVC-U ventilator blocks are added at centres along the ridge, and the joint between the ridge and the roof deck is sealed. Abutments are formed in a similar way to ridges, with a folded metal sheet that is fixed to the roof sheet on one side and is sealed against the adjacent masonry/concrete wall or parapet, typically with an additional metal flashing which projects from within the depth of the wall. Verges are formed with a metal clip which provide a closer to the edge sheet, with an additional metal coping set over this to provide a second line of defence against rainwater penetration.

Penetrations

Penetrations are formed in a way which forms a gutter around the opening in the roof. In some examples, rainwater runs off the roof into the gutter, at the edge of the upstand, which is set at the level of the adjacent roof sheets. Water is directed around the sides of the upstand and back onto the roof sheeting below. Penetrations in roofs are positioned so that the standing seam joints between roofing sheets are clear of the sides of the opening in the roof to ensure that rainwater can run freely around the sides of the upstand to the roof opening.

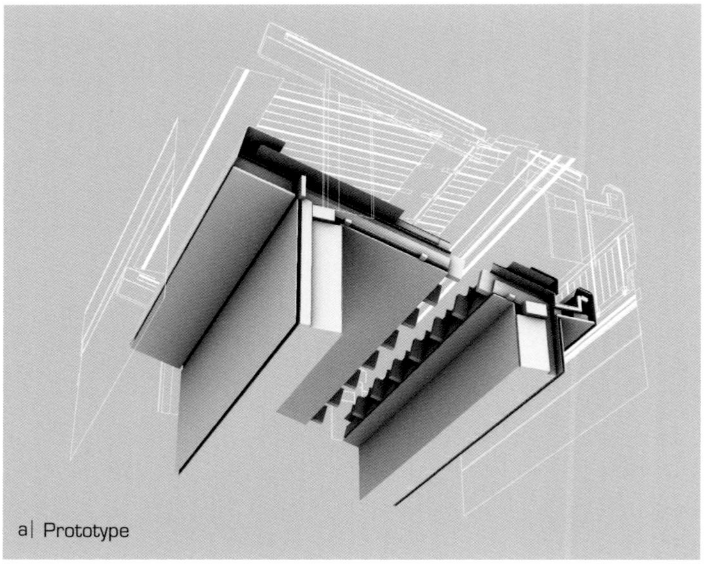

a| Prototype

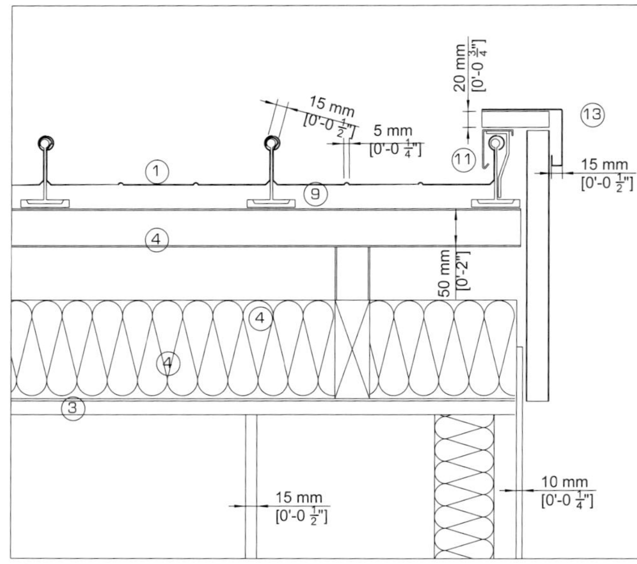

b| Vertical section. Verge

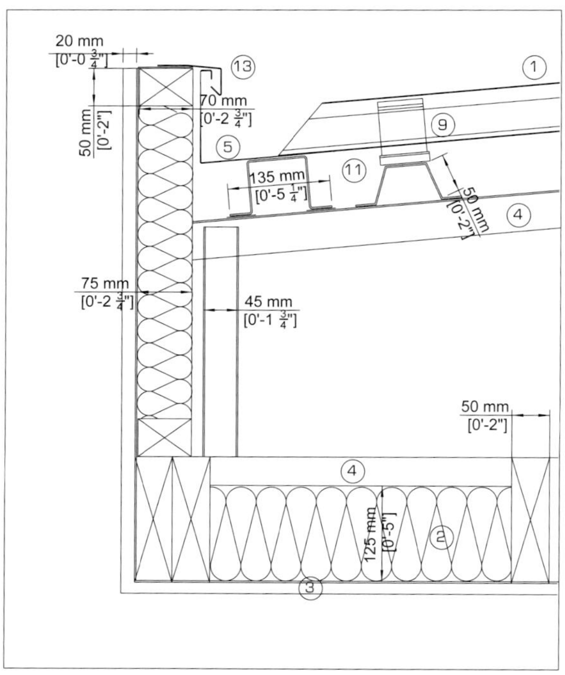

c| Vertical section. Penetration

Details:
Standing seam roof

1 Standing seam sheet
2 Fibre quilt thermal insulation
3 Vapour barrier
4 Timber beam/ joist
5 Folded metal gutter
6 Folded metal drip
7 Fascia board
8 External wall
9 Metal sheet fixing bracket
10 Soffit board
11 Metal supports
12 Ridge piece
13 Metal flashing
14 Parapet flashing

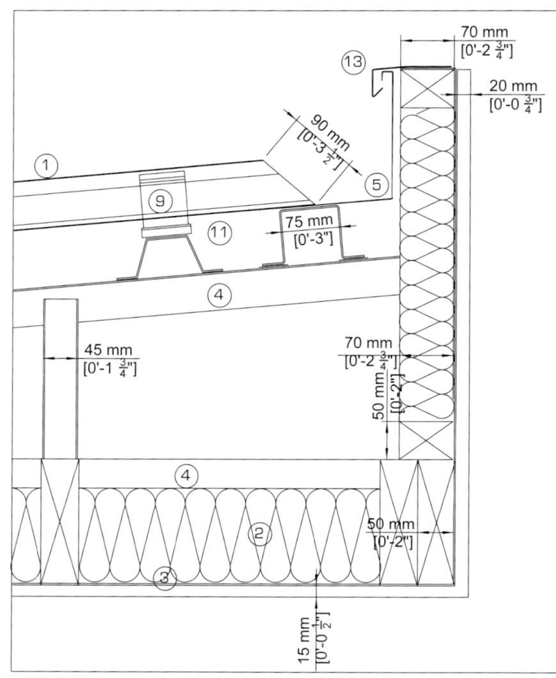

d| Vertical section. Penetration

Metal tiled roofs

Metal tiles are increasingly being used for roofs due to the flexibility of the roof pitch and tile lap that can be formed economically for individual projects. Copper and zinc are commonly used, with their characteristic patina that is suitable for both walls and roofs. Metal tiles are fixed in the same manner as clay tiles and slates, with battens and counter battens set onto a bitumen- or polymer-based waterproof underlay. Eaves and parapet gutters are formed in the same way as standing seam roofs, with ventilation slots provided to ensure that air can pass through the roof void. Metal tiles are either flat like slates, or alternatively are made from profiled sheet, which imitates the appearance of traditional tiled roofs made from curved or profiled tiles. Their appearance is very different from profiled metal sheet, with its continuous long lengths of sheeting. The short metal tiles, lapped over one another, can easily take up complex geometries with a pattern of joints and laps that clearly show the form of the roof.

Ridges and verges are usually made from folded metal sheet and are individually designed and fabricated for a building project. An advantage of metal sheet tiles over clay tiles is that tiles and edging pieces can be individually designed for each project while remaining economic, since metal can be folded without the need for moulds which are required for clay tiles and slates.

Details:
Tiled roof

15 | Metal tiles
16 | Softwood battens
17 | Vapour permeable membrane
18 | Gutter
19 | Softwood rafter
20 | Ventilation void
21 | Thermal insulation
22 | Vapour barrier
23 | Softwood joist
24 | Metal flashing
25 | Soffit board
26 | Fascia board
27 | External wall
28 | Ridge piece
29 | Ventilator

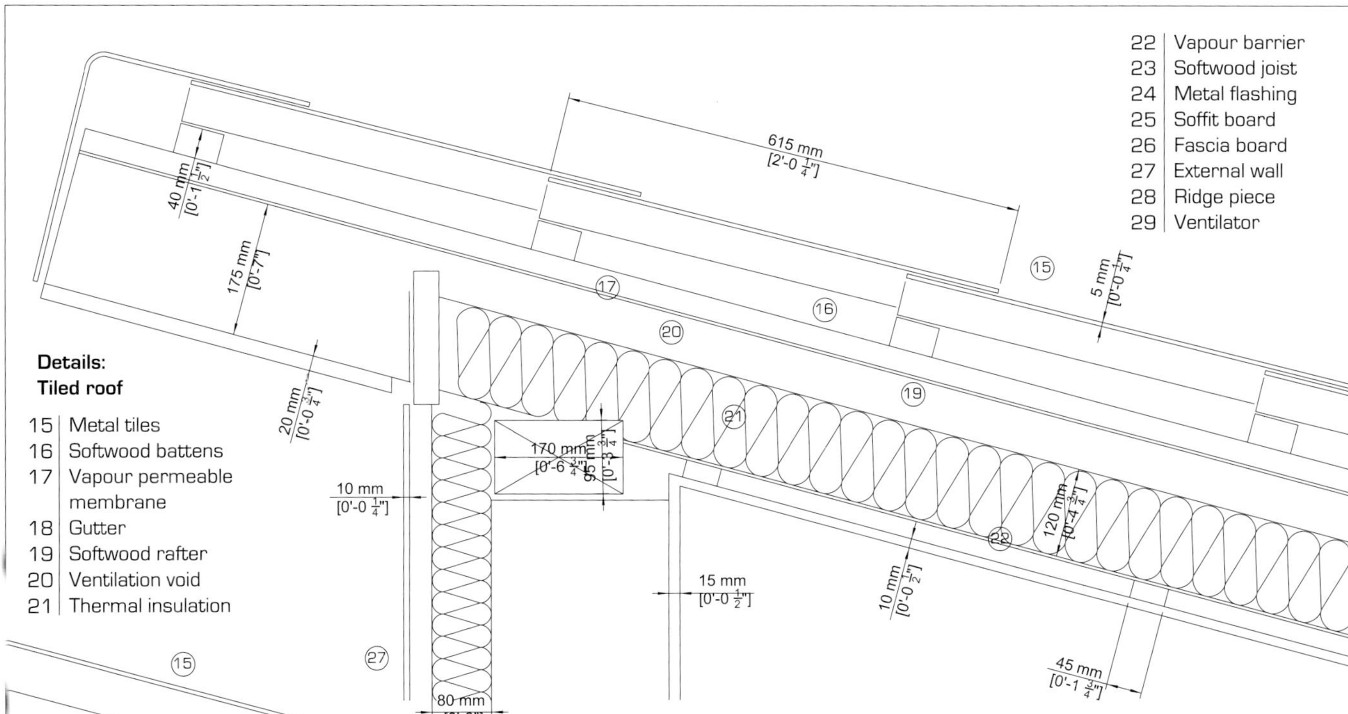

e| Vertical section. Monopitch ridge

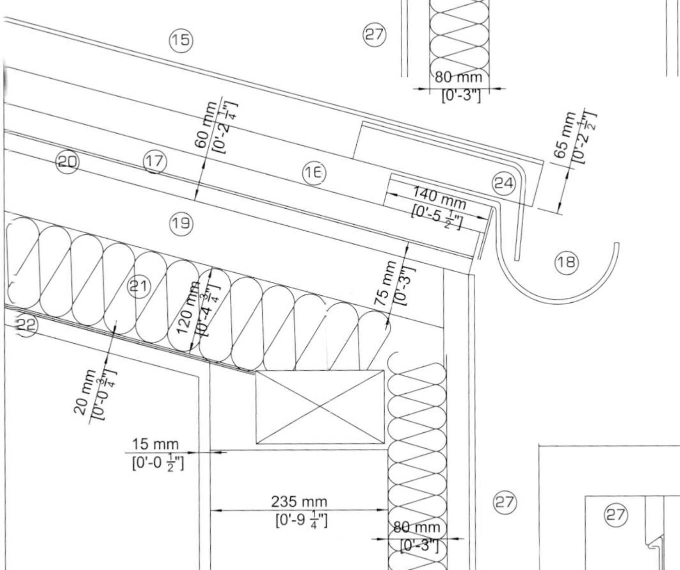

f | Vertical section. Eaves

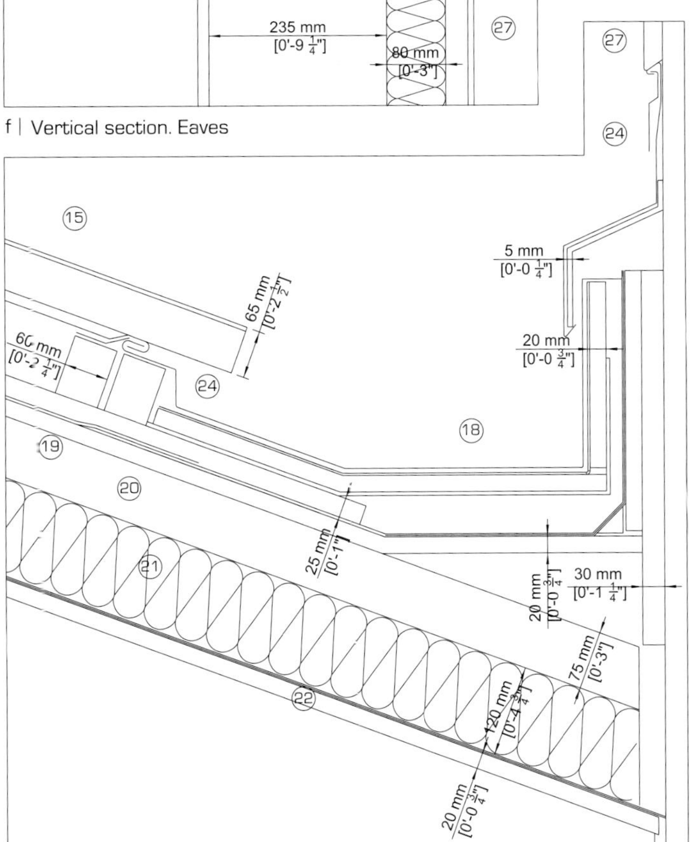

g| Vertical section. Valley gutter

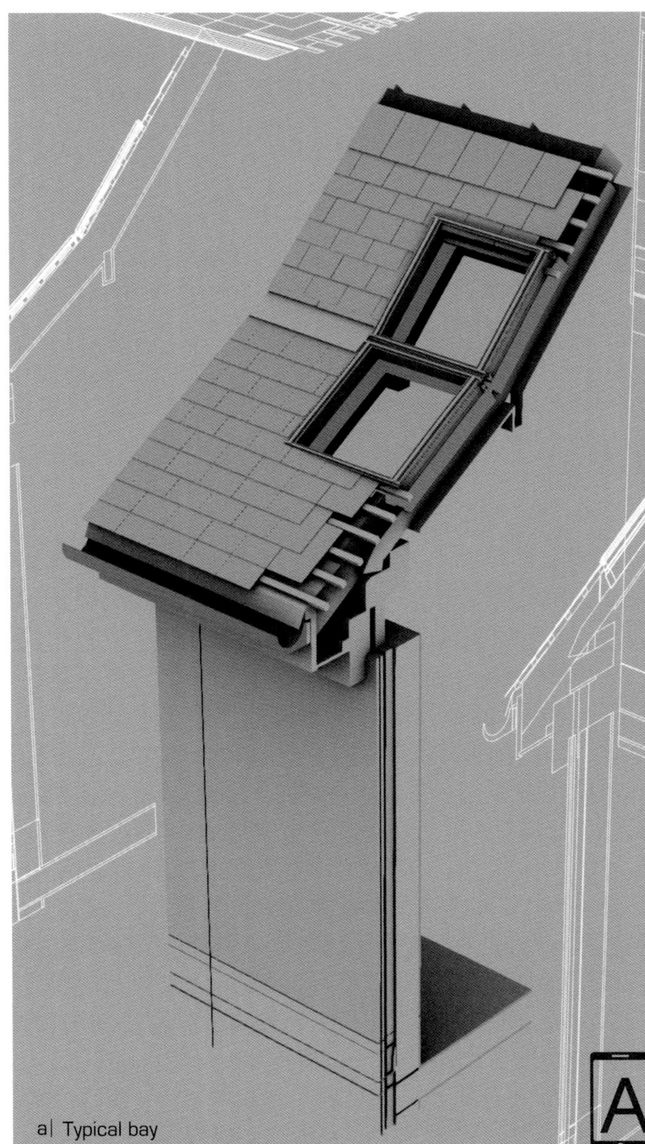
a| Typical bay

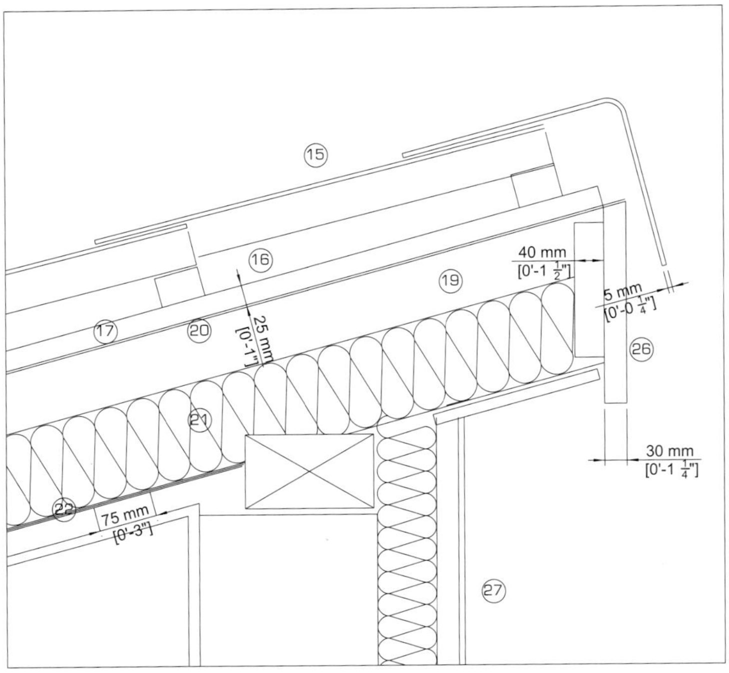

a| Vertical section. Monopitch ridge

b| Typical bay

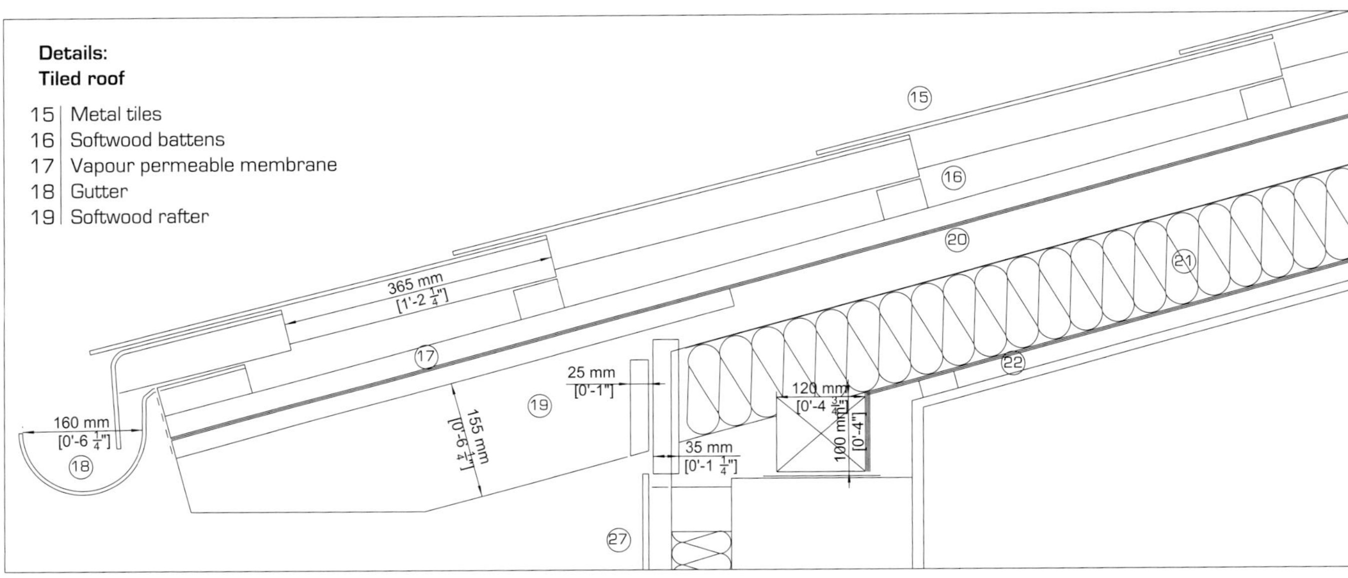

c| Vertical section. Eaves

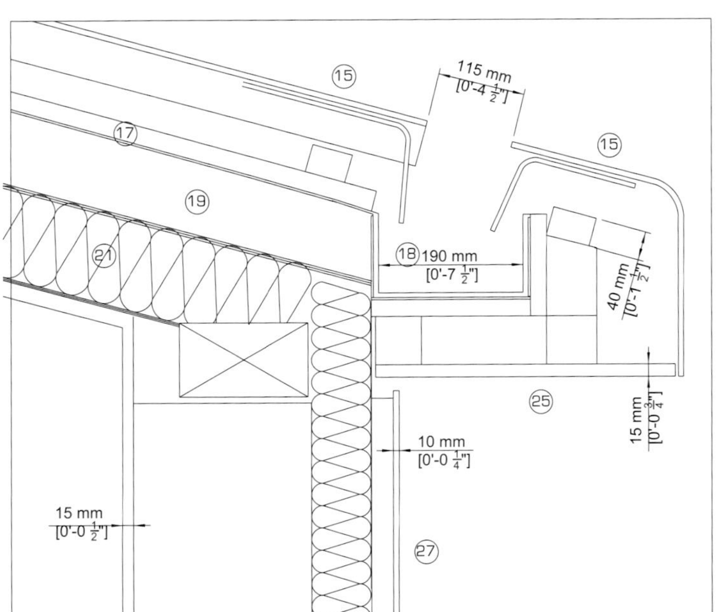

d| Vertical section. Eaves

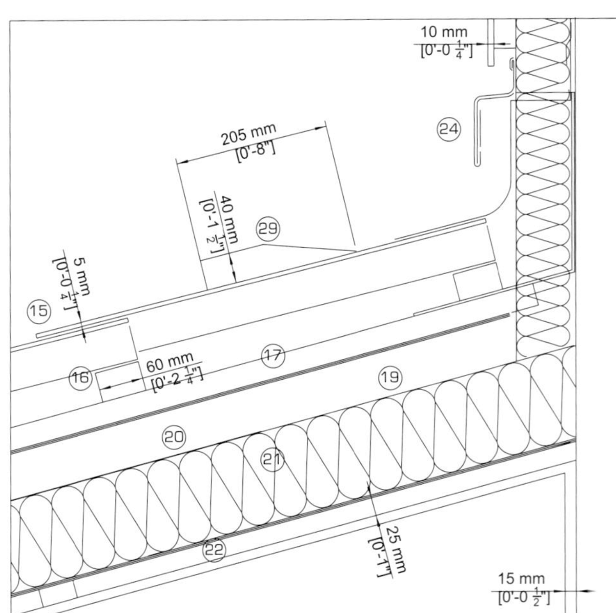

e| Vertical section. Abutment

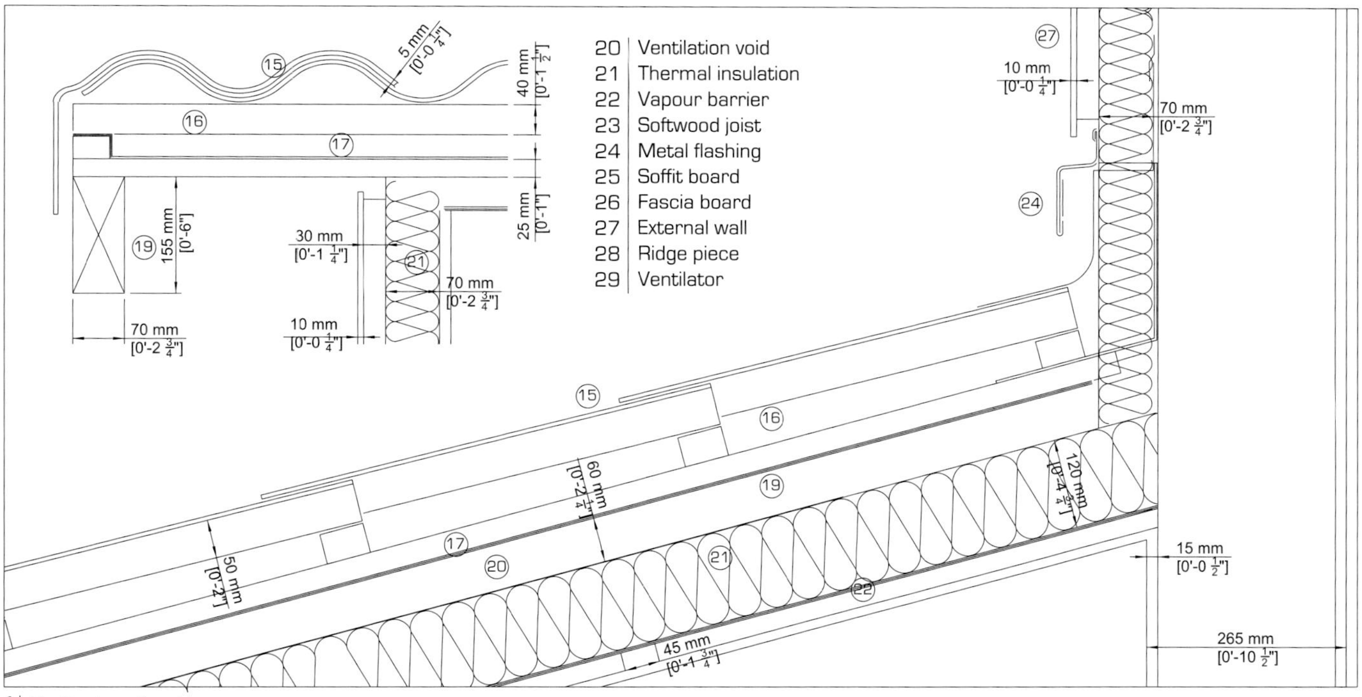

f | Vertical section. Verge

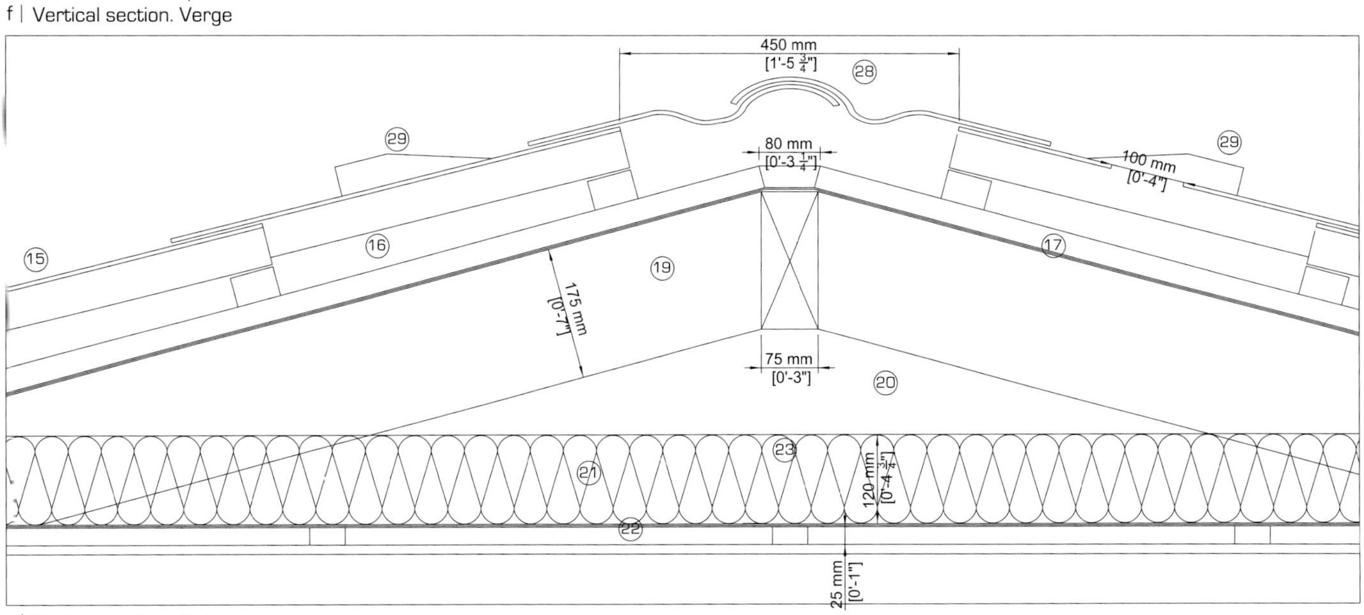

g | Vertical section. Ridge

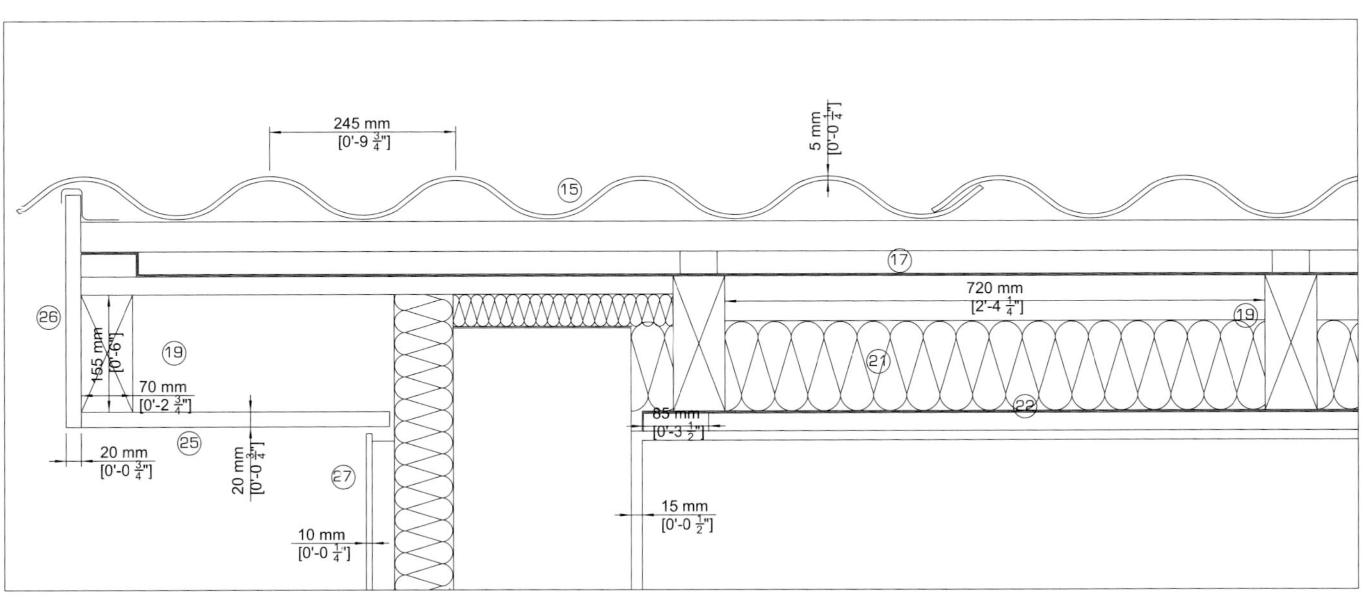

h | Vertical section. Verge

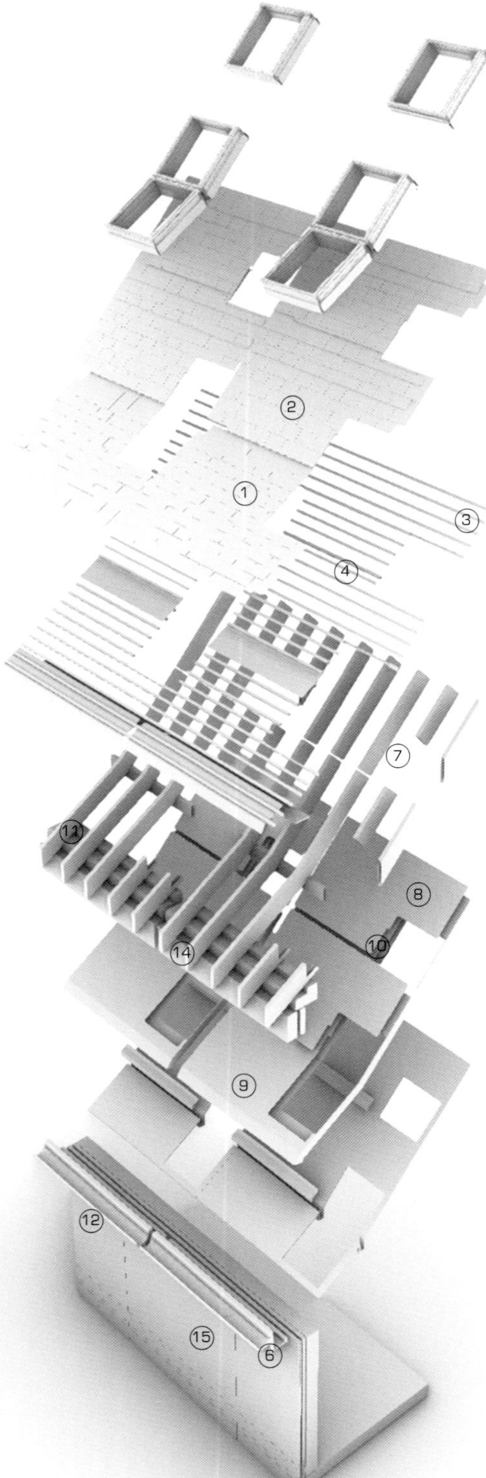

a | 3D exploded view showing tiles on pitched timber roof

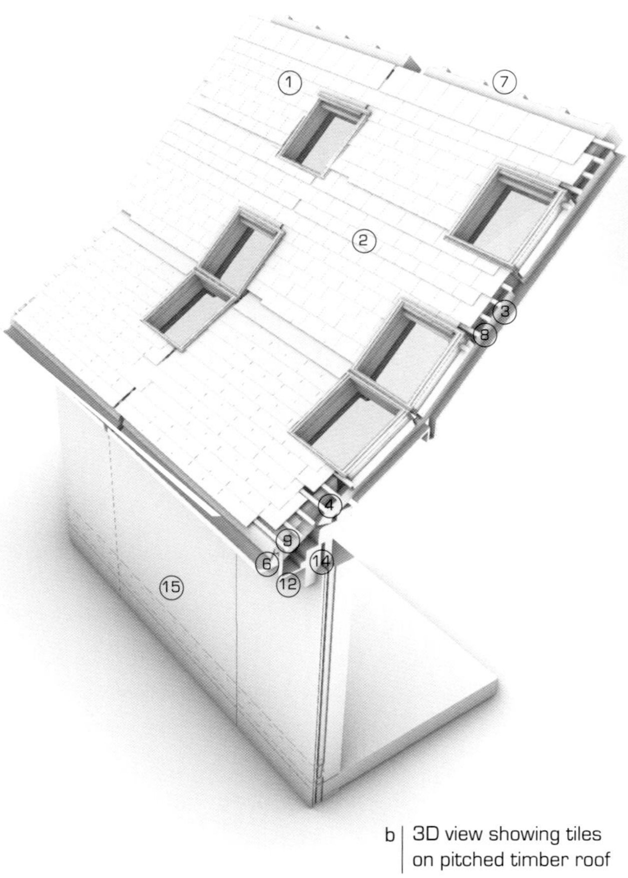

b | 3D view showing tiles on pitched timber roof

Details

1 | Plain tile
2 | Interlocking tile
3 | Softwood battens
4 | Softwood counter battens
5 | Roofing felt
6 | Gutter
7 | Softwood rafter
8 | Vapour-permeable membrane
9 | Thermal insulation

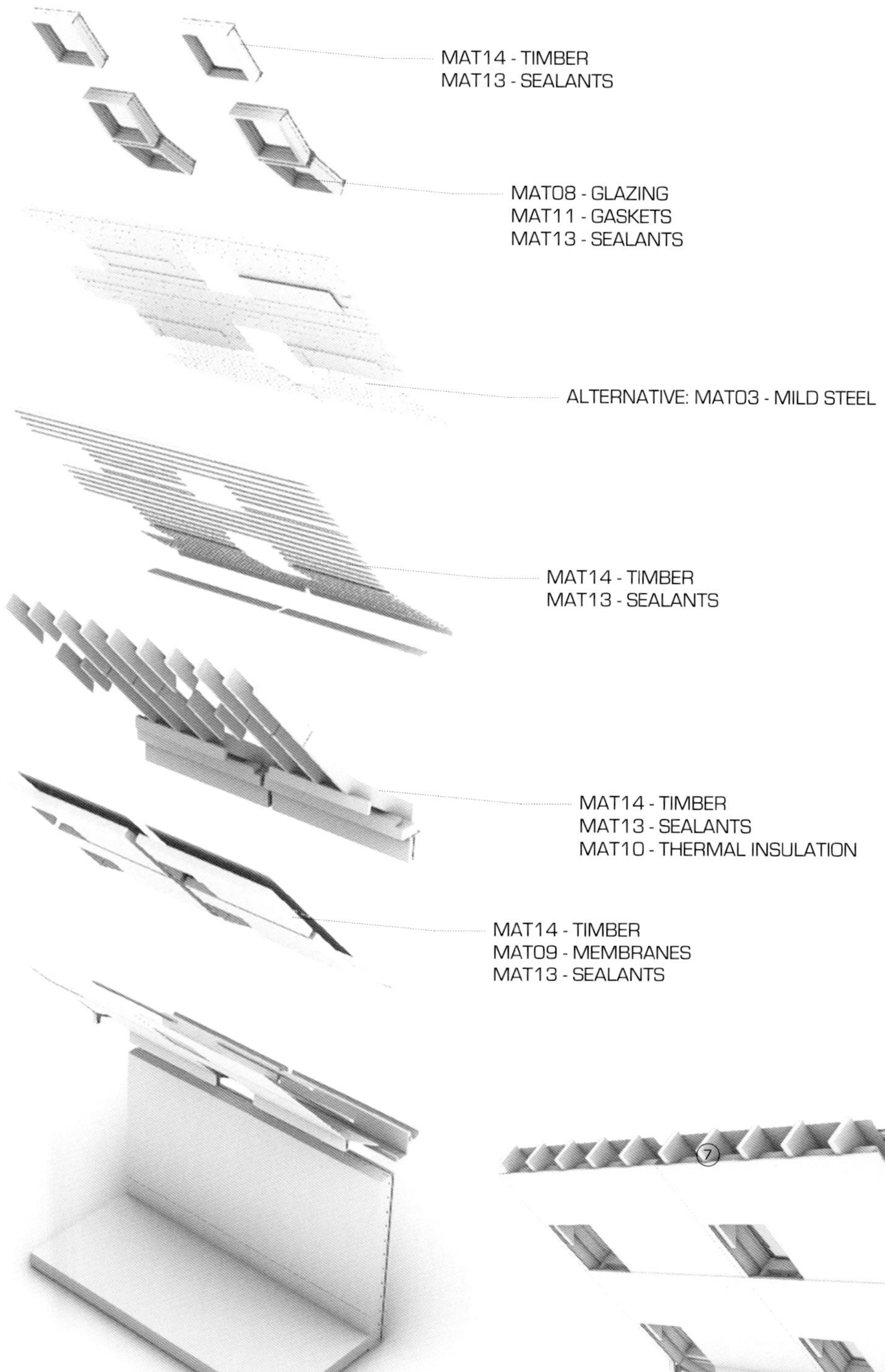

c | 3D exploded view of tiles on pitched timber roof components with material specification references

10 Vapour barrier
11 Softwood joist
12 Metal flashing
13 Ventilator
14 Fascia board
15 External wall
16 Ridge capping
17 Inner cavity leaf

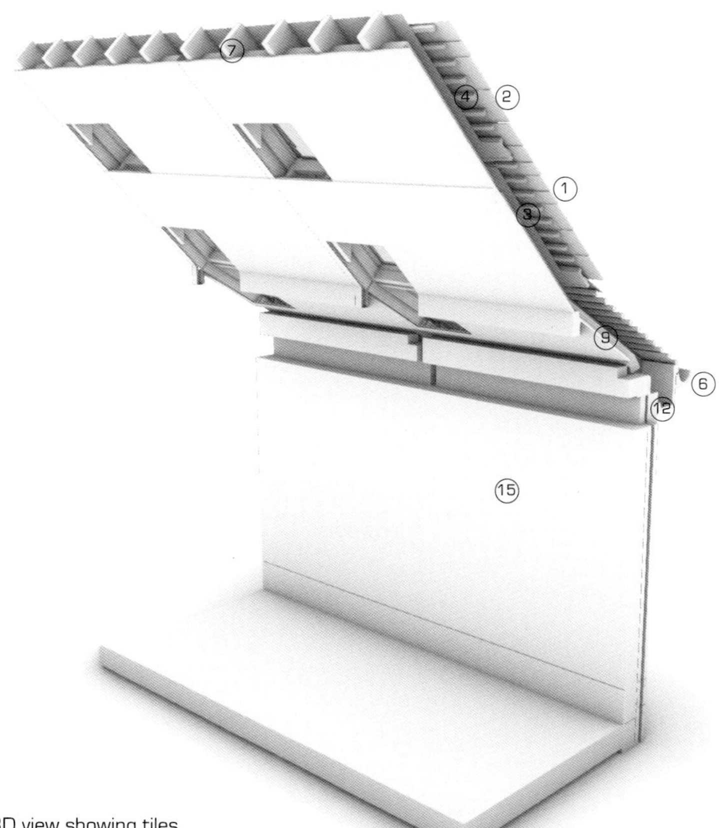

d | 3D view showing tiles on pitched timber roof

Example project of polymer facades
Heydar Aliyev Cultural Centre, Baku, Azerbaijan
Facade Engineer to Arabian Profiles: Newtecnic
Architect: Zaha Hadid Architects

POLYMER ROOFS

41 **GRP rooflights**

Eaves and upstands

Verges

Abutments

Sliding roof panels

42 **GRP panels and shells**

Smaller panels and shells

Larger panels and shells

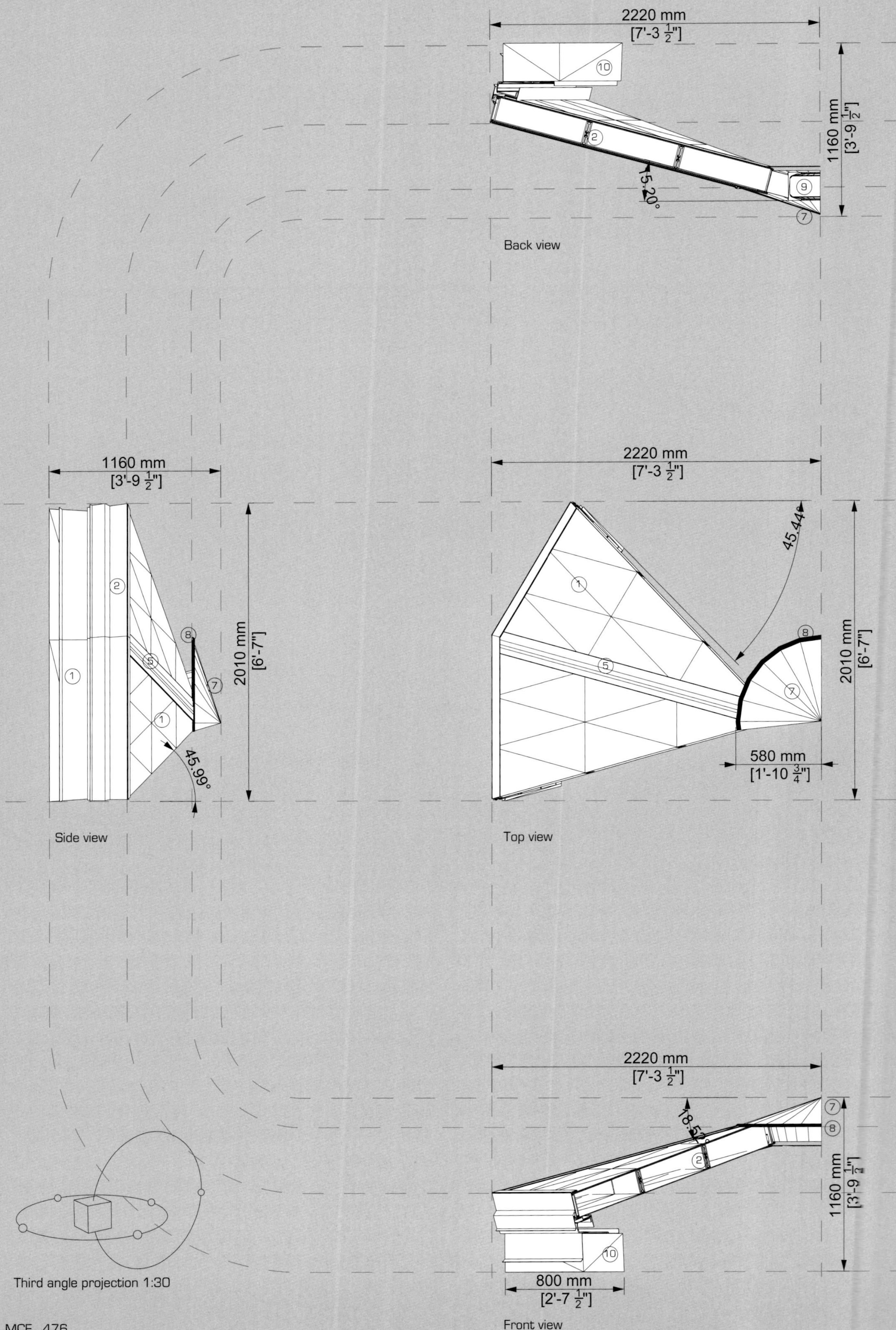
2220 mm
[7'-3 1/2"]
1160 mm
[3'-9 1/2"]
15.20°
Back view
1160 mm
[3'-9 1/2"]
2010 mm
[6'-7"]
45.99°
Side view
2220 mm
[7'-3 1/2"]
45.44°
2010 mm
[6'-7"]
580 mm
[1'-10 3/4"]
Top view
2220 mm
[7'-3 1/2"]
18.52°
1160 mm
[3'-9 1/2"]
800 mm
[2'-7 1/2"]
Front view
Third angle projection 1:30

Details

1 | Translucent and insulated GRP roof panel
2 | Thermally broken aluminium framing
3 | Inside
4 | Outside
5 | Pressure plate

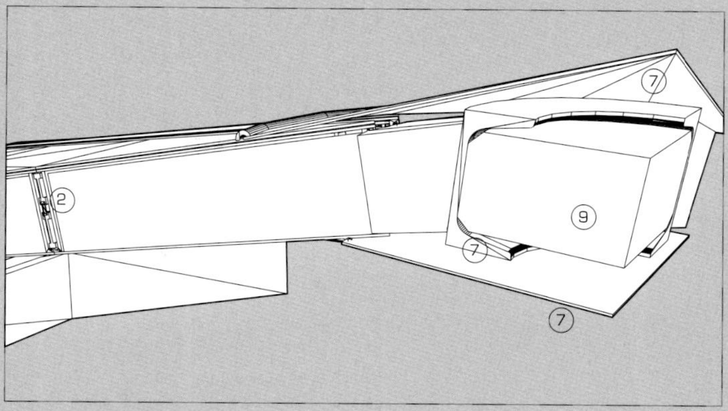

3D view of the assembly

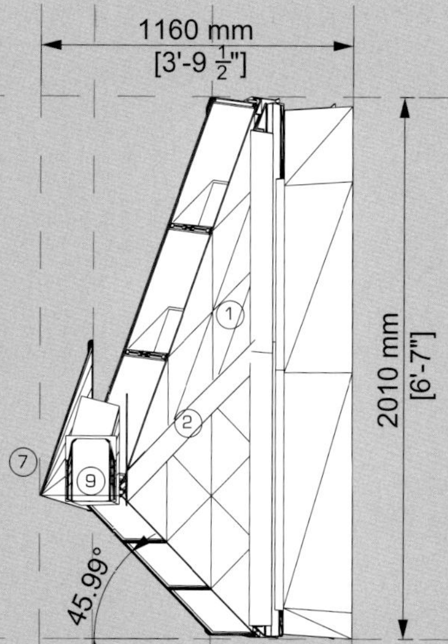

Side view

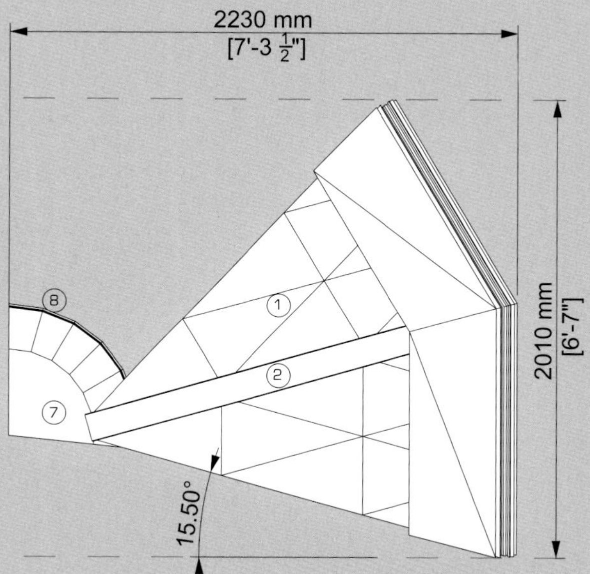

Bottom view

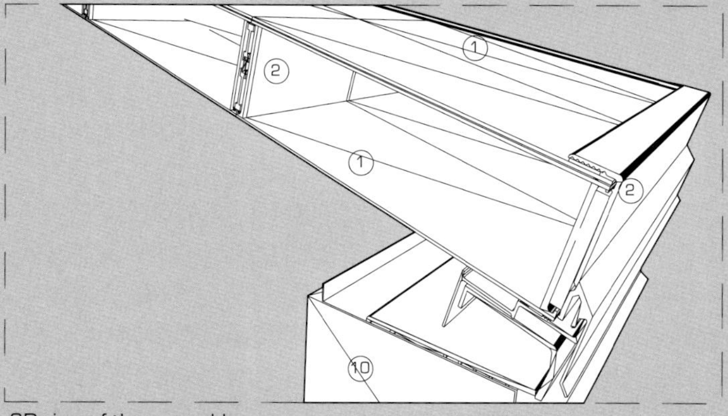

3D view of the assembly

6 | Adjacent masonry/ concrete wall
7 | Folded metal cover strip
8 | EPDM strip
9 | Supporting structure
10 | Concrete upstand

Details

1 | Translucent and insulated GRP roof panel
2 | Thermally broken aluminium framing
3 | Inside
4 | Outside
5 | Pressure plate
6 | Adjacent masonry/ concrete wall
7 | Folded metal cover strip
8 | EPDM strip
9 | Supporting structure
10 | Concrete upstand

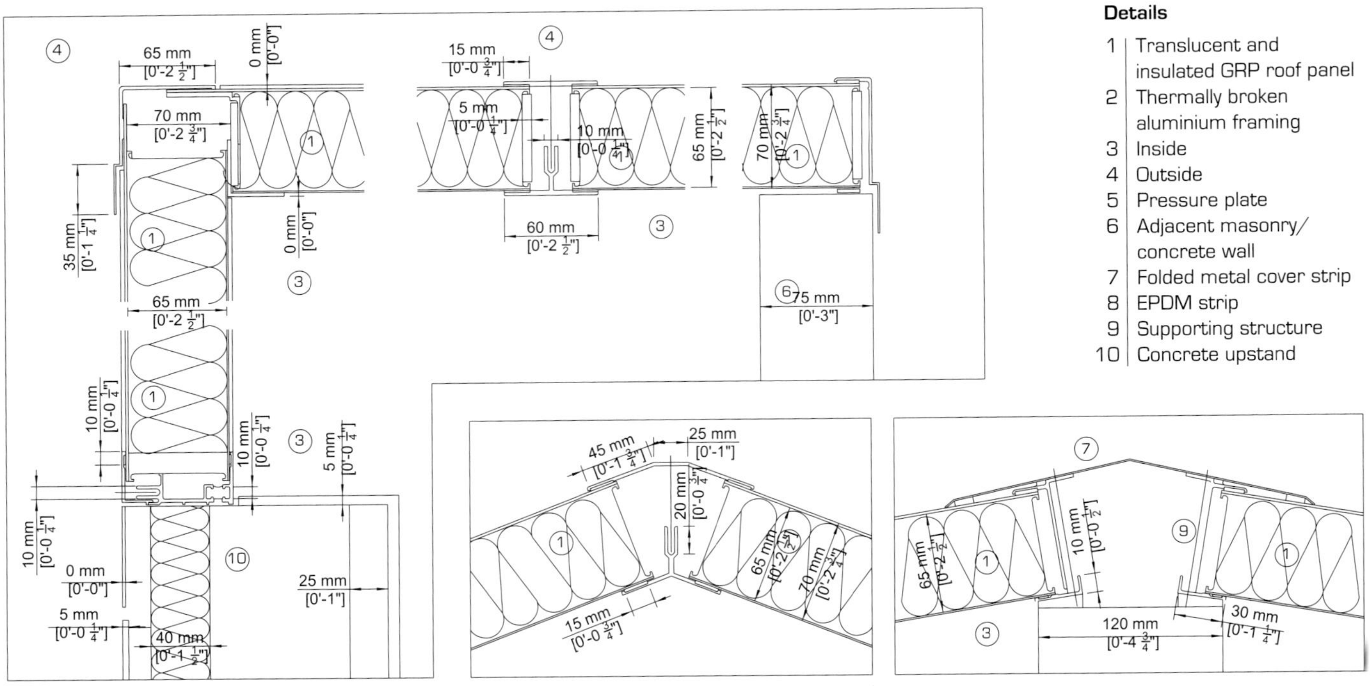

a| Vertical section. Small rooflight verge

b| Vertical sections. Small rooflight verge and eaves

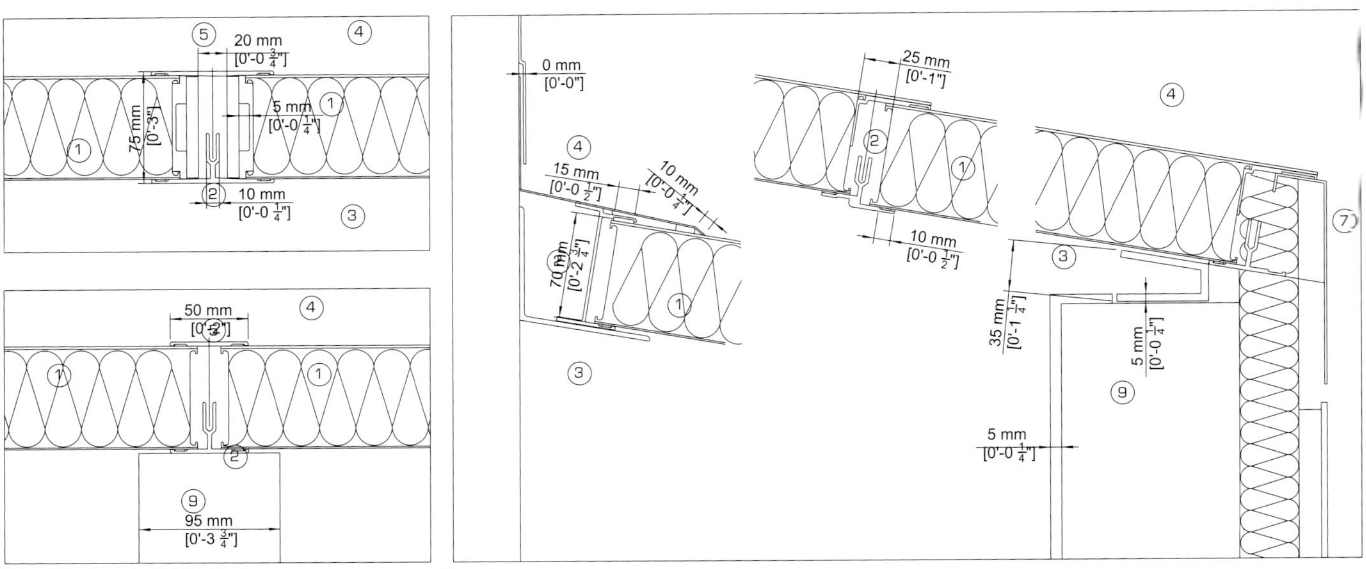

c| Vertical sections. Small rooflight panel to panel junction

Glass reinforced polyester (GRP) is used in the form of thermally insulated panels to form translucent rooflights (f) which are robust and economic when compared to an equivalent glazed rooflight with double glazed units. GRP rooflights use the principles of metal composite panels which are combined with greenhouse glazing framing to provide a lightweight, highly thermally insulated and economic rooflight. Where glazed rooflights sometimes need an additional layer of solar shading, or use solar performance glass to reduce the effects of the sun, GRP rooflights provide a more economic solution. The advantages of GRP over glass are its strength, lightness and flexibility, as well as the material's high resistance to impact damage. GRP is a composite material formed by reinforcing flexible fibreglass mat (or fibres) with thermosetting polyester resins that provide high tensile and compressive strengths. The material is not combustible, making it a suitable material for rooflights as well as an opaque roof cladding material. While GRP rooflight systems do not usually achieve the high levels of resistance to wind blown rain of internally drained and ventilated curtain walling systems, they are robust and economic, making them ideal where translucency is required rather than the transparency provided by glass panels.

Rooflights are made from GRP sheet which is bonded to an aluminium carrier frame around its edges. Thermal insulation set into the void is usually bonded to the outer GRP facing sheets to provide true composite action between the GRP skin and the insulated core. Like metal composite panels, GRP panels increasingly have a thermal break introduced into the framing to reduce the possibility of condensation forming on the underside of the panel in temperate climates as well as to improve the overall thermal insulation value of the rooflight. Thermal breaks are usually made from an extruded polymer that has a much lower thermal conductivity than aluminium, and are bonded to the extrusion in the manner of glazed curtain walling or are clipped to it and secured in place by self-tapping screws that hold the pressure plate in position (c).

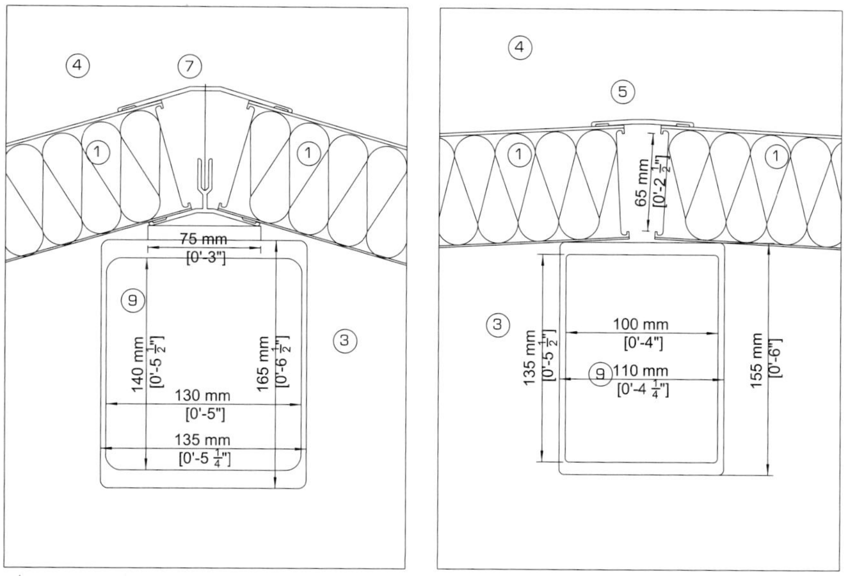

d | Vertical section. Large rooflight ridges

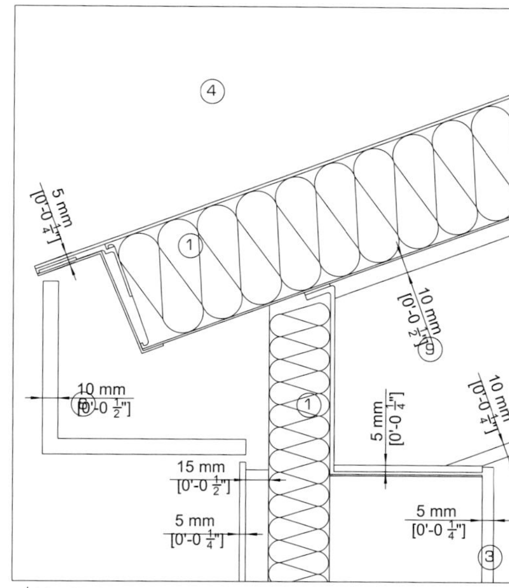

e | Vertical section. Large rooflight eaves

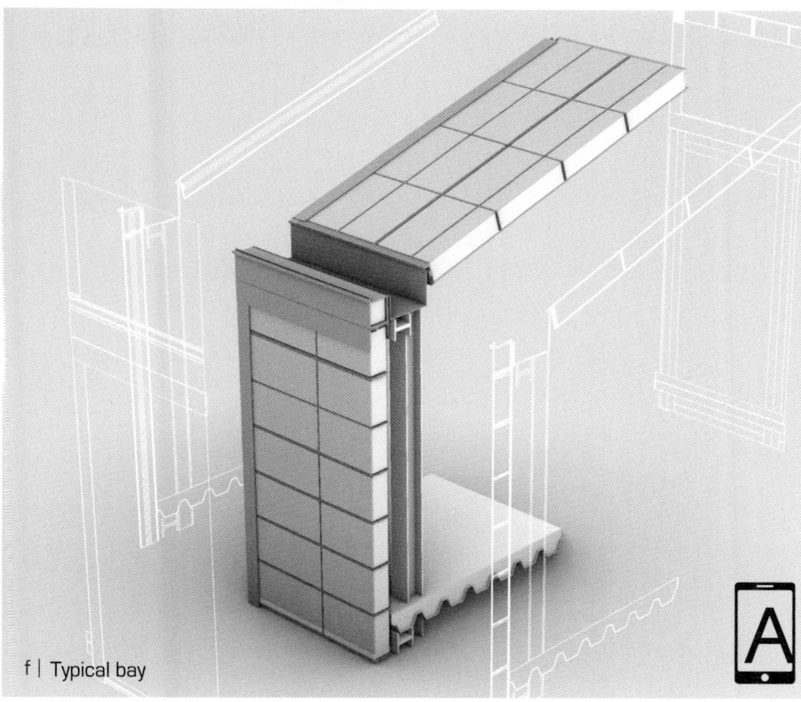

f | Typical bay

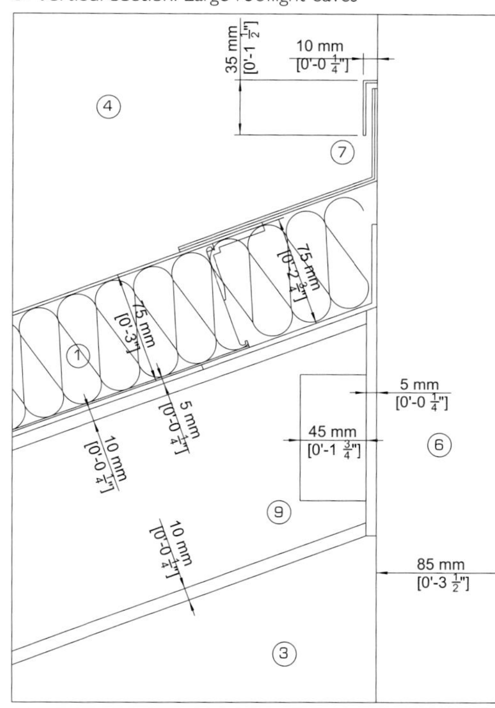

g | Vertical section. Large rooflight abutment

In small rooflights, up to around a 3000mm (10ft) span, GRP composite panels require no additional support, while those of greater span use an additional aluminium or steel frame beneath to support the composite panels over the greater span. Panel sizes vary with the proprietary system and with the individual rooflight design. Typical panel sizes range from around 400mm x 800mm (16in x 32in) to 800mm x 3000mm (32in x 120in). GRP rooflight panels use a lightweight framing system rather than a lapped junction or raised edges to providing a standing seam-type joint used in metal composite panels. These make their appearance more refined, since the framing is very visible in translucent GRP panels, unlike their metal equivalent. An extruded aluminium T-section forms the support to the panels on all four edges, with an extruded aluminium pressure plate being used to hold the panels in place on the outer face of the GRP panels (c). Most support frames are now internally drained and ventilated to provide a second line of defence against rainwater penetration. The outer seal is provided by an extruded EPDM gasket clipped into the aluminium extrusion. Proprietary tapes are also used, but are more dependent upon good workmanship on site than gaskets, which are fixed to the pressure plate in factory conditions.

Pressure plates for panel joints running down the slope are set over the joint between the panels in the manner of glazed curtain walling. Joints running across the slope sometimes have pressure plates with lapped joints in order to avoid water building up on the upper side of the joint and being unable to run over the joint. The lap is formed by setting an aluminium strip or extrusion under the bottom edge of the panel which laps over the top of the panel below (MCE_482, a). This combination of glazed curtain walling (or greenhouse glazing) techniques and composite metal panel techniques provides a method of construction for translucent rooflights which is visually lightweight with slender joint lines.

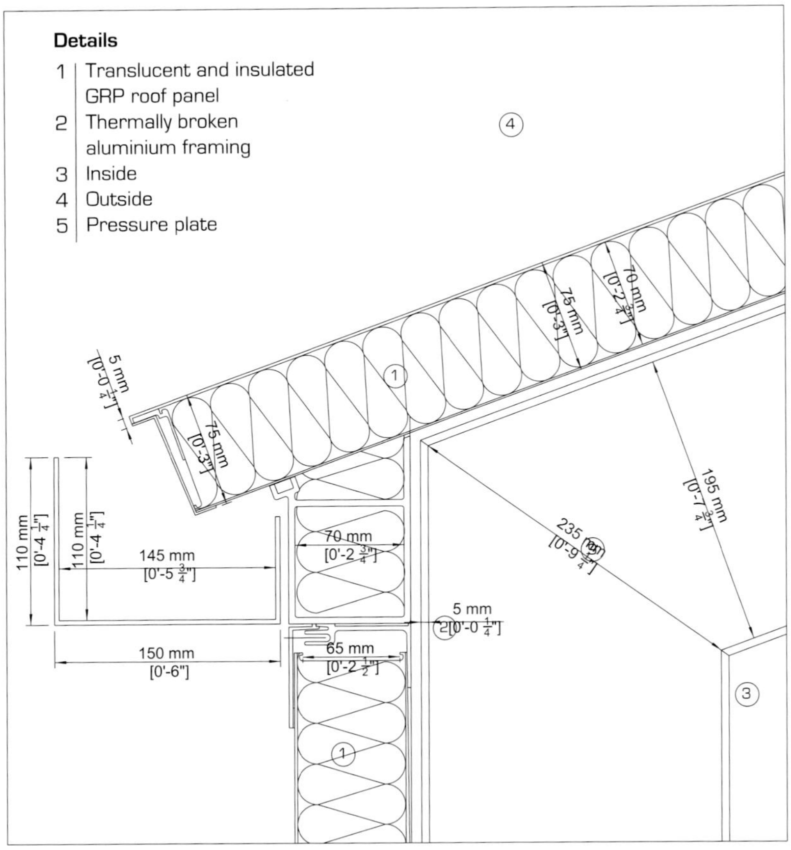

a| Vertical section. Large rooflight eaves

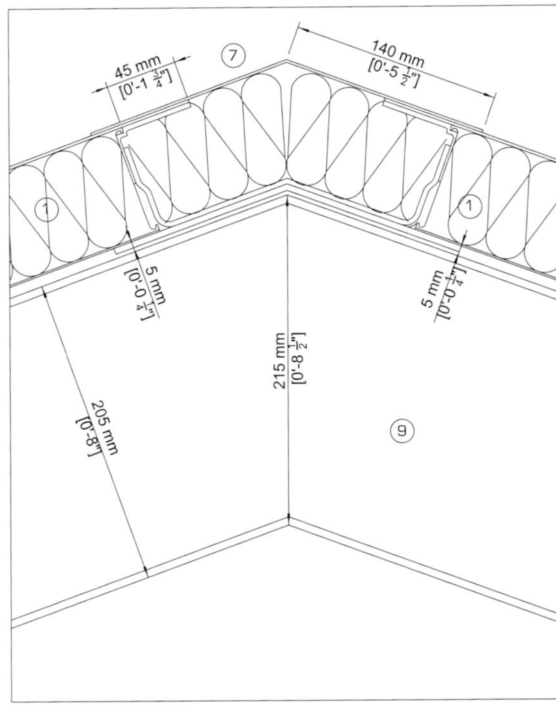

b| Vertical section. Large rooflight ridge

c| Vertical section. Large rooflight panel to panel junction

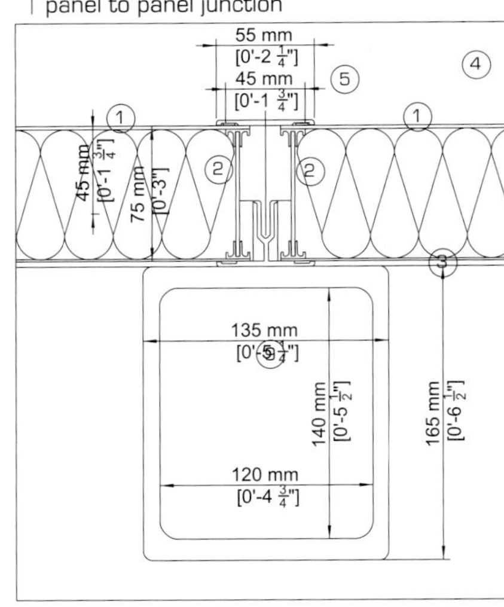

6 | Adjacent masonry/ concrete wall
7 | Folded metal cover strip
8 | EPDM strip
9 | Supporting structure
10 | Concrete upstand

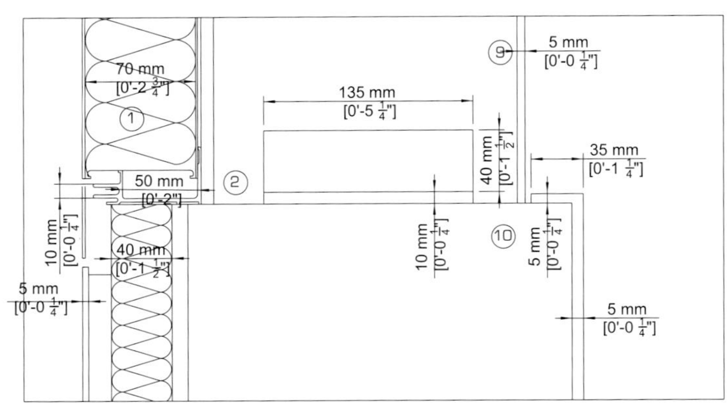

d| Vertical section. Large rooflight upstand at base

In addition to these standard junctions which form part of proprietary systems, panels are jointed with folded metal sheet, where unusual junctions are to be formed. A folded metal sheet is fixed over the top of the upper panels, to ensure that rainwater runs down the panel, and forms a drip where it is fixed to the top of the vertical panel. An inner metal sheet is bonded to the junction of the panels to provide an inner seal and vapour barrier. An alternative detail is to fill the void with thermal insulation in order to reduce the risk of condensation forming on the underside of the panel in temperate climates. This is typically used in a wider than typical joint where the provision of a thermal break is usually difficult to achieve. A second line of defence against rainwater penetration is provided by a membrane set immediately beneath the metal flashing rather than on the inner joint between panels, as is the case elsewhere, because the thermal insulation cannot be set within the drainage cavity where it would both impede the passage of the small amounts of water that would be drained in the internal cavity, as well as lose its thermal insulation properties due to the absorption of water.

A ridge flashing can be formed in the same way, with a metal sheet fixed to the upper face of the GRP panel frame and a membrane waterproofing layer beneath. Any rainwater that penetrates the outer seal is drained either at the ends of the ridge or down the joint between the panels that run down the slope of the roof (a).

Eaves and upstands

When a rooflight terminates in an eaves, an intermediary aluminium extrusion or folded sheet is used to form the junction (g). An outer EPDM or extruded silicone seal is used as an outer line of defence against rainwater penetration at the junction with the GRP roof panel. Drainage slots formed in the bottom of the aluminium closer piece drain away any water that passes through the outer seal. These drainage slots also take away any water to the outside from the drainage channels within the joints between panels that run down the slope of the roof. The GRP panel beneath the closer piece that forms a vertical

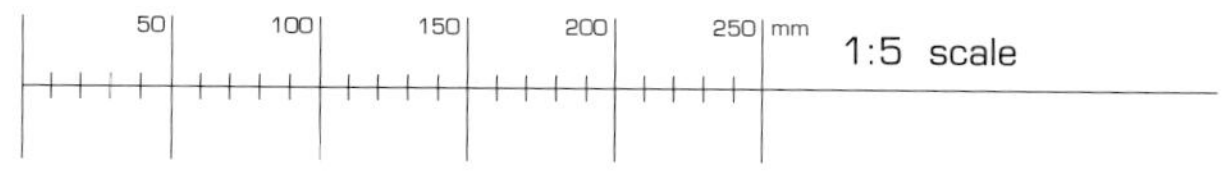

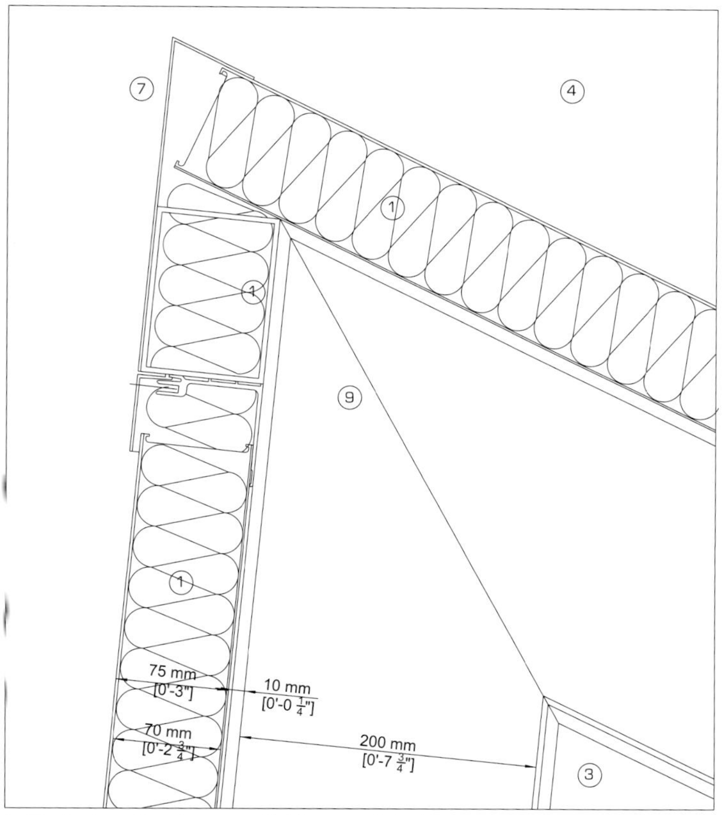

f Vertical section. Large rooflight monopitch ridge

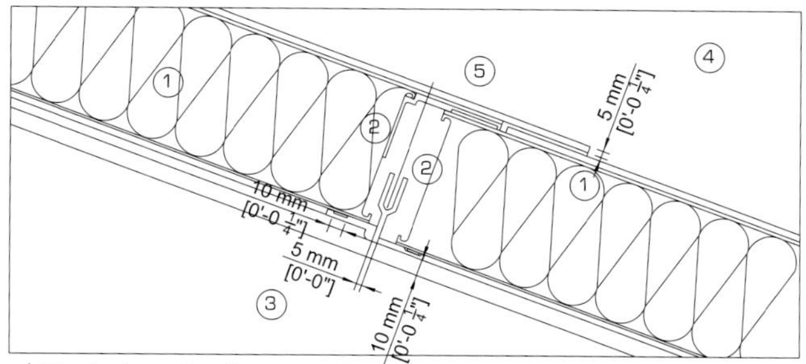

e | Vertical section. Large rooflight internal fold

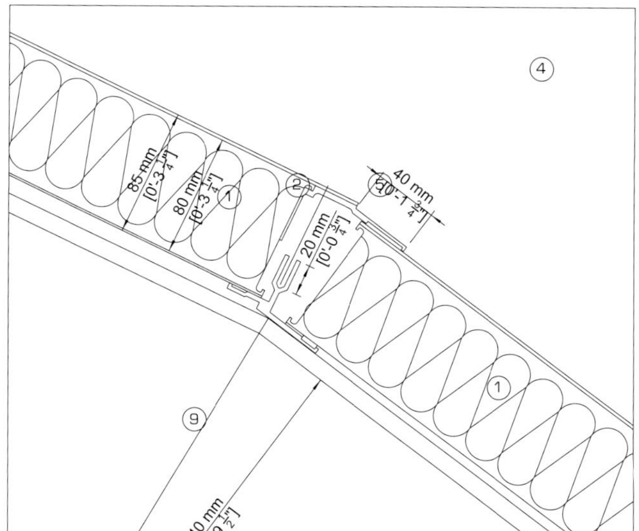

g | Vertical section. Large rooflight internal fold

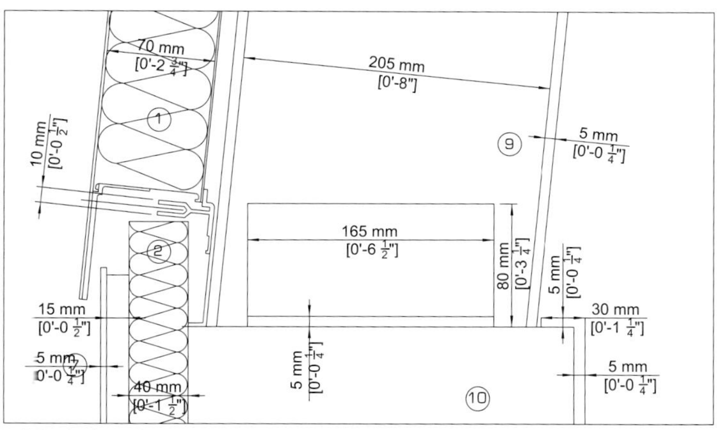

h | Vertical section. Large rooflight base

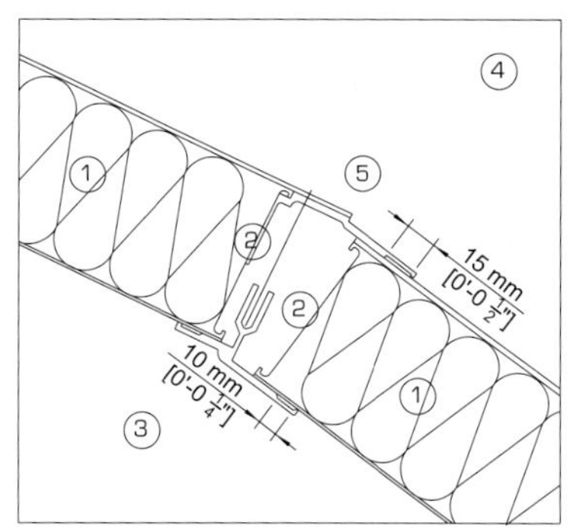

j | Vertical section. Large rooflight external fold

part of the rooflight is formed by setting the panel behind a vertical aluminium strip that forms a lapped joint over the top of the panel (a), avoiding the possibility of rainwater passing through the joint. The gap between the aluminium closer and the GRP panel is sealed with either an EPDM gasket, a proprietary tape, or a silicone sealant. A metal gutter is fixed to the metal closer if required, but this is usually exposed unless it forms part of a fascia, such as the curved eaves used in profiled metal roofing, for example. In smaller rooflights the rainwater typically runs off onto the surrounding area of flat roof.

Upstands are formed either with an aluminium closer piece, as used at eaves, or with a metal flashing, where the GRP panel is supported on an additional steel or aluminium supporting frame. A metal closer piece is used to provide a continuous edge support to the GRP panel as well as a weathertight seal (MCE_478, c). A waterproof membrane is bonded to the outside of the closer piece and is secured to the upstand below, typically formed in reinforced concrete. The closer piece is thermally insulated and is sealed on its inner face with a thin folded aluminium strip that serves both to retain the thermal insulation in place and as a vapour barrier. The visible width of the inner closer strip usually matches that of the adjacent aluminium joints between the panels and is visible within the building. The gap between the bottom of a GRP panel and the upstand beneath is sealed with a metal flashing set forward of the vertical face of the upstand. This allows water to drain out of the ventilated channels in the joints that form the second line of defence against rainwater penetration. A waterproof membrane, typically EPDM, is bonded to the bottom edge of the metal frame surrounding the GRP panel and is sealed against the waterproofing layer of the upstand, providing a continuous seal from the surrounding area of flat roof up to the rooflight. A thermal break in the framing of the GRP composite panel ensures a continuity in the thermal insulation from upstand to GRP rooflight.

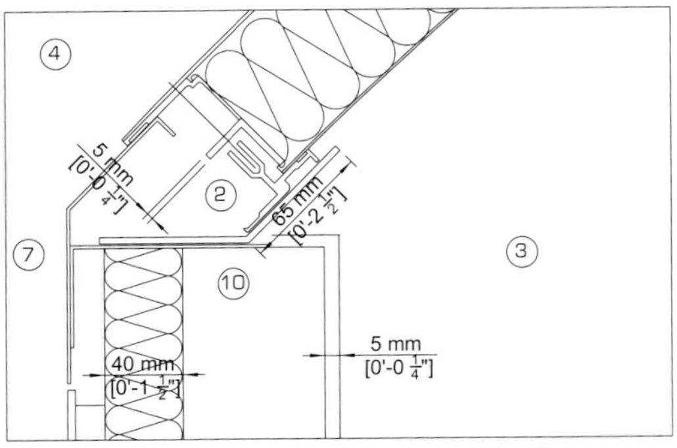

a | Vertical section. Small rooflight upstand at base

Details

1 | Translucent and insulated GRP roof panel
2 | Thermally broken aluminium framing
3 | Inside
4 | Outside
5 | Pressure plate

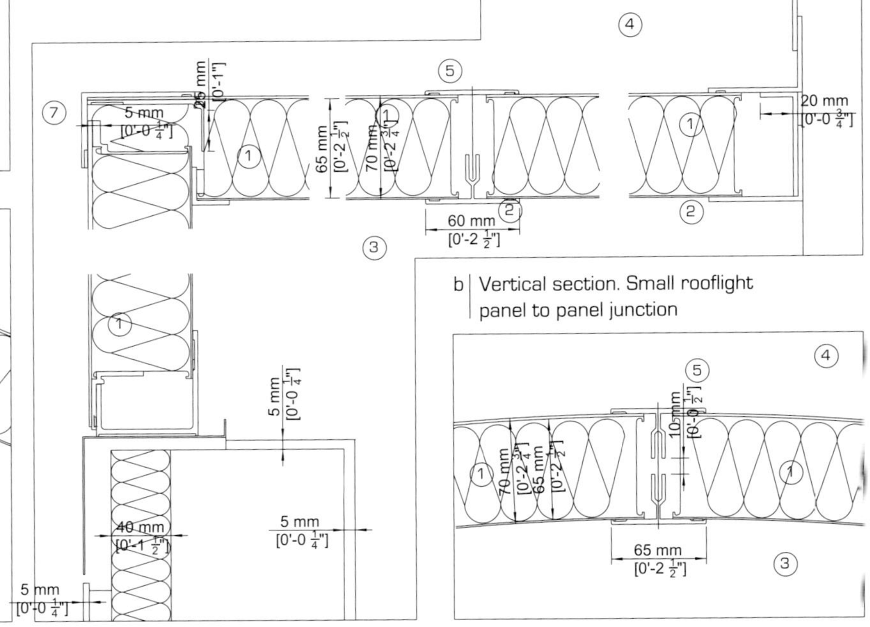

b | Vertical section. Small rooflight panel to panel junction

d | Vertical section. Abutment

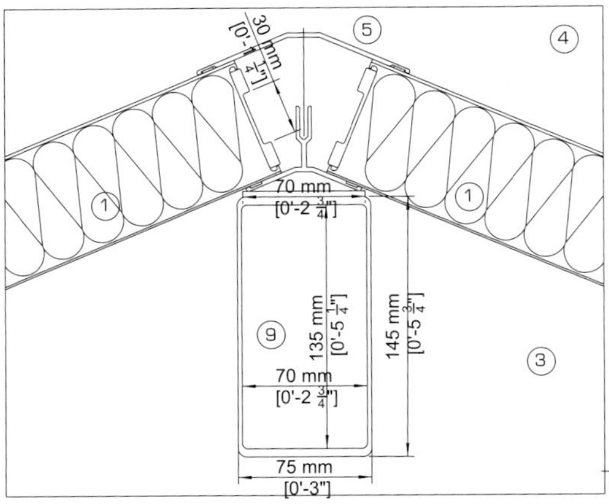

c | Vertical section. Small rooflight ridge

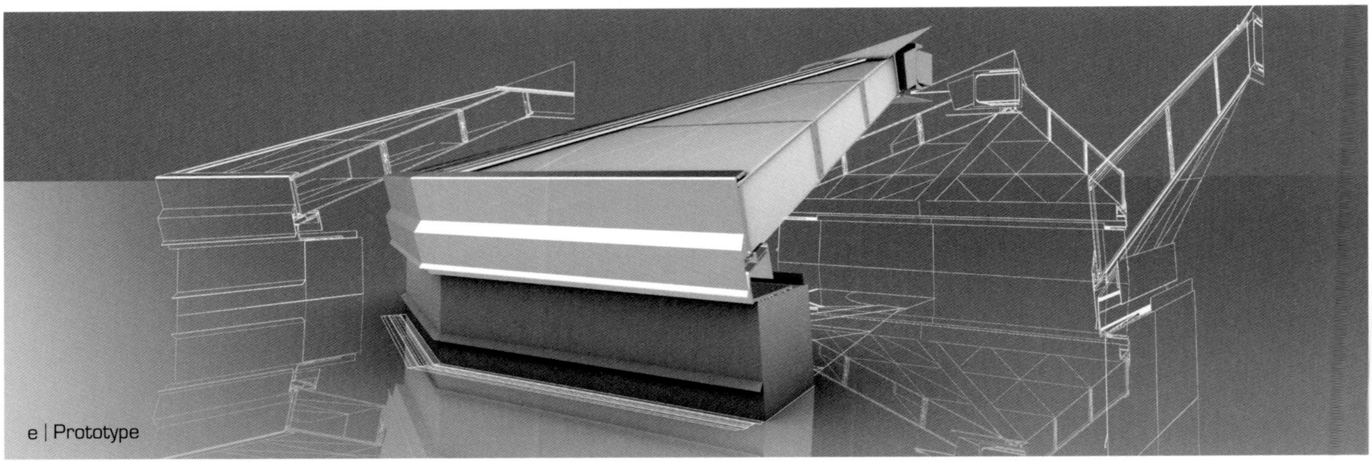

e | Prototype

Verges

The gable ends of sloping GRP rooflights are formed with an aluminium flashing that is bonded, or mechanically fixed and sealed, to the metal edge frame to the side of the GRP panel forming the sloping panel, and to the top of the triangular-shaped vertical end panel (j). The sloping panel can extend slightly forward of the vertical panels to give a thin edge to the roof. Alternatively the roof can terminate in a sharp edge, with a folded aluminium sheet or angle closing the gap between the panels. As with other panel-to-panel junctions, a waterproof membrane is set on the underside of the outer metal flashing as a second line of defence against rainwater penetration. The void between the panels is filled with mineral fibre quilt-type thermal insulation, which has the flexibility required to fill the irregular-shaped voids between GRP panels. An additional aluminium angle is used at the junction of the inner face of the panels to provide an additional seal and vapour barrier.

Abutments

Where a GRP rooflight is set against a vertical wall in another material, such as reinforced concrete or concrete block, a metal flashing is used. Beneath the metal flashing, a waterproof membrane is bonded to the edge of the GRP panel and is folded up the abutting wall to which it is bonded (MCE_479, g). Where wall and rooflight are not connected structurally, the membrane usually has an extra curve of material between the panel and the wall to allow for structural movement between wall and roof. The outer line of defence formed by the metal flashing is set onto the membrane and is folded up the vertical face of the wall. An additional metal flashing is used to cover the top edge of the metal flashing where significant structural movement is expected; the outer metal flashing being fixed into a continuous groove or horizontal joint line in the concrete or masonry wall to provide a weathertight seal to the top of the flashing.

1:5 scale

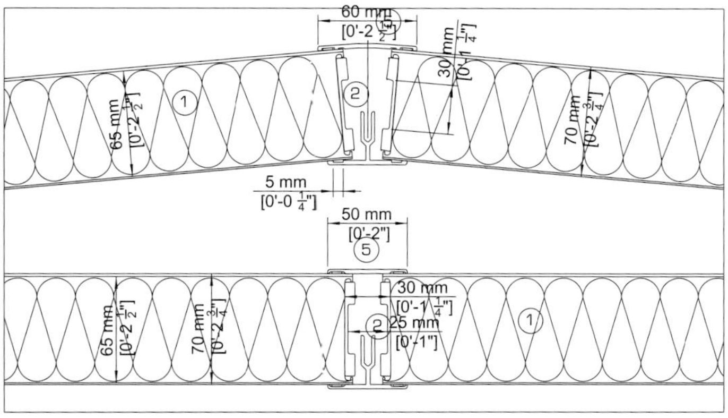

f | Vertical sections. Small rooflight panel to panel junction

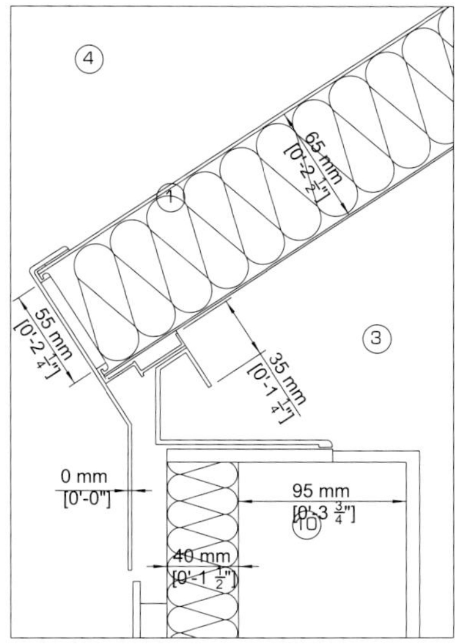

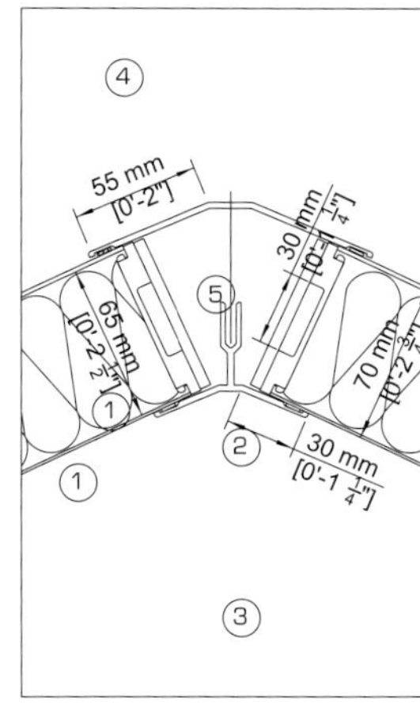

g | Vertical section. Small rooflight ridge

6 | Adjacent masonry/concrete wall
7 | Folded metal cover strip
8 | EPDM strip
9 | Supporting structure
10 | Concrete upstand

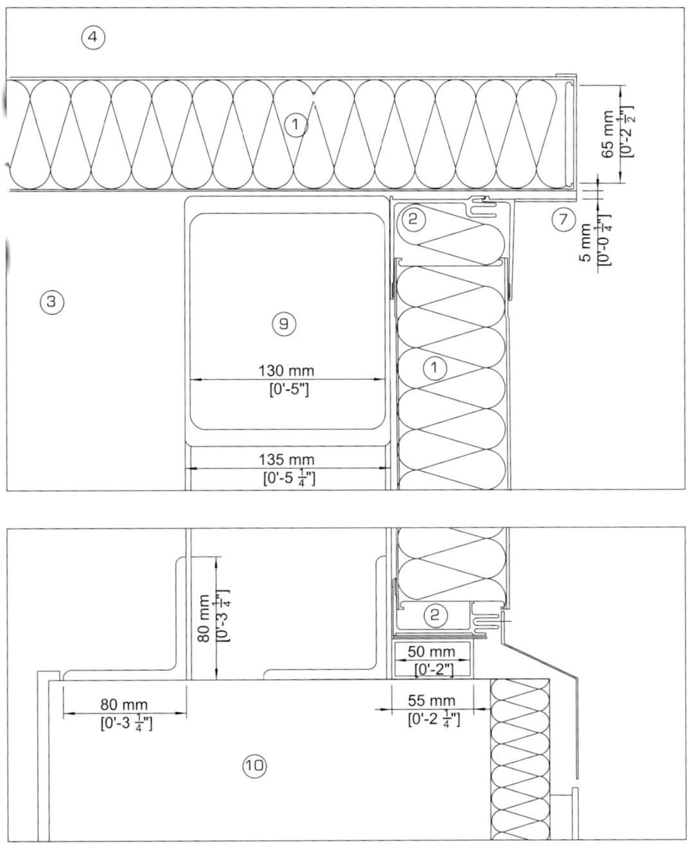

h | Vertical section. Verge

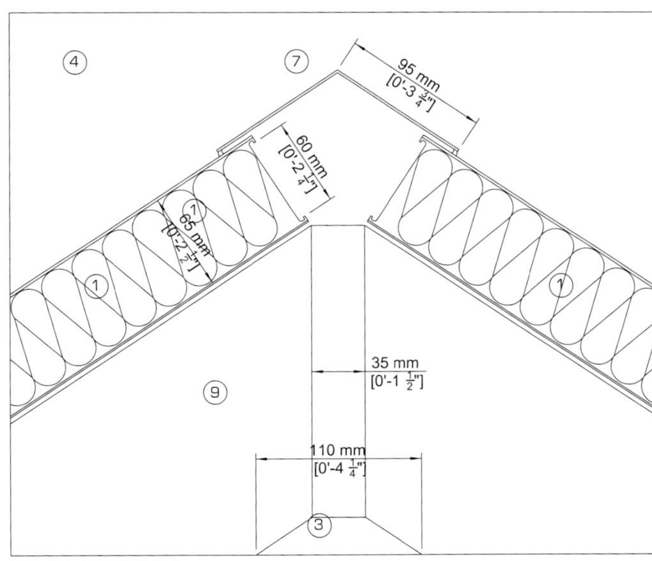

j | Vertical section. Ridge

1:5 scale

Sliding roof panels

The lightweight nature of GRP rooflight panels is being used increasingly in sliding panels that allow a rooflight to be opened at different times of the year. Typically up to around 40% of panels in a rooflight are opened by sliding them over adjacent fixed panels, in relatively modest electrically operated systems. Whereas larger-scale glass rooflight panels are difficult and expensive to make openable due to the weight of the glass and the complexity of their seals, GRP rooflight panels are easier to slide, with seals being provided by lapped joints with EPDM or extruded silicone seals in the manner of sliding doors. Opening panels are sealed to the levels of air infiltration associated with opening windows, which are usually lower than those of fixed glazed rooflights.

Panels can move either side to side in the manner of sliding doors, or vertically in the manner of sash windows. The same principles of fixing and waterproofing joints are used regardless of the direction of movement of the panels. For example, one side of the horizontally sliding panel would have an upstand, while the other three sides would be set into a slot. In the slot connection, the outer (upper) seal comprises EPDM or silicone-based gaskets, usually in a 'flipper' section that allows the aluminium framing to slide in and out of the housing on one side, and to slide on the other two sides. Any rainwater that penetrates the outer seal is drained away in a cavity within the rooflight frame, draining out through holes in the bottom of the aluminium profile onto the roof below. An inner air seal, either polymer foam based or similar to the outer gasket seals, is provided on the underside of the GRP panel. These general principles of sliding panels will no doubt be developed to suit more complex geometries of openable panels over the next ten years. The pop-out type sliding doors that move outward before sliding, together with hinged panels, will make this rooflight system more generally applicable to large-scale applications of all building types.

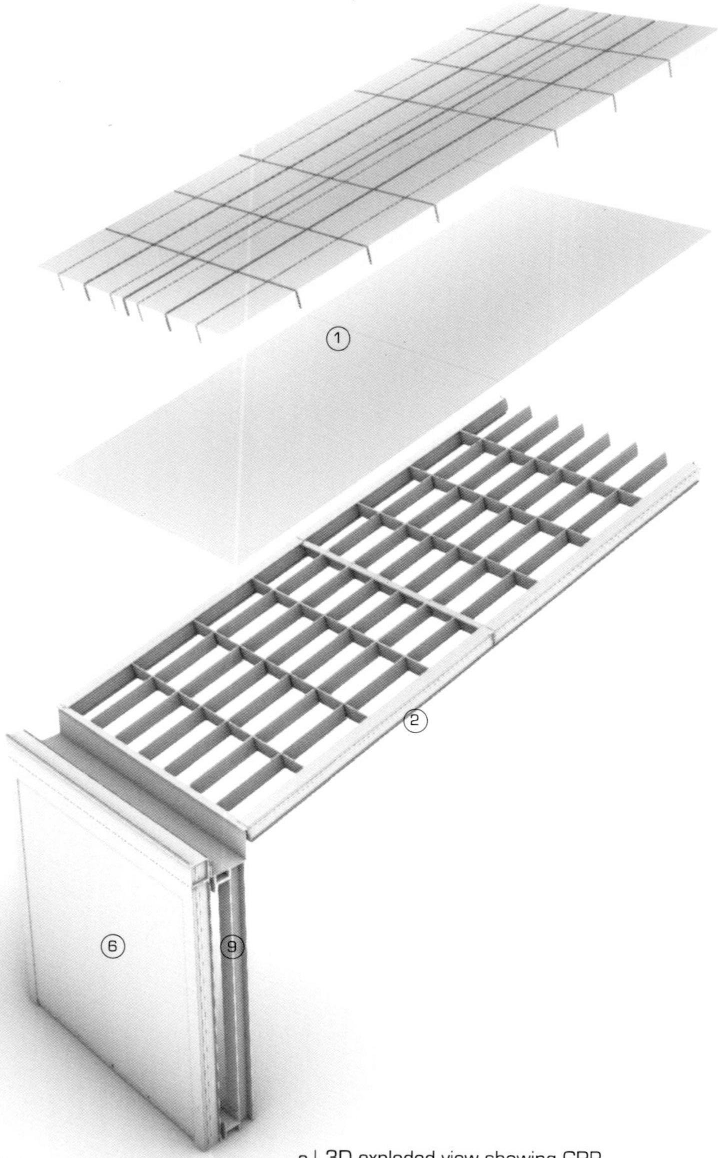

a | 3D exploded view showing GRP rooflights fixed to steel structure

b | 3D view showing GRP rooflights fixed to steel structure

Details

1 | Translucent and insulated GRP roof panel
2 | Thermally broken aluminium framing
3 | Inside
4 | Outside
5 | Pressure plate
6 | Adjacent masonry/concrete wall
7 | Folded metal cover strip
8 | EPDM strip
9 | Supporting structure
10 | Concrete upstand

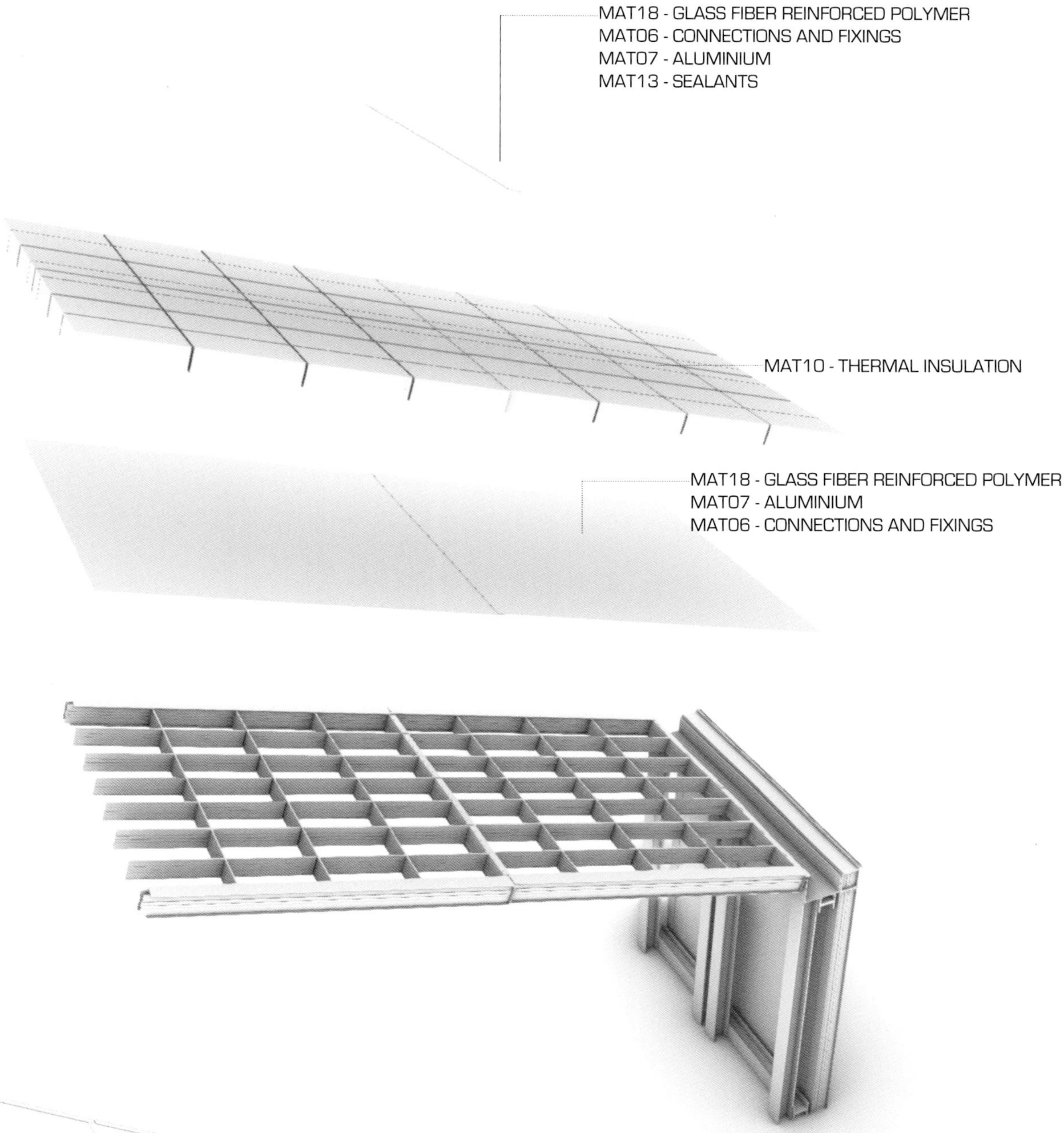

c | 3D exploded view of GRP rooflight components with material specification references

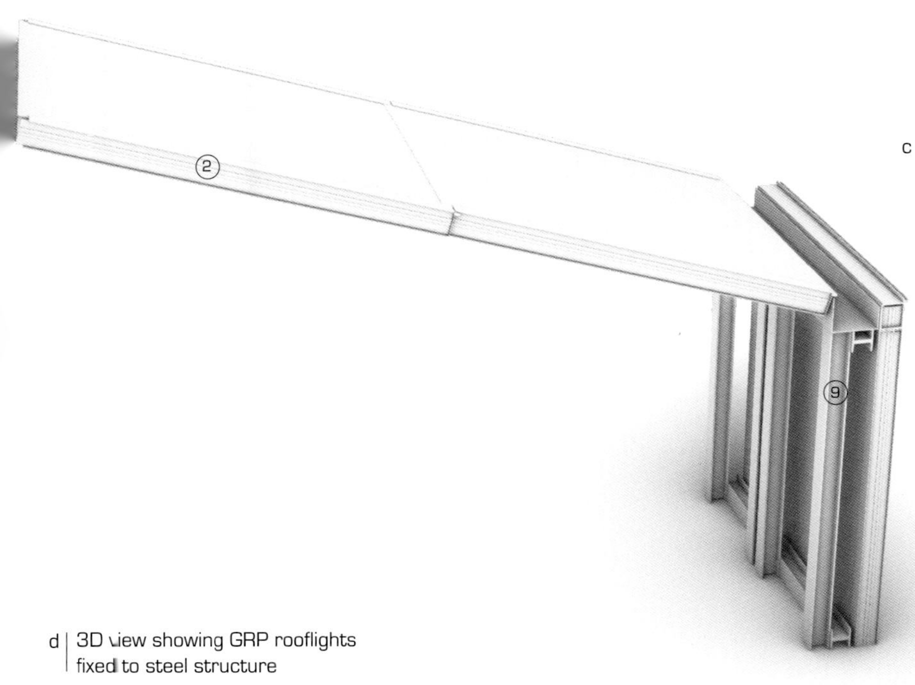

d | 3D view showing GRP rooflights fixed to steel structure

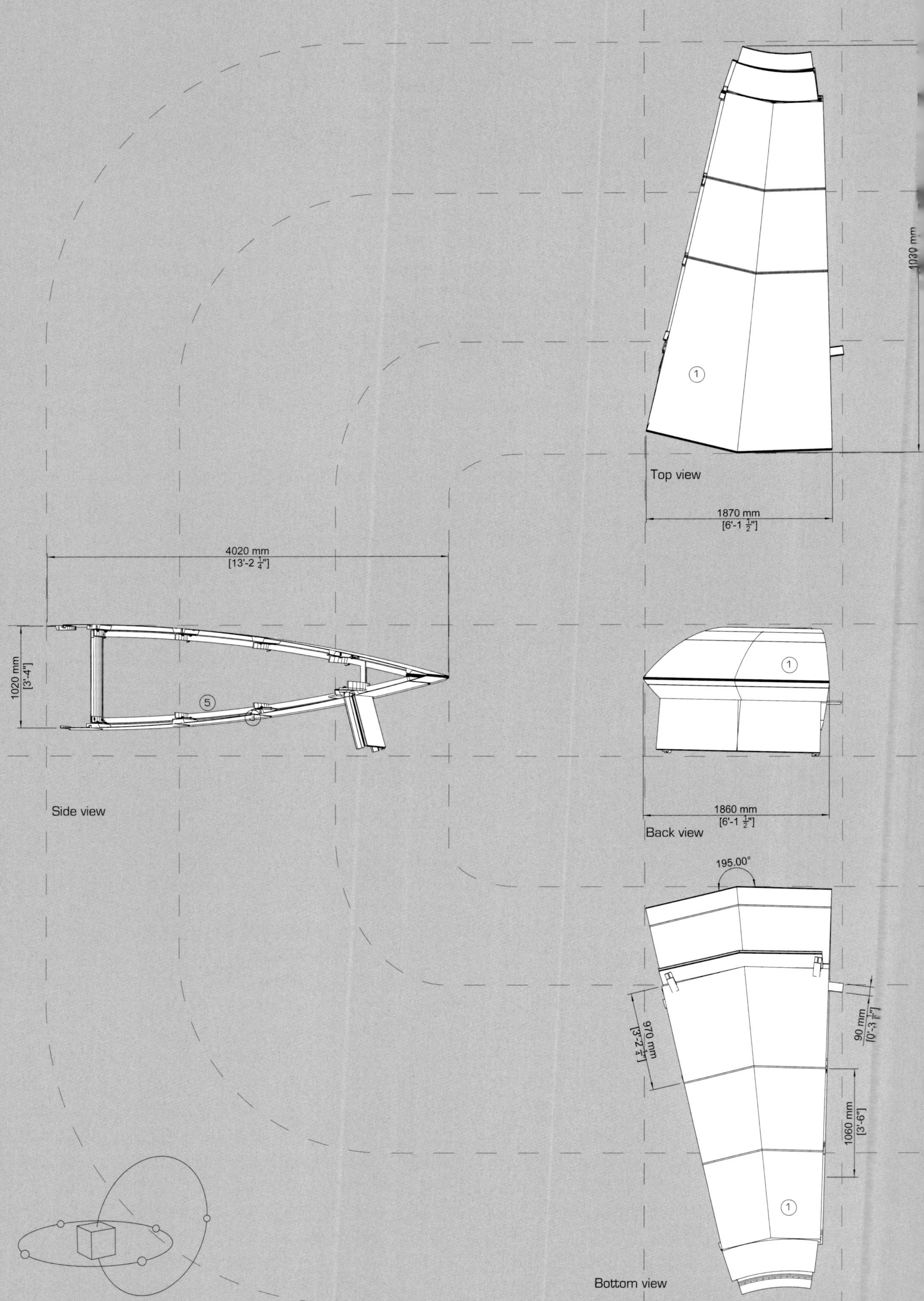

Third angle projection 1:30

Details

1 | GRP shell panel
2 | GRP outer cladding panel
3 | GRP structural rib
4 | GRP thin panel with honeycomb core

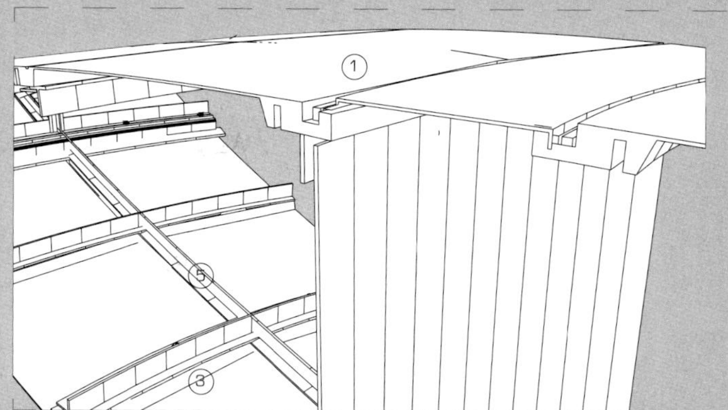

3D view of the assembly

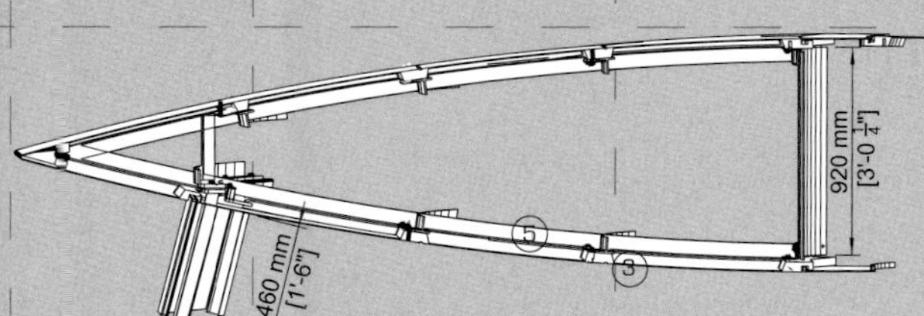

Side view

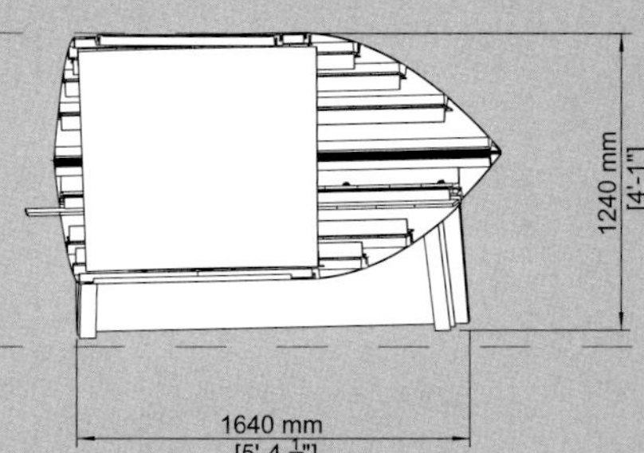

Front view

5 | Mild steel or aluminium frame
6 | Mild steel or aluminium truss
7 | GRP flashing
8 | Waterproofing membrane
9 | Metal fixing bracket
10 | Thermal insulation

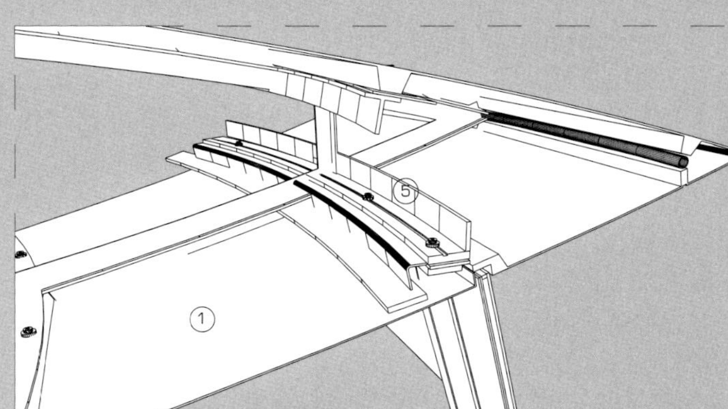

3D view of the assembly

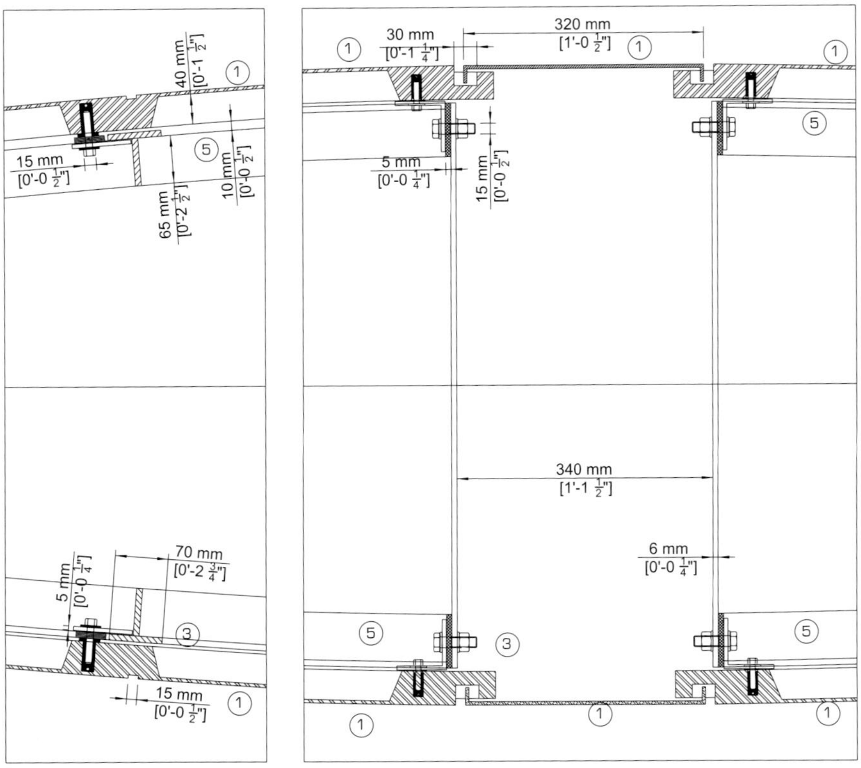

a| Vertical sections. Panel-to-panel junction

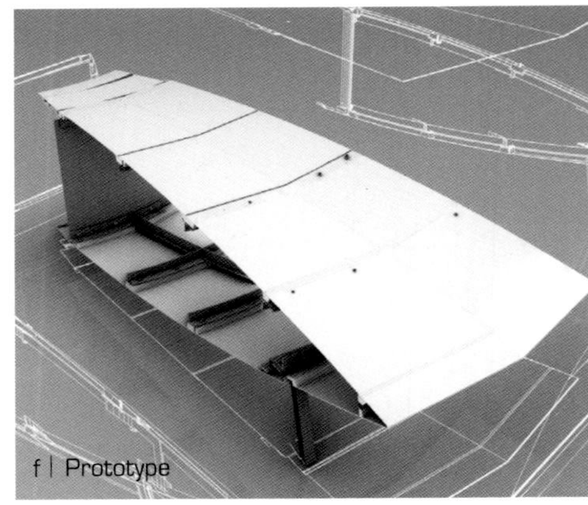

f | Prototype

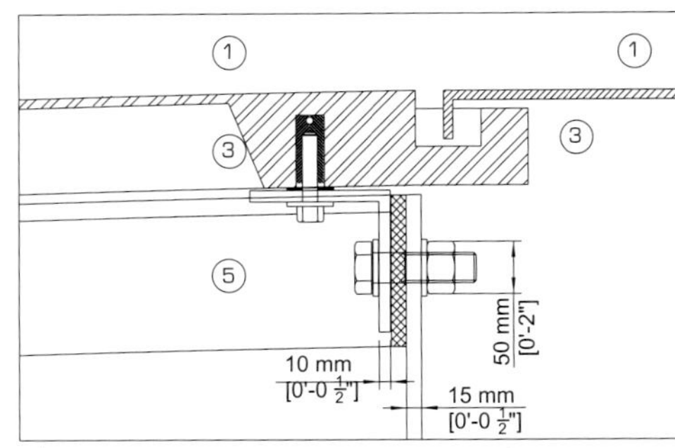

b| Vertical section. Panel-to-panel junction

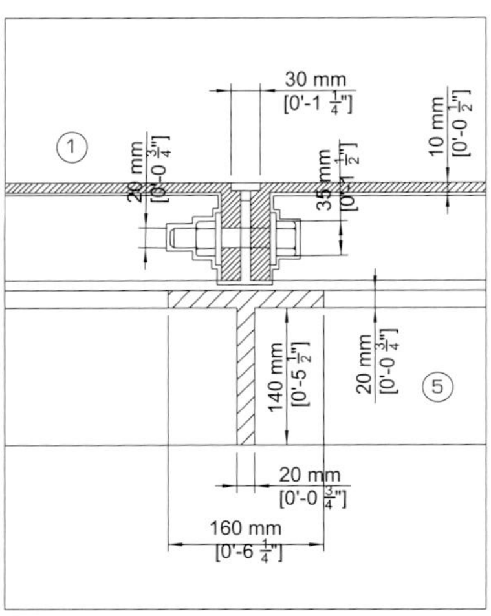

c| Vertical section. Panel-to-panel junction

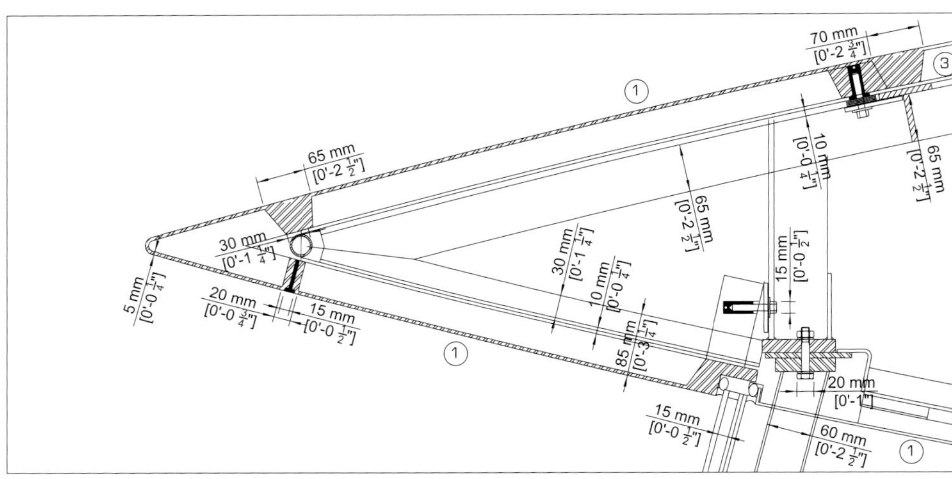

d| Vertical section. Perimeter

Where glass reinforced polyester (GRP) rooflights, discussed in the previous section on rooflights, are made as panels which are joined to form translucent rooflights, opaque GRP panels can be made as monolithic, self-supporting shells (e). These shells are usually made from panel segments which are brought to site and bolted together. The segment sizes of GRP shells are made in sizes which are suitable for transportation by road, usually set upright on a trailer. The shells can then be lifted by crane into place as a completed assembly, which makes them quite different from roof structures in other materials.

Smaller panels and shells

Smaller shells consist of a set of segmented panels which are bolted together to form a roof shell of approximately 7.0 metres in diameter. Some variations are supported by an additional frame (MCE_493, g), while others are self-supporting GRP shells.

GRP panels can be supported by a light metal frame beneath. The frame comprises steel or aluminium T-sections which are welded together to form a structure that supports the complete outer skin (MCE_491, c). The frame has curved members that radiate from the centre at the top to the edge and from the centre at the lowest point of the structure, back to the perimeter. The radiating 'spokes' of the wheel are held in place by T-sections that, in plan, form concentric circles. This 'bicycle wheel' form is supported near its perimeter by a metal ring beam that is set immediately above the glazing beneath the GRP roof (e). The ring beam is supported by posts that are fixed to the roof deck beneath.

The metal frame is clad in prefabricated GRP panels which are bolted to the support frame on their internal face in order to avoid visible fixings. Panels are made with an outer skin of GRP around 5mm thick, with an overall panel depth of around 45mm for the panels sizes shown of 3500mm long and 1800mm wide. The GRP

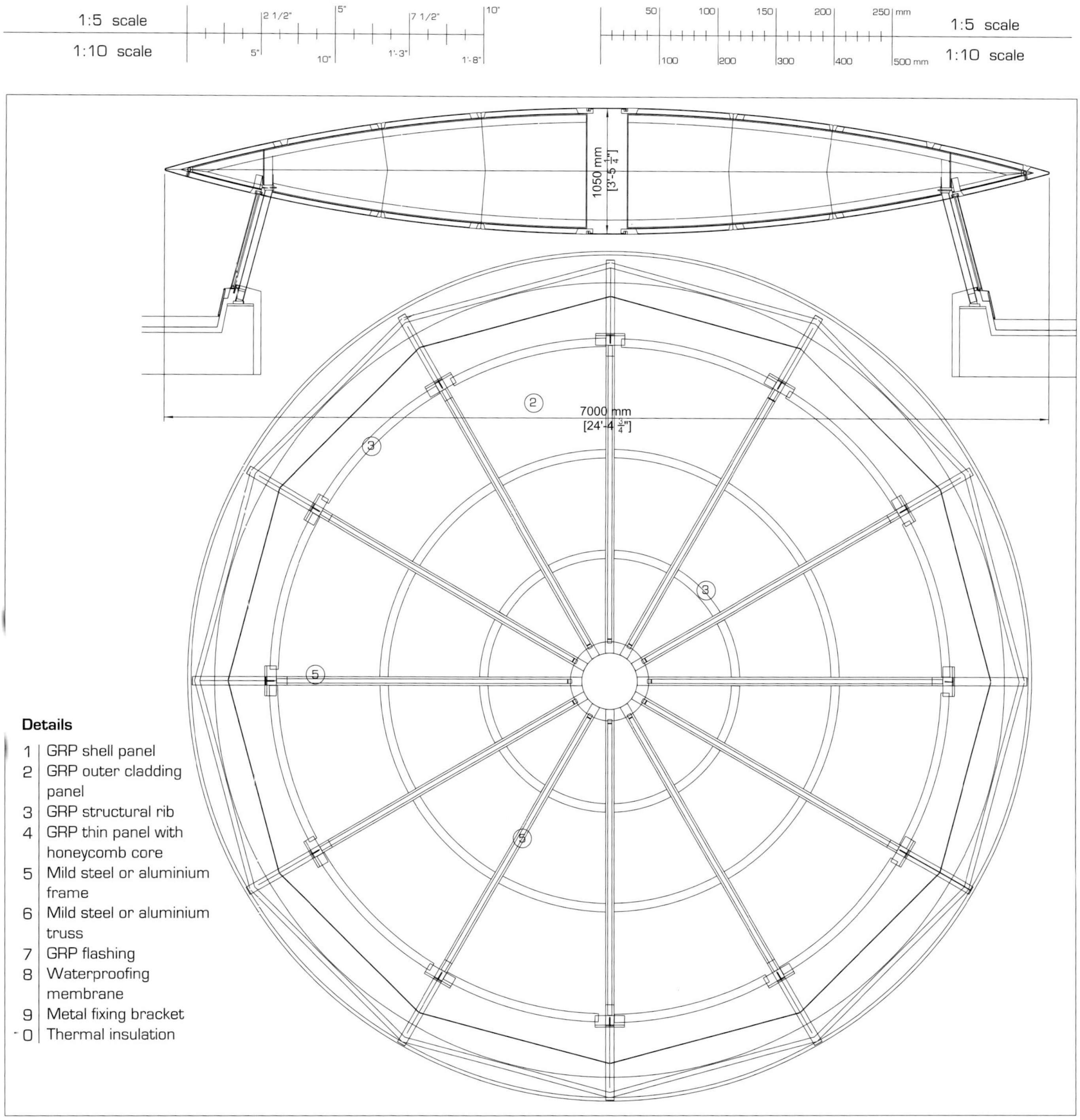

e | Vertical and horizontal section 1:50. Generic examples A1 and A2

panels are stiffened by concentric ribs, around 120mm wide. The long edges of panels do not require thick ribs, and these are around 10mm thick . Panels are secured with bolts which are fixed through the metal support frame into reinforcing ribs at the edges of the GRP panels. Joints between GRP panels are formed by butting panels up to one another and sealing the gap between the panels. The seal is formed in a continuous step profile on the long edges of each panel that creates a continuous groove at the joint between panels. The groove is filled with a lamination of glass fibre and resin to fill the groove to the level of the top of the panels. The external face of the GRP is then ground smooth, usually by a hand-held grinder, to achieve a uniform, smooth surface that conceals the joint lines. Finally, a paint finish is applied, usually as a spray, to give a smooth and reflective finish. Where pigments are applied to the top coat, or 'gel' coat, in the factory, a more limited range of colours is available. Thermal insulation is set on the underside of the shell, being bonded to the inner face of the GRP panels in order to achieve continuity of insulation.

Typical examples use glass fibre ribs that form part of the shell to provide integral structural stability to the shell. Panels are generally around 200mm deep, and are bolted together to form a self-supporting GRP shell. Ribs are made in solid GRP in order to allow them to be bolted together easily (MCE_494, a). Panels are joined and sealed on their external face. There are various methods for forming the top of the roof. Where panel segments converge at the top of the shell, a separate centre panel is used to create a smooth external finish. A central external panel avoids the difficulty of bringing up to 16 panels together at a single point, which would make it difficult to achieve a smooth transition from one side of the shell to the other. In the example shown, a shadow groove is set around the joint between the central panel and the segments in order to avoid any misalignment between segmented

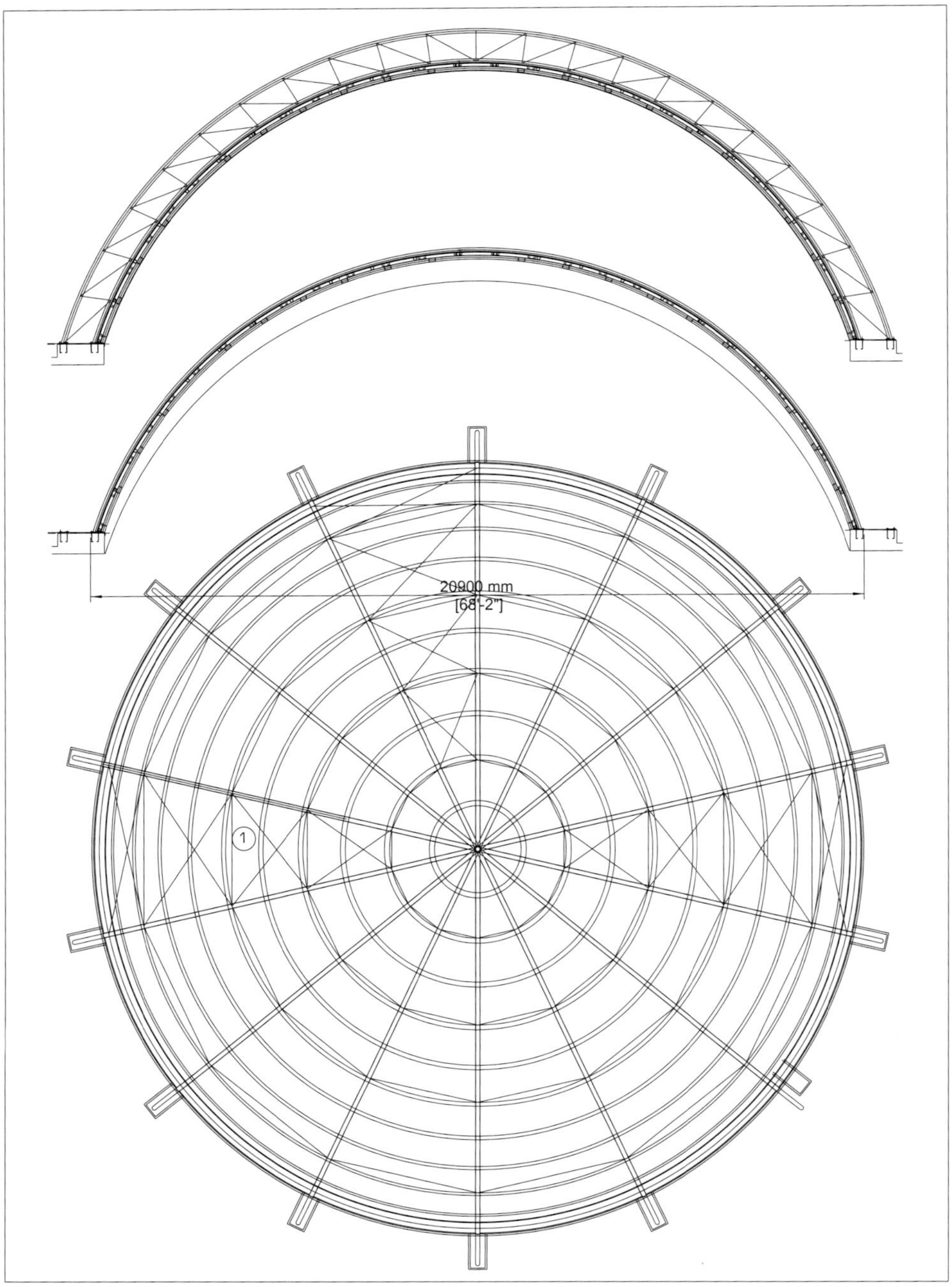

a | Vertical and horizontal section 1:200. Generic examples B1 and B2

b | Prototype

panels from being visible. The perimeter joint of the panel can be filled and sealed using the method described earlier, with additional grinding required on site to ensure a smooth finish.

The underside of the shell is mostly set internally behind curved glazed units installed near the perimeter of the roof. The internal part of the shell on its underside has the same panels with the same finish, but without any thermal insulation, which is typically set around the outer edge of the shell in order to keep temperatures within the void close to those within the building. Close to the edge of the shell, on the underside, the joint between the top panel which curves around the edges and the adjacent panel underneath has a groove formed along the joint to serve as a drip. Wind-blown rain will still be pushed along the underside of the soffit panel, but the drip reduces the amount of water that runs down to the glazing below.

The curved glazed units beneath the roof shell are set into a recessed slot at the panel joint. The recess avoids the possibility of a weakness in the seal that could result in using a butt joint between shell and glazing. The edge of the double glazed unit is sealed completely around the junction with the shell, typically with silicone.

Both the cladding panels and the self-supporting shell are made using the same manufacturing process. Panels are made in a mould, usually from a single segmented panel type to form a complete rooflight. Moulds are usually made from plywood to create the shape and are then finished in GRP to create the negative shape of the panel being formed. GRP panels are fabricated by first applying a release agent to the mould to allow the finished panel to be removed easily, then thermosetting polyester resins are applied to the face of the mould, with flexible fibreglass mat being laid into the resin, usually with rollers. The process of fabricating GRP panels is very labour-intensive, but requires no expen-

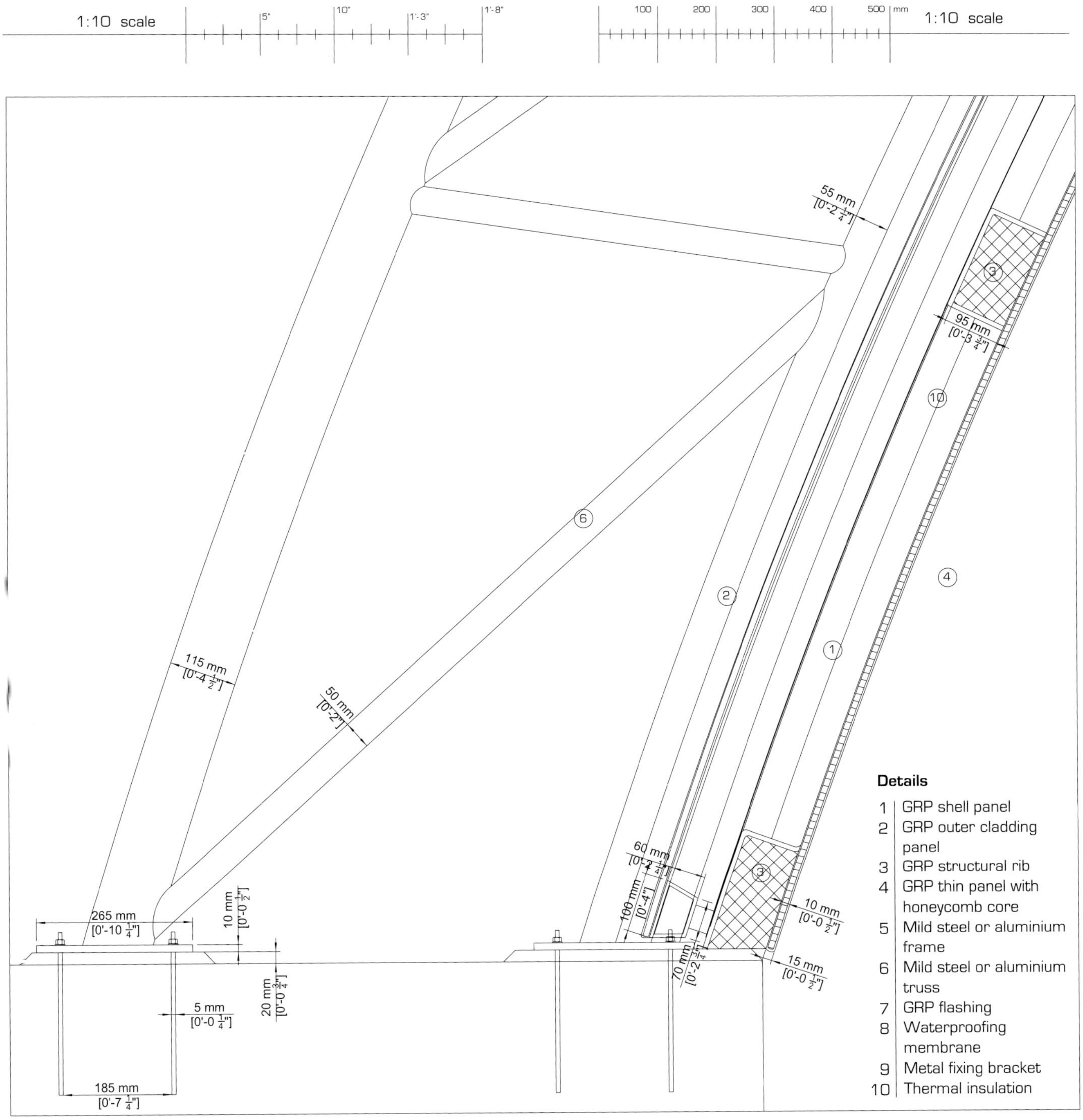

c | Vertical section. Junction at base

sive equipment, making panel production a craft-based technique rather than an industrial process. When the panels are released from the mould they are trimmed along their edges and ground smooth where necessary. An alternative method is to apply a mixture of resin and glass fibre particles as a spray directly into the mould. The mixture is applied to a thickness of 3mm to 5mm, depending on the panel size required.

Larger panels and shells

In many cases a hemispherical dome of 20 metre diameter is constructed as a self-supporting shell. The GRP shell can either be made from GRP only, or can use additional structural stability provided by an external steel or aluminium truss (MCE_493, g). The truss might be used to support a decorative outer screen or a lighting system to illuminate translucent panels, for example.

In many instances, a loadbearing inner shell, with integral GRP structural ribs, is clad with GRP panels that form a weathertight outer screen. The inner shell serves as the loadbearing structure, it being easier to install panels with a pre-finished internal surface since no further access is required for finishes on its internal face. Consequently, no scaffolding or access platforms are required inside the building to install and finish the rooflight. The outer skin of GRP cladding panels is fixed using a lightweight crane that lifts the outer segmented panels into position. The construction method can, of course, be reversed so that an outer shell with a factory-applied finish can be installed as a pre-assembled item that requires no further work externally. GRP cladding panels can then be applied to the internal face of the shell from inside the building, but this is the more difficult construction method.

The structural shell is usually assembled by bringing the panel segments to site individually, due to their size. The panels are then bolted

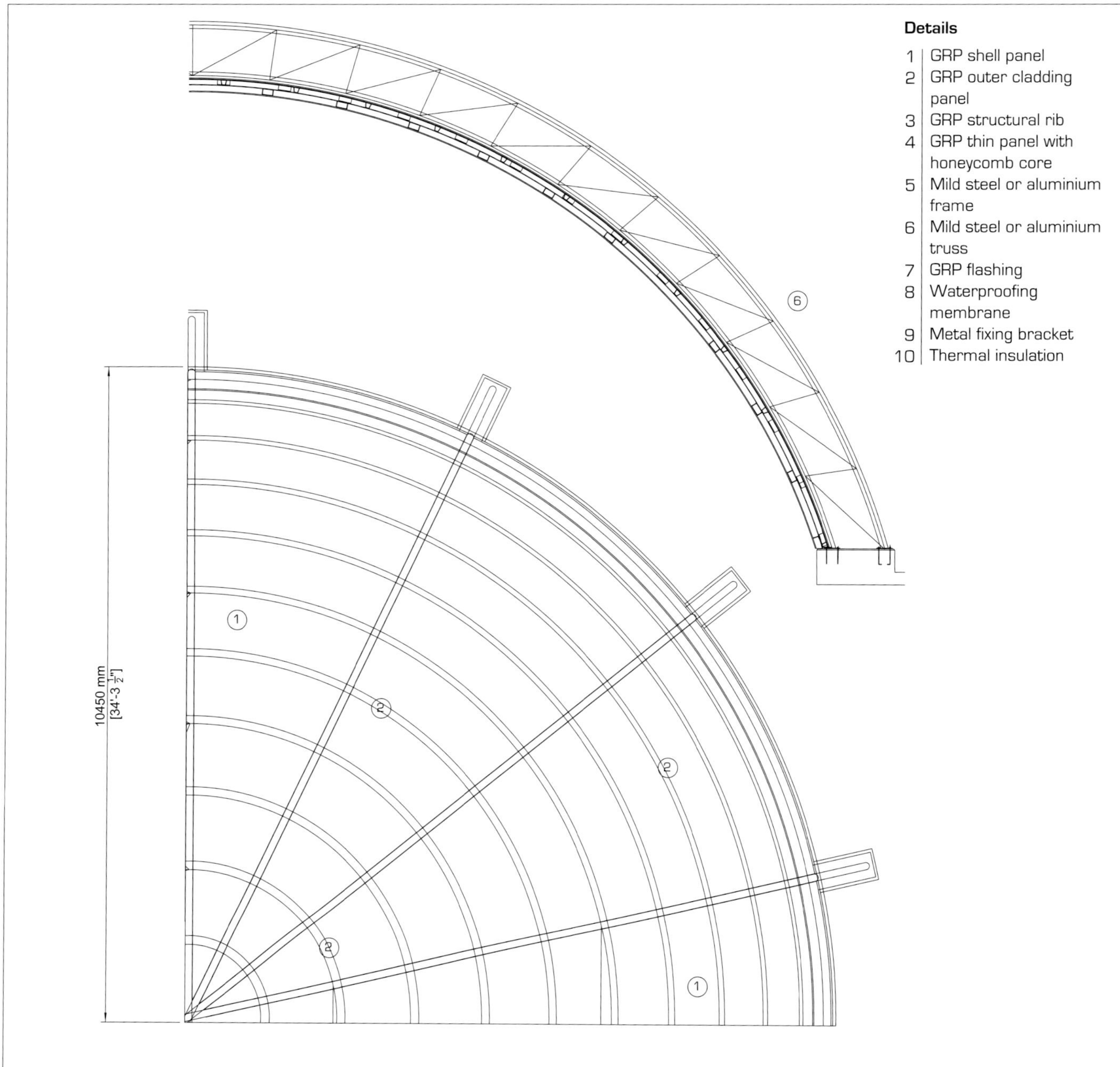

a | Vertical and horizontal section 1:200. Generic examples B1 and B2

together on site, on the ground. The joints are then formed on the finished face, either internal or external. Finally, the shell is lifted by crane into position on the roof as a complete item. The flexibility and lightweight nature of GRP makes this possible, which avoids the need for scaffolding and access equipment at roof level. An alternative method, where no space is available on site for assembly, is to create a platform beneath the shell inside the building, where the roof can be assembled in place. However, the sanding and grinding operations required to make the inner face of the shell smooth and continuous and ready for painting usually require the in-situ constructed solution to be in an enclosed space, to avoid GRP dust particles from spreading around the building.

The inner shell is generally made with an inner wall around 15mm thick made as a thin composite panel. The panel is formed with two outer skins of GRP, 2–3mm thick, with an inner core made from polypropylene-based honeycomb sheet. The honeycomb core provides a rigid reinforcing layer to the outer skins of GRP that give them a flat, smooth appearance. This inner skin is reinforced by structural ribs that form an integral part of the panel, and are set horizontally at approximately 1000mm centres. They are also made from GRP, filled with mineral fibre insulation or injected foam to provide thermal insulation, with overall section sizes of around 200mm wide x 300mm deep. The horizontally-set ribs span a maximum of around 450mm at the base of the shell, deceasing in size until they reach the top of the shell. GRP ribs are also set vertically on the edges of the panels, filled with thermal insulation but made as solid GRP where panels are fixed together with bolted connections (MCE_488, a).

The shell is thermally insulated with mineral fibre or foam-based insulation set in the voids between the ribs. The outer face of the shell is covered with a GRP skin that provides an outer seal against rainwater penetration. The layer is given a further protection and visual screen

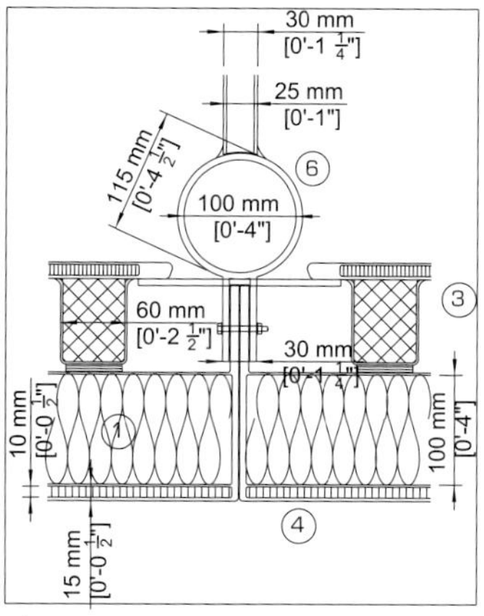

b | Horizontal section. Panel-to-panel junction

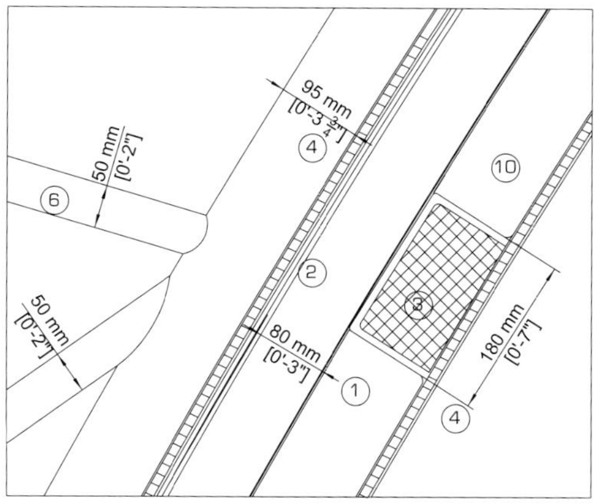

c | Vertical section. Panel-to-panel junction

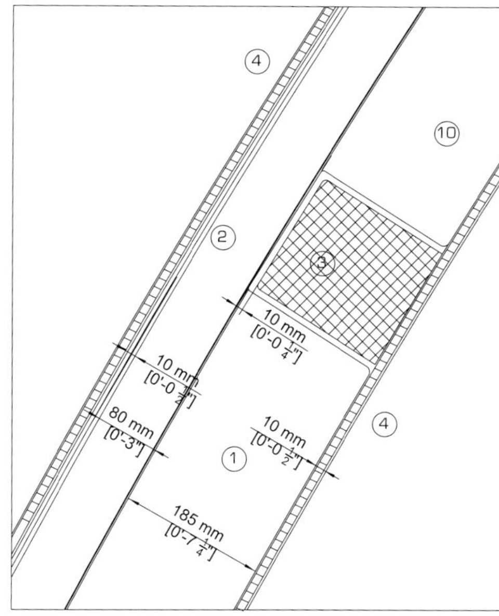

d | Vertical sections. Panel-to-panel junction

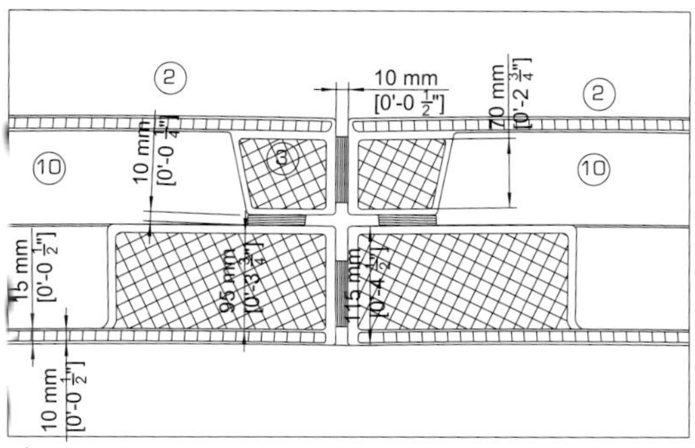

e | Horizontal section. Panel-to-panel junction

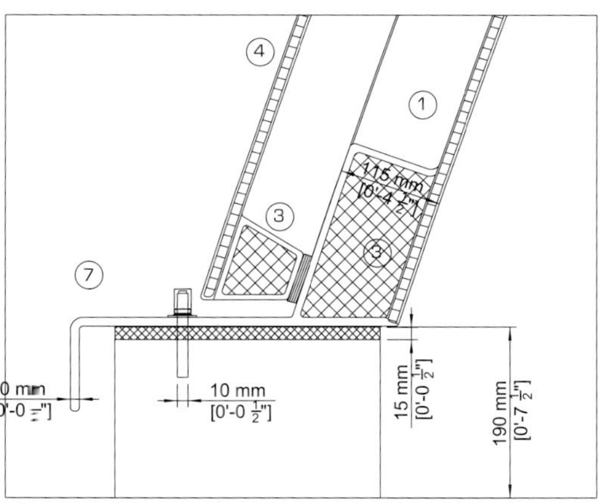

f | Vertical section. Junction at base

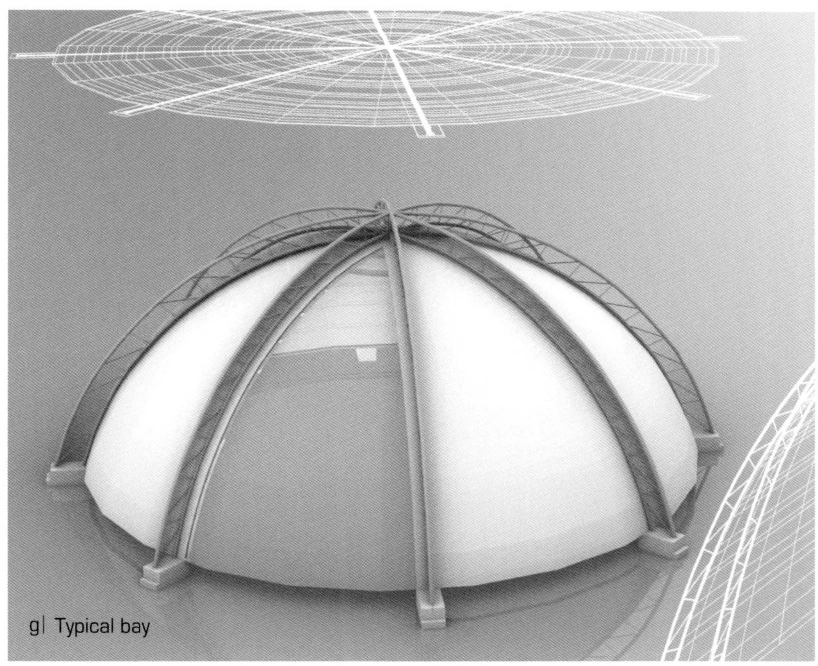

g | Typical bay

provided by outer cladding panels. Like the inner shell, the outer cladding is made from 15mm-thick composite panels, stiffened by GRP ribs around 100mm deep x 200mm wide. These outer segmented panels are bonded to the inner shell with resins or with silicone, and sealed with silicone at their joints to form a continuous outer skin. Small amounts of rainwater that might penetrate the outer line of defence are drained away down the outer GRP skin of the inner shell. While it is easier to provide a continuous smooth finish in a smaller shell, in larger examples it is more difficult to avoid visible surface irregularities on larger shells. For this reason, visible joints are usually preferred in larger GRP shells.

In generic examples, an additional metal truss in either mild steel or aluminium is fixed at the panel joints to provide structural stability (a). The GRP ribs within the panels forming the inner shell are reduced to 20mm thick ribs (a). A metal plate is set between the adjacent panels, with the joint being sealed with silicone. An outer skin of GRP cladding panels can provide an outer line of defence against rainwater penetration. Panels are mechanically fixed to a continuous metal plate forming part of the truss; the panel being sealed against the plate with silicone seal.

In many cases the junction at the base of the shell is treated as an upstand. This allows the shell to be set a minimum of 150mm above the adjacent roof level without the need for complex junctions between roof slab and shell. The base of the shell has a continuous GRP bonded flashing which also serves as a fixing plate onto the concrete upstand. The base plate is set on levelling shims to take up the construction tolerances in the height of the concrete upstand. The waterproofing membrane of the adjacent roof is taken up to the top of the upstand and is lapped under the GRP flashing. An additional waterproofing membrane may extend to the inside of the inner shell and up the internal face of the shell where it is protected by another GRP cover strip, concealed from view by internal finishes.

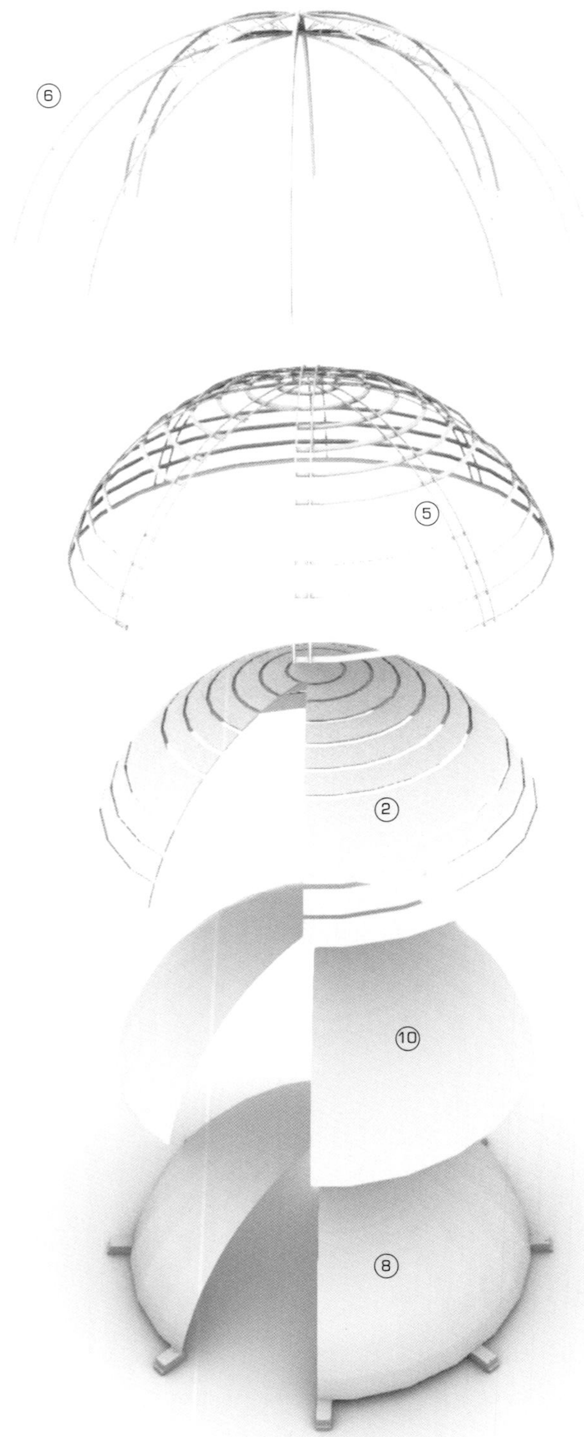

a | 3D exploded view showing GRP facade panel dome

Details

1 | GRP shell panel
2 | GRP outer cladding panel
3 | GRP structural rib
4 | GRP thin panel with honeycomb core
5 | Mild steel or aluminium frame

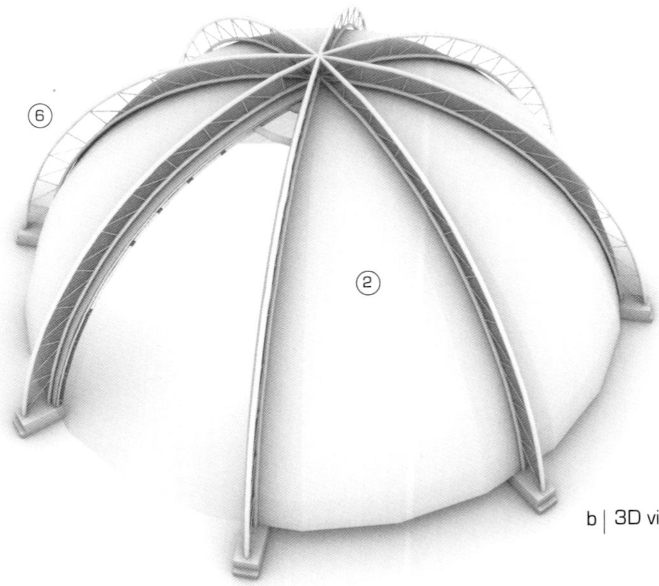

b | 3D view showing GRP facade panel dome

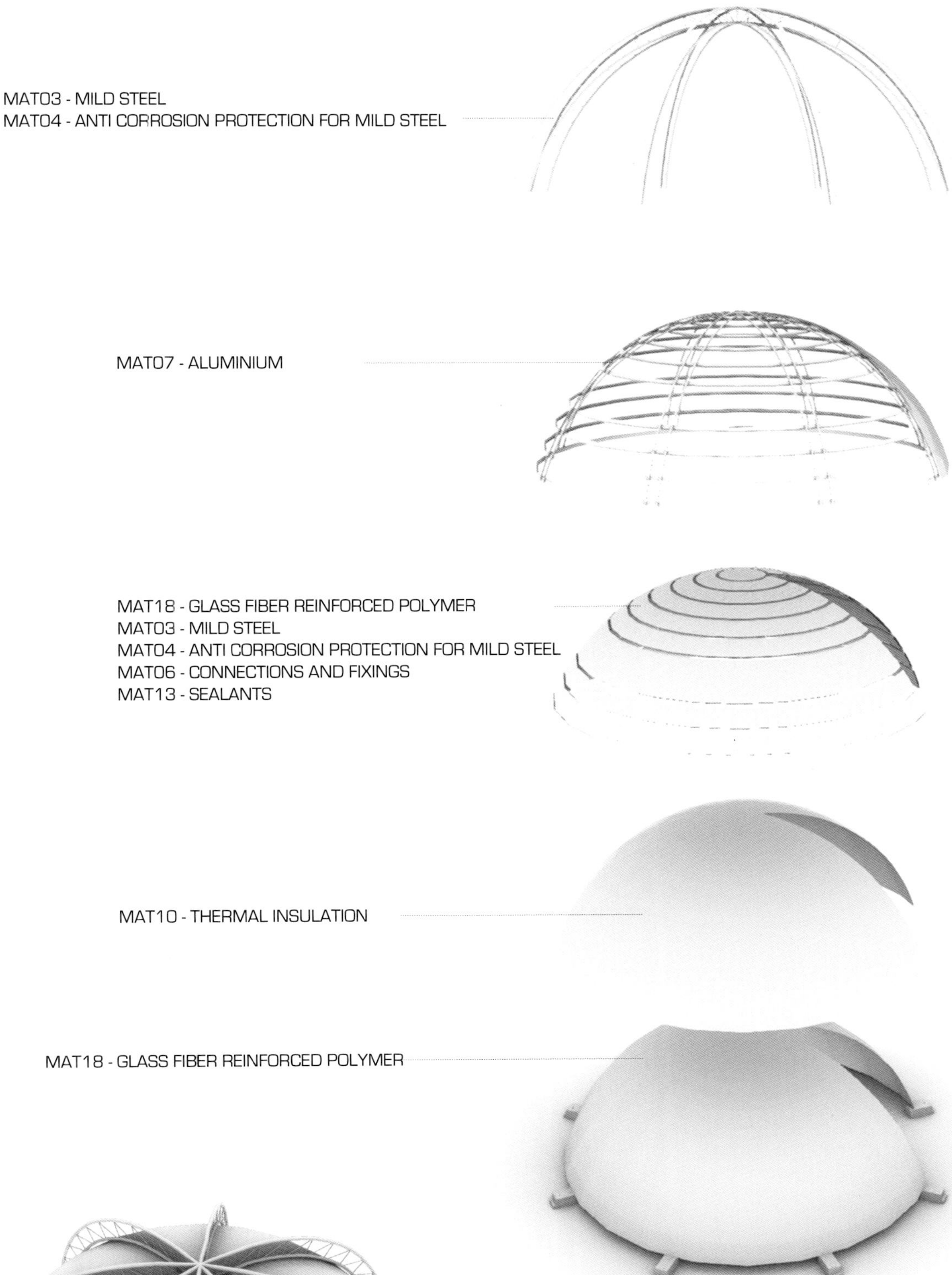

c | 3D exploded view of GRP facade panel dome components with material specification references

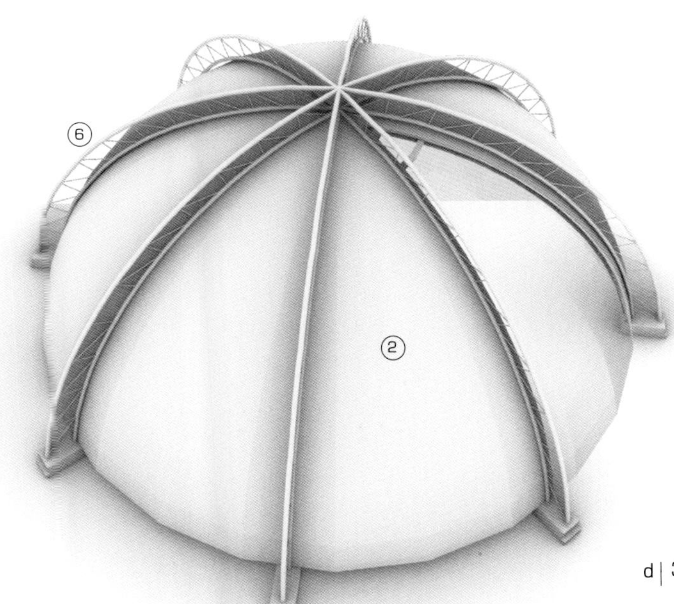

6 | Mild steel or aluminium truss
7 | GRP flashing
8 | Waterproofing membrane
9 | Metal fixing bracket
10 | Thermal insulation

d | 3D view showing GRP facade panel dome

Example project of fabric facades
Construction Laboratory
Facade Engineer: Newtecnic
Architect: Newtecnic

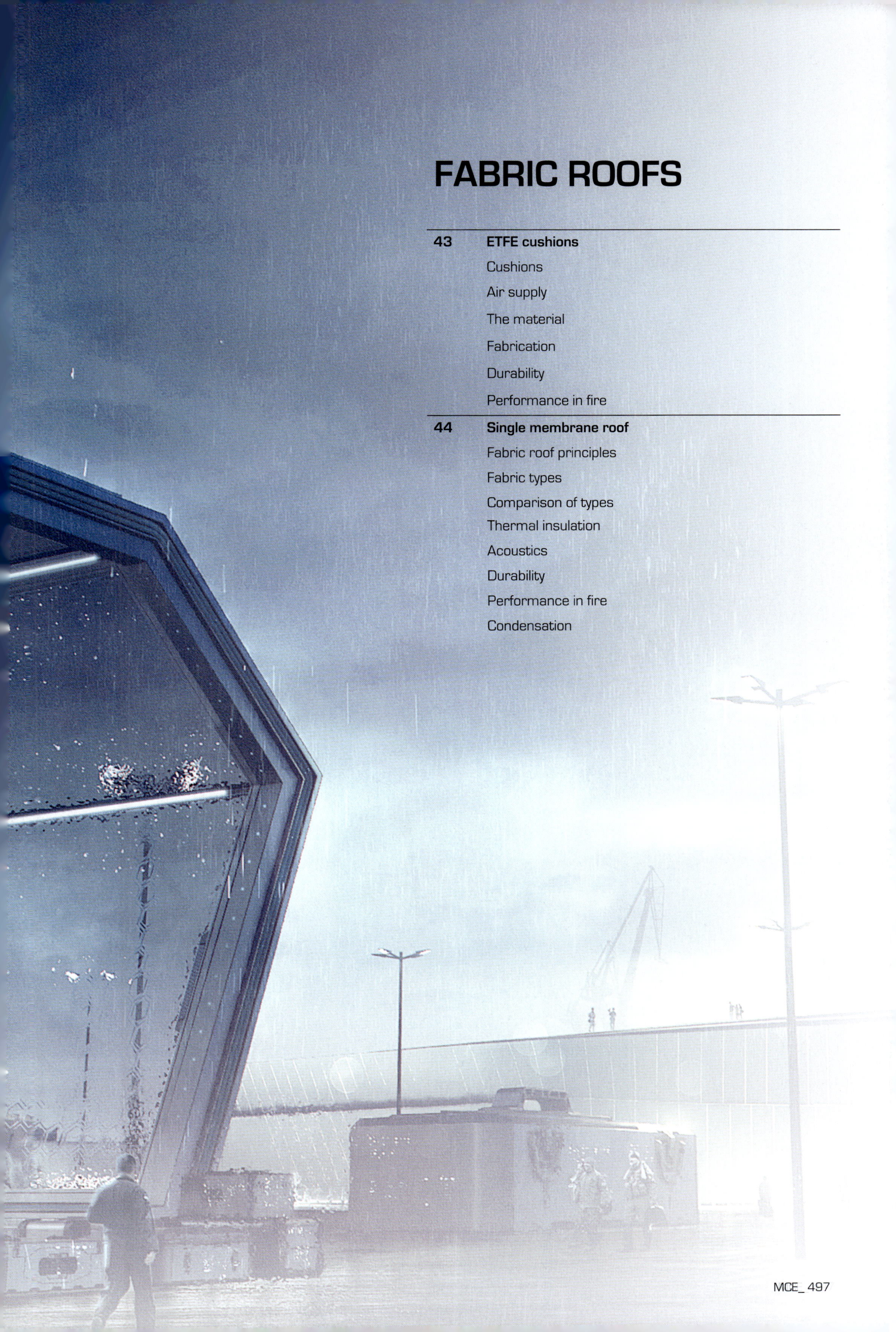

FABRIC ROOFS

43 **ETFE cushions**

- Cushions
- Air supply
- The material
- Fabrication
- Durability
- Performance in fire

44 **Single membrane roof**

- Fabric roof principles
- Fabric types
- Comparison of types
- Thermal insulation
- Acoustics
- Durability
- Performance in fire
- Condensation

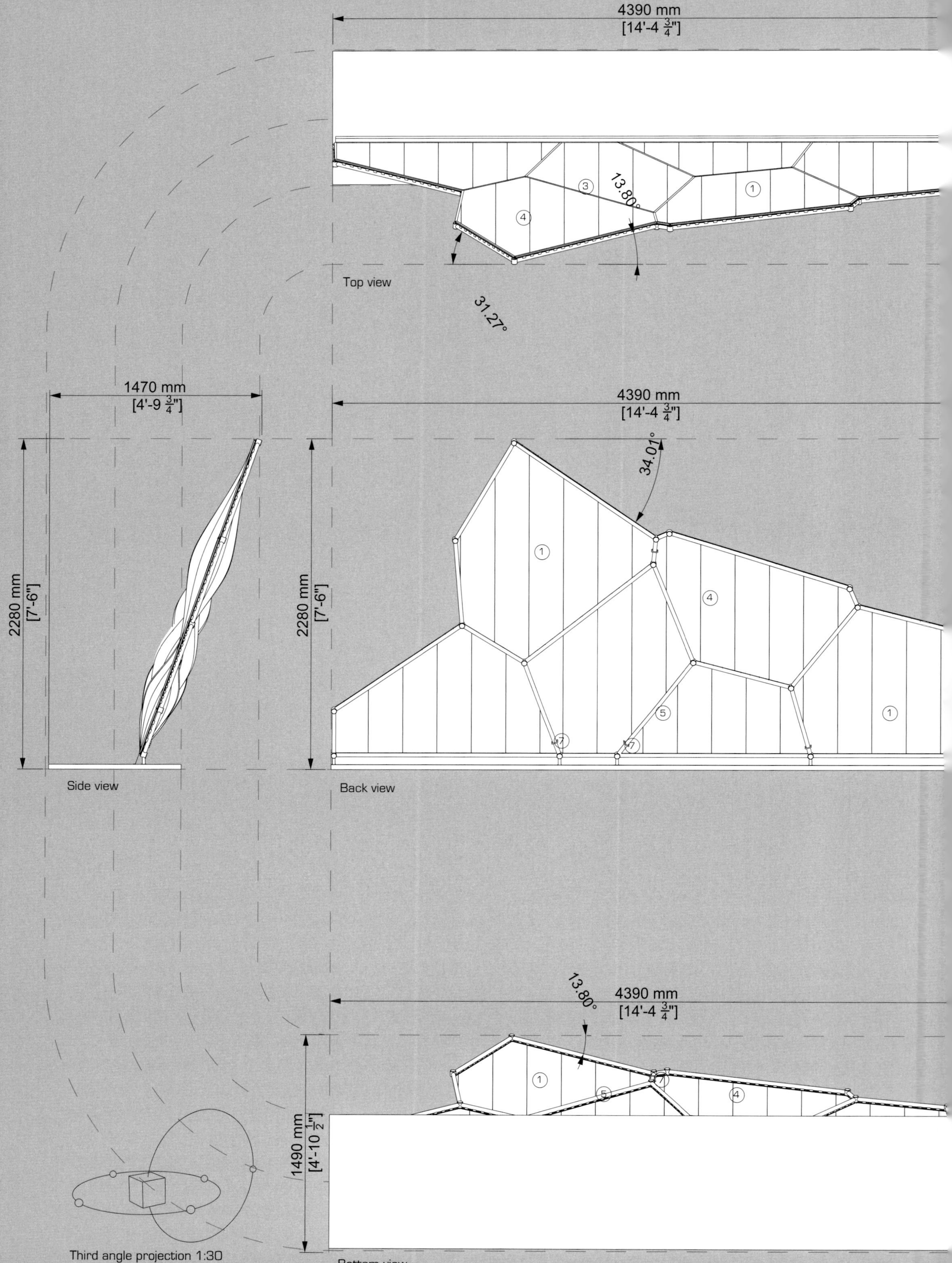

Third angle projection 1:30

Details

1 | ETFE cushion
2 | Extruded aluminium clamping plate
3 | Extruded aluminium retaining profile
4 | Plastic edge bead to fabric membrane
5 | Supporting structure

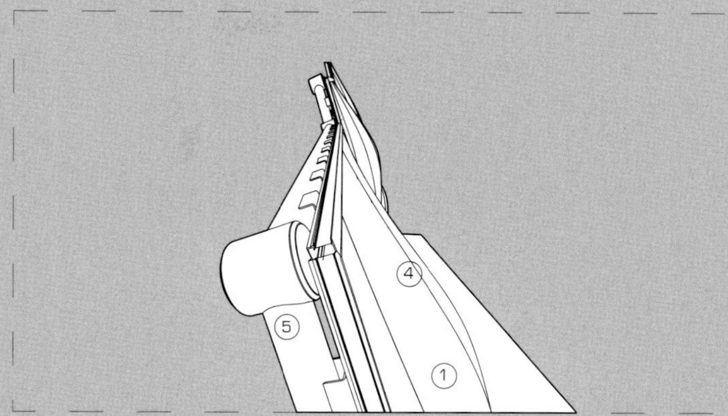

3D view of the assembly

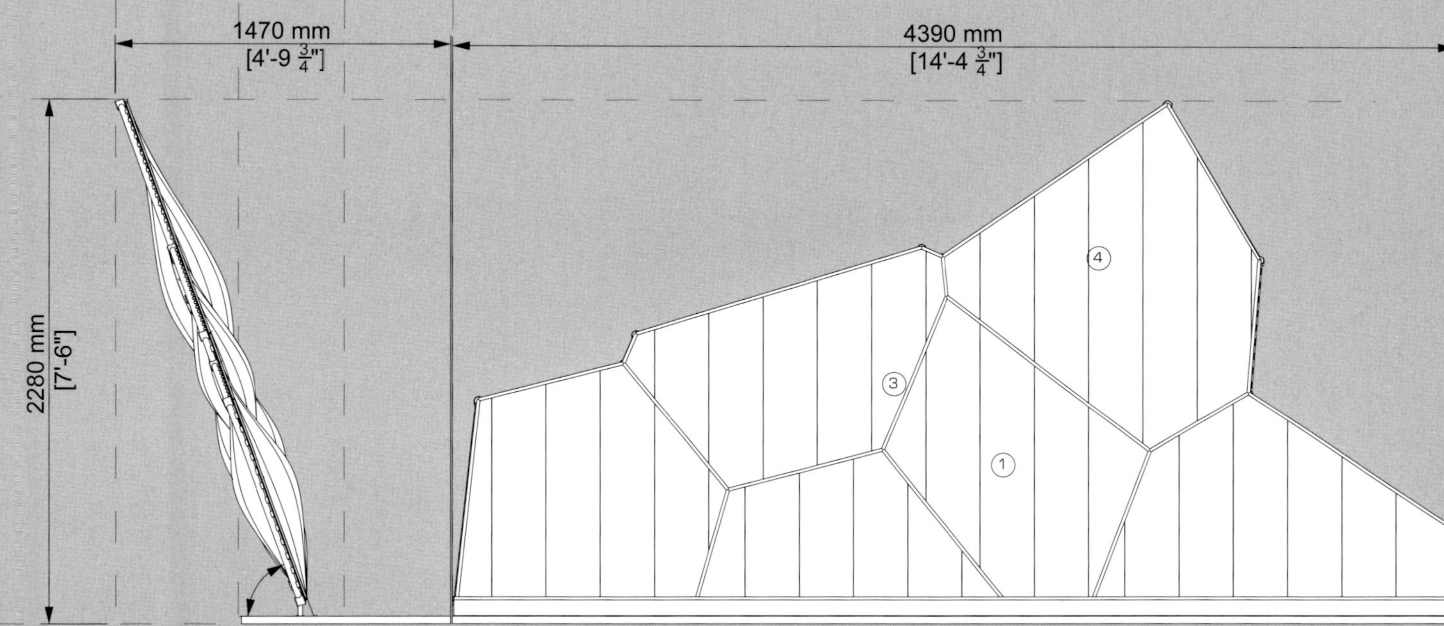

Side view

Front view

6 | Plastic air supply tube
7 | Main air supply tube
8 | Insulated metal lined gutter
9 | Metal flashing
10 | ETFE cushion with contours shown for clarity

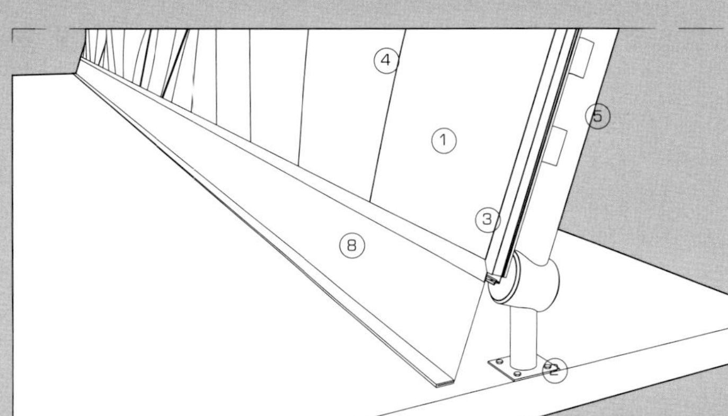

3D view of the assembly

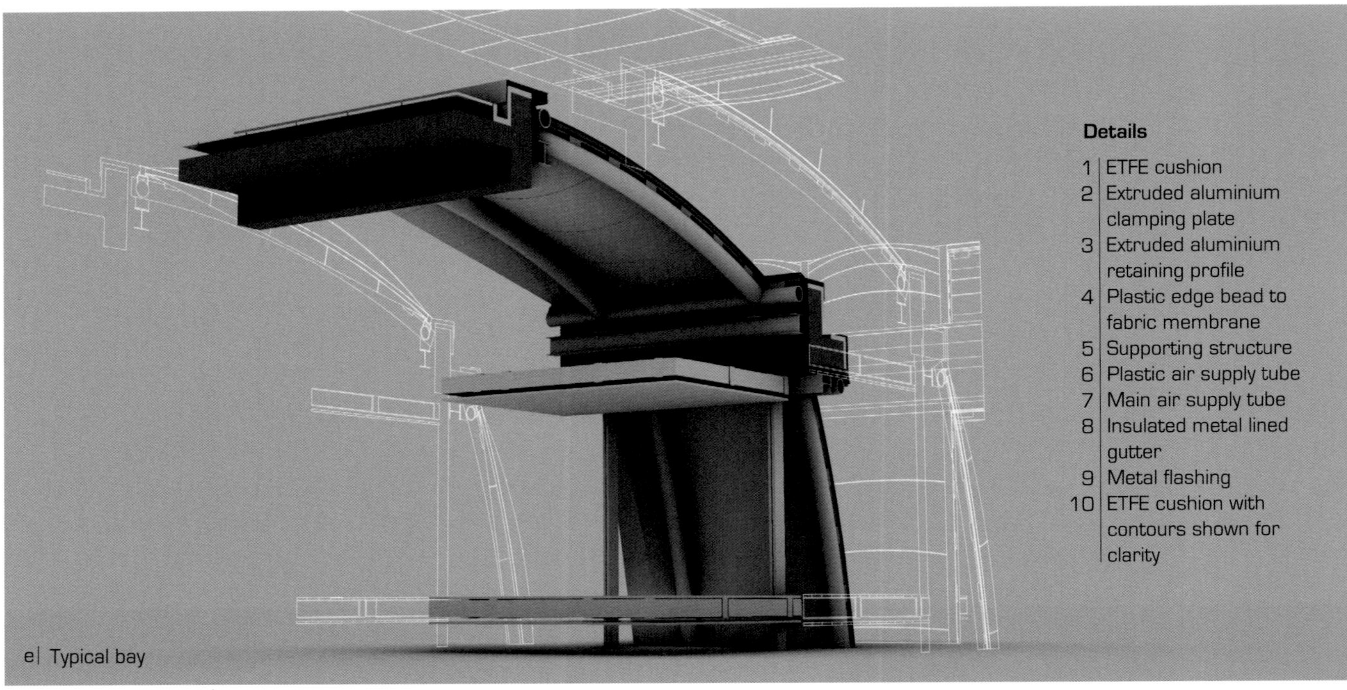

e| Typical bay

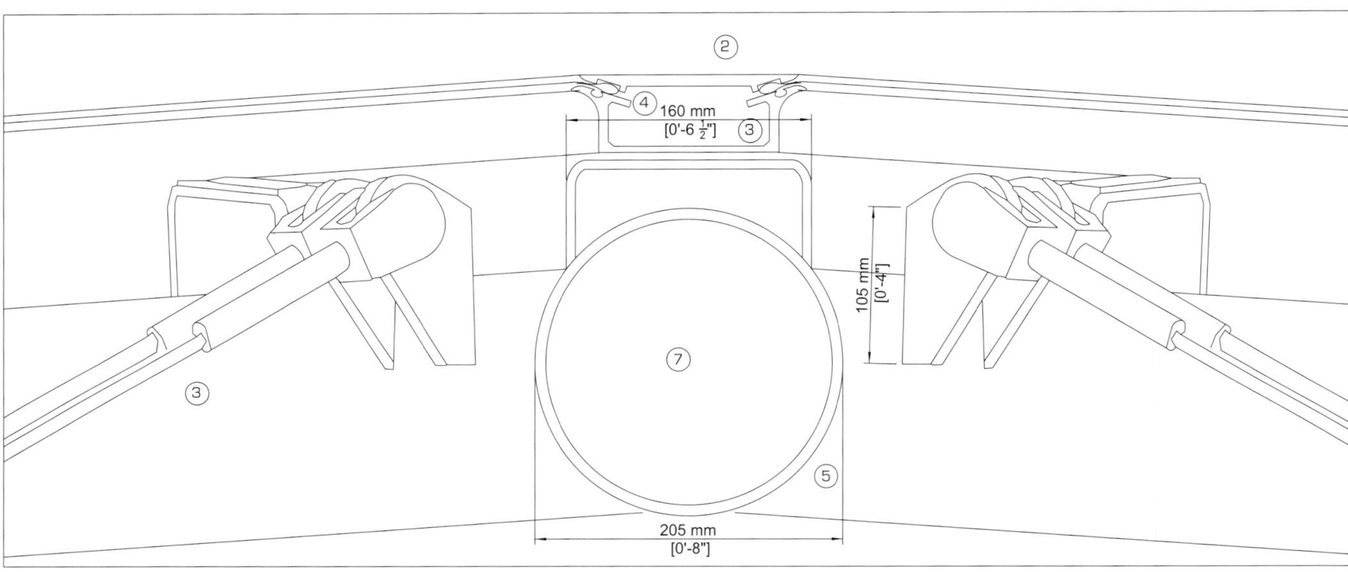

b| Vertical section. Junction of clamping plates

Fabrics used as roof membranes have the advantage of being light inweight, strong in tension and durable, and have the ability to be cut to different shapes and joined together economically, which is difficult to achieve so easily with metal and is very expensive to achieve with curved glass. Roof membrane fabrics are used in tension structures, either by stretching the material, or 'prestressing' the membrane between structural supports, or alternatively by supporting the material pneumatically in inflated structures. The use of fabric membranes in prestressed roofs is discussed in the next two sections on single-layer fabric membranes. This section considers ETFE cushions, also called 'pillows' or 'foils', which are the most common application of inflatable fabric roofs.

Although large-scale self-supporting inflatable roof structures are in use, particularly for covering sports stadiums, they remain structurally stable only while air is being supplied to the structure. If the air supply is interrupted, the complete roof structure deflates. In smaller-scale applications, air-filled cushions remain in place when the air supply fails or is switched off when used as non-loadbearing panels. This type, where ETFE sheet is used to make panels formed as air-filled 'cushions' (MCE_498, e), provides highly transparent, lightweight and resilient roofs that have thermal insulation values similar to those of double glazed units.

Cushions

ETFE cushions usually consist of a minimum of two layers of ETFE sheet which are set back to back to form a flat panel and are sealed at the edges (b). The void within the cushion-shaped panel is inflated with air to a pressure of between 200 Pascals and 700 Pascals, depending on the cushion size and the manufacturer's proprietary system, to provide structural stability to the panel. The increased air pressure stretches, or 'prestresses', the outer membranes, giving ETFE cushions their char-

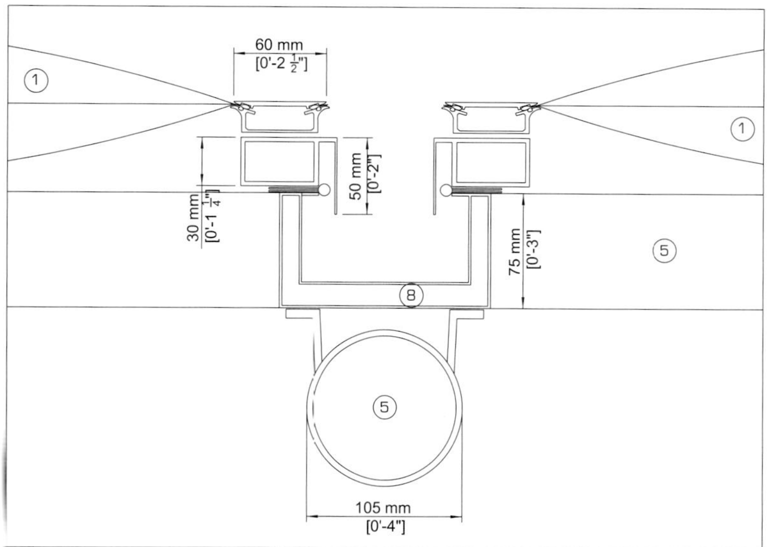

c | Vertical section. Junction of clamping plates

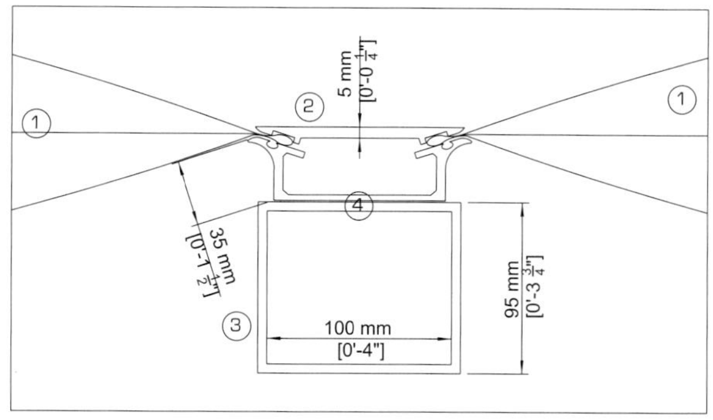

d | Vertical section. Opening smoke vents with insulated gutter

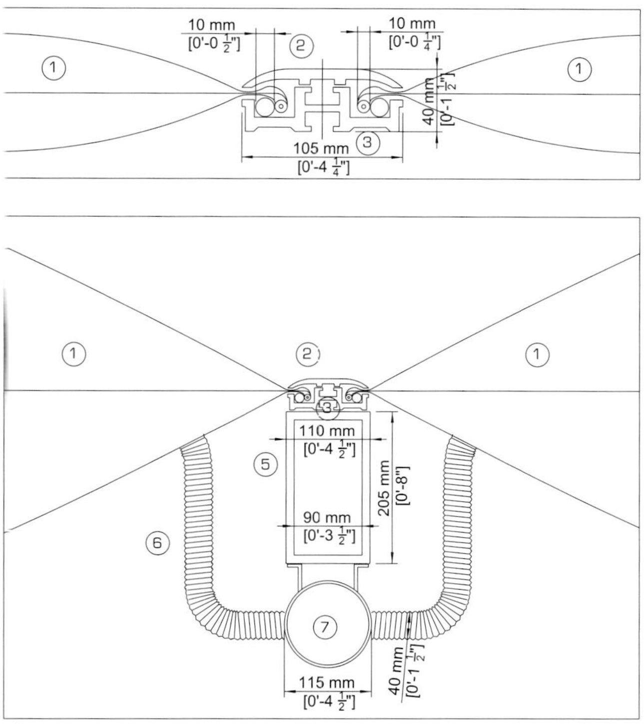

e | Vertical section. Clamping plate with air supply pipes

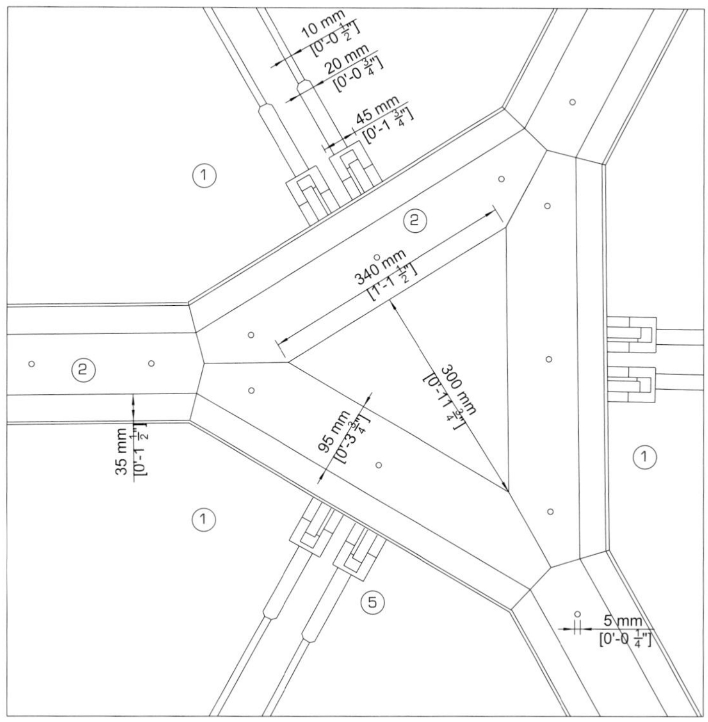

f | Horizontal section. Junction of clamping plates

acteristic curved shape. The cushions are held in place by clamps that form a frame around the cushions in the manner of glazed rooflights (c). The clamping frames are then supported by a mild steel structure formed typically as box sections or tubes (a). Cushions typically have three layers that form two chambers. The two chambers are linked by a hole formed in the middle (flat) membrane in order to allow air to pass to both chambers from a single air supply, and to ensure that the air pressure in both chambers remains equal. This three-layer cushion provides a U-value of around 2.0 W/m^{2}K, which is similar to a double glazed unit used in glass roofs. Cushions made from two layers of ETFE sheet are also used, but the thermal insulation performance is reduced considerably. Thermal insulation performance is reduced at the perimeter of the cushion where its thickness is reduced to a thin edge. Some cushions have thermal breaks at the panel perimeter to partially overcome this loss of thermal insulation. The overall U-value can be reduced by forming a wide gap between the two outer membranes, to provide a consistently wide air space within the cushion. Thermal insulation performance can also be improved by increasing the number of air chambers within the cushion by adding further layers of ETFE membrane.

Air supply

Air is supplied to cushions from rubber pipes or flexible plastic pipes that are connected to the underside of the cushion near the clamping assembly (d). Pipes are usually of around 25mm diameter, and are connected to a larger pipe that supplies the air to all the cushions either side of a single structural support. This main pipe is also usually made from plastic and can be concealed within the supporting structure, being only up to a diameter of around 60mm (b). The air supply, which maintains the air pressure within the cushions at a constant level, is supplied by electrically powered fans with air filters (to avoid the passage of dust). The humidity level of the air is usually controlled to avoid the possibility of condensation forming within the cushions. Once the

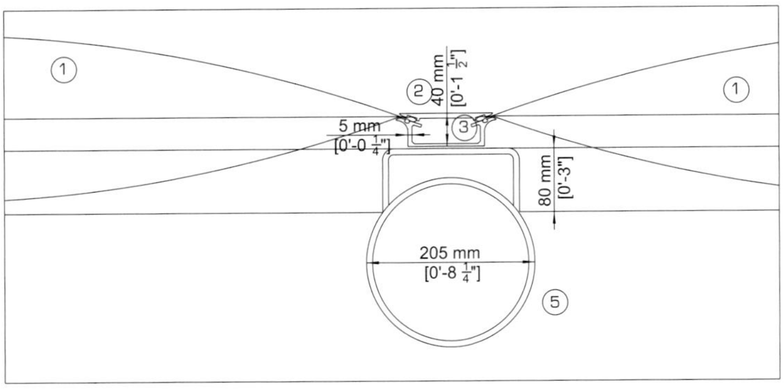

a| Vertical section. Junction of clamping plates

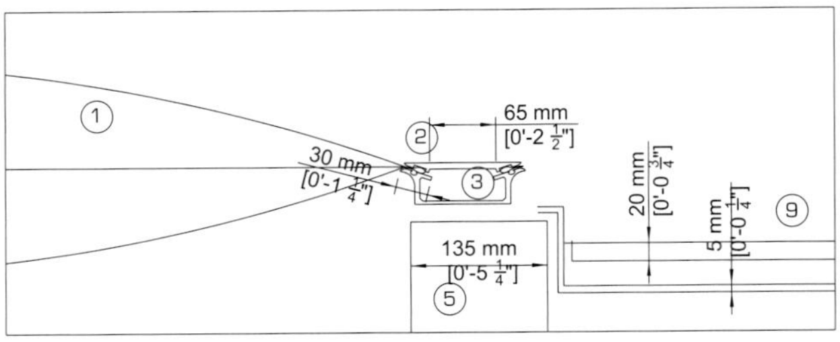

b| Vertical section. Junction with adjacent roof

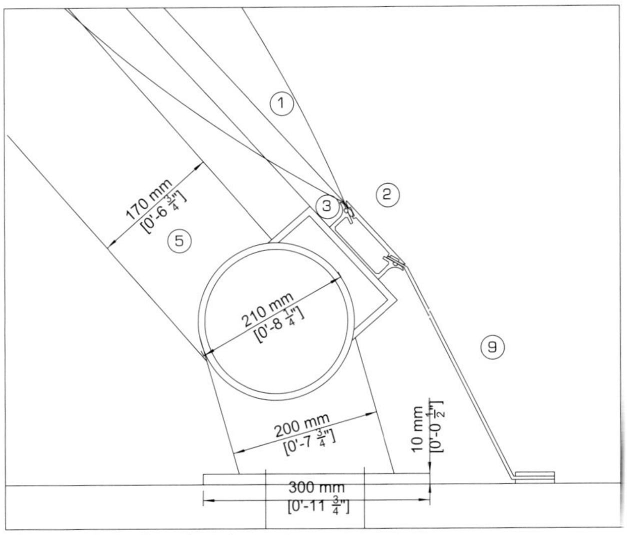

c| Vertical section. Junction at base

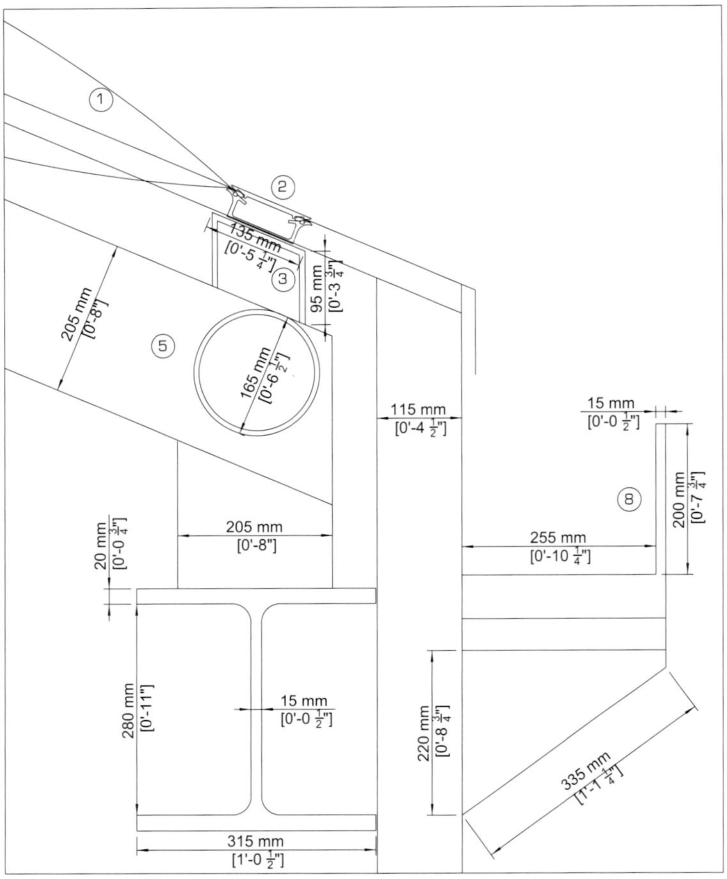

d| Vertical section. Junction of clamping plates

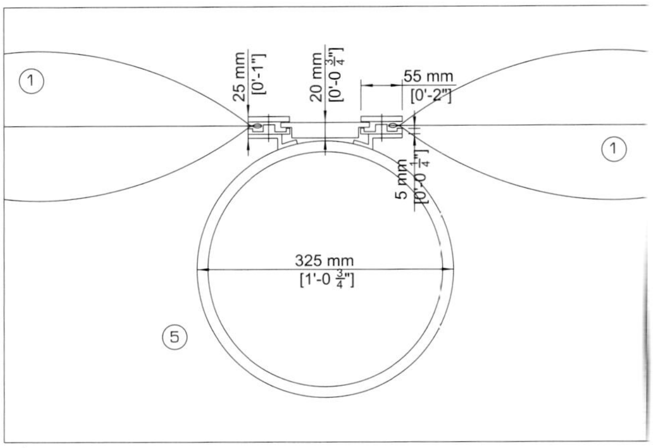

e| Vertical section. Junction of clamping plates.

ETFE cushions have been inflated, air is supplied to the cushions for only around 5–10 minutes per hour to compensate for loss of air pressure from leaks from cushions or from air supply pipes.

If air pressure within the cushion is lost as a result of damage to one of the outer membranes or from loss of air pressure in the supply pipe, the cushion deflates to its flattened shape. As wind pressures are applied to the external face of the cushion, the outer skin will deflect either inwards or outwards as a result of the positive or negative pressures. This does not usually cause damage to the cushions before the air supply is restored. Some manufacturers' systems have one-way valves to prevent loss of air pressure from the cushions back to the supply pipes.

Condensation on the underside, or internal face, of ETFE cushions is avoided by ensuring that ventilation levels within the building are sufficiently high to avoid this occurring, and that levels of relative humidity are suitably controlled. Where condensation is still a risk, condensation channels are fixed to the underside of the cushion clamping assembly (d), similar to that used for glazed rooflights. In most cases the temperature of the air supplied to the cushions is similar to that of the internal space immediately below the roof, so condensation does not usually occur on the underside of the cushion.

The material

ETFE (ethylene-tetra-flouro-ethylene) is a polymer similar to PTFE (marketed as Teflon) which is made by extrusion as a sheet material. Thicknesses vary but 0.2mm is a common thickness of material for ETFE cushions, allowing them to be very light in weight at around 350g/m² for this thickness of sheet. Heavier gauge sheet at 0.5mm thickness weighs around 1000g/m². Inner layers of ETFE sheet that provide separate chambers within the cushion are often made from 0.1mm thick sheet. The material is also used for its high level of transparency

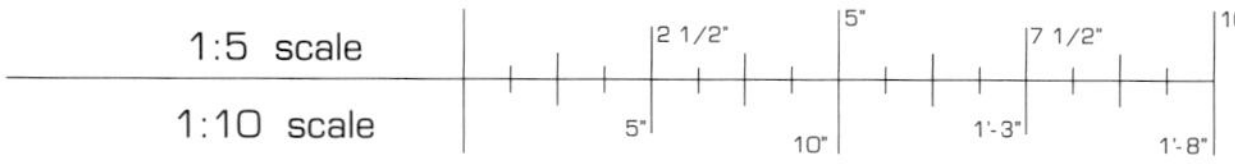

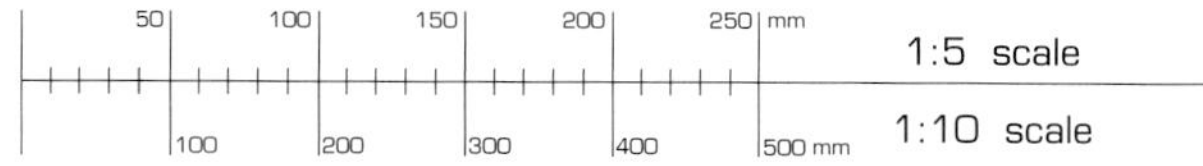

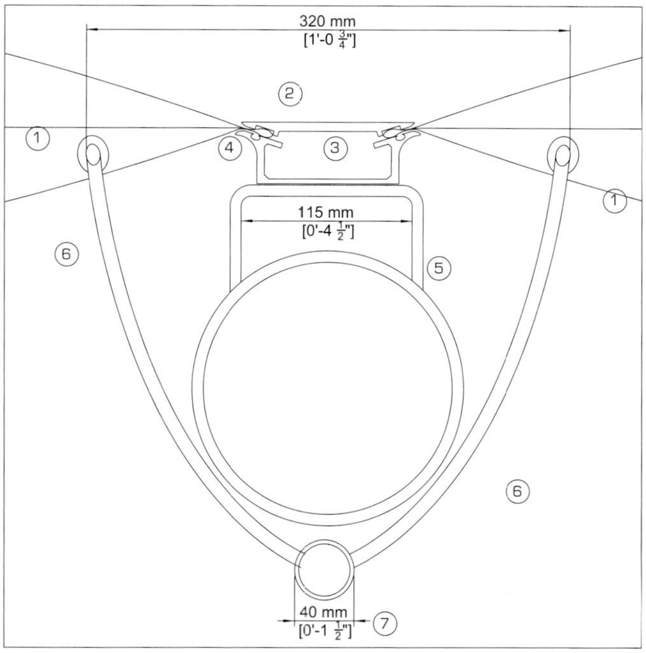

f | Vertical section. Junction of clamping plates

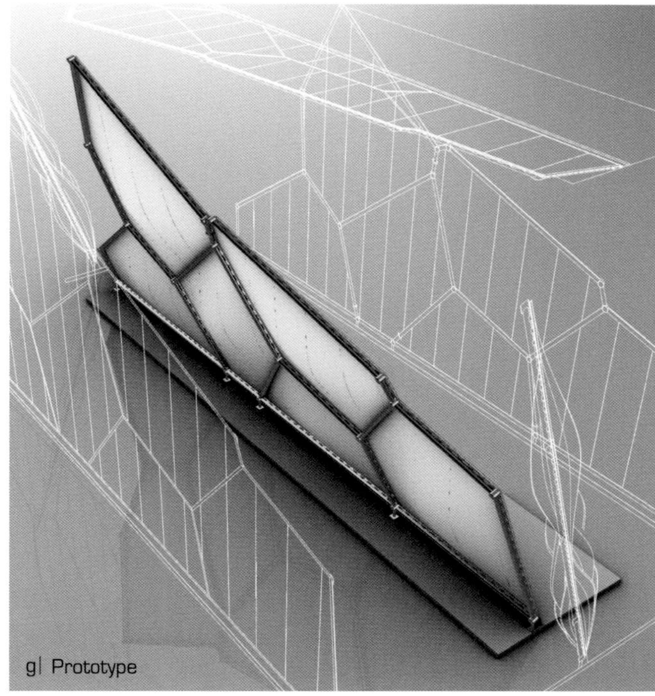

g | Prototype

Details

1 | ETFE cushion
2 | Extruded aluminium clamping plate
3 | Extruded aluminium retaining profile
4 | Plastic edge bead to fabric membrane
5 | Supporting structure
6 | Plastic air supply tube
7 | Main air supply tube
8 | Insulated metal lined gutter
9 | Metal flashing
10 | ETFE cushion with contours shown for clarity

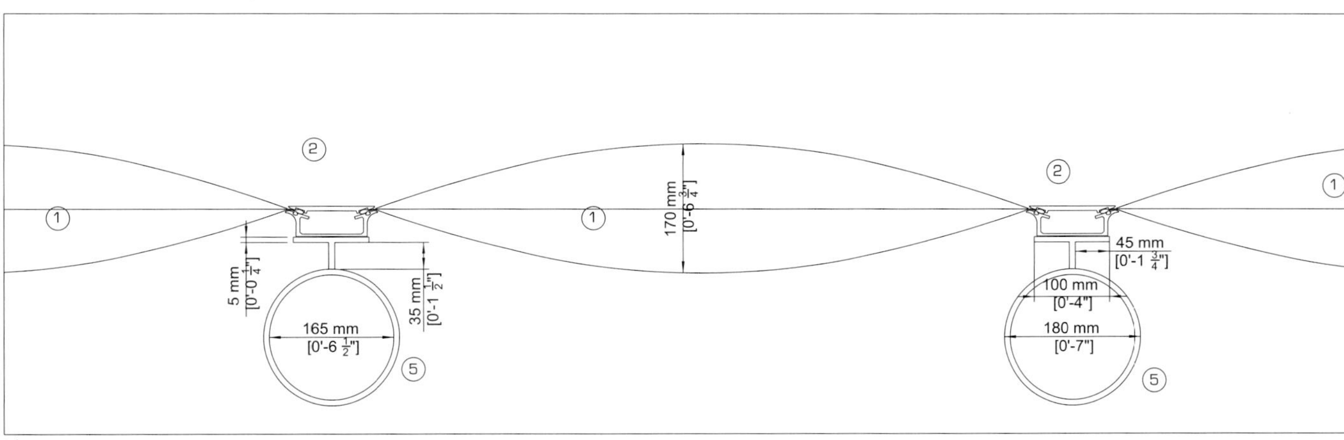

h | Vertical section. Junction of clamping plates

with 95% light transmission, and its durability when compared to other fabric materials, with a life expectancy of 25–35 years, based on visual criteria. In order to provide translucent areas of roof (or facade) using the same material, a white coloured translucent sheet is manufactured which provides around 40% light transmission. ETFE does not provide a barrier to the passage of UV light, making it ideal for use in buildings where extensive planting is displayed, though the translcent white sheet has greatly reduced UV light transmission. Solar shading can be provided by a pattern of dots printed onto the surface of the ETFE cushion, with a reflective silver colour being a popular choice, though other colours and patterns can be developed for individual projects. The printed dots on clear ETFE sheet reduce the light transmission to around 50–60%, but this can be increased further by printing dots on two faces of the airfilled cushion in areas of the roof where more shading is required. The amount of solar shading provided can be varied by allowing the middle layer to move as a result of changing the air pressure within the cushion. The middle layer moves either outwards or inwards to increase the overlap of the printed screen of dots that has the effect of varying the amount of solar shading provided.

The sound absorption of ETFE cushions is low, so that sound travels easily through the material. While this can be an advantage in noisy internal environments, it can be an inappropriate material if external noise is able to travel through the roof to internal spaces where a quiet environment is required. Cushions can also produce a drumming effect during rain resulting from the use of a thin stretched, membrane material.

Fabrication

ETFE cushions are usually fabricated in the workshop but can be assembled on site to suit project conditions. The material is manufactured in rolls of considerable length but of a width of around 1.5 metres. For this reason, ETFE cushions are often made with spans in widths

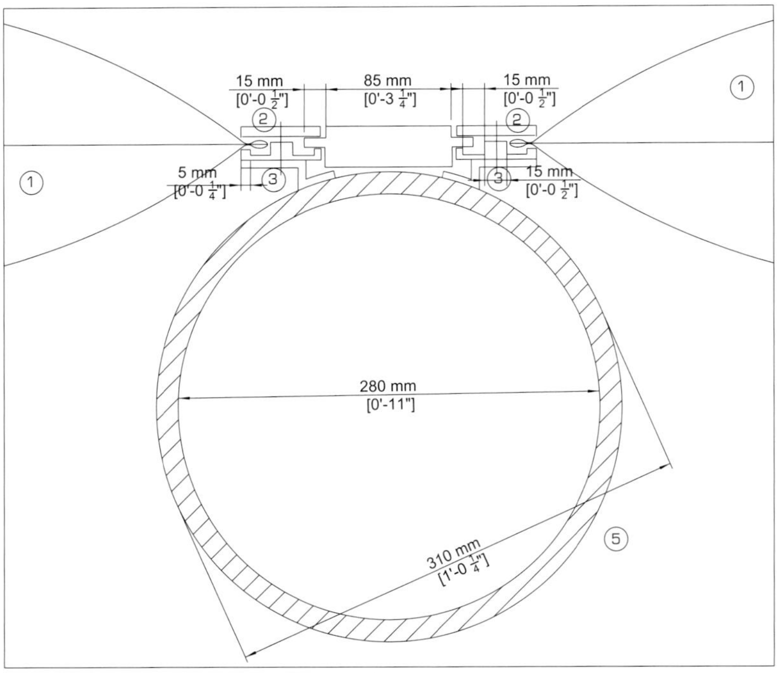

a| Vertical section. Junction of clamping plates

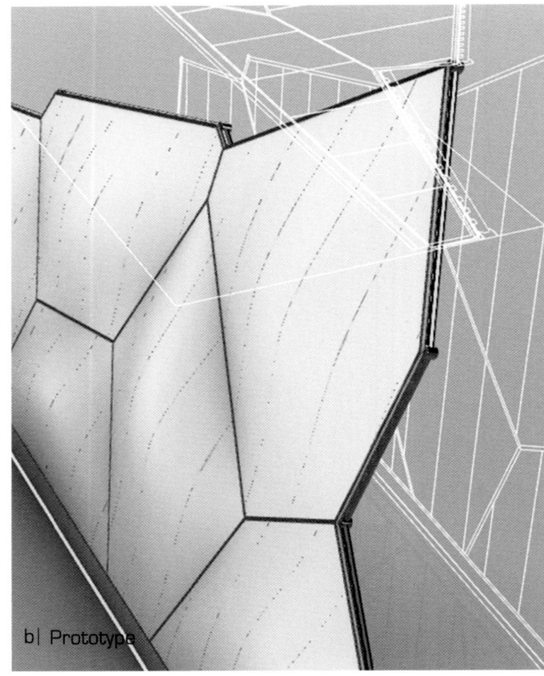

b| Prototype

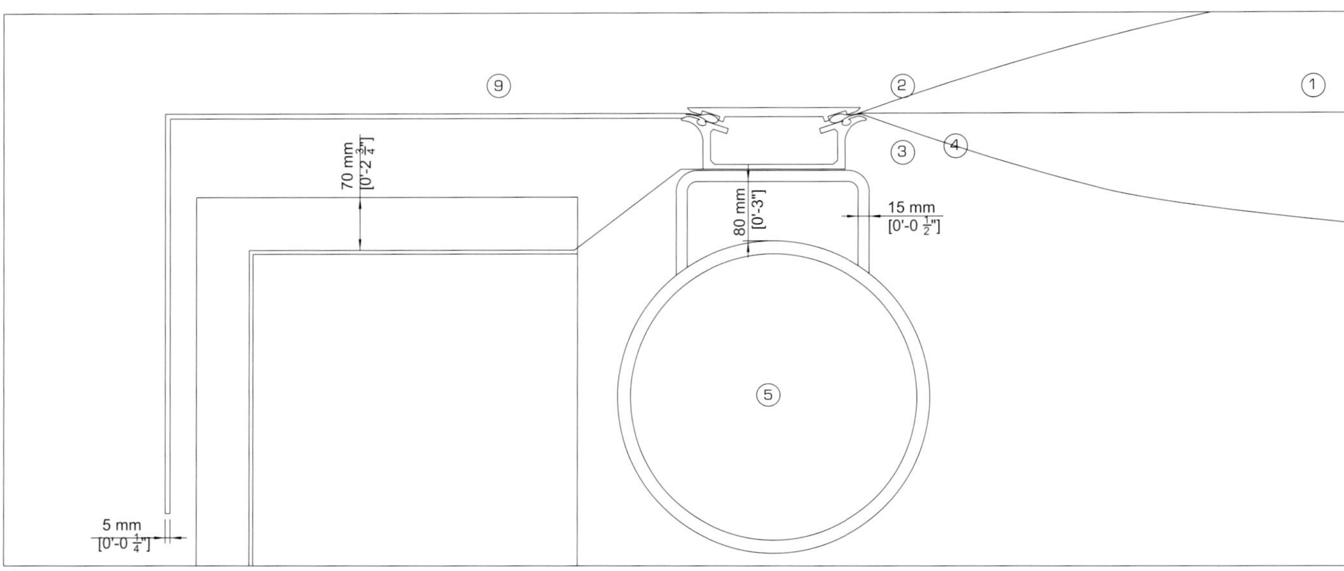

c| Vertical section. Junction with adjacent roof

of around 3.0 metres to 4.0 metres between clamped frames (h), in lengths up to between 15–30 metres, but lengths up to 60 metres have been used. Larger cushions are made by welding sheet together in the workshop to form wider sheets to form cushions that vary from the rectangular form created from a standard width of ETFE sheet. Cushions up to around 7.0 x 7.0 metres have been made in a variety of shapes from circles to hexagons using a hot welded process undertaken in the workshop. The welded seams are visible but are not visually striking when viewed from around the building. The width, or span, of the cushion has an effect on its depth, where the depth is typically around 1/5 the span of the cushion. Larger cushions sometimes use a net of connected cables to provide additional restraint.

The cushions have an edge bead, typically in plastic, which is used to retain the panel in its supporting frame. The cushion is set into a clamping frame, usually made from extruded aluminium, that holds the panel in place by clipping it into an aluminium profile, then holding it in place with an aluminium pressure plate that clamps the edges (c). The complete assembly usually performs in a similar way to the framing used for glazed roofs, with a drained and ventilated inner chamber that serves as a second line of defence against rainwater penetration (b). Any rainwater that passes through the outer clamping plate is drained away in grooves formed adjacent to the edges of the ETFE cushions, the water being drained back to the outside of the roof. The clamping assembly is typically around 100mm wide, which is wider than that used for glazed roofs, but less framing is required than is commonly used on glazed roofs.

The complete ETFE cushion assembly is supported by a structural frame beneath, with mild steel, aluminium and laminated timber all being used to suit the design. The clamped framing assembly is fixed to mild steel tubes by fixing the framing onto T-sections which are welded in short lengths to the mild steel tube (a). The aluminium framing spans

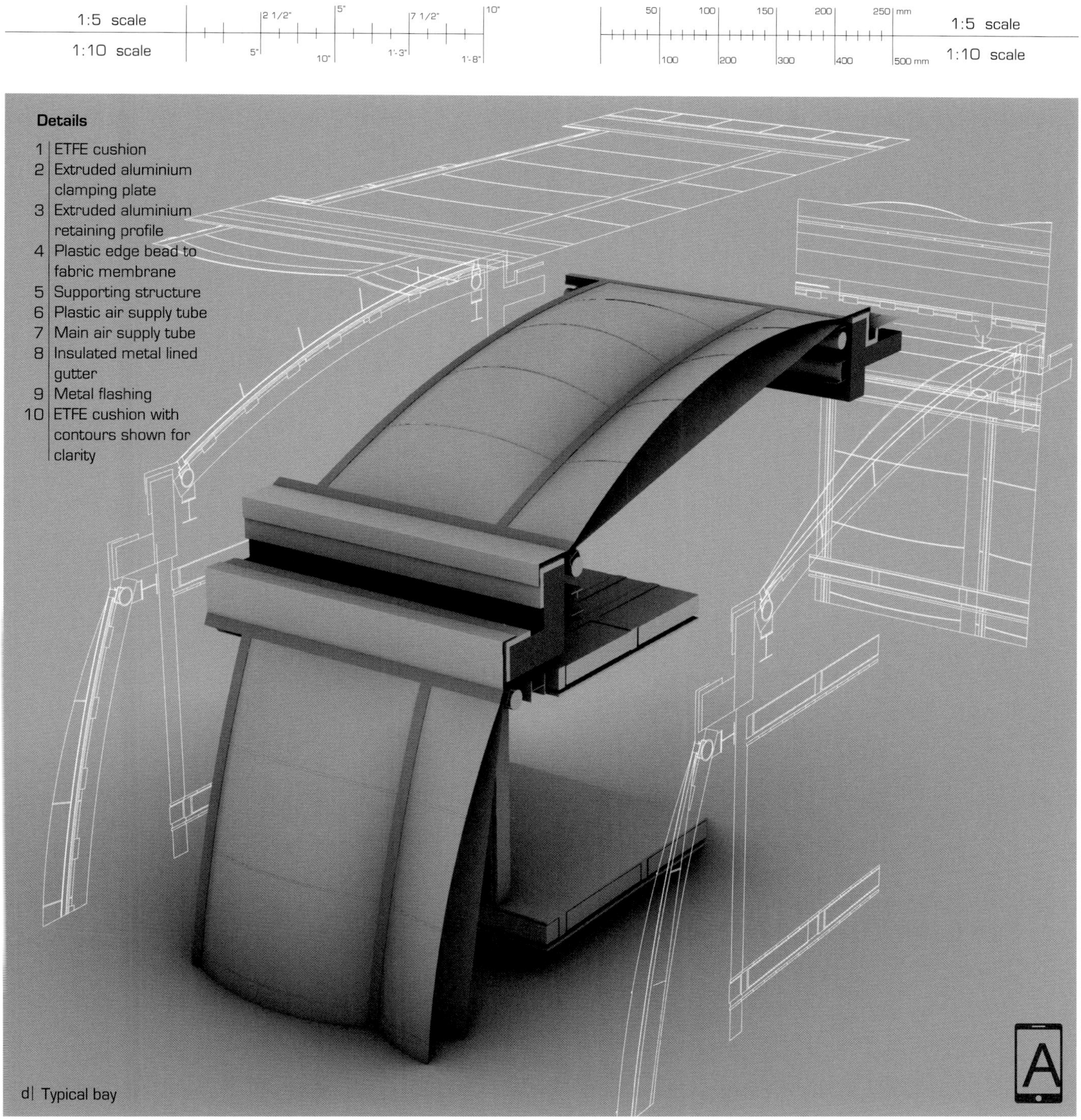

d| Typical bay

between the T-section supports to provide a visual separation between the ETFE cushions and the supporting steelwork.

Durability

The toughness of ETFE sheet is combined with a high resistance to tear. Damage by sharp objects puncturing an outer membrane does not spread easily into a larger tear. Birds can puncture the outer membrane, but they have great difficulty in coming to rest on the roof itself, except on the clamping plates, where wire is sometimes fitted to avoid providing any spaces for birds to stand. The material has fairly high resistance to surface fading from UV light, where there is a gradual loss of surface reflectivity. ETFE sheet is also highly resistant to attack from chemicals and from airborne pollution in urban areas. Its low level of surface friction ensures that cushions do not hold dirt and dust easily, making cushion roofs relatively easy to maintain. ETFE roofs are usually cleaned as a result of rain in temperate climates, though access for maintenance is required, usually provided by walking along the external clamping plates with cable assistance or from external structure. Roofs are designed so as to ensure that rainwater drains easily from the roof. Gutters are introduced on long span roofs between sets of clamping plates. Cushions are repaired by the use of ETFE tape, which is visible, or by complete replacement of the panel, depending largely on the visual requirements of the roof design.

Performance in fire

A major concern in the use of polymer materials for single-layer and multiple-layer fabric roofs is their performance in fire. ETFE sheet is not easily inflammable and will self-extinguish quickly under direct flame. Few burning fragments will fall below during a fire, as a result of the material melting rather than burning, with most of the burnt material being carried away in the rising hot air of a fire. ETFE sheet melts at around 275°C, forming holes in the fabric which allow the heat and

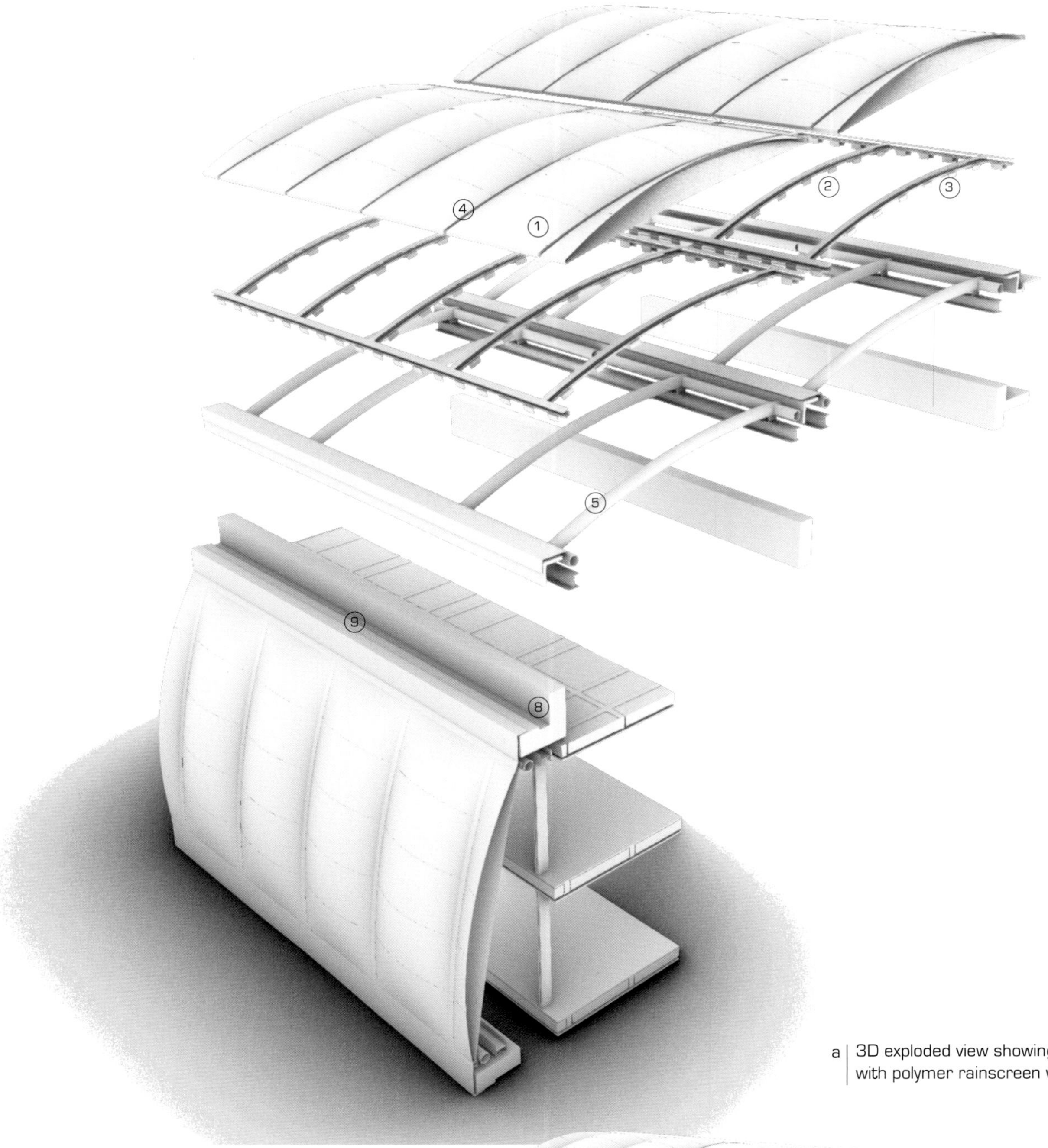

a | 3D exploded view showing ETFE roof with polymer rainscreen wall

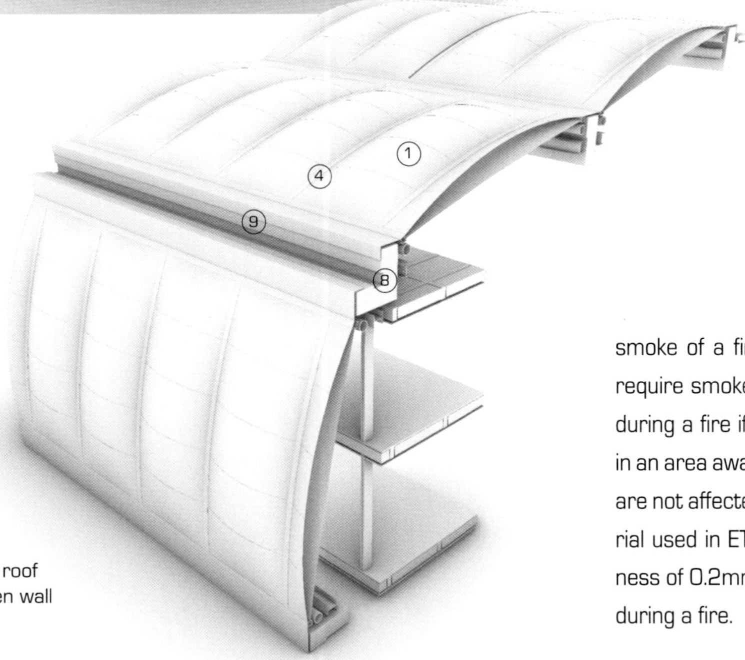

b | 3D view showing ETFE roof with polymer rainscreen wall

smoke of a fire to escape. However, some roofs still require smoke vents, since this may not always occur during a fire if the smoke and heat is being generated in an area away from the roof, where the ETFE cushions are not affected by the fire. The small amounts of material used in ETFE cushions, with an average wall thickness of 0.2mm, result in little material being deposited during a fire.

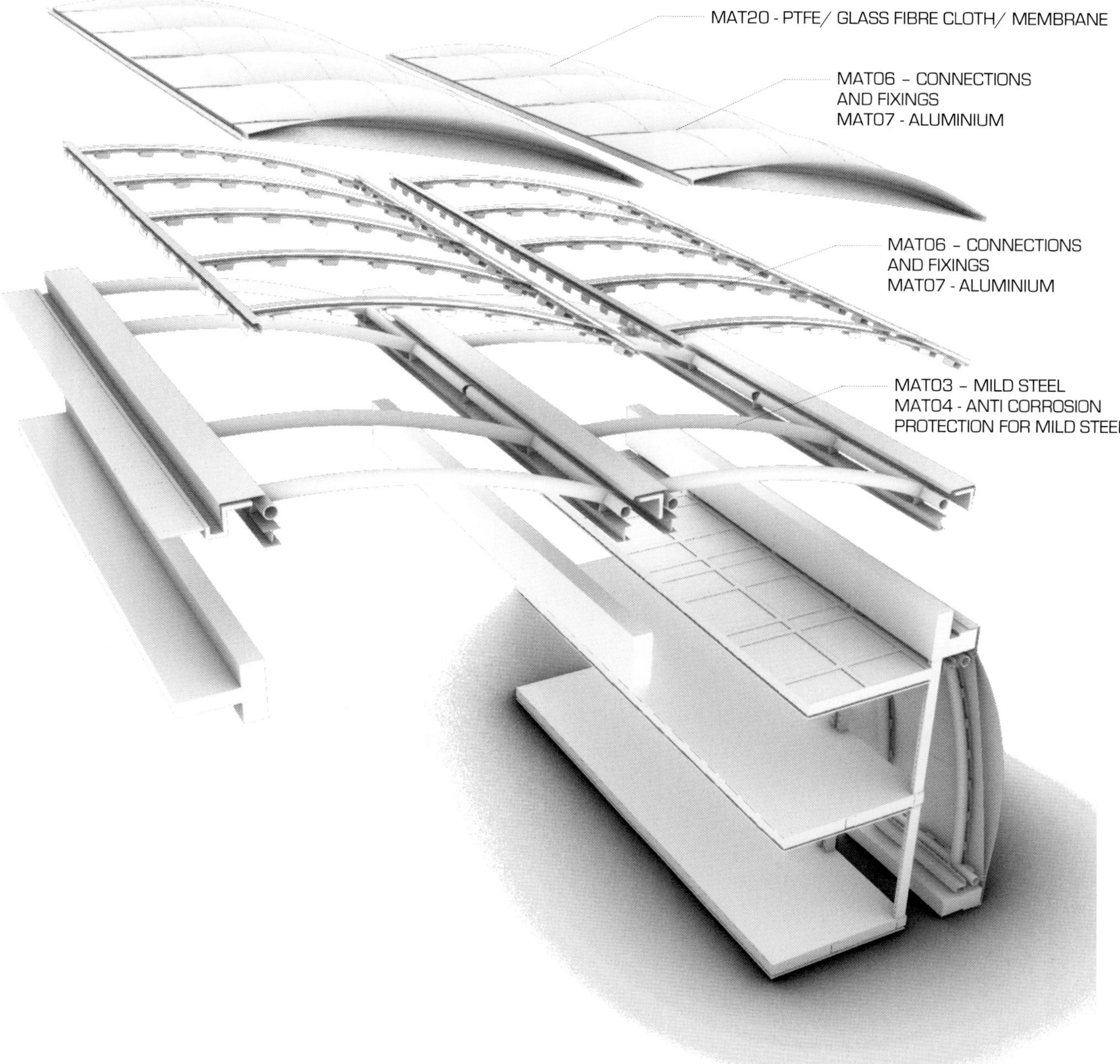

c | 3D exploded view of ETFE roof components with material specification references

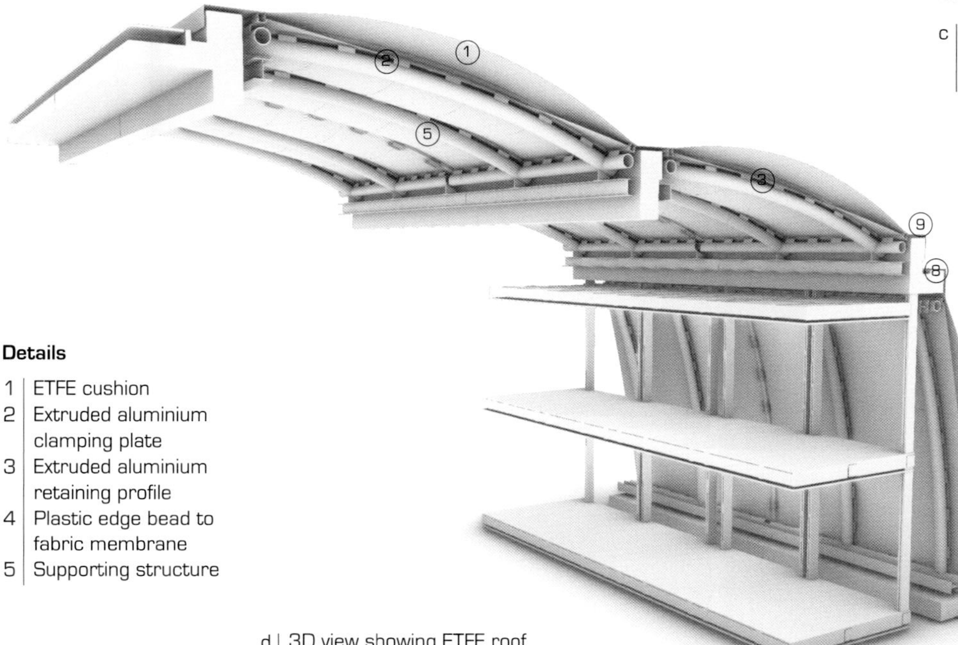

Details

1 | ETFE cushion
2 | Extruded aluminium clamping plate
3 | Extruded aluminium retaining profile
4 | Plastic edge bead to fabric membrane
5 | Supporting structure
6 | Plastic air supply tube
7 | Main air supply tube
8 | Insulated metal lined gutter
9 | Metal flashing
10 | ETFE cushion with contours shown for clarity

d | 3D view showing ETFE roof with polymer rainscreen wall

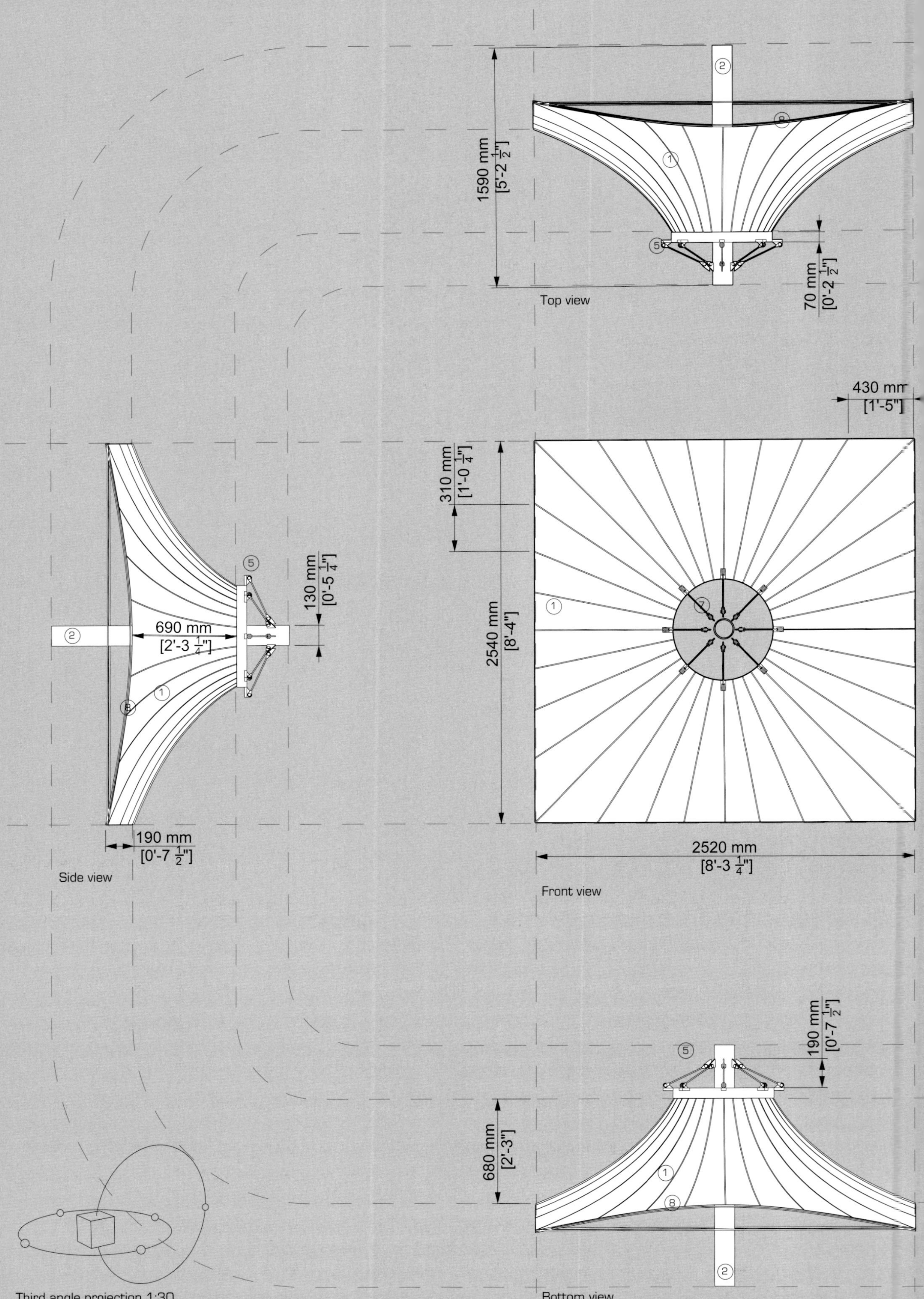

Third angle projection 1:30

Details

1 | Fabric membrane panel
2 | Supporting mild steel structure
3 | Extruded aluminium retaining profile
4 | Plastic edge bead to fabric membrane
5 | Extruded aluminium clamp assembly
6 | Stainless steel cable
7 | Stainless steel connector
8 | Membrane skirt
9 | Gutter formed by membrane skirt
10 | Mild steel ring support
11 | Fabric cover to close ring

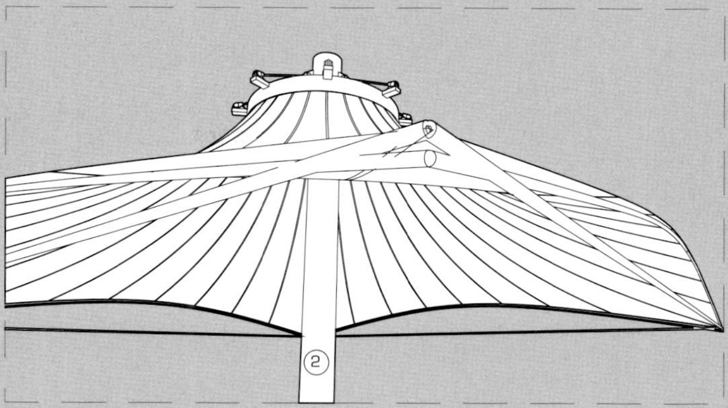

3D view of the assembly

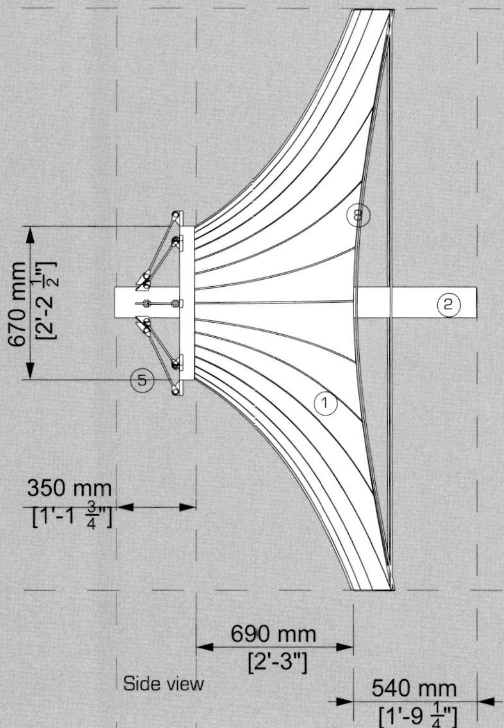

Side view

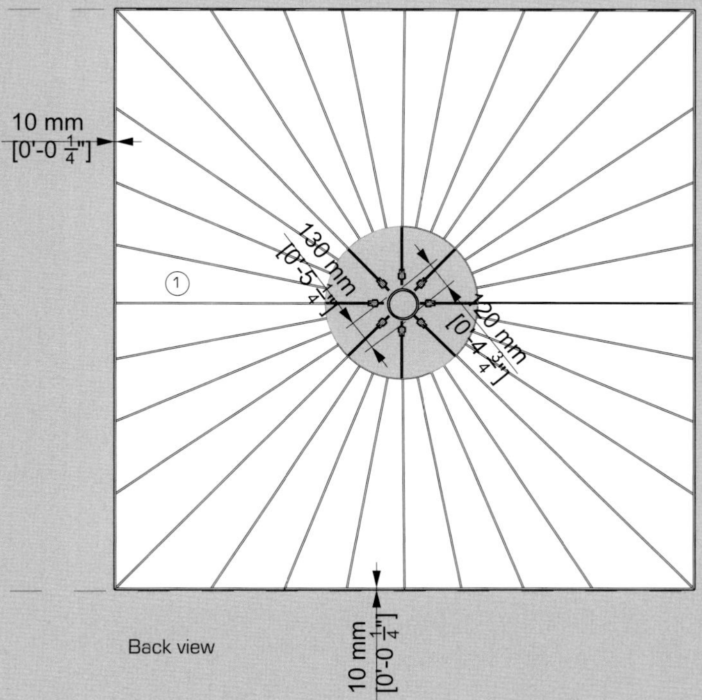

Back view

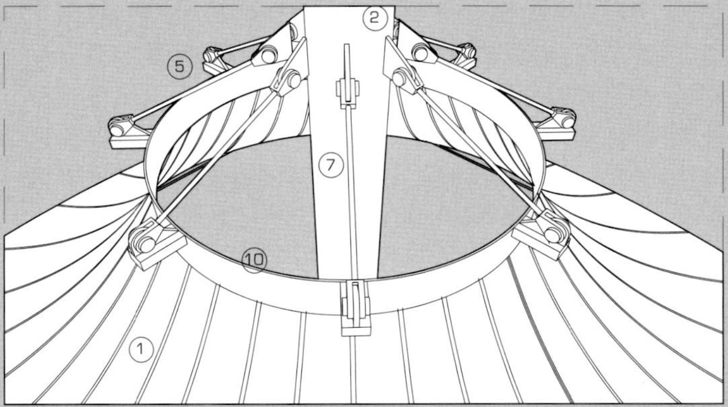

3D view of the assembly

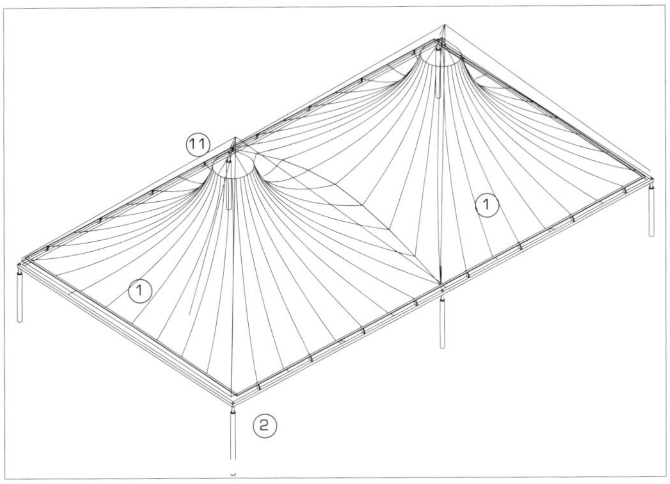

a| Isometric view of cone-shaped fabric roof

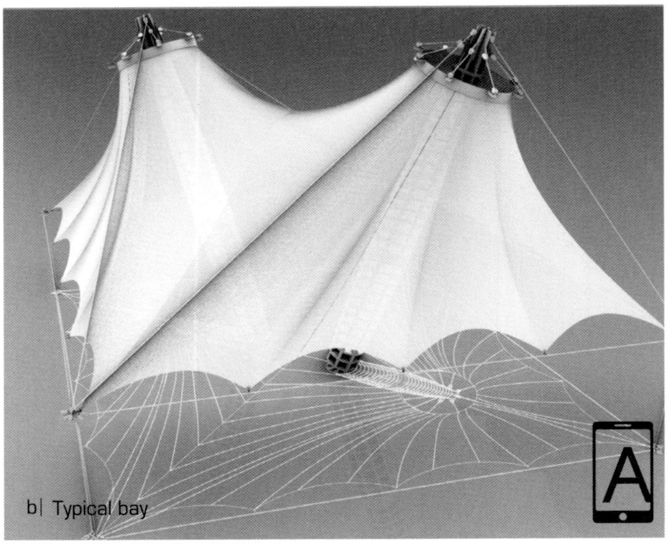

b| Typical bay

Details

1 | Fabric membrane panel
2 | Supporting mild steel structure
3 | Extruded aluminium retaining profile
4 | Plastic edge bead to fabric membrane
5 | Extruded aluminium clamp assembly
6 | Stainless steel cable
7 | Stainless steel connector
8 | Membrane skirt
9 | Gutter formed by membrane skirt
10 | Mild steel ring support
11 | Fabric cover to close ring

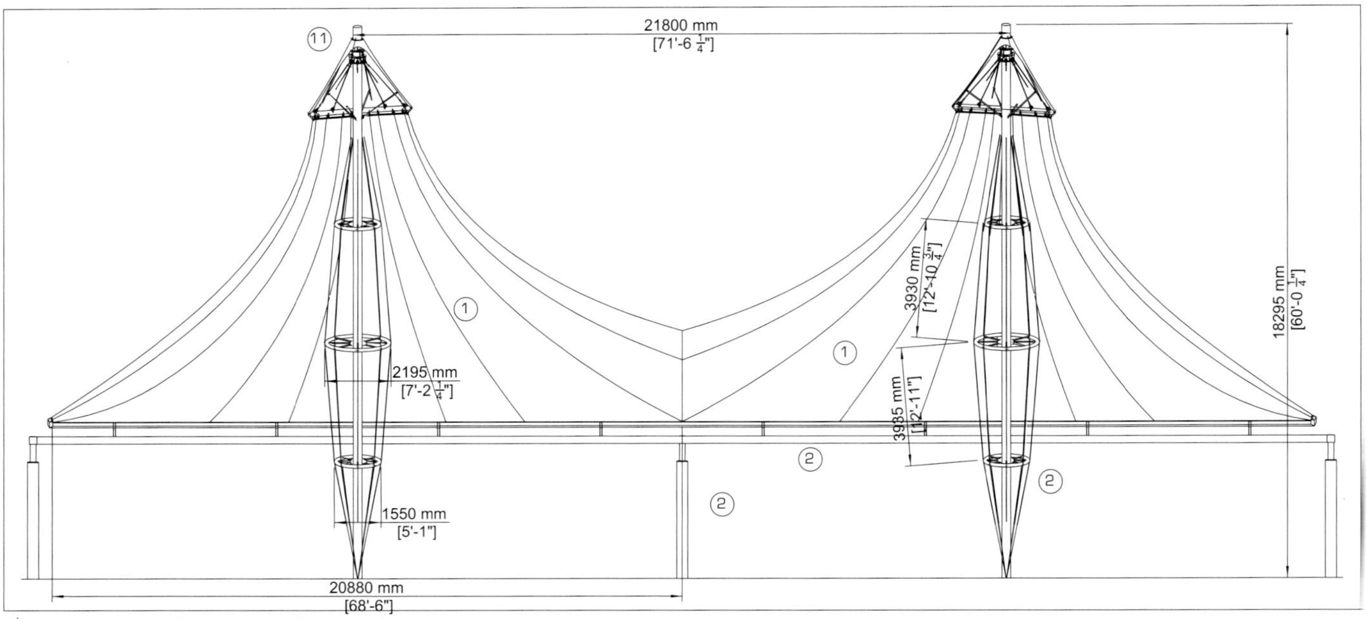

c| Vertical section. Cone-shaped fabric roof

The following two sections discuss the two most common shapes used for single-layer roofs: the 'cone-shaped' roof and the 'barrel-shaped' roof. A third type, based on the shape of a hyperbolic paraboloid, uses the same principles as those for these two types and is constructed using the same principles and same construction details. For that reason its specific geometry is not discussed here, but the construction principles described here can be applied equally to hyperbolic paraboloid forms.

The advantages of single-membrane fabric roofs are their smooth curves — typically with different and thin, sharp edges, in opposite directions — that provide translucent roofs that allow diffused daylight to pass through them. They use their curvature as a method of tensioning the membrane against a supporting structure, which is typically a mixture of mild steel tubes and stainless steel cables (MCE_509, d).

Fabric roof principles

Early examples of the fabric roof structures of 30 years ago were based, in part, on observations of how forces act in soap bubbles, where the soap film wall of the bubble takes up a minimum of surface area as a result of the surface tension of the wall of the bubble being evenly distributed. In a fabric roof, the membrane is structurally modelled so that the resultant form developed between architect and structural engineer distributes the tensile forces within the membrane without over-tensioning some parts and under-tensioning other parts of the membrane, with the resulting design resisting all load combinations in their different directions. This work is usually developed in the form of a computer model, either by specialist structural consultants or by manufacturers as part of the design development process. The minimal surface form must also be suitable for draining away rainwater (MCE_509, e), which forms another aspect of design development, together with the treatment of interfaces with adjacent areas of roof

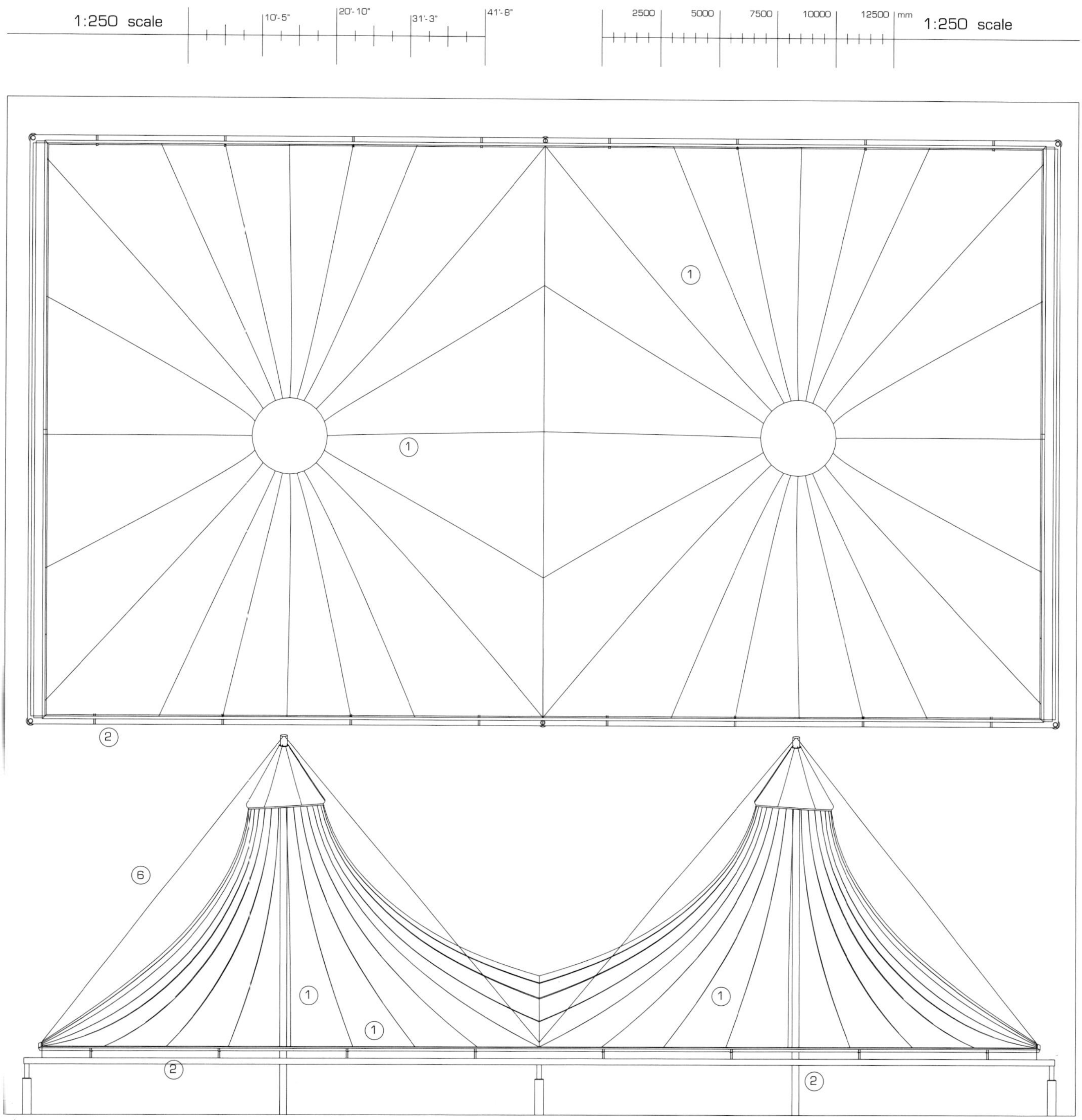

d | Horizontal section and elevation. Cone-shaped fabric roof

and external wall. The resulting form is designed to keep all parts of the fabric membrane in tension, not just from the supporting structure but from imposed loads, mainly wind loads. Wind pressures are resisted by re-distributing the forces within the fabric membrane. Any areas of the fabric roof that go into compression as a result of slackness in the membrane reveal themselves as creases in the material.

Both the cone-shaped example in this section and the barrel-shaped example of the following section make use of an internal steel structure that supports part of the membrane (MCE_509, d), to tension it in some areas, with roof edges where the material is held and tensioned, either at points or with continuous clamped fixings (c) similar to those described in the previous section on ETFE cushions. Generally speaking, the high points of the supporting structure take up the downloads from the membrane and imposed loads (mainly wind loads) and the lower points at the edges take up the effect of wind uplift. In shallow-sloped roof membranes, more of the structural loads are taken by the edges or points at the base, often resulting in large columns or posts being required at these points. The distribution of loads within the fabric roof design is revealed in the supporting structure, which can be as visually lightweight and elegant as the fabric membrane itself or can become visually heavy, which can detract from the intended lightweight effect of the membrane. Where roofs transfer forces to an adjoining structure, rather than contain the loads within their own supporting frame, the visual effect on the adjoining structure is balanced with the requirements of the membrane roof and its own frame.

Where imposed loads such as snow or sand can cause permanent stretching of the fabric membrane, the form of the roof and its associated slopes are made sufficiently steep to avoid creating areas or pockets on the fabric roof where they can collect.

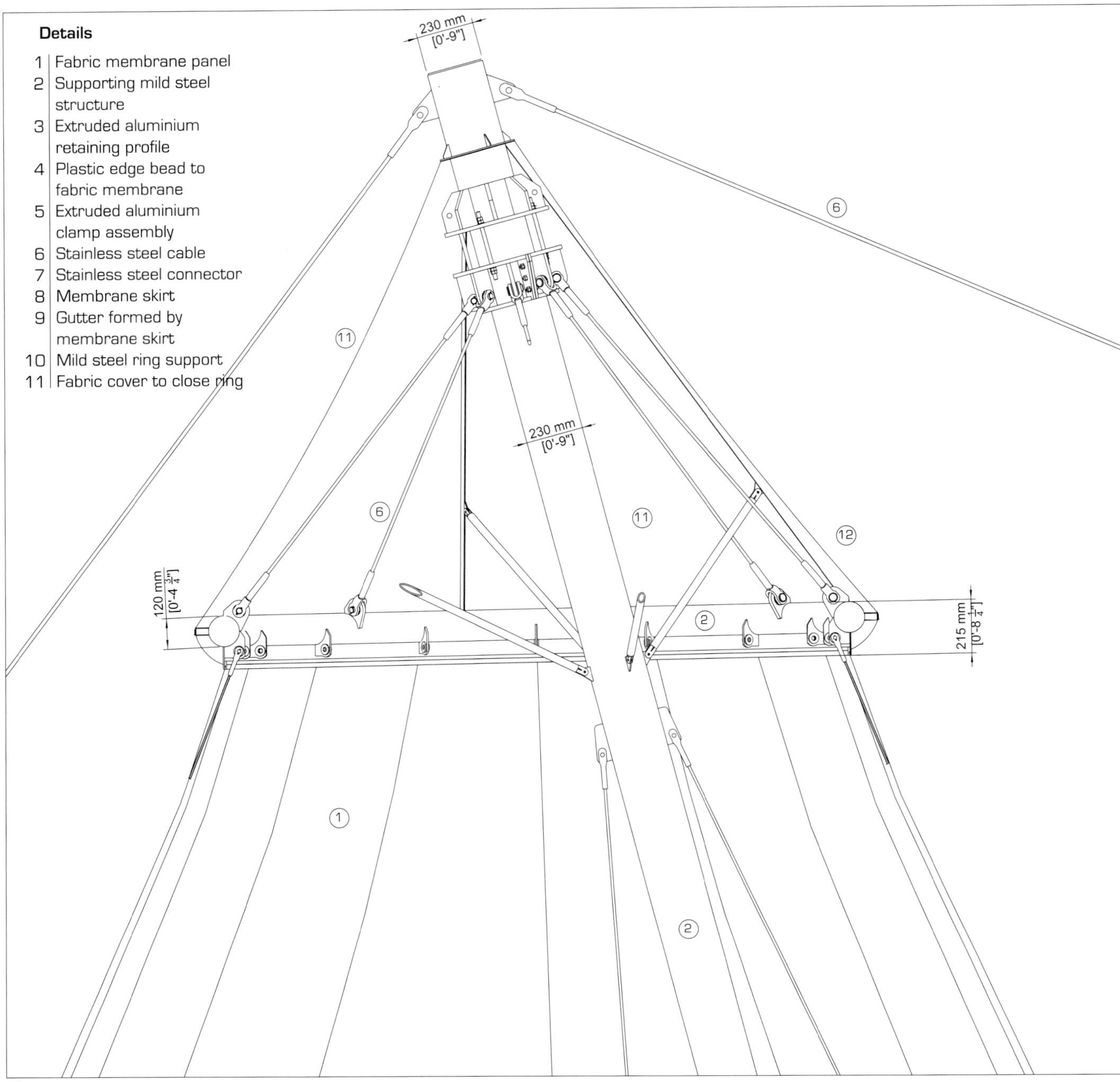

a| Vertical section. Clamping ring with fabric cover

Fabric types

The two most common fabrics used are PVC-coated polyester fabric and PTFE-coated glass fibre fabric. Both are woven cloth materials which are protected by coatings, usually applied on both sides. Other open-weave materials are used as solar shading only, and are manufactured without protective coatings. These are made from polymer threads, sometimes with a protective coating applied to the manufactured thread itself, to increase the life expectancy of the material. In all these woven materials the strength of the fabric can be different in the two directions in which the 'cloth' is woven. When selecting a material, the strength of the' warp' threads running the length of the material is compared to the 'weft' threads running the width. In most commonly used roof membranes, the tensile strength of the warp and weft directions are similar, but these need to be checked when the material type is chosen.

Most fabric roof materials imitate the appearance of natural canvas but this material is used only where its appearance and individual texture is considered to be the most important consideration. Natural canvas is less stable than synthetic fabrics when used in tension structures and is difficult to clean. A modified acrylic canvas material, with a similar texture to natural canvas, is sometimes used for its greater dimensional stability. Neither material is suitable for long span fabric roofs.

PVC/polyester fabrics are made from polyester cloth which is coated on both sides with a layer of PVC. The coating protects the fabric against the effects of rain and of UV light. The PVC coating is a mixture of PVC powder, softeners and plasticisers, UV stabilisers, pigments and fire retardants. An additional outer coat of lacquer slows down the effect in the PVC coating of becoming increasingly brittle with age, which results from the softeners within the material gradually moving to the surface of the PVC coating. The lacquer coating also slows down

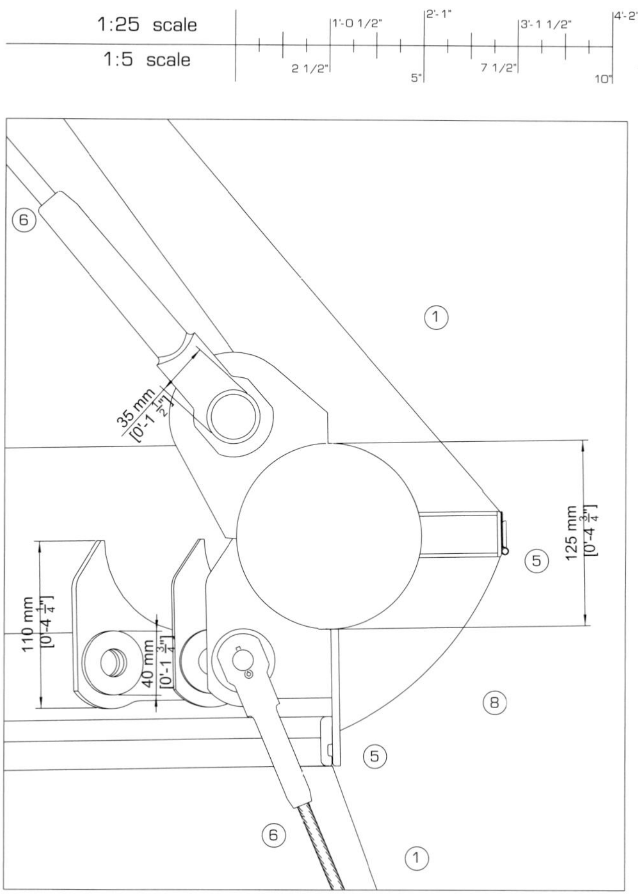

b| Vertical section. Base of roof with membrane skirt

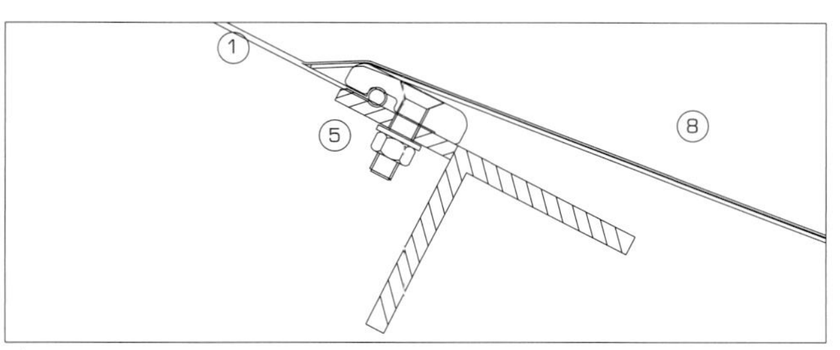

c| Vertical section. Clamping ring

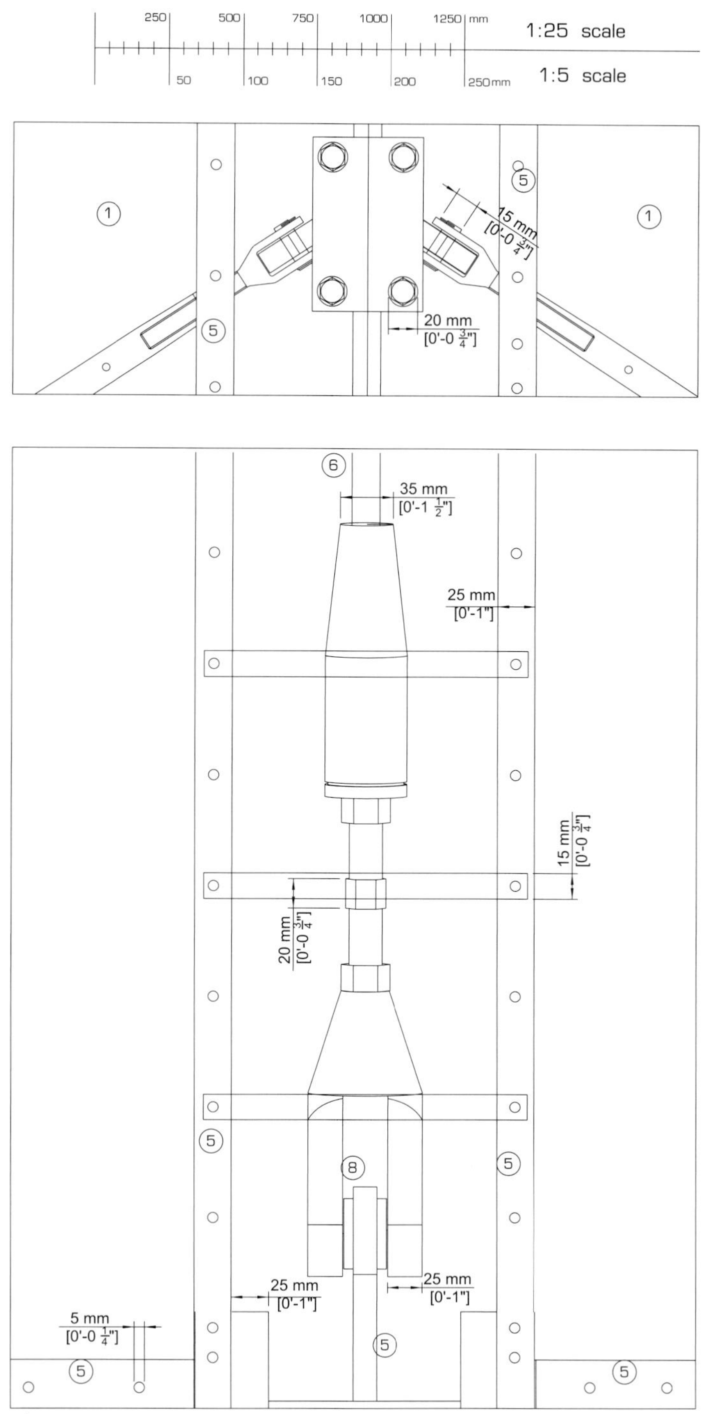

d| Horizontal section. Junction of adjacent fabric membranes

the fading effects of the colour pigments. A PVDF lacquer (a fluorinated polymer) is typically used, which also ensures that the surface has low surface friction so that it will attract little dust and dirt, and allows the membrane to be cleaned easily. Acrylic lacquers are also used. The typical weight of roof using this material is 500-800g/m². A PVC/polyester membrane roof will last around 15–25 years.

PTFE/glass fibre membranes are made from a glass fibre mat which is coated with a PTFE layer such as Teflon. As with PVC/polyester membranes, the coating protects the fabric from the effects of both the weather and UV light as well as forming a low-friction surface to reduce the collection of dirt and dust. Most dirt is washed away by rain, but some cleaning is needed using the same methods as for PVC/polyester roofs. Typically it weighs 800–1500g/m². The life expectancy of a PTFE/glass fibre membrane roof is longer than PVC/polyester membranes at around 30–40 years.

Comparison of types

Both PVC/polyester and PTFE/glass fibre have high tensile strength and high flexibility, making them very suitable for curved and double-curved roof membranes. They both have a light transmission of 5–20% depending on the thickness of membrane used, reflecting 75–80% of light. Neither will catch fire easily, and both resist the deteriorating effects of UV light, though PVC/polyester becomes increasingly brittle with age. Both have almost no acoustic performance and have poor thermal insulation performance when used as a single membrane roof. PVC/polyester has a greater range of colours readily available, while PTFE/glass fibre is usually white, the colour to which it bleaches naturally from its manufactured beige colour after a few months of being exposed to sunlight. Weld marks that occur during fabrication also disappear as a result of bleaching in sunlight. PTFE/glass fibre has lower surface friction than PVC/polyester, allowing the former to remain cleaner, while PVC/polyester requires cleaning more frequent-

a | Prototype

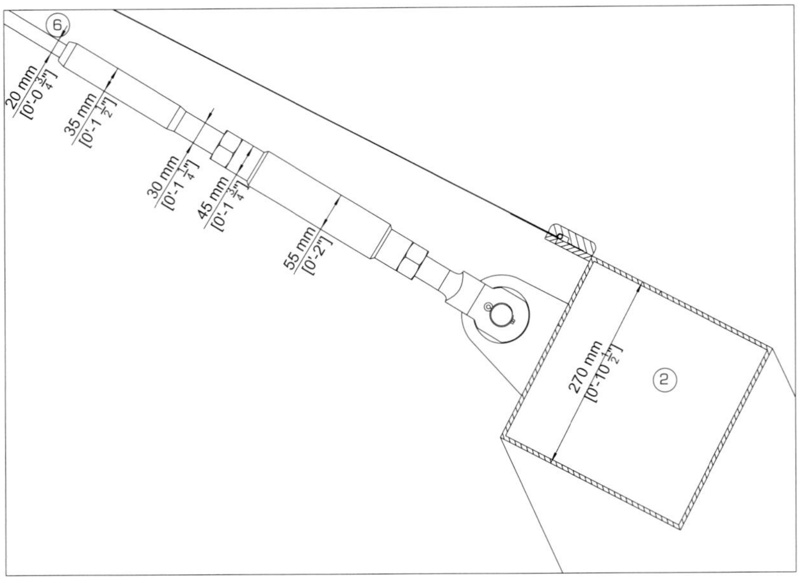

b | Vertical section. Base of fabric roof

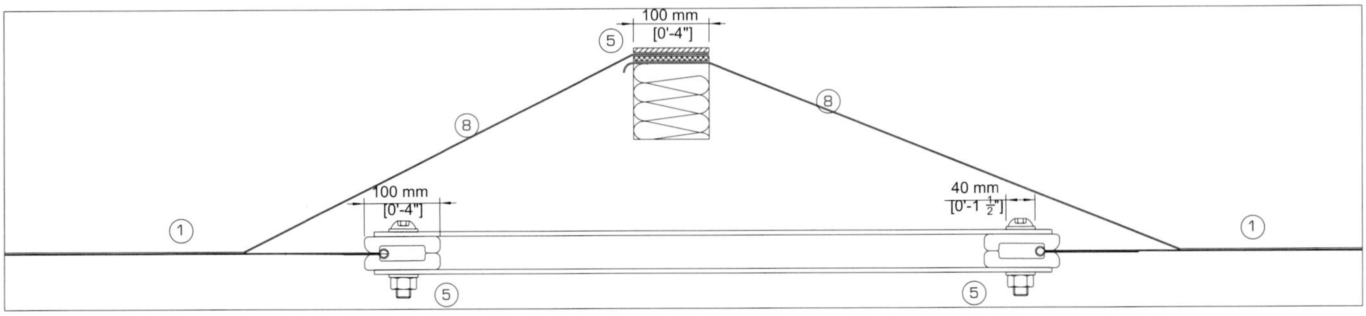

c | Vertical section. Junction of adjacent fabric membranes meeting at same angle

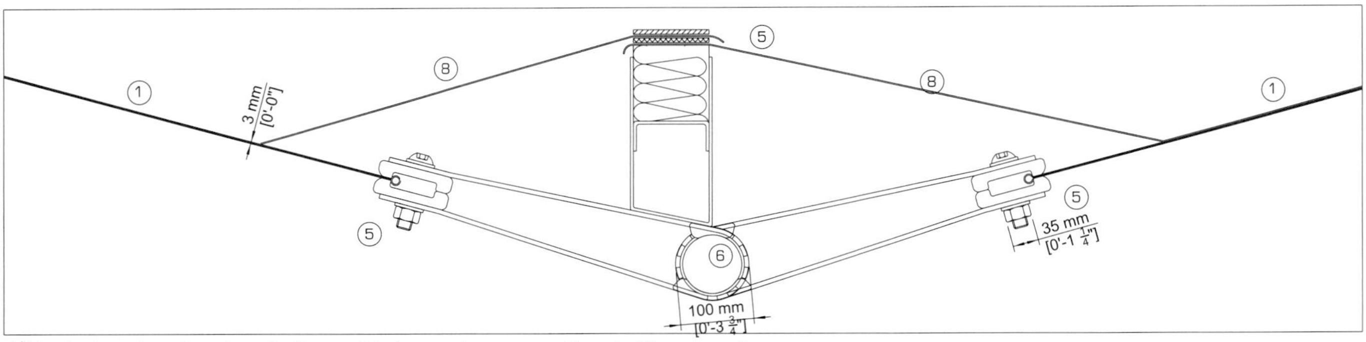

d | Vertical section. Junction of adjacent fabric membranes meeting at different angles

ly. PTFE/glass fibre requires greater care in transportation to site and erection than PVC/polyester, the latter being capable of being folded without damage to the membrane.

Thermal insulation

A single-layer membrane fabric roof in either PVC/polyester or PTFE/glass fibre typically has a U-value of around 6.0W/m²K. Where two layered membranes are used, with a minimum air gap of 200mm between the membranes, the U-value can be reduced to around 3.0 W/m²K. Double-layer membranes are less commonly used as they have a severely reduced light transmission, the translucency of the material being one of the main advantages of the material. Thermal insulation can be added to a double-layer membrane by using a translucent fibre-based insulation, as is used in fibre glass cladding panels discussed in the previous chapter. The insulation can be fixed to the inner face of either membrane, depending on how the roof void is ventilated. With the increasing importance of the role of thermal insulation in the reduction of energy use in buildings, the use of double-layer fabric membranes is set to develop considerably over the next ten years.

Suspension points

Suspension points at the top of a cone-shaped fabric roof are usually formed either by a metal ring, which is fixed back to a central mast by cables or cantilevered brackets (MCE_510, a) or, alternatively, by a 'palm tree' arrangement of projecting curved metal brackets which serve to tension the continuous membrane against its supporting mast set within the building.

The first option with a metal ring requires an additional membrane cover piece, while the second 'palm tree' option forms a continuous membrane with a smooth curved top, without any breaks in the continuity of the membrane. In the 'ring' solution, the membrane is clamped

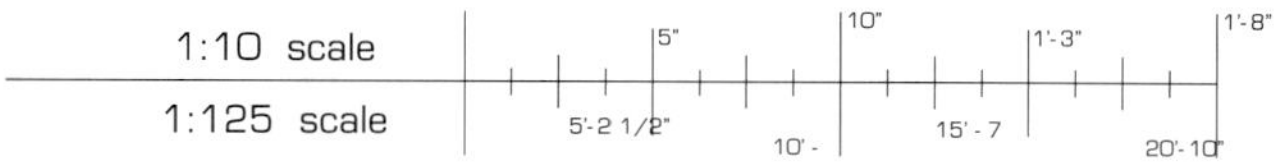

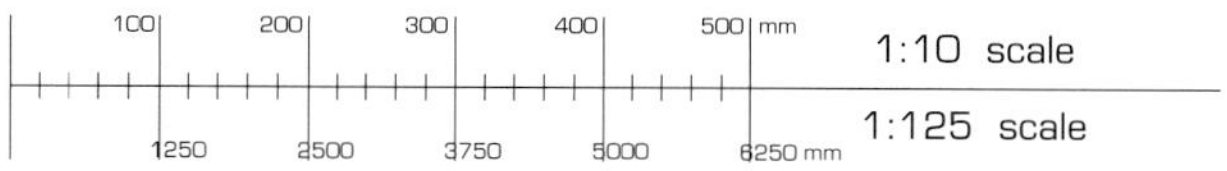

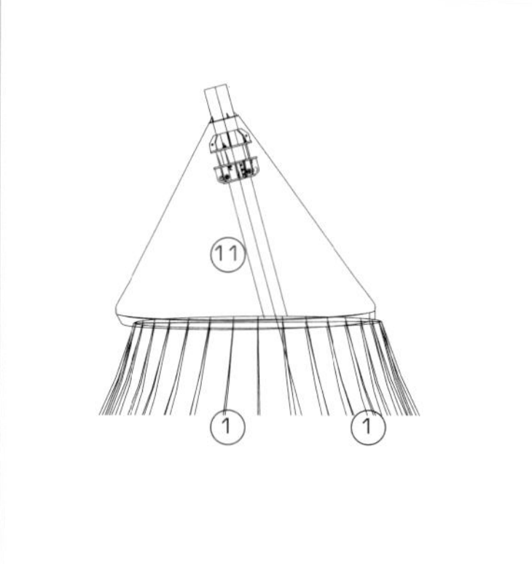

e | Elevation. Membrane cover

Details

1 | Fabric membrane panel
2 | Supporting mild steel structure
3 | Extruded aluminium retaining profile
4 | Plastic edge bead to fabric membrane
5 | Extruded aluminium clamp assembly
6 | Stainless steel cable
7 | Stainless steel connector
8 | Membrane skirt
9 | Gutter formed by membrane skirt
10 | Mild steel ring support
11 | Fabric cover to close ring

f | Prototype

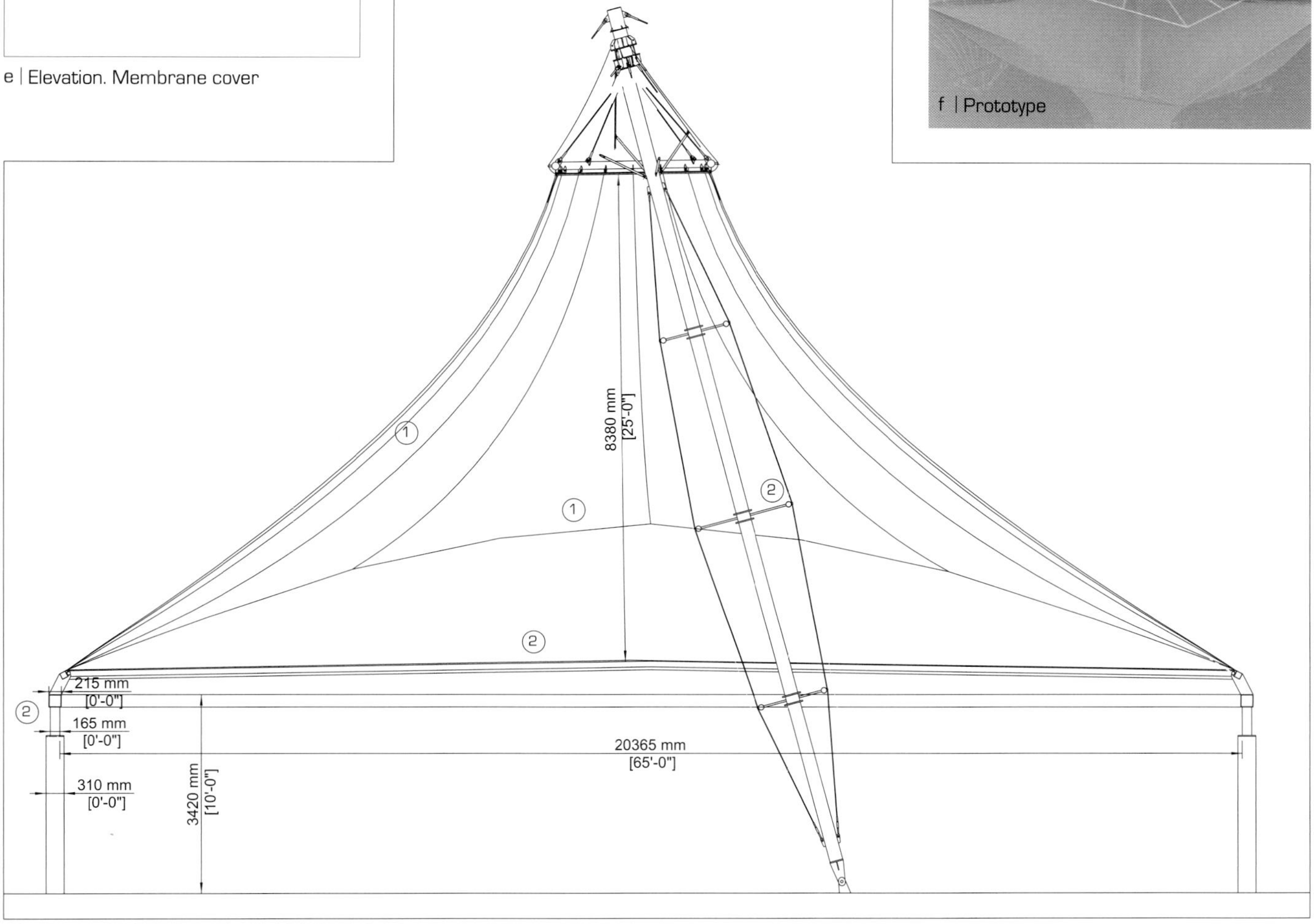

g | Vertical section. Cone-shaped fabric roof

between an inner ring and an outer ring fixed together with bolts (e). A second clamp is used to fix the cone-shaped membrane that covers the top of the ring. The top of the cone-shaped cover is either pulled over the top of the central mast or clamped around it (c). The ring is either freely suspended from cables, or is firmly fixed to the mast with cantilevered brackets to which the ring is secured.

In the 'palm tree' supported solution, cantilevered brackets with a curved shape are set out radially in order to create a smooth curved form on which the membrane is set. The brackets are usually aligned with joints between membrane panels.

Acoustics

Like the ETFE cushions discussed in the previous section, single-layer membranes provide no significant reduction of noise through the roof. A double-layer roof with an acoustic lining will provide some acoustic performance but will have the effect of losing most of the light transmission through the membranes. In addition, low-frequency sound is difficult to absorb due to the low mass of the cladding material. In common with ETFE cushion roofs, the roofs are almost transparent to sound emitted from within the building.

Durability

Fabric roofs are highly susceptible to damage from sharp objects. Small cuts in the membrane can be repaired with patches made from the same material which are glued into position. Larger tears are repaired with hot air welders, usually undertaken by the specialist contractor that installed the fabric roof. Large repairs are visible, and panels are replaced where visibility is the most important consideration. Since large tears can affect the overall structural performance of the membrane, the complete membrane is sometimes removed for another panel to be stitched or welded, and re-coated in the factory.

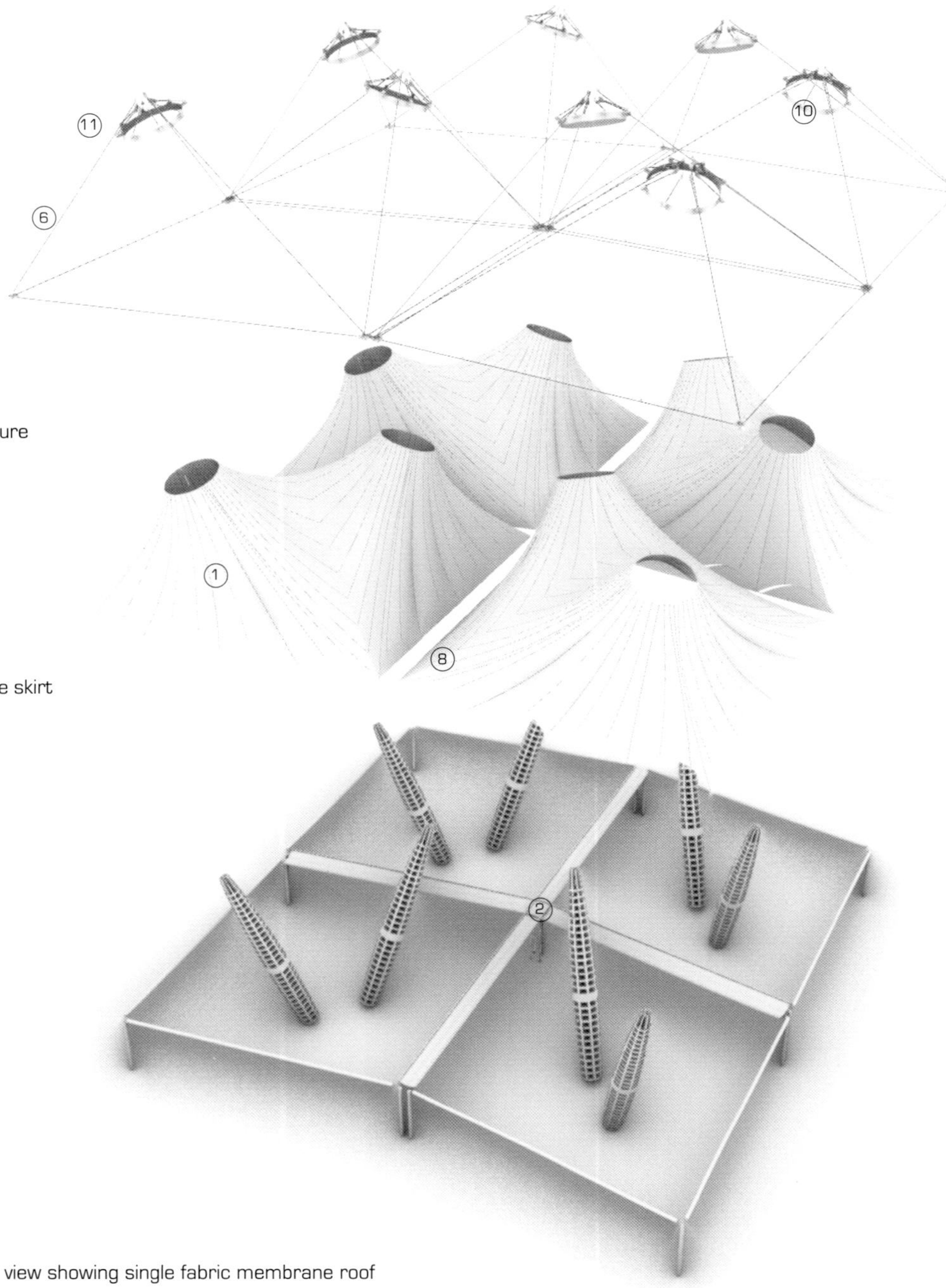

Details

1 | Fabric membrane panel
2 | Supporting mild steel structure
3 | Extruded aluminium retaining profile
4 | Plastic edge bead to fabric membrane
5 | Extruded aluminium clamp assembly
6 | Stainless steel cable
7 | Stainless steel connector
8 | Membrane skirt
9 | Gutter formed by membrane skirt
10 | Mild steel ring support
11 | Fabric cover to close ring

a | 3D exploded view showing single fabric membrane roof

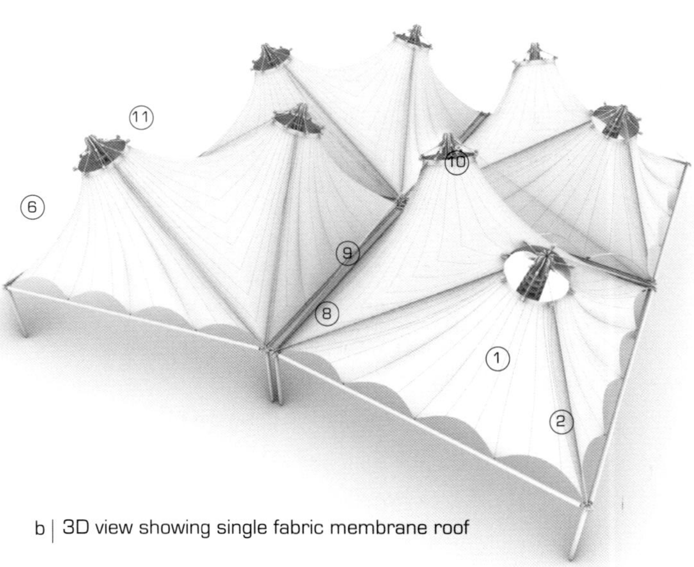

b | 3D view showing single fabric membrane roof

The outer surface of fabric roofs is cleaned with soft brushes that wash the membrane surface, the work being done typically from either a mobile platform or by rope access cleaners (abseilers) who are attached in harnesses and suspended from cables attached to a steel bracket anchor point on the top of the membrane roof. The anchor point forms part of the steel or timber supporting structure to the membrane roof. Cleaning is also an important consideration in areas of high humidity where there is a higher risk of mould forming on the surface of the fabric which can cause permanent staining. Regular cleaning prevents mould growth. PVC/polyester is more susceptible to mould growth than PTFE/glass fibre fabrics, essentially because the latter has lower surface friction.

Performance in fire

The performance of a membrane during a fire depends on both the fabric used and the stitching at the seams, where this joining method

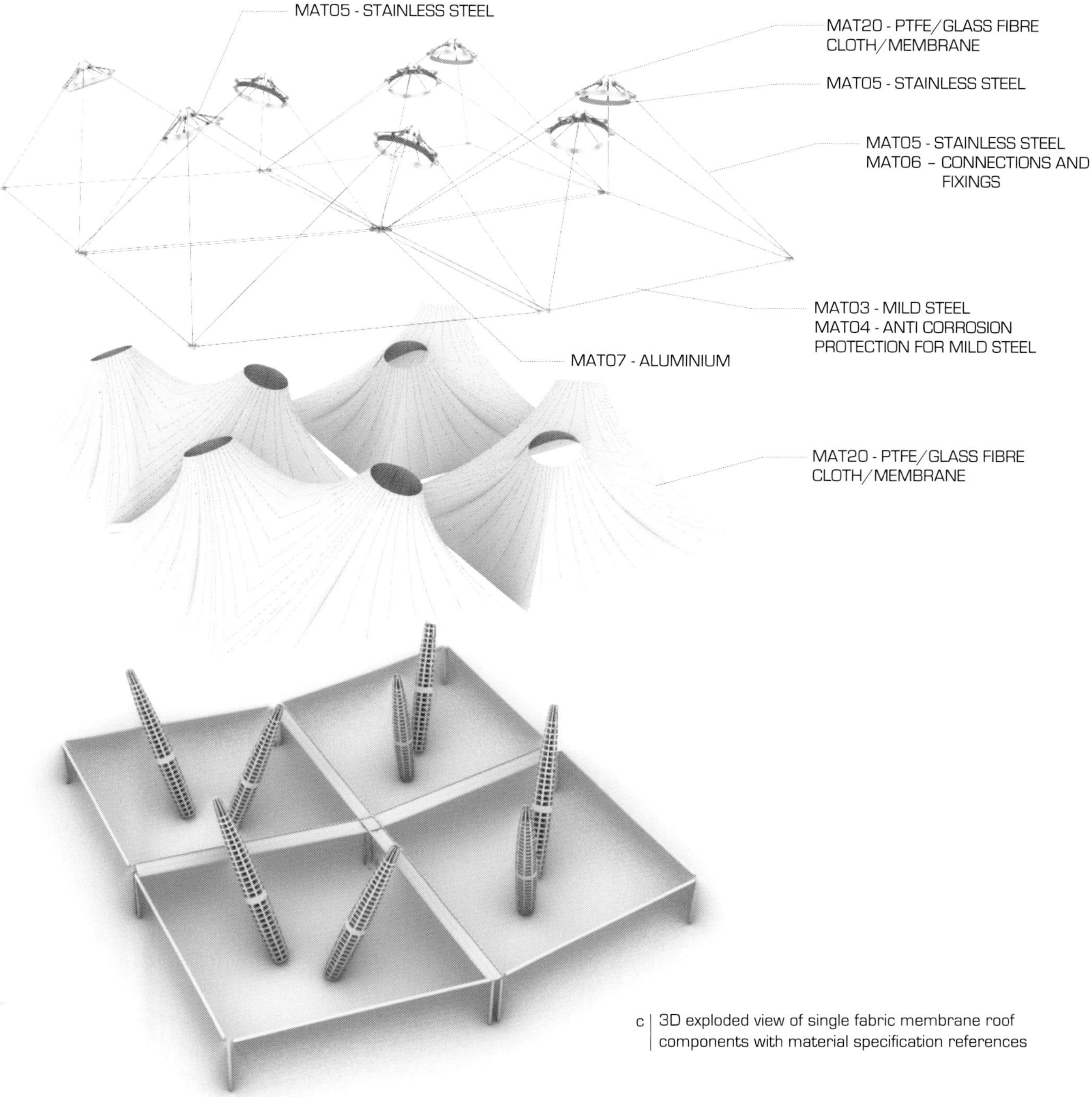

c | 3D exploded view of single fabric membrane roof components with material specification references

is used. Membranes lose their tension under high temperatures, with PVC/polyester stretching above 70–80°C, and PVC/polyester seams starting to peel at around 100°C. At 250°C the PVC melts, leaving holes in the membrane. PVC has fire retardants in the coating so that it self-extinguishes when the flame source is removed which results in few, if any, burning fragments dropping down from the roof. PTFE/glass fibre fabrics fail at around 1000°C, but seams will fail at a much lower temperature of around 270°C. With both materials, the failure of the membrane forms holes in the roof which allow heat and smoke to escape.

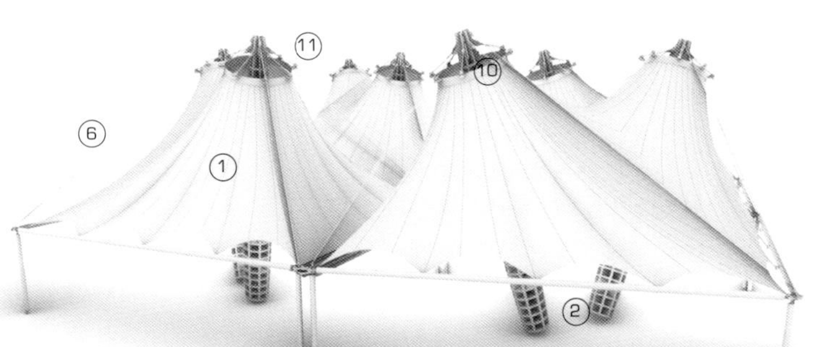

d | 3D view showing single fabric membrane roof

Condensation

Where condensation is likely to occur on the underside of the membrane in an internal space within the building, either a second membrane is added or, more commonly, increased ventilation is provided within the space below the roof.

A

13 Aluminium windows 178
- Windows in openings 180
- Window walls 184
- Composite windows 185

Authorship 520

B

Bibliography 520
BIM to fabrication 50
Bitumen-based sheet membranes 420
10 Bolt fixed glazed walls 148
- Support methods 150
- Bottom supported glazing 152
- Top hung glazing 153
- Corners 154
- Seals and interfaces 155

32 Bolt fixed glazing: small scale rooflights 380
- Generic structural support methods 383
- Supporting brackets 383
- Bolt fixings 384
- Arrangement of bolt fixings 385
- Glazed units 386

33 Bolt fixed glazing: large scale rooflights 390
- Base of glazed roof 393
- External and internal folds 394
- Small glazed rooflights 395
- Larger rooflights 396

34 Bolt fixed glazing: twin wall roofs 400
35 Bonded glass rooflights 410
- Generic conical rooflight 412
- Generic rectangular rooflight 413
- Generic monopitch rooflight 415
- Glass roof decks 415

18 Brick cavity walls 232
- Ground level 236
- Window and door openings 236
- Eaves and parapets 238

C

15 Cast in-situ / cast-in-place 200
- Parapets, drips and sills 203
- Finishes 204
- As-cast finish 205
- Washed finish 206
- Polished finish 207

25 Cladding panels, rainscreens – timber 306
- Timber boards 308
- Finishes 309
- Cladding panels and rainscreens 310
- Plywood sheets 313

24 Cladding the timber frame 296
- Timber frames 296
- Ground level 299
- Upper floors 301
- Corners 302
- Roof eaves and parapets 303

08 Clamped glazing 138
- Comparison with bolt fixed glazing 140
- Patch-plate glazing 141
- Clamped glazing 142
- Opaque glazing 144
- Sealing clamped glazing 144

Comparison of systems 16
28 Composite panels 338
- Single wall composite panels 340
- Twin wall panels 342
- Ridges, Verges 343
- Eaves, Parapets and valley gutters 344

03 Composite panels 76
- Parapets and sills 80
- Windows and door openings 80
- Developments 82
- Corners, Thermal bridges at sills 83

36 Concealed membrane roofs 422
- Materials 424
- Structural joints 426
- Parapet upstands 427
- Balustrades and plinths 428
- Rainwater outlets 428
- Penetrations for pipes and ducts 430

D

Design to protoype 6

E

43 ETFE Cushions 498
- Cushions 500
- Air supply 501
- The material 502
- Fabrication 503
- Durability, Performance in fire 505

37 Exposed membrane roofs 432
- Polymer-based membranes 435
- PVC membranes 436
- FPO (TPO) membranes 437
- Mechanically fixed method 437
- Bonded fixing method 438
- Parapets and upstands 439
- Ballasted roofs 439

F

Fabric roof systems 497
39 Flat roof: Bitumen-based sheet membranes 452
- The material 456
- Roof build-up 457
- Solar protection, Fixing methods 458
- Parapet upstands 459
- Junction with tiled roof 460
- Eaves and verges 461

G

11 Glass blocks 158
- Fixing glass blocks 161
- Support frames and walls 162
- Steel support frames 163
- Masonry, timber framed openings 164
- Cast glass channels 165

17 GRC cladding panels 220
- Individually supported panels 222
- Self-supporting stacked panels 224
- Parapets and sills 225
- Openings 226
- Sand-blasted finish and tooled finish 226

42 GRP panels and shells 486
- Smaller panels and shells 488
- Larger panels and shells 489

41 GRP rooflights 476
- Eaves and upstands 480
- Verges 482
- Abutments 482
- Sliding roof panels 483

L

06 Louvred screens 106
- Metal louvres 108
- Solar shading 110
- Glazed louvres 111
- Walkways 113

M

18 Masonry cavity walls: brick 232
- Ground level 236
- Window and door openings 236
- Eaves and parapets 238

19 Masonry cavity walls: concrete block 242
- Wall structures 245
- Ground level 245
- Openings in walls 247
- Eaves and parapets 248

20 Masonry stone cladding 252
- Fixings 255
- Cladding to precast concrete panels 256
- Joints, Closed joints, 257
- Movement joints 258
- Stone finishes 259

21 Masonry terracota rainscreens 262
- Manufacture of panels 264
- Corner pieces 265
- Fixing systems 266
- Panel sizes, Openings 268

05 Mesh screens 96
- Rigid mesh 98

Meshes flexible in one direction 99
Fully flexible mesh 100
Mesh used on curves 101
Perforated metal 102
30 Metal canopies 358
Bolt fixed panels 360
Fixed metal louvre canopies 360
Metal canopies 362
Electrically operated louvres 365
29 Metal rainscreens – roofs 348
Panel arrangement 351
Parapets 352
Monopitch ridges and verges 353
Roof geometry 354
Roof soffits 355
04 Metal rainscreens – walls 86
Materials, Fixing methods 89
Backing walls 90
Construction sequence 92
Window and door openings 92
Parapets and sills 93
26 Metal standing seam roofs 318
Site-based method 320
Prefabricated methods 321
Sealed and ventilated roofs 323
Roof openings 324
Ridges and valleys 324
Eaves and parapets 325

P

40 Pitched roof: Tiles 454
Standing seam cold roofs 466
Eaves and valley gutters 466
Ridges and abutments 467
Penetrations 467
Metal tiled roofs 468
38 Planted concrete roofs 442
System design 444
Planted roof components 445
Soil depth 446
Overflows, Roof junctions 447
Rainwater outlets 448
Balcony planters 448
22 Polymer-based cladding sealed panels 274
GRP panels 278
Polycarbonate cladding 280
23 Polymer-based cladding rainscreens 284
Flat polycarbonate sheet 286
Multi-wall polycarbonate sheet 287
Profiled polycarbonate sheet 288
Polymer-composite flat panels 289
UPVC board cladding 289
UPVC windows 290
GRP panels 291
Preliminary specification for inventory models 22
02 Profiled metal cladding – walls 66
Junctions 69
Parapets and gutters 70
Insulation and liner trays 71
Window and door openings 72
Developments 73
27 Profiled metal sheet – roofs 328
Profiled metal decks as substrates 330
Profiled metal roof sheeting 331
Sealed and ventilated methods 332
Twin skin construction 332
Ridges 333
Openings, Eaves and parapets 334
Ridges and valleys 335
Prototyping 44

R

Rainscreens:
Metal roofs 348
Metal walls 86
Masonry 262

S

01 Sheet metal 56
Fixing methods 58
Openings 59
Substrates and supporting walls 61
Corners, parapets and sills 62
31 Silicone-sealed glazing and rooflights 370
Silicone-sealed systems 372
Junctions 373
Use of capped profiles 375
Rooflights 376
44 Single membrane roof 508
Fabric roFof principles 510
Fabric types 512
Comparison of types 513
Thermal insulation 514
Suspension points 514
Acoustics, Durability 515
Performance in fire 516
Condensation 517
17 Small precast / GRC cladding panels 220
Individually supported panels 222
Self-supporting stacked panels 224
Parapets and sills 225
Openings 226
Sand-blasted and tooled finishes 226
12 Steel windows 168
Small-scale glazing 171
Large-scale glazing 173
07 Stick glazing 118
Comparison with unitised glazing 120
System assembly 121
Framing profiles, Corners 123
Opening lights 124
Parapets, sills and penetrations 124
Spandrel panels 125
19 Stone and blockwork cavity walls 242
Wall structures 245
Ground level 245
Openings in walls 247
Eaves and parapets 248
20 Stone cladding 252
Fixings 255
Cladding to precast concrete panels 256
Joints 257
Closed joints 257
Movement joints 258
Stone finishes 259
16 Storey height precast panels 210
Panel types 213
Thermal insulation 214
Joints 215
Acid etched finish 217

T

21 Terracotta rainscreens 262
Manufacture of panels 264
Corner pieces 265
Fixing systems 266
Panel sizes, Openings 268
25 Timber cladding panels and rainscreens 306
Timber boards 308
Finishes 309
Cladding panels and rainscreens 310
Plywood sheets 313
24 Timber framed cladding 296
Timber frames 298
Ground level 299
Upper floors 301
Corners 302
Roof eaves and parapets 303
14 Timber windows 188
Window walls 190
Window design 191
Windows in openings 194

U

08 Unitised glazing 128
Jointing panels 131
Opening lights 132
Corner panels, parapets and sills 133
Penetrations 134
Silicone-bonded glazing 134
UPVC windows 290

Authorship and bibliography

All contents shown in this book, including photographs, have been created entirely by Newtecnic . Drawings and models were created and rendered by the Newtecnic team. The book was designed by Yasmin Watts. The text is by Andrew Watts. The text and overall content of the book was updated with contributions from Howard Tee, Ross Kunchev, Andrew Watts and Yasmin Watts.

David Marold is Acquisitions Editor at Birkhäuser Verlag GmbH in Vienna. He has driven this book from a set of basic layouts to a completed book. He has a passion for books and their design, ranging from their wider content to the quality of print paper, to the overall reader experience.

Newtecnic have produced a number of volumes on contemporary building technology which provide reference material for engineering and architecture students, and professionals around the world. The Modern Construction Series is published by Birkhäuser Verlag GmbH. In addition, the Facades Technical Review, from RIBA Publications was published in spring 2007.

Newtecnic is an international world leader in the engineering design and architectural design of complex, highly ambitious construction projects and advanced building envelope systems.

The company undertakes the engineering and architectural design of building structures, facades and MEP installations in partnership with leading international developers and large contractors. Newtecnic also works on some projects with signature architects.

Founded in 2003, Newtecnic's design professionals team is completely and solely dedicated to the design and engineering of structures, façades and MEP. In partnership with the Engineering Departments of Cambridge University and the University of Houston. Newtecnic's R&D team analyses, develops, tests, validates and specifies new building products, technologies and methods. Newtecnic is certified for the ISO 9001 and ISO 27001 quality standards.

Newtecnic has offices in the US, the UK, Saudi Arabia, UAE and Hong Kong. The company is owned, directed and managed by long-established and experienced engineers and architects. Newtecnic are holders of *The Queen's Award for Enterprise* and is a *New Civil Engineer 100 Companies of the Year 2019.*

The building engineering design of Newtecnic is based on strong technical and visual concepts that are supported by analysis, modelling, geometry and coordination. Newtecnic's work ranges from conception and the development of design concepts when working alongside architects, to detailed design and engineering analysis for fabrication when working for contractors.

Newtecnic's ability to develop design solutions iteratively within short time frames is crucial in delivering complex building projects on time and budget. Newtecnic's design work is based on an intensive use of the latest digital tools, such as Catia, Rhino, Revit, Maya, Sofistik, SAP2000, Comsol Multiphysics and Phoenics. The use of these tools links design to digital manufacturing as used by building contractors and fabricators.

Newtecnic is grounded in the first principles of engineering design. The authorship suggests the following bibliography in order for the reader gain further knowledge of the technical aspects of the construction of building envelopes, within the context of architectural design. Each publication listed here is followed by the chapter headings in this book where this additional reading might be applied.

Metal walls

Heywood, M. D. (2006) 'Best Practice for the Specification and Installation of Metal Cladding and Secondary Steelwork', Steel Construction Institute.

01. Sheet metal
02. Profile metal cladding

Selves, N., Shanahan, J. and Sproul, C. (2004) 'Profiled Sheet Roofing and Cladding: A Guide to Good Practice', Taylor & Francis.

03. Composite panels

Davies, J.M. (2001) 'Lightweight Sandwich Construction', Oxford: Blackwell Science Ltd.

04. Metal rainscreens
05. Mesh screens
06. Louvre screens

Glass walls

Weller, B., Unnewehr, S., Tasche, S. and Harth, K. (2009) 'Glass in Building', Basel: Birkhauser.

07. Stick glazing

Patterson, M. (2011) 'Structural glass facades and enclosures', New Jersey: John Wiley & Sons.

07. Stick glazing

Wurm, J. (2007) 'Glass Structures: Design and Construction of Self-Supporting Skins', Basel: Birkhauser.

07. Stick glazing

Memari, A. M. (2013) Curtain Wall Systems: A primer. Reston Virginia: American Society of Civil Engineers.

08. Unitised glazing
09. Clamped glazing
10. Bolt fixed glazing
11. Glass blocks

Pennycook, B. (1987) 'Building with Glass Blocks', Doubleday Canada Ltd.

12. Steel windows
13. Aluminium and PVC-U windows
14. Timber windows

Concrete walls

Hsu, T.T.C and Mo, Y.L. (2010) 'Unified Theory of Concrete Structures', John Wiley & Sons.

15. Cast in situ
16. Storey height precast panels

Elliot, K. S. and Jolly, C. K. (2014) 'Multi-storey Precast Concrete Framed Structures', Oxford: Wiley – Blackwell.

17. Small precast/ GRC cladding

Valery, V. and Evgeny, V. (2013) 'Advanced Mechanics of Composite Materials and Structural Elements', Oxford: Kidlington.
15. Cast in situ
16. Storey height precast panels
17. Small precast/ GRC cladding

Masonry walls

Pacheco-Torgal, F. et al. (2014) 'Eco-efficient Masonry Bricks and Blocks:design, Properties and Durability', Cambridge: Elsevier.
18. Masonry cavity walls: brick
19. Masonry cavity walls: concrete block
20. Stone cladding

De Sousa Camposinhos, R. (2014) 'Stone cladding engineering', Springer.
21. Terracotta rainscreens

Polymer walls

Jeska, S. (2007) 'Transparent Plastics: Design and Technology', Springer Science and Business Media.
22. Polymer-based cladding sealed panels
23. Rainscreens

Timber walls

Herzog, T. (2012) 'Timber Construction Manual', Birkhauser.
24. Cladding the timber frame
25. Cladding panels and rainscreens

Davies, J.M. (2001) 'Lightweight Sandwich Construction', Oxford: Blackwell Science Ltd.
24. Cladding the timber frame
25. Cladding panels and rainscreens

Metal roofs

Oliver, M.S., Albon, J.M. and Garner, N.K. (1997) 'Coated Metal Roofing and Cladding', London: Thomas Telford.
26. Metal standing seam
27. Profiled metal sheet

Selves, N., Shanahan, J. and Sproul, C. (2004) 'Profiled Sheet Roofing and Cladding: A Guide to Good Practice', Taylor & Francis.
28. Composite panels

Davies, J.M. (2001) 'Lightweight Sandwich Construction', Oxford: Blackwell Science Ltd.
29. Rainscreens
30. Metal canopies

Glass roofs

Weller, B., Unnewehr, S., Tasche, S. and Harth, K. (2009) 'Glass in Building', Basel: Birkhauser.
31. Silicone-sealed glazing and rooflights
32. Bolt fixed glazing: small scale rooflights
33. Bolt fixed glazing: large scale rooflights
34. Bolt fixed glazing: twin wall roofs
35. Bonded glass rooflights

Wurm, J. (2007) 'Glass Structures: Design and Construction of Self-Supporting Skins', Basel: Birkhauser.
31. Silicone-sealed glazing and rooflights
32. Bolt fixed glazing: small scale rooflights
33. Bolt fixed glazing: large scale rooflights
34. Bolt fixed glazing: twin wall roofs
35. Bonded glass rooflights

Concrete roofs

Sedlbauer, K., Schunck, E., Barthel, R. and Kunzel, H. (2010) 'Flat Roof Construction Manual: Materials, Design, Applications', Basel: Birkhauser.
36. Concealed membrane
37. Exposed membrane
38. Planted

Luckett, K. (2009) 'Green Roof Construction and Maintenance', Mcgraw-Hill.
38. Planted

Timber roofs

Mindham, C.N. and Goss, R. (2016) 'Goss's Roofing Ready Reckoner: From Timberwork to Tiles: Including Metric Cutting and Sizing Tables for Timber Roof Members', Chichester, West Sussex: Wiley Blackwell.
39. Flat roof: bitumen based sheet membranes

Herzog, T. (2012) 'Timber Construction Manual', Birkhauser.
39. Flat roof: bitumen based sheet membranes

Sedlbauer, K., Schunck, E., Barthel, R. and Kunzel, H. (2010) 'Flat Roof Construction Manual: Materials, Design, Applications', Basel: Birkhauser.
40. Pitched roof: tiles

Polymer roofs

Zoghi, M. (2014) 'The International Handbook of FRP Composites in Civil Engineering', Boca Raton, FL: Taylor & Francis.
41. GRP rooflights
42. GRP panels and shells

Fabric roofs

Huntington, C. (2004) 'The Tensioned Fabric Roof', ASCE Press.
43. ETFE cushions
44. Single membrane roof

Rivera, R. (2014) 'Membrane structures: first steps towards form finding', San Juan, Puerto Rico: Tensile structures education.
43. ETFE cushions
44. Single membrane roof

Buchholdt, H. (1999) ' An Introduction to Cable Roof Structures'.
43. ETFE cushions
44. Single membrane roof

Author
Andrew Watts
London, England

Aquisitions Editor: David Marold, Birkhäuser Verlag, Vienna, Austria
Project and Production Editor: Angelika Heller, Birkhäuser Verlag, Vienna, Austria

Layout and Cover Design: Yasmin Watts, London, England
Cover image: Newtecnic Ltd
Proofreading: Alun Brown, Vienna, Austria
Printing and binding: Holzhausen Druck GmbH, Wolkersdorf, Austria

Library of Congress Control Number: 2018957756.

Bibliographic information published by the German National Library
The German National Library lists this publication in the Deutsche Nationalbibliografie; detailed bibliographic data are available on the Internet at http://dnb.dnb.de

ISBN 978-3-0356-1770-2 Hardcover edition
ISBN 978-3-0356-1769-6 Softcover edition
e-ISBN 978-3-0356-1781-8

P.O. Box 44, 4009 Basel, Switzerland
Part of Walter de Gruyter GmbH, Berlin/Boston

9876543 www.birkhauser.com

This book is augmented, so install the free Artivive app, and hold your smartphone over the marked 3D drawings within.
Depending on your internet connection, you may be charged by your internet service provider.

1. Install the Artivive app

2. Look for the Artivive images marked with the icon

3. Hold your smartphone over the image